PROCESS PIPING SYSTEMS

Edited by

David J. Deutsch

and

the Staff of Chemical Engineering

CHEMICAL ENGINEERING

McGraw-Hill Publications Co., New York, N.Y.

Printed in the United States of America.

Library of Congress Cataloging in Publication Data

Main entry under title:

Process piping systems.

Includes index.
1. Chemical plants—Pipe lines. I. Chemical engineering.
TP155.5.P764 660.2'83 80-13774
ISBN 0-07-606675-4 (paper)
ISBN 0-07-010706-8 (case)

CONTENTS

INTRODUCTION

As much as 30% of the total capital spending in a typical chemical-processing-industries (CPI) plant goes for valves, piping, fittings and allied components, claims one recent marketing study.* Accounting for 45% of the total U.S. consumption for this equipment, the CPI spent $4.2 billion in 1978, and the forecast calls for an increase of more than 50%, in real terms, by 1985, to $6.6 billion.

One purpose of this book is to try to ensure that all such money is well spent. By presenting the chemical engineer with practical information on piping systems — not only piping and valves, but insulation as well—this volume will aid in the proper design and optimum operation of this vital sector of any CPI plant. Further, a section on slurry-handling should prove useful as transport of coal, among other materials, becomes increasingly more commonplace in the synfuel era now dawning.

However, the emphasis of the book is—*as it should be*—on piping, the subject of its first section. A long chapter on piping design provides practical techniques that should prove valuable to all engineers involved in the design, construction, operation and maintenance of plants in the CPI. Practical information is the key, but also served up is a quick review of basic principles. The intent is to refresh and reinforce the reader's knowledge and to present many useful solving methods, so as to enable any engineer to readily get accurate answers to any piping problem.

The piping design chapter leads off with pipeline sizing information of a general nature. More-specific applications, such as piping for reboiler systems, or high-temperature piping, are then described. Useful properties of fluids and friction-factor equations provide reference material, and a number of shortcut calculation methods are offered. A brief chapter follows on selecting piping accessories, including welding fittings and flanges, and steam-hose couplings. Wrapping up the piping section is an overview on specifying piping materials of construction.

In similar manner, the second section, on valves, covers various selection and specification criteria. Specific information follows on butterfly valves, pressure-relieving devices and rupture disks, among others. Operation and maintenance problems are not overlooked — with articles on how to improve online control-valve performance and on the reconditioning of old valves.

Befitting the energy-conscious eighties, an insulation section focuses on this cost-saving area. Detailed are insulation for hot pipes, as well as cold systems, steam tracing methods, and even a means to insulate hot tanks with floating balls. Section IV explores slurry-handling topics—including piping systems, pumps and instrumentation.

The editors have presented a wealth of information in this one volume—touching most aspects of the design, selection, and operation and maintenance of piping systems. In some cases, the coverage of material overlaps; but in most cases, a number of different techniques are supplied to offer the chemical engineer the opportunity to choose the best practical means to solve a particular problem.

---

*"Valves, Piping and Piping Components in the Chemical Process Industries," Study #713, July 1979, Frost & Sullivan, New York.

Section I
PIPING

Designing plant piping

Typically, process piping costs 20% of the total investment in a process plant. Energy usage plus continued maintenance add additional costs. Clearly, sound sizing practices can have a substantial impact on overall plant economics.

Larry L. Simpson and Martin L. Weirick, Union Carbide Corp.

☐ Although the literature on pipe sizing is massive, reflecting its importance, sizing remains a somewhat intractable technology. Piping-flow phenomena are frequently complex and often not well understood. Nevertheless, engineers and technicians too often size piping by rules of thumb, ignoring sound sizing principles.

Before undertaking any sizing calculation, the designer first decides what basis to use. The decision is often a difficult one; it might be the economic trade-off between operating-energy cost and piping investment, the so-called economic or optimum size, or it might be peculiar to a particular process situation. The designer should avoid rules of thumb where possible.

Many of the tools that were the backbone of the sizing art ten years ago are now becoming obsolete as a result of improved calculation tools, and because of the trend toward metrication. Small card-programmable calculators, first introduced four years ago, have freed many designers from laborious calculations and tedious charts and nomographs associated with pipe sizing. These calculators have become highly sophisticated and are more suitable for line sizing than any other tool currently available, including large computers.

Line-sizing practice is also being influenced by the gradual shift toward metrication. This change causes us to rethink many of the practical parameters etched in our minds. It makes dimensional equations obsolete, forcing us to derive new equations or to think and work with dimensionally consistent equations. This article reflects this trend by incorporating dimensionally consistent equations. Dimensional equations will be so noted and most will be presented with multiple sets of English and metric units. [A conversion factor (g_c) may be needed in some of the pressure-drop equations, but such equations will be flagged with an asterisk.]

Sizing economics and limitations

When economical sizing is mentioned, some engineers immediately think of balancing energy and investment costs to minimize an annualized cost. It is better to think of economical sizing in a broader context, because there is usually *one* best and most economical size and that size may have nothing to do with energy consumption. For example, pipe sizes to and from fixed-pressure utility headers do not affect plant energy consumption. Yet in a given application there is a minimum size that will give satisfactory operation, and is therefore the most economical size. The term "economical sizing" is too ingrained in our literature to change, but the designer should constantly strive for economical sizes in the broader sense.

One rational approach to pipe sizing involves first determining if pipe size affects plant energy consumption. If so, optimum sizing should be given further consideration. If not, it should be totally ignored, and another rational basis selected. In either case, the following sizing limitations should be considered:

- Pressure-drop availability
- Two-phase flow patterns
- Drain-piping design
- Erosion corrosion
- Noise and cavitation
- Flow distribution
- Transient and vibration phenomena
- Piping expansion and support

Many techniques are available for estimating an optimum diameter or velocity. The more sophisticated techniques [*36*] contain correlated piping costs and complex analytical expressions for explicit estimation of diameter, given the cost factors, flowrate, and physical-property data. This is a good approach, but due to the complexity, average parameters are usually used and an economical velocity is selected from a table of velocity versus fluid density.

In 1974, the authors [*42*] developed an alternative method for economical sizing of piping in turbulent flow. Instead of estimating the pipe size that would minimize an annualized cost, we estimated the flowrate range for which a given size would function most economically. This approach has several advantages over the conventional: First, discrete sizes avoid rounding-off of the estimated diameter. Second, actual pipe costs can be used instead of correlated costs. Third, tables of economical flowrate ranges can be prepared to reflect the economic impact of piping complexity, materials of construction, and control-valve pressure drop. Finally, the economic tables can be easily updated.

Since pressure loss through piping and fittings is a

Originally published April 3, 1978

Economic piping ranges for Plant X, based on the W_{OPT} parameter, which reflects energy usage **Table I**

Location: Plant X
Piping configuration: Complex
Material: Carbon steel

Without control valve

Nominal pipe I.D. (in.)	Actual pipe I.D. (in.)	WOPT MIN.	WOPT MAX.
1.5	1.6100	0	1498
2.0	2.0670	1498	2461
3.0	3.1880	2461	5575
4.0	4.1880	5575	12302
6.0	6.2490	12302	22279
8.0	8.2490	22279	36865
10.0	10.3440	36865	61309
12.0	12.2500	61309	98122
14.0	13.5000	98122	112788
16.0	15.5000	112788	168791
18.0	17.5000	168791	212979
20.0	19.5000	212979	250563
24.0	23.5000	25,563	380367

With control valve

Nominal pipe I.D. (in.)	Actual pipe I.D. (in.)	WOPT MIN.	WOPT MAX.
1.5	1.6100	0	1187
2.0	2.0670	1187	1951
3.0	3.1880	1951	4420
4.0	4.1880	4420	9757
6.0	6.2490	9757	17670
8.0	8.2490	17670	29242
10.0	10.3440	29242	48637
12.0	12.2500	48637	77854
14.0	13.5000	77854	89483
16.0	15.5000	89483	133923
18.0	17.5000	133923	168995
20.0	19.5000	168995	198802
24.0	23.5000	198802	301828

Location: Plant X
Piping configuration: Complex
Material: Stainless steel

Without control valve

Nominal pipe I.D. (in.)	Actual pipe I.D. (in.)	WOPT MIN.	WOPT MAX.
1.5	1.6820	0	1683
2.0	2.1570	1683	2963
3.0	3.2600	2963	6402
4.0	4.2600	6402	13288
6.0	6.4070	13288	27899
8.0	8.4070	27899	54578
10.0	10.4820	54578	83761
12.0	12.4380	83761	151287
14.0	13.6240	151287	151612
16.0	15.6240	151612	197188
18.0	17.6240	197188	233701
20.0	19.5640	233701	276417
24.0	23.5000	276417	361967

With control valve

Nominal pipe I.D. (in.)	Actual pipe I.D. (in.)	WOPT MIN.	WOPT MAX.
1.5	1.6820	0	1334
2.0	2.1570	1334	2349
3.0	3.2600	2349	5077
4.0	4.2600	5077	10538
6.0	6.4070	10538	22130
8.0	8.4070	22130	43300
10.0	10.4820	43300	66457
12.0	12.4380	66457	120051
14.0	13.6240	120051	120298
16.0	15.6240	120298	156462
18.0	17.6240	156462	185433
20.0	19.5640	185433	219344
24.0	23.5000	219344	287209

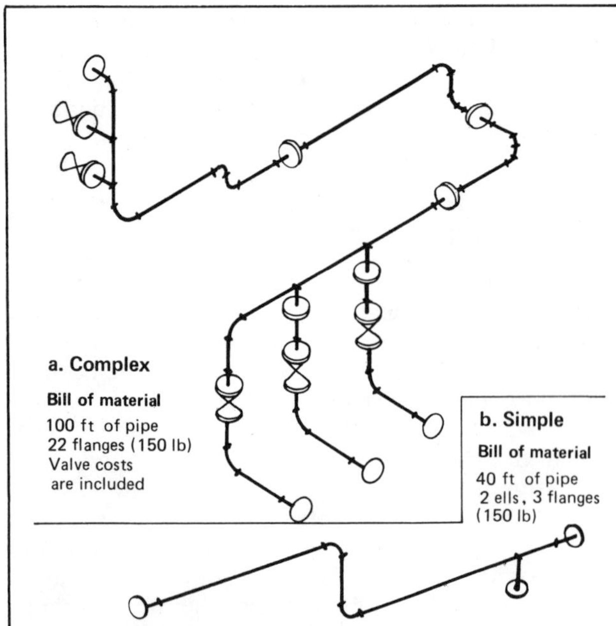

Piping configurations used as bases for cost estimates developed in the text **Fig. 1**

weak function of the kinematic viscosity of the fluid, we selected the kinematic viscosity of water as a basis, and developed a dimensional flowrate parameter

$$\left[W_{OPT} \equiv W/\rho^{2/3} \text{ (units of (lb/h)(lb/ft}^3)^{-2/3})\right]$$

that depends primarily on energy usage. The W_{OPT} parameter, correlated against the net present value of piping and pumping costs, allowed us to explore the economic impact of different configurations, materials of construction, and locations. Table I shows a typical set of economic W_{OPT} ranges at Plant X.

Several factors can affect economical line size:

Economic analysis—Plant startup assumed to be one year after the purchase of piping.

Piping configuration—Two different configurations—a complex one, Fig. 1(a), and a simple one, Fig. 1(b)—as bases for cost estimates.

Materials of construction—Cost estimates for two different material specifications prepared for each piping configuration.

Physical properties—A ten-fold increase in the kinematic viscosity will decrease the W_{OPT}(min)/W_{OPT}(max) numbers by 3 to 5%.

Control valves—In a controlled pipeline, it is necessary

(con't) **Table I**

Location:	Plant X			Location	Plant X		
Piping configuration:	Simple			Piping configuration:	Simple		
Material:	Carbon steel			Material:	Stainless steel		
Without control valve				**Without control valve**			
Nominal pipe I.D. (in.)	Actual pipe I.D. (in.)	WOPT MIN.	WOPT MAX.	Nominal pipe I.D. (in.)	Actual pipe I.D. (in.)	WOPT MIN.	WOPT MAX.
1.5	1.6100	0	997	1.5	1.6820	0	1186
2.0	2.0670	997	2621	2.0	2.1570	1186	2876
3.0	3.1880	2621	5086	3.0	3.2600	2876	5469
4.0	4.1880	5086	10494	4.0	4.2600	5469	10925
6.0	6.2490	10494	18065	6.0	6.4070	10925	22525
8.0	8.2490	18065	33788	8.0	8.4070	22525	47801
10.0	10.3440	33788	54967	10.0	10.4820	47801	69913
12.0	12.2500	54967	84489	12.0	12.4380	69913	114765
14.0	13.5000	84489	90062	14.0	13.6240	114765	125561
16.0	15.5000	90062	131294	16.0	15.6240	125561	155743
18.0	17.5000	131294	148068	18.0	17.6240	155743	187019
20.0	19.5000	148068	196713	20.0	19.5640	187019	218498
24.0	23.5000	196713	295847	24.0	23.5000	218498	306086
With control valve				**With control valve**			
1.5	1.6100	0	789	1.5	1.6820	0	940
2.0	2.0670	789	2077	2.0	2.1570	940	2279
3.0	3.1880	2077	4030	3.0	3.2600	2279	4335
4.0	4.1880	4030	8319	4.0	4.2600	4335	8661
6.0	6.2490	8319	14321	6.0	6.4070	8661	17858
8.0	8.2490	14321	26791	8.0	8.4070	17858	37911
10.0	10.3440	26791	43591	10.0	10.4820	37911	55451
12.0	12.2500	43591	67016	12.0	12.4380	55451	91042
14.0	13.5000	67016	71437	14.0	13.6240	91042	99606
16.0	15.5000	71437	104141	16.0	15.6240	99606	123548
18.0	17.5000	104141	117455	18.0	17.6240	123548	148359
20.0	19.5000	117455	156044	20.0	19.5640	148359	173331
24.0	23.5000	156044	234696	24.0	23.5000	173331	242829

for the control valve to take a considerable percentage of the total frictional pressure drop [*40*]. (The "with control valve" cases shown in Table I assume 50% of the total pressure drop occurs across the valve.)

Insulation—Insulation of pipelines tends to increase capital investment and decrease optimum pipe size. Insulating a line would tend to shift the W_{OPT}(min)/W_{OPT}(max) range upward for a particular line size. No examples of the effect have been included in the table.

Location—Separate sizing tables are prepared for each location to incorporate plant-to-plant variations in labor, material, energy, and overhead costs.

Inflation—Cost indices and/or revised costs are used to periodically update the tables.

Correction factors—The W_{OPT}(min)/W_{OPT}(max) tables are based on 8,000 h/yr operation and a 70% hydraulic-pump efficiency. The actual W_{OPT} parameter can be corrected for operating hours and pump efficiency by multiplying the W_{OPT} by:

$$\text{Hydraulic efficiency multiplier, } \left[\frac{70\%}{\%\ \text{Eff}}\right]^{1/3}$$

$$\text{and Operating time factor, } \left[\frac{\text{h/yr operation}}{8{,}000}\right]^{1/3}$$

Example of economical line-sizing

Economical sizing of a line used in batch operation is outlined below:

Flowrate	40,000 lb/h
Fluid properties	Non-corrosive hydrocarbon with a viscosity of 0.5 cP and a density of 51 lb/ft^3. (The flow will be turbulent.)
Operating h/yr	6,000 h/yr
Pump efficiency	≈70%
Control valve	None
Line description	Approximately 200 ft of line with 4 elbows, one shutoff valve and 2 tees.
Material of construction	Carbon steel with 300-lb flanges
Location	Plant X
Insulation	None
Initial W_{OPT}	$40{,}000/(51)^{2/3} = 2909$
Corrected W_{OPT}	

$$2909 \times \left(\frac{6000}{8000}\right)^{1/3} = 2640$$

At this point, engineering judgment must come into play. First, check the sizing figures in Table I that best represent the conditions. At Plant X (with carbon steel, simple configuration, and no control valve), $W_{OPT} = 2{,}640$ falls within the 3-in. nominal diameter range of $W_{OPT}(\text{min}) = 2{,}510$ and $W_{OPT}(\text{max}) = 5{,}430$. But note that:

■ The calculated $W_{OPT} = 2{,}640$ is close to the $W_{OPT}(\text{min}) = 2{,}510$ for 3-in. pipe.

■ The actual line configuration has fewer fittings and is simpler than the simple configuration used in the tabulation.

■ The actual line will require 300-lb flanges, but the simple configuration costs are based on 150-lb flange construction.

The simpler configuration and more expensive materials of construction tend to lower the $W_{OPT}(\text{min})/W_{OPT}(\text{max})$ values, so net present costs for the two-inch and the three-inch line are very similar. The two-inch line seems slightly more economical because of differences between the actual line and the one on which the table was based.

Use and misuse of sizing rules of thumb

There is a bewildering number of rules of thumb for selecting pipe size. They take the form of suggested bases such as "use 2 to 4 psi per 100 ft," or limitations such as "use 6 ft/s maximum velocity."

It is best to ignore such rules unless you understand the underlying reasons for them. Using the second example to illustrate the difficulty that can arise, consider the sizing of a Schedule 40 line for 70 gpm of water. In two-inch piping the water velocity would be 6.7 ft/s; in three-inch, it would be 3 ft/s.

The designer is left wondering what problems would be encountered if he selected the smaller and probably less expensive alternative. Often he won't want to take a chance, and will use the three-inch size because it satisfies the totally arbitrary rule he has learned. If he really knew the economics and the possible problems with velocities, he might realize that even 1.5-inch piping would be acceptable, with an installed cost of only one third that of the three-inch pipe. Moreover, situations arise where the 1.5-inch pipe would actually perform better.

Rules of thumb should be viewed only as guidelines or selection aids, not as inviolate laws. They represent the experience developed by other designers, making them no more than helpful references to past practice. In this context, the engineer should consider the following guidelines:

■ Select a sound sizing basis; avoid rules of thumb.

■ Consider special piping configurations or flow phenomena that might alter the selection.

■ Use short-cut pressure-drop estimating techniques for small piping only (less than six-inch), or when pressure drop is unimportant.

■ Avoid sizes not stocked in plant stores (e.g., $3\frac{1}{2}$ inch).

■ With small piping (e.g., one-inch), consider need for larger piping to ensure adequate support.

■ With turbulent flow, recognize that straight-pipe friction will usually dominate total frictional losses in small piping, and fitting losses will usually dominate the total in large piping.

■ Avoid the "equivalent length" concept for turbulent flow through large-diameter fittings and for all laminar-flow fittings.

■ Avoid the use of valves (except control valves) that differ from line size.

■ Use a design frictional pressure-drop approximately 1.3 times greater than the value calculated for new pipe, to allow for uncertainty of estimate and for pipe aging. Lower multipliers (<1.3) can be used with high-quality estimates in non-fouling services.

Turbulent hydraulics

Probably more than 90% of all chemical processing operations involve turbulent flow, and piping is no exception. Both single-phase and multiphase piping flows are usually at high Reynolds numbers where viscosity effects are much less important than inertial effects; consequently, many turbulent phenomena depend primarily on the density times square of velocity (ρV^2).

As with other processing equipment, piping must be integrated into the total system. Pressure changes must meet allowable limits, or these changes must be estimated, in order to specify other equipment. Thus, regardless of the sizing basis, the designer must work with flowrate, pressure drop, and diameter.

For convenience, think of piping hydraulics in terms of the Bernoulli equation:

$$\Delta P/\rho + \Delta(V^2)/2 + g\Delta Z + E = 0 \qquad (1)^{*\dagger}$$

Here, with the ΔP referring to the downstream minus upstream value, the E term, usually positive and representing the rate of irreversible conversion of mechanical energy into thermal energy or heat, is sometimes called *head loss, friction loss* or *frictional pressure drop*. Ignoring this factor would imply no energy usage in piping. The equation assumes a flat velocity profile, a steady-state operation, and an incompressible fluid. Expressed as frictional pressure drop for pipe and fittings:

$$-\Delta P_f \equiv \rho E = (fL/D + \Sigma K_i)\rho V^2/2 \qquad (2)^{*\ddagger}$$

This equation is particularly useful for analyzing plant piping because it divides the frictional loss into Reynolds number ($DV\rho/\mu$), dependent pipe friction (fL/D), and individual loss coefficients (K_i). The $\rho V^2/2$ term is a velocity head; K_i represents the number of velocity heads. With this approach, fittings can also be included in the calculations by adding their "hydraulic equivalent length" to the actual pipe length, remembering that many fittings have hydraulic behavior that approximates constant K_i values. Table III presents a compilation of K_i values.

Even with a very simple piping system such as that depicted in Fig. 2, the Bernoulli equation will cause confusion if not interpreted correctly. With 300 gpm of water flowing through 4-in. Schedule 40 pipe, the velocity amounts to 7.6 ft/s and the velocity head is 0.37

*Insert a g_c for use with English units.

†It can be shown that the $\Delta(V^2)/2$ term (Eq. 1) should be multiplied by 2 for laminar flow and by 1.06 for turbulent flow (from the "1/7 power rule").

‡Dimensional forms of Eq. 2 are given in Table II.

Dimensional forms for the equation that is used in calculating frictional pressure drop **Table II**

Pressure drop, Reynolds number, and velocity

Flowrate	$-\Delta P_f$ or ρH_f	N_{Re}	V	H_V
W	$\left(\frac{bfL}{D} + \Sigma K_i\right) \frac{aW^2}{\rho D^4}$	$\frac{cW}{\mu D}$	$\frac{dW}{\rho D^2}$	$\frac{\rho V^2}{e}$
Q	$\left(\frac{bfL}{D} + \Sigma K_i\right) \frac{a\rho Q^2}{D^4}$	$\frac{cQ\rho}{\mu D}$	$\frac{dQ}{D^2}$	$\frac{\rho V^2}{e}$

Friction factor

If $N_{Re} < 2{,}000$, $f = 64/N_{Re}$

If $N_{Re} \geqslant 2{,}000$, $f = [2 \log_{10} (0.27\, \epsilon/D + (7/N_{Re}^{\,0.9})]^{-2}$ *

*Ref. [*10*]

Units and constants

	Conventional units					**Metric units**			
					[Gas]				
$-\Delta P_f\ (H_f)$	psi	psi	(ft)	(ft)	inH_2O[60°F]	bar	bar	Pa	(m)
$W(Q)$	lb/h	(gpm)	lb/h	(gpm)	(acfm)	kg/s	(L/s)	kg/s	(m^3/s)
D	in.	in.	in.	in.	in.	mm	mm	m	m
ϵ	in.	in.	in.	in.	in.	mm	mm	m	m
L	ft	ft	ft	ft	ft	m	m	m	m
ρ	lb/ft³	lb/ft³	lb/ft³	lb/ft³	lb/ft³	kg/m³	kg/m³	kg/m³	kg/m³
μ	cP	cP	cP	cP	cP	mPa·s (cP)	mPa·s (cP)	Pa·s	Pa·s
V	ft/s	ft/s	ft/s	ft/s	ft/min	m/s	m/s	m/s	m/s
H_V	psi	psi	ft	ft	in. H_2O	bar	bar	Pa	m
a	2.799×10^{-7}	1.801×10^{-5}	4.031×10^{-5}	2.593×10^{-3}	0.02792	8.106×10^6	8.106	0.8106	0.08265
b	12	12	12	12	12	1,000	1,000	1	1
c	6.316	50.66	6.316	50.66	379.0	1.273×10^6	1,273	1.273	1.273
d	0.05093	0.4085	0.05093	0.4085	183.3	1.273×10^6	1,273	1.273	1.273
e	9,266	9,266	$64.35 \times \rho$	$64.35 \times \rho$	1.204×10^6	2×10^5	2×10^5	2	$19.61 \times \rho$

a,b,c,d,e	Constants	L	Pipe length	H_V	Velocity head
D	Pipe diameter	$-\Delta P_f$	Frictional pressure drop	W	Mass flowrate
f	Weisbach friction factor	Q	Volumetric flowrate	ϵ	Pipe roughness
H_f	Frictional head loss	N_{Re}	Reynolds number	μ	Fluid viscosity
K	Number of velocity heads	V	Velocity	ρ	Fluid density

psi, or 0.89 feet of water. Writing the Bernoulli equation between the tank and pump suction, one would conclude that the actual static head at the pump suction would be (using K_i values given in Table III):

$$6 - [(0.1 + 0.6 + 0.5 + 0.7 + 1)(0.89)] = 3.4 \text{ ft head}$$

where: valve $K_i = 0.1$; 2 elbows $= 0.6$; entrance $= 0.5$; straight pipe $= 0.7$ (assumed); velocity head $= 1$ ($\Delta V^2/2g$—from the Bernoulli equation).

Thus the head at the pump suction nozzle is 3.4 ft of water (about 1.4 psi) above that at the tank liquid level. A static-pressure or static-head instrument would measure such a value. Of the head change, 0.89 feet is expended in generating the 7.6 ft/s velocity and is the recoverable energy contained in the flowing stream. The Hydraulic Institute Standards [*25*] include this velocity head as additional head in the calculation of NPSHA (net positive suction head available). Since NPSHA represents the liquid head plus velocity head above the vapor pressure at the pump suction nozzle,

$$\text{NPSHA} = 3.4 + 0.89 = 4.3 \text{ feet of water}$$

The point being made regarding the Bernoulli velocity head is subtle and usually unimportant, but it does cause confusion and can lead to errors that can be avoided if the distinction between actual pressure drop and frictional loss is understood.

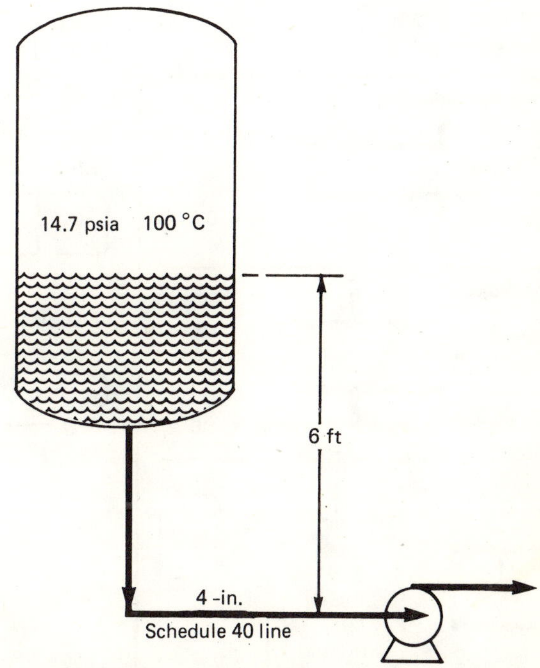

Pump suction piping used in NPSH example **Fig. 2**

G

(velocity heads) used for calculating the head losses through various fittings

ell

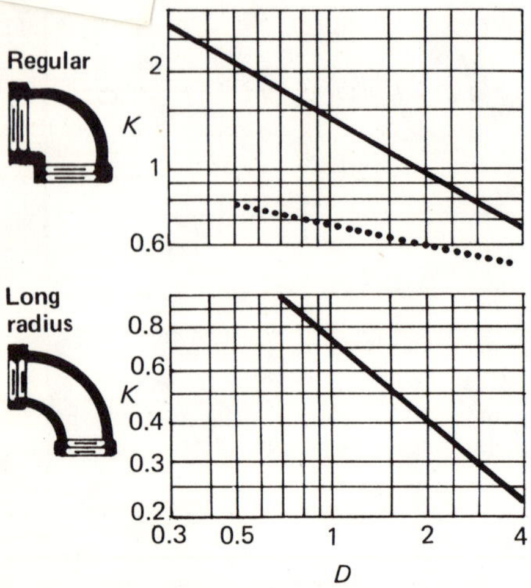

Flanged 90°ell

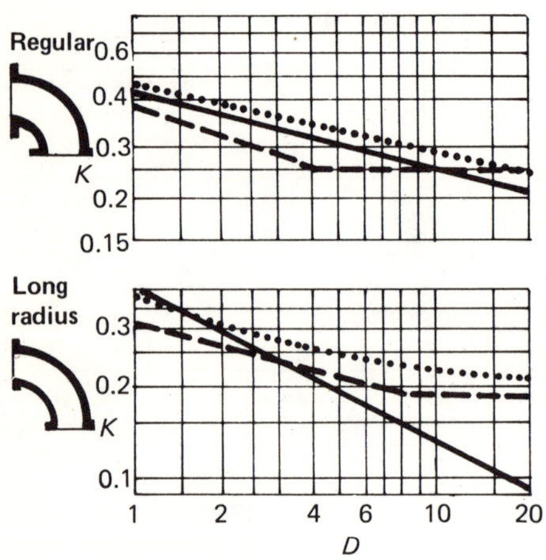

Square-edged inlet [24]

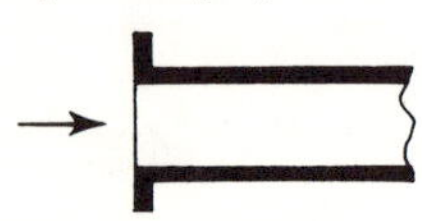

$K = 0.5$

Inward projecting pipe [24]

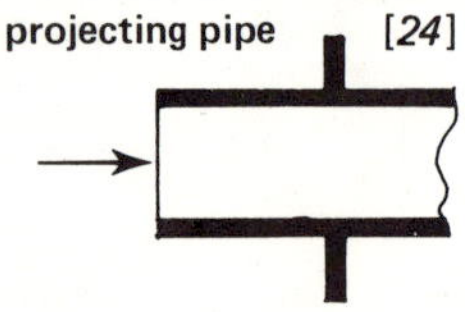

$K = 1.0$

Flush [12]

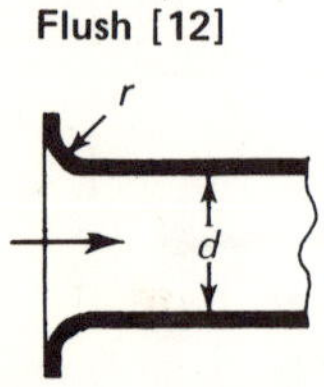

For K, see table

r/d	K
0.00*	0.5
0.02	0.28
0.04	0.24
0.06	0.15
0.10	0.09
0.15 & up	0.04

*Sharp-edged

Screwed tee

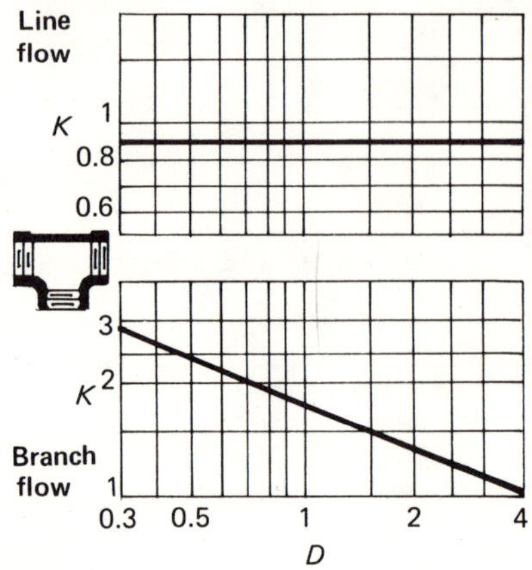

Flanged tee

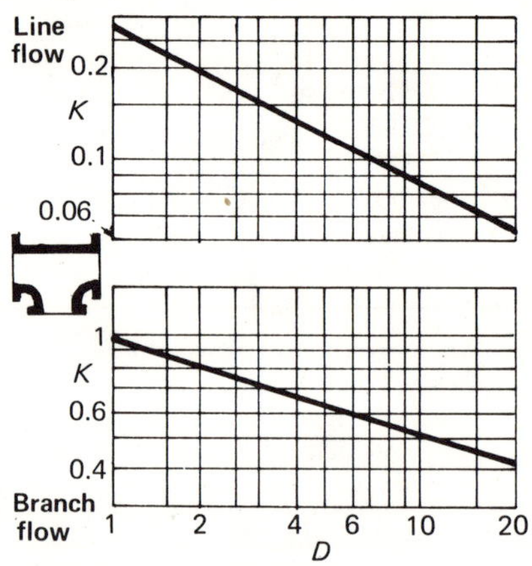

Contraction

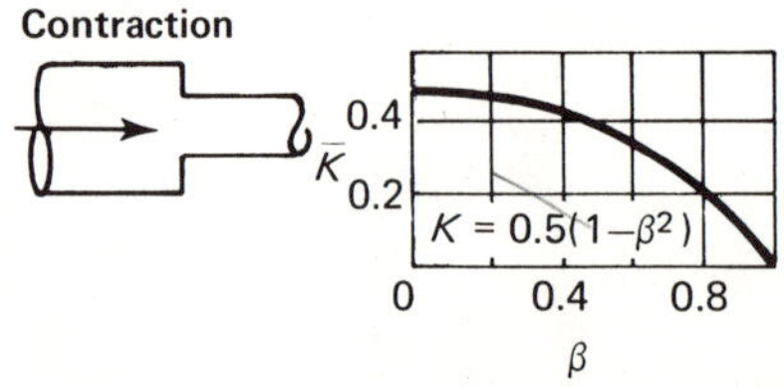

Orifice

$$K = 2.8\,(1-\beta^2)\left[\left(\frac{1}{\beta}\right)^4 - 1\right]$$

Enlargement

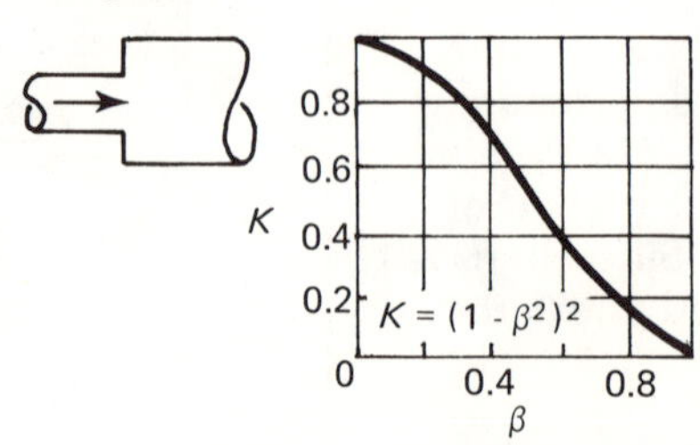

β = small diameter/large diameter
K based on velocity in smaller pipe·
based on main pipe for orifice. [40]

Globe valve

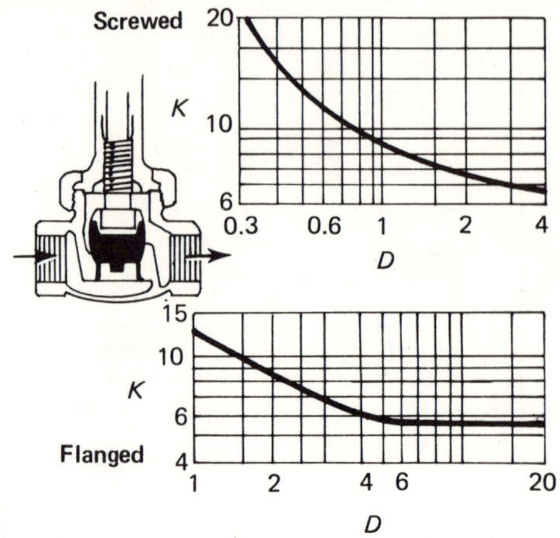

Gate valve

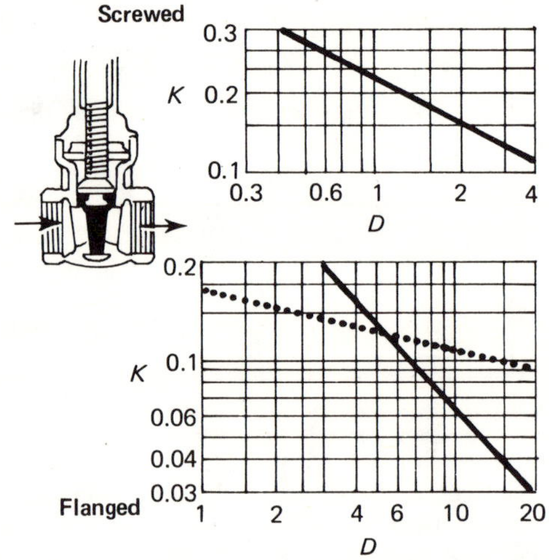

Regular screwed 45°ell

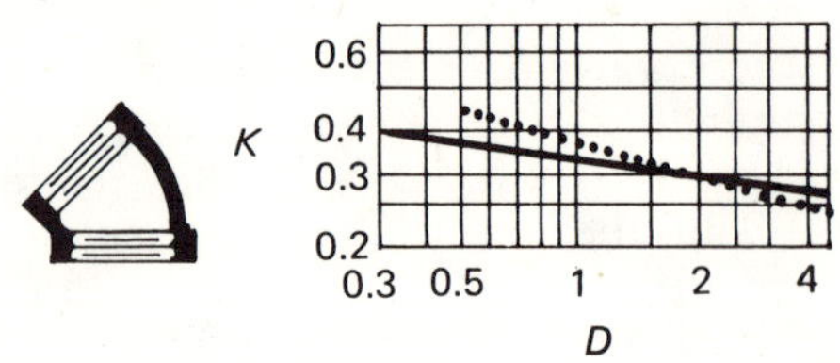

Long radius flanged 45°ell

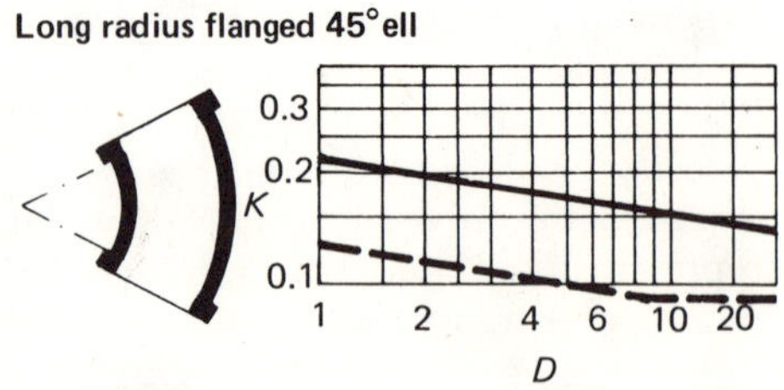

—————— Hydraulic institute [24]

·············· Crane [12]

— — — — — Miller, based on water at 6 ft/s [32]

Table III

Swing check valve

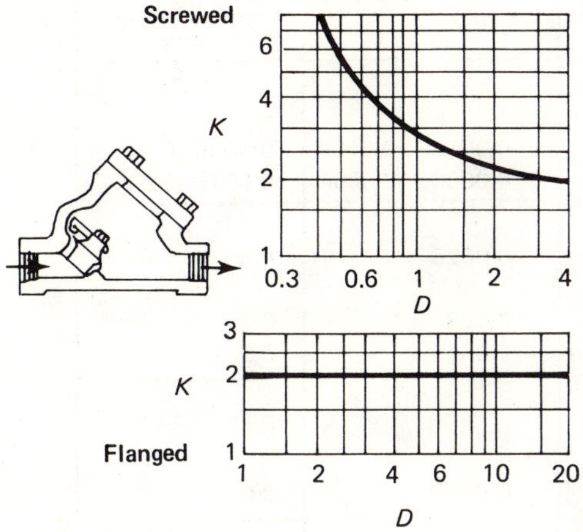

Angle valve

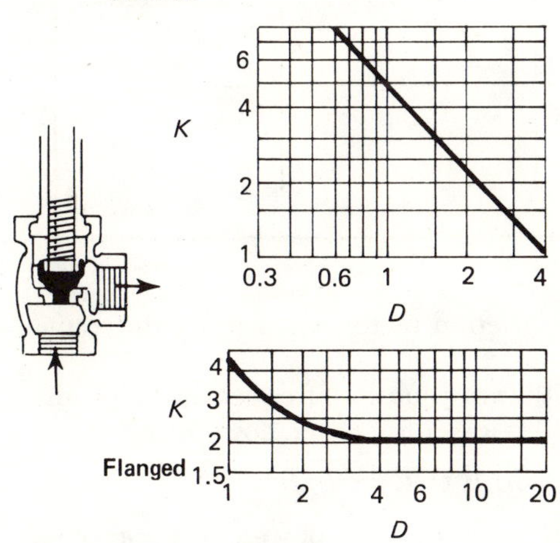

Screwed return bend

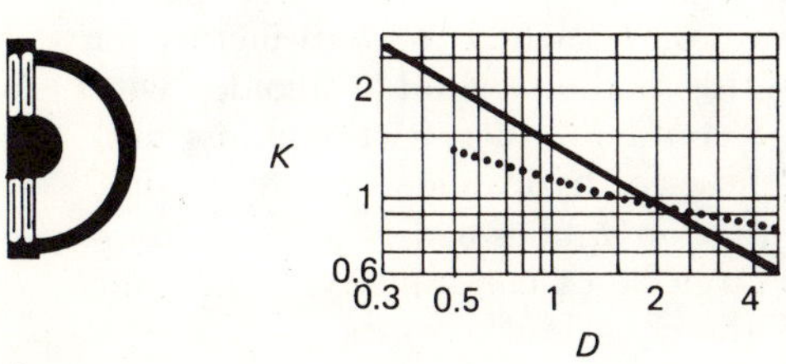

Flanged return bend

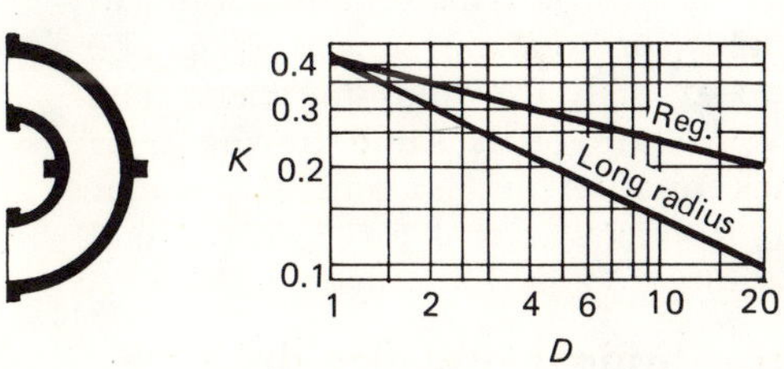

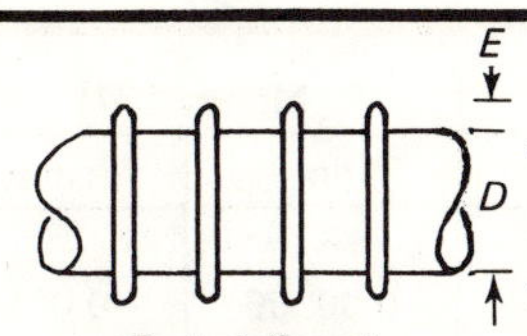

Typical dimensions

Nominal size (in.)	I.D. (in.)	$\frac{E}{D}$
1/2	0.520	0.250
3/4	0.750	0.275
1	1.000	0.187
1-1/2	1.500	0.200
2	2.000	0.161
3	2.962	0.143
4	4.000	0.081

Fanning friction factor for flexible metal hose [*14*]

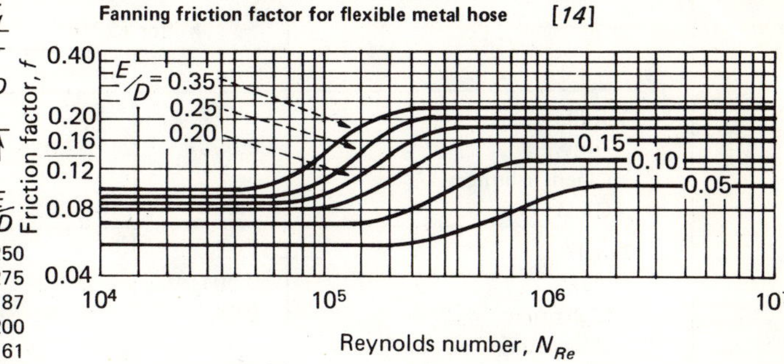

Head loss in conical diffusers

[*19*]

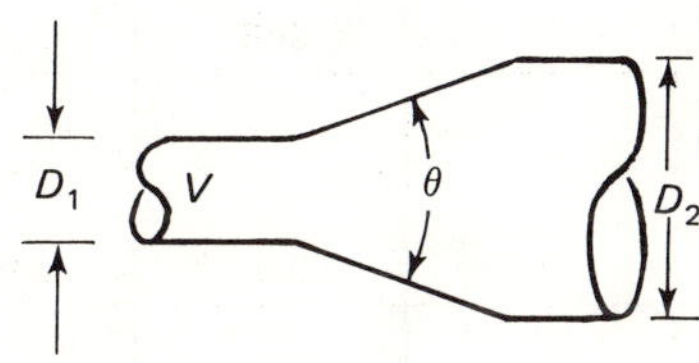

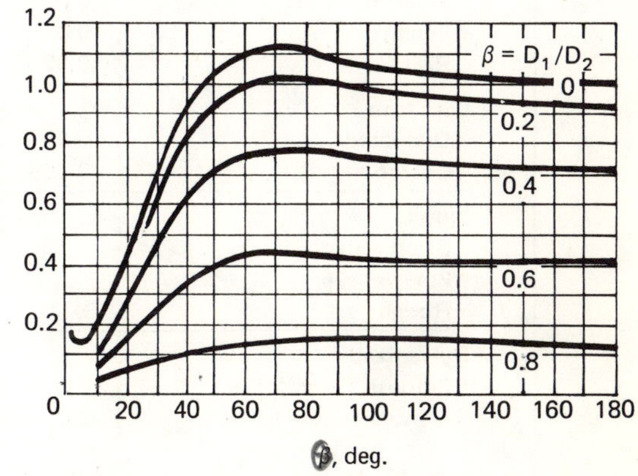

Head loss in circular miters

[*47*]

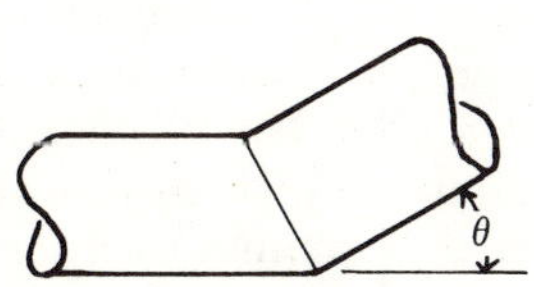

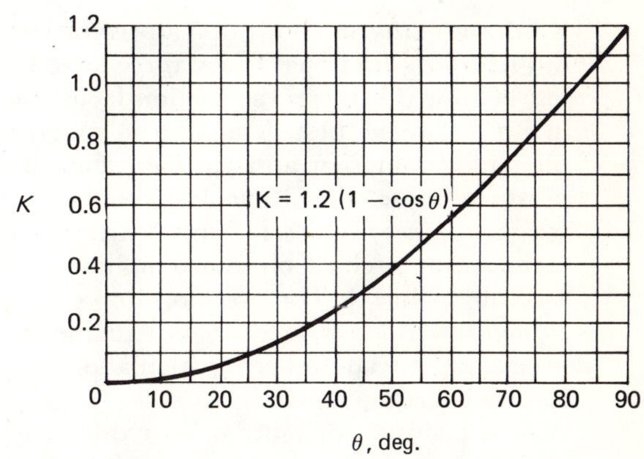

Plug cock valve [*19*]

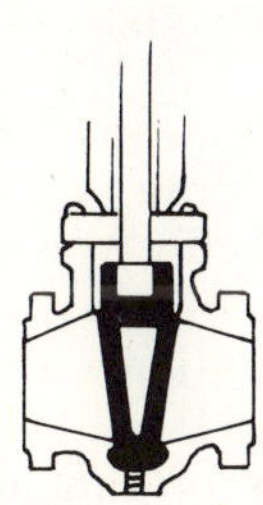

θ	K
5°	0.05
10°	0.29
20°	1.56
40°	17.3
60°	206.0

θ is angle between pipe axis and plug cock axis

Butterfly valve [*19*]

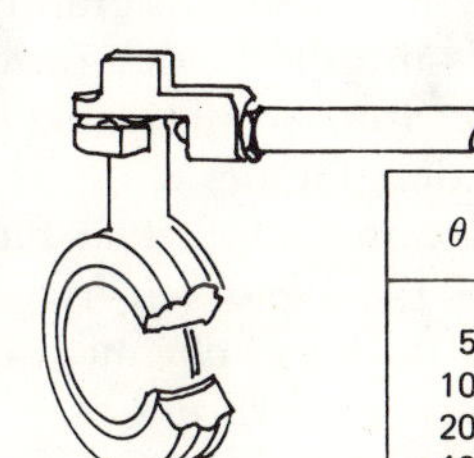

θ	K
5°	0.24
10°	0.52
20°	1.54
40°	10.8
60°	118.0

θ is angle between pipe axis and flapper plate

Surface roughness of various pipes and tubes **Table IV**

	Miller [32]		Moody [34]		Kutateladze [29]	
	in.	mm	in.	mm	in.	mm
Tubing						
Drawn	0.0001	0.0025	0.00006	0.0015		
Clean, seamless					0.00006- 0.0004	0.0015–0.01
Glass	0.0001	0.0025			0.00006–0.0004	0.0015–0.01
Steel						
New	0.001	0.025	0.0018	0.046	0.0025	0.06
Light rust	0.01	0.25				
Deaerated saturated steam					0.008	0.2
Condensate (heavy rust)					0.035	0.9
Concrete						
Smooth	0.001	0.025	0.012	0.3	0.03	0.8
Precast	0.01	0.25				
Rough	0.02	0.5	0.12	3.0	0.35	9.0
Cast Iron						
Uncoated	0.006	0.15	0.01	0.25	0.012	0.3
Coated	0.006	0.15	0.005	0.12		
Wood						
Birch veneer					0.001–0.002	0.025–0.05
Pine veneer					0.004	0.1
Rough			0.036	0.9		
Galvanized						
Smooth finish	0.001	0.025				
Normal finish	0.006	0.15	0.006	0.15		

Of the many methods developed for estimating liquid-handling frictional losses, three have withstood the test of time. Of these, the friction-factor method given in Eq. (2) is the best. It applies to rough and smooth pipe and to laminar and turbulent flow. To use it, one must estimate the Reynolds number and relative roughness (see Tables IV and V) and find a friction factor from a chart or from equations such as those given in Table II. The pressure drop can be estimated explicitly with this method, but diameter or flowrate necessitate a trial-and-error calculation, or the use of a special chart.

As a historic sidelight, at least three friction factors have been introduced since the 19th century. Weisbach can be credited with the velocity-squared resistance law [37], and his resistance coefficient became known as the Darcy-Weisbach friction factor. Using this same resistance coefficient, Blasius originated the "friction-factor vs. Reynolds-number" diagram that later became known as the Stanton diagram. However, the Stanton diagram friction factor amounts to one eighth of the Weisbach friction factor. The Fanning friction factor, one fourth of the Weisbach friction factor, is the most common one found in chemical engineering literature.

In engineering literature as a whole, the Weisbach friction factor has probably gained wider use because of the popularity of the Moody diagram [34]. Weisbach is used throughout this paper because it offers a slight computational advantage over the other coefficients in hydraulic calculations. In turbulent flow, its value will range from roughly 0.008 to 0.04. Often the designer must decipher the friction factor when using different literature sources.

The Hazen-Williams equation frequently helps in estimating hydraulic losses for water flow:

$$H_f = 10.4(Q/C)^{1.85}L/D^{4.87} \tag{3}$$

In this equation, H_f represents feet of water (at approximately 60°F); Q, in gpm; L, in feet; and D, in inches. The coefficient, C, varies from about 60 for corroded pipe to 150 for smooth tubing; a value of 130 is sometimes used for new, clean steel. Other values for specific applications are available elsewhere [6]. Satisfactory for many applications, this method should be avoided with high C values that could produce frictional-loss estimates lower than those for hydraulically smooth pipe.

The Manning equation applies widely to open channel flow [9] and to large closed conduits. With the same units given above, the equation becomes

$$H_f = 13.2n^2L\,Q^2/D^{16/3} \tag{4}$$

Manning's n varies from <0.01 for smooth tubing to roughly 0.04 for highly corroded pipe; $n = 0.009$ can be used for steel pipe at high Reynolds numbers. The equation suffers from many of the same drawbacks as the Hazen-Williams equation but, in a different form, it has proven a very convenient tool for open-channel hydraulics calculations.

Laminar and transitional pressure drop

At Reynolds numbers less than 2,000, liquids lose the turbulent characteristics usually associated with pipe

Nomenclature

a	Wave speed	P_o	Initial upstream pressure
C	Hazen-Williams flow coefficient	Q	Volumetric flowrate
D	Pipe diameter	N_{Re}	Reynolds number
f	Weisbach friction factor	t	Time
g	Gravitational acceleration, 9.807 m/s^2, 32.174 ft/s^2	V	Fluid velocity
g_c	Gravitational constant, 32.174 (lb/lb$_f$)(ft/s^2)	W	Mass-flowrate
H_f	frictional head loss	W_{OPT}	Flowrate parameter for economical pipe sizing
H_V	Velocity head	ΔZ	Change in elevation
K_i	Fitting loss in velocity heads	β	Ratio of small diameter to large diameter
L	Pipe length	ε	Pipe roughness
n	Manning's "n"	μ	Viscosity
NPSHA	Net positive suction head available	ρ	Density
P	Pressure		**Subscript**
$-\Delta P$	Pressure drop, initial minus final pressure	f	Frictional

flow. Velocity profiles become parabolic with essentially no radial mixing in this laminar regime. Pressure drop is estimated with an exact analytical equation called Poiseuille's Law:

$$-\Delta P_f = 32\mu VL/D^2 \tag{5}^*$$

Although Poiseuille cannot be credited with the equation, his pioneering experimental work discerned the essential features of laminar flow well before the equation was derived [*37*]. Note that pressure drop is independent of pipe roughness in laminar flow.

Eq. (5) can be readily transformed into the Reynolds-number vs. friction-factor relationship presented in Table II. However, in this case, the Table III fitting-loss coefficients are no longer applicable, nor is the "equivalent length" concept viable. Fitting-pressure-drop estimates from it would be worse than those from turbulent-velocity head coefficients.

Only a few laminar-fitting losses have been studied in depth. For contractions and expansions, the equations of Kaye and Rosen [*26*] and of Kays [*27*] are suggested.

$$\text{Contraction: } K = (1.32 + 159/N_{Re})(1 - \beta^2) \tag{6}$$

$$\text{Expansion: } K = 1 - 2.66\beta^2 + \beta^4 \tag{7}$$

where β is the usual diameter ratio and the K-values are based on the velocity in the smaller pipe.

Turbulent-flow head-loss coefficients can be used for orifices if the orifice Reynolds number exceeds 50, but they underestimate the loss at very low Reynolds numbers. For other fittings, use the Table II methods with the actual flow-path length and twice the turbulent-velocity head coefficients given in Table III.

Although maintaining laminar flow at a Reynolds number of 40,000 is possible, using an inlet exceptionally free of disturbances [*39*], the laminar-to-turbulent transition occurs in the 2,000 to 4,000 Reynolds-number range in industrial piping. In this region, random patterns of both laminar and turbulent flow coexist. At N_{Re} = 2,000, the flow will be mostly laminar; at N_{Re} = 4,000, mostly turbulent. Whereas friction-factor charts contain a discontinuity between laminar and turbulent flow, the Churchill equation [*11*] predicts continuous frictional behavior in this transitional or critical region.

*Insert a g_c for use with English units.

Designing drain piping

No simple cookbook procedure will suffice for designing drain piping, because many situations arise in practice and they must be approached separately.

Drain piping usually connects to an upstream pressure vessel and, even if the liquid level is controlled, gases and vapors can break through into the piping. Whirlpools, low liquid levels, flashing, and turbulence can contribute to the gas entrainment.

Whirlpools are highly unpredictable, being influenced not only by Coriolis forces but also by tank configuration, liquid viscosity, and the velocities of inflowing and outflowing liquids. Typically, a whirlpool forms when the liquid level above the outlet (in inches) is roughly equal to the fourth root of the drain flowrate (in gallons per hour). If liquid inflow to the tank stops, or if unusual conditions exist in the tank, the liquid level above the outlet can increase by a factor of ten or more.

An antiswirl baffle installed above or in the outlet nozzle will relieve any doubts about whether a whirlpool will form. The baffle can be a cross; size is not critical, so a width of about two nozzle diameters and a height of 0.5 to 1.0 nozzle diameter should suffice. A flat, circular plate, four nozzle diameters across and 0.5 nozzle diameter above the outlet, may be more effective than the cross, but it would prove impractical if the diameter of the nozzle were large compared to the tank diameter.

Even without a whirlpool, low liquid levels will cause gas entrainment in the drain piping, as discussed earlier [*40,41*]. Recently, McDuffie [*31*] reported that the Harleman equation [*40*] underestimates the critical level under some circumstances, but his conclusion seems largely based on data with a 3-mm drain line. Consequently, the Harleman equation seems the one to

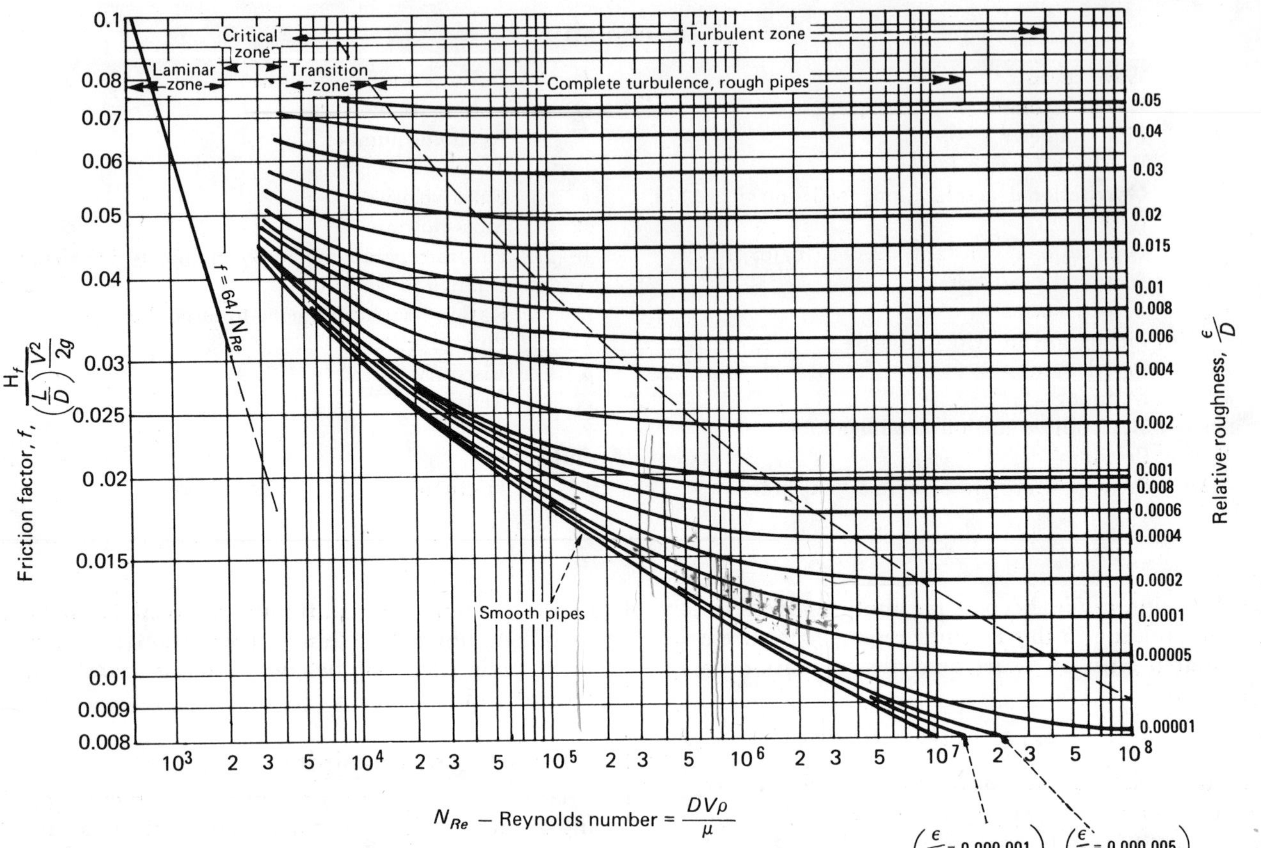

Friction factor diagram **Fig. 3**

use for design until the apparent discrepancy between the Kalinske [28] and McDuffie data gets resolved.

$$H = 0.688(Q/\sqrt{g})^{0.4} \quad (8)$$

When the tank liquid is at its bubble point, the drop in pressure that accompanies flow through a drain nozzle can cause local flashing or cavitation. Although little work has been reported [41] on the consequences of such flashing, it should be avoided in design, particularly in pump-suction piping. For a flush-mounted drain connection, a liquid level higher than three nozzle-velocity heads should avoid any problem.

Dumping a high-velocity stream, or one near the top of a vessel, onto a liquid surface will drive a substantial amount of gas into the liquid and even into the drain piping. High flowrate and low liquid level can create troublesome turbulence. All but the smallest bubbles usually have rise velocities above 0.5 ft/s, so superficial downflowing liquid velocities greater than this value will also aggravate the entrainment. Liquid velocities near the drain nozzle depend primarily on the distance from the nozzle; consequently, a convergence effect can cause entrainment of the bubbles.

Since flow patterns within drain piping represent a second facet of piping design, it is useful to distinguish between two regimes: *self-venting flow* and so-called *siphon flow*. Liquid in a self-venting line occupies a portion of the pipe cross-section, and gas moves freely with, or counter to, the liquid. Siphon flow refers to a liquid or two-phase downflow where pressure increases in the direction of flow.

Simpson's criterion [40] identifies self-venting flow in vertical piping:

$$V = 0.31\sqrt{gD} \quad (9)$$

At lower liquid velocities, small quantities of gases rise freely, counter to the liquid flow. At higher velocities, gases are pulled down with the liquid. As the liquid velocity increases, gas rate first increases, and then decreases as the frictional head approaches the elevation head. At the self-venting limit, long pipe-size bubbles are trapped inside the piping. Unless liquid, or a bend or slight obstruction, seals the bottom of the pipe, a falling film will develop and the transition to siphon flow will shift to

$$V \cong 2\sqrt{gD} \quad (10)$$

The self-venting criterion can be converted to useful dimensional sizing equations:

$$\text{English: } D = 0.92\,Q^{0.4} \quad (\text{in., gpm}) \quad (11)$$

$$\text{Metric: } D = 70\,Q^{0.4} \quad (\text{mm, L/s}) \quad (12)$$

Short, horizontal pipe will also become self-venting at this transition. However, the transition is very sensitive to pipe slope and somewhat sensitive to length. Open-channel-hydraulics technology must be used for in-depth analysis of horizontal piping.

The significance of entrainment in self-venting and siphon flows is illustrated in Fig. 4(a). According to Eq. (11), at flowrates less than 20 gpm for a 3-in line, the piping between Tanks A and B is self-venting and the entire system is at steady-state. At startup, the level builds in Tank A and in the seal until it reaches the top of the seal (5-ft level). The liquid enters Tank B and the flow is stable. Little or no vapor flows through the vent piping.

Beyond 20 gpm inflow rate, the Tank A level rises and a siphon develops in the downleg, filling it and increasing the flowrate to an estimated 294 gpm (13 ft/s), a value determined by the head and the piping resistance. This rate is much higher than the normal inflow rate, so the level quickly drops, draining Tank A. Eventually, the siphon will break and the whole cycle repeat.

In Figure 4(b) the drain piping will also be self-venting at rates less than 20 gpm. Its behavior will be similar to that in (a) except that a low volumetric flowrate of small vapor bubbles will be forced into Tank B and returned to Tank A through the vent piping. The liquid flow might also pulsate slightly. Once again, when the flowrate exceeds 20 gpm, the Fig. 3(b) drain line tries to fill and form a siphon. And, if the flowrate exceeds the 289 gpm that would balance the eight-foot elevation head with the frictional loss, the piping will fill with liquid but the tank level will be highly sensitive to the flowrate. More often than not, the flowrate is substantially less than the balanced value, pulling gas into the drain—decreasing the effective head and increasing the frictional head loss. The gas circulates back through the vent line and may cause a significant pressure drop in it.

This two-phase flow is often acceptable in plant piping, provided the gas flow does not disrupt the operation of Tank B. If the inflow to Tank A were 100 gpm, for example, with atmospheric air as the gas, the air rate would reach an estimated 25 cfm (neglecting vent-piping pressure drop). This estimate may be somewhat too high, since it is based on homogeneous flow, and the downflow could actually be more nearly annular.

Erosion corrosion

Of the many factors affecting line sizing, erosion corrosion probably causes the greatest uncertainty in selection of piping size. With meager data available on this important subject, so many variables that influence erosion corrosion rates, allowable velocities inevitably become a matter of judgment and experience.

A number of variables can influence erosion corrosion rates:

Surface chemistry—Highly reactive fluid-pipe wall combinations usually result in high erosion rates.

Metal properties—Hard metals are more resistant to attack than the soft aluminum, copper, and lead metals and alloys.

Velocity and turbulence—High velocities and associated turbulence, particularly at entrances, bends, tees, and valves, usually aggravate erosion; however, with corrosive surface-reaction products, increasing the velocity might reduce erosion [*18*].

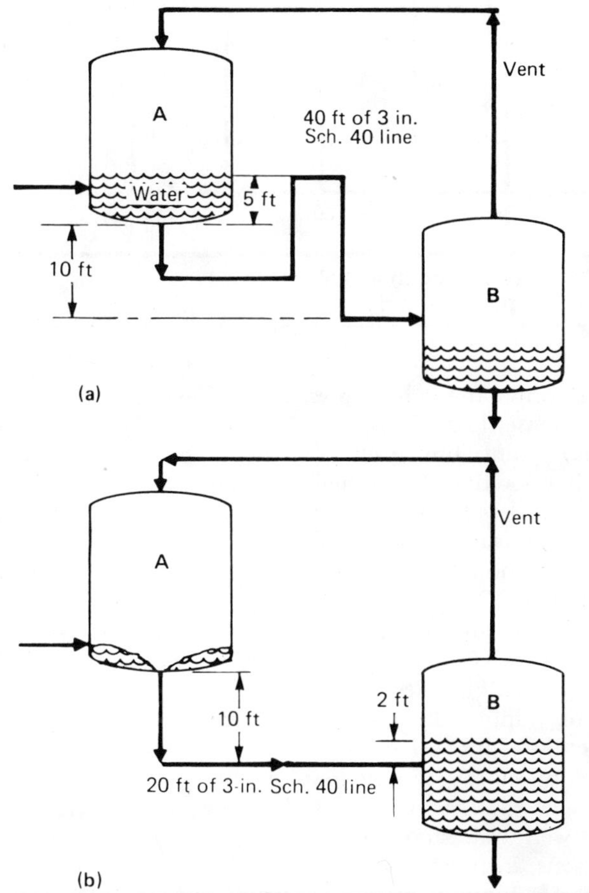

Configuration used in drain piping example **Fig. 4**

Cavitation—During the collapse of a bubble near a solid metallic surface, a liquid jet develops and hits the surface at such a high velocity that it causes erosion, regardless of the surface properties.

Impingement—Abrasive solids in slurries generate wear. As indicated by industry's continuing problems with two-phase steam-condensate piping, some mist flows also cause erosion. (A recent book by Moore and Sieverding [*33*] contains an excellent discussion of erosion in steel-condensate piping.)

Galvanic Effect—High velocities will increase galvanic attack [*18*].

In addition to the references mentioned above, plant materials engineers are a good source of erosion know-how, and can help in selecting allowable velocities.

Try to keep liquid velocities above 2 ft/s but under 25 ft/s. And if erosion corrosion is suspected, the upper velocity limit may have to be reduced—sometimes to as low as 4 ft/s, depending on the expected severity of attack. The 25 ft/s velocity range is achievable only in systems free of erosion corrosion.

Noise and cavitation

In contrast to piping gas, liquids handling rarely creates objectionable noise. When it does, the noise

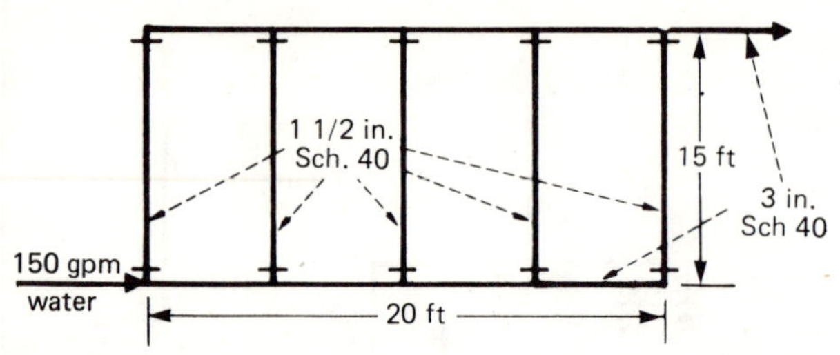

Flow distribution in a tank cooling tube manifold

Fig. 5

often originates in high-pressure let-down, cavitation, or two-phase flow. Velocities of 20 to 30 ft/s do result in noisy piping, but rarely does the designer specify such high velocities. As a rough guideline, a piping velocity-head less than 1.3 psi (0.9 bar) will avoid excessive noise (consult Seebold [*38*] for details).

Cavitation occurs all too frequently in chemical processing. The sudden collapse of vapor bubbles, with associated pressure pulsation, can cause cavitation in pumps, valves, and direct-contact steam heating. Since it can be highly destructive to piping and other equipment, piping design should strive to avoid it—frequently a difficult task.

Pumps often cavitate in spite of designer's best efforts to meet NPSH requirements of pumps. This behavior is not well understood, and could occur for a number of reasons, including:

■ High pump-suction velocities and long piping increase pressure fluctuations in the pump. Velocities less than 6 ft/s work best when NPSH requirements are difficult to meet.

■ Vacuum systems seem more prone to unpredictable cavitation than pressure systems.

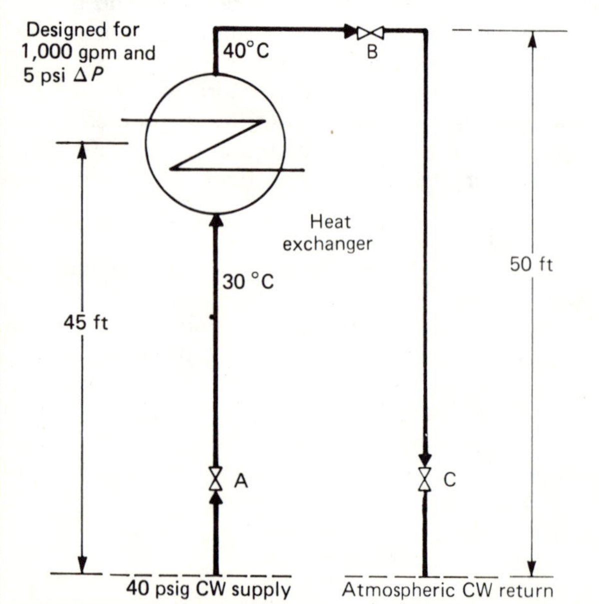

Flashing in elevated piping

Fig. 6

■ Vapor pressure is used in NPSH calculations when actually saturation pressure might be more appropriate. A better definition of vapor pressure is needed (e.g., pressure at which 2 volume-% of the gas is released). Be wary of dissolved gases [*45*].

■ A suction-piping elbow next to the pump aggravates cavitation, especially at high flow velocities. About five diameters or more of straight pipe between pump and elbow will minimize pressure disturbances near the elbow.

Valve cavitation can often be avoided by following manufacturers' recommendations regarding allowable pressure drop. Alternatively, a special low-cavitation valve can help solve the problem.

Flow distribution

When a fluid stream splits into several parallel streams, it often requires special precautions to ensure that each of the parallel lines carries an equal portion of the flow. Too small a manifold header will not give good distribution.

Whether it involves a sparger, a distributor pipe, air-cooler piping, or the tubeside inlet of a heat exchanger, flow distribution is related to pressure changes in the manifold piping and to the flow resistance of the parallel passages. Obviously, a 0.1-psi pressure change in a manifold would be unimportant if the parallel channels had an average pressure drop of 1 psi across them. This comparison of pressure changes is the essence of flow-distribution analysis.

Pressure changes in a manifold pipe can result from momentum and frictional effects. In a splitting manifold, the momentum-related pressure rises about 1.2 velocity heads along the manifold length, based on the velocity upstream of the first takeoff. In a combining manifold, the momentum effect drops the pressure by about 1.8 velocity heads in the direction of flow. (To remember the sign of the pressure change, note that increasing velocity reduces pressure and vice versa, as predicted by the Bernoulli equation.)

Frictional loss is approximately one-third of the value that would be obtained by calculating the loss on the assumption that the total stream flows through the total manifold.

Analysis of a distributor

A tank-cooling-tube manifold, shown schematically in Fig. 5, has inlet-flowrate of 150 gpm, resulting in a velocity of 6.51 ft/s and a velocity head of 0.285 psi. Frictional loss associated with this rate through 20 feet of 3-inch steel pipe amounts to 0.442 psi. With an assumed 1.5 velocity-head entrance/exit loss in the parallel 1.5-inch tubes, frictional pressure drop in them reaches 0.630 psi at an average flowrate of 150/5 = 30 gpm.

In this example, the difference between the pressure drop at the last and first 1.5-inch tube is:

$$\begin{aligned} &\text{Inlet manifold:} && [(1.2)(0.285) - 0.442/3] \\ &\text{Outlet manifold:} && +\ [(1.8)(0.285) + 0.442/3] \\ & && = 0.855 \text{ psi} \end{aligned}$$

Note that momentum effects add, and frictional effects cancel, for this co-current manifold.

Fouling of a 5/8-in. heat exchanger tube from siphon flashing **Fig. 7**

With maldistribution

$\equiv 100$ (high flow/low flow -1)

$\approx 100\,(\sqrt{(0.630 + 0.855/2)/(0.630 - 0.855/2)} - 1)$

$= 205\%$

This analysis indicates that the fluid distribution is poor. It would be improved by using larger manifold piping, smaller heat exchange tubes, an orifice in each tube, or a countercurrent configuration.

Transient and vibration phenomena

Whether the transients are caused by occasional plant outages, flow-control devices, two-phase phenomena, or by high-frequency-turbulence noise, fluid-transport operations are by nature unsteady processes. Even though most flow fluctuations are innocuous, many are worrisome, and a few are destructive. Discussed in a previous section, cavitation vibration is only one of many physical phenomena that can cause problems. A few others can create significant effects.

Vortex shedding—Movement of fluid past a blunt object (e.g., water past a thermowell) causes alternating vortices to be shed from each side of the object, generating a periodic force normal to the object. If the flow frequency—often predictable from the consistency of the dimensionless Strouhal number—closely approaches the natural mechanical frequency of the object, it can generate vibration, at a destructive amplitude [*5*]. Heat-exchanger tubes in cross-flow represent the most common examples of vortex shedding. But other structures are equally susceptible to such vibration, among them spargers and sparger supports, thermowells, distillation columns, and stacks.

Siphon flashing—Lifting a column of water more than 34 feet causes it to boil or flash. Not so well known is the fact that this same phenomenon frequently occurs in water piping and in process lines, and that it can lead to inoperable piping.

For example, assume that only one butterfly valve is installed in the Fig. 6 piping, and that it is located at point A. Upon startup, the valve opens, filling the piping and heat exchanger. And if the flowrate exceeds $Q = (7.981/0.92)^{2.5} = 222$ gpm, the downleg will also fill. At this siphon flowrate, and at higher rates, trouble develops. With 1,000 gpm, the total frictional pressure drop to the top of the outlet piping increases to approximately 5.8 psi plus the pressure drop across the valve. At a height of $Z = 14.7 \times 144/61.8 = 35$ ft above the cooling water return, the pressure is an estimated 0 psia. At 40°C, water boils at 1.1 psia, so piping at an elevation of >32 ft must create a two-phase mixture of liquid water and water vapor.

The water vapor collapses violently in the downleg and shakes the piping and associated support. This siphon-flashing can also leave deposits on the internal or external surfaces of heat-exchanger tubes, as illustrated in Fig. 7. The vacuum also interferes with sampling and with heat-exchanger venting. As a rule, heat exchanger outlet pressure should exceed the water-vapor pressure at the maximum tubewall temperature; better yet, the heat exchanger-outlet pressure referred to previously should be above atmospheric.

Condensing on cold surfaces—A subcooled liquid is being fed into a distillation column through a closed pipe sparger. The external piping is noisy and shakes perceptibly. While not commonplace, such events and subtle variations on them do cause equipment failures. The most likely explanation for this type of problem: the piping does not fill with liquid, so saturated vapors are pulled into it where they suddenly collapse due to an extremely low resistance to heat transfer. Water hammer and destructive forces ensue. To stop the water hammer, increase the exit velocity so that the piping fills completely with liquid.

Two-phase piping—Two-phase flow regimes encountered in industry tend to be somewhat pulsative. Hence, make the piping supports more substantial than those for single-phase flow. Also, avoid long, unsupported spans.

From a vibration standpoint, slug flow behaves worse than the other flow regimes. It can move large-diameter pipe several inches [*40*]. For a simple cure, use smaller piping—this avoids slug flow altogether.

Hydraulic transients—Rapidly closing a valve at the downstream end of a liquid-filled line will increase the pressure on the valve by some 30 to 50 psi for every ft/s change in the velocity. To reduce pressure buildup, close the valve slowly relative to the time for waves to move through the piping. Dependent on fluid properties and elasticity of the piping, wave speed of a liquid can vary from roughly 2,000 to 5,000 ft/s. Small quantities of free gas will lower the wavespeed to hundreds, or even to tens, of feet per second.

If a valve is closed in a time less than twice the

wave-transient time ($t < 2L/a$), the pressure rise at the valve increases to

$$P - P_o = -\rho a \Delta V,$$

where a is the wave speed or sonic velocity and P_o is the initial pressure at the upstream end of the pipe.

Although overpressure will be reduced if a larger pipe is used or a surge tank is installed, the most satisfactory solution may be to slow the valve closing. Several methods [*35,44*] can help for estimating the required closing time. The techniques range from simple charts to method-of-characteristics computer programs.

Regardless of how it occurs, any sudden change in velocity will cause a corresponding $-\rho a \Delta V$ change in pressure. In addition to the classic valve-closure problem, water hammer also may occur in many other situations, such as the startup of a pump in a long, closed pipeline.

A frozen check valve installed at the discharge of the same pump will change the velocity very suddenly if a power-failure-induced reverse flow frees the valve. And if the pump were reciprocating, there would be water hammer associated with the time-varying pump delivery rate.

References

1. Abramowitz, M. and Stegun, I. A., "Handbook of Mathematical Functions," National Bureau of Standards, Washington, D.C., 1964.
2. "American National Standard of Welded and Seamless Wrought Steel Pipe," (ANSI B37.10-1975), ASME, New York, 1975.
3. Baker, O., *Oil Gas J.*, pp. 156–167, Nov. 10, 1958.
4. Benenati, R. F., *Chem. Eng.*, pp. 201–206, Feb. 28, 1977.
5. Blevins, R. D., "Flow Induced Vibration," Van Nostrand Reinhold, New York, 1977.
6. "Cameron Hydraulic Data," Ingersoll-Rand Co., 1958.
7. Chisolm, D., NEL Report No. 535, National Engineering Laboratory, E. Kilbride, Glasgow, 1973.
8. Choe, W. G., "Flow Patterns and Pressure Drop in Horizontal Two-Phase Flow," PhD dissertation, University of Cincinnati, 1975.
9. Chow, V. T., "Open Channel Hydraulics," McGraw-Hill, New York, 1959.
10. Churchill, S. W., *AIChE J.*, Vol. 19, No. 2, p. 375, 1973.
11. Churchill, S. W., *Chem. Eng.*, pp. 91–92, Nov. 7, 1977.
12. Crane, "Flow of Fluids—Technical Paper No. 410," Crane Co., New York, 1976.
13. Croley, T. E., "Hydrologic and Hydraulic Computations on Small Programmable Calculators," Iowa Institute of Hydraulic Research, Iowa City, Iowa, 1977.
14. Daniels, C. M., and Cleveland, R., Jr., *Mach. Des.*, Nov. 25, 1965.
15. Dodge, D. W. and Metzner, A. B., *AIChE J.*, Vol. 5, No. 2, p. 189, 1959.
16. Dukler, A. E., Wicks, M., and Cleveland, R. G., *AIChE J.*, Vol. 10, No. 1, p. 38, 1964.
17. Dukler, A. E., Wicks, M., and Cleveland, R. G., *AIChE J.*, Vol. 10, No. 1, p. 44, 1964.
18. Fontana, M. G. and Greene, N. D., "Corrosion Engineering," McGraw-Hill, New York, 1967.
19. Gibson, A. H., "Hydraulics and Its Applications," 5th ed., Constable, London, 1952.
20. Gouse, S. W., "An Index to Two-Phase Gas-Liquid Flow Literature," M.I.T. Press, Cambridge, Mass., 1966.
21. Govier, G. W. and Aziz, K., "Flow of Complex Mixtures in Pipes," Van Nostrand Reinhold, New York, 1972.
22. Hewitt, G. F. and Hall-Taylor, N. S., "Annular Two-Phase Flow," Pergamon Press, Oxford, England, 1970.
23. Hughmark, G. A., *Chem. Eng. Progr.* pp. 62–65, April, 1962.
24. Hydraulic Institute "Pipe Friction Manual," 3rd Ed., Hydraulic Institute, New York, 1961.
25. "Hydraulic Institute Standards," 12th ed., Hydraulic Institute, New York, 1969.
26. Kaye, S. E. and Rosen, S. L., *AIChE J.*, Vol. 17, No. 157, p. 1269, 1976.
27. Kays, W. M., 1949 Heat Transfer and Fluid Dynamics Institute, June 22–24, 1949.
28. Kalinske, M., "Hydraulics of Vertical Drain and Overflow Pipes," Bulletin 26, University of Iowa Studies, 1939–1940.
29. Kutateladze, S. S., and Borishanskii, V. M., "A Concise Encyclopedia of Heat Transfer," Pergamon Press, New York, 1966. (Translated from Russian.)
30. Lockhart, R. W., and Martinelli, R. C., *Chem. Eng. Prog.*, pp. 39–48, Jan. 1949.

30a. Metzner, A. B. and Reed, J. C., *AIChE J.*, Vol. 1, No. 4, p. 434, 1955.

31. McDuffie, W. G., *AIChE J.*, Vol. 23, No. 1, p. 37, 1977.
32. Miller, D. S., "Internal Flow," British Hydromechanics Research Association, Cranfield, England, 1971.
33. Moore, M. J., and Sieverding, C. H., "Two Phase Steam Flow in Turbines and Separators," McGraw-Hill, New York, 1976.
34. Moody, L. F., Trans. ASME, Vol. 66, pp. 671–684, 1944.
35. Parmakian, J., "Water Hammer Analysis," Dover, New York, 1963.
36. Perry, R. H. and Chilton, C. H. "Chemical Engineer's Handbook," 5th ed., McGraw-Hill, New York, 1973.
37. Rouse, H., and Ince, S., "History of Hydraulics," Dover, New York, 1963.
38. Seebold, J. G., *Hydrocarbon Proc.*, pp. 189–191, Sept., 1973.
39. Schlichting, H., "Boundary-Layer Theory," 6th ed., McGraw-Hill, New York, 1968.
40. Simpson, L. L., *Chem. Eng.*, pp. 192–213, June 17, 1968.
41. Simpson, L. L., *Chem. Eng.*, pp. 167–181, Apr. 14, 1969.
42. Simpson, L. L., and Weirick, M. L., private communication, 1974.
43. Skelland, A. H. P., "Non-Newtonian Flow and Heat Transfer," Wiley, New York, 1967.
44. Streeter and Wylie, "Hydraulic Transients," McGraw-Hill, New York, 1967.
45. Taylor, I., "Two-Pump Applications Need Extra NPSH Available," presented at 1972 Fluids Engineering Conference, Polyphase Flow Forum, ASME, New York, 1972.
46. Users' Library Solutions—"Thermal and Transport Sciences," Hewlett Packard, 1977.
47. Vazonyi, A., Trans. ASME, Vol. 66, 177–183, 1944.
48. Wallis, G. B., "One-Dimensional Two-Phase Flow," McGraw-Hill, New York, 1969.
49. Weisman, J. and Choe, W. G., "Methods for Calculation of Pressure Drop in Cocurrent Gas-Liquid Flow," Presented at the Two-Phase Flow and Heat Transfer Symposium, Ft. Lauderdale, Florida 1976.
50. Weisman, J., Husain, A., and Harshe, B., "Two-Phase Pressure Drop Across Abrupt Area Changes and Restrictions, Presented at the Two-Phase Flow and Heat Transfer Symposium, Ft. Lauderdale, Florida, 1976.
51. Wohl, M. H., "Designing for Non-Newtonian Fluids Refresher Series," *Chem. Eng.*, Jan. 15, 1968–Apr. 26, 1968.

The authors

Larry L. Simpson is manager of heat transfer and fluid dynamics in the Engineering Department at Union Carbide's Technical Center, Box 8364, South Charleston WV 25303. His past work includes plant design and startup, and for the past eleven years, fluid dynamics specialization. He has written several articles on different aspects of this subject, including a chapter in the book "Turbulence in Mixing Operations." He has a B.S.Ch.E. from Purdue University and an M.S.Ch.E. from West Virginia University. He is a registered professional engineer in West Virginia and is active in AIChE.

Martin L. Weirick is an engineer in the Engineering Department at Union Carbide's Technical Center, South Charleston, West Virginia, where he specializes in fluid dynamics. He has a B.S. degree in chemical engineering from Virginia Polytechnic Institute and State University and is an associate member of the AIChE.

Preliminary pipeline sizing

Here is a compilation of formulas for determining pipeline internal diameter from typical or limiting fluid velocity, or from allowable pressure drop per hundred feet.

George R. Kent, Consultant

☐ When designing a process plant or unit, the designer needs a basis for sizing piping. Study layouts, which depend on reasonable pipe sizes, must be drawn before pipeline lengths are known. Therefore, many engineers have adopted rules that are independent of length, based on typical and limiting fluid velocities or allowable pressure drop/100 ft.

The formulas listed in Fig. 1 efficiently and directly size piping, by determining pipe internal diameter. In my experience, 80% of lines sized using these formulas did not require change after layouts were completed and more precise checks were made.

Typical velocity

Robert Kern [1] has presented tables listing typical liquid and gas velocities for different pipe sizes and fluid characteristics. These tables can be reduced to the following equations:

$$\text{for liquids } u = 5.6\,D^{0.304}\ \text{ft/s} \qquad (1)$$

$$\text{for gases } u = 43.6\,D^{0.45}/\rho^{0.16}\ \text{ft/s} \qquad (2)$$

For suction, drains or vents, Eq. (1) and (2) should be halved. Inserting these equations into the functional relationship between pipe internal diameter and fluid velocity,

$$D = 7.136(W/\rho u)^{1/2} \qquad (3)$$

yields a diameter corresponding to the typical velocity. The results of this operation are listed in Fig. 1.

Limiting velocity

The formulas for determining minimum internal pipe diameter for liquids have been developed from a table listing maximum velocities presented by Kern [1]. This table can be reduced to the following equation:

$$\text{for clear liquids } u = 48/\rho^{1/3} \qquad (4)$$

For erosive or corrosive liquids Eq. (4) is halved. Substituting into Eq. (3) gives the formulas shown in Fig. 1.

For gases, the maximum fluid flow attainable in a pipe is given by the sonic velocity:

$$u_s = (kgZRT)^{1/2} \qquad (5)$$

Substituting g = 32.2 and $R = 1{,}545/m$, yields:

$$u_s = 223(kZT/m)^{1/2} \qquad (6)$$

Liquids / **Gases**

Typical dia.

Liquids	Gases
$D = 2.607\left(\frac{W}{\rho}\right)^{0.434}$	$D = 1.065\,\frac{W^{0.408}}{\rho^{0.343}}$
$D = 3.522\left(\frac{W}{\rho}\right)^{0.434}$ ‡	$D = 1.414\,\frac{W^{0.408}}{\rho^{0.343}}$ ‡

Minimum dia.

Liquids	Gases
$D = 1.03\,\frac{W^{1/2}}{\rho^{1/3}}$ *	$D = 0.585\left(\frac{W}{\rho}\right)^{1/2}\left(\frac{m}{kZT}\right)^{1/4}$ *
$D = 1.457\,\frac{W^{1/2}}{\rho^{1/3}}$ **	$D = 0.827\left(\frac{W}{\rho}\right)^{1/2}\left(\frac{m}{kZT}\right)^{1/4}$ **

‡ Suction, drains or vents
* Clean fluid
** Erosive or corrosive fluid

Clean steel pipe in turbulent zone

$$f = 0.0185\left(\frac{\mu}{D\rho u}\right)^{0.16}$$

For known D

$$\frac{\Delta P}{100} = 13.29\,\frac{\mu^{0.16}W^{1.84}}{\rho D^{4.84}}$$

For known $\Delta P/100$

$$D = 1.706\left[\frac{\mu^{0.16}W^{1.84}}{\rho(\Delta P/100)}\right]^{0.207}$$

Economic pipe size in turbulent zone

$$D = 0.627\,\frac{W^{0.486}}{\rho^{0.342}}\,\mu^{0.0274}(C/K)^{0.171}$$

Useful formulas

$\rho = 62.4\ (\text{Sp gr})$	$\rho = \frac{Pm}{10.73\,ZT}$
$q = 2W/\text{Sp gr}$	$Q = 16.67\,\frac{W}{\rho}$

$$u = 51\,\frac{W}{\rho D^2}$$

$$D = 7.136\left(\frac{W}{\rho u}\right)^{1/2}$$

Formulas for sizing pipeline internal diameter **Fig. 1**

For turbulent flow, the average velocity can be approximated by two thirds of the maximum. This yields a limiting average clean-gas velocity of:

$$u = 148.7(kZT/m)^{1/2} \qquad (7)$$

For erosive or corrosive flow, Eq. (7) is halved. Inserting into Eq. (3) gives the formulas for minimum diameter as shown in Fig. 1.

Pressure drop

When allowable pressure drop is used as the basis for preliminary pipe sizing, we are confronted with determining a simple solution for friction factor. An excellent

Originally published September 25, 1978

Nomenclature

C	Equipment cost, \$/brake hp
D	Pipe internal dia., in.
g	Gravitational acceleration, 32.174 ft-lb mass/lb force-s^2
k	Specific heat ratio
K	Piping cost, \$/in./ft
m	Molecular weight
P	System pressure, psia
ΔP	Pressure drop, psi
q	Flowrate, gpm
Q	Flowrate, ft^3/min
R	Gas constant, 1,545 ft-lb force/(lb mole) °R
R_e	Reynolds number
Sp gr	Specific gravity
T	Rankine temperature, 460 + °F
u	Velocity, ft/s
W	Flowrate, 1000 lb/h
Z	Compressibility factor
μ	Viscosity, cP
ρ	Density, lb/ft^3

Subscripts

s	sonic
v	vapor

review is presented by R. P. Benedict [*2*]. S. W. Churchill [*3*] has developed a direct solution for friction factor that is excellent for precise checking of pipe sizes, but is too complex for preliminary sizing. A study of the Moody diagram in the turbulent region shows that friction factor for clean steel pipe may be represented by:

$$f = 0.0185(\mu/Du\rho)^{0.16} \tag{8}$$

Substituting this expression into the relationship between pressure drop and diameter presented by L. L. Simpson [*4*]:

$$\Delta P/100 = 0.518 f\rho u^2/D \tag{9}$$

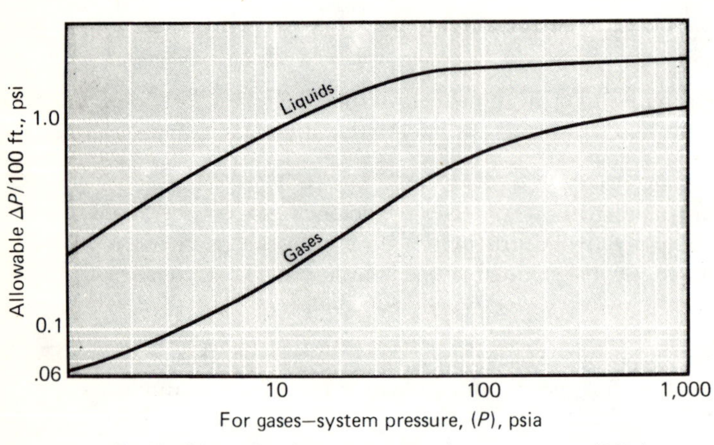

For $P > 1{,}000$, $\frac{\Delta P}{100} = 0.49P^{0.12}$; For $\frac{P}{P_V} > 1{,}000$, $\frac{\Delta P}{100} = 1.5\left(\frac{P}{P_V}\right)^{0.042}$

Allowable pressure drop for liquids and gases **Fig. 2**

yields an easy formula for preliminary sizing of pipe based on $\Delta P/100$ ft:

$$D = 1.706\left[\frac{\mu^{0.16}W^{1.84}}{\rho(\Delta P/100)}\right]^{0.207} \tag{10}$$

The reality of turbulent flow can be determined by the Reynolds number relationship:

$$R_e = 124Du\rho/\mu$$

A plot of allowable $\Delta P/100$ ft used for a number of plant designs, vs. system pressure for gases and the ratio of system pressure to vapor pressure for liquids, is found in Fig. 2. For suction, drains or vents, I suggest that 20% of the allowable pressure drop be used for either fluid.

Economic pipe size

The many detailed papers on determining economic pipe size boil down to higher pipe cost with increase in size, and lower pressurizing cost with decrease in system pressure. The summation of the two resulting curves points out the condition for minimum total cost.

Having made a generalized economic study of pressurizing versus piping, I found that pipe size reduced to:

$$D = \frac{1.37\, W^{1/2}}{\rho^{1/3}}(fC/K)^{1/6} \tag{11}$$

Using the friction factor from Eq. (7) and the general relationship of velocity, flowrate and internal pipe diameter, Eq. (3), the economic pipe size can be expressed by:

$$D = 0.627\frac{W^{0.486}}{\rho^{0.342}}\mu^{0.0274}\left(\frac{C}{K}\right)^{0.171} \tag{12}$$

Control valves

Where flow-control valves form part of a system, it is recommended that the pressure drop across the valve be not less than 30% of the total system pressure drop. This results in a valve pressure drop equal to 43% of the system drop, excluding the valve. To assure that the pipe in the control system is suitably sized, it should be capable of handling a maximum flowrate that is 21% above the design flowrate.

References

1. Kern, Robert, Useful Properties of Fluids, *Chem. Eng.*, Dec. 23, 1974, p. 58.
2. Benedict, Robert P., Friction in Pipe Flow, *Instrum. Control Syst.*, Dec. 1969.
3. Churchill, Stuart W., Friction Factor Equation Spans all Fluid Flow Regimes, *Chem. Eng.*, Nov. 7, 1977, p. 91.
4. Simpson, Larry L., Sizing Piping for Process Plants, *Chem. Eng.*, June 17, 1968, p. 192.

The author

George R. Kent, 44 Stearns St., Newton Centre, MA 02159, is a consultant who has spent 16 years in structural and civil engineering, 21 years in stress analysis and petrochemical engineering, and the last 11 years in the mechanical end of electric power generation. He has been the lead project engineer on five petrochemical projects, and project manager on six power-plant projects. He holds a B.S.C.E. from The Cooper Union, New York, and is a registered professional engineer in New York and Massachusetts.

How To Compute Pipe Size

Developing the basic relations for fluid flow leads to practical formulas for finding pipe size. Here are useful techniques for liquids or gases in pipelines under steady-flow conditions.

ROBERT KERN, Hoffman — La Roche Inc.*

Methods for sizing piping and piping components rely on units of measurement that are customarily found in manufacturers' literature and in handbooks.

In order to use these data, we must first understand the basic relations of fluid flow in piping systems. As we develop these relations, we will also apply them to practical formulas for sizing the components of such systems when handling liquids and vapors.

Euler's Derivation

Fig. 1a shows a reducing elbow in which an ideal fluid flows. (Ideal fluids have no internal or external resistance.) Let us investigate the forces acting on a partial mass, *dm*, of this ideal fluid, using the consistent units of English measurement. The pressure in the elbow is *p*.

The partial mass of fluid is enlarged in Fig. 1b to show its dimensions and the forces acting on it. The length of the partial volume is *dl*, and the area at both ends is *dA*.

The forces acting on this partial mass are F_1, F_2 and F_3. Parallel to its direction of motion, $F_1 = p\,dA$; and as this mass flows, one surface meets a somewhat higher pressure, $(p + dp)$, and $F_2 = (p + dp)\,dA$. The weight component, acting against the direction of flow, $F_3 = W \sin \alpha = \rho\, dl\, dA \sin \alpha$. (In an actual fluid, there is a force, F_s, which is the fluid's resistance acting against the direction of flow. For an ideal fluid, $F_s = 0$.)

Perpendicularly to the *X*-axis, the forces due to pressure, which act on the surface of the partial mass, cancel each other, i.e., $\Sigma F_y = 0$. The weight component of *dm* (i.e., $W \cos \alpha$), perpendicular to the direction of flow, is balanced with an equal reaction force on the pipe wall. If forces acting in the direction of flow are positive, and those acting against the flow are negative:

$$\Sigma F = F_1 - F_2 - F_3 = p\,dA - (p + dp)dA - \rho\,dA\,dl \sin \alpha \quad (1)$$

The difference in height between the end-points of

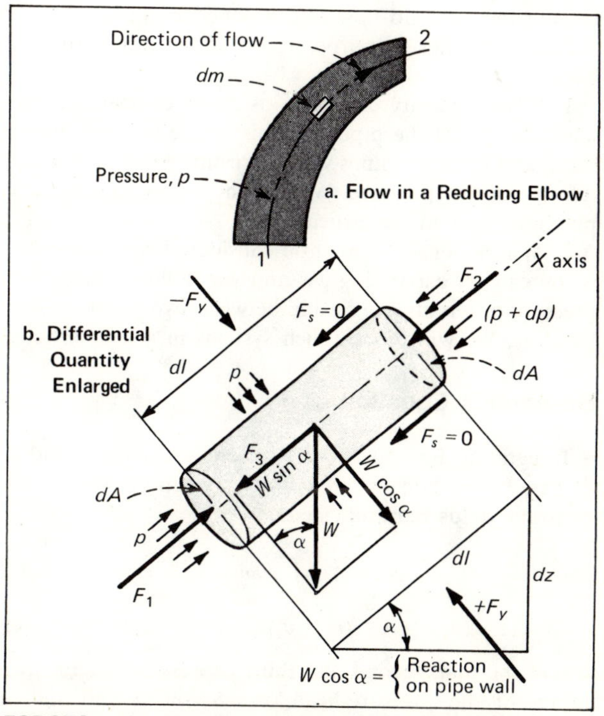

FORCES acting on differential mass in fluid—Fig. 1

*To meet your author, see p. 44.

Originally published January 6, 1975

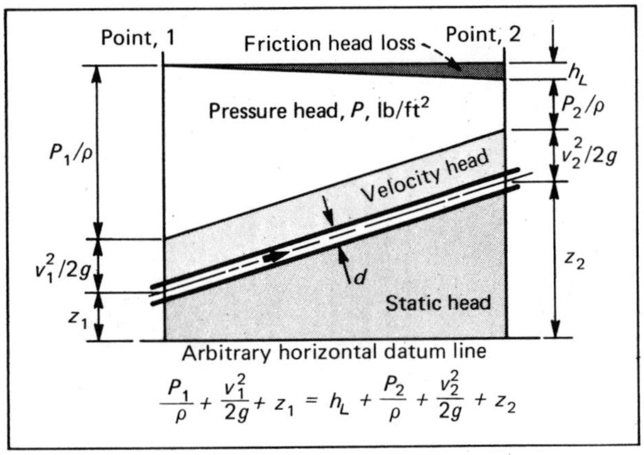

DISTRIBUTION of fluid energy in a pipeline—Fig. 2

length, dl, is $dz = dl \sin \alpha$, and Eq. (1) reduces to:

$$\Sigma F = -dp\, dA - \rho dA\, dz \qquad (2)$$

The resultant force acting on a body in the direction of motion is proportional to mass times acceleration: $\Sigma F = adm$; mass equals weight divided by the gravitational constant: $dm = W/g$; and acceleration is the velocity difference, dv, over a time span, dt: $a = dv/dt$. Consequently, $\Sigma F = (W/g)(dv/dt)$. Since the weight of fluid is its volume multiplied by density:

$$\Sigma F = dA\, dl(\rho/g)(dv/dt) \qquad (3)$$

Setting Eq. (2) and (3) equal to each other yields:

$$-dp\, dA - \rho\, dA\, dz = (\rho/g)dA\, dv(dl/dt) \qquad (4)$$

where dl/dt is velocity. After multiplying both sides of Eq. (4) by $-1/\rho\, dA$, we get:

$$(1/\rho)\, dp + (1/g)\, v\, dv + dz = 0 \qquad (5)$$

Eq. (5) is the general and fundamental differential relationship of fluid flow as developed by Euler.

For piping-design purposes, Eq. (5) is solved in two basic ways:

1. Where density can be considered constant in the whole length of the pipeline. This is true with all liquid lines, and those gas lines where pressure losses are small relative to line pressure. We will show how to size such pipelines later in this article.

2. Where density cannot be considered constant. This is the case of expanding gas and vapor flow, where the pressure differential is sizable between two points of the pipeline. We will discuss such systems in a later article.

Bernoulli's Equation

Integrating Eq. (5) between reference points 1 and 2 of Fig. 1a for fluids where the density can be considered constant yields Bernoulli's equation:

$$(1/\rho)\int_1^2 dp + (1/g)\int_1^2 v\, dv + \int_1^2 dz = 0$$

$$(1/\rho)(p_2 - p_1) + (v_2^2 - v_1^2)/2g + (z_2 - z_1) = 0 \qquad (6)$$

where the first, second and third expressions of Eq. (6) represent the pressure-head, velocity-head and static-head differences, respectively. Eq. (6) is used for investigating energy distributions or calculating pressure differentials between any two points in a pipeline.

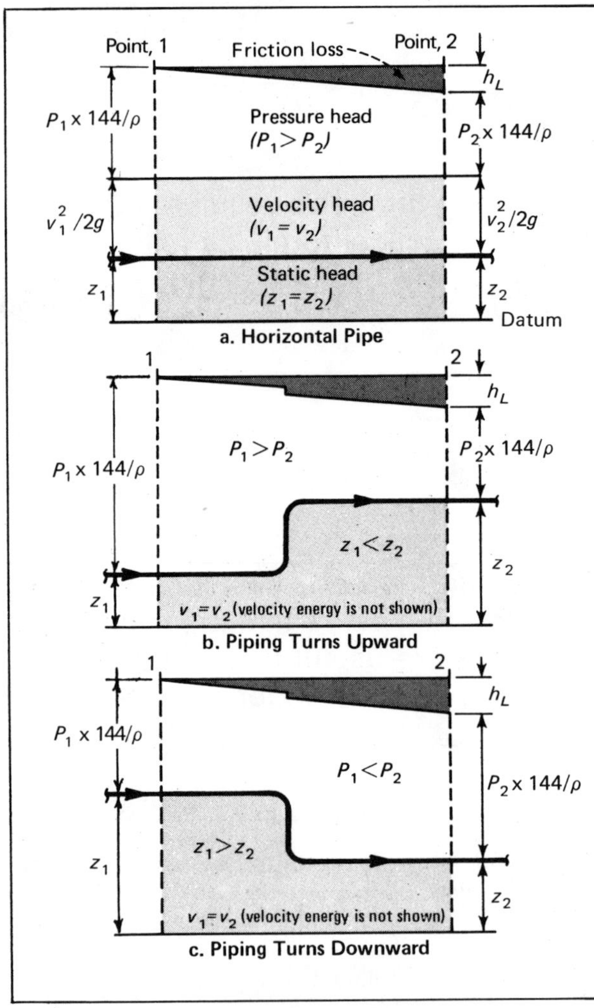

PIPE elevation affects energy distribution—Fig. 3

Fig. 2 graphically illustrates Bernoulli's energy distribution in a slanted pipeline, with two additional factors included: (1) there is a head loss, h_L, due to friction because we are now dealing with an actual fluid, and (2) in a pipeline with constant diameter, velocity usually does not change, i.e., $v_1 = v_2$. The first factor changes the righthand side of Eq. (6), and the second eliminates the component for the velocity-head difference.* Eq. (6) now becomes:

$$(1/\rho)(p_2 - p_1) + (z_2 - z_1) = h_L \qquad (7)$$

What is the significance of Eq. (7) from a piping designer's standpoint? Head loss, h_L, is consumed at the expense of pressure-head or static-head difference. The static-head difference can be positive or negative. For a negative static-head difference, the pressure-head difference must be greater than $h_L + (z_2 - z_1)$.

In practical design work, Eq. (7) can rarely be fulfilled as it stands, for two reasons:

1. To obtain a reasonable movement of fluid in a pipeline, the lefthand side of the equation must be greater than the resistance to flow.

* Do not use up velocity head for pipe resistance.

2. Commercial pipes are manufactured in increments of size (i.e., pipe diameter). Consequently, the calculated resistances for each size are also in increments for the same flow conditions. So, from a practical standpoint, Eq. (7) becomes:

$$(1/\rho)(p_2 - p_1) + (z_2 - z_1) > h_L \tag{8}$$

In design work, we calculate separately the lefthand and righthand side of Eq. (8). The sum of the left side must be positive and greater than the total resistance of a pipeline.

Resistance should be calculated for all alternative flow conditions. The flow direction is always from a higher pressure point to a lower one. If piping turns upward in the direction of flow, the static head reduces the pressure: $z_2 > z_1$. If piping turns downward in the direction of flow, the static head adds pressure to the fluid: $z_1 > z_2$. This is illustrated in Fig. 3. Pressure due to the static head of a gas is usually small and can be neglected in calculations.

The excess energy not used up for pipe resistance provides a safety margin and is finally lost at the end of a pipeline. In general:

$$\Delta P_a = \Delta P_p + \Delta P_x$$

where ΔP_a is the available pressure drop, ΔP_p is the pressure drop due to resistances of the pipe system, and ΔP_x is the excess pressure drop.

Flow of fluids through equipment and components in a pipe system also introduces additional resistances that have to be overcome by ΔP_a. To account for these resistances, the overall ΔP distribution can be:

$$\Delta P_a = \Delta P_r + \Delta P_c + \Delta P_q + \Delta P_x$$

where ΔP_r is pressure drop due to pipe resistance, ΔP_c is pressure drop due to resistances in pipe components, and ΔP_q is pressure drop due to resistances in equipment.

Sizing Liquid Lines

Let us apply Eq. (6) to an example: A large container is filled with liquid. Near the bottom of the container, an orifice is drilled or a piece of pipe is attached, as shown in Fig. 4. Velocity at the bottom opening is v_2. Because of the large difference in area between the bottom opening and the container's cross-section, the liquid level moves down very slowly, and its velocity, v_1, can be considered approximately zero. Pressure is atmospheric at the liquid surface and at the bottom outlet and, consequently, $P_2 - P_1 = 0$. For convenience, we will represent the elevation difference between the liquid surface and the bottom opening as $h_L = z_1 - z_2$. Taking these factors into account, the Bernoulli relation, Eq. (6), reduces to:

$$h_L = v_2^2/2g \tag{9}$$

When in motion, all actual fluids have internal resistance due to viscosity and friction between the fluid and the pipe wall and equipment parts. To take this into account, we will correct Eq. (9) by introducing a resistance coefficient, K, and letting $v = v_2$:

$$h_L = Kv^2/2g \tag{10}$$

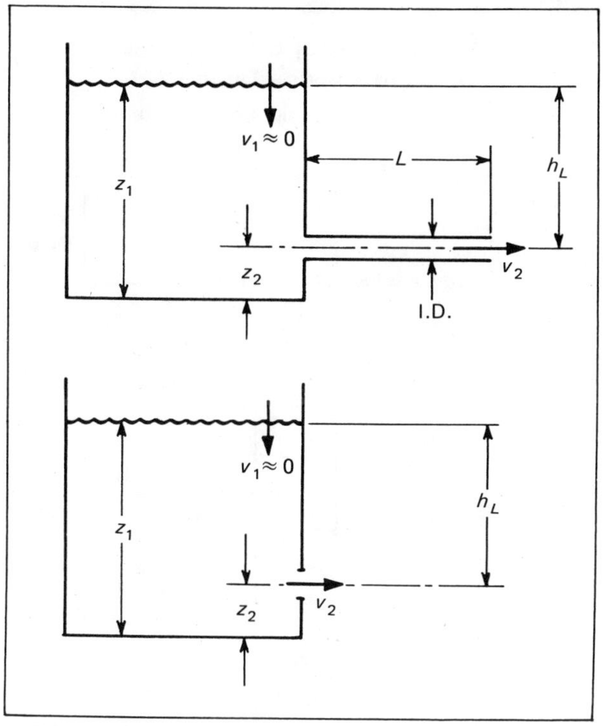

FLOW of liquid through pipe or orifice—Fig. 4

Eq. (10) is the Darcy equation and is the basic equation used for sizing piping and components. The velocity expression in Eq. (10) is: $v = C(2gh)^{1/2}$, where $C = (1/K)^{1/2}$.

Flow coefficients for piping and pipe components are obtained from experimental data. For restrictions such as orifices, the constant C is used; for control valves, the usual constant is C_v. When sizing pipes, the flow coefficient, K, is proportional to the friction factor, f, and pipe length, L, and is inversely proportional to the pipe diameter, D:

$$K = fL/D \tag{11}$$

Friction factor is a function of the Reynolds number and the relative roughness of the pipe wall.

Practical Formulas

We will now show how to derive sizing formulas in the convenient units used by designers, manufacturers and handbooks, by converting the previous relations from the consistent units of the English system of measurement.

We will begin by converting Eq. (10) to pressure drop, ΔP, in psi:

$$\Delta P = (h_L\rho)/144 \tag{12}$$

To get pressure drop, ΔP, in psi for pipelines, we combine Eq. (10) and (11):

$$\Delta P = (fL/D)(v^2/2g)(\rho/144) \tag{13}$$

If we want to express pipe diameter in inches, and

Resistances of Elbows, Tees and Bends—Table I

(Resistance in equivalent pipe length, ft)

Nominal Pipe Size, In	90° Elbows* Short Radius, $R = 1D$	90° Elbows* Long Radius $R = 1.5D$	90° Bends* $R = 5D$	90° Bends* $R = 10D$	Tee Flow-Through Branch	Tee Flow-Through
1½	4.5	3	2.5	4	8	3
2	5.25	3.5	3	5	11	3.5
2½	6	4	3.5	6	13	4
3	7.5	5	4	7.5	16	5
4	10.5	7	5.5	10	20	7
6	15	10	8.5	15	30	10
8	21	14	11	20	40	14
10	24	16	14	25	50	16
12	32	21	16	30	60	21
14	33	22	19	33	65	22
16	39	26	21	38	75	26
18	44	29	24	42	86	29
20	48	32	27	50	100	32
24	57	38	32	60	120	38

* For 45° elbows and bends, estimate 50% of tabulated values. For 180° returns, double the tabulated values.

velocity as a volumetric flowrate, Q, gpm, we substitute. $D = d/12$, and $v = 0.408(Q/d^2)$ into Eq. (13). These substitutions now yield:

$$\Delta P = 0.000216 f L \rho (Q^2/d^5) \qquad (14)$$

where ΔP is pressure drop, psi; f is friction factor; L is pipe length, ft; ρ is density lb/ft^3; Q is volumetric flowrate, gpm; and d is pipe diameter, in. Note that density and volumetric flowrate must be expressed at the flowing temperature.

If we want to express the pressure drop as the customary loss per 100 ft of pipe, we substitute $L = 100$ ft in Eq. (14), and get:

$$\Delta P_{100} = 0.0216 f \rho (Q^2/d^5), \text{ psi/100 ft} \qquad (15)$$

Eq. (15) can be expressed in terms of specific gravity, S, by substituting $\rho = 62.37S$:

$$\Delta P_{100} = 1.35 f S (Q^2/d^5), \text{ psi/100 ft} \qquad (16)$$

[Note that in establishing Eq. (16) the density of water, 62.37 lb/ft^3, is at 60°F. Hence, the specific gravity, S, in Eq. (16) must be the density of the liquid at flowing temperature compared to the density of water at 60°F.]

Eq. (15) and (16) are the most convenient for calculating unit losses in liquid lines. Values of d^5 are listed in manufacturers' catalogs.

Example 1—What is the pressure drop per 100 ft in a 6-in Schedule 40 (I.D. = 6.065 in, d^5 = 8,206 in^5) line for kerosene? Liquid conditions are: flowrate, Q_{60} = 900 gpm; density, ρ_{60} = 51 lb/ft^3; and temperature, t = 321°F.

Resistances to Flow for Various Types of Valves—Table II

(Resistance in equivalent pipe length, ft)

Nominal Pipe Size, In	Gate, Fully Open	Globe,* Fully Open, Bevel or Plug Seat 90°	Globe,* 60°	Globe,* 45°	Check Swing	Check Ball	Straight-Through Cock.[1]	Three-Way Cock[2] Straight-Through Flows	Three-Way Cock[2] Flow-Through Branch	Butterfly, Fully Open
1½	1.75	46	23	18	17	20	2.5	6	20	6
2	2.25	60	30	24	22	25	3.5	7.5	24	8
2½	2.75	70	38	30	27	30	4	9	30	10
3	3.5	90	45	38	35	38	5	12	36	12
4	4.5	120	60	48	45	50	6.5	15	48	15
6	6.5	175	88	72	65	75	10	22	70	23
8	9	230	120	95	90	100	13	30	95	27
10	12	280	150	130	120	130	16	38	120	35
12	14	320	170	145	140	150	19			40
14	15	380	190	160	150	170	20			45
16	17	420	220	180	170	190	22			50
18	18	480	250	205	180	210	24			58
20	20	530	290	240	200	240	27			64
24	32	630	330	270	250	290	33			78

* For partially closed globe valves, multiply tabulated values by 3 for three-quarters open, by 12 for one-half open, and by 70 for one-quarter open.
[1] With port area open. Port area = pipe area.
[2] Port area equals 80% of pipe area.

First, some data preparation is necessary, which is summarized as:

Specific gravity at 60°F: $S_{60} = 51/62.37 = 0.82$
Specific gravity at 321°F: $S_{321} = 0.72$ from Ref. 5a, on p. 93.
Density at 321°F: $\rho = 62.37(0.72) = 44.9$ lb/ft³
Expansion factor: $E = S_{60}/S = 0.82/0.72 = 1.14$
Flowrate at 321°F: $Q = Q_{60}/E = 900(1.14) = 1{,}026$ gpm
Viscosity: $\mu = 0.3$ cp from Ref. 5d, on p. 93.

We now calculate the Reynolds number at the flowing conditions:

$$N_{Re} = 50.6(\rho/\mu)(Q/d)$$

$$N_{Re} = 50.6(44.9/0.3)(1{,}026/6.065)$$

$$N_{Re} = 1.281 \times 10^6$$

The friction factor for this N_{Re} is found from Fig. 5 and 6 p. 92 of series, and hence, $f = 0.0154$ (Flow falls in the transitional turbulent zone.) Substituting the appropriate values into Eq. (16) yields:

$$\Delta P_{100} = 1.35(0.0154)(0.72)(1{,}026)^2/8{,}206$$

$$\Delta P_{100} = 1.92 \text{ psi}/100 \text{ ft}$$

Sizing Gas and Vapor Lines

To size gas or vapor lines, we must convert Eq. (14) and (15) to a weight flowrate by substituting $Q = 0.125(W/\rho)$ into them. This conversion yields:

$$\Delta P = 0.00000336L(f/\rho)(W^2/d^5) \qquad (17)$$

and for $L = 100$ ft, Eq. (17) becomes:

$$\Delta P_{100} = 0.000336(f/\rho)(W^2/d^5) \qquad (18)$$

where ΔP is pressure drop, psi, and ΔP_{100} is pressure drop, psi/100 ft.

Eq. (17) and (18) give acceptable results, providing that (1) the average gas density of flow is used: $\rho = (\rho_1 + \rho_2)/2$, where ρ_1 is the density at the beginning point and ρ_2 is the density at the terminating point in the pipe segment under investigation, and (2) $(P_1 - P_2) \leqq 0.4P_1$ because energy losses due to acceleration and density variations can be neglected up to this limit.

With longer pipelines where the total pressure drop is greater than 40% of the upstream pressure, computations are done by considering that the line is divided into segments, so that the pressure loss in each segment is less than 40% of the upstream pressure. Densities, of course, will be different in each segment.

If $P_1 - P_2 \leqq 0.1P_2$, average values of ρ need not be calculated. Either the downstream or upstream density can be used.

Example 2—What is the pressure drop per 100 ft in a 4-in Schedule 40 (I.D. = 4.026 in, $d^5 = 1{,}058$ in⁵) gas line for methane? Gas flowrate, $W = 10{,}750$ lb/h; molecular weight, $M = 16$; temperature, $t = 172$°F; pressure, $P = 127$ psig; and viscosity, $\mu = 0.0145$ cp.

To find the pressure drop, we must use Eq. (18); and in order to solve this equation, we must determine gas density at flowing conditions, and the friction factor.

For density, we place the appropriate values in the

Resistances of Eccentric and Concentric Reducers, And of Sudden Changes in Line Size—Table III

(Resistance in equivalent pipe length, ft)

Nominal Sizes, In		d_2 → d_1	d_1 → d_2
d_1	d_2		
¾	½	0.6	0.5
1	½	1.2	0.7
	¾	0.6	0.6
1½	¾	1.6	1.0
	1	1.2	0.9
2	1	2.2	1.3
	1½	1.3	1.3
3	1½	3.8	2.4
	2	2.7	2.3
4	2	5	3.2
	3	3	3
6	3	8	5
	4	4	4
8	4	12	7
	6	7	7
10	4	15	8
	6	14	9.5
	8	6	6
12	6	19	12
	8	14	12
	10	6.5	6.5
14	6	22	14
	8	22	14
	10	15	13
	12	6	6
16	8	27	17
	10	23	17
	12	15	15
	14	7	7
18	10	30	19
	12	23	19
	14	15	15
	16	4	4
20	12	30	23
	14	21	23
	16	13	13
	18	5	5
24	18	25	25
	20	12	12

Note: Add these equivalent lengths to the equivalent length of the smaller pipe and its components.

Resistances of Horizontal and Vertical Inlets and Outlets—Table IV

(Resistance in equivalent pipe length, ft)

Resistance Coefficient	$K = 1.0$	$K = 0.78$	$K = 0.5$	$K = 0.23$
Nominal Pipe Size, In	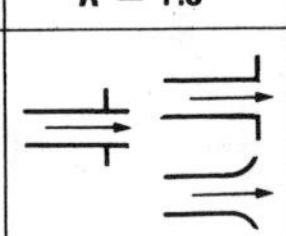	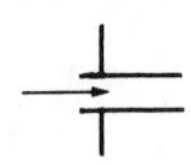	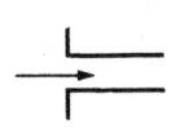	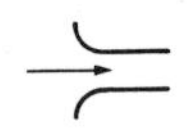
½	2	1.5	1	0.5
¾	3	2.5	1.5	0.75
1	4	3	2	1
1½	7	5.5	3.5	1.75
2	9	7	4.5	2.25
3	15	12	7.5	3.75
4	20	16	10	5
6	36	29	18	9
8	48	38	24	12
10	62	49	31	15
12	78	60	39	19
14	88	70	44	22
16	100	78	50	25
18	120	95	60	30
20	136	107	68	34
24	170	135	85	42

following relation, keeping in mind that pressure and temperature are in absolute units:

$$\rho = MP'/10.72Tz$$

$$\rho = \frac{16(127 + 14.7)}{10.72(460 + 172)1} = 0.334 \text{ lb/ft}^3$$

To establish the friction factor, f, we must determine the Reynolds number:

$$N_{Re} = 6.31W/d\mu$$

$$N_{Re} = (6.31)(10{,}750)/(4.026)(0.0145)$$

$$N_{Re} = 1.162 \times 10^6$$

For this value of the Reynolds number, we find that the friction factor is: $f = 0.0166$.

By substituting the appropriate values into Eq. (18), we obtain:

$$\Delta P_{100} = 0.000336(0.0166/0.334)[(10{,}750)^2/1{,}058]$$

$$\Delta P_{100} = 1.82 \text{ psi/100 ft}$$

Overall Pressure Loss

Most piping problems are solved by trial and error. A pipe size is selected, its resistance is calculated for a given set of flow conditions, and the overall pressure loss is computed. If the overall pressure loss is close to and less than the available pressure difference between two points in a pipeline, the selected size is accepted for the given flow conditions.

In pipeline calculations, it is convenient to obtain pressure loss per 100 ft of pipe, ΔP_{100}. Multiplying ΔP_{100} by the equivalent length of pipe and fittings (L, ft) between two points yields the overall pressure loss:

$$\Delta P = \Delta P_{100}(L/100), \text{ psi} \qquad (19)$$

The equivalent-pipe-length concept is the quickest and most convenient method for calculating overall pressure drop. Friction loss through a pipe component can be converted to an equivalent pipe length if the pressure drop between the piping ends is the same as between the two ends of the piping component. Sizes are assumed to be identical.

The equivalent length of fittings and valves, the exit length and the entrance length can be quickly obtained from Tables I to IV. These tables have been compiled from a nomograph published by Crane Co.

Example 3—What is the overall pressure loss between points 1 and 2 for a pipeline whose nominal size is 6 in, as sketched here?

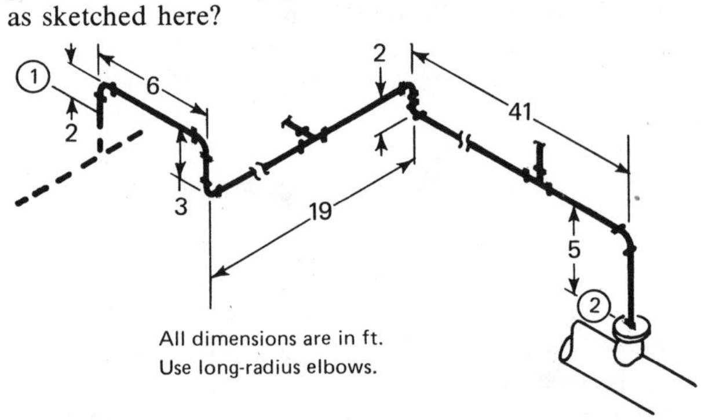

All dimensions are in ft.
Use long-radius elbows.

By the methods outlined previously, we have found that the pressure drop for the 100-foot length of the pipeline is $\Delta P_{100} = 1.92$ psi.

From Table I, we determine the equivalent pipe length for the 6-in long-radius elbow as 10 ft, and for flow through a 6-in tee as 10 ft. Hence, for this piping system, we get:

Actual length	78 ft
6 long-radius elbows:	$6 \times 10 = 60$ ft
2 flow-through tees:	$2 \times 10 = 20$ ft
Equivalent pipe length	158 ft

Using these data in Eq. (19), we find:

$$\Delta P = 1.92(158/100) = 3.04 \text{ psi}$$

The conversion between equivalent pipe length and the resistance coefficient, K, can be calculated from $K = fL/D$.

There is no shortcut to graphic piping design. Pipe configurations have to be drawn or sketched in considerable detail so that equivalent lengths of piping, fittings and other components contributing to the resistance of fluid flow can be accurately added.

How To Design Piping For Pump-Suction Conditions

Using the proper procedures to analyze process conditions, hydraulic relations and piping configurations will assure trouble-free installation and operation of process pumps.

ROBERT KERN, Hoffmann - La Roche Inc.*

Centrifugal pumps require vaporfree liquid flow in the suction line at the entrance to the impeller vanes to operate satisfactorily. Reciprocating pumps require such flow right after the inlet valve.

A centrifugal pump cannot pump vapor-liquid mixtures. A positive-displacement pump can pump liquids containing vapor but its volumetric efficiency will be greatly reduced—even if the weight percentage of vapor is a very small proportion of total flow.

The pressure gradient along the liquid path through a pump has characteristics similar to those of a control valve or orifice. Instead of a pressure loss, however, the downstream pressure will become much greater than the suction pressure, as sketched in Fig. 1. In both the pump-suction line and the pump, the minimum pressure should not drop below the vapor pressure of the liquid at the flowing temperature. System design and graphic piping design should satisfy this essential principle.

Net Positive Suction Head

If the pressure in a centrifugal pump falls below the vapor pressure of the liquid at the flowing temperature, bubbles form just before the inlet edge of the impeller vanes. Because of the rapid increase in pressure within the impeller, the bubbles then collapse. This phenomenon is called cavitation. It can reduce pump efficiency and cause noise, vibration, impeller wear and breakage.

To avoid cavitation, pump manufacturers require a positive suction pressure called net positive suction head, $NPSH_r$. $NPSH_r$ is the pressure above vapor pressure of the liquid being pumped, as measured at the pump-suction flange on a gage at the pump centerline. It is the pressure necessary to overcome friction and turbulence losses between the suction nozzle and the entrance to the impeller, and is usually given as head of water on pumps' head-capacity curves. $NPSH_r$ for a pump will increase with flowrate, and considerably increase with rotating speed of the impeller [1]:

$$(NPSH_r)_2 = (n_2/n_1)^2(NPSH_r)_1 \tag{1}$$

where $(NPSH_r)_2$ is the required net positive suction head if pump speed is increased to n_2, rpm. $(NPSH_r)_1$ and n_1

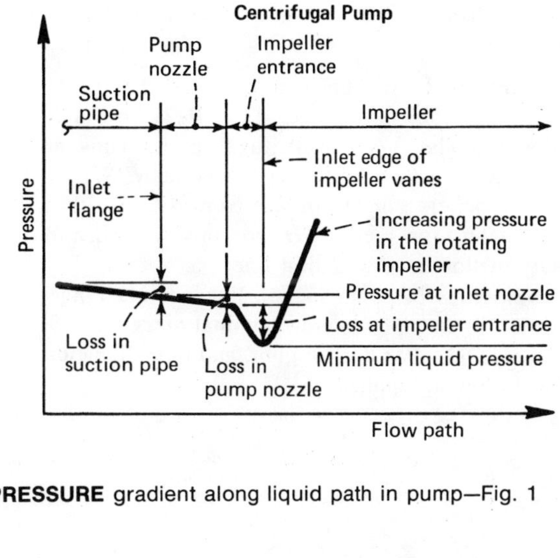

PRESSURE gradient along liquid path in pump—Fig. 1

*To meet your author, see p. 44.

Originally published April 28, 1975

are obtained from the pump's head-capacity curve. For flows below about 20 to 25% of rated capacity (near the maximum efficiency points), *NPSH* requirements can be higher than those for normal flow. This is especially true with high-capacity, low-head pumps.

Available net positive suction head, $NPSH_a$, is the potential energy of an installation, expressed in feet of liquid head, and can be calculated from:

$$(NPSH_a)(\rho)/144 = P \qquad (2)$$

where ρ = density of liquid at flowing temperature, lb/ft^3; and P = pressure above vapor pressure at the centerline of the suction flange, psi.

If we let $\rho = S\rho_w$, where ρ_w = density of water at 60°F, 62.37 lb/ft^3, and S = specific gravity of liquid at flowing temperature, Eq. (2) becomes:

$$NPSH_a = 2.31P/S \qquad (3)$$

Example Illustrates Design Procedures

A centrifugal pump having a 3-in suction inlet and a 2-in discharge will draw off 120 gpm of condensate water from an overhead suction drum, as sketched here:

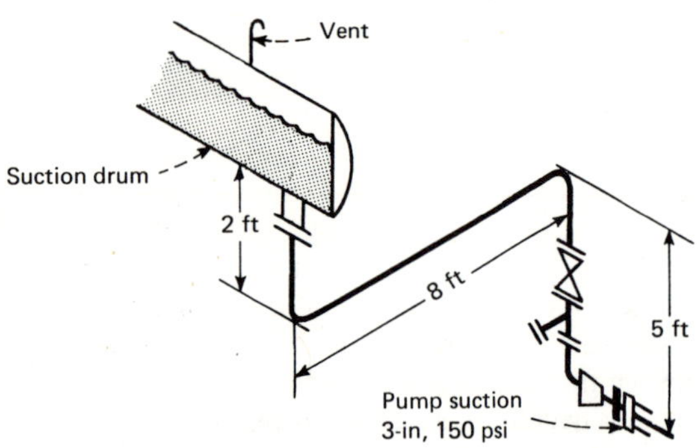

The suction line from the vessel is a 4-in Schedule 40 pipe, reduced to 3 in for the pump suction. The line contains three elbows, one strainer, one gate valve and one reducer. Flow data for condensate water are:

Flowrate	120 gpm
Density	60.67 lb/ft^3
Specific gravity	0.94
Suction drum pressure	14.7 psia
Temperature, subcooled	175°F
Temperature, saturated	212°F
Viscosity	0.35 cp

Additional information: The 4-in pipe has an inside diameter, d, of 4.026 in; hence, $d^2 = 16.21$ in^2, and $d^5 = 1{,}058$ in^5. The 3 × 2 centrifugal pump runs at 1,750 rpm, and at a flow of 120 gpm has an $NPSH_r$ = 5.5 ft, with impellers ranging in size from 4½ to 6½ in dia. Let us determine the $NPSH_a$ for this system, and the velocity of flow in the suction line.

Liquid velocities in pump-suction lines should range from 2 to 4 ft/s in pipes having nominal sizes from 3 to 10 in (see p. 89). For flow in gpm, and pipe diameter in in., the following applies:

$$v = 0.408(Q/d^2) \qquad (4)$$

$$v = 0.408(120/16.21) = 3.02 \text{ ft/s}$$

Therefore, the flow velocity in the 4-in suction line is well within the recommended values.

We determine the pressure loss in the 4-in suction line by first calculating the Reynolds number:

$$N_{Re} = 50.6(Q/d)(\rho/\mu) \qquad (5)$$

$$N_{Re} = 50.6(120/4.026)(60.67/0.35)$$

$$N_{Re} = 261{,}435$$

The friction factor, f, for this Reynolds number equals 0.018 for new, commercial steel pipe. Since the suction line in this problem is in an existing installation, we will increase the friction factor by 25% to allow for used pipe. Hence, $f = 0.018 \times 1.25 = 0.023$. (Friction factor from charts on p. 92.)

To determine the pressure drop as the unit loss per 100 ft, we will substitute into:

$$\Delta P_{100} = 0.0216 f\rho(Q^2/d^5) \qquad (6)$$

$$\Delta P_{100} = 0.0216(0.023)(60.67)(120)^2/1{,}058$$

$$\Delta P_{100} = 0.41 \text{ psi/100 ft}$$

We then find the equivalent length for the suction line and its fittings from data in Tables I to IV (pp. 22-24).

Pipe length	15 ft
Entrance loss to suction line	10 ft
3 Elbows: 3 × 10.5	31.5 ft
1 Strainer	40 ft
1 Gate valve	4.5 ft
1 Reducer	3 ft
Total equivalent length, L	104 ft

Therefore, the overall loss for the line and its fittings becomes:

$$\Delta P = \Delta P_{100}(L/100)$$

$$\Delta P = 0.41(104/100) = 0.43 \text{ psi}$$

We are now ready to perform the necessary computations and analyses for finding the available net positive suction head. These are detailed in Table I.

Meaning of Net Positive Suction Head

To ensure reliable pump operation, $NPSH_a$ should equal or exceed $NPSH_r$. Thus, process equipment must be elevated at least as high as the sum of the pump's $NPSH_r$, plus the friction-head losses in the suction line when saturated liquids are pumped.

At the suction nozzle, velocity head is also a positive energy component. However, this is not included in *NPSH* diagrams for pumps made in the U.S. Consequently, in calculations for $NPSH_a$, velocity head does not have to be considered.

According to the Hydraulic Institute [*2*], "The *NPSH* requirements of centrifugal pumps are normally determined on the basis of handling water. It is recognized that when pumping hydrocarbons, the *NPSH* to obtain satisfactory operation may be reduced for certain conditions." However, it is customary to use manufacturers' data for the minimum required *NPSH* when handling hydrocarbons having specific gravities less than 1. This means that (1) both available and required *NPSH* do not vary with the liquid's specific gravity, and (2) for a given

Calculations for Net Positive Suction Head—Table I

	Subcooled Liquid	Saturated Liquid
1. Pressure on liquid surface	14.7 psia	14.7 psia
2. Vapor pressure of liquid	−6.75 psia	−14.7 psia
3. Pressure available	7.95 psi*	0. psi*
4. Static head pressure (+) or suction lift (−), $(\text{ft} \times \rho)/144 = (7 \times 60.67)/144$	+2.95 psi*	+2.95 psi*
5. Line 3 + Line 4	10.9 psi*	2. psi*
6. Maximum equipment ΔP 0 psi		
7. Suction-line loss 0.43 psi		
8. Line 6 + Line 7	−0.43 psi	−0.43 psi
9. Available pressure at suction nozzle, P	10.47 psi*	2.52 psi*
10. Available $NPSH = 2.31(P/S)$	25.73 ft	6.19 ft
11. Available $NPSH = 6.19 > 5.5 =$ Required $NPSH$		

*These pressures are above the vapor pressure of liquid at pump-suction nozzle.

Notes: Density, ρ, and specific gravity, S, must be taken at flowing temperature.
Line 9 and Line 10 must be positive.
Line 11: Required *NPSH* from manufacturer's head-capacity curve.

centrifugal pump, one curve of required *NPSH* can be used with all normal specific gravities.

The effect of specific gravity can be explained as follows: Looking at the analogy of *NPSH* as a force that pushes liquid into a pump, and assuming a pump that has an $NPSH_r$ of 8 ft at a given capacity, we can show that a head of 8 ft of water equals 3.45 psi, while a head of 8 ft. of gasoline equals 2.60 psi. It would appear that the force pushing liquid into the pump decreases with specific gravity. However, the mass of liquid being accelerated has decreased by the same proportion [*3*].

When choosing between two pumps of equal capacity, the one having the higher $NPSH_r$ will cost less. But the less-expensive pump will be the economic choice only if its cost advantage is not offset by the expense of elevating the process equipment. Often, a vertical pump will be specified to bring equipment closer to grade elevation.

If a reciprocating pump is to transfer saturated liquid and if we desire maximum volumetric efficiency, we make analyses and calculations for *NPSH* conditions similar to those for centrifugal pumps. Resistance at the inlet valve of a reciprocating pump is greater than the entrance loss to an impeller. Due to pulsating liquid flow, we must also take into account increased line resistance.

Suction Lift in Centrifugal Pumps

With the arrangement shown in Fig. 2, we find that a centrifugal pump cannot pump saturated liquid. If we perform the necessary calculations, we would find a negative suction pressure, and available *NPSH* of zero.

Water at ambient temperature and atmospheric pressure can be lifted from about 15 to 22 ft (i.e., in Fig. 2, $H = 15$ to 22), depending on pump type and flowrate. A pressure-distribution scale for the pumping system is also shown in Fig. 2. With increasing temperature, the absolute vapor pressure of the liquid also increases. This will decrease $NPSH_a$ and dimension H in Fig. 2.

When suction lift is required, the end of the suction line usually has a foot valve and strainer to prevent liquid from draining out of the pump and suction line when the pump does not run. Some kind of priming arrangement should still be provided because dirt or wear may prevent a tight seal at the foot valve.

Four simple priming arrangements are shown in Fig.

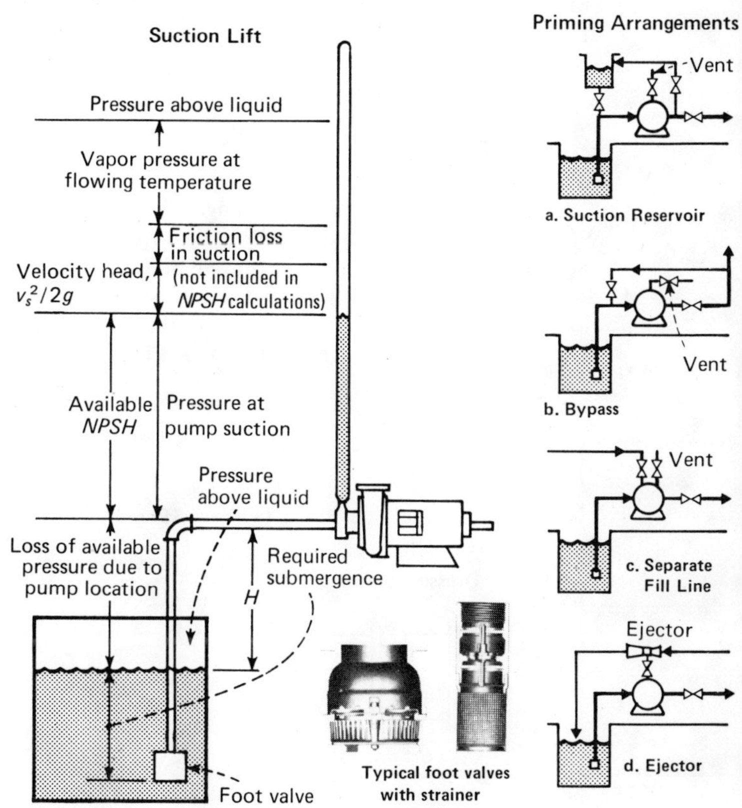

SUCTION lift requires priming the pump—Fig. 2

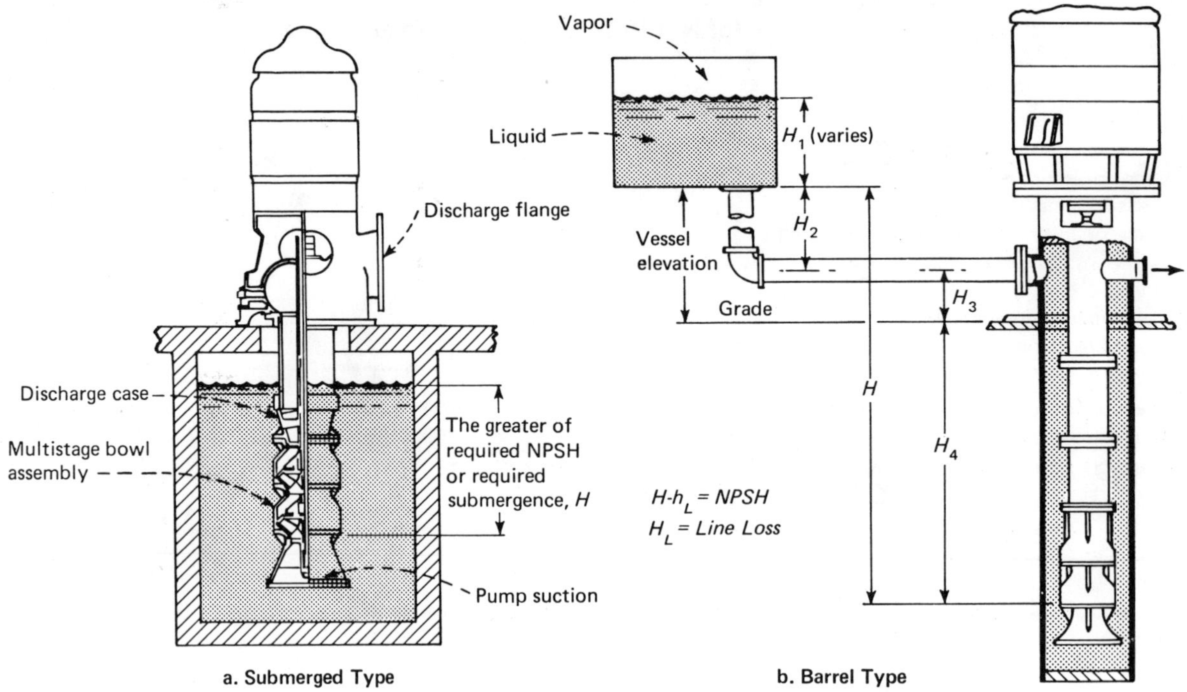

VERTICAL pumps need adequate submergence in addition to meeting or exceeding the required NPSH—Fig. 3

2. These consist of (a) a suction reservoir, (b) a bypass from the discharge to the suction line, (c) a separate fill line from another liquid source, and (d) an ejector to draw liquid into the pump. At large installations, a vacuum pump may replace the ejector. A pump-case vent is usually essential with priming systems except for the ejector or vacuum-pump arrangement.

Minimum Suction Head

A vortex is a swirling and funneling action in a liquid. When this occurs in a liquid between the liquid's surface and the drawoff nozzle in a vessel, air or vapor can be drawn into a pump. If the suction head of a pump is close to a minimum value, a vortex is most likely to form in pump installations such as shown in Fig. 3a and 4.

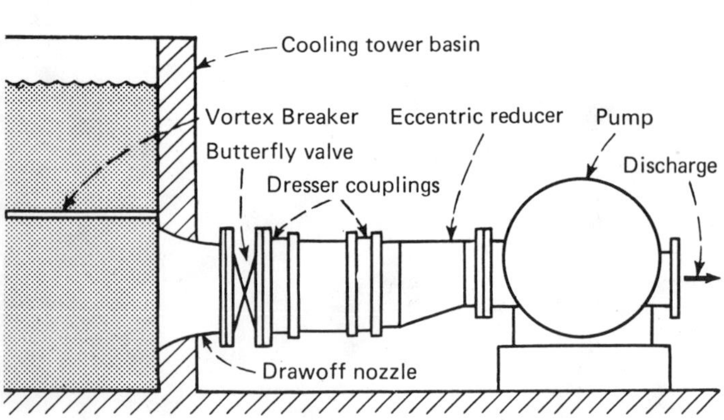

SUCTION piping for cooling-water pump—Fig. 4

To avoid vortex formation, we can take several precautions:

A submerged vertical pump, such as that in Fig. 3a, has no suction piping. Manufacturers of such a pump require a minimum submergence to prevent vortex formation. Greater submergence is needed if the pump has no inlet strainer.

Submergence and *NPSH* are not the same. It is possible to have sufficient submergence and insufficient *NPSH,* or vice versa. We must check a proposed installation for both adequate submergence and for *NPSH* equal to or greater than that required by the pump.

Submerged pumps should have an adequately sized suction sump. Baffles are needed if two or more pumps are in a row. The "Hydraulic Institute Standards" provide recommended practices for sump design [*4*].

A barrel-type vertical pump (Fig. 3b) usually takes suction from an elevated drum. The barrel of the pump serves as an annular suction pipe. With this configuration, vortex forming is unlikely. Because part of the static head for required *NPSH* is below grade, the suction vessel does not have to be elevated as high. The pump should be placed close to the vessel for simple suction piping. If dimension H in Fig. 3b must be maintained, dimensions H_2 and H_4 can be adjusted over a wide range.

An adequately sized drawoff nozzle will also help to prevent vortex formation. Hence, we size the drawoff nozzle so that liquid flow velocity ranges from 1 to 6 ft/s. In typical process applications, the drawoff nozzle is 2 to 6 in, and flow velocity ranges from 1 to 4 ft/s. For larger nozzles, flow velocity ranges from 3 to 6 ft/s.

Fig. 4 shows a typical example of the suction piping for a cooling-tower water pump. The water level in the cooling-tower basin determines the basin's height and cost. Consequently, a minimum water level is preferred, but the level must be sufficient to meet the pump's *NPSH* and suction resistance.

The drawoff nozzle in Fig. 4 is well rounded and tapered for least resistance. Pipe diameter is larger than the pump-suction nozzle, and piping is kept as simple and short as possible. The eccentric reducer is placed in the line with the flat side up in order to avoid an air pocket just before the pump entry. The butterfly valve in the line offers little resistance and has no air pocket in its housing. Because of the rigid-pipe system, expansion joints are necessary. In this type of arrangements, vortex breakers are essential.

Cooling-water pumps usually have double-inlet impellers. The "Hydraulic Institute Standards" provide the following recommendations for the suction piping [*5*]:

"Suction piping, for double-suction pumps, should not be installed so that there is an elbow close to the suction nozzle of the pump, except when the elbow is in vertical position. For example, an elbow that is bending either straight up or straight down is permissible, but one entering at an angle should not be used.

"There is always uneven flow in an elbow, and when it is installed in any position other than vertical, more liquid enters one side of the impeller than the other. This causes a reduction in capacity and efficiency, and a thrust that may heat the thrust bearing and cause rapid wear."

To transfer liquid from storage, inline pumps can be placed directly on the storage-tank nozzle. [According to the manufacturers, inline pumps (up to 100 gpm) can be mounted in any position.] Cradle-mounted pumps should be adjacent to storage tanks. Suction lines to diked tanks are usually long because pumps must be located outside the diked areas. Normally, a larger diameter pipe compensates for the longer runs. Gate valves in these lines should be preferably installed so that their stems are horizontal, thus eliminating the air pocket in the valve housing.

The liquid head in storage tanks varies; and if it falls below the required minimum, the pump will cavitate. A vertical pump can be used to pump a storage tank dry.

Close-coupled pumps can be space savers. For example, they can be mounted horizontally or vertically on steel or concrete columns.

A mechanical device known as a vortex breaker eliminates the possibility of air or vapor entering the pump. A vortex may arise when the *NPSH* is at a minimum value because the process-liquid surface and the pump-suction nozzle are close to each other, or the pump has no suction piping, or the piping is short and simple. Sometimes, a vortex may arise because the liquid level in the process vessel becomes very low. Typical vortex breakers and thier applications are shown in Fig. 5.

As a rule, we must keep process-pump suction lines as simple and short as possible. In many cases however, suction lines have considerable configurations because of vessel elevation, flexibility requirements of hot lines, and pump and standby locations. Suction lines should not have loops or pockets. Air or vapor can collect in a loop, and dirt can accumulate in a pocket.

When the liquid level in a suction vessel is higher than

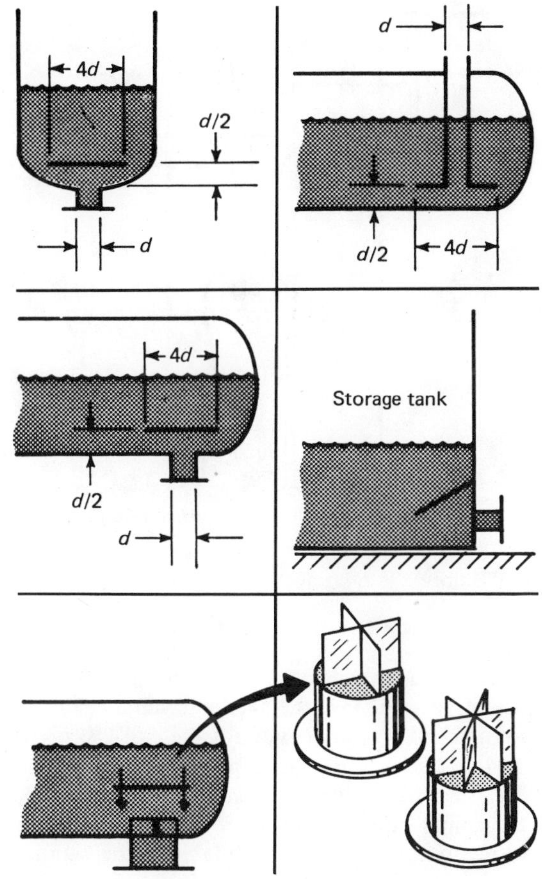

VORTEX breakers in typical process vessels—Fig. 5

the pump, the suction line has a block valve. A permanent or temporary strainer is usually placed between the block valve and pump in a straight portion of the suction line. A corner-type strainer, located at the pump-suction nozzle, should not be used. Its resistance can cause pressure drop at a location where it is least desirable.

Pumping Saturated Liquids

For subcooled liquids, the available pressure at the suction nozzle is usually more than adequate for *NPSH* and line losses. Consequently, suction vessels can be close to grade elevation, and the pipe configuration is not as critical (see Table I: *NPSH* calculations for subcooled liquid).

Many process pumps handle saturated liquids. Here, the suction vessel and drawoff nozzle are elevated to meet the *NPSH* requirement, plus a distance equal to the static height of the line losses (see Table I: *NPSH* calculations for saturated liquid).

We now repeat an important rule for the design engineer. At no point in the suction line should pressure fall (due to piping and components resistance) below the vapor pressure of the liquid at the flowing temperature. A horizontal pipe run immediately after the drawoff nozzle is undesirable because a gradual pressure reduction (in proportion to the length of horizontal pipe) can

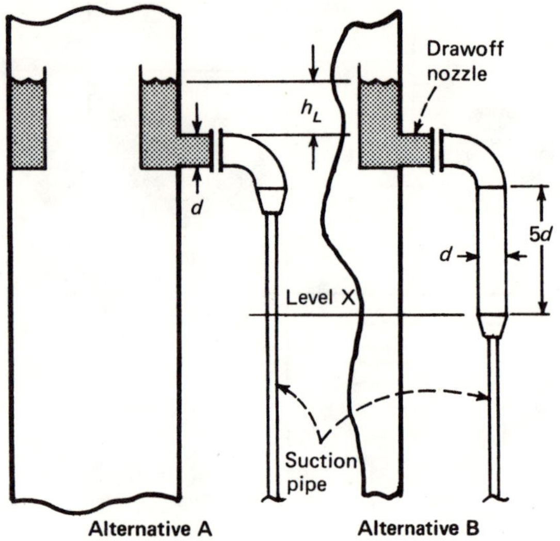

DRAWOFF nozzle is high above pump suction—Fig. 6

cause the liquid to vaporize. Also, if the liquid level in the suction vessel is low, the available static head might not be sufficient to compensate for the exit loss and the friction loss in the horizontal pipe section. This is especially true with light, volatile liquids.

A horizontal-pipe segment after a vertical drop is not as bad because static head is available at the beginning of the horizontal section. In a vertical segment of suction piping, the potential energy is usually much greater than the energy consumed by friction. Thus, when pumping hot liquids, a vertical pipe run after the drawoff nozzle is most desirable; and with light hydrocarbons, it can become essential.

To avoid vaporization at the vessel's drawoff nozzle, the minimum static head of liquid above the nozzle should be greater than the entrance loss to the suction pipe. Cavitation at the drawoff nozzle can cause undesirable pressure surges and inconsistent control conditions.

The exit resistance* of a nozzle can be expressed:

$$h_L = Kv^2/2g \qquad (7)$$

where h_L is liquid level above drawoff nozzle, ft; $K = 0.5$, the resistance coefficient* for a sharp-edged pipe entrance; v is flow velocity, ft/s; and g is gravitational constant, 32.2 ft/s^2.

Using a safety factor of four, the minimum liquid head above the drawoff nozzle will be:

$$h_L = 4Kv^2/2g \qquad (8)$$

Solving Eq. (8) for the flow velocity, v, for a sharp-edged pipe entrance, ($K = 0.5$), yields:

$$v = \sqrt{2gh_L/4K} = 5.67\sqrt{h_L}, \text{ ft/s} \qquad (9)$$

Eq. (9) is suitable for liquid levels up to 1.5 times the diameter of drawoff nozzle as measured from the top edge of the drawoff nozzle. Safety factor in Eq. (9) is justified because the liquid level can be agitated, or foam or emulsion can collect on the liquid surface. These reduce the reliable minimum estimated liquid level. Imperfections in fabrication and used pipe can also result in a resistance coefficient higher than $K = 0.5$.

To illustrate these principles, let us size the drawoff nozzle that is located high above the pump suction for a process vessel, as sketched in Fig. 6. Let us also determine the size of the vertical suction line. Flowrate through the drawoff nozzle will be 1,600 gpm, and h_L = 15 in. Size of the pump-suction nozzle is 8 in.

We calculate the flow velocity from Eq. (9):

$$v = 5.67\sqrt{15/12} = 6.34 \text{ ft/s}$$

Then, we solve Eq. (4) for d:

$$d = \sqrt{0.408(1{,}600)/6.34} = 10.14 \text{ in}$$

This is the calculated diameter for the drawoff nozzle. For practical reasons, the next larger size is chosen. Hence, the drawoff nozzle will have a 12-in nominal size.

If the drawoff nozzle is larger than a reasonable suction-pipe size, we can reduce the size (i.e., nominal diameter) of the vertical leg in Fig. 6. In this example, we can use one size larger for the suction pipe than the required 8-in pump-suction nozzle. To summarize:

Drawoff nozzle.	10.14 in (calculated)
Drawoff nozzle (nominal)	12 in (selected)
Suction pipe (nominal)	10 in (reduced)
Pump-suction nozzle (nominal) . .	8 in. (manufactured)

In Fig. 6, alternative designs are shown for the suction-line connection to the drawoff nozzle. Arrangement B has less resistance to flow up to level X than arrangement A. Alternative A is preferred in borderline cases where the calculated drawoff nozzle (10.14 in) is slightly larger than the determined size of the suction pipe (10 in). Alternative B is preferred if the calculated size of the drawoff nozzle (suppose for this example that d = 11.5 in) is close to the next nominal size.

Feed pressure to a centrifugal pump should not fluctuate because a sudden reduction of pressure in a hot liquid can release large quantities of vapor. To prevent flashing, a system for injecting cool liquid can be added to the suction piping. This might be necessary, for example, at a boiler feedwater pump.

Drawing vapor into a pump is likely to be a problem at greatly reduced flow (for example, at startup), or with fluctuating pump capacities. A vent line attached to the top of the pump housing at the first impeller inlet and connecting back to the vapor space of the suction drum may be necessary.

Cavitation may occur in large pumps when they operate below a minimum flowrate, as specified by pump manufacturers. If greatly reduced flow (for example, down to 20% of rated capacity) is expected, a minimum-flow bypass will have to be designed in the discharge line, with a throttling valve or restriction orifice, or both.

Sizing the Suction Pipe

Size of the suction pipe is usually one or two sizes larger than that of the pump's suction nozzle. A suction

*See pp. 21 and 26.

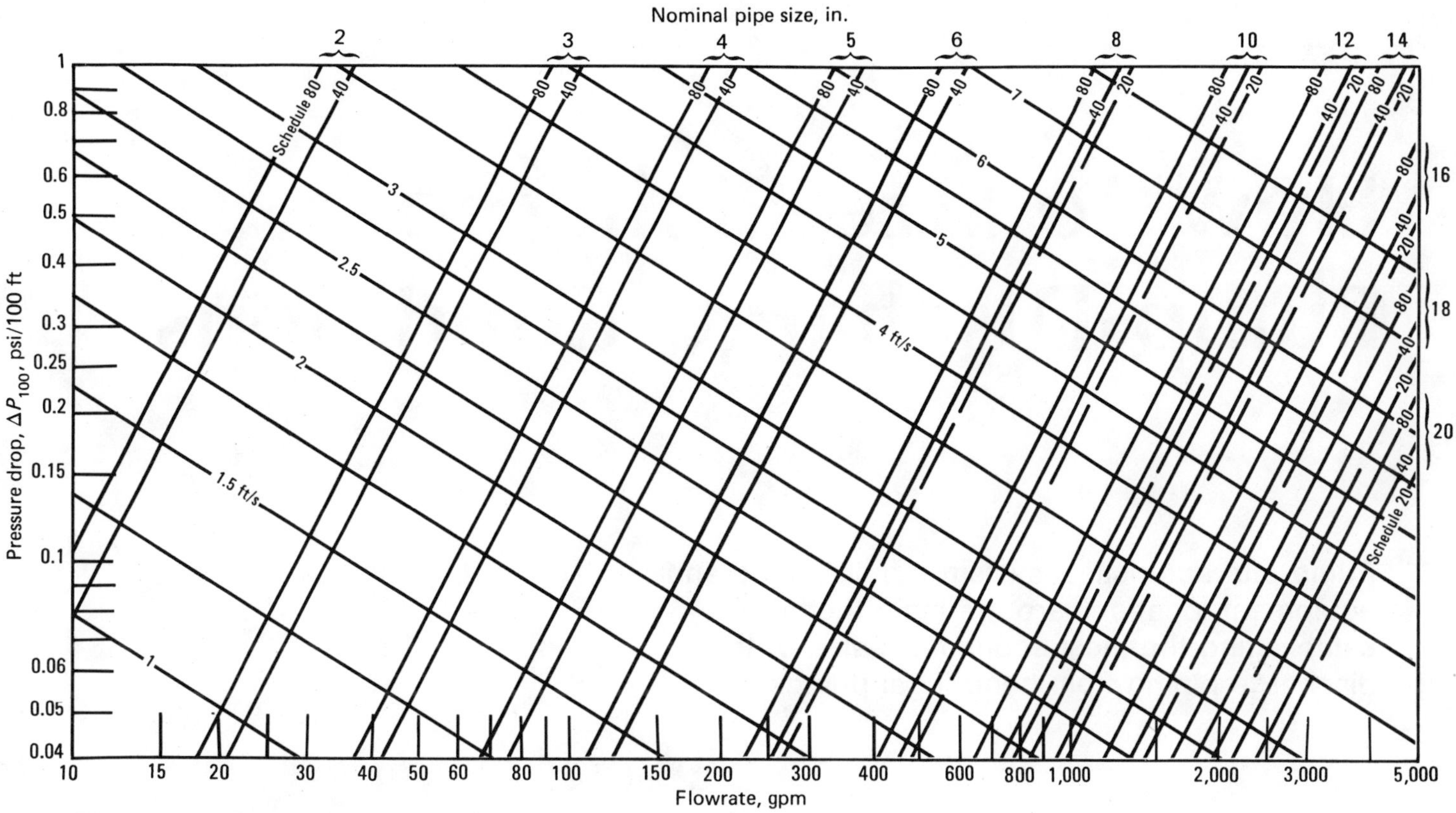

Note: ΔP_{100} values are for $S = 1$ and f for total turbulence. Multiply ΔP_{100} values by specific gravity, and correction for friction factor: f (actual) / f (total turbulence).

PRESSURE loss and flow velocity in process pipelines for totally turbulent flow and specific gravity equal to one—Fig. 7.

pipe smaller than the pump's suction nozzle should never be used. Total suction-line losses can be estimated between 1 to 3 ft of liquid head. Liquid velocities can be in the range of 2 to 8 ft/s; for pumping viscous liquids, 0.5 to 4 ft/s.

Fig. 7 provides a graphical solution of Darcy's equation:

$$\Delta P_{100} = 1.346 f S (Q^2/d^5) \tag{10}$$

for a specific gravity of $S = 1$, and friction factors, f, in the totally turbulent zone (see p. 93, Table VI). Also, in Eq. (10), Q is volume flowrate of liquid, gpm; and d is inside diameter of pipe, in.

Fig. 7 has been drawn for a practical range of sizes for suction pipes. Unit loss can be estimated as ranging from 0.05 to 0.5 psi/100 ft for saturated liquids, and from 0.5 to 1.0 psi/100 ft for subcooled liquids.

We enter the graph with the flowrate, intersect the line size, and read the unit pressure loss. At the intersection for pipe size and flowrate, we can also quickly read the flow velocity.

The unit-loss values, ΔP_{100}, obtained from Fig. 7 must be corrected if $S \neq 1$ and the friction factor falls in the transitional turbulent zone, or a corrected friction factor (such as for used water pipes) is considered. Hence:

$$\Delta P_{100}\ (\text{actual}) = \Delta P_{100}\ (\text{graph}) \times S \times \frac{f\ (\text{actual})}{f\ (\text{total turbulence})}$$

The actual friction factor may be obtained via the Reynolds number from charts on p. 92.

We can use Fig. 7 to check the calculated unit loss in the example for determining the *NPSH* in Table I. For a flow of 120 gpm in a 4-in, Schedule 40 pipe, Fig. 7 establishes the unit loss as 0.31 psi/100 ft, and velocity as 3 ft/s. Correcting for the specific gravity and friction factor, we find the actual unit loss as:

$$\Delta P_{100} = 0.31 \times 0.94 \times (0.023/0.0165)$$

$$\Delta P_{100} = 0.4 \text{ psi/100 ft}$$

The next article in this series will appear in the issue of May 26, 1975, and will deal with pump discharge piping.

References

1. "Hydraulic Institute Standards," 12th ed., p. 57, Hydraulic Institute, New York, 1969.
2. Ibid, p. 78.
3. "Cameron Pump Manual," Ingersoll-Rand Co., Phillipsburg, N.J.
4. "Hydraulic Institute Standards," 12th ed, pp. 87–92, Hydraulic Institute, New York, 1969.
5. Ibid, pp. 99–100.

How To Size Piping for Pump-Discharge Conditions

Pump characteristics, system requirements and equipment locations are the main factors in establishing the most-economical size of the discharge piping from centrifugal pumps.

ROBERT KERN, Hoffmann — La Roche Inc.*

Economy of both capital outlay and operating cost is the essential requirement when designing the discharge piping for a pump. This contrasts with the need for reliable operation when sizing the suction piping.

If the centrifugal pump receives saturated liquid on the suction side, the liquid becomes well subcooled in the discharge line, due to the increased pressure. This is one reason why filters, orifice runs, control valves, exchangers, and other flow restrictors are placed on the discharge side of process pumps.

In this article, we will analyze and compute the resistances to flow on the discharge side of the process pump. As a rule, we can readily select economical sizes for discharge pipe up to about 12-in dia. For larger diameters, we must often make a more-detailed cost comparison, in order to choose between alternative discharge-pipe designs.

Discharge Characteristics of Pumps

By reviewing typical pump data and knowing only the flowrate, we can readily indicate a reasonable design for the hydraulic system, regardless of discharge-pressure requirements of the pumped liquid.

Performance Curves—In Fig. 1, we find a composite rating chart for a series of standard chemical-process pumps, and the head-capacity curve for a specific pump. The performance curves for this pump show that it is suitable for handling flows from 150 to 300 gpm (close to the maximum efficiency points). Additional data are:

Flow, gpm	150 to 300
Impeller diameter, in	6 to 10
Power, hp	3 to 10
Efficiency, %	~58 to ~67
Total head, ft water	30 to 90

All these data were obtained from Fig. 1b for the pump having a 4-in suction nozzle and a 3-in discharge nozzle. The total head represents a differential pressure of 13 to 39 psi (when pumping water) between suction and discharge flanges.

For a flow of about 250 gpm, the suction-pipe size can be 6 in, the discharge pipe 4 in. The orifice run can be 4 or 6 in, the control-valve size 3 or 4 in; block valves and strainer should be line size. With these data, detailed calculations can be made, pump total-head requirement determined, and impeller diameter and motor size selected.

A centrifugal pump is a dynamic machine that can perform only on points on the head-capacity curve. Consequently, the sum of piping and components resistances during operation must fall exactly on the head-capacity curve. The relationship between the pump's head-capacity curve and pipe-system resistance is shown in Fig. 2. Resistance of the throttling valve should not be included as part of the system resistance. The gap between the head-capacity curve and the system-resistance curve is available for throttling resistance (i.e., control-valve ΔP). While pipe resistance increases with higher flowrate, throttling resistance decreases. Or, at higher

*To meet your author, see p. 44.

Originally published May 26, 1975

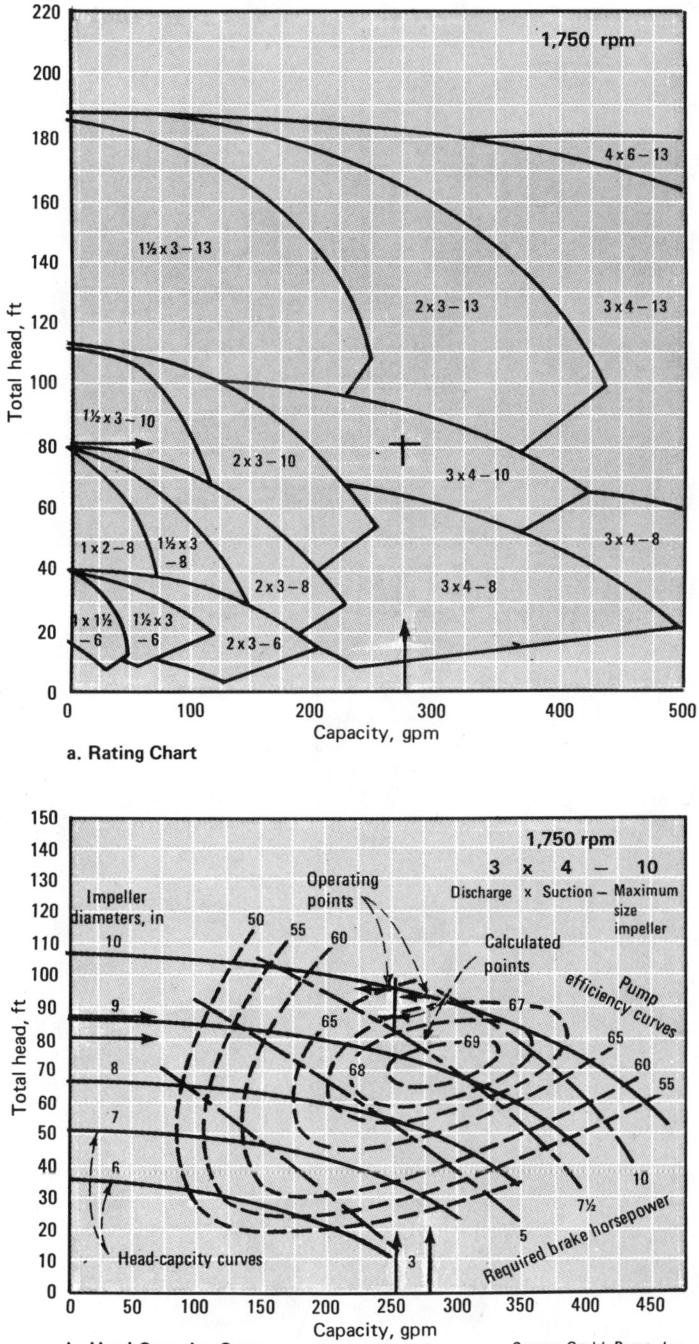

PERFORMANCE of a typical series of centrifugal pumps and detailed characteristics of the selected pump—Fig. 1

flowrates, the throttling valve has to open and pass larger flows with less resistance.

According to the Hydraulic Institute Standards [1], the velocity-head difference between pump nozzles should be included in the head-capacity curves. Some of the velocity energy converts to pressure energy if the discharge pipe is larger than the size of the pump's discharge nozzle. When computing the required total head, velocity head usually does not enter into the calculations, as in the total-head calculation for the example in this article.

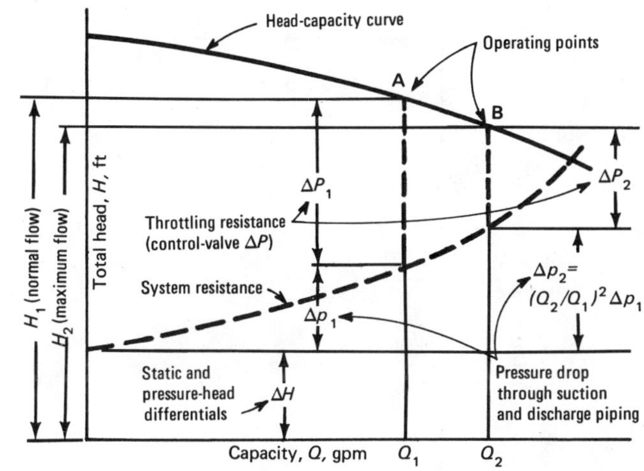

PUMP head and system resistance establish the available throttling resistance for control valve—Fig. 2

and system-resistance curves are reproduced as shown here:

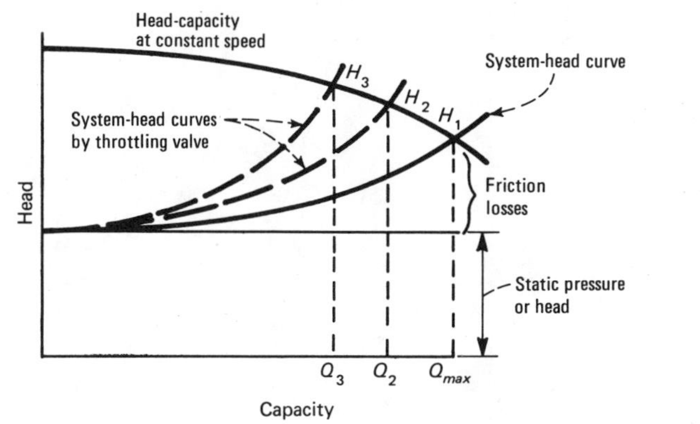

The curves in this drawing suggest that when the discharge is throttled, the system-resistance curve moves to another position. Not so! Only one system-resistance curve exists for a given discharge line. The varying ΔP of the control valve brings the system resistance up to a point on the head-capacity curve.

Head-capacity curves are drawn for pumping water. However, a centrifugal pump with a given impeller, speed and size will develop the same head in feet, no matter what the liquid or specific gravity—providing viscosities are normal. The static-head pressure at the pump nozzles will be higher with a heavier liquid, and the horsepower requirement will also increase. Brake-horsepower requirements for pumping water are superimposed on the head-capacity curve (see Fig. 1b). We multiply the horsepower required to pump water by the specific gravity of the liquid being pumped to obtain actual brake horsepower. Efficiency curves are also superimposed on the head-capacity curves (Fig. 1b). The ideal, normal operating point is in the maximum efficiency range.

When calculating total-head requirements at alternative capacities, it is useful to know the operating points

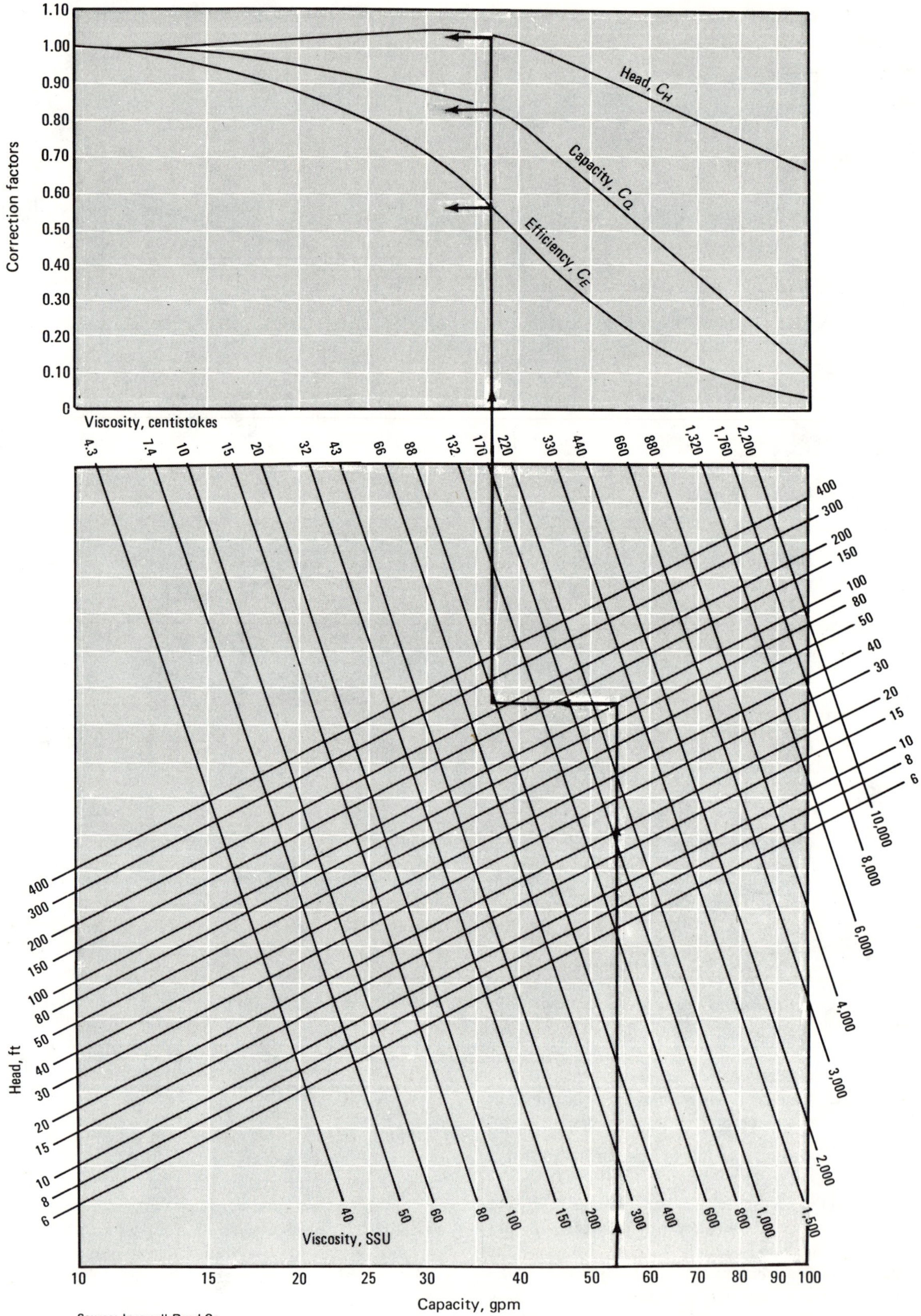

CORRECTION factors for viscous Newtonian liquids flowing through single-stage centrifugal pumps—Fig. 3

Nomenclature

β	Ratio of orifice bore to inside pipe diameter
β^2C	Orifice capacity coefficient (if $\beta = 0.7$, $\beta^2C = 0.339$)
C_v	Control-valve coefficient
C_{vc}	Calculated control-valve coefficient
d	Internal pipe dia., in
f	Friction factor
μ	Viscosity, cp
L	Equivalent pipe length, ft
N_{Re}	Reynolds number
Δp	Pressure loss, psi
ΔP_{100}	Unit loss, psi/100 ft
P	Pressure, psi
ΔP	Pressure difference, psi
Q	Flowrate, gpm
ρ	Density of liquid at flowing temperature, lb/ft³
S	Specific gravity at flowing temperature

on the head-capacity curve, so that the available control-valve pressure differentials can be more closely estimated.

If a pump has not been selected, we can still estimate the increase in total head for a given pump when capacity decreases by X% from a given capacity. This increase in total head can range from 0.5 to 1.0 multiplied by X%. A high value for this increase (selected from the range 0.5 to 1.0), say 0.85X%, is justified for three reasons:

1. The pump will not be undersized.
2. The head-capacity curve for a worn impeller will have a more pronounced decline with increasing flow as compared with a new impeller.
3. A somewhat increased control-valve ΔP will provide a more desirable plug position at a maximum flow.

Viscous Flow—A reduction in head, capacity and efficiency, and an increase in horsepower, will occur when pumping viscous liquids instead of water. (*NPSH* requirement does not change.) Fig. 3 shows a performance-correction chart for conventional, single-stage, centrifugal pumps handling viscous Newtonian liquids for capacities between 10 to 100 gpm. A similar chart is available for 100 to 1,000 gpm in the "Hydraulic Institute Standards" [*1*].

These charts should not be used for multiple-impeller, mixed-flow and axial-flow pumps, or for pumps with special hydraulic design. Manufacturers can supply viscosity-correction charts for their pumps.

To use Fig. 3, we select a pump from the manufacturer's head-capacity curves at optimum efficiency, and note the head, capacity and efficiency. We enter Fig. 3 with the same capacity. We intersect the head line, in feet, and move horizontally left or right to viscosity; then vertically up to intersect the efficiency-correction factor, C_E, capacity-correction factor, C_Q, and head-correction factor, C_H. An example is drawn in Fig. 3.

The corrected performance characteristics for viscous flow will be:

$$Q_w \times C_Q = Q_{vis}$$
$$H_w \times C_\theta = H_{vis}$$
$$E_w \times C_E = E_{vis}$$

where *Q, H* and *E* are capacity, total head, and efficiency, respectively, and where subscript *w* stands for water, and subscript *vis* stands for viscous liquid.

Two-Phase Flow—If the pressure drops below the vapor pressure of the liquid in the discharge line (for example, after a heat exchanger or a control valve), vaporization will occur. Friction resistance in the pipe increases considerably with the increased vapor content of the liquid. This section of the discharge pipe should be calculated for two-phase flow. If two-phase flow is expected in a pipeline between two pieces of process equipment, capital cost and operating cost can be reduced by locating this equipment side by side. Generally, lines with two-phase flow should be short.

Total-Head Calculations

We can use the data plotted in Fig. 4 for selecting a suitable size for the discharge piping of a centrifugal pump. The shaded zones in Fig. 4 will help in selecting economical sizes for alloy or carbon-steel piping in the case of electric-motor or steam-turbine drives.

In general, economical sizes for discharge pipes can be easily recognized if we list the unit losses for several adjacent pipe sizes. For example, the unit losses in a Schedule 40 pipe for a 300-gpm flow, as obtained from Fig. 4, yield:

Pipe Size, In	ΔP_{100}, Psi/100 Ft
3	8
4	1.9
6	0.23

In this case, the 4-in size is reasonable for the discharge pipe. The unit loss of a 3-in pipe will, most probably, increase utility cost to an extent that cannot be compensated for by lower capital costs of the pipeline and components. And the cost of a possibly larger pump and motor has not been taken into account. The unit loss of a 6-in pipe is far too low for an average process pump. Where the discharge line is very long, a 6-in line might be reasonable, such as the cooling-water supply header to a distant point.

An example will best illustrate the concepts for calculating the pump's differential pressure:

A pump moves a liquid having a density, ρ, of 50 lb/ft³ from an overhead drum to an elevated point on a column, as sketched here:

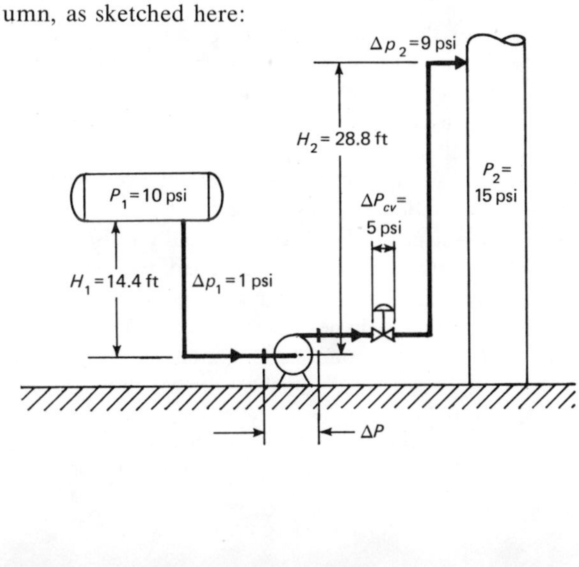

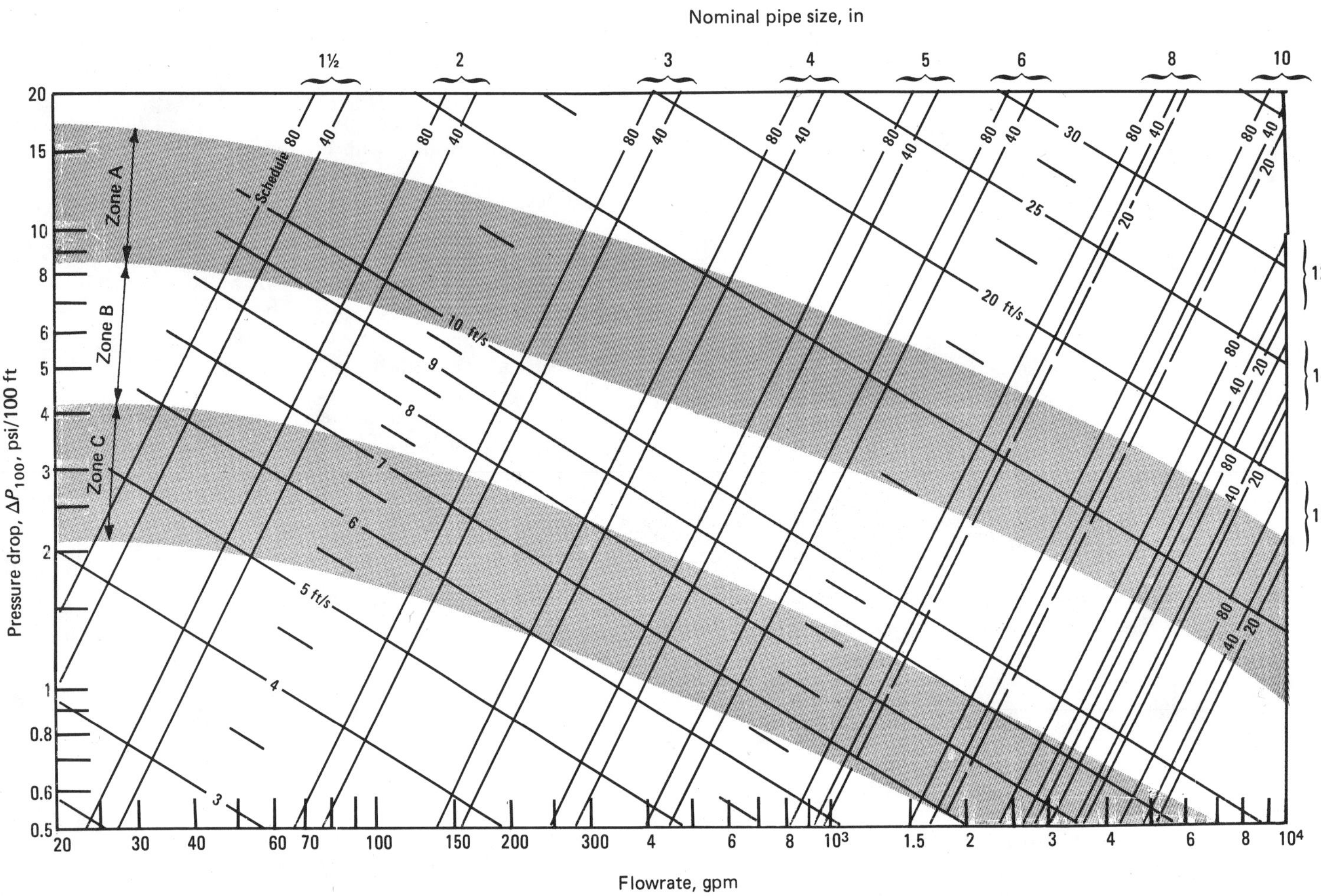

*Economical sizes for discharge pipe:

Zone A: Alloy-steel pipe, 2-yr payout.

Zone B: Carbon-steel pipe, electric-motor drive or 2-yr payout. Also, alloy-steel pipe, long payout.

Zone C: Carbon-steel pipe, turbine drive, or long payout.

Note: ΔP_{100} values are for $S = 1$ and f for total turbulence. Multiply ΔP_{100} values by specific gravity, and correction for friction factor: f(actual) /f(total turbulence).

UNIT pressure losses in new, commercial-steel pipe help in selecting suitable sizes for discharge piping—Fig. 4

Pressure in the drum, P_1, is 10 psi, and pressure in the column, P_2, is 15 psi. Pressure drop across the control valve, ΔP_{cv}, equals 5 psi; and the friction losses in the suction and discharge lines are $\Delta p_1 = 1$ psi and $\Delta p_2 = 9$ psi, respectively. Let us calculate the required total head for the pump, i.e., the differential pressure, ΔP, between the suction and discharge flanges.

Both suction head and discharge head consist of four components: static head, pressure head, velocity head, and piping and components resistance. Velocity-head differences normally do not have to be calculated because they are taken into account by U.S. pump manufacturers in their published head-capacity curves. Pressure and static head in a system usually remain constant with changing capacities. Suction-pipe, discharge-pipe and components resistances vary with flowrate (see Fig. 2). The analysis, therefore, becomes:

	Suction, Psi	Discharge, Psi
Vessel pressure	$P_1 = 10$	$P_2 = 15$
Static-head pressure	$P_{H1} = 5$	$P_{H2} = 10$
Friction loss	$\Delta p_1 = -1$	$\Delta p_2 = 9$
Total	14	34

We now subtract the total suction resistance from the discharge line backpressure to find the pump's differential pressure excluding the control valve: $34 - 14 = 20$ psi. Choosing a control-valve ΔP_{cv} of 5 psi, we must add this to the pump's differential pressure of 20 psi to obtain the total differential pressure across the pump flanges as

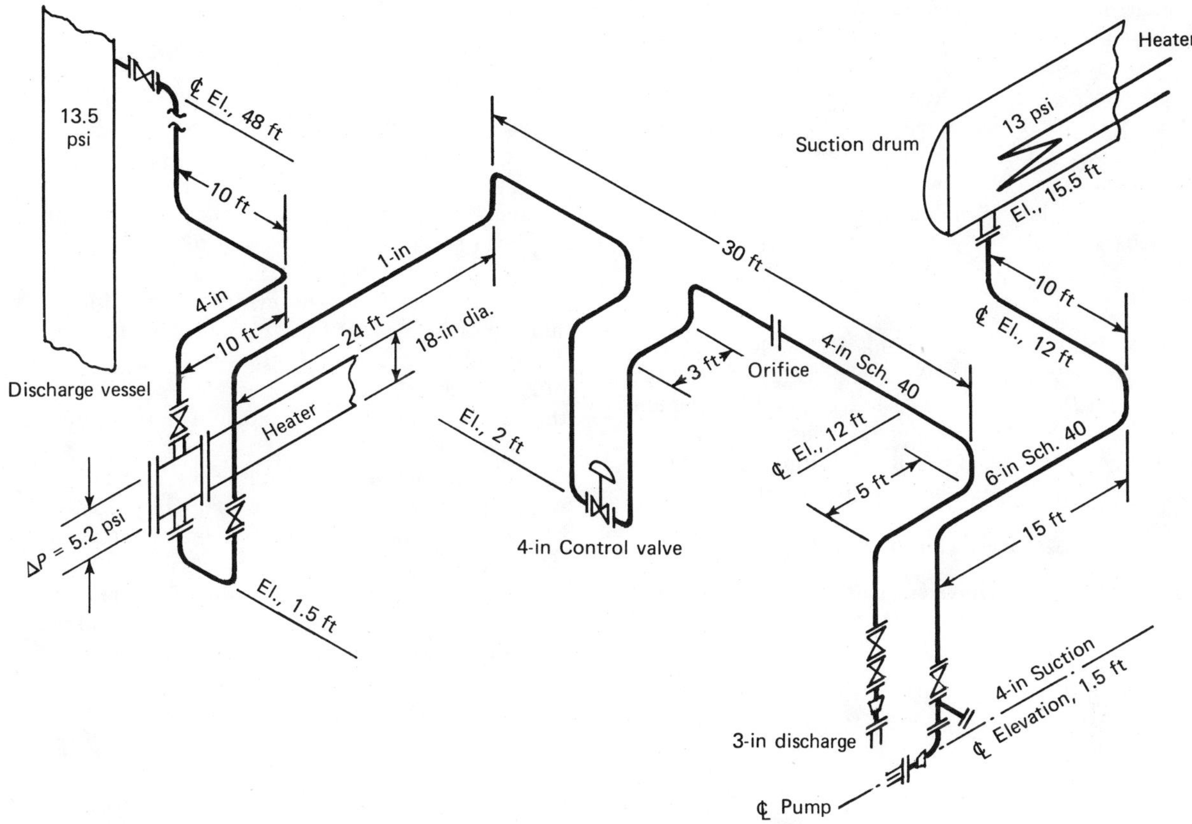

PIPING and equipment layout for the suction and discharge lines to the process pump as given for the problem—Fig. 5

25 psi, or expressed as head of liquid (the required total head):

$$H = (25)(144)/50 = 72 \text{ ft}$$

Example Illustrates Detailed Design

A centrifugal pump having a 4-in suction nozzle and a 3-in discharge nozzle will handle a gas oil at a normal flowrate of 250 gpm through a piping and components system, as drawn in Fig. 5. Allowing for a safety factor of 1.1, we find that the maximum flowrate will be 1.1(250) = 275 gpm. Specific gravity and density are: $S_{60} = 1.18$, and $\rho_{60} = 73.6$ lb/ft^3, respectively. At the flowing temperature of 555° F, $S = 1.04$ and $\rho = 64.87$ lb/ft^3. Viscosity of the gas oil is 0.6 cp. There is a flow control valve in the discharge line.

Let us calculate the total head on the pump when it is expected to operate at a normal flowrate and at a maximum flowrate.

We will begin the analysis by first determining the loss in the suction line and then that in the discharge line. Pertinent data for the Schedule 40 pipe are:

	Suction Line	Discharge Line
Nominal size, in	6	4
Inside dia., d, in	6.065	4.026
d^5	8,206	1,058

Suction-Line—We fine the loss in the suction line by initially calculating the Reynolds number from:

$$N_{Re} = 50.6(Q/d)(\rho/\mu)$$
$$N_{Re} = 50.6(250/6.065)(64.87/0.6) = 225{,}500$$

The friction factor, f, for this Reynolds number equals 0.0175 (from Fig. 6, p. 92). By substituting in the following equation, we obtain the unit loss:

$$\Delta P_{100} = 1.35 f S Q^2/d^5$$
$$\Delta P_{100} = 1.35(0.0175)(1.04)(250)^2/8{,}206$$
$$\Delta P_{100} = 0.19 \text{ psi}/100 \text{ ft}$$

We then find the equivalent length for the suction line and its fittings from data in Tables I to IV (pp. 22–24):

Pipe length	39 ft
5 Short-radius elbows	75 ft
1 Reducer	4 ft
1 Strainer	30 ft
1 Gate valve	6.5 ft
1 Inlet to pipe	18 ft
Total equivalent length, L,	172.5 ft

Hence, the overall loss for the line and its fittings at the normal flow of 250 gpm becomes;

$$\Delta P = \Delta P_{100}(L/100)$$
$$\Delta P = 0.19(172.5/100) = 0.33 \text{ psi at normal flow}$$
$$\Delta P = 0.33(250/275)^2 = 0.4 \text{ psi at maximum flow}$$

To find the pressure at the suction nozzle, we calculate the static-head pressure: (14)(64.87)/144 = 6.3 psi, and add this to the vessel pressure of 13 psi to get 19.3 psi. Since the pipe-friction loss at normal flow is 0.33 psi, the pressure at the suction nozzle becomes 19.3 − 0.33 = 18.97 psi. At maximum flow, the pressure at the suction nozzle is 19.3 − 0.4, or 18.9 psi.

Discharge-Line—We now perform similar computations to find the loss in the discharge line. Therefore, for the discharge line:

$$N_{Re} = 50.6(250/4.026)(64.87/0.6) = 340{,}000$$

The friction, factor, f, for this Reynolds number is 0.0178 from Fig. 6, p. 92. We now use Fig. 4 to find the unit loss, P_{100}, as 1.32 psi for totally turbulent conditions at the normal flow of 250 gpm. Since the flow for this Reynolds number is in the transitional region (see p. 92), we must correct the unit loss as follows:

$$\Delta P_{100} = 1.32(1.04)(0.0178/0.0165) = 1.48 \text{ psi/100 ft}$$

We then find the equivalent length for the discharge and its fittings from data in Part 2 of this series:

Pipe length	156 ft
20 Short-radius elbows	210 ft
4 Gate valves	18 ft
1 Reducer	3 ft
2 Exits	40 ft
1 Inlet	10 ft
Total equivalent length, L	437 ft

Hence, the overall loss for the line and its fittings at the normal flow of 250 gpm becomes:

$$\Delta P = \Delta P_{100}(L/100)$$
$$\Delta P = 1.48(437/100) = 6.47 \text{ psi}$$

And at maximum flow,

$$\Delta P = 6.47(275/250)^2 = 7.83 \text{ psi}$$

A detailed, specific and systematic procedure is shown in Table I for calculating the total head on the pump when it operates at a normal flowrate and at the maximum flowrate.

For the calculations presented in Table I, the following steps are recommended:

Lines 1 to 6 can be worked out simultaneously for normal and maximum flow. Lines 1 and 2: Discharge-vessel pressure and discharge static head do not change with alternative flow capacities. Lines 3, 4 and 5: Orifice ΔP, pipe-friction loss, and equipment-friction loss increase with capacity. Hence:

$$\Delta P_{max} = (Pump\ Safety\ Factor)^2 \Delta P_{normal}$$

Line 3: Orifice pressure drop depends on the manometer deflection and on the orifice diameter/pipe diameter ratio.*

The manometer deflection* for an orifice at a 250-gpm flow will be:

$$h_w = [0.176Q\sqrt{S}/(d_1{}^2\beta^2 C)]^2$$
$$h_w = [0.176(250)\sqrt{1.04}/(16.21 \times 0.339)]^2$$
$$h_w = 80.7 \text{ in, or } 6.73 \text{ ft}$$

*For details, see Part 3 of this series, *Chem. Eng.*, Feb. 3, 1975, p. 72–78.

For this deflection, we can use a 100- or 125-in-long manometer. The orifice ΔP_o will be:

$$\Delta P_o = 6.73(62.37)/144 = 2.92 \text{ psi}$$

With $\beta = 0.7$, the permanent loss will be 52% of the orifice pressure differential.*

$$\Delta P_o(\text{loss}) = 0.52(2.92) = 1.52 \text{ psi}$$

Line 4: Friction loss in the discharge (and suction) line has been previously computed as 6.47 psi. Line 5: Pressure drop through the exchanger (and other equipment) can be obtained from the manufacturers. Line 6 is the sum of Lines 3 to 5 at the normal flowrate. For a 10% greater flow and a pump safety factor of 1.1, resistance of the discharge line will increase by $(1.1)^2$. Line 7 is the subtotal of Lines 1, 2 and 6 at normal flow and at maximum flow.

At this point, we will continue the calculations at the maximum flowrate for reasons that will become evident.

Line 8: For the control valve to operate in an optimum range at normal flow, we usually consider the valve plug in a fully open position at maximum flow. This also gives a minimum pressure drop through the control valve†. A 4-in single-seat control valve has a valve coefficient† $C_v = 124$. And, with $C_{vc}/C_v = 1$:

$$\Delta P_{(min)} = \left[\frac{Q}{(C_{vc}/C_v)C_v}\right]^2 S$$

$$\Delta P_{(min)} = \left[\frac{275}{1 \times 124}\right]^2 1.04 = 5.12 \text{ psi}$$

Line 9 is the required discharge pressure, including control-valve ΔP at maximum flow. Suction-nozzle pres-

†For details, see p. 244.

Calculations for Total Head of Pump — Table I

Line	Variable		Normal Flow, Psi		Maximum Flow, Psi
1.	Discharge-vessel pressure		13.5		13.5
2.	Discharge static head		20.94		20.94
3.	Orifice ΔP_o	1.52			
4.	Discharge pipe-friction loss	6.47			
5.	Exchanger ΔP	5.2			
	[a]	+			
6.	Discharge-system resistance	→	13.19	× (1.1)² =	15.96 [b]
7.	Subtotals		− 47.63		50.40
8.	Control valve ΔP		10.71 [d]		+ 5.12 [c]
9.	Discharge pressure		+ 58.34		55.52
10.	Suction-nozzle pressure		18.97		− 18.9
11.	Pump differential pressure		39.37		36.62
12.	Required total head, $(144 \times \Delta P)/\rho$		87.4 [e] ft	←	81.3 ft
13.	Total head from head-capacity curve		95 ft		92 ft
14.	Total-head safety margin		7.6 ft		10.7 ft

[a]Other equipment resistance.
[b]Pump safety factor, $s.f.$ = 1.1
[c]With $C_{vc}/C_v = 1$.
[d]With C_{vc}/C_v = 0.5 to 0.8
[e]ΔH(normal flow) = ΔH (maximum flow) × 1.075 = 81.3 × 1.075 = 87.4

sure has been previously computed. By deducting the suction-inlet pressure from the required discharge pressure, we obtain the pump's differential pressure at maximum flow (Line 11). This is converted to the equivalent head (Line 12) at maximum flow by using the previously determined value of 36.62 psi from Table I:

$$\frac{144(\Delta P)}{\rho} = \frac{144(36.62)}{64.87} = 81.3 \text{ ft}$$

Let us now summarize some of these results. The pressure needed at the pump-discharge nozzle to overcome backpressure in the discharge line is the sum (Line 9 in Table I) of actual pressure in the discharge vessel, static-head lift up to the terminating nozzle (or liquid level in discharge drum), control-valve ΔP, and total discharge-pipe and equipment resistances. Pump differential pressure (Line 11) equals discharge pressure (Line 9) minus suction-nozzle pressure (Line 10).

We can now estimate the total head at normal flowrate. Total head will increase by an amount ranging from 0.5 to 1 of the percentage decrease in capacity. In this example, there is a 10% decrease in capacity. Hence, for a single-impeller pump, we will assume an increase in the total head of about 7.5% (i.e., 0.75 of 10%). The computed value is on Line 12 for the normal flow. Calculated values for the total head for this example are shown in Fig. 1b.

Suction-nozzle pressure at normal flow is on Line 10. Line 9 = Line 10 + Line 11.

The available pressure differential at normal flow for control-valve sizing (Line 8) equals discharge pressure (Line 9) minus the line backpressure without the control valve (Line 7). This ΔP should give a control-valve coefficient falling within the recommended ranges of $C_{vc}/C_v = 0.5$ to 0.8 for equal-percentage contoured plugs:

$$C_{vc} = Q\sqrt{S}/\sqrt{\Delta P} = 250(\sqrt{1.04})/\sqrt{10.71} = 77.9$$

For the selected 4-in control valve, $C_{vc}/C_v = 77.9/124$, or 0.63, which is acceptable.

For a normal flow of 250 gpm and 87.4 ft total head on the one hand, and a maximum flow of 275 gpm and 81.3 ft on the other, we can now select the pump, as shown in Fig. 1. Impeller diameter for the selected pump is 10 in, and a standard motor of 10 bhp is required. The motor will work with a better than 65% efficiency.

The calculated total-head points fall between the 9- and 10-in impellers. The pump will operate at 95-ft total head at 250 gpm, and at 92-ft at 275 gpm. The extra head (7.6 and 10.7 ft, respectively, here) provides a safety margin compensating for inaccuracies in the assessment of the flow-properties and line data. These additional pressure differentials can be absorbed by the control valve. Or, the block valve in the discharge line can be slightly closed to bring the operating point up to the head-capacity curve. Also, the motor will be able to drive the pump when the liquid is colder and specific gravities are greater than at operating conditions.

If a pump has not been selected and the head-capacity curve is not available, a safety margin of 5 to 15% can be estimated and added to the required total head of Line 12.

Economy of Pump Piping

As the data in Fig. 1b show, several sizes of impellers can be placed in one pump case. The cost difference between impeller sizes is negligible. Motor sizes are usually well determined. A difference in the cost of the pump and motor occurs for pumps falling in adjacent envelopes of the composite rating chart. In some borderline cases, it may be more economical to redesign the discharge piping for lower pressure drop rather than to invest in a larger pump and motor.

For economy in utility cost, the pump should work at its highest efficiency. High pump efficiency results in minimum horsepower input, and minimum wear and maintenance. High-efficiency pumps last longer, are quieter and vibrate less than low-efficiency ones [*2*].

Small pumps should not be oversized. The total of oversized small motors in a plant can add to substantial waste in energy usage.

For pipe diameters above 12 in, more than one size may be selected initially because capacity increments in large-size pipe are very close. Piping costs, of course, increase with diameter, while utility costs decrease because of smaller pipe and components resistances. The best size can be determined by adding the total cost of utilities over the period of capital payout to the cost of the mechanical and electrical installation. The lowest total cost calculated for a 2-, 5- or 10-yr amortization will give the most economical design.

A detailed investigation for the most economical pipe size is justified if line sizes are large, pipe configurations are long or complicated, or if the piping material is expensive. Pipe friction must contribute a major portion of the discharge pressure—otherwise, there will be little difference in total heads between alternative designs. Actual vessel pressures and static liquid heights usually cannot be altered.

For reciprocating pumps, the available *NPSH*, pump differential pressure, suction-line and discharge-line resistances cannot be calculated in the same way as for centrifugals. Because of pulsating flow, the minimum-pressure levels should not fall below the vapor pressure when saturated liquid is pumped. For identical flowrates, pressure losses in suction and discharge lines of reciprocating pumps are greater than in those of centrifugal pumps. These principles have been adequately discussed by Hattiangadi [3]. For information on power ratings, installation and operation of reciprocating pumps, consult the "Hydraulic Institute Standards" [*4*]. The Standards [*4*] also contain information on electric-motor-driven and steam-driven reciprocating pumps.

References

1. "Hydraulic Institute Standards," 12th ed., p. 81, Hydraulic Institute, New York, 1969.
2. Marischen, J. P., "Critical Centrifugal Pump Information," Ampco Metal Inc., Milwaukee, WI 53201.
3. Hattiangadi, U. S., Specifying Centrifugal and Reciprocating Pumps, *Chem. Eng.*, Feb. 23, 1970, pp. 101–108.
4. "Hydraulic Institute Standards," 12th ed. pp. 165–166, 181–182, 204–205, Hydraulic Institute, New York, 1969.

Pump Piping Design

Layout and sizing of pump piping and fittings are critically important to centrifugal pump performance. Suction and discharge piping present individual problems and must be designed with equal care.

ROBERT KERN, Hoffman-La Roche, Inc.

When designing suction piping, the engineer's main concern is reliable operation. In general, suction piping consists of a draw-off nozzle, one or two sections of horizontal and vertical pipe runs, a block valve, strainer and connecting elbow section before the final connection to the pump suction nozzle. Each of these parts should be designed and located to avoid vaporization or air pockets.

The lowest pressure point in a pump is just before the impeller vanes. If the pressure here falls below the vapor pressure of the liquid, vapor bubbles will form. Then the bubbles collapse because of the rapid pressure increase within the impeller. This phenomenon (the rapid formation and collapse of vapor bubbles) is called cavitation. It can reduce pump efficiency and cause noise, vibration, impeller wear and breakage.

Cavitation can be avoided by providing sufficient static head before the pump suction nozzle to satisfy net positive suction head requirements and to overcome friction losses in the line. A simple suction pipe configuration is essential.

Suction Line Layouts

Fig. 1 shows a typical example of suction piping for a cooling tower. Water level determines the cooling water basin height and its cost. Consequently, a minimum water level is preferred, but it must be sufficient for the pump's NPSH and suction pipe resistance.

The draw off nozzle in Fig. 1 is well rounded and tapered for least resistance. Pipe diameter is larger than the pump suction nozzle, and piping is kept simple and short as possible. The eccentric reducer is placed in the line flat side up to avoid an air pocket just before the pump entry.

The vortex breaker in the cooling water basin prevents funnel formation and air entry into the suction pipe. The butterfly valve in the line offers little resistance and has no air pocket in its housing. Because of the rigid pipe system, expansion joints are necessary. The bottom of the basin is designed to prevent excessive sediments from entering the pump. Screens are built into the wall between the basin and suction trough and a trench is provided for cleaning.

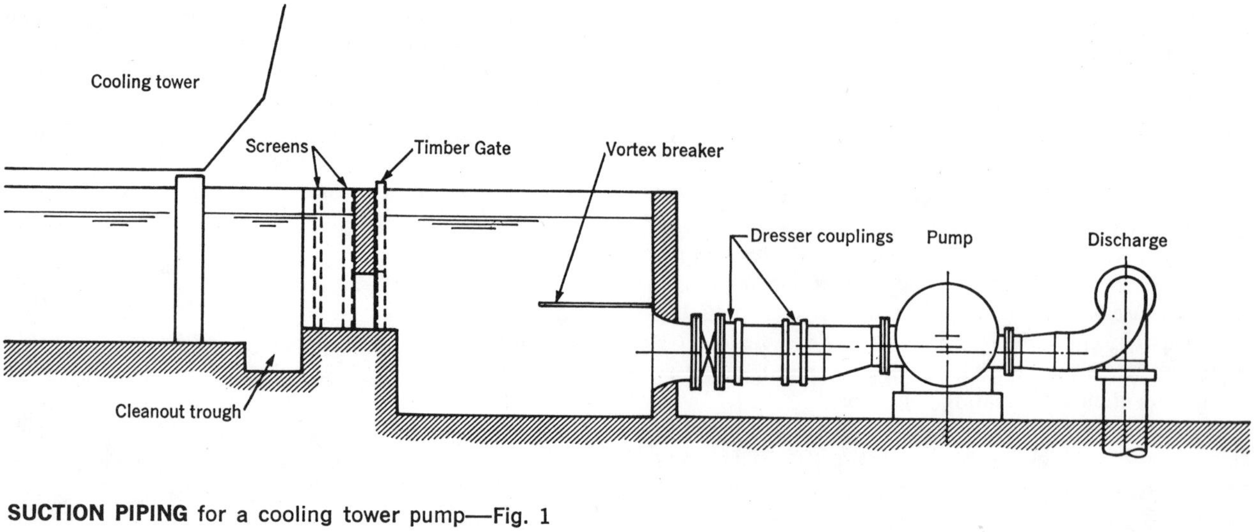

SUCTION PIPING for a cooling tower pump—Fig. 1

Originally published October 11, 1971

Fig. 2 illustrates the Hydraulic Institute recommendations for double inlet suction piping. "Suction piping, for double suction pumps, should not be installed so that there is an elbow close to the suction nozzle of the pump, except when the elbow is in a vertical position. For example, an elbow that is bending either straight up or straight down is permissible, but one entering at an angle should not be used.

"There is always uneven flow in an elbow and when it is installed in any position other than vertical, more liquid enters one side of the impeller than the other. This causes a reduction in capacity and efficiency, and a thrust that may heat the thrust bearing and cause rapid wear."[1]

For transfer of liquids from storage, in-line pumps can be placed directly on the storage tank nozzle. Cradle-mounted pumps should be adjacent to storage tanks. Suction lines to diked tanks are usually long because pumps must be located outside the diked areas. Normally, larger diameter pipe is used to compensate for longer pipe runs. Gate valves in these lines should be installed so that their stems are horizontal, thus eliminating the air pocket in the valve housing.

The head in storage tanks varies and if it falls below the design minimum the pump will cavitate. A vertical pump can be used for pumping a storage tank dry.

When the suction drum is elevated and the liquid is below its boiling point (subcooled), the available static head is usually more than adequate for NPSH and lines losses. Consequently, pipe configuration is not so critical. Valves can be placed in vertical pipe runs and pipe diameter can be the same as the pump suction nozzle. If the suction line is designed without loops or pockets, vapor and air collection are not likely.

Saturated Liquid Flow

Most process pump suction lines involve liquid at or near its boiling point (saturated) and a minimum static head. The suction vessel and draw-off nozzle are elevated to meet the NPSH requirement plus a distance equal to the static height of line losses. The important rule for the designer is that at no point in the line should pressure fall, due to pipe or component resistance, below the vapor pressure of the liquid.

A horizontal pipe run immediately after the draw-off nozzle is undesirable because a gradual pressure reduction (in proportion to the horizontal pipe length and friction) can cause the liquid to vaporize. Also, with a low liquid level in the tank, the available static head might not be sufficient to compensate for exit and horizontal pipe section losses. This is especially true with light, volatile liquids.

A horizontal pipe segment after a vertical drop is not as bad because static head pressure is available at the beginning of the horizontal section. In a vertical segment of suction piping, the potential energy

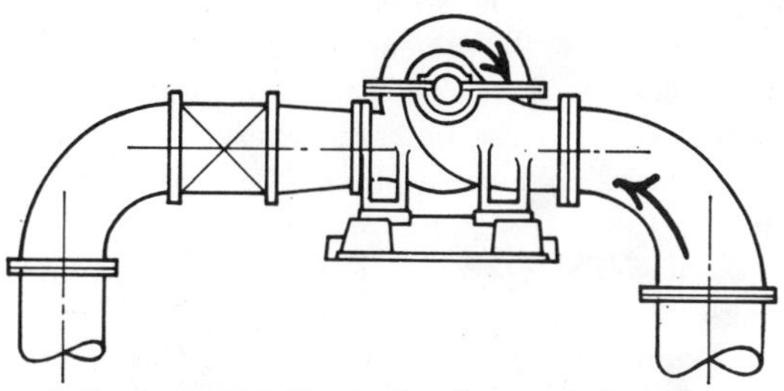

1. Use increaser and long radius elbow on discharge to conserve velocity head.
2. If elbow is necessary it should be of the long radius type.
3. Desirable to locate gate valve beyond increaser. Check valve when needed should be placed inside gate valve.
4. Discharge piping and suction piping should be supported close to the pump flange to prevent vibration and strain on pump casing.

DOUBLE INLET suction piping recommendations from the Hydraulic Institute—Fig. 2

is usually much greater than the energy consumed by friction. Thus, when pumping hot liquids, a vertical pipe run after the draw-off nozzle is most desirable and with light hydrocarbons it can become essential.

With saturated liquid and a minimum static head, suction piping is usually one or two sizes larger than the pump nozzle. Total suction line losses can be estimated at 2 to 3 ft. of liquid head. Fluid velocities are 4 to 8 ft./sec. except when pumping viscous liquids.

Suction Pipe Sizing

Data for sizing suction piping are given in Fig. 3. Pipe sizes should be selected in the shaded portion of the graph. Normally, the choice is limited to one size up to 12-in. dia., and two or three sizes at larger diameters.

The suction line size estimates in Fig. 3 are based on velocities that will give vapor-free flow. Larger piping is required for viscous liquids, and with subcooled water or hydrocarbons in elevated tanks the line can be a size smaller than shown. At small diameters, however, a size reduction might not be practical.

The static head loss for a specific fluid and piping configuration can be calculated from Darcy's equation and Moody's diagram.[2]

NPSH and Piping Design

As emphasized earlier in this article, reliable performance of a centrifugal pump depends primarily on having an adequate net positive section head. The following discussion covers NPSH from the standpoint of the piping systems designer.

NPSH is the pressure above vapor pressure of a liquid being pumped, as measured at the pump suction flange on a gage at the pump centerline. This is the pressure necessary to overcome friction and turbulence losses between the suction nozzle and the entrance to the impeller.

Required NPSH is specified by pump manufacturers, and is usually given as water head on head-capacity curves. It increases with pump capacity. Available NPSH is the potential energy of an installation, expressed in feet of liquid head. It can be calculated as follows:

Available NPSH $= 2.31\ (P/S)$

where

P pressure above vapor pressure at the centerline of the suction flange, psi.

S specific gravity of the flowing liquid

To ensure reliable pump operation when pumping liquids near their boiling point, the available NPSH should be higher than the required value. Practically, this means that process equipment must be elevated higher than that necessary for required NPSH plus friction losses in the suction line.

According to the Hydraulic Institute, "The NPSH requirements of centrifugal pumps are normally determined on the basis of handling water. It is recognized that when pumping hydrocarbons, the NPSH to obtain satisfactory operation may be reduced for certain applications."[1] It is customary in the petrochemicals industry to use manufacturers' data for minimum required NPSH when handling hydrocarbons with specific gravities less than one.

Several other points should be noted: 1) performance curves for American pumps include the velocity head and thus it does not have to be considered in calculations; 2) both available and required NPSH do not vary with the liquid's specific gravity; and 3) for a given centrifugal pump, one curve of required NPSH can be used with all normal specific gravities.

The effect of specific gravity can be explained as follows: "Looking at the analogy of NPSH as a force that pushes liquid into a pump, and assuming a pump that requires 8 ft. NPSH at a given capacity, it can be shown that 8 ft. of water is equal to 3.45 psi., while 8 ft. of gasoline is equal to only 2.60 psi. It would appear the force pushing liquid into the pump decreases with specific gravity. At the same time, however, the mass of liquid being accelerated has gone down by the same proportion."[3]

When choosing between two pumps of equal capacity, the one with the higher NPSH will cost less. But the less expensive pump will be the economic choice only if its cost advantage is not offset by the expense of elevating process equipment. Often a vertical pump will be specified to bring equipment closer to the grade level.

Mounting the Pump

Close-coupled pumps are versatile because they can be mounted horizontally or vertically. A simple arrangement for suction piping is shown in Fig. 4. Here, the deaerator supporting legs serve as suction piping for a vertically mounted pump. The suction nozzle is close to grade, while still satisfying the pump's NPSH requirement.

According to manufacturers, in-line pumps up to 100 gpm. can be mounted in any position. Suctions should be designed with a vertical pipe segment and not more than one elbow. Valves and strainers should be on or close to the draw-off nozzle.

Submerged vertical pumps have no suction piping and the pump housing serves as the suction nozzle.

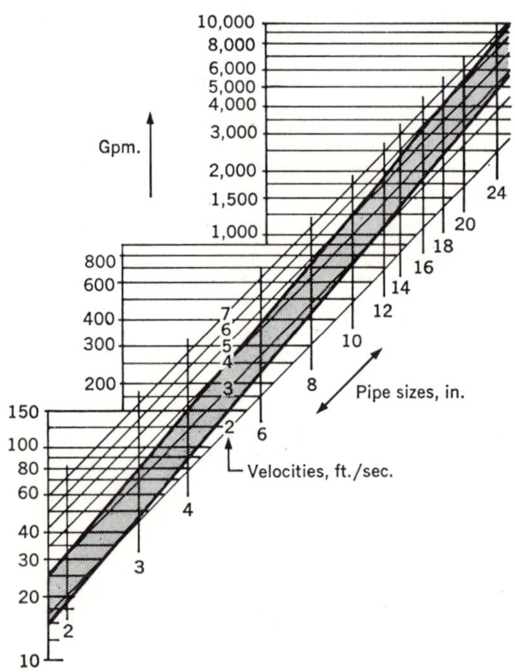

DATA for sizing suction piping—Fig. 3

CLOSE-COUPLED pumps can permit simple layouts for suction piping—Fig. 4

Manufacturers specify minimum submergence required to prevent vortex formation. Greater depth is needed if the pump does not have an inlet strainer. A proposed installation must be checked for adequate submergence and available NPSH equal to or greater than that required.

With barrel-type vertical pumps, part of the static head for required NPSH is below grade. As a result, the suction vessel does not have to be highly elevated. The pump should be close to the vessel for simple pipe layout, as illustrated in Fig. 5. If dimension H in the figure is fixed, H_2 and H_4 can be adjusted over a wide range.

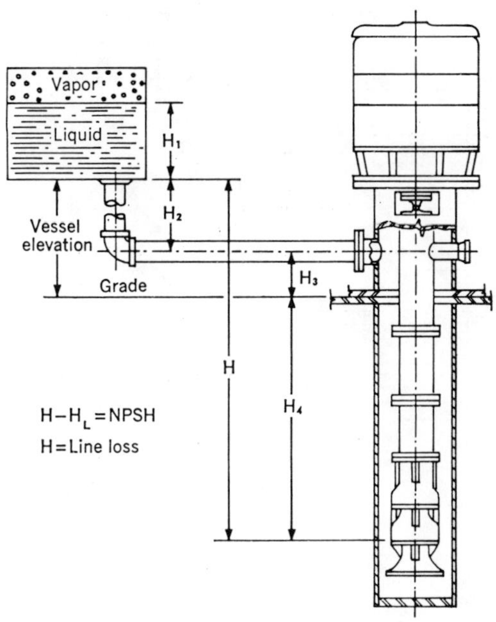

VERTICAL PUMP has part of its static head below grade—Fig. 5

Feed pressure to a centrifugal pump should not fluctuate because suddenly reduced pressure in a hot liquid can release large quantities of vapor. To prevent flashing, a system for injecting cool liquid can be added to the suction piping. This might be necessary, for example, at a boiler feed water pump.

Drawing vapor in a pump is likely to be a problem at greatly reduced flow (startup, for instance) or fluctuating pump capacities. A vent line attached to the top of the pump housing at the first impeller inlet and connecting back to the vapor space of the suction drum might be necessary.

There is danger of cavitation with large pumps when they operate below a minimum flow rate, which is specified by manufacturers. If greatly reduced flow (down to 20% of rated capacity, for example) is expected in the application, a minimum flow bypass will have to be designed in the discharge line, with a throttling valve or restrictive orifice.

When a suction lift is involved, the end of the suction line usually has a foot valve and strainer to prevent liquid from draining out of the pump when it is not running. Some kind of priming arrangement should still be provided since dirt or wear may prevent a tight seal at the valve.

Four simple priming arrangements are shown in Fig. 6. These consist of: A) a suction reservoir; B) a bypass from the discharge to the suction line; C) a separate fill line from another liquid source; and D) an ejector to draw liquid into the pump. At large installations, a vacuum pump may be used instead of the ejector. A pump-case vent is usually essential with priming systems (the arrangement in D is an exception).

Designs of self-priming pumps vary from maker to maker, but the principle is usually the same: a built-in reservoir keeps the pump flooded and serves as an air and vapor separator.

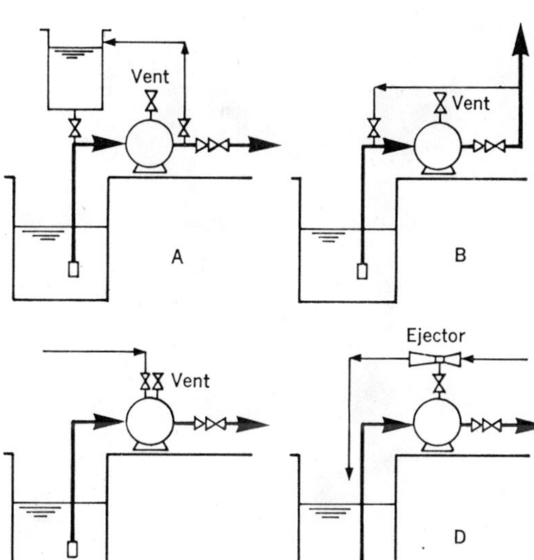

PRIMING methods for centrifugal pumps—Fig. 6

DISCHARGE PIPING

While at the pump suction the main concern is to provide vapor-free flow, at the discharge side piping economy is the principal consideration. It is necessary to reach an economic balance between line sizes, pipe configurations, power requirements and capital costs.

Pump discharge pipe sizes based on the most economical fluid velocities have been discussed previously.[4] The pipe selection chart in Fig. 7 is based on these velocities. Data are presented here for pipe sizes at various flow rates, including carbon steel and alloy steel construction at a payout of two years and a longer payout of five to 10 years. Note that longer payout times tend to favor larger pipe diameters.

Centrifugal pump discharge piping should not have a smaller diameter than the discharge nozzle of the pump. When pumping liquids that are at or near their boiling point, vaporization will occur if line pressure drops below the liquid's vapor pressure, as might happen after a control valve. In this case, friction resistance in the discharge pipe should be calculated as two-phase flow.[5] Otherwise, the pressure required at the pump discharge nozzle can be calculated in the usual manner.

System Economics

A detailed investigation of the most economical pipe sizes is justified if the configuration is long or complicated, if line sizes are large or if the piping material is expensive. The system should also be analyzed for the best arrangement if pipe friction is a major part of the required pump discharge pressure.

With diameters above 12 inches, more than one size may be selected initially because of the close capacity increments in large size pipe. Piping costs, of course, increase with diameter while utility costs decrease. The best size can be determined by adding the total cost of utilities over the period of capital payout to the capital cost of the installation. A similar comparison of sizes may be made with suction piping.

The total head produced by the pump may be increased by going to a larger impeller, which in turn requires a bigger motor for more horsepower. In some cases, it may be more economical to redesign the discharge piping for lower pressure drop rather than invest in a larger pump and motor.

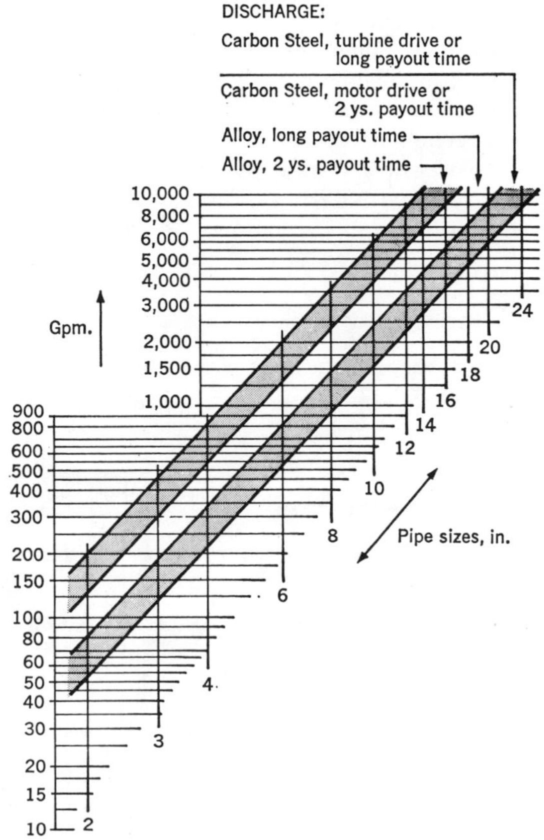

SIZE of discharge piping can be estimated—Fig. 7

The author

Robert Kern is head of the plant-layout section in the corporate engineering department of Hoffmann-La Roche Inc., Nutley, NJ 07110. He is a specialist in hydraulic-systems design, plant layout, piping design and economy. He is the author of a number of articles in these fields, and has taught several courses for the design of process piping, plant layout, graphic piping and flow systems, both in the U.S. and South America. Previously, he was associated with M. M. Kellogg Co. in England and the U.S. Mr. Kern has an M.S. in mechanical engineering from the Technical University of Budapest, and is a member of AIChE.

References

1. "Hydraulic Institute Standards," Hydraulic Institute, 122 E. 43rd St., New York, NY 10017.
2. "Flow of Fluid Through Valves, Fittings and Pipes," Technical Paper 410, Crane Co., 4100 S. Kenzie Ave., Chicago, Ill. 60632.
3. "Cameron Pump Manual," Ingersoll-Rand Co., Phillipsburg, NJ 08865.
4. Braca, R.M. and Happel, J., Economics of Pipe Sizing, *Chem. Eng.*, Jan. 1953, p. 180.
5. Kern, Robert, How to Size Process Piping for Two-Phase Flow, *Hydrocarbon Processing*, Oct. 1969, p. 105.

Piping around control valves

Control-valve piping should be arranged for ease of maintenance, and to ensure that solids are kept out of the valves.

John D. Constance, Consultant

☐ Control valves often have to be removed from pipelines for maintenance, so the piping around them has to be flexible enough to permit continued operations under such conditions.

Usually, the control valve is one line-size smaller than the main line itself. If it is the same as the line size, or more than two line sizes smaller, the system flow characteristics should be reexamined, and the control valve resized if necessary.

For the best setup from the operating and maintenance points of view, block valves should be installed, one on each side of the controller, together with a throttling-type globe valve for use as a bypass. During normal operation, the bypass valve remains closed, with flow taking place through the controller. Fig. 1a through l show various arrangements of piping around control valves, from those using small screwed-pipe connections, to flanged valves and pipe, and combinations of both.

The greatest source of trouble with control valve opera-

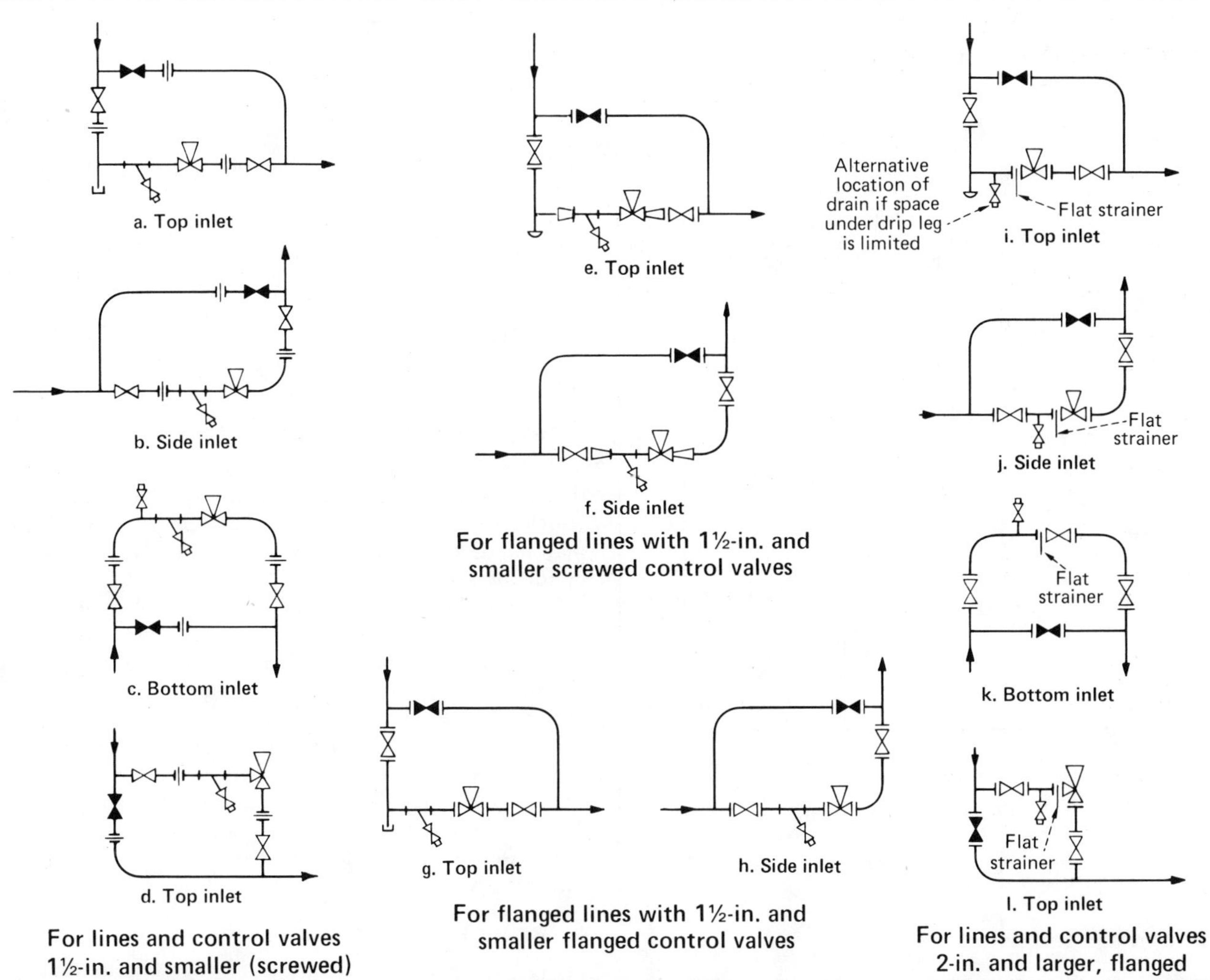

Control-valve bypasses — Fig. 1

Originally published November 5, 1979

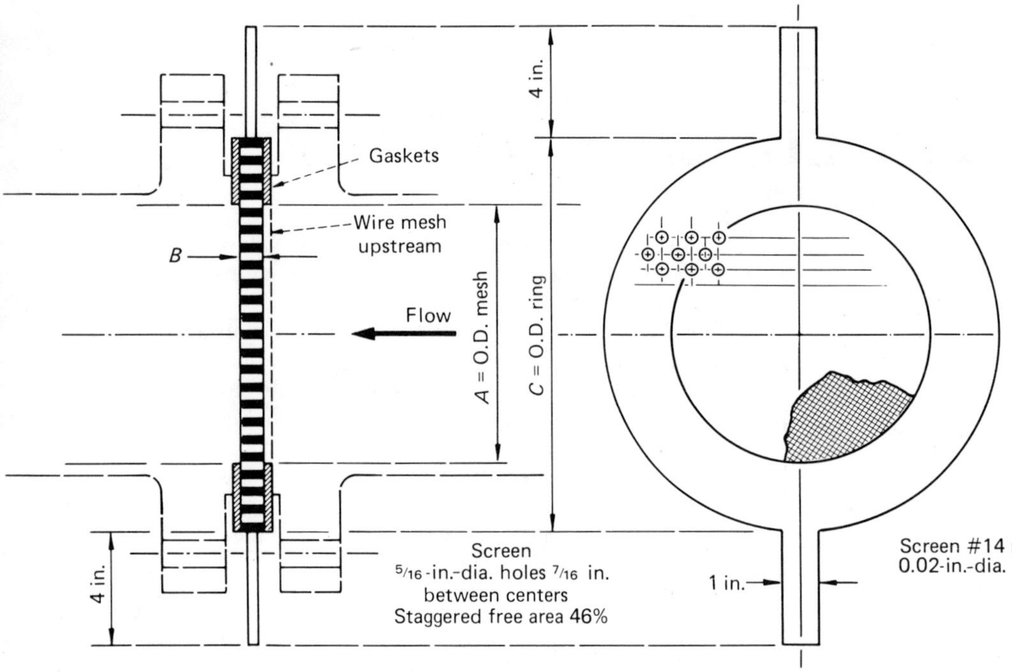

150 and 300 psi ANSI std. raised face

Pipe size, in.	A, in.	B, in.	C, in. 150 psi	C, in. 300 psi
2	$1^3/_4$	#10 USG	$3^3/_4$	4
$2^1/_2$	$2^1/_4$	#10 USG	$4^1/_2$	$4^3/_4$
3	$2^3/_4$	#10 USG	5	$5^1/_2$
4	$3^3/_4$	#10 USG	$6^1/_2$	$6^3/_4$
6	$5^3/_4$	$^3/_{16}$	$8^3/_4$	$9^1/_2$
8	$7^3/_4$	$^3/_{16}$	$10^5/_8$	$11^3/_4$
10	$9^3/_4$	$^1/_4$	13	$13^7/_8$
12	$11^3/_4$	$^1/_4$	$15^3/_4$	$16^1/_4$
14	13	$^1/_4$	$17^3/_8$	$18^3/_4$
16	15	$^1/_4$	$19^7/_8$	$20^7/_8$
18	17	$^1/_4$	$21^1/_4$	$23^1/_8$
20	19	$^1/_4$	$23^1/_2$	$25^3/_8$

Flat-type strainers for control valves — Fig. 2

Main line size, in.	Control valve size, *A*, in. 1		1½		2		3		4		6		8		10		12	
	B	*C*	*B*	*C*	*B*	*C*	*B*	*C*	*B*	*C*	*B*	*C*	*B*	*C*	*B*	*C*	*B*	*C*
1	1	1																
1½	1½	1½	1½	1½														
2	1½	2	2	2	2	2												
3			2	2	3	2	3	2										
4					3	3	4	3	4	3								
6							4	4	6	4	6	4						
8									6	6	8	6	8	6				
10											8	8	10	8	10	8		
12													10	10	12	10	12	10

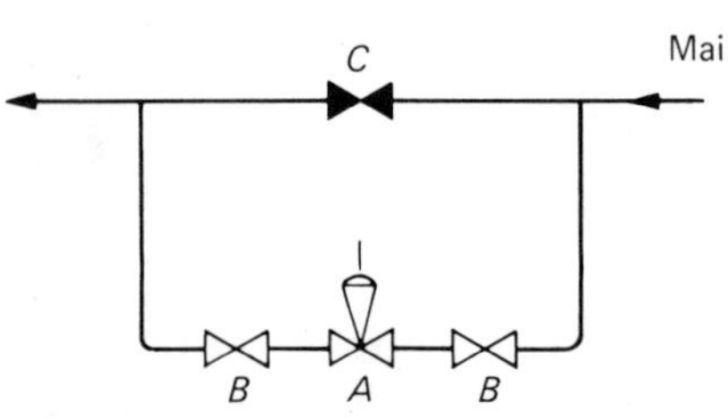

A = Control valve
B = Block
C = Bypass

Piping around control valves — Fig. 3

tion is from foreign matter in the pipeline. Therefore, strainers should be installed ahead of all control valves, with drip legs ahead of the strainers. Y-type strainers of reputable manufacture with 20-mesh screens should be used with all types of control valves $1^1/_2$ in. and smaller, as shown in Fig. 1a through h. Flat-type strainers with 14-mesh screening, as shown in Fig. 2, should be used with control valves 2 in. and larger, as shown in Fig. 1i through l.

Sizing block and bypass valves

The block valves should be full line size, except when the control valve is two or more sizes smaller than that, in which case the block valves should be one size smaller than line size. The block valve material and rating should be for the service specified.

Bypass valves and bypass lines at control valves should be main-line size for lines 2 in. and smaller. For lines 3 in. and larger, the bypass line and bypass valves should be one normal pipe size smaller than main-line size (except for gravity flow, where they should be line size).

Bypass valves should be globe valves in sizes up to and including 4 in.; above this size, they should be gates or, in special cases, gear-operated cocks where low pressure-drops in large lines are encountered. Fig. 3 provides a quick indication of block and bypass valve sizing for control valves from 1 in. through 12 in.

Swaging at control valves

With screwed control valves, a swage nipple should be screwed into the control valve body, with the unions at each side of the control valve (used for its removal) located at the large end of the swage nipple. No heavier or lighter swages than Sched. 80 should be used. Where valves are flanged and less than line size, the swaging should be immediately adjacent to the control valve.

Piping around control valves

The line piping adjacent to control valves should be arranged to provide flexibility for removing them. For example, the block valves on either side of a control valve should neither be at the same elevation as the control valve nor both in the same plane with it.

One block valve may be in the same horizontal plane as the control valve, and the other block valve in the vertical, with the bypass below the control valve. Or, the two block valves may be in the vertical plane, with the control valve horizontal, and with the bypass below the control valve.

Several arrangements are permissible, depending on the direction of flow, and elevation of the control valve with respect to the flow line. However, the piping around control valves should be arranged so that the lines need not be sprung to remove the control valve. Special attention should be given to the location of the control valve so that it is not pocketed in steam, air or gas flow and will not catch such foreign materials as oil scale, welding beads, and the like.

One-inch valved connections should be provided in the bottom of the line downstream of the first block valve and before the downstream block valve to serve as telltale indicators, to permit checking the tightness of the block valves should it be required that the control valve be bypassed for removal.

The piping around control valves should be self-supporting or should be permanently supported. This is required in order that when the control valve or block valve is removed, the lines should remain in place without the need of building scaffolding or providing other temporary supports.

Control-valve location

All control valves should be installed where they are accessible from equipment platforms and walkways, or at grade, so that they can be easily maintained. Sufficient clearance should be provided above and below control valves so that the bottom flange and plug (or the topwork and plug) may be removed, for repair or replacement, with the valve in place.

When vaporization occurs in control valves, they should be placed as close as possible to the destination of the steam, in order to reduce pressure drop, surging and excess lengths of large pipeline downstream of the control valve.

The author

John D. Constance is a consultant located at 625 Hudson Terrace, Cliffside Park, NJ 17010. He has had a long career as a design engineer, working for such companies as Ralph M. Parsons Co., Stone & Webster Engineering Co. and the M. W. Kellogg Co. He holds a B.S. in chemical engineering from New York University, and has taught at the Cooper Union (New York, N.Y.). He is a registered professional engineer in New York and New Jersey. An authority on engineering registration, he has written several books on the subject.

Piping Design for Two-Phase Flow

Here is the simplest and least frustrating method for sizing pipes when two-phase flow exists in the pipeline. And useful tips that exploit the inherent flexibility in the distribution of pressure losses in a piping system will let the designer get reasonable sizes for pipes and components.

ROBERT KERN, Hoffmann - La Roche Inc.*

Two-phase-flow theories and experiments have a threefold significance for the process-piping designer. It has been shown that:

1. If the vapor content of a liquid line increases, the friction loss is greater than the single-phase liquid pressure loss, and is greater than the pressure loss calculated with the average density.
2. For a given vapor-liquid ratio and associated physical properties, a characteristic flow pattern develops.
3. Between the various flow patterns, unit pressure losses can differ when comparing borderline cases.

Piping design for two-phase flow has been investigated by a great number of researchers through rational and empirical steps [*1,2*]. And limitations, generalizations and simplifications have been introduced for providing practical methods of design.

In this article, we will assume that two-phase flow is isothermal, turbulent in both the liquid and vapor phases, and steady (liquid and vapor move with the same velocity), and that pressure loss is not more than 10% of the absolute downstream pressure.

An often-asked question is: How accurate are two-phase-flow calculations? If actual pipelines closely resemble the experimental conditions, deviations are small. The application of correlations for two-phase flow to process-piping design is arbitrary. Experiments are usually done with small-diameter, straight, and relatively short pieces of horizontal or vertical pipe. Under laboratory conditions, flow patterns are kept constant and flow conditions consistent. However, most process piping probably has changing flow patterns in various segments of the line because of the three-dimensional pipe configurations in which one finds horizontal and vertical runs, elevation changes, offsets, branch connections, manifolds, pipe components, reducers and other restrictions. Sizable deviations can be expected in prediction of friction loss compared to actually measured values. Because of this, pipe runs for two-phase flow should be short and simple.

In this article, we present the simplest, and in its practical application the least frustrating, method of design from among the many available ones. Let us compute the resistance for two-phase flow in two main steps. These are:

Select a possible flow pattern by calculating the coordinates of a flow-region chart.

Determine unit pressure losses by calculating only the vapor-phase unit loss, corrected by an applicable correlation for two-phase flow.

Two-Phase-Flow Regions: Baker Parameters

We select two-phase-flow patterns [*3*] from Fig. 1. The borders of the various flow patterns in Fig. 1 are shown as lines. In reality, these borders have rather broad transition zones [*4*].

We can establish a particular flow region from the Baker parameters, B_x and B_y. From data supplied or usually available to the piping designer, the Baker parameters can be expressed as:

$$B_y = 2.16 W_v / A \sqrt{\rho_l \rho_v} \qquad (1)$$

B_y depends on the vapor-phase flowrate, vapor and liquid densities, and pipe size. The practical significance of pipe size is that by changing pipe diameters, the type

*To meet your author, see p. 44.

Originally published June 23, 1975

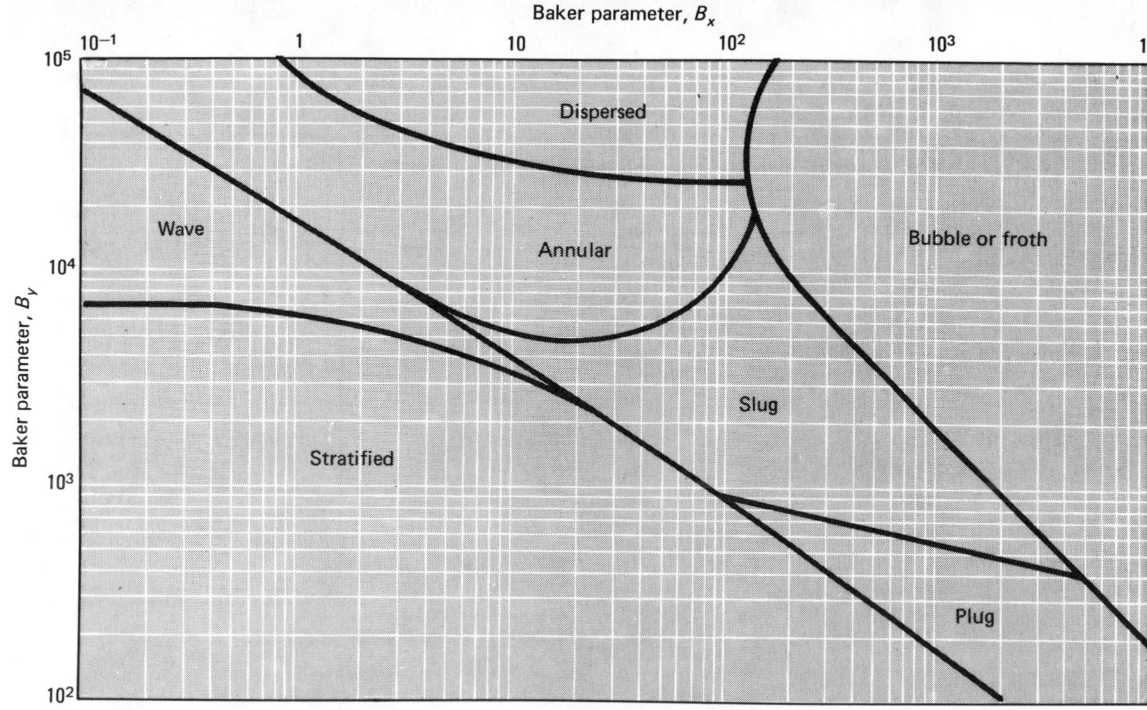

Two-Phase Flow Correlations						
Dispersed	**Bubble**	**Slug**	**Stratified**	**Wave**	**Plug**	**Annular**
Use Fig. 3. and Eq. (3)	$\phi = \frac{14.2\,X^{0.75}}{(W_l/A)^{0.1}}$	$\phi = \frac{1{,}190\,X^{0.815}}{(W_l/A)^{0.5}}$ Avoid slug flow	$\phi = \frac{15{,}400\,X}{(W_l/A)^{0.8}}$ Horizontal pipe	Use Fig. 5 and Eq. (9) and (10) Horizontal pipe	$\phi = \frac{27.315\,X^{0.855}}{(W_l/A)^{0.17}}$	$\phi = aX^b$ $a = 4.8 - 0.3125\,d$ $b = 0.343 - 0.021\,d$ d = I.D. of pipe, in For pipe 12-in and over, use $d = 10$.

Courtesy: Mr. Ovid Baker and The Oil and Gas Journal.

BAKER parameters determine type of two-phase flow, and appropriate two-phase-flow correlation sets unit loss—Fig. 1

of flow might also be changed, which in turn changes friction losses in the pipe.

$$B_x = 531\left(\frac{W_l}{W_v}\right)\left(\frac{\sqrt{\rho_l\rho_v}}{\rho_l^{2/3}}\right)\left(\frac{\mu_l^{1/3}}{\sigma_l}\right) \quad (2)$$

In Eq. (2), we can substitute the ratio of percent liquid to percent vapor for W_l/W_v, and

$$\sqrt{\rho_l\rho_v}/\rho_l^{2/3} = (\rho_v)^{0.5}/(\rho_l)^{0.166}$$

B_x depends on the weight-flow ratio and the physical properties of the liquid and vapor phases. Once calculated, B_x does not change with alternative pipe diameters. The position of the B_x line in Fig. 1 changes only if the liquid-vapor proportion changes and, to a much lesser extent, if the physical properties of the concurrently flowing liquid and vapor change. This can occur in long pipelines where relatively high friction losses reduce the pressure. Consequently, the vapor content of the mixture increases with a corresponding decrease in vapor density. The B_x line will shift somewhat to the left.

In Eq. (2), σ_l is the liquid-phase surface tension [*7,8*]. For the surface tension of water at various temperatures, see the chart by Yaws and Setty [*9*]. For the surface tension of paraffinic hydrocarbons and mixtures, consult the "Engineering Data Book" [*10*].

The intersection of B_x and B_y in Fig. 1 determines the flow region for the calculated liquid-vapor ratio and the physical properties of the liquid and vapor. With increasing vapor content, the point of intersection moves up and to the left.

Unit Losses for Two-Phase Flow

The calculations of unit losses for vapor-liquid mixtures are based on the method of Lockhart and Martinelli [*5*]. Only the essential relationships are repeated here. We will use these with the customary data for practical piping design. The general equation is:

$$\Delta p_{100(two\text{-}phase)} = \Delta p_{100(vapor)}\phi^2 \quad (3)$$

We calculate the pressure drop of the vapor phase by assuming that only vapor flows in the pipeline. We then correct the calculated vapor-phase unit loss with the correlations shown in Fig. 1. Most of these correlations result from experiments with large-scale, horizontal industrial piping [*4*].

The forms of the correlations (in Fig. 1) are identical:

$$\phi = aX^b \quad (4)$$

where a includes the vapor-phase flowrate and the pipe's cross-section, and b is a constant (in annular flow, only

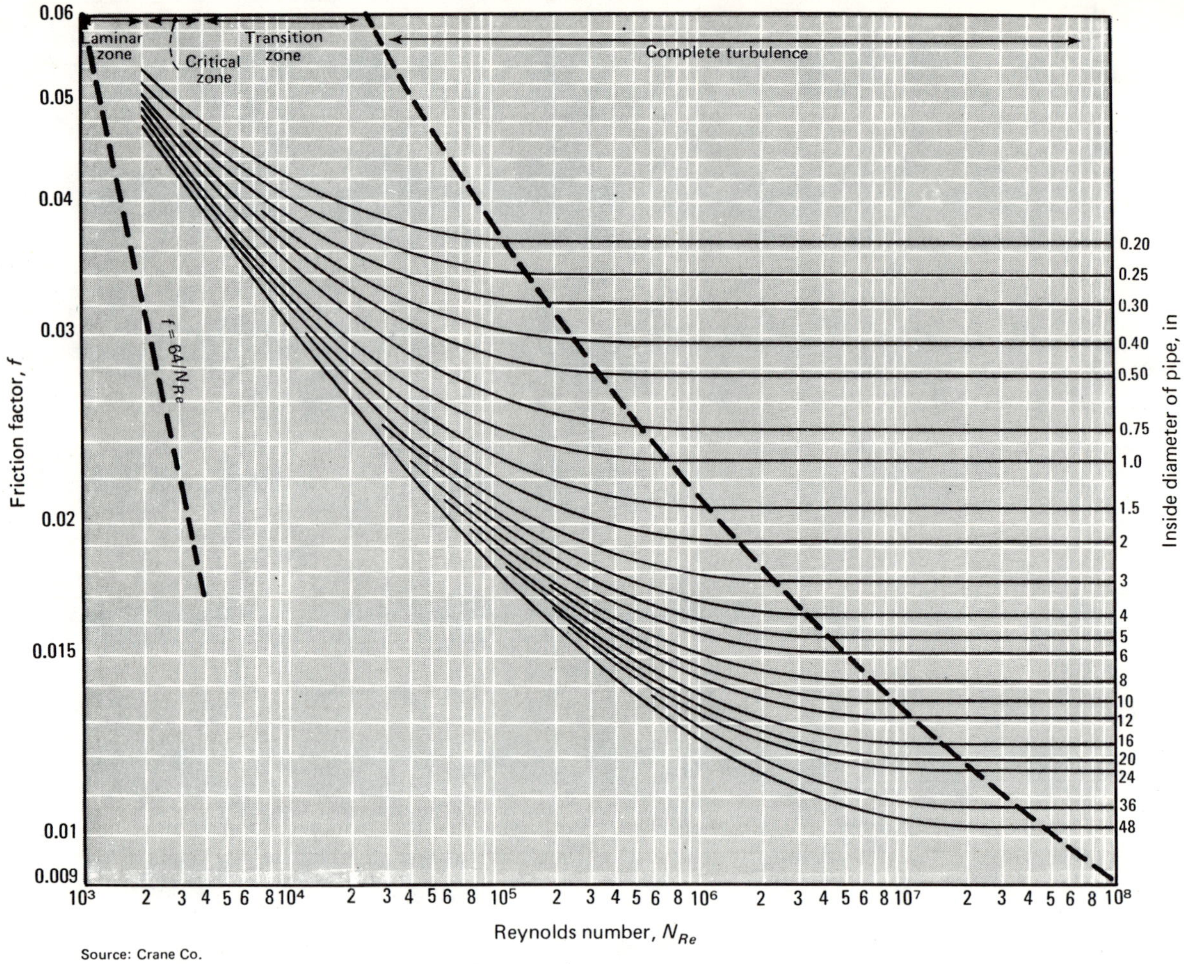

FRICTION FACTORS in new commercial-steel pipe for vapor-phase or liquid-phase flow—Fig. 2

pipe diameters appear as variants in a and b), and X is the Lockhart-Martinelli, two-phase-flow modulus:

$$X^2 = \Delta p_{100(liquid)}/\Delta p_{100(vapor)} \quad (5)$$

In Eq. (5), $\Delta p_{100(liquid)}$ is calculated by assuming only liquid flows in the pipe, and $\Delta p_{100(vapor)}$ by assuming only vapor flows in the same size pipe. The modulus, X^2, remains constant for one set of flow conditions and is independent of pipe size within two to three sequential diameters.

After inserting Darcy's equation in the numerator and denominator of Eq. (5) and simplifying, the two-phase-flow modulus becomes:

$$X^2 = (W_l/W_v)^2(\rho_v/\rho_l)(f_l/f_v) \quad (6)$$

where f_l is the liquid-phase and f_v the vapor-phase friction factor. The modulus can be obtained directly by calculating the liquid-phase and vapor-phase Reynolds numbers and using Fig. 2. (Alternatively, friction factors can be determined from Fig. 5 and 6 on p. 92).

Reynolds numbers are calculated separately for the vapor and liquid phases by using the same pipe diameter, corresponding flowrates and viscosities from:

$$N_{Re} = 6.31\,W/d\mu$$

A convenient form of Darcy's equation for unit pressure-loss calculations in pipelines for liquid or vapor is restated here:

$$\Delta p_{100} = 0.000336(f\,W^2)/d^5\rho \quad (7)$$

In Eq. (7), we use the same diameter for the liquid-phase and vapor-phase calculations, and the corresponding phase flowrate, density and friction factor. Also, in Eq. (7), we must use the Moody friction factors from Fig. 2.

A generalized form suggested by Blazius for the Lockhart-Martinelli relation expresses the friction factors for turbulent flow as:

$$f_l = 0.046/(N_{Re})_l^{0.2}$$
$$f_v = 0.046/(N_{Re})_v^{0.2}$$

Inserting these values and the equivalent relation for the Reynolds numbers into Eq. 6, and simplifying, yields:

$$X^2 = (W_l/W_v)^{1.8}(\rho_v/\rho_l)(\mu_l/\mu_v)^{0.2} \quad (8)$$

The Blazius friction-factor expression does not take into account the relative roughness of the pipe wall and the consequent relative reduction of the friction factor as the pipe diameter increases.

The widely used Moody friction factor (Fig. 2) will

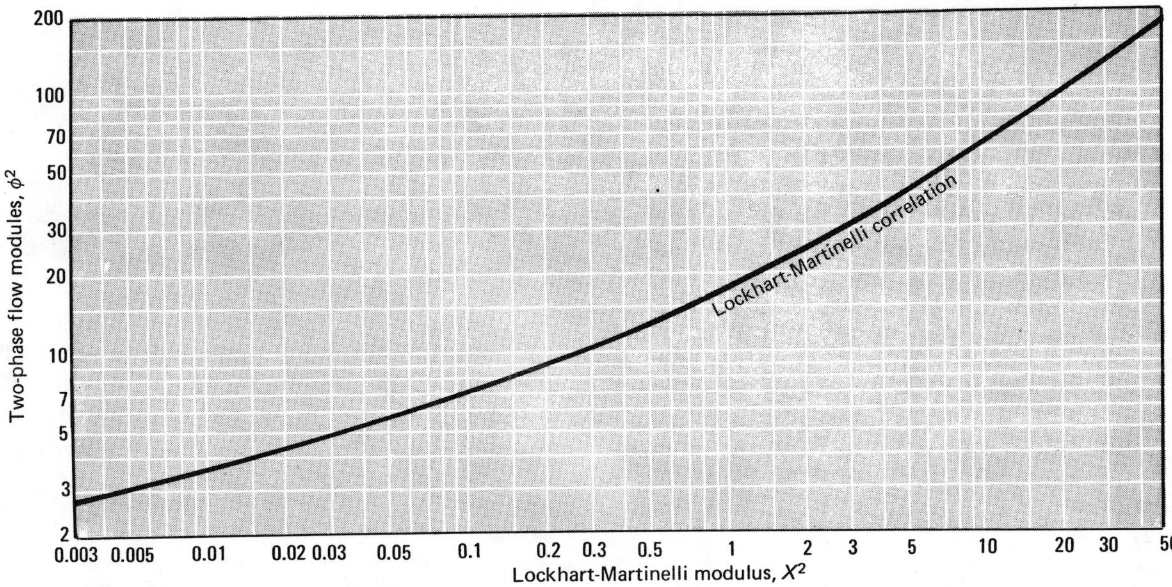

LOCKHART-MARTINELLI correlation relates vapor and liquid properties to establish two-phase flow modulus—Fig. 3

Nomenclature

A	Internal cross-section of pipe, ft²
D	Internal diameter of pipe, ft
d	Internal diameter of pipe, in
L	Equivalent length of pipe and fittings, ft
M	Molecular weight
W	Flowrate, lb/h
Δp	Pressure loss, psi
ΔP	Available pressure difference, psi
μ	Viscosity, cp
ρ	Density, lb/ft³
σ	Surface tension, dyne/cm
Dimensionless Quantities	
B_x	Baker parameter for Eq. (2)
B_y	Baker parameter for Eq. (1)
f	Friction factor for Eq. (7)
f_H	Huntington friction factor (Fig. 5)
H_x	Huntington correlation in Eq. (9)
N_{Re}	Reynolds number
X	Lockhart-Martinelli modulus
ϕ	Two-phase-flow modulus
Subscripts	
l	Refers to liquid
v	Refers to vapor
100	Refers to 100 ft of pipe

give somewhat smaller values for f_l/f_v, and consequently smaller unit losses than the Blazius form of f_l/f_v, depending on pipe size and liquid-vapor proportion.

As with all line-sizing procedures, pipe sizes must first be estimated. After selecting a pipe size, the flow-region coordinates can be calculated and the flow type determined from Fig. 1. After finding the vapor-phase unit loss and the applicable two-phase-flow correlation (Fig. 1), unit losses for two-phase flow can be calculated from Eq. (3).

Overall friction loss in the pipeline between two points will be:

$$\Delta p_{(two\text{-}phase)} = \Delta p_{100(two\text{-}phase)}(L/100)$$

where L is the equivalent length of pipe and fittings.

Dispersed Flow

When nearly all of the liquid is entrained as spray by the gas, the flow is dispersed. This type of flow can be expected when the vapor content is more than roughly 30% of the total weight flowrate. Some overhead-condenser and reboiler-return lines have dispersed flow.

The curve in Fig. 3 represents the Lockhart-Martinelli correlation for all pipe sizes and for turbulent-liquid and turbulent-vapor phases. Turbulence for this correlation is defined as $(N_{Re})_l \geqq 2{,}000$, and $(N_{Re})_v \geqq 2{,}000$.

Let us find the unit loss in a 12-in Schedule 40 line ($d = 11.938$ in, $d^5 = 242{,}470$ in^5, $D = 0.9965$ ft, $A = 0.7773$ ft^2) for the following flow data:

	Liquid	Vapor
Flow, W, lb/hr	300,000	350,000
Molecular weight, M	78.8	75.7
Density, ρ, lb/ft^3	33.5	2
Viscosity, μ, cp	0.1	0.01
Surface tension, σ, dyne/cm	5.7	—

We first calculate the Baker parameters from Eq. (1) and (2):

$$B_x = 531\left(\frac{300{,}000}{350{,}000}\right)\left(\frac{\sqrt{33.5(2)}}{33.5^{0.67}}\right)\left(\frac{0.1^{0.33}}{5.7}\right) = 29.3$$

$$B_y = \frac{2.16(350{,}000)}{0.7773\sqrt{33.52(2)}} = 118{,}800$$

We enter Fig. 1 with these coordinates, and find they intersect in the dispersed-flow region. Next, we calculate the modulus X^2 from Eq. (8):

$$x^2 = \left(\frac{300{,}000}{350{,}000}\right)^{1.8}\left(\frac{2}{33.5}\right)\left(\frac{0.1}{0.01}\right)^{0.2} = 0.072$$

From Fig. 3, we find $\phi^2 = 6.3$ for $X^2 = 0.072$.

We determine the vapor-phase unit loss from Eq. (7). To do so, we must find $(N_{Re})_v$ and f_v, as follows:

$$(N_{Re})_v = \frac{6.31(350{,}000)}{11.938(0.01)} = 18.5 \times 10^6$$

With this Reynolds number, we find $f_v = 0.013$ from Fig. 2. By substituting the appropriate values into Eq. (7), we find the vapor-phase unit loss:

$$\Delta p_{100(vapor)} = \frac{0.000336(0.013)(350{,}000)^2}{242{,}470(2)}$$

$$\Delta p_{100(vapor)} = 1.1 \text{ psi/100 ft}$$

And, from Eq. (3), we determine the dispersed-flow unit loss as:

$$\Delta p_{100(dispersed)} = 1.1 \times 6.3 = 6.93 \text{ psi/100 ft}$$

Bubble Flow

Bubble flow develops when vapor or gas bubbles, distributed in the liquid, move along at about the same velocity as the liquid. This type of flow can be expected when the vapor content is less than about 30% of the total weight-flowrate. The sizing procedure is similar to that for dispersed flow. The correlation for bubble flow in Fig. 1 should be used.

Slug Flow

Slug flow develops when waves of liquid are picked up periodically in the form of frothy slugs that move at a greater velocity than the average liquid velocity. Typically, this phenomenon may occur in a pocketed line between an overhead condenser at grade and an elevated reflux drum. At low velocities, liquid collects at the low point and periodically can become a slug in the line. With a sufficiently high velocity, the liquid phase can be carried through without developing slug flow.

Unit losses for slug flow in process piping are generally not calculated. In borderline cases, unit losses for slug flow are smaller than those for bubble or annular flow.

Slug flow causes pressure fluctuations in piping, which can upset process conditions and cause inconsistent instrument sensing. Slug flow can be avoided in process piping by:

■ Reducing line sizes to the minimum permitted by available pressure differentials.

■ Designing for parallel pipe runs that will increase flow capacity without increasing overall friction loss.

■ Using valved auxiliary pipe runs to regulate alternative flowrates and thus avoid slug flows.

■ Using a low-point effluent drain or bypass.

■ Arranging the pipe configurations to protect against slug flow. For example, in a pocketed line where liquid can collect, slug flow might develop.

Slug flow will not occur in a gravity-flow line. A hard-tee connection (i.e., flow through the branch) at a low point can provide sufficient turbulence for more-effective liquid carryover.

A diameter adjustment, coupled with gas injections, can also alter a slug-flow pattern to bubble or dispersed flow. Gas addition (used solely to avoid slug flows) can be expensive.

Let us determine a reasonable pipe size, downstream of the control valve, for the piping configuration sketched in Fig. 4. The available pressure difference, ΔP, is 10 psi, including that for the control valve. The two-phase flow data in the line after the control valve are:

	Liquid	Vapor
Flow, W, lb/h	59,033	9,336
Molecular weight, M	79.47	77.2
Density, ρ, lb/ft^3	31.2	1.85
Viscosity, μ, cp	0.11	0.0105
Surface tension, σ, dyne/cm	5.07	—

From Eq. (1) and (2), we calculate the Baker coordinates for several sizes of Schedule 40 pipe, including and larger than 4 in. Using the Baker coordinates, we establish the flow region from Fig. 1. The results are:

Pipe Size, In	Baker Coordinates, B_x	B_y	Flow Region
8	245	7,650	Bubble/Slug
6	245	13,250	Bubble
4	245	30,000	Bubble

Flow in the 8-in line falls on the borderline of slug

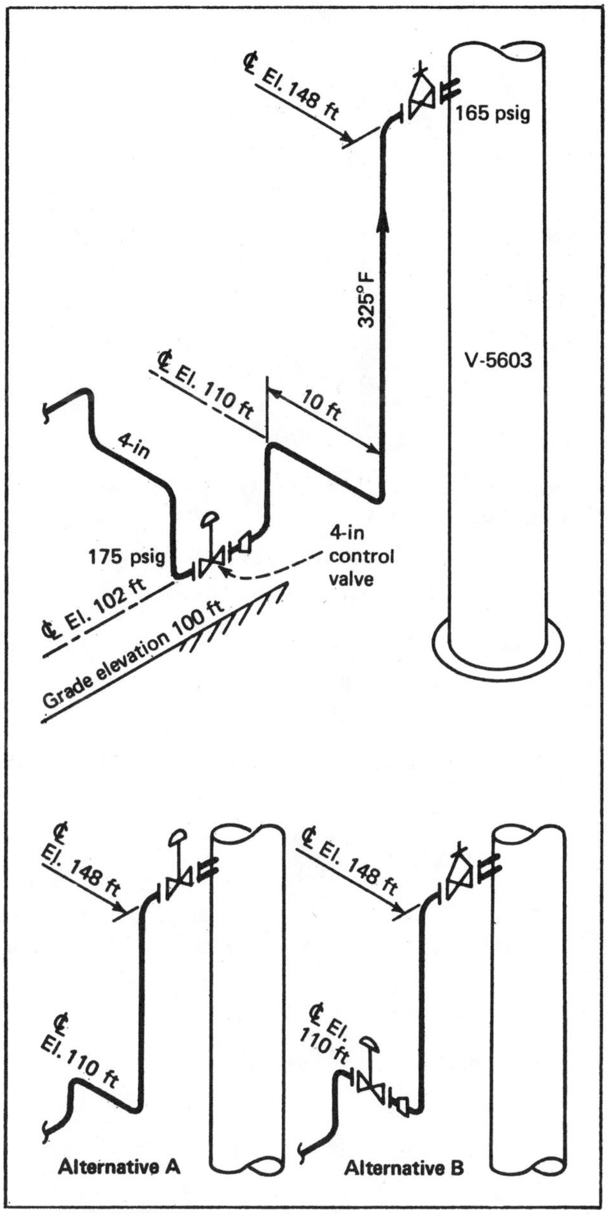

CONFIGURATIONS of piping for sample problem—Fig. 4

and bubble flow. In the 6-in line, we will assume both bubble and slug flow.

Using the correlations for the appropriate two-phase flow (see Fig. 1), we compute the unit losses and the overall friction losses, as summarized here:

Line Size, In	Two-Phase Δp_{100}, Psi	Equivalent Length, Ft	Total Δp, Psi	Flow Region
8	0.13	204	0.27	Slug
6	0.3	168	0.48	Slug
6	0.96	168	1.6	Bubble
4	6.9	132	9.1	Bubble

Reviewing the data in the preceding table, we find that a 6-in line is the optimum size. The 8-in line falls in the slug-flow region, which should be avoided. The total loss in the 4-in line of 9.1 psi plus the static-head backpressure of 3.15 psi (computed on the basis of two-phase density) is greater than the available ΔP of 10 psi. Consequently, the 4-in line cannot be used.

There is always an uncertainty in making reasonably accurate predictions of pressure loss, due to two possible types of flow. Vertical upflow and flashing in the line are affected proportionally by the decreasing static-head and friction-loss gradients. Overall friction can be assumed to range from 0.5 to 2 psi. In this example, if we add the static-head backpressure of 3.15 psi to the overall friction loss, we obtain a range of 4.85 to 6.35 psi for the control valve, i.e.:

$$\Delta P_{cv} = 10 - (2 + 3.15) \text{ to } 10 - (0.5 + 3.15)$$

A 4-in double-seated conventional control valve can accommodate this range within acceptable operating points.

Alternative locations for the control valve are also shown in Fig. 4. For alternative A, slug flow cannot develop. For alternative B, a shortened and self-draining pipeline improves the pipe configuration but at the expense of convenient access to the control valve. In all cases, there will be considerable turbulence after the control valve, which helps to provide slugfree liquid-phase carryover.

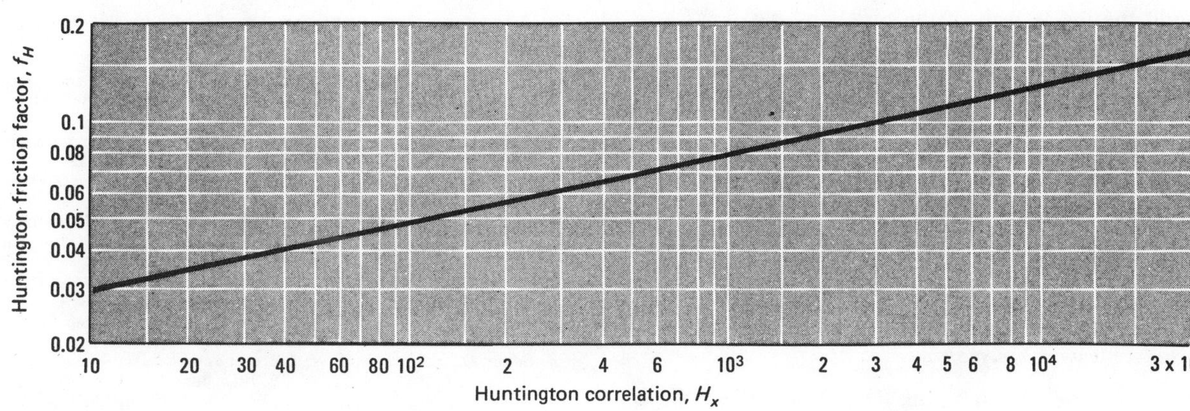

UNIT losses that occur during wave flow in pipelines for use with the Huntington correlation—Fig. 5

This example emphasizes two points. First, the majority of process-piping systems have an inherent flexibility in the pressure-loss distribution. Second, a control valve is able to operate within wide, available, pressure differentials. The process-piping designer can exploit these advantages for reasonable piping and components sizing.

There are a number of ways to adjust the pressure-loss distribution in a pipe system. Pipe sizes can be changed. A section of pipeline can be designed with increased or decreased pipe diameter. The amount of energy that can be lost at the terminating point of the pipeline can be varied. The static head of elevated vessels can be adjusted. Valve and orifice restrictions can be changed to consume more or less pressure differential. Within design limits, a pump can work against changing pipe resistances.

Stratified Flow

In horizontal lines, liquid flows along the bottom of the pipe, and the gas flows over a smooth gas-liquid interface. For stratified flow, a correlation is given in Fig. 1.

Wave flow is similar to stratified flow except the interface has waves moving in the direction of flow. A useful method for calculating the unit loss for wave flow is the Schneider-White-Huntington [6] correlation, H_x:

$$H_x = (W_l/W_v)/(\mu_l/\mu_v) \tag{9}$$

After computing H_x, we find the friction factor, f_H, from Fig. 5, and use it in the Darcy relation, Eq. (7), to determine the unit loss:

$$\Delta p_{100(wave)} = 0.000336 f_H (W_v)^2/d^5\rho_v \tag{10}$$

Plug flow develops when the interface is high in the pipe's cross-section. The waves congregate and touch the high point of the pipe wall. Here, plugs of liquid alternating with pockets of gas in the liquid close to the upper wall move along the horizontal line.

Gravity-flow lines have such types of flow. The initial, minimum pipe sizes can be selected with velocities typical of those for pump-suction lines.*

Annular Flow

Annular flow develops when liquid forms a film or ring around the inside wall of the pipe, and gas flows at a higher velocity as a central core.

The annular-flow relation in Fig. 1 correlates well with pipe sizes up to 10 in. For pipes larger than 10 in, use $d = 10$.

Let us find the unit loss in a 6-in, Schedule 40 line handling the following flows:

	Liquid	Vapor
Flow, W, lb/h	6,150	21,500
Density, ρ, lb/ft^3	52	1.92
Viscosity, μ, cp	0.1	0.01
Surface tension, σ, dyne/cm	6.25	—

Pipe data are: $d = 6.065$ in, $d^5 = 8{,}206$ in^5, and $A = 0.2$ ft.^{2}

*See pp. 30–32.

Using Eq. (1) and (2), we calculate the Baker parameters as:

$$B_x = \left(\frac{6{,}150}{21{,}500}\right)\left(\frac{\sqrt{52(1.92)}}{52^{0.67}}\right)\left(\frac{0.1^{0.33}}{6.25}\right) = 8$$

$$B_y = \frac{2.16(21{,}500)}{0.2\sqrt{52(1.92)}} = 23{,}200$$

With these coordinates, we find that the point of intersection in Fig. 1 falls in the annular-flow region.

From Eq. (8), we calculate the two-phase flow modulus, X^2:

$$X^2 = \left(\frac{6{,}150}{21{,}500}\right)^{1.8}\left(\frac{1.92}{52}\right)\left(\frac{0.1}{0.01}\right)^{0.2} = 0.0062$$

$$X = 0.0787$$

From Fig. 1, we find that the two-phase flow correlation for annular flow is: $\phi = aX^b$, where:

$$a = 4.8 - (0.3125 \times 6.065) = 2.9$$

$$b = 0.343 - (0.021 \times 6.065) = 0.216$$

Consequently:

$$\phi = 2.9(0.0787)^{0.216} = 1.67$$

$$\phi^2 = 2.80$$

To calculate the vapor-phase unit loss, we must find the Reynolds number:

$$(N_{Re})_v = \frac{6.31(21{,}500)}{6.065(0.01)} = 2.24 \times 10^6$$

With this Reynolds number, we find $f_v = 0.015$ from Fig. 2. Using this value of the friction factor in Eq. (7),

$$\Delta p_{100(vapor)} = \frac{0.000336(0.015)(21{,}500)^2}{8{,}206(1.92)}$$

$$\Delta p_{100(vapor)} = 0.15 \text{ psi/100 ft}$$

By substituting into Eq. (3), we now find the unit loss:

$$\Delta p_{100(annular)} = 0.15 \times 2.8 = 0.42 \text{ psi/100 ft}$$

References

1. DeGance, A. E. and Atherton, R. W., Chemical Engineering Aspects of Two-Phase Flow, *Chem. Eng.*, Mar. 23, 1970, pp. 135–139; Apr. 30, 1970, pp. 151–158; May 4, 1970, pp. 113–120; July 13, 1970, pp. 95–103; Aug. 10, 1970, pp. 119–126.
2. Baker, O., Brainerd, H. W., Coldren, C. L., Flanigan, O. and Welchen, J. K., "Gas-Liquid Flow in Pipelines: Design Manual," American Gas Assn., Arlington, VA 22209, Oct. 1970.
3. Alves, G. E., Co-Current Liquid-Gas Flow in Pipe Line Contactor, AIChE National Meeting, San Francisco, Sept. 14, 1953.
4. Baker, O., Simultaneous Flow of Oil and Gas, *Oil Gas J.*, July 26, 1954.
5. Lockhart, R. W. and Martinelli, R. C., Proposed Correlation of Data for Isothermal Two-Phase, Two-Component Flow in Pipe, *Chem. Eng. Progr.*, Jan. 1949.
6. Schneider, F. N., White, P. D. and Huntington, R. L., Correlation for Two-Phase Wave Flow, *Pipe Line Industry*, Oct. 1954.
7. "A.P.I. Technical Data Book—Petroleum Refining," Chap. 10, American Petroleum Institute, Washington, D.C. 20006.
8. Baker, O. and Swerdloff, W., Calculation of Surface Tension, *Oil Gas J.*, Nov. 21, 1955, Dec. 5, 12, 19, 1955, Jan. 2, 9, 1956.
9. Yaws, C. L. and Setty, H. S. N., Water and Hydrogen Peroxide, Fig. 7-2, *Chem. Eng.*, Dec. 23, 1974, p. 69.
10. "Engineering Data Book," 9th ed., pp. 16-27 to 16-33, and 16-39, Natural Gas Processors Assn., Tulsa, OK 74103, 1972.

How to design piping for reboiler systems

Interactions between hydraulic requirements and piping configurations require close attention to many fluid and mechanical details, in order to obtain the most efficient and economical distillation units.

Robert Kern, *Hoffmann - La Roche Inc.* *

☐ Familiarity with graphic piping design is an essential requirement for the designer of hydraulic systems. The accuracy of his calculations, predictions of flowrate and pressure differential, reliability of operation, and the economy of capital, energy, maintenance and operating costs depend to a great extent on pipe configurations and pipe components.

In these articles, we have recognized the importance of graphic piping design, and to a limited degree have presented its fundamentals. We will now evaluate the flow systems and piping design for a distillation column, which is a more integrated unit than the individual systems discussed in our earlier articles.

Layout for distillation columns

A process flow diagram of a typical distillation column with bottom pump, thermosyphon reboiler, overhead condenser, reflux drum and reflux pump are shown in Fig. 1 (F/1). The equipment components are located adjacent to each other in the actual plant. Also in F/1, we find elevation and plan drawings for the column. These show how the principal elements of a distillation column are usually integrated into an overall plant arrangement. Manholes face access roads (or access aisles at housed installations). Each manhole has a platform for maintenance. Valves and instruments are located above these platforms for convenient access.

For economy and easy support, piping should drop immediately upon leaving the tower nozzle, and run parallel, and as close as possible, to the tower itself. A vertical line lends itself as a suitable location for the straight run of an orifice. The horizontal elevations, after the lines leave their vertical run, are governed by the elevations of the main pipe rack. Lines that run directly to equipment at grade (more or less in the direction of the main pipe rack) often have the same elevation as the pipe bank.

Lines from tower nozzles below the pipe rack should approach the pipe bank roughly 2 ft below the pipe-rack elevation. The same elevation is used for those lines that run to pumps located below the pipe rack.

Pump-suction lines can also be arranged on this elevation. They should be as short as possible and run without loops or pockets. Pipelines, dropping from above the pipe-rack elevation, will approach the pipe bank roughly 2 ft higher than the elevation of the main pipe bank. This elevation is also used for steam lines to reboilers. These steam lines usually connect to the top of the headers to avoid excessive condensate drainage toward process equipment.

The plan view (F/1) of the tower shows the segments of its circumference allotted to piping, nozzles, manholes, platform brackets and ladders. Such a pattern usually leads to a well-organized arrangement for the process equipment and auxiliary components.

From a layout standpoint, it is preferable to have equal platform-bracket spacing, and the orientation of brackets lined up along the entire length of the tower. This will minimize interferences between the piping and structural members. According to OSHA, ladders between platforms should not be longer than 30 ft.

Area segments for piping going to equipment at grade are available between the ladders and on both sides of the manholes. Lines approaching the pipe rack

*To meet your author, see p. 44.

Originally published August 4, 1975

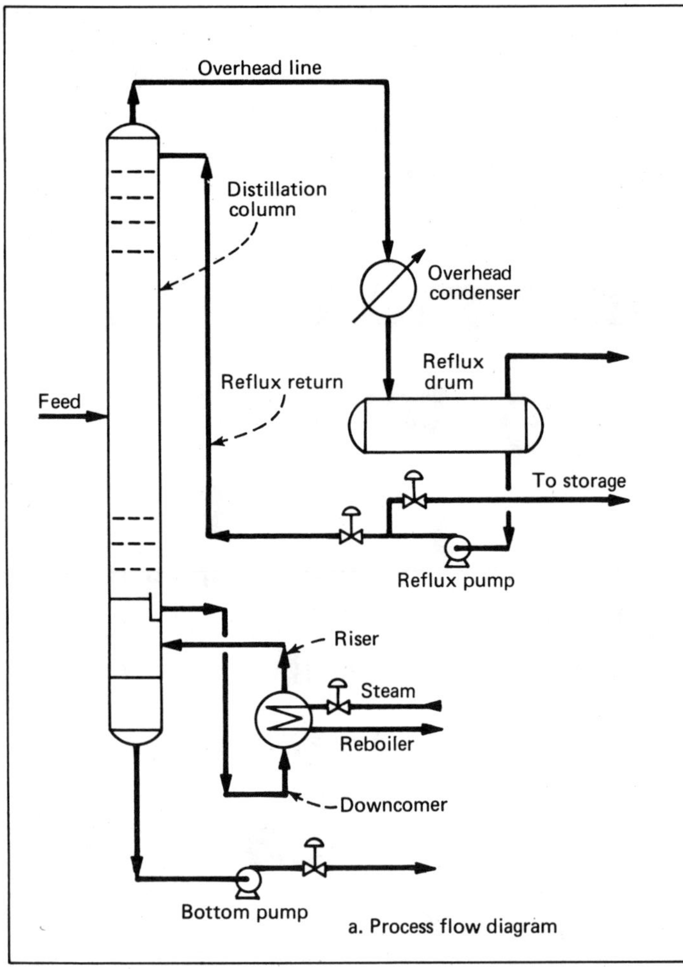

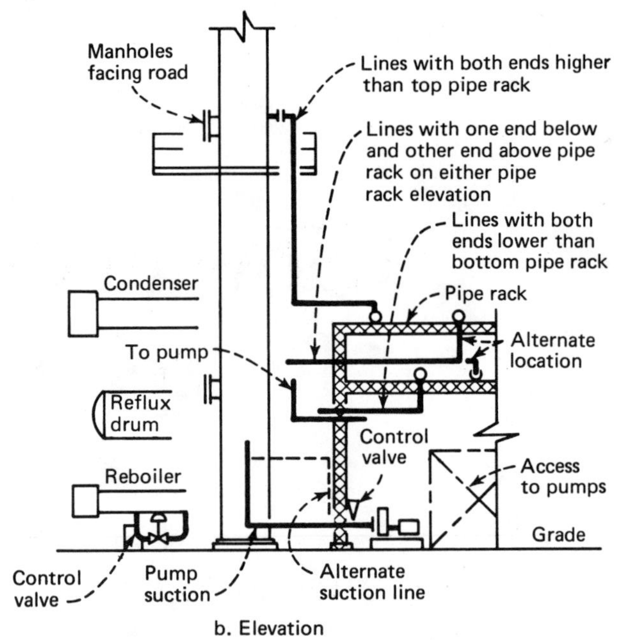

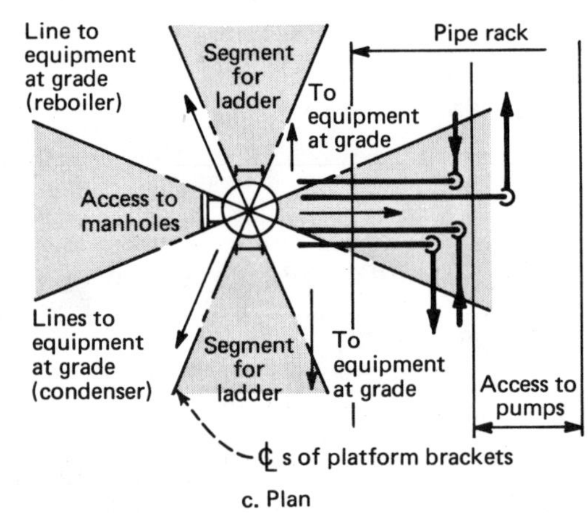

Piping the distillation column requires economy, ease of support and accessibility for good design. F1

can turn left or right, depending on the plant's overall arrangement.

On a distillation column, the largest lines are the overhead vapor line, and the reboiler downcomer and return. These lines should have the simplest and most direct configurations, to minimize pressure loss and cost.

During normal operation, the pump in Fig. 1 transports liquid at equilibrium. This means that the distillation column and reflux drum are elevated to satisfy NPSH (net positive suction head) requirements. The discharge lines often have two destinations. Total-head requirements should be designed and calculated so that operating points fall on the pump's head-capacity curve when pumping to an alternative destination. Alternative discharge lines can have equal capacity and alternative operation, or partial capacity with simultaneous operation. All alternatives should be investigated for the process pumps.

The design for a pumped reboiler circuit is similar to that of a reflux pump system. A bottom pump transports the liquid through an exchanger or fired heater and returns it to the distillation column. Close attention should be paid to possible two-phase flow in pipelines coming after the heater—especially when we want to locate the heater close to the column.

Inserted-type reboilers have no process piping. Larger-diameter towers can have one to four U-tube stub bundles inserted directly into the liquid space through tower nozzles, and extending across the tower diameter. Reboilers with small heat duties are usually designed as helical coils.

Reboiler arrangements

In horizontal thermosyphon reboilers, liquid flows from an elevated drum or tower bottom or tower-trapout boot through a downcomer pipe to the bottom of exchanger shell. The liquid is heated, leaves the reboiler in the return piping as a vapor-liquid mixture, and flows back to the tower or drum.

In vertical reboilers, heating usually occurs on the shellside. In horizontal reboilers, heating is on the tubeside. For a large evaporation rate (for example, 90% of total flow), a kettle-type reboiler is used.

Piping to horizontal reboilers is designed as simply and directly as possible within the limitations of thermal-expansion forces.

Symmetrical arrangements between the drawoff and

reboiler-inlet nozzles, as well as between the reboiler outlet and return connection on the tower, are preferred in order to have equal flow in the reboiler circuit. A nonsymmetrical piping configuration may also be accepted for a more-economical or more-flexible piping design.

Reboilers often have two outlets and two parallel-pipe segments. When sizing and arranging nonsymmetrical piping, an attempt should be made to equalize the resistance through both legs of the reboiler piping. More resistance in one leg produces a smaller flow than in the other. Hence, uneven heat distribution will occur in the reboiler—one segment of the riser will be hotter than the other.

At startup in reboilers having high, liquid drawoff nozzles, a gravity-flow bypass is usually provided from the tower's liquid space to a low point of the downcomer.

Valves are rarely included in reboiler piping, except when a standby reboiler is provided, or when two or three reboilers are used and operated at an extremely wide heat-capacity range. Some companies require line blinds to blank off the tower nozzles during shutdown, turnaround and maintenance.

The heating media (steam or a hot process stream) connect to the tubeside of horizontal reboilers. The inlet piping usually has a temperature-regulated control valve (with block valves and bypass globe valve, if required). This is normally arranged at grade near the reboiler's tubeside inlet.

Reboiler elevations

Most reboilers are at grade next to the tower, with centerline elevations of about 3 to 5.5 ft above ground level for exchangers about 1 to 3 ft dia. Exchangers at grade provide economical arrangements—valves and instruments are accessible, tube-bundle handling is convenient, and maintenance is easy. In this arrangement, the static heads are well determined between the exchanger's centerline and the drawoff and return nozzles on the tower. Vertical reboilers are usually supported on the distillation column itself.

Some reboilers have a condensate or liquid-holding pot located after the tubeside outlet, as shown in F/2. In such cases, the centerline elevation of the reboiler is somewhat higher than units that do not have these control vessels.

The arrangement in F/2a is a high-capacity steam trap. The top of the condenser pot should not be higher than the bottom of the exchanger shell, to avoid flooding the tubes with condensate and adversely affecting the exchanger's heat-transfer duty.

The arrangement in F/2b maintains a required condensate level in the reboiler, to provide for a wide range of heat-transfer control. Process conditions determine the precise relationship between the exchanger and the vertical condensate-control pot.

In F/3, we show an example where a reboiler has been elevated to meet the NPSH requirement of the centrifugal pump. The elevated reboiler, in turn, raises the tower because the minimum liquid level in the bottom of the tower must be higher than the liquid level in the exchanger. The elevation difference (dimen-

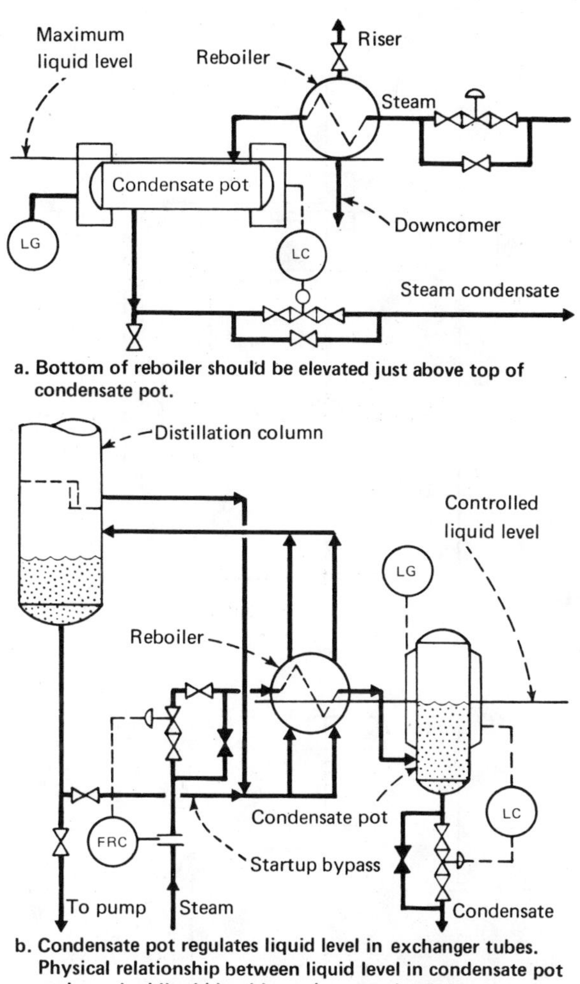

a. Bottom of reboiler should be elevated just above top of condensate pot.

b. Condensate pot regulates liquid level in exchanger tubes. Physical relationship between liquid level in condensate pot and required liquid level in exchanger tubes is important.

Process conditions determine equipment elevations. F2

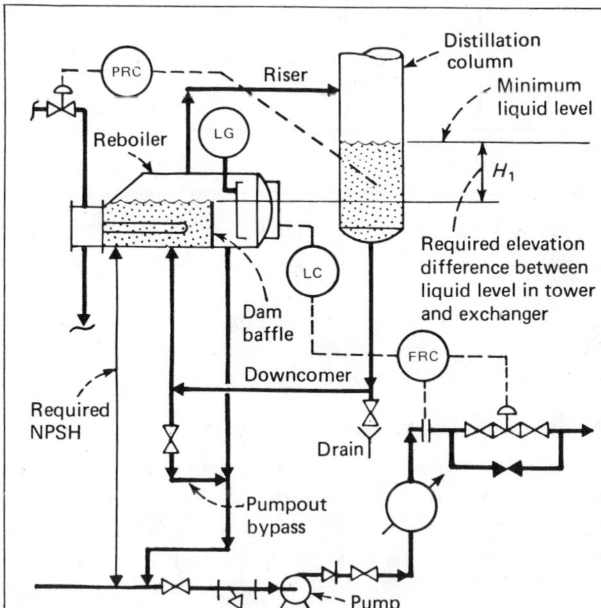

Reboiler raised to meet pump's NPSH; in turn, top of dam in reboiler elevates minimum liquid level in tower.

Reboiler raised to meet pump's NSPH requirement. F3

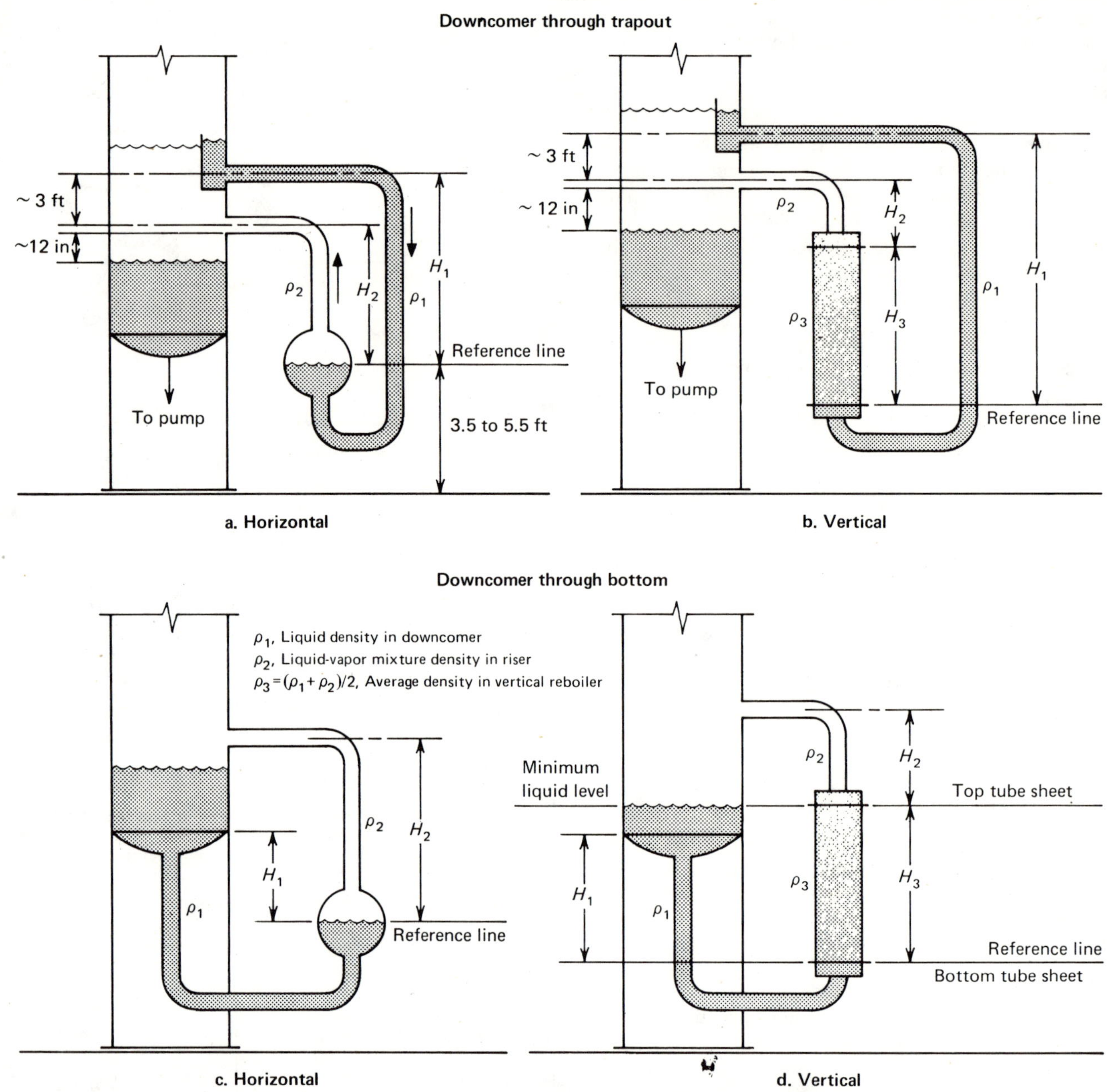

Reboilers having natural circulation rely on the difference in static head between the downcomer and riser. F4

sion H_1 in F/3) provides the positive static head for flow in the reboiler circuit, and overcomes friction losses in the exchanger, and downcomer and return lines.

Designing the reboiler system

Typical reboiler arrangements are shown schematically in F/4. In all cases, the vessel pressure is the same at the tower's outlet and return nozzles. Circulation is forced by the static-head difference between the liquid column in the downcomer and the vapor-liquid column in the riser. For convenience, reference lines are chosen at the exchanger's centerline for horizontal reboilers, and at the bottom tubesheet for vertical reboilers.

If P_1 is liquid pressure in the downcomer at the reference line, and P_2 is backpressure in the riser's vapor-liquid column, the pressure difference ($\Delta P = P_1 - P_2$) must overcome the exchanger and piping friction losses. Therefore, P_1 must be greater than P_2. If ρ_1 is the hot liquid density in the downcomer, then $\rho_1 H_1/144 = P_1$, psi. The backpressure, P_2, can have two alternative expressions:

1. For horizontal exchangers (see F/4a and F/4c):

$$P_2 = \rho_2 H_2/144, \text{ psi} \tag{1}$$

where:

$$\rho_2 = \frac{W}{\dfrac{W_l}{\rho_l} + \dfrac{W_v}{\rho_v}} = \frac{100}{\dfrac{\%\ Liquid}{\rho_l} + \dfrac{\%\ Vapor}{\rho_v}} \tag{2}$$

2. For vertical exchangers (see F/4b and F/4d):

$$P_2' = (\rho_2 H_2 + \rho_3 H_3)/144, \text{ psi} \quad (3)$$

where ρ_2 is again the mixture's density, as expressed by Eq. (2) for horizontal exchangers, and ρ_3 is the average density of liquid and liquid-vapor mixture in the reboiler:

$$\rho_3 = (\rho_1 + \rho_2)/2 \quad (4)$$

Eq. (4) provides a conservative estimate of the density gradient in vertical reboilers. Actual density will be less than that expressed by Eq. (4). In all equations, the units for ρ are lb/ft^3, and for H, ft. We also note that the vertical reboiler should be flooded. The maximum elevation of the top tubesheet should not be higher than the minimum liquid level in the tower.

Hydraulics in horizontal reboilers

In the following discussion, the hydraulic conditions only in horizontal exchangers will be developed. (The derivations are the same for vertical exchangers, except that P_2' will replace P_2.) For horizontal exchangers:

$$P_1 - P_2 = \Delta P = (1/144)(\rho_1 H_1 - \rho_2 H_2) \quad (5)$$

If a safety factor of 2 is introduced, then the available pressure difference for friction losses is halved, and:

$$\Delta P = (1/288)(\rho_1 H_1 - \rho_2 H_2) \quad (6)$$

The quantity $(H_1 - H_2)$ is usually 3 ft (see F/4a). Consequently, a minimum driving force of $\Delta P_{min} = (3/288)\rho_1 \approx 0.01\rho_1$ is always available at horizontal exchangers.

The maximum possible driving force depends on the elevation difference between the drawoff nozzle and exchanger centerline (dimension H_1) and on the total evaporation taking place in the reboiler. Neglecting the vapor-column backpressure in the return line, the maximum usable driving force is:

$$\Delta P_{max} = (H_1/288)\rho_1 \quad (7)$$

In most applications, the actual driving force is not much below this maximum. H_1 can range from 6 to 24 ft, depending on the size of the arrangement and on NPSH for a pump taking suction at the bottom of the tower. For these H_1 values:

$$\Delta P_{max} = (6/288)\rho_1 \text{ to } (24/288)\rho_1$$

$$\Delta P_{max} = 0.02\rho_1 \text{ to } 0.08\rho_1 \quad (8)$$

Thus, the driving force is reduced to a function of the downcomer liquid density at operating temperature. For example, if the piping geometry produces $H_1 = 12$ ft, and $\rho_1 = 50$ lb/ft^3 for kerosene:

$$\Delta P_{max} = (12/288)50 \approx 2.0 \text{ psi}$$

These simple relationships are useful when the evaporation rate is not known and line sizes have to be estimated. The available driving force will be near but less than ΔP_{max}.

ΔP_{max} as evaluated here is, of course, an extreme value taken at total evaporation. In reboilers, partial evaporation usually takes place, which will reduce ΔP_{max}. However, even if the driving force is assumed at the maximum value, any inaccuracy is well compensated for by the safety factor of 2, and by the necessity to use commercially available pipe sizes that are normally larger than calculated pipe diameters.

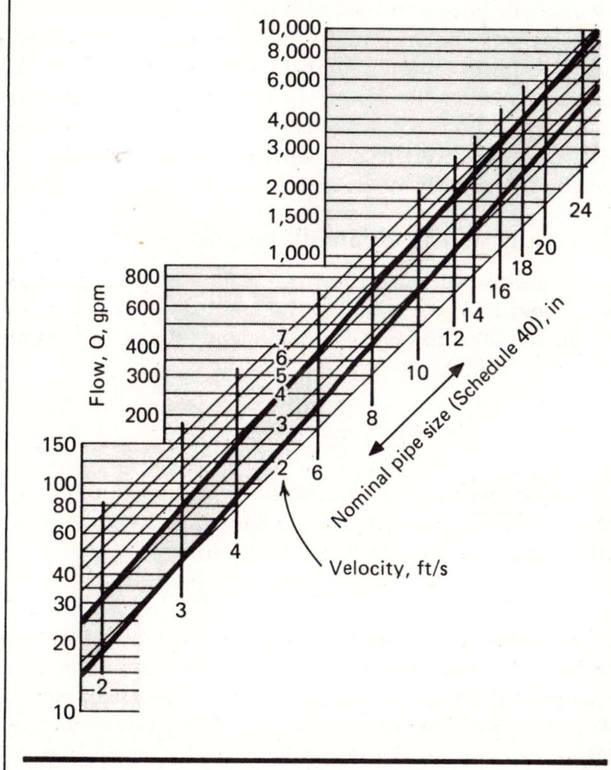

Shaded part of chart establishes size of downcomer. F/5

Friction losses in reboilers

The total friction losses in a thermal-circulating reboiler system must be smaller than the available driving force. The pressure loss caused by friction takes place in two main locations: in the exchanger itself, Δp_e, and in the piping, Δp_p. Hence:

$$\Delta p_e + \Delta p_p < \Delta P$$

Friction losses in reboilers, Δp_e, are generally given as 0.25 to 0.5 psi. (A note should indicate whether entrance and exit losses are included.) Unit losses in downcomers and risers are in fractions of 1 psi/100 ft.

Calculation procedures for liquid lines have been outlined in pp. 19–24 (Example 1); and for the two-phase flow risers in pp. 48–54 (Example 1). We calculate reboiler returns as dispersed flow, regardless of the intersection of the Baker parameters (see Fig. 1 on p. 49).

To avoid trial-and-error calculations, a selection chart for reboiler pipe size is presented in F/5. This chart is based on limiting velocities for flow in downcomers of 2 to 7 ft/s. We enter the graph with known liquid-flow quantities. We obtain downcomer pipe sizes from the shaded portion of the graph, and also find the corresponding flow velocities for computing the

Reynolds numbers. The riser can be assumed as one or two sizes larger than the downcomer pipe.

In vertical reboiler circuits, reboiler losses are greater, and pipe losses smaller, than in horizontal circuits. In this case, a safety factor of 1.25 applied to the driving force can be used. In kettle-type reboilers, evaporation rates are high. For these reboilers, a large-diameter return line is usually necessary.

Elevation of the drawoff nozzle

The minimum elevation, H_1, for the downcomer tower nozzle above that for the centerline of the horizontal reboiler may be found from Eq. (6), where $H_2 = H_1 - 3$, ft:

$$H_1 = \frac{288\Delta p - 3\rho_2}{\rho_1 - \rho_2} \tag{9}$$

The downcomer nozzle cannot be lower than H_1. Δp replaces ΔP in Eq. (6), and is the sum of the downcomer, riser and exchanger friction-losses:

$$\Delta p = \Delta p_d + \Delta p_r + \Delta p_e$$

The value of H_1 is useful when elevation adjustments are made to vessel heights during graphic piping design, or when the vessel can be located at a minimum elevation. The coefficient for ρ_2 in Eq. (9) is the elevation difference between the downcomer and riser nozzles. If this is other than 3 ft, the correct dimension should be inserted.

Many towers have a bottom drawoff pump. NPSH requirements usually cause the process vessel and the reboiler drawoff nozzle to be raised higher than that of the reboiler's minimum elevation. This increases the static head in the vertical legs, and also the driving force in the circuit. With increased tower height, it is worthwhile to check the reboiler circuit for a possible reduction in size of the liquid and return lines—especially where large-diameter lines are necessary. Uneconomical reboiler lines are just carelessly oversized or poorly routed.

Example demonstrates calculations

Let us size the reboiler lines for the kerosene distillation unit, as sketched in F/6. Flow data are:

	Downcomer	Riser	
	Liquid	**Liquid**	**Vapor**
Flowrate, W, lb/h	85,000	59,500	25,500
Density (hot), ρ, lb/ft³	36.7	36	1.31
Viscosity, μ, cp	0.6	0.5	0.01
Molecular weight, M	—	—	53

The flow data for the riser reflect that 70% of total flow is liquid and 30% vapor. We obtain the vapor density,* ρ_v, in the riser from:

$$\rho_v = MP'/10.72Tz$$

$$\rho_v = 53(181.7)/(10.72)(682) = 1.31 \text{ lb/ft}^3$$

We find the mixture density,* ρ_2, in the riser by substituting in Eq. (2):

$$\rho_2 = \frac{100}{(70/36) + (30/1.31)} = 4.05 \text{ lb/ft}^3$$

And, we calculate volume flowrate,* Q, from:

$$Q = W/500S$$

$$Q = 85{,}000/500(36.7/62.37) = 289 \text{ gpm}$$

We will begin by sizing the downcomer and then calculate the overall pressure loss in it. We follow up with similar computations for the riser—remembering that the line has vapor/liquid flow. The simplest part of this analysis is finding the preliminary size of the reboiler lines. The computations for checking whether these line sizes are adequate require considerable detail. We should also note that the reboiler has two inlets and two outlets (F/6). Consequently, we find full flow (100%) in each line from the column to the "tee" connection on the reboiler's inlet and outlet piping; but only 50% of total flow in each line segment after the tee. In the following procedures, we will see how this flow arrangement affects our design calculations.

Downcomer—For a liquid flow of 289 gpm, we find the pipe size for the downcomer as 6 in from the selection chart of F/5. (For a 6-in, Schedule 40 pipe, I.D. = 5.761 in, d^5 = 6,346 in⁵.)

In order to calculate the unit loss for the 100% flow section, we must calculate the Reynolds number from:

$$N_{Re} = 50.6(Q/d)(\rho/\mu)$$

$$N_{Re} = 50.6\left(\frac{289}{5.761}\right)\left(\frac{36.7}{0.6}\right) = 155{,}300$$

For this Reynolds number, we obtain the friction factor, f, as 0.0182 from the chart on p. 50, and calculate the unit loss from:

$$\Delta p_{100} = 0.0216 f\rho_1(Q^2/d^5)$$

$$\Delta p_{100} = 0.0216(0.0182)(36.7)[(289)^2/6{,}346]$$

$$\Delta p_{100} = 0.19 \text{ psi/100 ft}$$

We can then find the unit loss at 50% flow:

$$N_{Re} = 0.5(155{,}300) = 77{,}650;\ f = 0.022$$

$$\Delta p_{100}(50\%) = 0.19\left(\frac{144.5}{289}\right)^2\left(\frac{0.022}{0.0182}\right)$$

$$\Delta p_{100}(50\%) = 0.0574 \text{ psi/100 ft}$$

We now determine the equivalent length of pipe and fittings for each segment from tables pp. 19–24, as follows:

	Segment for 50% Flow, Ft	Segment for 100% Flow, Ft
Actual length	6	26
Entrance loss	—	18
Elbows*	10	20
Sharp tee	30	—
Exit loss	36	—
Total	82	64

*One elbow for 50% flow segment, 2 for 100% flow.

Overall pressure loss of the downcomer:

$$\Delta P = 0.19(64/100) + 0.057(82/100) = 0.167 \text{ psi}$$

*See pp. 86–93 for complete details for these calculations.

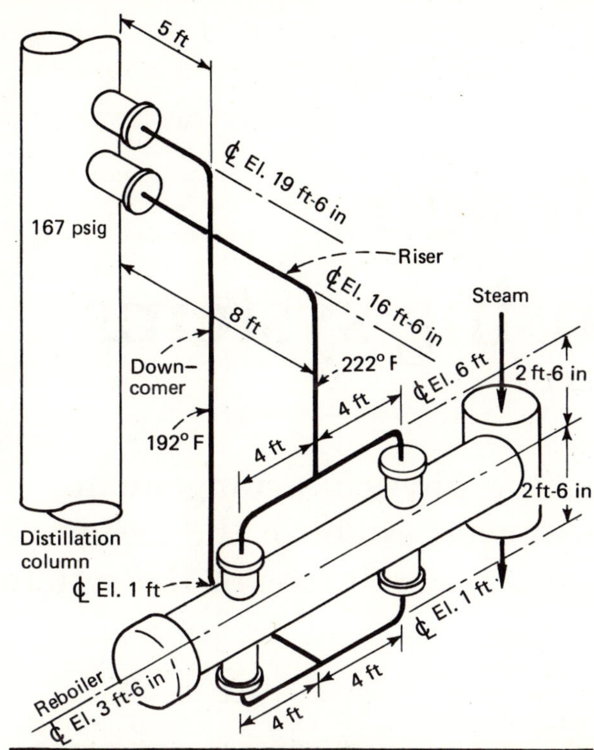

Reboiler flow relations and elevations for problem. F6

Riser—Since the downcomer is a 6-in pipe, let us use a riser having a nominal size of 8 in. (For an 8-in Schedule 40 pipe, I.D. = 7.981 in, $d^5 = 32{,}380$ in,[5] $A = 0.3474$ ft^2.)

In order to calculate the unit loss for the 100% flow section, we must calculate the vapor-phase Reynolds number from:

$$N_{Re} = 6.31 W_v / d\mu_v$$
$$N_{Re} = 6.31(25{,}500)/(7.981)(0.01)$$
$$N_{Re} = 2 \times 10^6$$

For this Reynolds number, we obtain a friction factor, f, of 0.014 from Fig. 2 on p. 50; and then calculate the vapor-phase unit loss from:

$$\Delta p_{100} = 0.000336(f/\rho_v)(W_v^2/d^5)$$
$$\Delta p_{100} = 0.000336(0.014/1.31)[(25{,}500)^2/32{,}380]$$
$$\Delta p_{100} = 0.072 \text{ psi}/100 \text{ ft}$$

Since the riser handles vapor/liquid flow, we must now determine the two-phase flow modulus* from:

$$X^2 = (W_l/W_v)^{1.8}(\rho_v/\rho_l)(\mu_l/\mu_v)^{0.2}$$
$$X^2 = \left(\frac{59{,}500}{25{,}500}\right)^{1.8}\left(\frac{1.31}{36}\right)\left(\frac{0.6}{0.01}\right)^{0.2} = 0.38$$

With this value of X^2, we find the two-phase flow modulus, θ^2, as 14 from Fig. 3 on p. 51.* We calculate the two-phase-flow unit loss* from:

$$\Delta p_{100} = 0.072 \times 14 = 1.01 \text{ psi}/100 \text{ ft}$$

*For complete details of these calculations, see pp. 48–54.

Let us estimate that the unit loss in the 50% flow section is one-third that in the 100% flow segment. Hence:

$$\Delta p_{100}(50\%) = 0.34 \text{ psi}/100 \text{ ft}$$

Again, we determine the equivalent length of pipe and fittings for each segment of the riser as follows:

	Segment for 50% Flow, Ft	Segment for 100% Flow, Ft
Actual length	6	18
1 Elbow	14	14
1 Sharp tee	24	—
1 Entrance loss	24	—
1 Exit loss	—	48
Total	68	80

Overall pressure loss of the riser:

$$\Delta P = 1.01(80/100) + 0.34(68/100) = 1.04 \text{ psi}$$

In summary, the total pressure loss is obtained by adding the loss in the downcomer, riser and reboiler:

6-in Downcomer	0.167 psi
8-in Riser	1.02 psi
Reboiler	0.35 psi
Total ΔP	1.557 psi

By substituting the appropriate values into Eq. (6) for this example, we determine the available pressure difference, as:

$$\Delta P = (1/288)[(36.7)(13.5) - (4.05)(10.5)] = 1.57 \text{ psi}$$

The available pressure difference of 1.57 psi is greater than the calculated pressure losses of 1.557 psi. Therefore, the design and sizes are acceptable.

Finally, we check the minimum elevation of the drawoff nozzle above the centerline of the reboiler by substituting into Eq. (9):

$$H_1 = \frac{288(1.535) - 3(4.05)}{31.7 - 4.05} = 15.55 \text{ ft}$$

Since the drawoff nozzle is actually 16.0 ft above the reboiler's centerline (F/6), the minimum value of 15.55 ft is acceptable.

How to design overhead condensing systems

The state of fluids in pipelines, and the physical arrangement of equipment around the distillation tower, establish the design parameters for meeting hydraulic and piping conditions of the system.

*Robert Kern, Hoffmann - La Roche Inc.**

☐ The state of fluid in the pipelines, and the hydraulic and thermal conditions in condensing systems of distillation columns, are the reverse of those in reboiler circuits.

The inlet line to condensers can carry superheated or saturated vapor, or dispersed vapor-liquid mixtures. Fluid is cooled in the exchanger, and partial or full condensation takes place. The condenser's outlet line can have stratified and dispersed two-phase flow, saturated liquid, or subcooled liquid. In addition, the flowing fluid can be a mixture of two substances. Thus, this type of condensation offers a wide range of classification from a thermodynamic standpoint [*1*].

In contrast, saturated liquid normally flows in the downcomer of reboilers. The liquid is vaporized while passing through the exchanger. The reboiler's outlet line carries a dispersed vapor-liquid mixture having a vapor content of 30 to 90% of total flow. (See p. 55 for more details.)

In this article, we will examine the hydraulics for the following:

1. Condensers with gravity-flow return lines.
2. Condensers with pumped-reflux lines.
3. Two-stage condensation.

Within these groups, hydraulic-design and piping-design conditions vary, depending on the state of fluid in the lines and the physical arrangement of the installation.

Vacuum technology has its own systems, equipment and terminology. Piping design of vacuum-condensing systems are outside the Euler-Bernoulli-Darcy theories and are not included in this article.

*To meet your author, see p. 44.

Originally published September 15, 1975

Gravity-flow reflux

Horizontal condensers—A condenser in gravity-flow arrangements is located above the level of the terminating point of the condenser's outlet line, as shown in F/1 and F/2. For the horizontal condenser in F/1b, vapor enters the exchanger at the top, and subcooled liquid leaves at the bottom. The looped-outlet pipe ensures a permanent liquid level in the condenser. This liquid level is controlled through the reflux branch and through the takeoff line to storage.

The static-head pressure difference, ΔP_s, between the vertical overhead line and the condenser's outlet line for the arrangements in F/1 can be written as:

$$\Delta P_s = (H/144)(\rho_1 - \rho_2) \tag{1}$$

ΔP_s must be equal to or greater than the sum of (1) the pipe-system resistance, Δp_p, between reference points A and B; (a) exchanger pressure drop, Δp_e; and (3) required pressure difference across the control valve, Δp_{cv}:

$$\Delta P_s \geqq \Delta p_p + \Delta p_e + \Delta p_{cv} \tag{2}$$

The required distance, H, between fractionator inlet and exchanger centerline can be calculated from Eq. (1) as:

$$H = (144\,\Delta P_s)/(\rho_1 - \rho_2) \tag{3}$$

The vapor column can be neglected by assuming $\rho_2 = 0$ in Eq. (1) and (3). All pressures are in psi; densities, ρ, in lb/ft^3; and dimensions, H, in ft.

As Eq. (3) shows, for a minimum of elevation difference between the top of the column and the exchanger centerline, the piping and components resistances must also be minimal.

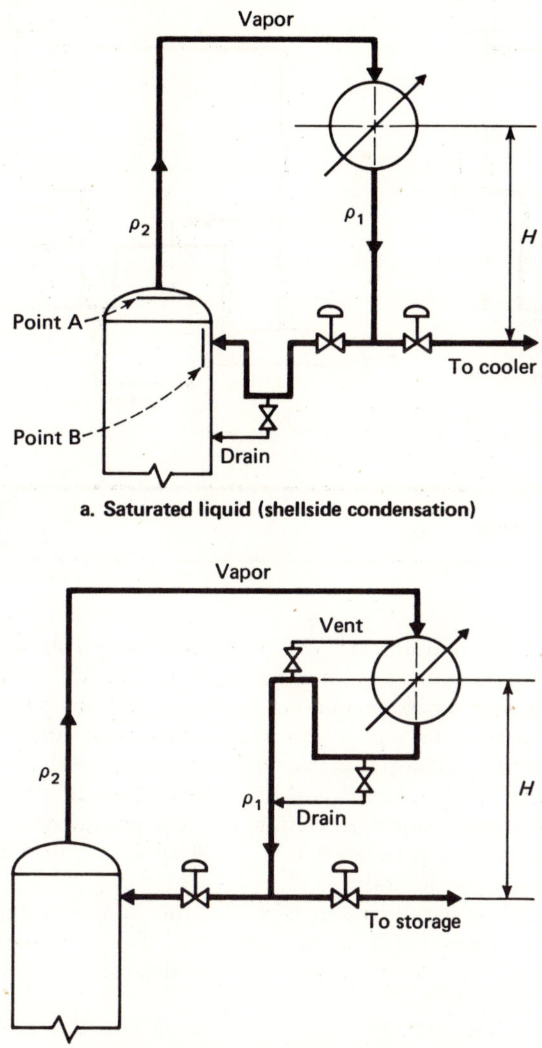

a. Saturated liquid (shellside condensation)

b. Subcooled liquid (shellside condensation)

Horizontal condensers with gravity-flow reflux F 1

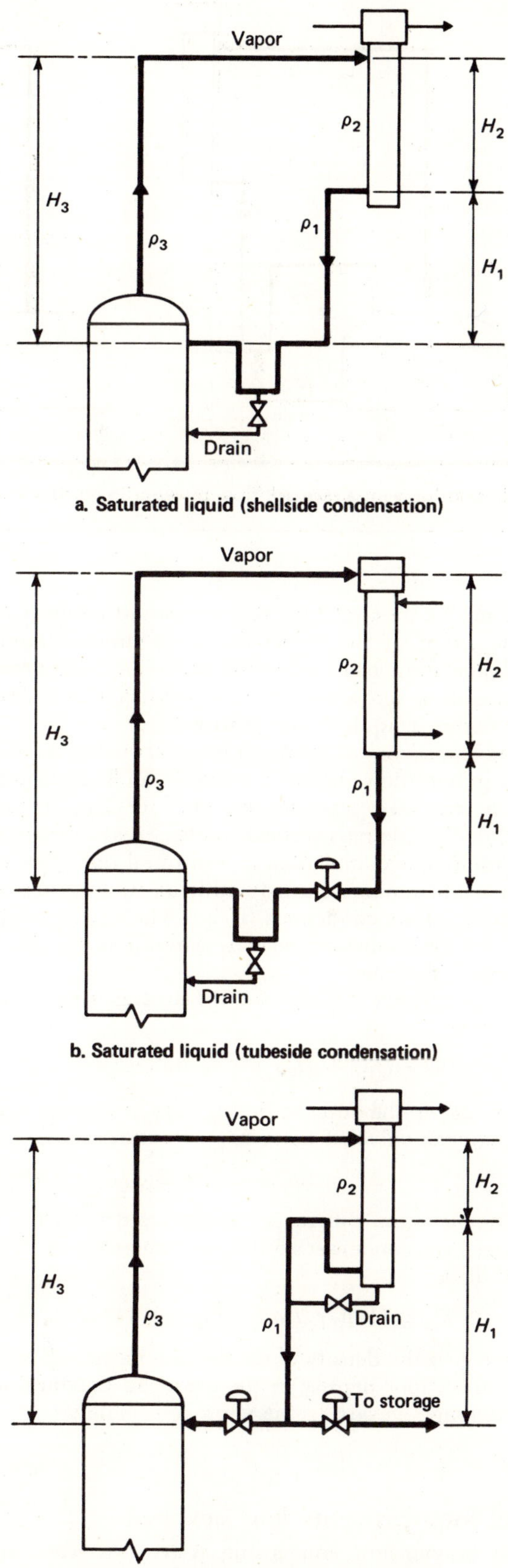

a. Saturated liquid (shellside condensation)

b. Saturated liquid (tubeside condensation)

c. Subcooled liquid (shellside condensation)

Vertical condensers with gravity-flow reflux condense vapors on the shellside or tubeside of exchanger F 2

Generally, in condensing systems, the unit loss in the piping is low—about a tenth or a hundredth of 1 psi/100 ft. Inlet and outlet resistances to process equipment usually take a considerable portion of the pipeline resistance and should not be ignored in the calculations. (Using three decimal places in the calculations is not unusual.)

In horizontal condensers, condensation takes place in the shell. This gives lower resistance than the tubeside. A baffle (or baffles) in the exchanger is in the horizontal plane through the exchanger's centerline. If necessary, two inlet and two outlet nozzles can halve the total flow, and reduce entrance and exit resistances considerably. In this case, the inlet and outlet piping should be symmetrical.

The subcooled liquid for the arrangement in F/1b can be drained or pumped directly to storage. The product stream for F/1a is usually directed through a cooler before storage.

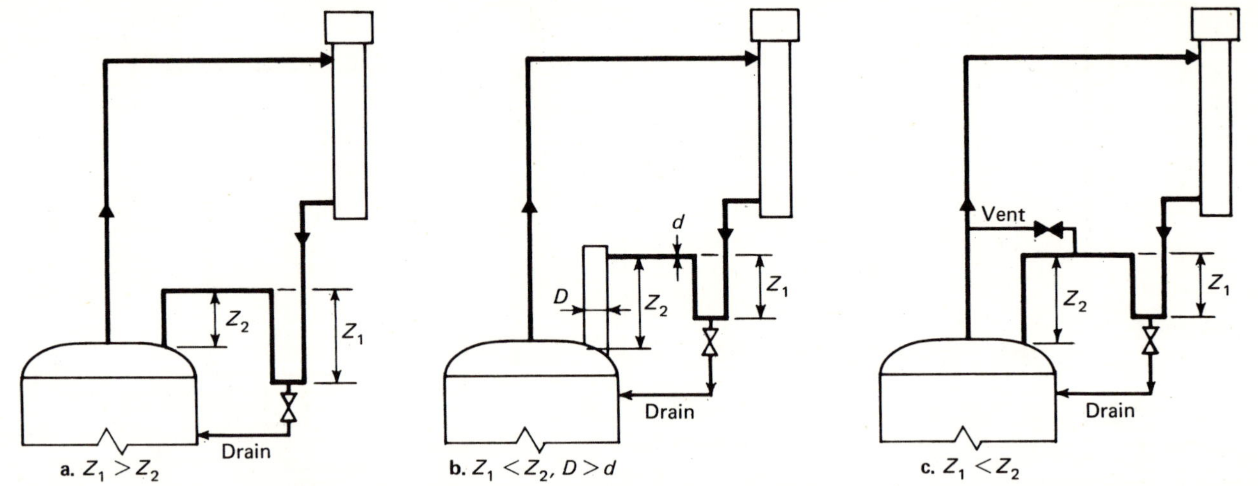

Seal loops prevent a reversed flow of vapor in gravity-flow condensing systems F3

Control valves in these systems should be located at a low point of the return line and product stream. Sufficient static head before the valve inlet will prevent vaporization across the valve. A product cooler should not receive a liquid-vapor mixture.

Vertical condensers—Arrangements for these condensers with gravity-flow outlets are shown in F/2. Condensation can take place in the shell (F/2a), or tubes (F/2b). A single-pass vertical condenser is more suitable for liquid subcooling than a horizontal one. The seal-loop height can be adjusted within a greater range than with horizontal condensers (F/2c). The required liquid level in the exchanger shell is determined by the exchanger's designer.

The hydraulic balance for the arrangements shown in F/2 is:

$$(1/144)(H_1\rho_1 + H_2\rho_2) \geqq (1/144)H_3\rho_3 + \Delta P \quad (4)$$

where ΔP is the sum of piping, Δp_p, exchanger, Δp_e, and control-valve (if any), Δp_{cv}, resistances:

$$\Delta P = \Delta p_p + \Delta p_e + \Delta p_{cv} \quad (5)$$

The elevation difference, as expressed from Eq. (4), between the condenser's outlet and the reflux inlet nozzle will be:

$$H_1 \geqq (1/\rho_1)(H_3\rho_3 - H_2\rho_2 + 144\,\Delta P) \quad (6)$$

where ρ_1 is the density of condensate in the reflux line, ρ_3 is the vapor density in the overhead line, and ρ_2 is the average density in the vertical exchanger:

$$\rho_2 = 0.5(\rho_1 + \rho_2) \quad (7)$$

Seal loop prevents flow reversal

In gravity-flow condensing systems, a seal loop is provided to prevent a reversed flow of vapor in the condenser's outlet line. This loop can be used for holding a liquid level in the condenser, as shown in F/1a and F/2c.

If the gravity-flow reflux line terminates in a vertical leg (Z_2 dimension in F/3), the piping design should be such as to prevent siphoning that can empty the seal. If dimension Z_1 is smaller than Z_2, and the pressure just before the seal loop and at the terminating point after the seal loop is identical (for example, with greatly reduced flow), liquid can be siphoned out of the seal; and intermittently, the condenser will not operate well. This can be prevented if Z_1 is designed to be longer than Z_2 (see F/3a).

For the arrangement in F/3b, the final vertical leg has a larger diameter than the gravity-flow reflux line. Again, this can prevent the siphoning of liquid from the seal loop.

Another arrangement (F/3c) has a closed vent line. This can be opened at reduced condensate flow to keep the seal loop filled with liquid. With this type of venting, the pressure difference across the vent valve should be zero. Therefore, it is essential to connect the endpoints of the vent line to locations where pressures are expected to be about equal.

Pumped-reflux arrangements

Typical overhead lines for hydrocarbon distillation columns are shown in F/4. Fluid circulation in the piping is the result of the thermosiphon effect in gravity-reflux condensation. For the systems shown in F/4, there is (and most of the time must be) a pressure difference between the top of the tower and reflux drum. The reflux drum has a pump, which returns the liquid to the top of the tower or sends it to storage.

In these arrangements (besides the sum of the static heads), actual pressure differences, ΔP_p, also enter into the calculations:

$$\Delta P_p = P_1 - P_2 \quad (8)$$

For the dimensions given in F/5, the static-head difference will be:

$$\Delta P_s = (1/144)(\rho_1 H_1 - \rho_2 H_2) \quad (9)$$

where ρ_1 is usually vapor density, and ρ_2 is vapor-liquid

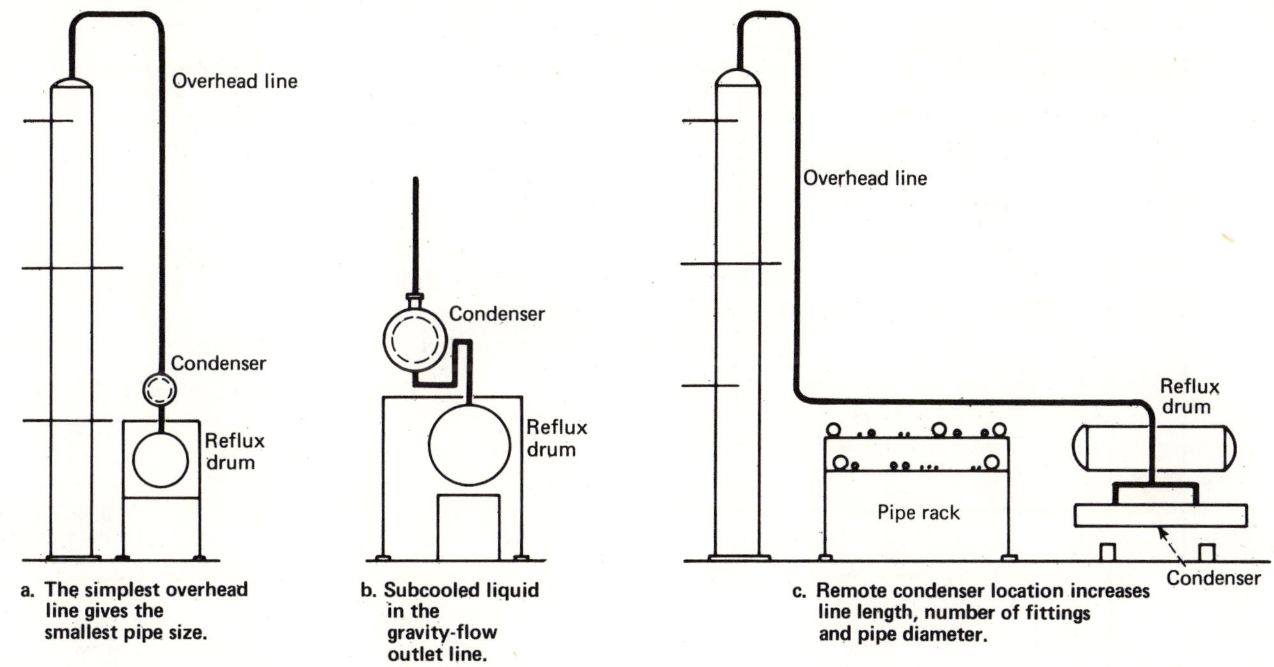

Overhead condensers for distillation columns F4

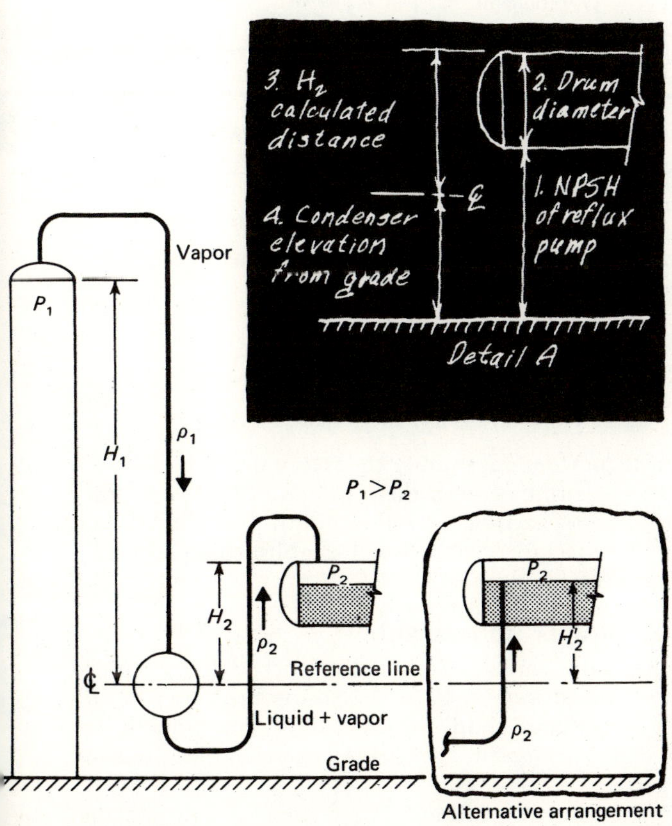

imensional relations for a condenser at grade F5

mixture (or liquid) density. The overall available ΔP is the sum of Eq. (8) and (9), or:

$$\Delta P = \Delta P_s + \Delta P_p \tag{10}$$

If the righthand side of Eq. (10) becomes negative, the condenser must be placed at an elevation closer to, or above, the reflux drum. A negative value indicates that the static-head backpressure ($\rho_2 H_2$) in Eq. (9) is greater than the sum of (1) pressure difference between the top of the tower and the reflux drum plus (2) the vapor static head ($\rho_1 H_1$) in the overhead line. The greater the condensation, the heavier the mixture becomes in the condenser's outlet line, which results in a greater backpressure. Of course, $\rho_2 H_2$ becomes positive when the line has a gravity-flow arrangement between the condenser outlet and the reflux-drum inlet, as shown in F/4a and F/4b.

Eq. (10) shows the driving force in the overhead system. This must be greater than the sum of piping, inlet and exit losses, ΔP_p, and exchanger resistance, Δp_e:

$$\Delta P > \Delta p_p + \Delta p_e \tag{11}$$

Δp_p usually is between 2 to 6 psi, and condenser Δp_e ranges from 0.5 to 5 psi.

The maximum possible condenser-centerline location below the reflux drum (dimension H_2 in F/5) can be calculated from the combination of Eq. (9), (10) and (11), to give:

$$(1/144)(\rho_1 H_1 - \rho_2 H_2) + \Delta P_p - (\Delta p_p + \Delta p_e) = 0 \tag{12}$$

As a safety factor, the positive static-head column pres-

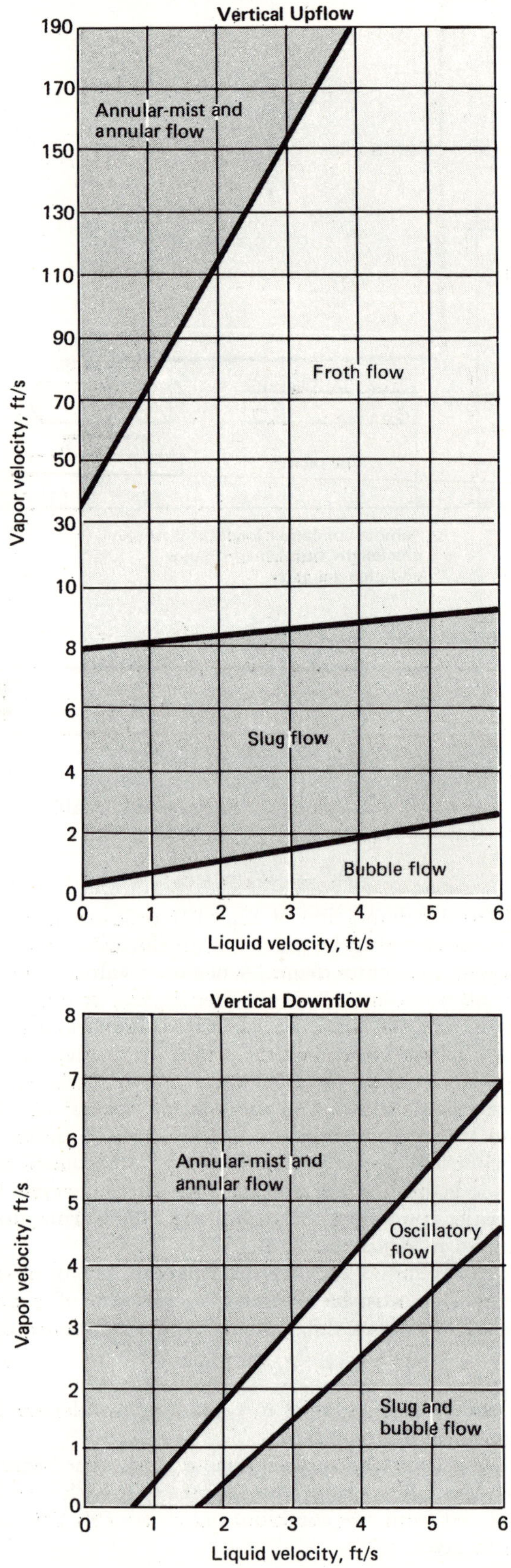

Liquid and vapor velocities determine type of flow F6

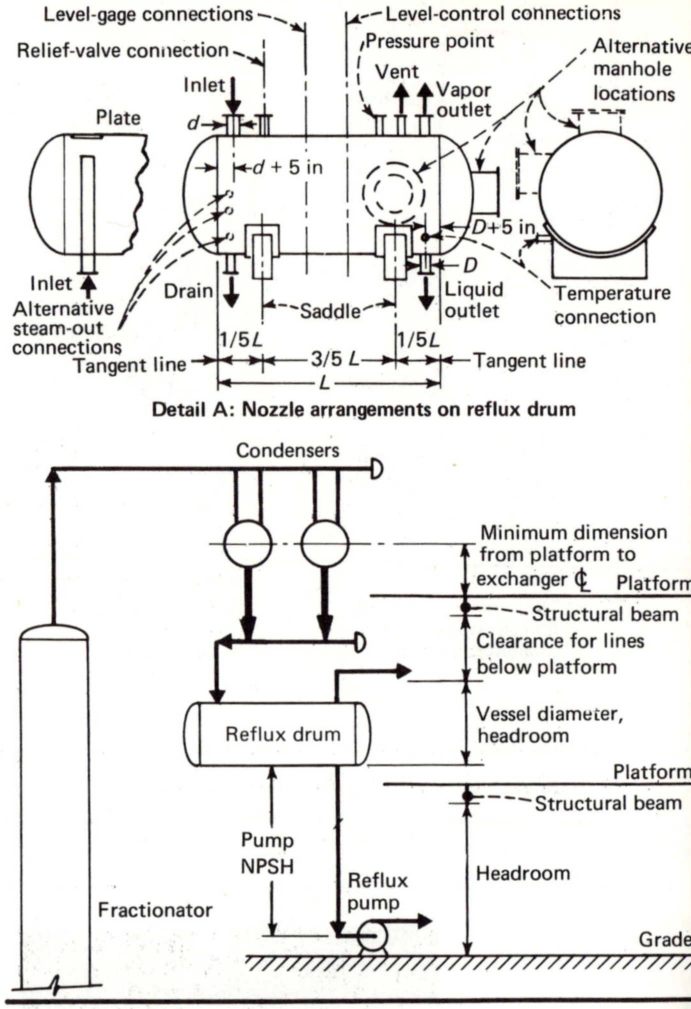

Elevated condensers and details of reflux drum F7

sure of the overhead vapor line can be neglected. Consider:

$$\rho_1 H_1 = 0.$$

Expressing H_2 from Eq. (12) in feet, we get:

$$H_2 = (144/\rho_2)(\Delta P_p - \Delta p_p - \Delta p_e) \qquad (13)$$

In layout design, usually the reflux drum is elevated first in accordance with the required NPSH (net positive suction head) of the reflux pump. The dimensions shown in Detail A of F/5 will establish the condenser elevation from grade.

Slug flow is undesirable

Slug flow can develop in the pocketed condenser-outlet line shown in F/5, depending on vapor-liquid proportion and fluid velocity. Slug flow should be avoided because it can cause undesirable pressure surges.

An empirical relation can be used to estimate the slug-flow region. If the velocity (calculated with two-phase density) in the pipeline is smaller than $(5\rho_1/\rho_v)^{1/2}$, slug flow is possible. The type of flow can also

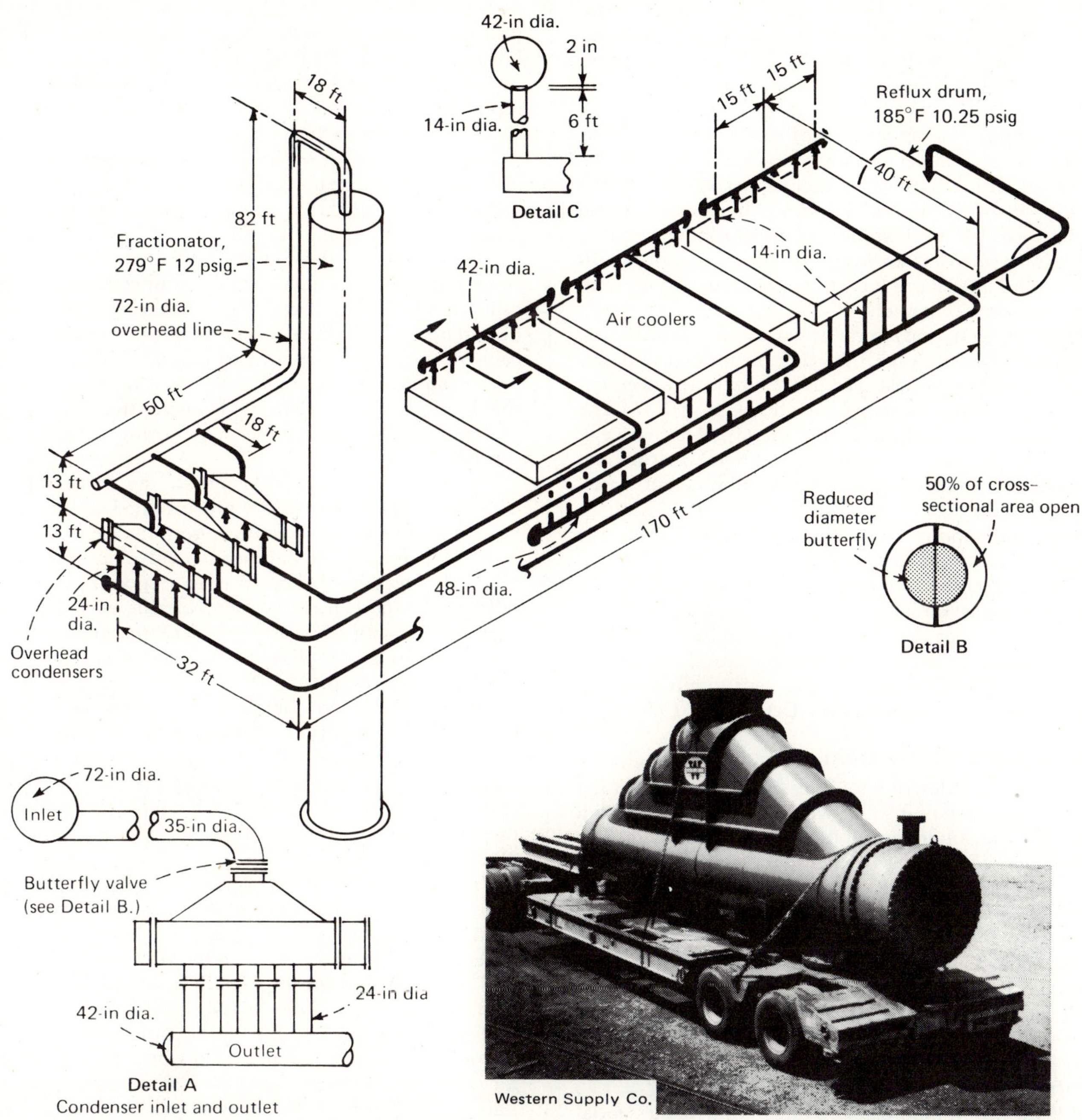

Two-stage condensing system has parallel flow through large surface condensers in series with air coolers F 8

be estimated by calculating the Baker parameters and using Fig. 1 p. 49. Flow-region maps such as are shown in F/6 are also available for vertical flows. To use these maps, the vapor-phase and liquid-phase velocities are calculated separately for the same pipe diameter. The intersection of these in the appropriate diagram of F/6 establishes the flow region.

The general criterion for selecting a suitable line size is that the pipe diameter must be sufficiently small to have the highest possible velocity, but large enough to stay within available pressure differential. The possibility of slug flow can be minimized by (a) increasing the pressure drop in the condenser's outlet line and reducing the resistance of the rest of the condensing system, (b) providing two parallel lines between condenser and reflux drum, (c) using valved alternative pipe runs for alternative flowrates, and (d) changing to a gravity-flow arrangement.

Equipment arrangement

In a gravity-flow arrangement, a structure is often required. Factors that influence layout design are shown in F/7. Below the elevated condensers, a platform is usually provided. Structural members, space for piping below the platform, and reflux-drum diameter or minimum headroom requirement determine the platform level for the reflux drum. Piping, NPSH or headroom will establish the distance between grade and the reflux-drum platform.

The piping arrangement of condensing systems should be as simple as possible. Each elbow in the overhead line represents a sizable percentage of the overall pipe-friction loss. However, lines can become longer because of flexibility considerations. Usually, there is no valve in an overhead line. Pressure loss in fittings and line length have to be compensated for with increased pipe diameter.

Piping design, and access to valves and instruments, also depend on how well nozzles are located and oriented on process vessels. The usual nozzle arrangement of a reflux drum is shown in Detail A of F/7. The condensate inlet is at one end of the top. Alternatively, a bottom inlet with a standpipe in the reflux drum can be designed, eliminating a couple of elbows and reducing pipe resistance. The pump drawoff nozzle is at the bottom and opposite end of the reflux drum. Level-gage and level-controller connections can be in the center. At the center, the liquid level is least agitated by the inlet and outlet streams. Pressure connections are usually in the vapor space, and temperature connections in the liquid space near the pump drawoff nozzle. Alternative manhole locations, and vent, drain and support connections are shown in Detail A of F/7.

Two-stage condensation

We will now describe a large, two-stage, condensing system, and in doing so we will find additional design ideas for the condenser piping.

Where cooling water is expensive, condensing systems can be designed for use with air coolers or with a combination of air coolers and water coolers in series. If close condensate-temperature control is required, the water coolers are located after the air coolers. For optimum heat transfer, a reversed sequence can be chosen.

A large-capacity condensing system handling the overhead from the primary fractionator of an ethylene unit is shown in F/8. The pressure difference between the fractionator and reflux drum is small, $\Delta P = 1.75$ psi. With this limited pressure differential, it is essential to have minimal resistance through exchangers, piping and pipe components.

Resistance of the surface condensers has been minimized by choosing three exchangers, each handling one-third of the total flow; and by providing a flared exchanger inlet, and four large outlets for each shell. Exchanger resistance is given as 0.12 psi, exclusive of inlet and outlet resistances. The photograph in F/8 shows one of the exchangers being delivered.

Air-cooler resistance has been minimized by using numerous single passes and large-diameter finned tubes. Air-cooler pressure loss is estimated as 0.7 psi, excluding nozzles.

Even with all these provisions, close to 50% of the available overall pressure difference has been consumed by the exchangers alone. About 30% of the overall ΔP has been taken up by the inlet and outlet resistances, leaving 20 to 25% of 1.75 psi for pipe resistance. This, and the rather long pipe configuration, mean very large pipe sizes. These are indicated in F/8. Velocities in these lines are less than 50 ft/s. Under these conditions, the piping system from the fractionator to the reflux drum has to be self-draining. Horizontal sections of the piping slope toward the reflux drum.

Saturated vapor flows in the 72-in-dia. overhead line. (Overhead lines are not usually insulated, and it is reasonable to assume that some condensation will take place in them.) Condensate, collected in the overhead line, can drain to the surface condensers through the three tangentially welded branch-connections (see Detail A in F/8). Quite likely, this condensed liquid is not distributed to the three surface condensers equally because of the nonsymmetrical arrangement of the branch lines. Considering the small amount of liquid, and that this liquid does not affect the heat-transfer duty of the exchangers, the nonsymmetrical arrangement is accepted.

For equal vapor distribution to the three surface condensers, butterfly valves are installed at the inlet flanges of the flared exchanger sections (see Detail B in F/8). The butterfly flap in each valve has a reduced diameter—leaving 50% of the cross-sectional area of the inlet nozzle permanently open. In this way, it is assumed that, with less resistance, a more-sensitive regulation can be accomplished than with a valve having a full-diameter flap.

Pipelines between the surface condensers and air coolers have two-phase dispersed flow. Because of the low velocity, elbows and tee connections, some liquid separation (with stratified flow) at the bottom of the lines is also assumed. (Under such conditions, and with these line sizes, it is difficult to assess how valid two-phase-flow theories are.)

The outlet lines of the surface condensers connect centrally to the inlet headers of the air coolers. Equal fluid distribution is assured by the relatively great difference in resistance between the very-low header loss and the high entrance resistance to the air cooler branch-connections. For equal distribution of the stratified liquid at the bottom of the pipe, each branch-connection extends about 2 in up into the horizontal air-cooler header (see Detail C in F/8). These extensions dam up the liquid in the bottom of the pipe, and the overflow facilitates equal liquid distribution across each air-cooler pass. Symmetrical piping at the outlet side of air coolers is not considered as essential as at the inlet side.

The next article in this series of piping design will appear in the issue of Oct. 13, 1975, and will cover calculation procedures for compressible fluid flow under isothermal and adiabatic conditions in pipelines and restrictions.

References

1. Kern, D. Q., "Process Heat Transfer," McGraw-Hill, New York, 1950.
2. Golan, L. P. and Stenning, A. H., Two-Phase Vertical Flow Maps, Joint Symposium on Fluid Mechanics and Measurement in Two-Phase Flow Systems, at University of Leeds, Leeds, England, Sept. 24–25, 1969; *Proc. Inst. Mech. Engrs. (London)*, Vol. 184, Part 3C (1970).
3. Cady, P. D., How To Stop Slug Flow in Condenser Outlet Piping, *Hydrocarbon Process.*, Sept. 1963.
4. Greene, J. L., Symmetrical Piping Arrangements Solve Two-Phase Flow Distribution Problems, *Hydrocarbon Process.*, Feb. 1967.

How to size piping and components as gas expands at flow conditions

An evaluation of isothermal and adiabatic conditions leads to practical formulas for establishing flowrate, pressure drop and the limiting velocity for gases through pipelines and restrictions.

Robert Kern, *Hoffmann - La Roche Inc.* *

☐ Flow of gases and vapors through pipelines, and piping components and other restrictions, is affected by the changing conditions of pressure and temperature.

Therefore, we must examine the nature of the flow, i.e., whether it is isothermal or adiabatic, and apply the appropriate flow relations to the piping design. Let us first review the principles for such isothermal and adiabatic flows, and then solve some typical problems.

Densities of gases and vapors vary with pressure and temperature. In isothermal flow (temperature is constant), density varies with pressure. In adiabatic flow, a decrease in temperature occurs when pressure decreases. This results in a density-increase correction. At high pressures and temperatures, the compressibility factor can be less than unity, and some increase in density will occur.

When calculating flowrate or pressure drop for piping design, pressure will have the dominant influence on density. As pressure decreases due to piping and components resistance, the gas expands and its velocity increases.

However, we may encounter a limitation. The velocity of flow in a pipeline of constant cross-section (or in the throat of a restriction) cannot go higher than the sonic or critical velocity. Even if the downstream pressure is lower than the pressure necessary to reach critical velocity, flowrate through a pipeline or restriction will not increase above that calculated at the critical velocity.

Originally published October 13, 1975

Useful fluid-flow relations

According to available data, the relation between the Reynolds number and friction factor is the same for liquid flow, steady gas flow, and expanding gas flow, up to the limit of sonic velocity [*1*]. In short lines having high pressure differentials, flow usually falls in the totally turbulent zone, and the friction factors in Table VI p. 93 can be used. At the beginning of long pipelines, transitional turbulence is also possible, with friction factors higher than at total turbulence. These friction factors for steel pipes can be obtained from Fig. 2 on p. 50.

Interpolation between the initial transitional-turbulent friction factor over a length of line, and the totally turbulent friction factor over the remaining portion of the line, can give a close estimate of the overall friction factor. This also means that the expressions for the resistance coefficient, K, head loss, h_L, and coefficients of restrictions, C, (derived for steady flow) can be used in the calculations of expanding gas flow.

Thermodynamic derivations assume that the ideal gas laws are applicable: i.e., $P\bar{V}$ = a constant for isothermal flow, and $P\bar{V}^k$ = a constant for adiabatic flow. While, theoretically, this is true only for frictionless flow, it makes little practical difference when calculating flowrate or pressure drop through pipelines and components.

*To meet your author, see p. 44.

For a given pressure differential, the mass discharge rate through a pipeline is greater for adiabatic flow than the rate for isothermal flow. The difference will never be more than 20%. Practically, there will be no difference if the pipeline is more than 1,000 pipe diameters long [2]. Actual flows are somewhere between isothermal and adiabatic conditions; and the difference in flowrate stays well below 20%—even for lines having lengths well under 1,000 pipe diameters.

In comparing the basic equations for adiabatic and isothermal flows, isothermal flow can be considered as a hypothetical adiabatic flow where the adiabatic exponent $k = 1$. This proposition is useful because isothermal derivations are simple and the solutions convenient compared with some of the lengthy and complicated procedures of the relations for adiabatic calculations.

Data for the adiabatic exponent, k, specific heats, c_p and c_v, and the gas constant, R, are given in handbooks, usually at some standard pressure and temperature. Increase in pressure and decrease in temperature tend to give higher k values and a lower compressibility factor, z. At high pressures and temperatures, z can be considered as a correction factor for the gas constant, R. These changes will make little difference in sizing piping and components, or estimating flowrates and pressure drop.

Elevation changes in pipelines have negligible effect on pressures, flowrates and pressure drops because gas densities are usually small. For restrictions in a pipeline or flows through a short pipeline, we must pay close attention to the design parameters in order to avoid overly high flow velocities. High velocities can cause vibrations or noise, or both.

Isothermal flow: flowrate through pipelines

Bernoulli's general equation for steady flow can be expressed graphically as:

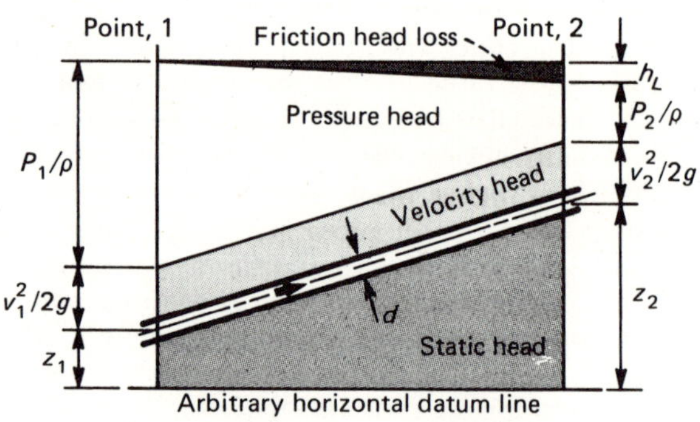

$$\frac{P_1}{\rho} + \frac{v_1^2}{2g} + z_1 = h_L + \frac{P_2}{\rho} + \frac{v_2^2}{2g} + z_2$$

These relationships can also be written as:

$$(1/\rho)(p_2 - p_1) + (1/2g)(v_2^2 - v_1^2) + (z_2 - z_1) + h_L = 0$$

For expanding-gas flow, $v_2 \neq v_1$; and with horizontal pipe, $z_2 = z_1$. Hence, the differential form of Bernoulli's equation can be written as:

$$-(1/\rho)\,dp = (1/g)v\,dv + h_L \qquad (1)$$

where the density, ρ, is not constant, and

$$h_L = Kv^2/2g \qquad (2)$$

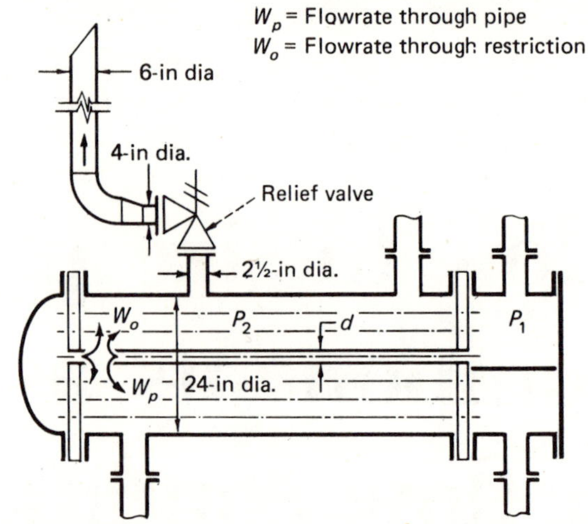

Flow through ruptured exchanger tube and tubesheet orifice

Nomenclature

A	Internal cross-section of pipe, ft^2
β	Ratio between restriction diameter and inside diameter
C	Mechanical flow coefficient
C_K	Kinetic-energy correction factor
d	Internal diameter of pipe, in
d_o	Diameter of restriction, in
D	Internal diameter of pipe, ft
f	Friction factor
g	Gravitational constant, 32.2 ft/s^2
h_L	Head loss, ft
k	Specific heat ratio
K	Resistance coefficient
L	Overall length of pipe, ft
M	Molecular weight
p	Absolute pressure, lb/ft^2
P	Absolute pressure, psia
ΔP	Pressure loss or pressure difference between two points, psi
ΔP_E	Pressure drop in steady flow, psi
ΔP_c	Critical pressure drop, psi.
R	Universal gas constant, 1,544/M
ρ	Fluid density, lb/ft^3
T	Absolute temperature, °R
v	Mean velocity of fluid, ft/s
v_c	Critical (or sonic) velocity, ft/s
$\bar{V}$	Specific volume, ft^3/lb
w	Weight flowrate, lb/s
W	Weight flowrate, lb/h
W_E	Weight flowrate for steady condition, lb/h
Y	New expansion factor
z	Compressibility factor

Subscripts

1	Upflow condition
2	Downflow condition
o	Orifice
p	Pipe
RV	Relief valve

Weight flowrate through the pipe is: $w = A\rho v$, or $w/A = \rho v =$ constant. Expressing this relation in differential form, and remembering that both density and velocity change along the pipeline:

$$\rho\, dv + v\, d\rho = 0$$

$$dv = -v\left(\frac{d\rho}{\rho}\right) \tag{3}$$

In expanding-gas flow, the pressure-and-density ratio stays constant: $p/\rho = dp/d\rho = p_1/\rho_1$. We can express this as two relationships:

$$\frac{d\rho}{\rho} = \frac{dp}{p} \tag{4}$$

$$\frac{1}{\rho} = \left(\frac{p_1}{\rho_1}\right)\frac{1}{p} \tag{5}$$

where p_1 and ρ_1 are the initial absolute pressure and density, respectively. Inserting Eq. (4) into Eq. (3) yields:

$$dv = -v(dp/p) \tag{6}$$

With Eq. (2) and Eq. (6), Eq. (1) becomes:

$$-dp = \frac{v^2\rho}{g}\left(\frac{K}{2} - \frac{dp}{p}\right) \tag{7}$$

Expressing velocity in terms of weight flowrate: $v = w/A\rho$; and inserting this into Eq. (7) yields:

$$-dp = \frac{1}{g}\left(\frac{w}{A}\right)^2\frac{1}{\rho}\left(\frac{K}{2} - \frac{dp}{p}\right) \tag{8}$$

Inserting Eq. (5) into Eq. (8), and multiplying both sides by $2p$, we get:

$$-2p\, dp = \left(\frac{w}{A}\right)^2\left(\frac{p_1}{\rho_1}\right)\left(\frac{1}{g}\right)\left(K - 2\frac{dp}{p}\right) \tag{9}$$

In Eq. (9), pressure is the only variant. Integrating Eq. (9) between two reference points in a pipeline:

$$-2\int_1^2 p\,dp = \left(\frac{w}{A}\right)^2\left(\frac{p_1}{\rho_1}\right)\left(\frac{1}{g}\right)\left(K - 2\int_1^2 \frac{dp}{p}\right)$$

$$p_1^2 - p_2^2 = \left(\frac{w}{A}\right)^2\left(\frac{p_1}{\rho_1}\right)\left(\frac{1}{g}\right)\left(K + 2\ln\frac{p_1}{p_2}\right) \tag{10}$$

Solving Eq. (10) for flowrate, w, lb/s, we get:

$$w = \sqrt{\left(\frac{A^2\rho_1 g}{K + 2\ln(p_1/p_2)}\right)\left(\frac{p_1^2 - p_2^2}{p_1}\right)} \tag{11}$$

We convert Eq. (11) to convenient units by letting: $W = 3{,}600w$, lb/h; $A = (d/12)^2(\pi/4)$, where d is in.; $p = 144P$, where P is psia; and $K = fL/D$. With these expressions, Eq. (11) now becomes:

$$W = 1{,}335.6d^2\sqrt{\rho_1/[K + 2\ln(P_1/P_2)]} \times \sqrt{(P_1^2 - P_2^2)/P_1} \tag{12}$$

Eq. (12) can be used to calculate flowrate through a pipe for expanding-gas flow under isothermal conditions, provided that upstream and downstream pressures are known.

Example 1—What is the flowrate of natural gas through a ruptured exchanger tube assuming (1) a complete break near the tubesheet, as shown in F/1, and (2) isothermal flow? (We will compare this calculation with that for adiabatic flow in Example 4 under the same flow data.)

Flow data for this example are:

Pressure in exchanger tubes, $P_1 = 1{,}100$ psig.
Relief-valve set pressure, $P_2 = 400$ psig, on shellside.
Gas temperature, 100°F.
Molecular weight, $M = 18.7$.
Compressibility factor, $z = 0.9$.
Exchanger tubes are $^3/_4$ in, Schedule 160 for which $d = 0.0512$ ft $= 0.614$ in; $d^2 = 0.377$ in^2; and O.D. $=$ 1.050 in.
Tube length $= 20$ ft.
Friction factor, $f = 0.026$ for complete turbulence.

First, we evaluate the upflow density, ρ_1, from:

$$\rho_1 = MP/10.72Tz$$

$$\rho_1 = \frac{18.7(1{,}110 + 14.7)}{10.72(460 + 100)0.9} = 3.89 \text{ lb/ft}^3$$

Then, we evaluate the total resistance as follows:

Tube resistance:

$K = fL/D = 0.026(20/0.0512)$	$= 10.15$
Tube entrance and exit resistance	$= 1.5$
Total resistance	11.65

For convenience, we will also evaluate the term: $2\ln(P_1/P_2)$ in Eq. (12) as:

$$2\ln(1{,}124.7/414.7) = 2$$

We can now calculate the flowrate, W, through the completely broken exchanger tube (when the relief-valve set pressure is 400 psig) by substituting into Eq. (12):

$$W = 1{,}335.6(0.377)\sqrt{\frac{3.89}{11.65 + 2}}\sqrt{\frac{(1{,}124.7)^2 - (414.7)^2}{1{,}124.7}}$$

$$W = 8{,}380 \text{ lb/h}$$

Isothermal flow: pipeline pressure drop

A convenient method for calculating pressure loss has been developed by others in a sequence of technical papers [3,4].

Replacing w/A by $\rho_1 v_1$ in Eq. (10), multiplying by $1/p_1^2$, and rearranging, we get:

$$1 - \left(\frac{p_2}{p_1}\right)^2 = \frac{\rho_1 v_1^2 K}{g p_1} + \left(\frac{\rho_1 v_1^2}{g p_1}\right) 2\ln\left(\frac{p_1}{p_2}\right) \tag{13}$$

In Eq. (13), we find:

$$2\ln(p_1/p_2) = -\ln(p_2/p_1)^2 \tag{14}$$

and we will let:

$$2\Delta p_E = \rho_1 v_1^2 K/g \tag{15}$$

Δp_E is the pressure loss in steady flow, expressed in consistent units. See Eq. (13) on p. 21. The coefficient of the logarithmic expression in Eq. (13) is:

$$(\rho_1 v_1^2)/g p_1 = v_1^2/v_c^2$$

because:

$$1/v_c^2 = \rho_1/g p_1 \tag{16}$$

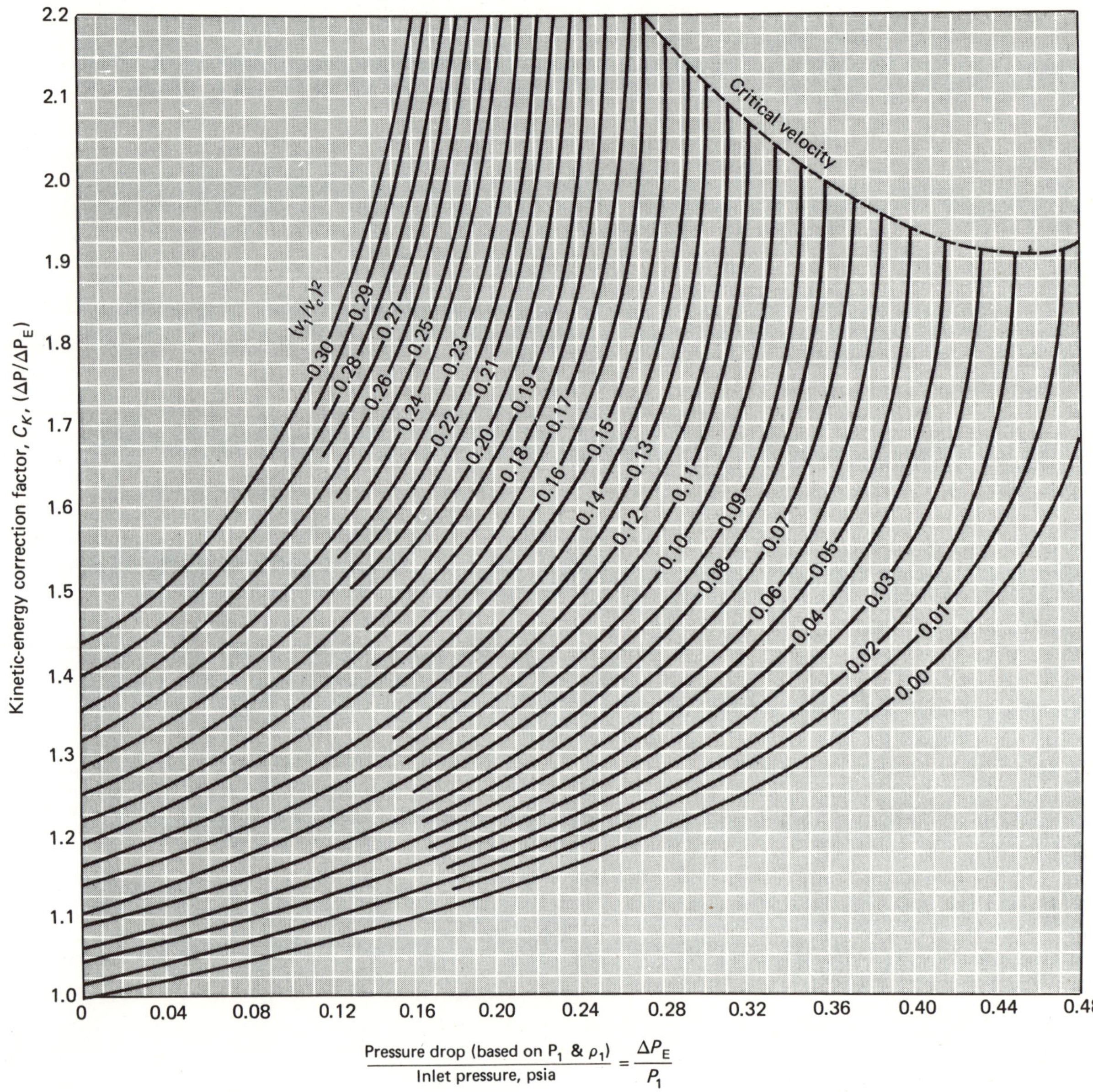

Source: ASME and M.W. Kellogg Co.

Kinetic energy correction for pressure drop for isothermal flow of gases F 2

Eq. (16) is the critical-velocity flow formula, expressed in consistent units.

By using Eq. (14), (15) and (16) in Eq. (13), and rearranging, we find:

$$1 - \left(\frac{p_2}{p_1}\right)^2 = \frac{2\Delta p_E}{p_1} - \left(\frac{v_1}{v_c}\right)^2 \ln\left(\frac{p_2}{p_1}\right)^2 \quad (17)$$

Chesler and Jesser [4] introduced the following proposition: To obtain pressure loss, Δp, during expanding-gas flow, multiply the steady-flow pressure loss, Δp_E, by a kinetic-energy correction factor, C_K:

$$p_1 - p_2 = \Delta p = \Delta p_E \times C_K \quad (18)$$

Eq. (18) can also be written as:

$$\frac{p_2}{p_1} = 1 - C_K\left(\frac{\Delta p_E}{p_1}\right) \quad (19)$$

Placing Eq. (19) into Eq. (17) (and because pressures are in ratios), we can replace all p and Δp by P, psia, and ΔP, psi, to get:

$$\frac{2\Delta P_E}{P_1} = 1 - \left[1 - C_K\left(\frac{\Delta P_E}{P_1}\right)\right]^2 + \left(\frac{v_1}{v_c}\right)^2 \ln\left[1 - C_K\left(\frac{\Delta P_E}{P_1}\right)\right]^2 \quad (20)$$

The original proposition of Eq. (18) now becomes:

$$\Delta P = \Delta P_E \times C_K \qquad (21)$$

where ΔP is the pressure loss for expanding-gas flow, psi, and ΔP_E, psi, is:

$$\Delta P_E = 0.00000336 fLW^2/d^5\rho_1 \qquad (22)$$

Eq. (22) was originally presented as Eq. (17) on p. 210.

In Eq. (20), v_1/v_c is the Mach number. The expression for the upstream gas velocity, v_1, with initial density, ρ_1, is:

$$v_1 = 0.0509W/d^2\rho_1 \qquad (23)$$

The sonic velocity, v_c, calculated with initial absolute pressure, P_1, and initial density, ρ_1, is:

$$v_c = 68\sqrt{P_1/\rho_1} \qquad (24)$$

The kinetic-energy correction factor, C_K, in Eq. (21) can be obtained from F/2, which is the graphical solution of Eq. (20). The dimensionless ratios of $\Delta P_E/P_1$, and corresponding values of $(v_1/v_c)^2$, can be conveniently calculated with upstream gas conditions. P_1 has dimensions of psia, and C_K is dimensionless.

Example 2—What is the overall pressure drop in an 800-ft-long, well-insulated, 12-in Schedule 30 pipeline (d = 12.09 in, d^2 = 146.17, d^5 = 258,304)? Flowrate, W, = 250,000 lb/h of carbon dioxide having a molecular weight, M, = 44. Pipe inlet conditions are: P_1 = 80 psia, T_1 = 600°R, and viscosity, μ, = 0.027 cp.

We calculate upflow gas density, ρ_1, from:

$$\rho_1 = MP_1/10.72T_1z$$

$$\rho_1 = \frac{44(80)}{10.72(600)1} = 0.547 \text{ lb/ft}^3$$

For steady flow (based on upstream conditions) with a friction factor of f = 0.013 (for a Reynolds number, N_{Re} = 4.87 × 10^6), the pressure loss is found by substituting in Eq. (22). Hence:

$$\Delta P_E = \frac{0.00000336(0.013)(800)(250{,}000)^2}{(258{,}304)(0.547)}$$

$$\Delta P_E = 15.46 \text{ psi}$$

To determine the value of C_K from F/2, we must now compute $\Delta P_E/P_1$ and $(v_1/v_c)^2$:

$$\Delta P_E/P_1 = 15.46/80 = 0.193$$

We substitute in Eq. (23) to find v_1, as:

$$v_1 = \frac{0.0509(250{,}000)}{(146.17)(0.547)} = 159.15 \text{ ft/s}$$

and now substitute in Eq. (24) to find v_c, as:

$$v_c = 68\sqrt{80/0.547} = 822.36 \text{ ft/s}$$

$$(v_1/v_c)^2 = (159.15/822.36)^2 = 0.0374$$

Using these values, we obtain C_K = 1.18 from F/2. Therefore, the overall pressure drop in this pipeline,* including kinetic energy losses, will be:

$$\Delta P = \Delta P_E \times C_K = 15.46 \times 1.18 = 18.24 \text{ psi}$$

*Designers often calculate expanding-gas flow initially with a roughly estimated average density and steady flow, and repeat the calculations to obtain greater arithmetic accuracy. In this example, if density is calculated with the median pressure of 71 psia, ΔP will be 17.41 psi as calculated with Eq. (22). In the case of sizable pressure loss, calculated with average density and as steady flow, the error of underestimating ΔP can be substantial.

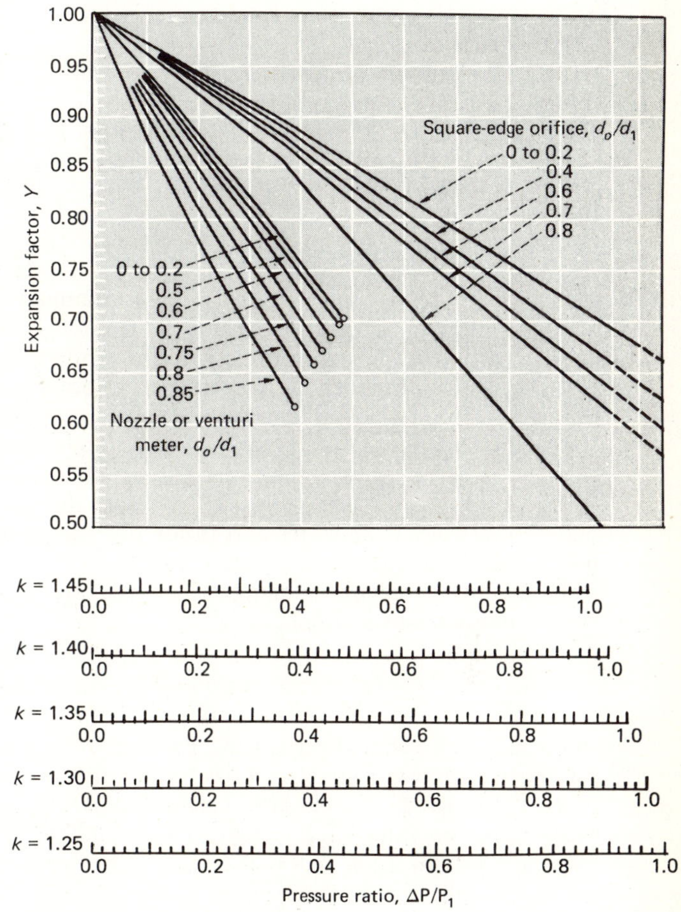

Source: Crane Co. and Ref. [7, 8].

Expansion factor for compressible flow through nozzles and orifices F3

Adiabatic flow

Adiabatic flow can be assumed for flows through an orifice, flow nozzle or a short pipeline. In addition to pressure change, there will be a change in temperature. This affects density, velocity and flowrate. Temperature change can be calculated from:

$$T_2 = T_1(P_2/P_1)^{(k-1)/k} = T_1(\rho_2/\rho_1)^{(k-1)}$$

The method of calculating flowrates through restrictions and short pipelines is identical to the relation expressed in Eq. (21). Without much elaboration, we will develop certain procedures in what follows. Detailed derivations are available in the literature [5,6,7].

Adiabatic flow through orifice

The mechanical principles are the same whether liquid or expanding gas is flowing. In addition, a change in the internal energy takes place in the expanding gas. This is reflected by multiplying Darcy's equation by a net expansion factor, Y:

$$W = W_E Y \tag{25}$$

where W_E is:

$$W_E = 157.66 C d_o \sqrt{h_L \rho_1^2} \tag{26}$$

Eq. (26) was presented as Eq. (5) in Part 3 [*14*].

We will modify this equation by substituting $h_L = 144\Delta P/\rho_1$, and then substitute the modified equation into Eq. (25):

$$W = W_o = 1{,}891 d_o^2 C Y \sqrt{\Delta P \rho_1} \tag{27}$$

where W or W_o is flowrate, lb/h; d_o is diameter of restriction, in; ρ_1 is upflow density, lb/ft^3; ΔP is the pressure difference between upflow and downflow, psi. C is the orifice or nozzle flow coefficient (from Fig. 1 and 4 in Part 3 [14]). Y can be obtained from F/3.

The net expansion factor or adiabatic flow coefficient, Y, is a function of the adiabatic exponent, $k = c_p/c_v$; absolute pressure ratio, $\Delta P/P_1$; and the ratio between restriction diameter and inside pipe diameter, $\beta = d_o/d_1$. P_1 is upstream pressure, psia.

To design a pressure-reducing orifice, we can calculate its diameter from Eq. (27) as:

$$d_o^2 = \frac{W_o}{1{,}891 C Y \sqrt{\Delta P \rho_1}}, \text{ in}^2 \tag{28}$$

where ΔP represents the required pressure reduction, psi. The restriction diameter, d_o, is usually small compared with pipe diameter. After calculating $\Delta P/P_1$, we can obtain the value of Y from F/3 by using the line for a nozzle ratio, $d_o/d_1 = 0$ to 0.2. We get the mechanical flow coefficient, C, from Fig. 1 in Part 3 [*14*].

Example 3—What is the flowrate through the tubesheet at the place of a completely ruptured exchanger tube? Data are the same as given in Example 1 and F/1.

We will treat the ruptured tube as an orifice for which $d_o = 0.614$ in, and $d_1 = 24$ in. Hence, $\beta = 0.614/24 = 0.026$. To find the net expansion factor, Y, we calculate:

$$\Delta P/P_1 = (1{,}124.7 - 414.7)/1{,}124.7 = 0.63$$

From F/3, we find $Y = 0.79$ for these values. Note that $k = 1.3$ (obtained from Table I in Part 1 [*12*]. The mechanical flow coefficient, C, = 0.62 from Fig. 1 of Part 3 [*14*]. Substituting the appropriate values into Eq. (27) and solving, we find:

$$W_o = 1{,}891(0.377)(0.62)(0.79)\sqrt{710(3.89)}$$

$$W_o = 18{,}351 \text{ lb/h}$$

Adiabatic flow through pipe

We will use the following equation previously developed as Eq. (17) on p. 23:

$$\Delta P = 0.00000336 L(f/\rho_1)(W_E^2/d^5) \tag{29}$$

We then let $(12fL)/d = K$, and solve Eq. (29) for flowrate W_E, lb/h:

$$W_E = 1{,}891 d^2 \sqrt{(\Delta P \rho_1)/K} \tag{30}$$

Using the proposition of Eq. (26), Eq. (30) becomes:

$$W = W_p = 1{,}891 d^2 Y \sqrt{(\Delta P \rho_1)/K} \tag{31}$$

where d is internal diameter of pipe, in; $\Delta P = P_1 - P_2$ is the pressure difference between the inlet and outlet points of the pipeline, psi; ρ_1 is inlet density, lb/ft^3. We obtain the net expansion factor, Y, from F/4.

Example 4—Let us now calculate the adiabatic flowrate of natural gas through the ruptured exchanger tube for the data given in Example 1.

To find the net expansion factor, Y, we evaluate:

$$\Delta P/P_1 = (1{,}124.7 - 414.7)/1{,}124.7 = 0.63$$
$$k = 1.3 \text{ (from Table I in Part 1 [12])}$$
$$K = 11.65 \text{ (calculated in Example 1)}$$

With these data, we enter F/4 and find $Y = 0.765$.

Inserting the appropriate values into Eq. (31) and solving yields:

$$W_p = 1{,}891(0.765)(0.377)\sqrt{710(3.89)/11.65}$$

$$W_p = 8{,}397 \text{ lb/h}$$

As this example demonstrates, the adiabatic flowrate differs very little from that of isothermal flow (8,380 lb/h in Example 1). The L/D ratio is $20/0.0512 = 390$, which is much less than the 1,000 pipe diameters previously discussed. As it should be, the adiabatic flowrate is greater than the isothermal flowrate.

The flowrate for sizing the relief valve, shown in F/1, will be the flow through the restriction (Example 3) plus the flow through the ruptured exchanger tube (Example 4):

$$W_{RV} = 18{,}351 + 8{,}397 = 26{,}748 \text{ lb/h}$$

This is also the flowrate through the relief-valve discharge line.

Critical pressure

We can also obtain the data for calculating the critical pressure drop, from F/3 and F/4 at the lower endpoint of each line. For example, in a flow nozzle where $\beta = 0.5$ and $k = 1.4$, F/3 gives a net expansion factor, $Y = 0.698$, and a pressure ratio, $\Delta P_c/P_1 = 0.47$. Therefore, the critical pressure drop is: $\Delta P_c = 0.47P_1$, and the critical downstream pressure at the point of sonic velocity will be $P_2 = P_1 - \Delta P_c$. The flowrate will not increase even if downstream pressure is decreased to less than P_2.

Example 5—Coke-oven gas is discharged to atmosphere through a 3-in Schedule 40 pipe ($D = 0.2557$ ft, $d^2 = 9.413$ in^2). The pipeline is 20-ft long and carries gas from a vessel that has a pressure of 125 psig. Gas properties are: molecular weight, $M = 12.18$; temperature, $t = 140$°F; $k = 1.4$; and $\rho_1 = 0.265$ lb/ft^3. What is the maximum flowrate through the line?

Actual pressure difference ΔP is 125 psi. Hence, $\Delta P/P_1 = 125/(125 + 14.7) = 0.895$. From Table VI in Part 1 [*12*], we obtain the friction factor, f, = 0.0178 for totally turbulent flow. We use this value to calculate pipeline resistance and total resistance as follows:

Pipeline resistance:

$K = fL/D = 0.0178(20)/0.2557$	= 1.4
Exit and entrance losses	= 1.5
Total resistance	2.9

We enter the chart in F/4 for $k = 1.4$ with these values and find their intersection falls beyond the limiting point of sonic velocity or critical pressure. Conse-

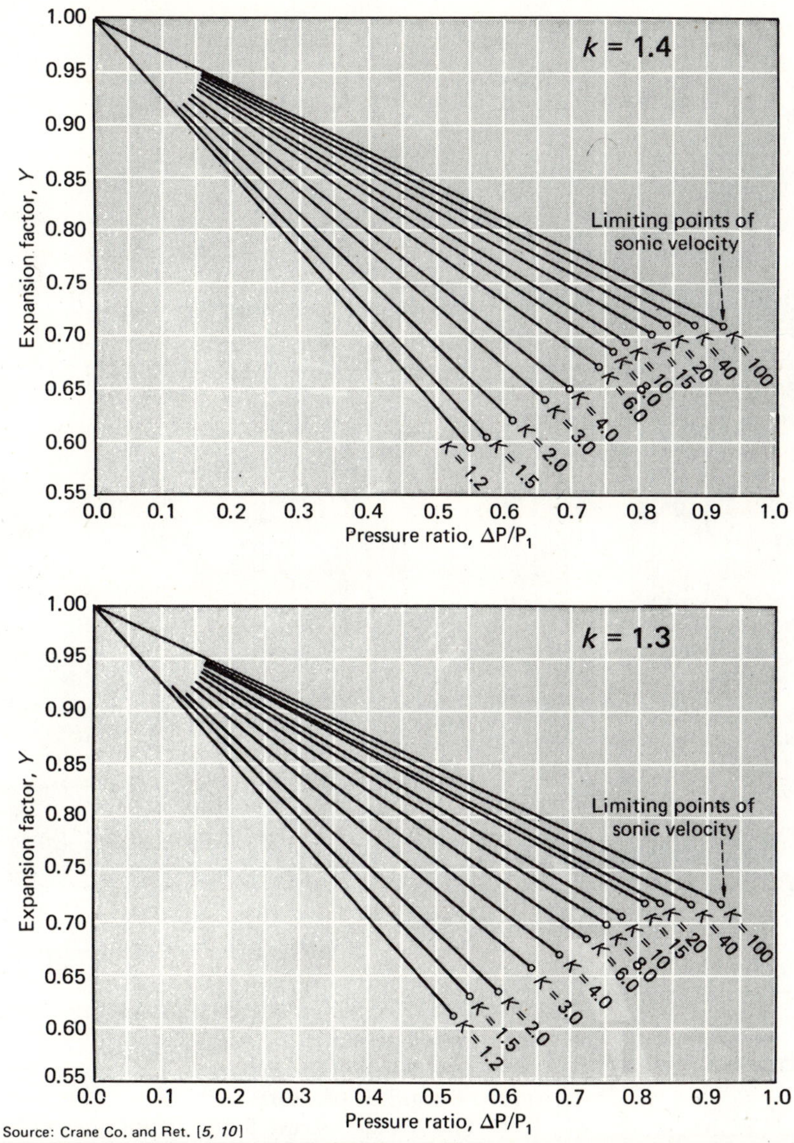

Limiting Factors for Sonic Velocity, $k = 1.4$†

K	$\Delta P/P_1$	Y
1.2	0.552	0.588
1.5	0.576	0.606
2.0	0.612	0.622
3	0.662	0.639
4	0.697	0.649
6	0.737	0.671
8	0.762	0.685
10	0.784	0.695
15	0.818	0.702
20	0.839	0.710
40	0.883	0.710
100	0.926	0.710

†k is approximately 1.4 for air, H_2, O_2, N_2, CO, NO and HCl.

Limiting Factors for Sonic Velocity, $k = 1.3$*

K	$\Delta P/P_1$	Y
1.2	0.525	0.612
1.5	0.550	0.631
2.0	0.593	0.635
3	0.642	0.658
4	0.678	0.670
6	0.722	0.685
8	0.750	0.698
10	0.773	0.705
15	0.807	0.718
20	0.831	0.718
40	0.877	0.718
100	0.920	0.718

*k is approximately 1.3 for CO_2, SO_2, water vapor, H_2S, NH_3, N_2O, Cl_2, CH_4, C_2H_2 and C_2H_4.

Source: Crane Co. and Ref. [*5, 10*]

Expansion factor for compressible flow through pipes to a larger flow area F4

quently, with a K value of 2.9, we must use the limiting factors for sonic velocity for which $\Delta P_c/P_1 = 0.66$ and $Y = 0.64$.

The critical pressure drop will be:

$$\Delta P_c = 0.66P_1 = 0.66(125 + 14.7) = 92.2$$

The critical downstream pressure, P_2, just before the end of the pipeline, is $125 - 92.2 = 32.8$ psig. We now calculate the maximum flowrate through the 3-in pipe by using Eq. (31):

$$W_p = 1{,}891(9.413)(0.64)\sqrt{92.2(0.265)/2.9}$$

$$W_p = 33{,}067 \text{ lb/h}$$

At the end of pipelines where sonic velocities can be expected, application of a silencer should be considered.

The next article in this series will appear to the issue of Nov. 10, 1975, and will deal with the major aspects of piping design for process plants.

References

1. Kennan, J. H., *J. Appl. Mech.,* Vol. 6, A11-20, A135-136 (1939).
2. Lapple, C. E., Isothermal and Adiabatic Flow of Compressible Fluids, *Trans. AIChE*, Vol. 39, 385-428 (1943).
3. Lobo, W. E., Fried, L. and Skaperdas, G. T., Pressure Drop in the Flow of Compressible Fluids, *Ind. Eng. Chem.,* Vol. 34, No. 7, 821-823 (1942).
4. Chesler, S. and Jesser, B. W., Pressure Drop for Flow of Compressible Fluids in Pipes, Appendix to Paper No. 51-SA-18, ASME, New York.
5. Dodge, B. F., "Chemical Engineering Thermodynamics," McGraw-Hill, New York, 1944.
6. Shapiro, A. H., "The Dynamics and Thermodynamics of Compressible Fluid Flow," Ronald Press, New York, 1954.
7. "Fluid Meters, Part I," ASME, New York.
8. Cunningham, R. G., Orifice Meters with Supercritical Compressible Flow, Paper No. 50-A-45, ASME, New York.
9. Smith, E. S., *Trans. ASME,* Vol. 52, HYD 52-7b, 89-109.
10. Flow of Fluid Through Valves, Fittings and Pipe, Technical Paper No. 410, Crane Co., Chicago.
11. Smith, B., Charts Used for Easier Pipe Sizing, *Hydrocarbon Process.,* May 1969.
12. Kern, R., *Chem. Eng.,* Feb. 3, 1975, p. 72.

Designing high-temperature piping

Hydrocarbon reformer photo: Howe Baker Engineers, Inc.

High-temperature operation introduces unique stress patterns in piping. This article describes the principal stresses involved and shows how to make the most efficient use of materials.

John D. Dawson, *Howe-Baker Engineers, Inc.*

☐ The most common criteria currently used in the U.S. for the design of hot piping in petrochemical plants are found in ANSI B31.3.* This code prescribes minimum requirements for the materials, design, fabrication, assembly, erection, inspection and testing of piping systems subject to pressure or vacuum, over a range of temperatures up to 1,500°F.

Allowable stress

Improperly designed piping may fail as a result of excessive stress levels generated during service. The "basic allowable stress" for a particular material at a particular temperature is based on several criteria, any of which may govern. In accordance with the Code, this basic allowable stress shall not exceed the lowest of the following:

- One third of the material's minimum tensile strength at room temperature.
- One third of the material's minimum tensile strength at design temperature.
- Two thirds of the material's minimum yield strength at room temperature.
- Two thirds of the material's minimum yield strength at the design temperature.
- The average stress for a creep rate of 0.01% per 1,000 h.
- 67% of the average stress for rupture at the end of 100,000 h.
- 80% of the minimum stress for rupture at the end of 100,000 h.

The various failure theories considered in selecting the above bases for allowable stress are beyond the scope of this discussion.

An exception to the fourth criterion above is that for austenitic stainless steels (generally the 300 series) and for some of the nickel alloys (such as Incoloy 800), when used at temperatures below 1,100°F, the limit may be as high as 90% of the minimum yield strength at the design temperature. However, this high allowable stress is not recommended for flanged joints or other applications where slight deformation could cause failure.

Principal stresses

For most piping design problems, where the pipe wall is thin (less than one tenth of the radius) in comparison

*ANSI B31.3. American National Standards Institute Code for Chemical Plant and Petroleum Refinery Piping, published by the American Soc. of Mechanical Engineers, 1976.

Originally published April 9, 1979

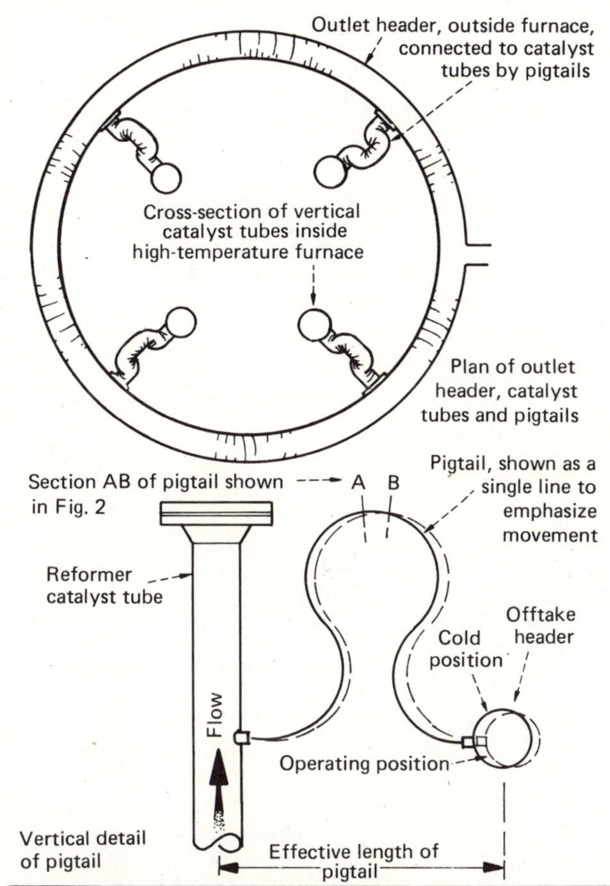

Catalyst tubes, outlet header and pigtails **Fig. 1**

with the other dimensions, we can assume that stresses in the pipe wall exist in only two directions, longitudinal and circumferential. This assumption is not valid for thick-wall pipe where radial stresses may become appreciable.

Longitudinal and circumferential "principal stresses" can be divided into two broad categories: primary stresses and secondary stresses.

Primary stresses are those that would not be relieved by local yielding of the material if the piping were stressed beyond the yield point. For example, internal pressure in a pipe causes longitudinal and circumferential tensile stresses to be generated. If the pressure is high enough, the yield strength of the material will be exceeded and the pipe will be deformed by bulging, which causes the pipe wall to become thinner, as well as larger in diameter. This will cause a further increase in stress, due to the internal pressure. Since stresses are not relieved by yielding of the material, the stresses caused by the pressure are said to be primary ones. Other causes of primary stresses are weight (of piping, contents and insulation) and, of course, external loads.

Secondary stresses are relieved by local yielding of the material in the highly stressed areas. For example, thermal stresses (primarily those caused by expansion of the metal in the pipe wall when the pipe is operated at a temperature higher than that at which it was installed) are relieved when the pipe material yields.

The B31.3 Code has different allowable design limits for primary stresses and secondary stre[...] that the limit for primary stresses will b[...] secondary stresses, since if primary [...] yield point, failure by rupture is ine[...] cases.

Due to the cylindrical shape of pipe, the [...] tensile stresses due to the internal pressure will be tw[...] great as the longitudinal tensile stresses. The minimum wall thickness for internal pressure will, therefore be determined by the allowable **circumferential stress.** It can be shown that the circumferential stress in a thin-wall cylinder under internal pressure is:

$$\text{Stress} = (\text{pressure} \times \text{radius})/\text{wall thickness}$$

which is basically the equation used to determine the minimum wall-thickness requirement in accordance with the Code. The basic allowable stress values tabulated in the Code may be used directly in the procedure for determining the required wall thickness.

In some piping systems, failures may occur as a result of excessive **longitudinal stress** levels generated by the combination of stresses from internal pressure, weight and other sustained loadings (primary stresses). Since the magnitude of combined stresses will vary according to the design of a particular system, the Code cannot present a simple equation to ensure a safe design, as it can in the case of wall thickness for internal pressure. Instead, the Code specifies that the sum of longitudinal stresses due to pressure, weight and other sustained loadings shall not exceed the basic allowable stress for the maximum metal temperature expected during the cycle under analysis.

Secondary stresses, on the other hand, have a higher allowable level because they are, in effect, self-limiting; that is, they will not cause immediate failure of the system as excessive primary stresses will.

Deformation and springing in cyclic loading

Under conditions of cyclic loading, there will be a maximum stress and a minimum stress (assume tensile stresses to be positive and compressive stresses to be negative) generated during each cycle. If these stresses stay within the elastic range for the material, no permanent deformation will occur. If, however, the combination of primary and secondary stresses exceeds the elastic limit, some local yielding (deformation) will occur in the highly stressed areas, and the piping system will acquire a "permanent set." If the system undergoes this plastic deformation on every cycle, failure will eventually occur. If a relatively few (say, less than 7,000) cycles are expected during the life of the system (as is the case for most plant piping), the allowable stress range will be greater than the basic allowable stress, and is calculated according to the equation given in Section 302.3.5 (d) of the Code.

$$S_A = 1.25\,S_c + 0.25\,S_h$$

where S_A = allowable-displacement-stress range; S_c = basic allowable stress, psi, at minimum (cold) metal temperature expected during the displacement cycle under analysis; and S_h = basic allowable stress, psi, at the maximum (hot) metal temperature expected.

This means that, in some cases, piping may be designed to undergo plastic deformation in service when the yield-

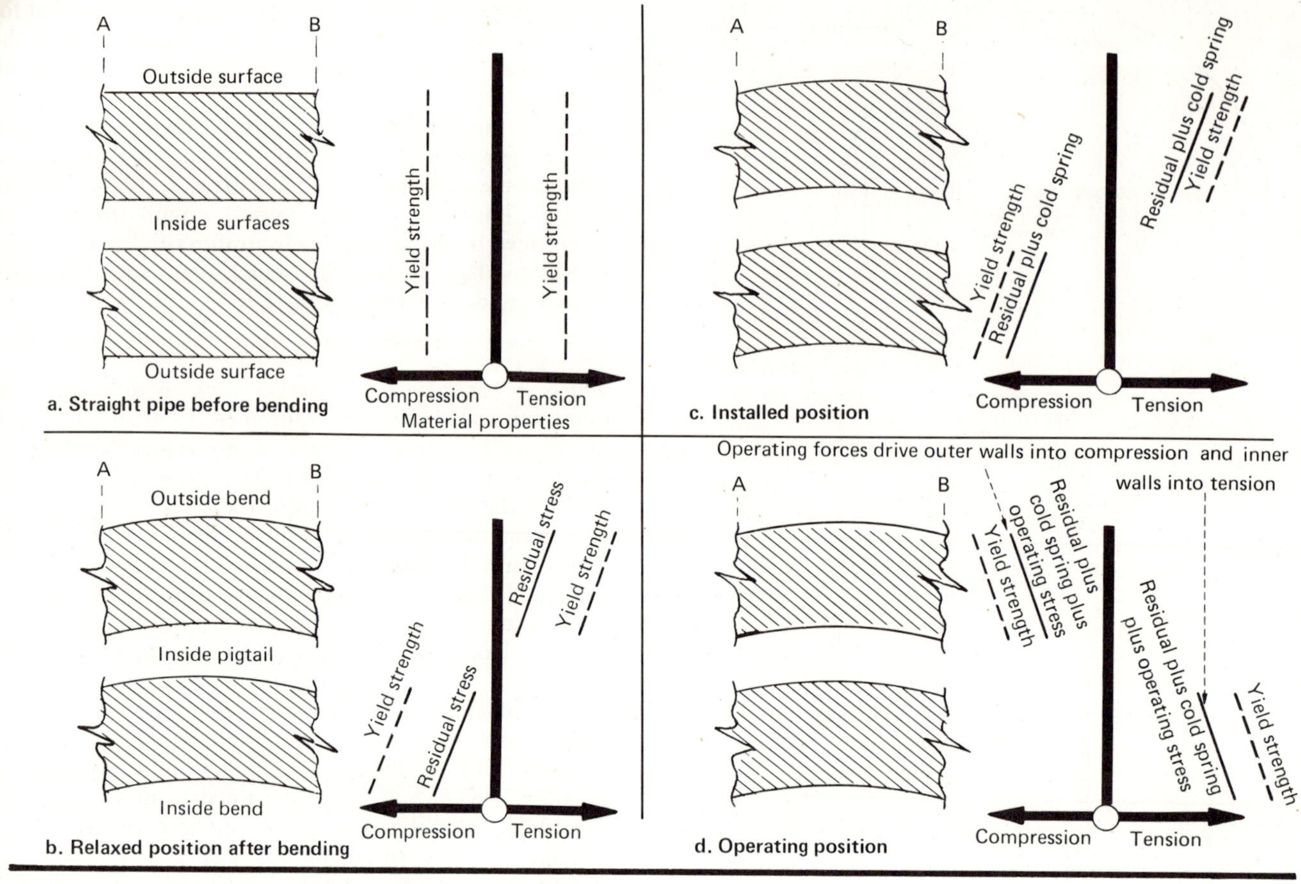

Cold-formed pigtail in various positions **Fig. 2**

ing is the result of self-limiting secondary stresses, such as those resulting from thermal expansion. (See Section 319.2.3 of the Code.) It is possible, of course, for failure to occur as a result of excessive yielding, and a properly designed system will not permit gross deformation to such an extent that failure or misalignment of piping guides and supports results.

One way that stresses resulting from thermal expansion can be offset is by **cold springing**. Cold spring is the intentional deformation of piping during assembly to produce a desired initial displacement, and stresses opposite to those expected in service.

Cold spring may be produced by "cutting short" or "cutting long" rather than by permanently deforming the system. This is the intentional fabrication of the piping to dimensions slightly different from the installed dimensions. The difference between the installed and fabricated dimensions is generally about half of the thermal expansion, but the arrangement of the particular system, the effects of weight, supports and allowable nozzle loadings or anchor reactions will determine the actual amounts of cold spring required.

A variation on cold springing is **self springing,** in which the piping is designed to undergo local yielding and consequent permanent deformation in the first cycle of operation. This leaves a residual offset stress that in subsequent cycles allows the piping to operate within the elastic limit.

Although self springing may eliminate additional yielding on subsequent cycles, the high stress (approaching the yield strength) may cause appreciable creep in the high-temperature range.

Expansion-loop example

The following example, although quite specialized, illustrates several important principles of design. The example concerns a steam/hydrocarbon reformer outlet-header, which is shown in Fig. 1. The reformer catalyst tubes run vertically through a high-temperature furnace. A circular offtake header rests on top of the cylindrical furnace and is connected to each of the catalyst tubes by a small-bore expansion-loop tube, otherwise known as a pigtail or omega because of its shape.

In service, the catalyst tube remains essentially stationary, while the circular outlet header expands radially. The amount of radial displacement depends on the material, temperature and circle radius. The pigtail also expands when heated, but since the straight-line distance between its ends is less than the radius of the header, the thermal growth of the pigtail is less than the displacement of the header. The result is that the pigtail is subjected to mechanical extension, as well as thermal expansion at operating temperature.

The pigtails are cold-formed from straight lengths of seamless pipe (Fig. 2a). In the forming operation, the material is stressed beyond its yield point, and permanent deformation (bending) occurs. When the forming-die forces are released, the pigtail has some "springback," but is left with residual stresses in the deformed (bent) areas—tensile on the outside of bends, and compressive on the inside (Fig. 2b). The minimum bend-radius for small pipe sizes is in the range of two to three pipe diameters. Tighter bends will result in excessive deformation.

High-nickel alloys, such as Incoloy 800, are subject to "strain-hardening," and as a result, the yield strength of the material increases as deformation takes place. Since the outer surface of the pipe undergoes more yielding than does the inside surface, the hardness will be greater on the outside surface, and the yield point will also be highest on the outside. The strain-hardening also decreases the ductility of the material; therefore, the outer surface is also the least ductile.

Under operating conditions, the pigtails are mechanically extended, with the result that new compressive stresses occur on the outside of bends, and tensile stresses occur on the inside. When added to the existing stresses in these locations, which are of the opposite direction, the new resultant operating stresses are less than they would have been had the material been in the annealed (unstressed) condition after forming.

The pigtails are compressed for installation, so that the tensile and compressive stresses in the outer and inner sides of the bends are greater than in the "relaxed" position (Fig. 2c). This allows a greater mechanical extension in the operating position without exceeding the yield strength, which would result in further deformation (Fig. 2d).

The design of the pigtail and header system is such that the residual stresses from the forming operation combine favorably with those arising from operating conditions, and the result is an efficient use of materials at minimum cost.

Thermal gradients

Thermal stresses are the result of differential expansion in a material. A high thermal gradient will, in general, cause more differential expansion and consequent thermal stresses than will a lower thermal gradient. Also, the differential expansion (and, therefore, the thermal stresses) in an irregularly-shaped area will be greater than that in an area of uniform cross-section, and so high thermal gradients are particularly to be avoided in areas of irregular cross-section. Other factors in stress design, such as sharp corners, which have a stress-concentrating effect, are even more critical in high-temperature systems than they are in lower-temperature applications, because they tend to aggravate the thermal gradient problem.

The thermal conductivity of stainless steels is less than that of carbon and low-alloy steels; therefore, high thermal gradients will occur more readily in stainless steels. The thermal expansion of stainless steel, however, is much greater (on the order of 40%) than that of carbon steel, so it is obvious that merely using a high-temperature material without regard to its different physical properties is not the solution to high-temperature design.

Allowable stress values at temperatures greater than 1,500°F are not tabulated in the Code, but some materials are suitable for use in higher-temperature systems, and appropriate allowable stress levels at these higher temperatures can be determined in accordance with Code paragraph 323.2.1. Two higher-temperature materials are the austenitic stainless steels and nickel-chrome alloys.

At these high temperatures, the allowable stress decreases rapidly with increases in temperature. For example, for Incoloy 800H, a nickel-iron-chrome alloy, an increase of 100°F in the design temperature results in a 40% reduction in the allowable stress.

Other considerations

In the critical design of high-temperature piping, proper specification and application of **insulation** is very important and is based on more than simply conservation of heat. After any maintenance work requiring insulation removal, the insulation should be carefully reinstalled, making sure that all large cracks are packed with loose fill. The thicknesses and types of insulation should not be changed arbitrarily, as they are a part of the overall design, and altering them without considering the rest of the system may decrease the reliability or life of the system. If the insulation is allowed to deteriorate or is not replaced when removed for maintenance, high thermal stresses can develop in the areas of discontinuity. These resultant secondary stresses may cause excessive distortion or possible failure.

Location of **welds** in high-temperature piping designs is also important, since residual stresses in and adjacent to the welds will combine with the operational stresses and may either improve or impair the design.

Digital-computer techniques have made feasible more detailed analysis of piping design and helped to ensure the most economical design consistent with the service requirements, design codes and good engineering judgment.

Proper **support and guiding** of any piping system is important, especially if it is for hot service, where thermal expansion is significant. If the system is designed to undergo local yielding or plastic flow, the proper guiding and support is not only important—it is absolutely essential. Improper or inadequate guides and supports will permit unintended deformation that may distribute forces and moments on anchor points in such a way that excessive stresses and consequent failure will result.

Many simplifying assumptions have been made in the foregoing discussion, and the graphs are intended to show directional trends only, not precise engineering data. For this reason, persons involved in designing systems of this type should refer to published design properties for appropriate criteria.

The author

John D. Dawson is a senior project engineer with Howe-Baker Engineers, Inc., Tyler, Tex., and is responsible for mechanical design and project management of steam/hydrocarbon reforming plants. Prior to joining Howe-Baker, he was with Texaco, Inc.'s Refining Dept. He is a registered professional engineer in several states and is a member of ASME.

Calculating line sizes for flashing steam-condensate

Flashing steam-condensate lines can be quickly sized, and flow velocities derived, using this method. Estimated pressure drops are conservative.

Richard P. Ruskin, *Arthur G. McKee & Co.*

☐ Condensate-return lines in refineries and chemical plants present a classic application for techniques that calculate pressure drop with two-phase flow. More than 20 correlations for such calculations are presented in the literature. Many of these have a theoretical basis, and most have constants derived from curve-fitting to experimental data. Several authors, including Paige [*8*], Kordyban [*6*], and Bankhoff [*2*], deal specifically with steam/water two-phase flow.

Two-phase-flow pressure-drop methods proposed by Baker [*1*], Lockhart and Martinelli [*7*], Chenoweth and Martin [*4*], and Dukler, et al. [*5*] are often referenced. Dukler's method has a semitheoretical basis and is accepted as the most accurate of available correlations by virtue of its close agreement to experimental data.

However, this and most of the other methods are rather difficult to apply, due both to the amount of physical data and to the lengthy computation required.

The accompanying figure provides a rapid estimate of the pressure drop of flashing condensate, along with a determination of fluid velocity. This figure is based on the simplifying assumption of a single homogeneous

Predicted pressure drops compared with Dukler

Flow rate lb/h	Pipe size	Saturation pressure psig	End pressure psig	Pressure drop psi/100 ft: Presented method	No slip	Dukler Constant slip	Horizontal flow pattern (baker chart)
1,000.	1½" SCH 80	600.	200.	0.28	0.22	0.25	Slug
100,000.	16" SCH 10	30.	0.	0.08.	0.068	0.034	Annular
100,000.	18" SCH 10	600.	0.	0.19	0.17	0.15	Annular
1,000.	1" SCH 80	400.	75.	7.4	4.6	4.5	Annular
1,000,000.	12" SCH 10	200.	100.	3.7	2.3	3.1	Dispersed
100.	3/4" SCH 80	400.	0.	2.5	2.2	1.8	Annular
1,000,000.	24" SCH 10	400.	200.	0.10	0.075	0.093	Annular
10,000.	6" SCH 40	100.	0.	0.22	0.19	0.11	Annular
1,000,000.	16" SCH 10	600.	75.	4.7	3.0	4.2	Dispersed
10,000.	3" SCH 40	150.	0.	9.4	6.1	4.8	Annular
10,000.	4" SCH 40	150.	0.	2.3	1.6	1.2	Annular
10,000.	6" SCH 40	150.	0.	0.27	0.23	0.15	Annular
10,000.	8" SCH 40	150.	0.	0.064	0.061	0.039	Wave

Originally published August 18, 1975

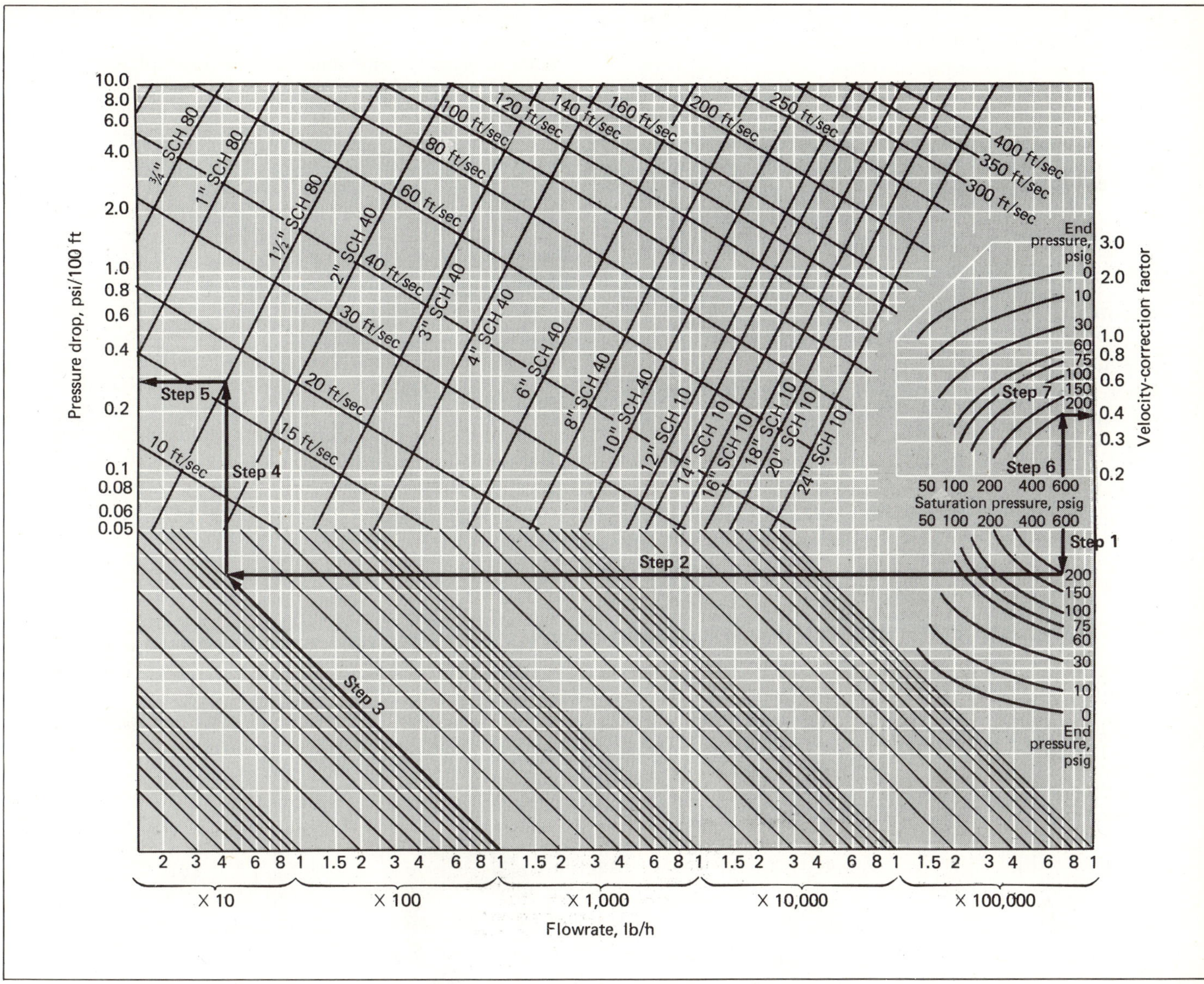

Nomenclature

d = pipe inside diameter, in.
f = friction factor, dimensionless
ΔP = pressure drop, psi/100 ft
V = velocity, ft/s
w = flowrate, lb/h
ρ = density, lb/cu ft

phase of fine liquid droplets dispersed in the flashed vapor. Pressure drop has been calculated accordingly, by Darcy's equation for single-phase flow:

$$\Delta P = 0.000336(fw^2)/(d^5\rho) \quad (1)$$

Steam-table data have been used to calculate the isenthalpic flash of liquid condensate from a saturation pressure to a lower end-pressure; and the average density of the resulting liquid-vapor mixture has been used as the assumed homogeneous fluid density, ρ.

Flows within the regime of the figure are characterized as either in complete turbulence or in the transition zone near complete turbulence. The friction factor, f, for Eq. (1) has been calculated by the relation:

$$f = 0.25[-\log(0.000486/d)]^{-2.0}$$

This relation is valid for complete turbulence flows in commercial steel and wrought-iron pipe.

Pressure drops for steam-condensate lines can be determined by assuming that the vapor-liquid mix throughout the lines is represented by the mix for conditions at the end-pressure. This assumption conforms to conditions typical of most actual condensate systems, since condensate lines are sized for low-pressure drop, with most flashing occurring across the steam trap or control valve at the entrance.

If the condensate line is to be sized for a considerable pressure drop, so that continuous flashing occurs throughout its length, end-conditions will be quite different from those immediately downstream of the trap. In such cases, an iterative calculation should be performed, involving a series of pressure-drop determinations across given incremental lengths.

This iteration is begun at the downstream-end pressure and worked back to the trap, taking into account the slightly higher pressure, and thus the changing liquid-vapor mix, in each successive upstream incremental pipe length. The calculation is complete when the total equivalent length for the incremental lengths equals the equivalent length between the trap and the end-pressure point. This operation can be performed using the figure.

Example

A reboiler is condensing 1,000 lb/h of steam at 600 psig and returning the condensate to a nearby condensate return header nominally at 200 psig. What size of condensate line will give a pressure drop of 1.0 psi/100 ft or less?

Step 1: Enter the figure at 600 psig below the insert near the right-hand side, and read down to the 200-psig end-pressure.

Step 2: Proceed left horizontally across the chart to the intersection, with:

Step 3: The 1,000-lb/h flowrate projected diagonally up from the bottom scale.

Step 4: Reading vertically up from this intersection, it can be seen that a 1-in line will produce more than the allowed pressure drop, so a $1\frac{1}{2}$-in size is chosen.

Step 5: Read left horizontally to a pressure drop of 0.28 psi/100 ft on the left-hand scale.

Step 6: Note the velocity given by this line as 16.5 ft/s, then proceed to the insert on the right, and read upward from 600 psig to 200 psig to find the velocity correction factor as 0.41.

Step 7: Multiply 0.41 by 16.5 to get a corrected velocity of 6.8 ft/s.

Accuracy

This rapid method provides pressure drops comparable to those computed by more-sophisticated techniques for two-phase flow. Thus, for the conditions of the above example, the Dukler no-slip method estimates 0.22 psi/100 ft and the Dukler constant-slip method estimates 0.25 psi/100 ft.

Comparisons for this and 12 other examples, representing all regimes of the chart, are shown in the table. Pressure drops predicted with the Dukler methods are consistently smaller than those estimated by the chart, with the Dukler no-slip method averaging 23% lower and the constant-slip method averaging 32% lower.

Results from the chart have also been compared to those calculated by a method suggested by Paige [*8*] and based on the work of Benjamin and Miller [*3*]. Paige's method assumed a homogeneous liquid-vapor mixture with no liquid holdup, and thus is similar in approach to the present method. However, Paige suggests calculation of the liquid-vapor mix based on an isentropic flash, whereas the present chart is based on an isenthalpic flash; and this is believed to be more representative of steam-condensate collecting systems.

For the example, the Paige method gives 0.26 psi/100 ft at the terminal pressure—and 0.25 psi/100 ft at a point 1,000 ft upstream of the terminal pressure, due to the slightly higher pressure, which suppresses flashing.

References

1. Baker, O, *Oil & Gas J.,* July 26, 1954.
2. Bankoff, S. G., *Trans. ASME,* Vol. C82, p. 265, 1960.
3. Benjamin, M. W., and Miller, J. G., *Trans. ASME,* Vol. 64, p. 657, 1942.
4. Chenoweth, J. M., and Martin, M. W., *Pet. Ref.,* Vol. 34, p. 151, 1955.
5. Dukler, A. E., Wicks, M., and Cleveland, R. G., *AIChE J.,* Vol. 10, p. 44, 1964.
6. Kordyban, E. S., *Trans. ASME,* Vol. D83, p. 613, 1961.
7. Lockhart, R. W., and Martinelli, R. C., *Chem. Eng. Prog.,* Vol. 45, p. 39, 1949.
8. Paige, P. M., *Chem. Eng.,* p. 159, Aug. 14, 1967.

The author

Richard P. Ruskin joined Arthur G. McKee & Co., Cleveland, Ohio, in 1971, and is currently a process engineer in the Petroleum and Chemicals Div. He holds a B.S. and M.S. in chemical engineering from Ohio University, is an associate member of AIChE and is a registered professional engineer in Ohio.

Designing steam transmission lines without steam traps

Here is proof that steam traps are not always needed, provided that certain conditions are met. The authors show how to design trapless transmission lines, with savings in equipment and energy.

Mileta Mikasinovic and *David R. Dautovich*, *Ontario Hydro, Toronto, Canada.*

☐ A steam transmission line carries energy from Point 1 to Point 2. This energy is a function of temperature, pressure and flowrate. Along the line, energy is lost through the pipe insulation and through steam traps. A design that would reduce the energy loss and the amount of required equipment would be highly desirable.

This article explains how to design steam transmission lines without steam traps, thereby saving the equipment cost of steam-trap stations and the energy normally lost through steam traps.

Design theory

The designer's primary concern is to ensure that steam conditions stay as close to the saturated line as possible. The steam state in the line changes according to the change in pressure due to a pressure drop, and the change in enthalpy due to heat loss through insulation. These changes of condition are plotted in Fig. 1. Point 1 is defined by P_1 and T_1 steam conditions. Due to the variability of such parameters as flowrate and ambient temperature, the designer should consider extreme conditions. Thus, P_2 would be defined by the minimum pressure drop produced by the minimum flowrate. Similarly, h_2 would be defined by the maximum heat loss produced by the lowest ambient temperature. Point 2 on the *h-s* diagram is defined by the above P_2 and h_2. If Point 2 is above or on the saturated steam line, no condensate is generated and steam traps are not required.

In some cases (small pressure drop, large heat loss), Point 2′ is below the saturated line, and some condensate is generated. The usual practice has been to provide trap stations to collect this condensate, and steam traps to remove it. However, current research [*3,4,6*] in two-phase flow demonstrates that the turbulent flow, produced by normal steam velocities and reasonable steam qualities, disperses any condensate into a fine mist equally distributed along the flow profile. The trap stations do not collect the condensate and, once again, the steam traps are not required. Fig. 2 shows that for velocities greater than 110 ft/s, and a steam fraction more than 98% by volume, the condensate normally generated in a transmission line will exist as a fine mist that cannot be collected by steam trap stations.

Operating the line

A few basic points should be followed when operating a steam line without steam traps. All lines must be sloped. If a line is long, several low points may be required. Globe valves are used on drains for each low point and for a drain at the end of the line. Since trap stations are not required, drain valves should be located as close to the line as possible to avoid freezing. Vents are placed at all high points. All vents and drains are opened up prior to warming up the line. Once steam is flowing from all vent valves, they are closed. As each drain valve begins to drain steam only, it is partially closed so that it may still bleed condensate if necessary. When full flow is established, all drain valves are shut. If the flow is shut down, all valves are opened until the pipe cools, and are then closed to isolate the line from the environment. The above procedure would be the same if steam traps were on the line.

Conclusion

It has been demonstrated that for steam transmission lines, steam traps are not required, provided that one of the following parameters is met:

1. Steam is saturated or superheated.
2. Steam velocity is greater than 110 ft/s, and the steam fraction more than 98% by volume.

Using the above design, the steam energy normally lost through traps is saved, along with the construction, maintenance and equipment costs for the traps, drip leg, strainers, etc., associated with each trap station.

Example

Design a steam line for transporting a minimum of 6.0×10^5 lb/h and a maximum of 8.0×10^5 lb/h saturated steam at 205 psig and 390°F. The line is 3,000 ft long, with eight 90-deg elbows and one gate valve. Ambient temperatures range from −40°F to 90°F.

Using a steam velocity for maximum flowrate of

Originally published March 14, 1977

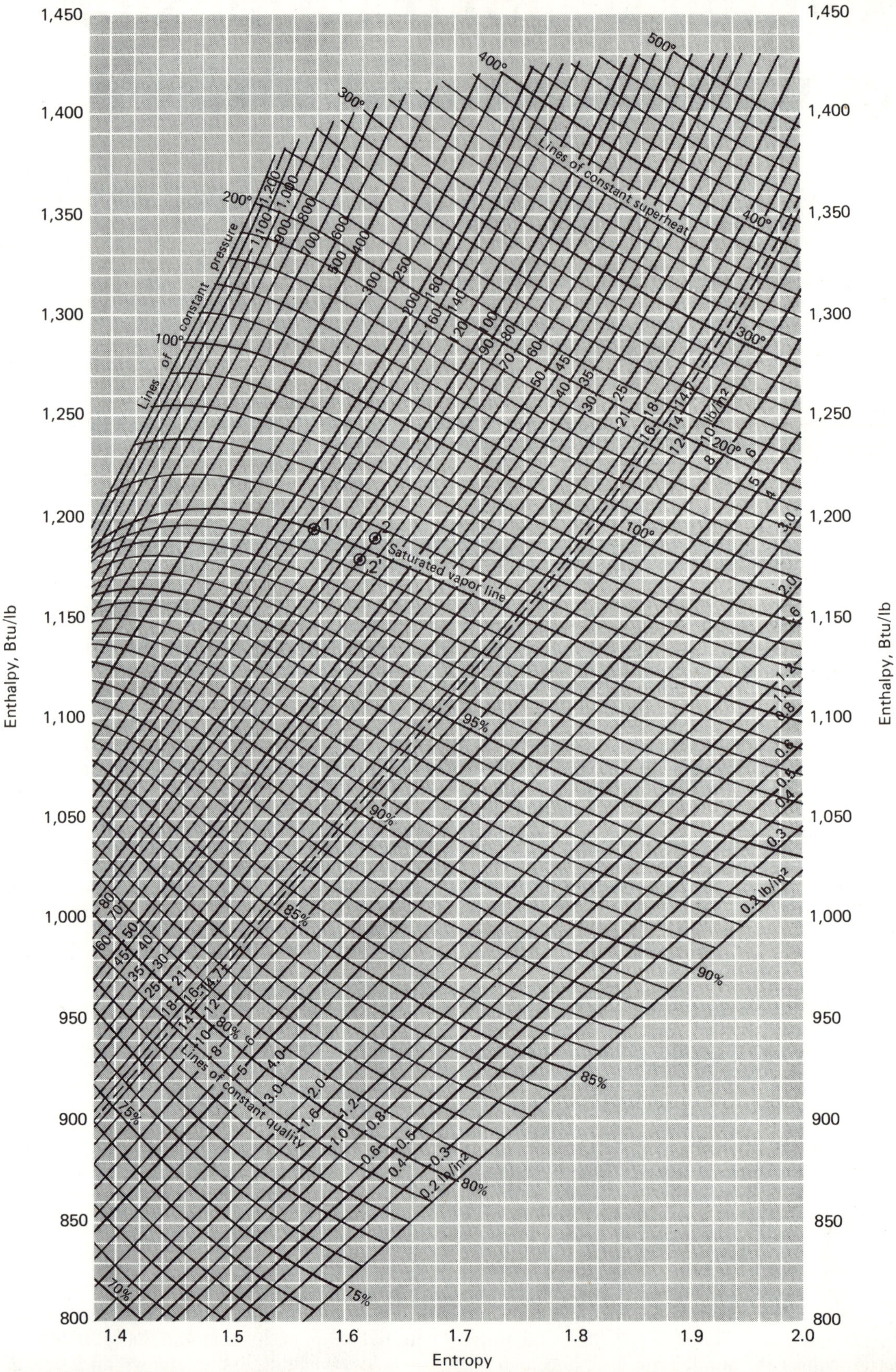

Enthalpy-entropy or Mollier diagram for steam

Fig. 1

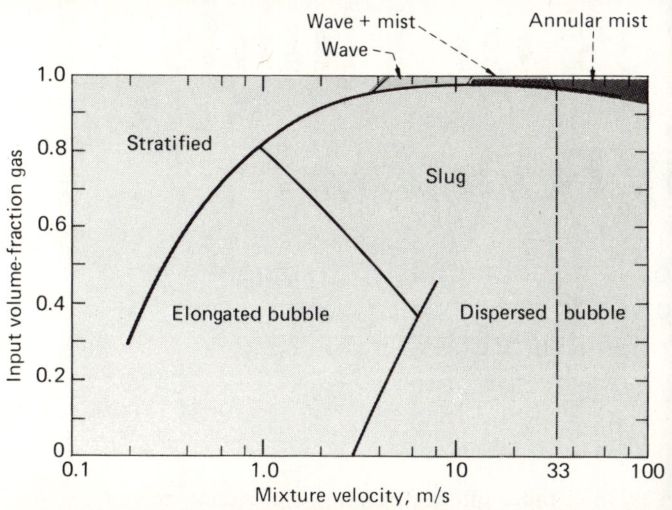

Flow-pattern map Fig. 2

160 ft/s to size the pipe:

$$D = 24 \text{ in., std. wt. (area} = 2.94 \text{ ft}^2)$$

The velocities are:

For maximum flow, 157.6 ft/s
For minimum flow, 118.0 ft/s

and the pressure drops and final pressures at Point 2 are:

For maximum flow, 23.3 lb/in^2 and 196.7 lb/in.^{2}
For minimum flow, 14.0 lb/in^2 and 206.0 lb/in.^{2}

Therefore, the velocities at Point 2 are:

For maximum flow, 173 ft/s
For minimum flow, 124 ft/s

The line will be covered with a 3-in.-thick insulation with a thermal conductivity of 0.48 (Btu-in)/(h) (ft^2)(°F).

The maximum heat loss at minimum ambient temperatures is:

$$Q_{max} = 1{,}452{,}931 \text{ Btu/h}$$

Therefore, the change in enthalpy becomes:

$$\Delta h_{max} = \frac{1{,}452{,}931}{6 \times 10^5} = 2.42 \text{ Btu/lb}$$

The minimum enthalpy at Point 2 is:

$$h_{2\,min} = h_1 - \Delta h_{max}$$
$$h_{2\,min} = 1{,}197.18 \text{ Btu/lb}$$

The minimum heat loss at maximum ambient temperatures is:

$$Q_{min} = 1{,}013{,}673 \text{ Btu/h}$$
$$\Delta h_{min} = \frac{1{,}013{,}673}{8 \times 10^5} = 1.26 \text{ Btu/lb}$$
$$h_{2\,max} = h_1 - \Delta h_{min}$$
$$h_{2max} = 1{,}198.34 \text{ Btu/lb}$$

Now we have two extreme states at Point 2:

(a) Minimum flow and lowest ambient temperature, where $P_{2\,max} = 206$ lb/in.^{2}; $h_{2\,min} = 1197.18$ Btu/lb.
(b) Maximum flow and highest ambient temperature, where $P_{2\,min} = 196.68$ lb/in^2; $h_{2\,max} = 1198.34$ Btu/lb.

The second condition is superheated and no condensate is generated. The first condition is below the saturated line and the amount of condensate and steam is:

$$2.22 \frac{\text{ft}^3 \text{ of dry steam}}{\text{lb of total mass}}$$

and $$.0000342 \frac{\text{ft}^3 \text{ of condensate}}{\text{lb of total mass}}$$

or .00154% condensate per volume
and 99.99846% dry steam per volume.

The velocity, 124 ft/s, $>$110 ft/s.

According to Fig. 2, we can see that the condensate will exist in the steam flow as a fine mist, and cannot be collected by steam traps.

References

1. Schwaigerer, S., "Rohrleitungen—Theorie und Praxis"; Springer Verlag (1967).
2. Simonson, J. R., "Engineering Heat Transfer," McGraw-Hill Co. Ltd., (1967).
3. Gorier, G. W., and Aziz, K., "The Flow of Complex Mixtures in Pipes," Van Nostrand Reinhold Co. (1972).
4. Tong, L. S., "Boiling Heat Transfer and Two-Phase Flow," John Wiley and Sons, Inc. (1967).
5. Shearer, C. J., and Hedderman, R. M., *Chem. Eng. Sci.*, Vol. 20, pp. 671–683, 1965.
6. Goldmann, K., Firstenberg, H., and Lombardi, C., Trans. ASME, *J. of Heat Transfer,* Ser. C, Vol. 83, pp. 158–162, 1961.

The authors

Mileta Mikasinovic is a design engineer at Ontario Hydro, 700 University Ave., Toronto, Canada. He holds a B.S. degree in mechanical engineering from the University of Belgrade. He has experience in the design and construction of mechanical equipment for the chemical and petrochemical industries, and is the author of a handbook of fluid flow published in Yugoslavia.

David R. Dautovich is a design engineer at Ontario Hydro, 700 University Ave., Toronto, Canada. Formerly employed by C-E Lummus, he has had experience in the design of mechanical equipment, pressure vessels and process piping in the petrochemical and thermal-generating fields. He holds a B.S. Sc. degree in mechanical engineering from the University of Waterloo, Ontario.

Useful Properties of Fluids for Piping Design

This up-to-date series will provide practical design techniques for piping; it should prove valuable to all engineers involved in the design, construction, operation and maintenance of plants in the chemical process industries.

ROBERT KERN, Hoffmann - La Roche Inc. *

It is important to recognize that piping often accounts for as much as one-third of the total plant cost, and that energy cost (among others) is directly related to the design of the piping and hydraulic systems.

The objectives of this series will be to:

- Give the simplest proven methods of design for obtaining accurate solutions to piping problems in the minimum number of steps.
- Provide basic design information and essential data.
- Show the design connections between piping and equipment in hydraulic systems.
- Explain, as required, the fundamental relationships of Euler, Bernoulli and Darcy to applied hydraulics.

Furthermore, specific references will be made to design data if not shown within these articles. These articles, the accompanying data and references will be useful to engineers just now entering this field as well to those who design piping and components as part of a larger job responsibility in process plants.

We will review the basic concepts of fluid properties in this first installment. The definitions, nomenclature, and design data covered here will be used throughout this series. Only Newtonian fluids will be considered.

Densities: Liquid

The density of a liquid is the weight of a unit volume at the temperature during measurement, and is usually expressed in lb/ft^3 [*5a*]. For example, the density of water is $\rho_{60w} = 62.37$ lb/ft^3 at 60°F.

Pressure has no practical effect on liquid density. However, increasing temperatures will cause liquids to expand. Process flowrates can be given at a basic 60°F temperature, coupled with an expansion factor at points where the temperature changes in a pipe system. This expansion factor is: $E = \rho_{60}/\rho$, where ρ is the density at flowing temperature. Hence, the volume flowrate, Q in gpm, at temperature will be: $Q = Q_{60}E$ where Q_{60} is volume flowrate at 60°F.

Specific gravity at standard temperature relates the density of any liquid at 60°F to that of water at 60°F:

$$S_{60} = \rho_{60l}/\rho_{60w} \tag{1}$$

A relation more often used gives the specific gravity of the liquid at flowing temperature, S, from:

$$S = \rho/\rho_{60w} \tag{2}$$

where ρ is the density of liquid at flowing temperature. If $S > 1$, the liquid is heavier than water at 60°F; and if $S < 1$, the liquid is lighter than water.

Densities: Vapor and Gas

A convenient equation for calculating vapor or gas densities is derived from the well-known gas law: $P\bar{V} = RTz$, where P is absolute pressure, lb/ft^2; $\bar{V}$ is specific volume, ft^3/lb; T is temperature, °R; R is the universal gas constant, (ft)(lb)/(lb)(°R); and z is a correction factor accounting for the nonideal behavior of a gas (usually, $z = 1$). Since $R = 1{,}544/M$, where M is the molecular weight; $P = 144P'$, where P' is the absolute pressure, psia; and $\bar{V} = 1/\rho$, ft^3/lb, the gas law can be rewritten as:

$$\frac{1}{\bar{V}} = \rho = \frac{P}{RTz} = \frac{144P'}{1{,}544T/M} \tag{3}$$

From Eq. (3), we can now find an expression for gas density, ρ, as:

$$\rho = \frac{MP'}{10.72Tz}, \text{ lb/ft}^3 \tag{4}$$

As Eq. (4) shows, gas densities depend on pressure and temperature. Hence, for purposes of calculation, a pipeline is separated into segments over which the pressure and temperature are essentially constant. By using this procedure, calculations for pipe size and pressure drop can be made for each segment over the length of the pipeline.

Specific volume is the reciprocal of density, $\bar{V} = 1/\rho$, ft^3/lb. At high temperatures and pressures, gases do not follow closely the ideal gas law, and $z \neq 1$. The numerical value of z can be obtained from Fig. 1 by calculating reduced pressure and reduced temperature.

Example 1—What is the compressibility factor of carbon dioxide at 450 psia and 300°F?

From Table I, we find that the critical pressure, P_c, and

*To meet your author, see p. 44.

Originally published December 23, 1974

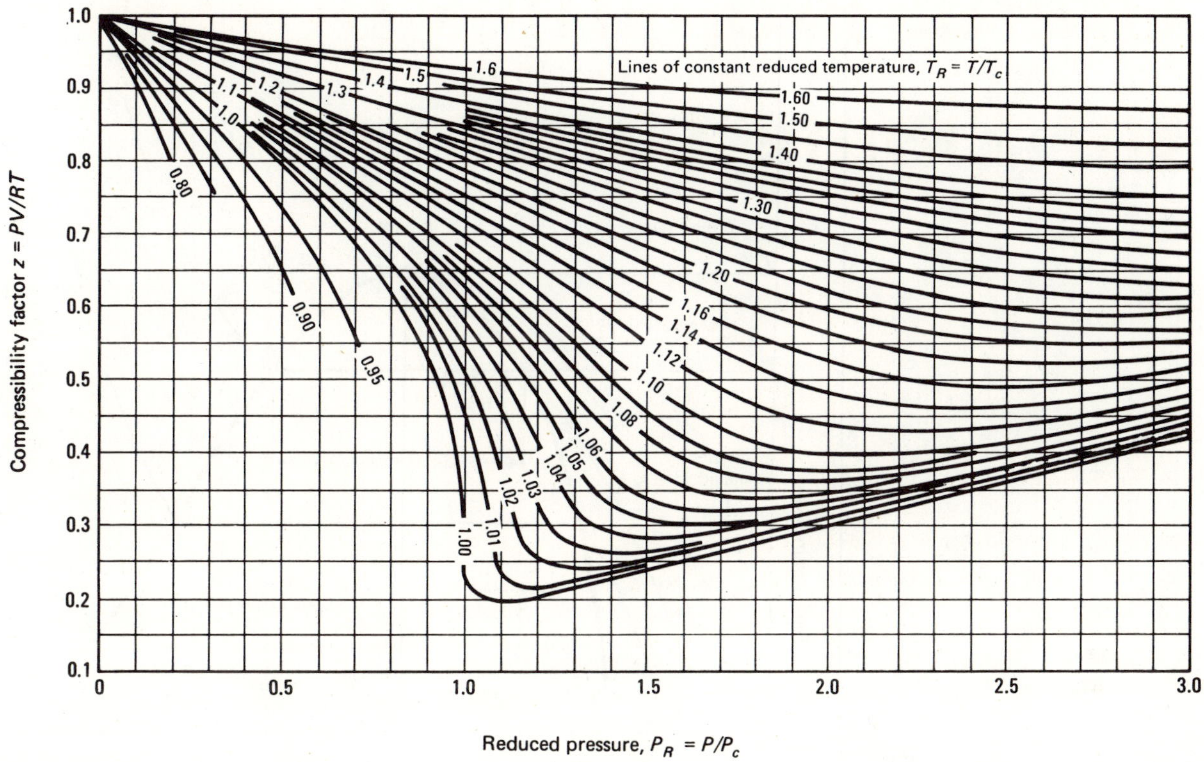

COMPRESSIBILITY factors for actual gases in relation to critical conditions for temperature and pressure—Fig. 1

critical temperature, T_c, of carbon dioxide are 1,073 psia and 548°R, respectively. We then calculate reduced pressure $P_R = P/P_c = 450/1{,}073 = 0.42$, and reduced temperature $T_R = T/T_c = 760/548 = 1.39$. Using Fig. 1, we find that $z = 0.95$ for these values.

Specific gravity of a vapor or gas, S_{60g}, relates the density of the gas at 60°F and 1 atm (14.7 psia), ρ_{60g}, to the density of air, ρ_{60a}, under the same conditions:

$$S_{60g} = \frac{\rho_{60g}}{\rho_{60a}} = \frac{M_g}{M_a} \quad (5)$$

Eq. (5) also shows that specific gravity of a gas can be obtained by dividing the molecular weight of the gas, M_g, by the molecular weight of air, M_a.

For determining the specific gravity of a gas, S_g, at the flowing temperature and pressure, the relation is:

$$S_g = \rho_g/\rho_{60a} \quad (6)$$

where ρ_g is the density of the gas at flowing temperature and pressure.

The density of air, ρ_{60a}, is 0.0764 lb/ft³ and the molecular weight, M_a, is 28.97. Densities and specific gravities of vapors and gases are listed in handbooks [*2,4,5a*].

Densities: Liquid-Vapor Mixtures

If a liquid-vapor mixture occupies 1 ft³ of volume, the weight of vapor, W_v, in the mixture, divided by the volume it occupies, V_v, gives its density: $\rho_v = W_v/V_v$.

Properties of Selected Gases and Vapors—Table I

Gas or Vapor	Molecular Weight	Specific Heat Ratio at 60°F, $k = c_p/c_v$	Critical Pressure, P_c, Psia	Critical Temperature, T_c, °R
Acetylene	26.04	1.24	905	557
Air	28.97	1.40	547	239
Ammonia	17.03	1.31	1,657	731
Benzene	78.11	1.12	714	1,013
Carbon dioxide	44.01	1.30	1,073	548
Carbon monoxide	28.01	1.40	510	242
Chlorine	70.91	1.36	1,119	751
Ethane	30.07	1.19	708	550
Ethyl alcohol	46.07	1.13	927	930
Ethyl chloride	64.52	1.19	764	829
Ethylene	28.05	1.24	742	510
Hydrogen	2.02	1.41	188	60
Methane	16.04	1.31	673	344
Methyl alcohol	32.04	1.20	1,157	924
Methyl chloride	50.49	1.20	968	750
Natural gas*	18.82	1.27	675	379
Nitrogen	28.02	1.40	492	228
Oxygen	32.00	1.40	730	278
Propane	44.09	1.13	617	666
Propylene	42.08	1.15	668	658
Water vapor	18.02	1.33	3,208	1,166

* Approximate values based on average composition.

Source: "Engineering Data Book—1957," 7th ed., Natural Gasoline Supply Men's Assn.

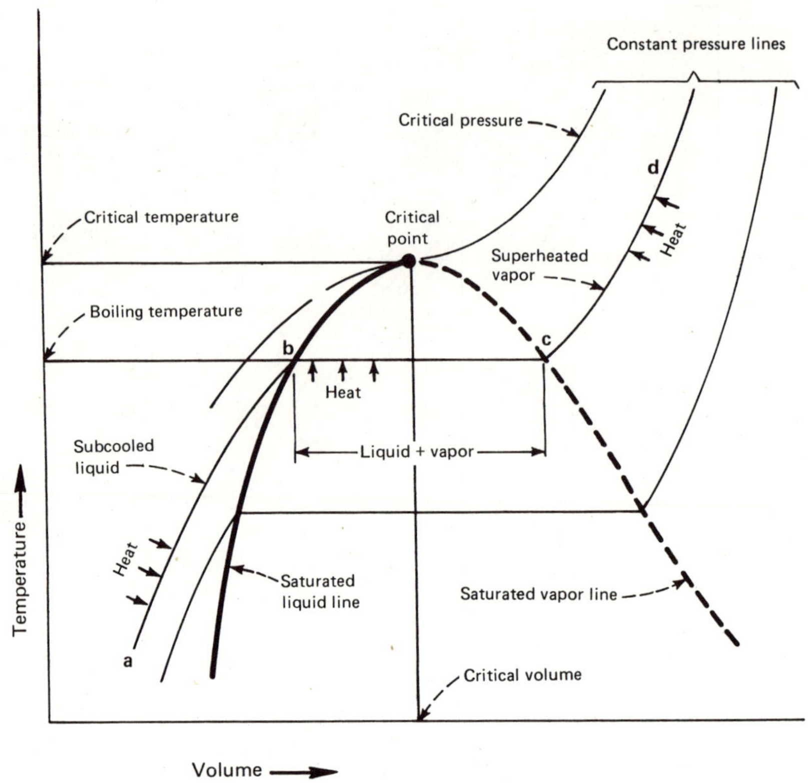

PHASE equilibria between liquid and vapor for a material undergoing heating or cooling in a closed system—Fig. 2

Similarly, for the liquid part: $\rho_l = W_l/V_l$. The mixture density will be: $\rho_{l+v} = W_{l+v}/V_{l+v}$. Since $V_{l+v} = V_l + V_v$, the mixture density becomes:

$$\rho_{l+v} = \frac{W_{l+v}}{(W_l/\rho_l) + (W_v/\rho_v)}, \text{ lb/ft}^3 \qquad (7)$$

In Eq. (7), W can represent the weight of fluid, weight flowrate, or percentage of each component in the mixture.

Example 2—Find the mixture density for a flow of water, W_l = 495 lb/h, and of steam, W_v = 5 lb/h, if the components are well mixed and flowing concurrently in a pipeline. The flowing fluid has a pressure of 110 psia and temperature of 344°F.

By substituting into Eq. (4), we find that steam density is:

$$\rho_v = \frac{18(110)}{10.72(344 + 460)1} = 0.23 \text{ lb/ft}^3$$

From a water properties table [*5a*], we find that the density of water at 344°F is 55.56 lb/ft.^{3}

By substituting the appropriate values into Eq. (7), we find that the mixture density is:

$$\rho_{l+v} = \frac{500}{(495/55.56) + (5/0.23)} = 16.32 \text{ lb/ft}^3$$

This example shows that even a small amount (about 1%) of vaporization greatly reduces liquid density. Hence, in designing flow systems, we must be aware that:

1. With a very small amount of vaporization, the decreased density significantly reduces the static head backpressure in a vertical pipe.
2. With a constant weight flowrate and a small amount of vaporization, the volume of flow greatly increases. In turn, this increases pipe resistance significantly. Such a condition is common in condensate lines.

Thermodynamic Properties

In routine calculations for piping and component sizing, thermodynamics is hardly involved. However, it is useful to recognize when physical change takes place in the flowing fluid. For example, if a liquid flows near its boiling point, even a little vaporization (usually due to pressure reduction) can increase piping and component resistances.

In order to understand the meanings of the terms subcooled and saturated as applied to a liquid, and the terms saturated and superheated as applied to a vapor, we will use the temperature-volume relations that are shown in Fig. 2.

When a liquid is heated at constant pressure, its temperature increases and its volume expands. (See line segment from a to b in Fig. 2.) On reaching the boiling temperature, the liquid becomes saturated. The liquid below the boiling point is called subcooled. More heat will gradually vaporize the liquid while its pressure and temperature remain constant but its volume increases

Typical Sections of Steam Tables—Table II

Saturated steam and saturated water

Pressure Absolute, P', Psia	Pressure Gage, P, Psig	Temperature, t, °F	Heat of the Liquid, Btu/Lb	Latent Heat of Evaporation, Btu/Lb
110.0	95.3	334.79	305.8	883.1
111.0	96.3	335.46	306.5	882.5
112.0	97.3	336.12	307.2	882.0
124.0	109.3	343.74	315.2	875.8
125.0	110.3	344.35	315.8	875.3
126.0	111.3	344.95	316.4	874.8
127.0	112.3	345.55	317.1	874.3

Superheated steam

Pressure Absolute, P', Psia	Pressure Gage, P, Psig	Saturated Temperature, t, °F		Total Temperature, °F			
				500	600	700	800
400.0	385.3	444.60	$\bar{V}$	1.2841	1.4763	1.6499	1.8151
			h_g	1,245.1	1,307.4	1,363.4	1,417.0
420.0	405.3	449.40	$\bar{V}$	1.2148	1.4007	1.5676	1.7258
			h_g	1,242.4	1,305.8	1,362.3	1,416.2
440.0	425.3	454.03	$\bar{V}$	1.1517	1.3319	1.4926	1.6445
			h_g	1,239.7	1,304.2	1,361.1	1,415.3

$\bar{V}$ is specific volume, ft³/lb; and h_g is total heat of steam, Btu/lb.

(Line b to c in Fig. 2). During this stage, both liquid and vapor phases are present. After sufficient heat absorption, all of the liquid vaporizes (Point c). At this point, the vapor is saturated. Further heating (Line c-d) will cause both the temperature and the volume of vapor to increase. Vapor above the saturation temperature is superheated.

At a higher constant pressure, the boiling temperature will be higher, and less heat will be required to vaporize the liquid. At the critical point (see Fig. 2), the densities of liquid and vapor become identical. Just below the critical temperature, the substance is considered liquid; just above, it is considered vapor.

The quantity of heat needed to vaporize a unit weight of saturated liquid at constant pressure (Line b-c in Fig. 2) is called the latent heat (usually expressed in Btu/lb).

Thermodynamic properties for various substances have been established and are available. Part of a typical page is shown in Table II, for saturated steam [*1,5b*].

Flashing Liquid

When a liquid is flowing near its saturation point (also called the equilibrium point or boiling point) in a pipeline, decreased pressure will cause vaporization. The greater the pressure difference, the greater the vaporization. When flashing occurs, pipe resistance cannot be calculated as for liquid flow. We will now have a two-phase flow problem. The quantity of vaporized liquid can be estimated by using the data of the thermodynamic properties of the liquid.

Typical Liquid Velocities in Steel Pipelines—Table III

Nominal Pipe Sizes, In.	2 or less	3 to 10	10 to 20
Liquid and Line	**Velocity, Ft/S**	**Velocity, Ft/S**	**Velocity, Ft/S**
Water			
Pump suction	1 to 2	2 to 4	3 to 6
Pump discharge (long)	2 to 3	3 to 5	4 to 7
Discharge leads (short)	4 to 9	5 to 12	8 to 14
Boiler feed	4 to 9	5 to 12	8 to 14
Drains	3 to 4	3 to 5	–
Sloped sewer	–	3 to 5	4 to 7
Hydrocarbon liquids			
(Normal viscosities)			
Pump suction	1.5 to 2.5	2 to 4	3 to 6
Discharge header (long)	2.5 to 3.5	3 to 5	4 to 7
Discharge leads (short)	4 to 9	5 to 12	8 to 15
Drains	3 to 4	3 to 5	–
Viscous oils			
Pump suction,			
Medium viscosity	–	1.5 to 3	2.5 to 5
Tar and fuel oils	–	0.4 to 0.75	0.5 to 1
Discharge (short)	–	3 to 5	4 to 6
Drains	1	1.5 to 3	–

Typical Velocities in Gas and Vapor Lines—Table IV

Nominal Pipe Size, In.	Saturated Steam or Saturated Vapor	Superheated Steam, Superheated Vapor, or Gas	
	Low Pressure	Medium Pressure	High Pressure
	Velocity, Ft/S	Velocity, Ft/S	Velocity, Ft/S
2 or less	45 to 100	40 to 80	30 to 60
3 to 4	50 to 110	45 to 90	35 to 70
6	60 to 120	50 to 120	45 to 90
8 to 10	65 to 125	80 to 160	65 to 125
12 to 14	70 to 130	100 to 190	80 to 145
16 to 18	75 to 135	110 to 210	90 to 160
20	80 to 140	120 to 220	100 to 170

Note: Within the above velocities and line-size ranges, (a) large lines can have higher velocities than smaller ones, and (b) short lines, and leads from headers, can have higher velocities than long lines and headers.

Equipment Lines	Velocity, Ft/S
Reboiler, downcomer (liquid)	.3 to 7
Reboiler, riser (liquid and vapor)	35 to 45
Overhead condenser	25 to 100
Two-phase flow	35 to 75
Compressor, suction	75 to 200
Compressor, discharge	100 to 250
Inlet, steam turbine	120 to 320
Inlet, gas turbine	150 to 350
Relief valve, discharge	$0.5v_c$*
Relief valve, entry point at silencer	v_c*

*Obtain sonic or critical velocity, v_c, from:

$v_c = 68\sqrt{k(P'/\rho)}$, ft/s. k from Table I.

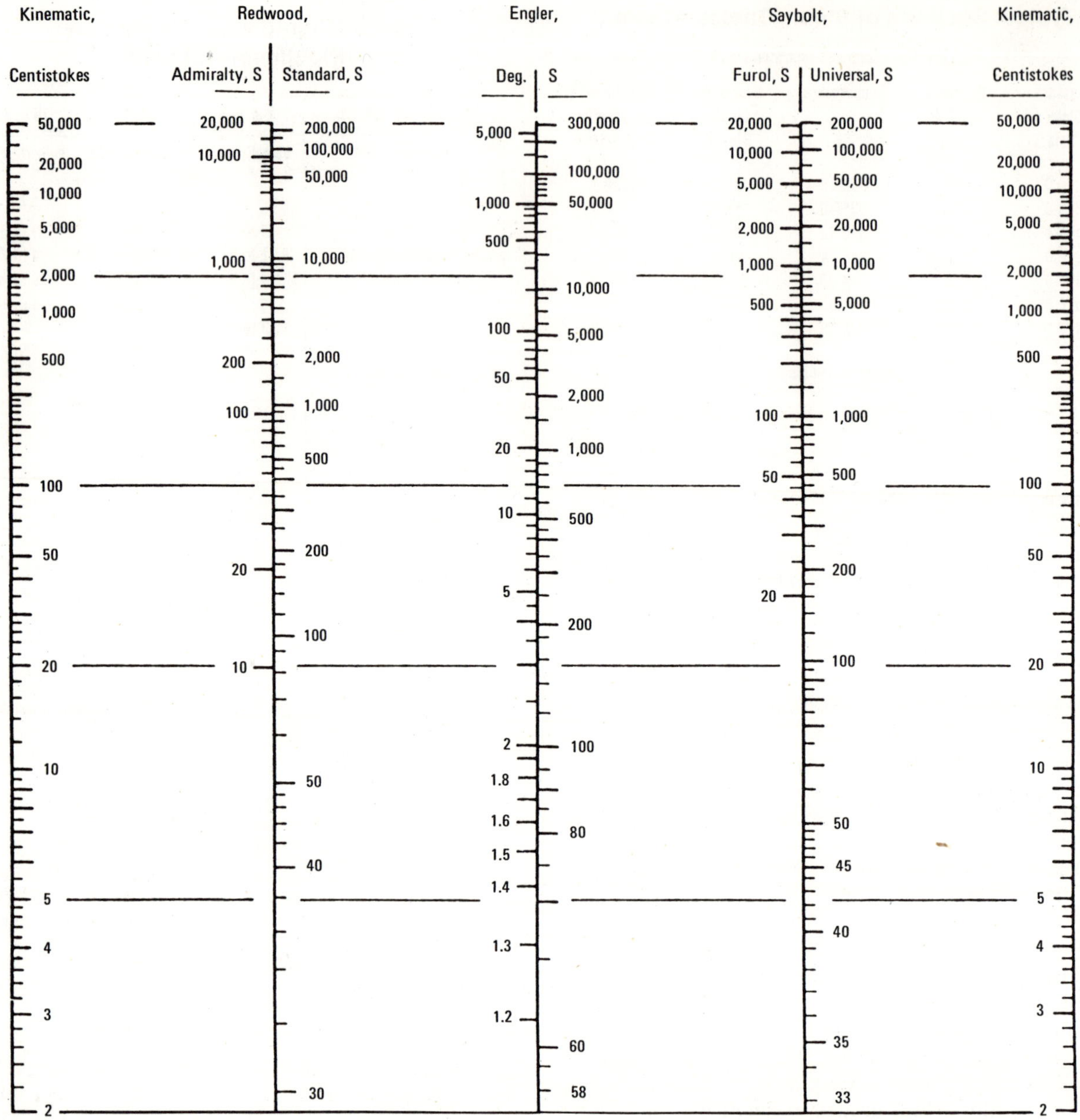

Read horizontally across chart to convert readings. Multiply readings in centistokes by specific gravity to obtain absolute viscosity in centipoises.

VISCOSITY conversion obtained conveniently by reading horizontally across the chart—Fig. 3

Example 3—Let us consider that condensed water is flowing at a rate $W = 500$ lb/h at 125 psia and 344°F. When pipe resistance reduces the pressure to 110 psia, how much water will flash to steam?

From Table II, we find that the heat content of 1 lb of water is 315.8 Btu at 125 psia, and 305.8 Btu at 110 psia. The difference is 10.0 Btu/lb, or a total of $10.0 \times 500 = 5{,}000$ Btu released when a pressure reduction of 15 psi (125 − 110) occurs. This amount of heat will vaporize some of the liquid.

We must now find the latent heat of water at the flowing conditions. From Table II, the latent heat at 110 psia is 883.1 Btu/lb, and at 125 psia is 875.3 Btu/lb. The average of the two values is 879.2 Btu/lb. The amount of heat released during the pressure reduction, divided by the average latent heat, will yield the amount of liquid vaporized: $5{,}000/879.2 = 5.69$ lb/h. Therefore, the quantities in two-phase flow will be: $W_v = 5.69$ lb/h saturated steam, and $W_l = 500 - 5.69 = 494.31$ lb/h saturated water.

Specific Heat

The quantity of heat absorbed by a unit weight of substance while its temperature is raised 1° is termed the specific heat. Customary units are in Btu/(lb)(°F). Spe-

Maximum Velocities To Prevent Erosion Or Corrosion of Pipe Wall—Table V

	Maximum Velocity, Ft/S
Liquid in carbon-steel pipe	
Phenolic water	3
Concentrated sulfuric acid	4
Cooling-tower water	12
Salt water	6
Calcium chloride brine	8
Caustic soda (>5% by volume)	4
Aqueous amine (mono- or diethanolamine)	10
Wet phenolic vapor	60
Liquid in plastic or rubber-lined pipe	10

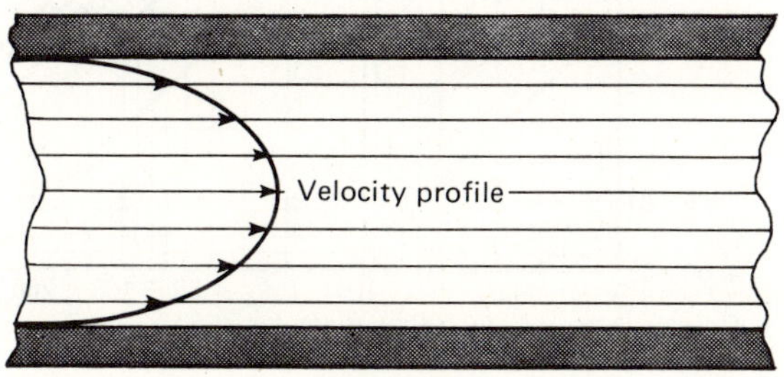

a. Viscous Flow

$N_{Re} \leqslant 2{,}000$
$f = 64/N_{Re}$

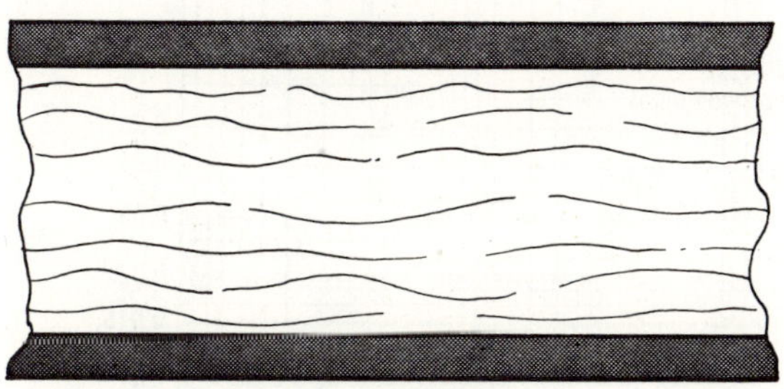

b. Transition Flow

$N_{Re} = 2{,}000$ to $4{,}000$
$f = 0.04$ to 0.052 for steel pipes.

C. Turbulent Flow

$N_{Re} > 4{,}000$
f from Fig. 5 and 6

FLOW patterns in a pipe containing water—Fig. 4

cific heat can be measured at constant volume, c_v, and at constant pressure, c_p. At constant pressure, $c_p \approx \Delta h/\Delta t$, where Δh is the number of Btu absorbed by 1 lb of substance over a temperature span of Δt°F.

At adiabatic gas flow (no heat exchange between the gas and its environment), the ideal-gas law becomes $P\bar{V}^k = RT$, where $k = c_p/c_v$. For various gases, the range of k is from 1.2 to 1.4, with few exceptions. In isothermal (constant temperature) flow, $k = 1$. The quantity k is called the specific heat ratio or adiabatic exponent. Data for c_p, c_v and k are available in engineering handbooks [*2,3,4,5c*].

Fluid in Motion

Velocity—A fluid moving in a pipeline has a parabolic velocity-distribution profile (see Fig. 4a). The average velocity is calculated at a given cross-section and steady flowrate: $v = q/A$, ft/s, where q is volume flowrate in ft^3/s, and A is cross-sectional area of the pipe in ft.^{2}

For liquid-flow calculations:

$$v = 0.408(Q/d^2) \qquad (8)$$

For vapor-flow or gas-flow calculations:

$$v = 0.0509\,W/(d_2\rho) \qquad (9)$$

where v is velocity, ft/s; Q is volume flowrate, gpm.; W is weight flowrate, lb/h; d is internal diameter of pipe, in; and ρ is gas density at flowing temperature and pressure, lb/ft.^{3}

The relationship between volume flowrate (Q, gpm) and weight flowrate (W, lb/h) is:

$$(Q\rho/7.48)60 = W = 8Q\rho$$
$$Q = 0.125(W/\rho)$$

Since specific gravity, S, is ρ/ρ_{60w}, we find that $W = 500QS$. In the case of water close to 60°F, $W = 500Q$, lb/h, with $\rho_{60w} = 62.37$ lb/ft.^{3}

The initial pipe diameter can be estimated by choosing a reasonable velocity for a specific type of pipeline. Thus, for liquid lines:

$$d^2 = 0.408(Q/v), \text{ in.}^2 \qquad (10)$$

and for vapor lines and gas lines:

$$d^2 = 0.509\,W/(v\rho), \text{ in.}^2 \qquad (11)$$

Table III gives practical velocities for liquid lines, and Table IV for vapor lines. Values of d and d^2 are tabulated in pipe manufacturers' catalogs [*5f*].

To prevent corrosion or erosion of the internal pipe wall, some chemicals should not exceed the maximum velocities listed in Table V. Only a few examples are listed in this table.

Viscosity—Viscosity is the measure of how easily the liquid or gas flows. It is a measure of the internal resistance of fluids. With increasing temperatures, liquid viscosity decreases and gas viscosity increases.

For measuring viscosity, many English and metric units have been developed. In this series, centipoise, cp, will be used. A convenient conversion scale between the various units of viscosity is shown in Fig. 3. Elaborate instruments are required for measuring absolute viscosity such as cp. Kinematic viscosity is simple to measure. The relation between kinematic and absolute is:

$$\nu = \mu/S \qquad (12)$$

where ν is viscosity in centistokes; μ is viscosity in cp; and S is specific gravity. Note that 1 Stoke has dimensions of cm^2/s.

The viscosity of water at 68°F is 1 cp. If data are not available, the following viscosities can be used: $\mu = 1$ cp for liquids similar to water, 0.1 cp for liquid-vapor mixtures, and 0.01 cp for gases or vapors [*5d*].

Reynolds Number and Friction Factors

Experiments and observations made in glass pipes have shown that several types of flow exist in a fluid stream; these depend on the velocity of the stream. Such flows can be observed by injecting fine colored streams into the main body of the liquid. At low velocities, the colored streams retain their sharpness, and flow is laminar (Fig. 4a). As fluid velocity increases, the fine streams begin to break up, and flow is now in a transition or critical zone (Fig. 4b). With a further increase in the flow velocity beyond the critical zone, flow becomes turbulent (Fig. 4c).

For estimating the type of flow in a pipeline, the Reynolds number, N_{Re}, is used. The Reynolds number

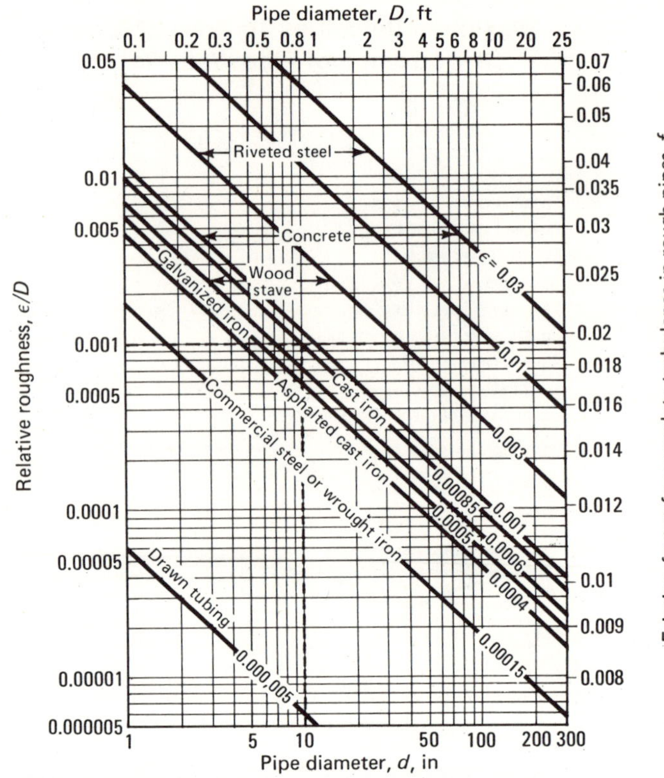

RELATIVE roughness and friction factor chart—Fig. 5

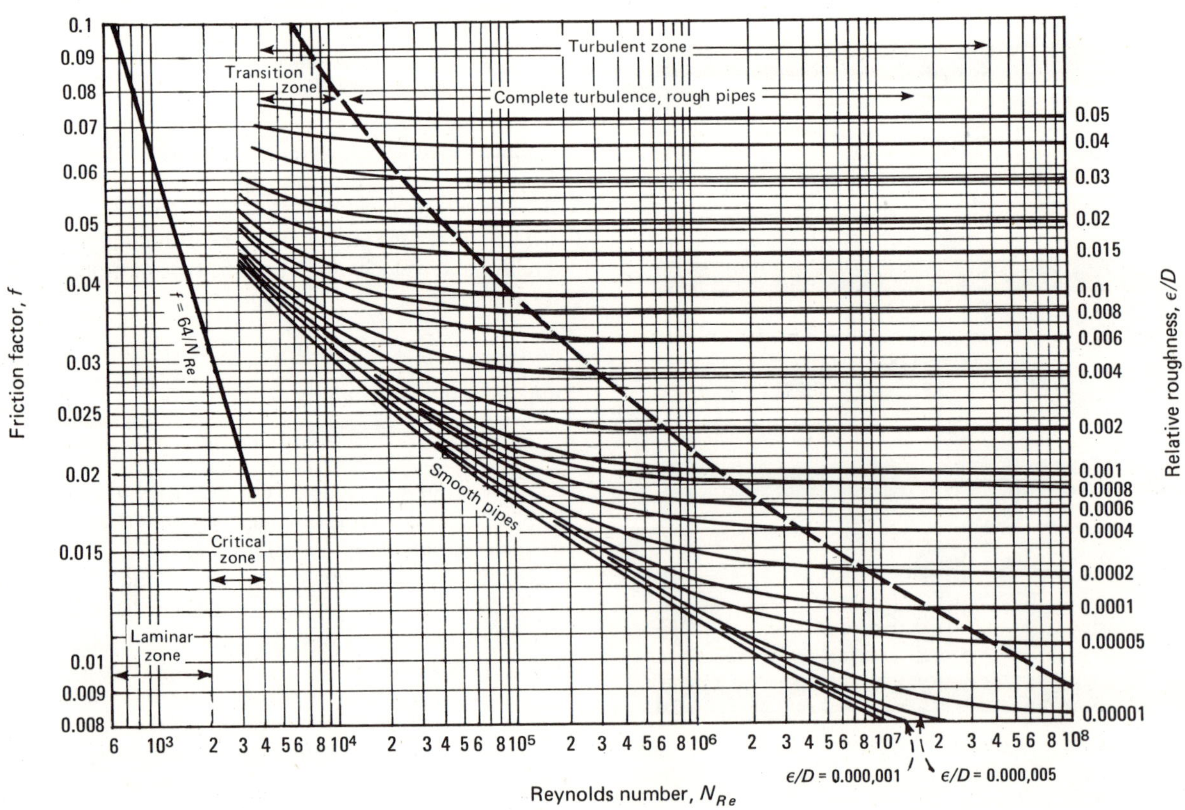

FRICTION factors for any type of commercial pipe under any condition of fluid flow—Fig. 6

is a dimensionless combination of pipe diameter, velocity, density and viscosity. $N_{Re} = Dv\rho/\mu_e$, where μ_e is the absolute viscosity in $lb_m/(ft\text{-}s)(ft^2)$.

Practical formulas for calculating N_{Re} are:

$$N_{Re} = 50.6(Q/d)(\rho/\mu) = 6.31\,W/d\mu \qquad (13)$$

where Q is volume flow, gpm; W is weight flow, lb/h; d is internal diameter of pipe, in.; ρ is density, lb/ft^3 and μ is viscosity, cp.

Resistance to fluid motion depends on the type of flow in the pipeline and on the roughness of the pipe wall. Relative roughness is ϵ/D where ϵ is the absolute roughness (i.e., the depth of the unevenness of the internal pipe wall), and D is the internal diameter of the pipe. Both ϵ and D should be measured with the same dimensional unit. Values for relative roughness can be obtained from Fig. 5, for various pipe diameters and materials.

The friction factor depends on the type of flow in the pipeline and the roughness of the pipe wall. It can be obtained from Fig. 6, for various flow conditions.

In the laminar zone, where $N_{Re} < 2{,}000$, the friction factor depends on the Reynolds number only. Hence, $f = 64/N_{Re}$. In the critical zone, where $2{,}000 \leqslant N_{Re} \leqslant 4{,}000$, the friction factor is unpredictable. Friction factors for this zone are given in Fig. 4b.

If the N_{Re} ranges from 4,000 to the dashed line in Fig. 6, the flow is in the transitional turbulent zone. Here the friction factor varies with the Reynolds number. The zone to the right of the dashed line is totally turbulent, and the friction factor remains constant with increasing Reynolds number.

Because glass and plastic materials have smooth pipe walls, the friction factors for them are independent of relative roughness or pipe diameter. Hence, there is only one line for these materials in Fig. 6.

For any given type of pipe material (for example, steel), a diagram of the Reynolds number vs. friction factor can be drawn [*5e*]. In this case, constant pipe-diameter lines will replace the constant relative-roughness lines of Fig. 6.

For steel pipes, the border line between transitional and total turbulence can be estimated. This is done by multiplying the nominal pipe size (in.) by 10^6 to get the borderline N_{Re}. Friction factors in the totally turbulent zone are listed in Table VI.

Example 4—Let us find the friction factor for a 4-in steel pipe. The estimated Reynolds number is then 4×10^6 for total turbulence. And from Table VI, the friction factor is $f = 0.0165$ for a 4-in pipe. If $N_{Re} < 4 \times 10^6$, the friction factor must be obtained from Fig. 5 and 6.

Friction factors obtained directly from tables and diagrams are used in calculations where the pipe material is new, and where the fluids (for example, light hydrocarbons) do not deteriorate the pipe wall. For fluids that corrode the pipe wall, form deposits, or cause scaling or erosion, the friction factor should be increased by a safety factor. For steam condensate, cooling water, salt water, the safety factor can be 1.25 to 1.5, depending on the size and the expected life of the installation.

The dimensional properties of pipe are published by manufacturers [*5f*]. For steel pipes, these are based on the recommendations of the American National Standards Institute.

Friction Factors for Total Turbulence—Table VI (In new commercial steel pipes)

Nominal Pipe Size, In.	Friction Factor f	Nominal Pipe Size, In.	Friction Factor, f
1½	0.0205	10	0.0136
2	0.0195	12	0.0132
3	0.0178	14	0.0125
4	0.0165	16	0.0122
5	0.016	18	0.012
6	0.0152	20	0.0118
8	0.0142	24	0.0116

The next article in this series will appear in the Jan. 6, 1975, issue and and will deal with the basics for sizing piping and components. Practical formulas will be given for liquid-line and vapor-line sizing for which the density of the fluid in the line is constant.

Acknowledgements

The charts and illustrations in this article have been kindly supplied by, and used with the permission of: American Petroleum Institute, Fig. 1; Dow Chemical Co., Fig. 3; Crane Co., Fig. 4; Crane Co. from data extracted from Friction Factors for Pipe Flow by L. F. Moody with the permission of the publisher, ASME, Fig. 5 and 6.

References

1. Keenan, J. H., Keyes, F. G., Hill, P. G. and Moore, J. G., "Steam Tables," Wiley, New York, 1969.
2. Gallant, R. W., "Physical Properties of Hydrocarbons," Vol. 1 and Vol. 2, Gulf Publishing Co., Houston, 1968 and 1971.
3. "A.P.I. Technical Data Book: Petroleum Refining," American Petroleum Institute, Washington, D.C.
4. Perry, R. H. and Chilton, C. H., "Chemical Engineers' Handbook," 5th ed., McGraw-Hill, New York, 1973.
5. "Flow of Fluids," Technical Paper No. 410, (a) p. A-6, physical properties of water; p. A-7, density of various liquids; p. A-8, physical properties of gases; p. A-10, weight density of air; (b) pp. A-12 to A-19, thermodynamic properties of steam; (c) p. A-8, physical properties of gases; p. A-9, specific heat ratios for steam; (d) p. A-2, viscosity of steam; p. A-3, viscosity of water and petroleum products; p. A-4, viscosity of various liquids; p. A-5, viscosity of gases and vapors; pp. B-3 to B-5, viscosity equivalents; (e) p. A-25, friction factor vs. Reynolds number diagram for steel pipes; (f) pp. B-16 to B-19, steel pipe data. (Crane Co., Chicago, IL 60632.)

Friction-factor equation spans all fluid-flow regimes

A single correlating equation relates pipe friction loss to Reynolds number and surface roughness for laminar, transitional and turbulent flow alike, thus making fluid-flow calculation simpler.

Stuart W. Churchill, *University of Pennsylvania*

☐ Equations are more convenient than tables or graphical correlations in computer-aided design and operation. A single equation that correlates the fluid-flow friction factor with all Reynolds numbers and all pipe-roughness ratios can be constructed from theoretical and correlating equations for the laminar, transition and fully developed turbulent regimes of flow, using a general model developed by Churchill and Usagi [*1*].

The friction factor considered herein is defined in terms of the shear stress on the pipe wall, as follows:

$$f \equiv \tau_w / \rho u^2 \qquad (1)$$

The shear stress at the wall can be related to the gradients in pressure, elevation and velocity by a force and momentum balance, such as:

$$\tau_w = \frac{D}{4}\left(-\frac{dP}{dL}\right) + \frac{g\rho D}{4}\left(-\frac{dh}{dL}\right) + \frac{w}{\pi D}\left(-\frac{du}{dL}\right) \qquad (2)$$

This equation is exact if the momentum-mean velocity is used in the right-hand derivative. However, the volumetric-mean velocity

$$u_b = 4w/\pi D^2 \rho \qquad (3)$$

provides a reasonably accurate approximation. We can then rewrite Eq. (2) more conveniently in terms of fluid density and mass flowrate by combining Eq. (1) and (2) and replacing u with w and ρ through Eq. (3), yielding:

$$f = \frac{\pi^2 D^5 \rho}{64 w^2}\left(-\frac{dP}{dL}\right) + \frac{\pi^2 D^5 \rho^2 g}{64 w^2}\left(-\frac{dh}{dL}\right) + \frac{D}{4\rho}\left(\frac{d\rho}{dL}\right) \qquad (4)$$

The friction factor defined by Eq. (1) and expanded in Eq. (4) is related to the commonly used friction factors of Fanning and Darcy, as follows:

$$f = f_F/2 = f_D/8 \qquad (5)$$

So, all equations in this article can be converted to the Fanning or Darcy friction factors, by multiplying the nonsubscripted f by 2 or 8, respectively.

The individual flow regimes

Laminar regime. For $Re < 2{,}100$, Poiseuille's law

$$f = 8/Re \qquad (6)$$

is applicable.

Transition regime. The various sets of experimental data for the transition regime between laminar and turbulent flow are quite scattered. Wilson and Azad [*2*] obtained a precise set of values for the friction factor in smooth pipe for $1{,}000 < Re < 500{,}000$ by numerical computations. Using their values, we derive an empirical equation for the central portion of the transition regime:

$$f = 7.1 \times 10^{-10} Re^2 \qquad (7)$$

Experimental data suggest that this expression is reasonably valid for rough, commercial pipe as well.

Turbulent regime in smooth pipe. Schlichting [3] and others have shown that the semitheoretical equation of Prandtl expressed in the form of

$$1/\sqrt{f} = 2.457 \ln(1.126\, Re\sqrt{f}\,) \qquad (8)$$

gives a good fit for experimental data on smooth pipe over the range $3{,}000 < Re < 3.4 \times 10^6$. They presume that Eq. (8) holds for even higher Re. For horizontal flow of a fluid of constant density,

$$Re\sqrt{f} = \frac{1}{2\mu}\sqrt{D^3 \rho\left(-\frac{dP}{dL}\right)} \qquad (9)$$

So Eq. (8) is convenient if the pressure gradient is specified.

However, Eq. (8) requires trial and error if instead the flowrate is specified. That difficulty is avoided by the Blasius equation:

$$f = 0.03955/\mathrm{Re}^{1/4} \qquad (10)$$

However, Eq. (10) is inaccurate for $Re > 10^5$. The expression

$$1/\sqrt{f} = 2.21 \ln(Re/7) \qquad (11)$$

suggested by Colebrook (see Churchill [*4*]) is also convenient if the flowrate is specified, and is essentially

Originally published November 7, 1977

Nomenclature

D	Diameter, m
f	Friction factor, $\tau_w/\rho u^2$
f_i	Friction factor given by Eq. (i)
f_D	Darcy friction factor, $8\tau_w/\rho u^2$
f_F	Fanning friction factor, $2\tau_w/\rho u^2$
g	Acceleration due to gravity, m/s^2
h	Elevation, m
L	Length of pipe, m
m	Arbitrary exponent
n	Arbitrary exponent
P	Pressure, Pa
Re	Reynolds number, $4w/\pi\mu D$
u	Velocity,
u_b	Volumetric-mean velocity, $4w/\pi D^2\rho$ (m/s)
w	Mass flowrate, kg/s
ϵ	Effective roughness, m
μ	Viscosity, Pa • s
ρ	Density, kg/m^3
τ_w	Shear stress, Pa

equivalent to Eq. (8) in accuracy. Eq. (8), (10) and (11) are applicable for $\epsilon Re\sqrt{f}/D < 5$ and $Re > 3{,}000$, where ϵ is the effective (or surface) roughness; this parameter has been tabulated in Perry's "Chemical Engineers' Handbook" and elsewhere for various kinds of pipe.

Fully developed turbulent flow in rough pipe. Nikuradse [5] determined the following asymptotic expression for very large Re in pipes having uniform artificial (laboratory-produced) roughness:

$$1/\sqrt{f} = 2.457 \ln (3.707D/\epsilon) \tag{12}$$

Eq. (12) also holds for rough, commercial pipe, and it is applicable for $\epsilon Re\sqrt{f}/D > 70$ and $Re > 10{,}000$.

Developing turbulent flow in rough and smooth pipe. Churchill [4] derived an expression for both developing and fully developed turbulent flow in both rough and smooth pipes by combining Eq. (8) and (12), as follows:

$$\frac{1}{\sqrt{f}} = 2.457 \ln\left(\frac{1}{0.888/Re\sqrt{f} + 0.27\epsilon/D}\right) \tag{13}$$

Eq. (11) was also combined with Eq. (12) to give an alternative expression:

$$\frac{1}{\sqrt{f}} = 2.457 \ln\left(\frac{1}{(7/Re)^{0.9} + 0.27\epsilon/D}\right) \tag{14}$$

Eq. (14) differs only slightly from Eq. (13).

Eq. (13) is essentially equivalent to the expression commonly used to construct the complete turbulent regime of the friction-factor chart that is reproduced in most textbooks and handbooks.

The full-range equation

The friction factors given by Eq. (7) and (14) can first be combined in the form of the Churchill-Usagi model to yield the following test expression for the transition and turbulent regimes:

$$f^n = f_7^n + f_{14}^n \tag{15}$$

Here the subscripts 7 and 14 indicate Eq. (7) and (14), respectively. The computed values of Wilson and Azad indicate that the best value of the arbitrary constant n is about -8, hence:

$$\frac{1}{f} = \left[\left(2.457 \ln\left(\frac{1}{(7/Re)^{0.9} + 0.27\epsilon/D}\right)\right)^{16} + (37{,}530/Re)^{16}\right]^{1/8} \tag{16}$$

Eq. (16) can in turn be combined with Eq. (6) in the form:

$$f^m = f_{16}^m + f_6^m \tag{17}$$

The computed values of Wilson and Azad suggest that, in this case, 12 is the best value for the arbitrary constant m, hence:

$$f = \left[\left(\frac{8}{Re}\right)^{12} + \frac{1}{(A+B)^{3/2}}\right]^{1/12} \tag{18}$$

where

$$A = \left[2.457 \ln \frac{1}{\left(\frac{7}{Re}\right)^{0.9} + \frac{0.27\,\epsilon}{D}}\right]^{16}$$

$$B = \left(\frac{37{,}530}{Re}\right)^{16}$$

Eq. (18) is valid for all Re and ϵ/D. A trial-and-error solution is necessary if the pressure drop rather than the flowrate is specified, but this situation would also occur if Eq. (13) were used rather than Eq. (14).

Conclusions

Eq. (18) is a convenient and accurate replacement for all of the friction-factor plots in the literature. The only uncertainty arises from the degree of accuracy of component Eq. (6), (7), (11) and (17), and from the experimental and theoretical values upon which they are based. The equation appears to be complicated but is actually suitable for calculations with even a hand-held computer.

This equation not only reproduces the friction-factor plot but also avoids interpolation and provides unique values in the transition region. These values are, of course, subject to some uncertainty, because of the physical instability inherent in this region.

References

1. Churchill, S. W., and Usagi, R., A General Expression for the Correlation of Rates of Transfer and Other Phenomena, *AIChE J.*, Vol. 18, No. 6, 1972, pp. 1121–1128.
2. Wilson, N. W., and Azad, R. S., A Continuous Prediction Method for Fully Developed Laminar, Transitional and Turbulent Flows in Pipes, *J. Aplied Mech.*, Vol. 42, 1975, pp. 51–54.
3. Schlichting, H., "Boundary Layer Theory," 4th ed., p. 515, McGraw-Hill, N.Y., 1960.
4. Churchill, S. W., Empirical Expressions for the Shear Stress in Turbulent Flow in Commercial Pipe, *AIChE J.*, Vol. 19, No. 2, 1973, pp. 375–376.
5. Nikuradse, J., Strömungsgesetze in rauhen Rohren, Ver. deutsch. Ing., Forschungsheft 361, 1933.

The author

Stuart W. Churchill is the Carl V.S. Patterson Professor of Chemical Engineering at the University of Pennsylvania, Philadelphia, PA 19174. He obtained bachelor's and doctoral degrees at the University of Michigan, gaining five years of industrial experience along the way. He served as Professor and Chairman of the Department of Chemical and Metallurgical Engineering at Michigan before moving to Penn. A past president and a fellow of AIChE, and a member of the National Academy of Engineering, he has done research mainly in heat transfer and combustion.

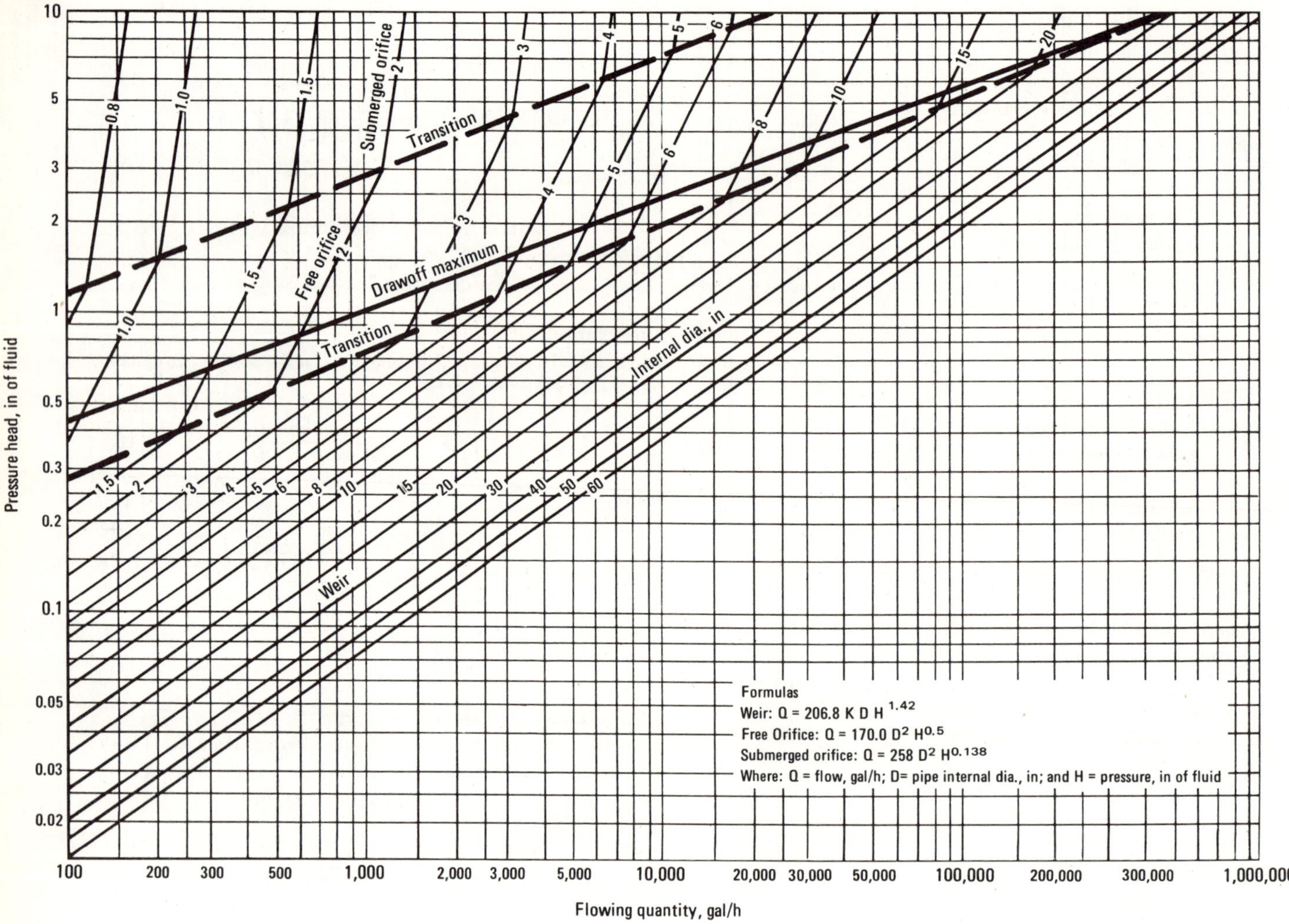

CONFIDENCE LINES FOR SIZING PLANT PIPING SYSTEMS

A. R. CHIMES, New York City

The separation between chemical engineering theory and practice is probably never greater than in the design of process-plant piping systems. The theory provides equations by Colebrook, Manning, Fanning, a dimensionless Reynolds No., coefficients for the roughness of pipe, equivalent lengths for elbows, valves, fittings, etc., and equations to balance pump horsepower against economical line sizes. Yet the careful application of all these

Originally published August 5, 1974

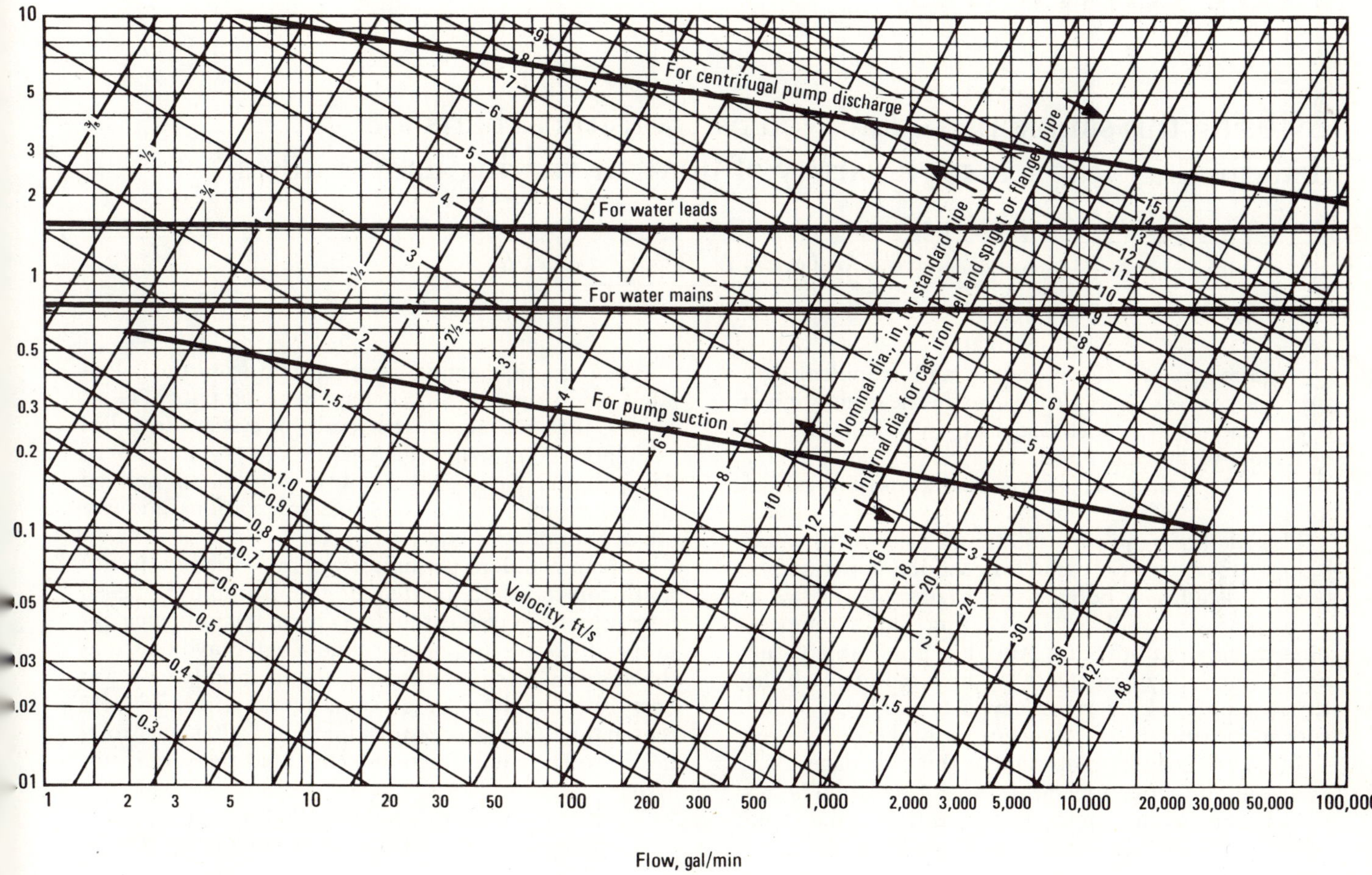

niceties to actual plant piping-design often borders on the ridiculous.

There is one simple reason: Process plant designers do not know the lengths of the sections of pipe they size. In order to know the length of a piping run, the plant must be laid out. Equipment must be located, pipe racks established, and the locations of individual piping runs assigned in the pipe racks. Yet a primary consideration for plant layout are the diameters and materials used for the piping.

Thus the piping must be sized for the plant to be laid out, while the layout must be complete for the calculation of piping pressure drops. A rigorous theoretical design would require that the entire physical plant design be iterative, i.e., with successive revisions of line sizes and layouts. This is impractical for all but 2–3 of the larger engineering companies, where a layout is assumed, based on past experience, and the piping sized according to this layout in a department manned by experienced operators and 10–20 specialists.

Otherwise, design engineers must follow past examples and a sense of judgment. The accompanying charts might be helpful for this. They show (Fig. 1) three types of flow down circular pipes—weir, free orifice, and submerged orifice, and (Fig. 2) pressure drops and velocities for water flowing through commercial pipe.

Superimposed on Fig. 1 in the free-orifice area is a "confidence line," giving approximate maximum sizes for drawoffs to stripping columns and reboilers, and for tray downcomers. Superimposed on the water-flow chart are 4 confidence lines, one each for pump suction, water mains, water leads, and centrifugal pump discharge. The pressure drops shown for these lines can be prorated to other than water by ratios of density or specific gravity, when pressure drops are necessary.

These two charts, their confidence lines, and the rule that pressure drop is proportional to diameter to the 5th power,* have been used to size hundreds of miles of piping in plants that were competitively priced and that have operated profitably for years.

*See Plant Notebook, *Chem. Eng.*, Nov. 27, 1972, p. 114. Also see "Piping Engineering," pp. 14–24, Pub. by Tube Turns Div. of Chemetron Corp., Louisville, Ky. (1969).

Streamlining head-loss calculations

By using this simple procedure, you can construct charts for determining head loss for a variety of fluids at varying temperatures.

Archie M. Larson, Jr., Gilbert Associates, Inc.

☐ A combination of calculation and graphical construction makes it possible to simplify the evaluation of $fv^2/2g$ and $v^2/2g$, which are used in calculating the head loss of flowing fluids. This can be done over a range of pipe sizes and temperatures, and for various fluids.

This method requires a minimum of detailed calculations. It is extremely useful for conducting parametric head-loss studies in liquid systems consisting of two or more pipe sizes and having varying liquid temperatures in different branches. Also, it is applicable to any liquid (separate charts must be prepared for each different liquid) as long as the Reynolds number, N_{Re}, lies within the transition or turbulent range (i.e., above 4,000), as defined by plots such as those on pages A-24 and A-25 of Crane Technical Paper No. 410. The nomenclature used herein is the same as that in the Crane paper except for terms specific to the method.

When care is taken in plotting data from the calculated points on 2 × 3-cycle log-log graph paper (11 × 17 in.), the accuracy of the results of system head-loss calculations (utilizing the plotted $fv^2/2g$ and $v^2/2g$ data) should compare favorably with results derived entirely from detailed slide-rule calculations and the use of plots of f vs. N_{Re}.

Although gallons per minute (gpm) is used as the abscissa for the data plotted in the example discussed later, any form of flowrate (e.g., lb/h, ft^3/min, etc., appropriate to the study being conducted) may be used as the abscissa. Further, although the method has not yet been employed for compressible fluids, it should yield reasonably accurate results, provided that the total head loss is a small fraction of the fluid pressure in the system.

The method is illustrated by the following example:

A piping system is estimated to potentially require

Calculation of flowrates in various pipe sizes (Table I)

Nominal pipe size, in	A, ft^2	Flowrate at 5 ft/sec. Q_5, gal/min	Flowrate at 5 ft/sec. Q_{15}, gal/min	D, ft	Dv 5 ft/sec	Dv 15 ft/sec
2	0.02230	52.4	157	0.1772	0.886	2.66
4	0.08840	198	595	–	–	–
6	0.2006	464	1,354	–	–	–
8	0.3474	781	2,340	–	–	–
12	0.7773	1,750	5,250	–	–	–
16	1.2272	2,760	8,280	1.250	6.25	18.75

Evaluation of Reynolds number at two temperatures (Table II)

Water Temp °F	Viscosity μ,* cp	μ_e, (lb)(ft)(s)	Density ρ, + lb/ft^3	ρ/μ_e, sec/ft^2	N_{Re}, $Dv\rho/\mu_e$ At 5 ft/s 2-in pipe	At 5 ft/s 16-in pipe	At 15 ft/s 2-in pipe	At 15 ft/s 16-in pipe
50	1.3077	8.78×10^{-4}	62.40	7.1×10^4	6.3×10^4	4.44×10^5	1.89×10^5	1.33×10^6
260	0.181	1.218×10^{-4}	58.60	4.81×10^5	4.26×10^5	3.01×10^6	1.28×10^6	9.05×10^6

*Perry, J. H., "Chemical Engineers' Handbook," 2nd ed., McGraw-Hill, Table 5, p. 874.
+Badger and McCabe, "Elements of Chemical Engineering," 2nd ed., McGraw-Hill, Appendix X.

Originally published October 27, 1975

Evaluation of friction factor, *f*, for various Reynolds numbers

Nominal pipe size, in	Fluid velocity, ft/s	Water temp., °F	N_{Re}	f*	$v^2/2g$†	$fv^2/2g$	Flowrate, gpm
2	5	50	6.3×10^4	0.0237	0.389	0.00923	52.4
	5	260	4.26×10^5	0.0199	0.389	0.00775	52.4
	15	50	1.89×10^5	0.0210	3.50	0.0735	157
	15	260	1.28×10^6	0.0191	3.50	0.0668	157
16	5	50	4.44×10^5	0.0149	0.389	0.00580	2,760
	5	260	3.01×10^6	0.0127	0.389	0.00494	2,760
	15	50	1.33×10^6	0.0133	3.50	0.0465	8,280
	15	260	9.05×10^6	0.0122	3.50	0.0427	8,280

*Crane Technical Paper 410; from plot, page A-25, based on N_{Re} calculated in Step 2.
†From step 1a
‡From columns 3 and 4 of Table I.

the utilization of six different nominal pipe sizes, as follows: 2, 4, 6, 8, 12 and 16 in. All of these will be Schedule 40, clean, commercial steel pipe. The temperature of the fluid flowing in different branches of the system is anticipated to range between 50 and 260°F. Therefore, the effect of temperature may be significant and should be included. The fluid flowing is clean water. The steps are as follows:

1. Calculate the value of $v^2/2g$ and flowrate, in gpm, in each different pipe size to be investigated, at 5 ft/s and 15 ft/s.

a. At 5 ft/s: $v^2/2g = 0.389$. At 15 ft/s: $v^2/2g = 3.50$

b. Q_x (flowrate, gpm, at fluid velocity x)

$Q_5 = (5)(60)(7.5)A = 2{,}250\ A$

$Q_{15} = (15)(60)(7.5)A = 6{,}750\ A$

2. Evaluate N_{Re} at 5 ft/s and 15 ft/s for water temperatures of 50 and 260°F (Table II).

3. Evaluate f for each of the eight Reynolds numbers, N_{Re}, calculated in Step 2, and calculate the corresponding values of $fv^2/2g$ for each (Table III).

4. Prepare log-log paper data chart and plot values of $fv^2/2g$, calculated in Step 3, above, vs. the corresponding flow rates (gpm in this example). Refer to Fig. 1.

a. Label abscissa (3 cycles) for flow rates from 10 to 10,000 gpm.*

b. Label ordinate (2 cycles) for values of $fv^2/2g$ from 0.001 to 0.1.*

c. Plot values of $fv^2/2g$ (Column 7, Table III) vs. corresponding flow rates (Column 8, Table III) for 2-in pipe size on chart.

d. Construct a straight line through the upper (50°F) set of data points plotted. The lines should extend above and below the data points.

e. Repeat step 4d for the lower (260°F) set of 2-in pipe data plotted.†

f. Construct vertical lines through each set of data points plotted (not shown in Fig. 1).
These should terminate slightly above and below the diagonal lines previously constructed.†

g. Repeat steps 4c thru 4f, above, for the 16-in. pipe size data in Columns 7 and 8 of Table III.

5. Prepare chart for plotting data for the remainder of pipe sizes to be investigated. Refer to Fig. 2.

a. Construct a dashed line between the two lowest data points plotted for the 2-in and the 16-in pipes (i.e., the 5 ft/s data at 260°F.) These lines should extend nearly across the width of the plot.

b. Repeat step 5a for the next higher set of data points for the 2-in and 16-in pipes (i.e., 5 ft/s at 50°F.) The set of lines so produced are labeled the "5 ft/s crossbars" on Fig. 2.*

c. Repeat step 5a for the next higher set of data points (i.e., 15 ft/s at 260°F.)

d. Repeat step 5b for the highest set of data points (i.e., 15 ft/s at 50°F.) The set of lines so produced are labeled the "15 ft/s crossbars" on Fig. 2.†

The chart is now ready for constructing the lines that enable evaluating $fv^2/2g$ vs. gpm for the remainder of the pipe sizes.

6. Construct the lines defining $fv^2/2g$ vs. gpm for

* If a different range of pipe sizes or fluids with physical properties different from water are under consideration, these ranges may differ. Also, for optimum accuracy, 2 × 3 cycle log-log paper is recommended. However, this may require the labeling of two or more sheets and calculation of additional points to be plotted on the different sheets if a wide range of pipe sizes is to be investigated.

† If the temperature of the flowing fluid is not a consideration, these steps are not necessary.

Nomenclature

A	Cross-section area of pipe, ft²
D	Internal diameter of pipe, ft²
h	Friction factor
g	Acceleration of gravity, 32.2 ft/s²
N_{Re}	Reynolds number (shown as *Re* in Crane Technical Paper No. 410)
Q_x	Flowrate, gal/min (at fluid velocity x, ft/s)
v	Mean velocity of flow, ft/s
μ	Absolute (dynamic) viscosity, cp
μ_e	Absolute viscosity, lb/(ft)(s)
ρ	Density of fluid, lb/ft³

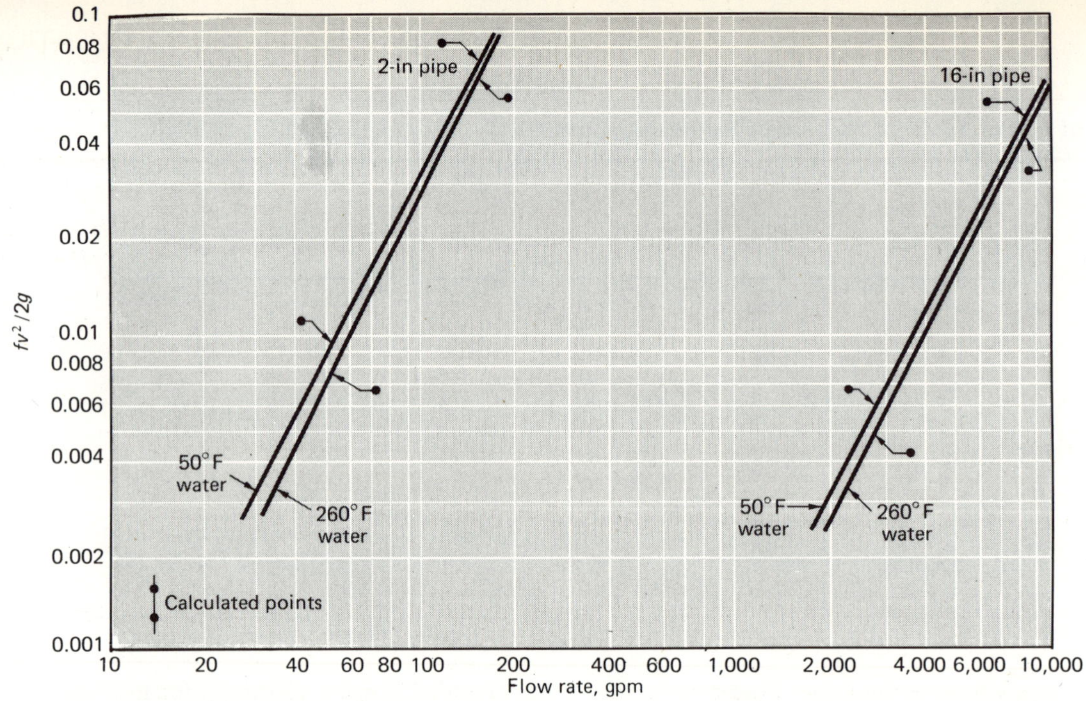

Graph as it appears following the completion of Step 4

F1

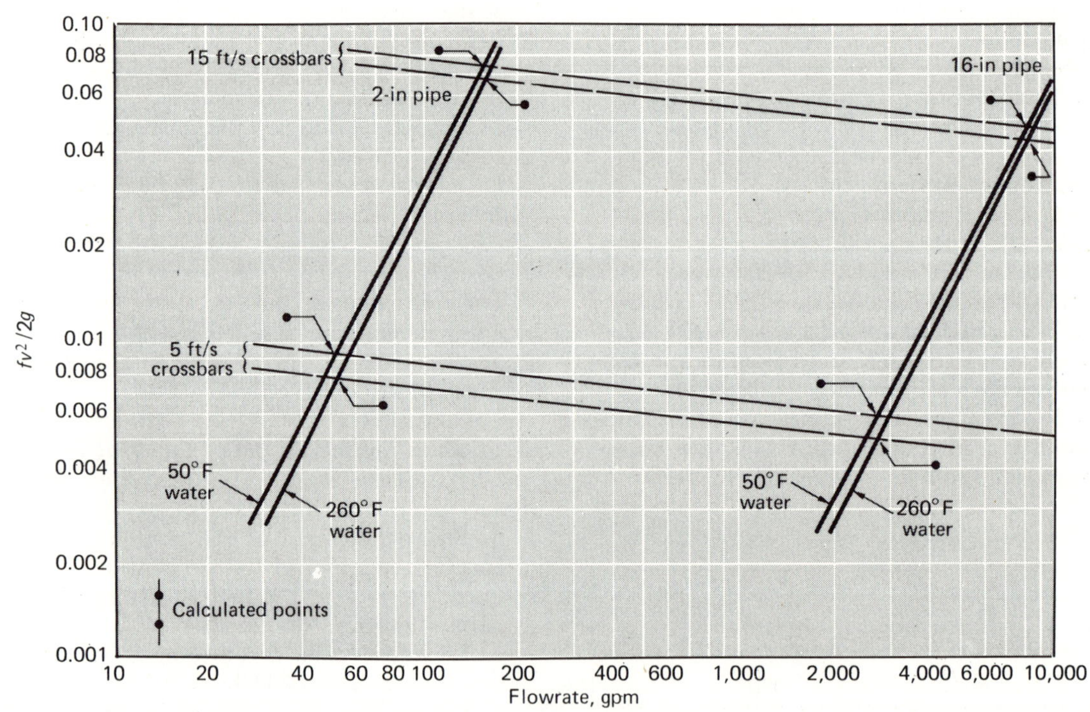

Steps 5a through 5d cover the plotting of the 5 ft/s and the 15 ft/s crossbars

F2

the remainder of the pipe sizes to be investigated. Refer to Fig. 3:

a. For 4-in pipe (refer to Column, 3, Table I.) Construct a short vertical line through the "5 ft/s crossbars" at 198 gpm.

b. For 4-in pipe (refer to Column 4, Table I.) Construct a short vertical line through the "15 ft/s crossbars" at 595 gpm.

c. Construct a straight line through the intersection of the vertical lines (steps 6a and 6b, above) with the 5' ft/s at 50°F and the 15 ft/s at 50°F crossbars. The line should extend above and below the data points.

d. Construct a line through the intersection of the vertical lines with the 5 ft/s at 260°F and 15 ft/s at 260F crossbars.*

e. Repeat steps 6a through 6d, above, for each of the

* If the temperature of the flowing fluid is not a consideration, this step is not necessary.

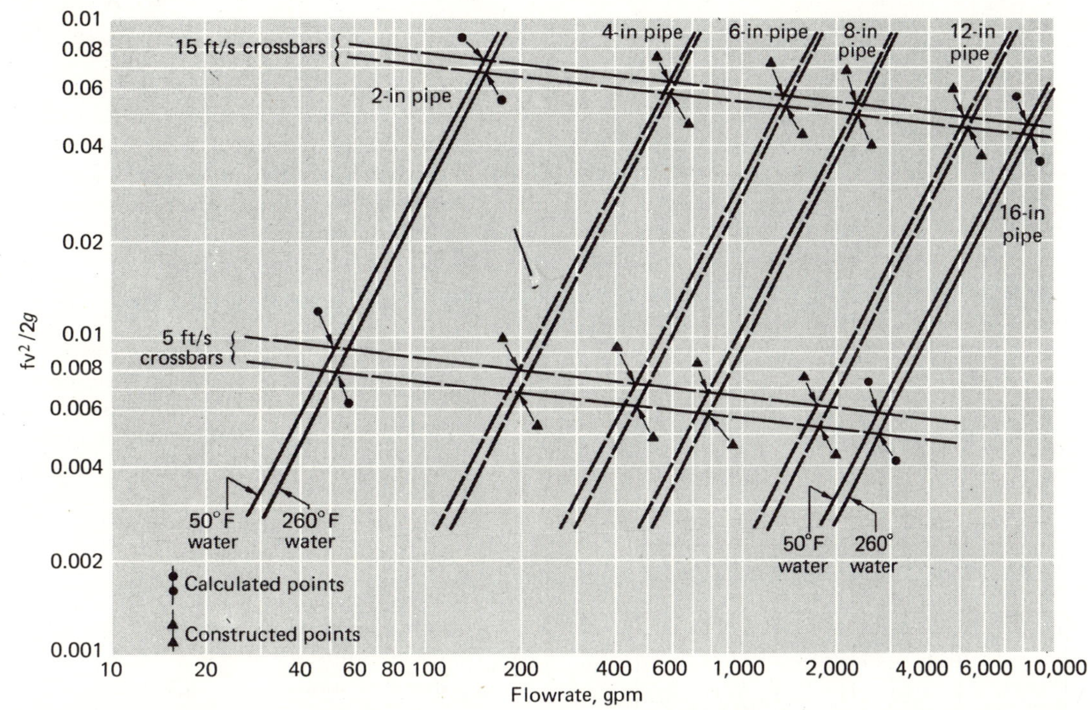

The graph is completed by performing the procedures of Steps 6a through 6e F3

remaining pipe sizes to be evaluated, using the appropriate gpm values from Columns 3 and 4 of Table I.

To construct values of $v^2/2g$ for rapid evaluation of contraction and expansion head losses by construction, utilize another sheet of log-log paper labeled similarly to the $fv^2/2g$ plot except that the values placed on the ordinate should range from 0.1 to 10. Construct horizontal lines across the chart at 0.389 (5 ft/s velocity) and at 3.50 (15 ft/s velocity.) Then, using the data from Columns 3 and 4 of Table I for each different pipe size, construct the sloping line that defines the range of values of $v^2/2g$ vs. gpm for that pipe size. Of course, one chart of $v^2/2g$ will suffice for *any fluid* flowing in the pipe sizes plotted.

It is further noted that log-log plots of v (as abscissa) vs gpm (as ordinate), or any other form of flowrate, are straight lines. Also, if the value of the pipe roughness factor is known, plots of $fv^2/2g$ vs. flow can be constructed by utilizing the plot on page A-24 of Crane Technical Paper 410 for pipes of materials other than clean, commercial steel.

When taking the effect of temperature into account (especially over as wide a range as in the example, data for one or two intermediate temperatures (e.g., at 120°F and 190°F) should be calculated for at least one 5 ft/s point since the effect of temperature on the value of $fv^2/2g$ is seldom linear. For the example problem, at a specific flowrate for a specific size pipe, nearly 60% of the reduction in the value of $fv^2/2g$ between 50°F and 260°F occurs when the water temperature reaches 120°F or ⅓ of the total temperature rise.

Get some log-log graph paper now and plot either the data tabulated herein or your own.

You will find this technique will save time and energy when you have many head-loss calculations to carry out.

Note:

Crane Technical Paper No. 410 is available from:

Crane Co., Advertising Dept., 300 Park Ave., New York, N.Y., 10022

Price is $8.00 including postage ($10.00 outside U.S., plus postage).

Check or money order must accompany order.

The author

Archie M. Larson, Jr., is a nuclear engineer at Gilbert Associates, P.O. Box 1498, Reading, PA 19603, where he is Mechanical Nuclear Project Engineer for Continuing Services for Metropolitan Edison Co.'s Three Mile Island Nuclear Station. His work has included operation, design and plant startup for fuel processing plants, test reactors and nuclear power plants. He received his B.S.Ch.E. from the University of Idaho, and is a member of the American Nuclear Soc.

Pipeline pressure drop—a C_v approach

*G. C. Shah**

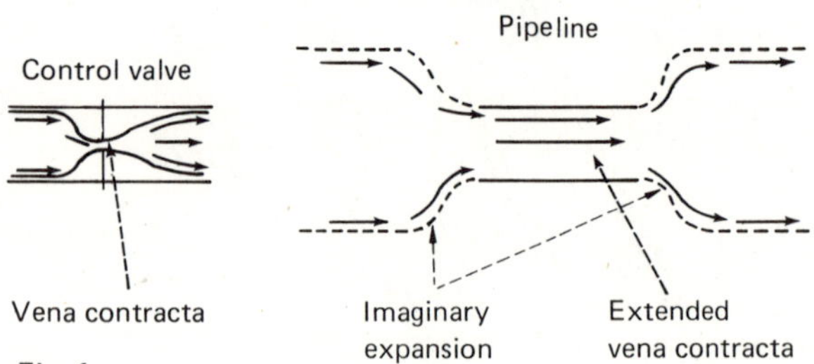

Fig. 1

☐ The pressure drop for liquid flow through a pipeline can be calculated using a method similar to that used for a control valve (or orifice). The analogy between these two flow situations is shown in Fig. 1.

Based on this analogy, the following equation is presented for calculating liquid flow through pipelines:

$$Q = fC_v\sqrt{\Delta p/\text{sp gr}}$$

where: Q is flowrate in gpm, C_v is the equivalent flow coefficient for a pipe, Δp is pressure drop in psi, and f is the flow-coefficient correction factor.

Equivalent flow coefficients for pipes have been derived by using Moody's friction factor chart.† Fig. 2 plots C_v as a function of pipe diameter and length. Fig. 3 establishes the correction factor.

Application limitations are:

1. Low-viscosity (centistokes nearly 1) liquids in horizontal pipes. For vertical pipes, an allowance for head difference must be made.
2. Pipes of moderate roughness, 1.5–2.0 × 10^{-3} in.
3. Not for laminar regions.

Example: Calculate the Δp for 100 gpm of water flowing through a 100-ft-long, 3-in.-I.D. steel pipe.

From Fig. 2: $C_v = 99.0$
From Fig. 3: $f = 0.915$

Substituting into the equation:

$$100 = (0.915)(99)\sqrt{\Delta p/1}$$
$$\Delta p = 1.2 \text{ psi}$$

*American Hoechst, 1120 NASA Road One, Houston, TX 77058.

†"Standard Handbook for Mechanical Engineers," 7th ed., McGraw-Hill, N.Y., 1967.

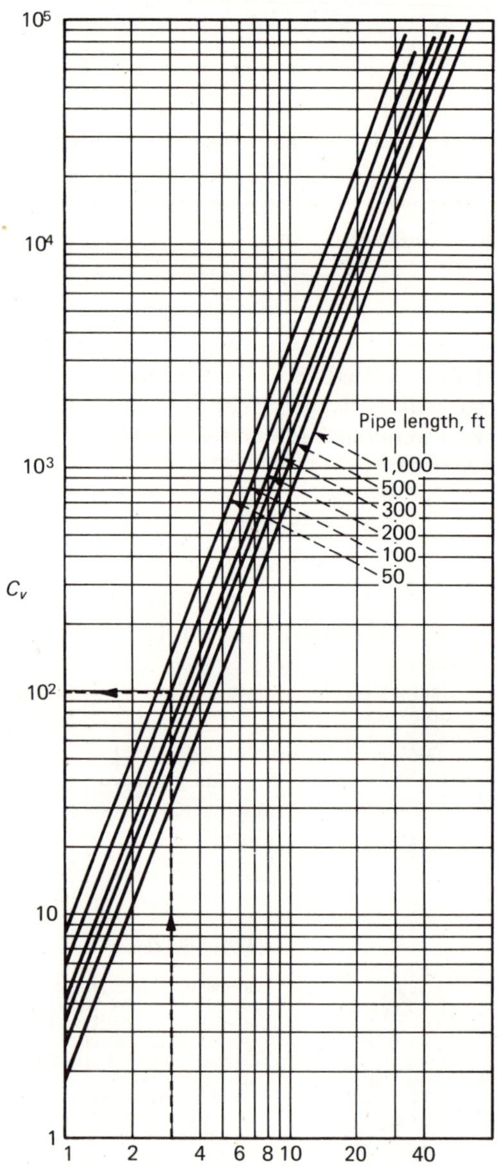

Fig. 2

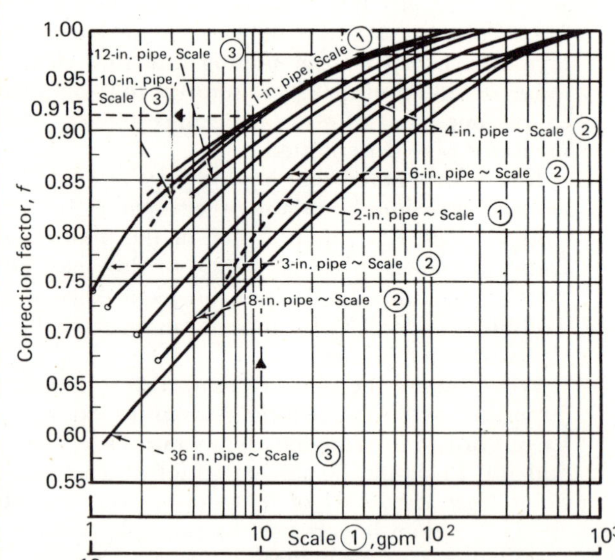

Fig.3

Originally published July 30, 1979

Pipe systems for process plants

Integration of process flows, equipment, instruments and controls with an economical and functioning piping system depends on the use of systematic procedures for plant layout and piping design.

*Robert Kern, Hoffmann - La Roche Inc.**

☐ The piping design of chemical process plants is a collective effort that depends on successfully dovetailing the activities of a great number of specialists within an engineering/design organization. The smoothness and effectiveness of this operation depend, to a great extent, on the hydraulic-systems designer.

Since piping design and associated activities are the most complex part of plant design, let us examine the major aspects for such design by answering the following:

1. What information is required for system and piping design?
2. How is this information used by the hydraulic-systems designer?
3. What information is produced by system and piping engineers for the design and construction of a process plant?

The information flow for these procedures is outlined in Fig. 1 (F/1). Whether one engineer, or an organization having specialized departments, develops much of this information makes no difference in the steps necessary for developing the piping and plant design.

Information Required

To design the hydraulic systems, we need the following kinds of information:

1. *Process Flow Diagram* (PFD) is the basic plan of an industrial plant. It shows how and what type of process equipment are used and interconnected with pipelines. It usually provides data on flow, material and energy balances; process flows (flowrates, specific gravities, molecular weights, pressures, temperatures and viscosities). Generally, the PFD presents essential process-design re-

*To meet your author, see p. 44.

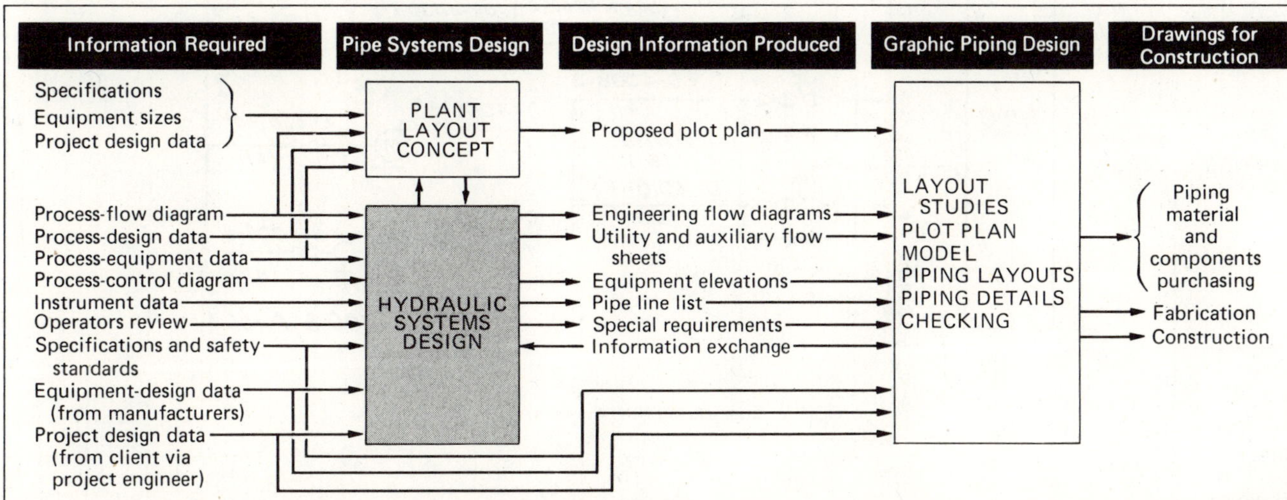

Sources and flows of information for the design of piping systems, layouts and details F 1

Originally published November 10, 1975

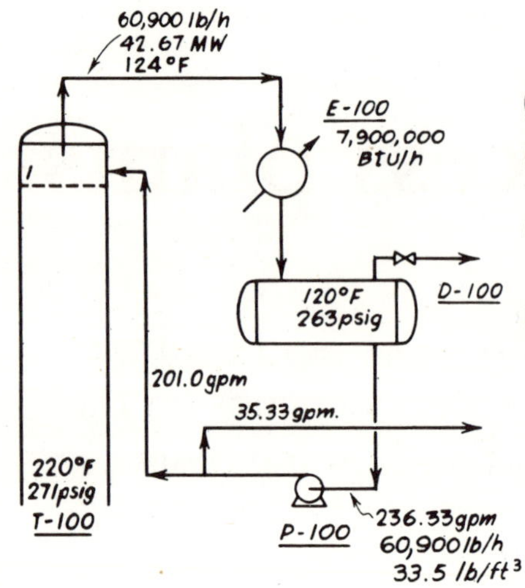

a. Process flow diagram

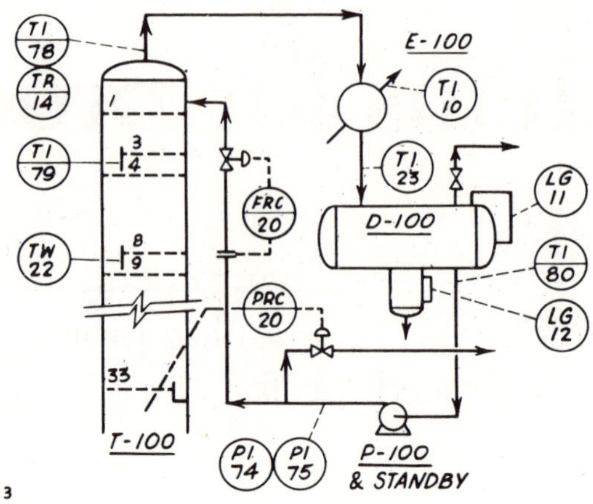

b. Process control diagram

Key:

Instrument Symbols

- FRC Flow recorder/controller
- LG Level gage
- PI Pressure indicator
- PRC Pressure recorder/controller
- TI Temperature indicator
- TR Temperature recorder
- TW Temperature well

Other Symbols

- CW Cold water
- DR Drain
- M Motor-driven pump
- O Oil
- RV Relief valve
- T Turbine-driven pump

On all pipelines, the coding indicates type of fluid flowing, line number and size, and materials specifications (A1, B1, etc.).

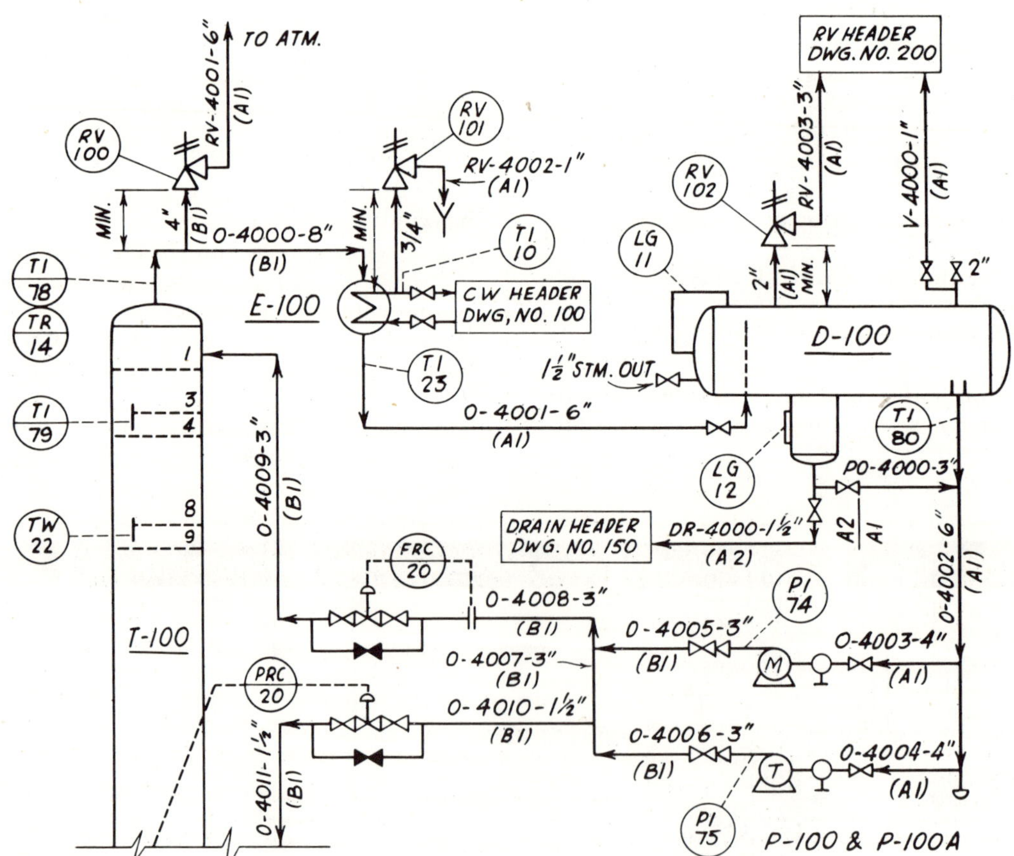

c. Engineering flow diagram

Flow diagrams are the basic documents for piping design

quirements. A process description also supplements the PFD. A typical portion of a PFD is shown as F/2a.

2. *Process Equipment Data* enumerate details about the equipment used, such as size and type; materials of construction; operating descriptions; manholes, handholes, and process, utility and instrument nozzles, and sizes; auxiliary equipment and devices, special requirements; and design, test and operating pressures and temperatures. References to manufacturers are also included on equipment data sheets. The PFD and process equipment data are usually worked out by process engineers.

3. *Process Control Diagram* (PCD) shows the instrumentation requirements of the process plant, such as locally-mounted and board-mounted instruments, and the connections between the elements of instrumentation loops. Instrumentation data sheets supplement the PCD and provide details of the instrumentation components, such as types of equipment, dimensions and manufacturers; and details of instrument locations at process vessels. A typical part of a PCD is shown as F/2b.

Instrument symbols identify, and a number system cross-references: (a) the PCD, (b) engineering flow diagrams, and (c) piping drawings.

4. *Piping Design Specifications* describe a company's basic requirements concerning piping and plant design. Specifications generally apply to:

- Development, presentation and interpretation of engineering flow diagrams; pipeline designation tables, and materials of construction.
- Design of process, utility and auxiliary piping; operating and safety conditions; allowable design pressures and temperatures; piping material, corrosion allowance, wall thicknesses, and insulation requirements; desirable unit losses; piping-and-components sizing methods. Valving is an important part of hydraulic-systems design. Table I (T/I) shows a typical page of a piping-design specification for valving.
- Mechanical plant-design requirements; clearance, access headroom, operation, maintenance, grading and paving requirements; economy and methods of plant layout and graphic piping design; mechanical handling requirements, and associated equipment.

It is essential for all engineers and designers to read and be familiar with a company's specifications.

5. *National Standards, Codes and Specifications* can be another information source for the piping designer. T/II lists the most important U.S. publications. Codes primarily outline design requirements and construction details for piping systems. Standards give piping, fittings and valve dimensions; materials, pressure and temperature ratings. Specifications describe materials for the manufacture of pipes and fittings.

6. *Project Design Data* provide site conditions, available cooling-water temperatures, available or required steam, brine, fuel oil, etc., data. Locations and distances from feed and product storage tanks are given. A site plan is included here, if necessary, with climatic conditions, average winter and summer temperatures, rain and snowfall quantities, seasonal temperature variations.

7. *Design Practices and Data Books* are another valuable source. These are compiled by contractors and owners to record useful data, preferred design methods and procedures, past operating experience affecting design; and descriptions of actual experience with processes, equipment and plants.

Manufacturers' catalogs and reports give information about the latest developments and available equipment. Design data in manufacturers' catalogs can be used at the preliminary stages of design. At the final stages, certified data and drawings of process equipment are required.

8. *Plot Plan* gives the physical relationships between process equipment in plan and elevation. With this drawing, the line length and assessments of pipe configuration can be made for pipe-sizing, control-valve and pump calculations. A conceptual or proposed plot plan can be developed from the PFD, process equipment data, and project design data.

Design of hydraulic systems

For a small plant, the process-flow, process-control and engineering-flow sheets can be produced by a single chemical engineer. The volume of work for designing a large chemical plant, however, requires an organization with departments of specialized skills.

Piping and Instrumentation flow diagrams (P&Is) are the basic documents for final plant layout and graphic piping design. The hydraulic-systems designer prepares the P&Is from documents available and supplied to him.

When piping drawings are checked, the P&Is should be complete in all minute details. P&Is should show all necessary notations and instructions so that the graphic piping designer can produce a reliable and economical piping layout. A section of a P&I is shown as F/2c.

The P&Is consolidate in one set of drawings the PFD and the PCD. P&Is show all piping interconnections between process equipment, components, pipelines and manifolds, utility and auxiliary piping connections. Line sizes are given, and locations of size and specification changes are indicated. All valves (including type) are shown with proper locations in the pipelines.

The piping-systems engineer should indicate to the piping-layout designer those lines that need special attention. These are:

a. *Minimum-length lines or circuits.* For example, length of a transfer line on one hand, and a closed compressor circuit on the other (noting where the compressor suction begins).

b. *Gravity-flow lines.* Most of these lines are readily recognized (for example, pump suctions or condenser outlets). Dimensions of seal loops in gravity-flow lines, sloping lines, and pocket-free gravity-flow lines should be clearly indicated on P&Is.

c. *Two-phase flow lines.* Since most layout designers are not interested in process data, these lines are not readily recognized. If the systems designer indicates the line where slug flow is possible, where vaporization takes place, or where two-phase flow exists, a more careful study, layout and piping design can be produced.

d. *Symmetrical pipelines.* These are required for even-flow distribution, and should be so indicated on the P&Is. These lines are not necessarily graphically sym-

Typical design specifications for valving

Block valves shall be provided in the water inlet and outlet of exchanger units when the exchangers are provided with process isolation valves for frequent inspection and cleaning during operation of the unit. In freezing climates, an antifreeze bypass from the inlet to the outlet piping (or other means) to prevent freezing shall be furnished.

Block valves shall be provided at the following locations in pump, turbine, and compressor piping:

- ☐ In suction and discharge piping to pumps.
- ☐ At the equipment in auxiliary piping for gland oil, flushing oil, and cooling water.
- ☐ At the header in overhead water-supply branches located outdoors in freezing climates.
- ☐ In all auxiliary piping when necessary to allow removal of the equipment during operation of the unit.

Block valves shall be provided at the following locations in steam piping:

- ☐ At equipment in steam piping to steam-driven equipment.
- ☐ Upstream of control valves that are spared (separate control valves furnished for standbys).
- ☐ At equipment in pressure exhaust-steam piping from steam-driven equipment.
- ☐ Adjacent to equipment in vacuum exhaust-steam piping from steam-driven equipment when the equipment may be shutdown for sustained periods during operation of the unit.

Utility systems shall be provided with valves at service connections. When a distribution system is used for water or for steam, a block valve shall be furnished at the header.

Block valves shall be provided and located as required for bleed and aeration systems.

A block valve shall be provided in the main line of fuel-oil and fuel-gas piping to fired heaters, to be located remote from the equipment, and be accessible for rapid operation in an emergency.

Block valves shall be provided for piping leaving and entering a building.

Two block valves with a bleeder or a resilient double-seated valve with body bleeder shall be provided in piping connecting systems in services in which contamination cannot be tolerated.

- ☐ When steam systems are connected to process services and are operated at a higher pressure than the process service, one block valve shall be furnished.

Valving of operating vent and drain connections shall be as follows:

- ☐ Services with ANSI ratings 600 psi or less: single block valve.
- ☐ Services with ANSI ratings 900 psi or over: single globe valve.
- ☐ Liquid hydrocarbon services with a vapor pressure over 65 psia at 100°F: two block valves.
- ☐ Caustic services: single globe (vent only) or lubricated-plug valve.

Valving at sample connections shall be as follows:

- ☐ Services with ANSI ratings 300 psi or less: single block valve.
- ☐ Services with ANSI ratings 400 psi over: single globe valve.
- ☐ Liquid hydrocarbon services with a vapor pressure over 65 psia at 100°F: two block valves.
- ☐ Caustic services: a single globe valve or lubricated-plug valve.

metrical, but should have identical resistances between similar functional points. An example: gas lines to a number of furnace burners.

e. *Special pipelines.* Some pipelines have special features and limitations that must be taken into account during plant layout and piping design. For example: glass piping having special fittings and support requirements; glass-lined, Teflon-lined or refractory-lined pipes, which usually have flanged fittings and limitations for line length; lines where solid deposits require special cleaning provisions; slurry lines (indicating where a vertical or near-vertical line is essential); jacketed and steam-traced lines; pipelines where the liquid carries an expensive catalyst (short and simple piping here can reduce initial and makeup catalyst quantities).

At the final stages of design, the P&Is are usually reviewed by the plant operators for adequacy in valving, instrumentation and pipe interconnections necessary for operating, startup and shutdown.

P&Is, layout studies and plot plans are simultaneously developed, and the final drawings released for piping layout and detailing.

Utility-Flow Diagrams are supply and return systems for steam, cooling water, process water, brine, condensate, air, nitrogen, etc. From the start of a project, these are produced as engineering flow diagrams, except occasionally for steam, when an elaborate system might require a steam-energy balance.

Since process piping interconnects equipment in a sequence, equipment arrangement usually does not influence its orientation. In contrast, utility-flow diagrams must be drawn to a plot plan, so that leads to equipment connect in a plot-arrangement sequence to utility headers. Otherwise, flowrates in these pipe sections cannot be realistically estimated.

Utility supply is controlled at sensing points in the process piping. Supply and return leads usually have valves at the terminating points on the process vessels. These valves, instruments and control loops are shown on the P&Is (for example, the steam-supply control valve to a reboiler). In addition, valves are often required at the header (as shown on the utility-flow sheets)—especially when a number of leads have to be provided.

Utility headers are usually sized with unit losses in decimals of 1 psi. Leads are sized with the available pressure difference between supply and return points at the header.

Flow Diagram Symbols. The piping and instrumentation symbols are the alphabet of all flow diagrams. Piping symbols usually represent the various valves and components in the line. Fittings are indicated where they have significance for process and piping design. A few examples are: a cap at the end of a header, a reducer or increaser at changes in line size, flanges where pipe sections have to be removed, and swing elbows where alternate tie-in connections are required.

Instrument symbols and interconnecting line designations have been issued by the Instrument Soc. of America [*1*]. These symbols indicate location of sensing elements; local and board-mounted instruments; pressure-, temperature-, flow- and level-control loops; and the instrumentation hardware in the pipeline.

If all the equipment, pipelines and symbols are shown on the overall and detailed design of a process having automatic and manual control, plant startup and shutdown can be understood and evaluated.

[Graphic piping-design symbols have been issued by ASME but are not widely accepted by the oil and chemical industries. Contractors and owners use their own piping symbols. These are identical to a great extent among the various companies.]

A Pipeline List is prepared for several reasons: to supply a complete record of pipeline data; to provide a checklist for material takeoff; to cross-reference piping details, plans, elevations and flow diagrams; to locate

National piping codes, standards and specifications

ANSI Code for Pressure Piping:

B31.1 – Power Piping
B31.2 – Fuel Gas Piping
B31.3 – Petroleum Refinery Piping
B31.4 – Oil Transportation Piping
B31.5 – Refrigeration Piping
B31.6 – Chemical Plant Piping (In Preparation)
B31.7 – Nuclear Power Piping
B31.8 – Gas Transmission and Distribution

ANSI Standards:

B16.5 – Steel Pipe Flanges
B16.9 – Steel Butt-Welding Fittings
B16.28 – Steel Short Radius Fittings
B36.10 – Welded and Seamless Steel Pipe
B36.19 – Stainless Steel Pipe

Issued by:

American National Standards Institute
1430 Broadway
New York, NY 10018

ASME Boiler and Pressure Vessel Code:

Section I – Power Boilers
Section II – Material Specifications
Section III – Nuclear Vessels
Section VIII – Unfired Pressure Vessels

Issued by:

The American Soc. of Mechanical Engineers
345 East 47th St.
New York, NY 10017

ASTM Specifications:

A72 – Wrought Iron Pipe
A106 – Carbon Steel Pipe
A155 – Pipe for High-Temperature Service
A182 – Stainless Steel and Alloy Steel Flanges
A234 – Carbon and Ferritic Alloy Steel Fittings
A312 – Stainless Steel Pipe
A333 – Low-Temperature Pipe
A335 – Ferritic Alloy Steel Pipe
A358 – Austenitic Chrome – Nickel Steel Pipe
A381 – Pipe for Transmission Service
A403 – Stainless Steel Fittings
A420 – Low-Temperature Fittings
B210 – Aluminum Pipe
B247 – Aluminum Flanges
B361 – Aluminum Fittings
B366 – Nickel and Nickel Alloy Fittings

Issued by:

American Soc. for Testing and Materials
1916 Race St.
Philadelphia, PA 19103

API Specifications:

5L – Line Pipe
5LX – High-Test Line Pipe
6A – Flanges

Issued by:

American Petroleum Institute
1801 K St., N.W.
Washington, DC 20006

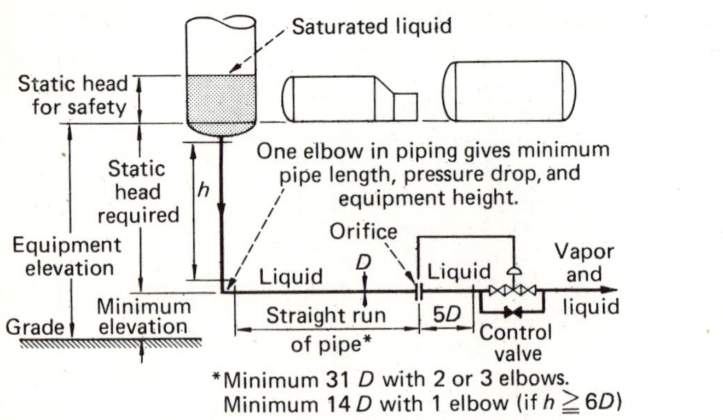

Static head before orifice prevents vaporization (F/3)

fabricated pipes on drawings; and to place pipelines in relation to the process equipment.

From the line list, the specialist in pipe support and flexibility can obtain temperature and wall-thickness data. The metallurgist can verify chosen materials of fabrication. Using temperature data, insulation thickness can be determined. For relief-valve sizing, the designer can find what fluid is flowing, and the alternate pressures in the line under normal, maximum, test and emergency conditions. From the list, the piping designer can find the appropriate P&I; and the construction foreman can locate a fabricated, delivered and identified line on one of the piping drawings.

Equipment elevations

The systems designer works out all equipment elevations, static-head pressure requirements, elevation differences between process equipment, and elevation differences between the starting and terminating points of a pipeline. This is also essential information for plant layout and piping design.

For unhoused process units, the most economical location for process equipment is usually at grade. The units seldom require a structure; construction is easy; most valves and instruments can be made accessible from grade; and operation and maintenance are convenient and safe.

Minimum skirt heights for towers vary from 3 to 6 ft for towers ranging from 2 to 12 ft dia., and having bottom temperatures of 100 to 400° F. Higher temperatures (or cryogenic temperatures) require the addition of 1 or 2 ft to these heights in order to avoid the transmission of unduly high (or low) temperatures to the foundation or supporting steel. Drums and exchangers are usually elevated 1 to 4 ft from grade to the bottom of the unit, depending on diameter, size and valving for the bottom-outlet piping.

When a pressure difference between two pieces of process equipment is not available, gravity flow is called for. The elevation difference between the equipment will be determined by pipe, fittings and valve resistances.

Gravity-flow lines can be sized for velocities similar to those for pump-suction or reboiler-downcomer lines (use F/5 in Part 9, *Chem. Eng.*, Aug. 4, 1975, p. 111).

To avoid vaporization in a hot liquid line, a positive static head can provide additional pressure to keep the liquid subcooled. An example of these conditions is shown in F/3. Flashing through the orifice (and control valve) can be avoided with a static liquid column ahead of the orifice. If the orifice is as close to the vessel as possible, and if the piping up to the control valve has only one elbow, we will need the minimum equipment elevation, orifice range and line length. To meet such conditions, we can elevate a tower, drum or exchanger.

When vapor and liquid have to be separated, the vapor-collecting drum is elevated, as shown in F/4. For example, the steam drum on a pyrolysis furnace (F/4a) collects the steam generated by the waste-heat boiler. A similar arrangement prevails for steam generated in a furnace-heating coil. In another example, the elevated steam drum (F/4b) prevents steam from collecting in top of reactor jacket, and thus avoids overheating it.

In F/5, we find a vapor-disengaging drum that is placed over the heaters. Here, the liquid level in the drum keeps the exchanger tubes flooded, and the vapor space in the drum prevents excessive liquid carryover in the vapor-outlet line. This drum must be elevated over the exchangers and can be mounted directly on the exchanger nozzles.

Elevation requirements for process vessels to meet pump NPSH, relative distances between reboilers and towers, and elevations for condensers and reflux drums have already been discussed. (See articles beginning on pp. 25, 55 and 62, respectively.)

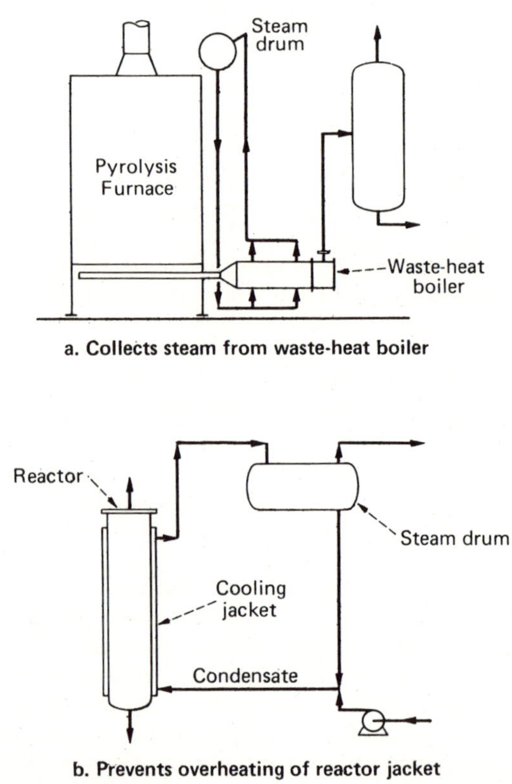

Elevated steam drums separate vapor/liquid mixture (F/4)

Elevation requirements should be noted on the P&Is, or on data sheets so that the layout and piping designers are fully aware of them. Similarly, if equipment is elevated for layout reasons, the process engineer should be notified.

Economy of systems design

Economy of piping and components depends primarily on the skill of the engineer. A rational approach to design development, systematic methods, and consistent data usage will yield far more reliable results than intuitive judgments and rule-of-thumb decisions. Even with rational methods, calculated pipe and components sizes end up oversized because piping and components are manufactured to standard increments. This provides a fairly safe and wide latitude in assembling a process-piping system. However, the selection of overly sized pipe, components and equipment is a waste of money. On the other hand, undersized systems waste energy, if they work at all. Undersizing often results in changes to design, fabrication and construction, and rework, all of which are expensive.

An important activity of a hydraulic-systems designer should be consultation with other specialists of plant design. There must be a full understanding between process and system designers.

When large and expensive piping is sized, when utility cost is high per unit loss, and when available pressure differentials are limited, the hydraulic-systems designer must obtain a dimensioned sketch from the graphic piping designer. From this sketch, the hydraulic-systems designer can suggest the elimination of bends and offsets for saving capital and utility costs.

Pipelines can be much longer than one may estimate from a plot plan. These lines can often be simplified after some discussion with interested specialists. It is also essential to obtain pump data from the pump specialist before calculating pump differentials, and to establish control-valve types with the instrument engineer before calculating their sizes.

For substantial capital-cost reduction, there is more opportunity during the process-development and design phases (and during the mechanical-design and plant-layout phases) than during the hydraulic-systems design phase.

Piping design

The piping designer has that exceptional and essential talent to be able to visualize equipment and interconnecting piping presented on flow diagrams as a three-dimensional plant, and to represent this on a drawing. Piping design is executed in three main steps: (1) layout studies, (2) piping plans and elevations, and (3) piping details (isometrics).

The layout designer coordinates specifications; process, project and equipment design data; and well-established design practices to produce a layout study that informs all disciplines of the physical features of a plant.

At this stage, the effective and economical use of space is determined. Structural-steel outlines are

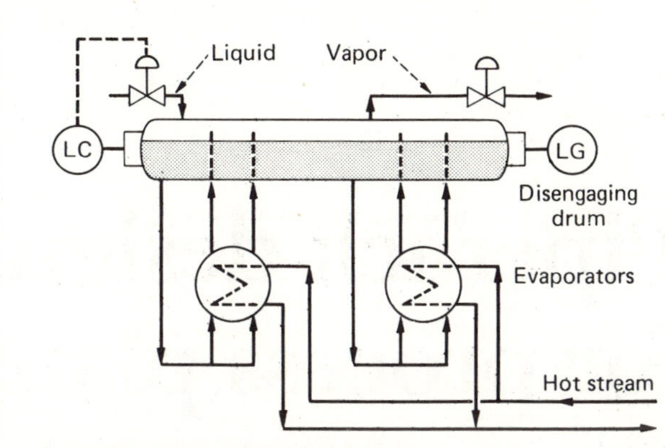

Disengaging drum is supported on exchanger nozzles F5

shown, large and critical pipelines are routed, space for main pipe banks is indicated, all equipment and major structures are located and dimensioned, and control-valve locations are given. Generally, all plant components that occupy space should be shown on a layout study in a well-organized manner. At this time, the most essential layout requirement is met—providing access space and access aisles for operating, maintaining and constructing the plant.

When released for detail design, layout studies should be presented in a way that all disciplines can work simultaneously and independently, and produce details that will not interfere with each other when the plant is built. At this point of design development, a final plot plan is drawn and, if required, a model is built.

The next step is the production of construction drawings—including piping plans and isometrics. This is a large-scale operation in contrast to the work of the single individual at the concept/design phase and, generally, to the two to four designers during the development of layout studies.

For effective piping design and detailing, all information presented on engineering-flow diagrams, design data, specifications, manufacturers' drawings, plot plants and structural drawings must be final. For example, during the layout-study phase, the addition of a pressure or temperature point in a pipeline will cost nothing. During piping detailing, the same change can cost as much as designing and detailing the pipeline, in addition to delays in making fabrication drawings.

References

1. Instrumentation Symbols and Identification, Standard ISA-S5.1, Instrument Soc. of America, Pittsburgh, Pa., 1968.
2. Heitner, I., Some Typical Flow Sheets and How They Grow, *Petrol. Refiner*, Oct. 1963.

Piperack design for process plants

Careful analyses of flow diagrams, plot plans, specifications and project data are essential to the layout of efficient and economical plant piperacks.

Robert Kern, *Hoffmann - La Roche Inc.* *

☐ The main arterial system of a process plant is in the yard piperack, which contains the long process lines for connecting distant equipment, and the lines for entering and leaving a unit. Utility headers, supplying steam air, gas and water to process equipment are also located in the yard, as well as relief headers and blowdown headers. Instrument lines and electrical-supply conduits are often supported on the yard steel.

We will provide a step-by-step evaluation for the layout design of piperacks for open (i.e., outdoors) chemical plants. Piperack design for housed (i.e., indoors) plants will be discussed in a later article.

Before attempting piperack layout, we must have the following information: specifications, project-design data, plot plans and flow diagrams. Let us summarize the details for each:

Specifications—Only a few items in the job specifications affect yard-piping design. Such items are: the minimum headroom between road surfaces and overhead pipelines or steel beams; access, headroom and handling requirements for equipment arranged under the yard; requirements for ladders, catwalks and platforms located in the yard; details affecting piping and structures, operating and safety requirements.

Project design data and site maps—These give required or existing conditions inside and outside the unit limits, such as the required location of cooling-water mains, whether below or above grade; location of furnaces, control and switch houses; location of utility and process lines entering and leaving unit limits; main pipe runs outside the unit; location of storage tanks relative to process units; location of blending, loading and fill-

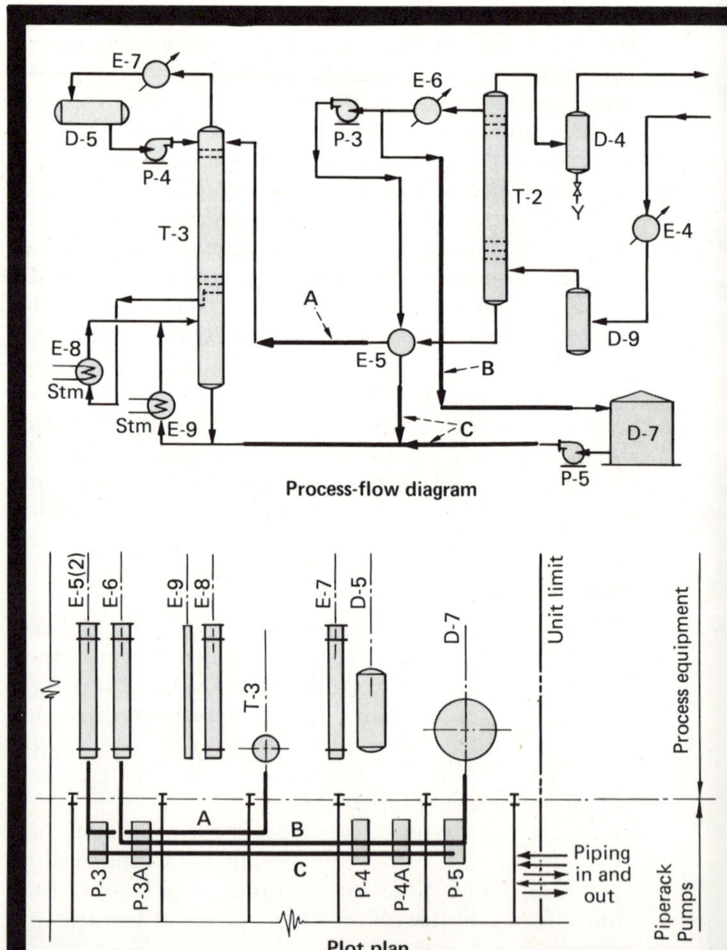

Process lines located in the piperack Fig. 1

*To meet your author, see p. 44.

Originally published January 30, 1978

ing stations, and cooling towers; variations in the site's grade level; safe distances between units, or between tankage, storage, loading, etc., facilities.

Plot plan—This shows the relationship between plant units, equipment, buildings and yard piping, and the position of incoming and outgoing lines. The plot plan also locates major structures, buildings and all equipment, and indicates roads crossing the yard or located under the piperack.

Flow diagrams—Process-flow diagrams show the essential lines that connect process equipment. Mechanical-flow diagrams (developed from process-flow diagrams) indicate the complete flow systems necessary for plant operation, and include pipe sizes, valving, manifolds, and all piping and instrumentation details. Utility-flow diagrams show the number and size of water, steam, condensate, gas, air, etc., headers, and all equipment supplied by these headers, with the necessary valving and piping details.

Fig. 1 shows portions of a process-flow diagram and a plot plan. From these two drawings, we can make an assessment regarding which portions of the process lines will be located in the yard and which lines will connect directly to nozzles on adjacent or nearby equipment. Heavy lines on the flow diagram indicate piping assumed to be located in the yard. These lines are also shown on the plot plan to give a visual idea of the yard-space requirement. A mechanical-flow diagram is similarly evaluated. The greater the number of lines included in the plot plan, the more accurate the estimate of the piperack width.

In addition to process lines, utility-flow diagrams will show individual service lines and utility headers. Utility mains generally run the whole length of the yard. These lines should also be taken into account when estimating space requirements. We can now proceed to the detailed analyses for yard piping.

Yard piping in plan

The plant layout determines the main-yard piping runs. Fig. 2 shows typical piperack layouts for various plant arrangements. Smaller plants usually have the simplest yard piping, as shown in Fig. 2a and 2b. In Fig. 2a, the process and utility lines enter and leave at the same end of the plot. Fig. 2b presents a frequently adopted layout, with utility lines entering at one end and process lines at the opposite end. Layout conditions sometimes result in an L-shaped yard (Fig. 2c).

In larger plants, yard piping will be more complicated, as shown in Fig. 2d, 2e and 2f. Fig. 2g shows the yard piping for a very large plant. This layout can be considered as a combination of several simpler yard-piping arrangements.

Of course, configuration of the piperack is not determined when doing the plant layout. The yard results from an overall plant arrangement, site conditions, customer's requirement and, above all, plant economy.

Pipelines in the piperack are classified as process lines, relief-line headers, utility lines and steam headers. Let us examine each type in detail.

Process lines interconnect nozzles on process equipment units more than 20 ft apart, as shown in Fig. 1. Product lines run from vessels, exchangers or, more often, from pumps to the unit limits, or to storage or header arrangements outside the unit limits. Crude or other charge and chemical lines enter the unit and usually run in the yard before connecting to exchangers, furnaces or to other process equipment.

Relief-line headers, individual relief lines, blowdown lines and flare lines should be self-draining from all relief-valve outlets to the knockout drum, flare stack, or to a point at the plant limit. A pocketed relief-line system is more expensive because an extra condensate pot is required with its associated instruments, valves and pumps. To eliminate pockets, some relief-line headers must be at a higher elevation. Such headers are placed above the main yard on a tee support attached to extended yard columns. For some noncondensing gas systems, self-drainage is not as important.

Utility lines in the yard can be classified as (1) utility headers serving equipment in the whole plant, and (2) utility lines either individually serving one or two equipment items, or a group of similar equipment (furnaces or compressors). The first category includes lines for low-pressure and high-pressure steam, steam condensate, plant air, and instrument air. The second category contains lines for boiler feedwater, fire steam, compressor-starting air, fuel oil, fuel gas, lubricating oil, inert gas, and chemical-treating.

Steam headers should drain to the steam separator for effective condensate collection. Branch connections are usually made at the top of the headers in order to avoid excessive condensate drainage to equipment.

Instrument lines and electrical cables are often supported in the yard, and extra space should be provided for them. The best instrument-line arrangement eliminates almost all elevation changes between the plant and the control room. This can easily be done when instrument lines are supported outside the yard column at a suitable elevation.

Line location in the piperack

Fig. 3a shows a one-level piperack. Regardless of service, heavy pipelines (very-large-diameter lines, or large lines full of liquid) are placed over or near the supporting columns. (A centrally loaded column and a smaller bending moment on the cross beam may yield a lighter structural design.) All process lines and relief lines are placed next to the heavy lines. Utility lines are in the central portion of the rack.

A general sequence for utility lines is also shown in Fig. 3a. The position of individual utility lines in the yard depends on the number and size of the branch connections. If a majority of similar-size branches connect to the header from the right, it is more economical to place the utility line in the right-half section.

From a support standpoint, it is advantageous to group hot lines that require expansion loops, as shown isometrically in Fig. 3a. The most commonly adopted design is to elevate the expansion loops horizontally over the yard, with the hottest and largest diameter line on the outside. Line guides, line stops, and anchor points are usually required along a hot line, somewhere in the piperack. Pipe-expansion forces at some of these points will affect the design of the supports.

Those process lines that connect equipment on the

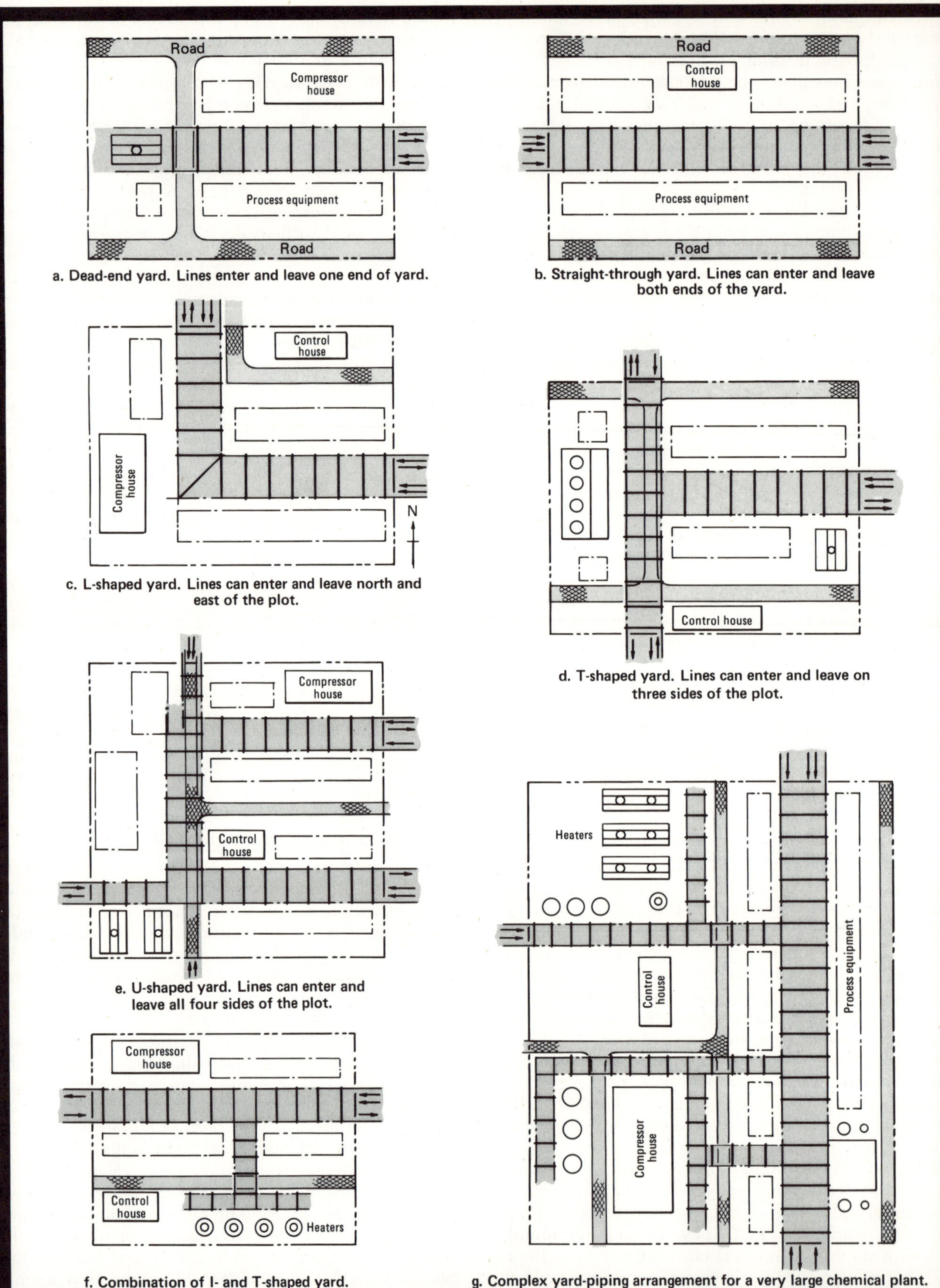

a. Dead-end yard. Lines enter and leave one end of yard.

b. Straight-through yard. Lines can enter and leave both ends of the yard.

c. L-shaped yard. Lines can enter and leave north and east of the plot.

d. T-shaped yard. Lines can enter and leave on three sides of the plot.

e. U-shaped yard. Lines can enter and leave all four sides of the plot.

f. Combination of I- and T-shaped yard.

g. Complex yard-piping arrangement for a very large chemical plant.

Piperack arrangements depend on plant layout — Fig. 2

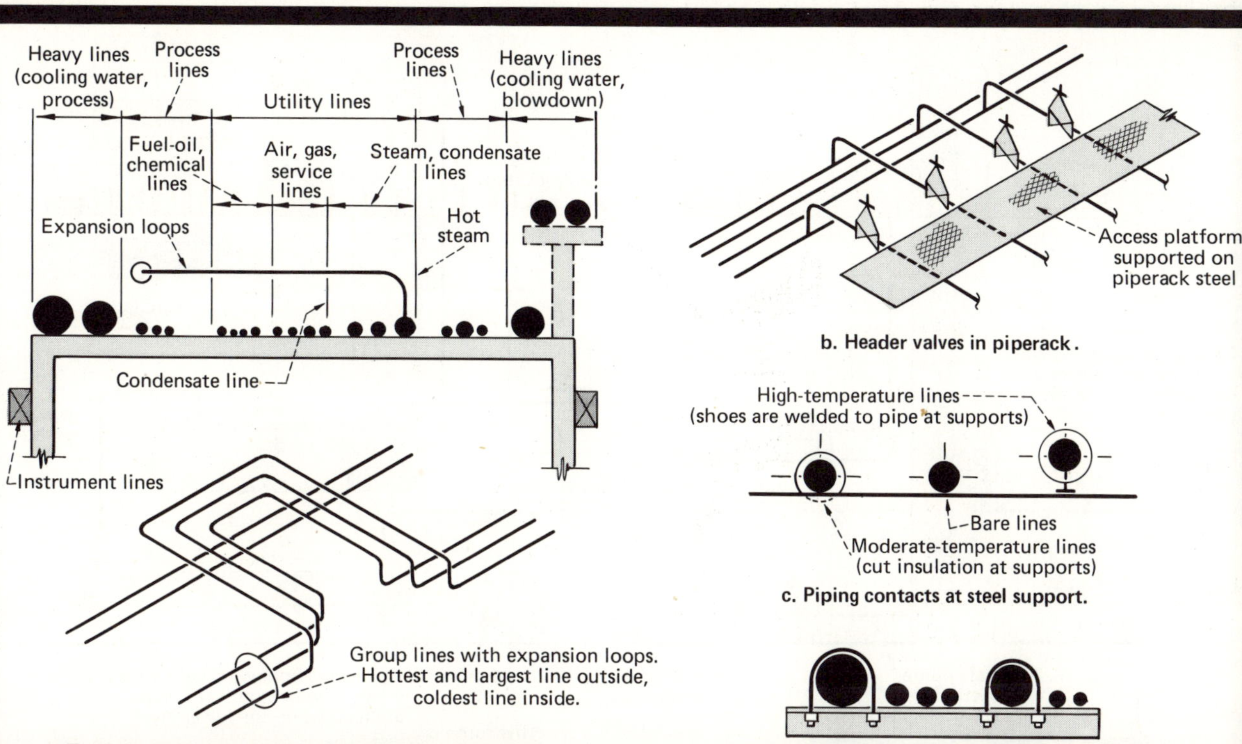

a. Typical pipeline distribution on a one-level piperack.

b. Header valves in piperack.

c. Piping contacts at steel support.

d. Large-diameter lines supporting small lines

Line locations and supports for pipelines located on piperacks **Fig. 3**

same side of the yard should be near the edges of the yard bank. Lines that connect equipment located on both sides of the yard should be closer to the utility lines, and can be placed on either side of the yard. The position of product lines is influenced by their routing after leaving the plant limit. Lines turning to the right should be on the righthand side of the yard; those turning left, on the lefthand side. Utility lines individually serving one or two equipment items should be on the same side of the yard as the equipment to which they connect.

If two elevations for the piperack are required to accommodate a large number of lines, all utility lines are usually placed in the top bank and all process lines in the bottom bank. Obviously, exceptions can always be made to the elevation of individual utility or process lines. Line sequence will be similar to that discussed for a one-level yard. Data for line spacing are usually available in contractors' and owners' design handbooks.

Valves at the headers in piperacks should be lined up, with handwheels accessible from platforms or walkways (Fig. 3b). Hot lines are usually provided with "shoes" and sliding plates. These normally elevate pipelines 4 in. higher than the supporting steel (Fig. 3c).

Small-diameter lines can be suspended from large ones, with the supports located approximately at the centers of the pipe span (Fig. 3d).

Piperack in elevation

Fig. 4 shows in cross-section typical yard piping for a process plant. The elevation is determined by the highest requirement for the following: (a) headroom over a main road, (b) headroom for access to equipment under the piperack, and (c) headroom under lines connecting the yard and equipment laterally outside the yard. Size of the supporting-steel beams should also be taken into account when considering headroom requirements between the piperack and grade.

Generally, process lines that connect two nozzles elevated higher than the top bank should be located in the top bank. Process lines having one end lower than the bottom yard elevation can run in the top or bottom bank. If both ends of a process line are lower than the bottom yard elevation, the line should be located in the bottom bank.

The elevation of a line can also be influenced by valves and instruments in the line. Often, a convenient access platform can be provided for valves arranged in the top bank. The preferred location of lines with orifice runs is near the edge of the yard, with orifice flanges near a yard column for convenient access with a portable ladder. Orifice runs, after pumps, can be located near supporting columns at 7-ft elevation. Control valves are usually near steel columns for convenient support.

When pumps are arranged under the piperack, one or two slots (i.e., open space) are often required along the yard. These slots are usually over the pump-discharge nozzle for process, steam and other utility lines connecting the appropriate header in the piperack to the pumps and drivers.

Local platforms and catwalks can also be supported on the piperack steel over yard and lateral piping.

Keeping Dimensions B and C (Fig. 4) to the smallest

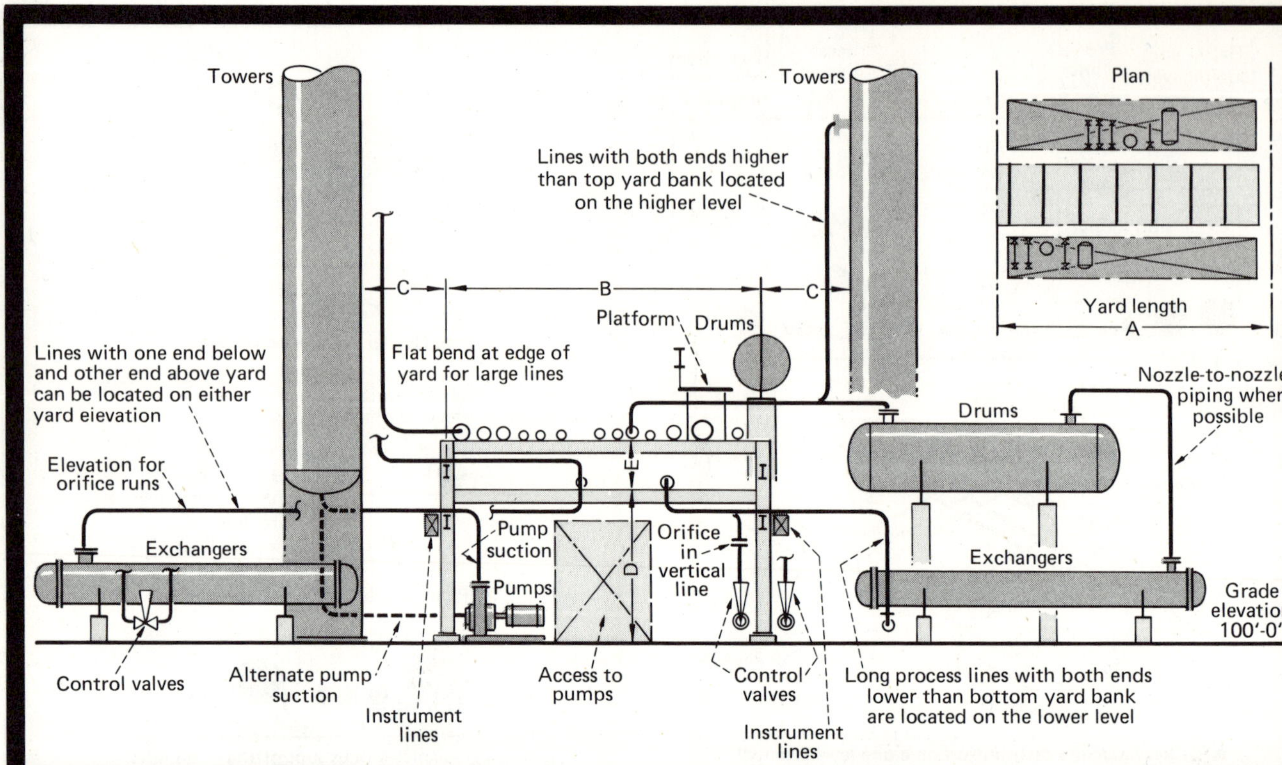

Piperack, piping and equipment relationships in a petrochemical plant **Fig. 4**

required sizes will minimize the length of pipe (a) between the yard and process equipment, and (b) for connecting equipment on opposite sides of the yard. Dimension C is usually 6 to 7 ft. Keeping Dimensions D and E at not more than the necessary yard height will minimize the length needed for vertical pipe runs.

A well-known rule in piping design states: Change elevation when changing direction. This applies to all lines connecting to yard piping. However, some large-diameter lines can make a flat turn when entering the yard. These lines should be placed at the edge of the yard. In any other location, such lines will require excessive space in the piperack (see Fig. 4).

Fig. 5a shows commonly used elevations for a piperack intersection. The 14-ft elevation of the lateral yard permits turning up or down at an intersection. It is important to elevate lateral pipebanks between two elevations of the main yard.

The difference in elevation between the main bank and laterally connecting pipelines is about 2 to $2\frac{1}{2}$ ft. This allows a difference of 4 to 5 ft between two main banks (Fig. 5a and 5b).

If a building (e.g., control house, pump house, etc.) is located under the main yard, piping elevations will be higher. Building clearances, roof pitches, steel structures and pipeline clearances will affect the height of the yard. Plant specifications determine elevations over roads and railways. Minimum clearances for cranes and mobile-handling facilities must be given.

The elevation difference is not required if a flat turn can be made within the yard (Fig. 5b). In this case, the line sequence must be identical before and after the turn. Varying line sequence in the two directions introduces an elevation difference, and an additional pipe elbow in each line.

Fig. 5c shows a yard-piping junction with an adjacent exchanger-supporting structure. Yard elevations are governed by headroom requirements over the access road. The top of the yard sets elevation of the first platform in the exchanger-supporting structure because lines must cross from below the first platform to the top bank of the yard.

All lines from the exchanger structure to the piperack drop along the structure. Space has been left open in the yard, adjacent to the exchanger structure, for lines that connect to the lower main yard or to pumps below the piperack. Process lines turning into the yard from the exchanger structure have been arranged at the highest yard elevation.

A number of vertical reflux drums have been arranged on the first level of the exchanger structure. All suction lines to pumps turn horizontally below the lower yard bank above headroom.

Length of piperack

The economy of yard piping primarily depends on the length of the piperack. Approaches to economical design are shown in Fig. 6. Process equipment arranged entirely on one side of a piperack will considerably increase its length (Fig. 6a). Process equipment placed on both sides of the piperack (Fig. 6b) will result in a piperack having half the length of the one in Fig. 6a. Facilities not closely associated with the piperack (such

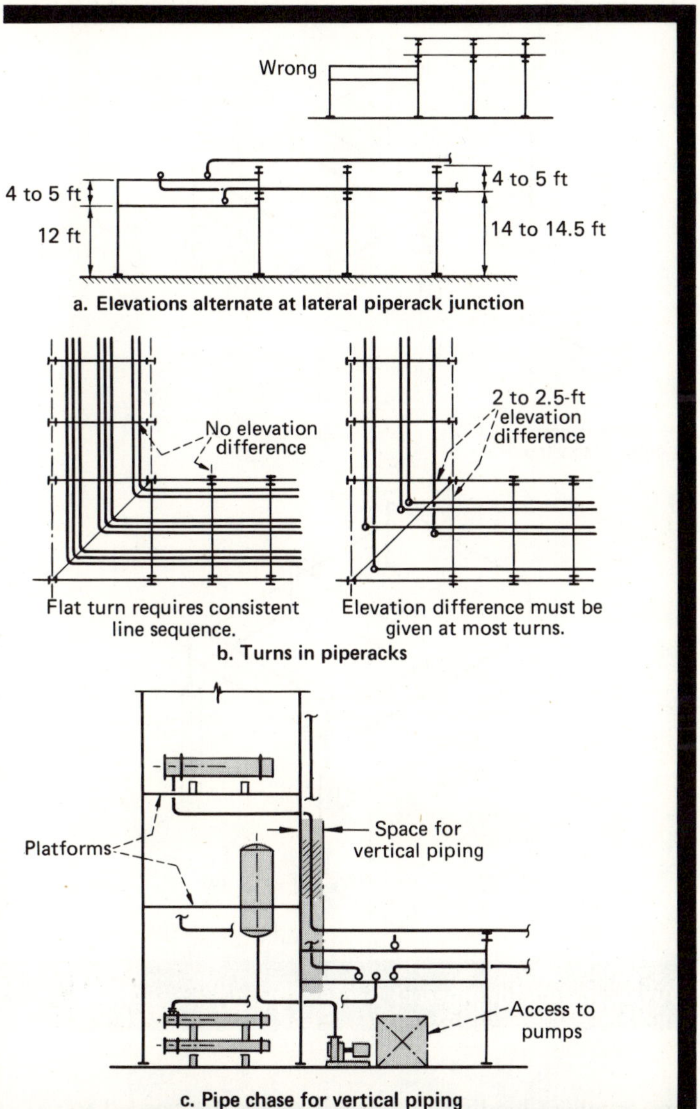

Change elevation when changing direction Fig. 5

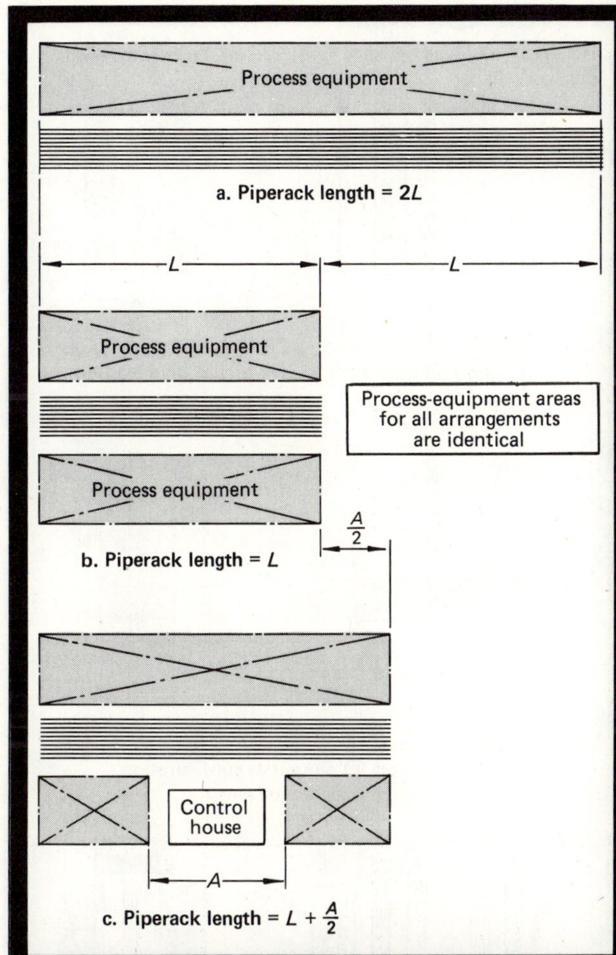

Rack and pipe lengths depend on arrangement Fig. 6

as control houses, treating buildings and storage areas) can waste valuable space along the piperack and increase its length (Fig. 6c).

Fig. 7 shows some actual examples of how a design engineer can minimize and simplify piperacks. The first example involves the future doubling of a tank farm. Of the two arrangements shown in Fig. 7a, the one on the right requires 50% less pipe length, and results in a delayed capital expenditure. A second example concerns the layout of a utility plant (Fig. 7b). Here, the preferred arrangement requires turning the unit 90° to bring it into alignment with the piperack, and thus shorten pipe length substantially. In the preferred arrangement, future piperack is available; while in the original, future piperack will require additional expenditure. The third example (Fig. 7c) relocates the control house. The alternate, preferred, combines the originally designed two piperacks into one.

The length of yard piping (Dimension A in Fig. 4) is governed by the number and size of equipment, structures and buildings arranged along both sides of the yard. On an average, about 10 ft of yard length are required for each piece of process equipment (exchanger, drum, tower, unhoused compressor, etc.) for petrochemical plants. With good layout practices, the cost of yard piping can be considerably reduced, by arranging the same number and size of equipment in a shorter length of yard. Under these conditions, a 7 to 8 ft average length of yard for each piece of process equipment is not unusual.

Equipment in pairs, stacked exchangers, exchangers under elevated runs, drums or exchangers supported on towers, two vessels combined into one, closely located towers with common platforms, drums supported on exchangers, drums supported on the piperack steel or concrete, and process equipment located under the piperack are only a few examples of what can help in shortening the yard. These arrangements, of course, shorten not only process lines connecting equipment but also the lines that pass through an area, and the utility headers serving an area.

Structural steel for piperack

Fig. 8 shows typical pipe-support bents. From a layout standpoint, cost depends on column spacing and span.

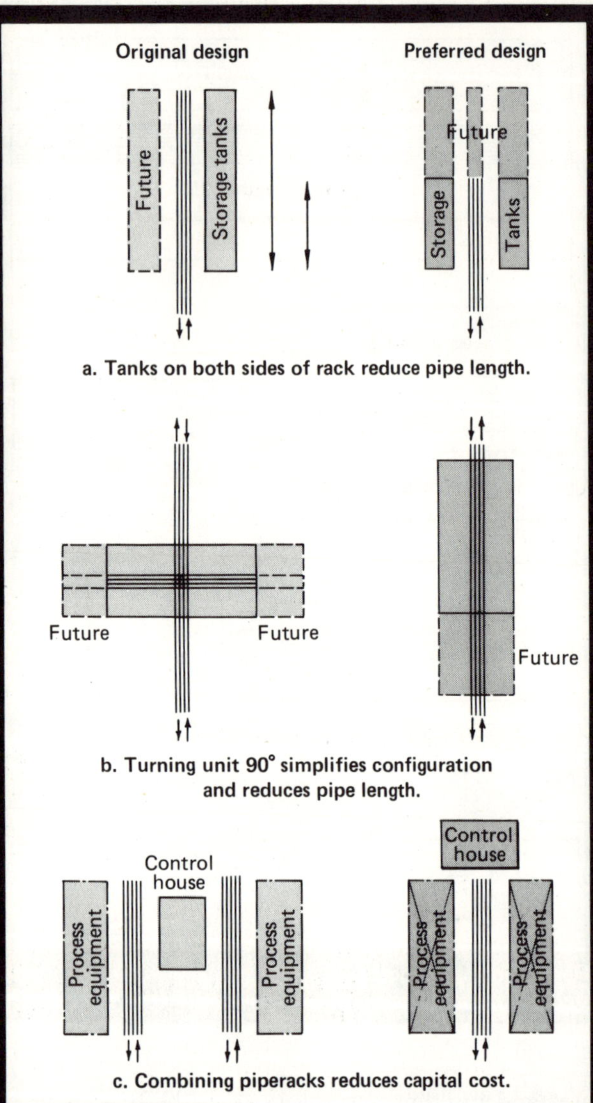

Examples of cost-reducing piperack layouts Fig. 7

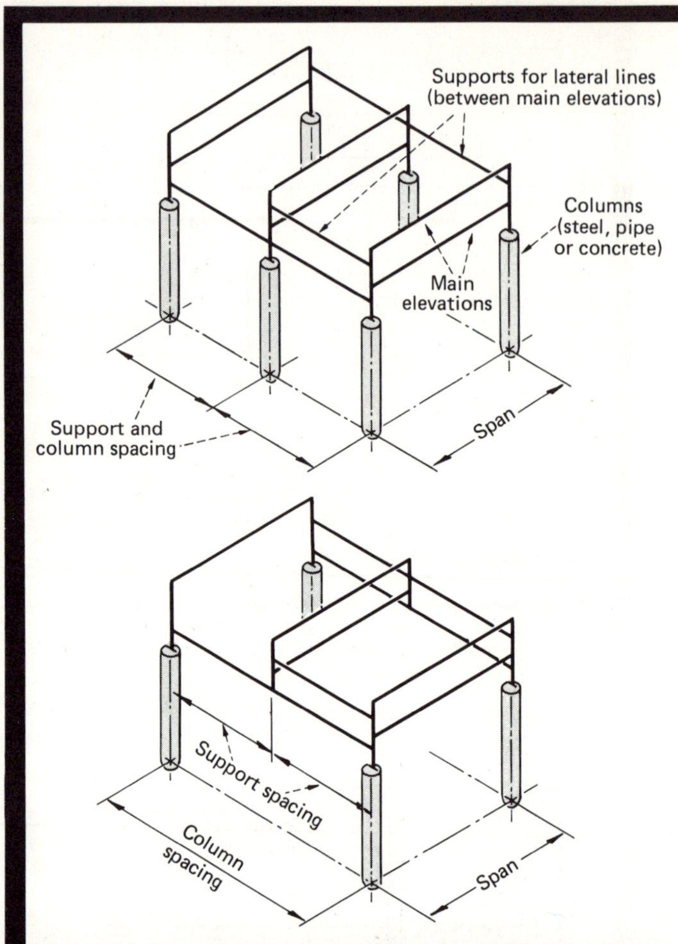

Arrangements for piperack bents Fig. 8

Spacing between bents for the pipe yard is about 16 to 24 ft. A few small-size pipes can be intermittently supported from larger lines, as shown in Fig. 3d. This can permit wider and more-economical bent spacing. In determining the bent spacing, consideration should be given to:

- Line sizes—Small-diameter lines must have support on shorter spacing than large-diameter ones.
- Span—Liquid-filled lines require shorter spans than gas-filled ones.
- Line temperature—Very-hot lines span shorter distances than cold ones of the same size and wall thickness.
- Insulation—Heavily insulated small-diameter lines operating under cold temperatures must be supported at relatively short intervals.
- Space requirement—Equipment at grade and equipment under the yard can sometimes influence the spacing between yard bents.

Wide spacing of yard steel is necessary at road crossings or where loading and access spaces are needed to process equipment.

The design manuals of contractors and owners usually contain tabulations of allowable spacing requirements for insulated, bare, liquid-filled and vapor-filled lines for three temperature ranges. The weights listed in these tabulations are for calculating pipe loads on yard steel and yard columns.

Width of the yard is influenced by the number of lines (including instrument and electrical lines), and space for future lines in the yard—or occasionally by space requirements for equipment under the yard.

The number of lines to be located in the yard can be estimated by routing the pipelines on a print of the plot plan, with the help of flow diagrams (such as shown in Fig. 1). By adding the number of lines in the densest section of the yard, we can obtain the total width of the yard from

$$W = fnS + A$$

where W = width of yard, ft; f = safety factor; n = number of lines up to 18 in. dia.; S = estimated average spacing between lines, ft; and A = additional required width, ft.

The safety factor, f, equals 1.5 if the lines have been laid out on the plot with the help of process-flow dia-

= means equal dimensions

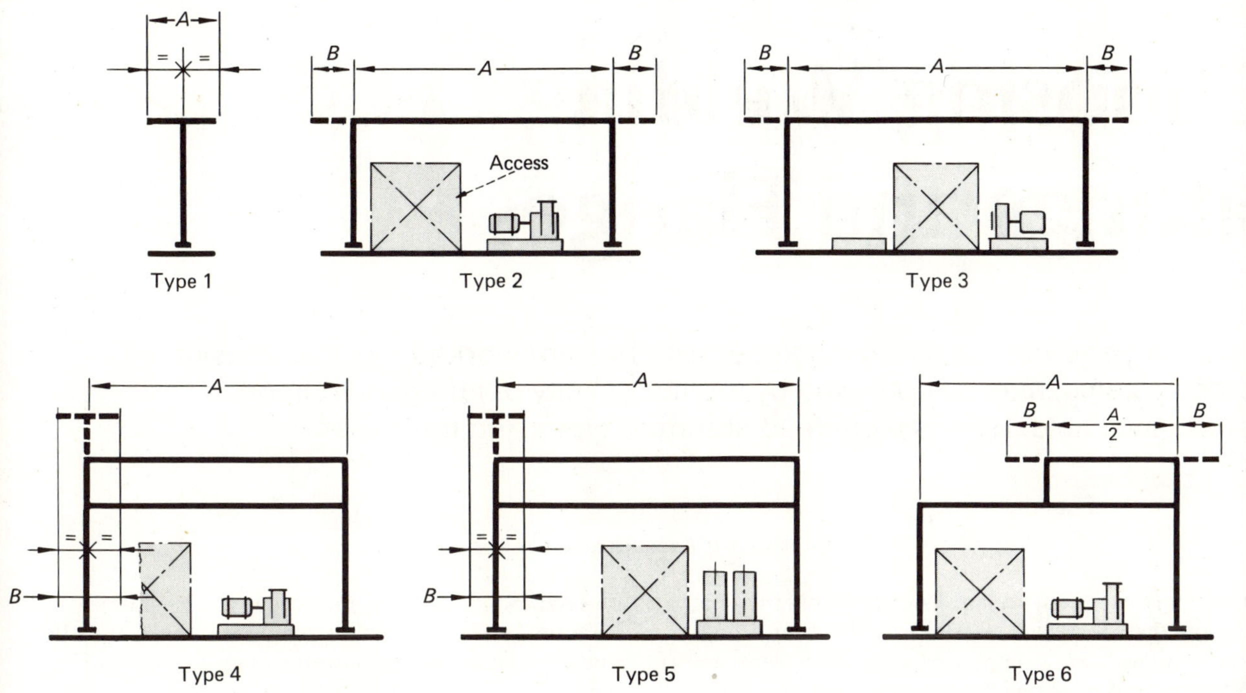

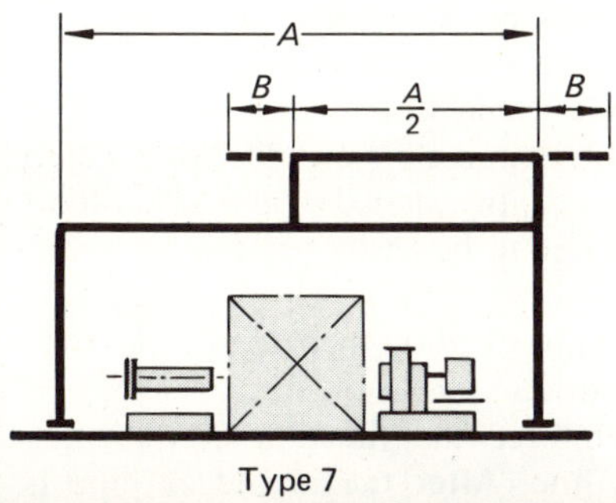

Total available width, W_t					
Without cantilever, ft	**With cantilever, ft**	**Dimension A, ft**	**Dimension B, ft**	**Elevations, No.**	**Type No.**
-	10	10	-	1	1
20-24	30-34	20-24	5	1	2
28-32	38-42	28-32	5	1	3
28-34	36-42	20-24	4	1½	6
39-47	45-53	20-24	6	2	4
40-44	48-52	28-32	4	1½	7
55-63	61-69	28-32	6	2	5

Dimensions for and types of piperacks **Fig. 9**

grams; $f = 1.2$ if the lines have been laid out with the help of fully detailed mechanical-flow diagrams.

Estimated spacing between lines, S, usually equals 1 ft; $S = 0.75$ ft if lines in the yard are smaller than 10 in.

Additional width, A, is required for (a) lines larger than 18 in., (b) future lines, (c) instrument lines (about 2 to 3 ft), (d) electrical cables (about 2 to 3 ft) if these are also supported on the yard steel, (e) one or two slots for pump-discharge and utility lines (about 1.5 or 3 ft).

Total width of yard, W, can be between 20 and 60 ft. If W is greater than 30 ft, two yard-banks will usually be required because the upper limit for the span of yard steel is 32 ft.

Space requirements for equipment plus access below the yard can also influence width of yard. For a single row of pumps with access aisles of 8 to 10 ft, a yard span of about 20 to 24 ft is required, depending on the length of the pumps. For a double row of pumps, a 28 to 32-ft span will be needed.

Fig. 9 shows typical yard-steel bents with dimensions. The total available yard width for each type of support is shown in the table. This tabulation can be used for selecting a type of yard support after the total required width has been estimated. The most commonly used yard supports are Types 2, 3, 4 and 5.

For economical design, an evaluation by a structural expert should be done. This expert can determine whether steel or concrete (or a combination of the two) is the most desirable choice, and what span and spacing will give the most economical arrangement.

Choosing Welding Fittings and Flanges

The young engineer, and any engineer who has not worked much on design of piping systems, may be confused by the multiplicity of fittings and flanges that are available. Here is a guide to standard types and their uses.

JOHN ASSINI, Babcock & Wilcox Co.

If you have ever designed an all-welded piping system, you know that the selection of exactly the right fittings can be a difficult and complex process. Unless you are actively engaged in piping design, you probably are not familiar with many of the fittings available. To help make the job of selection easier, this set of guidelines describes the various types of available fittings and flanges. and highlights applications for each type.

Available Welding Fittings

Long- and Short-Radius Elbows—The long-radius elbow offers minimum resistance to flow (without unduly sacrificing space) and is less expensive than the short-radius elbow.

Long-Tangent Elbows—These are special elbows that provide a straight length at the end to accommodate a slip-on flange.

Reducing Elbows—These are designed both to change direction and reduce pipe size; their use eliminates one fitting (a reducer) and lessens required welding by one-third. The gradual reduction in diameter through the arc of the reducing elbow provides maximum flow efficiency, reduces turbulence and erosion, and aids in maintaining pressure.

Returns—When a 180-degree change in direction is required, a 180-deg return fitting is recommended rather than two 90-deg elbows. Returns are easy to install and require only two welds (they are used mainly in heater coils and heat exchangers).

Tees—The straight tee is made with all three outlets of the same size—the "run" (the two outlets in line with each other) as well as the "branch" (the outlet at right angles to the other two). The reducing tee is manufactured with the branch outlet smaller than the run. thus giving the desired flow through the system.

Crosses—The four outlets on a straight cross are equal in dimension. The two branch outlets on a reducing cross are smaller than the run, for proper flow characteristics.

Reducers—These function to reduce a line to a smaller pipe size and to increase flow pressure. The eccentric reducer is manufactured with the reducing outlet off-center, allowing it to line up with one side of the inlet. The concentric design is less expensive and is made with both the inlet and outlet ends on a common center-line. (It can be used in most piping installations.)

Lap-Joint Stub Ends—Such fittings are used in lines that require a quick-disconnect system. Normally, stub ends are installed in pairs, two mated with two lap-joint flanges. This allows a line to be easily opened for cleaning or inspection.

Caps—A line may be blocked off by placing a cap over the end, and welding around the joint.

Saddles—These reinforce the junction of a 90-deg nipple or pipe to a line. After the nipple or pipe is welded on, the saddle is placed over the outlet and brought to rest on the header. A threaded plug is removed from the saddle before welding to permit the exit of gases formed during welding. Then the saddle is welded to the outlet and header, and the plug is replaced.

Shaped Nipples and Laterals—A shaped nipple is a piece of pipe with one end beveled and cut to shape to fit onto a matching or larger size pipe at either a 45 or 90-deg angle. Such fittings are inexpensive but can only be used in low-pressure systems and have little structural strength. Also, they do not meet the Pressure Vessel Code requirements unless they are reinforced.

Flanges

Welding-Neck Flange—Commonly referred to as the "high hub" flange, this is designed to transfer stresses to the pipe and is used in severe service applications involving high pressure, and subzero or elevated temperatures.

Slip-On Flange—This has a low hub, because the pipe slips into the flange prior to welding.

Originally published September 2, 1974

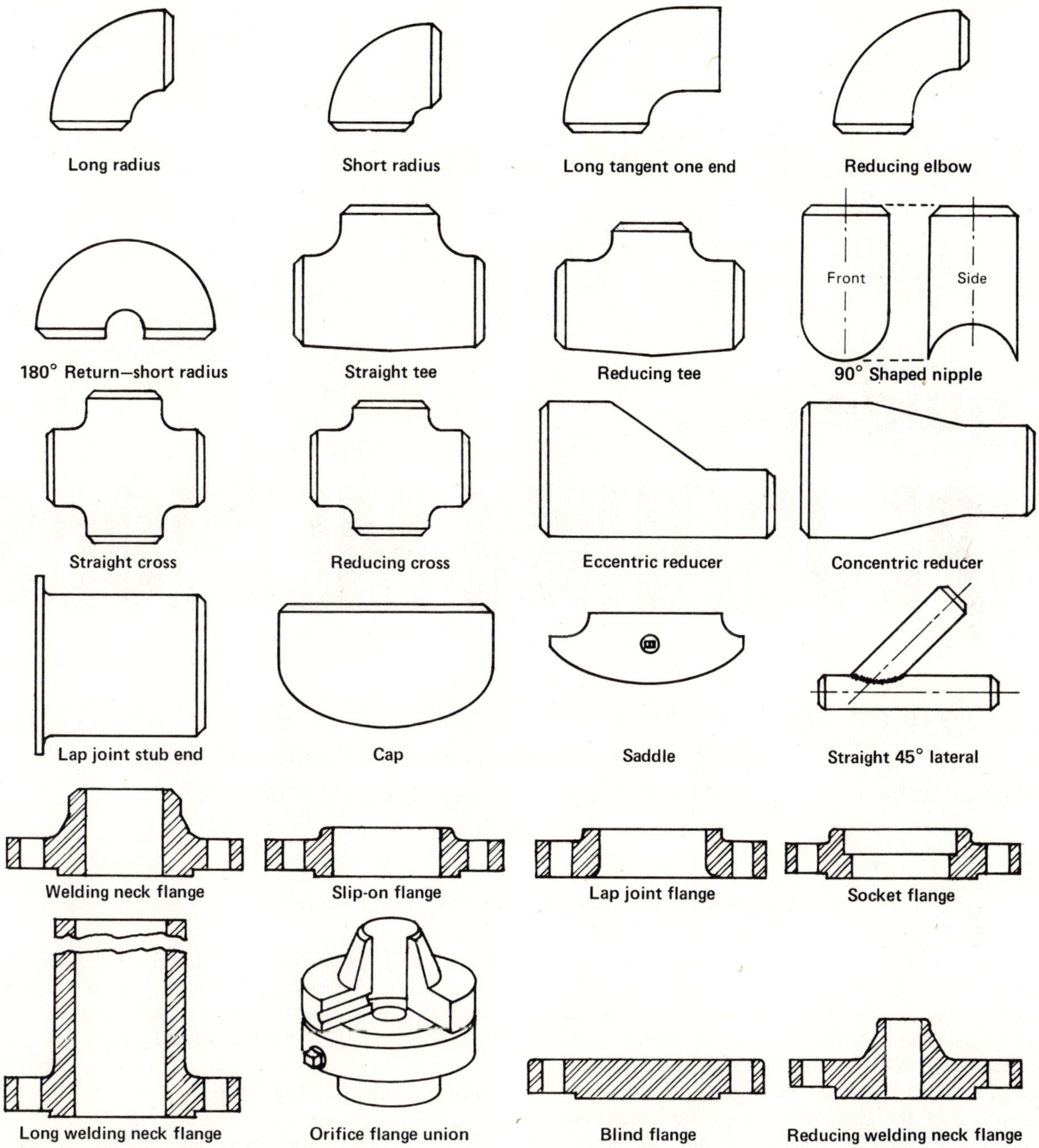

Lap-Joint Flange—This is practically identical to the slip-on flange except that it has a radius at the intersection of the bore and flange face. Normally, a lap-joint flange and a lap-joint stub end are mated together in an assembly system.

Socket Welding Flange—This type is similar to the slip-on flange, except that it has a bore and a counterbore dimension. Its applications include chemical-process, hydraulic, and steam-distribution lines.

Long Welding-Neck Flange—Sometimes referred to as a "nozzle," this is used as an outlet on pressure vessels and tanks. It has a long heavy wall, and a straight hub that is square-cut on the end.

Orifice Flange—Consisting of an orifice plate clamped between a pair of flanges, this is used to meter the flow of fluids through the pipeline.

Blind Flange—A blind flange has no bore. It is clamped to another flange to close off the end of a piping system or vessel opening.

Reducing Flange—This is used to reduce the line to a smaller pipe size, usually for increasing flow pressure through the system.

Meet the Author

John Assini is Manager of Fittings Sales for the Babcock & Wilcox Tubular Products Div., Beaver Falls, PA 15010. He is responsible for the nationwide marketing of welding fittings and flanges to the refining, process and distribution industries. He has worked in the manufacture of specialty steel tubular products since 1936.

A guide to...

PIPE JOINING METHODS

This survey covers current methods of joining metallic, nonmetallic, clad and lined pipe. It spotlights welding, soldering and fusion techniques, and the varieties of adhesives, fittings, flanges and couplings for connecting pipe.

MARX ISAACS and RICHARD B. SETTERLUND, Fluor Corp.

The threaded cast-iron fitting was for many years the standby for making mechanical pipe joints. It was superseded by the malleable iron fitting, which was stronger. The forged-steel screwed fitting, rated for as high as 6,000 psi. cold working pressure, was subsequently developed for the higher pressures and temperatures of the chemical process industries. For unusually high temperatures and pressures, the flanged cast-steel fitting came to be used with larger pipe sizes. Finally, after improvements in techniques, welding became the standard for making permanent, leakproof joints.

METALLIC PIPE

Welded joints now represent the ultimate in safety and reliability. In fact, where no leakage can be tolerated (such as in plants handling highly radioactive fluids) only welded joints are often specified.

All codes call for welding to be "qualified" by a procedure. Covered in the standard procedure are: base-metal specifications, filler metals, joint preparation, pipe position, welding processes, techniques, electrical details (current setting, electrode manipulation, etc.), preheating and interpass temperatures, and post-welding heat treatments. These requirements are listed in the ASME Boiler and Pressure Vessel Code, Section IX, "Welding Qualifications."

Joint Preparation and Design

The circumferential butt joint (Fig. 1) is the most common in welded-pipe systems. It is used in pipe-to-pipe, pipe-to-flange, pipe-to-valve, pipe-to-fittings, and even pipe-to-vessel-nozzle joints. Although all sizes of pipe can be butt joined, it is often more advantageous to join small pipe (up to 2 in. dia.) by means of fillet-welding, using socket-welding fittings.

The ends of heavier-walled pipe are usually prepared by machining or grinding. Carbon and low-alloy steel pipe of less than ½-in. wall thickness are also oxygen-cut and beveled. Basic ANSI (American National Standards Institute; formerly ASA, American Standards Assn.) welding bevels for all components (without backing ring, or with split ring) are shown in Fig. 2.

If the mismatching of the inside diameter of a

Originally published May 3, 1971

pipe being joined by welding exceeds the allowance of the code, the pipe wall must be machined or ground into the tolerance. (This can be costly if done in the field.) If, on the other hand, the machining diminishes the wall thickness too close to the minimum, the pipe wall must be built up with weld metal.

When pipes of unequal wall thickness are to be welded, the thicker wall should be smoothly tapered to its end. The length of the taper normally is four times the offset.

Fillet-welded joints require good penetration to the inside of the pipe because this type of weld is subjected to shearing and bending stresses. This is particularly important with the socket joint because the usual visual inspection often fails to reveal an unsound weld. Socket welds are not acceptable in piping systems exposed to nuclear or concentration-cell action. In such systems, butt welds are generally required for all pipe sizes, and the weld penetration must be complete.

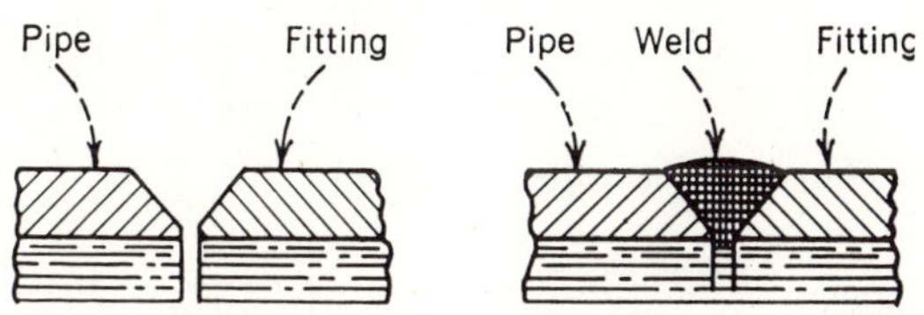

BUTT-WELD JOINT is common for all pipe sizes—Fig. 1

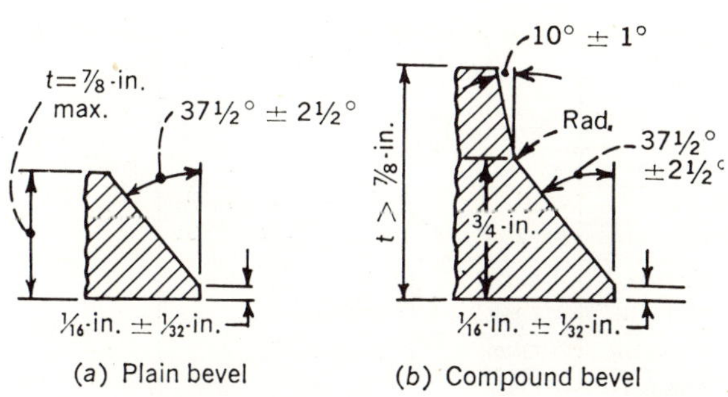

Nominal pipe wall thickness (t), in.	End preparation
Less than x*	Cut square or slightly chamfer, at manufacturer's option
x* to 7/8-in. included	Plain bevel as in (a) above
More than 7/8-in.	Compound bevel as in (b) above

x*=3/16-in. for carbon steel, ferritic alloy steel, or wrought iron; 1/8-in. for austenitic alloy steel.

BASIC BEVELS of ANSI Code for Pressure Piping—Fig. 2

Intersection Joints

Intersection joints (such as medium and large welded tees, laterals, wyes, and mitered elbows) are the most difficult to weld soundly. Chromium-molybdenum-steel, wrought-carbon-steel and carbon-molybdenum-steel fittings that conform to ASTM Specification A234 (Factory-Made Wrought Carbon-Steel and Ferritic Alloy Steel Welding Fittings) are preferred for assembling piping because their bursting strengths are equivalent to that of pipe of the same weight or schedule, and their installation requires only butt-welding. However, because economics often prohibit the use of tees, the "stub in" is generally specified.

Fig. 3 shows two methods of preparing pipes for joining at an angle of 90 deg. Because the opening in the lower header is large enough for the branch to be inserted, only the header opening requires beveling. In the upper header, the opening equals the inside diameter of the branch, and the latter only is beveled. This type of connection may have to be reinforced.

Pipes intersecting at an angle of 45 deg. or less are difficult to join with the penetration and soundness of weld deposit required for severe service conditions. Such intersection welds are not recommended for pressures above 125 psi. When possible, joint intersections, regardless of angle, should be welded in a shop because the pieces can be positioned flat.

Pipe joining has been simplified somewhat by the availability of factory-made welding nozzles, necks,

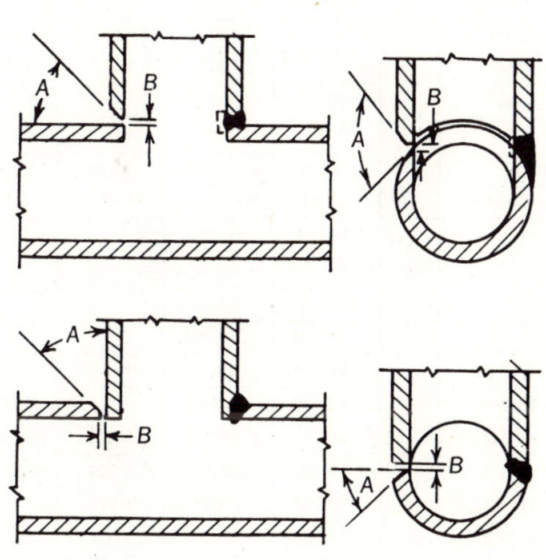

INTERSECTION JOINT of 90 deg. requires angle **A** be at least 45 deg., dimension **B** between 1/16-1/4 in.—Fig. 3

flanges and tees. The cutting, tacking and fitting required for welding branch connections have been considerably eliminated by the availability of extruded outlets having butt joints. However, some such fittings, especially laterals, should be selected with caution because they may not be sufficiently reinforced for full design pressure.

Selecting Welding Materials

Welding filler-metal is not simply a wire or coated electrode having the same properties as the metal being joined, but a product designed to provide a weld deposit equivalent in strength and corrosion resistance to the metal being joined. Additions or deletions of alloy are made to compensate for the interaction with the arc atmosphere and base-metal dilution and to prevent cracking.

Tables I and II include many welding materials used in the fabrication of ferrous and nonferrous piping. The P numbers in the ASME Code, Section IX, group together metals of similar welding characteristics. Since 1969, welding materials have been classified by the American Welding Soc. (AWS) rather than listed as ASTM standards. Unfortunately, the AWS standards (except for those on coated ferrous electrodes) are not followed as extensively as they should be.

Three welding-product forms are listed: coated electrodes, solid-wire filler metal and the newer composite electrodes; in the last, granular materials are encased in a metallic sheath. Coated electrodes provide their own shielding, but solid wires must be furnished with an external shielding gas (or flux in the case of submerged-arc welding). Composite wires, on the other hand, are designed specifically for welding with or without external gas shielding, although some may perform satisfactorily either way. Applications for composite wires, although still classified by AWS for carbon steel only, are increasing, with a self-shielding stainless wire the newest development.

Welding Materials and Heat Treatments for Ferrous Piping——Table I

Base Material	P Group No.	ASTM Specifications: Seamless Pipe	Plate	Flanges	Welding Fittings	Castings	AWS Specifications: Coated Electrodes (SMA)	Bare Wire (TIG, GMA)	Composite Wires (GMA)	Pre-heat Temperature, °F.	Pre-heat Treatment, °F.
Carbon Steel	1	A-53	A-285	A-181			[AWS A 5.1] E6010	[AWS A 5.18]	[AWS A 5.20]		
			A-515		A-234		E6011	E70S-2[1,2]	E70T-1[2]		
		A-106	A-516	A-105	WPA	A-216	E7020	E70S-3[1,2]	E70T-5[2,3]	200	1,100-1,200 F
		API-5L	A-283		or	WCB	E7018	E70S-5[2]	E70T-3[3]	(over ¼ in. thick)	(over ¾ in.)
			A-7		WPB		E7028				
Carbon Steel (for −50 F. service)		A-333 GR."1"	A-300 CL1	A-350 CL. LF1	A-420 WPL1	A-352 LCB	E7018	E70S-2[1,2]	E70T[2]		
Wrought iron	2	A-72	A-42				E6010	–	–	–	None
Carbon molybdenum	3	A-335-P1	A-204		A-234	A-217	E7018-A1	E70S-1B[1,2]	Available[2]	175	1,150-1,350 F
½ Cr – ½ Mo		A-335-P2	A-387A		WP1	WC1	[AWS A 5.5] E8018-B1	[No AWS Stds.] –	–		(over ½ in.)
1 Cr – ½ Mo	4	Z-335-P12	A-387B	[A-182] F12	A-234-WP12	A-217	–	–		300	1,325-1,375 F
1¼ Cr – ½ Mo		A-335-P11	A-387C	F11	A-234-WP11	WC6	E8018-B2	ER515[1,2]	Available[2]		
2¼ Cr – 1 Mo	5	[A-335] P22	A-387D	F22	A-234-WP22	A-217-WC9	E9018-B3	ER521[1,2]	Available[2]	350	1,325-1,375 F
5 Cr – ½ Mo		P5	A-357	F5	A-234-WP5	A-217-C5	[AWS A 5.4] E502-15/16	[AWS A 5.9] ER502[1]	Available[2]		
7 Cr – ½ Mo		P7		F7			E7CR-15/16	ER7CR[1]	Available[2]		
9 Cr – 1 Mo		P9		F9		A-217-C12	E505-15/16	ER505[1]	Available[2]		
TP. 410	6	A-268	[A-240] GR. 410	A-182	A-182	A-351	E410-15/16	ER410[1]	Available	400	1,400-1,450 F
TP. 430 modified		TP. 410	GR. 430A	F6	F6	CA15	E430-15/16	ER430[1]	[No AWS Standards]		
TP. 405	7	[A-268] TP. 405	[A-240] GR. 405				E309-15/16	ER309[1]		100	None
TP. 410S		TP. 410S	GR. 410S				E309-15/16	ER309[1]			
TP. 430		TP. 430	GR. 430B				E309-15/16	ER309[1]			
Stainless TP. 304	8	[A-312] TP. 304	[A-240] GR. 304	[A-182] F-304	[A-403] WP-304	[A-351] CF8	E308-15/16	ER308[1]	Available[1]	None	None
TP. 304L		TP. 304L	GR. 304L	F-304L	WP. 304L	CF3	E308L-15/16	ER308L[1]	Available[1]		
TP. 309		TP. 309	GR. 309		WP. 309	CH20	E309-15/16	ER309[1]			
TP. 310		TP. 310	GR. 310	F-310	WP. 310	CK20	E310-15/16	ER310[1]			
TP. 316		TP. 316	GR. 316	F-316	WP. 316	CF8M	E316-15/16	ER316[1]	Available[1]		
TP. 316L		TP. 316L	GR. 316L	F-316L	WP. 316L	CF3M	E316L-15/16	ER316L[1]	Available[1]		
TP. 321		TP. 321	GR. 321	F-321	WP. 321		E347-15/16	ER347[1]			
TP. 347		TP. 347	GR. 347	F-347	WP. 347	CF8C	E347-15/16	ER347[1]	Available[1]		
3½ nickel	9	A-333 GR.3	A-203 GR.D	A-350 LF3	A-420WPL3	A-352-LC3	A5.5 E8018-C2	Available[1]	Available[2]	200	1,100-1,150 F
9 nickel	10	A-333 GR.8	A-353				A5.11ENiCrFe-3	A5.14 ERNiCr-3	–	200	None

[1]Used primarily with inert gases. [2]Used with CO_2 or CO_2 containing mixtures. [3]Used without shielding.

Welding Processes

Manual, shielded, metal-arc (SMA) welding with coated electrodes is still the most accepted method of welding, although the automatic submerged-arc and gas-metal-arc (GMA) processes may now be more common in fabrication shops. Tungsten-inert-gas (TIG) and metal-inert-gas (MIG) welding are the only methods for joining titanium and aluminum. Both also have numerous other applications. Oxyacetylene welding continues to be used, although it is seldom seen in major shops or at field-fabrication sites.

Manual, shielded, metal-arc welding is applied to almost all metals used in piping systems. The equipment is comparatively simple, compact, portable and safe. The welding machine is a.c. or d.c. powered, but can also be driven by a gasoline or diesel engine.

Most welding is done with electrodes that correspond to AWS A5.1 (Table I); of these, the most popular is the versatile E6010 electrode, which, because of its excellent penetration and fluid characteristics, makes a sound root or stringer pass. The low-hydrogen electrode, such as the E7015, E7016 or E7018, is good for low-temperature service or for preventing "underbead" cracking in some materials. Even when a low-hydrogen electrode is necessary, the E6010 may be used for the root-pass welding.

However, the E6010 is not the proper choice when a.c. power is used, or the pipe can be rotated, or a smooth appearance is desired, as when welding very-light-gage material, or in fillet welding; in the latter case, titania-coated electrodes, such as the E6012 or E6013, are recommended.

Nonferrous electrodes, and electrodes that correspond to AWS A5.4, are most often chosen for making fill passes over one or more tungsten-inert-gas stringer passes—because of the difficulty of making a satisfactory root pass with these electrodes, and because the alloys of these electrodes are most often used in more-critical applications that call for a smooth root-pass to reduce stress concentration.

To keep down cost, the largest electrode consistent with sound welding practice is generally chosen. An important exception is in low-temperature applications, such as on P-9 or P-10 materials, because small stringer beads are required in order to meet the specified impact values. When welding these materials, the fabricator should hold to the heat input specified in the qualification procedure for verifying impact strength. Slag entrapment, the most common weld defect, also becomes more prevalent with larger electrodes.

Tungsten-Inert-Gas Welding

Also known as "heliarc," tungsten-inert-gas welding is used in many forms and on all metals that can be welded. It derives its name from the fact that an electric arc is drawn between the work and a nonconsumable tungsten electrode shielded with an inert gas. It is used both with and without a filler metal. Examples of welding without filler metal are with light-gage 5S or 10S stainless steel piping whose edges are cut flat and abutted, and in the fusion of consumable insert rings—even in these cases, filler metal must often be added manually into the arc.

This welding method is most often chosen for root-pass welding of alloy, or otherwise critical, piping. The two most common methods are the "open butt" techniques and another in which consumable inserts of various designs are used. Each

Welding Materials for Nonferrous Piping——Table II

Base Material	P Group No.	ASTM Specifications: Seamless Pipe	ASTM Specifications: Plate	ASTM Specifications: Welding Fittings	AWS Specifications: Coated Electrodes (SMA)	AWS Specifications: Bare Wire* (TIG, GMA)
Pure Aluminum AA1100 ASTM 990A	21		B-209	B-361 WP3003	Not used	ER1100
AA3003 ASTM MIA		B-241				ER4043
Aluminum with Mg AA5154 ASTM Gr. 40A	22	B-235	B-209	B-361 WP5154		AWS A5.10 ER5154
Aluminum with silicon AA6061 ASTM G511A	23	B-235	B-209	B-361 WP6061		ER4043
Aluminum with high 51 AA5083 ASTM GM41A	24		B-209	B-361 WP5083		ER5183
70-30 copper-nickel	34	B-111	B-171		A 5.6 ECuNi	A 5.6 ERCuNi
Nickel	41	B-161	B-162	B-366 WPN	ENi-3	ERNi-3
Monel	42	B-165	B-127	B-366 WPNC	ENiCu-7	ERNiCu-7
Inconel	43	B-167	B-168	B-366 WPNC1	AWS A5.11 ENiCrFe-3	AWS A5.14 ERNiCrFe-5
Hastelloy B	44	Case 1323	B-333	B-366	ENiMo-1	ERNiMo-4
Hastelloy C		Case 1324	B-334	B-366	ENiMo-2	ERNiMo-5
Incoloy	45	B-407	B-409	B-366	ENiCrFe-2	ERNiCr-3
Titanum	–	B-337 GR-2	B-265 GR-2	B-363 GR-2	Not used	ERTi-2

*With inert gas shielding.

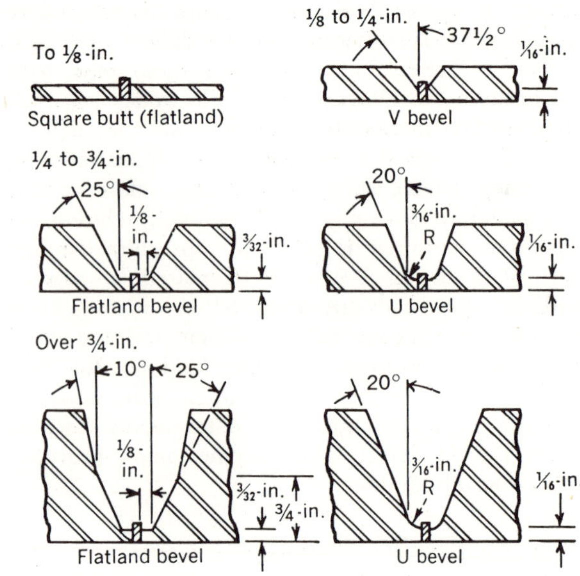

FLATLAND, U BEVEL are best for TIG welding—Fig. 4

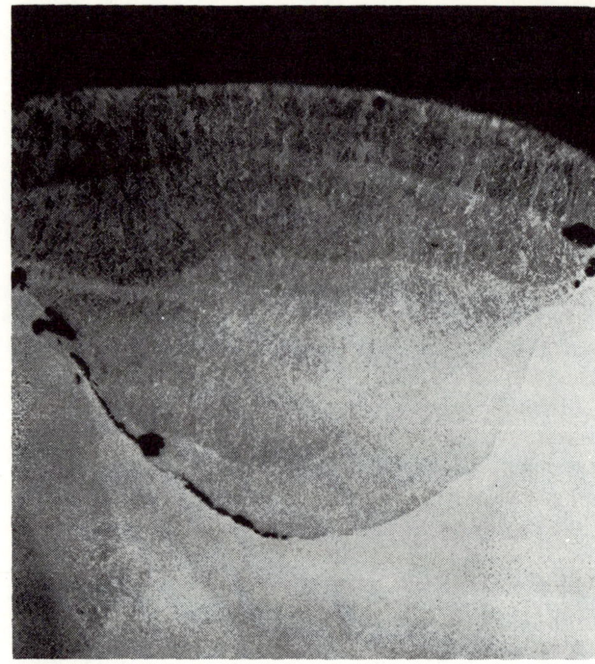

COLD LAP is the result of incomplete fusion—Fig. 5

has its merits and, when done by a skilled welder, both can meet the most exacting standards. The open-butt method, in which the filler metal is added manually, demands more skill but is nevertheless the method most favored in the petroleum refining and petrochemical industries. Joints are either the standard ANSI bevel, the U bevel, or the extended-land bevel (Fig. 4).

Alloys of P-5 and higher groupings require protection from oxidation during welding, which is often accomplished by purging the inside of the pipe with argon or nitrogen.

Clean surfaces are important in tungsten-inert-gas welding because the process does not include a flux to float off impurities. When aluminum or titanium is welded, the surfaces and filler metal often must be treated chemically to remove oxides.

Gas-Metal-Arc Welding

In gas-metal-arc welding, filler metal is fed continuously into the gas-shielded arc. At one time, argon or argon-helium was the shielding gas. The most important application of this process was the welding of aluminum.

Earlier attempts to apply the process to the welding of steels were not very successful because of the high cost of the shielding gas. However, with carbon dioxide or mixtures of carbon dioxide and other gases, and improvement in the electrical system, the process has been increasingly used in recent years for fabricating pipe.

Gas-metal-arc welding may be used for all-position root- and cover-pass welding, up to a maximum current of about 250 amp. With higher currents, welding is restricted to the downhand position, in which case the wires are often composites.

A typical procedure for horizontal, fixed, carbon-steel welding calls for a fairly wide (3/32 in.) weld gap with a 0.035-in. E60S-2 filler metal, and carbon dioxide shielding. The root pass is started from the top of the pipe and run downward, with subsequent passes made upward from the bottom.

Most problems with this process arise from poor training of the welder and incorrect selection of AWS filler wire. A serious problem has been "cold lap" (Fig. 5). Because of earlier difficulties of this type, some major oil companies have placed restrictions on gas-metal-arc welding.

Submerged-Arc Welding

Automatic submerged-arc welding is widely used in shops for joining carbon steel, chromium-molybdenum steel, and stainless steel piping. It is generally suitable for the circumferential welding of 6-in.-dia. and larger pipe whose geometry allows rotation. It requires a relatively large capital investment because the equipment includes 400 to 1,200-amp. d.c. generators or a.c. transformers; work positioners; and facilities for supporting and positioning the welding head. But, once the equipment has been installed, submerged-arc welding is the cheapest process.

Backing rings are sometimes used with groove welds. If the fitting-up is poor, the first weld pass on the backing ring should be made by manual, shielded metal-arc or inert-gas tungsten-arc welding. Sometimes it is necessary to deposit one or more shielded metal-arc weld passes over an inert-gas tungsten-arc root pass, before submerged-arc welding is started, to prevent burning-through by the submerged arc. These steps must be carefully performed because this process allows little, if any, flexibility to compensate for irregularities in the preparation or fitting-up of joints.

The interpass temperature tends to increase when heavy-walled pipe is butt-welded. Intermittent welding, or air or water cooling, may be needed on stainless steel piping to prevent an excessive rise in interpass temperature.

All submerged-arc fluxes contain some alloying ingredients; however, for greatest uniformity, the major alloying elements should be in the wire. Excessive alloy pickup in the weld may be caused by welding variables when highly alloyed submerged-arc-welding fluxes are used.

Oxyacetylene Welding

Widely employed about 25 years ago, oxyacetylene welding is now limited to small-sized steel piping. Consequently, qualified oxyacetylene welders for erecting field piping are often difficult to find. Because the equipment required is less expensive and more portable than that for shielded metal-arc welding, oxyacetylene welding is an inexpensive means of joining pipe when arc welding is not possible because of lack of electrical power, or when the cost of electrically powered equipment would be prohibitive. Weld metal is deposited as in manual shielded metal-arc welding for the various pipe positions. Joint design and the cleaning procedure are also similar. However, the layers of deposited weld metal are usually thicker.

Brazing and Soft Soldering

Brazing differs from welding in that the parent material is not melted.

When the filler is not placed beforehand into a groove machined into the end of the fitting, brazing is done by hand feeding the filler alloy in the form of a wire or strip.

Copper-silver-phosphorus or copper-phosphorus alloys are generally used with bronze or brass fittings. The brazing alloy should be of the silver-copper-cadmium-zinc type for malleable fittings with ferrous piping, for which brass filler metal (40% Zn-60% Cu) of the type regularly used in braze welding is also applicable.

The brazing area of the joint must first be thoroughly cleaned with steel wool, emery cloth, or a wire brush. Next, brazing flux is applied to both the fitting and the pipe. Afterwards, the assembled pipe and fitting is heated with an oxyacetylene flame or any other heat source capable of melting and flowing the brazing alloy.

Stainless Steel Joined With Epoxy Cements

For several years, epoxy cements have been used for joining light-walled stainless steel piping. Joints of all sizes can be made with regular bell-end or long-tangented socket fittings. Such joints are limited by temperature and vibration restrictions. Because of the speed at which these connections can be made, considerable savings are possible.

Welding Ferrous Materials

Table I lists common welding materials and thermal treatments for ferrous piping. In most cases, welding materials are designed so that their major chemical elements match those of the base material. However, there are several exceptions: (1) ferritic stainless steels corresponding to the P-7 group are most often welded with high-alloy stainless or Inconel-type electrodes, but this piping material is seldom used; (2) titanium-stabilized Type 321 stainless steel is welded with columbium-stabilized filler metal, Type 347, because titanium cannot be transferred through the welding arc; and (3) Inconel-type filler metal must be used when welding 9 nickel steel according to ASME Code Case 1308 in order to meet the required impact tests at temperatures down to −320 F.

Preheating is required for materials in the P-3 through P-6 groups because they all air-harden to varying degrees—and preheating, by tending to slow down the cooling rate, creates a microstructure less prone to cracking. The preheating temperatures in Table I are from the ANSI code B31.3 for pressure piping. The power piping code, as well as many company specifications, call for higher preheating temperatures. Preheating carbon steel is mandatory only for boiler-code piping, but for practical reasons it should be performed for all piping fabrication. Cracking or excessive hardness may result when the higher carbon grades of P-1 materials (Table I) are joined into heavy sections without first having been preheated.

Post-weld heat-treatment requirements become progressively more stringent as the material numbers go from 1 to 6 (Table I). Temperature requirements may vary because of differing codes and specifications. The treatment may also be applied for process considerations: for example, carbon steels are often stress relieved to prevent cracking by hot caustic; and austenitic stainless steels may be stress relieved (usually at between 1,550 and 1,650 F.) as an additional safeguard against stress-corrosion cracking. This heat treatment should only be applied to stabilized or low-carbon-grade steels, otherwise the corrosion resistance of the weld will be impaired by carbide precipitation at grain boundaries.

Post-weld heat treatment is performed at construction sites with exothermic kits, or by induction or resistance methods. The once common practice of stress relieving with direct-firing gas ring-burners is seldom used today for code piping because the process is difficult to control and flame impingement is a problem.

A mixture of aluminum powder, metal oxides,

refractories and binders reacts inside insulating sleeves when the treatment is done with exothermic materials. The exothermic and insulating materials are packaged into kits tailored for a particular material and joint geometry. These kits are usually made up directly from piping drawings for a particular job by the manufacturer. Kits of approximately 15,000 different designs are currently available. These kits should not be used with P-9, 3½ Ni materials, when precise control is needed for meeting the impact strength required at low temperature.

Exothermic stress relieving may be used only on code piping covered by ANSI 31.3 because the petroleum piping code allows evaluation based on hardness readings. Piping covered by the boiler code must be held at a specific temperature for a definite length of time, which is why electrical heating methods are necessary.

Electrical-resistance stress relieving has been developed considerably over the past decade. Reusable kits with sintered-alumina-encased electrical elements and ceramic-fiber insulation that can be clamped onto weld joints are now available. Because of the time this method saves, electrical-resistance post-weld heat treatment has been increasingly used on field piping.

Induction stress relieving is less used than electrical resistance or exothermic methods in refineries because of the cost and large size of the equipment. Induction stress relieving develops heat within the metal by means of magnetic fields set up by induction coils wrapped around the weld joint. This method is extensively used on conventional and nuclear-power piping. As with electrical resistance, the temperature can be closely controlled.

The Petroleum Refinery Piping Code, ANSI B31.3, allows an important exclusion in the post-weld heat treatment of carbon-molybdenum and chromium-molybdenum welds: i.e., the frequently used socket and seal welds—if made with electrodes that do not harden in air (such as stainless steel, E309; Inconel type, ENiCrFe; or low-carbon matching varieties, E8018-B2L)—on P-3 and P-4 or higher materials do not require post-weld treatment.

Because the zone heated by the welding will be hard regardless of the filler metal used on P-4 through P-6 materials, it is prudent to heat-treat when hardened metal is likely to be sensitive to its environment. Most companies require post-weld heat treatment of socket welds having a chromium content of 5% or higher.

Welding Nonferrous Metals

The nonferrous alloys in Table II do not require post-weld heat treatment.

Aluminum alloys are welded exclusively by the gas-tungsten-arc and metal-inert-gas processes. The most common materials for piping are pure aluminum, P-21, and the higher strength P-23 silicon alloys. The P-21 material must be welded with pure wire that matches the base metal's corrosion-resistance properties. Selecting wires for the more common P-23 materials is much less restricted.

Although aluminum presents only minor problems in piping fabrication, when radiographic quality is needed, a higher degree of cleanliness of base metal, wire and equipment is desirable, along with good gas shielding. In some shop practices, a stainless steel snap-ring is placed under the weld bevel; after the weld is complete, the snap-ring is pulled out the end of the pipe.

Copper and copper alloys, most often in tubing form, are brazed rather than welded. However, red and yellow brass can be satisfactorily oxyacetylene-welded in all positions, and red brass can be welded by the tungsten-arc inert-gas process. Copper-nickel alloys, P-34, may be welded with coated electrodes, as well as by inert-gas processes.

The nickel and nickel-base alloys in Table II can be welded by nearly all the processes already mentioned. It is important that all welding and heating be done on material completely free of foreign material. These alloys are susceptible to embrittlement by lead, sulfur, phosphorus, and some low-melting-point metals that may be present in lubricants, paints, marking crayons or inks, shop dirt or processing chemicals.

The Hastelloy alloys, and P-44 in particular, must be welded with small stringer beads at a low interpass temperature. This is necessary both to preserve the corrosion resistance of the materials and to prevent cracking. Because of the latter problem, these alloys are most often welded in a qualified shop, and field joining is avoided if possible.

When solution-annealed Incoloy, P-45, is to be joined for high-temperature service, additional precautions must be observed. First, the weld selected should match the creep strength of the base material (by choosing ENiCrFe-2 coated electrodes rather than the newer ENiCrFe-3 variety). Second, all weld slag must be removed, because traces of it have been shown to promote corrosion at high temperatures, particularly in a reducing or sulfurous environment. Sand- or grit-blasting is most often specified.

Most titanium in piping systems is commercially pure Grade 2 variety. This material is now welded routinely in qualified shops. Titanium, when heated over 1,000 F., becomes embrittled if exposed to oxygen, nitrogen, hydrogen and other gases. It is important, therefore, that the weld metal and heated zones be well shielded with inert gas and that all materials be meticulously clean. Contaminated metal may be inspected by observation of discoloration or by hardness measurement.

Welding Dissimilar Metals

When welding metals of dissimilar composition, the filler metal must be compatible with both metals being joined.

One of the more common combinations is between austenitic stainless steels, P-8, and carbon or low-alloy steels, which are conveniently welded by either highly alloyed stainless steel, E309, or nickel-base, ENiCrFe, filler metals. Problems are likely to arise with the normal 18-8 electrodes because ferritic metal dilution causes the filler metal to become hardened and crack-prone.

Welds between different grades of low-alloy steels, or between carbon and low-alloy steels, are also common. Because of the similarities of these materials, the choice of filler metal is not critical; and some fabricators match the properties of the lower alloy, and others those of the higher. In either case, the preheating and post-weld treatment requirements of the higher alloy should be followed.

Some metals differ so greatly that joining them by welding is not practical. For example, in cryogenic systems that necessitate joining aluminum and stainless steels, either mechanical joints or bimetallic transition sections are employed.

Bimetallic Piping

Internally clad piping represents a form of dissimilar-metal welding. This type of piping is sometimes used to conserve expensive corrosion resistant materials or to improve reliability, and sometimes for thermal conductivity characteristics.

Bimetallic piping is of two general types: that made from clad plate with a specific cladding thickness, usually in the range of 3/16 to 3/8-in.; and piping with a thin electroplated or diffused metallic lining.

Internally clad piping generally has some form of stainless steel in the inner layer, with a carbon steel or low-alloy steel for backing. Such piping is seldom of a diameter large enough to allow welding from the inside, so it is almost always done from the outside. The root and initial passes, when backing-metal dilution is not likely, are always made with filler metal that matches the cladding. When environmental conditions permit, the weld is completed with a metal that is compatible with the initial stainless steel deposits and the carbon steel or low-alloy steel backing. The fill passes can be made with either high-alloy stainless steel or Inconel-type deposits; Inconel offers the advantage of more closely matching the thermal expansion characteristics of the base material. When an exact match is desired, it is necessary to put down a "butter" pass of Inconel prior to filling the joint with carbon or low-alloy weld metal. With this method, the first carbon or low-alloy fill pass will harden because of alloy pickup. However, this is not a significant drawback in many applications, particularly in heavy-wall-pipe fabrication. A typical joint is shown in Fig. 6.

Aluminized piping, which has been successfully welded by means of special techniques, has an internal aluminum-rich zone, 0.010 to 0.015 in. thick, which is applied by either a dipping or gaseous-diffusion process. Welding is sometimes done simply by heliarc-fusing the root pass without filler-metal addition. The purpose of this technique, which is used in refineries, is to pick up enough aluminum from adjoining areas to make the root pass corrosion-resistant. It has two disadvantages: (1) the root pass, because of the aluminum pickup, is brittle and subject to cracking; and (2) the aluminum content of the weld metal exposed to the environment is less than that of the adjacent material, and thus not equivalent in cor-

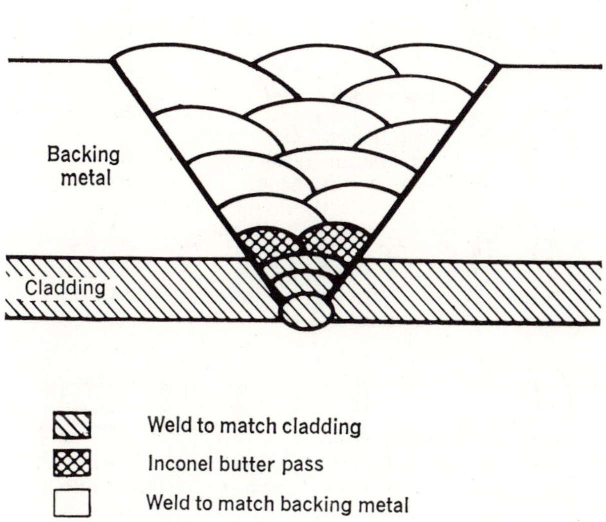

CLAD PIPE requires passes of three materials—Fig. 6

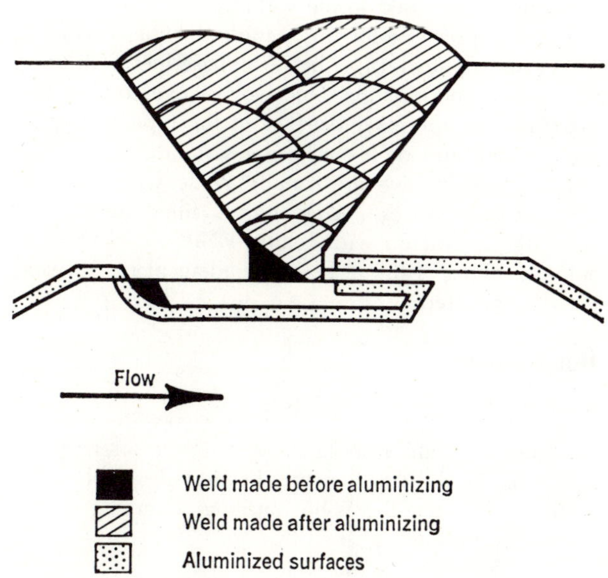

ALUMINIZED PIPE is welded with backing ring—Fig. 7

T. D. Williamson, Inc.

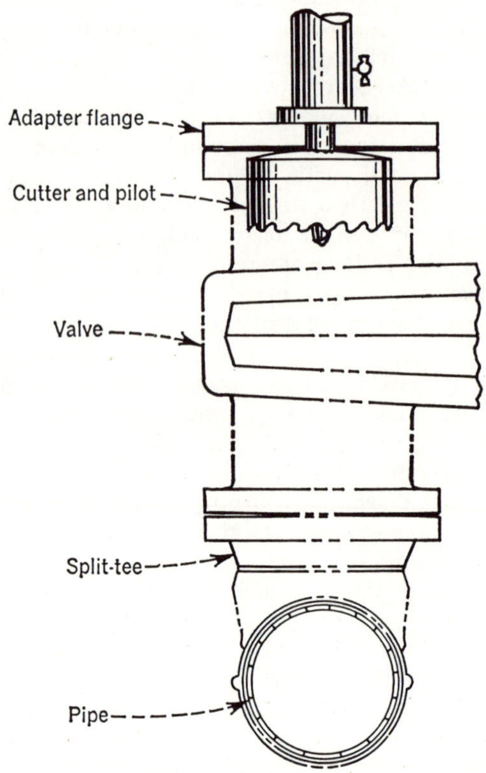

ONSTREAM PIPE can be intersected by hot tap—Fig. 8

rosion resistance. Because of these objections, a joint arrangement with a backing ring is commonly used (Fig. 7). In this design, the backing ring is attached to the upstream side of the joint prior to the aluminizing, and the beveled surfaces are masked so that only bare steel is exposed to the welding arc.

Special precautions are required when welding nickel-plated steel pipe because base-metal dilution often is not permissible. One way of overcoming this is to "butter" the inside surface of the pipe ends with pure nickel, and then machine a standard bevel. The initial passes, which are usually made with the gas-tungsten-arc process using nickel filler metal, are normally completed with coated ENiCrFe electrodes. Some cracking of the nickel plate adjacent to the weld is to be expected.

Hot Tapping

It is often economically advantageous to make a connection to an onstream line without interrupting operation or shutting down a header. A widely used method for doing this is hot tapping, a service that is usually contracted.

The procedure is as follows: (1) an accurately made branch fitting is carefully aligned and welded onto the pipeline, with all the welding done on the outside of the fitting because weld metal and slag can damage the cutter teeth of the tapping machine (Fig. 8); (2) a flange is welded to the fitting for permanent attachment of a valve; (3) the valve is opened and the cutter of the tapping machine is lowered through it to the pipe surface; (4) after the cutting operation, the cutter, holding the plug that has been cut from the pipe, is withdrawn through the valve, which is then closed; the connection can then be made to the free end of the valve.

An alternate method for making a connection without shutting down is more often used to isolate a portion of pipeline for maintenance. A bypass is installed across the part of the line on which the new connection is to be made, and stopples are placed to plug both ends of the bypassed section.

FITTINGS, FLANGES AND COUPLINGS

Of the standard fittings made from the major piping materials, the most common are of cast iron, malleable iron, brass, copper, cast-steel and wrought-steel. Cast and wrought fittings are also made from other major nonferrous piping materials.

Butt-Weld Fittings

Butt-weld fittings are furnished to the ASTM specifications for welding fittings shown in Tables I and II. The fittings normally have a heavier wall thickness than the pipe to which they are joined, for two

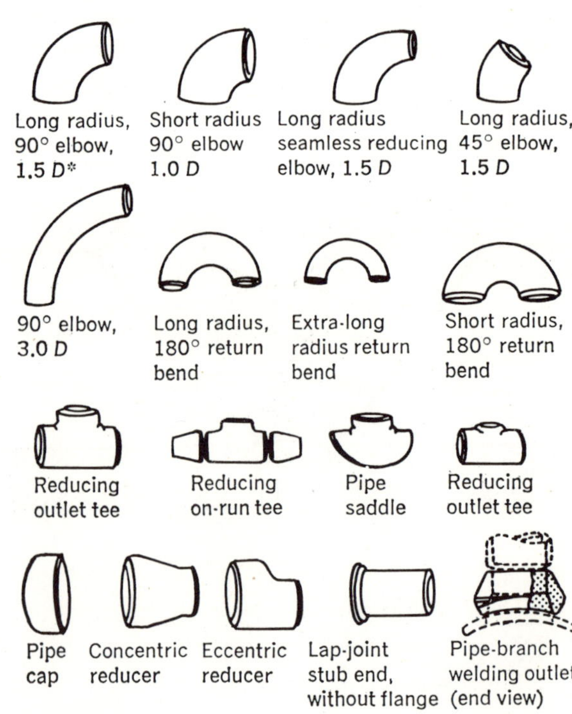

*The 1.5 D designation means that the radius of curvature is 1.5 times the fitting's nominal-dia. size of the fitting

BUTT-WELD FITTINGS make full-strength joints—Fig. 9

reasons: (1) external reinforcement is needed to give the fitting an allowable working pressure fully equal to that of pipe (the necessary reinforcement has been established by the Mfrs. Standards Soc. to conform to ANSI piping code requirements; however, the end of the fitting is furnished turned, to match the outside diameter of the pipe side); and (2) butt-weld fittings are not stocked to the full range of pipe schedules available (therefore, for example, a standard-weight, Schedule 40, taper-bored fitting is provided for Schedule 30 piping). Fig. 9 shows various designs of seamless butt-weld fittings.

When very-thick-wall fittings (over X-strong) or fittings to nonstandard dimensions are to be used, the designer must provide the supplier with the pressure and temperature at which the fittings are to serve, as well as end-preparation requirements.

Very-thin-wall fittings, 5S and 10S, are not reinforced and have pressure ratings substantially less than those of straight pipe of the same thicknesses. The design criteria for these fittings are mechanical strength, economy, and resistance to vacuum, with pressure seldom a critical consideration. For example, a 3-in., Schedule 10S, Type 304 pipe has an allowable working pressure of 925 psig. at 300 F., yet the matching fitting is rated at only 210 psig. by the established ratings of MSS-SP-43.

Socket-Welding Fittings

Forged socket-welding fittings are generally used in sizes from ½ to 2 in. but are available up to 4 in.

Installing a permanent, leakproof connection by socket welding is relatively simple because the fitting is slipped over the squared-off end of the pipe (Fig. 10). Because the fitting is self-supporting on the socket, aligning the pipe is easy. Drippings of welding metal cannot get inside the pipe to restrict flow.

Socket-welding fittings are available in carbon steel, carbon-molybdenum, chromium-molybdenum and stainless steels. Welding-rod materials must be compatible. The classes of these fittings, which are made in accordance with ANSI B16.11, are: 3,000 CWP (cold working pressure), for Schedule 80 extra strong pipe, and 6,000 CWP, for Schedule 160 and double extra strong pipe.

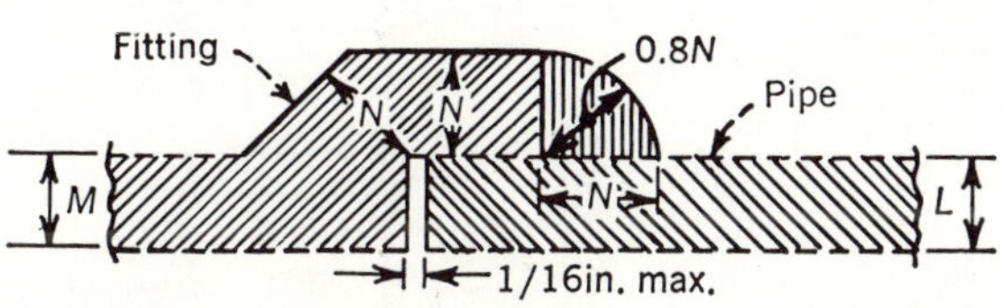

PENETRATION is the key to a good socket weld—Fig. 10

Forged Branch Fittings

Three common self-reinforced forged branch attachments bear the trade names Weldolets, Sweepolets and Latrolets. Weldolets and Sweepolets are both 90-deg. fittings. The Weldolet must be reinforced to meet ASME and ANSI codes. It has integral reinforcement but with a sweeping portion tying in horizontally (in longitudinal cross-section) with the run pipe by means of an insert butt-weld. Therefore, its edge is as thick as the pipe into which it is welded.

The Latrolet, which is available in sizes through 12 in., is a 45-deg. forged connection attached in a manner similar to the Weldolet. It makes possible inexpensive 45-deg. branch connections when 45-deg. laterals would not be used because of the cost or the poor quality of field-fabricated laterals. Unreinforced factory-made 45-deg. laterals have allowable pressure ratings of only 40% of the pipe from which they are made. However, heavier laterals may be substituted with the weld ends taper-bored to match the pipe dimensions for full working pressure.

Screwed and Bolted Joints

Smooth, clean threads are essential if screwed joints are to go together easily and remain pressure tight. The lip angle, sometimes called the "rake" or "hook," is a critical feature of a pipe-threading die. The angle between the front face of the chaser and the actual cutting edge varies with the material being threaded: not over 16 deg. for wrought iron; 15 to 20 deg. for ordinary Bessemer steel; at least 25 deg. for open-hearth and basic steel; and very small for brass. A rough thread and metal tearing can result from an improper lip angle or the absence of cutting oil.

Because pipe threads are imperfect, a compound is usually applied, both to lubricate and to ensure tightness. A light lubricant, such as linseed oil, is sometimes sufficient for well-made threads. White or red lead is generally used in plumbing work. A paste containing oil and a filling material is applied to steam-piping joints. A compound of molybdenum disulfide and mineral oil is often used on screwed oil-piping. A mixture of litharge and glyptol, or glycerine, satisfactorily tightens joints against liquids that leak easily, such as oils, and is often preferred on screwed stainless-steel joints. Thread tape is economical, easy to handle and quickly applied. Joints made with tape are more easily taken apart.

Piping that must be disassembled for maintenance is usually joined by bolted flanges. Forged-steel flanges are available in welding-neck, slip-on, lap-joint, threaded, blind, socket, and reducing types. Unless otherwise ordered, flanges usually conform to the following ASTM Specifications: 150 and 300 lb., A181 Grade I; 400 and 600 lb., A105 Grade I; and 900 to 2,500 lb., A105 Grade II.

Flanges are screwed, welded or lapped onto pipe, and the facing of the contact surface can be plain, serrated, grooved for ring joints, or ground and lapped

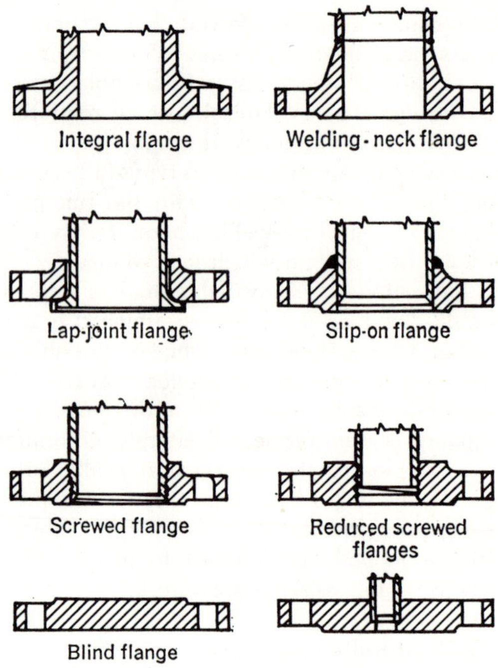

STANDARD steel-pipe flanges per ANSI B16e—Fig. 11

for metal-to-metal contact. Common flanges that comply with ANSI codes for pipeline connections are shown in Fig. 11 and 12.

Integral flanges have been cast onto fittings, valves, pumps and other equipment.

Welding-neck flanges gradually taper toward the welding end. They are used for a wide range of severe operating conditions, including shock and vibration.

Lap-joint flanges, whose lap equals the pipe wall thickness, are used in joining expensive nonferrous metallic piping, usually with short stub ends that are welded on and bolted between with suitable gaskets. They can be rotated for easy alignment of connecting piping.

Slip-on flanges are preferred by some designers because the pipe does not have to be cut too accurately. Their disadvantages are that they require double-fillet welds, front and back, are weaker than welding-neck flanges, and have the objectionable feature of an inside pocket between the pipe and flange into which liquid or sediment can settle.

Screwed-pipe flanges are restricted to low temperatures and pressures and to smaller sizes. Cast-iron screwed flanges can be used only below 400 F. on air, steam and gas lines, and below 300 F. on oil lines.

Long welding-neck flanges come in lengths of 9 to 12 in. for connections on pressure vessels, heat exchangers and instruments.

Socket-welding flanges are approximately equivalent to the slip-on type. They are usually limited to 2-in. and smaller sizes. They are unsuitable for services in which the crevice at the pipe base of the socket is likely to promote corrosion or product contamination.

Rolled-on flanges are most often used on steel piping systems having walls too thin for threading, and for economic reasons. Roll-on joints are limited to the terminal fittings of oil-refinery heaters. Such piping is installed in stainless steel raised-face-type inserts that have serrations for rollling in the pipe ends with an expanding tool. These inserts are part of the flanges, which are bolted together with a gasket.

Reducing flanges, both slip-on and threaded types, are for reducing line size without resorting to threaded or welded reducers.

Ring or plate flanges are for large openings on low-pressure equipment fabricated from steel plates.

Orifice flanges, slip-on, weld-neck and threaded types, are for mounting orifice plates.

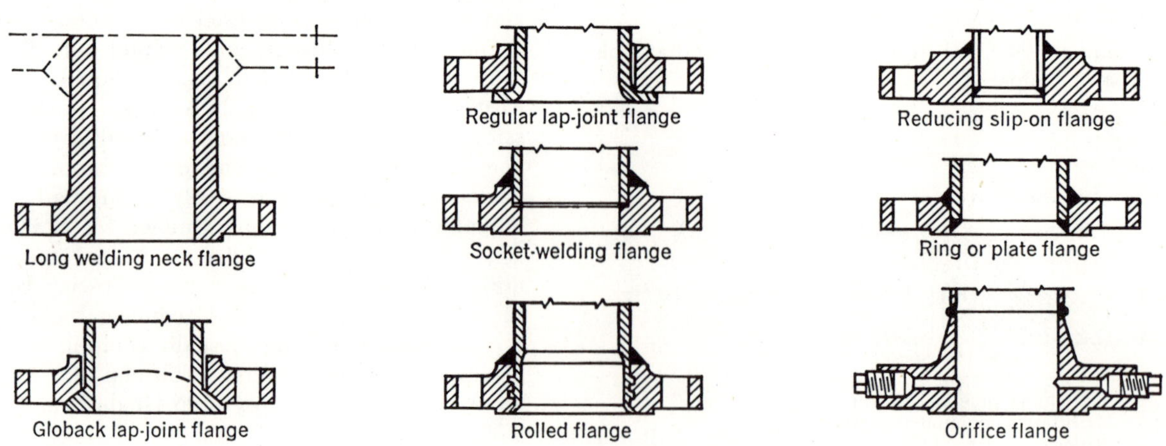

SEMISTANDARD flanges have been designed for special services, such as connecting thin-wall pipe—Fig. 12

Flange Specifications

Flanges from ½ to 24 in. are rated for 150, 300, 400, 600, 900 and 1,500 lb., and from ½ to 12 in. for 2,500 lb. Catalog tables also indicate outside diameter, bolt circle, number of bolts required, and recommended bolt lengths, as well as pressure-temperature service ratings.

Fig. 13 shows common types of flange facings. Because of the small contact area, the ring-type facing can provide a tight joint at low bolting loads, which results in lower flange stresses. Although numerous special facings are available for specific services, it is generally more economical to use a standard facing.

Experience is usually the best guide in selecting flange facing. For temperatures up to 220 F., flat-face joints with full-face asbestos gaskets are satisfactory. Serrated raised-face joints with asbestos-composition gaskets are generally used for temperatures up to 750 F. Facings providing a high contact pressure for a given bolt load, such as tongue-and-groove and ring joints, are often used for high temperatures and pressures. For most services, an equally successful joint can be made with a profile-serrated metal gasket contacting the flange facing, which can be plain male-to-male raised-face.

The extent to which a gasket must flow to achieve an impervious seal depends importantly on the surface finish. Bolting that adequately provides a satisfactory seal for a smooth contact surface may be inadequate for a rough surface. A smooth tool finish is most frequently put on steel and cast-iron pipe flanges. Steel flanges often have a serrated finish, particularly if they are to be used with an asbestos-composition, ring-type gasket with a wide contact area, such as is available on raised facings. The serrated finish usually consists of 1/64-in.-deep spirals on concentric grooves, 32/in. A flange surface is usually lapped smooth for metal gaskets.

It is considerably more expensive to make a tight joint by grinding and lapping contacting faces than by using a gasket of a softer material. Pressure from the tightening of bolts causes a gasket to flow into the minor machining imperfections to produce a tight seal. Gasket materials for various applications are listed in Table III.

Asbestos gaskets are normally plain and have smooth surfaces. Semimetallic gaskets, which combine a metal and soft material, depend on the metal to withstand temperature, pressure and attack by fluids, and on the soft material for resilience. Gaskets are available in a wide variety of designs and cross-sectional shapes. In general, the object of the design is to retain the advantages of a metal gasket while minimizing the bolting load required for a tight seal.

Fig. 14 shows the general types of flanged joints that take gaskets. For raised-face flanges, which are the most widely used, the flat-ring gasket can be of paper, cloth, rubber or composition asbestos for moderate services. The spiral metallic gasket is used for operations at or near the flange pressure rating, and in or near possible flash-fire areas, to prevent gasket blowouts.

Metallic asbestos-filled gaskets, which require a relatively high contact-pressure to effect a seal, are used with raised-face or small or large tongue-and-groove-joint flanges. The metal-jacketed gasket is confined in the groove, which prevents it from slipping during assembly or disassembly of the joint, and eliminates blowouts during operation.

Next in popularity to the raised-face type is the ring-type-joint flange, particularly in oil refin-

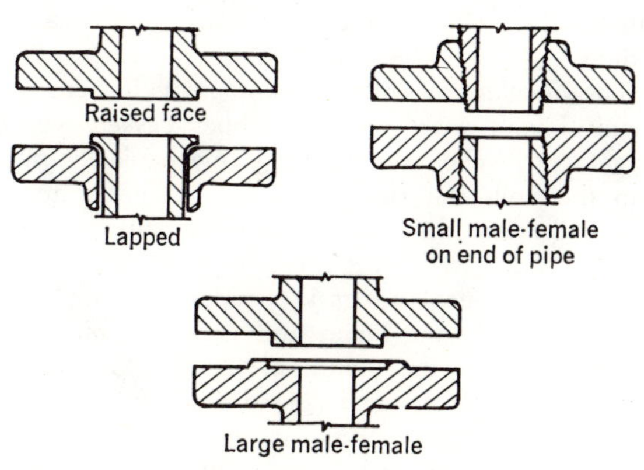

SERVICE largely determines choice of facing—Fig. 13

Gasket Materials for Different Services——Table III

Fluid	Application	Gasket Material
Steam (high pressure)	To 1,000 F.	Spiral-wound composition asbestos
	To 1,000 F.	Steel, corrugated or plain
	To 1,000 F.	Monel, corrugated or plain
	To 1,000 F.	Hydrogen-annealed furniture iron
	To 1,000 F.	Stainless steel 12 to 14% chromium, corrugated
	To 1,000 F.	Ingot iron, special ring-type joint
	To 750 F.	Composition asbestos, spiral-wound
	To 600 F.	Woven asbestos, metal asbestos
	To 600 F.	Copper, corrugated or plain
Steam (low pressure)	To 220 F.	Red rubber, wire inserted
Water	Hot, medium, and high pressures	Black rubber, red rubber, wire inserted
	Hot, low pressures	Brown rubber, cloth inserted
	Hot	Composition asbestos
Water	Cold	Red rubber, wire inserted
	Cold	Black rubber
	Cold	Soft rubber
	Cold	Asbestos
	Cold	Brown rubber, cloth inserted
Oils (hot)	To 750 F.	Composition asbestos
	To 1,000 F.	Ingot iron, special ring-type joint
Oils (cold)	To 212 F.	Cork-fiber
	To 300 F.	Neoprene composition asbestos
Air	To 750 F.	Composition asbestos
	To 220 F.	Red rubber
	To 1,000 F.	Spiral-wound composition asbestos
Gas	To 1,000 F.	Asbestos, metallic
	To 750 F.	Composition asbestos
	To 600 F.	Woven asbestos
	To 220 F.	Red rubber
Acids	Varies	Sheet lead or alloy steel
	Hot or cold mineral acids	Composition blue asbestos
		Woven blue asbestos
Ammonia	To 1,000 F.	Asbestos, metallic
	To 700 F.	Composition asbestos
	Weak solutions	Red rubber
	Hot	Thin asbestos
	Cold	Sheet lead

eries and chemical plants. It is applicable for high pressures, requires low sealing force, and is not subject to leakage with sudden temperature changes. However, the inside gap between the flange facings may be an objectionable feature in some chemical processes because it might contribute to the contamination of product. In some applications, this space is filled with plastic-filler ring-gaskets. Oval ring gaskets are more often used than the octagonal type.

Carbon and low-alloy steel bolts are used for service temperatures up to about 700 F. Common types include carbon-steel grades 1018, 1037, 1040 and 1045; in alloy steel, they are often made of 8640 grade, followed by the 4140 and 4340 grades. In low-temperature service, bolting materials must have specific chemical compositions and meet impact test requirements. The grades commonly used for low temperatures are 2317, 4140, 4142, 4145, 4340, 3-1/2% Ni steel and stainless steel Types 303, 304, 321 and 347. For temperatures much over 900 F., special precipitation-hardened bolts conforming to ASTM A-453 are required for strength and oxidation resistance.

The following ASTM specifications cover bolting materials: A-307 for low-carbon-steel externally and internally threaded fasteners; A-449 for quenched and tempered steel bolts and studs; A-354 for quenched and tempered alloy-steel bolts and studs with suitable nuts; A-320-63 for alloy-steel bolting materials in low-temperature service; A-193 Gr B7 for alloy-steel bolting materials in high-temperature service; and A-453 Gr 660 for high-strength, high-temperature bolting materials.

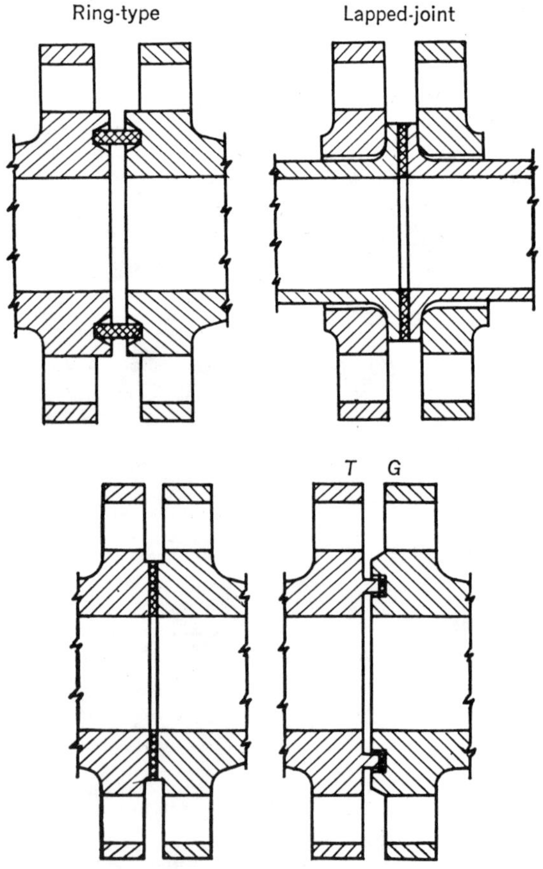

GASKET flows into surface to give tight joint—Fig. 14

Joining Cast-Iron Pipe

Bell-and-spigot joints for cast-iron pipe can be sealed with lead and oakum, sulfur compounds, cement, or rubber (including neoprene). For water-supply systems, the prevalent joint sealer consists of lead and oakum. Low-pressure distribution systems for manufactured wet gas depend on oakum with a lead or cement backing. To ensure tightness, the oakum must be kept wet by either the moisture in the conveyed fluid or, for example, in a dry gas system, by an impregnating compound in the lines.

Although bell-and-spigot joints have been found satisfactory in water-supply systems operating at pressures in excess of 200 psig., they are not recommended for gas pressures above 10 psig. For fire-water lines and other high-pressure water applications, bell-and-spigot pipe and fittings are available with bolting lugs.

Mechanical (gland type) joints, a modification of the bell-and-spigot joint, are used in low- and intermediate-pressure gas-distribution systems, particularly those for natural gas or dry manufactured gas. Mechanical joints also see service in water lines likely to suffer excessive vibration, and in aboveground water and sewage piping.

The *mechanical lock joint* is similar to the gland-type mechanical joint, except that the spigot end of the pipe is grooved or recessed to grip a gasket. If a sag or lateral thrust in the pipeline is likely to cause a joint to come apart, a mechanical joint with a self-locking feature is used to resist end-pull. Lock joints are used aboveground in the process industries, on bridges and trestles, and in unusually loose or marshy soils. With the locking feature on the spigot rather than on the bell, this type of piping can be used with the regular line of mechanical-joint fittings.

The *mechanical roll-on joint* is a low-cost mechanical joint with a round rubber gasket over the spigot end. When the spigot is mechanically pulled into the bell, the ring is placed in the bottom of the bell. On the outside of the rubber gasket, braided jute, wedged behind a projected ridge in the bell, confines the gasket under pressure in the joint. To retain the hemp and gasket, a bituminous compound seals the mouth of the bell. Either bell-and-spigot or mechanical (gland type) fittings can be used with this type of pipe.

The *mechanical screw-gland joint* contains a coarse-threaded screw gland drawn up by a spanner wrench that compresses a standard rubber or composition packing gasket. A lead ring inserted in the bell ahead of the gasket will seal off the contents of the line from the gasket. This joint can be used on lines carrying water, gas, oil and other fluids at considerable pressure.

The *ball-and-socket joint* provides considerable flexibility, such as for river crossings. Cast-iron pipe is available with ball-and-socket joints of either the lead-and-oakum or the mechanical gland type. A positive stop against disengagement is a feature of all designs of ball-and-socket joints. Angular deflection of as much as 15 deg. can be tolerated without leakage. This type of joint is suitable for water, sewage, oil, air and gas under considerable pressure. Pipe supplied for ball-and-socket joints can be used with either bell-and-spigot or mechanical (gland type) fittings, but the integral ball on the spigot end of some designs has to be cut off before the pipe can be inserted into a regular ball.

The *universal pipe joint,* a cast-iron pipe joint, has a machined, tapered seat that obviates the need for either caulking or a compression gasket. After the joint is pulled up snugly with the two bolts, the nuts are backed off slightly to allow the lock washers to give enough to prevent overstressing the lugs or socket. The universal joint is used to some extent on 4 to 24-in.-dia. piping in underground water-supply systems, but is not recommended for gas service. It does not permit much angular displacement or expansion.

The *compression-sleeve coupling* is widely known by the tradenames Dayton and Dresser couplings. It is used extensively with plain-end pipe of either cast iron or steel in both underground and above-ground piping for air, gas, water and oil. This type of joint must be anchored or braced solidly at dead ends or turns to prevent the line from pulling apart. Compression couplings and fittings with screwed packing glands are available for small-diameter cast-iron or steel pipe. To allow for expansion, compression couplings instead of welding are used on a certain percentage of the joints in welded transmission lines for oil or gas if significant changes in temperature are anticipated.

A *split coupling* (known as the Victaulic Coupling) is used with either cast-iron or steel pipes that have been grooved near their ends, so that the coupling can grip them to prevent disengagement. This coupling comes in two segments in sizes from ¾ to 14 in., and in four or six segments in larger sizes. These are also attached to grooved-end fittings. This joint can be used in underground or above-ground service with almost any liquid or gas, if the gasket material is properly selected. It has the advantages of considerable angular displacement and some expansion; a gasket that seats tighter under pressure; a positive stop against disengagement; and a design that permits rapid erection or disassembly. One disadvantage, particularly with lightweight pipe, is the need for the grooving at the pipe ends to be nearly as deep as taper threads. However, this objection can be overcome by banding or welding on heavier grooved nipples.

The snap-on Victaulic couplings save time and money in making and breaking pipe joints. The snap joints are hinged together in one assembly; there are no loose parts or bolts.

Another split-ring coupling, widely known by its tradename Grayloc, is not used on cast-iron pipe but is designed for higher-pressure applications. Its hub, the equivalent of a flange, is usually welded to the pipe. The clamps correspond to the flange bolts and the seal ring to the gasket. Bolts retain the clamp segments.

The clamp, which is designed to match the hub in strength, normally carries all of the load imposed by the fluid pressure, as well as all axial or bending loads that the pipe can transmit, such as those from thermal stress and vibration. The standard design has two bolts on each side, so that there is 100% backup should one bolt fail.

The seal ring is made of high-strength material, usually 17-4 PH, and is designed to tighten under internal pressure. In most cases, the seal ring is coated for more effective sealing and to prevent galling during makeup. The coating depends on the service environment—the stock coatings being: cadmium plating for carbon-steel rings, silver plating for stainless-steel rings, and Teflon, molybdenum sulfide and indium-graphite for other applications.

The hub is usually made of the same material as the pipe and can be of almost any metal that can be machined, including carbon steel, chromium-molybdenum steels, all austenitic stainless steels, aluminum, Monel, Inconel, Hastelloy, titanium and zirconium.

The Grip-Seal joint also cannot be used on cast-iron pipe, because the pressure required to seal the joint would fracture cast iron. Pipe cut to length is inserted into the loosely assembled fitting until the pipe ends rest against the Teflon seal. The union nut is then tightened on the body threads, and this slides the collets forward on the pipe until they meet the seal. The chamfered collets act as wedges between pipe and fitting; the greater the axial load on the pipe, the tighter the fitting grips.

NONMETALLIC PIPE

Catalogs represent the best sources of information on nonmetallic fittings and piping and on methods of joining nonmetallic pipe because comprehensive literature is not available on these subjects.

In many cases, nonmetallic systems can be installed with much less field work than is required by welded or flanged metallic systems, and sometimes more economically. However, there are disadvantages. For example, cementing or welding plastic piping in the field is practical and usually simple, but human error, possibly due to limited experience with joining plastics, can frequently offset the cost advantage. Failure to clean surfaces properly before joining, or trying to weld under wet conditions, are common errors that lead to leaks.

Thermoplastic Piping

Thermoplastic piping is joined without difficulty by solvent welding (cementing) and hot-gas (thermal) welding. Socket joints are generally solvent welded, and socket-weld fittings are readily available. Cement is applied to the pipe end and to the socket of the fitting. The joint should set for at least 10 hr. and will develop full strength in 24 hr.

Air, nitrogen, oxygen or carbon dioxide is passed through the heating torch in hot-gas welding. Compressed air is most widely used. With polyvinyl chloride (PVC) joints, oxygen produces the strongest and carbon dioxide the weakest joints. On polyethylene piping (PE), however, oxygen causes surface oxidation and results in poor-quality welds, so an inert gas or nitrogen is recommended.

One of the two types of heating torch is electrically heated; the other has a gas flame to heat the coil through which the welding gas flows. In the latter, the flowrate of the welding gas through the torch determines the welding temperature. In general, the flowrate corresponds to a gas pressure of 5 to 10 psig. Lowering the flowrate increases the temperature, which is indicated by a thermometer held about ¼ in. from the tip of the torch.

With a "nonrigid" plastic, the filler rod should be pressed against the piping at a forward angle of 45 deg. When joining "rigid" plastics, the rod should be perpendicular to the joint surface. Properly made butt joints should have 60 to 90% efficiency. Fig. 15 illustrates typical welds for butt and fillet joints that require filler rods.

Flanged joints should connect plastic pipe to metallic pipe.

Plastic piping is generally available in iron-pipe sizes. Fittings are made mainly by injection molding and are available in the same variety and types as steel fittings.

Polyvinyl chloride (PVC) pipe can be cut to length before threading with a hack saw or pipe cutter. The ends of the pipe can be beveled with a hand file and the interior deburred with a regular tool or knife. The pipe is held in a pipe vise when it is threaded because saw-toothed jaws would mar the surface.

Threaded PVC pipe and fittings should only be assembled with strap wrenches and not with Stillson wrenches. Joints should be made hand-tight, then given only an additional one-quarter to one-half turn. More force is likely to stretch or distort the pipe. Teflon thread-lubricant is recommended as a compound.

Polyethylene (PE) piping procedures are essentially the same as for PVC piping.

The difficulties of joining polypropylene (PP) pipe promise to be largely overcome by an electrical thermal-fusion method recently developed. This technique depends on a small coil of polypropylene-insulated resistance wire that is slipped over the male end of the joint. The parts are fitted together cold and held with an adjustable clamp. Current is applied by means of leads that trail to the outside. The wire heats, melting the polypropylene and the two adjacent surfaces merge. On cooling, the joint becomes a strong, solid connection, with the coil permanently embedded. If the joint should leak during testing, the coil can be reheated. While the interface is molten, the connection can be realigned or adjusted as necessary.

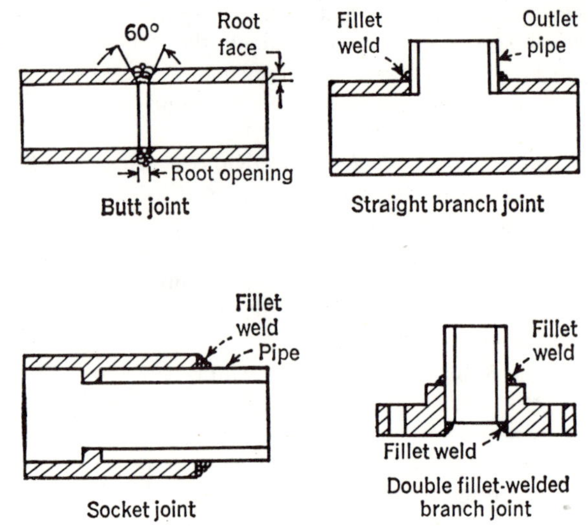

BUTT and fillet joints offer 90% efficiency—Fig. 15

Cellulose-acetate-butyrate (CAB) piping, is usually joined by solvent welding. Socket-weld fittings are readily available. Joints are generally made by screwed connections or thermosetting cements.

Thermosetting Plastic Piping

Thermosetting-plastic piping is chiefly made of two resins: epoxy and polyester, reinforced with glass or blue-asbestos fiber (fiberglass reinforced is the popular FRP piping). Piping of thermosetting resin is joined by "welding," flanges, couplings, and a wide range of standard fittings.

A joint is "welded" by butting together ends of pipes and wrapping it with layers of fiberglass impregnated with catalyzed resin. This is an economical and permanent method of joining pressure and drain pipe.

Flanges are either formed on the pipe at the factory, or welded to the pipe by means of a stub end. Dimensions are the same as those of standard 150-psi. steel flanges, except for thickness. Gaskets should be full face and ⅛-in. thick.

Standard screwed and socket fittings are also available. To provide improved strength and sealing, some fittings have tapered socket joints. Screwed connections are extensively used in the oil industry, with welded joints preferred in chemical plants.

Concrete, Cement and Cement-Lined Pipe

Nonreinforced concrete pipe is used for conveying sewage, industrial waste and storm water. Its joints may or may not be sealed with rubber gaskets. Rubber-gasketed joints for standard-strength, bell-and-spigot, nonreinforced-concrete sewer pipe are covered by ASTM Specification C443. Pipe sizes for conveying irrigation water range from 6 to 24 in. with water head pressures of 30 ft. or higher, depending on pipe diameter.

Reinforced concrete pipe ranges in sizes from 12 to 144 in. and is available with either tongue-and-groove or bell-and-spigot joints. Its joints also may or may not be sealed with rubber gaskets. If rubber-gasketed, these joints must also conform to ASTM Specification C443.

Reinforced and prestressed-concrete pressure pipe is used for water-distribution systems, ocean outfalls and sewer mains. Sizes range from 12 in. through 108 in. for pressures up to 600 psig.

For high pressures, reinforced-concrete pipe is manufactured with a watertight, flexible-expansion, bell-and-spigot type of joint with joint surfaces formed by steel rings in the ends of the pipe. For moderate-pressure service, joints are made watertight with rubber gaskets instead of steel rings. In the non-cylinder-type, longitudinal reinforcement extends through the pipe to join the rings. In the cylinder-type, the rings are welded to the ends of the cylinder to form a continuous watertight membrane throughout the length of the pipe.

The rubber gasket is placed in a rectangular groove on the spigot end of the pipe. It is compressed into the groove by the flared portion of the bell as the sections of pipe are pushed together, and thereby confined on all four sides. The joint rings, or concrete surfaces, are shaped and sized to make the joints self-centering, even without a rubber gasket. The space between the ends of the pipes is filled with cement mortar, premolded asphalt, or other jointing compounds, both inside and outside of the joint. This is done to further protect the steel joint rings and not for joint watertightness, which depends solely on the rubber gasket.

All types of concrete pipe can be trapped under pressure for joining onto threaded or flanged connections. Elbows, reducers, tees, wyes and closures are standard items; special fittings can be supplied for specific requirements.

Asbestos-Cement Pipe

Asbestos-cement pipe is made from asbestos fiber, portland cement and silica flour. Because of its strength, corrosion resistance and high-flow characteristics, it is one of the most widely used pipings for conveying water and sewage.

For jointing, pipe ends are machined square, and coupling areas are lathed to close tolerances. Sleeves larger in inside diameter than the piping are machined to fit over the ends of the pipe and to enclose two rubber gaskets, one on each pipe end, which act as seals. The joint is made by placing the rubber rings in the coupling grooves, then inserting the pipe ends into the coupling to seal the joint. Because couplings differ, asbestos-cement adapters are available. The drilling and tapping of asbestos-cement pipe for service connections can be done with standard equipment. Factory-fitted tapped couplings are also available.

Cement-Lined Ferrous Pipe

Cement-lined cast-iron pipe is widely used for ambient-temperature water lines; cement-lined steel pipe sees much service where corrosion is a problem. The petroleum industry relies on cement-lined pipe extensively for oil-field pipelines and casings, because of its resistance to salt water, hydrogen sulfide and other corrosive materials. It is also used for handling brine in salt works, in coal mines for conveying highly corrosive sulfur water, in paper and pulp mill lines for conveying dilute acids and other corrosive liquids, and in process plants for carrying water and other liquids that must not be contaminated by rust.

Jointing is by screwed seal rings, which prevent the corrosive liquid from contacting the steel, and by flanges. Some piping assemblies involving welding of the steel joints are prefabricated. Field joining of such assemblies is usually done with flanged ends. After these ends are welded, the joints are filled with mortar.

Clay Pipe

Vitrified and unglazed clay pipe is used to carry sewage, industrial wastes and storm water. It is available in sizes from 4 to 36 in. Nearly all vitrified clay pipe is supplied with a factory-fabricated compression-type joint. The clay-pipe industry furnishes three types of compression joints that are classified as follows in ASTM Specification C425:

- Type I joints have the same resilient material on the spigot and in the bell of the pipe, although the material of either end (bell or spigot) may differ in hardness. By controlled design, these preformed ends make up a positive closure.
- Type II joints use different resilient materials but in a manner similar to that in Type I, except that the material in either end may differ both in hardness and resilience.
- Type III joints depend on a gasket or compression ring of a resilient material having a controlled shape that is compressed within the annular space to form a closing seal. The gasket or ring may or

may not be attached to one of the joining faces prior to socketing.

Vinyl chloride plastisols, polyurethanes, polyester resins, and natural or synthetic rubber are the resilient materials used in manufacturing joints for clay pipe.

Glass Pipe

The chemical, food, beverage and pharmaceutical industries rely extensively on glass piping and fittings. In addition to the standard fittings, special fittings, such as sweep elbows, U-bends, reducing tees and crosses, and eccentric reducers, are also available.

Joints are usually made with metal flanges cushioned from the glass by molded asbestos inserts. When the flange bolts are tightened with a torque wrench, an interface gasket of Teflon or other suitable material (depending on resistance to the fluid conveyed) is gripped between the pipe ends. This prevents the fluid from contacting the metal flanges.

Wood Pipe

Wood-stave pipe of Douglas-fir or redwood staves finds extensive use, mainly in the western U.S., in large diameters for water transportation in hydraulic power developments, municipal water supplies, mining, irrigation and other applications. Hot, cold or acid water can be carried and the wood can be either creosoted or untreated. Douglas fir and redwood are selected because of their availability and resistance to decay.

Three types of wood piping are available:

Continuous-stave pipe, which is made of staves milled accurately to true radial planes and to correct curvatures on both the inside and outside faces. This is available in sizes from 10 in. to 16 in. and for pressures up to 350-ft. static head; the staves are assembled into a circle and banded together with steel rods.

Machine-banded stave pipe, whose staves, milled similarly and assembled in sections, comes in two varieties, either wound with wire or with flat banding; the former is supplied in sizes from 2 to 24 in. for pressures up to 400-ft. static head, the latter from 2 to 48 in. for the same pressure range.

Bored pipe, which is manufactured from straight-grained stock having outside dimensions from 4x4 in. to 12x12 in., by boring through the center inside diameters from 1½ to 6 in.

Wood-Lined Steel Pipe

Steel pipe lined with wood staves has been successfully used for handling chemical solutions under pressures exceeding the limits of banded wood pipe. This type of piping is extensively used in pulp and paper mills, in mining operations to transport acid waters under high pressure, and in the chemical process industries for carrying acid solutions and gases.

Wood-lined steel pipe is supplied with standard flanged joints and is easy to install. Cutting to length and flanging in the field, if necessary, are done simply by sawing the pipe to the required length and welding the flange in place. Fittings for this kind of pipe—whether they are standard, such as ells or tees, or of a special nature, are also lined with wood staves.

Acknowledgement

The author wishes to thank Rip Weaver, Alfred J. Smith, Norman S. Davis and C. E. Frost, Piping Design Section, Fluor Corp.'s Houston Div., for their assistance in supplying reference materials for the preparation of this article and for reviewing the manuscript.

References

1. Canham, W. G. and Hagerman, J. R., Reduce Piping Connection Costs, *Hydrocarbon Proc.,* May 1970.
2. Cherrington, D. C. and Ciuffreda, A. R., Piping Design for Hydrogen Service, *Hydrocarbon Proc.,* May 1967.
3. Dudych, D., Ball Joints Boon to Cramped Areas, *Oil & Gas J.,* Aug. 8, 1966.
4. King, R. C., and Crocker, S., "Piping Handbook," Fifth Ed., McGraw-Hill, New York, 1967.
5. Linnert, G. E., "Welding Metallurgy—Vol. II," American Welding Soc., 1966.
6. Masek, J. A., Metallic Piping, *Chem. Eng.,* June 17, 1968, p. 215.
7. Neal, E. J., Combination of CO_2 and Stick Welding Attains Code Quality, *Oil & Gas J.,* July 17, 1967.
8. Palo, J., Exothermics for Pipe Annealing, *Southern Eng.,* Dec., 1963.
9. Sosnin, H. A., The Joining of Light-Wall Stainless Steel Piping, *Welding J.,* Oct., 1967.
10. Ward, J. R., Lined-Pipe Systems, *Chem. Eng.,* June 17, 1968, p. 238.
11. Weaver, R., "Process Piping Drafting," Gulf Pub. Co., Houston, Tex., 1970.
12. Wright, E. E., Nonmetallic Pipe: Promise and Problems, *Chem. Eng.,* June 17, 1968, p. 230.
13. "Alonized Steels," Alon Corp.
14. "Fabrication of Hastelloy Alloys," Stellite Div., Cabot Corp.
15. "Handbook of Stress Relieving," Cooperheat Corp.
16. "Joining the Huntington Alloys," Huntington Div., International Nickel Co.
17. "Microwire Welding," Hobart Technical Center.
18. 'Second-Chance' Plastic Pipe Connections, *Chem. Proc.,* July 1967.

Meet the Authors

◀**Marx Isaacs** is a process engineer and technical writer/editor with Fluor Corp.'s Houston Div. After graduating from Tulane University with a B.S. degree in chemical engineering, he joined Shell Oil Co. as a chemist and technologist. He was for three years the editor of **Corrosion,** a publication of the National Assn. of Corrosion Engineers, and is now secretary of the South Texas Section of AIChE and editor of **Intercom,** a newsletter published by the Soc. for Technical Communication.

Richard B. Setterlund is a metallurgical/welding engineer with▶ Fluor Corp.'s Houston Div. (4620 N. Braeswood, Houston, Tex. 77035). Before joining Fluor, he was a metallurgist with C. F. Braun & Co., Magnaflux Corp., Aerojet-General Corp. and Bechtel Corp. He is a registered engineer and a member of National Assn. of Corrosion Engineers and American Welding Soc.

Steam-hose couplings

Properly installed and maintained couplings are essential to steam-hose safety. Here are the techniques you need to know, plus a list of steam-hose safety tips.

Dave Carriker, *Gates Rubber Co.*

☐ Unless properly installed, a steam hose cannot be safely operated. Because most steam-hose failures occur at the hose ends, maintenance crews must use the proper couplings, carefully installed, to protect the ends.

Before outlining the correct hose-coupling procedure, let us consider some helpful installation hints:

- Use only interlocking, clamp-type couplings, two- or four-bolt, for all kinds of steam hose. With such couplings, the bolts can be tightened periodically as the rubber permanently compresses from the effects of heat.
- Couple the hose to a length of pipe that will separate it from the steam source, thus minimizing heat damage.
- Install a pressure gage and a positive shutoff valve between the pressure-reducing valve (or steam source) and the hose.
- Remember that although steam hose is flexible a severe bend will weaken it.
- If hose must be used at an acute angle to the steam source, make the angle with a pipe fitting.
- Provide a permanent rack upon which portable steam hose can be coiled when it is not in service.
- Support permanent steam-hose lines throughout their entire lengths by hose trays or channels.

Installing couplings

Here is the procedure for installing couplings on steam hose:

Cut the hose at a 90-deg angle to the longitudinal axis. The ideal way is to use a power, circular, hose-cutting saw. However, a hacksaw can be used—one that has a blade with 32 teeth per inch.

When manually cutting steam hose, use a vise with jaw protectors of lead, nylon or copper. These protect the hose cover from the serrations on the jaw faces. When tightening the vise, be sure not to collapse the hose. This can kink the reinforcement and permanently damage the hose.

Another caution: Be sure to watch for and remove sharp ends of the wire-braid reinforcement protruding from the end of the hose.

Assembling the coupling begins with installation of the stem. For easy insertion, slightly bevel the inside of the hose at a 45-deg angle. This is especially necessary when (Photos continue overleaf, text continues on p. 140).

Properly stored steam-hose, off the floor and away from the source of heat

Steam hose that has failed through heat aging (note cracks in cover). Also, coupling has been tightened to the maximum (front and side lugs have bottomed out)

Originally published January 29, 1979

Steam-hose safety

1. Always wear asbestos gloves, rubber boots, and an apron that reaches below the boot tops. Protect eyes with goggles.
2. Make sure no one is in the way when steam hose is being used.
3. The hose should be grasped firmly at, or close to, the nozzle so it will not whip.
4. The nozzle should always be pointed toward the floor before the valve is opened. Never open the valve when the nozzle is lying free on the floor.
5. If a steam-water mixture is being used, always turn the water on first; then turn on the steam. When closing the valves, shut off the steam first; then turn off the water.
6. Store steam hose properly after use, in a location protected from adverse environmental elements. Wipe off the hose, if possible, and coil it on a support well away from any direct heat source. Never leave a steam hose lying unattended on the floor where it can be run over and damaged by crushing. Coiling it on a support also eliminates a tripping hazard.
7. Do not use steam hose for any other fluids.

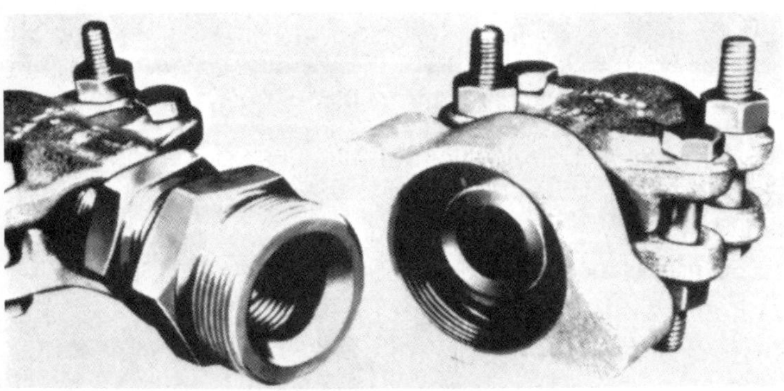

Ground-joint couplings do not need washers

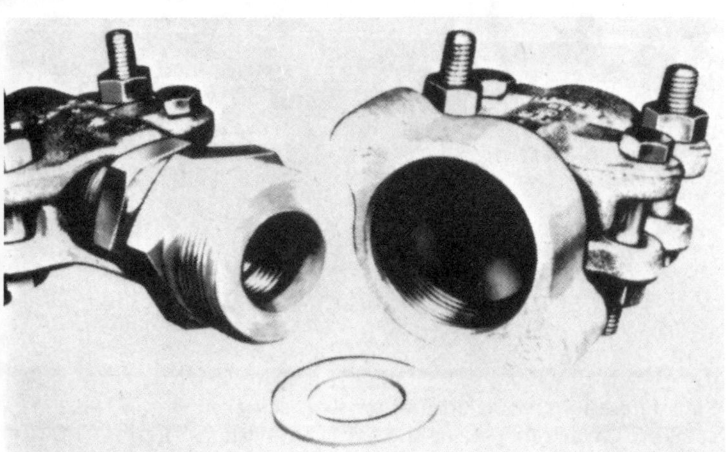

Washer-joint coupling, showing replaceable washer

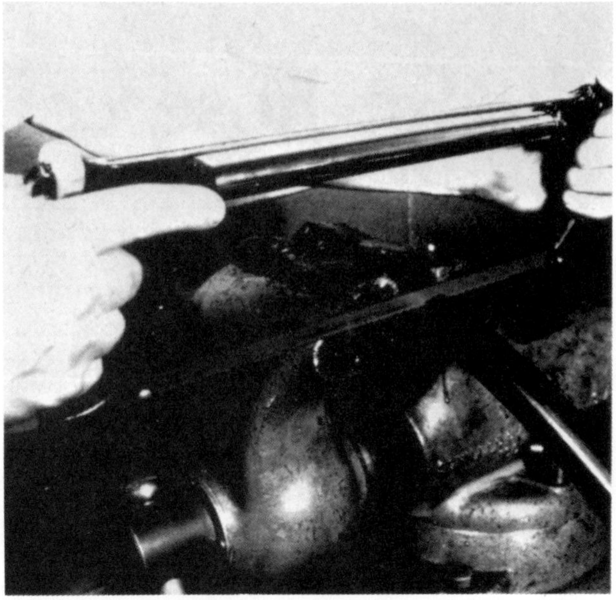

Cutting hose with a 32-teeth/in. hacksaw. Note vise jaw-covers that protect hose from possible damage by serrations on face of jaws

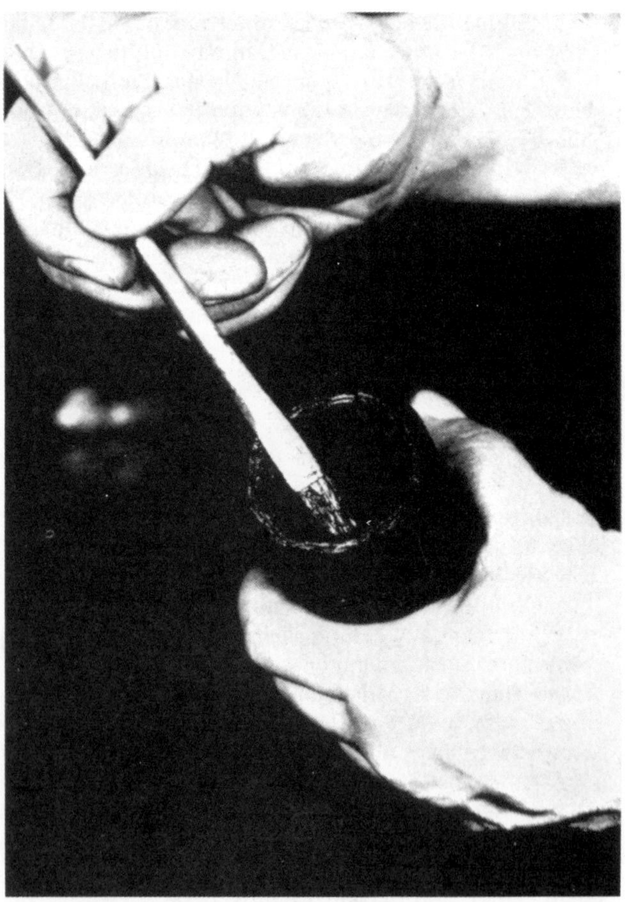

Use a recommended lubricant on both the hose and the coupling stem before inserting stem

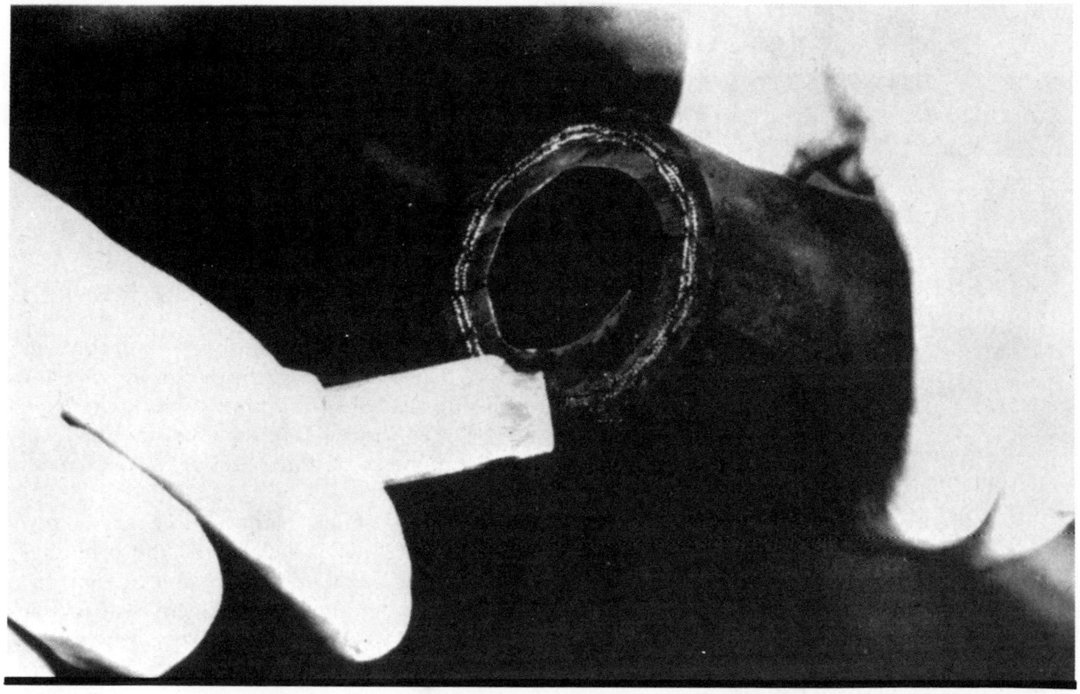

With stiff or large-diameter hose, a 45-deg bevel will help when inserting the coupling stem

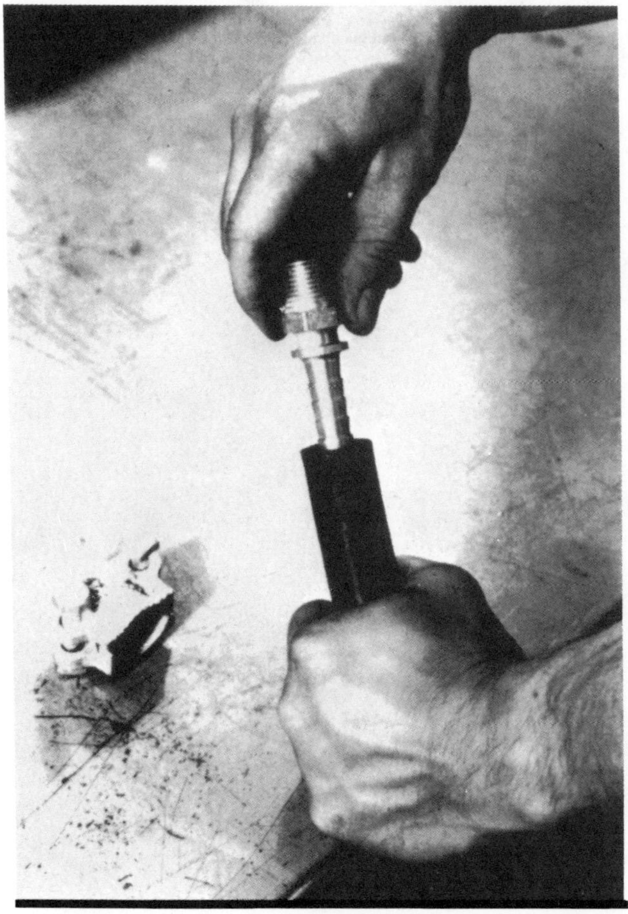

Push stem straight into hose. Alternatively, place stem in vise and push hose on stem

Clamp ready to be tightened by a wrench. Front lugs are aligned on stem. One bolt is up, one is down

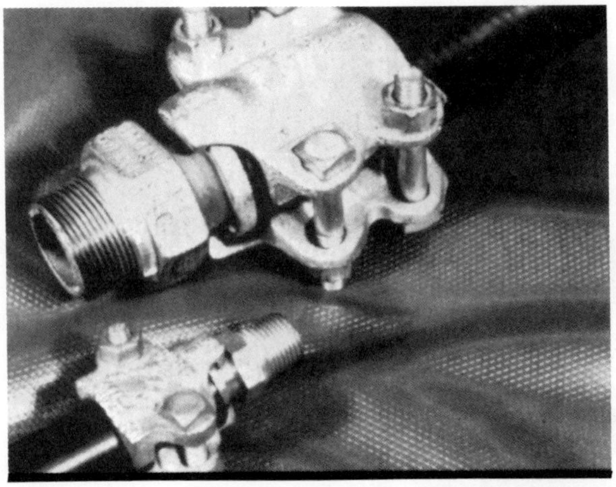

Two- and four-bolt clamp couplings, tightened and ready for use

working with a stiff hose that tends to neck down at the end.

When inserting the stem, it may be necessary to use a lubricant. Water should be satisfactory, but if more lubricity is needed, water-plus-soap, an alcohol-glycerin-water mix, or a commercial hose-lubricant can be used. Never use oil or petroleum-base products, because they may attack the tube stock and cover.

It is very important that the clamp and stem fit the hose properly—use accurate I.D. and O.D. measurements to get the proper fit.

Clamp couplings are recommended because, as steam hose is used, the hose cover and tube will compress and take a set, due to the heating and cooling cycles. The clamp allows for routine tightening, so that slack can be taken up, preventing steam leaks. The couplings—either ground joint or washer joint—interlock to provide the necessary holding power.

When assembling the clamp over the stem and hose, insert the clamp bolts in opposing directions. The bolt head should fit into the shoulder on the clamp, to facilitate tightening with one wrench.

Finger-tighten the nuts and carefully position the lugs on the clamp so that they line up in the grooves on the stem. Wrench-tighten bolts uniformly and make sure there is a gap between the lugs and the grooves. The front lugs should not bottom out, and there should be overlap on the side lugs.

In summary, the chemical-plant maintenance crew should select the right hose, install it and the couplings correctly, operate the hose with due precaution, and maintain it correctly. Steam-hose replacement costs will be minimized and the plant will be a safer place in which to work.

The author

Dave Carriker is a coupling engineer in the Hose Technical Laboratory, Hose Lab Unit 25, The Gates Rubber Co., 999 South Broadway, Denver, CO 80217. He is an expert on industrial hose-coupling applications, recommends couplings for specific applications and evaluates new couplings as they appear on the market. He also supervises testing of couplings and hoses in sizes from 3/16 in. to 24 in. O.D. He holds a B.S. in mechanical engineering from the University of Denver.

The corrosive connection

Flexible connectors are frequently used during the manufacture and shipping of corrosive fluids. Here is what you need to know about selecting the right type.

Glen Chase, Phillips Co.

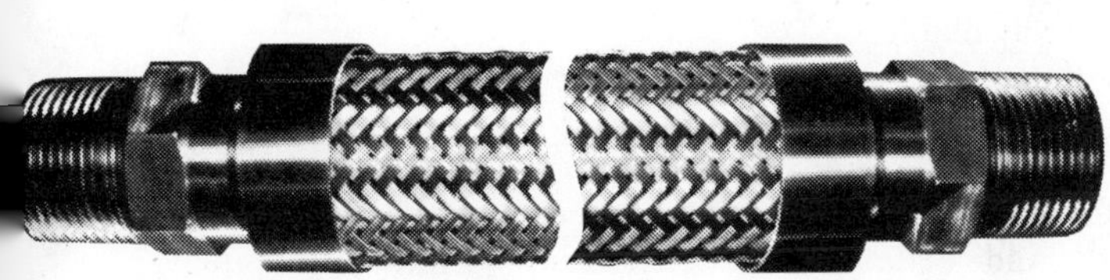

Stainless steel flexible metal hose with male pipe thread at each end — Fig. 1

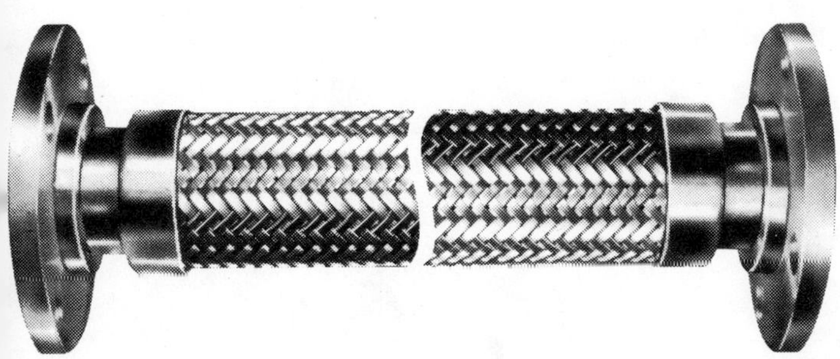

Stainless steel flexible metal hose with stainless steel flanges at each end — Fig. 2

Stainless steel jacketed flexible metal hose — Fig. 3

Originally published September 10, 1979

☐ During the manufacture and distribution of corrosive liquids, it is often necessary for them to pass through flexible connections. For example, a flexible connection is required to transfer the product from the process to the shipping container—tank truck, tank car, barge, plastic or glass carboys.

The choice as to the type of flexible connection varies according to the corrosivity of the product. The least expensive is rubber hose, provided that the manufacturer can guarantee its life and chemical resistance. Metal hose, 300-series stainless steels, bronze, steel and Monel are available. Among the plastics, Teflon hoses will handle the widest spectrum of corrosives.

Later, we will discuss an alternative to the flexible connection—a loading arm.

The rubber connection

The rubber hose is the least expensive method of liquid transfer and has the greatest flexibility. Generally there are two types: discharge hose for use under pressure only, and suction-and-discharge hose, which has wire reinforcement so that the hose will not collapse under vacuum. The rubber industry offers a rather large number of types, each having chemical resistance to certain liquids, concentrations and temperatures.

Hose elastomer-liners include gum and natural rubber, Hypalon, butyl, ethylene-propylene, Viton and cross-linked polyethylene. The manufacturer's recommended maximum pressure, temperature and concentration should not be exceeded. Heat will vulcanize and harden the carcass; excessive pressure and concentration can perforate it. Rubber also undergoes age deterioration.

The metal-hose connection

The flexible-metal-hose industry offers four basic alloys: bronze, Monel, steel and stainless steel. The stainless steel alloys include Types 304, 304L, 316, 316L, 321 and 347.

The hoses are constructed with an inner core of flexible metal—tubing that is made flexible by corrugating a thin-wall (max. 0.025 in.) metal tube. The application of braid over the corrugations and the welding of the fittings on each end make the tube into a pressure-containing unit. The smaller sizes have the highest pressure ratings; pressure ratings decrease as the size increases. Hoses are generally available from 3/16 thru 16 in., with specials being as large as 30 in.

To increase the pressure rating, additional braid can be used. The upper temperature limit for stainless-steel hose usually is considered to be 1,500°F. Ratings are reduced on increase of temperature as well as by other factors such as shock loading and rapid continuous flexing.

Flexible hoses are also used as vibration isolators and

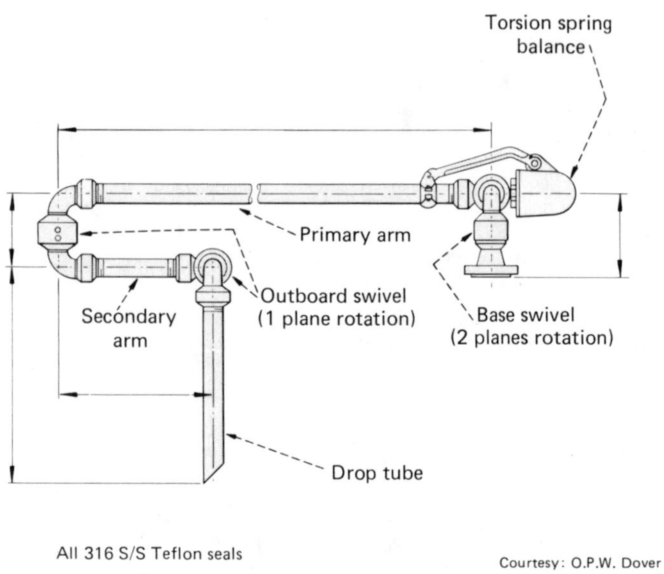

Loading arm assembly **Fig. 4**

off-line expansion compensators. Corrosion charts will aid in the selection of the grade of stainless steel to use for a specific corrosive liquid or gas.

Fig. 1 shows a standard hose with male pipe threads at each end. Fig. 2 is an example of a unit with flanged ends. If the hose is jacketed, as in Fig. 3, it will keep the product hot during transfer.

The jacketed unit is constructed of a hose within a hose. Fittings at each end of the jacket are used to introduce steam or hot water to keep the product at pumping temperature.

Monel or Teflon hoses, required for chlorine transfer, are covered later in this article.

The Teflon hose connection

Hoses of Teflon can handle the widest spectrum of corrosives. There are only two classes of chemicals known to attack Teflon—the molten alkali metals such as sodium and potassium, and the fluorochemicals such as chlorine triflouride, oxygen diflouride and flourine gas. If we eliminate these, a Teflon hose can handle any corrosive that is compatible with the end fittings.

The hoses are constructed of FEP and TFE Teflons, with smooth-liner and convoluted-liner designs. The smooth-hose TFE units are in sizes 1/8 thru 1 1/2 in., with stainless steel or bronze braid. The fittings are generally permanently attached, but are, in some cases, reusable.

FEP straight-wall hoses are constructed by lining stainless-steel/asbestos interlock hose. Sizes from 1 through 8 in. are available.

All of the straight-wall hoses are limited by the critical bend-radius. If this is exceeded, a kink will weaken the hose and may result in leaking.

Corrugated and braided metal hose can also be Teflon-lined, and Teflon-lined rubber hose is available in some sizes. The convoluted Teflon design is available from 1/2 thru 4 in. with stainless steel braid, and has an excellent bend radius so that the hose is less susceptible to damage from kinking.

The loading-arm connection

The liquid loader has been available for many years, and is used extensively in the loading of tank trucks and tank cars. These units are extremely rugged and versatile,

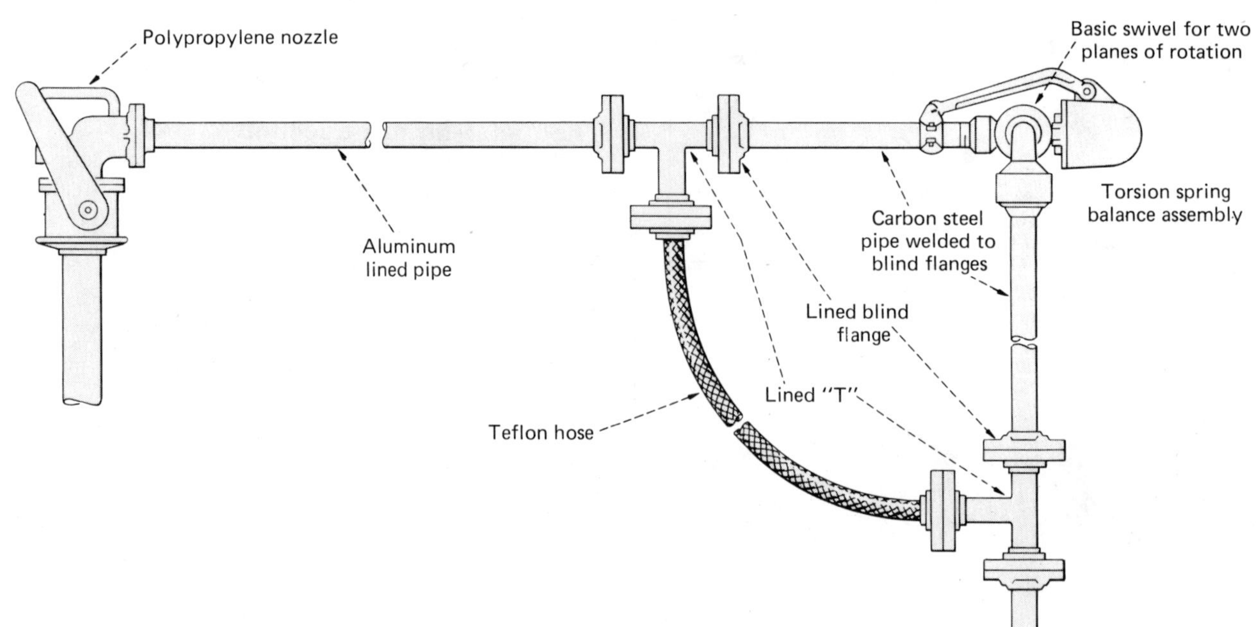

Lined loading arm (Teflon, Kynar, polypropylene) **Fig. 5**

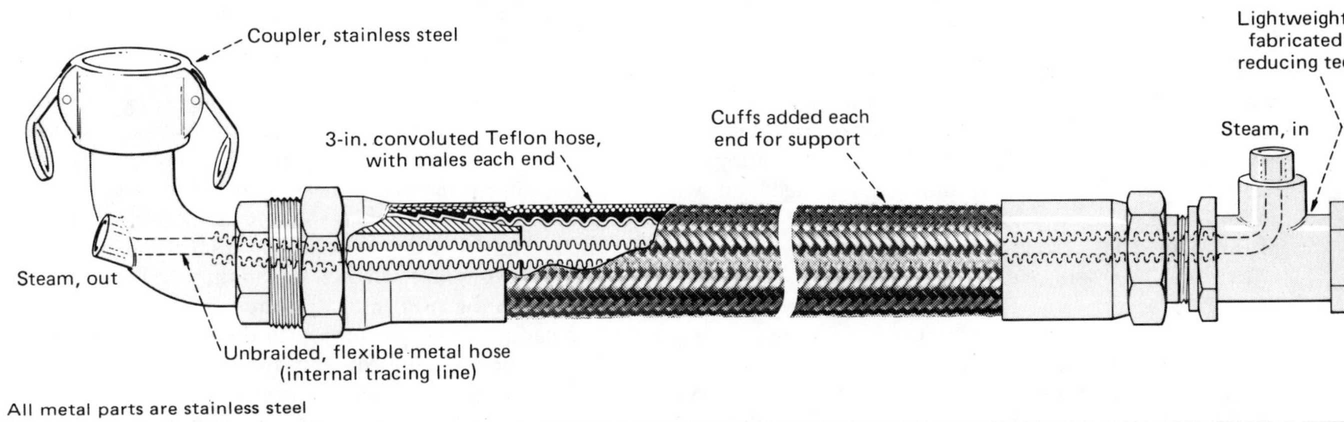

Sulfur connection with internal tracing to keep sulfur liquid during transfer **Fig. 6**

and are constructed to be easily handled by one person.

The materials of construction include aluminum, bronze, carbon steel and Type 316 stainless steel. Fig. 4 illustrates one of the common designs, but other configurations and accessories are available to suit requirements. Vapor-recovery units also are available.

Loading-arm pipes can be jacketed if heat is required to maintain the product temperature. However, the swivel joints cannot be jacketed. Where metals are not suitable for the corrosive liquid, plastic-lined piping can be used. Although the flexibility of the system is reduced, a swivel joint bypassed with a hose of Teflon has been used successfully. (See Fig. 5.)

Price comparison

To sample the pricing differences, consider a piece of hose 2 in. dia. × 10 ft long with Type 316 stainless steel male pipe threads at each end. The chemical-rubber hose using a cross-linked-polyethylene inner tube would be the least expensive.

The same size metal hose in Type 316 stainless steel with single braid and 316 stainless-steel male ends costs about 25% more. Convoluted Teflon with stainless-steel braid and 316 stainless-steel male ends is about 80% more than the polyethylene-lined rubber. And a stainless-steel loading-arm assembly with a ten-foot reach costs approximately 800% more.

The chlorine connection

One of the most difficult liquids to handle through a flexible hose is chlorine. The only hose recognized by the Chlorine Institute is made of Monel. This hose consists of a Monel inner core, two Monel braids and Schedule 80 Monel male ends. Great skill is required in the proper welding of the fittings.

The standard TFE/stainless-steel-braided hose has been tried and found to be unacceptable. As chlorine passes through the hose under pressure, it effuses through the Teflon tube in extremely small quantities. However, moisture in the air combining with the effused chlorine can corrode the braid, with subsequent hose failure.

Teflon-hose manufacturers have developed special types for chlorine service. One company uses the smooth hose design; another, a convoluted innercore design. The inner core of both consists of TFE Teflon.

The braid in each case is Kynar monofilament, double braided, and the fittings are Schedule 80, Monel, male National Pipe Thread. Pressure ratings are 375 psi working pressure at temperatures from −20°F to 120°F. The hose has a minimum burst pressure of 1875 psi, resulting in a safety factor of 5. The units are factory-assembled and proof-pressure-tested to 750 psi before shipment.

In each case, the manufacturer uses a swaging system to secure the Monel male ends to the hose. The smooth-bore type of hose uses a Kynar internal spring to prevent kinking of the inner core. The convoluted design requires no inner support to prevent kinking or tube collapse. The advantage of the Teflon hose is that it is non-corroding and non-aging in this extremely rough service.

The molten-sulfur connection

Transfer of molten sulfur requires a special type of hose or loading arm, because it must be heated. Sulfur, to remain at a flowable viscosity, must be held at or close to 290°F. When the temperature is below 250°F, solidification results, and in the range of 315 to 320°F, the viscosity increases beyond the pumping range.

To maintain the proper pumping temperature, the connection requires either steam jacketing, internal or external steam tracing, or electrical tracing.

One solution in the past has been to use a jacketed metal hose and to introduce 50-psi steam to the jacket. At one installation, a 3-in. metal hose with a 5-in. jacket was used. This did the job but was extremely bulky—it required at least three persons to hook up the system. This hose was replaced by a 3-in. convoluted Teflon hose having swaged-on carbon-steel male fittings. On one end, a 3-in. 90-deg.-elbow camcoupler was attached, and the other end had a lightweight reducing tee for connecting to the pump. This hose was internally traced through the bull of the tee using ½-in. flexible stainless-steel unbraided hose (suitable at 50 psi) and coming out through the side of the camcoupler (see Fig. 6). Only one person was required to hook up this system.

Sulfuric acid unloading

A recent installation of Teflon convoluted hose was made to unload barge quantities of sulfuric acid. This hose was double-braided stainless steel with overall Type 304 stainless-steel interlocking armor. The end fittings are stainless-steel Teflon-coated stub ends swaged on with carbon-steel (Van Stone) backup flanges. The bend radius of this 4-in.-I.D. hose is 20 in.

Previously, a 4-in. rubber hose had been used, which developed a leak and sprayed acid all over the barge before the system could be secured. (There was some suspicion of age deterioration.)

Hose of Teflon does not deteriorate with age. With armor to protect the braid, and with proper handling, this hose should last a goodly number of years. If a leak should develop, the armor will dissipate the spray pattern and lessen the possibility of personnel injury.

Maintenance of the connection

Preventive maintenance of hose is based on proper installation and handling. Most hose failures occur at the fitting connection. A horizontal line connection to the hose is best made at a 90-deg. elbow so that there is not a sharp bend at the fitting. If the hose must run out horizontally—and will sag—12 to 24 inches of interlock armor "cuffs" will alleviate the sharp bend at the fitting, spreading the load and reducing the failure rate.

In connecting hoses to the system, there should be no twist induced in the hose. Unions or Van Stone flanges, properly used, will eliminate this possibility. With long hoses, both flexible metal and Teflon with external braid will last longer with a chafe sleeve. The chafe sleeve can be a spring guard, interlock armor, a rubber cover, or plastic film.

Periodic testing at 1.5 times the operating pressure will probably forewarn of impending flaws and help to avoid in-service failures. Each plant will have to set its own standards for test type and frequency for each kind of hose being used. Remember that metal hoses are subject to metal fatigue over a period of time, and straight-wall Teflon hoses are subject to kinking and should be restricted to the proper bend radius.

Loading arms are subject to the corrosive action of the liquids being handled, and swivel-joint seals should be checked and replaced periodically.

The author

Glen Chase is a sales representative for Phillips Co., P.O. Box 2008, South Hackensack, NJ 07606, telephone (201) 488-3008. Phillips Co. is a manufacturers' representative and fabricator. Formerly, the author worked for Wright Aeronautical (a div. of Curtiss-Wright), Gates Rubber Co., King and Sheppard, Yarway Corp., and Ferris Engineering. He holds a bachelor's degree in mechanical engineering from Virginia Polytechnic Institute.

Corrosion and piping materials in the CPI

The tremendous growth of the CPI continues to highlight need for improved service from materials used in industrial piping and related equipment. Optimum material selection in piping-system design for corrosive environments can present very complex problems involving both (1) behavior of metals and alloys in contact with corrosive fluids, and (2) basic corrosion theory.

Thomas C. Bauman and *Leslie T. Overstreet,* Ralph M. Parsons Co.

☐ No single, economical, corrosion-resistant material suits all conditions. So, selecting the optimum piping material requires a detailed, accurate profile of the operating environment. The data should encompass temperature ranges, pressure levels, and information on process variables, such as composition and concentration of reactants, chemical additives, occurrence of condensation or vaporization, etc.

A knowledgeable appraisal of possible interplay between environmental factors (temperature, pressure, process) and material performance variables (mechanical, metallurgical, corrosion) can assist in identifying potential corrosion problems in a given piping system. Material suppliers, manufacturers' catalogs, applicable codes, technical publications, related plant experience, and laboratory testing programs provide useful information regarding material recommendations for transporting a particular fluid. Piping designed for handling industrial liquids will generally conform to the ANSI Pressure Piping Code (B31.3) for metallic materials.

However, the final selection of materials usually depends on service experience—in conjunction with an established economic standard based on initial cost, reliability, and maintenance requirements. Several major factors play a vital role in the design of a satisfactory piping system: (1) material selection and suitability, (2) method of fabrication and joining, (3) product availability and completeness, and (4) current and accurate cost comparisons.

Theory of corrosion

Corrosion, the deterioration of a material by chemical or electrochemical reaction with its environment, generally relates to metallic components. For nonmetals, some measurable change in physical or mechanical properties must define performance criteria.

Corrosion—a difficult and complex problem to analyze, due to the number of variables involved—represents one of the most serious destructive agents confronting the engineer. In designing a piping system, the engineer must not only meet strength requirements, but also prevent serious deterioration of material as a result of exposure to its environment. In order to accomplish this, he must understand the mechanism, nature, control, and manifestation of corrosion.

Probably one of the most familiar forms of corrosion occurs by direct chemical attack, where the metal enters into a chemical reaction with other elements such as oxygen, chlorine, etc. to form a nonmetallic compound. Corrosion cells exist on the surface of a metal or alloy in contact with a corrosive medium. This surface consists of anodic and cathodic areas. At the anodic regions, metal ions leave the metal and enter into the solution, thus "dissolving" the surface.

Metals attacked by this method generally exhibit an etched appearance, or may darken by the presence of compounds which form. This type of corrosion can result in the creation of a protective, or passive, film. Examples include rust on iron, the oxide layer on aluminum, and "passivation" of stainless steel by immersion in nitric acid. If they remain intact, films formed from this type of action, either naturally or chemically, will generally protect the metal or at least retard further degradation. However, high stream velocities, vibration, and thermal shock can all have adverse effects on film continuity. The pickling of pipe to remove mill scale illustrates controlled corrosion by direct attack.

Galvanic action, or bimetal corrosion, exemplifies another common mechanism. Two dissimilar metals establish an electrical potential when they come in contact, or are connected by an electrical pathway, in the presence of a conducting solution (electrolyte). The difference in potential between the two metals causes

Originally published April 3, 1978

the more active one (anode) to go into solution by forming positive ions. The relative tendency for a metal to deteriorate this way follows the order shown in the electromotive series of metals with respect to hydrogen or in a galvanic series for a particular environment, such as seawater. Tables I and II.

The metal closest to the anodic (active) end will corrode when two metals are placed in contact. Additionally, the farther apart in the series, the greater the tendency to corrode, all other factors being equal. However, two metals of different composition in a solution—such as salt water—with no electrical connection, may both be attacked by direct chemical action. Conductivity, composition, temperature, hydrogen formation, degree of polarization, and flow velocity of the electrolyte also become important factors affecting corrosion rate.

In bolted joints, such as pipe flanges, the relative sizes of the anodic and cathodic areas become important. An anodic area that is small in relation to the cathodic area accelerates the corrosion. Where flanges and bolts are dissimilar metals, the bolting metal should be cathodic to the flange metal.

Corrosion can also occur when a concentration cell forms on the surface of a metal exposed to a solution varying in composition or concentration. The most important concentration cells are those which are generated by changes in the amount of dissolved oxygen. In actual service, concentration cells are likely to form when the initial corrosion forms discontinuous layers of corrosion products on the metal surface. In such a case, precipitation of rust can establish an oxygen concentration cell, with the shielded portion under the rust acting as the anode, and the adjacent rust itself as the cathode.

This occurs more readily in a quiescent solution than in a moving one, which carries the corrosion products away as they are formed. A break in the continuity of a solid coating can also promote a similar situation by providing a void into which oxygen does not diffuse readily. This creates a concentration differential with the surface environment and results in pitting. Pitting is a form of concentration-cell corrosion where the accumulation of non-adherent products forms an electrolytic cell with the base metal. Pit formation during the corrosion process is readily recognizable because of the small localized areas of metal where reaction takes place to produce the pits of different depth and size.

Intergranular corrosion occurs when material in the grain boundaries of certain alloys is less resistant to the corroding agent than the grains themselves. Penetration of the corroding media follows grain boundaries and destroys the bond between grains. This action can seriously affect the strength of an alloy. Intergranular attack is difficult to detect by visual surface examination of the metal, although microscopic examination will readily reveal such penetration along the grain boundaries. Certain grades of stainless steel are known to be highly susceptible to this type of corrosion. This type of attack does, however, also occur in other alloys.

Another phenomenon that can significantly affect the rate of corrosion is erosion. Pipeline corrosion-erosion occurs when action of liquid stream or its particulate matter mechanically erodes the protective corrosion-product film responsible for passivity. This effect increases with increasing velocity, up to some maximum rate, and varies with smoothness of the metal surface, degree of turbulence, and presence of impurities.

The means of reducing corrosion include: (1) Use of corrosion-resistant metals and alloys, (2) protective coatings such as paint, electroplating, etc., and (3) cathodic protection by the use of a sacrificial metal higher in the electropotential series.

Corrosive fluids and environments

Water

Since corrosion by water is quite commonplace, its effects may be much less spectacular than those produced by other fluids; however, water—fresh or salt—is the most widespread corrosion medium encountered in industrial service.

The term "fresh water" implies any well, river, pond, or lake water not within the classification of salt water. In the case of a municipal water supply, it may have undergone chlorination, aeration, or filtration to render it potable. If destined for use as boiler feed, it will probably have been treated to control scale formation. In some instances, such treatments can significantly reduce the corrosive character of water. Interestingly, the natural scale, which reduces heat transfer and chokes pipeline cross-sections, may actually be beneficial—in thin deposits—in view of its ability to limit corrosion.

Most natural waters have pH values close to the neutral point; while they may cause some corrosion by hydrogen evolution, most of the corrosion they create will proceed by oxygen depolarization. Depolarization occurs where oxygen diffuses to the metal surface and combines with the hydrogen that has previously evolved in the formation of positively charged ions, making the cathode surface available again for hydrogen deposition. Fresh waters generally contain enough dissolved oxygen to continue causing this kind of corrosion in the absence of protective coatings.

The term "salt water" applies to seawater and to other natural or brackish waters used for condenser cooling and other industrial purposes. Salt waters usually contain, in addition to the usual calcium and magnesium bicarbonates, substantial concentrations of chlorides and sulfates of those metals and of sodium and potassium as well. The readily soluble salts of these latter metals do not tend to precipitate out on metal surfaces in the form of protective coatings.

In addition, the chlorides in salt water appear to interfere with the formation of protective films by other constituents in the solution. They can diffuse through, and cause corrosion beneath, protective films, and so are usually more corrosive than fresh waters.

Salt-water corrosion frequently takes the form of pitting. Salt water can induce local failures of the natural oxide coatings on metal surfaces, reducing their ability to resist corrosion. Copper alloys, used extensively in salt-water service, owe their usefulness to the fact that they are not subject to this type of attack.

Because of the usual tendency toward pitting which characterizes salt-water corrosion, thin-wall piping for this service must be made from material which not only has satisfactory resistance to general corrosion but also corrodes uniformly without pitting.

Acids

In corrosion terminology, an acid can be defined as a substance which, when dissolved in water, increases the hydrogen-ion concentration, thus reducing the pH to some value less than 7.0.

The mineral, or inorganic, acids listed here fall into three groups (Table III).

Under normal conditions of temperature and pressure, the members of Group I exist as gases. They ordinarily are used in the form of water solutions of those gaseous compounds. Since the gases tend to escape from solution at elevated temperatures, handling these hot acids may involve a vapor phase as well as the liquid phase.

Hydrochloric acid, an important industrial reagent, enters into the manufacture of pharmaceuticals and many chemicals in the paper, sugar, leather, and textile industries. Hydrofluoric acid finds wide use as an etching agent in the manufacture of frosted glass, as a pickling solution in the metal industry, in the purification of beet sugar, and as an antiseptic agent in the manufacture of brewed and distilled liquids. The concentrations commonly encountered include 30%, 48%, and 60% HCl in the commercial grades.

The acids in Group II form when their anhydrides dissolve in water.

Carbonic, sulfurous, and sulfuric acid show comparatively slight solubility of their respective anhydrides in water at ordinary temperatures and pressures. Problems in handling them may involve both liquid and vapor phases.

Carbonic acid, encountered in the manufacture of dry ice, though rather unstable and a weak acid, can cause troublesome corrosion. It is present in all carbonated beverages.

Chromic acid, an important source of chromium in metal-plating baths, is also used in printing and lithographers inks, paints, rubber pigments, colored glasses and enamels, and in the tanning industry.

Nitric acid is an essential raw material for the manufacture of explosives, pyrotechnics, fertilizers, and dyes, and in metallurgical processes. Most purposes require a grade containing 90% HNO_3 or more. Common laboratory grades contain about 68% HNO_3 by weight.

Phosphoric acid plays a critical role in the manufacture of phosphate chemicals, ethylene gas, dental cements, incandescent-light filaments, rubber, and textiles, and as a substitute for citric and tartaric acid in the production of jellies and soft drinks. It also finds application in rust-preventive coatings for iron and other metals.

Sulfuric acid, the world's most widely used manufactured chemical, finds major uses in the fertilizer, petroleum, chemical, metallurgical, paint, explosives, and textile industries. Handling sulfuric acid presents many problems, because its manufacture and many of its industrial processes involve broad ranges of concentration at temperatures up to its boiling point. Common grades include 66%, 78%, 93% and 98% H_2SO_4, as well as oleum (>100% H_2SO_4).

The members of Group III have rather indefinite compositions, but they contain varying percentages of one or more acids from Groups I and II.

Many processes using nitric acid employ it in the form of mixtures with sulfuric acid, the latter acting initially as a carrier for the nitric acid and later also as an absorbent for the water formed in the nitration reaction.

The term "sludge" acid usually applies to waste acids resulting from treatment of petroleum oils with sulfuric acids of various strengths. In addition, piping often carries alkaline solutions that follow the acid treatment.

Alkalies

When dissolved in water, alkali chemicals increase the hydroxyl ion concentration, thus raising pH above 7.0. Chemicals that yield alkaline solutions are either water-soluble or water-insoluble. The hydroxides of sodium, potassium, and barium are examples of the

Electromotive series with respect to hydrogen — **Table I**

Electrode reaction				Standard oxidation potential, E°, V, 25°C*
Li	=	$Li^+ + e^-$		3.05
K	=	$K^+ + e^-$		2.93
Ca	=	$Ca^{++} + 2e^-$		2.87
Na	=	$Na^+ + e^-$		2.71
Mg	=	$Mg^{++} + 2e^-$		2.37
Be	=	$Be^{++} + 2e^-$		1.85
U	=	$U^{+3} + 3e^-$		1.80
Hf	=	$Hf^{+4} + 4e^-$		1.70
Al	=	$Al^{+3} + 3e^-$		1.66
Ti	=	$Ti^{++} + 2e^-$		1.63
Zr	=	$Zr^{+4} + 4e^-$		1.53
Mn	=	$Mn^{++} + 2e^-$		1.18
Nb	=	$Nb^{+3} + 3e^-$	ca.	1.1
Zn	=	$Zn^{++} + 2e^-$		0.763
Cr	=	$Cr^{+3} + 3e^-$		0.74
Ga	=	$Ga^{+3} + 3e^-$		0.53
Fe	=	$Fe^{++} + 2e^-$		0.440
Cd	=	$Cd^{++} + 2e^-$		0.403
In	=	$In^{+3} + 3e^-$		0.342
Tl	=	$Tl^+ + e^-$		0.336
Co	=	$Co^{++} + 2e^-$		0.277
Ni	=	$Ni^{++} + 2e^-$		0.250
Mo	=	$Mo^{+3} + 3e^-$	ca.	0.2
Sn	=	$Sn^{++} + 2e^-$		0.136
Pb	=	$Pb^{++} + 2e^-$		0.126
H_2	=	$2H^+ + 2e^-$		0.000
Cu	=	$Cu^{++} + 2e^-$		−0.337
Cu	=	$Cu^+ + e^-$		−0.521
2Hg	=	$Hg_2^{++} + 2e^-$		−0.789
Ag	=	$Ag^+ + e^-$		−0.800
Pd	=	$Pd^{++} + 2e^-$		−0.987
Hg	=	$Hg^{++} + 2e^-$		−0.854
Pt	=	$Pt^{++} + 2e^-$	ca.	−1.2
Au	=	$Au^{+3} + 3e^-$		−1.50

*Standard reduction potentials, ϕ°, have the opposite sign.
From: Uhlig, H.H., *"Corrosion and Corrosion Control,"* 2nd ed., p. 29.

Galvanic series of metals in seawater — **Table II**

Active (Read down)

- Magnesium
- Magnesium alloys
- Zinc
- Aluminum 5052H
- Aluminum 3004
- Aluminum 3003
- Aluminum 1100
- Aluminum 6053T
- Alclad
- Cadmium
- Aluminum 2017T
- Aluminum 2024T
- Mild steel
- Wrought iron
- Cast iron
- Ni-Resist
- 13% Chromium stainless steel, type 410 (active)
- 50-50 lead-tin solder
- 18-8 stainless steel, type 304 (active)
- 18-8, 3% Mo stainless steel, type 316 (active)
- Lead
- Tin

(cont. at top of next column)

- Muntz metal
- Manganese bronze
- Naval brass
- Nickel (active)
- 76% Ni-16% Cr-7% Fe (Inconel) (active)
- Yellow brass
- Aluminum bronze
- Red brass
- Copper
- Silicon bronze
- 5% Zn-20% Ni, bal. Cu (Ambrac)
- 70% Cu-30% Ni
- 88% Cu-2% Zn-10% Sn (composition G-bronze)
- 88% Cu-3% Zn-6.5% Sn-1.5% Pb (comp. M-bronze)
- Nickel (passive)
- 76% Ni-16% Cr-7% Fe (Inconel) (passive)
- 70% Ni-30% Cu (Monel)
- Titanium
- 18-8 stainless steel, type 304 (passive)
- 18-8, 3% Mo stainless steel, type 316 (passive)

Noble (Read up)

From: Clark, D. and Varney, W., *"Physical Metallurgy for Engineers"*, 2nd ed., 1962, p. 301

water-soluble groups. The water-insoluble group includes hydroxides of calcium and magnesium, which form heavy suspensions in water.

Perhaps the greatest single threat to metals from alkali substances is caustic embrittlement, often characterized by intergranular attack. Temperature control, stress relief of steel, and the presence of sulfate ions in the process stream, will all tend to inhibit this type of corrosion.

Salt solutions

Numerous solutions of metal salts are encountered in the process industries. Many of them lie within the narrow neutral pH zone, others are strongly alkaline (high pH), and still others fall on the acid (low) side of the pH range. No hard and fast rules cover the choice of materials for handling salt solutions. Solutions that are normally acid or neutral may sometimes be treated to render them alkaline. Typical examples of salts from each of these groups are shown in Table IV.

Solvents and oils

The term "solvents" covers a large number of organic fluids. The choice of piping materials used in their manufacture and transport is determined basically by the chemicals present in the solvent compound. Representative solvents are listed in Table V.

Oils fall into two main groups, excluding petroleum products—animal oils and vegetable oils. Examples of each are listed in Table VI. These products rarely introduce serious corrosion problems. In many cases, ordinary grades of steel pipe work entirely satisfactorily.

Food products

The number of food products now being processed industrially has become so vast, and the processing methods so diverse, that it is difficult to give detailed facts regarding their behavior toward metals. However, in all cases, equipment must not contaminate the process fluid to any degree that might either endanger consumer health or affect flavor.

The corrosion that food products cause often results from dissolved carbon dioxide (carbonic acid) or organic acids, which foods characteristically contain. Also, in special cases, small amounts of mineral acids (e.g., phosphoric acid) can form. To facilitate proper material selection for food process piping, the engineer should specify the product being processed and the stage of the process. Type 316 stainless steel has gained wide acceptance for piping and other food-processing equipment.

Metallic piping systems

Iron and steel

This basic category includes iron, cast irons, and carbon steel (but not stainless steel) used for piping. In general, these materials do not differ greatly in their corrosion resistance. It is usually other factors that determine choices among them: physical properties, cost, availability, and personal preference. While there are exceptions to the above statement, the applications for which the carbon and low-alloy steels were developed generally do not involve corrosion resistance as a primary consideration.

Three groups of inorganic (mineral) acids — **Table III**

Group I (Halogen acids)	Group II (Oxy acids)	Group III (Mixtures)
Hydrobromic (HBr)	Carbonic (H_2CO_3)	Mixed acids
Hydrochloric (HCl)	Chromic ($H_2Cr_3O_7$)	Sludge acids
Hydrofluoric (HF)	Nitric (HNO_3)	
	Phosphoric (H_3PO_4)	
	Sulfuric (H_2SO_4)	
	Sulfurous (H_2SO_3)	

Typical examples of salts — **Table IV**

Acid salts	Alkaline salts	Neutral salts
Aluminum sulfate	Sodium carbonate	Calcium chloride
Sodium bisulfate	Sodium bicarbonate	Calcium nitrate
Zinc sulfate	Trisodium phosphate	Calcium sulfate
Aluminum chloride	Sodium silicate	Magnesium chloride
Ferric chloride	Sodium cyanides	Magnesium nitrate
Ferrous chloride	Potassium sulfides	Magnesium sulfate
Stannic chloride	Barium sulfides	Disodium phosphate
Stannous chloride		Barium chloride

In most situations, their serviceability depends on the fact that following the initial corrosive attack, protective films form that tend to reduce the corrosion rate to less than 20 mils/yr. With few exceptions, these "natural" protective films do not form on the surface of iron or steel when in contact with dilute acid solutions. But, in neutral and alkaline solutions, protective coatings do form readily on the surface, since the hydrogen-ion concentration has not risen high enough to promote rapid corrosion. Corrosion that does occur is controlled by the concentration of dissolved oxygen present and the ease with which it can diffuse from the body of the solution to the metal surface. Over the pH range of 4 to 10, corrosion rate remains constant in aerated water.

When controlled only by diffusion of oxygen, corrosion rate at a given concentration will approximately double for every 30°C rise in temperature. Where attended by hydrogen evolution, the rate more than doubles for the same temperature increase (e.g., iron-corroding in hydrochloric acid will double for every 10°C rise).

In spite of its limitations, carbon steel remains the most widely used material in the process industries for metallic piping systems. In the absence of corrosive liquids and special processing, its range can extend from approximately 20°F to about 1,000°F. Much present-day steel-pipe usage conforms to API 5L or ASTM A53 Grade B. These materials are most economical, and have code-specified carbon content ($<$0.3%) for good weldability.

For service above 750°F, the preferred silicon-killed grades (ASTM A106) provide homogeneity and improved mechanical properties. Most piping used to transport fluids is made from steel produced in open-hearth furnaces or oxygen converters. These grades of steel provide excellent chemical composition and residual-element control. Evidence indicates that impurities and lack of homogeneity have some influence on the manner in which ferrous materials corrode.

Stainless steel

The three main classes of stainless steels, designated in accordance with their metallurgical structure, include martensitic, ferritic, and austenitic. Nickel-bearing austenitic types have the highest general corrosion resistance, and higher-nickel alloys are more resistant than lower-nickel compositions. Stainless steels of primary interest include: Types 304, 316, 321, and 347, and special alloys such as Carpenter 20Cb3. Pipe and pipe-fittings are available in all these alloys.

Type 304 has a wide range of corrosion resistance, but the lowest overall resistance. It can suffer from intergranular corrosion due to carbide precipitation caused by welding. "L" grades, limiting the carbon content to $<$0.035%, help reduce this tendency toward intergranular corrosion.

Type 316 resists corrosion, in general, better than the 304 grade, with one exception: with oxidizing acids such as nitric. However, due to its molybdenum content, it resists pitting better and has excellent resistance to fatty acids at high temperatures. This makes it popular for handling pharmaceuticals and food products, to avoid excessive contamination by metals.

Types 347 and 321 resemble the 304 grade, except for the addition of columbium to the former and titanium to the latter, to further reduce intergranular corrosion. In general, they show corrosion resistance very similar to that of Type 304.

The Carpenter 20Cb3 alloy offers improved corrosion resistance in handling sulfuric acid. Other grades of stainless steel do not resist attack by dilute H_2SO_4 as effectively.

Stainless steels perform best under fully aerated or oxidizing conditions, which favor the passive state. When handling chemicals, keeping the alloy surface clean and free from contamination helps eliminate concentration cells that might cause pitting. Physical properties of commonly encountered stainless steels are shown in Table VII.

In general, stainless steels resist corrosion by:

- Nitric acid.
- Dilute sulfuric acid at room temperature, if aerated.
- Most organic acids, including food acids.
- Alkalies, except under stress in hot concentrated caustic.

They do not resist:

- Dilute or concentrated HCl and HF.
- Oxidizing chlorides.
- Some organic acids, including oxalic, formic, and lactic acids.

Aluminum

Aluminum pipe is available in various alloy compositions that show very little difference in actual corrosion resistance. Commonly used alloys include: 3003, 5052, 6061, and 6063. Aluminum piping serves widely in the cryogenic field since the values of its tensile and yield properties increase, and ductility improves, at lower temperatures. Some room temperature properties are listed in Table VIII.

The average thermal expansion amounts to 1.5 inches per 100 feet of pipe per 100°F rise, and adequate allowance should be made for it. Aluminum is light in weight and has good resistance to atmospheric corrosion. Aluminum also resists:

- NH_4OH (hot or cold).
- Acetic acid (hot or cold).
- Citric, tartaric, and malic acids.
- Fatty acids.
- Nitric acid, 80% up to 120°F.
- Distilled water.
- Sulfur and its compounds.

It does not, however, resist the effects of:

- Acids, such as H_2SO_4, HF, and HCl.
- Strong alkalies.
- Mercury and its salts.
- Chlorinated solvents.
- Seawater (subject to pitting).

Copper and its alloys

Copper pipe is generally made of pure, electrolytically refined copper, deoxidized with phosphorus. Its applications range from water (hot and cold), steam, fuel oil, and lubricants, to many uses in the food industry. Pure-copper pipe—joined by means of threading,

Solvents commonly used in the process industries — **Table V**

Alcohols	Chlorinated solvents
Amyl alcohol	Carbon tetrachloride
Butyl alcohol	Chloroform
Ethyl alcohol	Ethylene dichloride
Ethylene glycol	
Glycerine	
Methyl alcohol	
Petroleum solvents	**Miscellaneous solvents**
Naphtha	Acetone
Gasoline	Ethers
Kerosene	Ethyl sulfate
Oleum spirits	Turpentine

soldering, or brazing—has virtually complete immunity to stress-corrosion cracking. However, alloys containing phosphorus or over 15% zinc may be susceptible. Also, copper is sensitive to corrosion, by high-velocity water or other aqueous solutions, by impingement attack. The rate of corrosion increases with the dissolved-oxygen content.

Copper will resist corrosion by:

- Seawater.
- Atmospheric exposure.
- Deaerated, dilute H_2SO_4, H_3PO_4, and other non-oxidizing acids.

It will not resist:

- Oxidizing acids, e.g., HNO_3.
- NH_4OH plus O_2 (which can cause stress-corrosion cracking).
- Hydrogen sulfide, sulfur and its other compounds.

Copper/zinc alloys (brasses) show improved physical properties and resist impingement attack better than copper alone. Therefore they find wide use as condenser tubes. However, brasses can corrode by "dezincification," a selective attack that leaches the zinc out of the copper base-metal, leaving a porous sponge-like structure. This becomes more likely as zinc content increases, and in differential aeration or high-velocity conditions. Inhibited grades of brass containing arsenic or antimony are more resistant to dezincification. Brasses may crack when exposed to ammonia and its compounds.

Nickel and nickel-based alloys

Nickel piping is available as commercially pure wrought nickel and in a low-carbon version (<0.02% C).

Nickel resists:

- Hot and cold alkalies.
- Dilute nonoxidizing organic and inorganic acids.
- Oxidation in air at elevated temperatures.

It does not resist effects of:

- Oxidizing acids (e.g., HNO_3).
- Oxidizing salts (e.g., $FeCl_3$).
- Aerated ammonium hydroxide.
- Seawater.
- Sulfur or sulfur-reducing environments (>600°F).

When alloyed with copper, nickel shows improved resistance under reducing conditions. The 70% Ni, 30% Cu alloy (Monel) resists the effects of high-velocity seawater, boiling sulfuric acid in concentrations <20%, and unaerated HF at all concentrations and at temperatures up to boiling. Its resistance to alkalies is good, except for hot concentrated caustic.

Inconel, a nickel-iron-chromium alloy, has become widely used in the food industry because of its resistance to organic acids. Immune to chloride-ion stress-corrosion cracking, it may crack in caustic or weak sulfurous acid. It performs ideally for handling alkaline sulfur solutions in the manufacture of paper.

Hastelloy signifies a line of commercial alloys containing nickel, molybednum, and chromium with smaller amounts of iron. The B- and C-designated alloys enjoy the widest use. Hastelloy B exhibits excellent resistance against reducing chemicals (e.g., HCl). It resists boiling sulfuric acid up to 60%, but does not resist oxidizing agents such as HNO_3. Hastelloy C resists wet chlorine gas, hypochlorite solutions, and chlorine dioxide solutions. It has exceptional resistance against strong oxidizing solutions, and does not pit in seawater. Both alloys have a tensile strength of about 100,000 psi and a yield strength of about 45,000 psi.

Hastelloy pipe can be joined by welding, threading, or flanging. Care must be taken not to induce stresses during installation, by properly aligning and supporting the structure.

Special metals and alloys

Special metals include titanium, tantalum, and zirconium. Titanium, generally considered an exotic metal, resists the effects of oxidizing media in the presence of chloride ion, aqua regia at room temperature, and wet chlorine. It resists nitric acid at elevated temperatures better than stainless steels do, and shows marked resistance to pitting and crevice corrosion in seawater. Titanium cannot handle hydrofluoric acid or dry chlorinated compounds, however.

Zirconium exhibits general corrosion resistance to most oxidizing media and reducing media in the absence of chlorides. This broad range of corrosion resistance results from formation of a tough inert-oxide surface film. The commercial grade of zirconium used in the chemical industry contains up to 2.5% hafium.

In brief, it resists:

- All concentrations of alkalies up to their boiling point.

Common commercial animal and vegetable oils — **Table VI**

Animal-oil group:	Vegetable-oil group
Cod liver	Castor oil
Lard	Coconut
Sperm	Cottonseed
Neatsfoot	Linseed
Tallow	Palm
	Peanut
	Soybean oil

- Hydrochloric acid.
- Nitric acid.
- Sulfuric acid < about 70% (boiling).
- Phosphoric acid < about 55% (boiling).

Zirconium will corrode in the presence of:

- Wet chlorine.
- Hydrofluoric acid.
- Aqua regia.
- Oxalic acid (boiling).

Tantalum is limited by its high cost, to use only in extremely corrosive conditions. For economy, most tantalum piping consists of thin-wall tubing inside of carbon-steel pipe. Though attacked by alkalies (causing embrittlement), hydrofluoric acid, and fuming sulfuric, tantalum does, however, resist hydrochloric acid, nitric acid, and sulfuric (except fuming) at all concentrations up to the boiling point. Physical properties of these special metals are given in Table IX.

ASTM specifications for metallic pipe

The American Soc. for Testing and Materials establishes certain minimum mechanical standards for piping. These requirements appear in the appropriate ASTM specification (listed in Table X). Table XI shows typical chemical compositions of the various metals used in piping.

Non-metallic piping systems

Plastics have found wide application in chemical process plants because of their excellent resistance to a wide range of chemicals. They are often used at lower installed cost for many conditions which would otherwise require very expensive metals. Although particularly effective against inorganic substances, they often cannot compete with metals in their resistance to organic solvents.

Each piping system bears the name of the resin from which it is manufactured. Proper formulation, compounding, and method of manufacture can improve some properties of the base resin, but usually at the expense of other properties.

The mechanism of attack for plastics differs from that for metals, which corrode by electrochemical reaction. Plastics are attacked by solvation or chemical reaction. Solvation involves penetration of a corrosive agent into the plastic, causing softening, swelling, and ultimate failure. Plastics usually resist attack completely or they deteriorate rapidly, generally offering excellent resistance against most dilute acids, alkalis, and salt solutions.

Various sources can provide information on plastics and other nonmetallics. These include "Resins and Molding Compounds" ("Modern Plastics Encyclopedia"), "Plastics and Rubber" (*Materials in Design Engineering*), and Perry's "Chemical Engineers' Handbook." A knowledge of manufacturing methods can help the engineer to understand the properties and inherent cost factors important in using plastic. Much pipe is made by casting or molding, although extrusion has become quite common. Production of pipe is usually highly automated.

The engineer must check suitability of a given resin for maximum operating pressure and temperature (internal and external). Temperature is particularly important, and almost any increase or decrease can result in significant property and behavior changes. All plastics gradually elongate or "creep" when subjected to sustained loads for long periods of time, even at ambient temperatures. The need for adequate support becomes important when specifying plastic pipe. Current codes have recently expanded to include non-metallic piping. The engineer should also make use of manufacturers information and actual service experience when selecting a plastic material.

Polyvinyl chloride (PVC), generally stronger and more rigid than other thermoplastic materials, exhibits excellent resistance to hydrochloric acid but is susceptible to damage by ketones, aromatics, and some chlorinated hydrocarbons. It finds extensive use in process- and potable-water piping.

Most reinforced polyester pipe uses glass fibers as the reinforcing agent. Some chemicals, however, such as hydrofluoric acid and fluorides, will attack the glass fibers. These instances call for the use of nylon, Dacron, or asbestos. The average commercial polyester pipe will contain 70% resin and 30% reinforcing material after manufacture.

In general, there are five basic types of polyester resins, including:

1. Isophthalic types with a maximum operating temperature of 140 to 160°F. (These resins are used in the handling of foodstuffs and potable water.)
2. Bisphenol-A isophthalic types (up to 180°F).
3. Hydrogenated bisphenol-A types. (These resist

Typical properties of some common stainless steels — Table VII

	Stainless steel type number				
	304	316	321	347	20Cb3
Ultimate strength, psi	85,000	90,000	80,000	95,000	85,000
Yield strength, psi	35,000	40,000	40,000	40,000	35,000
Rockwell hardness	B-80	B-85		B-85	B-85

Typical physical properties of aluminum alloys — Table VIII

	3003	5052	6061	6063
Tensile strength, psi	17,000	41,000	45,000	35,000
Yield strength, psi	8,000	36,000	40,000	31,000

Typical physical properties of special metals — Table IX

	Titanium	Zirconium	Tantalum
Tensile strength, psi	50,000	60,000	30–40,000
Yield strength, psi	40,000	35,000	20–30,000

alkaline as well as strong mineral or organic acids at temperatures up to 200°F.)

4. Chlorinated resins (usable at 250°F).

5. Fumarate bisphenol-A resins (allowable working temperature to 285°F).

Polypropylene, the lightest in weight of any of the thermoplastic materials, has a wider range of chemical resistance than polyethylene. It is used widely for salt-water disposal and in handling sulfur-bearing compounds. Not prone to environmental stress-cracking like polyethylene, it serves at temperatures up to approximately 200°F. Polypropylene is attacked by fuming sulfuric acid and liquid halogens, and swells rapidly in chlorinated solvents.

Other non-metallic materials of interest include impervious graphite, Transite, and clay. Impervious graphite—for handling hydrofluoric, sulfuric, and nitric acids—has some temperature and concentration limits. Also, it is unsuitable for use with free bromine, fluorine, and iodine, as well as chromic acid and sulfur trioxide. Normally, its porosity makes graphite unsuitable for use as a piping material; however, various resins used as impregnants will produce an impervious material. The pipe manufacturer should know what corrosives are to be handled, so he can provide the proper impregnant.

"Transite," the brand name for pipe produced from a mixture of asbestos fibers, silica, and cement, serves extensively in corrosive soils and, because of its light weight, as duct material in the transport of corrosive vapors.

Clay finds wide use in the conveyance of sewage, industrial wastes, and storm water. It is generally available in diameters ranging from about four inches up to three feet.

ASTM specifications for metallic pipe — **Table X**

Material	ASTM Specification
Carbon steel	A53 (Welded and seamless) A106 (Seamless pipe) A105 (Forgings) A181 (Forged flanges) A234 (Fittings)
Type 304, 304L Type 316, 316L Type 347	A312 (Seamless and welded) A182 (Forged flanges) A403 (Fittings)
20Cb3	B464 (Seamless)
Hastelloy B	B335 (Rod)
Hastelloy C	B336 (Rod)
Nickel 200, 201	B161 (Seamless pipe) B160 (Flanges) B366 (Fittings)
Monel 400, Inconel 600	B167 (Seamless pipe) B166 (Flanges) B366 (Fittings)
Aluminum 3003, 6061, 6063	B241 (Seamless pipe) B235 (Pipe, modified wall) B247, B361 (Flanges)
Copper	B43 (Seamless pipe)
Red brass	B43 (Seamless pipe)
Titanium	B337 (Seamless pipe) B363 (Fittings)
Zirconium	B353 (Tubing)

Lined-piping systems

The specification of lined piping systems for corrosive service is rapidly increasing. These include glass, plastics, elastomers, and various metals. Lined pipe systems usually rely on carbon steel as the main structural component. Its ease of fabrication and low cost permit use of carbon steel with most linings and manufacturing techniques.

Polytetrafluorethylene resin is virtually inert to all commercially available chemicals and solvents except molten alkali metals, and fluorine. As with other plastic linings, TFE is susceptible to permeation by gases or vapors. For example, use of too thin a TFE liner inside steel pipe for HCl service will allow acid vapors to permeate the liner and attack the steel. It is also limited in service under negative pressure conditions, which tend to collapse the liner. TFE can stand temperatures up to 500°F.

Fluorinated ethylene/propylene resin (FEP), is a companion resin to TFE, with about the same corrosion resistance but with an upper temperature limit of only 400°F.

Glass-lined steel pipe has one of the broadest ranges of corrosion resistance of any of the lined-piping systems with the exception of TFE and FEP. It resists all acids up to 350°F, excluding hydrofluoric acid. Its smooth surface improves product flow. But it has poor impact and thermal-shock resistance, requiring care in handling during installation and maintenance.

Rubber-lined pipe is perhaps the oldest of the lined-pipe systems. Natural rubber has excellent resistance to acid and alkaline solutions except those of oxidizing agents like nitric and chromic acids, and to animal, vegetable and mineral oils. Rubber linings are usually limited to continuous-process temperatures of about 200°F. Soft rubber has better abrasion resistance. The selection of a rubber lining should include consultation with the manufacturer, since each does his own compounding and therefore best knows the performance of his product.

Epoxy-lined pipe, produced by casting the resin onto a Transite pipe wall, produces a continuous protective film on the internal bore. These systems find application underground, when conveying corrosive fluids under acid soil conditions.

Tin-lined pipe, produced by applying a $\frac{1}{16}$-inch layer of tin to the inside of brass pipe, safely handles distilled and demineralized water, maintaining the purity of the water for drinking and scientific use.

"Kynar" designates a vinylidene fluoride resin which exhibits excellent chemical resistance against most corrosive agents and organic compounds, including acids, alkalies, strong oxidizers, and halogens. However, oleum attacks it at room temperature.

"Penton" chlorinated polyether resin serves widely in

higher-temperature ranges for handling corrosives and organic materials which other plastic linings cannot handle.

"Saran" polyvinylidene chloride liner material, used extensively in the plating industry, also enters into food processing and the handling of potable water.

Pipe-system economics

Many factors affect the final selection of a piping system. Initial investigation will usually show that several materials of construction seem acceptable candidates. The material cost in its fabricated form is important; however, the true cost appears only after installation of the system. Installed cost depends on many factors, including: material cost, size, pipe schedule, system complexity, joint make-up, fabrication techniques, and labor rates.

Selection of the piping material should keep future maintenance costs and downtime to a minimum. Estimated life expectancy of the operation must also be considered. Metallic piping materials usually have known corrosion rates; this permits specification of the least expensive material based on expected length of service. Criteria for a piping system having been established, a list of potential materials reduces to those which meet the requirements; this leads to a comparison of *relative* installed cost, which is needed for the most economic decision.

In order to give the reader a relative comparison for the installed cost of pipe systems in 1970, see Ref. 5, pp.

Chemical compositions of metallic piping materials **Table XI**

Material	C, %	Cr, %	Cu, %	Fe, %	Mn, %	Mo, %	Ni, %	Si, %	Other Elements, %
Admiralty metal	—	—	70	—	—	—	—	—	Zn, 29; Sn, 1
Aluminum 2S	—	—	—	—	—	—	—	—	Al, 99.5+
Aluminum 3S	—	—	0.1	0.5	1.25	—	—	0.5	Al, 97.7
Brass, admiralty	—	—	71	—	—	—	—	—	An, 28; Sn, 1
Brass, aluminum	—	—	76	—	—	—	—	—	Al, 2; Zn, 22
Bronze, silicon	—	—	96	—	—	—	—	3	
Carpenter 20 alloy	—	20	4	44	—	3	29	—	
Copper, arsenical	—	—	99.9+	—	—	—	—	—	As, 0.04
Croloy 2¼ alloy	0.15	2.25	—	96.1	—	1.0	—	0.5	
Cupro-nickel, 70-30	—	—	69.3	0.75	—	—	30	—	
Cupro-nickel, 90—10	—	—	88.8	1.25	—	—	10	—	
Durimet 20 alloy	0.07	20	3.5	45.7	—	1.75	29	—	
Duriron	0.85	—	—	84.7	—	—	—	14.5	
Hastelloy B alloy	0.1	—	—	5	—	28	62	—	
Hastelloy C alloy	0.1	16	—	5	—	16	55	—	W, 4
Incoloy alloy 800	0.04	21	0.3	45.3	1	—	32	0.4	
Inconel alloy 600	0.04	15.8	0.1	7	0.20	—	76	20	
Iron, cast	3.4	—	—	94.3	0.5	—	—	1.8	
Lead	—	—	—	—	—	—	—	—	Pb, 99.95
Monel alloy 400	0.12	—	31.5	1.4	1	—	66	0.15	
Monel alloy K-500	0.15	—	29.5	1.0	0.60	—	65	0.15	
Nickel 200	0.06	—	0.05	0.15	0.25	—	99.4	—	
Nickel silver, 18%	—	—	65	—	—	—	18	—	Zn, 17
NI-Hard alloy	3.4	1.5	—	89.5	0.5	—	4.5	0.6	
NI-Resist, Type I	2.8	2.5	6.5	69.6	1.3	—	15.5	1.8	
NI-Resist, Type II	2.8	2.5	—	71.9	1.0	—	20.0	1.8	
Steel, carbon-molybdenum	0.15	—	—	99.4	—	0.5	—	—	
Steel, low-carbon, electric-resistance-welded, ASTM A587-68	0.1	—	—	99.5	0.4	—	—	—	P, 0.01; S, 0.02; Al, 0.05
Steel, mild (SAE 1020)	0.2	—	—	99.1	0.45	—	—	0.25	
Steel, Ni-Cr (SAE 3140)	0.4	0.65	—	96.7	0.8	—	1.25	0.25	
Steel, stainless									
AISI Type 30, or ASTM A312-64, Gr. TP-304	0.06	18	—	72	—	—	10	—	
AISI type 316, or ASTM A312-64, Gr. TP-316	0.07	18	—	67.4	—	2.5	12	—	
AISI Type 347	0.07	18	—	70.2	—	—	11	—	Cb, 0.7
Tantalum	—	—	—	—	—	—	—	—	Ta, 99.9+
Titanium (Ti-50A)	0.08 max.	—	—	0.20 max.	—	—	—	—	N, 0.05 max.; H, 0.015 max.
Worthite alloy	0.07	20	1.75	46.7	—	3	25	3.5	

Ref: Masek, J. A., *Chemical Engineering*, June 17, 1968, Table I, p.216

347 and 349, in which the authors studied several mixtures of pipe and fittings. (When comparing costs, note that technology and competitive situations change constantly, causing wide fluctuations in prices. However, the relative differences generally remain the same.

The information obtained from such sources should be used only as a screening tool in estimating installed costs for corrosion-resistant piping. The final choice must depend upon detailed economic studies, including capitalized cost for the specific piping system under consideration. It is important here to use the most current information available.

The selection of piping material for use in a particular corrosive environment is not a simple matter. The best material is not always the most desirable, because factors other than corrosion will influence, and control, the final selection.

In some instances, specific physical properties may control the selection. For example, one might tolerate some sacrifice in corrosion resistance in order to obtain higher strength. In other cases, the best material might not be readily available, indicating a compromise between desirability and need. In many situations, complete resistance to corrosion is not really required, and the most desirable material will usually be available at moderate cost along with reasonable life. When plant conditions or processes differ from what is considered normal for a particular fluid, the material normally specified may also require change.

Many apparently extraneous factors can exert a tremendous influence on the progress of corrosion. It is often helpful to know something about the installation itself, such as through study of its layout or through previous experience. The opportunity to examine corroded pipe, valves, or fittings can aid materially in making effective recommendations in response to corrosion problems.

Factors which are not always obvious often affect behavior of metals and alloys in contact with corrosive fluids. The questions listed below may serve as a general guide for the type of information which often helps in evaluating the corrosive environment when selecting materials for piping.

1. What is the name or composition of the corrosive fluid?
2. What is the concentration (specific gravity, pH, etc.)?
3. What is the operating temperature and pressure?
4. Is the fluid a water solution or an organic fluid in which water may be present at any time?
5. Are there substances other than water, such as oil, solids, etc.?
6. Is air present, or is the opportunity for air-leaks high?
7. Is the flow intermittent or continuous?
8. Is the system ever flushed, rinsed, or drained out?
9. Is a slight amount of corrosion objectionable from a contamination standpoint?
10. Has any specific trouble or problem been experienced with certain materials?
11. What materials are being, or have been, used for valves and fittings?
12. What were the comparative lives of the materials used?
13. Are there special fabrication, handling, or installation problems?
14. What is the estimated installed cost or capitalized cost for the projected life of the piping?

References

1. "Properties and Selection of Metals," 8th ed., Metals Handbook, Vol. 1, American Soc. for Metals, 1961.
2. "Steel Piping, Tubing, and Fittings," 1977 ASTM Std., Pt. 1., American Soc. for Testing and Materials, Philadelphia, Pa.
3. Perry, J. H., and Chilton, C. H., "Chemical Engineers' Handbook," 5th ed., McGraw-Hill, New York, 1973.
4. Newell, H. D. and Rutherford, J. J. B., "Ferrous Metal Tubing in Oil Refining Processes," Third World Petroleum Congress (The Hague).
5. Schweitzer, P. A., "Handbook of Corrosion Resistant Piping," Industrial Press Inc., N.Y.
6. Evans, L. S., "Selecting Engineering Materials for Chemical and Process Plant," Wiley, New York, 1974.
7. "Petroleum Refinery Piping," ANSI B31.3, American National Standards Inst., New York, 1974.
8. Masek, J. A., Metallic Piping, *Chem. Eng.*, June 17, 1968.
9. NACE: Basic Corrosion Course, National Assn. of Corrosion Engineers, Houston, Tex.
10. Uhlig, H. H., "Corrosion and Corrosion Control," 2nd ed., Wiley & Sons, New York, 1971.
11. Gleekman, L. W. and Ayres, J. M., Plastic Lined Steel Pipe for the Process Industries, *Metal Prog.*, Jan. 1970.
12. Lochman, W. J., "Introduction to Corrosion Control," NACE Western States Corrosion Seminar, 1970.
13. Evans, W. R., "The Corrosion and Oxidation of Metals," Edward Arnold Pub., London, 1960.
14. Long, W. P. and Montrone, E. D., Economics, *Chem. Eng.* Deskbook, Oct. 12, 1970.
15. Bauman, Thomas C., Steel Selection and Usage for Refinery Operations, *J. Metals* (AIME), Aug. 1970.
16. Clark, D. and Varney, W., "Physical Metallurgy for Engineers," 24 ed., Van Nostrand, 1962.
17. Polar, J. P., "A Guide to Corrosion Resistance," Climax Molybdenum Co.
18. Short Course on Process Industry Corrosion, Ohio State University and National Assn. of Corrosion Engineers, 1960.

The authors

Thomas C. Bauman is an engineer with The Ralph M. Parsons Co., Pasadena, CA 91124. As a member of his company's Materials Application Group, he assists in resolving problems related to material selection and corrosion in chemical-process units. He holds a degree in metallurgical engineering from California Polytechnic State University and has completed graduate study in the field of energy systems. A Registered Professional Engineer in California, he holds memberships in the American Soc. for Metals, American Welding Soc., American Soc. of Mechanical Engineers and the National Assn. of Corrosion Engineers.

Leslie T. Overstreet is a principal materials application engineer with The Ralph M. Parsons Co. His responsibilities include material selection and failure analysis. He has several years of nuclear power experience, along with seventeen years of work in the petrochemical field. Mr. Overstreet has a B.S. degree from the University of California (Berkeley) and holds professional engineer licenses in corrosion and metallurgy. He belongs to the American Soc. for Metals, American Soc. of Mechanical Engineers and the National Assn. of Corrosion Engineers.

Tips on specifying FRP piping

Growth in the use of fiberglass-reinforced plastic piping has outpaced the adoption of uniform design standards.

L. S. Surtees, Kerr-McGee Corp., and P. Rooney, The Ralph M. Parsons Co.

☐ With the use of fiberglass-reinforced plastics for corrosive services rising as much as 12% per year, more and more engineers are being confronted with writing specifications for the purchase of FRP piping.

This article will describe one way to write a conservative specification for piping components handling fluids under pressure (up to 150 psig) in chemical and hydrometallurgical processes. Such description is needed because there is a lack of suitable standards around which a user's specification can be written for filament-wound pipe. Failure to recognize this lack has contributed to the horror stories sometimes recited about FRP piping system failures.

Existing standards—mostly for hand layup

Historically, the standard FRP piping has been hand-layup, custom-fabricated. However, technological advancement has resulted in wider use of the lower-cost filament-wound piping.

If you are a buyer of a piping system using hand-layup piping, you can specify that the pipe, fittings and flanges be made in accordance with NBS Standard PS-15-69 [*1*] and be reasonably confident that you will be buying a quality product. Unfortunately, there is no comparable standard that the buyer of filament-wound piping can specify. Such standards as do exist include ASTM D2996-71 [*2*] and ASTM D2517-73 [*3*]. From the user's point of view, these are deficient in that they are limited to pipe sizes under 12 in., oriented toward suppliers' (not buyers') needs, and unsuited for some process situations. The industry is working on more-suitable standards, but, in the meantime, the chemical or metallurgical process user of filament-wound pipe is forced to tailor-make his own specification.

There are many ways that filament-wound pipe can be manufactured and assembled. However, where the severity of the service conditions warrants, the most reliable and trouble-free installation will be obtained by selecting pipe that has an inner, 100-mil SPI (Soc. of the Plastics Industry) corrosion barrier, contact-molded fittings (with stub-ends on the flanges), and butt-and-wrap joints. Some references to support this approach include Mallinson [*4,5*] and Webster [*6*]. Of course, milder operating conditions (low corrosivity and pressure) and smaller diameters may allow some relaxation of the requirement for liner thickness, and may permit other than contact-molded fittings. But the following pertinent details should be included in any specification for all normal process conditions.

Even though the pipe is filament-wound, you should specify that the vendor conform to certain portions of PS-15-69 that apply to resin makeup and structural integrity. These include Sections 1.1, 3.1.1, 3.2.–3.3.5, 3.3.7.1, 3.3.8, 3.3.9, 3.5.1., 3.5.2., 3.5.4.–3.5.7.2, 3.5.8., 3.5.10 and 4.1–4.3.6.

The pipe wall

Filament-wound pipe should consist of (1) a corrosion barrier of "C" glass surfacing-veil made up of 90% resin and 10% glass and (2) an anti-wicking interior layer of chopped mat containing 75% resin and 25% glass. Over this should be a filament-wound outer layer consisting of no more than 40% resin and 60% minimum glass content. Thickness and construction of the corrosion barrier and anti-wicking layer should follow Sections 3.3.1 and 3.3.2 of PS-15-69. Thickness and construction of the outer layer should satisfy structural and operational demands on the pipe.

Fittings and joints

Hand layup (contact-molded) fittings, made per PS-15-69, should be specified for all piping over 12 in. dia. Flanges should have stub-ends. For piping smaller than 12 in., specifications can be relaxed to permit filament-wound fittings; however, for critical services, retention of the stub-ended hand-layup flanges is recommended.

Cheaper, molded adhesive-joint flanges and fittings are available, but should be used mostly for atmospheric lines such as vents and tank overflows.

For versatility in design, you will want to specify some hand-layup fittings, using various combinations of flanges and plain ends.

Shop and field joints. Specify butt-and-wrap joints. Permit the cheaper adhesive socket-type joints only on piping 12 in. dia. or smaller and then, preferably, only in low-pressure service. Butt-and-wrap joints should conform to the requirements of PS-15-69, Sections 3.3.5 and 3.3.6. However, change the last sentence of Section 3.3.5 to read "The interior of all accessible joints shall also be sealed by covering with not less than 0.100 in. of reinforced resin-rich surface as described in 3.3.1 and 3.3.2." For economical design, you will want to minimize the use of flanged joints.

Dimensions. Require that vendors submit certified data defining pipe outside diameter and wall thickness, together with tolerances. Advise them that they will be required to manufacture to these dimensions.

Originally published November 21, 1977

Tolerances with respect to lengths, roundness, flange-face flatness and orientation should be as you specify, or referenced to applicable sections of PS-15-69—for example, Section 3.5.1 defines pipe inside dimensions; Section 3.5.5, fittings; Section 3.5.5.1, elbows.

Mechanical properties. The best approach is to specify the working pressure requirements for the pipe, fittings and flanges (25, 50, 75, 100, 125 or 150 psig) and require that individual pieces (or assemblies) be hydrotested at 1.5 times the working pressure rating. If you want hydrotesting done in the vendor's shop on every piece or on portions of the production run, be sure to state so in your specification.

The design safety factor for the pressure rating of the pipe should be 10 (i.e., design pressure is 10 times the hoop tensile strength). In order for your piping system designers to make necessary calculations, specify that the vendor submit for each pipe size and pressure rating calculated data on minimum (a) hoop tensile-stress, psi; (b) longitudinal tensile strength, psi; and (c) longitudinal tensile modulus, (in.^{2}) lb/in. The vendor should also be prepared to submit actual failure-test data that confirm his design calculations.

Match resin and glass to the service

Resins are best specified from your own experience or test data. Otherwise, provide enough data to the piping vendor and resin supplier to allow them to define the best resin for your service. Also, permit only one resin to be used throughout the pipe wall.

Fillers should be allowed only to the extent required for viscosity control and fire retardancy. They should be limited to a maximum of 5 weight percent and should not interfere with the operator's ability to visually inspect the pipe.

Glass should be specified as a commercial grade matched with a coupling agent compatible with the resin.

Pigmentation. If you want the pipe to be pigmented, specify so and name the color. But remember that pigmentation will interfere with visual inspection for pipe quality. As an alternative to using pigmentation for ultraviolet protection, specify 0.25 wt% of an inhibitor such as UV No. 9 [*4*].

Valves

Ball, globe and check valves should be specified with flat-face flanges. If you are using wafer-type butterfly valves, make the vendor guarantee that the pipe flanges are strong enough to take the required bolting torques for these valves without breaking or cracking and without using spacers between flanges.

Workmanship and quality control

Inside surfaces must be free of cracks, dents, crazing or other defects. Joint sealing surfaces should have no defects (such as cracks or pinholes) that will affect the integrity of the joints. Exterior surfaces need to be sealed with resin so that no unwetted glass is showing. If the outer surface of the pipe is to be exposed to a corrosive environment, ask the vendor to meet Section 3.3.1 of PS-15-69.

Testing. Specify exactly what tests you want run. For normal services, you may only want to run hydrotests in the vendor's shop on a sample of the production, or, in the shop or field, on completed shop assemblies. For more-critical services, you will want to require verification of hoop strength, longitudinal tensile strength, and stiffness on a certain proportion of the order. (Such testing will normally involve extra charges.)

Inspection. Reserve the right to inspect pipe manufacture at any time. Indicate to the vendor that you will verify appearance, dimensions, hardness, resin/glass ratios and (where specified) mechanical properties and hydrotesting. Then do it.

Field training. It is essential that you establish that the vendor will supply experienced, qualified personnel to conduct training sessions in field-assembly of FRP pipe for your own field people, followed by a period of onsite supervision to ensure that the training is being properly applied. You will also want followup field visits by the vendor's personnel.

Other factors

Somewhere in your negotiations, you will want to agree on maximum shipping lengths (usually 40 ft), fob point (make the vendor responsible for getting the pipe to your plant in good condition), and engineering assistance to your piping designers. Although many vendors are not equipped to provide this assistance, with diligent investigation you can pinpoint those having sufficient engineering talent to give you needed advice on special problems of piping system design, span lengths, expansion-joint requirements, isolation from vibration, and protection against shock loading.

If you incorporate these considerations into your filament-wound FRP pipe specification, and if your field crews use proper care in installing the pipe, you will have an excellent chance of ending up with a high-quality FRP piping system.

References

1. NBS Voluntary Product Standard PS-15-69, "Custom-Contact-Molded Reinforced-Polyester Chemical-Resistant Process Equipment," Natl. Bureau of Standards, Washington, D.C.
2. ASTM D2996-71, "Standard Specification for Filament-Wound Reinforced Thermosetting Resin Pipe," American Soc. for Testing and Materials, Phila.
3. ASTM D2517-73, "Standard Specification for Reinforced Thermosetting Plastic Gas Pressure Pipe and Fittings," American Soc. for Testing and Materials, Phila.
4. Mallinson, J. H., "Chemical Plant Design with Reinforced Plastics," McGraw-Hill, New York, 1969.
5. Mallinson, J. H., "Reinforced Plastic Failure—Causes and Cures, The Zero Maintenance Concept," paper presented at the seminar "Managing Corrosion Problems with Plastics," San Antonio, Tex., Nov. 3–7, 1975.
6. Webster, R. M., "Safety in the Use of Fiberglass Reinforced Plastic Equipment for Chemical Service; Problem Solving with Plastics," NACE, Houston, 1971.

The authors

L. S. Surtees is manager of materials engineering, Technology Div., Kerr-McGee Corp., Oklahoma City, OK 73102. Responsible for materials selection companywide, he has recently been involved in specifying the reinforced plastics used for construction of a new, 1.3-million tons/yr soda ash plant at Trona, Calif. A registered professional engineer in California, he is active in the Natl. Assn. of Corrosion Engineers, AIChE, and American Soc. for Metals. He holds a B.S. in chemical engineering from the University of California at Berkeley.

P. Rooney is principal process engineer, Parsons-Jurden Div., The Ralph M. Parsons Co., Pasadena, CA 91124, directing engineering and procurement of equipment for part of the Kerr-McGee soda-ash plant. He earned his B.S. in chemical engineering from the University of Saskatchewan and participated in a special University of Indiana executive management program. A member of AIChE and the Chemical Institute of Canada, he is a registered professional engineer in California.

Polybutylene— a new thermoplastic for industrial piping

Capable of withstanding a wider range of temperatures and pressures than other thermoplastics, this recently offered material boasts a property mix that suits it for handling hot water, gases, aqueous slurries and corrosive solutions in both aboveground and underground pipelines.

Jack Shannon, *Project Engineer, Polymer Div., Witco Chemical Corp.*

☐ Polybutylene resin* carries the highest design-stress rating of any flexible, thermoplastic piping material and has proved itself in pressure applications up to 200°F. Pipe made from this material has shown excellent resistance to stress cracking and abrasion when exposed to aqueous slurries, and a general resistance to attack by severe chemical environments such as hot acids and bases; it also has good liquid-flow characteristics and resists scale buildup. Other properties favorable to industrial applications are high flexibility, low creep under load, heat fusibility, and light weight.

Polybutylene pipe comes in sizes up to 24 in dia. and can be produced on standard extrusion equipment, using either pressure or vacuum sizing techniques. In addition, sizes up to 48 in are under development. Piping systems are currently in service in the chemical-process, power-generation and mining industries.

Operating range—temperature and pressure

Polyethylene has already established itself as a versatile piping material but has been limited by its temperature and pressure capabilities. Polybutylene, on the other hand, offers improved temperature and pressure properties, while retaining most of the mechanical characteristics of polyethylene.

In continuous-pressure applications, polybutylene pipe withstands temperatures up to 200°F; in unpressured and intermittently pressured systems, the material can tolerate service temperatures as high as 220°F. Polybutylene pipe and components carry a recommended hydrostatic design-stress rating of 1,000 psi at 73°F, and 500 psi at 180°F. Since the allowable working-pressure decreases with temperature, it must be adjusted with a "de-rating factor" at high temperatures. Maximum-allowable internal pressures at various temperatures are given as a function of pipe series, or Standard Dimension Ratio (SDR)* in Fig. 1.

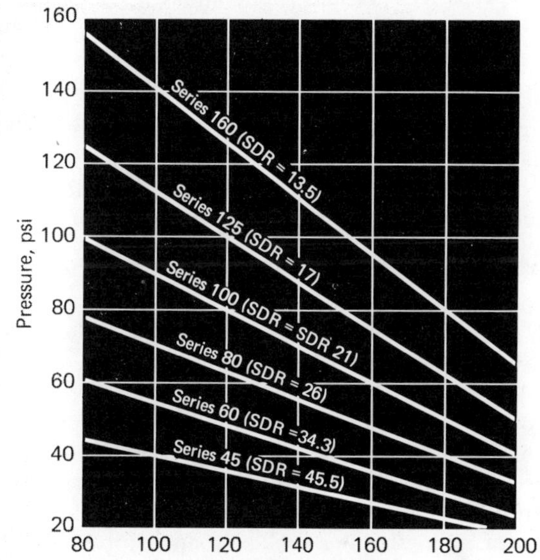

Maximum-allowable internal hydraulic-pressure for polybutylene pipe at various temperatures **Fig. 1**

Creep resistance

In many instances, the poor creep resistance of some thermoplastics has limited their use in piping applications, particularly at elevated temperatures. Polybutylene, however, has excellent creep resistance over a broad temperature range, even at very high stress levels. Test data indicate that polybutylene creeps less than 12% under a stress of 1,000 psi at 85°C for 2,000 h, and less than 4% under 1,000-psi stress at room temperature for 10,000 h. Further evidence of polybutylene's resistance to creep may be seen from the long-

* Manufactured by Witco Chemical Corp. as WITRON polybutylene resin.

† The Standard Dimension Ratio of thermoplastic pipe is defined as the ratio of the outer diameter to the wall thickness.

Originally published August 10, 1978

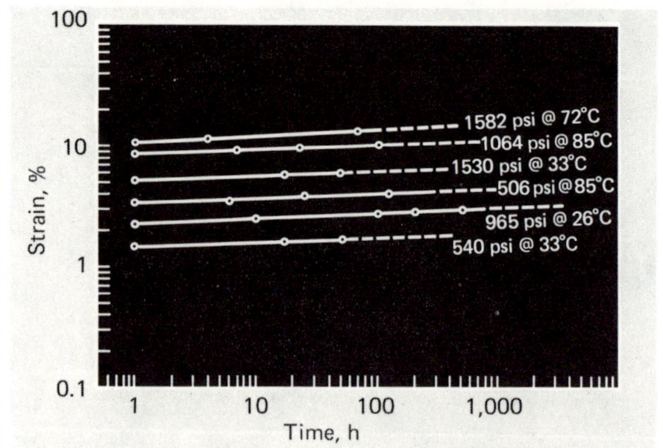

Tensile creep of polybutylene pipe (0.4 melt index, type II) at various temperatures **Fig. 2**

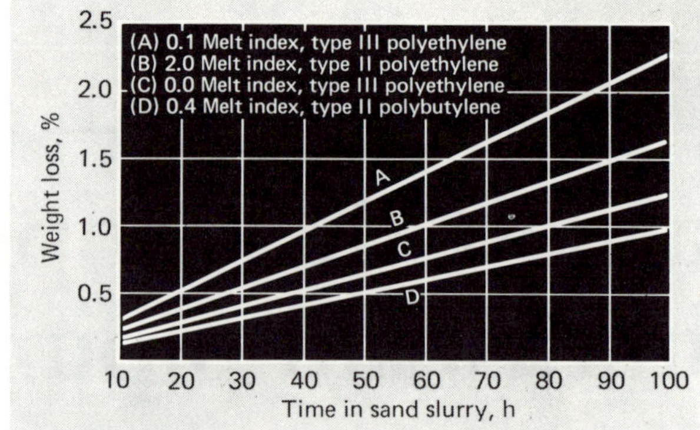

Slurry abrasion resistance: exposure time vs. weight loss **Fig. 3**

term stress curve, which shows less than 15% change in strain at room temperature after 100,000 h. After an initial deformation, the change in strain with time for polybutylene is slight. Its creep characteristics under a variety of test conditions are shown in Fig. 2.

Abrasion and stress-cracking resistance

The ability to transport slurries of abrasive materials such as gypsum, fly ash and phosphate rock without excessive wear is one of the outstanding features of polybutylene piping. During laboratory sand-slurry tests, the performance of polybutylene exceeded even that of some high-density polyethylenes (Fig. 3). Field installations handling abrasive slurries have reinforced these test results. A 4.5-in pipe (SDR = 13.5), handling a 55% slurry of 20-mesh silica rock at room temperature, held up for 23 mo. Moreover, in an application involving phosphoric acid and abrasive fines at 200°F and 50 psi, polybutylene pipe outlasted both glass-lined asbestos-cement pipe and rubber-lined steel pipe.

Polybutylene, because of its unique crystalline morphology, has no known susceptibility to attack by environmental stress-cracking agents. (When tested for 15,000 h under conditions of ASTM-D-1693, type II polybutylenes of 0.4 and 2.0 melt index sustained no failures due to stress-cracking.) In this regard, it is superior to other polyolefin piping materials.

Flexibility

The combination of high-strength per unit wall-thickness and low stiffness-modulus accounts for the high flexibility of polybutylene pipe. In general, the bending radii of polybutylene pipes are somewhat less than polyethylene ones of the same wall thickness. However, the reduced wall thickness needed for polybutylene for a given operating pressure permits up to three times greater flexibility than with some high-density polyethylenes. This flexibility allows the pipe to conform to ground contours and reduces the need for fittings to produce gradual changes of direction.

Light weight

The weight of polybutylene pipe ranges anywhere from one-fourth to one-tenth that of conventional non-plastic piping materials and is 25% lighter than polyethylene pipe of the same diameter and pressure rating. The reduced weight translates into easier handling and assembly and eliminates the need for installation with large, expensive equipment. In practice, long lengths of polybutylene pipe have been handled with a single, small "cherry picker" crane.

Underground piping systems

The ability of a flexible-piping system to withstand earth loading and superimposed loads depends upon the lateral support pressure of the soil on the pipe. The pipe itself contributes relatively little bending strength. Consequently, its ability to support loads must derive from passive pressures induced at the sides of the pipe as the pipe deforms outward against the soil. For this reason, proper compaction during burial—as per ASTM* guidelines—is essential for good performance of underground flexible pipelines.

To illustrate the effect of soil resistance on pipe deflection, let us consider two soil conditions: loose compaction and dense compaction. Fig. 4 and 5 give the amount of deflection at different applied earth loads for a series of SDR's and both soil conditions. It is accepted engineering practice to limit the pipe deflection to no more than 5%. Under this limitation, it is evident from Fig. 4 that with loose compaction (modulus of soil resistance = 300 psi) the allowable depth of cover for certain thin-walled pipe is severely restricted.

However, when denser soil compaction is employed (modulus of soil resistance = 700 psi; Fig. 5), there is no danger of exceeding the deflection limits for the pipe series shown.

There are other factors to consider in burial conditions, such as superimposed live loads resulting from traffic conditions over the installation, but these tend to be less critical than the actual burial conditions.

Resistance to collapse

The resistance of a flexible pipe to collapse due to external hydrostatic pressure must be considered whenever the pipe is to be immersed (as with a river or lake

* "Underground Installation of Thermoplastic Sewer Pipe," ASTM D-2321-75, American Soc. for Testing and Materials, Phila., Pa., 1975.

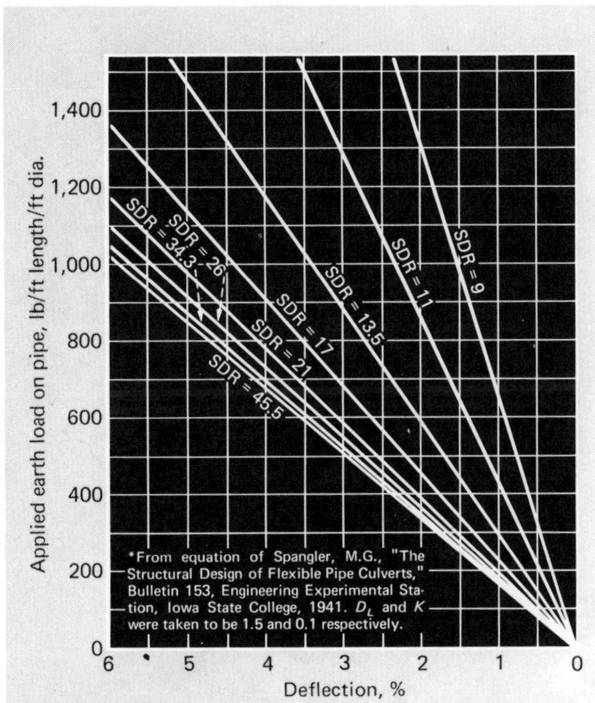

Underground pipe deflection at various applied earth loads* (modulus of soil resistance = 300 psi) **Fig. 4**

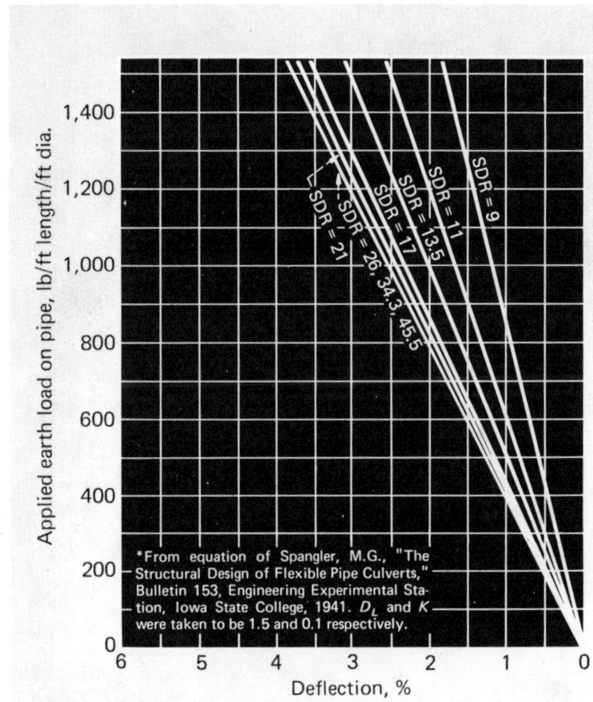

Underground pipe deflection at various applied earth loads* (modulus of soil resistance = 700 psi) **Fig. 5**

crossing), buried in hydrostatic soil, or placed under vacuum. Fig. 6 summarizes calculated collapse pressures for polybutylene pipe under short- and long-term conditions. It should be noted that these curves are based on round, unconstrained pipe at room temperature. For other conditions, corrections must be made. Furthermore, for design purposes, a safety factor of at least 2.0 should be applied to the external hydrostatic pressure.

Assembly and installation

Polybutylene industrial pipe is well suited for a variety of installation techniques, including the aboveground, direct-burial, insert-renewal and subaqueous types.

Pipe made from polybutylene can be joined by standard thermal-fusion welding techniques to provide socket- or butt-welded joints. Flexible piping systems can be coupled to other piping systems through the use of butt-welded stub ends with metal-flanged backup rings.

Butt fusion can be accomplished easily with a portable fusion machine, which provides the alignment jigs, the trimming tool to shave and square up the pipe-ends, and the heating plate. The fusion technique involves heating the adjoining ends of two lengths of pipe until the plastic melts, and then forcing the melted ends together. Melted polymer at the ends of the pipe flows together to produce a joint that is permanent, reliable, and as strong as the pipe itself. No specialized skills are required to operate the equipment, and joining can be performed at the jobsite under most ambient conditions.

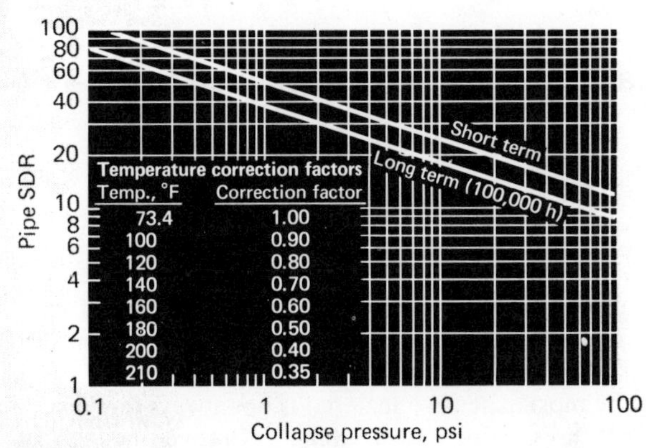

Hydrostatic collapse pressures for unconstrained, round polybutylene pipe **Fig. 6**

The author

Jack Shannon is a project engineer for the Polymer Div. of Witco Chemical Corp., 291 Fairfield Ave., Fairfield, NJ 07006. He has been engaged in application development and technical service for polybutylene pipe resins at Witco since 1974. Previously, he worked in polymer development positions at two major chemical companies for eleven years. A graduate of Rensselaer Polytechnic Institute, he holds a B.S. in chemical engineering.

Selecting Steel Tubing For the CPI

The chemical process industries consume a great deal of steel tubing, but the best material is not always chosen for the job. Here is how to choose correctly.

WILLIAM C. MACK, The Babcock & Wilcox Co.

Miles of steel tubing are used each year in the process industries. Achieving the best possible balance of engineering properties and economics, however, requires a most careful evaluation.

The most expensive material is not always the best. In many cases, carbon steel tubing has proved better than the more expensive stainlesses, based on strength, corrosion rate, and retubing and maintenance costs.

Cheap Materials Have Useful Properties

For example, in sulfuric acid solutions, all steels can suffer substantial chemical attack. In some of these applications, common stainless steels do not provide significant improvements in corrosion resistance over carbon steels. Concentrated sulfuric acid attacks carbon steel, but the corrosion products adhere very tightly to the base metal and prevent continued attack. As a result, carbon steel is normally used to contain concentrated sulfuric acid at ambient temperatures.

In another case, the less expensive of two stainless grades was the choice. A chemical and food processing plant that operated its equipment at high temperatures required the oxidation resistance of a stainless grade. But a ferritic (Fe-Cr) stainless worked better than the more expensive austenitic (Fe-Cr-Ni) stainless. A green corrosion product of nickel compounds from the austenitic grade objectionably discolored the process fluid. The less-expensive ferritic stainless does not contain nickel.

A less-expensive ferritic stainless was also the choice in a flue application, because the sulfur compounds in the flue gas attacked the nickel in the austenitic stainless. Austenitic stainlesses are normally favored in these flue-gas applications because of their high-temperature strength.

In many cases where corrosion rate is the primary factor, it may be more economical to accept the higher corrosion rate of a less-expensive material and compensate for it with a heavier tube-wall thickness.

Most-Expensive Material Often Economical

In other cases, where corrosion products can cause damage to other areas of the processing system, a corro-

Originally published November 26, 1973

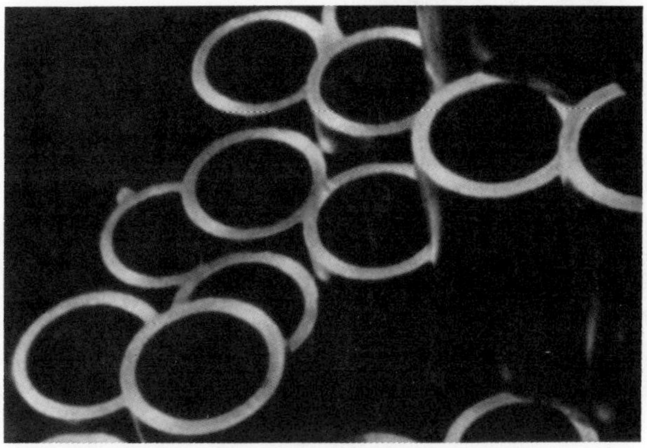

STAINLESS STEEL tubing (left) or carbon steel (right)? The choice depends on the process and on the economics.

sion-resistant tube, though of high cost, will be more economical than a cheaper material. An expensive material can often provide substantial savings in maintenance and retubing costs.

In one sulfuric acid plant, for example, carbon steel was used in a *vapor* portion of the system. Corrosion was somewhat rapid and the tubing had to be replaced periodically. By switching to B&W Croloy 2Si, a chrome-moly steel with silicon, at a moderate additional cost, the customer was able to use the tubing approximately twice as long, with resulting savings in retubing costs.

In oil refineries, naphthenic acid corrosion is sometimes a problem. Although a chrome-moly steel may seem a good choice for the temperature range of the application, the chrome-moly steels can be attacked rapidly by the naphthenic acid. A switch to stainless, although much more expensive, can provide substantial savings.

Maintenance considerations vary. One company had been using a chrome-moly steel in a heat exchanger with satisfactory tube life. But when the unit finally needed retubing, the procedure was extremely costly, since other parts of the system had to be removed to get at the heat exchanger. A change to stainless increased the time between retubings, and more than offset the increased cost of the stainless retubing.

Choosing the Most Economic Tubing

Making the right economic choice of tube material includes a number of interrelated factors—many of them specific to a particular process. But there are some basics:

Heat is removed or added in most processes. Heat-transfer characteristics of the tubing are therefore a critical consideration.

Corrosion places limits on effective tubing strength and therefore tube life. Corrosion products can also decrease heat transfer, which in turn decreases heat-exchanger efficiency.

Tube size is a function of strength, corrosion allowances and heat-transfer characteristics of the tube.

Maintenance is often a function of corrosion.

Testing of candidate materials under laboratory conditions to determine susceptibility to corrosion can sometimes give erroneous indications. It is difficult to duplicate the many variables outside the system, such as fluid velocity, turbulence and temperature. For this reason, the effects of corrosion are best determined by in-plant evaluations if at all possible.

Outline for Materials Selection

The following outline can help in making an economic evaluation of candidate materials. A cardinal rule is that no material should be eliminated from consideration on the basis of price alone.

- List all tube materials suitable for the process.
- For each of these materials, list sizes (diameter and wall thickness) and quantities, based on corrosion rates, strength and heat transfer.
- Estimate the cost of the tubing of each material.
- Estimate tube life for each material.
- Determine whether intended unit life is a factor—will the process tubing under consideration be phased out after a given time? If so, a material that corresponds to this unit life should be given special attention.
- Calculate retubing and/or maintenance costs for each material.
- Compile and compare total costs.
- Eliminate all but the most economical materials.
- Work the remaining materials through the process unit design with a complete economic evaluation.
- Select the material that yields the best return on investment.

At the process-unit-design stage of the evaluation, consultation with the tube mill may add valuable information regarding the problems of materials used with particular processes. In any case, the evaluation should include any possible modifications of the process, probabilities of increasing energy and labor costs, and other variables likely to influence the economics of the tubing material.

Meet the Author

William C. Mack is a Senior Metallurgist, Technical Services, Tubular Products Div., The Babcock & Wilcox Co., Beaver Falls, PA 15010, where he engages in research on use of tubular products in corrosive environments and at elevated temperatures. He has a B.S. in metallurgical engineering from Pennsylvania State University and an M.S. in the same subject from Michigan Technical University. He is a member of American Soc. for Metals, American Soc. for Testing and Materials as well as the American Institute of Mining, Metallurgical and Petroleum Engineers.

Pipe and equipment lined with elastomers and non-fluoropolymers

Many corrosive materials can be transported and processed in lined pipe and equipment, particularly at low temperatures and pressures. Fluorocarbons are covered in the next article—this one considers the other linings.

Lewis W. Gleekman, Materials and Chemical Engineering Services, Inc.

☐ An engineer once defined the difference between a paint and a protective coating by saying that lipstick is a paint and Chapstick is a protective coating. In this article, a differentiation must be made between a protective coating and a lining. This distinction cannot be made on the basis of either one being thinner or thicker than the other, since protective coatings and linings each have thickness ranges that overlap.

Thickness for thickness, linings are more free from discontinuities (pinholes, blowholes, cracks, etc.) than coatings, though linings will have more discrete discontinuities at edges because of size limitations. A coating, on the other hand, is an organic polymer usually applied from solution (by any means, such as brush, roller, spray or dipping) that is transformed to a solid by any one or a combination of solvent evaporation, oxidation or catalysis.

In contrast, a lining is applied to a substrate in solid form as a sheet; the sheet may previously have been extruded, molded, calendered, etc. The lining is usually a thermoplastic organic-based compound or mixture of compounds, though, conventionally, natural polymeric materials such as rubber are included as linings, as are other elastomers. Solids in powder form (applied by fluidized bed or by spraying) that are fused to form an integral film most closely resemble protective coatings.

Materials for piping and related fittings

For many years, rubber and other elastomers have been used as lining materials for pipe and fittings (such as elbows, tees, valves, etc.). These materials are still widely employed for this purpose, though in many cases thermoplastics such as Saran and polypropylene (along with the fluorocarbons that are discussed elsewhere) now command a large part of the lined-pipe market. There is an ASTM specification, F492-77, that deals with polypropylene-lined ferrous pipe and fittings (F491-77 deals with polyvinylidene fluoride—Kynar-lined, and F423-75 with polytetrafluoroethylene—Teflon-lined). All products from the three major U.S. suppliers conform to these standards.

The ASTM standards are full of specifications and recommended practices for thermoplastic materials in various forms, including unlined pipe and fittings. In the "Handbook of Corrosion Resistant Piping," by Philip A. Schweitzer (Industrial Press, New York, 1969), one finds reference not only to plastic-lined pipe but also to rubber-lined and a variety of other piping materials. In addition, there is a rather comprehensive corrosion-resistance chart directed both to metallic materials and to a wide variety of plastics materials and other nonmetallics. The most comprehensive summation of resistance of materials of construction to common and uncommon industrial chemicals in various combinations is still that found in the National Assn. of Corrosion Engineers' "Corrosion Data Survey," Sect. 1 for metals and Sect. 2 for nonmetals.

There are some limitations with plastic-lined pipe, specifically involving size. Obviously, lined pipe has a relatively high pressure rating, due to the support offered to the lining by the steel exterior. This is one of the great advantages offered by lined pipe. In addition, steel offers mechanical strength to the lined pipe similar to that of steel pipe itself.

However, one must face the fact that plastic-lined steel pipe commonly is available only up to 12-in. dia. with lengths not greater than 20 ft, has an appreciable weight factor and requires joining by flanges. At one time, 40-ft lengths were offered but these have generally been withdrawn. Penton (a chlorinated polyether) was once available as a lining but it too has been taken off the market. On occasion, plastic-lined pipe can be ob-

Originally published November 20, 1978

tained in 14 and 16-in. dia., though these sizes are not standard. Plastic-lined pipe is, in effect, an off-the-shelf material, relatively readily available within its size and material limitations.

Rubber-lined pipe, on the other hand, is usually a custom material, made to order for a specific customer and frequently for a specific service. This means that large-diameter pipe (up to 8 ft or more) can be made with rubber lining, frequently cured in place. One therefore has to balance the availability and somewhat broader corrosion resistance of plastic-lined pipe with the larger sizes and longer lengths of rubber-lined pipe.

Rubber-lined pipe

Natural rubber, particularly, can be cured or vulcanized to various hardnesses ranging from soft to what is frequently called ebonite, indicating extremely hard material. The curing can be done under pressure in an autoclave or, depending on compounding, with atmospheric steam, or it can be done chemically with carbon disulfide. The pipe can be its own "autoclave," depending on its wall thickness and end-closures for the necessary steam pressure (really, necessary curing temperature). Generally, the harder the rubber, the less permeable it is to various solutions and, therefore, the more chemically resistant the material. However, with hardness goes a loss of flexibility, and, therefore, the ability of the material to expand and contract as a function of operating temperature is reduced. For this reason, rubber formulators over the years have offered duplex rubber, where the soft rubber is attached to the substrate to offer flexibility, and the hard rubber (calendered originally to the soft rubber) offers chemical resistance. In addition, there are those who offer triplex, usually soft-hard-soft, where the soft layer exposed to the environment offers the additional value of abrasion resistance, while the hard sandwiched layer offers chemical resistance.

The unfortunate problem with natural rubber is the temperature limitation—most soft rubber compounds have an upper limit of 140°F for continuous exposure; for hard rubber the upper limit is 180°F. It is to extend the temperature resistance, as well as chemical resistance, that various synthetic-rubber and elastomeric materials have been developed. Notable among these are Buna S (GR-S), nitrile or Buna N (GR-M), Hypalon (chlorosulfonated polyethylene) and Thiokol. In addition, polymers of ethylene and propylene (EPDM) have been developed with elastomeric properties.

Like natural rubber, each of these may be compounded in several ways for maximum resistance to specific chemical exposure. The compounds chosen depend upon the stated chemical and temperature, environmental conditions and the experience of the rubber compounder and formulator. Adhesion of the rubber lining to the steel pipe, irrespective of the rubber formulation, is an obvious must. This requires the usual considerations of surface preparation and cleanliness.

On small-diameter pipe, inspection is relatively critical but difficult, so that one has to rely upon the degree of care shown by the rubber-lining firm. Handling techniques, particularly those involving the longitudinal skive joint, again are critical. So also is the flanging, which in the case of rubber-lined pipe is generally full face. Detailing of the pipe for fabrication of the steel substrate is important since once the rubber has been cured, it is extremely difficult to obtain adhesion of additional rubber to the cured material.

Elastomer liners being installed in pipe fittings **Fig. 1**

The Rubber Manufacturers Assn. (RMA) has a specification for protective linings that covers fabrication, tolerances and inspection, among other items. Fig. 1, 2 and 3 are views of 8-in. rubber-lined (3-ply) steel pipe that had been in both 49% alum and 40% $FeCl_3$ service at a sewage treatment plant where a leak occurred in 1,000-ft-length of pipe after 6 weeks' service. Fig. 1 is a good flange and 20-ft length of pipe, while Fig. 2 shows where the rubber lining has puckered just inside the flange. Fig. 3 also shows puckered rubber. Remote TV viewing of the pipe, while still in place, showed blisters and delamination at many locations. Failure of the rubber lining was traced to questionable application techniques and poor inspection before installation.

Plastic-lined pipe

In contrast, plastic-lined pipe, as made by its three major manufacturers in the U.S. (there are several non-U.S. manufacturers whose fabricating techniques differ from those in the U.S.) usually involves the extrusion of the plastic liner in the form of tubes, and the insertion of that liner in the previously prepared steel-pipe section.

Different companies have an assortment of techniques to achieve a bond between the plastic pipe and the steel; for this reason, the plastic-liner thickness may vary from company to company, though always meeting the requirements of the ASTM standard for polypropylene. The reason for establishing a bond is that most thermoplastic materials have a coefficient of expansion approximately ten times greater than that of steel. It is because of this expansion difference that

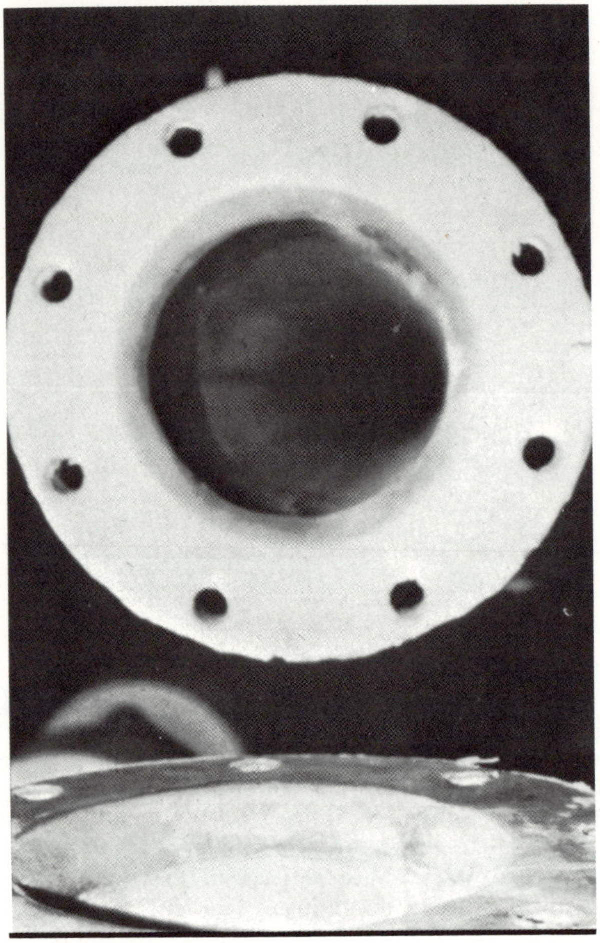

A good rubber-lined flange and pipe Fig. 2

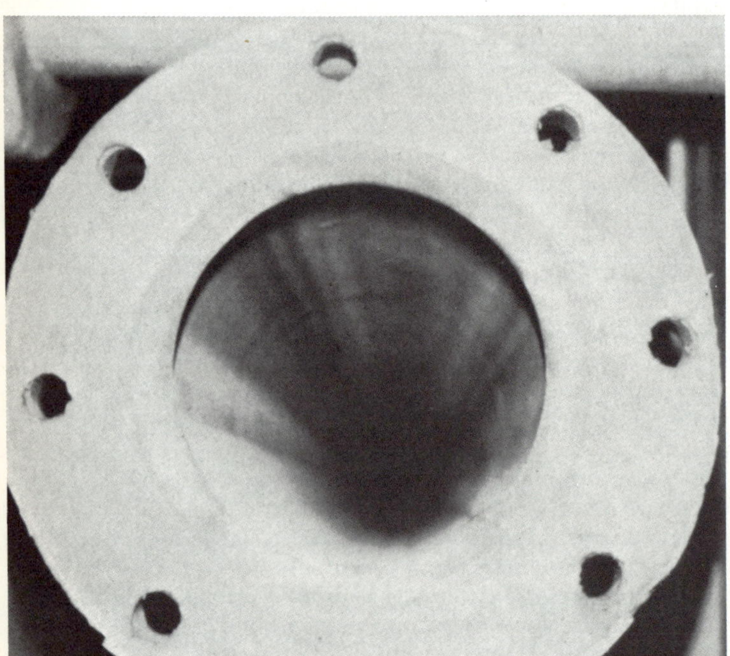

Here lining has puckered just inside of flange Fig. 3

thermoplastic-lined pipes have definite upper-temperature limitations.

Flanging in the shop is usually done in the form of a raised flange face, rather than a full-face flange, frequently with Van Stone flanges for ease in assembly. Several companies have developed field flanging techniques that allow a piece of pipe, usually 8 in. or less, to be cut to a shorter length, with a field flange being subsequently made that is integral with the plastic liner. There is only one U.S. company that offers Saran (polyvinylidene chloride) lined pipe, and that is Dow Chemical; several valve manufacturers offer Saran-lined (molded) valve bodies, along with polypropylene and fluorinated polymers.

One of the advantages of thermoplastic materials is that they can be softened by heat (theoretically any number of times) and when soft, can be readily deformed or formed. The thickness of steel pipe governs the pressure rating of the lined pipe. This is true also of the pipe fitting, where most frequently the plastic liner is introduced by injection molding.

Vacuum service can sometimes cause problems with the lining by separating it from the substrate. Here, proper design and fabrication of the lining and the steel pipe is rather critical. Most lined steel pipes have vent holes to allow air trapped between the liner and the steel to escape. By virtue of the support of the steel pipe, the wall thickness of the plastic component can be reduced for a given high operating pressure compared with the same plastic alone.

On occasion, attention must be given to the variation in corrosion properties of molded-plastic liners versus extruded or calendered material of the same basic composition; this is frequently a function of additives that must be used to enhance extrusion or calendering properties. It is important, therefore, to test the plastic in the form in which it will ultimately be used.

This means that evaluation over and beyond the tables of data (either supplied by the manufacturers or found in standard reference sources) should be done by actual testing of fittings as well as straight lengths.

Heating of lined pipe (either electric or steam tracing) can be successful but must be carefully done within certain well-established parameters. Long-lasting, maintenance-free exterior protective coatings on lined pipe are also available, usually on special order.

Competing materials

Obviously, lined pipe, both rubber and thermoplastic, has competing materials. These can run the gamut of corrosion-resistant metals such as nickel-base alloys, titanium or zirconium, where the choice of material is a function of the temperature pressure, corrodents, and economics. In addition, reinforced plastics compete in some cases with lined metal pipe, both the machine-made filament-wound materials and the hand-layup FRP materials involving polyesters, vinyl esters, furans, epoxies, etc. Not to be forgotten are the molded, random-oriented materials usually using furan or phenolic resins. Corrosion-resistance information is available from the three major manufacturers of lined pipe, as well as from the aforementioned Schweitzer's handbook and the NACE Corrosion Data Survey. Other standard handbooks, such as Perry's "Chemical Engineers' Handbook," also offer information on the subject.

One recent development in lined pipe is the use of polysulfone with a fiberglass wrap, now available up to 6-in. dia. in 10 or 20-ft lengths. This pipe handles a pressure of 150 psi and temperatures up to 250°F. Because of the wide resistance of polysulfone to acids, alkalis, salts and certain organics, this material is reported to offer economic competition to the fluorinated-plastic-lined pipe, especially in terms of installed cost. The polysulfone-FRP pipe is lighter, size for size, and involves fewer flanges.

The ultimate in lining comes with thin-gage metallic materials in steel, as for example, titanium, the high-nickel alloys, zirconium, tantalum. However, in keeping with the earlier definitions of linings, these materials will not be discussed, other than with the caveat that one considering such linings for pipe should review the experiences with "Strubing"—a great idea that had a few too many bugs that were too expensive to work out. "Strubing" was a process developed during the late 1950s that involved the reduction of a piece of alloy (or other corrosion-resistant-metal) pipe to the form of thin-wall tubing. This was done by compressing a length of pipe to a flat, and then rolling the flat so as to reduce the wall thickness and elongate the material. (Obviously, the pipe required good ductility.)

Ultimately, the thin-wall, flat tubing, with one end sealed, was drawn through a piece of steel pipe. Then, air pressure was applied to the open end of the flattened tube, causing it to expand into a cylinder against the interior of the steel pipe. If all things were done correctly, including provision for the two flat projections 180-deg apart (corresponding to the edges of the rolled material), the inflated tubing would fit snugly inside of the steel pipe and thus constitute a liner. Unfortunately, because of various design and manufacturing difficulties, things did not always work out correctly.

For larger-diameter ductwork, there is always the possibility of clad higher-alloy linings, either metallurgically roll-clad or explosive-clad.

Fig. 4 is a view of a sectioned polypropylene-lined pipe fitting where the operation of molding the plastic caused shrinkage of the polypropylene from the steel wall and subsequent cracking under hydrostatic testing even before the pipe was in service. Also, the specifications for this particular job called for 150-lb steel fittings in terms of the drilling, the pressure rating and face-to-face dimensions. Unfortunately, the manufacturer used 125-lb cast-iron fittings, which caused a great many problems in the installation. This problem, fortunately, does not happen too often.

Materials for lining vessels

A vessel (tank, reactor, heat-exchanger head, evaporator, etc.) differs materially from a pipe in terms of the *L/D* ratio (length/dia.), and a vessel does not have the advantage of the lining being mechanically attached (as for pipe) or internally molded (as for fittings). More particularly, vessels are frequently of an unsymmetrical and complex design that requires a much more involved assembly of the lining material than do pipe and fittings. It is for this reason that the use of thermoplastic and elastomeric lining materials in steel equipment depends very largely both on the adhesive that attaches the lining to the steel and the ability of the lining to conform to the vessel contour.

Add to this the mechanical requirements (thermal expansion, and vibrational or flexural movement of the lining relative to the steel), plus chemical requirements of the lining, and one realizes that lining steel vessels is no routine activity. Many mistakes have been made over the years in applying "new" lining materials to steel vessels; these mistakes have caused many techniques and even some materials to fall by the wayside.

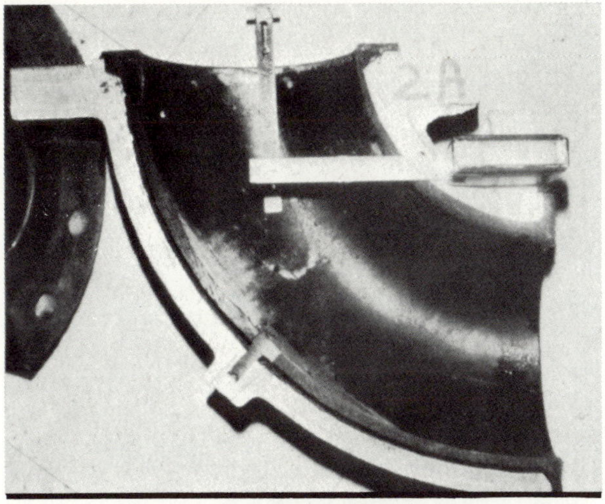

Section of polypropylene-lined fitting where lining has shrunk from wall and cracked **Fig. 4**

Steel tank being bricked up with two courses of acid-resistant brick over rubber lining **Fig. 5**

Rubber-lined vessel readily for shipping **Fig. 6**

In choosing a liner, first and foremost to consider is the chemical resistance of the lining to the environment at the maximum operating temperature. Obviously, if one or more constituents of the environment can permeate the lining and get to either the adhesive or the steel substrate or both, with related failure of one or both, then the lining material offers little promise of a long maintenance-free life.

If the lining material does not expand and contract to the same degree as the steel substrate, then the lining will become detached in time, and can fail due to its inability to withstand hydrostatic pressure from the process liquid on the one hand or expansion from entrapped air between the lining and the steel on the other. Therefore, the thermal expansion properties of the plastic lining materials must always be kept in mind.

At one time, a duplex PVC (polyvinylchloride) lining was made available, consisting of a rigid layer facing the environment and a flexibilized layer attached to the steel substrate. While this was fine in theory, it did not withstand the rigors of many industrial service conditions and has essentially disappeared.

Unlike elastomers, thermoplastic materials used for linings do not need to be cured. (To maximize their chemical and physical properties, elastomers must be in a fully cured state.)

Perhaps one of the meanest environments to contain has been the effluent from a chlorine emergency-disposal tower. Before the days of stringent discharge requirements on plant wastes, many companies circulated a 5% caustic solution through a packed tower to handle any emergency discharges of chlorine from chlorine cells or chlorine processing, including such things as stiff gases from transfer lines. This meant that the caustic solution increased in absorbed chlorine, with greater and greater amounts of sodium hypochlorite being formed, sometimes in great surges that resulted in extensive increases in temperature due to the exotherm. That caustic or hypochlorite was then dumped and a fresh charge of cold caustic added.

Historically, after various rubber compounds failed as the lining for the steel receiver tank (neoprene was excellent for the caustic but not suitable for the hypochlorite, and compounds suitable for hypochlorite were not suitable for caustic), PVC-lined tanks were tried. Here, again, the extremes in temperature and the permeation of the hot bleach caused failures. Even fiberglass-reinforced plastic, using the best of the chemically resistant resins, could not take the full gamut of fluctuating conditions. Acid-resistant brick with resin mortars had its limitation because resin mortars that were good for caustic did not resist hypochlorite.

It should not be assumed that all substrates that receive a protective lining are steel. Concrete has been lined at times, as has fiberglass-reinforced plastic. However, if the corrosive conditions are so severe that failure of the substrate will occur if the lining is permeated, then that substrate and lining should not be used in the first place. If the process conditions cannot be changed, one should consider using high-alloy or higher-resistance metal construction, in the form perhaps of a thin lining. The successful use of light-gage tantalum as a lining on a high-temperature, high-pressure vessel employed in the Kel-Chlor process has been described in the literature.

Rubber linings for vessels have been well proven (within temperature limitations of the elastomer), particularly where abrasion resistance is required along with chemical resistance. It should be recognized that, as with rubber lining of pipe, rubber linings in vessels have many joints that, if not properly cured, can present an edge where the adhesive is exposed; if not manufactured correctly, this can result in delamination at exposed edges.

A case in point occurred when a chlorobutyl layer was calendered, via a tie-gum rubber, to a butyl lining, for an evaporator in the production of 52% phosphoric acid. The chlorobutyl layer was designed to produce an extended life over the conventional butyl material in view of the high operating temperature (225°F) and the presence of trace amounts of fluorides and sulfuric acid. Unfortunately, in at least one case, the tie-gum rubber that was exposed at various edges was attacked, resulting in pieces of the chlorobutyl layer becoming free and ultimately plugging up a heat exchanger.

Designing substrates

There are many "do's" and "don'ts" with regard to the design of steel substrates that are to receive a lining, whether it be plastic or rubber. The reader is referred to various committee activities of NACE, particularly T6A-29.

The use of acidproof brick with resin mortar has been common for extending the service conditions of the rubber or plastic lining, since the common rule of thumb says that there is approximately a 15-deg F drop per inch of shale or fire-clay brick (10-deg F for carbon brick). This allows a process fluid to operate at a higher

Corrosion-resistant properties of common nonfluorinated plastics and elastomers (within temperature limits of each material)

	Strong oxidizing acids and NaOCl	Non-oxidizing acids HCl, H_3PO_4	Alkalies	Mild organics (oils and higher-molecular weight compounds)	Strong organics, especially solvents and chlorinated compounds	Inorganic salts with aeration
Plastics						
Saran	Fair	Good	Good	Good	Fair	Good
Polypropylene	Poor	Good	Good	Good	Poor	Good
ABS	Fair	Good	Good	Good	Poor	Good
PVC	Good	Good	Good	Good	Poor	Good
Elastomers						
National rubber						
Soft	Poor	Fair	Fair	Poor	Poor	Good
Hard	Poor	Good	Fair	Poor	Poor	Good
Neoprene	Fair	Good	Good	Good	Fair	Good
Nitrile	Fair	Good	Poor	Good	Fair	Good
Butyl	Good	Good	Fair	Poor	Poor	Good
Chlorobutyl	Good	Good	Fair	Good	Poor	Good
Hypalon	Good	Good	Fair	Fair	Poor	Good
EPDM	Good	Good	Good	Fair	Poor	Good

temperature since the one or two-course acidproof brick with its temperature drop means that the temperature against the interface of the lining and the brick will be within the limits of operation of the lining, usually 150°F maximum.

"A Handbook of Chemically Resistant Masonry" by W. Lee Sheppard (CCRM Inc., Havertown, Pa., 1977) has a section dealing in some length with membranes for use under chemically resistant masonry. There is also a good deal of information devoted to design of vessels, as well as other considerations involving these materials.

Fig. 5 shows a view of a steel tank being bricked up with two courses of acid-resistant brick and resin mortar over a rubber lining previously applied to the steel. Note that all brick joints are staggered from one course to the other horizontally and vertically, that the mortar joints are thin (not exceeding 1/16 in.), and that all bricks are saw-cut, rather than hammer-severed, for proper fitup.

The rubber lining shows a butt joint at the manway to the tank; this lining had previously been spark-tested to establish freedom from pinholes, and proper continuity. In this particular vessel, rather than use a 1-in.-thick ceramic sleeve for protection of the rubber manway (thus reducing a 30-in.-dia. steel nozzle to less than 28 in. dia.), a 16-gage titanium nozzle liner was installed, set in the resin mortar.

In retrospect, nozzle liners of this type, particularly made of titanium, in a solution of 6% propylene chlorohydrin plus 3% HCl at 180°F, proved to be susceptible to crevice corrosion and suffered early failure, allowing the solution to attack the rubber and then the steel, with an inevitable major leak. Currently, such a nozzle liner would be made from the titanium–1% nickel alloy with its known crevice-corrosion resistance.

The use of ABS (acrylonitrile-butadiene-styrene; a popular tradename being Krelastic) has been common, applied in sheets that are thermally formed, cemented to the steel and having all joints covered with a butt strap thermally welded to the sheet on either side. Some vessels are shop fabricated up to 12 ft dia. (this being the common maximum for transportation over the road or by rail) with an ABS or PVC or polypropylene liner.

Essentially, the laminate of thermoplastic material is formed (made flexible by heat) on a mandrel, and subsequently overwrapped with polyester-fiberglass while still on the mandrel. Attention must be paid to the fact that the thermoplastic lining materials usually are not flexibilized; a flexibilized lining loses chemical resistance (e.g., PVC Type II in contrast to PVC Type I). Because of the somewhat "nonstick" nature of polypropylene, adhesives are generally not suitable, unless the polypropylene is glass-backed during its fabrication as a sheet.

Once a vessel is lined, repairs or revisions by welding are impossible. Mechanical damage must be avoided. Fig. 6 shows a lined vessel with typical warnings.

The table summarizes the relative corrosion-resistant properties of the linings previously discussed.

The author

Lewis W. Gleekman is President and Technical Director of Materials and Chemical Engineering Services, Inc., P.O. Box 346, Southfield, MI 48037, phone (313) 356-4047, a consulting firm specializing in chemical, corrosion, materials and metallurgical engineering. He holds a B.Ch.E. from The Cooper Union, New York, and M.S. and Ph.D. degrees in chemical engineering from the State University of Iowa. He is a registered professional engineer in Michigan, Louisiana and California and is an active member of AIChE and National Assn. of Corrosion Engineers (Accredited Corrosion Specialist) and is a fellow of the American Soc. of Metals.

Fluorocarbon-lined pipe and equipment

The fluorocarbon plastics have extremely high corrosion resistance, and a higher upper-temperature limit than most other plastics. These qualities make them the choice for many challenging applications.

Roy V. Hughson, Associate Editor

☐ The fluorocarbon resins are widely used in the chemical process industries (CPI) because they include some of the most corrosion-resistant materials known.

Although some process equipment is made from solid fluorocarbon resins (plug valves, rotary pumps), the materials' most common use is as lining materials.

The first of the fluorocarbon resins was discovered serendipitously, at Du Pont in 1938. It was polytetrafluoroethylene (PTFE), and its first application in chemical processing was in the Oak Ridge gaseous diffusion plant that produced uranium-235 for the first atomic bomb. This plant had to process large quantities of uranium hexafluoride (UF_6), which though extremely corrosive to most materials was withstood by PTFE.

After the war, Du Pont introduced the material on a commercial scale as Teflon. Later, as other fluorocarbon resins were developed, Du Pont retained the use of the name Teflon for all of them, and distinguished PTFE as Teflon TFE. PTFE is now made by several companies, under their own tradenames.

PTFE resists almost all chemicals except fluorine and related oxidizing chemicals such as chlorine trifluoride, and violent reducing agents such as metallic sodium. No solvent is known for the resin. Its upper service-temperature limit is 260°C (500°F).

A fluorocarbon that is sometimes used as a lining material is polychlorotrifluoroethylene (PCTFE), which is marketed by Allied Chemical Corp. under the tradename of Halar. It is somewhat less resistant to chemical attack than TFE, and has an upper temperature limit of about 150°C (300°F).

Another fluorocarbon resin, of wide utility as a lining material, is fluorinated ethylene propylene (FEP). It has a chemical resistance comparable to PTFE, but has the advantage of being melt-processable. A disadvantage is that its service temperature limit is about 200°C (400°F), which is somewhat lower than that of PTFE.

Since it is a truly thermoplastic material, FEP is useful for making complex molded shapes, such as are required in valves, pumps and fittings.

Polyvinylidene difluoride (PVDF or PVF_2), which is sold by Pennwalt Corp. as Kynar, has a somewhat lower degree of chemical resistance than PTFE, FEP or PFA, and also has a lower upper-temperature limit—about 150°C (300°F). See Table I [*1*] for its chemical resistance.

A modified copolymer of ethylene and tetrafluoroethylene was introduced by Du Pont in 1970 under the tradename of Tefzel. Its advantages are increased resistance to abrasion, and greater mechanical toughness than straight TFE (Tefzel contains over 75% TFE by weight). Its disadvantage is that it has somewhat lower corrosion resistance, being attacked by strong oxidizing acids (e.g., nitric), organic bases (such as amines) and sulfonic acids—particularly when these materials are at high concentrations and near their boiling points. Even so, it is exceeded in resistance only by PTFE, FEP, and PFA (described below).

Tefzel is more used for solid molded shapes such as column packings than as a lining, but it has been used as a lining for diaphragm valves (to replace a glass lining), as well as for lining pumps.

One of the most recently introduced materials in the fluorocarbon family is PFA (perfluoralkoxy) resin, which was commercialized in 1972. It has a chemical resistance equivalent to PTFE, and similar upper-temperature limits (260°C, or 500°F), but is melt-processable.

In addition to corrosion resistance, the fluorocarbons have a unique "non-stick" property that makes them ideally suited to applications where buildup on the inner wall of the pipe would be detrimental, as in the processing of foods, drugs and pharmaceuticals, and certain fine chemicals. Of course, this same non-stick property is what causes problems in bonding fluorocarbon linings to substrates.

Originally published November 20, 1978

Fluorocarbon-lined pipe

When pipe is lined with fluorocarbon, the liner is not bonded to the pipe. No adhesives are available, and no bonding techniques have been developed that will withstand the severe service conditions for which fluorocarbon-resin-lined pipes are usually employed. The standard pipe construction used consists of the fluorocarbon liner in a vented housing.

Liners are generally made by molding or extruding the fluorocarbon tube. One manufacturer, however, builds up the liner by wrapping an extruded and calendered fluorocarbon tape upon a mandrel and then sintering the resulting assembly into a homogeneous, stress-relieved tube. These liners do not have to be locked into the steel supporting tube, and the manufacturer claims [*2*]: lowest permeability to gases and liquids; up to twice the strength at the collar area; and more-uniform physical characteristics.

Manufacturing lined pipe

The most common method of producing lined pipe involves placing the liner under a lengthwise (longitudinal) stress so that its diameter is reduced. While under stress, it is introduced into the steel supporting pipe, and the stress is relaxed. The fluoroplastic tube then decreases in length and increases in diameter, producing a snug fit between the tube and pipe.

A fluoroplastic tube longer than the pipe is used, so that the ends of the tube can be flared over the face of the pipe flange. This has the effect of producing a gasket that is integral with the tube. When mating flanges are bolted together, there is a continuity of plastic from one pipe section to the next, so that the steel pipe is nowhere exposed to the chemicals carried within it.

Another method of lining is to insert a plastic tube into an oversize steel pipe that has a roughened inner surface. This oversize pipe (with the liner in place) is passed through swaging rolls that reduce the outside diameter of the steel pipe to standard size. This squeezes the steel down on the fluorocarbon liner, and the rough inner surface of the steel is mechanically "locked" to the outside of the liner.

Flanges are then affixed to the steel pipe, and the liner (which, again, is longer than the steel support) is heated and flared over the flanges.

When used within its temperature limitations, such a liner will not move within the pipe.

Venting is required because of possible permeation of the pipe's contents through the plastic. Like other plastics, the fluorocarbons are subject to permeation by gases and vapors at elevated temperatures, though the rate is lower than for most other plastics.

At low temperatures, rates are so low for the fluoroplastics that the phenomenon can safely be ignored. However, these materials are frequently used at higher temperatures because the nonfluorinated plastics are seldom usable above 120°C (250°F).

At higher temperatures, permeation might cause a pressure buildup between the liner and the wall of an unvented supporting pipe, causing collapse of the liner tube. An alternative to venting is to have the liner so securely bonded to the support that vapor accumula-

Pipe tees having linings of PFA fluoropolymer

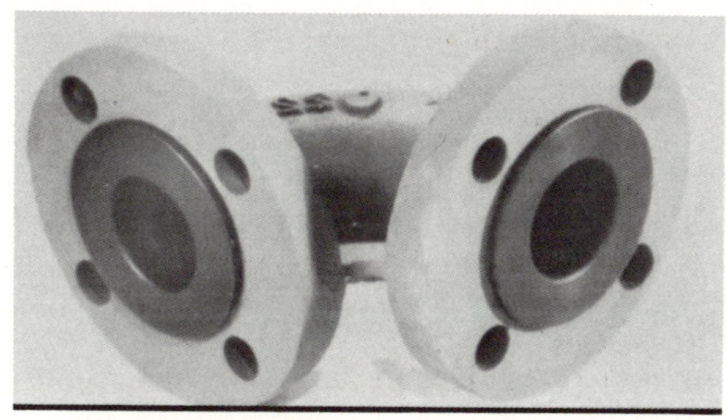

Pipe elbow lined with PVDF

Cutaway view of pipe with a PVDF liner

End of pipe lined with PVDF
Note how liner forms a self-gasket

Chemical resistance of PVDF in a variety of corrosive materials

Chemical	Temperature °C/°F 21/70	49/120	77/170	100/212	110/230	121/250	135/275
Acids							
Acetic, 50%	+	+	+	+	?	–	–
Acetic, glacial	+	?	–	–	–	–	–
Benzoic	+	+	+	+	+	?	–
Benzene sulfonic	+	+	?	–	–	–	–
Chlorosulfonic	–	–	–	–	–	–	–
Chromic, 50%	+	+	+	?	–	–	–
Citric	+	+	+	+	+	+	+
Formic	+	+	+	+	+	+	?
Hydrobromic, 50%	+	+	+	+	+	+	+
Hydrochloric, 10%	+	+	+	+	+	+	+
Hydrochloric, 35%	+	+	+	+	+	+	+
Hydrofluoric, 30%	+	+	+	+	+	+	?
Hydrofluoric, 70%	+	+	+	+	?	–	–
Nitric, 10%	+	+	+	?	–	–	–
Nitric, 50%	+	+	?	–	–	–	–
Nitric, fuming	–	–	–	–	–	–	–
Oxalic	+	+	?	–	–	–	–
Phenol, 10%	+	+	+	?	–	–	–
Phenol, 100%	+	+	?	–	–	–	–
Phthalic	+	+	+	+	?	–	–
Phosphoric, 30%	+	+	+	+	+	+	+
Phosphoric, 85%	+	+	+	+	+	?	–
Succinic	+	+	+	+	+	+	?
Sulfuric, 50%	+	+	+	+	+	?	–
Sulfuric, 85%	+	+	+	?	–	–	–
Sulfuric, 98%	+	+	?	–	–	–	–
Sulfuric, fuming	–	–	–	–	–	–	–
Bases							
Ammonium hydroxide, 30%	+	+	+	+	?	–	–
Aniline	+	+	?	–	–	–	–
Barium hydroxide	+	+	+	+	+	+	+
Calcium hydroxide	+	+	+	+	+	+	+
Hexamethylene diamine	–	–	–	–	–	–	–
Magnesium hydroxide	+	+	+	+	+	+	+
Propyl amine	–	–	–	–	–	–	–
Sodium hydroxide	+	+	+	+	+	+	+
Sodium hydroxide, 10%	+	+	+	?	–	–	–
Sodium hydroxide, 50%	+	+	?	–	–	–	–
Halogens							
Bromine,	+	+	+	?	–	–	–
Chlorine, liquid	+	+	+	+	?	–	–
Iodine, liquid	+	+	+	?	–	–	–
Oxidants							
Benzoyl peroxide	+	+	+	?	–	–	–
Chlorine dioxide, 15%	+	+	+	+	?	–	–
Hydrogen peroxide, 30%	+	+	+	+	?	–	–
Nitrogen dioxide	+	+	+	?	–	–	–
Ozone	+	+	+	+	+	?	–
Potassium chlorate	+	+	+	+	?	–	–
Potassium permanganate	+	+	+	+	+	+	?
Sodium hypochlorite, 17%	+	+	+	+	?	–	–
Sulfur dioxide	+	+	+	?	–	–	–

Chemical	Temperature °C/°F 21/70	49/120	77/170	100/212	110/230	121/250	135/275
Aliphatic hydrocarbons							
Acetylene	+	+	+	+	+	+	?
Butadiene	+	+	+	+	+	+	?
Butylene	+	+	+	+	+	+	+
Gasoline	+	+	+	+	+	+	+
Kerosene	+	+	+	+	+	+	+
Hexane	+	+	+	+	+	+	+
Mineral oil	+	+	+	+	+	+	+
Naph	+	+	+	+	+	+	+
Aromatic hydrocarbons							
Benzene	+	+	+	?	–	–	–
Naphthaline	+	+	+	+	?	–	–
Toluene	+	+	+	?	–	–	–
Halogenated hydrocarbons							
Allyl chloride	+	+	+	+	?	–	–
Carbon tetrachloride	+	+	+	+	+	+	+
Chlorobenzene	+	+	+	?	–	–	–
Chloroform	+	+	+	?	–	–	–
Dichloroethylene	+	+	+	+	+	?	–
Ethylene bromide	+	+	+	+	+	?	–
Refrigerant 113	+	+	+	+	?	–	–
Oxygenated solvents and esters							
Acetone, 10%	+	+	?	–	–	–	–
Acetone, 100%	–	–	–	–	–	–	–
Acetophenone	–	–	–	–	–	–	–
Dimethyl formamide	–	–	–	–	–	–	–
Ethyl ether	+	+	?	–	–	–	–
Ethyl acetate	–	–	–	–	–	–	–
Ethylene oxide	+	+	+	+	?	–	–
Ethylene glycol	+	+	+	+	+	+	+
Glycerine	+	+	+	+	+	+	+
Methyl cellosolve	+	+	+	+	+	+	+
Methyl ethyl ketone	–	–	–	–	–	–	–
Triethyl phosphate	–	–	–	–	–	–	–
Salts							
Ammonium nitrate	+	+	+	+	+	+	+
Calcium phosphate	+	+	+	+	+	+	+
Calcium sulfate	+	+	+	+	+	+	+
Ferrous chloride	+	+	+	+	+	+	+
Sodium acetate	+	+	+	+	+	+	+
Sodium chlorate	+	+	+	+	+	+	+
Sodium chloride	+	+	+	+	+	+	+
Gases							
Ammonia, anhydrous	–	–	–	–	–	–	–
Carbon dioxide	+	+	+	+	+	+	+
Hydrogen	+	+	+	+	+	+	+
Hydrogen sulfide	+	+	+	+	+	+	+
Methane	+	+	+	+	+	+	+

Key + = Satisfactory
? = Caution, requires testing
– = Unsatisfactory

tion is prevented, but this may not always be possible.

Since the vent holes are important in preventing collapse of the liner, they must never be covered by paint or, particularly, by insulation. Vent extensions are available if needed, to extend through insulation. An added use for vents is that they provide an early warning of failure if the lining should accidentally be cut or otherwise damaged. They also reveal leaks during hydrostatic testing.

Piping tips

The following hints on lined piping are taken from the *Journal of Teflon*, No. 27 [*3*], and from Ref. [*4*].

Dimensions—Pipe and fittings are available flanged only. The length of pipe is measured face-to-face from the fluorocarbon lining facing the flanges (i.e., the self-gasket), not face-to-face from the steel flanges themselves.

Spacers—Cumulative discrepancies in length in the assembled system may be corrected with spacers.

Gasket-face protective disks—The disks used to protect the gasket face must be kept in place during shipping and handling. They should be removed only when the assembly is ready to be bolted into position. If removed for inspection purposes, they should be immediately replaced. These precautionary measures are necessary because the gasket face may be damaged if it should come into contact with a sharp object.

Lap-joint flanges—Pipe ends for connection to equipment nozzles, or other fixed flanges, should be specified as lap-joint (i.e., rotatable) flanges to facilitate bolt-hole alignment.

Gaskets—No additional gaskets are necessary between gasket faces on lined pipe or fittings. If an assembly is to be bolted to a flange face of another material (metal, glass, carbon, ceramic, etc.), an additional envelope gasket of fluorocarbon should be used to protect the assembly.

Bolting torques—Get recommended bolt torques from manufacturers. Overtorquing of the bolts when sealing gasket faces will deform the faces and result in a poor seal. Note also that threads must be lubricated and washers must be used for correct torque readings. If torque wrenches are not available, bolts must be only snug-tightened with a wrench no longer than 8 in. If leakage occurs at any connection after the system has been cycled at elevated operating temperature, bolts should be retorqued at room temperature. No further retorquing should be necessary.

Vent holes—As previously noted, if the piping system is equipped with vent holes, never plug them up.

Vacuum—Piping and equipment should be installed so as to avoid excessive vacuum from the barometric-leg effect of draining lines in closed systems. Likewise, vacuum resulting from condensing vapors should be avoided by proper valving or by vacuum-relief devices.

Pipeline tracing—When heating of pipelines is required, the tracing system should be designed for uniform application of heat around the periphery of the pipe, especially in the smaller sizes (4 in. and below). High-temperature, localized straight-line tracing can cause failure of liners by local buckling. It is desirable to have heat-tracing elements near the bottom of horizon-

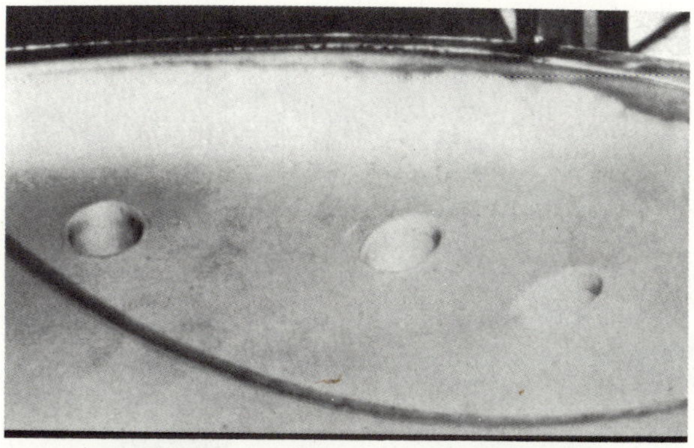

Inside of dished head of 500-gal vessel. FEP film was vacuum formed to shape, and FRP then applied

500-gal vessel during fabrication. FEP liner can be seen through manholes and on flanges

500-gal fiberglass-reinforced plastic vessel lined with FEP film

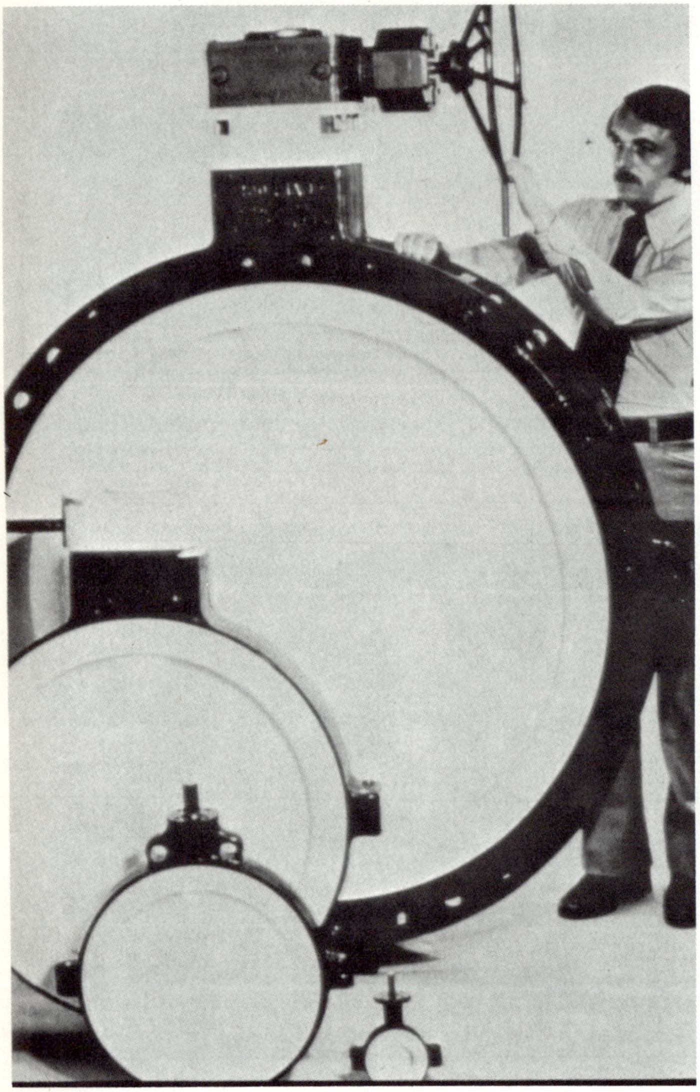

Butterfly valves up to 42 in. are available lined with FEP

Injection-molded FEP liner is held to valve mechanically by slots, keys and grooves

Body of a diaphragm valve that is lined with Tefzel

tal pipe, and to use oversize insulation to take advantage of thermal convection to help distribute the heat.

Fluorocarbon-lined plastic pipe

Instead of using a steel (or other metal) pipe for support, fluoropolymer linings can also be used with fiberglass-reinforced plastic pipe. These composite structures are used where the outside of the pipe may be exposed to corrosive environments that would attack a steel pipe.

Standards

Standards for the following fluorocarbon-lined metal pipe and fittings are available:

PTFE—ASTM F423-75, Polytetrafluoroethylene (PTFE) Plastic-Lined Ferrous Metal Pipe and Fittings.

FEP—ASTM F546-77, Perfluoro (Ethylene-Propylene) Copolymer (FEP) Plastic-Lined Ferrous Metal Pipe and Fittings.

PVDF—ASTM F491-77, Poly (Vinylidene Fluoride) (PVDF) Plastic-Lined Ferrous Metal Pipe and Fittings.

These standards include tests to be applied by the manufacturer to determine whether or not the product meets the standards. These standards differ from the previous ones in that they include tests for vacuum rating. Users' specifications for plastic-lined pipe and fittings should require that the manufacturer qualify his product according to the standards.

Lined equipment

Valves, pumps, vessels, heat exchangers, distillation columns, scrubbers, expansion joints, etc., are all available with fluorocarbon linings.

Valves

TFE and FEP resins have been used for years for lining butterfly, diaphragm, plug and ball valves. Also available are check valves, drain valves, an instrument control valve and a clamp valve. The clamp valve operates by a mechanism similar to the pinchcock used on laboratory equipment. A special mechanism is used to constrict a fluorocarbon tube, thereby shutting off flow. This valve

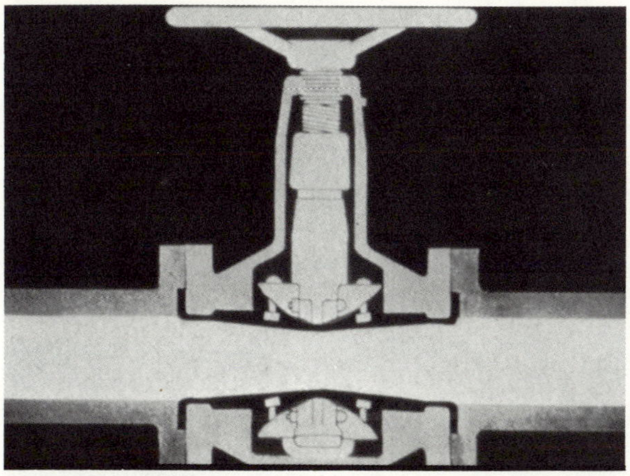

Diagram of TFE clamp valve described in text

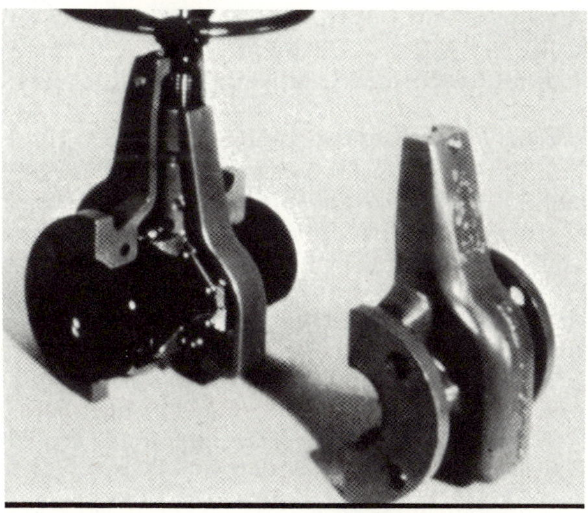

Cutaway of actual TFE clamp valve

offers a full pipeline-size opening for flow (when open), no stem seals, and is rated from full vacuum to 150 psi (1,000 kPa) at 350°F (170°C).

As mentioned earlier, a Tefzel-lined diaphragm valve with a TFE diaphragm has recently been made available. In one application [*5*], such a valve showed no sign of corrosion or abrasion after use for $1\frac{1}{2}$ yr throttling waste organic and mineral acids containing abrasive particles. Stainless-steel gate valves had lasted only 6 mo in this service.

In another application, these lined valves handle a solvent mixture containing chloroform, acetone, alcohol, and sometimes hydrochloric and abrasive ascorbic acids at temperatures up to 140°F and 40 psi (270 kPa). Only glass-lined valves have otherwise been considered suitable for this service.

Pumps

Centrifugal pumps are available lined with either FEP, TFE or PFA. Linings are transfer molded into split metal housings.

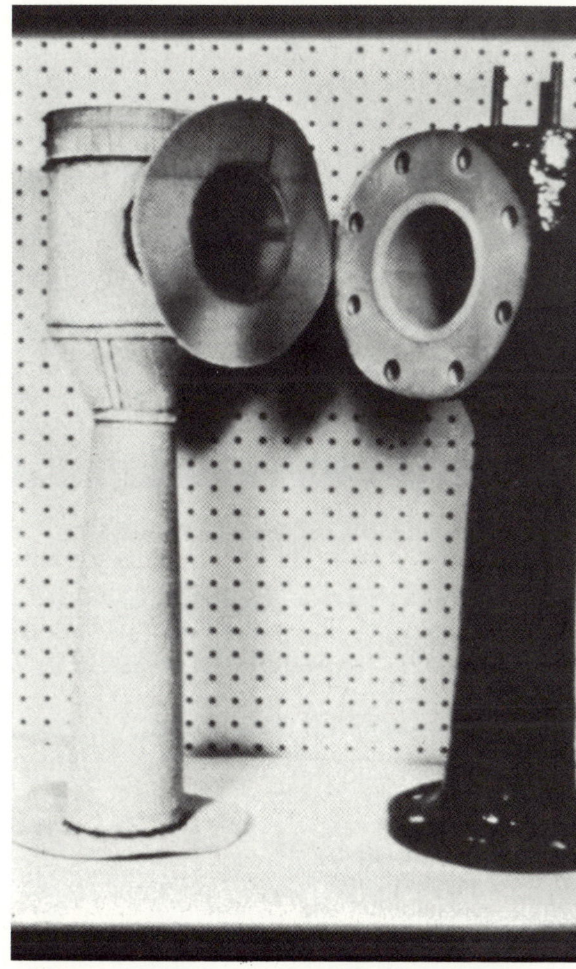

FRP steam scrubber before (left) and after (right) being lined with FEP (which is bonded to FRP)

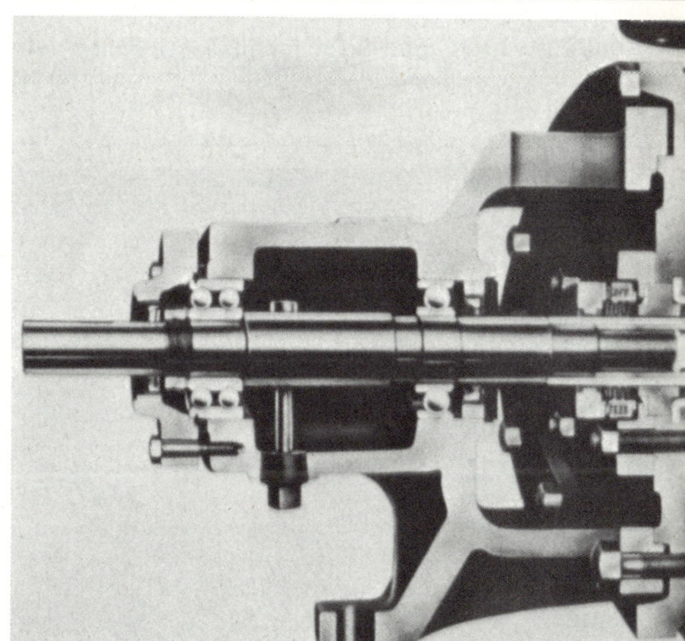

Centrifugal pump lined with PTFE fluorocarbon

Process equipment shells and heads lined with FEP

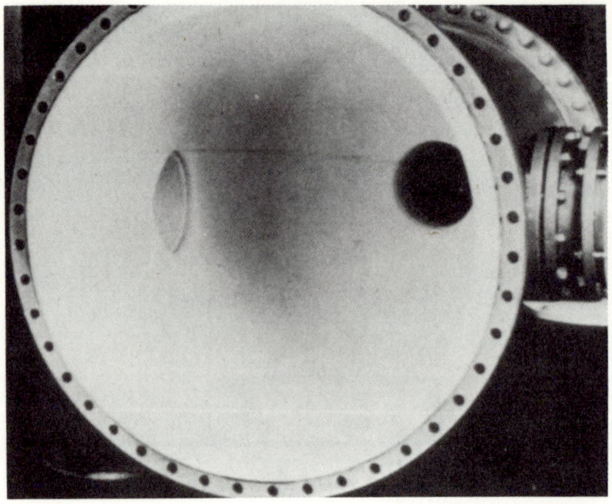

PTFE-lined column. Linings are available to ¼-in. thick

The following is a typical list of chemicals that have been handled by TFE-lined centrifugal pumps, in service for at least 5 yr [*5*]:

Materials handled	Temperatures °F (°C)	Head, ft.
Bromate, bromine, HF, H_2SO_4, H_2O	100 (40)	84
Methylene chloride, free Cl_2	150 (40)	110
Toluene sulfonyl chloride	75 (25)	74
Sulfuric acid, 98%	200 (95)	80
Sulfuric acid, 35%	190 (90)	83
HNO_3, 38% and HF, 3%	140 (60)	86

Some other applications of lined pumps:

An open-impeller FEP-lined pump has been in use in a chlorocarbons pilot plant, handling 82% sulfuric acid at 285°F (140°C).

A PFA-lined centrifugal pump has handled nitro-sulfuric acid, monochlorobenzene and various organic solvents at temperatures up to 160°F (70°C). The pump has operated intermittently for over a year.

Tanks, columns, etc.

Equipment such as tanks, distillation columns, scrubbers and the like are normally lined with sheets of PFA, FEP, PVDF or PCTFE. Standard sheet thicknesses are 40, 60 and 90 mils. For most applications, the lining is prefabricated and then fitted into the shell. The lining is then flared over the faces of any body or nozzle flanges.

Manufacturers have perfected techniques for making seams in the sheets of fluorocarbon resin—seams as strong as the sheet itself can be made by melt-bonding and thermoplastic welding. Vacuum forming is used to shape the sheet to fit dished heads.

Laminated linings

One manufacturer [*6*] uses a patented process in which glass or carbon-fiber reinforcement is combined

FEP-lined vessel showing welded seams

FEP vessel-lining being formed over flanges

FEP-lined expansion joint has been successful in paper industry

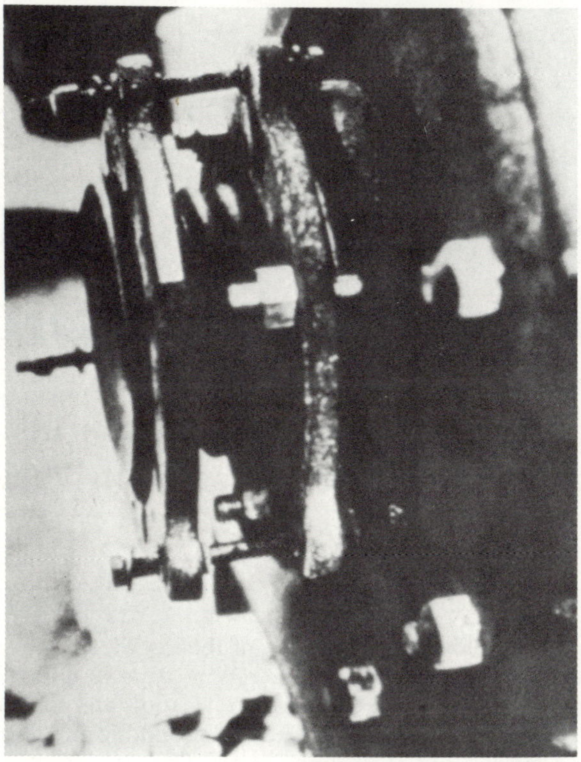

Bellows lined with PTFE installed in plant

with a fluorocarbon to provide a monolithic reinforced-fluoropolymer lining that is fused directly to a metal substrate. The fused-on lining is claimed not to separate from the substrate under normal conditions and can, therefore, be used in a vacuum up to the rated temperature of the lining material.

Using this technique, the interior or exterior of complex shapes, such as column internals, agitators, fans and impellers, can be provided with an appropriate plastic lining. (The purpose of the glass backing—versions of which are available from other manufacturers—is to provide a surface that will take adhesives.)

These laminated linings are said to be relatively good in heat transfer as compared with glass linings. A standard laminate thickness (55 mils nominal) is claimed to be comparable in heat transfer to the normal glass lining, which varies from 40 to 90 mils in a given piece of equipment.

This manufacturer is prepared to line any equipment that can be shipped over the road—normally up to 13 ft dia. and 50 ft long.

Expansion joints

The same manufacturer [7] offers TFE-lined expansion joints having 2, 3 or 5 convolutions, in sizes from 1 to 8 in. The joints having 2 and 3 convolutions will withstand pressures to 150 psi (1,000 kPa) and temperatures to 450°F (230°C). They absorb shock, correct pipe misalignment, and serve as sound insulators. They fit both metal and glass piping systems.

The units having 5 convolutions are used for lower-pressure service where additional flexibility is needed—as in weigh-tank applications.

Acknowledgments

Major suppliers of information for this article include: The Carborundum Co., Dow Chemical U.S.A., E. I. du Pont de Nemours & Co., Pennwalt Corp. and Plastic Omnium, Inc.

The lower three photographs on p. 125 were furnished by Pennwalt; the remainder by Du Pont.

References

1. Kynar: A Versatile Engineering Resin With a Unique Balance of Physical and Chemical Properties for a Wide Range of Applications, Pennwalt Corp., Plastics Dept., 3 Parkway, Philadelphia, PA 19012, Brochure No. PL-148-678-10M-WI.
2. TFE Lined Pipe and Fittings, Plastic Omnium Inc., 70 Shawmut Rd., Canton, MA 02021, Catalog P101, 1978.
3. Atkinson, H. E., Pipe Lined With "Teflon": Here's the How and Why, *J. of Teflon,* Reprint No. 27.
4. Atkinson, Harvey E., Fluoroplastic Linings for Corrosive Service, *Chem. Eng.,* Dec. 25, 1972, p. 76.
5. Pillsbury, Ross D., Jr., Du Pont Fluoroplastics for the Chemical Processing Industry, Paper No. 86, presented to the National Inst. of Corrosion Engineers meeting Corrosion/77, San Francisco, Calif., Mar. 14–18, 1977.
6. Carborundum Lined Chemical Equipment, The Carborundum Co., Protective Plastics Div., 117 State Rd., Avondale, PA 19311 (brochure).
7. TFE Expansion Joints, The Carborundum Co., Protective Plastics Div., 117 State Rd., Avondale, PA 19311, 1978.
8. Teflon Fluorocarbon Resins: Mechanical Design Data, E. I. du Pont de Nemours & Co., Plastic Products and Resins Dept., Fluorocarbons Div., Wilmington, DE 19898. Bulletin No. E-05561.
9. "Teflon" Film Protects Process Equipment, *J. of Teflon,* Vol. 13, No. 2 (1972), Reprint No. 48.
10. Corrosion Protection With Du Pont Fluoroplastics. E. I. du Pont de Nemours & Co., Wilmington, DE 19898, Bulletin No. E-08571.

The author

Roy V. Hughson is an associate editor of CHEMICAL ENGINEERING. He holds a B.Ch.E. from New York University and is a member of AIChE and National Assn. of Science Writers.

Installed cost of corrosion-resistant piping—1978

Here is a comparison of the installed cost of 26 different materials of construction in various sizes and pipe schedules.

James Yamartino, *Dow Chemical U.S.A.*

☐ The following comparison of the installed costs of a variety of corrosion-resistant piping systems will help engineers determine the most likely candidates for solving a particular problem. Detailed economic and corrosion studies can then pinpoint the final choice.

The installed cost of any piping system depends primarily upon:

1. Type of pipe.
2. Material costs.
3. Size of pipe.
4. Pipe-wall thickness (schedule).
5. Complexity of piping system.
6. Method of joint makeup.
7. Fabrication and installation techniques.
8. Area congestion and height above grade.
9. Labor rates.
10. Labor productivity.

1. Type of pipe

Cost comparisons are given for 26 different materials of construction, in 2-, 4- and 6-in. nominal pipe sizes. However, it should be noted that certain materials may not be readily available in all diameters.

Only piping systems commonly found in large-scale chemical processing plants have been included. Hence, no solid-plastic pipe has been considered, although fiberglass-reinforced plastic (FRP) piping is shown.

2. Material costs

The material costs used here are based on manufacturers' list prices in effect during the third quarter of 1978.

No discounts or other price incentives were included in the calculations, because they differ from company to company, and often depend on the quantity of piping ordered.

The cost listings also ignore the effects of prices that are higher than list (owing to short-term shortages). Hence, material costs may have to be adjusted upward or downward to reflect above-list or discount prices. Also, in making this compilation, it was common to find wide variations in prices for the same type of product, depending upon the manufacturer.

3. Size of pipe

Installed-cost comparisons can fluctuate widely, depending on pipe size. That is, the most economical choice in 2-in. pipe may not be most economical for 6-in. To bracket the range of sizes most often found in process piping, 2-, 4- and 6-in.-dia. (nominal) pipe is compared for all materials of construction. Caution is urged when extrapolating costs to other sizes.

4. Pipe-wall thickness (schedule)

Although Schedule 40 pipe is most common, it is improper to compare high-alloy piping in only this wall-thickness to other piping systems, since thinner-wall metallic piping is often satisfactory for many applications. In general, this study provides data for both Schedule 40 and Schedule 5 alloy piping systems. These two bracket the range encountered in most corrosion-resistant process piping. Cost interpolations between these two extremes are fairly valid.

5. Complexity of piping system

Comparison of the installed cost of different corrosion-resistant piping systems hinges heavily on the systems' complexity. The most economical choice for a simple straight-run pipeline may not be the most economical for a highly complex installation consisting mainly of short runs with numerous fittings and valves.

Consequently, cost comparisons have been based on two separate piping layouts. One consists of a 1,000-ft straight run of piping, and includes no fittings. This might be considered representative of a long, interplant transfer line. This system is assumed to be erected 20 ft above grade on existing supports.

The other layout consists of a moderately complex system that is representative of process piping found in a typical indoor installation. It was assumed that all

Originally published November 20, 1978

process vessels, heat exchangers, pumps, etc., were already in place, as were sufficient support points to adequately hold the piping. The complex layout consists of 500 ft of piping and includes 56 90-deg. ells, 8 45-deg. ells and 22 tees.* Labor charges for installing 30 valves were also included, but not valve costs.

An additional factor considered in selecting the complex layout was the mix between standard 10-ft spool pieces and shorter lengths. In certain of the nonmetallic systems, the pipe is supplied by the manufacturer in cut-to-order lengths. In such cases, a premium is often charged for cutting and preparing pipe under 10 ft long. Also, when using flanged connections, the length of a pipe spool affects its cost.

An analysis of several large installations showed that 80% of the spool pieces were cut to order, and 20% were standard 10-ft lengths. Hence, in the systems joined by flanges, the following makeup was used:

20 10-ft spool pieces	200 ft
80 cut-to-order pieces	300 ft
Total	500 ft

6. Method of joint makeup

Joint makeup has a substantial effect on piping costs. Some systems require flanged connections; others, such as glass pipe, need special couplings. With metallic systems, a choice is often available among welded, flanged or screwed connections. An unrealistic choice of the joining technique can easily bias a cost study.

An analysis of a large number of jobs on typical indoor chemical processes revealed that about 70% of all alloy joints were flanged, with the balance being welded. On long runs, such as are often found on transfer lines, the ratio was usually reversed—i.e., 70% welded and 30% flanged.

It was felt that the 70% flanged-connection figure might be somewhat high, especially in applications where process streams are relatively clean, seldom plug and therefore do not require disassembly for cleanout. Consequently, for the purpose of the comparison study, an approximate 50%-welded and 50%-flanged connection ratio was assumed.

This breakdown presumes that all connections to process equipment, pumps and valves are flanged. In general, it also enables a flange to be placed about every 15 ft in the piping arrangement. This appears to be a reasonable spacing to allow for ease in handling shop-fabricated piping subassemblies. The final system assumed includes 130 joints (53%) welded and 114 joints (47%) flanged.

In the 1,000-ft transfer line, the longest available lengths were chosen for the lined-piping systems. For some installations, this meant a flange spacing of 10 ft; on others, 20 ft. For the metallic systems, a flange was assumed every 40 ft to allow for ease of installation and to provide access for cleanout.

All flanged connections are ductile iron, 150-psi rating. For metal pipe, lap-joint flanges are assumed.

7. Fabrication and installation techniques

Fabrication technique has a large effect on installed costs. Availability of shop-fabrication facilities can affect the installed cost of a project by 20% or more.

Complex layout—For the 500-ft complex layout, costs for alloy and common metallic piping systems were calculated based on three fabrication techniques:

1. Shop welding, using welding stubs for the flanged connections, followed by field installation.
2. Shop welding and machine-formed Van Stone-type lap joints, followed by field installation.
3. Field fabrication with welding stubs, followed by field installation.

The rest of the nonmetallic systems studied—such as glass and lined-piping systems—were assumed to be supplied from the manufacturer in cut-to-order lengths that required only field installation.

Straight-run layout—For the 1,000-ft straight-run layout, costs were based on field-fabricated butt welding and welding stubs; for a simple layout, this is the most commonly encountered technique.

8. Area congestion; height above grade

Both the degree of area congestion and the elevation of piping above grade have a bearing on the total installed cost. This study assumes new-construction conditions, where there is little (if any) interference from existing conduits, piping, etc. The installation area was assumed to be nonhazardous and without restrictions on field welding.

The 1,000-ft straight-run pipeline was assumed to be installed on existing supports, 20 ft above grade.

The 500-ft complex system was designed to be placed in a three-level plant, which is typical of most installations. Thus, some piping was at floor or grade level, some near the ceiling, and some in between.

9. Labor rates

The cost of labor has a major impact on any installed-cost study. Most large, new piping installations, and most major additions to existing plants, are installed by contractors rather than by regular plant employees. Consequently, a contractor-labor figure is assumed for all labor. For this study, it is taken to be $23/h and includes direct wages and fringe benefits, contractor overhead and profit, and is representative of labor rates in effect in the upper Midwest during the third quarter of 1978.

10. Labor productivity

All labor costs used are based on actual hours required for installation, at 80% labor efficiency. Labor

(Text continues on p. 190)

*The assumptions employed in this study were also the ones used in the two previous articles in this series: Installed Cost of Corrosion-Resistant Piping, by Stephen P. Marshall and J. Lee Brandt, *Chem. Eng.*, Aug. 23, 1971, p. 68; and 1974 Installed Cost of Corrosion-Resistant Piping by the same two authors, *Chem. Eng.*, Oct. 28, 1974, p. 94.

Cost of 2-in-dia., 500-ft-long, complex piping systems, by material of construction — **Table I**

Tubing material	Cost of piping and flanges, $	Fittings and welding stubs $	Cost of hanger mate-rials, $	Equip-ment rental, bolts, nuts, $	Sand-blast, paint mate-rials, $	Time to sand-blast and paint, h	Field instal-lation labor man-h	Shop or field fabri-cation, man-h	Total installed cost, $	Cost/ 100 ft piping, $	Cost ratio	Fabrication method
Carbon steel,		506						158	11,234	2,247	0.82	Machine lap joint
Sch. 40	1,596	683	140	362	212	40	168	171	11,710	2,342	0.86	Weld stub, shop
		683						256	13,665	2,733	1.00	Weld stub, field
304 S.S., Sch. 40		3,013						228	18,759	3,752	1.37	Machine lap joint
	5,396	4,113	140	778	140	8	168	256	20,503	4,101	1.50	Weld stub, shop
		4,113						384	23,447	4,689	1.72	Weld stub, field
304 S.S., Sch. 5		2,220						208	15,092	3,018	1.10	Machine lap joint
	3,046	3,268	140	714	140	8	168	231	16,669	3,334	1.22	Weld stub, shop
		3,268						347	19,337	3,867	1.42	Weld stub, field
316 S.S., Sch. 40		3,704						249	21,359	4,272	1.56	Machine lap joint
	6,746	5,017	140	854	140	8	168	281	23,408	4,682	1.71	Weld stub, shop
		5,017						422	26,651	5,330	1.95	Weld stub, field
316 S.S., Sch. 5		2,743						226	16,753	3,351	1.23	Machine lap joint
	3,696	4,004	140	788	140	8	168	253	18,635	3,727	1.36	Weld stub, shop
		4,004						380	21,556	4,311	1.58	Weld stub, field
Nickel 200,		12,876						269	38,220	7,644	2.80	Machine lap joint
Sch. 40	13,761	18,307	140	1,068	140	8	168	306	44,502	8,900	3.26	Weld stub, shop
		18,307						459	48,021	9,604	3.51	Weld stub, field
Nickel 200,		10,219						244	31,513	6,303	2.31	Machine lap joint
Sch. 5	10,371	15,331	140	983	140	8	168	276	37,361	7,472	2.73	Weld stub, shop
		15,331						414	40,535	8,107	2.97	Weld stub, field
Alloy 20,		8,287						248	30,547	6,109	2.24	Machine lap joint
Sch. 40	10,947	11,275	140	1,281	140	8	168	281	34,294	6,859	2.51	Weld stub, shop
		11,275						422	37,537	7,507	2.75	Weld stub, field
Alloy 20,		5,324						226	21,900	4,380	1.60	Machine lap joint
Sch. 5	5,933	7,243	140	1,117	140	8	168	255	24,486	4,897	1.79	Weld stub, shop
		7,243						380	27,361	5,472	2.00	Weld stub, field
Monel 400,		10,514						269	33,855	6,771	2.48	Machine lap joint
Sch. 40	11,709	14,348	140	1,117	140	8	168	306	38,540	7,708	2.82	Weld stub, shop
		14,348						459	42,059	8,412	3.08	Weld stub, field
Monel 400,		9,258						244	28,057	5,611	2.05	Machine lap joint
Sch. 5	7,791	13,902	140	1,068	140	8	168	276	33,437	6,687	2.45	Weld stub, shop
		13,902						414	36,611	7,322	2.68	Weld stub, field
Inconel 600,		17,612						268	45,143	9,029	3.30	Machine lap joint
Sch. 40	15,915	24,129	140	1,124	140	8	168	306	52,534	10,507	3.84	Weld stub, shop
		24,129						459	56,053	11,211	4.10	Weld stub, field
Inconel 600,		15,067						244	38,355	7,671	2.81	Machine lap joint
Sch. 5	12,302	20,608	140	1,046	140	8	168	276	44,632	8,926	3.27	Weld stub, shop
		20,608						414	47,806	9,561	3.50	Weld stub, field
Hastelloy B,		0						0	0	0	0.00	Machine lap joint
Sch. 40	32,884	32,545	140	926	140	8	168	336	78,411	15,682	5.74	Weld stub, shop
		32,545						454	81,125	16,225	5.94	Weld stub, field
Hastelloy B,		0						0	0	0	0.00	Machine lap joint
Sch. 5	13,935	29,960	140	919	140	8	168	303	56,111	11,222	4.11	Weld stub, shop
		29,960						455	59,607	11,921	4.36	Weld stub, field
Hastelloy		0						0	0	0	0.00	Machine lap joint
C-276,	29,094	32,085	140	926	140	8	168	336	74,161	14,832	5.43	Weld stub, shop
Sch. 40		32,085						454	76,875	15,375	5.63	Weld stub, field
Hastelloy		0						0	0	0	0.00	Machine lap joint
C-276,	11,769	28,742	140	919	140	8	168	303	52,727	10,545	3.86	Weld stub, shop
Sch. 5		28,742						455	56,223	11,245	4.11	Weld stub, field
Zirconium,		15,528						294	63,945	12,789	4.68	Machine lap joint
Sch. 40	33,775	21,126	140	1,068	140	8	276	336	70,509	14,102	5.16	Weld stub, shop
		21,126						454	73,223	14,645	5.36	Weld stub, field
Zirconium,		14,998						266	49,199	9,840	3.60	Machine lap joint
Sch. 5	21,061	20,406	140	854	140	8	248	303	55,458	11,092	4.06	Weld stub, shop
		20,406						455	58,954	11,791	4.31	Weld stub, field
Titanium,		8,715						294	46,113	9,223	3.37	Machine lap joint
Sch. 40	22,756	11,704	140	1,068	140	8	276	336	50,068	10,014	3.66	Weld stub, shop
		11,704						454	52,782	10,556	3.86	Weld stub, field

Cost of 2-in-dia., 500-ft-long, complex piping systems, by material of construction **Table I (Continued)**

Tubing material	Cost of piping and flanges, $	Fittings and welding stubs $	Cost of hanger materials, $	Equipment rental, bolts, nuts, $	Sand-blast, paint materials, $	Time to sand-blast and paint, h	Field installation labor man-h	Shop or field fabrication, man-h	Total installed cost, $	Cost/ 100 ft piping, $	Cost ratio	Fabrication method
Titanium,		8,075						266	32,971	6,594	2.41	Machine lap joint
Sch. 5	11,756	11,980	140	854	140	8	248	303	37,727	7,545	2.76	Weld stub, shop
		11,980						455	41,223	8,245	3.02	Weld stub, field
Aluminum,		2,512						228	14,397	2,879	1.05	Machine lap joint
Sch. 40	1,821	4,053	140	492	140	8	168	256	16,582	3,316	1.21	Weld stub, shop
		4,053						384	19,526	3,905	1.43	Weld stub, field
Glass Pipe	7,035	2,189	169	128	47	3	386	Purchased fabricated	18,515	3,703	1.35	Purchased fabricated
Armored Glass Pipe	13,864	6,263	169	128	47	3	386		29,418	5,884	2.15	
Glass-lined steel, Sch. 40	23,428	18,282	177	926	212	40	276		50,293	10,059	3.68	
Fiberglass reinforced vinylester	5,458	6,154	177	226	47	3	168		15,995	3,199	1.17	
Fiberglass reinforced polyester	5,028	5,729	177	226	47	3	168		15,140	3,028	1.11	
Epoxy fiberglass pipe	4,615	6,531	177	226	47	3	168		15,529	3,106	1.14	
Rubber-lined steel, Sch. 40	6,023	6,930	177	128	212	40	276	Purchased fabricated	20,738	4,148	1.52	Purchased fabricated
Tantalum lined steel, Sch. 40	70,140	45,312	140	1,068	212	40	276		124,140	24,828	9.08	
Tantalum lined steel, Sch. 5	53,240	34,394	140	854	212	40	248		95,464	19,093	6.99	
Saran-lined steel	6,250	5,526	177	128	212	40	276		19,561	3,912	1.43	
Polypropylene lined steel	6,632	5,815	177	128	212	40	276		20,232	4,046	1.48	
Kynar-lined steel	10,460	7,404	177	128	212	40	276		25,649	5,130	1.88	
FEP Teflon-lined steel	12,850	8,990	177	128	212	40	276		29,625	5,925	2.17	
PTFE Teflon-lined steel	13,159	11,938	177	128	212	40	276		32,882	6,576	2.41	
PFA Teflon lined steel	14,000	10,326	177	128	212	40	276		32,111	6,422	2.35	

Meaning of abbreviations: Sch.=Schedule; S.S.=stainless steel; F.R.P.=glass-fiber-reinforced plastic
Penton=chlorinated polyester; Kynar=polyvinylidene fluoride; TFE Teflon=tetrafluoroethylene;
FEP Teflon=fluorinated ethylene-propylene; Saran=polyvinylidene chloride.
Cost ratio=(Cost of listed system) + (Cost of Schedule 40, carbon-steel piping system, field-fabricated using welding stubs).

Cost of 4-in-dia., 500-ft-long, complex piping systems, by material of construction **Table II**

Tubing material	Cost of piping and flanges, $	Fittings and welding stubs $	Cost of hanger mate-rials, $	Equip-ment rental, bolts, nuts, $	Sand-blast, paint mate-rials, $	Time to sand-blast and paint, h	Field instal-lation labor man-h	Shop or field fabri-cation, man-h	Total installed cost, $	Cost/ 100 ft piping, $	Cost ratio	Fabrication method
Carbon steel,		1,168						267	20,798	4,160	0.82	Machine lap joint
Sch. 40	3,441	1,523	226	697	316	68	315	303	21,981	4,396	0.86	Weld stub, shop
		1,523						454	25,454	5,091	1.00	Weld stub, field
304 S.S.,		10,940						391	44,807	8,961	1.76	Machine lap joint
Sch. 40	15,166	14,437	226	1,855	175	9	315	455	49,776	9,955	1.96	Weld stub, shop
		14,437						733	56,170	11,234	2.21	Weld stub, field
304 S.S.,		6,958						354	31,585	6,317	1.24	Machine lap joint
Sch. 5	6,916	9,947	226	1,716	175	9	315	410	35,862	7,172	1.41	Weld stub, shop
		9,947						615	40,577	8,115	1.59	Weld stub, field
316 S.S.,		13,461						428	52,572	10,514	2.07	Machine lap joint
Sch. 40	19,341	17,650	226	2,073	175	9	315	500	58,417	11,683	2.30	Weld stub, shop
		17,650						750	64,167	12,833	2.52	Weld stub, field
316 S.S.,		8,678						388	35,863	7,173	1.41	Machine lap joint
Sch. 5	8,541	12,274	226	1,867	175	9	315	451	40,908	8,182	1.61	Weld stub, shop
		12,274						727	47,256	9,451	1.86	Weld stub, field
Nickel 200,		48,603						465	106,825	21,365	4.20	Machine lap joint
Sch. 40	37,086	66,126	226	2,588	175	9	315	545	126,188	25,238	4.96	Weld stub, shop
		66,126						823	132,582	26,516	5.21	Weld stub, field
Nickel 200,		30,707						421	69,569	13,914	2.73	Machine lap joint
Sch. 5	18,990	41,778	226	2,336	175	9	315	491	82,250	16,450	3.23	Weld stub, shop
		41,778						747	88,138	17,628	3.46	Weld stub, field
Alloy 20,		27,108						428	83,724	16,745	3.29	Machine lap joint
Sch. 40	35,808	36,882	226	3,111	175	9	315	500	95,154	19,031	3.74	Weld stub, shop
		36,882						750	100,904	20,181	3.96	Weld stub, field
Alloy 20,		18,844						388	55,496	11,099	2.18	Machine lap joint
Sch. 5	17,074	25,638	226	2,801	175	9	315	451	63,739	12,748	2.50	Weld stub, shop
		25,638						727	70,087	14,017	2.75	Weld stub, field
Monel 400,		34,529						465	83,229	16,646	3.27	Machine lap joint
Sch. 40	27,490	46,978	226	2,662	175	9	315	545	97,518	19,504	3.83	Weld stub, shop
		46,978						823	103,912	20,782	4.08	Weld stub, field
Monel 400,		23,923						421	58,671	11,734	2.30	Machine lap joint
Sch. 5	14,795	32,549	226	2,417	175	9	315	491	68,907	13,781	2.71	Weld stub, shop
		32,549						747	74,795	14,959	2.94	Weld stub, field
Inconel 600,		56,165						465	109,896	21,979	4.32	Machine lap joint
Sch. 40	32,380	76,415	226	2,803	175	9	315	545	131,986	26,397	5.19	Weld stub, shop
		76,415						823	138,380	27,676	5.44	Weld stub, field
Inconel 600,		31,390						421	74,489	14,898	2.93	Mach lap joint
Sch. 5	23,170	42,708	226	2,393	175	9	315	491	87,417	17,483	3.43	Weld stub, shop
		42,708						747	93,305	18,661	3.67	Weld stub, field
Hastelloy B,		0						0	0	0	0.00	Machine lap joint
Sch. 40	76,689	86,973	226	1,851	175	9	315	599	187,143	37,429	7.35	Weld stub, shop
		86,973						899	194,043	38,809	7.62	Weld stub, field
Hastelloy B,		0						0	0	0	0.00	Machine lap joint
Sch. 5	30,670	58,488	226	1,712	175	9	315	540	111,143	22,229	4.37	Weld stub, shop
		58,488						810	117,353	23,471	4.61	Weld stub, field
Hastelloy		0						0	0	0	0.00	Machine lap joint
C-276,	67,486	82,306	226	1,851	175	9	315	599	173,273	34,655	6.81	Weld stub, shop
Sch. 40		82,306						899	180,173	36,035	7.08	Weld stub, field
Hastelloy		0						0	0	0	0.00	Machine lap joint
C-276,	26,880	59,220	226	1,712	175	9	315	540	108,085	21,617	4.25	Weld stub, shop
Sch. 5		59,220						810	114,295	22,859	4.49	Weld stub, field
Zirconium,		47,003						509	175,025	35,005	6.88	Machine lap joint
Sch. 40	101,021	63,950	226	2,588	175	9	526	599	194,042	38,808	7.62	Weld stub, shop
		63,950						899	200,942	40,188	7.89	Weld stub, field
Zirconium,		45,310						461	133,095	26,619	5.23	Machine lap joint
Sch. 5	63,625	61,646	226	2,070	175	9	473	540	151,248	30,250	5.94	Weld stub, shop
		61,646						810	157,458	31,492	6.19	Weld stub, field
Titanium,		30,923						509	126,490	25,298	4.97	Machine lap joint
Sch. 40	68,566	37,065	226	2,588	175	9	526	599	134,702	26,940	5.29	Weld stub, shop
		37,065						899	141,602	28,320	5.56	Weld stub, field
Titanium,		20,662						461	70,688	14,138	2.78	Machine lap joint
Sch. 5	25,866	24,567	226	2,070	175	9	473	540	76,410	15,282	3.00	Weld stub, shop
		24,567						810	82,620	16,524	3.25	Weld stub, field

Cost of 4-in-dia., 500-ft-long, complex piping systems, by material of construction — **Table II (continued)**

Tubing material	Cost of piping and flanges, $	Fittings and welding stubs $	Cost of hanger materials, $	Equipment rental, bolts, nuts, $	Sand-blast, paint materials, $	Time to sand-blast and paint, h	Field installation labor man-h	Shop or field fabrication, man-h	Total installed cost, $	Cost/ 100 ft piping, $	Cost ratio	Fabrication method
Aluminum,		5,523						391	27,518	5,504	1.08	Machine lap joint
Sch. 40	4,341	8,612	226	808	175	9	315	455	32,079	6,416	1.26	Weld stub, shop
		8,612						682	37,300	7,460	1.47	Weld stub, field
Glass pipe	18,781	5,130	323	214	59	3	533	Purchased fabricated	36,835	7,367	1.45	Purchased fabricated
Armored glass pipe	35,310	12,280	323	214	59	3	533	↓	60,514	12,103	2.38	↓
Glass-lined steel, Sch. 40	42,853	24,869	214	1,851	316	68	526	↓	83,765	16,753	3.29	↓
Fiberglass reinforced vinylester	8,585	9,682	214	394	59	3	315	↓	26,248	5,250	1.03	↓
Fiberglass reinforced polyester	7,635	8,609	214	394	59	3	315	↓	24,225	4,845	0.95	↓
Epoxy fiberglass pipe	7,823	10,339	214	394	59	3	315	↓	26,143	5,229	1.03	↓
Rubber-lined steel, Sch. 40	12,314	10,847	214	310	316	68	526	↓	37,663	7,533	1.48	↓
Tantalum lined steel, Sch. 40	211,834	138,618	226	2,588	316	68	526	↓	367,244	73,449	14.43	↓
Tantalum lined steel, Sch. 5	118,350	77,446	226	2,070	316	68	473	Purchased fabricated	210,851	42,170	8.28	Purchased fabricated
Saran-lined steel	11,924	9,620	214	310	316	68	526	↓	36,046	7,209	1.42	↓
Polypropylene lined steel	13,086	10,634	214	310	316	68	526	↓	38,222	7,644	1.50	↓
Kynar-lined steel	21,510	15,176	214	310	316	68	526	↓	51,188	10,238	2.01	↓
FEP Teflon lined steel	28,200	19,624	214	310	316	68	526	↓	62,326	12,465	2.45	↓
PTFE Teflon lined steel	27,168	23,226	214	310	316	68	526	↓	64,896	12,979	2.55	↓
PFA Teflon lined steel	31,585	22,145	214	310	316	68	526	↓	68,232	13,646	2.68	↓

Cost of 6-in-dia., 500-ft-long, complex piping systems, by material of construction — **Table III**

Tubing material	Cost of piping and flanges, $	Fittings and welding stubs $	Cost of hanger materials, $	Equipment rental, bolts, nuts, $	Sand-blast, paint materials, $	Time to sand-blast and paint, h	Field installation labor man-h	Shop or field fabrication, man-h	Total installed cost, $	Cost/ 100 ft piping, $	Cost ratio	Fabrication method
Carbon steel,		3,111						429	34,371	6,874	0.80	Machine lap joint
Sch. 40	5,713	4,265	317	2,028	386	90	473	501	37,181	7,436	0.87	Weld stub, shop
		4,265						750	42,908	8,582	1.00	Weld stub, field
304 S.S., Sch. 40		25,764						634	83,391	16,678	1.94	Machine lap joint
	28,138	35,225	317	3,269	212	10	473	751	95,543	19,109	2.23	Weld stub, shop
		35,225						1,127	104,191	20,838	2.43	Weld stub, field
304 S.S., Sch. 5		14,810						572	54,540	10,908	1.27	Machine lap joint
	12,638	20,490	317	2,298	212	10	473	676	62,612	12,522	1.46	Weld stub, shop
		20,490						1,014	70,386	14,077	1.64	Weld stub, field
316 S.S., Sch. 40		31,883						696	96,248	19,250	2.24	Machine lap joint
	33,138	43,193	317	3,631	212	10	473	826	110,598	22,120	2.58	Weld stub, shop
		43,193						1,239	120,097	24,019	2.80	Weld stub, field

Cost of 6-in-dia., 500-ft-long, complex piping systems, by material of construction **Table III (continued)**

Tubing material	Cost of piping and flanges, $	Fittings and welding stubs $	Cost of hanger materials, $	Equipment rental, bolts, nuts, $	Sandblast, paint materials, $	Time to sandblast and paint, h	Field installation labor man-h	Shop or field fabrication, man-h	Total installed cost, $	Cost/ 100 ft piping, $	Cost ratio	Fabrication method
316 S.S., Sch. 5		18,325						628	62,550	12,510	1.46	Machine lap joint
	15,638	25,095	317	2,505	212	10	473	743	71,965	14,393	1.68	Weld stub, shop
		25,095						1,220	82,936	16,587	1.93	Weld stub, field
Nickel 200,		93,881						757	192,596	38,519	4.49	Machine lap joint
Sch. 40	65,163	127,730	317	4,503	212	10	473	901	229,757	45,951	5.35	Weld stub, shop
		127,730						1,357	240,245	48,049	5.60	Weld stub, field
Nickel 200,		61,242						682	122,987	24,597	2.87	Machine lap joint
Sch. 5	31,290	83,322	317	3,131	212	10	473	810	148,011	29,602	3.45	Weld stub, shop
		83,322						1,215	157,326	31,465	3.67	Weld stub, field
Alloy 20,		53,763						695	157,849	31,570	3.68	Machine lap joint
Sch. 40	71,017	73,148	317	5,446	212	10	473	826	180,247	36,049	4.20	Weld stub, shop
		73,148						1,239	189,746	37,949	4.42	Weld stub, field
Alloy 20,		33,131						627	97,283	19,457	2.27	Machine lap joint
Sch. 5	34,336	45,076	317	3,757	212	10	473	743	111,896	22,379	2.61	Weld stub, shop
		45,076						1,220	122,867	24,573	2.86	Weld stub, field
Monel 400,		76,644						757	161,533	32,307	3.76	Machine lap joint
Sch. 40	51,133	104,277	317	4,707	212	10	473	901	192,478	38,496	4.49	Weld stub, shop
		104,277						1,357	202,966	40,593	4.73	Weld stub, field
Monel 400,		42,821						682	97,724	19,545	2.28	Machine lap joint
Sch. 5	24,205	58,260	317	3,374	212	10	473	810	116,107	23,221	2.71	Weld stub, shop
		58,260						1,215	125,422	25,084	2.92	Weld stub, field
Inconel 600,		107,269						757	207,376	41,475	4.83	Machine lap joint
Sch. 40	66,184	145,945	317	4,874	212	10	473	901	249,364	49,873	5.81	Weld stub, shop
		145,945						1,357	259,852	51,970	6.06	Weld stub, field
Inconel 600,		80,428						682	153,735	30,747	3.58	Machine lap joint
Sch. 5	42,747	109,426	317	3,236	212	10	473	810	185,677	37,135	4.33	Weld stub, shop
		109,426						1,215	194,992	38,998	4.54	Weld stub, field
Hastelloy B,		0						0	0	0	0.00	Machine lap joint
Sch. 40	128,594	132,829	317	2,574	212	10	473	991	298,428	59,686	6.96	Weld stub, shop
		132,829						1,436	308,663	61,733	7.19	Weld stub, field
Hastelloy B,		0						0	0	0	0.00	Machine lap joint
Sch. 5	52,257	102,815	317	2,807	212	10	473	891	190,010	38,002	4.43	Weld stub, shop
		102,815						1,336	200,245	40,049	4.67	Weld stub, field
Hastelloy		0						0	0	0	0.00	Machine lap joint
C-276,	112,352	116,052	317	2,574	212	10	473	991	265,409	53,082	6.19	Weld stub, shop
Sch. 40		116,052						1,436	275,644	55,129	6.42	Weld stub, field
Hastelloy		0						0	0	0	0.00	Machine lap joint
C-276,	45,760	97,248	317	2,807	212	10	473	891	177,946	35,589	4.15	Weld stub, shop
Sch. 5		97,248						1,336	188,181	37,636	4.39	Weld stub, field
Zirconium,		76,132						831	283,183	56,637	6.60	Machine lap joint
Sch. 40	163,624	103,581	317	3,131	212	10	888	991	314,312	62,862	7.33	Weld stub, shop
		103,581						1,436	324,547	64,909	7.56	Weld stub, field
Zirconium,		67,982						749	202,312	40,462	4.72	Machine lap joint
Sch. 5	95,463	92,492	317	2,504	212	10	799	891	230,088	46,018	5.36	Weld stub, shop
		92,492						1,336	240,323	48,065	5.60	Weld stub, field
Titanium,		66,171						831	239,413	47,883	5.58	Machine lap joint
Sch. 40	129,815	76,516	317	3,131	212	10	888	991	253,438	50,688	5.91	Weld stub, shop
		76,516						1,436	263,673	52,735	6.15	Weld stub, field
Titanium,		42,520						749	123,225	24,645	2.87	Machine lap joint
Sch. 5	41,838	47,728	317	2,504	212	10	799	891	131,699	26,340	3.07	Weld stub, shop
		47,728						1,336	141,934	28,387	3.31	Weld stub, field
Aluminum,		10,384						634	45,244	9,049	1.05	Machine lap joint
Sch. 40	7,763	15,308	317	877	212	10	473	751	52,859	10,572	1.23	Weld stub, shop
		15,308						1,126	61,484	12,297	1.43	Weld stub, field
Glass pipe	29,973	8,190	915	413	71	3	888	Purchased fabricated	60,055	12,011	1.40	Purchased fabricated
Armored glass pipe	58,073	24,737	915	413	71	3	888	↓	104,702	20,940	2.44	↓
Glass-lined steel, Sch. 40	69,995	39,795	425	2,574	386	90	888	↓	135,669	27,134	3.16	↓

Cost of 6-in-dia., 500-ft-long, complex piping systems, by material of construction — **Table III (continued)**

Tubing material	Cost of piping and flanges, $	Fittings and welding stubs $	Cost of hanger mate-rials, $	Equip-ment rental, bolts, nuts, $	Sand-blast, paint mate-rials, $	Time to sand-blast and paint, h	Field instal-lation labor man-h	Shop or field fabri-cation, man-h	Total installed cost, $	Cost/ 100 ft piping, $	Cost ratio	Fabrication method
Fiberglass reinforced vinylester	13,488	15,210	425	668	71	3	473	Purchased fabricated	40,810	8,162	0.95	Purchased fabricated
Fiberglass reinforced polyester	10,616	11,972	425	671	71	3	473	↓	34,703	6,941	0.81	↓
Epoxy fiberglass pipe	10,859	15,893	425	668	71	3	473	↓	38,864	7,773	1.03	↓
Rubber-lined steel, Sch. 40	20,642	19,752	425	413	386	90	888	↓	64,112	12,822	1.49	↓
Tantalum lined steel, Sch. 40	409,961	268,268	317	3,131	386	90	888	↓	704,557	140,911	16.42	↓
Tantalum lined steel, Sch. 5	208,959	136,738	317	2,504	386	90	798	↓	369,328	73,866	8.61	↓
Saran-lined steel	19,497	17,623	424	413	386	90	888	↓	60,837	12,167	1.42	↓
Polypropylene lined steel	20,146	18,578	424	413	386	90	888	↓	62,441	12,488	1.46	↓
Kynar-lined steel	33,119	33,037	424	413	386	90	888	↓	89,873	17,975	2.09	↓
FEP teflon lined steel	51,500	39,986	424	413	386	90	888	↓	115,203	23,041	2.68	↓
PTFE teflon lined steel	46,128	46,407	424	413	386	90	888	↓	116,252	23,250	2.71	↓
PFA teflon lined steel	56,685	45,537	424	413	386	90	888	↓	125,939	25,188	2.94	↓

Cost of 2-in-dia., 1,000-ft-long, straight-run piping systems, by material of construction — **Table IV**

Tubing material	Cost of piping and flanges $	Cost of hanger mate-rials, $	Equip-ment rental, sup-plies $	Sand-blast, paint mate-rials, $	Time to sand-blast and paint, h	Field fabri-cation labor, man-h	Field instal-lation labor, man-h	Total installed cost, $	Cost/ 100 ft piping, $	Cost ratio	Fabrication method
Carbon steel, Sch. 40	1,900	181	3,883	140	94	26	281	15,327	1,533	1.00	Field-fabricated. (butt welded and welding stubs)
304 S.S., Sch. 40	10,150	181	6,217	56	42	39	390	27,437	2,744	1.79	↓
304 S.S., Sch. 5	5,413	181	6,221	56	42	36	360	21,945	2,195	1.43	↓
316 S.S., Sch. 40	13,000	181	7,383	56	42	47	452	33,063	3,306	2.16	↓
316 S.S., Sch. 5	6,863	181	5,503	56	42	42	407	23,896	2,390	1.56	↓
Nickel 200, Sch. 40	29,185	181	8,036	56	42	52	494	50,982	5,098	3.33	↓
Nickel 200, Sch. 5	20,750	181	7,394	56	42	47	443	40,617	4,062	2.65	↓
Alloy 20, Sch. 40	22,572	181	11,693	56	42	78	598	51,016	5,102	3.33	↓
Alloy 20, Sch. 5	12,382	181	10,007	56	42	117	439	36,380	3,638	2.37	↓
Monel 400, Sch. 40	27,079	181	7,596	56	42	52	494	48,436	4,844	3.16	↓
Monel 400, Sch. 5	18,890	181	8,224	56	42	47	443	39,587	3,959	2.58	↓

Cost of 2-in-dia., 1,000-ft-long, straight-run piping systems, by material of construction **Table IV (continued)**

Tubing material	Cost of piping and flanges $	Cost of hanger materials, $	Equipment rental, supplies $	Sandblast, paint materials, $	Time to sandblast and paint, h	Field fabrication labor, man-h	Field installation labor, man-h	Total installed cost, $	Cost/ 100 ft piping, $	Cost ratio	Fabrication method
Inconel 600, Sch. 40	33,428	181	7,845	56	42	52	494	55,034	5,503	3.59	Field-fabricated (Butt-welded and welding stubs)
Inconel 600, Sch. 5	25,269	181	8,475	56	42	47	443	46,217	4,622	3.02	
Hastelloy B, Sch. 40	70,398	181	20,716	56	42	200	663	112,166	11,217	7.32	
Hastelloy B, Sch. 5	32,028	181	17,832	56	42	180	513	67,002	6,700	4.37	
Hastelloy C-276, Sch. 40	62,743	181	20,716	56	42	200	663	104,511	10,451	6.82	
Hastelloy C-276, Sch. 5	28,845	181	17,832	56	42	180	513	63,819	6,382	4.16	
Zirconium, Sch. 40	70,064	181	8,036	56	42	52	230	85,789	8,579	5.60	
Zirconium, Sch. 5	44,501	181	6,429	56	42	47	207	57,975	5,798	3.78	
Titanium, Sch. 40	43,770	181	8,036	56	42	52	230	59,495	5,950	3.88	
Titanium, Sch. 5	27,972	181	6,429	56	42	47	207	41,446	4,145	2.70	
Aluminum, Sch. 40	3,310	181	5,838	56	42	39	360	19,528	1,953	1.27	
Glass pipe	6,246	361	2,150	18	14	Purchased fabricated	190	13,467	1,347	0.88	Purchased fabricated
Armored glass pipe	17,121	361	2,150	18	14		190	24,342	2,434	1.59	
Glass-lined steel, Sch. 40	33,700	361	3,218	140	94		187	43,882	4,388	2.86	
Fiberglass reinforced vinylester	6,924	541	4,469	18	14		330	19,864	1,986	1.30	
Fiberglass reinforced polyester	5,755	541	4,469	18	14		330	18,695	1,870	1.22	
Epoxy fiberglass pipe	6,639	541	4,469	18	14		330	19,579	1,958	128	
Rubber-lined steel, Sch. 40	9,270	181	2,732	106	90		140	17,579	1,758	1.15	
Tantalum lined steel, Sch. 40	142,754	181	3,218	106	90		140	151,549	15,155	9.89	
Tantalum lined steel, Sch. 5	107,939	181	2,574	106	90		126	115,768	11,577	7.55	
Saran-lined steel	8,810	361	3,218	140	94		187	18,992	1,899	1.24	
Polypropylene lined steel	7,155	181	2,732	106	90		140	15,464	1,546	1.01	
Kynar-lined steel	13,965	181	2,732	106	90		140	22,274	2,227	1.45	
FEP Teflon lined steel	19,175	181	2,732	106	90		140	27,484	2,748	1.79	
PTFE Teflon lined steel	22,306	361	3,218	140	94		187	32,488	3,249	2.12	
PFA Teflon lined steel	20,425	181	2,732	106	90		140	28,734	2,873	1.87	

Cost of 4-in-dia., 1,000-ft-long, straight-run piping systems, by material of construction **Table V**

Tubing material	Cost of piping and flanges $	Cost of hanger materials, $	Equipment rental, supplies $	Sandblast, paint materials, $	Time to sandblast and paint, h	Field fabrication labor, man-h	Field installation labor, man-h	Total installed cost, $	Cost/ 100 ft piping, $	Cost ratio	Fabrication method
Carbon steel, Sch. 40	4,625	226	5,878	240	162	54	372	24,493	2,449	1.00	Field-fabricated (Butt welded and welding stubs)
304 S.S., Sch.40	30,288	226	9,033	85	48	81	493	53,938	5,394	2.20	↓
304 S.S., Sch. 5	13,430	226	8,700	85	48	73	457	35,735	3,574	1.46	
316 S.S., Sch. 40	39,125	226	11,253	85	48	97	558	66,858	6,686	2.73	
316 S.S., Sch. 5	17,108	226	10,117	85	48	88	493	42,003	4,200	1.71	
Nickel 200, Sch. 40	77,240	226	11,500	85	48	108	614	106,761	10,676	4.36	
Nickel 200, Sch. 5	37,980	226	10,398	85	48	98	553	64,766	6,477	2.64	
Alloy 20, Sch. 40	71,960	226	20,906	85	48	162	735	114,912	11,491	4.69	
Alloy 20, Sch. 5	35,839	226	17,999	85	48	243	652	75,838	7,584	3.10	
Monel 400, Sch. 40	59,350	226	9,553	85	48	108	614	86,924	8,692	3.55	
Monel 400, Sch. 5	29,590	226	12,718	85	48	98	553	58,696	5,870	2.40	
Inconel 600, Sch. 40	78,973	226	16,255	85	48	108	614	113,249	11,325	4.62	
Inconel 600, Sch. 5	54,284	226	13,047	85	48	98	553	83,719	8,372	3.42	
Hastelloy B, Sch. 40	158,388	226	46,872	85	48	400	850	235,425	23,543	9.61	
Hastelloy B, Sch. 5	69,139	226	35,758	85	48	360	765	132,187	13,219	5.40	
Hastelloy C-276, Sch. 40	139,555	226	46,872	85	48	400	850	216,592	21,659	8.84	
Hastelloy C-276, Sch. 5	61,695	226	35,758	85	48	360	765	124,743	12,474	5.09	
Zirconium, Sch. 40	199,543	226	11,500	85	48	108	376	223,590	22,359	9.13	
Zirconium, Sch. 5	129,903	226	9,199	85	48	98	338	150,545	15,055	6.15	
Titanium, Sch. 40	113,942	226	11,500	85	48	108	376	137,989	13,799	5.63	
Titanium, Sch. 5	58,541	226	9,199	85	48	98	338	79,183	7,918	3.23	
Aluminum, Sch. 40	8,350	226	7,734	85	48	81	420	29,022	2,902	1.18	↓
Glass pipe	15,970	450	3,478	28	16	Purchased fabricated	296	27,102	2,710	1.11	Purchased fabricated
Armored glass pipe	44,408	450	3,478	28	16	↓	296	55,540	5,554	2.27	↓
Glass-lined steel, Sch. 40	58,813	450	5,299	240	162		279	74,945	7,495	3.06	
Fiberglass reinforced vinylester	10,241	676	5,321	28	16	↓	486	27,812	2,781	1.14	↓

Cost of 4-in-dia., 1,000-ft-long, straight-run piping systems, by material of construction **Table V (continued)**

Tubing material	Cost of piping and flanges $	Cost of hanger mate-rials, $	Equip-ment rental, sup-plies $	Sand-blast, paint mate-rials, $	Time to sand-blast and paint, h	Field fabri-cation labor, man-h	Field instal-lation labor, man-h	Total installed cost, $	Cost/ 100 ft piping, $	Cost ratio	Fabrication method
Fiberglass reinforced polyester	8,055	676	5,321	28	16	Purchased fabricated	486	25,626	2,563	1.05	Purchased fabricated
Epoxy fiberglass pipe	11,677	676	5,321	28	16	↓	486	29,248	2,925	1.19	↓
Rubber-lined steel, Sch. 40	19,395	226	4,037	212	156	↓	220	32,518	3,252	1.33	↓
Tantalum lined steel, Sch. 40	433,481	226	5,299	212	156	↓	220	447,866	44,787	18.29	↓
Tantalum lined steel, Sch. 5	240,822	226	4,239	212	156	↓	198	253,641	25,364	10.36	↓
Saran-lined steel	16,872	450	5,299	240	162	↓	279	33,004	3,300	1.35	↓
Polypropylene lined steel	14,136	226	4,037	212	156	↓	220	27,259	2,726	1.11	↓
Kynar-lined steel	30,751	226	4,037	212	156	↓	220	43,874	4,387	1.79	↓
FEP Teflon lined steel	43,350	226	4,037	212	156	↓	220	56,473	5,647	2.31	↓
PTFE Teflon lined steel	47,242	450	5,299	240	162	↓	279	63,374	6,337	2.59	↓
PFA Teflon lined steel	48,493	226	4,037	212	156	↓	220	61,616	6,162	2.52	↓

Cost of 6-in-dia., 1,000-ft-long, straight-run piping systems, by material of construction **Table VI**

Tubing material	Cost of piping and flanges $	Cost of hanger mate-rials, $	Equip-ment rental, sup-plies $	Sand-blast, paint mate-rials, $	Time to sand-blast and paint, h	Field fabri-cation labor, man-h	Field instal-lation labor, man-h	Total installed cost, $	Cost/ 100 ft piping, $	Cost ratio	Fabrication method
Carbon steel, Sch. 40	8,151	450	11,553	399	211	75	649	42,058	4,206	1.00	Field fabricated (Butt welded and welding stubs)
304 S.S., Sch. 40	58,850	450	17,899	112	54	113	883	101,461	10,146	2.41	↓
304 S.S., Sch. 5	25,188	450	14,735	112	54	111	442	54,446	5,445	1.29	↓
316 S.S., Sch. 40	70,188	450	21,172	112	54	130	1,023	119,683	11,968	2.85	↓
316 S.S., Sch. 5	31,956	450	18,176	112	54	117	921	75,810	7,581	1.80	↓
Nickel 200, Sch. 40	137,179	450	23,305	112	54	150	1,116	191,406	19,141	4.55	↓
Nickel 200, Sch. 5	62,580	450	16,904	112	54	135	1,005	107,508	10,751	2.56	↓
Alloy 20, Sch. 40	137,542	450	40,636	112	54	225	1,354	216,299	21,630	5.14	↓
Alloy 20, Sch. 5	72,034	450	34,076	112	54	338	1,219	143,725	14,373	3.42	↓
Monel 400, Sch. 40	106,938	450	29,687	112	54	150	1,116	167,547	16,755	3.98	↓
Monel 400, Sch. 5	48,410	450	23,977	112	54	135	1,005	100,411	10,041	2.39	↓

Cost of 6-in-dia., 1,000-ft-long, straight-run piping systems, by material of construction **Table VI (continued)**

Tubing material	Cost of piping and flanges $	Cost of hanger materials, $	Equipment rental, supplies $	Sandblast, paint materials, $	Time to sandblast and paint, h	Field fabrication labor, man-h	Field installation labor, man-h	Total installed cost, $	Cost/100 ft piping, $	Cost ratio	Fabrication method
Inconel 600, Sch. 40	154,486	450	31,205	112	54	150	1,116	216,613	21,661	5.15	Field fabricated (Butt welded and welding stubs
Inconel 600, Sch. 5	100,823	450	24,996	112	54	135	1,005	153,843	15,384	3.66	
Hastelloy B, Sch. 40	264,513	450	95,651	112	54	640	1,485	410,843	41,084	9.77	
Hastelloy B, Sch. 5	118,614	450	72,186	112	54	580	1,345	236,879	23,688	5.63	
Hastelloy C-276, Sch. 40	230,473	450	95,651	112	54	640	1,485	376,803	37,680	8.96	
Hastelloy C-276, Sch. 5	104,588	450	72,186	112	54	580	1,348	222,922	22,292	5.30	
Zirconium, Sch. 40	322,220	450	23,305	112	54	150	525	362,854	36,285	8.63	
Zirconium, Sch. 5	194,502	450	18,643	112	54	135	473	228,933	22,893	5.44	
Titanium, Sch. 40	215,493	450	23,305	112	54	150	525	256,127	25,613	6.09	
Titanium, Sch. 5	63,280	450	18,643	112	54	135	473	97,711	9,771	2.32	
Aluminum, Sch. 40	14,906	450	12,699	112	54	115	650	47,004	4,700	1.12	
Glass pipe	26,285	901	4,744	38	18	Purchased fabricated	444	42,594	4,259	1.01	Purchased fabricated
Armored glass pipe	73,229	901	4,744	38	18		444	89,538	8,954	2.13	
Glass-lined steel, Sch. 40	99,201	901	7,100	400	211		381	121,218	12,122	2.88	
Fiberglass reinforced vinylester	15,213	1,352	8,769	38	18		645	40,621	4,062	0.97	
Fiberglass reinforced polyester	13,956	1,352	8,769	38	18		645	39,364	3,936	0.94	
Epoxy fiberglass pipe	16,310	1,352	8,789	38	18		645	41,718	4,172	0.99	
Rubber-lined steel, Sch. 40	30,813	450	6,236	352	200		325	49,926	4,993	1.19	
Tantalum lined steel, Sch. 40	827,474	450	7,100	352	200		325	847,451	84,745	20.15	
Tantalum lined steel, Sch. 5	419,248	450	5,680	352	200		293	437,069	43,707	10.39	
Saran-lined steel	28,299	902	7,100	400	211		381	50,317	5,032	1.20	
Polypropylene lined steel	24,406	450	6,236	352	200		325	43,519	4,352	1.03	
Kynar-lined steel	52,978	450	6,236	352	200		325	72,091	7,209	1.71	
FEP Teflon lined steel	82,000	450	6,236	352	200		325	101,113	10,111	2.40	
PTFE Teflon lined steel	87,084	902	7,100	400	211		381	109,102	10,910	2.59	
PFA Teflon lined steel	89,318	450	6,236	352	200		325	108,431	10,843	2.58	

Cost ranking and cost ratios, in ascending order, of 500-ft complex piping systems **Table VII**

Rank	2-in. dia.		4-in. dia.		6-in. dia.	
	Piping material	Cost ratio	Piping material	Cost ratio	Piping material	Cost ratio
1	Carbon steel, Sch. 40	1.00	Fiberglass reinforced polyester	0.95	Fiberglass reinforced polyester	0.81
2	Fiberglass reinforced polyester	1.11	Carbon steel, Sch. 40	1.00	Epoxy fiberglass pipe	0.91
3	Epoxy fiberglass pipe	1.14	Epoxy fiberglass pipe	1.03	Fiberglass reinforced vinylester	0.95
4	Fiberglass reinforced vinylester	1.17	Fiberglass reinforced vinylester	1.03	Carbon steel, Sch. 40	1.00
5	Glass pipe	1.35	Saran-lined steel	1.42	Glass pipe	1.40
6	304 S.S., Sch. 5	1.42	Glass pipe	1.45	Saran-lined steel	1.42
7	Aluminum, Sch. 40	1.43	Aluminum, Sch. 40	1.47	Aluminum, Sch. 40	1.43
8	Saran-lined steel	1.43	Rubber-lined steel, Sch. 40	1.48	Polypropylene lined steel	1.46
9	Polypropylene-lined steel	1.48	Polypropylene lined steel	1.50	Rubber-lined steel, Sch. 40	1.49
10	Rubber-lined steel, Sch. 40	1.52	304 S.S., Sch. 5	1.59	304 S.S., Sch. 5	1.64
11	316 S.S., Sch. 5	1.58	316 S.S., Sch. 5	1.86	316 S.S., Sch. 5	1.93
12	304 S.S., Sch. 40	1.72	Kynar-lined steel	2.01	Kynar-lined steel	2.09
13	Kynar-lined steel	1.88	304 S.S., Sch. 40	2.21	304 S.S., Sch. 40	2.43
14	316 S.S., Sch. 40	1.95	Armored glass pipe	2.38	Armored glass pipe	2.44
15	Alloy 20, Sch. 5	2.00	FEP Teflon lined steel	2.45	FEP Teflon lined steel	2.68
16	Armored glass pipe	2.15	316 S.S., Sch. 40	2.52	PTFE Teflon lined steel	2.71
17	FEP Teflon-lined steel	2.17	PTFE Teflon lined steel	2.55	316 S.S., Sch. 40	2.80
18	PFA Teflon-lined steel	2.35	PFA Teflon lined steel	2.68	Alloy 20, Sch. 5	2.86
19	PTFE Teflon-lined steel	2.41	Alloy 20, Sch. 5	2.75	Monel 400, Sch. 5	2.92
20	Monel 400, Sch. 5	2.68	Monel 400, Sch. 5	2.94	PFA Teflon lined steel	2.94
21	Alloy 20, Sch. 40	2.75	Titanium, Sch. 5	3.25	Glass-lined steel, Sch. 40	3.16
22	Nickel 200, Sch. 5	2.97	Glass-lined steel, Sch. 40	3.29	Titanium, Sch. 5	3.31
23	Titanium, Sch. 5	3.02	Nickel 200, Sch. 5	3.46	Nickel 200, Sch. 5	3.67
24	Monel 400, Sch. 40	3.08	Inconel 600, Sch. 5	3.67	Hastelloy C-276, Sch. 5	4.39
25	Inconel 600, Sch. 5	3.50	Alloy 20, Sch. 40	3.96	Alloy 20, Sch. 40	4.42
26	Nickel 200, Sch. 40	3.51	Monel 400, Sch. 40	4.08	Inconel 600, Sch. 5	4.54
27	Glass-lined steel, Sch. 40	3.68	Hastelloy C-276, Sch. 5	4.49	Hastelloy B, Sch. 5	4.67
28	Titanium, Sch. 40	3.86	Hastelloy B, Sch. 5	4.61	Monel 400, Sch. 40	4.73
29	Inconel 600, Sch. 40	4.10	Nickel 200, Sch. 40	5.21	Zirconium, Sch. 5	5.60
30	Hastelloy C-276, Sch. 5	4.11	Inconel 600, Sch. 40	5.44	Nickel 200, Sch. 40	5.60
31	Zirconium, Sch. 5	4.31	Titanium, Sch. 40	5.56	Inconel 600, Sch. 40	6.06
32	Hastelloy B, Sch. 5	4.36	Zirconium, Sch. 5	6.19	Titanium, Sch. 40	6.15
33	Zirconium, Sch. 40	5.36	Hastelloy C-276, Sch. 40	7.08	Hastelloy C-276, Sch. 40	6.42
34	Hastelloy C-276, Sch. 40	5.63	Hastelloy B, Sch. 40	7.62	Hastelloy B, Sch. 40	7.19
35	Hastelloy B, Sch. 40	5.94	Zirconium, Sch. 40	7.89	Zirconium, Sch. 40	7.56
36	Tantalum lined steel, Sch. 5	6.99	Tantalum lined steel, Sch. 5	8.28	Tantalum lined steel, Sch. 5	8.61
37	Tantalum lined steel, Sch. 40	9.08	Tantalum lined steel, Sch. 40	14.43	Tantalum lined steel, Sch. 40	16.42

Cost ranking and cost ratios, in ascending order, of 1,000-ft straight-run piping systems **Table VIII**

Rank	2-in. dia. Piping material	Cost ratio	4-in. dia. Piping material	Cost ratio	6-in. dia. Piping material	Cost ratio
1	Glass pipe	0.88	Carbon steel, Sch. 40	1.00	Fiberglass reinforced polyester	0.94
2	Carbon steel, Sch. 40	1.00	Fiberglass reinforced polyester	1.05	Fiberglass reinforced vinylester	0.97
3	Polypropylene lined steel	1.01	Polypropylene lined steel	1.11	Epoxy fiber-glass pipe	0.99
4	Rubber-lined steel, Sch. 40	1.15	Glass pipe	1.11	Carbon steel, Sch. 40	1.00
5	Fiberglass reinforced polyester	1.22	Fiberglass reinforced polyester	1.14	Glass pipe	1.01
6	Saran-lined steel	1.24	Aluminum, Sch. 40	1.18	Polypropylene lined steel	1.03
7	Aluminum, Sch. 40	1.27	Epoxy fiber-glass pipe	1.19	Aluminum, Sch. 40	1.12
8	Epoxy fiberglass pipe	1.28	Rubber-lined steel, Sch. 40	1.33	Rubber-lined steel, Sch. 40	1.19
9	Fiberglass reinforced vinylester	1.30	Saran-lined steel	1.35	Saran-lined steel	1.20
10	304 S.S., Sch. 5	1.43	304 S.S., Sch. 5	1.46	304 S.S., Sch. 5	1.29
11	Kynar-lined steel	1.45	316 S.S., Sch. 5	1.71	Kynar-lined steel	1.71
12	316 S.S., Sch. 5	1.56	Kynar-lined steel	1.79	316 S.S., Sch. 5	1.80
13	Armored glass pipe	1.59	304 S.S., Sch. 40	2.20	Armored glass pipe	2.13
14	304 S.S., Sch. 40	1.79	Armored glass pipe	2.27	Titanium, Sch. 5	2.32
15	FEP Teflon lined steel	1.79	FEP Teflon lined steel	2.31	Monel 400, Sch. 5	2.39
16	PFA Teflon lined steel	1.87	Monel 400, Sch. 5	2.40	FEP Teflon lined steel	2.40
17	PTFE Teflon lined steel	2.12	PFA Teflon lined steel	2.52	304 S.S., Sch. 40	2.41
18	316 S.S., Sch. 40	2.16	PTFE Teflon lined steel	2.59	Nickel 200, Sch. 5	2.56
19	Alloy 20, Sch. 5	2.37	Nickel 200, Sch. 5	2.64	PFA Teflon lined steel	2.58
20	Monel 400, Sch. 5	2.58	316 S.S., Sch. 40	2.73	PTFE Teflon lined steel	2.59
21	Nickel 200, Sch. 5	2.65	Glass-lined steel, Sch. 40	3.06	316 S.S., Sch. 40	2.85
22	Titanium, Sch. 5	2.70	Alloy 20, Sch. 5	3.10	Glass-lined steel, Sch. 40	2.88
23	Glass-lined steel, Sch. 40	2.86	Titanium, Sch. 5	3.23	Alloy 20, Sch. 5	3.42
24	Inconel 600, Sch. 5	3.02	Inconel 600, Sch. 5	3.42	Inconel 600, Sch. 5	3.66
25	Monel 400, Sch. 40	3.16	Monel 400, Sch. 40	3.55	Monel 400, Sch. 40	3.98
26	Nickel 200, Sch. 40	3.33	Nickel 200, Sch. 40	4.36	Nickel 200, Sch. 40	4.55
27	Alloy 20, Sch. 40	3.33	Inconel 600, Sch. 40	4.62	Alloy 20, Sch. 40	5.14
28	Inconel 600, Sch. 40	3.59	Alloy 20, Sch. 40	4.69	Inconel 600, Sch. 40	5.15
29	Zirconium, Sch. 5	3.78	Hastelloy C-276, Sch. 5	5.09	Hastelloy C-276, Sch. 5	5.30
30	Titanium, Sch. 40	3.88	Hastelloy B, Sch. 5	5.40	Zirconium, Sch. 5	5.44
31	Hastelloy C-276, Sch. 5	4.16	Titanium, Sch. 40	5.63	Hastelloy B, Sch. 5	5.63
32	Hastelloy B, Sch. 5	4.37	Zirconium, Sch. 5	6.15	Titanium, Sch. 40	6.09
33	Zirconium, Sch. 40	5.60	Hastelloy C-276, Sch. 40	8.84	Zirconium, Sch. 40	8.63
34	Hastelloy C-276, Sch. 40	6.82	Zirconium, Sch. 40	9.13	Hastelloy C-276 Sch. 40	8.96
35	Hastelloy B, Sch. 40	7.32	Hastelloy B, Sch. 40	9.61	Hastelloy B, Sch. 40	9.77
36	Tantalum lined steel, Sch. 5	7.55	Tantalum lined steel, Sch. 5	10.36	Tantalum lined steel, Sch. 5	10.39
37	Tantalum lined steel, Sch. 40	9.89	Tantalum lined steel, Sch. 40	18.29	Tantalum lined steel, Sch. 40	20.15

efficiency can fluctuate within a range of 55 to 85% or wider, depending on local conditions. It may be affected by the number of skilled craftsmen available in the area, local economic conditions, number and size of concurrent construction projects, amount of supervision, etc. Although the 80% figure might be considered high, it is the efficiency level historically experienced in the upper Midwest.

For most-accurate results, the figures for both hourly labor rate and labor efficiency should be adjusted to reflect local conditions.

Effects of major assumptions

Most of the cost-determining factors assumed here do not deviate substantially from those used in previously published installed-cost studies.* However, this study does deviate from past practice in two respects: the breakdown between welded and flanged connections in the metallic systems, and the assumed labor rate used in calculating fabrication and erection costs.

Most prior studies assumed a much higher ratio of welded to flanged connection than the 50:50 ratio used here. As a result, the installed costs of metallic systems in these prior studies tend to be somewhat lower. But, based on surveys of existing plants the 50:50 ratio appears to be somewhat on the conservative side for complex systems.

Even if flanged connections were limited to those used for connecting piping to process equipment, pumps and valves, the ratio of welded to flanged connections would be in the neighborhood of 75:25. Such a ratio, while constituting a departure from the 50:50 ratio actually used in the calculations, would most likely not have a drastic effect on the total installed cost of the metallic systems.

Most prior studies have used labor rates lower than the $23/h used here. The effect of using a recent labor rate is to escalate, to varying degrees, the total installed cost of all of the piping systems studied. The ratio of fabrication-and-erection costs to material costs is generally higher in alloy systems than in many nonmetallic ones, because the nonmetallic systems are often supplied by the manufacturer in cut-to-length spool pieces that require only field assembly. Fabrication costs of such systems are chiefly reflected in material costs.

Because of this, the indicated labor rate tends to elevate the installed costs of alloy systems somewhat more, proportionately, than those of other piping systems. It is definitely felt, however, that the choice of a more-or-less current labor figure results in installed-cost data that more accurately reflect the true situation in the majority of new piping installations.

*The diagram of this system appears in Installed Cost of Corrosion-Resistant Piping, by Stephen P. Marshall and J. Lee Brandt, *Chem. Eng.*, Aug. 23, 1971, p. 74; and 1974 Installed Cost of Corrosion-Resistant Piping, by the same authors, *Chem. Eng.*, Oct. 28, 1974, p. 106.

Observations

This study shows the dramatic effect that labor costs have on the total installed costs of a piping system. Thus, piping-system comparisons must be made on the basis of total installed costs rather than material costs. This is most obvious in the case of many of the alloys, where materials costs may represent less than half of the total installed-system cost.

On the other hand, many of the lined and other nonalloy piping materials yield lower installed costs (despite higher materials costs) because they can be purchased ready for field installation.

The cost data also reveal that the use of machine-formed lap joints can result in savings up to 20% or more for an alloy system, when compared with costs for the identical system installed using a field-fabricated welding-stub technique. Costs for alloy systems installed by using a shop-fabricated welding-stub technique are intermediate between the two.

In view of the history of steadily rising labor costs, much greater emphasis will likely be seen on making use of the most efficient fabrication methods available. Techniques such as bending and machine forming of lap joints will doubtless gain favor. Also, in line with this trend, there will be a growing advantage for those piping systems that can be supplied, by the manufacturer, ready for field installation.

Since industrial trends seem to be leading toward larger-diameter piping systems, we plan for 1979 an update of these installed-cost data, which will include both 8- and 10-in.-dia. pipe. This will better reflect the range of sizes encountered in process piping.

Conclusions

These cost data point out the need to consider all factors—pipe schedule, diameter, layout complexity, fabrication technique, etc.—to obtain a valid cost estimate for corrosion-resistant piping. Very few general conclusions can be made concerning relative installed costs. Each case should be considered individually. The data in this study should be taken only as a guide, and used basically as a screening device.

The author

James Yamartino is a chemical engineer in the Plastic Lined Pipe Plant, Central Div., Dow Chemical U.S.A., P.O. Box 815, Bay City, MI 48707, where his activities include research and development as well as customer service on plastic-lined pipe. He holds a B.S. in chemical engineering from the University of Massachusetts (Amherst, Mass.), and is a member of AIChE.

Section II
VALVES

Selecting and specifying valves for new plants

Valves make up 20 to 30% of the piping cost for a plant, depending on the process; and the cost of a given type and size of valve can vary 100%, depending on its construction. Thus the selection of valves is extremely important to the economics, as well as the operation, of process plants. Here is how one engineering design company decides on a valve type, the cost versus construction features, the materials of construction, and so forth, for competitive and trouble-free plant design.

Arkadie Pikulik, *Scientific Design Co.*

Photo inserts: Susan R. Horowitz

Originally published September 13, 1976

☐ Engineering companies select and specify valves through the coordinated efforts of the process, project and engineering design divisions.

The process engineers assume the overall responsibility for establishing the process-design parameters of temperature, pressure, flow, startup and shutdown procedures, and so forth. The project engineering group assumes responsibility for mechanical design performance. The design division assumes responsibility for detailed design and specifications for the entire plant.

Thus, in our company, the process division specifies the basic materials of construction necessary for the process fluids; the project division determines the need and location of valves, as well as the types of valve to be used; and the design division prepares the valve specifications to satisfy the requirements of the project divisions.

These valve specifications are most generally included with a piping material specification, which is prepared for the process and utility streams being handled in a plant and takes into account maximum operating pressures, temperatures, materials of construction, corrosion allowances, and any other factors that influence the selection of the material of construction.

The piping material specifications include all pertinent design data regarding fittings, flanges, pipes and valves to be used for a given service. Thus, for example, a piping material specification describing a carbon-steel material suitable for water at ambient temperature and 150-lb pressure will also include a list of suitable valves (Table I).

For the lone engineer who does not have the facilities of an engineering contractor or a company engineering group to provide piping material specifications, the procedure is the same, except that the burden of deciding which valve is acceptable lies solely with the engineer. As an aid to valve selection, that engineer should prepare a valve-classification list indicating as a minimum the following information (see Table II):

- Type of valve, sizes, manufacturers and manufacturer model number.
- Pressure and temperature ratings of the valves.
- Materials of construction.
- Packing material.
- Valve trim material, including stem, seat-ring and disk.
- Manufacturer's recommended service.

When new valve models are introduced, this list should be updated.

If the engineer finds that preparing and updating a valve classification list that covers all manufacturers is too time-consuming, the list might be limited to only a few manufacturers for each type of valve, provided that this limited listing does not preclude the engineer from obtaining competitive quotes from other manufacturers, once the basic specifications of the valves have been established.

Valves are ordinarily employed for two basic functions: shutoff and throttling services. Those valves used for isolating equipment, instruments and piping specialty items (strainers, steam traps, inline filters, etc.) for maintenance purposes are referred to as block valves (shutoff valves). In addition, block valves are used in piping manifolds for the purpose of diverting streams to various locations as desired. In essence, any valve that is either fully open or fully closed during plant operation may be regarded as a block valve. Block valves are usually pipeline size and have a port sized comparable to the pipe bore.

In calculating the pressure drop across a block valve, the usual procedure is to use an equivalent length of pipe, as specified by the manufacturer. Pressure-drop calculations for individual valves are *not normally* carried out, since the manufacturers' data have over the years proved valid (see Table III).

Manually operated valves whose purpose is to regulate flow and/or pressure are referred to as throttling valves. Selecting the size of a throttling valve will depend on the valve's flow coefficient, C_V. The rating is determined by flow tests for a given pressure drop across the valve's body and may be obtained from valve manufacturers for valves in throttling services. The engineer, after computing the required valve coefficient for the indicated service by use of the equation $Q = C_V(\Delta P/G)^{1/2}$, can properly select the size of the throttling valve by using the manufacturer's C_V ratings. (Q = flowrate, C_V = flow coefficient, ΔP = pressure drop between valve inlet and outlet, G = specific gravity of the fluid).

Valve selection

It is difficult to imagine a chemical plant, oil refinery, food-processing plant, drug-manufacturing unit, dairy, and so on, without valves. Valves are so extensively used and so familiar that they are at times treated casually, taken for granted.

As technology advances and plant capacities grow larger, however, valves have grown in size and cost; and it becomes increasingly important to be other than casual in selecting them.

Valve selection involves many factors, and it is best to cross-reference the key ones into a pattern that facilitates the selection. As a minimum, the following basic features should be considered; Type of valve, materials of construction, pressure and temperature ratings, packing material, cost and availability.

Type of valve

At the outset, the valve type will depend on the function it is to perform, whether shutoff, throttling, or non-reverse flow. These functions should be determined only after careful consideration of the requirements of the unit and the system for which the valve is intended.

Since there are several types of valves available for each function, it is also necessary to determine the service conditions in which the valve is to be used. And it is of primary importance to know the chemical and

Piping material specification and list of suitable valves — **Table I**

Service: Cooling, utility and fire water above grade
Primary flange: 125-lb cast iron, full face and 150-lb carbon steel, full face
Pressure-temperature rating: ANSI B16.1 — 1967; up to 150F maximum
Corrosion allowance: 0.05 in minimum

Item	Size, in	Schedule or rating	Description	Material specification
Pipe	½ – 1 ½	Sch. 80	Seamless, carbon steel	ASTM A106, Gr. B
	2 – 6	Sch. 40	Seamless, carbon steel	ASTM A53, Gr. B
	8 – 12	Sch. 20	Electric resistance welded, carbon steel	
	14 – 34	Sch. 10	Electric resistance welded, carbon steel	
	26 – 36	¼ in	Electric fusion welded, carbon steel. Wall thickness = ¼ in	ASTM A134; Plate-ASTM A283, Gr., C
Flanges				
Line	½ – 1 ½	150 lb	Threaded, carbon steel ., full face	ASTM A105
	2 – 24	150 lb	Welding neck, carbon steel., full face	
	26 – 36	*	Slip-on, carbon steel., full face	
Reducing	½ – 1 ½	150 lb	Threaded, carbon steel., full face	
	2 – 24	150 lb	Slip-on, carbon steel., full face	
Orifice	1 ½ – 36	300 lb	Welding neck, carbon steel., raised face	
Fittings				
Threaded	½ – 1 ½	300 lb	Banded malleable iron	ASTM A197
	½ – 1 ½	300 lb	Union, malleable iron., brass seat	
	½ – 1 ½	3,000 lb	Couplings, carbon steel	ASTM A105
	½ – 1 ½		Solid bound-head plugs and hex-head bushings, carbon steel.	
Butt welding	2 – 6	Sch. 40	Seamless, carbon steel	ASTM A234, Gr. WPB
	8 – 12	Sch. 20		
	14 – 24	Standard		
	26 – 36	¼-in wall	Seamless or welded, carbon steel	
	18 – 36	Same as pipe	Miter elbows	Same as pipe
Branch connections	½ – 1 ½		Straight or reducing tees	
	2		Full size: Tees, Reducing sizes: nozzle weld	
	3 – 36		Full and reducing: nozzle weld	
Pipe joints	½ – 1 ½		Couplings	
	2 – 36		Butt weld	
Gaskets	All	G-20	Asbestos full face	
Bolting	All		Alloy-steel studs	ASTM A193, Gr., B7
	All		Carbon-steel hex nuts	ASTM A194, Gr., 2

Valves	Size in	Type	Standard valve number	Dimensional standard
Threaded ends	½ – 1 ½	Gate	VC600A†	††
		Globe	VC610A	
		Check	VC620A	
Flanged ends	1	Gate	VC105A	††
	1 ½	Gate	VB105B	
	1 – 1 ½	Check	VC125A	
	2	Gate	VB105B	
	2 – 8	Globe	VB115A	
	2	Check	VB125A	
Wafer	3 – 6	Butterfly	VB155A	††
	8 – 24	Butterfly	VB155AG	
	26 – 36	Butterfly	VB155BG	
	3 – 36	Check	VB135A	

*In sizes 26 in and larger, flanges shall be 175 lb at nozzles and 150 lb at valves

†These valve numbers are given by way of example and are used in conjunction with Table II.

††Dimensional standards as developed to meet particular design requirements. For this reason, none are shown.

Valve classification list — **Table II**

Cast-iron, flanged, gate valves (125 lb, full-face flange)

All valves are solid wedge, bolted bonnet, outside screw and yoke, bronze trim, standard flange finish and handwheel operated unless otherwise noted.

Valve No.	Packing	Size range, in	Variations and/or remarks	Typical manufacturer
VB 105A	Asbestos	1½ – 48		Crane 465½
VB 105AG	Asbestos	8 – 48	Gear	Crane 465½
VB 105B	Asbestos	1½ – 4	Clamp bonnet	Crane 487½
VB 105C	Asbestos	1 – 4	Clamp bonnet, inside screw, rising stem	Crane 490½
VB 105D	Asbestos	2 – 48	Nonrising stem	Crane 461
VB 105E	Asbestos	4 – 14	Inside screw, nonrising stem. U.L. standard 175 lb. Indicator post.	Fairbanks 0433
VB 105F	Asbestos	4 – 14	Inside screw, nonrising stem. Double-disk, parallel seat; provided with automatic drain and indicator post.	American Darling 52 (modified)
VB 105G	Asbestos	2 – 12	Inside screw, nonrising stem. Indicator post.	Crane 461
VB 105H	Asbestos	2½ – 14	Double-disk. Parallel seat. U.L. approved.	Kennedy 68
VB 105K	Asbestos	2 -- 36	Double-disk. Inside screw, nonrising stem. American Water Works Assn. (AWWA). No handwheel. Provided with 2-in-square standard AWWA operating nut.	Kennedy 561X
VB 105L	Asbestos	4 – 12	175-lb bronze-mounted. Double-disk, parallel renewable seats, indicator-post flange. UA and FM approved, with indicator post.	Kennedy 701X

physical characteristics of the fluids being handled. In summary, consideration must be given to:

Valve function

- Shutoff valves, which are commonly referred to as block valves or on-off service valves.
- Throttling valves.
- Non-reverse-flow valves.

Type of service

- Liquids.
- Gases.
- Liquids with gases.
- Liquids with solids.
- Gases with solids.
- Flashing vapors from a liquid stream, due to a decrease in system pressure.
- Corrosive or non-corrosive.
- Erosive or non-erosive.

Once the function and type of service have been established, a valve type according to its construction can be selected by referring to the valve classification list described earlier.

In that list, the general valve functions are merely guides as to the best or most common use of a specific construction-type of valve. Often, more than one type of construction is suited to a specific function.

The principal features and most common uses of the various types of valves for shutoff service are:

Gate valves—Minimum resistance to line flow. Used full open or full closed. Infrequent operation.

Plug valves—Tight shutoff. Full open or closed.

Ball valves—Flow path is unobstructed. Used for viscous fluids and slurries. Positive shutoff. Used full open or full closed.

Butterfly valves—Mainly used for on-off and throttling services for large flows of gases and liquids at low pressures. Straight-through, open-disk design prevents any buildup of solids and yields low pressure drop.

The principal features and uses of valves for throttling service are:

Globe valves—Infrequent operation. Positive shutoff. Seat generally placed parallel to the direction of flow, providing substantial resistance and pressure drop.

Needle valves—Needle valves are basically globe valves that have tapered needle-like plugs, fitting with precision in their seats. Accurate throttling of small flows is possible because the orifice formed between the tapered plug and its corresponding tapered seat can be varied in very narrow and precise increments.

Y-valves—Y-valves are globe valves that offer the same straight-way passage and unobstructed flow that

Equivalent length in pipe diameters of various valves **Table III**

		Valve type and operating conditions			Equivalent length in pipe diameters, L/D
Globe valves	Stem perpendicular to run	No obstruction in flat, bevel, or plug-type seat		Fully open	340
		With wing or pin-guided disk		Fully open	450
	Y-Pattern	No obstruction in flat, bevel, or plug-type seat			
		With stem 60° from run of pipeline		Fully open	175
		With stem 45° from run of pipeline		Fully open	145
Angle valves		No obstruction in flat, bevel, or plug-type seat		Fully open	145
		With wing or pin-guided disk		Fully open	200
Gate valves	Wedge, disk, Double-disk or plug disk			Fully open	13
				¾ open	35
				½ open	160
				¼ open	900
	Pulp stock			Fully open	17
				¾ open	50
				½ open	260
				¼ open	1,200
Pipeline gate, ball, and plug valves				Fully open	3*
Check valves	Conventional swing		0.5†	Fully open	135
	Clearway swing		0.5†	Fully open	50
	Globe lift or stop: stem perpendicular to run or Y-Pattern		2.0†	Fully open	Same as globe
	Angle lift or stop		2.0†	Fully open	Same as angle
	Inline ball		2.5 vertical and 0.25 horizontal†	Fully open	150
Foot valves with strainer		With poppet lift-type disk	0.3†	Fully open	420
		With leather-hinged disk	0.4†	Fully open	75
Butterfly valves (8-in, and larger)				Fully open	40

*Exact equivalent length is equal to the length between flange faces or welding ends.
†Minimum calculated pressure drop (psi) across valve to provide sufficient flow to lift disk fully.

the gate valve does. Advantage is a lower pressure drop across the valve than the conventional globe valve.

Angle valves—Essentially the same as globe valves. Fundamental difference between the two is that the fluid flow through an angle valve makes a 90° turn.

Butterfly valves—Mainly used for on-off and throttling services for large flows of gases and liquids at low pressures (150 psig down to vacuum). Straight-through, open-disk design prevents any buildup of nonsticking solids and offers low pressure drop.

Non-reverse-flow valves have automatic reaction to pressure changes to prevent reversals of flow.

A special assortment of valve patterns is available for handling coarse or fine slurries. Most-common patterns are angle, flush bottom, plug, ball and diaphragm valves, and pinch valves. These are designed with minimum resistance to flow and are often lined with special alloys for corrosion/erosion resistance.

Materials of Construction

Having established the valve function and selected the valve type, the engineer must give consideration to materials of construction suited for the service in which the valve will be used. All valve parts in contact with the fluid should be suitably corrosion-resistant.

In selecting corrosion-resistant materials of construction, the engineer should use as a guide the valve manufacturers' listings of recommended materials for the various services encountered, as well as published data. Where such information appears inadequate, corrosion data may have to be obtained by performing corrosion tests in the laboratory. In general, unless the engineer is dealing with a first-of-a-kind process, there should be no difficulty in establishing the materials of construction, based on the amount of information that has thus far been published.

However, the published data should not be taken as final for the media they cover, because other conditions present in actual service may affect the corrosion rate and must be considered. For example: the presence of dissolved salts, of process contaminants and of different process compounds, fluid aeration, high fluid velocities, the presence of abrasives, the occurrence of cavitation and/or flashing, varying temperatures and concentrations, and so forth. The effect of these cannot be completely determined, except where data are available from an identical operating unit or system. Accordingly, while published corrosion data will prove valid in most cases, absolute certainty can be had only through actual experience.

When there is no previous experience, and no directly applicable data, the engineer must rely on a logical review and analysis of the data available for similar fluid compositions and conditions. This imposes a certain risk, which may be weighed against the increased cost of using a more reliable material of construction. If the cost of the valve is significant and the service requirement critical, then a material-testing program should be carried out with or without the assistance of the manufacturer.

Pressure and temperature ratings

Once the maximum operating pressures and temperatures have been identified, the engineer will be able to determine the appropriate pressure rating required for the valve. In this regard, the engineer should check the selection against manufacturers' lists for their pressure and temperature ratings, to assure that they conform.

As a guide toward pressure-class selection, the pressure ratings for a number of available materials are given for threaded, socket weld-end and flanged valves in Table IV. These ratings are based on ANSI standards. For materials other than those shown in the table, the engineer should refer either to ANSI standards or to the valve manufacturer's literature that applies to the service.

Packing material

It is just as important to select the proper valve-packing material as it is to select the proper materials of construction for the intended service. An improper selection of packing could lead to a leaky valve and thus cause a shutdown for replacement. Furthermore, if the fluid leaking through the valve is toxic or flammable, a disastrous situation could occur, resulting in plant damage and injury to personnel. Examples of such packings for various applications and the corresponding temperature limits of those packings are listed in Table V.

Such costly shutdowns and hazards are inexcusable and can be easily avoided. In selecting valve-packing material, it is recommended that the engineer refer to literature as published by valve and gasket manufacturers and to current technical publications, and that the selected packing be compatible with the stream being handled.

Furthermore, the shape or form in which the packing material is supplied must be compatible with the mechanical design of the valve. Certain packing materials require high compression, whereas certain valves are too flimsy or coarse to allow application of such high-compression forces. Alternately, the high compression pressures required by the construction features of a valve may cause certain packing materials to cold-flow. And other incompatible packing materials may cause the valve stem to gall.

Cost and availability

Often, more than one type of valve can be found to do a specific task. When all things are equal regarding materials of construction, performance, pressure and temperature ratings and availability, the least costly valve should, of course, be selected.

Having determined the type of valve best suited for the service, the engineer next must consider its availability and cost. It is not wise to select a valve that will not be delivered on time or be reasonably priced. Availability and cost data must be obtained from suppliers and/or manufacturers. The valve industry is cyclic in nature; today's price will not necessarily be tomorrow's price, nor was it necessarily yesterday's price. The price per valve also depends on the total number of valves to be placed on order at the time. Discounts are the usual practice in the valve industries, and the greater the number of valves the lower the unit cost.

However, a lower unit cost is often negated because of the effect a large order will have on the ability of a manufacturer to supply the valves on time. There are no guidelines as to cost and availability of any specific type of valve other than what can be discovered when soliciting bids from several suppliers and/or manufacturers.

Evaluation

When the function, type, materials of construction, packing, etc., are determined, the requirements plus the characteristics of suitable valves are complete, and bids can be solicited from suppliers and manufacturers. And when these bids have been received, an evaluation should be performed. Critical data, such as manufacturer's name, cost, delivery time, model no., materials of construction and packing are listed for comparison. Experience indicates that there will be a wide variation in costs, delivery time, and quality. Generally, higher quality valves are more expensive, but delivery time and costs do not usually fall into any definite pattern.

Although higher quality valves obviously hold a higher potential for fulfilling their requirements, there is no sure way of establishing the minimum quality that is necessary. The engineer will usually be forced to search for less expensive valve which, while not of equal quality to another valve, will still be acceptable. A carefully prepared bid comparison list will help the engineer in making such decisions; and experience will bring confidence. However, experience is not in itself reason to eliminate or shortcut the bid comparison.

Once the comparison has produced first and second-choice recommendations, the engineer should take one additional step. This step should be emphasized. Especially in the case of large valves, exotic materials or a critical service, the engineer should thoroughly review and analyze the details of the selected valve to be sure that it will carry out the function that has been assigned to it.

For example, a gate valve may have been chosen to close off a major process stream. In an emergency, it is important that this gate valve be closed as quickly as possible. Accordingly, a review of the details of selected valves may show that the number of turns required to

Valves commonly available from U.S. sources for plants in the chemical process industries (Valve size corresponds to nominal pipe diameter in inches) — Table IV

Gate valves

Pressure rating, psi

Material	125	150	175	200	250	300	400	600	900	1,500	2,500
Stainless steel	–	½ – 24	–	¼ – 2	–	½ – 24	–	½ – 24	–	–	–
Cast iron	2 – 48	½ – 4	2½ – 14	–	2 – 16	–	–	–	–	–	–
Ductile iron	–	2 – 24	–	–	–	–	–	–	–	–	–
Bronze	¼ – 3	⅛ –3	–	⅛ – 3	–	⅛ – 3	⅛ – 3	–	–	–	–
Cast steel	–	2 – 48	–	–	–	2 – 30	4 – 16	½ – 24	3 – 24	1 – 24	2½ – 24
Forged steel	–	–	–	–	–	½ – 3	–	½ – 2	–	–	–

Globe valves

Materials	125	150	200	250	300	400	600	900	1,500	2,500	4,500
Stainless steel	–	½ – 24	¼ – 2	–	½ – 24	–	¼ – 24	–	¼ – 3	¼ – 2	–
Cast iron	2 – 10	–	–	2 – 8	–	–	–	–	–	–	–
Ductile iron	–	2 – 24	–	–	2 – 6	–	–	–	–	–	–
Bronze	⅛ – 3	⅛ – 6	¼ – 3	–	¼ – 3	¼ – 3	–	–	–	–	–
Cast steel	–	2 – 14	–	–	2 – 16	4 – 12	2 – 18	3 – 24	1 – 18	2½ – 24	–
Forged steel	–	–	–	–	½ – 2	–	¼ – 2	–	¼ – 4	½ – 4	½ – 4

Angle valves

Material	125	150	200	250	300	400	600	900	1,500	2,500
Stainless steel	–	½ – 6	¼ – 2	–	½ – 2	–	–	–	¼ – 3	¼ – 2
Cast iron	2 – 10	–	–	2 – 8	–	–	–	–	–	–
Ductile iron	–	2 – 24	–	–	2 – 6	–	–	–	–	–
Bronze	⅛ – 3	⅛ – 6	¼ – 3	–	¼ – 3	¼ – 3	–	–	–	–
Cast steel	–	2 – 14	–	–	2 – 16	4 – 12	2 – 8	3 – 8	1 – 18	2½ – 24

Plug valves (lubricated)

Material	125	150	250	300	400	600	900	1,500	2,500
Stainless steel	–	1 – 4	–	½ – 4	–	–	–	–	–
Cast iron	½ – 16	–	½ – 24	–	–	–	–	–	–
Cast steel	–	1 – 36	–	½–24	4 – 26	½ – 26	2 – 20	½ – 16	½ – 16

Ball valves

Material	125	150	200	300	400	600	900	1,000	1,500	2,500
Stainless steel	–	½ – 14	–	½ – 14	–	¼ – 3	–	1 – 2½	½ – 2	1 – 1¼
Ductile iron	–	½ – 12	–	–	–	1 – 2½	–	–	–	–
Bronze	¼ – 2	¼ – 14	½ – 14	–	¼ – 3	¼ – ¾	–	–	–	–
Cast steel	–	½ – 16	–	½ – 16	–	1½ – 8	½ – 1	–	½ – 2	–
Forged steel	–	–	–	–	–	¼ – 2	–	–	¼ – 2	–

close the valve are critical and result in rejecting the valve otherwise listed as first choice.

Furthermore, if this final comparison of the requirements of selected valves indicates that the type originally intended for the application is not suitable, it may be necessary to re-specify the valve and solicit new bids.

Industry standards

We have mentioned the term quality in connection with bid evaluation, and elsewhere. It becomes necessary to define the term more precisely. Quality, in this context, has to do with industry standards.

The valve-manufacturing industry is an old and firmly established one, both in the U.S. and elsewhere. Development of most of the valves in current use was carried out many decades ago, and, as with so many other manufactured items, standards have been established to assure a uniformity among the various manufacturers (see list, Table VI).

In specifying valves, it is normal practice to state that the valve shall comply with the material, design, manufacture, testing and inspection requirements of an appropriate standard. This assures a certain quality within the uniformity established by the industry. However, it is possible to provide better than this assured quality. Forgings may be used instead of castings, for example, and components may be heavier. Also, the workmanship may be considerably better than that required by the standards. Quality deals with above-

Valves commonly available from U.S. sources for plants in the chemical process industries (cont'd)
(Valve size corresponds to nominal pipe diameter in inches) **Table IV**

Check valves, swing

Material	125	150	175	200	250	300	400	600	900	1,500
	Pressure rating, psi									
Stainless steel	–	½ – 24	–	¼ – 24	–	½ – 24	–	½ – 24	–	–
Cast iron	2 – 24	–	2½ – 12	–	2 – 12	–	–	–	–	–
Ductile iron	–	2 – 16	–	–	–	2 – 12	–	–	–	–
Bronze	⅛ – 3	¼ – 4	–	¼ – 3	–	¼ – 3	¼ – 3	–	–	–
Cast steel	–	2 – 24	–	–	–	2 – 20	4 – 16	½ – 16	3 – 10	1½ – 14

Check valves, tilting disk

Material	125	150	250	300	600	900	1,500	2,500	4,500
Cast iron	2 – 72	–	2 – 48	–	–	–	–	–	–
Cast steel	–	2 – 72	–	2 – 36	2 – 30	3 – 24	2 – 24	3 – 24	6 – 10

Check valves, lift

Material	125	150	200	300	600	900	1,500	2,500
Stainless steel	–	–	–	–	¼ – 2	–	–	–
Cast iron	2 – 6	½ – 2	–	–	–	–	–	–
Bronze	⅛ – 3	⅛ – 2	¼ – 2	¼ – 2	–	–	–	–
Cast steel	–	–	–	3 – 12	2½ – 19	2½ – 24	2½ – 24	2½ – 24

Butterfly valves

Material	125	150	300
	Pressure rating, psi		
Stainless steel	–	3 – 16	3 – 16
Cast iron	–	2 – 24	–
Ductile iron	–	2 – 24	–
Cast steel	–	3 – 30	3 – 16

Diaphragm valves

Material	125	150	175	200	300	400	600
	Pressure rating, psi						
Stainless steel	–	2½ – 4	1½ – 2	½ – 1¼	–	–	–
Cast iron	5 – 6	2½ – 4	1½ – 2	½ – 1¼	–	–	–
Ductile iron	–	2½ – 4	–	–	1½ – 2	–	½ – 1¼
Bronze	5 – 6	2½ – 4	–	1½ – 2	½ – 1¼	–	–
Cast steel	–	2½ – 4	1½ – 2	½ – 1¼	–	–	–

Plug valves (nonlubricated)

Material	150	300
Stainless steel	¼ – 12	¼ – 12
Ductile iron	¼ – 12	¼ – 12
Cast steel	¼ – 12	¼ – 12

Miscellaneous valves

Valve type	Material	150	300	600	900	1,500	2,500	4,500
Flush bottom	Stainless-steel	½–10	1 – 10	–	–	–	–	–
Y-pattern	Stainless-steel	½–2	½–6	½–10	–	–	–	–
Needle	Stainless-steel	–	–	–	–	¼–2	¼–2	¼–1½
Needle	Bronze	–	⅛–¾	–	–	–	–	–
Needle	Forged steel	–	–	–	–	¼–2	¼–2	¼–1½
Stop check	Cast steel	–	3–12	2½–18	2½–24	2–24	2½–24	–
Blowoff	Forged steel	–	1½–2½	1½–2½	1½–2½	1½–2½	–	–

standard features that a manufacturer may offer in a particular valve; all manufacturers should be expected to satisfy the minimum standards.

How valves are procured

Procedures vary from one company to another; but the essential functions of valve procurement must always be performed, and these should be revealed by a description of our company's procedures.

Our process engineering group initiates plant design activities by issuing process flow sheets indicating the materials of construction and critical instrumentation for both the process and utility (air, water, etc.) systems. Once these data are issued, the hub of the project activities center around the project engineers, who prepare a Piping Service Index that describes in detail the piping and valves to be used for the project.

All the fluid process and utility streams are separately listed in this index according to their maximum/minimum operating pressures and temperatures, materials of construction, corrosion allowances and gasket materials. After the index is reviewed and approved by the process engineers, it is forwarded to the piping materials group, who prepare the Piping Materials Specifications.

The specifications are in most cases standardized classifications of general services, as needed to satisfy the process data. Thus a standardized specification for "Cooling, Utility, and Fire Water, Above Grade" is illustrated in Table I. The result is a complete Piping

Packing materials for process valves in various services **Table V**

Packing material	Form	Used for	Temperature
Flexible, all-metallic	Spiral wrapping. Thin ribbons of soft babbit foil	Valve stem	Up to 450 F
Flexible metallic packings (aluminum)	Spiral wrapping. Thin ribbons of soft annealed aluminum foil loosely around a small core of pure dry asbestos	Hot oil valves, diphenyl valves	Up to 1,000 F
Flexible metallic packing (copper)	Soft annealed copper foil loosely around a small core of pure dry asbestos	Hot oil valves, diphenyl valves	Up to 1,000 F
Long-fiber pure asbestos and fine lubricating graphite (nonmetallic)	Graphite special long-fiber asbestos binder	Extreme resilience	Up to 750 F
Closely braided asbestos yarn; top jacket reinforced with Inconel wire; core: long fiber asbestos	Spools, die-formed rings	High-temperature valves	Up to 1,200 F
Pure asbestos yarn with an Inconel wire insert around a resilient asbestos core impregnated with graphite	Spool form, die formed	Valve stem for air, water steam, mineral oil	Stuffing—box temperature up to 1,200 F
Twisted long-fiber Canadian asbestos	Spool form, die formed	Valves handling, high & low-pressure steam	Up to 500 F
Asbestos, graphite and oilproof binder	Spool form, die formed	Shutoff valves	Up to 550 F
Solid, braided TFE	Coil, spool, ring	Valve shaft for highly corrosive service	−100 F to 500F
Braided asbestos with complete impregnation of TFE	Coil, spool, ring	Valve stems in mild chemical or solvent service	−100 F to 600 F
Braided of high quality wire-inserted asbestos over a loose core of graphite and asbestos	Coils, spools	Valve stems, steam, air, mineral oil	Up to 1,200 F
Braided of high quality wire-inserted asbestos over a loose core of graphite	Coils, spools	Stainless-steel valve stems, air, steam, water	Up to 1,200 F
Braided of long-fiber Canadian-asbestos yarn each strand impregnated with heat-resistant lubricant	Coils, spools	Valves for steam, air, gas and mild chemicals	Up to 550 F
Long-fiber Canadian asbestos yarn, each strand treated with a synthetic oilproof binder and impregnated with dry graphite.	Coils, spools	Refinery valves	To 750 F
Braided/overbraided, wire inserted, white asbestos packing impregnated with a heat-resistant lubricant	Coils, spools	Valve stems, for valves handling steam, air, gas cresylic acid	Up to 750 F
Braided white-asbestos yarn impregnated with TFE suspensoid	Coils, spools	Valve stems	− 100 F to 600 F
Braided of bleached TFE multifilament yarn	Spools, coils	Valve stems for highly corrosive liquids	− 12 F to + 500 F
Braided TFE multifilament yarn impregnated with TFE suspensoid	Spools, coils	Valve stems for corrosive chemicals, solvents, gases	−120 F to 600 F
Asbestos jacket, braided over a dry-lubricated plastic core of asbestos graphite and elastomers	Spools and coils	Valve stems, for valves handling superheated steam, hot gases	Up to 850 F

Standards and specifications for valves and piping	Table VI

ANSI Standards

B16.1	— Cast Iron Pipe Flanges and Flanged Fittings (25, 125, 250 and 200 lb)
B16.5	— Steel Pipe Flanges, Flanged Valves and Fittings (150, 300, 400, 600, 900, 1,500 and 2,500 lb)
B16.10	— Face-toFace and End-to—End Dimensions of Ferrous Valves
B16.11	— Forged Steel Fittings (Socket-Welding and Threaded)
B21	— Nonmetallic Gaskets for Pipe Flanges
B31.3	— Petroleum Refinery Piping

Issued by:

American National Standards Institute
1430 Broadway
New York, NY 10018

API Specifications

598	— Valve Inspeciton and Test
600	— Steel Gate Valves
602	— Compact Design Carbon-Steel Gate Valves for Refinery Use
603	— 150 lb, Lightwall, Corrosion-Resistant Gate Valve for Refinery Use
604	— Flanged Nodular-Iron Gate and Plug Valves for Refinery Use

Issued by:

American Petroleum Institute
1801 K St., N.W.
Washington, DC 20006

ASTM Specifications

E23	— Notched-Bar Impact Testing of Metallic Materials
E165	— Liquid Penetrant Inspection

Issued by:

American Soc. for Testing and Materials
1916 Race St.
Philadelphia, PA 19103

MSS Standards

SP25	— Standard Marking System for Valves, Fittings, Flanges and Unions
SP42	— MSS 150-lb Corrosion-Resistant Cast Flanged-Valves, Flanges and Flanged-Fittings Standard
SP53	— Quality Standard for Steel Castings for Valves, Flanges and Fittings, and Other Piping Components
SP54	— Radiographic Quality Standard for Steel Castings for Valves, Flanges and Fittings, and Other Piping Components
SP55	— Quality Standard for Steel Castings for Valves, Flanges and Fittings, and Other Piping Components.
SP61	— Hydrostatic Testing of Steel Valves
SP67	— Butterfly Valves
SP72	— Ball Valves with Flanged or Butt-Welding Ends for General Service

Issue by:

Manufacturers Standardization Soc. of the Valve and Fittings Industry
1815 N. Ft. Myer Dr.
Arlington, VA 22209

Service Index plus a set of Piping Material Specifications that are customized for the design and engineering work to be done on the project.

After the specifications and index are reviewed and approved by the process and project engineers, they are combined with similar documents prepared in much the same manner by other engineering groups and bound into one volume called The Project Design Basis. This volume is submitted to the client. On the client's approval, the procedures and standards set forth there are binding to both parties, defining both the work to be performed and the materials to be used in designing and building the plant.

The project engineers next prepare and issue piping and instrument flow diagrams (also called engineering flow diagrams) for the plant. These show schematically all the piping, plus all of the valves and instruments for the process and utility systems. It is possible at this point to count all the valves and prepare an estimated bulk take-off (from the drawings) for piping materials (pipe, fittings, valves, etc.) for the purposes of estimating the capital cost of the plant or soliciting bids from supplier/manufacturers.

The piping design group normally prepares a model of the plant. Models have been extremely popular for the simple reason that they are easy to review and modify. Their alternative is a set plan-view and elevation-view drawings, which are considerably more difficult for the process and project engineers to review and modify.

Meanwhile a plot plan and equipment arrangement drawing for the plant has been prepared by a specialized layout group. In the piping design it is important to have arrangement drawings that show the elevations and locations of the major equipment items at the beginnings and ends of piping runs.

As the piping model advances and areas or sections of the plant are finalized, isometric drawings are prepared and issued for all lines 1½ in and over. Each of these drawings has an itemized material list on it. These lists are summarized at scheduled intervals to update the total figures, which are issued by the piping group.

This information together with the Piping Material Specification makes up the package that the purchasing group uses to solicit bids from suppliers and manufacturers, for purposes of bid evaluation. In instances where special features have been requested, or the valve is critical, members of the piping design group are consulted for the bid evaluation.

Once the bid evaluation is prepared, it is forwarded to the project group for the project manager's approval. Once approved, the purchase order is placed with the supplier.

Inspection

Because of industrywide standardization, such normal inspection as is carried out for equipment items like heat exchangers, pressure vessels, pumps, etc., is not generally done for valves. Over the years, valve manufacturers have demonstrated that they can control the standard quality to which equipment is made in their respective shops.

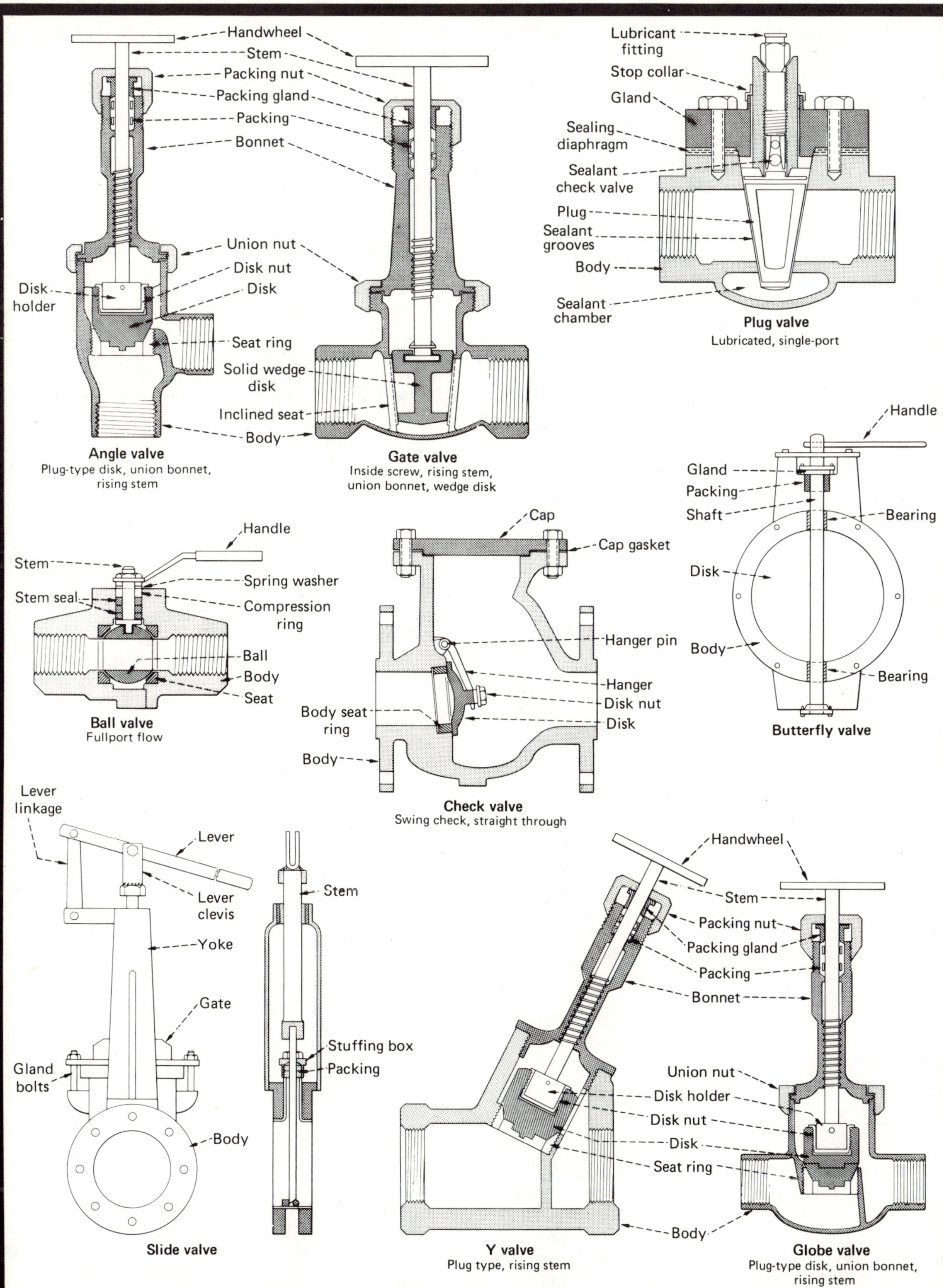

Assembled valves show the many components of the various types of valves used by the chemical process industries

Valves that are standard will be as nearly identical as is possible to mass-produce. Occasionally, a supplier will send valves with the wrong ends, or flanges will have the wrong face-finish. However, such instances are exceptions.

In the case of valves that are nonstandard, by contrast, inspections should be carried out at all key steps in the design and manufacture of the valve. Valves in highly critical service (that is, high pressures, high temperature, etc.) should also be inspected.

Fundamentals of valve construction

We have been describing the selection of standard valves for conventional services. "Standards" have been described as pre-designed or pre-calculated answers to routine engineering applications. However, the chemical process industries are rapidly developing more and more sophisticated processes; today they are handling, in large-scale equipment, chemical compounds and mixtures that only yesterday were confined to the laboratory because of their excessive reactivity. Moreover, these highly reactive compounds and mixtures are required to be handled at elevated temperatures and pressures in the large-scale plants.

The nonstandard valve

So what advice can we give to the chemical engineer who has to provide valves for a service in which any of the standard materials of construction are known to be unsatisfactory?

After repeatedly facing this problem during the past 25 years, we can provide some general guidelines, but no firm rules, on the subject:

- By and large, the reputable valve manufacturers are quite conservative. They are reluctant to use materials with which they have little or no experience. Nevertheless, it is important to place the eventual fabrication responsibility with a reputable manufacturer.
- The chemical process engineer, who knows the ramifications of the stream for which the valves will be provided and who is familiar with all the corrosion data taken on that stream, must go back to fundamentals, learning all the mechanical engineering considerations that go into the design of the valve in order to foresee all the problems that might arise and possibly to take a hand in their solution. For example, valves require differential hardness between the moving part and the stationary part (preferably the valve seat should be harder than its globe, gate or needle). Unless corrosion data for a material covers its behavior at different hardnesses, such data will be of little use to the valve manufacturer.
- The manufacturer's pressure and temperature rating for valves constructed of nonstandard materials should be carefully reviewed prior to the selection, in order to be certain that the ratings are consistent with the maximum operating pressure and temperatures of the services indicated.
- Get in touch with the expert in the valve manufacturer's organization—with the engineer who is actually familiar with the design of valves and who thoroughly knows the evolution of the design history of valves. Alternatively, such experts should be hired as consultants.
- Pump and compressor manufacturers, who have used the proposed novel material in the design of their

Titanium valves

Gate valves made of titanium should currently be considered nonstandard. Ball valves through 6-in. are the most common choice for an all-titanium valve (with liner), based on price and availability. Although large titanium gate valves have been in use for a decade or more, the number of firms with experience in manufacturing these valves is quite small when compared to those making valves of carbon steel or stainless steel.

Recently, we have been involved in several process plants that required a substantial number of titanium valves. In talking to the manufacturers, we found that they recommended different treatments to raise the hardness of the titanium seat rings. But none had enough experience to assure that their procedures were reliable and reproducible, since their quality control reverted back to casting foundries where design details are rarely dealt with.

The valve industry is highly competent in policing itself regarding the manufacture of standard valves. But for nonstandard valves there simply are no criteria. It is the buyer's responsibility to work with the manufacturer to first define the problems and/or conflicting opinions, and then to work with the manufacturer to find solutions. One must not simply assume that the responsibility rests solely with the manufacturer.

Specifically, as regards the titanium valves, we were inclined to accept the hardening procedure of one of the manufacturers. However, further discussions with that manufacturer revealed that in hardening the titanium, the relative concentration of one of the chemical components of the metal changed by a small percentage.

Titanium had originally been selected as the material of construction because of its corrosion resistance. What effect (if any) a change in the chemical composition of the metal would have on its resistivity, no one was in a position to say with certainty. Because of the critical service of the titanium valves and their high cost, the buyer (Scientific Design) arranged for laboratory tests, so as to establish the needed corrosion data.

This type of cooperation between buyer and manufacturer is essential when a nonstandard material or type of valve is required.

equipment, are an invaluable source of information, particularly when it comes to the design of compatible O-rings, seals and gaskets.

■ When using standard materials of construction (carbon steel, stainless steel, brass, etc.) the hydraulics inside the valve rarely affect the service; however, the proposed non-standard material may be prone to failure by abrasion or vibration, thus the hydraulics inside the valve may have to be closely investigated.

■ The cost of valves requiring non-standard materials may be reduced by lining or coating the valve internally. In these instances, the metallic valve materials being protected would not be resistant to corrosion by the contained fluids. Therefore, particular care must be exercised. Any flaw in the lining or the seals that will permit the contained fluids to contact the valve metal would invariably cause valve failure. Coatings often cannot be applied perfectly to the complicated internal shapes of the valves. Also, coatings often are not impervious to the diffusion of some fluids present in the process stream. Linings, on the other hand, may get detached from the support metal because of the severe hydraulic, mechanical and temperatures-reversal conditions to which the valve may be subjected. Lined valves frequently have seals, such as O-ring seals, to separate the protected sections from the unprotected sections. It is important that the seals are so located that their failure can be detected before disastrous corrosion has taken place.

Construction types

These rules for nonstandard valve-specification have often been applied in the past to develop valves currently offered as standard items. Engineering companies and the engineering and service departments of large corporations have little trouble in using such standard items, because the scope of their work allows for many identical or similar installations.

However, the lone engineer faced with selecting valves for a unique process or plant expansion is not in such a comfortable situation. He finds that, for lack of an experienced organization, he is somewhere between the extremes of completely novel versus all-standard applications.

Our recommendation for such engineers is to emphasize the rule for the chemical process engineer—learn the fundamentals of valve construction. As an aid toward that end, here follow more-detailed descriptions of the following valves: gate, globe, butterfly, plug, ball, needle, angle, pinch, slide, ram-type drain, Y-valve, flush-bottom tank, check and specialty. See the illustration on page 203 for a comparison of the constructions of these types.

Gate valves

The gate valve outnumbers the other types of valve in those services where uninterrupted flow and little pressure drop is required. Throttling services are not recommended for gate valves, since the gate and seal have a tendency to erode rapidly when restricting flow and producing turbulence with the gate in the partly open position.

When the valve is wide open, the gate is lifted entirely out of the flow passage so that the fluid flows straight through a passage that has essentially the same diameter as the piping.

The main characteristics of gate-valve services include: fully closed nonthrottling service, infrequent operation, and minimum resistance to flow.

The principal structural elements in a gate valve are, as shown in Fig. 3A, the handwheel, stem, bonnet, fluid-control element, seats and body. These valves are available with stem assemblies of the following types:

■ Inside-screw nonrising stem, which is advantageous where headroom is limited.

■ Outside-screw rising stem (OS&Y), which requires more headroom but avoids exposing the threads to the process fluids.

■ Inside-screw rising stem, which exposes the stem threads to the process fluid and is thus not to be used with corrosive fluids.

The following bonnet types are generally available with gate valves:

■ Screwed-in and screwed-on bonnets for smaller valves and low-pressure services.

■ Union bonnets for smaller valves where frequent maintenance is necessary.

■ Flanged and bolted bonnets for larger valves and high-pressure, high-temperature service.

■ Clamped bonnets for moderate-pressure valves that need frequent cleaning.

■ Pressure-sealed bonnets for high-pressure and high-temperature service.

■ Lip-sealed bonnets for high-pressure and high-temperature service.

Gate valve permits straight-through flow

■ Breech-lock bonnets for high-pressure and high-temperature service.

The following fluid-control elements are generally available with gate valves:

■ Solid or single-wedge disk with tapered valve seats, for oil, gas, air, slurries and heavy liquids.

■ Flexible wedges (the disk is solid only through the middle, with both seating surfaces flexible) for fluctuating temperatures and pressures.

■ Split-wedge disk (a ball-and-socket design where two back-to-back disks can adjust to both seating surfaces, so each disk moves independently for good sealing) for noncondensing gases, liquids at normal temperatures, and corrosive fluids, all at low pressures.

■ Double-disk (parallel-disk) valves (operating parallel to the body seats, the disks are forced apart by spreaders or wedges to force the disks against the seating surface) for noncondensing gases.

Gate-valve seats are provided either integral with the valve body or in a ring type of construction. For high-temperature service, seat rings are pressed into position and seal-welded to the valve body.

Leakage of gate valves can occur at: both ends where the valve is joined to the piping (when the valve is open), at the joint between bonnet and body, at the stem, and downstream of the control element when the valve is in the closed position. Seals can be provided to prevent leakage to the outside, or downstream when the valve is closed. Such seals can be metal-to-metal, metal in contact with a resilient material, or metal in contact with a resilient insert that is located in the face of the metal.

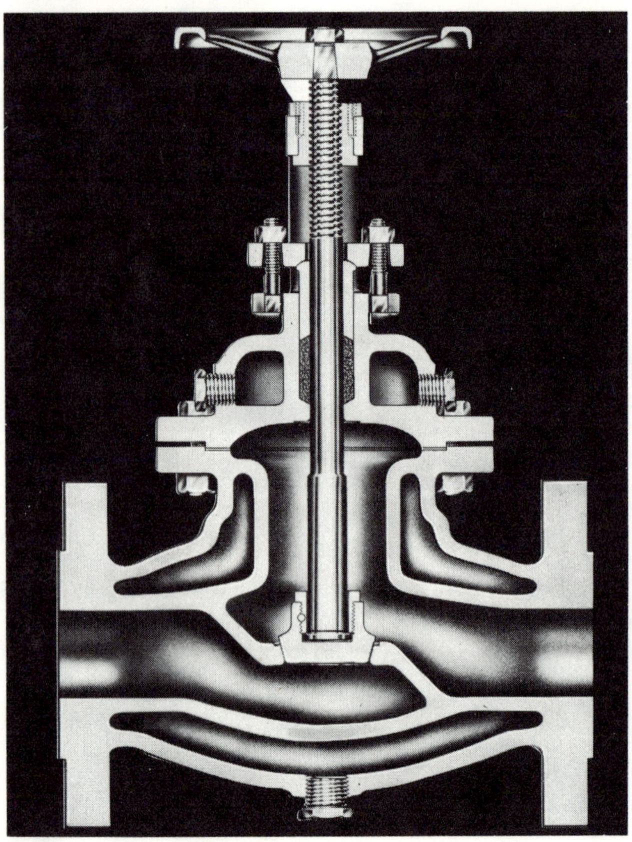

Globe valve is suited to flow control

Veteran valves

During the early 1960s, the project manager for a U.S. engineering company took up residence in Omuta, a small town on Kyusho, the southernmost island of Japan, where his company was assisting in the construction of a process plant it had designed and engineered.

The surrounding country was devoted to farming for rice, wheat, tea and oranges on terraces built into the slopes of the mountains. Both the terraces and the flat plains at the base of the slopes were irrigated, and the entire area was honeycombed with irrigation ditches and aquaducts, bringing and distributing water gathered in the high mountains.

From the view behind his house, the U.S. resident manager was struck by a sparse grove of thin, straight, regularly-spaced poles that spread out over the fields. As time passed, he saw that a Japanese workman would approach these poles at certain intervals and appear to be inspecting and brushing or painting them. Our resident manager undertook to walk across the fields to get a closer look, and discovered that the poles were threaded iron rods—the stems of gate valves regulating the water in the irrigation ditches.

Later, the resident manager chanced to meet and talk with the Japanese maintenance man, and from him learned that the irrigation ditches had been installed in 1904 using wooden slide valves, which had been replaced by cast iron valves in 1924. The maintenance man was the third generation to work at servicing the system and its valves, brushing and greasing the stems periodically. The maintenance man could not recall a single valve that had ever needed replacement or repair.

The stuffing box is the most common method of sealing the stem—having a gland flange to exert sealing pressure on the gland, and packing to prevent leakage. Also, stems can be sealed by means of applying side-injected grease to a lantern ring. The stem may be provided with fresh packing and/or fresh O-rings as needed. Materials include graphite-asbestos and TFE-impregnated asbestos for the stuffing box.

Globe valves

Globe valves are used either to stop or to regulate the liquid flow but are mainly used to do the latter. The change of direction of the fluid flow (two right-angle turns) through the valve causes turbulence and pressure drop. This turbulence causes the seat life to be limited.

The main characteristics of globe-valve services include: frequent operation, throttling to control the flow to any desired degree, positive shutoff for gases and air, and tolerably high line resistance and pressure drop.

The main structural elements that are typical of a globe valve are: handwheel, stem, bonnet, seats, disk, body.

Stem assemblies of the following types are generally available:

- Inside-screw rising stem, not for lines handling corrosive material because the stem threads are only partially protected.
- Outside-screw rising stem (OS&Y).
- Sliding stem, for quick opening and quick closing.

The following bonnet types are generally available:

- Screwed-in bonnet and screwed-on bonnet, for smaller valves and low pressure and low temperature.
- Union bonnet, for smaller valves and where frequent dismantling is necessary.
- Flanged (bolted) bonnet, for larger valves, and/or high pressures and temperatures. Bonnet gaskets seal the body-bonnet joint.
- Pressure-sealed bonnet, for high-pressure and high-temperature service.
- Breech-lock bonnet, for high-pressure and high-temperature service.

Globe valves are available with the following fluid-control elements:

- Composition disk. In this case the disk has a flat face that is pressed against a flat, annular, metal seating-surface. Not recommended for steam, gasoline and other liquids, this disk provides positive shutoff for gases and air.
- Metal disk. Here there is line contact between a tapered or spherical seating surface and a conical seat. Not recommended for throttling service, but provides positive shutoff for liquids. This type of valve is desirable where deposits could build up on seats.
- Plug-type disk. This has a wider seating contact, because of a long, tapered configuration, which enables the plug-type disk to handle throttling services; but it offers minimum resistance to erosion and corrosion.

Globe-valve seats are provided as integrally cast or as replaceable seat-rings that are screwed in or similarly attached. The leakage points of globe valves are identical to those for gate valves. Preventing leakage downstream of the control element is the primary difference between the two. For manually operated globe valves having as fluid-control element a disk or plug that fits up with a metallic seat ring, the disk may be either all metal or of a resilient material. When the pressure is not too high, a resilient seal is preferable, since it means pressing a metal surface against an elastomer surface. If a particle is trapped between such surfaces, there is not as much danger of breaking the seal. For stem packing, long-lasting TFC-impregnated asbestos is used for temperatures of 450°F or less. Asbestos and graphite with a suitable binder are used for valves rated over 450°.

Globe valves are available in a wide range of materials of construction: bronze, all iron, cast iron, forged steel, cast steel, stainless steel, brass, and corrosion-resistent alloys. The ends of the valve bodies (these may be flanged, butt welded, screwed) ordinarily range from ⅛ to 30 in.

Butterfly valves

Butterfly valves are one of the oldest types of valves known. They are simple, lightweight and low in cost. Maintenance costs are also low since there are a minimum number of moving parts. Butterfly valves are mainly used for both on-off and throttling services involving large flows of gases and liquids at relatively low pressures.

The straight-through, open design prevents any buildup of solids and provides low pressure drop. Operation is easy and quick, with a handle. It is possible to bring the valve from fully open to fully closed very quickly. Flow regulation is carried out by the valve disk sealing against a seat.

The principal characteristics of butterfly-valve services include: fully opened, fully closed or throttling, frequent operation, positive shutoff for gases or liquids, and low pressure drop.

The main structural elements of a butterfly valve are the shaft, the flow-control disk and the body casing. There are three major body types:

- Wafer type (lug type). This valve is simply held between two pipe flanges by bolts joining the flanges and passing through holes in the valve body's casing.
- Flanged type. This valve has flanged ends that join to the pipe flanges.
- Screw-on type. This valve is screwed directly onto the pipe.

Flow through a butterfly valve is controlled by a disk having about the same diameter as the connecting piping. Through this disk passes a shaft—the stem—which is supported at both ends by the body casing and is secured to the disk either by bolts or pins or by broaching the upper end of the disk's bore to a squared stem. Ro-

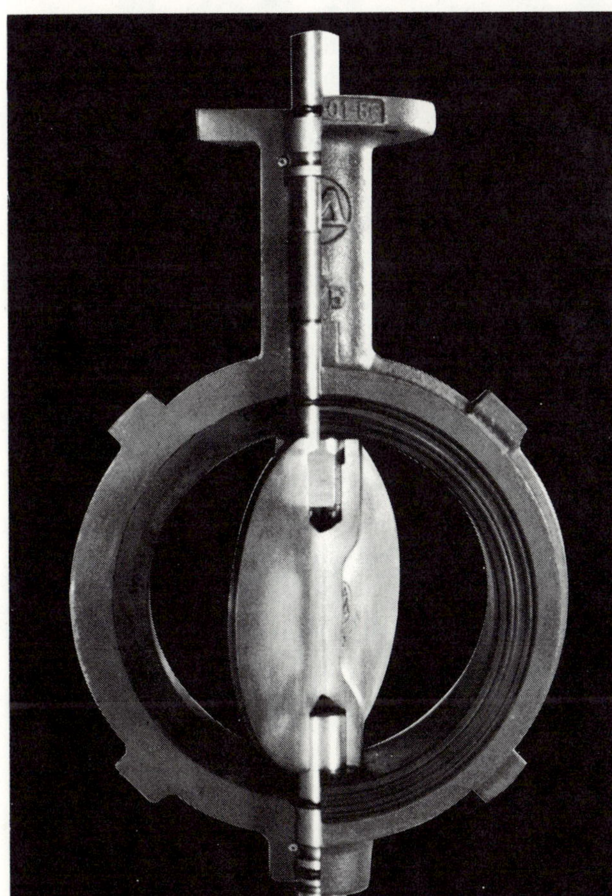

Butterfly valve is simple and light-weight

tating the stem (and disk) 90° fully opens or closes the valve. Throttling is done by moving the disk to intermediate positions, where it is held by a locking device.

To ensure a proper positioning, the stem is held firmly at both end-supports by bushings, which should be sealed to prevent any contact with corrosive media. Usually, the seals consist of a conventional stuffing box of O-rings.

According to their seal and leakproof characteristics, butterfly valves are classified into two major groups:

In the first group, the seats against which the disk closes are of metal, making the valve well-suited for handling semi-solids, since abrasive materials cannot cause any damage to these seats.

The second group employs resilient O-ring seals around the disks to provide a leakproof closure. In this latter type of valve, the materials used as seats are buna-N, Viton, natural rubber, TRE, Hypalon, etc. Such tight-shutoff butterfly valves are limited in temperature by their seat and seal materials.

Butterfly valves are available in a wide range of metals. Metal disks are also available in a wide range of linings, such as TRE, buna-N, Kynar, neoprene and Hypalon, in sizes ranging from 2 in to 15 ft.

Plug valves

Plug valves, like gate valves, are used mainly for on-off and nonthrottling services. Since the flow through the valve is smooth and uninterrupted, there is little turbulence within the valve, therefore the pressure drop across the valve is low. The main advantages of the plug valve are: quick action, simple operation, minimum installation space, and tight shutoff (with tapered plug).

There are two major types of plug valves: lubricated (to prevent leakage between plug surface and body seat and reduce friction during turning) and nonlubricated (which has a plug coating that eliminates the need for lubrication).

The main service areas of the plug valves include fully opened or fully closed nonthrottling service, minimum resistance to flow, frequent operation, low pressure drop. The basic components of plug valves are: body, plug and cover.

Plug valve offers quick action, tight shutoff

The two main categories of plug valves are: straight-through and multiport. *Straight-through*—The straight-through plug has a conical or a cylindrical form. There are different port designs for this plug:

- Full round port. This has a full-bore opening in plug and body.
- Rectangular port. This has full-area-sized ports generally rectangular, with minimum opening 70% of the line size.
- Venturi port. This has round or rectangular openings having reduced area, with venturi flow through the body.
- Diamond port. This has a plug opening that is diamond-shaped.

In the lubricated type, the lubricant is injected under pressure to prevent fluid leakage between the plug face and body. In addition, this reduces friction when turning the plug.

The plug is provided with grooves that permit the lubricant to make a seal. The lubricant pressure lifts the plug, permitting easy operation.

The advantage of lubricated plug valves is quick operation. They have a limited temperature range, depending on the lubricant used.

Nonlubricated plug valves are available in either (1) lift or (2) elastomer-sleeve or plug-coating type. The lift-type valve is operated by mechanically lifting the plug to permit easy rotation. The elastomer sleeve or plug coating (TFE) is inlaid (or coated) on the metal body completely surrounding the plug. TFE is durable and inert to all but a very few chemicals; having a low friction coefficient, it is self-lubricating.

The main advantages of nonlubricated plug valves are absolute shutoff, quick operation, no lubricant problem, and wide temperature range.

There are two main types of covers (glands), which correspond to the bonnets of gate and globe valves.

- Screwed cover: for smaller sizes and low-pressure and -temperature services, where shock and vibration are not encountered.
- Flanged (bolted) cover, for larger sizes and high-pressure and -temperature services.

Plug valves are available in a wide range of materials of construction, including cast iron, ductile iron, steel, stainless steel, bronze, nickel, brass, PVC, and corrosion-resistant alloys. Sizes range from ¼ in to 30 in.

Plug valves come with screwed, flanged and welded ends.

Multiport—Plug valves are easily adapted to multiport construction. Multiport valves have ports arranged so that when the plug is turned from one position to another, the channels previously in connection will be entirely cut off before new channels begin to open. The use of multiport plug valves simplifies the piping. A single multiport valve can be used in place of two, three or four straight-through valves. Multiport valves provide shutoff against working pressure only in the positive direction (pressure inside the valve forces the plug against the port). These valves do not shut off differential pressures when the line pressure is in the negative direction. (The pressure stream will try to get into the valve.) All other characteristics are like the straight-through plug valve.

Ball valve

Ball valves are basically modified plug valves. Although they have been available for a long time, their use was limited because of metal-to-metal seating, which could not provide bubble-tight sealing. Advances in plastics have permitted the metallic seats to be replaced with modern plastomer and elastomer seats.

The ball has a port opening that joins with the body in the open position. Ball valves are mainly used for on-off services; they are not satisfactory for throttling services. They are quick to operate, easy to maintain, require no lubrication, give tight sealing with low torque, and exhibit a pressure drop that is a function of the port size selected.

The ball valve is limited to those temperatures and pressures allowed by the seat material. When the valve is closed, some fluid is trapped between the seats and in the hole of the ball, which is undesirable in many cases.

Ball valves are not limited to a particular fluid material. They are used for: steam, water, oil, gas, air, corrosive fluids, slurries, and dry powdered materials.

The main structural elements of ball valves are the body, the seat and the ball.

There are two main body types of ball valves; top entry and split body. Top entry; the ball and seats are fitted through the top. Split body; the ball and seats are fitted from the ends.

Ball valves come in the full, venturi and reduced-area ports. The full port is equal to the inside diameter of the pipe. The venturi port has reduced port areas, with venturi flow through the body. The reduced port has a smaller bore than the inside diameter of the pipe.

The most commonly used seat materials of ball valves are TFE, nylon, buna-N and neoprene. Their usage is temperature-limited. Graphite seats have been developed for operation up to 1,000°F.

The stem in a ball valve is not normally fastened to the ball. Usually, a slot is cut in the ball, and the end of the stem is shaped to fit the slot accordingly and permits the ball to be turned. The stem is sealed with O-rings or conventional packing with TFE, TRE-filled or TRE-impregnated material held in place by a compression ring.

Ball valves are manufactured in a wide range of materials: cast iron, ductile iron, bronze, aluminum, carbon steel, stainless steel, brass, titanium, zirconium, tantalum and many corrosion-resistant alloys and plastics. The ordinary sizes range from ¼ in to 36 in.

The body ends are usually socket-welded, flanged, or threaded. There is also the possibility of changing the end-pieces in various materials of construction.

Like plug valves, ball valves can be provided as multiport valves. This type of ball valve can be used in place of two or three straight-through valves, which simplifies piping and cuts cost.

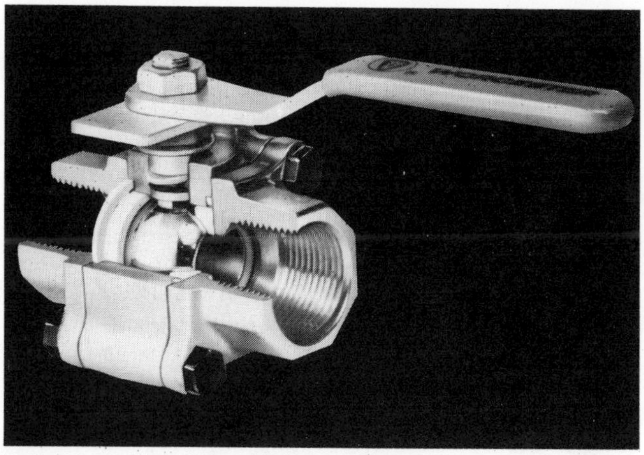

Ball valve offers tight seal for low torque

Needle valves

Needle valves are basically globe valves that have tapered needle-like plugs fitting precisely in their seats. Their stems rotate and move outward when they are opened. Accurate throttling of small flows is possible because of the variable orifice formed between the tapered plug and the correspondingly tapered seat. Generally used as instrument valves or in hydraulic fluid service, they are not used for high temperatures.

Materials of construction are usually bronze, stainless steel, brass and alloys. Ends are usually threaded. Sizes run ⅛ in to 1 in. Cheaper materials of construction are not generally used because the close-tolerance machining makes labor the controlling cost factor.

Angle valves

Angle valves are basically globe valves having inlet and outlet connections at right angles. They are mainly used for throttling services and offer less resistance to flow than globe valves. The stem rotates and moves outward when opened.

The overall construction elements of an angle valve are basically the same for the stem, disk, and seat-rings design as the globe valve. The axis of the stem is in line with one of the ends.

The right-angle form of the body eliminates the use of an elbow, since the flow on the inlet side is at a right angle to the flow on the outlet side. Materials of construction (and size ranges) are about the same as globe valves: bronze, cast iron, all iron, forged steel, Monel, cast steel, stainless steel, PVC, polypropylene, Penton, impervious graphite.

Pinch valves

Pinch valves are the cheapest and simplest in design of any valve type. They can be used for on-off service, or for throttling service that can vary from 10% to 95% of rated flow capacity. The main characteristics of pinch-valve services are: on-off and throttling (with low maintenance cost), low pressure drop, moderate temperature, and slurry flow.

Since the liquid is isolated from the metal parts by rubber or synthetic tubes, corrosive fluids can be controlled very well. The principle is that of clamping a flexible tube with a pinching mechanism. Since pressure drops through these valves are low, the valves are suitable for slurries and liquids having large amounts of suspended matter.

Although some pinch valves are claimed to be good for high-vacuum service, the ordinary pinch valve is not recommended for this because the flexible tube tends to collapse.

The main structural elements of pinch valves are the valve body and a pinching mechanism. The valve body is a molded-sleeve section of flexible material such as rubber. The flexible sleeve is equipped with flanged and clamp ends for pipe coupling. The molded bodies have pressure and temperature limitations. The pinching mechanism can be operated with a hand wheel, a chain wheel, or a hydraulically- or electrically-operated device.

Pinch valves are manufactured in a wide range of materials, including: rubber, Hypalon, neoprene, urethane, rubber, butyl, silicone, buna-S, and Viton A.

Slide valves

Slide valves, which are generally used for controlling low-pressure liquids and gases, are not intended for services where absolute tight closure is required. Due to their unrestricted bore, they may be used in services where liquids or gases passing through the valve contain a high percentage of solids, pulp stock, or free-flowing granular materials.

The flow control element consists of one or two disks which slide between parallel body seats, without making use of any disk-spreading mechanism. In the closed position, fluid pressure against the downstream surface of the gate (disk) forces it against the body seat, ensuring a tight joint between the downstream disk-surface and its seat. In the open position, there is very little obstruction to the flow since the disk is completely removed; and the pressure drop is minimal. The material of construction of the slide valves include low-chrome alloys, steel and stainless steel. The sizes range from 2 in to 75 in.

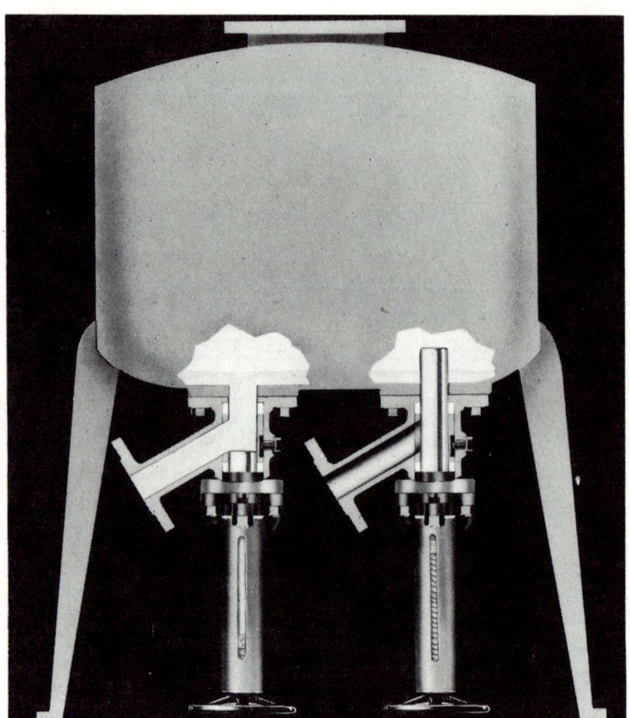

Ram-type drain valve avoids solids buildup

Flush-bottom tank valves

Flush-bottom valves are used to empty tanks and other vessels. There are two major types, both designed to have a seal and disk similar to globe valves. The disk of one type rises into the tank, whereas that of the other type lowers into the valve. The main parts are disk, stem and body.

The disk and seat arrangement is similar to the metal-to-metal seating of a globe valve, whether the disk moves away from its seat into the tank or into the valve. The stem construction is characterized as outside screw, rising (or lowering) stem, and the stem is equipped with a stuffing box and packing to prevent leakage. The bonnet is bolted.

When the tank nozzle is flanged and larger than the valve, adapters can be used to fit the valve to the nozzle.

Ram-type drain valve

Another type of flush-bottom valve is the ram-type drain valve, where the piston or ram extends into the tank when in the closed position and prevents the accumulation of materials that could plug the outlet of the tank or vessel. In the open position, the piston is drawn back into the bonnet, and the fluid in the bottom of the tank has an unobstructed flowpath.

The piston moving back upward into the tank pushes aside any material that may cause clogging. Because of the close tolerance between the piston and surrounding parts, it is possible to use this valve on low-temperature services.

The main construction parts of this valve type are the piston or ram, the stem, and the body. The piston or ram is kept tight by means of two resilient packing rings through which the piston moves. The stem of a ram-type drain valve is usually outside-screw, rising for smaller sizes, and outside-screw, nonrising for larger sizes. The body construction resembles the flush-bottom tank valve. Both valve types can be made in any cast metal, and they are usually flange-ended. The sizes range from 1 in to 6 in.

Diaphragm valves

Diaphragm valves are used for on-off and throttling services. They perform a wide and important range of services in liquid flow. In diaphragm valves, the diaphragm isolates the liquid being handled from the operating mechanism. Fluid cannot get to any working parts to cause corrosion and ultimate service failure.

When the valve is opened, the diaphragm is lifted out of the flowpath and the fluid has a smooth, streamlined flow. When the valve is closed, the diaphragm is tightly seated against a weir or circular area at the bottom of the valve.

The main applications of diaphragm valves involve low operating pressures and slurries that would clog or corrode the working parts of most other types of valves.

Diaphragm valves do not require stem packing.

Their life expectancy depends on the pressure, temperature and the frequency of opening and closing.

The main structural elements of a diaphragm valve are body, bonnet, and flexible diaphragm. The two general body types are the straight-through and the weir, or Saunders, type.

The straight-through valve has the same inside diameter and shape as the pipeline. To operate this type of valve, long strokes are necessary. This limits use of this valve, because there are few plastics that can bear the long strokes.

The weir or Saunders type is better for throttling and gives a tight shutoff as well. A shorter stroke will operate the valve and this permits the use of harder diaphragm materials such as Teflon.

Another type employs a plug and diaphragm molded into one unit. The open position, gives a straight-line flow; in the closed position, the plug seats on the bottom of the valve body, with a diaphragm shield connected to the plug and the top of the flow passage.

Stems in diaphragm valves are not rotating—the diaphragm is simply moved up or down with the help of a compression piston, which in turn is moved either with a lever arm or rotating stem. Usually, there is no packing, and this causes fewer maintenance problems. However, for toxic and dangerous chemicals, a bonnet is provided and sealed with a Teflon or other packing.

Diaphragm valves are manufactured in a wide range of materials, including cast iron, ductile iron, cast steel, stainless steel, and corrosion-resistant alloys. Linings may be materials such as virgin REP-Teflon, which can operate up to 400°F and is noncontaminating.

The valve ends can be either screwed, flanged, butt welded, socket welded, or sanitary male-threaded. Ordinary sizes range from ⅛ in to 24 in.

Y-valves

Y-valves, a modification of the globe valve, offer the straightway passage of a gate valve. The seat orifice is at approximately a 45-deg. angle to the flowpath. Because of this, a smoother flowpath is realized, similar to that of a gate valve, having a lower pressure drop across the valve than the conventional globe valve, and a good throttling ability.

Overall construction elements of a Y-valve are the stem, disk, and seat ring, as in the globe valve. Materials of construction and size ranges are about the same as for globe valves. Any specification for a conventional globe valve can be met by a Y-pattern valve.

Check valves

Check valves are self-contained and designed to prevent reversal of flow direction through a line. The pressure of the flowing fluid opens the valve, and the weight of the check mechanism and any reversal of the flow closes it. The disks and any associated moving parts may be in a constant state of movement if the velocity force is not sufficient to hold them in the wide-open, stable position.

There are different types of check valves, the selection of a particular type depending on temperature, pressure drop available, and cleanliness of the fluid.

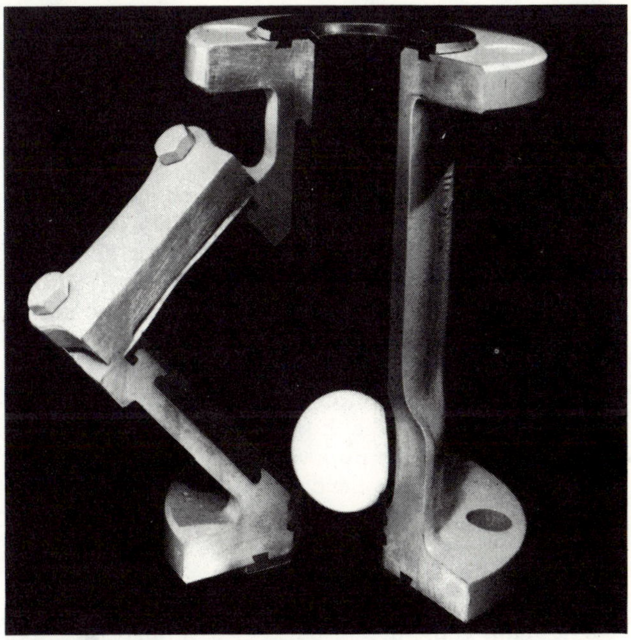

Ball-check valve distributes wear over ball

The swing-check valve opens with the pressure of the line, where the flow in the normal direction will cause the disk to swing away from the seat. It closes when the pressure decreases and reaches zero—in which case, the disk is held against the seat ring by its weight, or by external weights connected to a shaft extended through the check valve body.

The swing-check valves are used on low fluid velocities where flow reversals are infrequent; they are employed in systems along with gate valves. The main service characteristics of swing-check valves are minimum resistance to flow, low-velocity services, and infrequent change of direction.

The main construction parts of swing-check valves are the body, disk, swing pin, and cap. There are two major body design types: Y-pattern and straight-through pattern.

Y-pattern valves have an opening in line with the seat, which is integral with the body; this makes it possible to regrind metal-to-metal seated valves.

The straight-through swing-check valves contain a disk hinged at the top, so that the seating surface is at a slight angle. This permits the disk to swing open at lower pressures. The seat rings are replaceable.

Two disk types that are employed in swing-check valves are the metallic and the composition. Metallic disks are available in Y and straight-through patterns. Composition disks are better for flows that contain foreign particles. They provide a quieter operation and can seat tightly even when there are small particles in the line. At low pressures, positive shutoff is better accomplished with composition disks.

Cap construction types are: screwed cap, the most economical and the simplest design, and bolted cap, where a cap gasket seals the body-cap joints.

When faster reaction to flow reversal is necessary, certain check valves can be equipped with an outside lever and weight. This will provide a rapid closure of the disk and reduce the possibility of surge and shock. The lever and weight are adjusted so that the disk will open at the slightest pressure, or will not open until a desired pressure is attained.

Swing-check valves are available in a wide range of materials: bronze, all iron, cast iron, forged steel, Monel, cast steel and stainless steel. The ends can be screwed, flanged or welded.

Swing-check valves can be installed in horizontal lines, or in vertical lines with flow upward.

The tilting-disk check valve is similar to the swing-check valve. Fluid resistance is low because of the straight-through design of the valve. The tilting-check valve essentially consists of a cylindrical housing having a pivoted disk. The disk lifts easily off the seat to open the valve to permit flow. Disk stops built into the body serve to position the disk for optimum flow when fully open. As flow decreases, the disk starts to close by tilting at an increasing angle to the flowpath. The tilting-disc check valve has less pressure drop at low velocities and more pressure drop at high velocities.

The main construction parts of a tilting-disk check valve are the disk, the pivot shaft (rod) and the body.

The pivot point of the disk is located just above its center and is offset from the plane of the body. The design decreases the travel distance of the disk; and the closing force is reduced, which in turn minimizes the slam. When high reverse-pressure heads are expected, valves can be equipped with a special disk-restraining dashpot to control closure.

These valves are available either with a soft seal-ring or a metal-to-metal seal. The larger valves also have an insert valve seal.

Tilting-check valves are available in different materials of construction, including carbon steel, cast iron, stainless steel, aluminum, bronze and corrosion-resistant alloys. The ends can be flanged. Tilting-disk check valves can be installed in horizontal lines and in vertical lines having upward flow.

Lift-check valves

In the lift-check type of check valve a disk or ball is raised within guides from a seat by the pressure of upward fluid flow. When flow stops or reverses, the disk is reseated by gravity. Some lift-check valves utilize spring loading for positive seating.

In general, lift-check valves require relatively high pressure drops. They have an internal construction similar to globe valves. Their service characteristics include: frequent change of direction, increased flow resistance, and prevention of backflow. They are used with globe and angle valves.

The main construction elements of the lift-check valves are: disk, cap, body, seat and guide.

In the lift-check valves, there are two major types of disk construction, a composition disk and a regrindable disk. The composition-type disk is good for tight shut-off services and is equipped with a holder to maintain proper alignment. Regrindable disks are guided for alignment. These disks are usually metallic and are commonly furnished in steel. There are three body patterns: horizontal, angle and vertical. Design of these valve types is very similar to the globe valves, including the seat and disk configurations. Lift-check valves have three different cap constructions: screwed cap, which is usually used where lower pressures are encountered; union cap, which is preferred for frequent valve dismantling; and bolted cap, which is used on flanged lift-check valves.

Lift-check valves are built in a wide range of materials: bronze, all iron, cast iron, forged steel, Monel, stainless steel, and many plastic materials. The ends may be threaded or flanged.

Lift-check valves may be installed in horizontal or vertical lines, with the flow upward.

Ball-check valves

Similar to the lift-check, this valve has the guided disk replaced by a ball, which can rotate freely, dis-

A man to write home about

A U.S. engineering company was committed to purchasing equiment and materials in Spain for a plant to be located there. Although valve manufacturers from outside Spain offered bids, the orders were awarded to only Spanish firms strictly on the basis of price, despite reservations concerning delivery dates and quality. This left the outside firms feeling they had been treated unfairly.

This feeling was compounded, as time passed, for it became apparent that the deliveries of the larger valves, particularly those of stainless steel, would fall behind schedule. Valves were pulled out of the committed orders and new orders placed with the bidders from outside of Spain, but with the provision that the delivery dates of the original bids should be kept—a sort of trading that did not lend itself to making friends.

The situation reached a climax when the project was 80–85% complete. Two deliveries of Spanish valves had to be rejected because the castings were simply not acceptable. About two hundred valves were involved. Frantic telephone calls brought little or no encouragement that replacement valves could be found.

Just when the situation looked darkest, the salesman for the manufacturer who had objected most strongly to the bid restrictions reacted sympathetically. He personally delivered a first batch of valves to the Madrid residence of the project manager—at 1:30 a.m. With the help of the night porter and the valve salesman, the project manager loaded them into the trunk of his car and at 4:00 a.m. was on his way to the construction site.

Over the next two weeks, that salesman kept valves arriving at the construction site from just about everywhere. He gave his complete attention to the order and did not stop until the entire 200-odd valves had been delivered. The U.S. engineers expected a sizeable increase in price for this type of service; but much to their surprise, the price was exactly the same as that made on the original bid.

tributing wear evenly over the surface. Ball-check valves are limited to smaller sizes and are used for viscous and coking service.

Piston-check valves

Very similar to lift-check valves, these are equipped with a dashpot consisting of a piston and cylinder that provide a damping or cushioning effect during operation. Flow characteristics and installation are the same as for the lift-check valve.

Butterfly-check valves

Butterfly-check valves have a seating arrangement very similar to butterfly valves, so they are often used in conjunction with these. The main service characteristics are: minimum resistance to flow, frequent change of direction, and use in lines equipped with butterfly valves. Butterfly check valves may be installed horizontal or vertical, with vertical flow either up or down.

Specialty valves

Almost all piping installations require special-duty valves, which are designed for only one particular service. Two such valves are the globe-and-angle stop-check valve, and the blowoff valve. Both are mainly used in steam services.

Stop-check valves (globe and angle): Stop-check valves combine the function of a stop valve and check valve by providing positive shutoff and preventing reverse flow. These valves have a floating disk, which will lift under flow conditions with the force of boiler pressure. They are mainly used in steam-power plants for steam and feedwater services. The main uses of stop-check valves include: preventing backflow of steam from the main steam header, assisting in bringing a boiler into service after shutdown, assisting in cutting out a boiler when firing is stopped, and acting as a "safety first" valve by preventing backflow of steam from the header.

Examples of how different specifications for several types of valve affect price — **Table VII**

Plug valves (bottom entry)

Specification	1	2	3
Size, in	4	4	4
Rating, psi	400	200	200
Body material	Semisteel	Semisteel	Semisteel
Ends	Threaded	Flanged	Flanged
Plug	Lubricated	Lubricated	Lubricated
Port	Rectangular	Round	Diamond
Gasket	TFE	TFE	TFE
Price, $	320	435	226

Ball valves (top entry)

Specification	1	2	3
Size, in	6	6	4
Rating, psi	300	300	150
Body material	Carbon steel	Stainless steel	Ductile iron
Ends	Flanged	Butt welded	Flanged
Trim	Stainless steel, Type 316	Monel	Alloy 20
Ball	ASTM A-216	ASTM A-351	ASTM A-216
Seat ring	Reinforced TFE	Carbon/graphite	TFE
O-ring	Viton A	Viton A	Viton A
Price, $	2,107	2,805	621

Diaphragm valves

Specification	1	2	3
Size, in	3	3	3
Diaphragm material	Polyethylene	Viton	Solid TFE
Body material	Stainless steel	Cast steel	Ductile iron
Ends	Screwed	Flanged	Flanged
Stem	Extended stem	Outside stem and yoke	Sliding stem
Price, $	355	496	274

Swing-check valves

Specification	1	2	3
Size, in	10	10	10
Rating, psi	150	150	150
Body material	Cast steel	Cast steel	Stainless steel
Ends	Flanged	Butt welded	Flanged
Cap	Bolted	Bolted	Bolted
Trim	Monel	18% Cr-8% Ni-2½% Mo	Stainless steel
Price, $	2,949	2,677	3,971

Gate valves

Specification	1	2	3
Size, in	2	2	2
Rating, psi	150	150	150
Body material	Cast steel	Forged carbon steel	Stainless steel
Ends	Screwed and socket weld	Flanged	Flanged
Bonnet	Bolted	Bolted forged carbon steel	Bolted stainless steel
Stem	Outside screw and yoke (OS&Y)	OS&Y	OS&Y
Trim	Exelloy	Monel	Stainless steel
Packing	Pure asbestos yarn, Inconel wire insert around resilent asbestos	Pure asbestos yarn, Inconel wire insert around resilent asbestos	Blue asbestos
Price, $	305	310	268

There are two main body patterns: straight-through and angle. The angle design, with inlet and outlet at right angles, permits installation where an elbow might normally be used.

In both patterns, the stem is used for securing the disk in the seated position. The seat should be temperature-resistant, hard-faced, and precision-ground, to permit the disk to fit it for tightness. The seat is not restrictively fastened; thus, if forward flow is interrupted, the seat will act as a backward-flow check.

Materials of construction of stop-check valves include: carbon steel, stainless steel, forged steel, cast steel, and alloy steel. Ends can be socket welded, butt welded, flanged, or threaded. Sizes range ¼ to 24 in.

Blowoff valves: Blowoff valves are modified globe valves, mainly used in high-pressure steam service, to purge the boiler periodically in order to maintain a satisfactory concentration.

Blowoff valves are frequently used in combination of a straight-way and angle valve, but two similar valves or some other type of valve may be used as well. This arrangement ensures the proper pressure drop through the valves to prevent wire drawing. One valve is used for blowing and the other for sealing.

The valves are installed back to back, without a spacer so as to resist severe strains and violent temperature changes at the disk. The guide rings and the seats must be of hard facing. Usually, the seat-ring is welded on integral seats of Stellite. The stem is accurately guided, and a long shank extends above the packing to protect the threaded section. Stellite-faced disks have little play for self-alignment or tight sealing. Packing is especially selected for specified blowoff service.

When using two valves in series, both should be opened quickly and fully to increase the life of working parts. The valves should never blow down when partly open. The blowoff valves are produced with flanged and welded ends. Sizes usually range from 1 to 2½ in.

The engineer has two reasons for considering the costs of valves before a bid analysis. First, the differences in cost between various valve types should play a part in the choice of types to meet different services. Second, the cost of valves contributes to the overall capital cost estimates required for budgeting and project control.

Costs by type and construction: Since it is necessary to specify valves for the various plant services long before the number of valves is counted or bids are solicited, it is usually not feasible to have accurate cost data for those specifications. Thus the choices between globe versus angle, bolted bonnet versus union bonnet, or metal disk versus plug-type disk are frequently made months before the valve suppliers have had an opportunity to bid on the valves. In these choices, the engineer should compare prices for the different types, as given by a single manufacturer. Otherwise, price variations among manufacturers may confuse the choice.

Table VII compares the effects of some valve-construction features on common valve types. However, a careful specification requires an extensive analysis of manufacturers' list prices that are given for the different types.

Estimating valve costs: The custom among engineering firms has been to estimate valves along with piping costs in a code of accounts similar to that shown for an ethylene oxide plant in Table VIII. Because of this, the amount invested in valves frequently goes unknown; and even valve suppliers have difficulty knowing their potential markets.

However, we have found that, as the capacities of process plants increase, the contribution of valves to the total investment grows faster than other bulk-ordered materials. It becomes increasingly important to estimate valves separately, like pumps and heat exchangers—particularly specialty valves.

One guide to such estimation will be found in Tables VIII and IX. In Table X, the total piping cost shown (19.0% to 20.9%) is first divided into materials and labor. Then these material costs are divided into valves and pipe/supports. From this, we learn that the valves for an ethylene oxide plant can be expected to vary from 4.5% to 6.1% of the piping cost. And these valves are further divided into 3 in and over, versus 2½ in and under, for carbon steel and stainless steel.

Quite likely, the individual engineer will be interested in estimating valves for other than ethylene oxide

Breakdown of capital cost for ethylene oxide plants — Table VIII

For plants operating at a design capacity of:	50%	100%	200%
Item	%*	%*	%*
Fired heaters	1.6	1.3	1.1
Compressors and fans	13.93	14.3	14.5
Converters and towers	18.32	21.5	22.3
Heat exchangers	7.26	8.8	9.7
Tanks and bins	0.82	0.7	0.55
Pumps	2.6	3.1	3.4
Separating and drying equipment	0.07	0.09	0.11
Cooling tower	0.31	0.35	0.38
Foundations	2.2	2.1	1.96
Piping (including valves)	19.0	19.8	20.9
Buildings, structural steel and fireproofing	4.53	3.58	4.1
Instruments	6.71	5.8	3.2
Insulation	1.9	2.0	2.1
Electrical	1.3	1.16	1.0
Equipment errection	2.25	2.52	2.7
Painting	0.7	0.7	0.6
Construction supervision	3.6	3.2	2.6
Total engineering costs	12.9	9.0	8.8
	100.0	100.0	100.0

*Allocation of various capital costs expressed as a percentage of the total installed-platn capital cost.

Distribution of costs in piping-and-valve account for ethylene-oxide plants — **Table IX**

Maximum distribution for costs of plant:

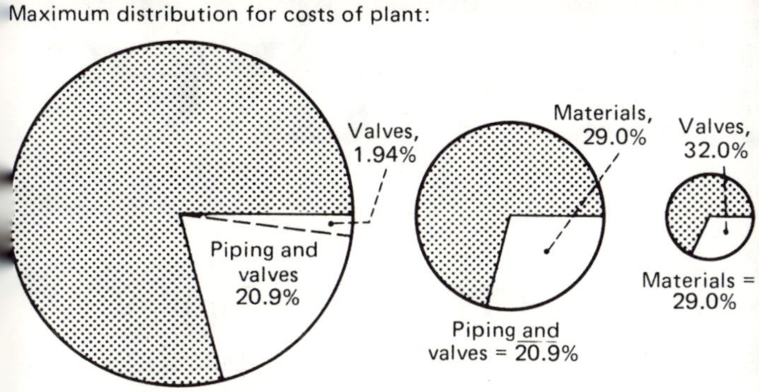

Minimum distribution for costs of plant:

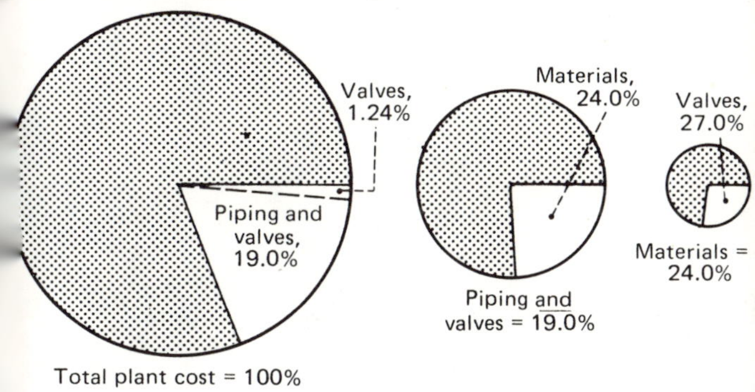

	Maximum, %	Minimum, %
ɔing account from Table VIII As percent of total installed-plant apital cost)	20.9	19.0
ɪterials in piping account Breakdown of piping costs into ɿaterials and labor. Where materials re 29.0%, labor is 71%; where mate- als are 24%, labor is 76%)	29.0	24.0
lve cost in materials account 3reakdown of materials cost into alves, and pipe and supports. Where alves represent 32.0%, pipe and ıpports are 68%; where valves re- resent 27%, pipe and supports are 3.0%)	32.0	27.0
ɪve cost in piping account As percent of total installed-plant ıpital cost)	1.94	1.24

ve cost by size and materials of construction:

	Carbon steel	Stainless steel
in, and under	26.0%	12.5%
, and over	42.9%	18.6%

Cost of valves for typical process plants — **Table X**

Plant	Capacity	Valve cost, %*
Maleic anhydride†	60 × 10^6 lb/yr	32.3
Propylene oxide†	446 × 10^6 lb/yr	28.1
Propylene oxide†	1,000 × 10^6 lb/yr	28.84
Terephthalic acid	500 × 10^6 lb/yr	30.9
Ethylene oxide	99 × 10^6 lb/yr	29.58
Petroleum refinery†	100,000 bbl/d	29.46
Antibiotics	15.5 × 10^6 units/yr	27.1
Fertilizer	1,100 tons/d	24.3
Polyvinyl chloride	–	20.7
Brewery	6 × 10^6 bbl/yr	19.3
Instant coffee	8 × 10^6 lb/yr	18.7

*Valve cost expressed as percent of total piping cost. See Table IX.
†Grass-roots construction.

plants. Equally accurate data for other processes are hard to come by; but Table X presents the valve costs as a percentage of total piping costs for a wide range of plants, from instant-coffee units to a grass-roots petroleum refinery. These data should be helpful.

Acknowledgements

The following manufacturers supplied literature that was used for this report: Cooper Stainless & Alloy Valves, Houston, Tex.; Crane Co., New York, N.Y.; Fabri-Valve, Portland, Ore.; Crane Packing Co., Morton Grove, Ill.; Posi-Seal International, North Stonington, Conn.; Rockwell International Flow Control Div., Pittsburgh, Pa.; Smith Valve Corp., Worcester, Mass.; Velan Valve Corp., Plattsburg, N.Y.; Walworth Co., Valley Forge, Pa.; Whitey Co., Cleveland, O.; Yarway Corp., Blue Bell, Pa.

The following manufacturers supplied information for this report: ACF Industries, W-K-M Valve Div., Houston, Tex.; Andale Co., Lansdale, Pa.; Combination Pump Valve Co., Philadelphia, Pa.; Corchester Controls Distributor Products Div., West Boylston, Mass.; Dover Corp., OPW Div., Cincinnati, O.; Dow Chemical USA, Midland, Mich.; Durion Co. Inc., Dayton, O.; Fisher Controls Co., Marshalltown, Ia.; Garlock Inc. Plastics Div. Camden, N.J.; Hills-McCanna Co., Carpentersville, Ill.; Jamesbury Corp., Worcester, Mass.; Jordan Valve Div. Richards Industries Inc., Cincinnati, O.; Kamyr Inc., Glens Falls, N.Y.; Masoneialn International Inc., Norwood, Mass.; Mosser Industries, Inc., Allentown, Pa.; Ogontz Controls Co., Willow Grove, Pa.; The William Powell Co., Cincinnati, O.; Resistoflex Corp., Roseland, N.J.; Stockham Valves and Fittings, Birmingham, Ala.; Strahman Valves, Inc., Florham Park, N.J.; TRW Mission Manufacturing Co., Houston, Tex.; Tufline Div. of Xomox Corp., Cincinnati, O.

The author

Arkadie Pikulik is a Sales Executive at Scientific Design Company, Inc., Two Park Ave., New York, N.Y. 10016, where he has been employed for twenty-two years at various positions, including pilot plant supervisor, process engineer, project engineer, project manager, project director, construction manager, and startup supervisor. He holds a B.Ch.E. from Brooklyn Polytechnic Institute.

The following members of Scientific Design Co. have shared in the preparation of this report: Bryan Flint (specification details), Manfred Gans (conceptual guidance), Arthur J. Hermes (capital cost data), Hagop Isnar (data research and writing assistance), M. George Tashjan (critical review), and Walter Van Pelt (pricing data).

Selecting valves

Plants in the chemical process industries (CPI) consist mainly of vessels, pumps, piping and valves. Each standard kind of valve has characteristics that make it best for a particular application. Here is the information you need to know about which valve to put where.

John T. Wier, Hudson Engineering Corp.

☐ Valves and piping make up the largest single investment of the various components of a hydrocarbon-processing plant. They represent approximately 22% of the total capital expenditure for materials and equipment. Many components that receive considerable attention represent a much lower investment—for example, pumps about 4%; compressors, 4.5%; drivers, 6.5%; and furnaces, 4%. After valves and piping, the next largest investment component is vessels, at about 15.4% [*1*].

Valve selection

A designer must consider many things when selecting a valve. Although the degree of importance assigned to each will vary, the first order of priority is generally valve function. Is the valve to be a block valve (on-off service only); is it for modulation (throttling) or diverting; is it to prevent backflow—or perhaps some combination of these? Valve function more than any other single criterion limits the choice; for purposes of this article, valve function will be categorized into four groupings*:

1. On-off service
   Gate valves
   Plug valves
   Ball valves
2. Throttling service
   Globe valves
   Butterfly valves
   Diaphragm valves
   Pinch valves
3. Prevention of backflow
   Check valves
4. Miscellaneous
   Control valves, solenoid valves, etc. (These are generally regarded as items of instrumentation and will not be discussed in this article.)

The chemical process industries employ valves in a wide variety of flowing media from granular solids to industrial wastes. In general, the most important characteristics to be considered are viscosity, corrosiveness, and abrasiveness; however, a designer must also consider the process parameters—i.e., any predictable short-term upset conditions. Other special circumstances must be recognized, such as handling more than one fluid with the same valve, or potentially high pressures resulting from a fluid being entrapped in a valve body (upon valve closure), which then vaporizes due to heat gain.

Pressure drop in valves may account for a substantial portion of overall friction losses in a system. Valve selections that result in a minimum pressure drop and still meet other requirements are obviously best. Today, the energy crisis gives greater impetus to the minimization of pressure drop than has ever before been true in the U.S.

*This classification refers to general use and does not cover special designs of types that may have application outside the categories in which they are listed. The balance of this article will discuss the different types of manual and self-operating valves.

Originally published November 24, 1975

for the CPI

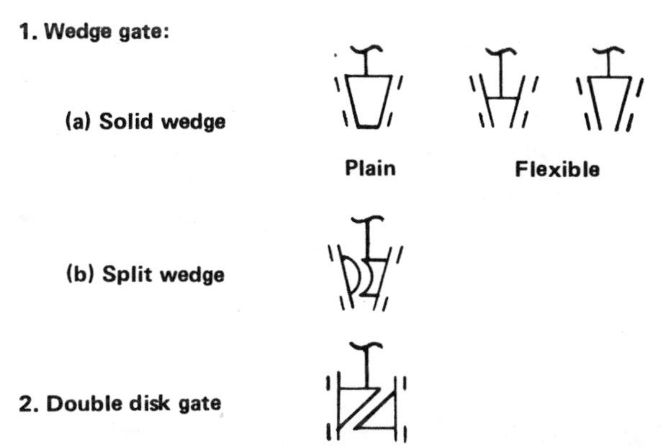

Gate valve flow-control mechanisms F1

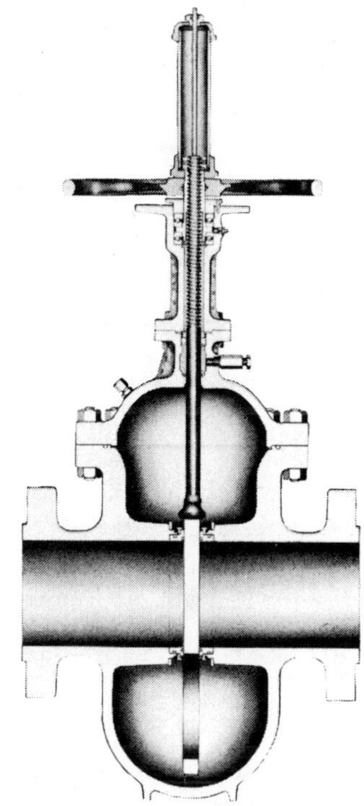

Full bore through conduit pipeline gate valve F2

Operating conditions—the coincident pressure and temperature at which a valve must function—frequently limit the choice of valves. Valves dependent on synthetic seals or nonmetallic materials of construction may have to be excluded because of high process temperatures. Conversely, low operating temperatures often limit the choice of valves to those available in suitable alloys.

Materials of construction must be compatible with all other factors involved in valve selection. Valve bodies, linings, and trims are available in a wide variety of materials to meet almost any service. Frequently, it is necessary to consider the body material separately from the trim (i.e., the other internal wetted parts) in order to optimize the valve design from an economic standpoint.

Valves made entirely from thermoplastic resins, and metallic valves with plastic-lined wetted parts, have become common in corrosive services.

When compared with gray cast iron, ductile iron both withstands higher pressures and temperatures, and also offers superior shock resistance. Since ductile iron is less expensive and has a higher corrosion resistance to some fluids than does steel, it has replaced the latter in many instances.

Valve materials of construction are sometimes tied to certain sizes. Valves for water, oil, gas, air, etc., usually are manufactured from brass or bronze in smaller sizes—whereas in sizes 4 in. and larger, iron and steel bodies are usually used.

GATE VALVES

In the on-off category, the gate valve far outranks the others in percentage of valves in operation. Even so, it has limitations. The gate valve does not lend itself to close control of flow because a disproportionate percentage of flow change occurs near shutoff—at high velocity. Nor is it intended for use in throttling service, because the gate and seat have a tendency to erode rapidly in any position other than fully opened or fully closed. When the valve is slightly opened in a throttling service, the disk and seat are subjected to "wire drawing" and erosion, ultimately precluding tight shutoff.

When fully opened, most gate valves allow straight-through flow in a passage that has essentially the same diameter as does the associated piping. While there are variations to this basic theme, the gate valve generally imposes a smaller pressure drop in the fluid-flow system than does any other style valve.

Flow-control mechanisms

There are several reasons why gate valves far outrank others in usage, but perhaps the most significant is that many variations are produced for specialized application. These valves may be generally classified by the type of fluid-control element used. One such classification, given in API Standard 600-1973, shows two major divisions, as indicated in F/1 (Fig. 1).

The most common gate valve is the plain solid-wedge, inclined-seat type. As operating temperatures (and pressures) have increased over the years, galling of

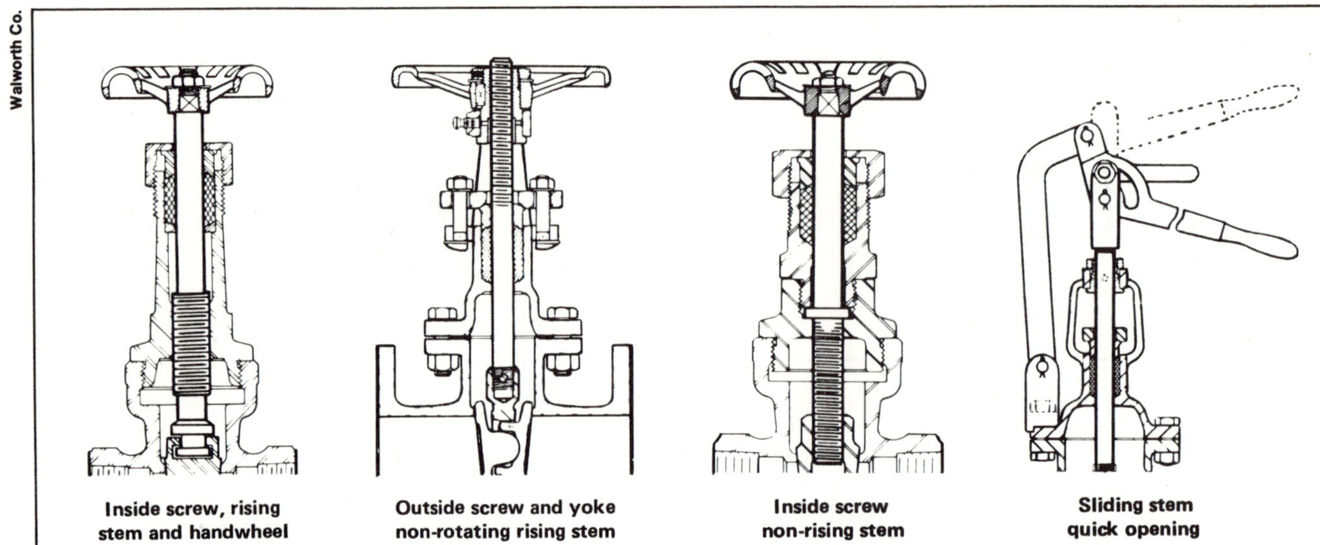

Variations in gate-valve stem mechanisms F3

the solid wedge against its inclined seat has become a problem. One solution is the "flexible" wedge, which tends to accommodate misaligned seats and thereby minimizes galling of the sealing surfaces. The wedge is "H" shaped and obtains flexibility from its geometry.

In the split wedge (or double disk) ball-and-socket type, the two disks are in contact with each other through a ball-and-socket joint. Since the disks are able to rotate independently, they accommodate changes in seat angles, providing good sealing and longer life. The split-wedge valve is most often used in low-pressure corrosive services in conjunction with integral seats.

The double-disk gate valve consists of two disks that are forced apart against parallel seats at the point of closure by a mechanical spreader device. This provides tight sealing without the assistance of fluid pressure and automatically compensates for angular misalignment of the seats. Additionally, it maintains a tight closure during the longitudinal shrinkage of the valve body on cooling.

Liquid- or gas-transmission pipelines employ a family of gate valves that permit the "pigging" of the line. The internal diameter of these valves must accommodate scrapers or product-separation spheres, and must be equal to or greater than the attached piping. The valves are identified as full bore through conduit (F/2).

Moving the fluid-control elements

An important part of the design of all valves is the means of moving the fluid-control element. In gate valves, this is accomplished by a stem that extends from the fluid-control element to the outside of the valve (F/3). Numerous techniques are employed to make the stem move the gate in and out of the flowpath. Rising-stem valves offer the advantage of easy visual indication of the amount of valve opening. Outside screw and yoke (OS&Y) construction removes the stem threads from contact with the flowing medium, and is recommended for high-temperature or corrosive-service applications. Valves intended for corrosive external environments including marine service or offshore platform service often employ inside-screw valves in order to protect the stem threads.

Large gate-valves require an excessive number of turns of a handwheel to open or close them. Gear units are available that reduce the torque required to move the gate, but if rapid cycling is required, the only acceptable solution is a power actuator.

Sealing methods

In gate valves, there are four places where sealing is required. Three are for prevention of leakage to the outside, and the fourth is to restrict leakage of the flowing medium (usually downstream) when the valve is closed.

The seals against leakage to the outside are at the end-connections, at the bonnet-to-body joint, and around the stem. The end-connection seals are identical with those for the joining of other pipe fittings, and are generally detailed in the piping specifications. The bonnet-to-body joint is a part of the valve design and must be suitable for the process conditions (F/4). The bonnet often is designed to facilitate entry to the removable internal components, and generally contains the stem seal. High-pressure, high-temperature valves in steam service frequently are of the "pressure-seal bonnet" design. Internal line pressure is utilized to make the body-to-bonnet-joint tight. Once tight, no adjustment is required. Stem sealing is more difficult than the other two seals because a moving surface is involved. Most gate-valve designs incorporate external stem-packing adjustment features, and often the stem seal can be repacked under pressure because of a back-seating of the wedge against the bonnet when the valve is fully open.

The seal to restrict leakage of the flowing medium downstream when the valve is closed depends on a

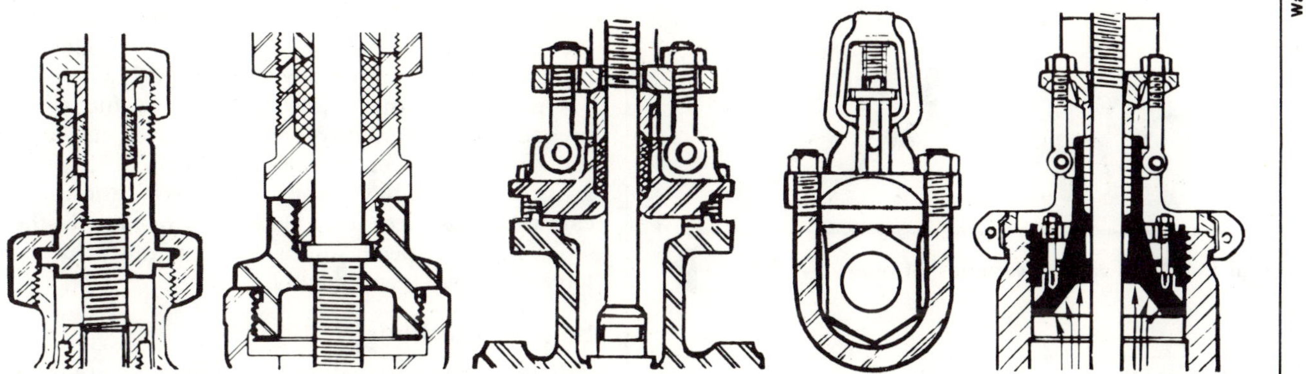

Bonnet variations in various gate-valve types

F 4

tight-fitting closure of the wedge against a seat. The selection of a wedge-to-seat design must take into account pressure and temperature changes. Such changes may result in strains being imposed by the connected piping, with resultant distortion or misalignment of the sealing surfaces.

There are two basic types of flow-sealing in gate valves—metal-to-metal contact, and metal in contact with metal that contains a resilient-material insert. A metal-to-metal seal provides the greatest mechanical strength but has the disadvantage of potential seizing and galling. The wedge and seats of many gate valves are highly machined at the factory and are then lapped for a close fit. This tight fit is very difficult to maintain, and when the valve is placed in service, some leakage must be anticipated after a few valve cycles. Most manufacturers of metal-to-metal-seal wedge gate valves design the materials of construction to achieve a differential hardness between the wedge and the seats to minimize seizing or galling.

A metal-to-metal seal in which one of the surfaces has a resilient insert provides both a primary resilient seal and a secondary metal-to-metal seal. This construction is limited in application to temperatures and pressures that do not adversely affect the resilient material. A number of user companies report a high degree of success with gate valves containing primary resilient seals (with secondary metal-to-metal seals) on both the upstream and downstream sides of the closure in double-block-and-bleed applications. These valves are used where leakage of either seal would result in product contamination, and are used in conjunction with a bleed valve between the seats. When the main valve is closed, the bleed valve is opened and any leakage is immediately detected.

Physical space requirements and weights

The gate valve offers a wide latitude in mechanisms for moving the fluid control element. It also has the disadvantages of being heavy and requiring a lot of space for installation. If an OS&Y rising-stem valve is selected, the designer's job is not complete until he considers the dimension from the valve centerline to the top of the stem with the valve open. Often another style of valve has to be chosen in order to reduce the space requirement. Since gate valves are sensitive to loads imposed by connected piping, adequate supports must be incorporated in the piping systems.

PLUG VALVES

The plug valve, a refinement of the simple cock, is the oldest of the valve family. Almost-intact, bronze, plug-cocks (reportedly used in aqueduct systems) have been found in the ruins of Pompeii, and were in use as long ago as AD 79. Modern plug valves, like gate valves, are primarily intended for on-off service; however, some owner companies have used plug valves in mild throttling services for many years with good success.

The basic components of the plug valve are the body, plug and cover. The plug is either tapered or cylindrical, and contains a passage through which the fluid flows. In the open position, the bore in the plug connects the inlet and outlet ends of the valve, providing straight-line flow.

Plug valves are available in round-opening, regular, venturi and short patterns (F/5). The round-opening valve has a full-bore round opening in both the plug and the body. The regular-pattern valve also has essentially full-area seat ports that are generally rectangular in shape.

The venturi-pattern offers reduced-area round or rectangular seat ports and a body that approaches a venturi. The short-pattern valve has full or reduced-area seat ports that are generally rectangular, and offers the same face-to-face dimensions as gate valves of the same size and pressure class.

An important feature of the plug valve is its easy

Rockwell Mfg. Co.

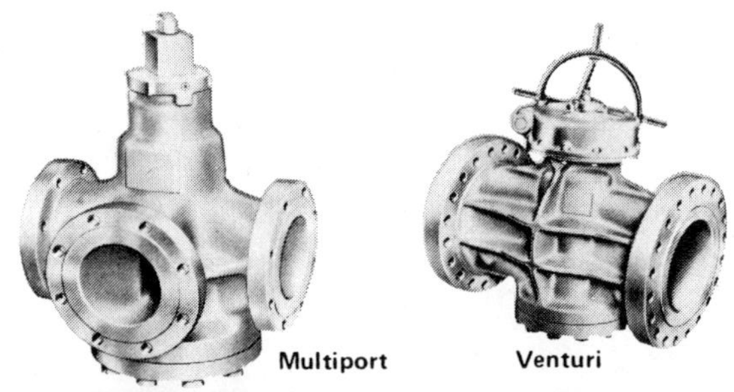

Multiport and venturi varieties of plug valves F/5

Walworth Co.

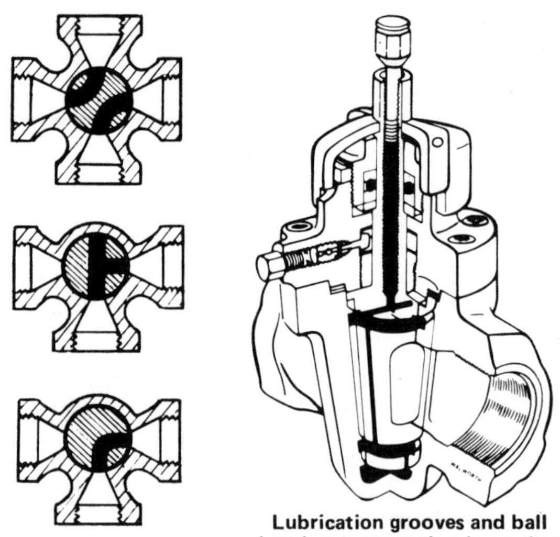

Flow patterns in multiport plug valves F/6

adaptation to multiport construction (F/6). Three- and four-way plug valves are widely used throughout the chemical process industries. They offer considerable economies in that the valve operates conveniently in many installations to simplify connected piping. A multiport valve may be used in lieu of several single-port valves, simultaneously reducing the number of fittings.

Full-area plug valves result in pressure drops comparable to full open gates, while reduced-area plugs vary widely in fluid friction losses. Designers are cautioned not to overlook pressure drops in the reduced-area plug valves, and particularly in the multiport patterns.

Moving the fluid control element

In single-port valves, the plug rotates through 90 deg from full-open to full-closed positions, and the valve is referred to as a quarter-turn valve. Multiport valves may rotate as much as 270 deg, depending on the number and arrangement of the ports.

In smaller sizes, the plug is rotated by means of a lever operator that is externally attached to the stem. The lever is generally L-shaped, but can be provided as a T shape when chain operation from below is required. In single-port valves, stops are usually an integral part of the cover, and the position of the valve port can be readily determined visually. The lever is perpendicular to the line of flow when the valve is closed, and is parallel to the line of flow when the valve is open.

Operating torque increases with size and with line pressure, and manually operated valves employ gear units when a lever becomes impractical. Most operating companies have proprietary specifications to indicate those sizes requiring gear operators, but in the absence of such specifications, valve manufacturers will have their own recommendations.

Sealing methods

Points requiring sealing in a plug valve are the same as for gate valves. The seals for prevention of leakage to the outside use the same principles as those for gate valves. Stem seals may be made with O-rings, or by means of injecting fluid packing into an annular groove, or through special devices. Seals for the prevention of leakage downstream are unique to the plug valve.

There are two general types of plug valves—lubricated (F/7) and nonlubricated (F/8). In the lubricated type, the lubricant is injected into the valve under pressure to prevent leakage between the plug face and body seat. The lubrication takes advantage of Pascal's law, which states that "a unit pressure applied to the fluid contained in a sealed vessel is transmitted uniformly to all areas of the confining surfaces of the fluid with undiminished force, thus multiplying the force many times, depending on the area of the interior of the vessel." Because of the "multiplying of force," the lubricant also serves to lift the plug off its seat to reduce the initial turning effort, as well as to reduce friction and wear between these surfaces during the turning. Following long periods of no operation, the ability to "pump" the plug off its seat is quite helpful in reducing the breakaway torque. The lubricated plug-valve seals only on the downstream seat, but is probably the most leakfree seal available when kept properly lubricated.

One nonlubricated plug valve uses resilient primary seals against one or both seats. Through a cam action, perpendicular expansion and contraction of the seating segments is accomplished. A bottom access plate is provided, making it possible to maintain the valve in the line by handling relatively small parts. One model is extensively used for double-block-and-bleed applications, and body bleed taps are provided between the seats.

The nonlubricated valve may be either a lift type, an eccentric type, or may have either an elastomer sleeve or a plug coating that eliminates the need to lubricate the space between the plug and seat. The sleeve-lined plug valve (F/8) finds extensive application in the chemical process industries in corrosive fluid service. A plastic sleeve of TFE fluorocarbon completely surrounds the plug and is retained and locked in place by the metal body. This results in a continuous primary seal being maintained between the sleeve and the plug at all times, both while rotating the plug and in either the open or closed position. The TFE sleeve is durable and is essentially inert to all but a few rarely encountered fluids. It also has a low coefficient of friction and therefore tends to be self-lubricating. The valve also contains a TFE diaphragm and thrust disk, completely eliminating metal-to-metal contact.

Physical space requirements and weights

The plug valve offers a compact on-off device requiring a minimum of physical space. Because of its compact size, it weighs less than gate valves or globe valves.

BALL VALVES

Flow-control mechanisms

Although ball valves have been available for some time, they have only found widespread acceptance in the chemical process industries within the past ten years. This has been due both to advances in elastomer and plastic technology and the development of machine tools capable of mass producing the valve balls. Early designs were metal-to-metal seating and were not "bubble tight." The crude packing and metallic seats of the early types have been replaced with materials such as fluorinated polymers and nylon, and the balls currently produced are much better than those in earlier designs.

The ball valve is basically an adaptation of the plug valve. Instead of a plug, it has a ball with a hole through one axis to connect the inlet and outlet ports of the body. In the open position, the flow is straight-through, whereas turning the ball 90 deg effects shutoff. Ball valves come in venturi, reduced and full port patterns (F/9). Pressure drop is a function of the port selected; in the full port pattern, it is appreciably the same as a gate valve of the same size and rating. As with the plug valve, ball valves may often be obtained in multiport patterns, effecting savings in valving and other piping components.

Moving the fluid control element

The techniques for moving the ball are generally comparable to those used in the plug valve. Early designs had stem-to-ball attachments that did not mechanically prevent the stem from exiting through the valve cover when under pressure, and it is still advisable to check this point. The ball valve, like the plug, is a quarter-turn valve. Smaller sizes are turned with a lever, and larger sizes use gear units. The sizes requiring gearing are more variable in ball valves than in the plug valves, and the manufacturer should be consulted.

One style of ball valve has a unique device for rotation of the ball (F/10). The stem goes through a packing area, has a spiral slot, and then is threaded (similar to a gate valve). The threaded portion is driven up or down by a yoke nut that is turned by a handwheel. A stem guide in the valve bonnet engages the spiral slot in the stem, and as the stem is moved up or down the movement is translated into a rotary motion of the ball. Thus the valve can be thought of as a "rising-stem ball valve," and a closure indicator is included to show the position of the ball from full open to full closed. The spiral slot in the stem is designed to effect a mechanical energizing of the ball against its seat in the closed position through a "cam" action, and at the same time eliminate friction between the ball and its seat in both the opening and closing cycles.

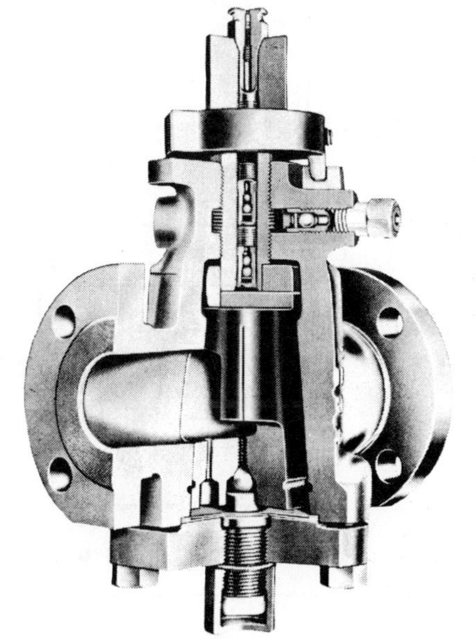

Rockwell Mfg. Co.

Lubricated plug valve (bolted bottom cover type) F7

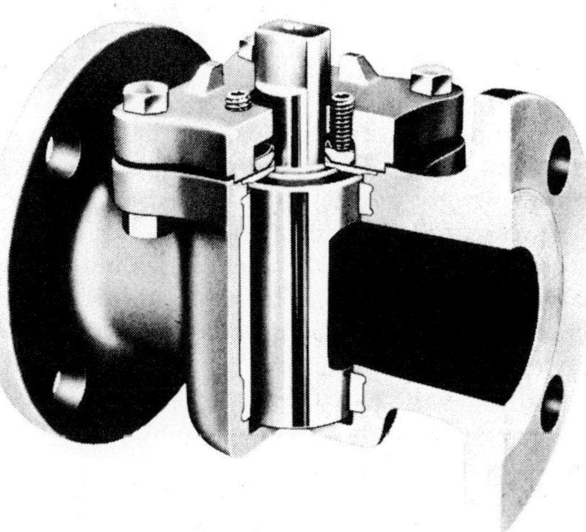

XOMAX Corp.

Sleeve-lined nonlubricated plug valve F8

Sealing methods

In sealing for prevention of leakage to the outside, ball valves employ stem seals ranging from O-rings to conventional packing. Sealing of the end-connections follows piping specifications and is identical with seals used for the joining of other pipe and fittings.

Ball valves come in top-entry, split-body end-entry, and completely welded-up, body styles. Additionally, ball valves are classified as floating ball and trunnion mounted. In the floating ball valve, line pressure

Walworth Co.

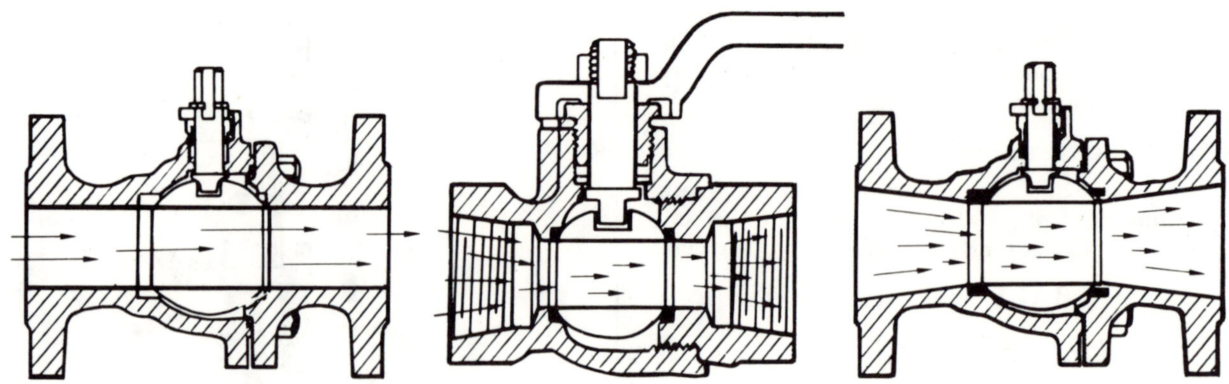

Ball valves: full and, reduced port and venturi types

F9

pushes a freely floating ball into the downstream seat. As line pressure increases, the effectiveness of the seal also increases; however, low differential pressures present problems. To effect a seal at low differential pressures, the plastic seats are often precompressed during assembly. The operating torque in floating ball valves increases with valve size and with differential pressure, and the size and pressure class range is therefore limited.

In trunnion-mounted ball valves, the position of the ball is fixed by top and bottom guides with bearings, and the seats are moved against the ball by line pressure. Each seat moves independently, and most designs incorporate spring loading to effect sealing at low differential pressures (F/11). The seats are free to rotate or are forcibly rotated a small amount each time the valve is cycled, to distribute the seat wear. Trunnion-mounted ball valves are frequently selected for double-block-and-bleed application in conjunction with bleed valves between the seats.

In either the floating-ball or the trunnion-mounted types, the seat rings may require sealing against the body in order to prevent leakage. In top-entry patterns, seals are required between the valve body and the bonnet or cover. In split-body end-entry patterns, seals are required for each body piece against its mating piece.

Ball valves offer quick quarter-turn on-off operation, are easy to maintain, require no lubrication, and provide tight sealing with relatively low torque requirements. They are limited in application because of pressure- and temperature-resisting properties of the seat and seal materials. Even though lubrication is not essential, some ball valve designs incorporate lubrication ports and external lubrication fittings. Where tight shutoff is imperative, and periodic valve maintenance is impossible, the lubrication feature assures the ability to effect a seal under emergency conditions.

Physical space requirements and weights

Ball valves are compact and offer appreciable weight savings over gate valves. So they find wide application where space is at a premium. They are the standard valves on most offshore well and production platforms, and are gaining in usage in petroleum refineries and chemical process plants.

GLOBE VALVES

Flow control mechanism

The designation "globe valve" encompasses a large number of designs including the hand-operated and power-actuated types used for automatic control. The common feature of all is the internal construction, consisting of a disk or plug moving within the valve body and mating with a seat to effect closure (F/12). Globe valves are normally used for throttling applications, and the design usually incorporates a tortuous flowpath—often a 90-deg change in the direction of flow. Most designs are unidirectional; a flow arrow is indicated on the valve body. (The flow is usually brought in under the disk or plug.) Pressure drop is often high in a globe valve, and in order to minimize the loss, many manufacturers offer Y-pattern and angle-pattern valves.

Moving the fluid control element

The techniques for moving the fluid control element of globe valves are essentially the same as those of the gate valve family. For illustrations, see F/13.

Sealing methods

Sealing problems and their solutions for globe valves are identical to those for gates with the exception of the seal to restrict leakage downstream.

Manually operated globe valves contain either a disk or plug that mates with a metallic seat ring. The disk may be either all metal or may have a resilient insert. A resilient seal involves pressing a metal surface against what is usually a rubber or plastic surface. Where the service is not severe or the pressure is not high, this type of seal provides tighter shutoff and is particularly desirable for fluids containing solid particles. When a par-

Orbit Valve Co.

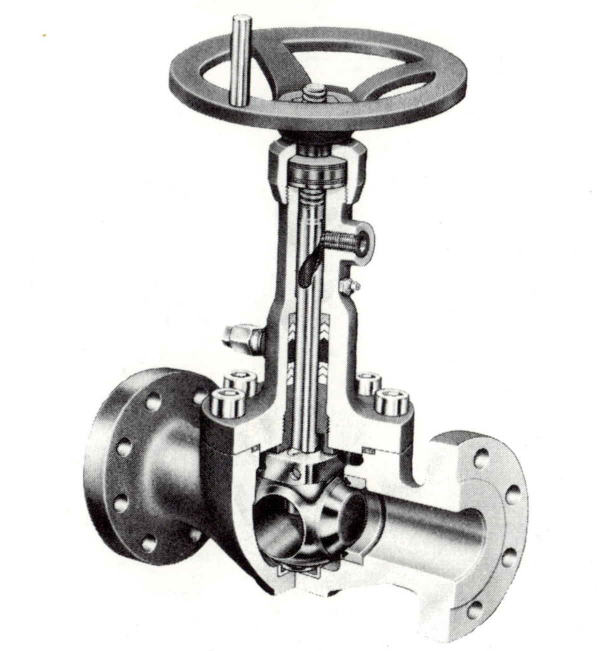

"Rising stem" ball valve has unique actuator F10

Cameron Valve Co.

Trunnion-mounted ball valve with rotating seats F11

Walworth Co.

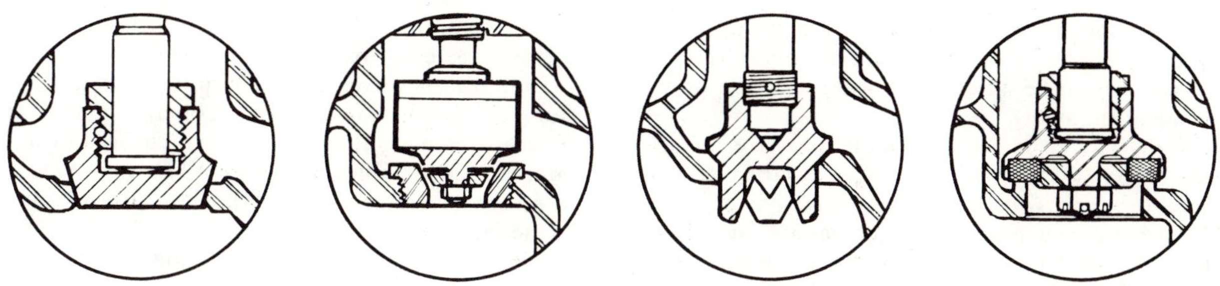

Globe-valve disks of various designs F12

ticle is trapped between the sealing surfaces, it is depressed into the soft surface and thus may not interfere with the sealing. The resilient seal is not as suitable for throttling as is metal.

Metal disks have a tapered or spherical seating surface with line contact on a conical seat. Seats may be integral with the valve body or may be replaceable (or otherwise renewable). One variation of a conventional globe valve incorporates a "Vee port" throttling disk, usually in conjunction with a vernier-type closure indicator. These valves find limited application and are for close regulation of flow, with the ability to obtain a given flow on a recurring basis through use of the closure indicator.

Physical space requirements

Globe valves have essentially the same space requirements and weights as gate valves. While not as sensitive as gates to loads imposed by connected piping, globe valves should also be adequately supported.

BUTTERFLY VALVES

The design principle of the butterfly valve is the same as that of a damper in a stovepipe. The disk or flow-control element has about the same diameter as the connected piping. Butterfly valves are made in the "swing through" style, or with seats for shutoff. In the shutoff valve, the disk is rotated against a seat that generally incorporates an elastomeric seal. Low-working-pressure valves often are completely lined with rubber or other material; in closing, the disk compresses the lining around the entire 360 deg. A number of engineers feel that when the seal of the disk against its lining is directly in line with the stem, it is very difficult to maintain the seal in the area of the stem. In other words, if the seal of the disk is in line with a projection

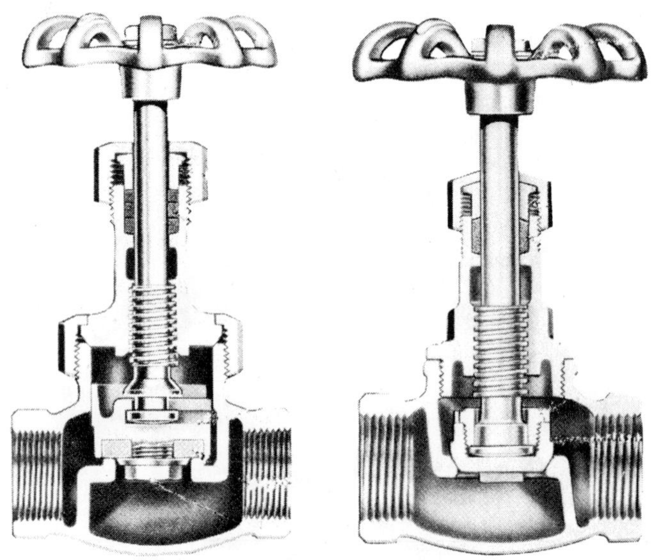

Globe valve, screwed bonnet, rising stem

Globe valve, union bonnet, rising stem—composition disk

Screwed-and union-bonnet styles of globe valves F13

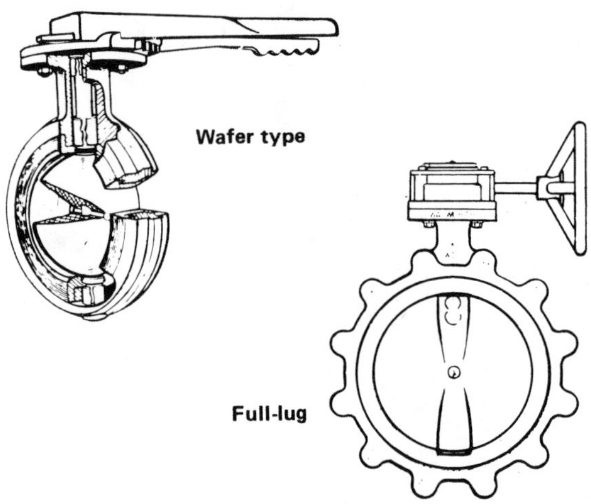

Butterfly valves in wafer and lug styles F14

of the centerline of the stem, problems may result in the disk seal. Because of this, some manufacturers cause the seal of the disk to be offset (from the projection of the centerline of the stem).

Butterfly valve bodies in small sizes are of the threaded type that is screwed directly onto the connecting pipe. In larger sizes, the valves are designed for installation between a pair of pipe flanges and are classified as wafer style or lug style. Where the wafer valve body is used, the bolts span the body. The lug style was developed to facilitate dead-end service. Such a valve body is equipped with lugs that are usually drilled and tapped. Bolts can be installed from either side, and thus the valve can be left in the line when piping on one side is removed. (F/14).

Butterfly valves are good for on-off or throttling service, and cause low fluid-friction loss. They are especially suited for large flows of gases or liquids at relatively low pressures. Such valves are a good selection for slurries or liquids with large amounts of suspended matter and do not permit sediment buildup. They are fast acting, since one quarter-turn of the stem will change the disk position from fully opened to fully closed. The lined valves are limited to pressures not usually in excess of 150 psi, and all butterflies are limited in temperature by their seat and seal materials. Valves that rely on the elastomeric seal are available in pressure classes, including ANSI 600-16. Valves for high-temperature service are available with metal-to-metal seats, but are not tight-shutoff valves.

DIAPHRAGM VALVES

A number of diaphragm-valve body designs are available; they fit two general classifications—the weir (or Saunders-patent, F/15), and the straight-through type. The Saunders valve is the most widely used because it offers both tight shutoff and a short stroke that permits the use of harder (and less flexible) diaphragm materials such as Teflon. The straight-through pattern is limited because there are few elastomers (and even fewer plastics) available that are sufficiently flexible to tolerate the long stroke that is needed.

Diaphragm valves do not require stem packing because the diaphragm isolates the operating mechanisms from the flowing fluid. The valve consists of a body, bonnet, and flexible diaphragm. The stem is attached to a compressor component that, in turn, is connected to the diaphragm.

When the valve is opened, the diaphragm is moved out of the flow passage, and when closed the diaphragm is tightly seated against a weir or a contoured area at the bottom of the valve body. Pressure drop in the weir-style valve is roughly equivalent to that experienced in a globe valve; like the globe valve, the diaphragm valve is frequently used in throttling service.

The diaphragm valve is good for viscous flowing media, slurries, or corrosive fluids. Many solutions or slurries that would clog, corrode or gum up the moving parts of other valves pass through a diaphragm valve with no problems. In extremely corrosive applications, diaphragm valve bodies are lined with elastomers, plastics or glass. In nuclear power plants, stainless steel bodies are used extensively.

CHECK VALVES

As the name implies, check valves check (or prevent) the reversal of flow in piping systems. Automatic in operation, they are kept open by the pressure of the flowing fluid. Closure is accomplished through the weight of the check mechanism or by the backpressure of a flow reversal.

Check valves are available in swing, tilting-disk, lift (disk, piston, or ball), or stop-check designs (F/16).

Conventional swing-check valves contain a flapper

ITT Grinnell

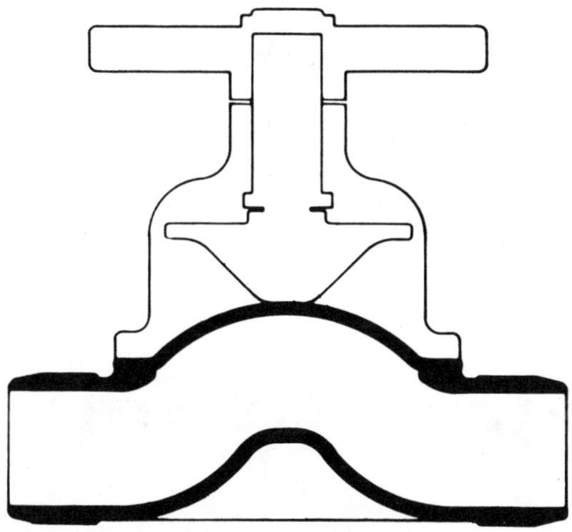

Weir-style (Saunders patent) diaphragm valve F 15

Walworth Co.

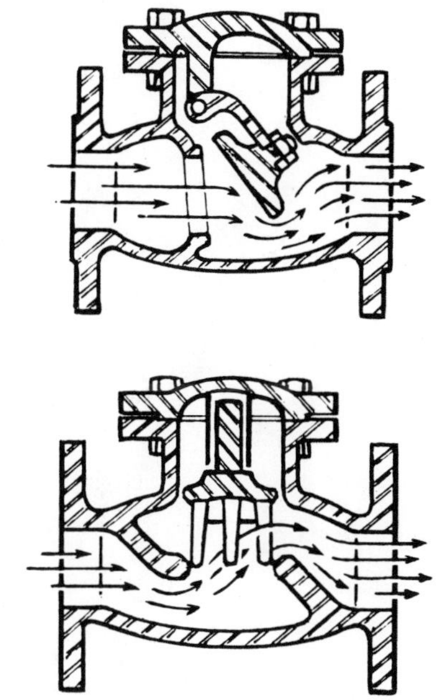

Check valves of swing-check and lift-check types F 16

that is hinged at the top, resulting in relatively low pressure drop. The flapper may have a composition disk when solid particles are present in the fluid, when noise is objectionable, or when tighter closure is required on shutoff. Water hammer or shock pressures can be reduced by attaching an external lever and weight to cause faster closing, but pressure drop will be increased.

Tilting-disk check valves keep fluid resistance low because of their straight-through design. The pivot point is selected to distribute pressure to provide quick closing on flow reversal, and is considered "non-slam." (Where desirable to slow down closure, an external dashpot can be provided.) The flapper floats at full flow and begins to close by tilting at an increasing angle to the flowpath as the flow decreases. The tilting disk has less pressure-drop at low velocities and more at high velocities than a comparable swing check.

Recent valves designs have departed from conventional flanged bodies and now come with wafer bodies for installation between pipe flanges. Wafer style bodies are available with single flappers eccentrically mounted, resulting in a tilting-disk effect, or with two flappers hinged in the middle. The two-flapper style is generally provided with springs to assist in closing quickly in order to minimize shock on closure. Pressure drop in either style of wafer-body check valve is usually greater in small sizes than a swing check, but lower in the larger sizes. Both wafer-style and flanged-body swing check valves should be installed either in horizontal lines or in vertical upflow lines—never in vertical downflow service.

The disk or ball in a lift check valve is raised, within guides, from the seat by upward-flow fluid pressure. When flow reverses, the disk or ball is forced back onto the seat by backflow and gravity. The piston-type lift check is essentially a disk with a dashpot consisting of a piston and cylinder that provides a cushioning effect upon flow reversal. The valve is suitable for services where frequent flow reversals occur, and is perhaps the only successful check valve for reciprocating applications. Pressure drop is greater than in other check valves.

Stop-check valves are usually produced in globe or angle styles, with a stem that is used either to tightly seat or to restrict the lift of the disk. The stem is not attached to the disk, and cannot be used to open the valve. The stop-check valve is used on steam lines from boilers more than in any other application.

Foot valves are specialty check valves used at the bottom of a suction line on a horizontal pump. They maintain the prime on such pumps and usually incorporate strainers to keep foreign matter from entering.

References

1. Hydrocarbon Processing Market Data, 1967, Gulf Pub. Co., p.4 and p.11.
2. From paper prepared by Robert B. McKee, Sun Oil Co., presented in a panel session during the 29th Annual Petroleum Mechanical Engineering Conf., Sept., 1974.
3. Based on a speech presented by L. E. Upp of Daniel Industries, Houston, Tex., in a panel session during the 29th Annual Petroleum Mechanical Engineering Conf., Sept., 1974.

The author

John T. Wier is manager of engineering services for Hudson Engineering Corp., Box 36100, Houston, TX 77036, where he is responsible for pressure vessels, design engineering and drafting services and the material takeoff department. He has a B.S. in mechanical engineering from Texas A and M University, is a member of ANSI B31.3 Petroleum Refinery Piping Code Committee, and is a registered professional engineer in the state of Texas.

How to Select

Over the last decade, high-performance valves have evolved from advances in materials of construction coupled with improved designs for bodies, seats and seals. Here are tips for selection and sizing.

DONALD W. BEAN, Jamesbury Corp.

This article emphasizes high-performance valves, those of advanced design made from critically selected components. Of course, there is still a demand in the chemical process industries (CPI) for "ordinary valves for ordinary service." Yet the ever-increasing complexity—and size—of new plants calls for valving that comes as close as possible to being failure-proof under extreme operating conditions. The primary need of users, and the goal of manufacturers, is repeatability over an extended period with little or no maintenance.

It is tempting to claim that valve manufacturers foresaw the need for new and improved components for the CPI as these industries grew in size and complexity. The fact is that the reverse was true. Manufacturers were literally forced to learn how to communicate with end-users. This meant not only a basic change in attitude and philosophy on the part of management, but also a restructuring of engineering, sales and marketing activities.

Today's valve salesman, for example, is the antithesis of the type of salesman who was prepared only to quote price and delivery. He must now have a thorough understanding of his customers' processes and must be able to pass on information about the performance records of valves used in specific applications. He also has to know case histories of failures of products made by his own company as well as by competitors, and must counsel against misapplications so that failures are not repeated.

VALVE SELECTION

Valve selection involves many decisions quite aside from simply choosing one type or style over another. Users have learned that the most important criterion is system performance. They are looking for optimum productivity, yield, quality and plant safety, with the assurance of an absolute minimum of downtime. They must also consider limitations imposed by the new environmental codes.

Until recently, valve selection was often a hit-or-miss procedure carried out according to a few long-established tenets: gate and plug valves are good for on-off service; globe types are best for throttling; butterfly valves are useful as dampers if working pressures are low. Today, any of these valves may be equally suitable

Originally published February 5, 1973

High-Performance Valves

for a particular job, thanks chiefly to advances in materials of construction.

Some selection decisions can be made by referring to manufacturers' specification sheets. If the application has limitations on weight, available space, operating torque, pressure drop, or other mechanical factors, these data will appear in the specification sheets. The most important decision, however, will be the choice of materials of construction for body and trim, and seat material. Of course, valve sizing and actuator sizing are also part of the selection process.

The fluid passing through a valve can affect the body and trim either by erosion or corrosion. Chemical corrosion by halogens can be particularly severe; chlorides, for example, attack and penetrate the natural protective film on stainless steel. Most valve manufacturers publish corrosion guides to aid in materials selection.

Three less common kinds of corrosion can also occur.

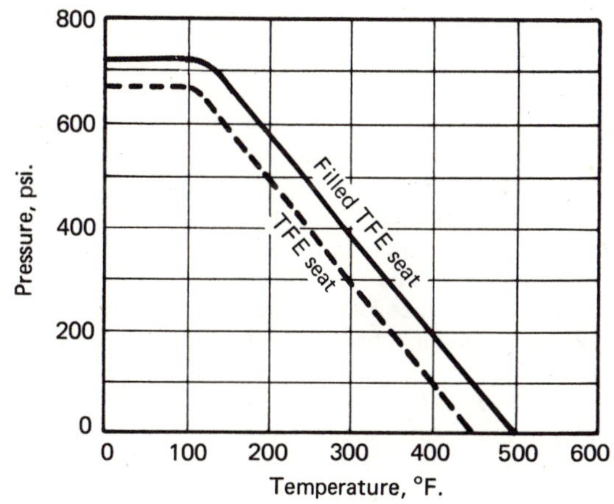

SEAT-RATING charts simplify material selection—Fig. 1

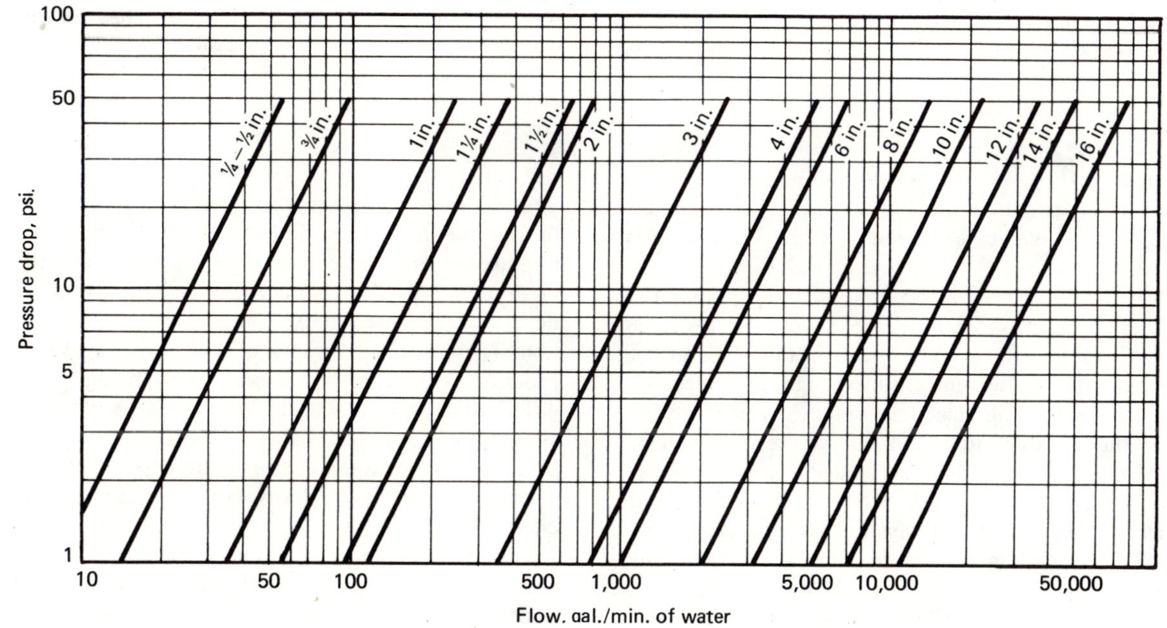

TYPICAL graph for valve sizing—Fig. 2

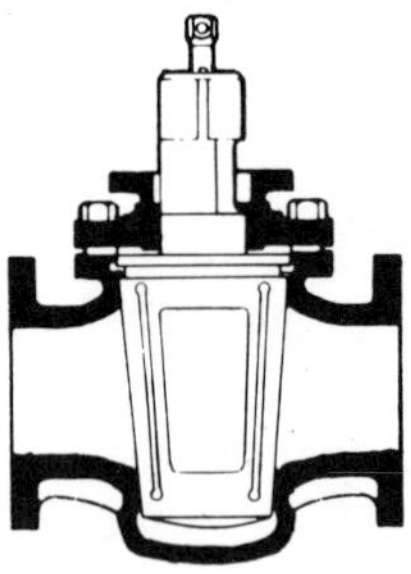

Plug Valves

A plug valve was used in the mechanism of an Egyptian water clock made prior to 2000 B.C. Its simple, no-nonsense design has been preserved through the ages, probably because it still offers rapid on-off service and quarter-turn operation, at minimal cost. The basic limitations of plug valves are problems with torque and sealing. Earlier forms of the plug used nonlubricated, metal-to-metal seals, but galling and sticking caused problems. Even so, plugs of this kind are still used, though in small sizes. More-modern versions come in sizes up to 34 in., with pressure ratings up to 10,000 psi.

Two standard types of plug valves are available. One is lubricated, which solves the friction problem. But this design is limited by the chemical compatibility of the grease and by the constant attention required to keep the valve lubricated. Plugs lined with elastomer, usually tetrafluoroethylene (TFE), are becoming standard, though even this design experiences high torque. Some elastomer-lined plugs use a spring to force the plug down; this design is limited by the compatibility of the spring material.

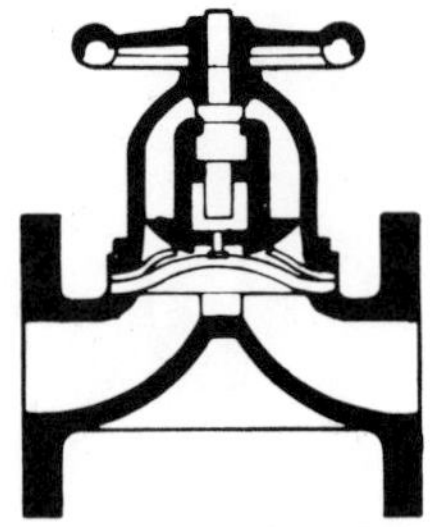

Diaphragm Valves

Particularly suitable for handling slurries, the diaphragm valve is essentially an on-off device, since its throttling characteristics are poor at low flowrates. The principal advantage of this valve is that since it consists basically of only a body, a bonnet and a flexible diaphragm, the stem seal is eliminated. Its major drawback lies in the diaphragm material. The more chemically inert the material, such as TFE, the less flexible it is; the more flexible the material, such as rubber, the poorer its chemical compatibility.

Two less-serious disadvantages are inherent in diaphragm valves. They are multiturn devices, and the most common type (the weir) may leak after media build up on its back side or after erosion wears down the weir.

Gate Valves

The one distinct advantage of gate valves is that they offer little restriction to flow in the fully open position. However, due to the effects of dynamic flow on the unsupported wedge, gates cannot be used effectively for throttling. Chattering induced by the velocity of flowing media against a partially opened wedge, plus repeated cycling at line pressure, creates dragging on the downstream side that, combined with erosion, causes seat-galling and subsequent leakage.

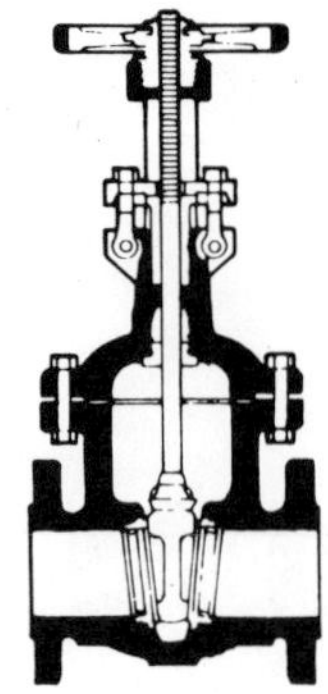

During full-open operation when handling slurries, media tend to build up in the seat cavity, preventing full closure. Also, frequent repacking is necessary since gates are multiturn valves, and wear on steam seals can be severe. Still another item contributing to high maintenance costs is periodic remachining of seats.

Several modified designs are available that deal with some of these limitations. The most common is the split-wedge gate, in which the sealing structure is a split disk. Sealing depends on the force applied to the wheel, or handle, rather than on upstream pressure.

Globe Valves

The globe valve is used mostly for throttling service because it can introduce repeatable pressure drop across a broad range of pressures and temperatures. However, it has low capacity and limited seat life due to turbulence.

The first of these is galvanic corrosion, which can be serious when two dissimilar metals in a valve are far apart in the galvanic series and a strong electrolyte is present.

Intergranular corrosion occurs at the grain boundaries of austenitic stainless steels (all 300-series stainless steels are austenitic) when they are heated in the range of 850 F. to 1,650 F. In this temperature range, often reached during the annealing stage, chromium carbide precipitates at the grain boundaries, making these areas susceptible to corrosive attack. Stainless steels can be desensitized by heating to 1,800 F. or higher, followed by rapid quenching to prevent the carbides from precipitating. Lowering the carbon content of stainless steels permits welding without the danger of carbide precipitation.

Another kind of corrosion, stress-corrosion cracking, sometimes occurs as the result of either residual internal stresses or external stresses in a piping system. Annealing after fabrication will relieve stresses induced by cold-forming or quenching. Sound design and good pipefitting are the only antidotes for piping that is not aligned or supported properly, or for systems that are subject to flexing, vibrations, and excessive expansion and contraction. These applied stresses, as well as residual stresses, produce areas particularly susceptible to halogen salts.

The seat material should be resilient, with sufficient strength to seal repeatedly against a specific shutoff pressure and temperature, taking into account the media's chemical nature and abrasiveness. Standard seat-rating charts, such as the one shown in Fig. 1, simplify seat-selection considerably. The charts include notes on the chemical or abrasive effects of various media. A typical note might read, "Special nylon seats cannot be used in oxygen service."

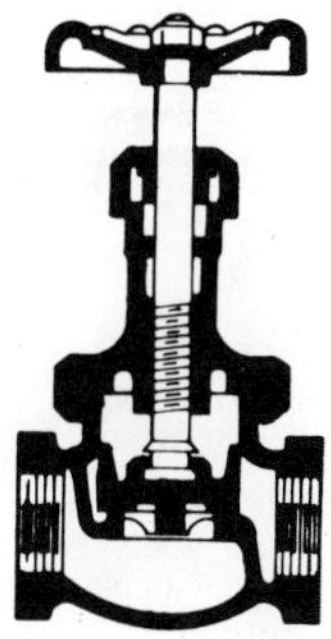

Globes are costly to maintain because the seats are usually metal-to-metal, although seats made from elastomeric materials are now being offered. These limitations explain to a large degree why globe valves are a poor choice for slurry service.

Butterfly Valves

Together with the ball valve, the butterfly valve has shared a large proportion of the research and development effort aimed at adapting the newer materials of construction, particularly elastomers and plastomers, to valve bodies, seats and seals. The effort has produced what amounts to a whole new line of valves. Butterfly valves are simple, small, light and low cost. Their straight-through, open design minimizes sediment buildup and gives low pressure-drop. An added advantage is that they are quarter-turn devices. Latest designs are suited for handling large flows of liquids, gases or slurries, and are available in sizes from 2 in. to several feet.

Some designs have hard seats with resilient O-ring seals around the disk; others may have a soft seat and offset

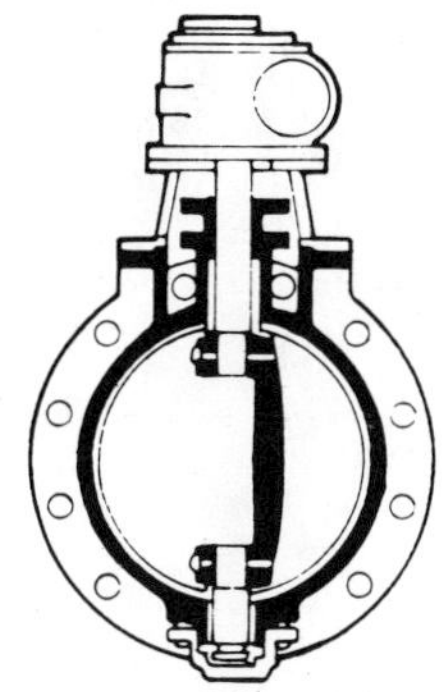

disk. Design is determined by service requirements, which may call for temperatures up to 1,000 F. or down to –32 F., tight shutoff at $1x10^{-5}$ torr, and pressures in specialized applications to 1,500 psi. Some sealing and torque problems still remain, however, and it is difficult to throttle a butterfly valve between the 60-degree-open and the full-open positions.

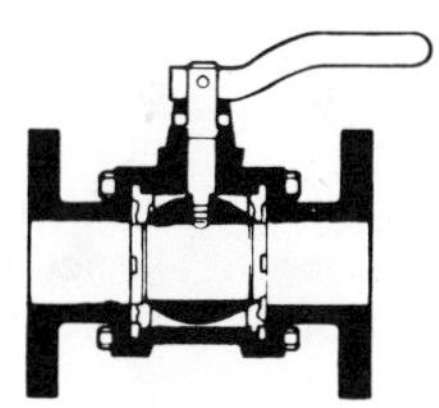

Ball Valves

The ball valve has inherent advantages of low operating torque, good throttling characteristics and high flow-capacity. This design became prominent in the high-performance-valve field when modern plastomer and elastomer technology provided dependable and repeatable seat and seal materials. For example, tetrafluoroethylene has the chemical and mechanical properties to operate over a wide range of temperature, pressure and fluid properties.

A structural seal for a ball valve is illustrated in the sketches. The seat and lip on the right can be compared to a beam section where the thin cantilever

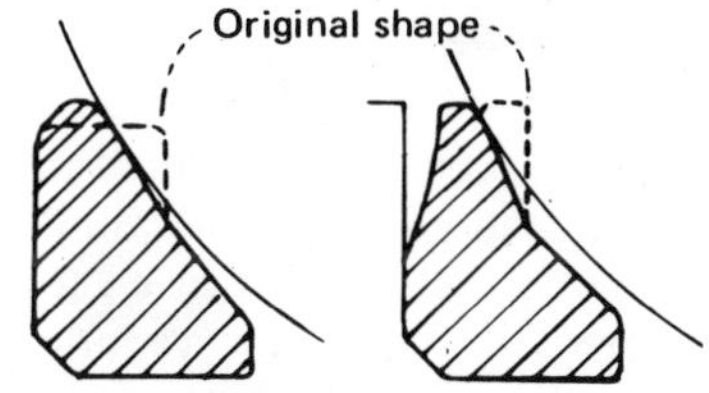

lip is held by the stronger structure of the outer ring. Because TFE is an elastomer, deflection of this ring structure produces a spring effect that maintains the original shape of the seat under stress. As applied to the ball valve, the cantilever concept explains why the structure, when pressurized, will deflect and return. In addition, the spring action automatically compensates for wear, and the lip shape (which makes line contact between ball and seat) ensures minimum torque for operation. The lip also acts as a wiper to clean the ball surface with every operation of the valve.

While the cantilever design gives large sealing force on the downstream seat, there are also smaller forces on the upstream seat which, in effect, provide a double seal. At lower pressures, and in vacuum, this becomes a great advantage. Limitations of ball valves are usually related to the structural strength of polymeric seat materials. Seat ratings should never be exceeded without a specific recommendation from the manufacturer.

Valve-Sizing Tips

Valve sizing is related to the pressure drop through a valve at a preselected flowrate. Experience with stop-valves has taught that the pressure drop must not be too high. An undersized stop-valve means that the system will either require a larger-than-necessary pump to drive it at the required flowrate or, if the system is already designed, that it will not have the required flow capacity. On the other hand, oversizing a stop-valve never does any harm, except possibly in terms of initial capital outlay.

This is not true for control valves. It is essential that a control valve be neither undersized nor oversized. Undersizing will cause insufficient flow capacity in the system. Oversizing can lead to an installed characteristic that provides inadequate rangeability or controllability, or can damage the valve when it is run too close to the closed position.

In sizing a control valve, consideration must be given to the position of the valve at minimum expected flowrate. Under no circumstances should a valve run for any length of time at less than 10% open. A rule of thumb is that a control valve should take 30 to 50% of the system's total pressure drop at maximum flowrate; it is undesirable for a control valve to take less than 10% of the total pressure drop at maximum flow.

The designer of the system should balance line size and velocity against the power cost of the pressure drop. Fluid velocity, which is related to erosion and fluid viscosity, is usually in the range of 5 to 15 ft./sec. The important pressure drop for sizing is that which occurs when the valve is in the open position, or, in the case of throttling, at the maximum, minimum and normally used positions. Once the pressure drop is determined, valve size is arrived at by calculating the valve constant or flow factor, (C_v).

In ordinary practice, the valve constant is obtained by actual test, using an open valve with a preselected pressure drop across it. The experimental values obtained in this way are used to prepare charts (Fig. 2), which most manufacturers can supply.

Even the experts are wary about relying solely on calculation to arrive at valve constants. Not only are different formulas required for liquid, gas and steam service, but special correction factors must be applied in each case. Cavitation, for instance, limits the reliance that can be placed on the liquid sizing formula. Whenever line pressure falls below the vapor pressure of the liquid flowing through it, flashing takes place. If pressure then returns to a level higher than vapor pressure, cavitation occurs, manifesting itself as a pinging noise that increases to a constant banging. At this point, no additional flow will pass through a valve no matter how much the inlet pressure is increased. There are several mathematical techniques to predict cavitation, as well as rules-of-thumb, which are spelled out in sizing literature offered by valve manufacturers.

Actuator sizing has become increasingly important as automation gradually replaces manual operation. As the term implies, actuator sizing consists of matching the operating torque in nonrising stem valves with the torque output of the actuator. Several factors contribute to valve operating torque, chief of which are the coefficient of friction of the sealing materials, the cycling frequency, and the nature of the fluid. Other factors are line pressure, valve design, and the relative smoothness of the sealing surfaces.

Actuator torque output depends on the unit's design and varies with the power input. Pneumatic, double-acting actuators have outputs that vary directly with input air pressure from 40 to 150 psi. Electrical actuators are

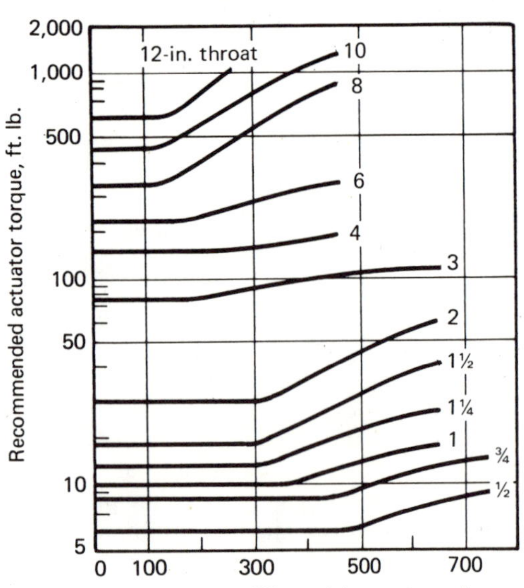

ACTUATOR TORQUE at various pressures—Fig. 3

Typical Valve Costs*—Table I

Size, In.	Gate	Globe	Ball	TFE-Seated Butterfly	Rubber-Lined Butterfly
1/2	$ 35.40	$ 33.00	$ 58.00		
1	$ 51.80	$ 43.00	$ 85.00		
2	$ 96.00	$ 96.00	$ 172.00		$ 33.60
4	$ 334.00	$423.00	$ 400.00	$ 218.00	$ 75.00
8	$1130.00		$1100.00	$ 542.00	$175.00
12			$3300.00	$1002.00	$295.00

*Flanged, 150 lb., 316 stainless steel

rated at slightly less than rated voltage; the variation in torque to rated voltage is not significant. Fig. 3 is a graph showing recommended actuator torques.

Costs and Warranties

In the effort to keep pace with user demands, manufacturers have greatly increased research and development staffs, 100% testing has become routine, and the cost of a single special test may reach $10,000. But volume production of even specialized items brings these costs to reasonable level. The trend is to modular valve design to obtain maximum flexibility in manufacturer's and user's inventories. Also, new seats have been designed for easy replacement by production people. All of these innovations contribute to reducing costs.

Table I represents typical selling prices for flanged, 150-lb., 316 stainless steel valves of several types. Where blanks occur, there has been little demand for that type in a particular size. These prices serve only to illustrate the relative cost of different valve types.

Most manufacturers hold regular training sessions and seminars for customers. Certification is another service. A certified valve (it costs from $25 to $200 for certification) is pretested and guaranteed to perform according to a preselected set of specifications. In many critical applications, customers feel that this form of insurance is well worth its cost.

Even for noncertified, standard, run-of-the-mill valves that may malfunction or fail, an implicit guarantee is now the rule. It is in a manufacturer's self-interest not only to replace a valve without question, but to make sure that the same sort of failure does not happen twice.

Meet the Author

Donald W. Bean is corporate products manager of Jamesbury Corp., Worcester, MA 01605. He has been with the company for the past 15 years in various positions in engineering, technical sales and marketing. Mr. Bean is a chemical engineering graduate of Worcester Polytechnical Institute.

Valve Selection

The first step in valve selection is to understand the capabilities of various types. Here is a rundown on design and operation of globe, gate, butterfly, plug, ball, check, needle and pressure relief valves.

AUGUST BRODGESELL, Crawford & Russell, Inc.

Valves by their nature present a restriction to fluid flow and therefore a pressure drop is always associated with flow through a valve. The pressure reduction is caused by frictional energy losses in the process fluid. Since the valve acts as an energy absorber in relation to the process, it must be adequate not only from the standpoint of containing the fluid under static conditions of pressure, temperature, corrosiveness, etc., but also under dynamic conditions of velocity, pressure drop, erosion, etc.

The type of service together with the operating conditions will determine the type of valve required. Generally, some valve designs are better suited for on-off or shutoff service, while others are designed for throttling.

Globe Valves

The designation "globe valve" encompasses a large number of designs including hand operated and power actuated types used for automatic control. The common feature of these valves is their internal construction, which consists of a disc or plug reciprocating within the valve body and mating with the seat on closure (Fig. 1).

Manually operated globe valves contain either a disc or plug that mates with a metallic seat ring. The disc may be either all metal or have a resilient insert. Metal discs have a tapered or spherical seating surface with line contact on the conical seat. Discs with resilient inserts have a flat seating surface and mate against a similar surface on the seat. These inserts provide tight shut-off but are poorly suited for throttling. Metal discs with hardened surfaces can provide some cleaning action on closure. Bronze as a seat material should be used with caution since it is easily damaged by foreign matter.

Seats are either integral with the body or replaceable screw in types. Because the flow path through a conventional globe valve is rather tortuous, these valves have relatively high pressure drop. The Y-

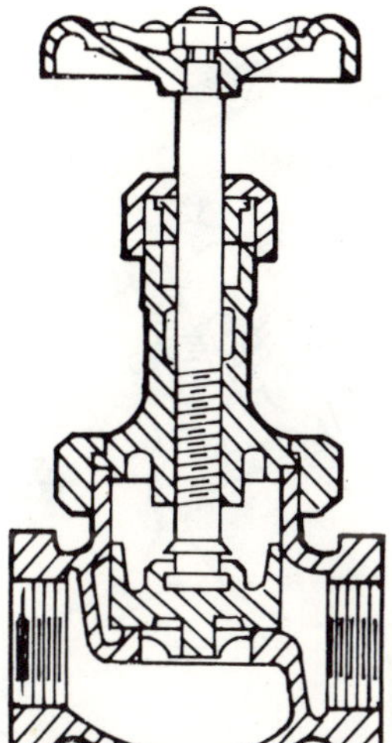

GLOBE VALVE with metal seat—Fig. 1

Originally published October 11, 1971

pattern globe in Fig. 2 is similar to the conventional design except that it seats at an angle relative to the valve centerline. The flow path is contoured and produces less pressure drop than in the conventional globe.

Globe Body Control Valves

Globe valves designed for automatic process control differ somewhat from hand actuated types. Linear motion of the stem is produced directly by the automatic actuator rather than through a threaded stem. Single or double seating plugs may be used to provide the desired flow versus lift characteristic. The seat and plug are usually stainless steel and can be provided with hardened facings for high pressure drop or other erosive services. Generally, operating pressure drops in excess of about 150 psi. warrant consideration of hardened trim. (See the following article for more information on control valve characteristics.)

A single-seated top guided valve has a low mass plug, which means its resonant frequency is high, and it is therefore less sensitive to vibration than valves with more massive plugs. The single-seated valve offers tighter shutoff than the double seated type, but has less flow capacity. Lapping of plug and seat is recommended for best tightness. The valve plug is unbalanced so that high capacity actuators are required, particularly on large valves or those with a high pressure drop.

Installations with flow under the plug are preferred for dynamic stability. Top and bottom guiding offers the advantage of an invertible body and a bottom cleanout flange. Inverting the valve body reverses the valve action; that is the valve closes up instead of down.

As mentioned before, the double-seated globe valve offers somewhat higher capacity than the same size single-seated valve. The valve plug is partially balanced because static pressures acting on the upper and lower valve plug tend to cancel each other so that less actuator force is required. However, double-seated valves have relatively high leakage and cannot duplicate the shutoff qualities of a single-seated valve.

Valve designs discussed up to this point do not solve the problem of seat ring replacement. The screwed in seat ring is quite rugged but can become a problem when replacement is necessary, particularly after prolonged service. The cage trim valve offers a ready solution. The seat ring is clamped between the valve and bonnet and can be easily removed once the valve is disassembled. Alternate designs clamp the seat between body and bottom flange, which permits removal with the bonnet in place.

The split body valve shown in Fig. 3 offers quick change valve trim with added advantages because the valve body is made in two pieces. The valve is adaptable to slip on flanges, which can be a cost saving with expensive alloys. In addition, one body will accommodate several pressure ratings by addition of appropriate flanges. Also, the lower body half can be set at 90 deg. with respect to the upper half if required by the piping configuration.

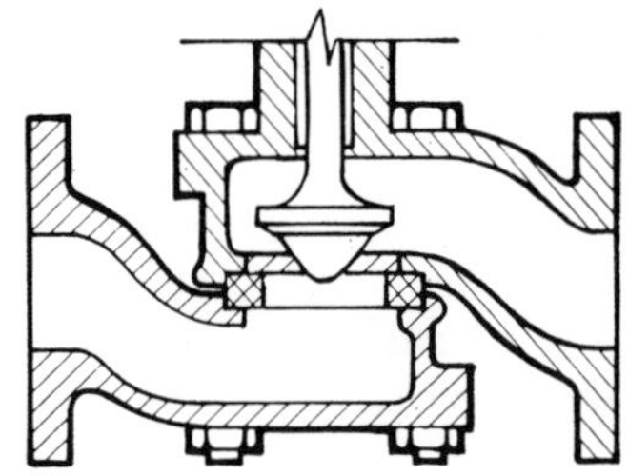

SPLIT BODY valve is easy to service and change seats—Fig. 3

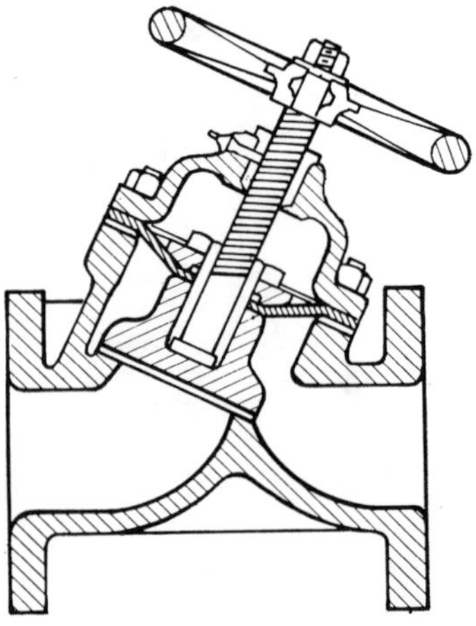

GLOBE VALVE with Y-body and diaphragm-sealed bonnet—Fig. 2

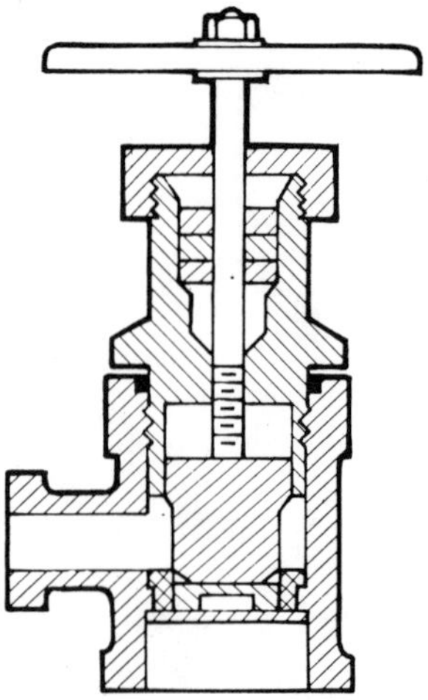

ANGLE VALVE is a special configuration of globe body—Fig. 4

Angle Valves

Angle valves are a special configuration of the globe body form. The body has inlet and outlet connections at right angles, with the stem axis in line with one of the connections. Generally, hand operated angle valves are used to accommodate a special piping configuration or to allow drainage of the valve body. However, for some applications—such as erosive situations and coking hydrocarbon service—the use of angle valves is indicated.

A typical angle body valve is shown in Fig. 4. Note the flow path that allows unobstructed passage of the fluid and drainage of the downstream side. Angle valves have lower pressure drop than comparable globe valves.

Angle body control valves have been used for high pressure drop services with flow over the seat. This arrangement, however, results in a substantial pressure reduction load on the downstream piping. In addition, the valve plug tends to be unstable when operating close to the seat. For these reasons, angle valves are less frequently installed for general pressure reducing service. Also, angle valves—particularly the sweep flow designs—have high pressure recovery so that the valves are very likely to cavitate in liquid service. Split body construction provides easy access to the valve trim for maintenance.

Valve Selection Guide

Valve Type	Size Range, in.	Pressure Rating, psi.	Temperature Rating, F.	Materials of Construction	Service
Globe	½ to 30	to 2,500	to 1,000	Bronze, iron, steel, stainless steel and special alloys.	Throttling & shut-off on clean fluids.
Angle	⅛ to 10	to 2,500	to 1,000	Bronze, iron, steel, stainless steel and special alloys.	Throttling & shut-off on clean, viscous or slurry services.
Gate	½ to 48 (larger in some designs)	to 2,500	to 1,800	Bronze, iron, steel, stainless steel and special alloys.	Shut-off (limited throttling) clean & slurry services.
Butterfly	2 to several feet	to 2,000 (limited pressure drop)	to 2,000 (limited to lower temperature if liners or soft seats are used)	Castable or machinable materials. Liners can be plastic, rubber or ceramic.	Throttling (shut-off only with special designs or seats) clean & slurry services.
Plug	to 30	to 5,000	to 600	Iron, steel, stainless steel & various alloys. Complete rubber or plastic lined valves are available.	Shut-off (throttling with some designs)
Ball	⅛ to 42	to 10,000	cryogenic to 1,000	Iron, steel, brass, bronze, stainless steel, plastic and special alloys for nuclear applications, complete plastic lined valves.	Throttling & Shut-off clean, viscous & slurry services.
Relief	½ thru 6 (inlet)	to 10,000	cryogenic to 1,000	Iron, bronze, steel, stainless steel, nickel steel & special alloys.	Pressure limiting.
Needle	⅛ to 1	to 10,000	cryogenic to 500	Bronze, iron, steel, stainless steel.	Fine flow throttling & shut-off clean services.
Check	⅛ to 24	to 10,000	to 1,200	Bronze, iron, steel, stainless steel and special alloys.	Prevent flow reversal (special designs prevent excess flow)

Gate Valves

Gate valves consist of a disc reciprocating in the body. Generally, the gate valve is used for shutoff rather than throttling service. Throttling is possible near the valve seat; however, the disc erodes fairly rapidly with the valve partially open. When fully open these valves have a pressure drop equivalent to a pipe section.

Gate valves are supplied with rising or non-rising stems. A rising stem requires more head room, but the threads are not exposed to the process. On non-rising stem gate valves, it is difficult to determine the degree of valve opening.

The single wedge disc is usually solid with tapered valve seats. This construction is shown in Fig. 5. To overcome misalignment and dimensional temperature changes, flexible discs are also available. In this case, the disc is solid only through the center so that some movement of the two faces relative to each other is possible.

In double disc gate valves the two seating surfaces can move relative to each other. This gives good shutoff even with misaligned seats or different seat angles. In one design, the two discs are held together by a ball and socket, which permits movement of the discs when they mate against tapered seats. Parallel seating, double disc designs use spreaders or wedges to force the discs against the seating surface. Seat wear is minimized because the discs contact the seat without sliding motion.

Another design, with soft inserts in the discs and vent or drain connections in the valve body, provides tight shutoff and permits the valve to be used for double block and bleed applications. However, this type is limited to temperatures of less than 600 F.

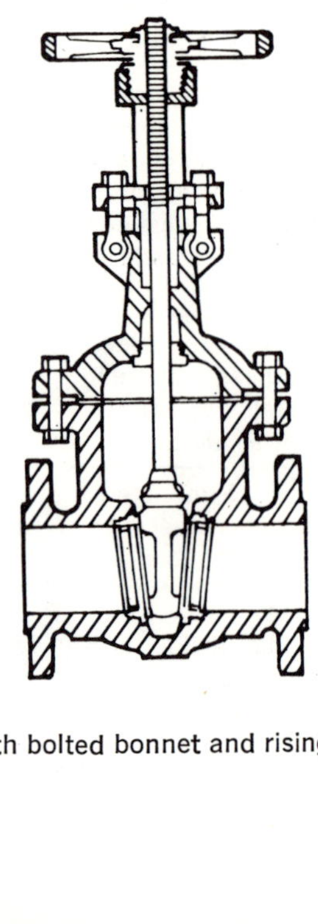

GATE VALVE with bolted bonnet and rising stem—Fig. 5

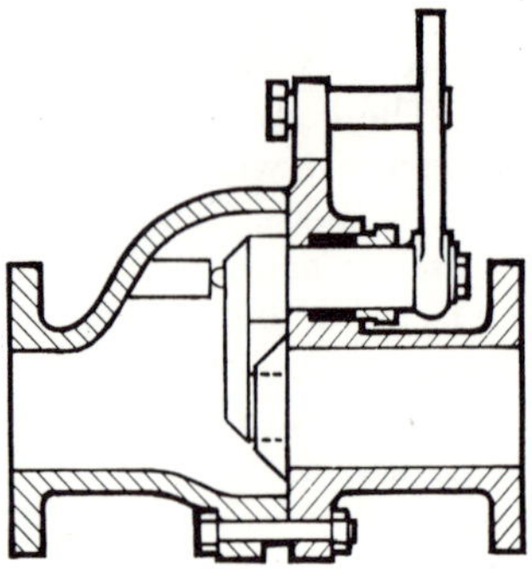

GATE VALVE with sliding disc is used on slurries—Fig. 6

The gate valve shown in Fig. 6 is designed for slurry service. The disc is positioned over the valve port by a lever arm. This motion causes any material obstructing the disc's path to be sheared off while simultaneously cleaning and grinding the seating surface for tight shutoff.

Sliding gate valves are used on low pressure clean or slurry services. The gate is a single or double disc moving within a wafer type body. Shutoff is not achieved by wedging action, but rather by the pressure of the process fluid pushing the gate against the downstream seat. The seat can be provided with a soft insert for better closure. The valve is manufactured in large sizes up to several feet in diameter; however, pressure ratings are limited to a few hundred psi. The main advantages of this design are low weight, low pressure drop and economy.

The conduit type gate valve consists of a single parallel faced disc and seat rings that are flush with the fluid passage. The flow path is a straight pipe section so that no solids can accumulate or settle out, and the valve is therefore well suited for slurry or coking applications. Shutoff is obtained by the force of the upstream pressure holding the disc against the seat, which may be either metal to metal or grease sealed.

Butterfly Valves

Butterfly valves are probably one of the oldest types still in use. Present designs, however, have expanded their application to services with high pressure drop and tight shutoff requirements. Low weight and low cost contribute to the popularity of the butterfly valve. Its straight-through flow pattern offers advantages of high capacity and self-cleaning action for handling slurries.

Essentially, the butterfly valve consists of a disc (also called a vane, flapper, or blade), a shaft and body with packing and bearings for sealing and support. The two body forms available are the solid ring or wafer type and the spool type shown in Fig. 7.

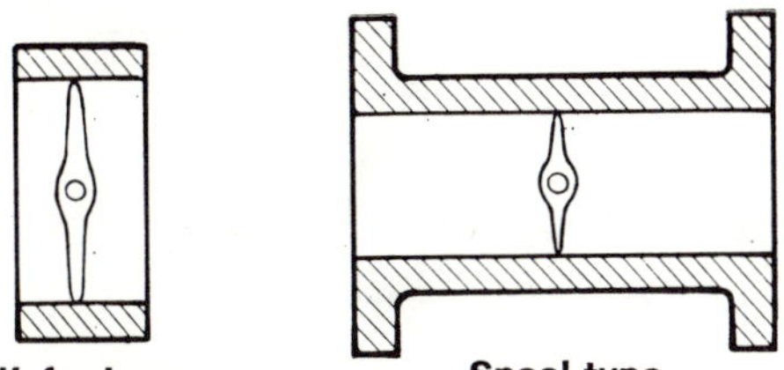

BUTTERFLY VALVES have two basic types of body designs—Fig. 7

The solid ring body is bolted directly between pipe flanges and thus requires very little space. The shaft is a continuous rod supporting the disc. Shaft and disc thickness are determined by the maximum pressure drop required.

Design of the butterfly disc is of major importance. Circular types close with the disc parallel to the body diameter. Elliptical disc valves close 10 to 15 deg. off the body diameter. This design is more expensive but provides tighter shutoff. Circular discs are normally made to rotate 360 deg. and some clearance is required between the disc and body. External stops limit vane rotation.

Disc and shaft thickness, which limit the flow area, are determined by the pressure drop across the valve and the torque required to operate the valve; the higher the pressure drop the higher the torque requirement. Low pressure drop, low torque valves (also called "light pattern") use a thin disc and shaft for maximum flow capacity. "Heavy pattern" valves are designed for higher pressure drops and have a thicker, reinforced disc and a thicker shaft to handle the large torque. The net effect is reduced flow area and capacity with the valve fully open.

Tight Shutoff

Several modifications have been introduced to overcome the basic weakness of the conventional swing through butterfly valve. One of these mentioned previously is the angle seating valve. This provides a seating surface that greatly reduces leakage because the clearances between disc and valve body of the swing through design are eliminated. A resilient facing on the seating surface gives further improvement in valve tightness.

The T-ring seal is another means of insuring bubble tight shutoff. An elastic ring in the valve body is pressurized with air when the disc is closed. The air pressure forces the ring to contact the disc, thus forming a leak tight seal. The T-ring contacts the disc only when it is in the fully closed position. The seal can be made of almost any commercially available elastomer that is suitable for the process.

The piston ring seal provides tight shutoff at temperatures beyond the limits of resilient liners or rings. These rings, acting akin to the piston ring in a gas engine cylinder, eliminate leakage around the circumference of the disc by contacting the valve body.

For applications where the rangeability required exceeds the capabilities of a conventional butterfly, multiple disc valves can be used. The larger outer disc remains closed until the inner disc is fully open.

Plug Valves

The plug cock consists of a cylinder or plug inserted into a matching cavity in the valve body. In this form it was used in ancient times on water. Presently the plug cock is common for on-off control and variations are used on a variety of services for both on-off and throttling.

The most widely used plug valve is the tapered or parallel plug type. Transverse ports in the plug are matched against similar openings in the body. A quarter turn moves the valve from full open to closed. When full open, the valve offers almost no obstruction to flow so that pressure drop is at a minimum.

Lubricated plug valves are used extensively in the oil industry. Lubricant is forced by a gun through a check valve and ducts in the plug to the seating surface. It aids in providing tight shutoff for the metal seat and inhibits corrosion. Where product contamination cannot be tolerated, non-lubricated valves as shown in Fig. 8 must be used. The plug rotates within a Teflon sleeve that prevents sticking and also acts as a seal.

In addition to the two-way valves discussed above, multiple port arrangements are possible for special applications such as flow diversion or blending.

V-port valves are used on slurry or pulp service in the chemical and paper industries. The body openings and plug are V shaped so that the resultant flow area is diamond shaped. On closure, the shearing action between plug and seat cleans the seating surface. This fact, plus the free flow path, allow the valve to be

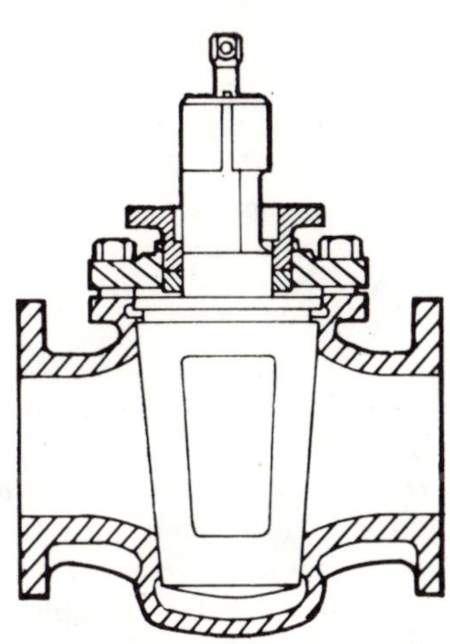

NON-LUBRICATED plug valve avoids product contamination—Fig. 8

used on fibrous slurries. High rangeability and precise throttling control near the closed position are claimed for this design. The seat-to-port clearance remains constant throughout the throttling range while the V-orifice retains its shape.

As opposed to the V-port design, the eccentric plug valve does not retain constant plug to seat clearance. In the full open position, the plug retracts into the valve body. When closed, the plug moves over the seat and is wedged against the seating surface. Various corrosion resistant liners can be provided that extend applications without undue increase in cost. However, hard solids may damage the liner.

The eccentric spherical segment type is designed primarily for automatic process control in applications similar to those of globe body control valves. The plug is a spherical segment and closes without rubbing by means of an offset center with respect to the valve stem. Motion of the stem rotates the spherical segment and simultaneously moves the spherical segment forward into the seat.

The valve characteristic is almost linear but an equal percentage characteristic can be achieved with a positioner and cam. The body is designed to be bolted between pipe flanges. Advantages of this valve include low weight, high capacity and low actuating force. In addition, capacity changes are simplified because only the seat ring is changed to obtain reduced area trim.

Ball Valves

The conventional ball valve, as the name implies, has a spherical plug that controls fluid flow. The two basic types are the rotary motion ball valve and the lift or ball-in-cage type. The ball valve provides not only tight shutoff but also good flow characteristics approaching those of an equal percentage valve. For this reason, it is often applied as a combination throttling and shutoff valve. Due to low friction forces, ball valves have been known to open gradually after prolonged exposure to vibration.

The most common body form is the two-way globe body. The spherical plug has a passage that aligns with body ports in the open position. A quarter turn of the plug closes the valve. To achieve tight shutoff, resilient seat rings are embedded in the body. Tetrafluoroethylene plastics are the common seat ring materials because of their low friction coefficient and resilience.

These resins are applicable up to about 350 F. For higher temperatures the manufacturer should be consulted.

The seats on ball valves are free to rotate, or are forcibly rotated a small amount each time the valve is operated to distribute the wear. For ease of seat replacement and cleaning, the assembly is sometimes made as a cartridge that can be removed without disturbing the pipe connections. Alternately, the body is made in two or three sections that can be disassembled for access to the valve internals.

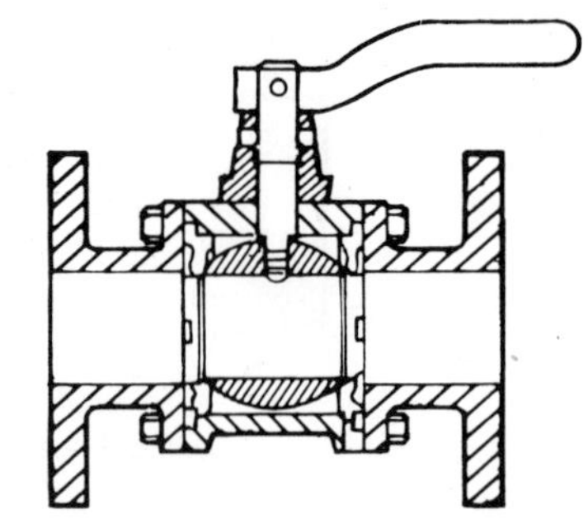

FULL-PORTED ball valve of fire-safe design—**Fig. 9**

A typical ball valve of fire safe design is shown in Fig. 9. Note the secondary seating surface formed by the metal lip, which provides a seal should the soft seat be destroyed by external heat. When handling toxic or other hazardous materials this feature is imperative.

Ball valves do not require lubrication and operate with a minimum of torque. Most commonly, the ball is free floating and sealing is accomplished by upstream pressure that pushes the ball against the seat ring. The stem is made with a square end that engages on a recessed hole in the ball. Alternately, the location of the ball is fixed by top and bottom guides with ball bearings and the seats are spring loaded. Since the plug is subject to wear, a design that lifts the ball off the seat before rotation is recommended where dead shutoff and long operating life are required. Abrasive fluids can be handled this way.

Ball valves can be provided with three or four-way plugs. In some installations this can represent a considerable savings in fittings, valves and space since one unit can replace two or three flow through valves. This arrangement may also reduce maintenance costs.

Ball And Cage Valves

This valve consists of a venturi-ported body with seat rings; a ball to affect closure; and a cage with a stem to position the ball relative to the port. In the closed position, the ball seats against the outlet port and is held by wedging of the cage and upstream line pressure. As the valve is opened, two of four inclined surfaces in the cage lift the ball out of the seat and roll it over the edge of the seat ring. Further opening causes the ball to roll over the two inclines to the center of the cage, where it rests on all four surfaces. The ball remains in this position while the valve is open, held in place by the Bernoulli effect of the flowing fluid.

The cage is designed to offer minimum obstruction in the open position, and it can lift the ball completely out of the flow path. The shape of the opening at the valve seat is a crescent that grows larger as the ball is lifted away by the cage. The flow characteristic is almost linear and permits throttling over a wide range.

Valve tightness is maintained over a long period of time since the rolling action of the ball gives a new seating surface after each operation. The upstream seat ring is not utilized and the body can be reversed to use this "spare ring" should the other become worn. Opening and closing forces are much less than for a comparable globe valve; however, valve travel is at least equal to the throat diameter since the ball must be lifted out of the flow path.

The venturi design of this valve reduces turbulence and offers some advantages when handling cavitating or flashing liquids. If cavitation is present, it will occur in the outlet side beyond the seat ring, where hardened surfacing or replaceable liners can be used effectively. The expanding discharge of the valve contains flashing liquids effectively.

Characterized ball valves are a modification in which only a portion of the plug is utilized. The essential difference between manufacturers' products is the shaping of the leading edge of the ball. Parabolic, V-notch and U-notch are available with essentially the same flow characteristic. The valve is applicable to fibrous and solids slurries as well as clean service. In the full open position, the spherical segment is rotated into the upper portion of the body out of the flow path. For throttling, the ball is rotated into the flow path. As the ball rotates and slides against the valve seat surface there is a wedging and shearing effect that tends to clean the seat.

Metallic or resilient seat rings can be supplied, including various hardened finishes for parts exposed to erosion. Valve bodies are designed for insertion between pipe flanges. When the seal and retainer ring are held in place by the pipe flange, the seal can be damaged by improper tightening of the flange bolts. Deformation of the asymmetrical flow control element under load can cause leakage at the shaft seal or distort the shaft. Pressure ratings are lower than for conventional ball valves. In terms of advantages, this valve offers a very wide throttling range, low weight, good shutoff and versatility, since one valve body will accommodate several pressure ratings.

Check Valves

Check valves are self contained and act automatically to prevent flow reversal. Selection of a particular type depends on operating temperature and pressure, cleanliness of the process fluid, pressure drop available, and to a lesser extent by the limitations of the piping configuration.

The swing check valve consists of a hinged disc over a valve port as shown in Fig. 10. Under no-flow conditions, the disc is held against the seat ring by gravity or weights attached to external levers. Flow in the normal direction (from left to right in Fig. 10) will cause the disc to swing away from the seat. On reversal, the disc is pushed against the seat and held in place by differential pressure. This valve depends on gravity for its operation and the installation must take this fact into consideration. Normally the valve is installed horizontally, but vertical up flowing situations are possible.

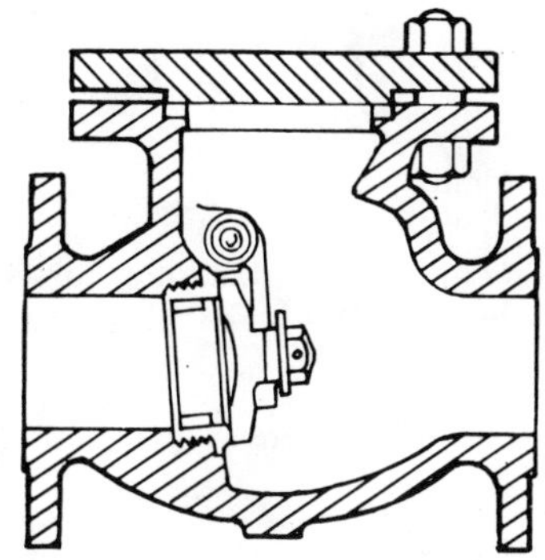

SWING CHECK valve prevents flow reversal—Fig. 10

Swing checks are used on low fluid velocities where reversals are infrequent. With manual overrides, they can be used on the discharge side of pressure relief valves. When several relief valves are piped to a common header, the check valve will prevent back flow while maintenance is done on the pressure safety valve.

The split disc swing check is a varient of the above design. The body is a solid ring for bolting between pipe flanges. The two disc halves are hinged on a pin and spring loaded to hold them closed at no flow. The valve does not depend on gravity for operation, thus allowing more flexibility in installation. It is suitable for situations with frequent flow reversals since, unlike the swing check, it will not slam shut and cause pressure shocks.

Lift check valves, usually of the globe body form, depend on gravity for operation. They are designed for either horizontal or vertical installation, but the two designs are not interchangeable. Under forward flow, a piston is lifted off the seat by the fluid while being retained in the valve by guides. On reverse flow, the piston is forced against the seat to block further flow. Some lift check valves utilize spring loading to assure positive seating. In any case, lift check valves require relatively high pressure drops. They are used on higher pressure services and in smaller pipe sizes than swing checks.

A wafer body for bolting between pipe flanges is also available in another variation of the lift check design. The spring-loaded valve disc is guided by bushings and installation in any position is possible. Lift check valves are better suited than swing checks for services where flow reversals are frequent because the piston is cushioned to reduce water hammer.

The ball check is similar to the lift check except that the guided piston or disc is replaced by a ball, which is free to rotate so that wear is distributed evenly over its surface. The globe body check shown in Fig. 11 must be installed horizontally, while the through flow design can also be used vertically. Ball checks are limited to small sizes, 6 in. or less, and are

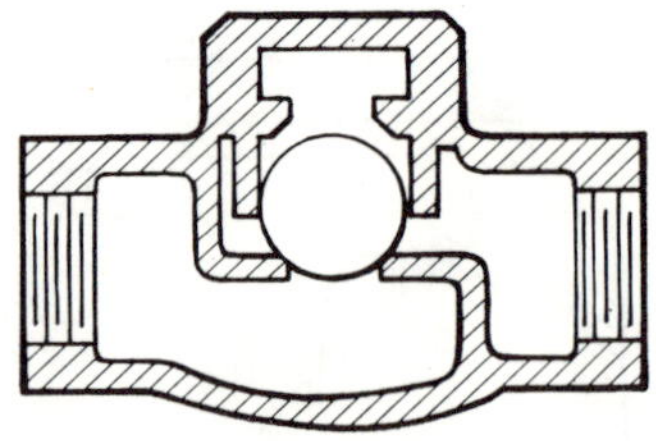

BALL CHECK valves generally give trouble-free service —Fig. 11

preferred for viscous and coking services where their operation is more trouble free than that of other types.

Excess-flow check valves are in-line devices that act automatically to prevent high forward flow rates. They are spring actuated to hold the valve disc open. On high flows that might occur with pipe rupture, the fluid force is sufficient to overcome the spring and the valves closes. A bleed port in the valve disc allows upstream and downstream pressures to equalize so that the valve can reopen.

This check valve is applied on high pressure liquid storage tanks to prevent accidental spills. Where a rupture disc is installed below a pressure relief valve, the excess flow check is used to maintain atmospheric pressure between the rupture disc and relief valve.

Various flow through check valves are also available in small sizes. Generally, these are applicable to clean services with pressures to 10,000 psi. and operating temperatures to 300 F. These valves use spring loading of the check that allows installation in any attitude. Their metal-to-metal seat is backed by a resilient seal to provide bubble tight shutoff.

Pressure Relief Valves

Pressure relief is required whenever the design pressure of a system may exceed what the system is capable of handling. The pressure relief valve, because of its simplicity and self-actuation, is probably the most reliable device providing shutoff after an over-pressure condition. The design pressure of the equipment the pressure relief valve is intended to protect will determine its set pressure. For further details on set pressure determination, the reader should refer to the appropriate ASME code.

In order to avoid ambiguity, several definitions are necessary:

▶ **Relief Valve:** An automatic pressure relieving device actuated by static pressure on the upstream side. The valve opens in proportion to the increase over opening pressure and is used mainly on liquid service.

▶ **Safety Valve:** An automatic pressure relieving device actuated by static pressure on the upstream side. The valve opens very quickly (pop action) and is used mainly on steam and vapor service.

▶ **Safety Relief Valve:** An automatic pressure relieving device that can be used as a relief or safety valve.

▶ **Pressure Relief Valve:** A generic term encompassing relief, safety, and safety relief valves.

▶ **Conventional Safety Relief Valve:** A safety relief valve with the bonnet vented either to atmosphere or the downstream side. Valve set pressure, reseating pressure and capacity are influenced by the back pressure.

▶ **Balanced Safety Relief Valve:** A safety relief valve with the bonnet vented to atmosphere. The effects of back pressure on valve performance are minimized.

▶ **Blowdown:** Difference between set and reseating pressures of a pressure relief valve expressed as a percent of set pressure or in psi.

The function of the safety relief valve is to sense a pressure build up in the system and to provide a flow path for material out of the system. Pressure build up is sensed by a force-balance arrangement that consists of process pressure acting on a fixed area and opposed by a spring or weight. Weight actuated valves are no longer used in process plants, except where very low pressures settings are required.

The direct, spring-opposed pressure relief valve usually has an angle body. The inlet connection is suitable for the upstream temperature and pressure, while the outlet side and bonnet are designed for lower pressures. The outlet side is normally larger than the inlet to allow for expansion of the flowing medium. The inlet port contains the valve disc and the seat, which may be a bushing, semi-nozzle, or the full-nozzle shown in Fig. 12.

The disc is most commonly directly spring loaded with the force transmitted to the disc through the stem. Process valves are usually top guided, while disc (bottom) guiding is common on boiler and liquid relief valves. The bushing is utilized on bottom guided valves and the semi-nozzle on cast iron valves. These seats are screwed into the body. Full-nozzles, which are used on steel valves, incorporate seat, orifice and

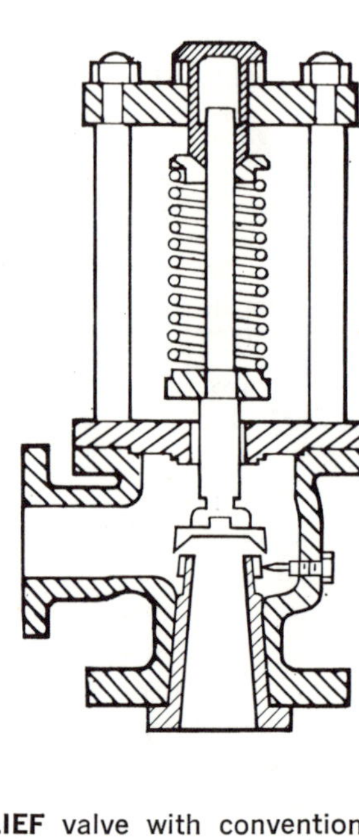

SAFETY RELIEF valve with conventional full nozzle—Fig. 12

flange facing in a single unit. There is an adjustable ring on the nozzle and a deflecting lip on the disc.

The nozzle converges at the orifice to give high velocity at this point. The nozzle ring provides a secondary orifice, whose setting determines the pressure at which the valve re-seats. The deflecting lip on the disc recovers some of the kinetic energy as static pressure, which results in the pop action of the safety relief valve. Relief valves do not have this lip and thus open in proportion to overpressure. Since some force is required to open the valve fully, safety valves reach full capacity at 3% above set pressure; relief valves, on the other hand, do not reach full capacity until 25% above set pressure.

The set pressure of the relief valve is determined by the spring and by adjusting the spring compression bolt for correct force. Springs are grouped by ranges so that the coils allow space for full valve travel, and the spring force does not exceed the force of the flowing medium when the valve is fully open.

Set pressures may be varied up to 10% from the factory settings. The spring is housed in the valve bonnet when the fluid must not be allowed to escape to the atmosphere. Steam valves usually have a yoke in place of the bonnet and the spring is exposed.

Back pressure on the outlet of the relief valve may either increase or decrease set pressure depending on valve design. If back pressure is constant, its effects can be compensated. Since the effect of back pressure is tied to valve design, however, compensation by increasing or decreasing spring set pressure should not be attempted without consulting the manufacturer.

Balanced Valves

Balanced disc-vented piston and bellows type pressure relief valves minimize the effect of back pressure. In the conventional relief valve, back pressure adds to or subtracts from the spring force. Balanced valves are designed to cancel back pressure forces acting on the disc. They are also used for flashing service to minimize the effect of vapor formation on valve operation.

For small variations in back pressure, the conventional pressure relief valve can be used without detrimental effects. Since back pressure does effect the lift of the conventional valve, increased back pressure variations will reduce capacity. For this reason, the conventional valve should never be used when back pressure is expected to vary more than 10% of set pressure.

The pilot operated valve is also a force-balanced design, but the forces are reduced so that closer set point control is possible. Adjustment is made in the pilot valve, which vents or admits process pressure to the top of the valve piston or disc. Since high nozzle velocities are not required for pop action, these valves provide higher capacity in the same body size than direct spring-loaded valves. Blowdown can be set as low as 1%. However, pilot operated safety valves have not found wide acceptance because they contain many moving parts with small clearances.

Special Considerations

Valves in polymer or other viscous and coking applications need to be protected from plugging. Rupture discs are usually mounted below the relief valve and flush with the inside of the pressure vessel. They are set at the valve set pressure and the piping between valve and rupture disc is vented to atmosphere through an excess flow check valve. There are also flush seating valve designs where the seat extends into the vessel nozzle so there are no dead-ended cavities between the valve seat and pressure vessel. For fluids with high freezing points, completely steam jacketed pressure relief valves are available.

Piping between the equipment and pressure relief valve must be sized adequately. Since the safety relief valve responds to static pressure, subsequent pressure loss in the pipe may be sufficient to close the valve. Under these conditions it is possible for the valve to cycle between open and closed (chatter) and to sustain seat damage. Common practice is to limit inlet piping pressure loss to 3% of set pressure. Valve chatter can also be caused by oversizing because a substantial fraction of full flow is required to hold the valve open. Pilot operated valves are less subject to chatter than conventional ones.

A test gag is used to hold a pressure relief valve closed during pressure testing of the equipment. This is typically a bolt that screws into the valve cap against the stem. Levers are used for manual opening of the relief valve, either to check valve operation or to clean the seat area. Local or remote indicators for valve operation, bellows failure or pilot valve failure are available.

Needle Valves

Needle valves give accurate throttling of small flows. They are therefore available in smaller sizes with a variety of designs for special applications. The valve consists of a tapered or characterized plug reciprocating in the body. Throttling occurs because the annular orifice formed by the plug and seat varies with plug position. The body and bonnet for process service have the form of globe valves.

Needle valves may be used to control actuating air or hydraulic fluids. Combination check-needle valves permit controlled flow in one direction or free flow in one direction and controlled flow in the opposite direction. Speed control of air cylinder actuators is a common application. Micrometer-type stem adjustments with a locking device can be provided for precise flow setting. Needle valves are also used on instrument leads to reduce pressure pulsations.

Meet the Author

August Brodgesell is senior instrument engineer for Crawford & Russell, Inc., Stamford, CT 06904. He has been with the company since 1966 and previously was employed by M. W. Kellogg Co. for 10 years. Mr. Brodgesell holds a B.S. degree in electrical engineering from Brooklyn Polytechnic Institute and is a senior member of the Instrument Society of America.

Control Valves In Process Plants

For proper performance in any piping system, here are the design relations, sizing formulas and installation procedures for selecting and using control valves for fluids.

ROBERT KERN, Hoffmann - La Roche Inc.*

Control valves are the basic regulatory devices in any process operation handling fluid streams. Hence, we must be thoroughly familiar with the different types of these valves and their flow characteristics. This enables us to meet process conditions, and to ensure proper installation in the fluid system.

Major Types of Control Valves

In the following brief discussions, only the general features of each control valve are given. For complete details about a specific control valve, consult the manufacturers' literature.

One major group of control valves resembles the globe valve (Fig. 1). In place of a handwheel, an actuator moves the valve stem and plug, thereby opening and closing the valve. The usual actuator is an air-operated device whose housing contains a diaphragm that separates it into two compartments. The diaphragm (and attached valve stem) is balanced in its position by a spring on one side and air pressure on the other. In flow control, the air pressure changes in response to a signal resulting from the measurement of the differential pressure across an orifice or other flow-sensing element.

The single-ported control valve (Fig. 1) finds use where tight shutoff is required in addition to flow control. The double-ported control valve (Fig. 1) has two seat rings with two plugs on a common stem. This is a higher capacity valve than the single-seated one of the same size. With hard seat rings and high temperatures, the double-seated valve cannot shut off tightly. The valve accessories, shown in Fig. 2, allow for various operating functions and conditions.

*To meet your author, see p. 44.

Originally published April 14, 1975

In recent years, a second group of control valves has received wide acceptance. In these types, the actuator rotates a butterfly flap, plug or disk around its axis (Fig. 3). Size for size, these valves usually have higher capacities and less flow resistance than the contoured-plug valves. Generally, control valves with rotating axes are suitable for a wide range of flow-control applications.

Characteristics of Valve Plugs

The valve plug can be disk type, solid contoured or ported. Flow-control characteristics depend on the shape or cavities of the plug. The three basic types of plug and their flow characteristics are:

■ *Quick Opening*—A single-disk (for high temperatures) or a double-disk (for low temperatures) plug is used for total shutoff or opening. A disk-type plug has linear flow characteristics and short stem movement.

■ *Linear Flow*—A plug has linear flow characteristics when the flowrate through the valve is proportional to the lift.

■ *Equal Percentage*—A plug has equal-percentage characteristics if at any plug position, the same percentage of change in flow takes place for the same amount of plug movement. The percentage of change is related to the flowrate just before the plug is moved, as shown in Fig. 4.

Most plug characteristics are somewhere near or between those described. Manufacturers provide diagrams similar to Fig. 4 for each valve.

A plug having linear-flow characteristics is commonly specified for liquid-level control. The equal-percentage plug is used for pressure or flow control; or where only a small percentage of the overall pressure differential is

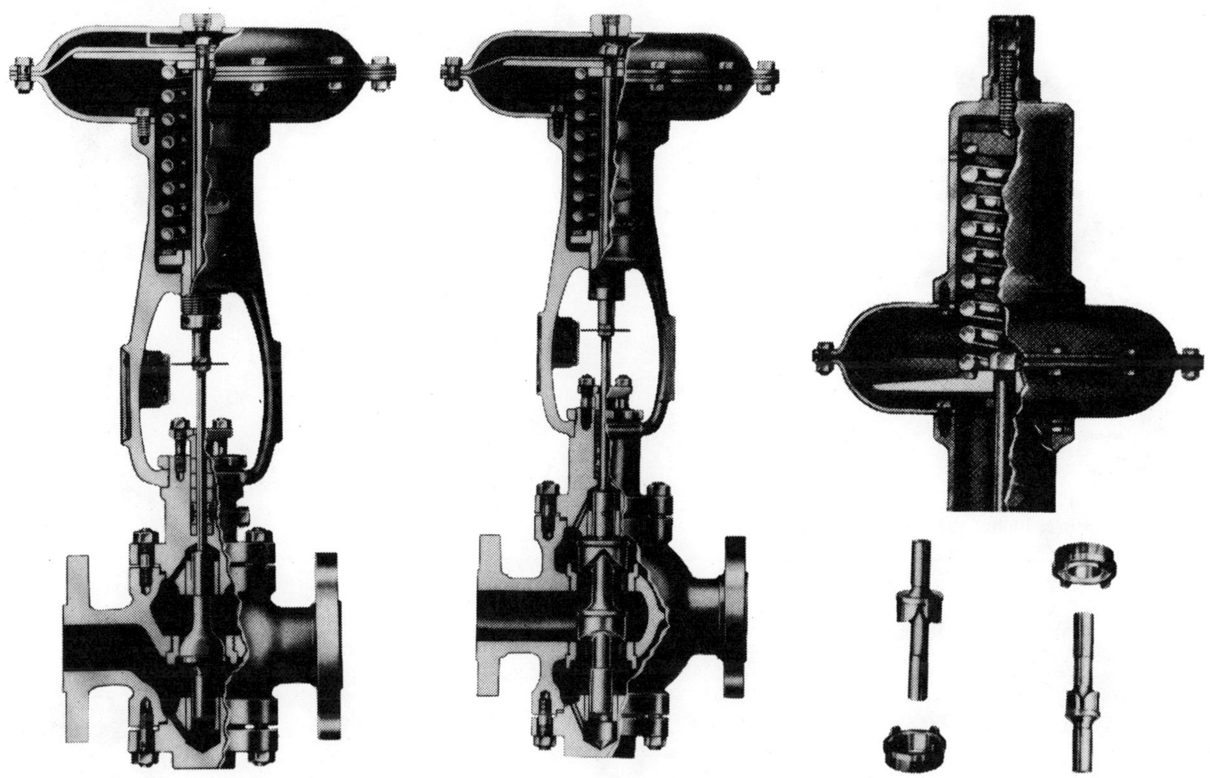

Single-Seat (Equal-percentage contoured plug, fails closed) **Double-Seat** (Equal-percentage ported plug, fails open) **Alternative Actuator and Plugs**

CONTROL valves handle many types of process fluids, and are actuated by air in response to a process signal—Fig. 1

available; or where pressure drop across the control valve varies greatly.

The modified parabolic-flow characteristic falls between the linear and equal-percentage characteristics. This type of plug (usually V-port) finds use where the major part of the system pressure drop is available for control.

Actuators (also called operators or valve positioners) lift the valve stem and plug above its seat, or move the plug in the seat cylinder. Butterfly or ball-type control valves have the actuators side-mounted because the actuator stem rotates the valve axle. Plug characteristics can be influenced by the linkage between actuator stem and valve axle.

The valve housing and the operator's yoke are separate pieces. Hence, after a valve is installed, the operator can be rotated around the valve stem or valve axle, relative to the valve body. This enables a convenient position to be chosen for the actuator, in order to provide access to operating points on the valve.

Hydraulic, mechanical and piston operators are also available.

Safety Requirements

Without air pressure in the pneumatic actuator, the valve can be in closed or open position. These alternative

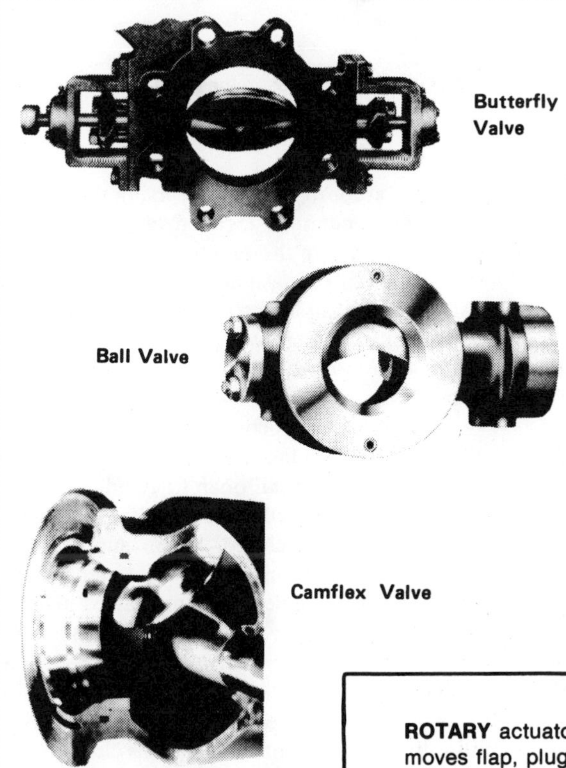

ROTARY actuator moves flap, plug or disk—Fig. 3

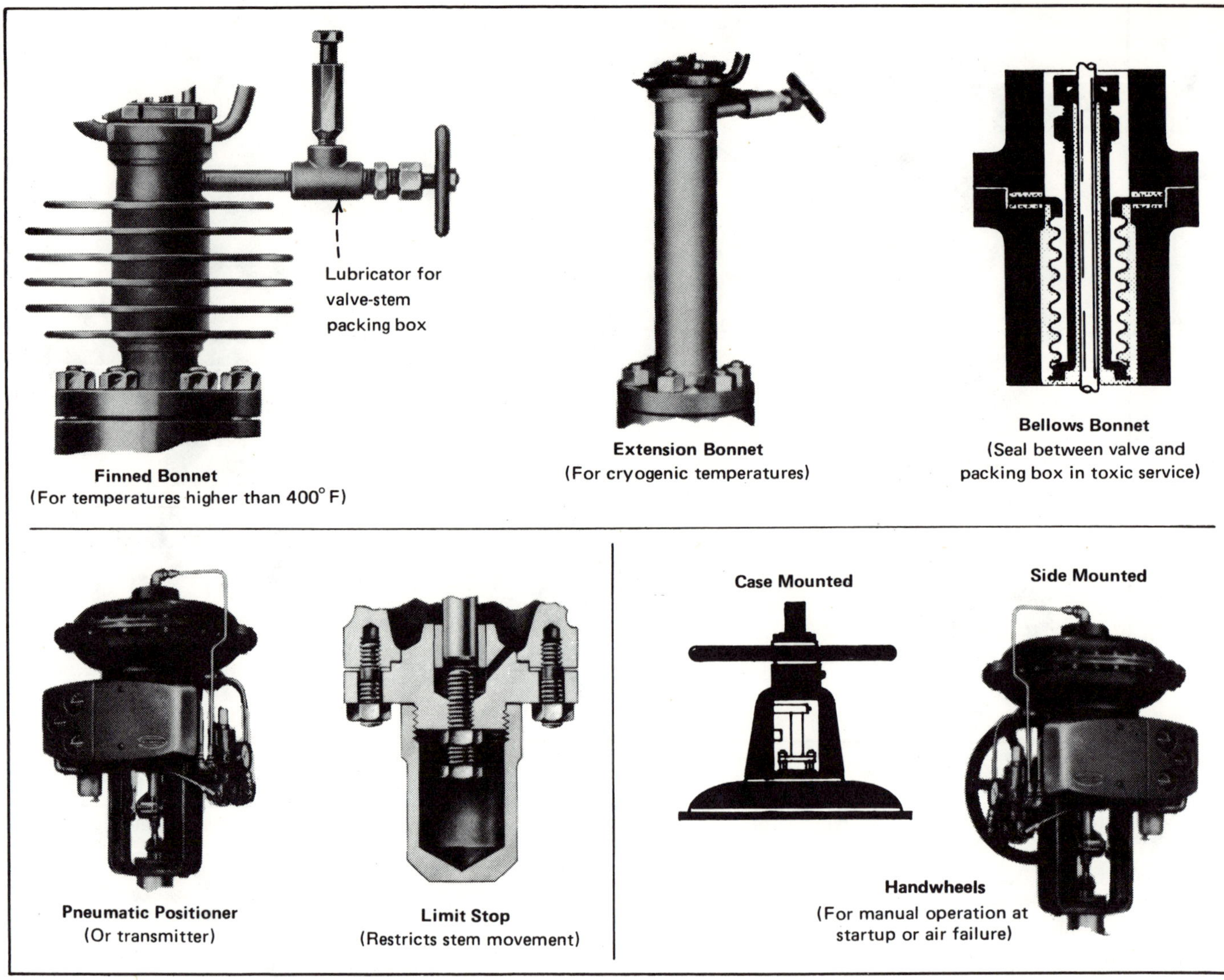

ACCESSORIES extend usefulness of control valves by providing for extreme and unusual conditions—Fig. 2

positions are accomplished by reversing the seat ring and plug, or by reversing the location of the actuator spring from below to above the diaphragm (Fig. 1).

One concern of the designer is to select valves that will fail-safe in the event of instrument-air failure. In principle, a control valve fails safe if temperature and pressure of the process system do not increase after the control valve becomes inactive.

For example, fuel-oil control valves to heater burners should fail closed. At the same time, feed to heater tubes (in most cases) should fail open to avoid overheating the furnace tubes. The feed-control valve to fractionating columns usually fails closed. Steam supply to reboiler fails closed. Reflux-drum vapor outlet and reflux pump-discharge valves fail open. Control valves in minimum-flow bypass lines at centrifugal-pump discharge lines, compressor bypass lines, and reciprocating-machine bypass lines fail open.

Reactors are protected under controlled conditions, and usually the feed-control valve fails closed. Generally, a designer of flow systems should consult process, instrumentation and equipment engineers when deciding on fail-safe positions for control valves so as to assure orderly shutdown procedures.

Capacity Coefficients of Valves

Valve flow coefficient, C_v, depends on the internal dimensions of the valve and the smoothness of surfaces. Tests made by manufacturers (using water or air at predetermined pressure difference) establish C_v values. Manufacturers give the following definition:

$$C_v = Q(\sqrt{S}/\sqrt{\Delta P})$$

C_v is a capacity index indicating the flow of 60°F water in gpm, which will pass through the completely open valve under a pressure difference of 1 psi between the inlet and outlet flanges. Obviously, if $S = 1$ and $\Delta P = 1$ psi, then $C_v = Q$.

Capacity indexes for the butterfly valve are also given at two throttling positions of the flap, in addition to the fully open position.

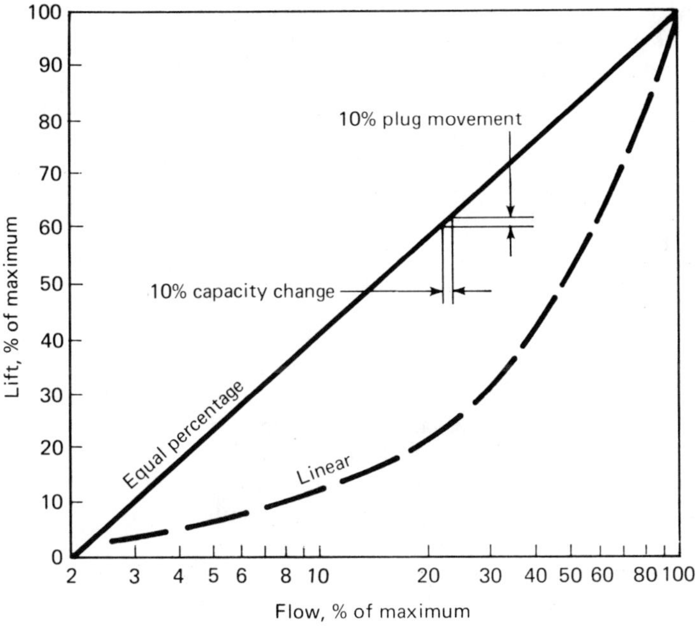

FLOW characteristics of ported or contoured plugs—Fig. 4

Nomenclature

C_f	Critical flow factor for line-size valve
C_{fr}	Critical flow factor for valve between pipe reducers
C_v	Capacity coefficient for control valve in fully open position
C_{vc}	Calculated coefficient for control valve
D/d	Ratio between larger pipe dia. to smaller pipe dia.
E	Expansion factor, ρ_{60}/ρ
k	Ratio of specific heats
M	Molecular weight
P	Absolute pressure, psia
P'	Absolute pressure, psia
P_c	Critical pressure, psia
ΔP	Differential pressure, psi
P_v	Vapor pressure of liquid at flowing temperature, psia
Q	Volume flowrate, gpm
R	Correction factor for control valve between pipe reducers
S	Specific gravity of liquid, ρ/ρ_{60w}
S_{60}	Specific gravity of liquid at 60°F
T	Absolute temperature, °R
v_s	Sonic velocity, ft/s
W	Weight flowrate, lb/h
μ	Viscosity, cp
ρ	Density of fluid at flowing condition, lb/ft³
ρ_{60}	Density of fluid at 60°F, lb/ft³
ρ_{60w}	Density of water at 60°F, 62.37 lb/ft³
Subscripts	
1	Upstream condition
2	Downstream condition

Control-valve coefficients for single- and double-seated valves are given in Table I.

Calculated Flow Coefficient, C_{vc}—When sizing control valves, a flow coefficient is calculated with normal design flowrate in gpm from:

$$C_{vc} = Q(\sqrt{S}/\sqrt{\Delta P})$$

Then a valve is selected whose capacity index, C_v, exceeds C_{vc}. For a good range of control, the capacity index should fall between 1.25 to 2 times the calculated flow coefficient, or:

$$C_{vc}/C_v = 0.5 \text{ to } 0.8$$

This is an optimum range for linear and percentage-contoured plugs. Some valves have a wider optimum range. All valves will operate below and above these C_{vc}/C_v ratios, but the plug will be closer to the fully open or fully closed position. Under these conditions, we lose the important advantage of having wide flexibility in controllable flow-capacity range, and this may limit operability of the process.

High velocities across the valve orifice can wear out the plug and seat, especially if temperature is also high or when abrasive fluid is present.

Critical Flow Factor, C_f—The pressure gradient across a control valve is shown in Fig. 5. For liquids, the flow can be considered subcritical if the vapor pressure of the liquid will not get higher than the lowest pressure point across the control valve. (Vapor pressure is the pressure at which the liquid begins to vaporize at its flowing temperature. Tables of thermodynamic properties of liquids give corresponding saturated-liquid pressures and temperatures.)

If the vapor pressure falls between the ranges of A and B (see Fig. 5), vaporization or cavitation will occur in the control valve. If the vapor pressure nears the downstream pressure, P_2, cavitation can be suspected. Cavitation can cause rapid wear of valve plug and seat as well as vibration and noise. If the vapor pressure falls between upstream and downstream pressures, P_1 and P_2, vaporization can occur. In this case, there will be two-phase flow in the pipeline after the control valve. If the vapor pressure is higher than the inlet pressure, P_1, the control valve receives two-phase flow; and additional vaporization can be considered across the valve. For this condition, diameter of the downstream pipe will usually be larger than the upstream pipe.

The criteria for subcritical and critical flows in liquids are, respectively:

$$\Delta P < C_f^2(\Delta P_s) \quad (1)$$

$$\Delta P \geqq C_f^2(\Delta P_s) \quad (2)$$

where: $$\Delta P_s = P_1 - (0.96 - 0.28\sqrt{P_1/P_c})P_v \quad (3)$$

and P_c is the critical pressure, psia.

For simplicity: $\Delta P_s = P_1 - P_v$, provided that $P_v < 0.5P_1$.

The sizing formula for critical flow is:

$$C_{vc} = (Q/C_f)(\sqrt{S}/\sqrt{\Delta P_s}) \quad (4)$$

We will use a simplified version of Eq. (4) later in this article.

One example of subcritical flow is that occurring in a control valve located in the discharge line from a centrifugal pump. Critical flow can occur across a pressure-

Flow Coefficients for Control Valves—Table I

	Flow Coefficient, C_v	
Size, In	Single-Seat*	Double-Seat*
3/4	—	8
1	9	12
1 1/4	14	18
1 1/2	21	28
2	36	48
2 1/2	54	72
3	75	110
4	124	195
6	270	450
8	480	750
10	750	1,160
12	1,080	1,620
14	1,470	2,000
16	1,920	2,560

*These values have been obtained for Masoneilan 10,000-series (either equal-percentage or V-port) plug valves having full-capacity trim, but also apply to similar valves of other manufacturers [2].

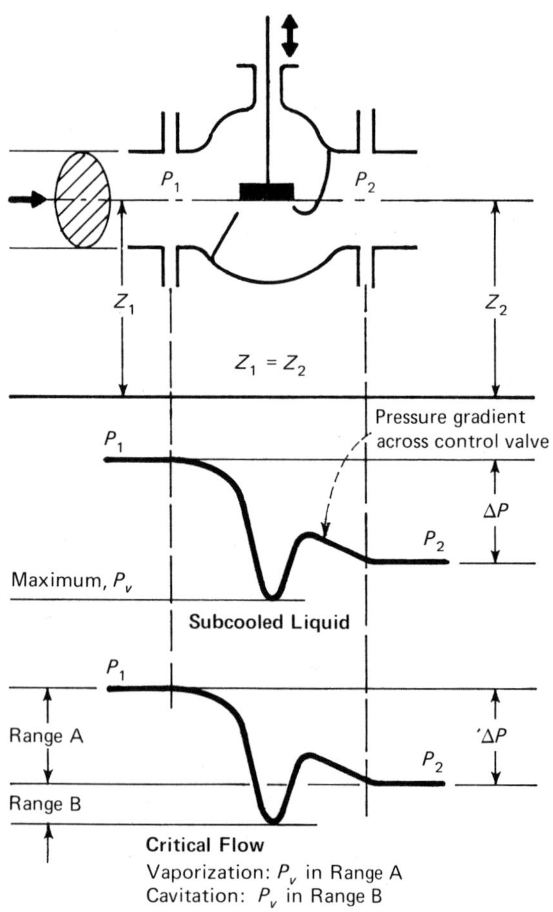

PRESSURES during liquid flow in a control valve—Fig. 5

reducing valve where the upstream liquid condition is close to the boiling point.

For gases, critical flow is assumed when gas velocity reaches the sonic velocity:

$$v_s = 68\sqrt{k(P'/\rho)},\ \text{ft/s} \tag{5}$$

Sonic velocity should be avoided because it can cause noise and vibration.

The criteria for subcritical and critical flows in gases are, respectively:

$$\Delta P < 0.5C_f^2P_1 \tag{6}$$

$$\Delta P \geqq 0.5C_f^2P_1 \tag{7}$$

Critical flow can be avoided by reducing the pressure drop across the valve, by relocating the valve in the flow system, or by choosing a valve with a high C_f value.

The critical flow factor, C_f, is a dimensionless number, which depends on the valve type [6]. C_f is the ratio between the control-valve coefficient under critical conditions and the flow coefficient as published in manufacturers' literature.

Valve Between Pipe Reducers—Flow capacity of a control valve placed between pipe reducers is slightly decreased. In subcritical flow, this is accounted for by a correction factor, R. In critical flow, the correction factor is C_{fr}, which replaces C_f in the calculations. R and C_{fr} also depend on the ratio between pipe size and valve size. C_f, C_{fr} and R have values smaller than 1. Numerical values for the valves shown in Fig. 1 are listed in Table II.

Let us now summarize a number of formulas for sizing control valves for liquid and gas services under different flow conditions [1].

Liquid Service

Subcritical Flow—For a liquid flowing well below its saturation temperature in the turbulent zone, with viscosity close to that of water, and sizes of the pipe and control valve identical, the calculated control-valve coefficient is:

$$C_{vc} = Q\sqrt{S}/\sqrt{\Delta P} \tag{8}$$

where the specific gravity, S, and flowrate, Q gpm, are taken at the flowing temperature; and $\Delta P = P_1 - P_2$.

For minimum pressure drop at the fully open plug position, C_v replaces C_{vc}:

$$\Delta P_{(min)} = (Q/C_v)^2S,\ \text{psi} \tag{9}$$

If we are interested in the pressure drop at a selected plug position between $C_{vc}/C_v = 0.5$ to 0.8, a convenient expression is:

$$\Delta P = \left[\frac{Q}{(C_{vc}/C_v)C_v}\right]^2 S,\ \text{psi} \tag{10}$$

where C_v is taken from the manufacturer's catalog, and C_{vc}/C_v is the selected plug position. (The methods of Eq. (9) and (10) can also be adapted to vapor flow.)

The calculated flow coefficient for laminar or viscous flow is:

$$C_{vc} = 0.072\sqrt[3]{(\mu Q/\Delta P)^2} \tag{11}$$

Critical Flow—If the valve and piping are the same

Correction Factors for Control-Valve Flow Coefficient—Table II

		Single-Seat*		Double-Seat*	
Condition	Factor	Equal-Percentage	V-Port	Equal-Percentage	V-Port
Critical flow Line size control valve	C_f	0.98† or 0.85‡	0.98	0.90	0.98
Critical flow (Control valve between pipe reducers)	C_{fr}	0.86	0.94	0.86	0.94
Subcritical flow, $D/d = 1.5$	R	0.96			
Subcritical flow, $D/d = 2$ (Control valve between pipe reducers)	R	0.94			

*These values have been obtained for Masoneilan 10,000-series plug valves having full-capacity trim, but also apply to similar valves of other manufacturers [2].
†Factor for flow to open.
‡Factor for flow to close.

size, the simplified calculated control-valve coefficient becomes:

$$C_{vc} = (Q/C_f)(\sqrt{S}/\sqrt{P_1 - P_v}) \quad (12)$$

provided $P_v \leqq 0.5P_1$.

Gas, Steam and Vapor Service

The calculated control-valve coefficient for subcritical flow will be:

$$C_{vc} = \frac{W}{11.65\sqrt{\Delta P(P_1 + P_2)\rho_1}} \quad (13)$$

where $\Delta P = P_1 - P_2$, provided that $\Delta P < 0.5C_f^2P_1$.
For critical flow when $\Delta P \geqq 0.5C_f^2P_1$:

$$C_{vc} = \frac{W}{10.13C_fP_1\sqrt{\rho_1}} \quad (14)$$

If the valve is located between pipe reducers, multiply the righthand side of Eq. (8), (11) and (13) by $(1/R)$; and Eq. (9) and (10) by $(1/R^2)$. Replace C_f with C_{fr} in Eq. (12) and (14).

These corrections can be neglected if the capacity of the selected control valve at normal flow gives a coefficient ratio, C_{vc}/C_v, well within 0.5 to 0.8. The operating position of the valve plug will perhaps not be identical to the calculated position, but this will not change valve or pipe size. Also, in sizing valves for critical flow, make sure that the plug will not operate close to its seat.

Two-Phase Flow

For well-mixed liquid and inert gas in turbulent flow with no additional vaporization, the following applies:

$$C_{vc} = \frac{W}{44.8\sqrt{\Delta P(\rho_1 + \rho_2)}} \quad (15)$$

where ρ_1 and ρ_2 are the upstream and downstream two-phase densities, respectively.

When saturated liquid enters the valve (i.e., $P_1 = P_v$), or saturated liquid and its saturated vapor flow concurrently (i.e., $P_v > P_1$), additional vaporization of the liquid can be assumed inside the control valve. For this condition:

$$C_{vc} = \frac{W}{63.3\sqrt{\Delta P\rho_1}} \quad (16)$$

where the maximum $\Delta P = 0.5C_f^2P_1$. (For calculating the densities in two-phase flow, see pp. 87–88.)

Example Illustrates Computations

Let us size the control valves for handling a flow of 113,000 lb/h (348 gpm) of liquid ammonia in each of

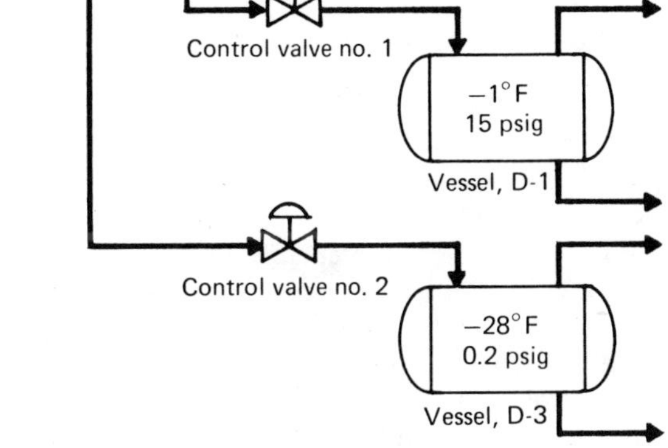

FLOW relations for sizing control valves—Fig. 6

two lines for the system sketched in Fig. 6. The three vessels are located side by side (i.e., all are at the same elevation). Physical property data for liquid ammonia are: $S_{60} = 0.615$, $E = 0.92$, and $M = 17$. Therefore, $S = 0.615 \times 0.92 = 0.566$. Pressure and temperature in each vessel, and corresponding thermodynamic properties, are:

Vessel	D-1	D-2	D-3
Temperature, °F	−1	−9	−28
Pressure, psig	15	225	0.2
Heat content, Btu/lb	41.8	33.2	12.8
Latent heat, Btu/lb	569.6	575.6	589.3

Control Valve No. 1—As liquid ammonia flows from vessel D-2 to D-1, its heat content increases, and liquid in the pipeline after the control valve is subcooled. (Pipelines before and after the control valve can be sized for liquid flow.)

Due to the large pressure difference between vessels D-2 and D-1, cavitation is possible in the control valve. Hence, we can consider the liquid to be in critical flow, and estimate the maximum vapor pressure in the valve from $P_v = 0.5P_1$, or:

$$P_v = 0.5(225 + 14.6) = 119.8 \text{ psia}$$

We will assume that a single-seat valve having a $C_f = 0.98$ (see Table II) will prove adequate by substituting the appropriate values into Eq. (12) to find:

$$C_{vc} = (348/0.98)(\sqrt{0.566}/\sqrt{239.6 - 119.8}) = 24.4$$

From Table I, we establish that a 2-in single-seat V-port control valve having a flow coefficient of 36 may be adequate. We then check the ratio $C_{vc}/C_v = 24.4/36 = 0.68$, which falls well within the desired range of 0.5 to 0.8.

The 2-in lines (before and after the control valve) are relatively short, and when handling 348 gpm will have a small pressure loss. Consequently, pipe resistance will have practically no effect on size of the control valve.

Control Valve No. 2—As liquid ammonia flows from vessel D-2 to D-3, its heat content decreases. Heat is released in the liquid, and as the liquid flows across the valve, vaporization will occur. The actual pressure drop is approximately 225 psig. This is greater than the maximum pressure drop, as determined from $\Delta P = 0.5C_f^2P_1$, or:

$$\Delta P = 0.5(0.98)^2(239.7) = 115 \text{ psia}$$

Substituting the appropriate values into Eq. (16) yields:

$$C_{vc} = \frac{113{,}000}{63.3\sqrt{115(35.2)}} = 28.0$$

We may choose a $2\frac{1}{2}$-in single-seated control valve whose C_v is 54 (Table I). The coefficient ratio will be $28.0/54 = 0.52$, which is acceptable. (Note: a low C_{vc}/C_v value was aimed for in this control valve because C_{vc} was calculated with liquid density. It is not unreasonable to use a two-phase downstream density taken at the outlet of the control valve. Flashing and vapor density are then calculated with the critical downflow pressure. A much higher C_{vc} will result, and a high C_{cv}/C_v can be accepted.)

Because of the subzero temperatures, an extension bonnet can be specified as an accessory to the control valve (see Fig. 2).

In flowing from vessel D-2 to D-3, the liquid releases $33.2 - 12.8 = 20.4$ Btu/lb of liquid, or a total of $20.4(113{,}000) = 2.3 \times 10^6$ Btu. The amount of vapor flashed with this heat is $2{,}300{,}000/582.5 = 3{,}950$ lb/h. This leaves $113{,}000 - 3{,}950 = 109{,}050$ lb/h liquid. These quantities can be used for calculating the two-phase flow resistance of downstream pipe.

Operating Conditions

Control valves are usually the same size or one size smaller than the upstream pipe size, never larger. Control valves are much smaller than line size when high pressure differentials have to be absorbed.

Control valves can accommodate a wide range of capacities and pressure differentials. Flowrates and process conditions are usually well determined for piping and components sizing. When sizing control valves, verify alternative capacities, periodically changing capacities and the related pressure differentials. Control over an extremely wide capacity range might require two control valves in parallel, one for the high flowrates, the other for the low ones. In borderline cases, or for a future increase in capacity, a larger valve body with reduced trim might be desirable.

In most instances, pressure differentials are part of the entire resistance of the piping system. Where an overall pressure differential is determined (for example, between two process vessels), one-third of the overall pressure drop can be attributed to the control valve, and two-thirds to friction losses in piping and equipment. At high pressure differentials, most of the pressure drop will be absorbed by the control valve. When pressure differentials must be minimized, the control valve should be line size, such as in steam feedlines to turbines.

Butterfly valves operate with very little pressure drop (decimals of 1 psi). They are usually suitable in compressor-discharge lines and cooling-water supply lines. However, under throttling conditions, the butterfly valve's coefficient decreases considerably. The coefficient is 50% at 72° position compared with the fully open (90°) position, and 33% at 60° setting.

At centrifugal pumps, pipe resistance n the discharge line (including that of any equipment in this stream) is usually known. An additional 25 to 50% of the discharge pipe resistance can be added for the control valve. With two control valves in series, there is double the amount of additional resistance. For a control valve installed in long discharge lines or in a system with high resistance and relatively small flow changes, the pressure drop across the valve can be 15 to 25% of total system resistance.

A control valve (except butterfly) can only regulate flow by absorbing and giving up pressure drop to the system. Economy in operation of control valves dictates lower pressure drops. However, the valve's capacity and range of control decrease rapidly with lower available pressure differentials.

Changes in the specific gravity, or inaccurate density

Flow Coefficients for Hand-Operated Throttling Valves—Table III

Jenkins Bros.

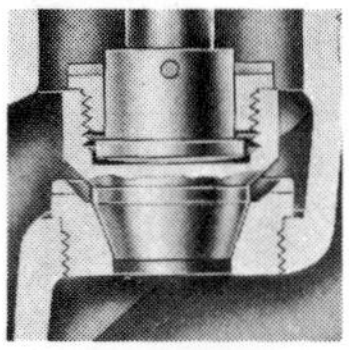

No. 576P

No. 556

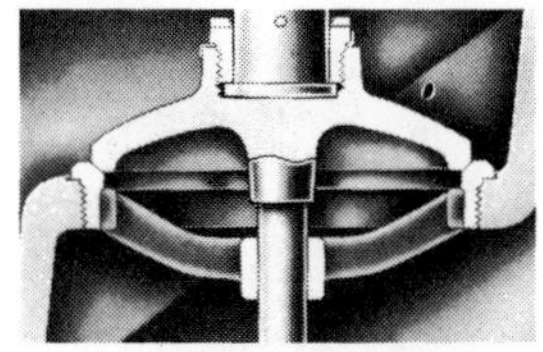

No. 1040

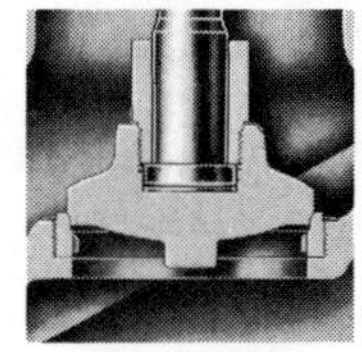

No. 1042

No. 1046

Bronze Globe Valves (Threaded)			Steel Globe Valves (Flanged)		
Size, In	Flow Coefficient, C_v For Valves No. 546P-150 Psi No. 556P-200 Psi No. 576P-300 Psi	Flow Coefficient, C_v For Valves No. 556-200 Psi No. 576-300 Psi	Size, In	Flow Coefficient, C_v For Valve No. 1040-150 Psi	Flow Coefficient, C_v For Valves No. 1042-300 Psi No. 1046-600 Psi
1/4	0.9	1.2	2	46	55
1/2	2	4.2	2 1/2	72	90
3/4	5	8.6	3	105	130
1	10	14.5	4	200	235
1 1/2	24	29.5	6	400	400
2	41	49	8	720	720

Note: Flow coefficients have been obtained for valves manufactured by Jenkins Bros., but also apply to similar valves of other manufacturers.

estimates, will have minor effect on valve capacity. These are small values—square-root functions of the calculated flow coefficient.

When critical flow occurs in the liquid, the piping after the control valve (and bypass valve) should be carefully sized. Vaporization increases pipe resistance considerably. To stay within reasonable velocities when vaporization occurs across the control valve, the downstream piping and block valve will often be larger in size than the upstream pipe size.

In some cases of saturated liquid flow, vaporization in and after the control valve can be avoided by providing a static head of liquid upstream of the valve. This should be noted on the engineering flow diagram.

At high pressures, high temperatures, or large pressure differentials, the control valve should not operate close to its seat. High velocities can wear the plug and seat. This causes inaccurate flow control, and leakage when the valve shuts off.

Bypassing the Control Valve

A bypass is usually provided for control valves smaller than 2 in., in lethal and high-viscosity services, in handling liquids containing abrasive solids, in boiler feedwater service, and in high (over 100 psi) pressure-reducing steam service.

For consistency in piping design, the flow coefficient for the bypass valve should be about the same as that for the control valve. Table III lists the flow coefficients for some of one manufacturer's globe valves. Because of various seat-and-plug designs, valve coefficients are not the same for comparable globe valves made by different manufacturers.

We find by comparing the data in Table III for globe valves with the flow coefficients for double-seated control valves in Table I that the bypass valve and control valve can be the same size. For single-seated control valves, the bypass globe valve can be one size smaller than the control valve. We can size bypass globe valves or manually operated throttling valves in the same way as control valves provided that flow coefficients are available.

Piping the Control Valve

The best position for a control valve is with the stem vertically upward. A control valve will operate in angular, horizontal or vertically downward position. Neither piping designers nor operators accept these positions. Large angle-control valves are an exception; a horizontal position for them can be most practical.

A single control valve without block valves and bypass is usually sufficient in clean-fluid service; or where parallel equipment containing control valves is installed with block valves located at pipe headers. Where dirty fluid or solid particles can be occasionally expected, a temporary or permanent strainer is installed, upstream of the control valve. Single control valves have handwheel operators.

Most piping specifications call for control valves to be

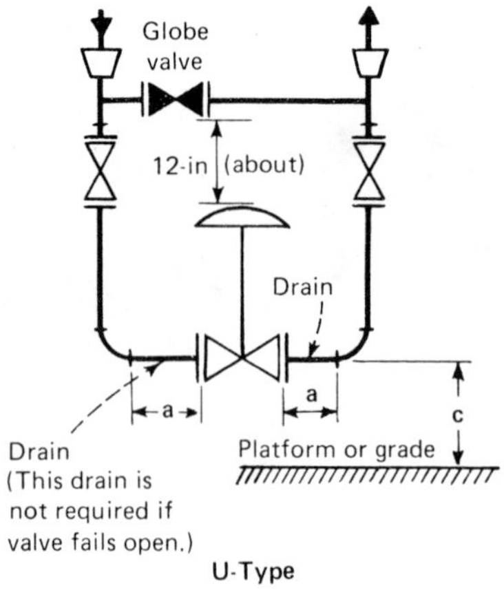

Clearance Dimensions

Pipe or Control Valve Size, In	a, In	b, In
2 to 4	9	12
6 to 8	10	15

Control Valve Size, In	c, In
2	24
3	30
4	36
6	45
8	48

Note: Clearances apply to all manifolds or bypasses.

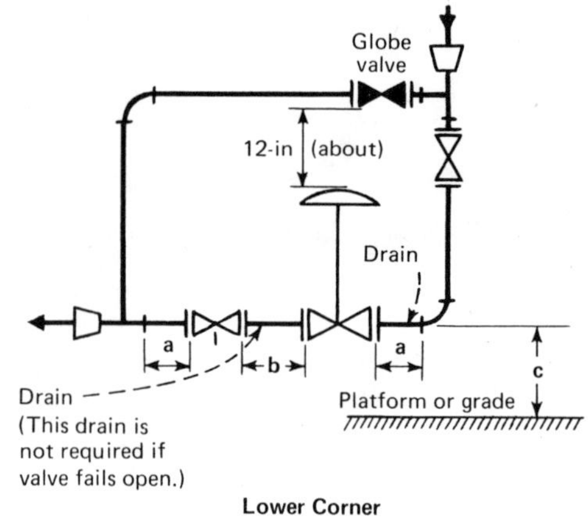

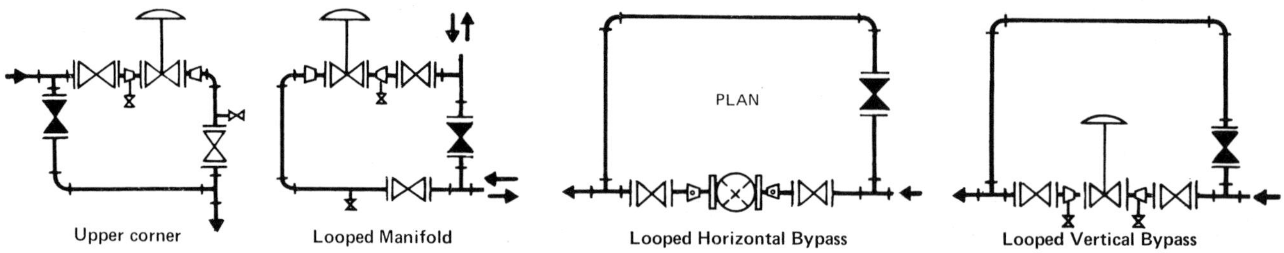

MANIFOLDS and bypasses for installing control valves into the process piping require proper clearances and drains—Fig. 7

located above grade or platform elevation, and at the edge of accessways, except for those valves that have to be located in self-draining pipelines. For example, a control valve placed in an overhead gravity-flow slurry line.

For inplace maintenance, clearance space is required below and above the valve for removing the seat, plug, actuator cover, spring and yoke. Estimated clearance requirements are shown in Fig. 7. Dimensions of control valves are given by manufacturers [*2,4*].

If flow conditions permit, manifolds for the control valve that are smaller in size than the main piping will prove economical. Typical standard manifolds are shown in Fig. 7 [*10*]. The U-type is chosen when the inlet and outlet flows approach the control valve from an elevation higher than that of the valve. The corner type is used when flow is from a high point to a low point, or the reverse. The looped-bypass type serves horizontal flows near grade. A looped-corner bypass can bring a control valve over the operating platform. For economical support, control-valve manifolds should be located near structural columns.

For pressure-relieving and draining a control-valve manifold, provide drain valves or plugs at low points. One drain point is required if the control valve fails open. Drains on each side of the control valve are needed if it fails closed. In saturated-steam flow, one or two steam traps are advisable at the low points of a pocketed control-valve manifold.

The automatic control valve is part of an instrumentation system. Sensing points for flow, pressure, temperature and level should be close to the control valve, as should the transmitter. Instrument wiring and tubing connect these elements. Air lines run from the transmitter to the diaphragm housing, and from the transmitter to the instrument-air header.

Level controllers usually have gage-glass companions. It is convenient for the plant operator to see the gage glasses from the control-valve manifold when operating the control-valve handwheel or the bypass globe valve.

References

1. "Handbook for Control Valve Sizing," Masoneilan International, Inc. Norwood, MA 02062.
2. Dimensions—Masoneilan Control Valves and Auxiliary Equipment, Masoneilan International, Inc., Norwood, MA 02062.
3. "Valve Sizing," Catalog 10, Fisher Controls Co., Marshalltown, IA 50158.
4. Fisher Control Valve Dimensions, Bulletin 1-100, Fisher Controls Co., Marshalltown, IA 50158.
5. Boger, H. W., Recent Trends in Sizing Control Valves, 23rd Annual Symposium on Instrumentation for the Process Industries, Texas A&M University, College Station, TX 77843, 1968.
6. Baumann, H. D., The Introduction of Critical Flow Factor for Valve Sizing, *ISA* (*Instr. Soc. Am.*) *Trans.*, Apr. 1963.
7. Baumann, H. D., Effect of Pipe Reducers on Valve Capacity, *Instr. Control Systems*, Dec. 1967.
8. Boger, H. W., Sizing Control Valves for Flashing Service, *Instr. Control Systems*, Jan. 1970.
9. Boger, H. W., Flow Characteristics for Control Valve Installations, *ISA* (*Instr. Soc. Am*) *J.*, Oct. 1966.
10. Hutchison, J. W. (Ed.), "ISA Handbook of Control Valves," Instrument Soc. of America, Pittsburgh, 1971.

How to select liquid-flow control valves

A chemical process plant employs numerous power-actuated throttling valves to control flow from less than a drop per minute to many thousands of gpm, and pressure drops as low as a few inches of water column to thousands of psi.

James A. Carey and *Donn Hammitt*, *Fisher Controls Co.*

☐ The range of valve applications includes a myriad of differing valve designs, from standard cage-style globe valves to massive, special designs specifically for high-pressure extremes. Choosing the right valve from the mix available might, at first, seem a bewildering task. Yet experience shows that the selection process requires close adherence to only a few fairly simple procedures.

Types of valves

Selecting the right valve for an application can be made easier by first reviewing the four basic styles of throttling-control valves: cage-style globe valves, ball valves, eccentric disk valves, and butterfly valves.

Globe valves are the "standard" of the control valve world. However, the modern cage-style valve (Fig. 1) has largely displaced the top-and-bottom-guided globe valve, either single- or double-ported. Notable in cage-style valve design is its quick-change trim consisting of the valve plug, cage, and separate seat ring.

The cage is a multi-purpose component. It provides valve plug guiding and retains the seat ring in the body, while the shape of the openings in the cage wall determines the flow characteristic of the valve. In contrast to the top-and-bottom-guided valve, the cage-guided valve gives much more stable operation. The cage offers massive valve-plug guiding for stability in the region where maximum pressure drop occurs.

The cage-style valve can meet most application requirements, since there is a wide range of trim styles from which to choose. Balanced, unbalanced, elastomer-seated, restricted or full-size trim are a few of the many options. Any trim configuration is interchangeable within a single valve body, giving the user a saving in purchase price and lowering his parts-inventory needs.

Originally published April 3, 1978

The globe valve, typically available in sizes to 16-inch, comes in most castable alloys with screwed, flanged, or weld-end connections. It meets standard face-to-face dimensions and is available in ANSI pressure-rating designations through Class 2500.

When it is compared to other valve styles, however, some limitations of the globe valve become apparent: (1) a size limitation, normally 16 inches, (2) lower capacity when compared to an equal size line-of-sight valve such as the ball or butterfly, and at times, (3) overall expense, especially in the larger sizes.

Although many applications are likely to remain the

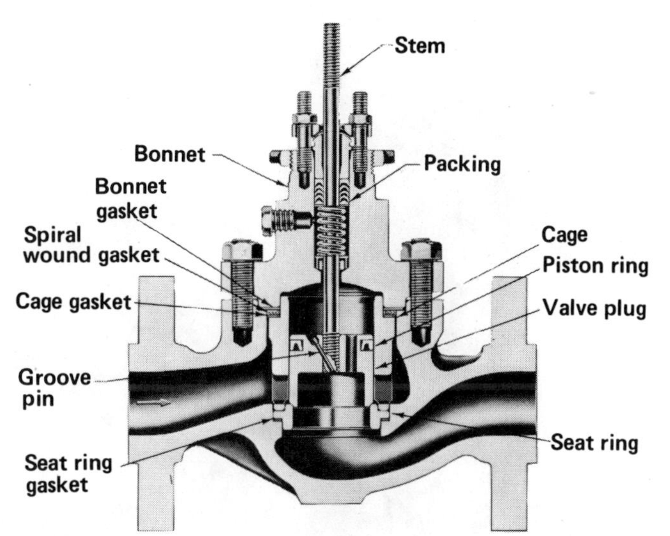

Globe valve with cage-style trim — Fig. 1

Characterized ball valve for throttling operations Fig. 2

domain of globe valves, either a ball- or butterfly-valve may prove the better choice because of higher capacity, simpler and more compact design, lighter weight, and more attractive cost.

There are several types of throttling-ball valves. In addition to the standard full ball (either full bore or reduced bore), there are designs that use a characterized, partial ball (Fig. 2). Designed originally for pulp-stock applications, these now serve in all industries. In another design, a hybrid between a ball valve and a high-performance butterfly valve, a disk pivots from the closed position to a position out of the flow stream.

Most ball valves use elastomeric or fluoroplastic ball seals, and exhibit good shutoff capabilities. Flexible metal seals are becoming standard for high temperatures, and rigid metal seals are increasingly available. The greatest progress in recent years has been made with tetrafluoroethylene (TFE) seals, which survive both relatively high temperatures and corrosive fluids.

Ball valves are classified as high-recovery valves, meaning that outlet pressure recovers to a level close to inlet pressure—higher than the pressure recovery through a globe valve. In other words, they require less pressure drop to produce a given flowrate. But ball valves are more limited in allowable pressure drop and temperature than globe valves.

ANSI Class 600, usually the top body-rating available in a ball valve, is a flangeless design that fits between line flanges. This means that tensile stresses due to piping are transmitted through the flange bolts, not the valve body.

In those applications where either a globe valve or a ball valve will serve, the ball valve can cost as little as half the price of a globe valve.

Another rotary-shaft style gaining in popularity is the high-performance butterfly valve, using a disk with its axis offset from the centerline of the valve (Fig. 3). The result is an eccentric motion of the disk as the valve strokes. An important design feature is that the disk contacts its seal for only a few degrees of rotation at valve closure. This tends to reduce seal wear and prevents permanent seal deformation. Also, because the seal does not rub against the disk when the valve is throttling, friction is low and less operating torque is required. As with a ball valve, this eccentric-design butterfly valve comes in either elastomer or metal seal construction.

This valve style is normally available in sizes through 24 inches, with ANSI Class ratings of 150, 300 or 600. It fits between line flanges, and requires very little space. The comparative costs of eccentric-disk and ball valves vary with size, but on the average an eccentric disk costs less. The cost differential becomes significant at sizes above 6 inches.

The fourth basic type of throttling-control valve, the standard butterfly valve (Fig. 4), is the most economical of all designs on a cost-per-flow-capacity basis.

For general applications not requiring low leakage rates, butterfly valves with swing-through disks are recommended. These valves can handle high inlet pressures and pressure drops over a wide range of

A high-performance rotary-shaft valve Fig. 3

Butterfly valve with fluorocarbon lining and coated disk, to withstand corrosive liquids Fig. 4

temperatures. The swing-through design, with clearance between the body-bore and disk, allows differential thermal expansion of the disk and body without seizing of the disk in the body.

Where low leakage rates are required, piston-ring designs can provide rates approximately one-third those of valves with swing-through disks. They can control high inlet pressures over a wide range of temperatures. The piston ring, retained on the disk periphery, also allows differential thermal expansion of the disk and body.

Valves with either an elastomer T-ring seat or a fully lined body-bore can provide tight shutoff. Also, in fully lined butterfly valves the liner provides total isolation of the body from the process fluid. This permits use of low-cost body materials for corrosive-fluid service.

Butterfly-valve sizes and styles cover a wide range of flow rates, inlet pressures, pressure drops, and shutoff requirements. Sizes range above 100 inches, with body ratings from ANSI Class 125 through Class 2500.

Table I shows principal characteristics of popular designs for globe and rotary-shaft valves.

Pressure considerations

The four major types of control valves discussed above account for the majority of control-valve applications. To choose a specific construction from the options obtainable first requires a look at the pressure conditions the valve must withstand.

Most commercially available control valves meet the standards of the American National Standards Institute (ANSI). These codes establish a pressure and temperature rating for a valve body, based on its material of construction. The first step then in the selection process is to determine which valve style or styles meet the pressure rating of the application. Some valves do not have pressure ratings above ANSI Class 600, while others are not available in relatively low pressure ratings, such as ANSI Class 125.

Selection of the best valve configuration for a particular use depends also on the maximum pressure drop expected. Although a quick study of manufacturers' literature will reveal precise pressure-drop capabilities, it is generally safe to expect that globe valves will control higher pressure-drops than rotary-shaft valves. Cage-guided globe valves serve successfully for pressure drops as high as 50,000 psi. Cage-guided designs are generally used where no other type of valve-plug guiding is satisfactory. Other globe valves also give relatively good pressure-drop capabilities, although some designs are intended for applications where pressure drops do not exceed 100 psi.

The pressure-drop capability of ball, butterfly, and other rotary-shaft valves depends largely on (1) the support of the throttling member, (2) shaft-bearing sizes and locations, and (3) the sealing-member design. Consequently, pressure drop capabilities vary considerably among common rotary-shaft valves. For example, trunnion-mounted ball valves can handle very high pressure drops, while various designs of floating-ball

Pertinent characteristics of common nonproprietary control valves — **Table I**

	Globe valves				Rotary-shaft valves			
Characteristic	Cage guided	Split body	Flangeless	Top-and-bottom guided	Lined butterfly	High performance butterfly†	Characterized ball valve¶	"Floating" ball valve**
ISI pressure classes	125 to 2,500	150 to 600	150 to 300	125 to 2,500	125 to 300	150 to 600	125 to 600	125 to 300
ual size range, in.	1/2 to 16	1/2 to 8	1/2 to 4	1/2 to 24	2 to 24	2 to 42	1 to 24	1/2 to 12
mperature range, °F	Cryogenic to 1,500	−100 to 450	−100 to 450	50 to 800	−20 to 300	Cryogenic to 1,500	Cryogenic to 1,000	−50 to 400
ssure drop, psi	to 14,000	300	740	3,000	200	1,440	up to 1,440	up to 720
acity relative to 3-in.	100*	75	50	90	250	200	240	350
rchase cost relative to 3-in.	100*	70	70	110	50	65	80	80
ative maintenance cost	Low	Moderate	Moderate	High	Highest	High	High	High
w characteristic	Any	Any	Any	Any	Equal percentage	Linear	Modified equal %	Modified equal %
ngeability‡	30 to 50:1	30 to 50:1	30 to 50:1	30 to 50:1	100:1	75:1	200 to 300:1	100:1
oical application	Steam letdown; high ΔP flow control; general application	Corrosive service	Corrosive service	Flow control; general applications	Low ΔP general service; corrosives; HVAC applications	General service; steam; water; hydrocarbons; natural gas dehydration	General service; pulp and erosive slurries; dry chlorine service; superheated steam	General service; hydrocarbon; water; saturated steam
ge trend	Level	Down	Down	Down, sharply	Up	Up	Up	Up

end
alues are compared with the globe cage-guided valve having a base of 100.
angeability is the ratio of maximum C_V to the minimum controllable C_V
gh performance butterfly valve is an offset disk valve which is trunnion supported and suitable for high pressure drop.
haracterized ball valve has a partial ball supported by trunnions.
"Floating" ball valve has a full or solid ball supported by seals.

Suggested service temperature limits for elastomeric valve components **Table II**

Material	Limits, °F
Natural rubber	−60 to +160
Neoprene	−40 to +175
Nitrile	−20 to +200
Polyurethane	−40 to +200
Butyl	−20 to +300
Ethylene propylene	−40 to +300
Tetrafluoroethylene (TFE)	0 to +450
Silicone	−65 to +400

valves are usually limited to drops of 300 to 400 psi or less. Special butterfly-valve designs accommodate pressure drops well in excess of 1000 psi, but the common rubber-lined butterfly valve ordinarily operates up to 100 or 200 psi.

Flowing fluid

The flowing fluid must enter into selection of the valve configuration. For example, fibrous or erosive slurries are difficult to handle at best; specifying valves designed for these services give optimum control and service life.

Another example is corrosive service. Exotic alloys to withstand corrosive liquids are often not readily available in cast form; therefore, TFE-lined bodies or valves made from bar stock might provide the answer. Fortunately, most applications involve relatively non-corrosive fluids at reasonable pressures and temperatures. Therefore, cast iron and cast carbon steel are the most commonly used valve-body materials. For further information, see the article "Corrosion and Piping Materials in the CPI in this Deskbook.

Temperature limitations

Temperature is another major operational consideration. Valves that rely on elastomeric materials for seals or for bearing surfaces cannot handle temperatures much higher than 300 or 400°F. Temperature ranges in Table II suggest limits within which elastomers will function adequately. Dynamic forces imposed on the materials also enter into consideration, since, in many cases, tear strength and other physical properties decrease rapidly as temperature increases.

Cryogenic service demands valves with low cool-down mass, short face-to-face dimensions for ease of installation, and a way to keep the valve packing out of the cold-box area. Extremely high-temperature service usually means a special body and valve-trim configuration and special materials of construction.

Shutoff capability

Besides temperature, other secondary criteria help determine the valve trim—the so-called "innards" of the valve. Often neglected, but very important to the selection of an economical control-valve package, is the degree to which the valve must shut off. This degree of shutoff can range from a relatively large amount of seat leakage in the closed position (such as one might expect in a swing-through butterfly valve) to leakages of less than 1 bubble per minute (the standard for soft-seated globe control valves). ANSI standard B16-104 defines the industry-recognized standard levels of leakage for control valves.

Stringent shutoff specifications usually add to the cost of a valve. Assume a 4-inch globe valve operating at temperatures around 600°F, with a pressure drop in the shutoff position of 1,000 psi. If leakages of 0.5%, or 0.1% of maximum valve flow, can be tolerated, the net price might be around $1,500, complete with a pneumatic actuator. If, on the other hand, the leakage has to be held at the level of 1 or 2 bubbles per minute, the price could easily exceed $3,000. Know what leak rates can be tolerated, and then specify to that level. Otherwise, expect to pay for oversized actuators or special valve construction.

Flow characteristics

Although thoroughly explored in innumerable technical discussions, valve flow characteristics remains one of the least-understood factors in selecting valve trim. The three most common characteristics—linear, equal-percentage, and quick-opening—are inherent; i.e., they measure the inherent capacity of a valve with a constant pressure drop across it as the valve plug is traveled.

■ The linear characteristic is self-explanatory; the valve capacity varies linearly with the travel of the valve plug.

■ The quick-opening characteristic is linear for the first portion of the travel, and very little gain in capacity is experienced thereafter. This is the type of characteristic that is found in a poppet type valve where no throttling surface is exposed to the fluid flow to modulate capacity in the early stages of control valve travel.

■ An equal-percentage characteristic increases flow capacity by the same percentage for each equal increment of travel. For example, assume capacity increases by 10% in the first 0.1-inch of travel; it will increase another 10% in the next 0.1-inch, and so on until fully opened. This characteristic exhibits a straight line on a semilogarithmic plot; it is exponential, not linear.

As its main objective, valve-flow characterization can show how the gain of the valve varies to compensate for changes in process gain with changing loads. The gain of the valve indicates the sensitivity of its output (flow) to changes in the input (valve travel). A high-gain valve exhibits a large change in flow for a small change in the valve position.

Thus, (1) a quick-opening characteristic has a high gain in the initial travel of the valve plug and then a low gain in the upper portions of the travel, (2) the linear valve characteristic maintains a constant gain throughout the entire travel, and (3) the equal-percentage and related modified-parabolic characteristics show low valve gains in the low-travel regions but, as the travel increases, valve gain increases sharply.

The flow characteristics discussed so far are described as inherent, and are those observed when pressure drop across the valve is constant. Flow through the valve, however, is influenced by more than just the flow area. Variations in pressure drop across the valve also cause changes in the fluid flow, even at constant valve area. In laboratory testing, holding pressure drop constant produces flow that is strictly a function of valve travel. In actual applications, the flow/travel relationship that results is referred to as the installed flow characteristic—the characteristic obtained in service where pressure drop varies with flow and with other system changes.

Pressure drop across the valve also influences the amount of flow change that occurs as a result of a travel change. Consider a linear valve in a system where pressure drop across the valve increases with load flow. At low load flows in correspondingly low travels, the pressure drop will also be low. As the load flow increases, the pressure drop also increases. Plotting flow-data points shows that, even though the inherent flow characteristic of this valve is linear, the installed flow characteristic more nearly approaches equal-percentage.

Conversely, if the pressure drop changes with flow in such a manner that, as the load flow increases, the pressure drop across the valve decreases, then the installed flow characteristic will appear more like a quick-opening characteristic. To establish the necessary installed characteristic requires a dynamic analysis of the system, so that the valve gain will compensate properly for gain variations in the process.

Reliable guidelines that help in selecting the proper valve flow characteristic are tabulated in Table III. The optimum method for selecting a characteristic involves a dynamic analysis of the control system. For those wishing to pursue the subject farther, more information may be found in the bibliography at the end of this article.

If a good dynamic analysis of the system is available, or if there is sufficient time to make one, the valve characteristic that is best suited to a particular control loop can be selected without reliance on the guidelines listed in Table 3. However, without a dynamic analysis, trouble can usually be avoided by following the previous suggestions, although optimum control may not be reached.

Long experience and numerous analyses show that it is better to pick an equal-percentage characteristic when in doubt. Using a linear characteristic where an equal-percentage one would have been better often leads to an unstable system. However, the reverse situation seldom causes instability. Fortunately, an equal-percentage characteristic is inherent in most of the common ball and butterfly valves.

A recently revived theory from the late 1950s holds that a "universal" characteristic exists between linear and equal-percentage. The fit of a universal characteristic to a flow system depends on (1) the percentage of total system pressure-drop designated for the valve, (2) the range in load experienced, (3) the gain of the control loop as load varies, and (4) the change in total system loss as a function of load. It is possible that this characteristic suits some applications, but in most instances it is not likely to give control as good as that provided by a properly selected linear or equal-percentage characteristic.

Flow-characteristic selection guidelines **Table III**

Control factor	Conditions encountered	Characteristic to use
Liquid level	Pressure drop increases by 2 to1 ratio, or more, as flow increases	Quick-opening
	Any other	Linear
Pressure	Liquid	Equal-percentage
	Compressible fluid:	
	Fast system—low volume downstream (generally, less than 10 ft of pipe); pressure increases quickly	Equal-percentage
	Slow system—generally, more than 100 ft of pipe downstream	Linear
	If pressure drop varies by 5-to-1 ratio, or more, for either a fast or slow system	Equal-percentage
Flow	Measuring element in series with valve	Linear
	Measuring element in bypass:	
	Linear measuring device	Linear
	Square-root measuring device	Equal-percentage
	Flow range small; pressure-drop at valve changes greatly	Equal-percentage

Rangeability of control valves

The word rangeability, used alone, is not definitive. Any discussion of it must address itself to inherent rangeability and installed rangeability.

Inherent rangeability can best be defined as "the ratio of maximum to minimum flow-coefficients between which the gain of the valve does not exceed the specified gain by some stated ratio." This definition must, of course, be further qualified by the limitation that in no case can the inherent rangeability exceed 100 divided by the percent of leakage.

As an example, if a double-ported or balanced type of valve with a leakage of approximately 0.5% could throttle down to the point of seat contact, the inherent rangeability would be 200. If the valve had tight shutoff (zero leakage), and again could throttle down to the point of seat contact, the inherent rangeability would be infinite. Unfortunately, it is not possible to achieve accurate control down to the point of seat contact, due to the necessary tolerances between parts. (This has a significant effect on the rangeability as the valve plug approaches the seat.)

The point of minimum controllable flow most commonly comes at the point of clearance flow, a flow that would be achieved through the dimensional clearance between the valve plug and cage. Since the clearance flow can be as much as 2% of the total flow, this would indicate an inherent rangeability of 50.

Installed rangeability can be defined as "the ratio of maximum to minimum flow over which the valve, as installed, will provide satisfactory control." Most people feel that the installed rangeability of a valve can be established from the inherent characteristic curve plus a knowledge of the range of actual service conditions. This is rarely the case and, in fact, applies only where the limiting factor is leakage flow.

The following points help make a reasonably accurate estimate of installed rangeability:

- The required valve gain over the intended flow range.
- The inherent flow characteristic of the valve.
- The maximum permissible ratio of actual to required gain.
- Stem force characteristics in the low-flow region.
- Actuator stiffness.

At this point, it should be obvious that there are some serious limitations in trying to estimate installed rangeability. To do this with any degree of confidence requires considerable time, along with access to information not readily available. Although a complete rangeability analysis can seldom be justified, and comprehensive guidelines are not feasible, there are a few rules that may help in avoiding rangeability problems:

1. Whenever possible, select the flow characteristic that best matches system requirements.
2. Do not expect installed rangeability to exceed that calculated from leakage flow.
3. Operating a globe-style balanced valve below 0.050 inches lift requires that spring rate (pounds per inch) of the actuator be substantially greater than pressure drop (psi), or negative gradient problems may possibly arise.

Packing materials

Unless otherwise specified, a control valve comes equipped with packing the manufacturer considers standard for the application. Sometimes the manufacturer's past experience and knowledge of the strong and weak points of available packing materials leads him to recommend a packing other than that specified.

Basic packing materials in widespread use today include asbestos, graphite, and TFE. Asbestos is low in cost and stable to about 800°F, but tends to give a high-friction seal.

Graphite packings may also have higher friction. Graphite is stable to 3,000°F in nonoxidizing service, but practical limits are 740°F for oxidizing service and 1,200°F for nonoxidizing service. It has high thermal conductivity and long service life, but requires higher gland pressures for leak-free operation. Graphite is impervious to a wide range of liquids, including sulfuric acid (up to 95% acid at 160°F), sodium hydroxide at all concentrations, and almost all inorganic compounds. Strong oxidizers, however, are not always compatible. Sodium chlorite, sodium hypochlorite, bromine, chlorine, and iodine will attack graphite packings above room temperature.

TFE materials, more chemically inert, are suitable for use with strong oxidizers. They also have very low friction properties. They require a smooth stem finish (4 to 6 micro-inches RMS) and will leak if stem or packing surface is damaged. Temperatures should range between −40 and 450°F.

To produce optimum properties, various materials can be mixed with each other. Two good examples of materials that are not suitable as packing when used alone, but which result in better packing when used as impregnants, are soft metals and fiber glass. Some soft metals provide better retainment and can provide cathodic protection in systems where stem or packing-box corrosion is a problem. Substances such as molybdenum disulfide strengthen packing materials, and are often used in TFE packings.

To achieve the most sensitive actuator response to an input signal requires the lowest possible packing friction without leakage. Accordingly, packing suppliers offer a number of packing compositions designed for low friction performance:

- Braided asbestos impregnated with TFE, molded into split rings.
- Asbestos fiber, lead wool, flake graphite, metal particles, and neoprene binder.
- Braided, preshrunk Teflon-fiber yarn impregnated with TFE.
- Molded rings of TFE, glass fibers, and compounding material.
- Square-braided blue African asbestos impregnated with TFE.

All packings require some periodic maintenance. Although spring-loaded TFE packing requires less maintenance than others, slight leakage should be expected. "Zero leakage" calls for a bellows seal with leak-off connection and backup packing. Usually this expense cannot be warranted except for hazardous fluids or extremely precious materials.

Lubricants can be used to help reduce friction. Lantern-ring spacers strategically placed on the stem between packing rings allow lubricant access to the stem; lubricant can be added at any time. Lubricant composition is optional, but it must be compatible with both controlled fluid and packing material. For moderately high and low temperatures, silicone fluids are widely used. Above 500°F, however, silicones "coke," or oxidize, and often become more of a problem than an aid in obtaining low friction or a tight seal.

Valve sizing

With the valve configuration and flow characteristic selected, the control valve can be sized. Sizing is exactly what the name implies; it is the technique of determining the valve size *best* suited to control the process. Obviously, a valve too small will not pass the required amount of flow. A valve too large will be unnecessarily expensive and may create trim wear and control problems at very low increments of travel.

Several sizing techniques are available, some in a form designed for calculation by hand-held programmable calculators. This permits convenient calculation of required valve capacity, based on maximum and minimum flow rates of the system and the pressure drop at each condition. For liquids, this capacity is expressed by the coefficient C_v, numerically equal to the number of U.S. gallons of water at 60°F that will flow through the valve in one minute under a 1 psi differential. C_g and C_s coefficients were developed for gases and steam to correct for compressibility effects and critical flow limitations at increased pressure drops.

(Some manufacturers do not use the C_g and C_s coefficients, but employ a conversion method utilizing the liquid sizing coefficient C_v.)

With the manufacturer's sizing technique, it takes only a few minutes to determine the flow coefficient, check for flashing or cavitating conditions in the case of liquid service, or critical flow for gases, and then select a valve with the proper capacity from a sizing chart. Most manufacturers have a logical progression of capacities to aid in selecting valve sizes. Restricted trim is available for applications where future demands for capacities might be greater than today's, or where the designer wishes to avoid a situation which could see an 8-inch line swaged down to a 2-inch valve.

Cavitation

There are three categories of liquid flow through a control valve: cavitating, noncavitating, and flashing. Each can be of concern to the valve specifier, because of their potential for mechanical valve damage and excessive valve noise.

For noncavitating and flashing liquid flow, laboratory tests and field experience show that noise levels are quite low, and so are generally not a problem. Also, proper valve design and material selection can eliminate mechanical damage from these two threats. (As an example, erosion damage caused by flashing liquid can be reduced or eliminated by specifying erosion-resistant valve body and trim materials.)

Cavitating flow, on the other hand, can cause considerable noise, and cavitation damage can make a valve useless. Cavitation must, therefore, receive major consideration in selecting valves for liquid flow.

Cavitation refers to the formation and subsequent collapse of vapor bubbles in the flowing liquid stream. As the liquid flows through the orifice of a control valve, its velocity increases while its static pressure decreases. In many applications, the increase in velocity causes pressure in the control valve to drop below the vapor pressure of the liquid, and vapor bubbles are formed. As the fluid moves downstream within the control valve into a larger flow area, the velocity decreases, with resulting pressure recovery. When the static pressure exceeds the vapor pressure of the liquid, implosions of the vapor bubbles occur, generating extremely high-pressure shock waves that hammer against the valve outlet and piping. Pressures in these collapsing cavities, reported as high as 500,000 psi in magnitude, can cause severe and rapid damage to the valve and piping.

The equation used to determine the maximum pressure drop effective in producing flow through globe valves can also be used to indicate when significant cavitation can occur. Minor cavitation will occur at a slightly lower pressure differential than that predicted by the equation, but should produce negligible damage in most instances. This formula is:

$$\Delta P = K_m(P_1 - r_c P_v)$$

where ΔP = Maximum allowable pressure drop; K_m = Valve-recovered coefficient; P_1 = Inlet pressure (psia); r_c = Critical pressure ratio; P_v = Vapor pressure of the fluid at the inlet condition (psia).

For high-recovery rotary valves, the equation becomes

$$\Delta P_c = K_c(P_1 - P_v),$$

where K_c is a cavitation index.

Solving cavitation problems begins with either controlling the cavitation process or, ideally, eliminating cavitation altogether. Before "anti-cavitation" valve trims appeared, several techniques served—and some still are serving—with various degrees of success.

Erosion-resistant body material, along with hardened valve trim, became a common technique. A sacrificial member, either in the valve itself or in the downstream piping, was another attempt to extend valve or piping life. An example is the "flow-down" angle valve, with a hardened liner in the downstream part of the valve body. The success of this method depends on the energy levels involved in pressure drop and flowrate. If both are high, the sacrificial member may soon disappear.

Another approach to preventing cavitation involves sharing the overall pressure drop, either between valves in series or between one valve and a breakdown orifice device or capillary tubing. If the series method is the choice, complicated control problems may occur, and the initial equipment cost may be prohibitive. The drawback of the downstream-orifice method is that it applies to only a narrow range of flow, and therefore has little application in throttling control.

Strategic location of the valve along the flow path can give effective and economical cavitation control. For example, locating the valve next to a tank and allowing it to flash into the tank eliminates cavitation problems that could occur if the valve were installed elsewhere in the system.

All these methods have limited applicability or limited design life—perhaps both. In the final analysis, the requirement is usually for a single valve/trim combination that will either control the location of cavitation and prevent damage, or eliminate the potential for cavitation altogether.

Pressure drop and flowrate are usually the chief factors in choosing the best valve/trim combination where cavitation is possible. The higher these two factors, the more sophisticated must be the valve/trim package.

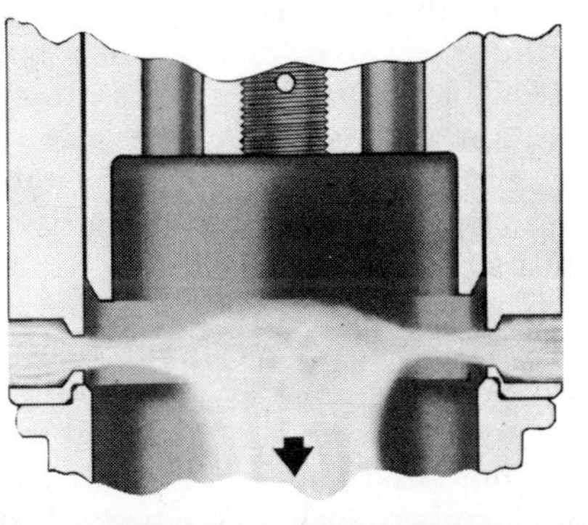

Cage-trim holes form a flow cushion that prevents cavitating liquids from contacting metal surfaces Fig. 5

Criticality of the application and economics quickly enter the picture, so it is advantageous to have several valve/trim styles available, so the best and most economical trim can be matched to the system application.

For applications with low pressure drops and relatively low flowrates, a cage with knife-edge orifices in the cage wall (Fig. 5) was one of the first valve/trim answers. As the valve plug moves away from the seat, more orifices open for the fluid. Each hole's jet aims at the center, to meet the diametrically opposite jet. A fluid cushion surrounds the central core of cavitating liquid, preventing contact of imploding bubbles with metal seating surfaces.

For applications with high pressure drops (1,000 to 3,000 psi), the high energy levels call for a valve/trim combination that eliminates cavitation. If a single control valve serves, the valve trim must "absorb" the pressure drop in stages, so that flowing pressure remains above the vapor pressure of the liquid. Concentric, cylindrical sections, each containing specially drilled orifices, are designed into valves with this style trim (Fig. 6). In operation, each section stages the pressure drop, and the number of stages required depends on the inlet pressure and the total pressure drop across the valve.

This cage design can also be characterized for applications where pressure drop across the valve decreases with increasing valve-plug travel. The characterization gives pressure-staging plus cavitation-prevention at low plug-lifts where it is needed. The number of stages becomes progressively lower as the pressure drop becomes smaller at high travel. Many other designs can prevent cavitation. Most work on some variation of a multiple-stage or tortuous-path principle to break the overall pressure drop into many smaller ones.

Excessive valve noise

Pressures, flows, and temperatures of many processes have increased over the years with a resulting increase in the operating noise level of valves controlling compressible flow. The response by control-valve manufacturers has been to develop noise-reducing valve trims and high-capacity valve configurations.

Understanding the major sources of control-valve noise is the first step to understanding today's noise-abatement technology.

One noise source, the mechanical vibration of valve components, results from either random pressure fluctuations within the valve body, or fluid impingement upon the movable or flexible parts. A more common type of vibration, the lateral movement of a valve plug against its guiding surfaces, produces sound with a frequency of less than 1,500 Hz, often described as a metallic rattling.

Another type of vibration is a valve component resonating at its natural frequency. This generates a single-pitched tone at a frequency of 3,000 to 7,000 Hz. Valve components susceptible to natural-frequency vibration are contoured valve plugs with hollow skirts, and flexible members such as the metal seal ring of a ball valve.

With both types of vibration, noise is usually a secondary concern, since it may be a warning of valve failure. Resonant vibration produces high levels of stress which can fatigue the part. Fortunately, most modern control valves have been designed very conservatively—sturdy, rugged, and well guided—and they suppress vibration to assure a long service life. Vibration-caused noise is a minor consideration today.

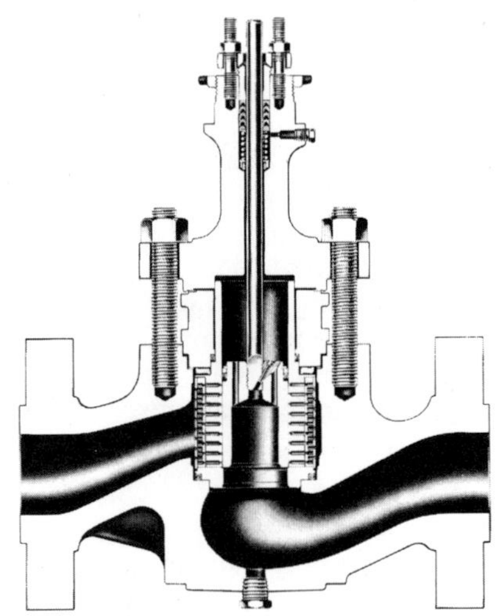

Valve with trim for ΔP of 1,000 to 3,000 psi. Fig. 6

Actuator

Usually, the final step in valve specification is selecting the valve actuator. An integral part of every automatic control loop, it provides the muscle, or motive force, required to position the final control element. And, since stability and operability of the loop hinge on satisfactory actuator performance, the actuator must be able to take control of the many varying static and dynamic forces created by the valve.

For the many available valve styles, there are four basic actuator types suitable for throttling-control: 1. Spring-and-diaphragm; 2. Pneumatic piston; 3. Electric motor; 4. Hydraulic or electro-hydraulic.

Diaphragm actuators

The pneumatic spring-and-diaphragm actuator, a very common, extremely simple design (Fig. 7), offers low cost and high reliability. Diaphragm actuators normally operate with air-supply ranges of 3 to 15 psi or 6 to 30 psi. Because of this, they are often suitable for throttling service using direct instrument signals. Available designs offer either adjustable springs or wide spring selections, for tailoring the actuator to the application. Spring-and-diaphragm actuators have few moving parts that can contribute to failure and, therefore, offer extreme reliability. If they fail, maintenance is simple.

The overwhelming advantage of spring-and-diaphragm actuators is the ever-present provision for fail-safe action. As air is loaded on the actuator casing, the diaphragm moves the valve and compresses the spring.

Pneumatic spring-and-diaphragm actuator **Fig. 7**

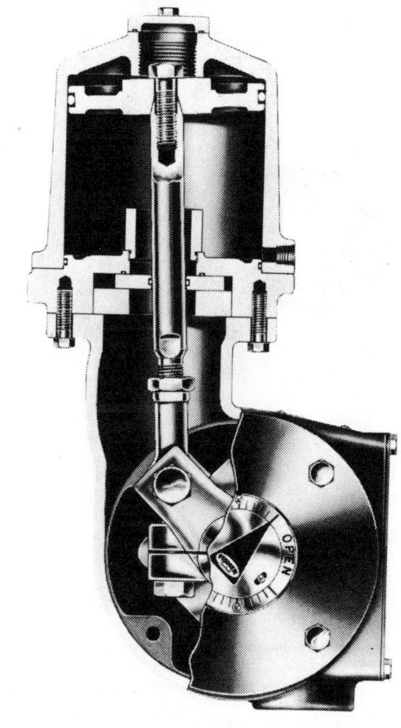

Pneumatic piston actuator for a rotary-shaft valve **Fig. 8**

Stored energy in the spring acts to move the valve back to its original position as air is removed from the casing. In the event of signal-pressure loss to the instrument or actuator, the spring moves the valve to its initial (fail-safe) position. Spring-and-diaphragm actuator designs offer either fail-open or fail-closed action on loss of signal pressure.

The primary drawback of a spring-and-diaphragm actuator is its relatively limited capability. Much of the thrust created by the diaphragm is taken up by the spring and does not result in output. The spring-and-diaphragm actuator ceases to be cost-effective for requirements in excess of approximately 2,000 lb thrust or 5,000 in.-lb torque. Unless there are severe overriding factors, use of diaphragm actuators above this level can be costly. It is simply not economical to build and use diaphragm actuators in this thrust range because the size, weight, and cost grow out of proportion.

Piston actuators

Higher-force requirements than those provided by spring-and-diaphragm actuators, call for one of the other three types previously mentioned. Pneumatic-piston actuators provide the next economical force-output range for operation of automatic control valves. Piston actuators normally work with supply pressures between 50 and 150 psi. Although some feature spring return, this construction has limited capabilities.

Piston actuators used in throttling service must be furnished with double-acting positioners which will simultaneously load and unload opposite sides of the pistons (Fig. 8), causing travel toward the lower-pressure side. The positioner senses the motion of the piston, and when the required position is reached, the positioner equalizes the opposing pressures on the piston, creating equilibrium.

The pneumatic-piston actuator is an excellent choice when compact, high-thrust units are required. It can also serve very effectively where varying service conditions require a wide range of output forces. Constructed primarily of metallic parts with few elastomers, piston actuators adapt readily to situations involving high ambient temperatures or humidities.

The main disadvantages of piston actuators are the high supply pressures required, the requirement for positioners when used in throttling service, and the lack of built-in fail-safe systems. As mentioned, piston actuators can be equipped with spring-return options, but addition of springs limits the construction to much the same force outputs as the diaphragm actuator. The only alternatives to springs are pneumatic-trip systems to move the piston actuator to its fail-safe condition. While these systems prove highly reliable, they add to overall system complexity, maintenance and cost.

Other available double-acting, high-pressure pneumatic actuators use vanes or rubber bladders to create the output thrust or torque directly.

Electric actuators

Electric-motor actuators, useful in many process situations, usually consist of motors with gear trains, and are available in a wide range of torque outputs. They offer a prime advantage in remote installations where no other power source is available.

Electric actuators are economical for normal application in smaller size ranges only. Larger units generally operate slowly and weigh considerably more than their pneumatic counterparts. At present, no economically

Electro-hydraulic actuator provides high thrust Fig. 9

feasible high-thrust electric actuators are available with fail-safe action other than lock-in-last-position. Throttling versions of electric-motor actuators have limitations in both capability and availability. In continuous, closed-loop applications requiring frequent changes to control valve position, the electric actuator may not be suitable, primarily because of its limited duty cycle.

Hydraulic and electro-hydraulic actuators

Electro-hydraulic actuators (Fig. 9) are electric actuators in which a motor pumps oil at high pressures to a piston, which in turn creates the output force. An electro-hydraulic actuator is an excellent choice for throttling service, because of its high stiffness (resistance to changing valve-body forces) and its compatibility with analog signals. Most electro-hydraulic actuators can provide high-thrust outputs (often up to 10,000 lbs). They are, however, handicapped by high initial cost, complexity, and size.

Hydraulic actuators, essentially the same as the electro-hydraulics, differ principally in gaining their power from an external pumping unit. A central hydraulic pumping unit can supply hydraulic fluid under considerable pressure, sometimes up to 3,000 psi. Actuator control is accomplished through a servo amplifier and a system of hydraulic valves. This system can provide the ultimate in actuator performance: exceptional stiffness, fast stroking speeds, very high thrust (often up to 50,000 lb), and excellent dynamic-response characteristics. But the price tag for all this performance is quite high.

Actuator selection

Selecting an actuator for power-operated valves requires looking at both performance and economics. The control valve can perform its function only as well as the actuator withstands the forces exerted on it. Also, an actuator can represent a significant portion of the total package price, especially when used with a small control valve. Careful selection, though, can lead to significant savings. As mentioned before, the wide range of actuator types and sizes might make the selection process seem highly complex. It is not. With a few simple rules in mind, knowledge of fundamental process needs can make the selection process very simple.

Several major actuator characteristics can be of help in making a selection:

Power source—The available source of power at the location of the valve can point directly to what type of actuator to choose. Typically, valve actuators are powered either by compressed air or electricity. (Some designs, however, use water pressure, hydraulic fluid, or even pipeline pressure.) The majority of actuators now use compressed air, and operate at supply pressures from 15 psi to 200 psi.

Since most plants have readily available electricity and compressed air, actuator selection depends on the convenience and cost of furnishing either power source to the actuator location. Other considerations include reliability and maintenance requirements of the power system (and their effects on subsequent valve operation), and provision for backup power to critical plant loops.

Fail-safe characteristics—Although the overall reliability of power sources is high, many processes demand specific valve actions if the power source fails. Fail-safe systems incorporated into many actuator designs automatically shut down the process, thereby preventing possible loss of product in case of power failure. Some systems store energy either mechanically in springs or pneumatically in volume tanks or hydraulic accumulators. Failure of actuator power triggers fail-safe systems to drive the valves to the required position and then hold this position until resumption of normal operation.

Available designs allow choice of valve-failure mode: failing open, failing closed, or holding in the last position. Many actuator systems, such as the spring-and-diaphragm, incorporate failure modes at no extra cost. In other designs, they may be optional.

Actuator capability—An actuator must have sufficient torque or thrust for the specific application. In some cases, torque requirements can dictate actuator type as well as power-supply requirements. For instance, large valves which require a high amount of torque, or thrust, may be limited to electric or electro-hydraulic actuators, because of unavailability of pneumatic actuators with sufficient capability. Conversely, electro-hydraulic or hydraulic actuators would represent a poor choice for a valve with very low force requirement.

Matching actuator capability with valve-body requirements is best left to the control-valve manufacturer. Although the sizing is not difficult, the great variety of designs on the market and the ready availability of vendor expertise (normally at no cost) make detailed knowledge of the procedures unnecessary.

Valve positioners

Pneumatic valve positioners transduce an instrument signal to a valve position, rather than using the air signal to act on the actuator directly. Several studies have shown that positioners are often used where pneumatic amplifiers (volume boosters) would be a better choice. Further, in many cases, better control can be achieved through use of neither an amplifier nor a positioner.

The principal reasons for selecting either a positioner or a volume booster are:

- A split range is required.
- A maximum loading pressure greater than the instrument signal is desired.
- The best possible control is desired. (Examples might include fast recovery from disturbances, or minimization of overshoot.)

Selection of positioner or pneumatic amplifier is loosely related to process dynamics, but not to valve size, unbalance, packing friction, or transmission-line length:

Process type	Positioner or pneumatic amplifier
"Slow" processes Most thermal systems, reactors, liquid-level control, and some low-pressure large-volume gas processes.	Fast positioner, no input restriction. Large-port 1-to-1 pneumatic amplifier may be included in positioner loop for the large actuators.
"Fast" processes Liquid-pressure, small volume gas-pressure and flow processes.	Standard (1/8-in. port) or high-capacity pneumatic amplifier commensurate with actuator size and stroking-speed requirements. Pressure gain of amplifier may be 1 or higher, depending on actuator-loading pressure. Positioner generally not recommended. If used, input restriction should be on the smaller actuators, and may or may not be desirable on larger-sized actuators.

Control functions

The functions of actuator control most clearly define the options available on the actuator-selection process. They include signal type, signal range, ambient temperature, vibration level, operating speed and frequency, and the quality of control required.

Signal groups are generally grouped as being either two-position (on/off) or analog (throttling). On/off actuators are controlled by two-position electric, electro-pneumatic or pneumatic switches. This is the simplest type of automatic control, and therefore the least restricted by actuator hardware.

Throttling actuators have considerably higher technological demands put on them, from the standpoints of both compatibility and performance. The throttling actuator receives its input from electronic or pneumatic instruments which measure the control-process variable. The actuator must then move the final control element accurately and timely, in response to the instrument signal, to ensure effective control. Compatibility with instrument signals is inherent in many types of actuators, and it can be obtained with add-on equipment.

Stroking speed, vibration, and temperature resistance may also be critical to the application. Stroking speed is generally not critical, but flexibility to adjust it is desirable. Fast stroking speeds can be detrimental on liquid loops, because of the possibility of water hammer and its subsequent damage to valve components.

In many cases it is desirable to operate the actuator manually for startup or emergency situations. Most power actuators can be equipped with optional handwheels to allow this.

Actuator weight, when combined with the weight of the control valve, may necessitate support structures for the assembly. Proper selection of compact, light-weight actuators can eliminate this unnecessary expense. Often, economics is an overriding factor in actuator selection. The total cost of an actuator includes not only its initial cost but also the operating and maintenance expenses over its total lifetime. These costs vary widely, but easily lend themselves to qualitative judgments.

A simple actuator with few moving parts is easier to service and will generally cause fewer problems; in addition, maintenance personnel understand them, and can work comfortably with them. An actuator made specifically for a control valve eliminates the chance for a costly mismatch. An actuator manufactured by the valve vendor and shipped with the valve will eliminate separate charges for mounting, and assures easier coordination of spare-parts procurement. Inventory of spare parts, a significant hidden cost, can be minimized by selecting actuators designed with commonality of parts in mind.

Spring-and-diaphragm actuators generally cost less than piston actuators of comparable quality. Part of this saving results from using instrument output-air directly, eliminating the need for positioners or boosters, in many cases. The inherent provision for fail-safe action in the diaphragm actuator is also an important consideration.

If a diaphragm actuator cannot be used for the application, the next best choice would be a pneumatic-piston actuator. This type offers a good combination of high thrust and relatively low initial cost, and the simplicity and maintainability characteristic of pneumatic actuators.

In choosing the type of actuator, the fundamental requirement is to understand the application. Knowing control signal, operating mode, power source available, thrust or torque required, manual-operation requirements, and fail-safe position can make decisions easier. Consider the simplicity, maintainability, and lifetime cost of actuator selected. Safety is another factor not to be overlooked.

The spring-and-diaphragm actuator is the most popular, versatile, and economical type. Consider it first. If the inherent limitations eliminate it, then consider pistons, and then electrics or electro-hydraulics, keeping in mind the capabilities and limitations of each (Table IV).

Installation tips

The consideration of line size vs. valve size involves primarily the strength of the valve in relation to the strength of the adjacent piping. The two rules of thumb

Summary of actuator advantages and disadvantages Table IV

Pro	Con
Spring-and-diaphragm	
Low cost	Limited torque availability
Simplicity	Limited temperature range
Inherent failsafe action	Inflexibility to changing service conditions
Low supply-pressure requirement	
Adjustability	
Maintainability	
Ability to throttle sans positioner	
Fast stroking speeds possible	
Pneumatic piston	
High torque capability	Failsafe requires accessories
Compactness	Positioner required for throttling
Light weight	Higher cost
Adaptable to high ambient temperatures	High supply-pressure requirement
Adaptable to varying valve-torque requirements	
Fast stroking-speed possible	
Relatively high actuator stiffness	
Electric motor	
Compact	High cost/torque ratio
Suitable for remote applications	Lack of failsafe action
	Limited throttling ability
	Slow stroking-speed
	Lack of adjustability
Electro-hydraulic/hydraulic	
High torque	High cost
Very high actuator stiffness	Complexity
Excellent throttling stiffness	Large size and weight
Fast stroking-speed	Failsafe action requires accessories

commonly used in this regard are: valve size not less than one half pipe size; and valve size not less than two sizes below line size. Either rule may be used with the assurance of safety.

Control valves are often installed with little thought as to how the piping arrangement immediately adjacent to the valve will affect its performance. Process-control engineers carefully size each valve and specify suitable valve-plug characteristics, but then the actual installation layout is turned over to piping designers who may give little thought to the control functions of the valve. Allowance for pressure drop through the adjacent block valves and piping is often neglected.

It is regrettable that in many instances the care used in specifying control valves in the engineering stages of a project, and the efforts of manufacturers to provide precise valve characteristics and capacities, are partially nullified by poor installation practices. The following general recommendation will help to assure optimum control-valve performance:

1. Avoid arrangements which amount to a nozzle-effect into the inlet of a control valve, as this will alter the valve characteristic adversely.

2. Avoid close-coupled inlet-block valves reduced from line size, since they affect both characteristics and capacity.

3. If a block-and-bypass arrangement must be used around a control valve, the valve is best placed in a straight-through section. Avoid the tortuous manifolds sometimes provided for control valves to achieve accessibility. They may have an adverse affect on valve characteristic and capacity.

4. Avoid sharp turns close to the valve inlet. Straight-line sections into and out of the control valve should, as closely as possible, simulate the piping required to originally establish flow capacity and characteristic test data. These test data are based on straight sections of approximately 5 nominal pipe diameters downstream of the valve and 12 diameters upstream for reducers and expanders; 13 pipe diameters upstream to the outlet of a tank; 18 pipe diameters upstream for elbow fittings oriented in the same plane; and 30 pipe diameters upstream for elbows oriented in a different plane.

While these recommendations tend to be ideal in nature, following them, and taking care in selecting the piping arrangements around the control valve, can help in getting the best performance from valve selection.

References

1. Control Valve Handbook, 2nd ed., Fisher Controls Co., Marshalltown, Iowa, 1977.
2. Coughanowr, L. B., and Koppel, L. D., "Process System Analysis and Control," McGraw-Hill, New York, 1965.
3. Harriot, P., "Process Control," McGraw-Hill, New York, 1964.
4. Lloyd, S. G., and Anderson, G. D., "Industrial Process Control," Fisher Controls Co., Marshalltown, Iowa, 1971.
5. Lloyd, S. G., Generalized Control Theory, TM-8, Fisher Controls Co., Marshalltown, Iowa.
6. Murrill, P. W., "Automatic Control of Processes," International Textbook Co., Scranton, Pa., 1967.
7. Schuder, C. B., The Dynamics of Level and Pressure Control, TM-7, Fisher Controls Co., Marshalltown, Iowa.
8. Schuder, C. B., "Control Valve Characteristics," *Instr. Control Syst.,* Mar. 1967.

The authors

James (Jack) A. Carey is assistant manager of the Chemical Sales Group at Fisher Controls Co., Marshalltown, IA 50158. He has also held positions in the refining, petrochemical, and pulp and paper sales areas. He holds a B.S. in industrial technology from Texas A&M University.

Donn Hammitt is Product Manager, Control Valves, within the marketing organization of Fisher Controls Co., where he has had both marketing and valve-design experience. He holds a B.S.M.E. degree from Iowa State University and an M.S.M.E. degree from the University of Nebraska. He is the Fisher representative to the Valve Manufacturers Assn.

Control Valves

The control valve is a critical element in pumped liquid systems. A knowledge of flow characteristics, both actual and theoretical, is essential for proper sizing. Optimum system design requires that process and instrument engineers work together closely.

ORVAL P. LOVETT, JR., E. I. du Pont de Nemours & Co.

The use of a control valve implies that the process has an automatic control system of some sort. It may be a level, flow, temperature, pressure or other type. It will usually involve the typical elements of a closed loop, feedback control system consisting of sensor, transmitter, controller, control valve and the process itself. Selection of the control valve requires an understanding of all of these factors, especially the process.

A valve selected as optimum for a level control process might not be the best selection for a flow control system. Also, the best valve for one flow control system might not be optimum for a system utilizing a different primary element or flow measurement means. This article discusses the various components of a control valve, especially as they are influenced by the total control system.

Control valves are used in many applications including liquid flow control, gas pressure reduction, steam flow to heaters, etc. This discussion is limited to liquid pumped systems and covers body and actuator styles, body and innervalve sizing, allocation of pressure drop to the control valve, inherent and installed valve flow characteristics, and materials of construction. It emphasizes the systems approach to pump, piping and control valve sizing.

A control valve consists of two major parts: body and actuator (Fig. 1). The body sub-assembly contains the flowing fluid, while the actuator's function is to respond to the automatic controller's signal and move the inner valve to cause variation in fluid flow.

Originally published October 11, 1971

Valve Body

The control valve body contains the fluid between pipe and connections. It must therefore serve as a pressure vessel and is subject to the same conditions of temperature, pressure and corrosion as the balance of the piping system.

Valve bodies are available in a wide variety of styles. They may be broadly classified as those with

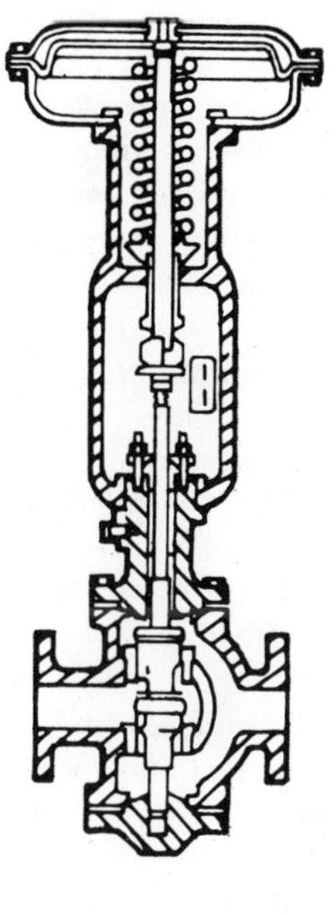

CONTROL VALVE with diaphragm actuator—Fig. 1

linear stem motion and those with rotary stem motion. The former include single seated globe style, double seated globe, split body globe, 3-way, angle body, diaphragm type valves, and many variations of these. Valves with rotary stem motion include the buttterfly, ball valve, and their variations.

Many other special forms of control valve bodies are available for applications in very small flows, cryogenics, noise reduction, etc. Reference sources for information include manufacturer's literature and handbooks.[1, 6, 7]

While the control valve body serves primarily as the containment vessel for the fluid, it must also house the inner valve mechanism and mechanically support the actuator and accessories. It contains moving parts that may wear, corrode, erode, etc. Thus it is important that the body can be removed from the piping system or that its interior be accessible for maintenance. Usually the body has flanged end connections. Flange ratings to match the piping system specifications are available.

Valve body face-to-face dimensions follow uniform standards. Instrument Society of America Standard RP4.1 applies to valve sizes up through 8 in. while Fluid Controls Institute Standard FCI 65-2 includes sizes 10 in. through 16 in. They both include 125 through 600 lb. ratings. Another face-to-face dimensional standard sometimes used is American National Standards Institute B16.10, "Face to Face and End to End Dimensions of Ferrous Valves." When installing control valves it is necessary to know which dimensional standard applies.

Control valves are also available with screwed and welded ends. Flangeless end connections are available from some manufacturers. In this design, the end of the valve matches the raised face section of the pipeline mating flange to provide a gasket seating surface. The valve is held between a pair of pipeline flanges with long stud bolts. Economy results from the reduced amount of material in the valve body. The face-to-face dimensional standards don't apply for this type, however.

Butterfly valves are economical in the larger sizes. They are generally used for low pressure drop and high flow rates. Ball valves including their many variations, are also used as throttling control valves.

Materials of Construction

Materials of construction are important in a control valve body and its trim components. Those parts in contact with the fluid must, of course, be compatible with that fluid from the standpoint of corrosion resistance. Valve bodies are available in all castable metals and alloys. Plastics and linings are also widely used. Manufacturers and their literature form excellent source material for references on materials of construction. There are other excellent references as well.[2, 10]

Corrosion is not the only factor of concern. Cavitation within the valve body and trim can have a very damaging effect. It usually occurs together with high pressure drop, although it actually can happen whenever the fluid vapor pressure is approached within the valve body. This may take place at the point of maximum fluid velocity near the seat of the valve. In pumped systems, where pump cavitation is also to be avoided, and where control valve pressure drop is low, valve cavitation is normally not a problem.

Usually the inner valve is made of a material that is more corrosion resistant and harder than the body. This minimizes effects of cavitation at the valve seat, permits construction with close tolerances to reduce seat leakage, and also aids in stem guiding. This subject is thoroughly treated in Ref. 3.

Valve Actuators

The control valve actuator responds to a signal from an automatic controller and moves the inner valve. It is the muscle of control or the power amplifier between controller and fluid flow. There are two basic pneumatic actuator designs. One uses a spring and diaphragm construction; the other is a piston within a cylinder. There are other kinds, as well as many variations of the basic pneumatic types.[4]

The spring and diaphragm actuator is usually less expensive than the piston type, especially where a valve positioner is not required. A positioner is an accessory with a motion feedback mechanism from actuator stem travel, a sending system from the controller output, and air output to the actuator. Much has been written on positioners.[5]

It is outside the scope of this article to include a complete treatment of positioners. They are sometimes necessary in cases of long transmission lines between controller and valve, to reduce valve hysteresis, to utilize flow characterized cams, to operate pressure balanced cylinder actuators and to provide an interface between electronic control systems and pneumatic valves. There are alternate means of accomplishing many of these functions but the valve positioner is popular and used frequently.

The Inner Valve

The functions of the inner valve include: 1) provide a variable restriction within the valve body to cause changes in fluid flow; 2) characterize the flow with respect to travel; and 3) provide some degree of flow shutoff when in the fully closed position. To accomplish these functions, there are more design variations of inner valve parts than there are in valve bodies.

A variable restriction is obtained in two general ways, which are comparable to the valve body classifications of linear and rotary motion. In both cases there is a relationship between motion and flow that is called "Flow Characteristic." This is a general term and should always be designated as either the inherent flow characteristic or installed flow character-

Inherent flow characteristic is that produced with a constant pressure drop across the valve. Three commonly used inherent flow characteristics are linear, equal percentage and quick-opening. These are illustrated in Fig. 2. Mathematical descriptions and theoretical curves can be found in Ref. 8.

Installed flow characteristic is the characteristic when the pressure drop across the valve varies as dictated by flow and related conditions in the system.[9] It is the factor of ultimate interest to users and is determined by measuring pressure drop variation as a function of valve travel. The physical make-up of the system in which the valve is installed determines this relationship, as described in the following section.

Pump and Valve System

A pumped liquid flow system can have many configurations. A typical arrangement in Fig. 3 includes a suction tank, pump, control valve, discharge tank and connecting piping. The suction tank is at pressure P_S and has an elevation head Z_1. Piping from tank to control valve has a pressure drop ΔP_{L1}. The pump has a characteristic flow vs head curve. The control valve has a pressure drop ΔP_V, and the piping after the control valve has a pressure drop ΔP_{L2}. The discharge vessel has a pressure P_D and an elevation head Z_2. Some of these factors vary with flow rate, while others are fixed and independent of flow rate.

These relationships are seen by a plot of pressure vs. flow in Fig. 4. The elevation head Z_2-Z_1 is constant and for this process we will assume that the pressure

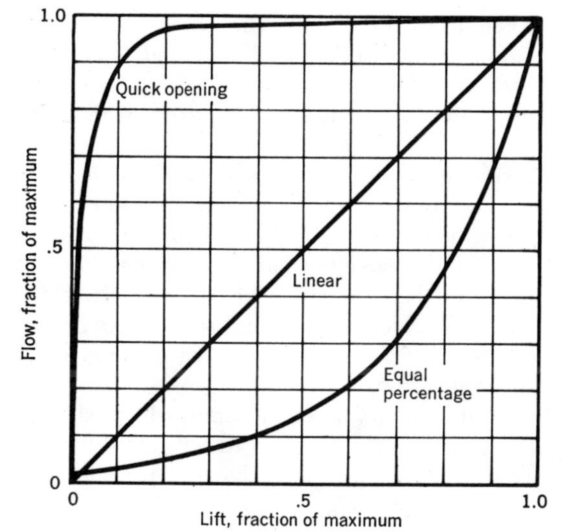

THREE INHERENT flow characteristics of valves—Fig. 2

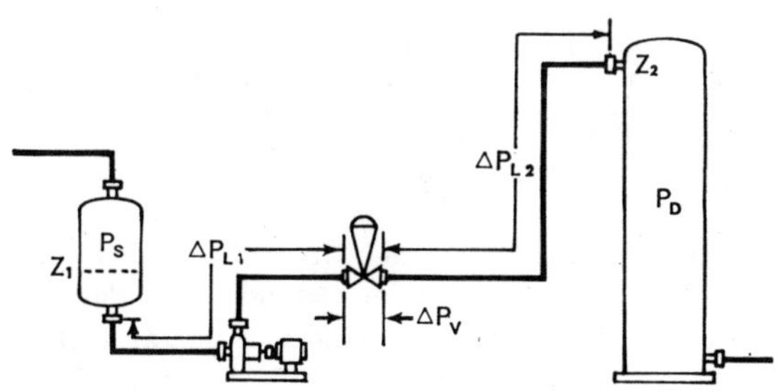

TYPICAL LIQUID flow system—Fig. 3

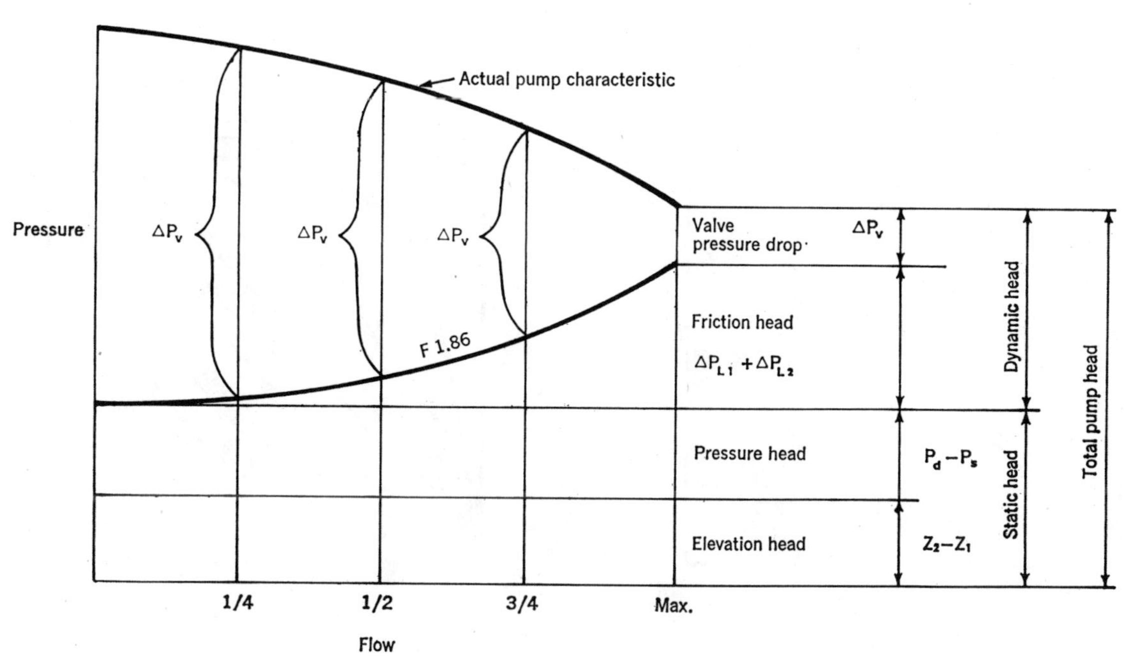

PRESSURE relationships in pumped liquid system—Fig. 4

head P_D-P_S is also constant. (If it varied, it is clear that it could be so plotted.) The pump characteristic curve and pipeline friction head is plotted. The control valve pressure drop is then obtained by subtracting the sum of the various pressure heads from the pump curve.

The situation shown is a typical case where the control valve is sized for maximum flow rate with a relatively small pressure drop. When the system is turned down, the valve must absorb considerably more pressure drop than that for which it was sized. The installed flow characteristic and the inherent flow characteristic are different due to the varying control valve pressure drop.

Fig. 4 also shows a relationship which shall be called the valve pressure drop ratio, P_R. It is the ratio of valve pressure drop to total dynamic pressure drop.

$$P_R = \Delta P_V / (\Delta P_V + \Delta P_{L1} + \Delta P_{L2})$$

The static head is not a factor here. The valve pressure drop ratio is a very important consideration in the design of the pump and piping system, and in control valve sizing.

Calculations of the installed flow characteristics for a linear valve are presented in Fig. 5. Actual and theoretical inherent flow characteristic for an equal percentage design are plotted in Fig. 6. In each figure, several values of pressure drop ratio are shown. It is apparent that the higher this value, the less the distortion. By use of these figures and a calculation of the valve pressure drop ratio, one may see at a glance the installed flow characteristic for his system.

Fig. 6 illustrates another reality when dealing with flow characteristics. This is the distinction between theoretical and actual characteristics. The theoretical characteristic may be described mathematically. The actual inherent flow characteristic is obtained by testing the valve after manufacture. It may match the theoretical within ±5%, or may be off by 40 to 50%. The values in Fig. 6 are from a specific case history and should not be construed as typical for all valves. They do serve to illustrate the degree of accuracy one can expect.

Allocation of Pressure Drop

A subject of utmost importance in the design of pumped liquid systems is the allocation of pressure drop to the control valve. At one extreme is the desire to have minimum distortion of the inherent flow characteristic. Maximum pressure drop in the valve accomplishes that objective. However, pumping against a higher than necessary head increases costs. Thus from the economics standpoint, it is desirable to operate at the lowest possible value.

The true answer to the optimum pressure drop can be found by a dynamic analysis of a complete system. If the engineer knows the type of control, control tolerances, pump and piping system characteristics, he can measure or calculate time constants and predict the optimum control valve flow characteristic (gain). Short of performing a complete dynamic analysis, the experience gained from previous systems provides some helpful rules.

The problem almost disappears in those systems where there is essentially no turndown. With almost any smooth characteristic the change in valve gain is small. Also, if turndown requirements are greater but

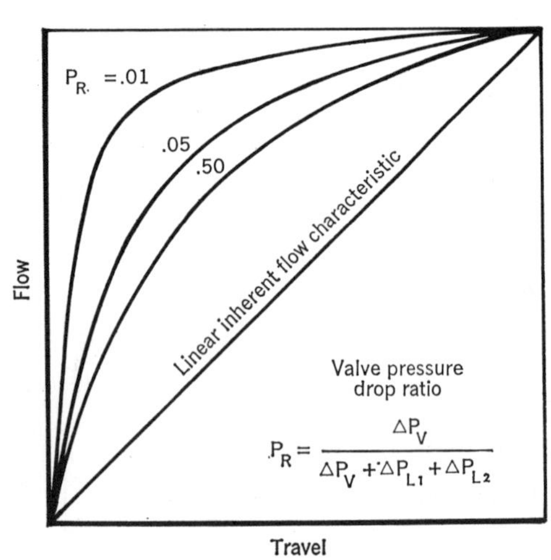

INSTALLED characteristics for a linear valve—Fig. 5

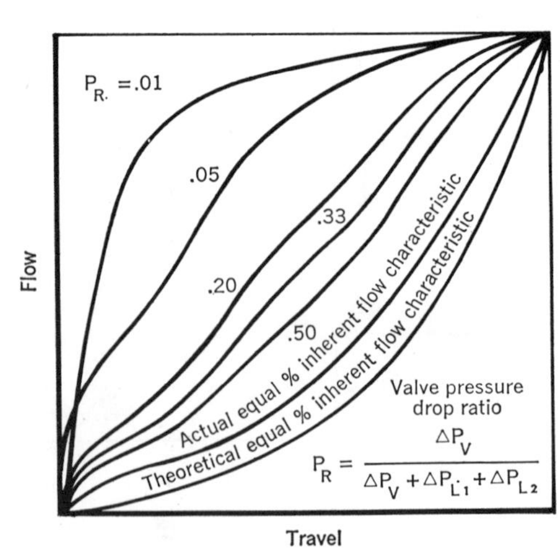

ACTUAL AND THEORETICAL flow characteristics—Fig. 6

done infrequently, or if for other reasons it is possible to retune the controllers, there is no problem. Retuning will permit good controllability at all flow settings.

In both of these cases any of the commonly available flow characteristics would be satisfactory. It would not matter appreciably how much the installed flow characteristic varied from the theoretical. The pressure drop ratio could be as low as 0.05. Practical considerations, which include the accuracy of calculating pipeline pressure losses and predicting pump curves, are the only factors preventing a P_R of 0.01. The problem become one of valve size prediction. A small error could result in a completely wrong size.

The above situation is not a typical one. Most commonly it is necessary to cover a wide range of flow rates (large turndown) without the opportunity for retuning the controller. It is in such cases that the actual installed flow characteristic becomes important.

The first step is to determine what flow characteristic is optimum. To illustrate the method, we will assume that a linear characteristic is desired. This is a reasonable expectation in many systems involving level and pressure control. Inspection of Fig. 6 reveals that the curve at $P_R = 0.20$ is reasonably linear from about 20 to 80%. This is the range of operation of a properly sized control valve. Thus a valve with 20% of the system drop at the maximum flow condition and having an equal percentage flow characteristic should be selected.

Control Valve Sizing

Having obtained the control valve's pressure drop allocation from pump head available, the engineer can proceed to size the valve. The other factors involved are flow rate and liquid specific gravity. The valve size coefficient, C_V, may be calculated by the equation:

$$C_V - \text{GPM}\sqrt{\frac{\text{Sp. gr.}}{\Delta p}}$$

This applies for most liquid sizing problems. There are correction coefficients for piping geometry, choked flow and Reynolds number. Turbulent flow usually exists in a pumping system and the valve pressure drop is likely to be low. Thus the piping geometry factor (effect of reducers) is the only item that is significant. This correction may also be calculated as part of the piping system friction loss if valve size is approximately known, or if the piping losses and valve size are recalculated after a trial computation. ISA Committee SP39 continues to work on refinements of the equation.

There are cautions to observe in sizing. Experience teaches that when control valve size is wrong, it is mostly in the direction of too large a valve. This can easily happen where there is uncertainty in piping and equipment pressure losses. There is a tendency to "be safe" and attribute more loss to the piping than might actually occur.

The pump application engineer selects a pump head to be adequate for the pipe losses, static head considerations and valve pressure drop. He may wish to add a little more head "to be safe" and then still more in order to match an available standard impeller. The net result is considerable excess head. Fortunately, the control valve is able to accommodate the excess pressure drop, but does so by considerably reducing its travel. This reduces the rangeability of the valve and it may not be able to throttle the flow at lower throughputs. These problems become especially troublesome at start-up when low throughputs are frequently encountered.

The solution is for all persons involved to avoid hidden safety factors in sizing calculations. Instead of adding a pressure drop safety factor, it is preferable to check the control valve percentage opening at various conditions. If an equal percentage valve is selected and sized for 80% travel at flowsheet maximum, it can pass 2.2 times maximum flow at the wide open position. Taking advantage of such flexibility and sizing valves "neater" can result in significant cost savings.

Pumps, piping systems, equipment and control valves should be selected as a complete system to obtain optimum performance and lowest total cost. In particular, the control valve size, flow characteristic, materials of construction and hardware configuration cannot be optimized without considering the complete system. Process and instrument engineers must cooperate and communicate fully during the course of the job to achieve this objective.

References

1. "ISA Handbook of Control Valves," J. W. Hutchison (Ed.), Instrument Society of America, Pittsburgh, Pa., Chap. 3.
2. Ibid., Chap. 9.
3. Ibid., Chap. 4.
4. Ibid., Chap. 7.
5. Ibid., Chap. 8.
6. "Final Control Elements," Chester S. Beard, Chilton Book Co., Philadelphia.
7. "Instrument Engineers Handbook," Volume II, Chap. I, B. G. Liptak (Ed.), Chilton Book Co., Philadelphia, 1969.
8. ASME Standard 112, "Diaphragm Actuated Control Valve Terminology."
9. Lovett, O. P., "Valve Flow Characteristic," ISA Paper 46.1.63.
10. Engineering Materials Deskbook, *Chem. Eng.*, Oct. 12, 1970.

Meet the Author

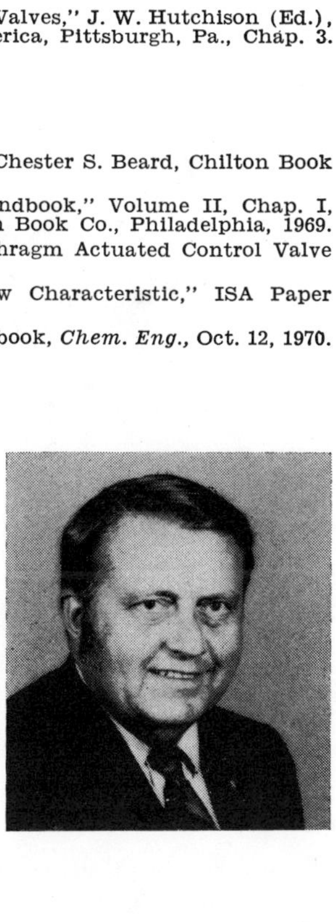

Orval P. Lovett, Jr., is instrumentation section supervisor in the design division of the engineering department, E. I. du Pont de Nemours & Co., Louviers Bldg., Wilmington, DE 19898. He is a civil engineering graduate of Drexel University and has been with Du Pont for the past 20 years. Mr. Lovett is chairman of ISA's Standards & Practices Committee SP32 and control valve sizing director of the Process Measurement and Control Div. of ISA. He has also been president of the ISA Wilmington Section.

Special Control Valves

High-velocity fluid flow through conventional control valves causes erosion, vibration and noise in the valve and pipeline. Here is an analysis of how to handle these problems, and an evaluation of some new control valves and other devices that can reduce these problems or eliminate them at the source.

J. B. ARANT, E. I. du Pont de Nemours & Co.

The conventional plug-and-orifice control valve is subject to severe maintenance and operational problems when forced to operate under conditions of high fluid velocity. This is also true of other valve configurations, such as ball and butterfly.

High velocity is the big problem in control valves, such as the angle valve shown in Fig. 1. It results from the conversion of pressure head to velocity head as the fluid undergoes a pressure drop while flowing through the valve orifice.

In liquid services, the problem arises when the velocity through the valve trim exceeds about 300 ft./sec. In gas or vapor services, it arises when the velocity exceeds about Mach 0.4. At these velocities, one or a combination of the following problems occurs: (1) erosion, (2) erratic control, (3) vibration, (4) shutoff leakage and (5) noise. Let us examine each of these problems in detail in order to define their parameters.

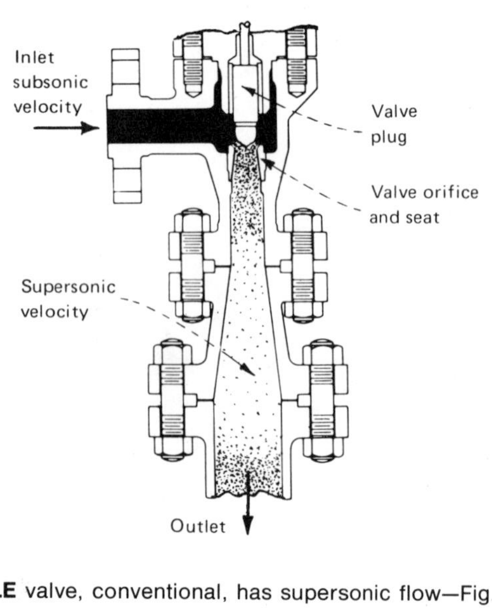

ANGLE valve, conventional, has supersonic flow—Fig. 1

Originally published March 6, 1972

Erosion

Erosion can result from liquid flow, by high-velocity "washing" of the metal surface whereby the protective film is continuously removed and the metal wears away. It also occurs in fluids containing solids particles such as scale, catalyst fines and liquid droplets, whereby the metal surface is abraded away.

In some cases, a special type of surface damage is caused by cavitation. While the exact mechanism of cavitation is complex and not completely understood, the following is generally accepted:

As the pressure head of a liquid is converted to velocity head through the valve trim, it will reach a zone of maximum velocity and minimum pressure called the vena contracta. If this minimum pressure is below the vapor pressure or gas-solution pressure of the liquid, localized flashing and bubble formation will occur. If this were as far as the phenomenon would go, there would be no problem because the bubbles do not cause damage. However, since pressure recovery occurs downstream in the valve, the bubbles quickly collapse in an implosion. The resultant energy dissipation can impose concentrated surface stresses on the order of 100,000 psi. or more. These act like a myriad of small, high-energy hammer blows that will chip metal particles from the surface to form the easily recognized "sand blasted" or "termite hole" damage. Even hardened-trim materials cannot withstand these stresses for long.

One type of valve design that has been used for cavitation situations is the "swiss cheese" or "flash flow" trim. While extremely helpful, this trim does not always solve the problem.

Reduce Noise and Vibration

Generally, clean dry gases do not cause erosion problems, but the presence of liquid droplets or solids can cause severe erosion in a relatively short time. This is highly accelerated when the valve is leaking or operating near the seat. Such erosion is called wiredrawing.

Steam is particularly notorious for wiredrawing since it is often in the saturated or wet condition. Even dry, superheated steam is highly erosive, though we would expect it to behave like any dry gas. The explanation for this is a reverse situation to cavitation, with liquid droplets formed in the vapor (Fig. 2).

Here, inlet pressure, P_1, drops in isentropic throttling to the pressure in the valve throat, P_3, where it is below the saturation temperature. This causes water droplets to form, and creates the same erosion situation as does wet steam. As pressure recovery takes place, the water returns to the vapor state, and dry steam leaves the valve at pressure P_2. Though reduction in steam pressure seems to be isenthalpic, actually it is not. The high velocities generated, the water droplets formed and the high fluid temperature, all combine to play havoc with the valve trim. It is not uncommon for a valve used in reducing 1,500 psig. steam to the 300 to 500-psig. level to be destroyed internally in a matter of days.

Erratic Control

Erratic control can result from excessive velocity, which may be due to one or a combination of factors. Extremely high velocities will produce turbulence that causes erratic side-loadings on the stem and plug. Changes in the relationship between velocity and static pressure across the plug face can seriously alter the force-balance relationship in the vertical direction. These situations can:

1. Alter the concentric plug-to-seat relationship, which will vary the flow characteristic.
2. Cause high-frequency vibration and wear, which will destroy the inner-valve components.
3. Create vertical instability in the valve operator, which will produce vertical oscillations.
4. Cause variable flashing or cavitation in liquids, which will give poor control.

Vibration

Vibration is often the result of high velocity both within the valve and in the piping. Such vibration will cause valve failure and, worse, may lead to piping failure.

Pipe vibrations, downstream of a high-drop gas-control valve, with a measured frequency on the order of 2,500 Hz. and an amplitude of 100 mils (1 mil = 0.001 in.) have been personally observed. Needless to say, this caused line failure. Attempts to dampen the vibration with an overwrap of lead sheeting resulted in destruction of the sheeting in less than 24 hr. Correction required complete redesign of the line's configuration, an increase in wall thickness of the pipe, bracing of the pipe, and lead-sheet wrap. The correct answer would have been to reduce or eliminate the high-velocity energy from the valve, which was the vibration source.

Shutoff Leakage

Loss of shutoff capability is the end-result of several of the problems previously mentioned. Erosion, by whatever means, will destroy the seating and shutoff surfaces. Vibration may peen and distort the plug and seat. In many cases, it is a cannibalizing situation, since the more the valve leaks, the worse it leaks. And, high velocity often limits the ability to use a soft elastomer

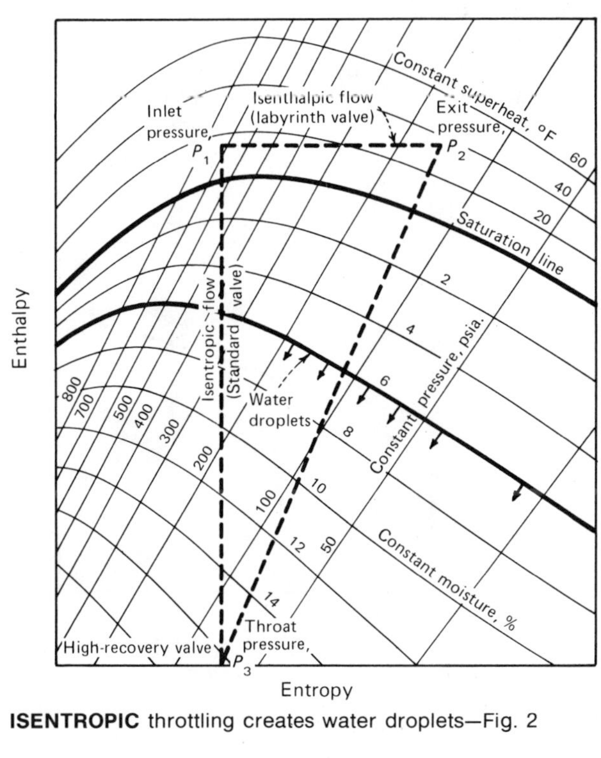

ISENTROPIC throttling creates water droplets—Fig. 2

or Teflon® seat to effect tight shutoff because of inability to retain these seals mechanically.

Noise in Control Valves

Noise and velocity go together. There are three main sources of control-valve noise: mechanical vibration, cavitation, and aerodynamic action.

Mechanical vibration is caused by turbulence generated by velocity and/or large massflows, and is usually unpredictable. It is most common with double-seated valves, but single-seated valves are not immune. Noise levels are usually low and in the 50 to 1,500-Hz. range. Frequencies above this are usually due to excitation of the valve's internals at natural resonance. This resonance is often very noisy and may be up in the 90-dbA range, with a frequency up to 7,000 Hz. This may be acceptable as noise, but is very likely to lead to unacceptable valve failures.

Cavitation can be noisy depending upon the degree present. Cavitation noise will increase with increasing pressure drop to some maximum value and then decrease to some lower magnitude. This is due to the progressive reduction in the downstream pressure recovery, which causes bubble collapse. Equations for predicting cavitation and for calculating the attendant noise are available in one vendor's valve handbook.[1] Noise levels during cavitation are generally in the 90 to 100-dbA range, but can be higher.

Aerodynamic noise is the commonest and worst source of valve noise. It is caused by fluid turbulence and shock waves due to high velocity or massflow of gas, and can be readily predicted as to occurrence and magnitude.[1,2]

While Mach 1.0 velocity (or sonic flow) guarantees noise, high-noise levels can be generated when Mach number is as low as 0.4 and there are relatively low pressure drops where large massflows are handled (Fig. 3).

However, the noise that affects us most is that generated by sonic or near-sonic flow. As an example of the tremendous power encountered, noise levels up to 140 dbA have been measured near valves. Natural-gas letdown from 4,000 psig. to 1,000 psig. generates about 5,000 w. of sound pressure level. About 4,950 w. will be absorbed by the valve body and pipe wall. The 50 w. escaping will yield about 110 dbA at 3-ft. distance. If the letdown had been a free jet or vent, the sound pressure level would have been greater than 150 dbA. For comparison, the noise level of a jet aircraft as it leaves the terminal is about 117 dbA.

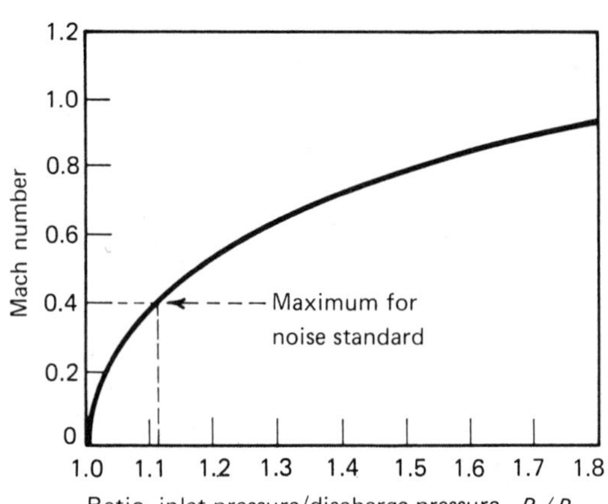

SONIC or near-sonic flow causes noise—Fig. 3

Anyone who has been around high-noise-level valves knows how intolerable this can be. Temporary symptoms may be dizziness, nausea, headache and deafness. Prolonged exposure can give permanent hearing damage. Even at a distance, such noise can be extremely wearing to the nerves and temper, and can cause self-generated safety hazards.

Noise pollution is drawing more attention from federal, state and local authorities, and increasingly stringent regulations concern not only the worker exposed to noise, but also the effects of noise on nearby residential areas. It behooves industry to recognize the impact of this trend and to bend every reasonable effort to eliminate or alleviate such pollution.

Valve noise is one of the major offenders in many plants. Its elimination is possible with current technology and is a big step in meeting noise standards. A guideline that is becoming accepted as a valve-noise standard is that the sound pressure level should be equal to, or less than, 85 dbA when measured 3 ft. from the source. While this is not exactly quiet, it is acceptable for hearing conversation. This criterion normally requires that gas or vapor velocity be limited to less than Mach 0.4 and, preferably, no more than Mach 0.3.

Much work has been done in predicting aerodynamic noise, and appropriate information is readily available.[1,2] A quick and approximate way to start is with the equation: $P_1C_v = X$, where P_1 is the upstream pressure, psia., and C_v is the valve-capacity coefficient at maximum service flowrate. If X equals or is less than 1,000, the maximum noise level is about 90 dbA. If X has a value of more than 1,000, more-definitive calculations are required in order to assess the problem. If the noise prediction is 90 to 100 dbA, there are several approaches possible. These are based primarily on path treatment rather than on source treatment.

1. Evaluate effect of distance. Maybe the valve's location on top of a structure or a pipe bridge indicates there would be no problem for personnel at their normal working locations. This can be checked by using the equation for reduction in sound pressure level (SPL):

$$SPL_{distance} = SPL_{source} - 10\log\left(\frac{\text{distance in feet}}{3}\right)$$

2. Increase wall thickness. Doubling the pipe wall thickness reduces escaping sound by 5 dbA. Adding one inch or more of acoustical insulation will reduce sound level another 10 dbA. However, keep in mind that sound travels down a pipeline, and increasing the pipe wall thickness and/or applying acoustical insulation may be an expensive solution.

3. Use a silencer. This can also be expensive and is not always effective. Inlet velocity must be subsonic and the silencer should not be forced to act as a pressure

reducer. The silencer itself can act as a sound propagator if the shell-wall thickness is too light and vibrates.

4. Try to limit exit velocity at valve so that it is equal to or less than Mach 0.3. Even if we expand to a larger diameter of pipe, exit turbulence will concentrate at the expander and generate noise. It is a common fallacy that we can eliminate noise by reducing the exit velocity at the valve by pipe enlargement.

If the noise-prediction calculation is equal to or greater than 100 dbA, then we have a problem of the worst order. The only recourse is to go to source treatment.

Resolving the Problems

Until recently, all we could do was to install one of a variety of hardened-trim materials, strengthen the stems and guides, and hope for the best. We might try using an angle-, Y-, or S-body configuration, or a different trim type, or reversing the body, etc. All we were doing was to try to live with the problem and stretch out the service life of the valve. If any of our changes increased valve life, we counted it as a success and it passed from the realm of an engineering problem to a maintenance one. It then became a way of life, accepted as such, and often we denied we had a problem anymore.

In the last three or four years, some new control-valve designs and other devices have come on the market. None is an absolute cure-all, although one comes close.

In general, any special design can be more expensive than a standard control valve by a factor of two to five. For larger valves or for higher temperatures and pressures, the cost difference between custom-built conventional valves and the special designs may be very modest. There are many benefits to be gained from special designs, such as reduced maintenance costs, increased service life, improved control, and reduced noise levels. We have many cases where the entire cost of the special valve was absorbed by elimination of vent mufflers or line silencers or support steel, by reduced piping complexity, and vastly reduced space requirements. The point is, you can not balance valve cost against valve cost.

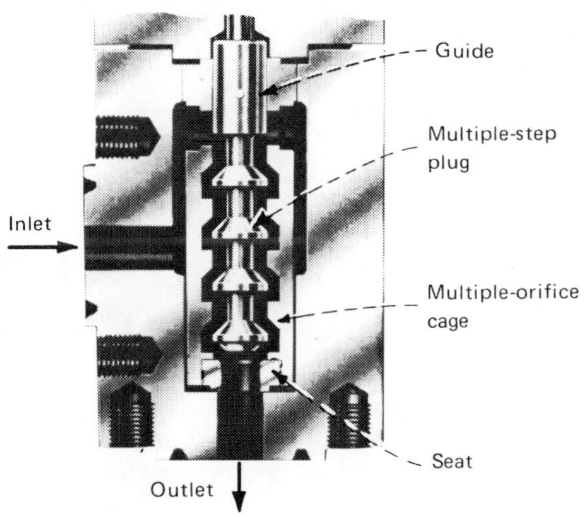

STAGED-TRIM valve distributes pressure drop—Fig. 4

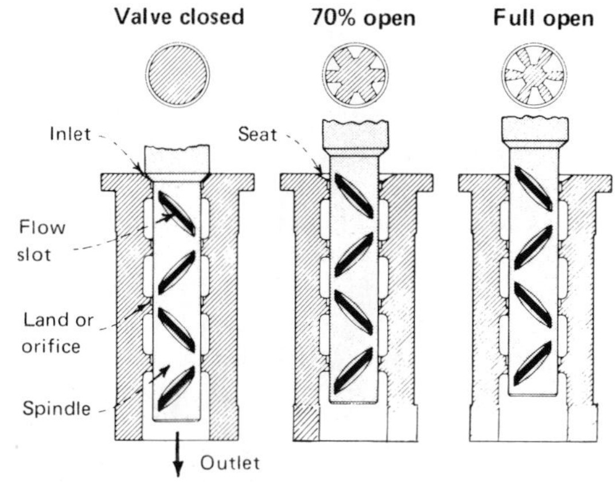

MILLED slots on plug provide staging—Fig. 5

Multiple-Stage Design

The first design is the staged trim (Fig. 4), which is a plug-and-orifice valve in a series or multistage configuration. There are several varieties available from various manufacturers for liquid service. One manufacturer also offers a version for gas or vapor service.

The virtue of the staged trim is that instead of taking the total pressure drop in one "fell swoop," it is distributed across the several plugs and orifices, or stages. These vary from two to a maximum of about six.

For best results, the pressure drop per stage should be limited to about 600 to 700 psi. Under these conditions, the velocity through each stage is considerably lowered. However, each stage still acts as a high-recovery valve, which generally means higher than desirable throat (or trim) velocities due to the efficiency of the nozzle-type trim. Therefore, staged valves are not immune to erosion or cavitation, although these are usually reduced in magnitude. Such valves are generally limited in size to 2 or 3 in. maximum, using pneumatic operators. This is due to power limitations as a result of their unbalanced or semibalanced design. They can be subject to vertical instability in the larger sizes. Flow coefficients, C_v, are on the order of one-third to one-half those of an equivalent-size standard control valve, and turndown, or flow-control rangeability, is much lower. These valves are large and massive for their pipe size and capacity, due to body-design requirements to contain the inner valve.

One manufacturer makes an interesting variation of the staged trim for liquids (Fig. 5). In this design, the spindle or plug is a straight cylinder, and the flow passages are obtained by milled slots or flutes that are cut at a 45-deg. angle to the spindle's vertical axis. As the spindle moves out of its corresponding land or orifice in each stage, its projected slot area is increasingly exposed and the flow-capacity area developed (Fig. 5). Further, the slots in each stage are angled 90 deg. to the preceding and following stages, so that flow direction from stage to stage is reversed. This aids in dissipating additional energy due to increased turbulence from the flow reversals. The efficiency of this reversal in making a low-recovery valve decreases as the valve plug moves toward the closed position, and the valve then behaves as a conventional staged-trim valve. In general, it has the same basic limitations as other staged-trim valves.

These various liquid-service staged-trim valves are commonly applied in high-pressure-drop services such as

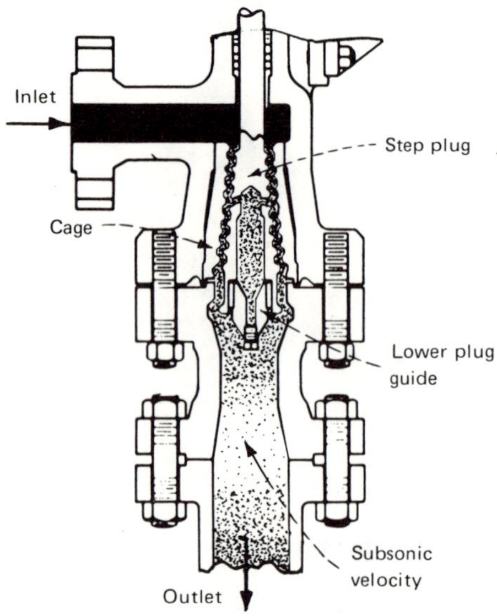

STAGED-TRIM valve for gas or vapor service—Fig. 6

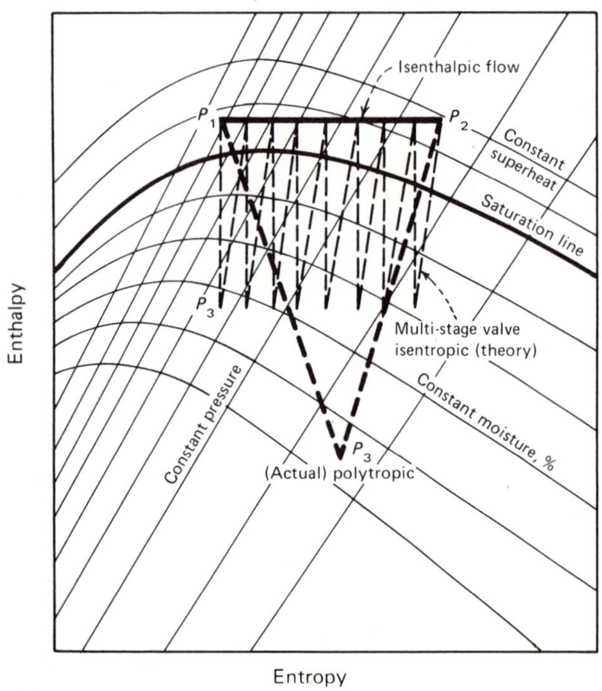

THROTTLING may be polytropic with Lanco trim—Fig. 7

boiler-feedwater-pump bypass and recirculation, desuperheater spray water, boiler-feedwater startup, and oil-separator levels. They can be considered for other services within their operating capabilities and size availability.

Only one staged-trim valve for gas and vapor service is manufactured (Fig. 6). This is the "Lo-Db," sometimes referred to as the "Lanco valve." Though in use in Europe, to the writer's knowledge none is in service in the U.S. The main differences between this valve and regular liquid-service staged-trim valves are:

1. The step plug has a Christmas-tree shape. The circumference and thus the flow area is gradually increased toward the discharge. This helps to compensate for the increase in specific volume as the pressure decreases, and helps to limit the increase in velocity buildup.

2. The step plug and mating-body lands have a semi-labyrinth pattern that provides some changes in fluid direction (similar to the liquid valve described previously). This can result in additional velocity-head loss.

The manufacturer states that the valve uses the principle of adiabatic flow with friction. The throttling process itself is adiabatic. The unique difference in a staged-trim vs. a plug-and-orifice trim is the degree of isenthalpic throttling achieved. The desired isenthalpic effect is that obtained in a long friction tube, like a capillary, but this is impractical. A short tube can be used if it is made to turn often and quickly.

A close look at the Lanco trim raises a question as to its efficiency as a friction device. It is probably more effective, in this respect, in the first portion of the plug's travel; and after that, it is doubtful if the fluid even knows the turns are there. What probably exists is multi-stage throttling, with each stage being isentropic initially. As the valve opens and the turns become less effective, the valve probably becomes polytropic (Fig. 7). In-service pictures have shown it heavily coated with frost and ice, which tends to confirm this analysis. The valve's effectiveness in noise reduction is more probably due to the geometry of the passage size coupled with limitation in the velocity buildup that is inherent in the trim shape.

These staged-trim valves are even larger and more massive for their pipe size and capacity than the liquid-service versions, due to the body-design requirement to contain the inner-valve geometry. To be effective, they are limited to a maximum pressure reduction of 4 to 1. Above this ratio, it is necessary to add restrictors in series to impose enough backpressure to limit the ratio to 4 to 1, or less. The necessity for fixed restrictors in this situation imposes limitations on flow turndown. This is because the restrictors lose their effectiveness as the flow decreases and quickly absorb available pressure drop as flow increases.

The valve has a very short stroke, and even though it has a partial hydrostatic-balance system, it must be operated with a ratio-multiplication lever system to obtain the necessary power and position sensitivity.

The manufacturer of the Lo-Db valve also supplies a unique device called the "Lo-Db Expansion Plate." This is a constant-area special-restriction orifice device that will handle part of the pressure drop in the system. One or more units in series can be used as needed (Fig. 8).

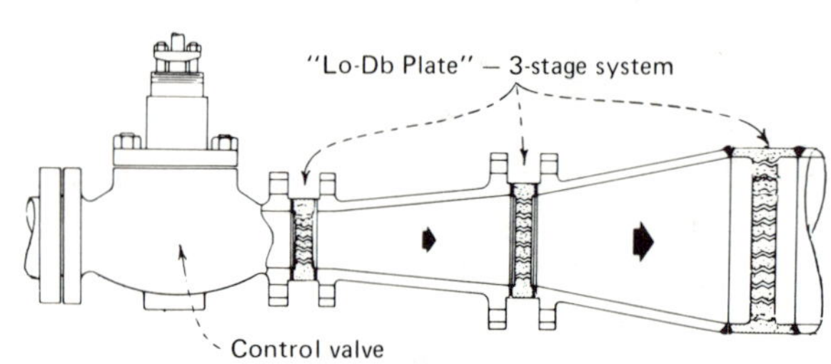

EXPANSION plates are multiple-orifice devices—Fig. 8

Such devices have a multiple-path geometry with some change in the flow-direction path. They not only act as restrictors, but also produce much lower noise levels than a larger single-path equivalent restrictor. Sound pressure level can be reduced as much as 20 dbA. The devices are effective only over a limited flow range. The vendor's guideline is that if $P_1C_v \leq 5{,}000$, these can be evaluated for throttling services. Above 5,000, they should only be used for on-off services such as free vents or system dumping. They can be an economical answer to some noise-throttling problems, either in conjunction with standard plug-and-orifice valves or to extend the pressure-ratio range of the Lo-Db valve.

Multiple-Path Design

The second design available is a conventional valve fitted with a special trim called the "Whisper Trim." This trim is for use with gas or vapor, and was developed by experimental and empirical means. The trim operates on the principle that a lower noise level is generated by a given massflow passing through a number of small areas rather than one large area. In other words, it functions in a manner similar to the Lo-Db plate except the multiple-path geometry is different, and the design allows flow modulation. The size and shape of the flow slots used for the multiple-path system have been developed by experimental testing (Fig. 9). The trim fits into the body like a conventional cage trim, with the plug position exposing the slot-area flowpath (Fig. 9).

In general, these valves have a capacity reduction of about 30 to 40% compared to the same size standard valve. It may be necessary to use oversized bodies to reduce exit velocities to acceptable levels and to gain optimum noise reduction. The reduction in sound pressure level obtained in optimum situations is a maximum of 20 dbA, but more typically is 10–15 dbA. This occurs at approximate sonic pressure drop. It is less effective at both lower and higher pressure ratios. Therefore, these valves should be considered only for moderate-noise-reduction and near-sonic pressure-drop applications.

Labyrinth-Disk Design

The third design is perhaps the most unique of all—a new approach to valve design, technology and operation. Called the "Drag Trim" valve, it is the most versatile valve currently available for severe-service and noise-reduction applications. It can be used for liquid, gas, vapor, or a combination of these, and is essentially immune to erosion, cavitation, wiredrawing, vibration and shutoff leakage. It does not generate noise and can be designed to give any level desired. It is a constant-impedance and true-velocity-limiting valve with actual isenthalpic letdown flow.

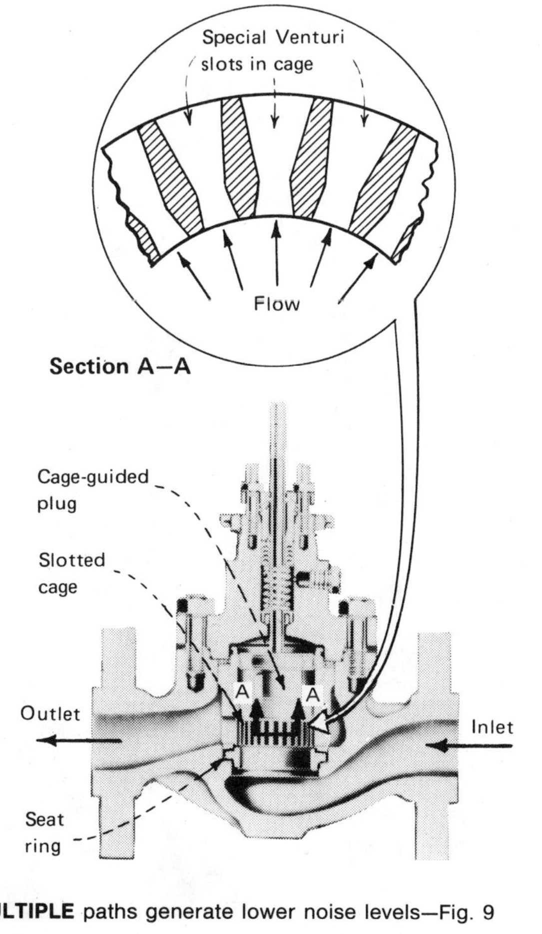

MULTIPLE paths generate lower noise levels—Fig. 9

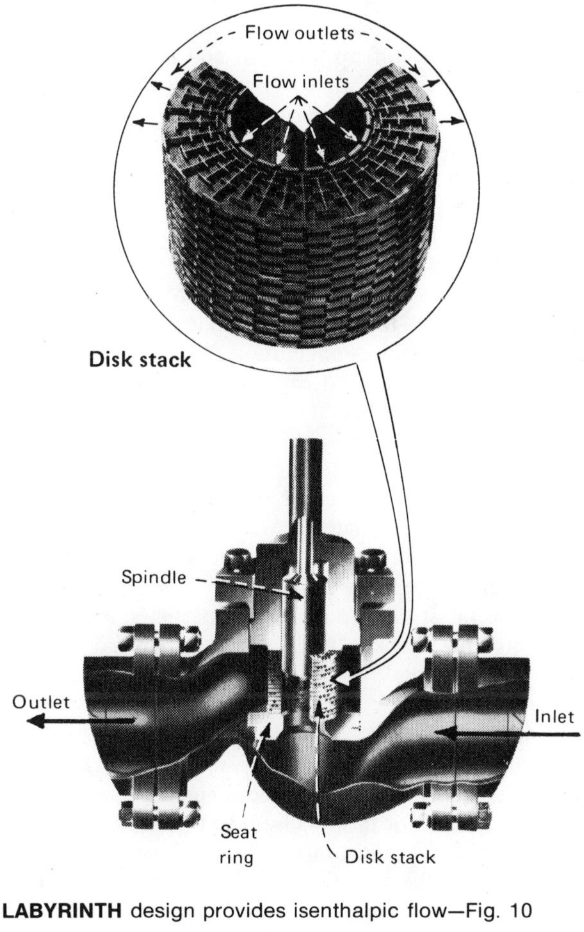

LABYRINTH design provides isenthalpic flow—Fig. 10

Basically, this valve consists of a stack of round disks with a hole in the center and a cylindrical spindle or plug in the hole. The individual disks are designed with a number of inlet passages, and each passage is a labyrinth that is actually a short friction tube consisting of a number of right-angle turns in series (Fig. 10). The number and shape of these turns are calculated and designed to be sufficient to dissipate the total required valve-drop at a velocity roughly equal to that at the inlet. Each disk has a specific flow capacity, so the required number of disks are made up in a "stack" to give the desired flow.

In general, the inlet area is about equivalent to the inlet-pipe area. Stack-discharge area is that necessary to handle the required flow, specific volume and desired exit-velocity.

The complete disk-stack is assembled in the valve body, similar to a conventional cage trim (Fig. 10). The spindle or plug positioning by the operator determines the exposed flow-area. This is inherently a linear-characteristic valve. It is relatively easy to characterize to other requirements by altering the disk-stack design and the configuration of the disks in the stack, or by using a cam positioner, or both.

In liquid service, the passage dimensions are the same from inlet to exit. In liquid flow with flashing, two-phase flow, gas flow and vapor flow, the passages increase in flow cross-sectional area after each turn. This accommodates the changes in specific volume while maintaining complete velocity control independent of pressure ratios. Another feature is the ability to provide continuous tight shutoff with the elimination of seat erosion. Bubble-tight shutoff up to about 500 F. with ultrahigh differential pressure can be obtained with an encapsulated Teflon® seat. Other designs are available for higher temperatures.

While the Lo-Db staged-trim design is limited to maximum pressures ratios on the order of 4 to 1 to achieve noise reduction, the Drag Trim valve is virtually unlimited in this respect. Services are in operation at 400 to 1, with valve-noise levels well under 85 dbA at a 3-ft. distance. Further, the design concept lends itself to a relatively efficient body configuration for the inner valve at any given pipe size and capacity. Since the disk-stack design has a greater height compared to the diameter, the long spindle stroke and ability to characterize disks can give turndowns of 100 to 1, or better.

However, utopia is not here yet. These valves are sensitive to line junk such as weld beads, slag, scale and rock. The disk stack does look like a good filter but it is not very susceptible to plugging from normal solids such as catalyst fines, dusts, slurry particles, etc. Generally, anything that can go in a passage will go through. In most cases, the passage size is larger than we might expect.

Line junk can cause jamming in the close-fitting spindle in the stack, where the flow direction is from the center hole to the disk's outside diameter. This orientation is required for flashing liquids, two-phase flow, gas, and vapor. For liquids, this problem is essentially eliminated because the flow direction is from the disk's outside diameter to the center hole. Here, the stack acts as its own filter and will tolerate a reasonable amount of debris without much effect on capacity or operation because there are two or three entrances into each disk passage. In any case, screens of adequate size and strength installed ahead of the valve are highly recommended. Of course, there may be situations where the system is known to be clean and free of line junk.

These valves have been used in the services previously mentioned, and in such tough applications as urea, diamine, ammonia, methanol and polyethylene letdown. Other applications include: high-pressure, steam, hot-process-gas, and natural-gas letdown and venting; and light-duty services such as moderate-pressure natural gas, steam and air at high massflow rates.

At this time, no other design can operate under high-velocity situations with such a degree of immunity to noise and destructive internal vibrations. For example, in a high-pressure liquid-letdown service, a custom-design angle-pattern plug-and-orifice valve with tungsten carbide trim has a service life on the order of 250 hr. A Drag Trim valve in the same service has had better than 6,000 hr. of operation to date with no problems.

Conclusion

There are several special-control valve designs or special devices for the tough, severe-service applications and noise problems. Each has its areas of usefulness and its limitations. Selection will depend upon a number of factors such as type of fluid, static and pressure-drop conditions, temperature, corrosion, and relative costs to benefits. All of these devices being specialized designs require more engineering evaluation of the application, the mechanical design and the metallurgy than is usually the case with conventional control valves that are used in many processes.

Illustration Credits

The illustrations in this article were contributed by the following firms: Fig. 10, Control Components Inc., Los Alamitos, CA 90720; Fig. 9, Fisher Controls Co., Marshalltown, IA 50158; Fig. 1, 4, 6, 8, Masoneilan International, Inc., Norwood, MA 02062; Fig. 5, Yarway Corp., Blue Bell, PA 19422.

References

1. Bulletin No. 340E, Masoneilan International, Inc., Norwood, MA 02062.
2. Bulletin TM-24, Fisher Controls Co., Marshalltown, IA 50158.

Meet the Author

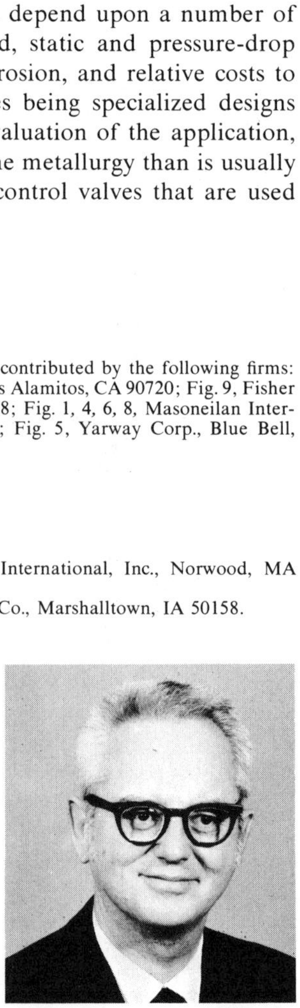

J. B. Arant is a senior instrument specialist in the Design Div., Engineering Dept. of E. I. du Pont de Nemours & Co., Wilmington, DE 19898. He is presently responsible for instrument design for plant projects and is involved with special control-valve problems. He joined Du Pont in 1950 at the Sabine River Works in Orange, Tex., as a division engineer in the instrument department. He has a B.S. in chemical engineering from the University of Texas and is a registered professional engineer in Texas.

Butterfly valves for flow of process fluids

Butterfly valves combine on/off and throttling-control characteristics. Compact design and wide availability of elastomers and high-alloy trim materials extend uses to many process media.

Brad E. Bertrem, *Norris Div., Dover Corp.*

☐ Butterfly valves find use in the chemical process industries for liquid, gas and solids handling.

While butterfly valves are excellent control valves, they are more commonly used as shutoff (i.e., block) valves. These valves provide positive "drop-tight" shutoff in almost any process applications, including steam, air, gases, liquids, slurries and solids.

The basic operation of a butterfly valve is simple, requiring only a 90° rotation of the disk to fully open the valve from a closed position. Simplicity of operation has led to increased use of automated butterfly valves for remote operation and for control.

Due to the linear-flow characteristic of butterfly valves between 20° and 70° opening, they are particularly adaptable to control applications (see Fig. 1). Butterfly valves are highly efficient control valves when compared to globe-type control valves. The velocity head of the flow stream is not lost as the fluid flows in a streamlined manner around the valve disk. Flow through the globe valve's restricted seat area and around the plug causes high pressure drop.

The quality of control obtained from a valve is dependent upon the ratio of the pressure drop across the valve to that of the entire system. Consequently, the butterfly valve is ideally suited for control applications in which minimum pressure loss is desired.

Butterfly control valves can be equipped to fail with the disk automatically opening or closing, through the use of fail-safe devices (see Fig. 2). These may be springs, reserve-air tanks, or mechanical systems that are activated when the loss of power supply to the valve actuator occurs. Warning devices can be added to the valve assembly to signal failure or to shut down pumps for safety and more-convenient operation.

Pressure and temperature limits

A primary function of most butterfly valves is to achieve positive shutoff. The emergence of new elastomers allows the valve designer to obtained improved resistance to process media. A great deal has been learned about elastomer compounding and molding, resulting in dependable butterfly-valve designs that provide shutoff up to 720 psi.

Butterfly valves are lightweight. Lighter than gate, plug or ball valves, they offer considerable advantage during installation because they can be handled without the use of a crane up through approximately 10-in. size. Smaller pipe hangers and supports will suffice for the piping system using the lighter valve.

Due to the minimum amount of materials in butterfly valves, they are economical in both original and maintenance costs. This advantage is amplified when

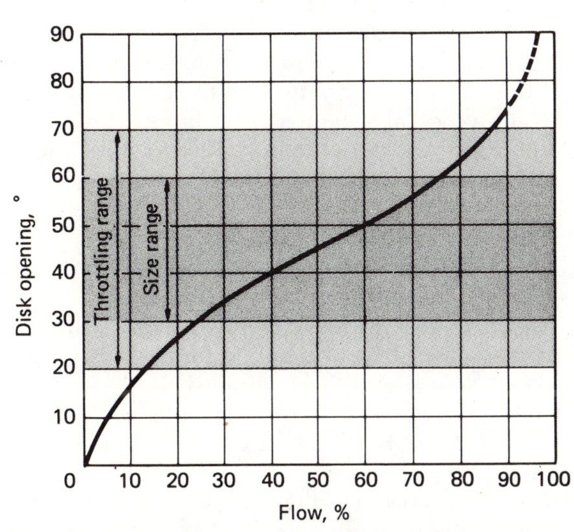

Butterfly valves for throttling control have linear characteristics for over 70% of the flow range **Fig. 1**

Originally published December 20, 1976

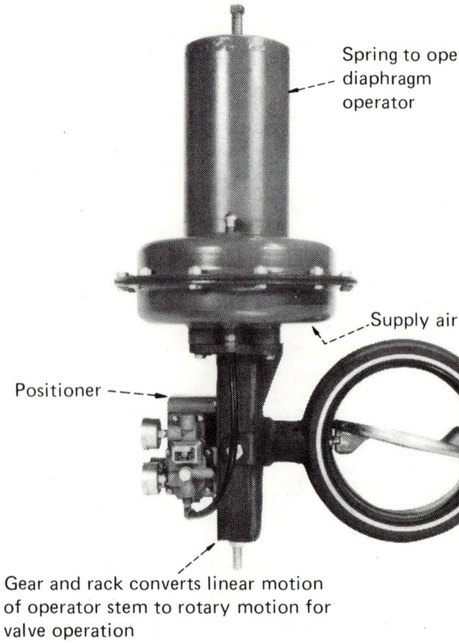

Fail-safe diaphragm operator with positioner Fig. 2

premium material trims are considered, such as titanium, Hastelloy, Teflon or Viton. Most butterfly valves have "dry back" construction that prevents process media from reaching the valve body. This minimizes the amount of premium material required to handle corrosives. Other valve types require high-alloy bodies as well as high-alloy trims if they are to be compatible with the flow stream. The butterfly valve can be equipped with carbon-steel or ductile-iron bodies, with only the internal wetted parts made from alloy material.

In flow control, the advantages of the butterfly valve include ease of maintenance with no special tools or equipment required and no need for special plant shutdown. The wafer-style butterfly valve offers very simple installation. All that is required is to put in a cradle of bolts in the lower bolt pattern of the flanges and insert the wafer-type valve between the flanges until it rests on the cradle. Bolts are then tightened until the end-seals are secure and metal-to-metal contact occurs. No gaskets or special seals are needed for end-seals on most butterfly valves.

If butterfly valves are properly selected with regard to metals and elastomers, they will provide bubble-tight service life with a minimum of maintenance. The movement of the disk across the seat acts as a squeegee, providing a seat-cleaning action. This is particularly true when the butterfly valve is installed with its shafts horizontal. Most butterfly valves are permanently lubricated.

With the proper selection of metals and elastomers, butterfly valves are capable of operating from –65 to 450°F. With a properly selected trim, the butterfly valve will handle any medium that will flow through a pipeline within the above pressure and temperature limitations.

Valve trim is critical

Various media temperatures and pressures require the correct valve configuration and trim for each service. The following information should be reviewed in selecting butterfly valves:

- Pipeline size.
- Type of fluid that valve will handle.
- Viscosity and specific gravity of fluid.
- Minimum and maximum temperatures. Average temperature.
- Minimum and maximum line pressures. Normal pressure.
- Minimum and/or maximum flowrates. Normal flow.
- Pressure drop at maximum and minimum flows.
- Flange type. Disk clearance.
- Shutoff conditions.
- Will valve be subject to free discharge?

The wrong valve can cause problems

Valves must be compatible with other equipment such as pumps, strainers, etc., in the system. The right valve with poorly chosen trim materials can deteriorate. Evaluation of the job to be done and choice of accessories are important.

Metal-to-metal swing-disk valves are primarily used for throttling control and are intentionally designed to have low leakage in the closed position. This is done to minimize rubber torque and to eliminate seating torque in order to give improved throttling control when using actuators. Since the butterfly valve has a fairly linear characteristic between approximately 20° and 70° open, an inexpensive linear actuator will provide the desired control mode.

Metal-to-metal or swing-through butterfly valves are generally used in those applications where high velocity and a large variation of flow or pressure drop is anticipated (Fig. 3a). The process engineer selects the proper valve size by using the values of the flow coefficient as published by each manufacturer. (See flow-coefficient curve in Fig. 1; this curve indicates that the butterfly valve is basically an equal-percentage valve between the 20° and 60° opening.)

It is standard practice to size a butterfly valve so that the minimum flow occurs when the valve is approximately 20° open, and the maximum flow when the valve is approximately 60° open. Within these guidelines, butterfly valves give excellent linearity for throttling control. The butterfly valve is a more efficient valve than most other types because approach velocity of the fluid is not lost. Hence, butterfly valves are ideally suited for process systems requiring a valve that produces a minimum pressure loss. This factor must be considered when using a butterfly control valve, just as it is when using other types of control valves. Pipe fittings adjacent to valves affect performance characteristics, and their effects should be considered when selecting precision-control butterfly valves.

Elastomer-lined butterfly valves are the most popular style (Fig. 3b). They are available in three basic configurations: (a) vertical-centered disk, (b) vertical-offset disk and (c) angle disk, as shown in Fig. 4. Ver-

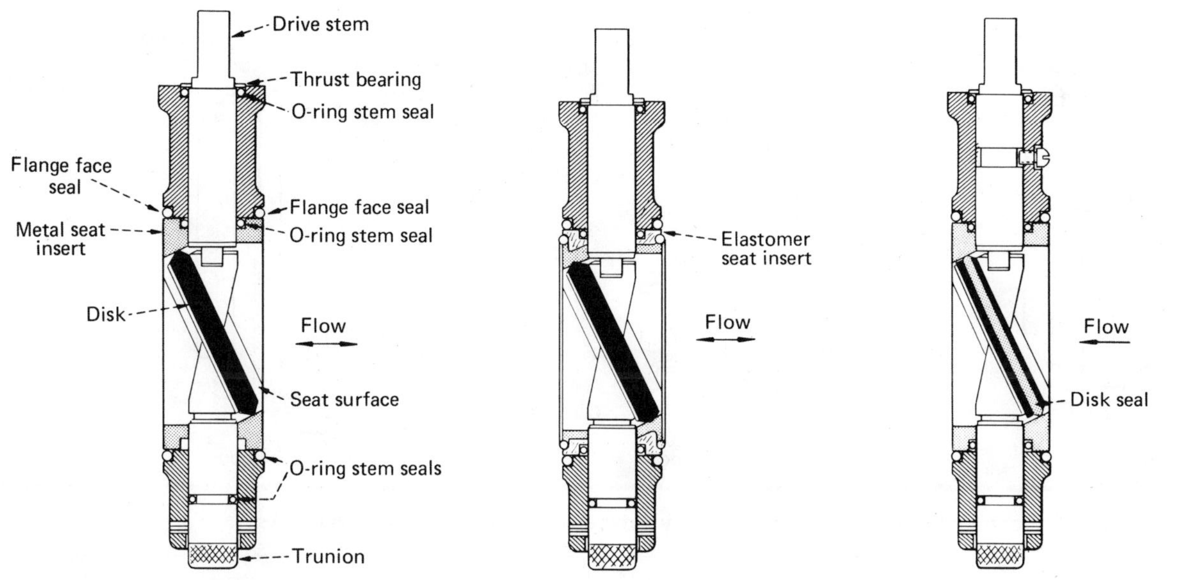

a. Metal-to-metal throttling valve b. Resilient-lined shutoff valve c. Limited-elastomer shutoff valve

Seat-and-seal materials of construction establish the process applications for butterfly valves **Fig. 3**

tical-centered disk valves usually have a lower cycle life due to the scrubbing and compression set that takes place on the seat at the disk-boss area. Vertical-offset disks eliminate this problem but have reduced port diameters, which means a less-efficient valve. Angle-disk valves eliminate both of these problems.

The resilient-lined butterfly valve has the basic advantage of shielding process media from the valve body. The resilient liner or seat can be molded from a variety of elastomers so as to be compatible with almost any process medium. The resilient-lined butterfly valve is a positive-shutoff valve. The seal is obtained by penetrating a metal disk into the elastomer seat until a unit pressure is built up in the elastomer that exceeds the pressure the valve is intended to hold. Since the disk cycles into and out of the elastomer seat, it is self-cleaning and requires little or no lubrication. Most resilient-lined butterfly valves are equipped with replaceable

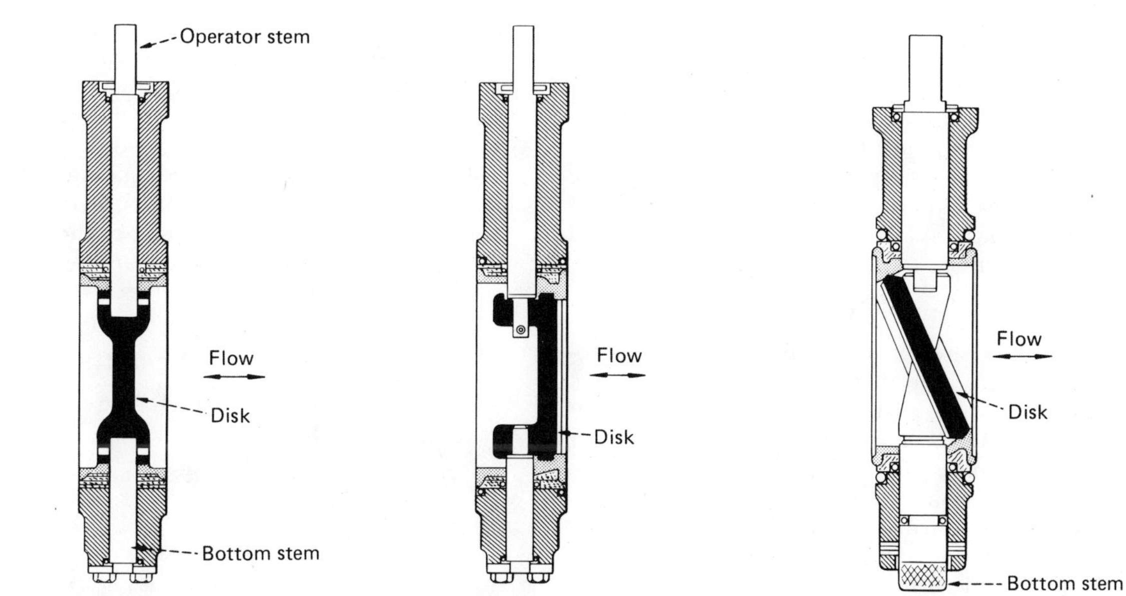

a. Vertical-centered disk b. Vertical-offset disk c. Angle disk

Elastomer-lined butterfly valves for handling process fluids have three basic configurations **Fig. 4**

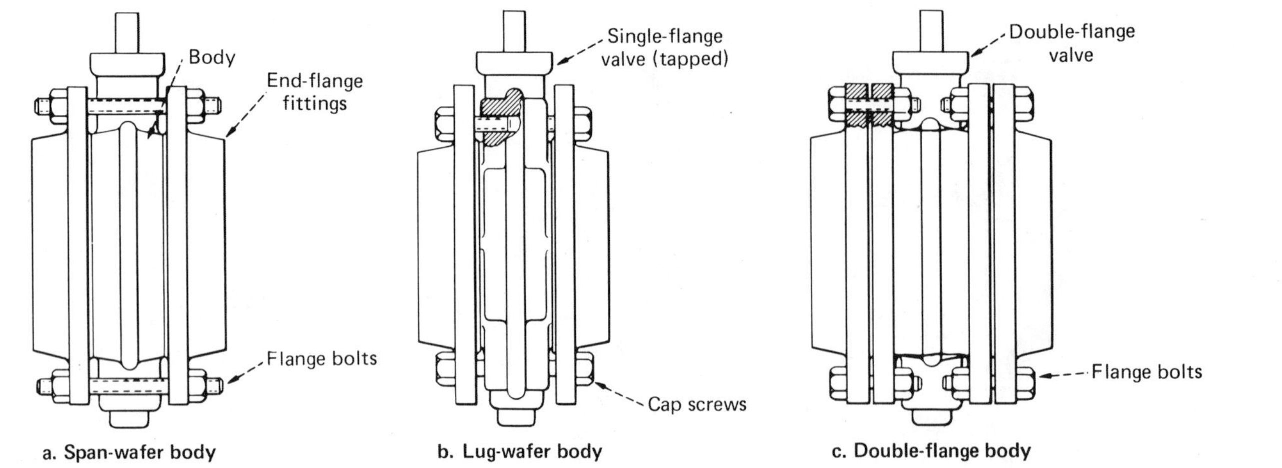

Body styles and end-connections of a typical butterfly valve meet application requirements **Fig. 5**

seats having a hard backup ring to prevent seat collapse. This ring also facilitates seat installation and removal.

The limited-elastomer sealing-type butterfly valve (Fig. 3c) is also a positive-shutoff valve. It offers application advantages over a resilient-lined butterfly valve by minimizing the amount of elastomer required for the seal. Limited-elastomer-sealed valves have a wider temperature-application range (-65 to +450° F), and are capable of containing higher pressure differentials.

End-connections for butterfly valves

The slip-in wafer-style valve is available in span, lug, single-flange and double-flange configurations (Fig. 5). The most popular one is the span type, due to the simplicity of slipping it between flanges in a cradle of bolts. One set of bolts secures it, with no special alignment being needed. The lug-style unit may also be inserted but it must be lined up, and cap screws inserted from each side and correctly tightened.

The wafer-style, narrow, face-to-face valve has the widest use in process plants because it offers many advantages when designing a piping system. Among these are space savings, light weight, lighter pipe hangers and supports, reliability, variety of trim materials, and variety of valve types.

Applications for butterfly valves

Butterfly valves have been used in the chemical process industries for many years for water-distribution systems, cooling towers and air lines; as isolation or intake valves for pumps; and in foam or water lines for fire-fighting systems.

Improved performance and wider trim selection have boosted the butterfly valves into more-demanding applications. They are now used for gas-feedstock regulation, and as condenser-water inlet and outlet valves. With special trims, they can be found in various applications such as waste and sewage treatment, waste-combustion control, and in digesters and centrifuges for process separations. Applications in corrosive media include hot ammonium nitrate, salt solutions, caustic cleaning solutions, and steam.

Butterfly valves perform well in dry-process services such as for powdered polymers and resins; as discharge valves for hoppers and dryers; and in pneumatic-conveyor lines and dust-recovery systems.

In the pulp-and-paper industry, butterfly valves find use in pulp lines, black- and white-liquor control, raw water, ammonia, caustic soda, spent nitric acid, washed bleached stock and vacuum piping. To handle such materials, valves made of Illium, with Nitronic 50 stainless and titanium trims—along with Viton, Teflon or Hypalon seals—are often used.

Butterfly valves are often found in plant-ventilating systems and in pollution-control systems, due to their ease of remote operation.

The butterfly valve has the advantage of combining in one valve the block-valve (i.e., on/off) function with the throttling-control function. This may require a positioner to assure that proper valve control is obtained, but it does allow for combining the two functions into one valve. Also, the single wafer-disk that pivots in the body may easily be operated manually or automatically. The use of geared operators, levers, electric-motor drives, air-cylinder or hydraulic-cylinder operators, or diaphragm operators is quite common for local or remote operation of butterfly valves.

The author

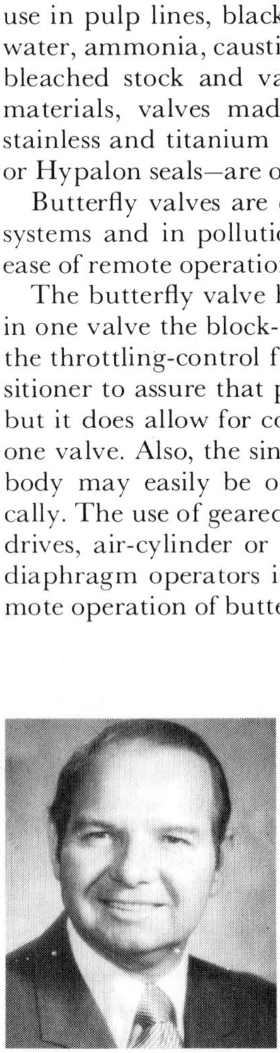

Brad E. Bertrem is vice president for butterfly valve operations in Dover Corp.'s Norris Div., P. O. Box 1739, Tulsa, OK 74101. He joined Norris in 1956, and has been responsible for sales, engineering and manufacturing of butterfly valves. He achieved his present position on Jan. 1, 1976. Mr. Bertrem has a B.S. in mechanical engineering from the University of Tulsa. He has been active in establishing standards for butterfly valves, serving on committees in the refinery and production groups of the American Petroleum Institute, ASME, Oklahoma Soc. of Professional Engineers, American Soc. for Testing and Materials, and Manufacturers Standardization Soc.

Manually operated valves

Valve selection is extremely important to the economics and the operation of many kinds of process plants. Valve catalogs—showing innumerable designs of valves, many or all of which may seem suitable for a particular application—can be bewildering. How, then, to choose the right valve?

Arkadie Pikulik, *Scientific Design Co., Inc.*

☐ There are countless different types of valves available—each with its own particular uses. A valve starts off designed for particular functions, and when used properly will give good service through a long life. In practice, however, valves are often misused, with varying consequences.

About 50% of industrial valves are used for on/off service, 40% for throttling and 10% for the prevention of back-flow. There are many ways in which valves control flow, and these offer varying degrees of accuracy.

Valve operation

Whether a valve will be operated by hand or by means of a driver (motor, air-piston, etc.) has no bearing on the type of valve selected for the application. Drivers can be operated by actuating them manually, or with an automatic signal from a sensing device. This article discusses the varieties of manually operated valves, and their accessory equipment. Process designers should understand completely the construction features of manually operated valves that they specify, and should be aware of the alternatives offered to the process industries by manufacturers or valve designers.

Manually operated valves may require special devices such as gears, chains and pulleys, and other accessories to aid the operator. They also may require added features, because of size, unavailability of a particular type of valve, difficulty of operation because of valve location, etc. However, selection of valve and associate devices sometimes depends purely on economics.

Valve functions can be defined as follows:

1. On/off service.
2. Throttling service.
3. Prevention of reverse flow, or back-flow.
4. Pressure control.
5. Special functions.
   a. Directing flow.
   b. Sampling service.
   c. Limiting flows.
   d. Sealing vessel or tank outlets.
   e. Other.

Valve details

Valves have a basic function of controlling the flow of fluid by inserting an object—a flow-control element—to

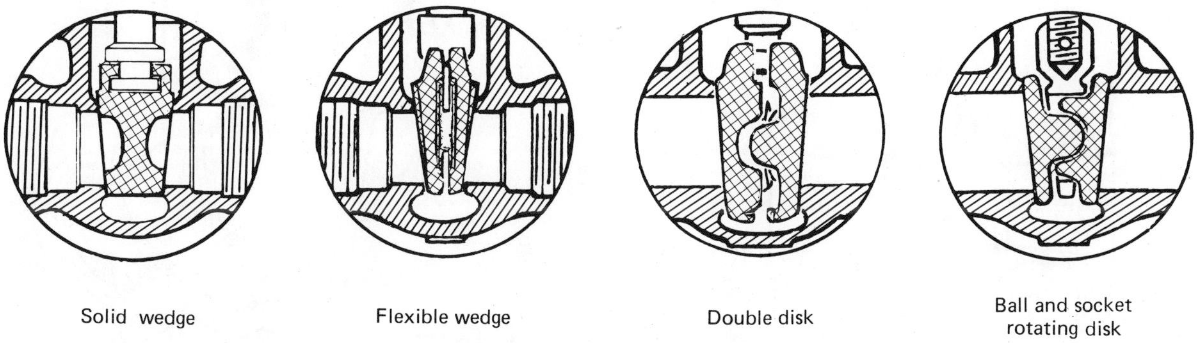

Some of the various designs of disks that effect the closure in gate valves **Fig. 1**

Originally published April 3, 1978

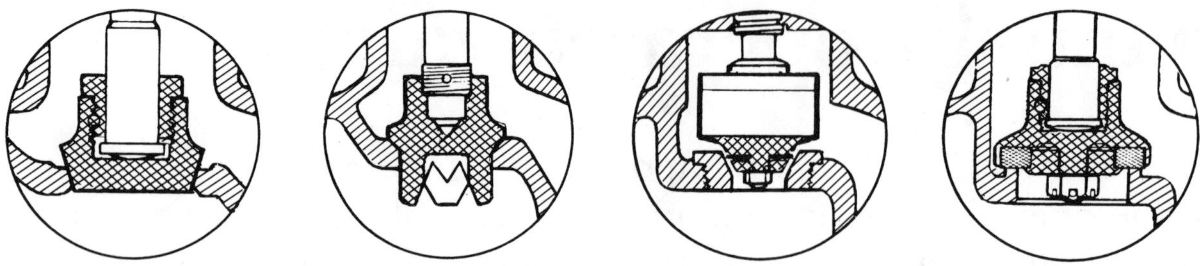

Plug type disk | V-port type plug disk | Dual-seal wafer disk and renewable seat | Composition disk

Plug valve sealing disks are designed for both throttling and tight seal **Fig. 2**

divert, restrict, or regulate flow. The mechanism for adjusting the flow-control element, to permit control of the rate of flow through the valve body, is a critical feature. While considering this mechanism, designers for various valve manufacturers have advanced a number of methods to isolate the mechanism that regulates movement of the flow-control element from the fluid being handled in the valve.

Valve selection requires consideration of three basic and critical details: (1) the flow control element, (2) the regulating mechanism, and (3) the seal to contain the fluid within the valve. In addition to these three important design aspects, the process-design engineer must consider such features as mechanical strength, materials of construction, dimensional arrangement, and types of end-connections.

Flow-control elements

Four basic methods are employed to control flow:

1. Move a disk or plug into or against an orifice, as in globe, angle, Y, and needle valves.
2. Slide a flat disk (which may be split), cylindrical or spherical surface across an orifice, as in gate, plug, ball, slide, and piston valves.
3. Rotate a disk or ellipse about a shaft extending across the diameter of a circular casing, as in butterfly valves and dampers.
4. Move a flexible material into the flow passage, as in diaphragm and pinch valves.

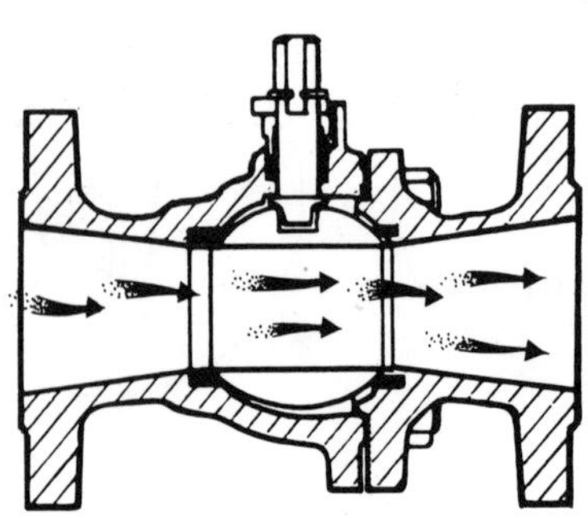

An open ball valve allows unimpeded flow **Fig. 3**

Movement of the flow-control element

The opening through which fluid can pass is varied by movement of the flow-control element. A stem is connected to this element and either rotates or slides it (sometimes both) to control flow. In most types of valves, the stem extends outside the valve body.

Stem rotates—Found on non-rising stem gate valves, as well as ball, butterfly and plug valves, and cocks.

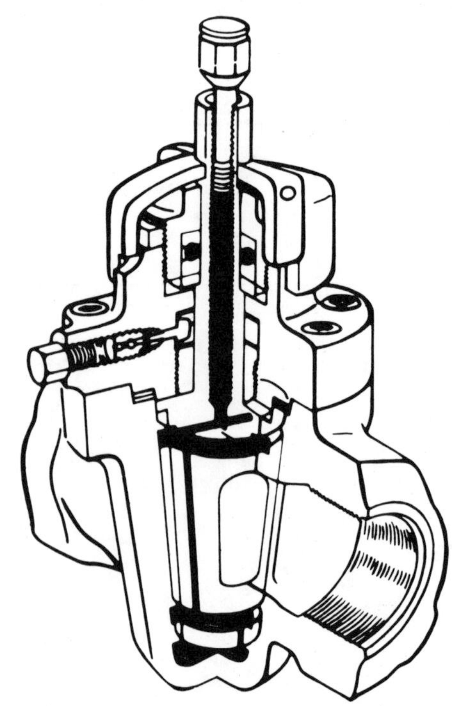

Lubricated plug valve **Fig. 4**

Stem moves endwise without rotation—Found in outside stem and yoke (OS&Y) gate valves, quick-opening gate valves, globe and diaphragm valves, and slide, piston, and sleeve valves.

Stem both rotates and moves endwise—Found in globe, angle, Y, and needle valves; rising-stem gate valves; lift-type plug valves; diaphragm, and pinch valves.

Leakage

There are three types of leakage:

- Process fluid escapes downstream, past flow-control element in closed position. Identified as "flow seal" leakage.
- Process fluid escapes to the outside of the valve, from around the stem and from the joints (bonnet) with the body. Identified as either stem-seal or bonnet-seal leakage.
- Air leaks into the valve body and to the process medium under vacuum.

Flow seals

The valve design must provide a tight seal between the flow-control element and the valve seat. Pressure and temperature changes should not misalign the sealing or seating surfaces. The most common types of seals include:

Metal to metal—Permits a strong seal, but metal seizing and galling may occur.

Metal to resilient material—This provides a tighter shutoff and is recommended for fluids containing solid particles and for medium-pressure services.

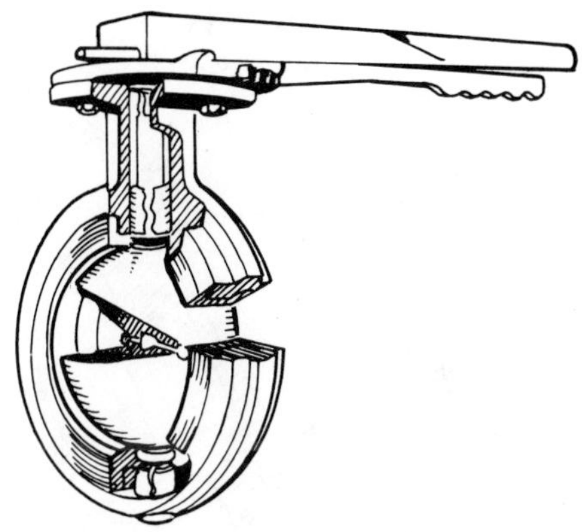

Butterfly valve with lever operator **Fig. 5**

Metal to resilient material contained in metal—Provides good seal and can be used for relatively high-pressure services.

Stem seals

The most common type of stem sealing is a stuffing box containing flexible packing material such as graphite/asbestos, TFE/asbestos, etc. TFE (tetrafluoroethylene) is of particular importance for corrosive ap-

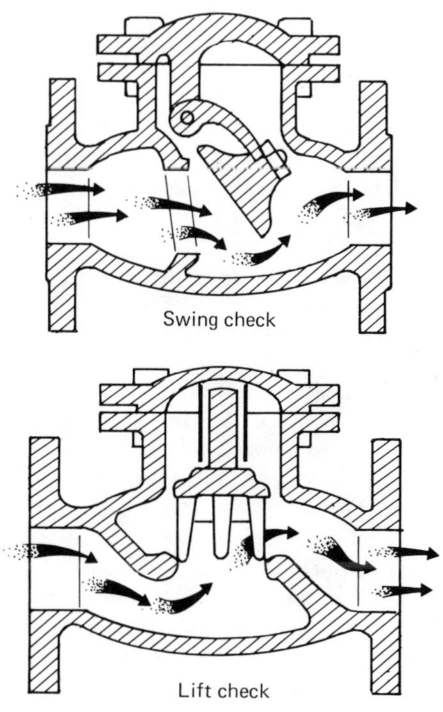

Check valves prevent reverse of fluid flow **Fig. 6**

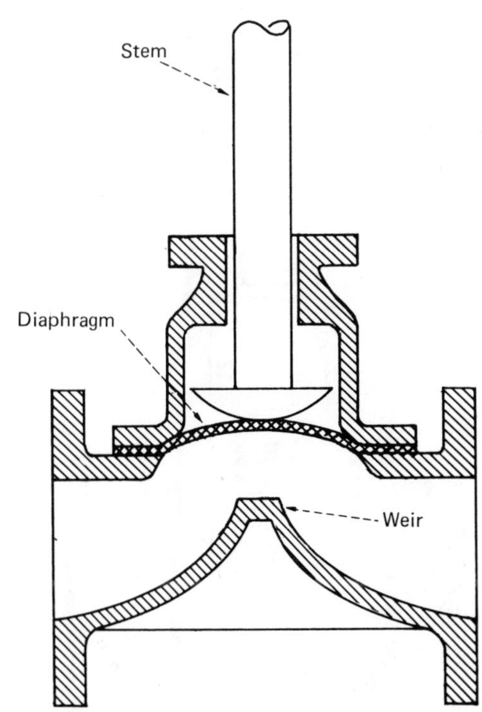

Weir type of diaphragm valve **Fig. 7**

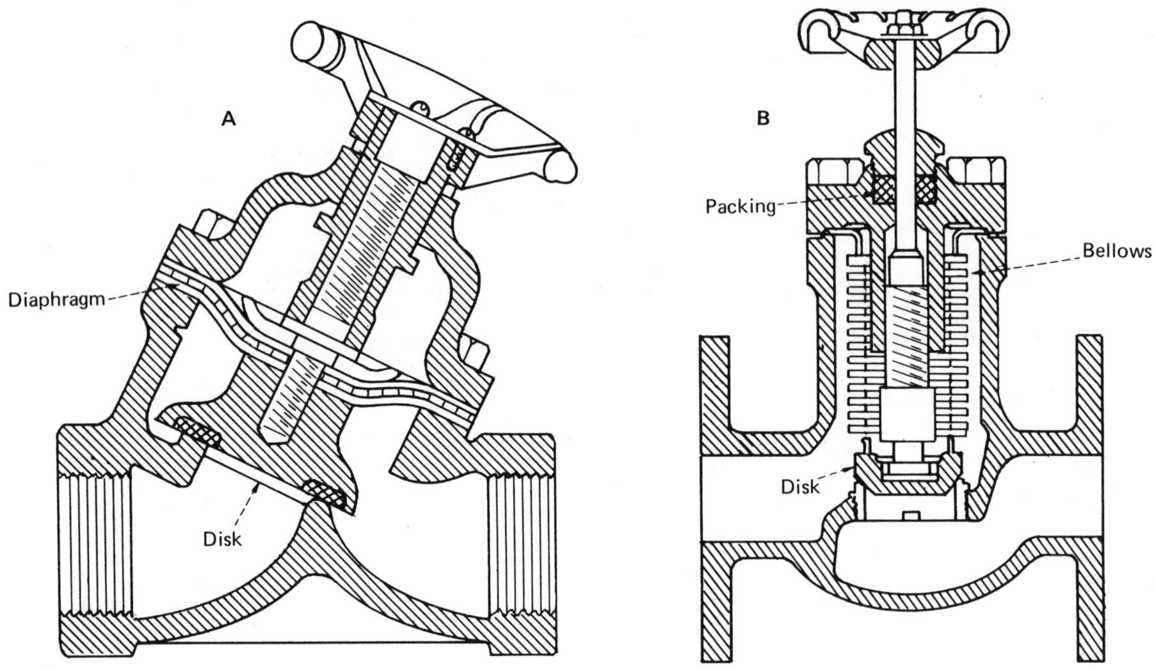

Globe valve with diaphragm bonnet seal (left) and one with bellows bonnet seal (right) **Fig. 8**

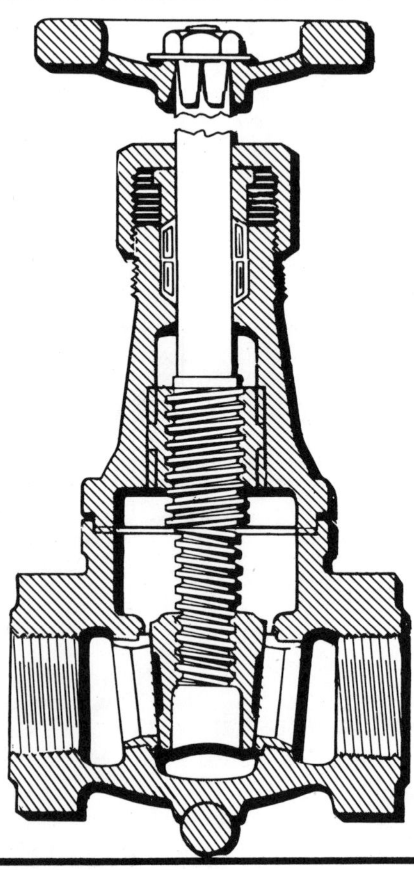

U-bolt bonnet is used mainly on smaller valves **Fig. 9**

plication. The packing may be solid, braided, or a loose fill of granulated TFE, asbestos fibers and TFE, and other compositions. Packing designs include square, wedge, and Chevron rings and O-rings. In order to retain fluid pressure inside the valve, it is necessary to compress the packing, by means of a compression device, or gland, which forces and compacts the packing into the stuffing box and against the stem.

Tightening up on the gland from time to time helps maintain adequate packing compression to prevent leakage. Invariably, if a valve has not been operated for a period of time, it will leak through the packing, when reactivated.

Occasionally, a spacer, or lantern ring, is placed in the stuffing box to separate lower and upper sections of the packing, as a means of introducing a lubricant or inert sealant.

When no leakage to the outside can be tolerated, a packless type of valve is called for. One design employs a flexible diaphragm between the bonnet and the body. A compressor component pushes the diaphragm into the flow path (Fig. 7).

A globe valve with a diaphragm, as shown in Fig. 8A, isolates working parts from the process fluid.

Another type of packless valve employs a metallic bellows (Fig. 8B). This construction, desirable for high-vacuum service, includes a stuffing box, in case of failure of the bellows.

Bonnet seals

The bonnet—the component of a valve that provides a closure for the valve body—gives access to the valve

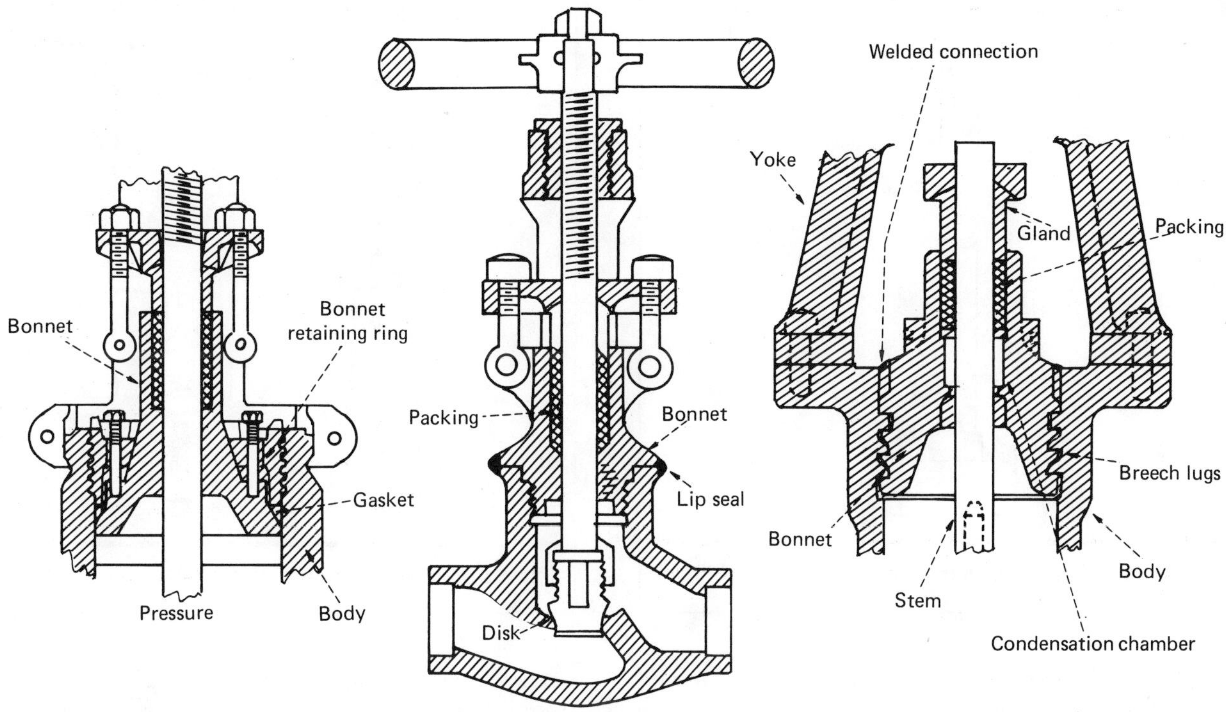

Pressure-sealed bonnet (left), lip-sealed bonnet (center), and breech-lock joint (right) **Fig. 10**

seat and flow-control element. For the purposes of repair and/or replacement, the bonnet is removed from the valve body in the following ways:

Screwed-on (and screwed-in) bonnet—These designs, the simplest and most economical, serve in low-pressure operations and for small valves. Disadvantage: the seal between the body and bonnet may loosen during normal valve operation or may even unscrew accidentally. Recommended for service with minimum of shock and vibration, and infrequent operation.

Union bonnet—This provides an easy, quick method of coupling and uncoupling the bonnet from the valve body. The bonnet has a loose nut, or union-bonnet ring, which screws onto the valve body. With all parts in compression and held firmly in place, distortion is unlikely. Allows easy access for repairs. Recommended for smaller valves.

Flanged bonnet—This type of bonnet construction, generally for larger valves, works well for handling corrosive media and high temperatures or pressures. Like a flanged piping joint, the bonnet flange bolts to a similar body-flange, using a suitable gasket between the faces. The flange joint employed can be of flat-face, male-and-female, or tongue-and-groove and ring-joint design.

U-bolt bonnet—In the smaller range of valves, for chemical service in particular, a U-bolt, that passes around the body, secures the bonnet (Fig. 9). This construction provides a strong connection.

Pressure-sealed bonnet—One of several types of seals available for high-temperature and high-pressure service, this construction results in lighter-weight design than the conventional flanged bonnet. The pressure-sealed bonnet uses the fluid pressure in the line to make the seal effective. This technique involves fitting the bonnet into the body, topped by a ring gasket or ring seal retained in the body, either by a segmented ring fitted into a groove in the body or by a retaining ring screwed into the body. The initial seal is created by lifting the bonnet against the gasket or seal ring with bolts. In service, the internal fluid pressure acts on the underside of the bonnet, forcing it against the gasket or seal ring. Increasing fluid pressure increases the tightness of the seal (Fig. 10).

Lip-sealed bonnet and breech-lock bonnet—These bonnet seals are for high-pressure high-temperature service. Both designs require welding the bonnet to the valve body. Chipping off the welded joints permits disassembly of the valve without removing it from the line.

Accessories for manual operation

Most valves operate simply by means of a handwheel or lever supplied with the valve. However, valve designers have developed alternative methods of operating valves either manually or by automatic devices. This article discusses only manual devices (the others are covered in the article on control valves).

Various accessories can be supplied to fit most types of valves, to meet three principal needs:

1. To make valve operation easy, when valve location is inaccessible.

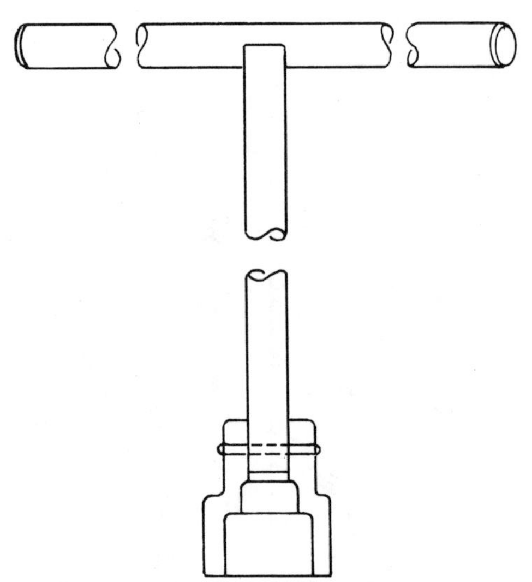

T-wrench, used as valve stem extension Fig. 11

2. To provide more turning leverage when the handwheel is not large enough.

3. To slow the closing and opening rates.

Stem extension

It is not always possible or practical to place valves within reach (e.g., underground water main valves), so valves are supplied with extension stems for remote operation. Stem extensions—wheel, key, socket wrench, or floor stand—can be of any required length. The extension unit usually consists of a cold-rolled steel rod and a coupling device designed for attaching to the valve stem.

For longer extensions, the rod is replaced with a heavy steel pipe, securely fastened to a wheel end-plug and a lower extension plug. A square-hole wheel or operating nut (2 in. square) fits the tapered square tip of the rod or wheel end-plug (Fig. 12). Extra-long stem-extension units require support.

Normally, the operating end of the extension is 2 inches square, permitting the use of a 2-in. square-socket wrench or any other suitable 2-in.-square

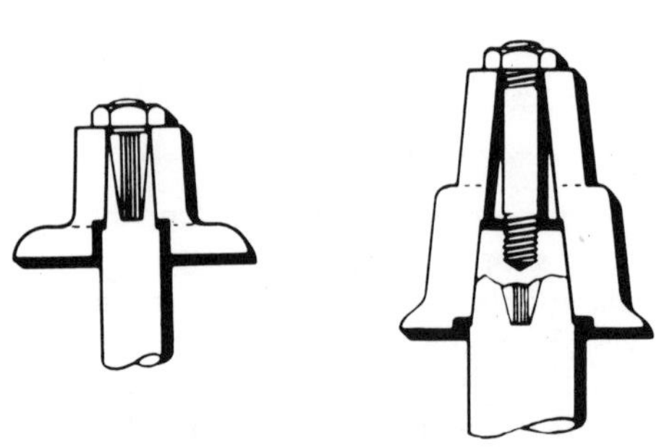

Square nut used to adapt valve stem to wrench Fig. 12

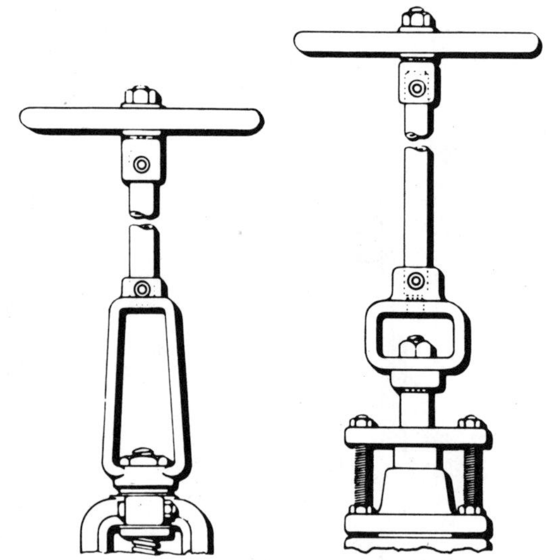

Extension units for lengthening valve stems Fig. 13

wrench, such as the type known as a T-wrench, illustrated in Fig. 11.

Fig. 13 shows examples of extension units for rising-stem and non-rising-stem valves. There are also flexible shafts—combinations of steel rods and universal joints—that permit the operator to work the valve from points not in line with the valve stem.

Floor stands

Floor stands help in operating gate, globe, and other types of valves installed in pipe trenches or other inaccessible locations. Floor stands are connected to stem-

Floor stands for operating inaccessible valves Fig. 14

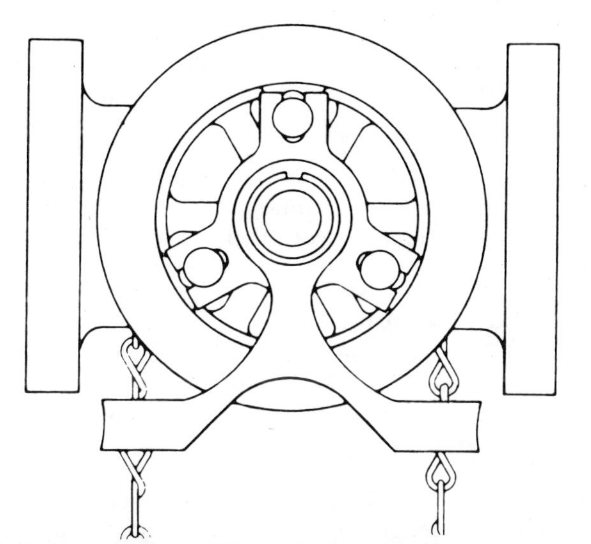

Chain wheels for valves in high places Fig. 15

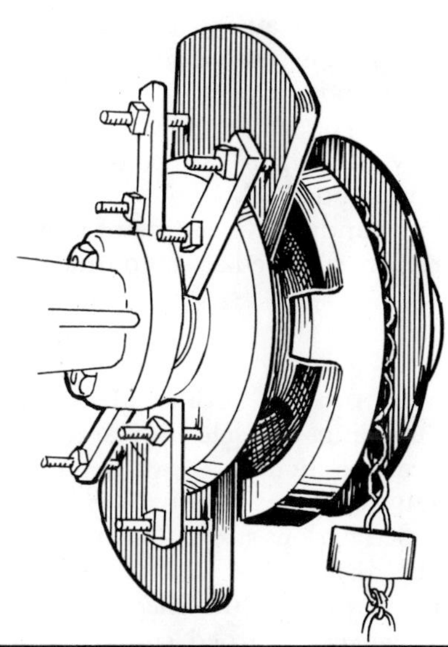

Hammerblow chain wheel eliminates reduction gears Fig. 16

extension with couplings. Fig. 14 shows two basic types: with indicator attachment showing the position of the flow control element, and a floor stand without the indicator.

Chain-wheel operators

Wheel operators permit use of overhead valves in either vertical or horizontal pipelines. These operators are easily attached to the rim, or to the spokes, of the handwheel. Chain wheels (Fig. 15) require a strong enough valve stem to withstand the extra weight and added pull. Some chain wheels have guides to prevent the chain from slipping off. There are also chain wheels designed to replace the handwheel.

Hammerblow chain-wheel

The hammerblow chain-wheel aids in opening and closing all standard hand-operated valves, with or without extension stems. It eliminates the need for reduction gears on large valves. Operations people find it convenient, as it requires neither valve wrenches nor additional help, in most cases, for operation. (Fig. 16). Also, emergency valves fitted with this device can be opened or closed without delay where lines are exposed to the danger of fires.

The operation of hammerblow chain-wheels compares to a hammer striking an anvil, with the anvil clamped to the handwheel or bolted to an adapter replacing the handwheel of the valve. This rotary-

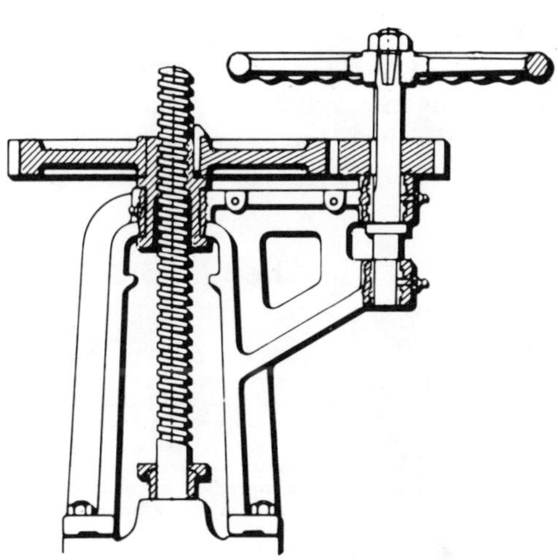

OS&Y spur-gear-head operator for large valves Fig. 17

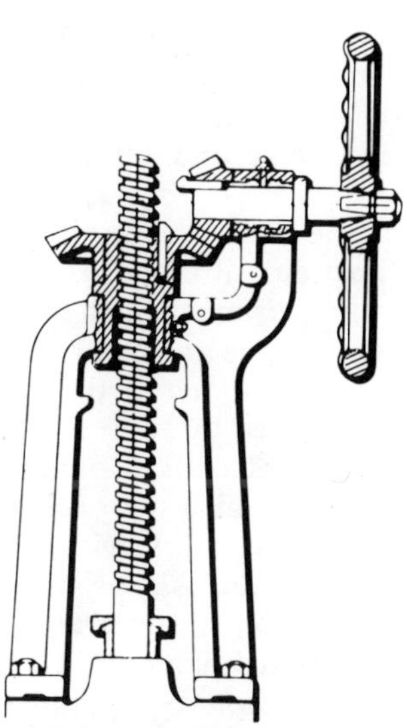

OS&Y bevel-gear head operator Fig. 18

impact action transfers to the elevating mechanism that actuates the valve gate, (causing a positive closing or an effective unseating.) The user should specify that these devices not spark upon impact.

Fig. 16 shows a type where the valve handwheel has been replaced by an adapter. A type is also available to clamp on to the handwheel.

Hammerblow handwheels, mostly of large diameters (24 in. or more), are supplied as standard equipment on some valves.

Gear operators

Gear operators provide additional mechanical advantages for opening and closing large valves. The operator unit can be mounted directly on the valve or worked remotely by means of extension stems. Designers have developed various styles of gears for industry in general (Fig. 17, 18).

The gearing normally furnished opens the valve when the handwheel is turned counterclockwise, but can be furnished to open on clockwise rotation. Covers for protecting the gears are also available.

Other accessories

Lock-shields—To prevent unauthorized tampering, valves can be ordered with operating nuts and without handwheels. These nuts are covered by lock-shields so that only a special T-wrench "key" can operate the valves.

Position indicators—Indicators are used on non-rising stem valves to show the relative position of the flow-control device. An indicator travels up and down in the slot as the valve opens or closes.

Pressure and temperature ratings

Knowing maximum operating pressures and temperatures, the engineer can determine the pressure rating required, in selecting a valve. The selection should be checked against manufacturers' lists for their pressure and temperature ratings, to assure that they conform.

Packing material

It is just as important to select the proper valve-packing material as to select the proper materials of construction for the intended service. Improper packing can lead to a leaky valve and cause a shutdown for replacement.

Toxic or flammable fluid leaking through the valve could create a disastrous situation, resulting in plant damage and injury to personnel. Such costly shutdowns and hazards are inexcusable and can be easily avoided. In selecting valve-packing material, the engineer should refer to literature published by valve and gasket manufacturers, and to current technical publications.

The shape or form in which the packing material is supplied must also be compatible with the mechanical design of the valve. Certain packing materials require high compression, but some valves are too flimsy or coarse to apply such high-compression forces. Conversely, the high compression required by the construction features of a valve may cause cold-flow in certain packing materials. Further, incompatible packing materials can cause the valve stem to gall.

Materials of construction

In selecting corrosion-resistant materials of construction, the engineer should be guided by the valve manufacturers' listings of recommended materials for the various services encountered, as well as by published corrosion data. Where such information appears to be inadequate, corrosion testing may have to be done in the laboratory. In general, unless the engineer is dealing with a first-of-a-kind process, there should be no difficulty in establishing the proper materials.

Cost and availability

Often, more than one type of valve can be found to handle a specific task. When all things are equal—regarding materials of construction, performance, pressure and temperature ratings, and availability—the basis of selection comes down to cost.

Having determined the type of valve best suited for a specific job, the engineer next must consider its availability and cost. It is not wise to select a valve that cannot be delivered on time or that is not reasonably priced. Availability and cost data must be obtained from suppliers and/or manufacturers. Concerning cost, the valve industry is cyclic in nature. Today's price will not necessarily be tomorrow's price, nor was it necessarily yesterday's price. The price per valve also depends on the number of valves ordered at the time. Quantity discounting is the usual practice.

It should be remembered, however, that a lower unit price can lead to higher overall costs, because of the effect a large order sometimes has on a manufacturer's ability to supply the valves on time. There are no hard-and-fast guidelines as to cost and availability of any specific type of valve—other than what can be discovered when soliciting bids from several suppliers.

Acknowledgements and illustration credits

The following companies supplied information and/or illustrative material for this article. Numbers in parentheses are figure numbers of illustrations supplied by these companies:

American Darling Valve and Manufacturing Div. (American Cast Iron Pipe Co.), Birmingham, AL 35202; Duriron Co., Dayton, OH 45401; Roto Hammer Co., Tulsa, OK 74107 (16); Stockham Valves & Fittings, Birmingham, AL 35212 (9, 11, 12, 13, 15, 17, 18); Strahman Valves, Inc., Florham Park NJ 07932; Walworth Co., Bala Cynwyd, PA 19004 (1, 2, 3, 4, 6, 14).

The author

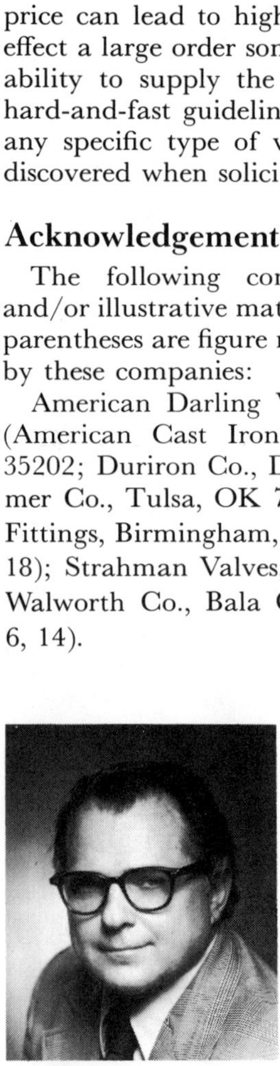

Arkadie Pikulik is a sales executive at Scientific Design Company, Inc., Two Park Ave., New York, NY 10016, where he has been employed for twenty-four years at various positions, including pilot plant supervisor, process engineer, project engineer, project manager, project director, construction manager and start-up supervisor. He holds a B.Ch.E. from Brooklyn Polytechnic Institute.

Pressure-relief valves for process plants

Relieving overpressures in process systems depends on whether a gas, vapor or liquid phase is vented. Design criteria for relief valves, fluid systems, and relief-valve piping will provide adequate capacities to assure compliance with the various codes.

Robert Kern, *Hoffmann - La Roche Inc.* *

☐ Pressure-relieving valves prevent undesired overpressurizing of process equipment and piping. Such valves operate automatically at a predetermined pressure to vent fluid and relieve the overpressure.

In this article, we will emphasize the operating characteristics and sizing of relief valves and their interrelation to system and piping design. For detailed definitions of the types of overpressure protection devices, pressure levels and required relief capacities, we can consult the literature [*1,2,3*]. It is also essential for us to know the requirements of the ASME boiler and pressure vessel codes [*4*] and the recommended practices of API [*3*].

Commercial relief valves

A pressure-relieving valve usually consists of an angle-type body having an inlet flange at the bottom and an outlet flange at the side (Fig. 1). The inlet flange is designed for the inlet-pressure and inlet-temperature ratings. The larger, outlet, flange usually has a lower pressure rating. In process relief valves, the inlet flange houses a conical nozzle whose small-diameter end serves

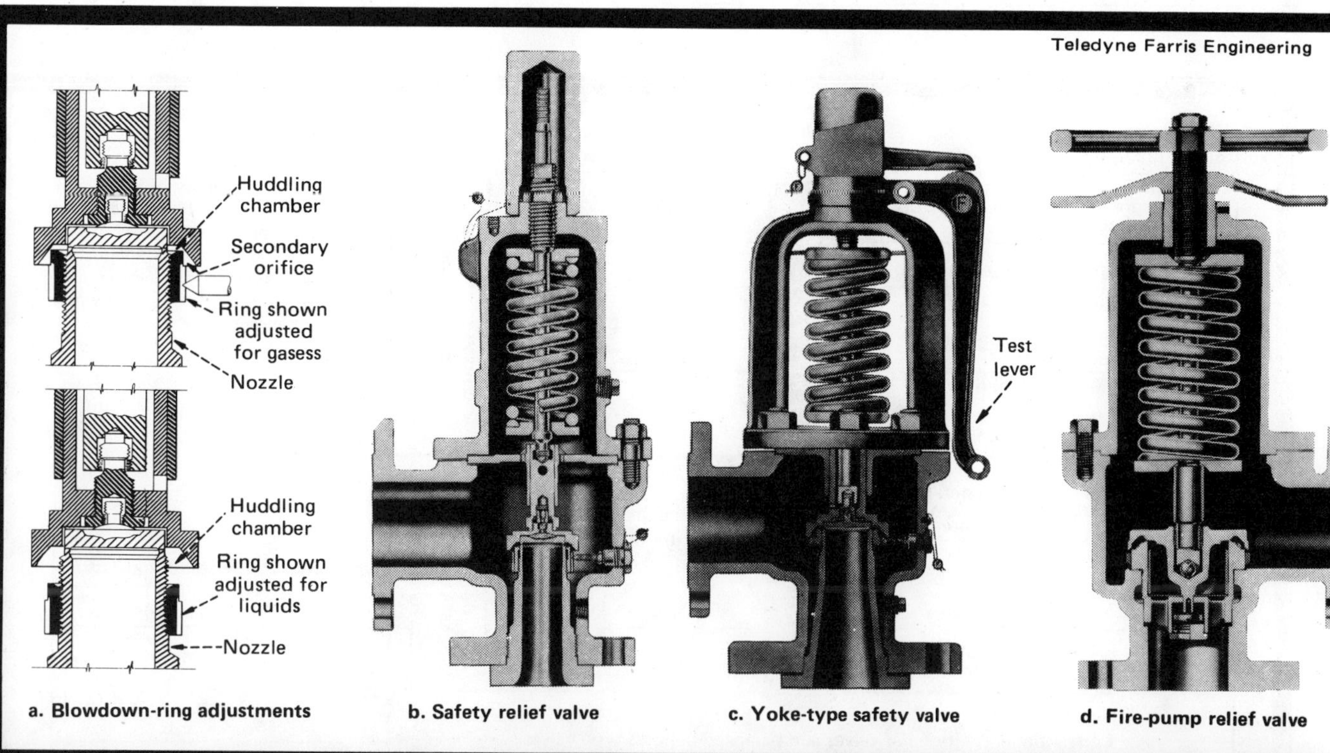

Construction details for typical pressure-relief valves for use by the chemical process industries Fig. 1

*To meet your author, see p. 44.

Originally published February 28, 1977

Nomenclature

A	Required valve-orifice area, in.^{2}
k	Adiabatic exponent (specific heat ratio, c_p/c_v)
M	Molecular weight
P	Absolute relieving pressure ($P_s + P_o + 14.7$), psia
P_b	Backpressure on discharge side, psi
P_d	Differential set pressure ($P_s - P_b$), psi
P_o	Overpressure, psi
P_s	Valve set pressure, psi
Q	Flowrate across the valve, gpm
ρ	Fluid density at flowing temperature, lb/ft^3
S	Specific gravity of liquid at flowing temperature
T	Absolute temperature, °R
W	Weight flowrate across valve, lb/h
z	Gas compressibility factor

Capacity-correction factors

C	Gas- or vapor-flow constant, from Table II
K	Valve coefficient of discharge
K_b	Gas- or vapor-flow correction factor for constant backpressure, from Table I
K_p	Liquid capacity-correction factor for overpressures less than 25%, from Fig. 4a.
K_{sh}	Steam-superheat correction factor, from Table III
K_u	Liquid-viscosity correction factor
K_v	Gas- or vapor-correction factor for variable backpressure (for balanced-bellows valves only), from Fig. 7
K_w	Liquid flow factor for variable backpressure (for balanced-bellows valves only), from Fig. 4b.

Subscripts

l	Liquid
s	Steam
v	Vapor

as the valve seat on which a spring-loaded disk rests. Such safety-relief valves can be used for liquid and vapor services (Fig. 1b).

The spring load on the disk offsets the inlet pressure. To avoid leakage (simmering) across the valve seat during normal operation in liquid or vapor service, the spring is set 10% above the normal operating pressure. This set pressure, P_s, can be adjusted with a screw above the spring. To avoid tampering with the set pressure, the screw is covered with a threaded cap. When an alternate spring setting is desired, the adjustment should not be more than ±10% of the manufacturer's setting up to 250 psi, and ±5% above 250 psi.

The spring can be housed in a bonnet. A closed bonnet is used when the discharged fluid must be confined within the valve body and discharge piping. This bonnet is sized for the outlet pressure rating of the valve. Yoke-type valves expose the spring and are usually selected for air and steam services. According to the ASME Code [4], safety valves in steam service must have test levers (Fig. 1c).

The disk with vanes under it will be guided in the inlet port (i.e., bottom guided). With a stem above it, the disk will be guided in a sleeve in the bonnet or yoke (i.e., top guided). Liquid relief valves sometimes have bottom guides (Fig. 1d). Process- and boiler-relief valves are usually top guided (Fig. 1b).

The following terminology applies to pressure relief valves:

- Relief valve—primarily for liquid service.
- Safety valve—for steam, gas and vapor service.
- Safety relief valve—for liquid or vapor service.

Liquid-relief valves begin to open when the internal pressure reaches set pressure. Pressure will rise gradually

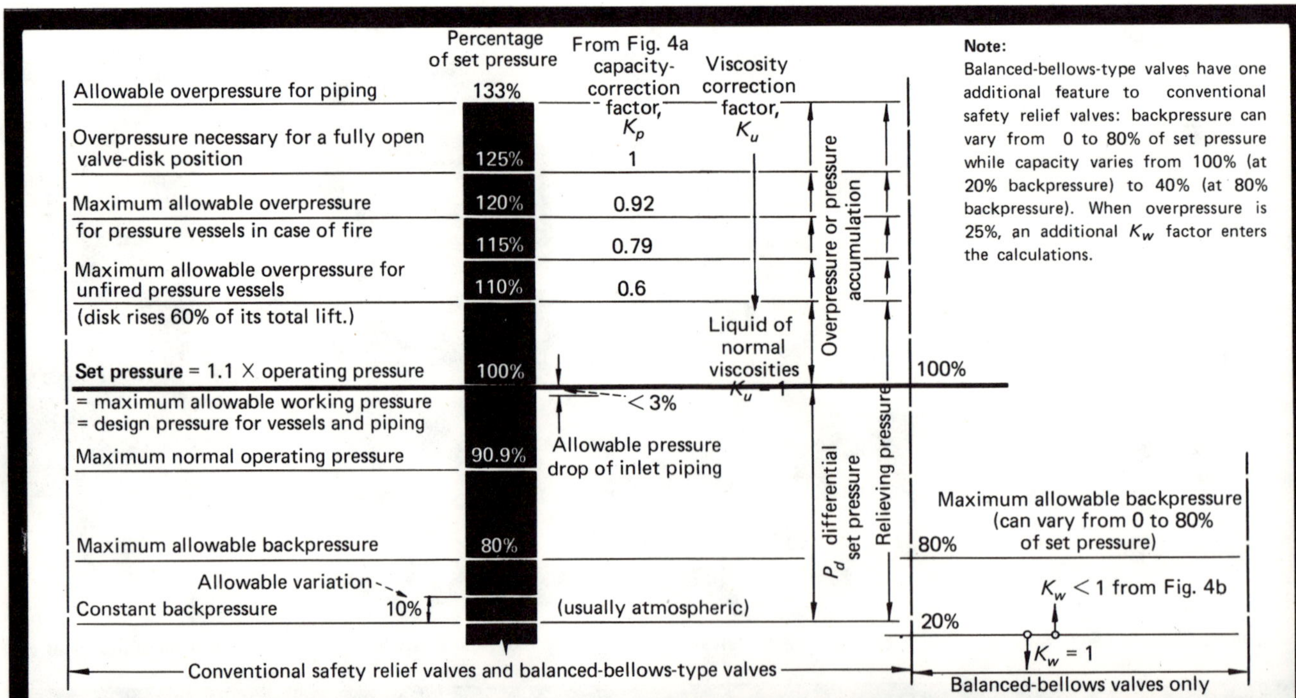

Liquid service: relative pressure levels for relief valves as related to the valve set pressure — Fig. 2

until it is 10 to 33% (depending on the service) over the set pressure. The valve reaches full capacity at 25% overpressure. Relieving pressure is defined as set pressure plus overpressure. Fig. 2 and 3 show the pressure relationships and definitions for conventional and balanced-bellows relief valves.

In steam, air, gas and vapor services, the safety valve pops open when pressure reaches the set pressure. Pressure will continue to rise—usually 3 to 33% above the set pressure. After relieving (called blowdown), the disk reseats at about 4% below set pressure.

In gas and vapor valves, static pressure opens the disk, and a dynamic force keeps it open. This force is created by the increasing fluid velocity in the conical nozzle below the valve disk and the bell-shaped underside of the disk, called the huddling chamber (Fig. 1a). The huddling chamber diverts the gas flow. The velocity, the mass of gas, and the diversion are proportional to the force that keeps the valve seat open.

A flow ranging from 25 to 30% of maximum valve capacity is needed to keep the valve seat open. A lesser flow might cause frequent opening and closing (i.e., chattering) of the valve.

In safety relief valves, an adjustable blowdown ring protrudes into the huddling chamber (Fig. 1a). At a high position for the ring, the pop action is sharper and blowdown longer than at a lower position. In liquid service, the blowdown ring should not interfere with operation of the valve. For this reason, the blowdown ring is placed at its lowest position (Fig. 1a).

Special relief valves and devices

The sentinel valve is a small pop valve. It warns the operator of excessive pressure buildup. The diaphragm relief valve is applied to corrosive services. Below a 1-in. size, the relief valve usually has threaded connections. Small relief valves protect against pressure buildup in liquid thermal expansion.

A pilot-operated relief valve is used where set pressure is very close to normal operating pressure. This type of valve allows a lower design pressure to be specified for high-pressure services and can save material and manufacturing costs.

The rupture disk is a thin saucer-shaped plate between flanges [7]. The thickness of the plate and the strength of its material is designed so that it breaks through at an exact predetermined pressure. Rupture disks give a tight seal against toxic, corrosive or flammable liquids and gases and are economical. They can be used upstream of a conventional relief valve. However, spare units must be at hand for replacement. For continuous operation, parallel installations have to be provided, with isolating valves that must be locked open during operation.

Sizing relief valves for liquid service

To size relief valves, we calculate a required valve orifice (A, in.^{2}) and select the next higher available orifice from manufacturers' catalogs. A summary of valve-sizing formulas for liquid services follows:

Piping—The derivation for a sizing formula is based on Bernoulli's equation (see p. 20). A convenient basic expression including the flow coefficient is:

$$A = \frac{Q\sqrt{S}}{27.2\sqrt{P_d}} \tag{1}$$

where A = required valve orifice, in.^{2}; Q = flowrate at

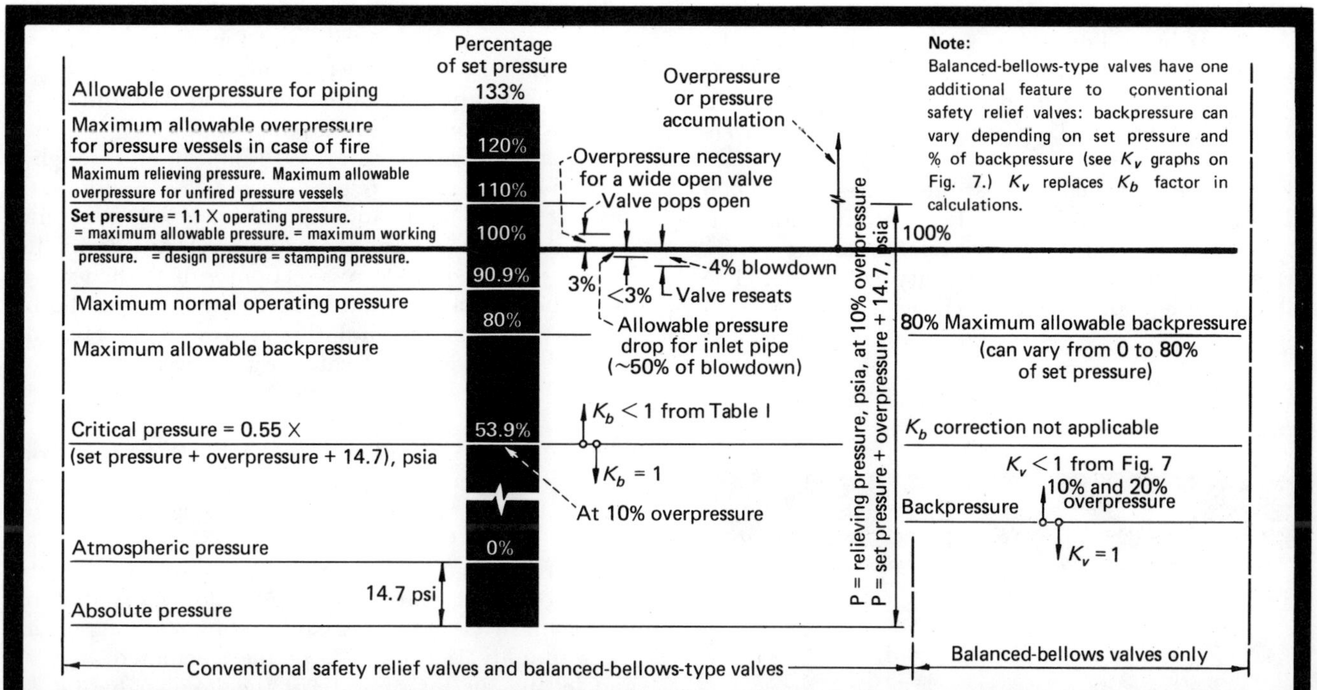

Vapor and gas service: relative pressure levels for safety valves as related to valve set pressure — Fig. 3

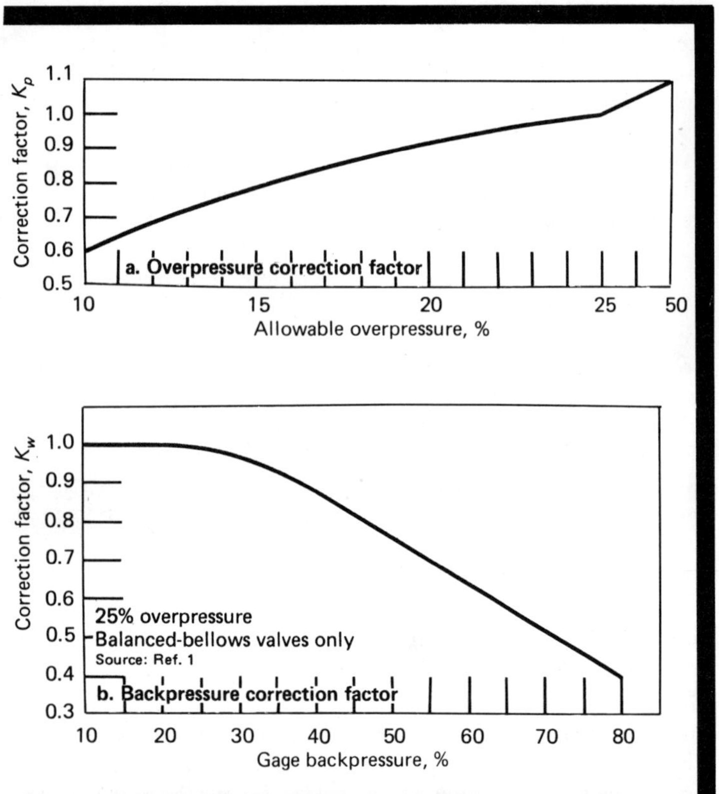

Correction factors for sizing relief valves in liquid service **Fig. 4**

flowing temperature, gpm; and S = specific gravity of liquid at flowing temperature. P_d is the differential set pressure, psi. It is the valve set-pressure (P_s, psi) minus backpressure on the outlet side (P_b, psi), or $P_d = P_s - P_b$. For relief valves discharging to atmosphere, $P_d = P_s$.

A relief valve in liquid service is fully open when the pressure reaches $1.25P_s$, i.e., 25% overpressure. Eq. (1) is suitable for calculating relief valves for overpressure in piping. Codes allow 33% pressure accumulation (overpressure) for piping.

Viscous liquid—When the viscosity is less than 2,000 SSU (Saybolt Seconds Universal), a viscosity correction is applied to Eq. (1). It is advisable to assume a 25% overpressure in addition to viscosity correction. The sizing formula, Eq. (1), now becomes:

$$A = \frac{Q\sqrt{S}}{27.2\sqrt{P_d}\,K_u} \qquad (2)$$

Constant backpressure sizing factor for conventional valves (vapor or gas service) **Table I**

Absolute backpressure, %	Constant backpressure factor, K_b
55	1.0
60	0.995
65	0.975
70	0.945
75	0.9
80	0.845

Source: Ref. 1

Capacity-correction factors for viscosity, K_u, at various Reynolds numbers,* N_{Re}, are:

N_{Re}	K_u	N_{Re}	K_u
5,000	0.98	100	0.59
3,000	0.97	50	0.41
2,000	0.96	40	0.35
1,500	0.95	30	0.28
1,000	0.91	20	0.19
500	0.89	10	0.11

In many cases, the use of steam tracing or a steam-jacketed valve body will prevent too high viscosities, or solidification of material in the valve body.

Pressure vessels—Pressure-vessel codes do not allow 25% overpressure when the valve relieves. Consequently, a capacity-correction factor, K_p, is required in Eq. (1) for overpressures less than 25%. The sizing formula for Reynolds numbers greater than 2,000, and viscosities over 2,000 SSU, is:

$$A = \frac{Q\sqrt{S}}{27.2\sqrt{P_d}\,K_p} \qquad (3)$$

Capacity-correction factors for other than 25% overpressure are:

Overpressure, %	K_p	Disk position, % open
33	1.03 (use $K_p = 1$)	100
25	1	100
20	0.92	92
15	0.79	79
10	0.6	60

Correction for under 10% of overpressure is not recommended. In the rare case of relieving viscous liquid, we multiply the righthand side of Eq. (3) by $(1/K_u)$.

Maximum allowable overpressure for pressure vessels in case of fire is 20% ($1.2P_s$) and for unfired vessels, is 10% ($1.1P_s$). For other overpressures, see Fig. 4a. Fig. 2 shows the relative pressure levels, and a summary of correction factors for liquid-relief valves.

Saturated liquids—As saturated liquid flows through a restriction and the pressure decreases, an adiabatic flashing takes place in and right after the valve opening. Up to the sonic velocity, the sizing recommendation [6] is: Calculate required cross-section with flashed-vapor quantities, A_v; and with the remaining saturated-liquid quantity, calculate the liquid cross-section, A_l. The sum gives the required relief-valve orifice area:

$$A = A_l + A_v$$

where A_l can be calculated from Eq. (3), and A_v from Eq. (6), for conventional safety-relief valves. To calculate the areas for a balanced-bellows-type valve, use Eq. (4) and Eq. (9)

According to the literature, this method oversizes the valve. To avoid chattering because flow capacities are too small, two relief valves can be chosen having orifice areas roughly 1/3 and 2/3 of the calculated area, A, and letting the set pressure of the small valve be 3% higher than that of the large one.

*For calculating Reynolds numbers and viscosity conversions, see Fig. 3 p. 90 and Eq. (13) p. 93.

A comparison between control valves and relief valves for flow and sizing can be misleading. When a control valve opens, flow increases with decreasing pressure differential. In contrast, when a relief valve opens, flow increases with increasing pressure differential. Pressure differentials at control valves are usually a fraction of overall pressure drop compared with the high pressure drops at relief valves. Also, relief-valve design takes into account a volume increase at the outlet side, with a generous valve housing and larger outlet flange than the inlet.

Piping after the relief-valve discharge nozzle is often swaged up to accommodate flashing liquid or expanding gas, and to avoid excessive velocities or excessive buildup of backpressure.

Balanced bellows—Let us assume that a conventional relief valve operates with a backpressure of between 10 to 80% of set pressure. The valve disk is pressed to its seat by the spring force plus a force equal to the seat-ring area times the backpressure. If the backpressure varies, builds up, or changes from time to time, the relief valve will not open at the desired set pressure. This cannot be allowed.

If the valve spring is sealed from the outlet chamber by a bellows whose diameter is identical to the seat-ring diameter, the backpressure will not affect set pressure. The sketches in Fig. 5 show these conditions.

Balanced-area bellows are used in case of buildup, intermittently-changing or permanent backpressures. In addition, the bellows can be used to separate the spring from a corrosive fluid flowing through the valve. A balanced-bellows valve must be used when the variation in backpressure exceeds 10% of the set pressure.

When sizing a balanced-bellows valve, the righthand side of Eq. (3) is multiplied by a backpressure correction factor, $1/K_w$:

$$A = \frac{Q\sqrt{S}}{27.7\sqrt{P_d}K_pK_w} \quad (4)$$

where K_p values are taken from Fig. 4a, and K_w can be obtained from Fig. 4b by entering the graph with the percentage of gage backpressure, P_{bg}, which is:

$$P_{bg} = (P_b/P_s)100 \quad (5)$$

In Eq. (5), P_b and P_s are given in psig. For viscous flow, we multiply the righthand side of Eq. (4) by $1/K_u$.

Sizing valves for gas and vapor services

There is a difference in the operation of relieving valves for liquid and vapor services. For liquid-pressure relief, an overpressure of 25% is necessary for a fully open valve. For vapor service, the valve seat pops open at 3%, or less, overpressure. Consequently, an overpressure correction for vapor valves is not required.

A convenient sizing formula for relieving vapor, gas and air is:

$$A = \frac{W\sqrt{Tz}}{CKK_bP\sqrt{M}} \quad (6)$$

where A is required valve-orifice area, in.^{2}; W is flowrate, lb/h; z is compressibility factor* corresponding to

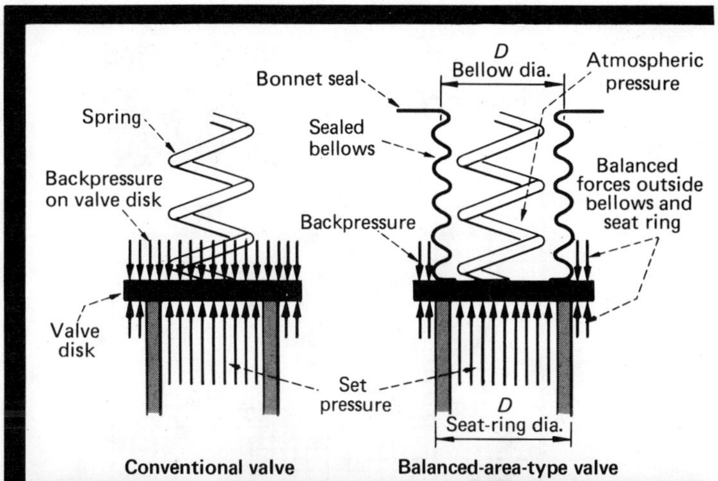

How backpressure acts on valve disks in conventional and bellows relief valves — Fig. 5

the absolute relieving pressure, P, psia, and absolute flowing temperature, T, °R. (Note: $P = P_s + P_o + 14.7$, and $T = °F + 460$.) M is average molecular weight of vapor or gas. K is valve coefficient of discharge determined by manufacturers, for example, $K = 0.975$ [*1*].

Set-pressure plus overpressure values recommended by the ASME code [*4*] are:

$P_s + P_o = 1.1P_s$ for unfired pressure vessels
$P_s + P_o = 1.2P_s$ for vessels in case of fire
$P_s + P_o = 1.33P_s$ for piping

Also in Eq. (6), K_b is the constant-backpressure sizing factor that is a function of the percentage of absolute backpressure, P_{ba}:

$$P_{ba} = [P_b/(P_s + P_o + 14.7)]100 \quad (7)$$

Gas or vapor flow constants — Table II

Specific heat ratio, k	Constant, C	Specific heat ratio, k	Constant, C	Specific heat ratio, k	Constant, C
1.00	315	1.26	343	1.52	366
1.02	318	1.28	345	1.54	368
1.04	320	1.30	347	1.56	369
1.06	322	1.32	349	1.58	371
1.08	324	1.34	351	1.60	372
1.10	327	1.36	352	1.62	374
1.12	329	1.38	354	1.64	376
1.14	331	1.40	356	1.66	377
1.16	333	1.42	358	1.68	379
1.18	335	1.44	359	1.70	380
1.20	337	1.46	361	2.00	400
1.22	339	1.48	363	2.20	412
1.24	341	1.50	364		

Source: Ref. 1

*See p. 87 Fig 1

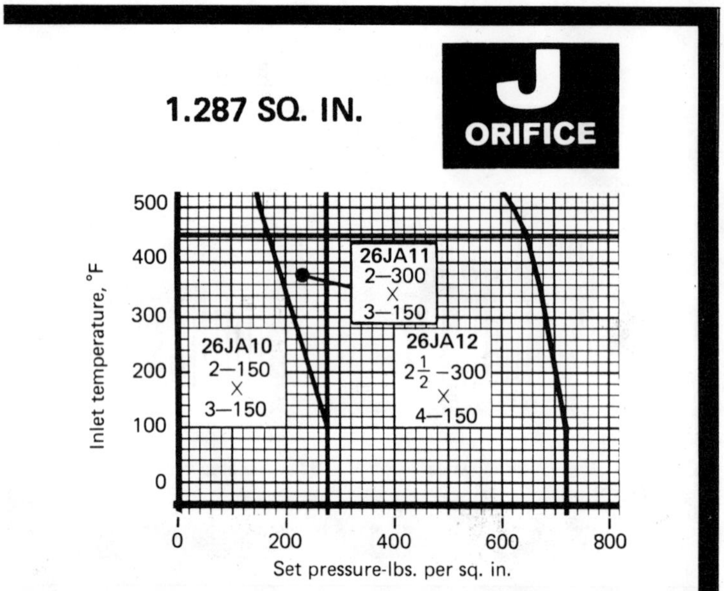

Portion of typical selection chart from a manufacturer's catalog

Fig. 6

where P_b is backpressure, psia; P_s is set pressure and P_o is overpressure, both in psig. K_b values are given in Table I. $K_b = 1$ when back pressure is less than the absolute relieving pressure $P_b < 0.55(P_s + P_o + 14.7)$.

The gas or vapor-flow constant, C, is a function of the adiabatic exponent $k = c_p/c_v$:

$$C = 520\sqrt{k\left(\frac{2}{k+1}\right)^{(k+1)/(k-1)}} \tag{8}$$

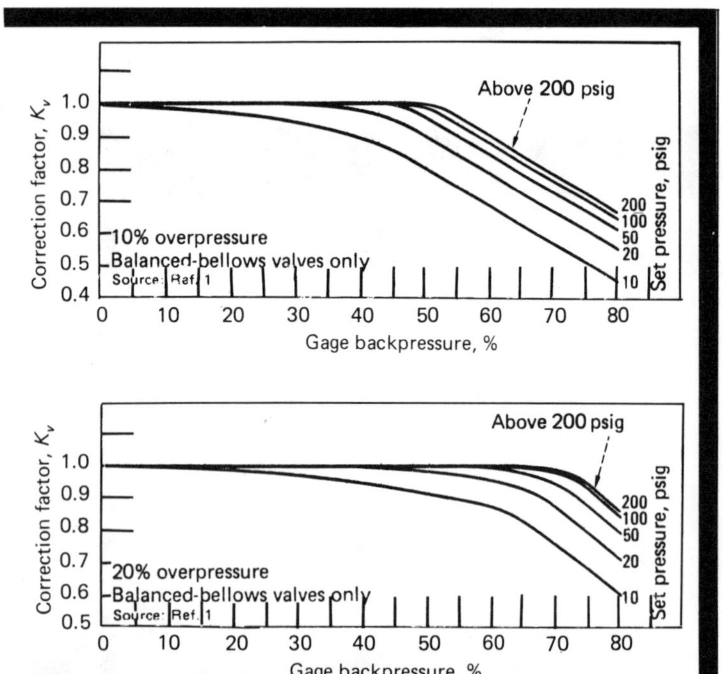

Backpressure sizing factors for relief valves in vapor or gas service

Fig. 7

Corresponding values of k and C, calculated with Eq. (8), are given in Table II. If k is not known, the assumption of $k = 1$ will give a conservative result, and $C = 315$. Fig. 3 shows the relative pressure levels for safety valves, with the summary of correction factors.

Example—Let us size a safety valve discharging to atmosphere ($P_b = 14.7$ psia). The following flow data apply: $W = 26{,}748$ lb/h; set pressure, P_s, = 400 psig; gas temperature = 100°F; molecular weight = 18.7; compressibility factor, z, = 0.9; and $k = 1.3$.

With $k = 1.3$, we find that $C = 347$ from Table II. Since backpressure is below 55% of absolute relieving pressure, $K_b = 1$. We calculate the required valve-orifice area, A, by substituting into Eq. (6):

$$A = \frac{26{,}748\sqrt{560(0.9)}}{347(0.975)(1)(454.7)\sqrt{18.7}} = 0.9 \text{ in.}^2$$

From manufacturers' catalogs, we select the safety valve having an orifice area closest to and larger than the calculated value. A section of a typical page from such a catalog [1] is shown in Fig. 6. Hence, for a set pressure of 400 psig and inlet temperature of 100°F, the valve selected will be one having a 2½-in.-dia., 300-lb flange inlet; a 4-in.-dia., 150-lb flange outlet; and an orifice area of 1.287 in.² Of course, if a different orifice area is required, the appropriate selection chart must be used.

In this example, the difference between calculated orifice area (0.9 in.²) and available area (1.287 in.²) is rather large. Alternatively, two safety valves can be selected—one with an orifice area of 0.307 in.² and the other of 0.785 in.² This gives an available orifice area of 1.092 in.², which is much closer to the calculated quantity. The smaller valve can be set to relieve at a pressure 3% higher than the larger one.

Balanced-bellows valves—Backpressure affects the rated capacity of a valve. Consequently, a correction is necessary. The correction factor, K_v, for gas and vapor is equal to the valve capacity with backpressure, divided by the rated capacity without backpressure. K_v can be obtained from the graphs of Fig. 7 by using the percentage of gage backpressure, P_{bg}, as computed from Eq. (5).

The sizing formula for balanced-bellows valves with constant or variable backpressure in gas and vapor service is the same as Eq. (6) except that K_b is replaced by K_v:

$$A = \frac{W\sqrt{Tz}}{CKK_vP\sqrt{M}} \tag{9}$$

An example of variable backpressure is when a safety valve discharges to a low-pressure process system.

Saturated and superheated steam

Because the molecular weight of steam is known, and steam density and saturation point are a function of temperature and pressure, Eq. (6) can be simplified and modified by a superheated-steam correction factor. The sizing formula for steam-relief valves is:

$$A = \frac{W_s}{51.5KPK_bK_{sh}} \tag{10}$$

where W_s = steam flowrate, lb/h; $K = 0.975$; $P = (P_s + P_o + 14.7)$ = relieving pressure, psia. K_b values are given in Table I. $K_b = 1$ when backpressure is below 55% of absolute relieving pressure. $K_{sh} = 1$ for saturated steam. Correction factors for superheated steam are given in Table III.

The ASME Boiler Code [*4*] allows only 3% pressure accumulation (i.e., $P_s + P_o = 1.03P_s$) except for steam generators having a size greater than 500 ft^2 and fitted with two safety valves. The second safety valve can be at 3% higher set-pressure than the first, and can relieve 6% higher than the first valve's set-pressure. See the ASME code [*4*] for specific details.

Relief-valve piping

A safety relief valve (RV) must open and operate at its design conditions. These include sizing, selecting and specifying the RV, and also the location of the valve on the equipment and the piping connected to it.

A relief valve is usually positioned with stem vertical, inlet nozzle down, and outlet nozzle horizontal. If an RV is connected to a horizontal nozzle or pipe, it should have a long-radius elbow before the inlet nozzle of the RV. On horizontal vessels, the relief valve is usually connected on the top. On vertical vessels, a side connection should be radial.

Inlet and outlet piping—The minimum pipe size to the inlet of the relief valve must be the same as the inlet diameter of the RV. Resistance of the inlet pipe should not be more than 2 to 3% of set pressure (50% blowdown). Excessive inlet-pipe resistance can also cause chattering. The inlet piping should be short, simple and obstruction-free. Relief valves requiring frequent maintenance (for example, in viscous or corrosive services) can have standby relief valves with block valves. The block valve should have a full pipe-area opening and should be locked in a completely open position. A strict valve-closing procedure is required by the unfired-pressure-vessel code. Block valves can be interlocked—when one is closed, the other must open. For corrosive fluids, the RV can be protected by a balanced bellows, or by a rupture disk before the relief-valve inlet.

When an RV opens, a blowout thrust, *F*, is imposed at the centerline of outlet pipe in the direction opposite to the outlet. This causes a bending moment, *M*, and a shear force at the point where the relief-valve nozzle connects to the vessel, as shown:

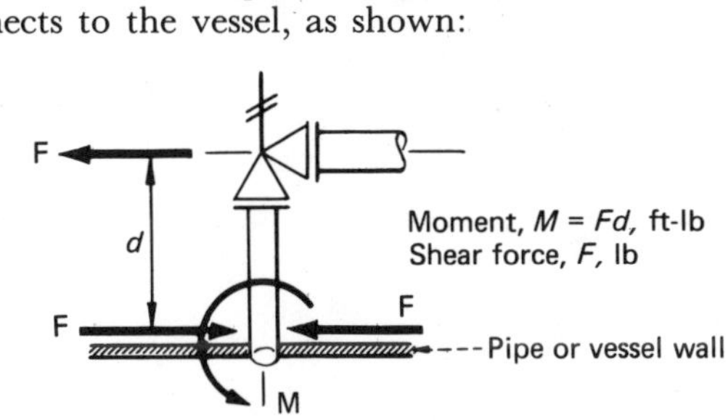

Force *F* imposes a bending moment and shear at the vessel or pipe wall. The bending moment increases in proportion to dimension *d*. To minimize the bending moment, the RV should be as close to the vessel or pipe as possible. If dimension *d* has an appreciable length, it may be necessary to reinforce the vessel wall, or support the relief valve.

A pipeline will withstand a bending moment if the RV outlet is parallel with it. At a perpendicular outlet, the bending moment tends to twist the pipeline.

Generally, if the inlet pipe of the RV is 3 in. or greater, and has even the smallest configuration, an investigation for pipe stress, load and support is in order. In many cases, the relief valve must withstand considerable thermal and mechanical loads during a hot discharge in addition to internal pressure and possible vibration. Dead loading or expansion loading of the piping should not add to the unavoidable relieving strain during RV operation.

Open relief-valve discharge

The location of a relief valve depends on its service and discharge functions. With an open discharge, a liquid RV can be located at the liquid space of a vessel. The discharge outlet turns down and is usually piped, without a loop or pocket, to a drain point. Relief valves, for gas and vapor, with open discharge are usually located at the high point of the vapor space in a vessel or pipeline. The discharge turns up and terminates about 10 ft above grade or platform. To avoid rain, snow or condensate accumulation in the discharge line, a bent-down tip or loose cap can be provided. A weep hole at the low point of a discharge line will drain condensate out of the line. A liquid static head cannot be allowed in the discharge line because it will increase pressure on the outside of the valve disk.

When the possibility of igniting the discharged gases

Sizing factors for superheated steam — **Table III**

Set pressure psig	Saturated temp. °F	Total temperature, °F							
		320	400	500	600	700	800	900	1,000
10	240	1.0	0.96	0.91	0.87	0.83	0.80	0.77	0.74
20	259	1.0	0.98	0.92	0.88	0.84	0.81	0.78	0.75
40	287	1.0	0.99	0.93	0.89	0.85	0.81	0.78	0.75
60	308	–	0.99	0.94	0.90	0.86	0.82	0.79	0.76
80	324	–	0.99	0.95	0.90	0.86	0.82	0.79	0.77
100	338	–	0.99	0.96	0.90	0.87	0.82	0.80	0.77
120	350	–	0.99	0.97	0.92	0.87	0.83	0.80	0.77
140	361	–	0.99	0.97	0.92	0.87	0.83	0.80	0.77
160	370	–	1.0	0.97	0.92	0.87	0.83	0.80	0.77
180	379	–	1.0	0.97	0.92	0.87	0.83	0.80	0.77
200	388	–	1.0	0.98	0.92	0.87	0.83	0.80	0.77
220	396	–	1.0	0.98	0.92	0.87	0.83	0.80	0.77
240	403	–	1.0	0.98	0.93	0.87	0.84	0.80	0.77
260	409	–	–	0.99	0.93	0.88	0.84	0.80	0.77
280	416	–	–	0.99	0.94	0.88	0.84	0.80	0.77
300	422	–	–	0.99	0.94	0.89	0.84	0.81	0.78
350	433	–	–	0.99	0.94	0.89	0.85	0.81	0.78
400	448	–	–	0.99	0.95	0.89	0.85	0.81	0.78
500	470	–	–	0.99	0.95	0.89	0.85	0.81	0.78
600	489	–	–	1.0	0.97	0.90	0.85	0.81	0.78
800	520	–	–	1.0	0.97	0.90	0.85	0.81	0.78
1,000	546	–	–	–	0.98	0.90	0.85	0.81	0.78

Source: Ref. 1

exists, a connection for snuffing steam can be installed at the low point of the RV discharge. The steam-activating valve must be accessible from a safe distance, and the RV discharge line should not be pocketed.

Safety valves and vents in a process building usually discharge to the atmosphere through a vent box. The condensate-collecting trough for the vent box usually is drained to a process sewer at grade.

For subcooled liquids, the outlet-pipe diameter can be the same size as the RV outlet. When saturated liquid flows, an increase of one size in pipe diameter is generally necessary. Vapors and gases expand after flowing through the valve seat. Here, fluid velocities in the discharge line should stay well below the sonic velocity if attainable with an increase in line size. Near-sonic or higher-than-sonic velocities at the end of the discharge line might require use of a silencer. A silencer or a long discharge-line can impose a built-up backpressure in the line. Overall pressure drop in the discharge line should be less than 10% of set pressure.

Many hazards may arise with open-discharge lines. The discharged liquid, vapor or gas can be ignited, and cause fire, explosion and personal injury. Vapor condensation can be corrosive to the surrounding structures and equipment. Toxic gases can pollute the air; flashing liquid can evaporate over a wide area and its odor can be unpleasant; noise can create problems.

Closed-discharge systems

For avoiding hazards, or where large quantities of discharge are expected, or where the liquid or gas must be reclaimed, a closed-discharge system is usually designed.

A closed system consists of the individual leads from the discharge flange of the safety relief valves to the collecting header. The leads connect to the header at the top. The whole piping arrangement, from the relief-valve outlet to the terminating point of the header, should be self-draining. When a low-point pocket in a relief line is unavoidable, a liquid collecting pot can be provided with level gages, controls and pumpout arrangement.

The number or combination of systems can be influenced by: client preference, economy, overall relieving systems in a chemical-plant complex, and available spare capacity in existing relieving headers. Reasons for the possible separation of relieving systems (beside the obvious one of separate systems for liquids and vapors) are:

■ Fluid temperatures. Materials of construction, specifications, supports and disposal can differ between a cold and hot relieving system.

■ Piping material. Because of temperature and corrosiveness, some systems require alloy piping. It can be more economical to keep such systems separate from others, in which carbon steel can be used.

■ Viscosities. Relief-valve discharges of normal and heavy viscosities can be separated. Relief valves and pipelines in viscous services usually need more maintenance, possibly steam tracing, clean-out openings in piping, steam-out connections, spare relief valves and block valves than do lines and equipment in normal viscous service.

■ Condensing-vapor systems. These need liquid-removal provisions, and sloped gravity-flow arrangements. Noncondensing gases do not. However, a common system can be designed with the requirements of the condensing system.

■ High pressure. When frequent high-pressure discharge can be expected, there is an advantage to separate it from other low-pressure discharges in closed relieving systems.

A closed header for relief valves is sized as a collecting system where the flowrates in the pipelines vary and are intermittent. A flow diagram of the relieving system should be drawn with RV leads in geographically realistic sequence. A liquid-collecting header can be sized as a gravity-flow line. With gas lines, the basic requirement is that a discharging RV should not impose excessive backpressure in the header.

By assuming simultaneous discharges from large safety valves, or from grouped or closely located valves, and by choosing unit losses in fractions of 1 psi, the initial line-size selection can be made. Jenett [*5*] has summarized procedures on how to calculate vent-line sizes and backpressures.

All relief valves, in liquid or vapor service, should be located above the collecting header, with self-draining leads. The valves should be conveniently accessible from a platform.

A closed relief-valve header can terminate at a low-pressure or high-pressure blowdown drum, a quench drum or pit, smokeless flare, process sewer, or suitable low-pressure process line or equipment.

References

1. Safety and Relief Valves, Farris Engineering Corp., No. FE336, Palisades Park, N.J.
2. Weber, C. G., Safety Relief Valve Sizing, Farris Engineering Corp., Palisades Park, N.J.
3. Recommended Practice for the Design and Installation of Pressure Relieving Systems in Refineries, Part I: Design; Part II: Installation, API RP-520, American Petroleum Institute, Wash., D.C., 1960.
4. ASME Boiler and Pressure Vessel Code, Section I: Fired Pressure Vessels, Piping, Valves and Fittings, pp. 31–40; Safety Valves and Safety Relief Valves, pp. 40–44; Section VIII: Unfired Pressure Vessels, Pressure Relief Devices, pp. 49–55, American Soc. of Mechanical Engineers, New York, 1965.
5. Jenett, E., Design Considerations for Pressure-Relieving Systems, *Chem. Eng.,* July 8, 1963; Components of Pressure-Relieving Systems, *Chem. Eng.,* Aug. 19, 1963; How To Calculate Back Pressure in Vent Lines, *Chem. Eng.,* Sept. 2, 1963.
6. Driskell, L. R., Piping of Pressure-Relieving Devices, *Petrol. Refiner,* July 1960.
7. Safety With Rupture Discs, Continental Disc Corp., Riverside, Mo.
8. Anderson, F. E., Pressure relieving devices, *Chem. Eng.,* May 24, 1976.
9. Kern, R., *Chem. Eng.,* Jan. 6, 1975, p. 116, Eq. (6).
10. Kern, R., *Chem. Eng.,* Dec. 23, 1974, p. 62.

Pressure-relieving devices

Although codes are explicit about relief-system sizing and general mechanical design, the engineer is left pretty much on his own when it comes to choosing specific devices for specific services. Here is some advice, based on experience, on how to choose this equipment for process operations.

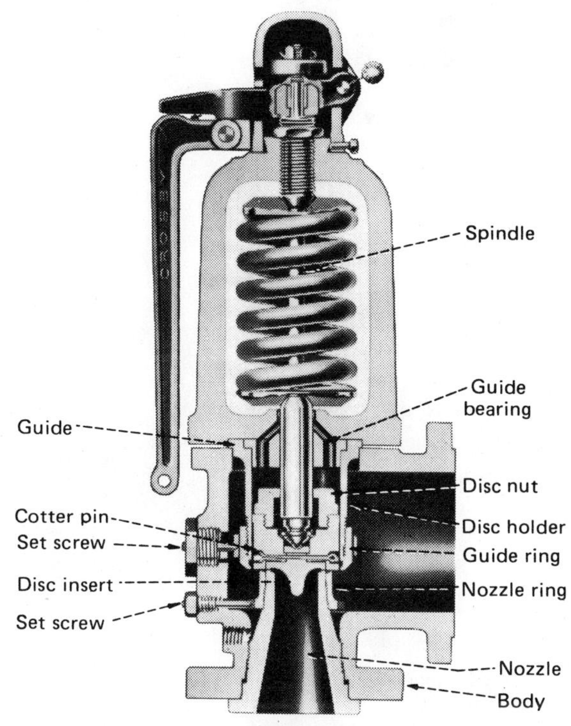

Safety valve is for boilers and steam **Fig. 1**

Source: Crosby Valve & Gage Co.

Floyd E. Anderson, *Fluor Engineers and Constructors, Inc.*

☐ Process plants are replete with equipment that is vulnerable to surges in pressure.

Fortunately, procedures for coping with pressure surges have been included in many codes. At the very least, these codes represent sound engineering practice. The American Petroleum Institute's API RP 520, Recommended Practice for the Design and Installation of Pressure Relieving Systems in Refineries, Part I—Design, is excellent for determining relief loads; and though it applies particularly to petroleum refineries, it has relevance for all types of process plants. Similarly, the relief loads for power boilers and superheaters are based on the maximum forced combustion of a furnace, as described in Appendix A-12 and section PG-67 of the American Soc. of Mechanical Engineers' Code, Section I, Power Boilers. Capacity conversions for these loadings may be made according to Part UA-280, Appendix J, Section VIII, Div. 1, of the ASME Code. And the general design of safety devices for all types of pressure vessels is prescribed in Parts UG-125 to UG-134 of Section VIII, Div. 1.

However, this leaves the engineer with the problem of selecting a relief device to suit a specific situation. Various types of devices are available. While no single one is suitable for all services, each is ideally suited for some particular application.

Parts UG-125–134 provide guidelines to which most of the commercially available relief devices are designed. Pressure-relieving devices are broadly separated into two groups: (1) valves and (2) rupture disks. The valves are spring-loaded, unless activated by a fail-safe type of pilot (Fig. 6); and if they are used for steam or air, they are equipped with a lever to open the valve when the contained pressure is no more than 75% of the relief pressure. Neither their seat nor their disks should be made of cast iron. And if the design is such that liquid could collect on the discharge side of the disk, the valves should be equipped with a drain.

Furthermore, these valves are broken down into (a) safety valves and (b) relief valves. A safety valve is: "An automatic pressure-relieving device actuated by the static pressure upstream of the valve, and characterized by full-opening pop action. It is used for gas or vapor service." A relief valve is: "An automatic pressure-relieving device actuated by the static pressure upstream of the valve, which opens further with the increase in pressure over the operating pressure. It is used primarily for liquid service."

Rupture disks may be substituted for safety valves as protection against corrosion and to avoid leaks. Thus, a rupture disk may be installed between a vessel and a spring-loaded valve, as long as there is a pressure gage, free vent or suitable indicator for rupture or leakage. When used, a rupture disk should have a connection at least as large as the relief area of the valve's disk; it should have an actual bursting pressure guaran-

Originally published May 24, 1976

Source: Crosby Valve & Gage Co.

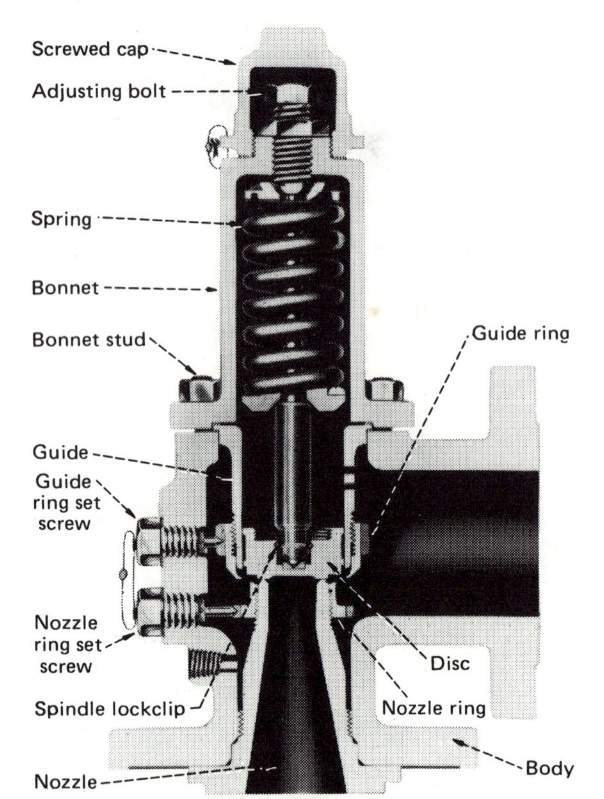

Full-nozzle safety-relief valve Fig. 2

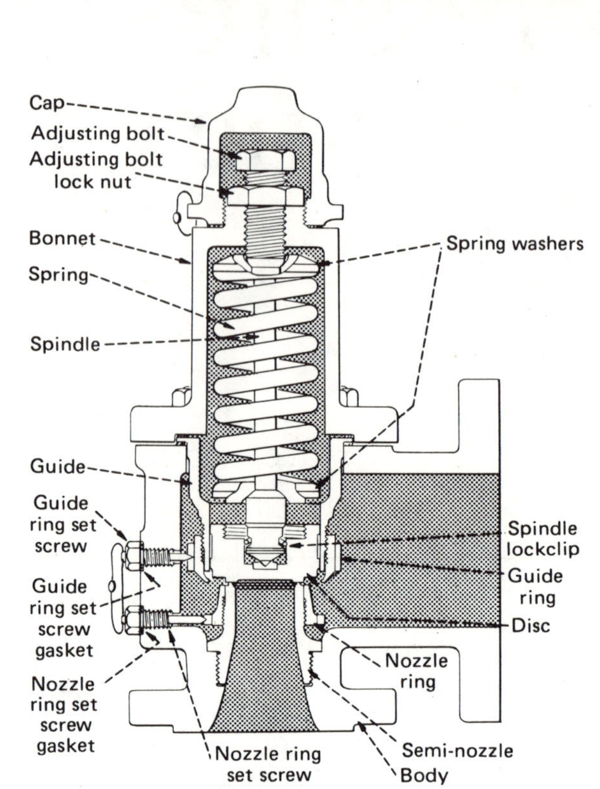

Semi-nozzle safety-relief valve Fig. 3

teed to within 5% of the relief pressure; and the opening after rupture must be able to handle the capacity of the valve without interference.

Safety valves

In industry, the term "safety valve" most generally refers to the type used in steam service on power boilers. Such a valve generally has the following features: flanged or weld-end inlet connections, full or semi nozzle, exposed spring, and lifting lever (Fig. 1). Safety valves used in superheated-steam service over 450°F must have bodies, bonnets and spindles of carbon steel or better; their springs must be fully exposed.

Full-nozzle valves (Fig. 2) have raised-face or ring-joint flanged connections. The base of the nozzle forms the raised-face portion of the flange. Only the nozzle and disk are in contact with the fluid when the valve is closed. Nozzles and disks are normally stainless steel. Bodies of full-nozzle valves are usually carbon steel or alloy steel, depending on the service temperature. Semi-nozzle valves (Fig. 3) have weld-end or flanged connections either raised or flat-face. The nozzle is part of the flange.

The flat-face valve finds little application in modern process plants because its cast iron body does not conform to most piping specifications.

Relief valves

In industry, the term "relief valve" is commonly used to refer to every type of pressure-relieving device except rupture disks. Strictly speaking, it should designate a valve designed for, and used for, liquid service. Most relief valves are small and have threaded (NPT) connections. These are described as base nozzle (Fig. 4), which means the inlet nozzle is not an insert but is merely a hole bored through the base part of the body.

All relief valves have closed bonnets. One type, which is a combination pressure regulator and relief valve, is used primarily in lube oil systems for pumps and compressors. Small bronze relief valves, some with resilient seats, are used for liquid thermal expansion of cooling water on shell-and-tube heat exchangers.

Relief valves are used on the discharge of positive-displacement pumps for thermal expansion of liquid in lines that can be blocked in and are exposed to solar radiation or other heat sources. These valves are generally not suitable for polymer service because the polymer tends to set up and freeze or plug the valve.

Spring materials for closed-bonnet valves are usually carbon steel in services below 450°F. Tungston alloy springs are required above this temperature. Springs are available in stainless steels. Various types of corrosion-resistant coatings are available.

Safety-relief valves

The ASME defines a safety-relief valve as: "An automatic pressure-actuated relieving device suitable for use either as a safety valve or relief valve, depending on application." A safety-relief valve can, therefore, possess all of the features of both types, except that it always has a closed bonnet (Fig. 3). It can be used in steam service on power boilers, but it must be tested and certified by ASME for this service, and it must carry the ASME stamp for power boilers. A safety-relief valve is mandatory on a high-temperature water boiler, but may not be used on a superheater, where a safety valve is required.

The safety-relief valve's widest application is on unfired pressure vessels conforming to the ASME Code. It is also used on the discharge of positive-displacement pumps and compressors, for thermal expansion of liquids or gases, and for general steam or air service. For steam or air usage, the lifting lever is mandatory.

Like the relief valve, the safety valve should not be used in polymer service unless the inlet is isolated from the fluid by a rupture disk. When used in coking service, a steam purge should be used at the inlet. Full-nozzle safety-relief valves are available with bellows to isolate the disk from variable or built-up backpressure.

The bellows-type pressure-balanced valve has at least two distinct advantages over conventional unbalanced valves. It provides a constant set-pressure even though the valve must discharge into a system where the pressure varies; and it permits discharge into a system where the variable or built-up backpressure may exceed the 10% limit normally imposed on conventional valves.

Pressure-balanced valves have been used with variable backpressure up to 80% of the set pressure; however, a capacity correction must be made whenever the backpressure exceeds approximately 15% for a valve in liquid service or 20% for a valve in vapor service. The charts (Fig. 10) present typical capacity correction-factors for bellows-type valves, but the valve manufacturer should be consulted for this information on specific applications.

Pilot-operated valves

Pilot-operated valves, which are safety-relief valves, should not be used with fluids such as slurries or polymers, since these tend to plug the pilot (Fig. 6). These valves are excellent for high-pressure gas service at temperatures below 300°F, but should not be used at higher temperatures because the O-ring construction of the valve has temperature limitations of about 350°F.

The devices consist of two valves, the main and the pilot. The main valve employs a piston with a larger cross-sectional area on the discharge (upper) side than on the inlet side. The spring-loaded pilot valve (which opens at relief pressure) loads the upper side of this piston with gas or liquid at the same pressure as the process. Therefore, this piston has the same pressure on the top as on the bottom; and the larger area on top develops greater force to drive the piston downward to its seat. Thus, the valve is seated tighter as the operating pressure increases; and the valve can be set at pressures only slightly above the operating pressure and not simmer.

Pilot-operated valves have at least three other good points:

1. With proper piping, they can be operated remotely for manual blowdown of the system.
2. By mounting the pilot close to the protected vessel, the main valve can be piped a considerable distance

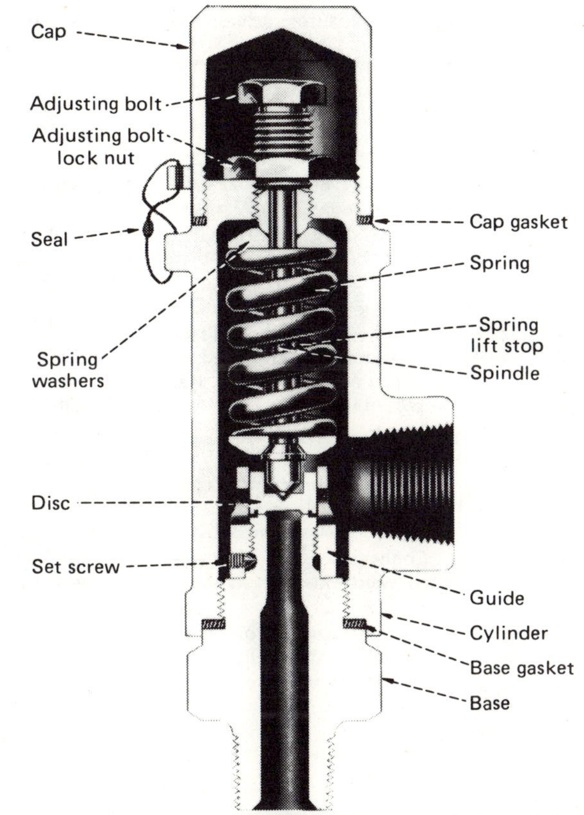

Relief valve is for pump-discharge lines — Fig. 4

Source: Crosby Valve & Gage Co.

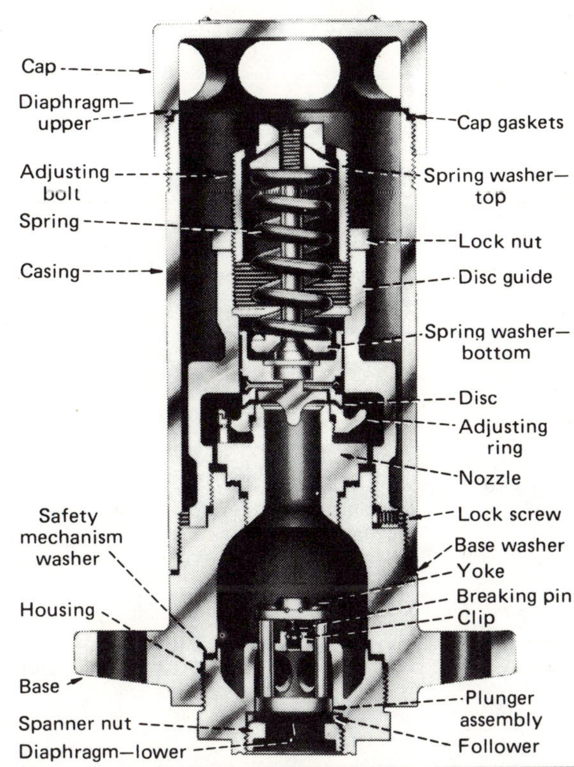

Chlorine-service relief avoids corrosion — Fig. 5

Source: Crosby Valve & Gage Co.

Here are the general features, advantages, disadvantages and typical uses for pressure relieving devices

Type	General	Advantages	Disadvantages	Recommended service
Safety valve See Fig. 1	Available in full nozzle or semi-nozzle design. Has exposed spring and lifting lever.	Open bonnet isolates spring from process temperature. Reseats after relieving.	For steam service only. Not available with balanced bellows; however, does not normally discharge into closed systems.	ASME power boilers and general steam service.
Full nozzle safety-relief valve See Fig. 2	Nozzle is the raised-face portion of flange. Used for liquid and vapor relief. Has closed bonnet.	Body of valve completely isolated from process fluid when not relieving. Available with balanced bellows. Reseats after relieving.	Limited to raised-face connections. Not good for polymer service. Maximum inlet pressure approximately 10,000 psig.	ASME unfired pressure vessels. Pump and compressor discharge. Water boilers.
Semi-nozzle safety-relief valve See Fig. 3	Used for liquid and vapor relief. Nozzle consists of insert screwed into base. Connections can be flanged, weld-ended or screwed.	Cheaper construction than full nozzle. Available in flat face for mating with cast-iron flanges. Available in weld-end connections.	Not good for polymers. Maximum inlet pressure about 1,500 psig. Not available with balanced bellows.	ASME unfired pressure vessels. Pump and compressor discharge. Water boilers.
Base-nozzle relief valve See Fig. 4	Nozzle formed from base of valve. Typically have screwed connections, but also available flanged or weld-ended. Fully open at 25% overpressure. Has closed bonnet.	Available in small-sizes. Low cost. Reseat after relieving. Suitable for toxic materials.	Not good for polymers. Limited to about 2,000-psig inlet. Not available with bellows. Not suitable for service on steam boilers.	Pump discharge. Thermal relief of lines and heat exchangers. Water heaters.
Chlorine service relief See Fig. 5	Internals protected by two diaphragms. One diaphragm isolates valve from process fluid, second diaphragm seals out ambient air.	Can be used in very corrosive services. Inlet diaphragm ruptures only after breaking-pin severs. Disc reseats after relieving.	Not available as balanced valve. Limited to approximately 375-psig inlet pressure.	Chlorine and other corrosive and toxic services. Tank cars and storage vessels.
Pilot-operated safety-relief valve See Fig. 6	Consists of two valves. Pilot valve controls the main valve.	Can withstand high inlet pressure. Can be set to relieve near operating pressure. Can be remotely operated for manual depressuring.	Not suitable for temperatures above 350°F. Not recommended for dirty services, slurries, polymers.	ASME unfired pressure vessels, particularly clean high-pressure gas service. Gas pipeline and reciprocating compressors.
Rupture disc See Fig. 7	Consists of frangible disc secured between flanges of other holders. Two principal types: prebulged, reverse buckling.	No moving parts to freeze or plug. Good for slurries, polymers, toxic materials. No leakage. Can handle very large capacities. Can be used in very high pressures. Fast response—can relieve explosions, detonations.	Subject to stress fatigue. Entire system depressures when disc ruptures and unit must shut down.	ASME unfired pressure vessels. Primary relieving device for slurries, polymers, or in series with relief valves for any of the above. Should be used in series with relief valves for toxic substances. High pressure services, internal explosions, detonations.
Tank vent See Fig. 8	Usually is dual function: combination pressure and vacuum relief. Generally, weight-loaded pallet type.	Can be set close to operating pressure. Very sensitive. Typical settings are in ounces or fractional parts thereof.	Limited to low-pressure service. Construction materials can be a problem when used in chemical service.	API RP 2000 aboveground petroleum storage tanks. Can be used for any atmospheric pressure storage tanks.
Atmospheric-type relief valve See Fig. 9	Two basic types: weight-loaded pallet, multiport. Designed for low-pressure steam service.	Provides large relieving capacity at low cost.	Highly specialized nature of valve tends to limit aplication to low-pressure steam service.	Surface condensers on outlet of condensing steam turbines. Generally low-pressure steam service where large quantities must be relieved.

Source: Anderson, Greenwood & Co.

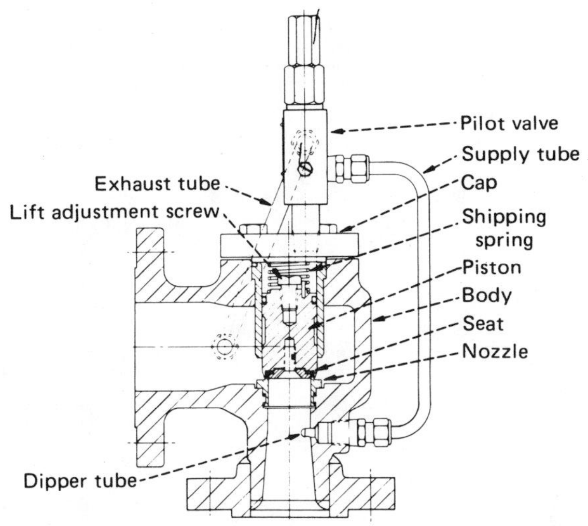

Pilot-operated safety-relief valve **Fig. 6**

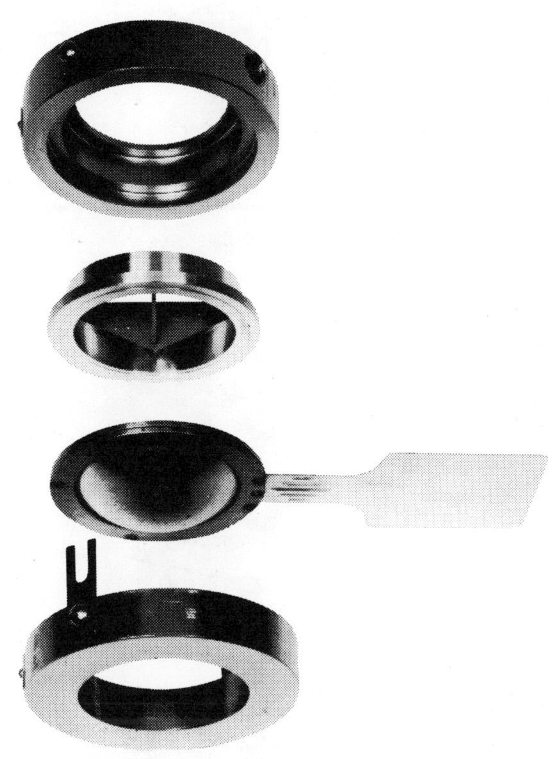

Rupture disc relieves explosions **Fig. 7**

away and will not chatter, since the pilot valve is not exposed to the effects of piping friction loss.

3. Pilot-operated valves are designed to withstand very high backpressures of up to 90% of set pressure when supplied with modified outlet flanges, and this makes them ideal for protection of multistage reciprocating compressors.

Rupture disks

A rupture disk consists of a suitable holder and a metal disk that is designed to burst at the relief pressure. Rupture disks can be used to protect vessels with very high design pressures far beyond the capacity of relief valves. Rupture disks are ideal for large capacities; and the entire head of a vessel can be designed as a rupture disk.

There are two basic types prebulged (which is conventional) and reverse buckling (Fig. 7). One disadvantage of the prebulged type is a required set pressure about 1.5 times the operating pressure, due to a susceptibility to stress fatigue. This means that the protected vessel must be designed for a higher pressure. Also, the popping pressure of the prebulged type can be erratic.

By contrast, the reverse-buckling type can be set to within 1.1 times the operating pressure. Its accuracy makes it ideal for protecting the inlet of relief valves in dirty or polymer service.

The most serious disadvantage of rupture disks is their inability to reseat. Rupture causes the protected system to depressure, and the process unit must be shut down to replace the disk. Therefore, rupture disks are unsafe for toxic fluids, and relief valves in series after a rupture disk should be seriously considered for such materials.

Tank venting

Atmospheric-pressure storage tanks may be protected with special devices designed to relieve at very small positive or negative pressures. Typically, these devices are combination pressure- and vacuum-relieving valves, which normally utilize weight-loaded pallets (Fig. 8). The pressure setting must be established on the safe side of the tank's design pressure so that the valve will be fully relieved at that pressure. The manufacturer of the valve should be consulted in determining the proper setting. Typical settings might be -1 in of water-column of vacuum and $+0.25$ psig positive pressure, for a storage tank with a design rating of -2 in of water vacuum and 0.5 psig. The applicable code is API Standard 2000, Venting Atmospheric and Low Pressure Storage Tanks.

Approved pilot-operated relief valves may also be used to protect low-pressure storage tanks. These valves reach their rated capacity essentially at set pressure. Hence, they can be set closer to the design pressure than can the pallet-type valve.

Atmospheric-type relief valves

Surface condensers associated with condensing-steam turbines operate at a vacuum. Loss of cooling water to the surface condenser results in the loss of ability to condense the steam. Hence, there is an immediate need to vent a large quantity of steam. Special, large-capacity relief valves have been designed for this service. There are two basic types: the weight-loaded pallet (Fig. 9), and the multiport. The valves are set to relieve at or slightly above atmospheric pressure.

These valves may be sized in accordance with the Standards of the Heat Exchange Institute—Barometric and Low Level Jet Condensers—Fig. B-14, page 17.

Source: Varec Inc.

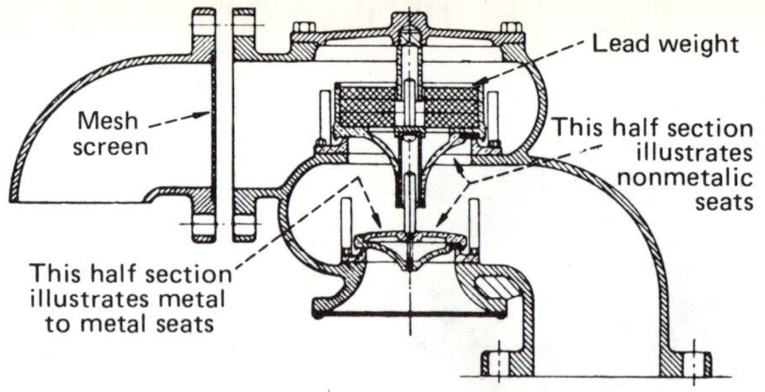

Tank vent for atmospheric storage tanks **Fig. 8**

Source: Atwood & Morrill Co.

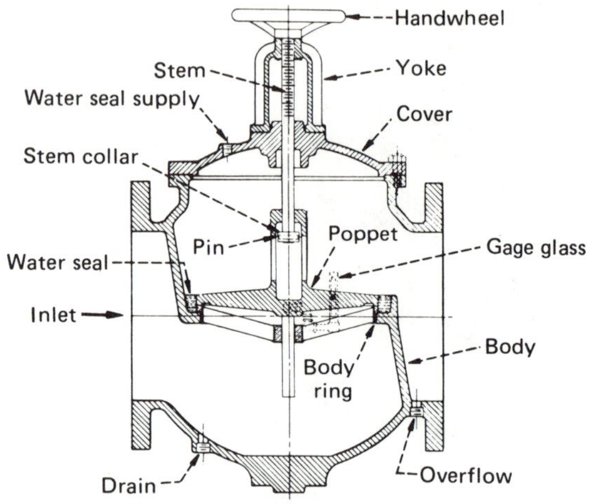

Atmospheric relief for surface condensers **Fig. 9**

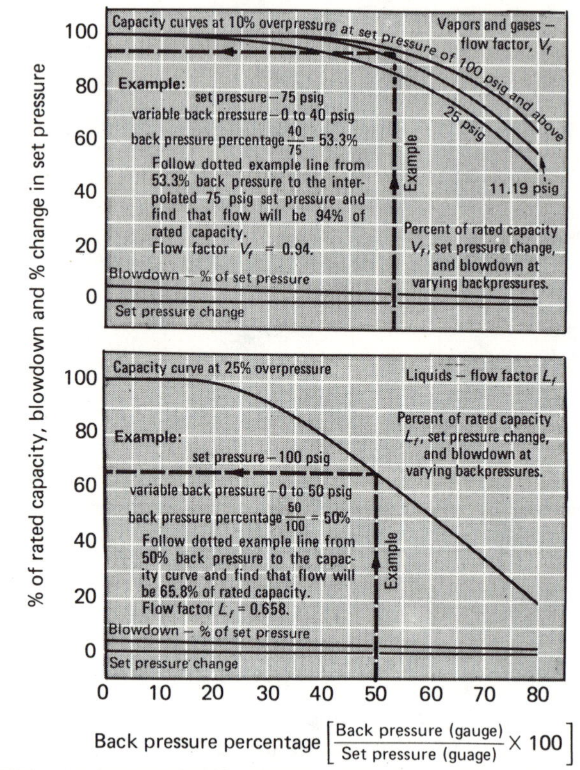

Performance characteristics for bellows type pressure balanced valves **Fig. 10**

Special applications

Reciprocating pumps generate pressure pulses that tend to lift the disk of a relief valve and cause the valve to leak. Thus, relief valves on reciprocating pumps should be set at least 25% above the operating pressure.

Safety-relief valves on reciprocating compressors should be set at least 15% above the operating pressure. Pilot-operated valves can be set closer.

Definitions

Overpressure: Pressure-relieving devices do not attain full capacity at the pressure at which they are set to open. For example, a relief valve in liquid service is not fully open until the inlet pressure is 25% above the set pressure. This increase is called overpressure.

Accumulation: The ASME Boiler and Pressure Vessel Code permits the maximum relieving pressure to exceed the vessel design pressure by certain specified limits. This margin is called accumulation, and is expressed as a percentage of the vessel design pressure. Typical values of permissible accumulation are:

- ASME Power Boilers 6%
- ASME unfired pressure vessels 10%
- ASME pressure vessels exposed to fire 20%
- Pressure piping and pump-discharge piping 25%
- Atmospheric storage tanks 0%

Blowdown is the difference between the set pressure and the automatic reseating pressure of a relief valve, expressed as a percentage of the set pressure. Normally, blowdown is 4% below the set pressure.

Rules and philosophy

Besides adhering to the appropriate codes, relief systems should conform to principles evolved from common sense and experience. Frequently, a relief valve must protect equipment subjected to overpressure from a number of unrelated causes. A fractionating column, for example, could be subject to overpressure from: external fire, blocked discharge, loss of reflux, power failure, cooling-water failure, or instrument failure. The relief valve should be sized for each of the separate conditions, and the selected size should be large enough to handle the largest capacity.

In determining relief loads, one should not assume double jeopardy; two separate and unrelated events—for example, power failure and blocked discharge—are not assumed to occur simultaneously.

No credit may be taken for standby equipment that is not normally operating, or for shutdown devices.

A control valve open at the time of the upset may be assumed to continue to function, unless the condition that caused the upset would also cause the valve to fail.

Wherever fuel may be present, it is necessary to protect a pressure vessel with a safety-relief valve large enough to handle the vapor generated by an external fire. Credit may be taken for the insulation if it is fireproof and installed in a manner to effectively resist being dislodged by the fire or fire-fighting apparatus.

Equations for sizing valves

Although the applicable code should be used for design calculations, the following equations, adapted from API RP 520, may be used for sizing the valves. Equations used by the various manufacturers may vary slightly, but are acceptable; and the specialized slide rules distributed by some of the manufacturers are sufficiently accurate for most sizing problems.

Liquid service:

$$A = \frac{gpm\sqrt{G}}{27.2K_pK_wK_v\sqrt{p - p_b}}$$

where:

A = effective discharge area of nozzle, in^2

gpm = flowrate, gal/min

K_p = capacity correction factor due to overpressure = 1.00 for 25% overpressure, dimensionless.

K_w = capacity correction factor due to backpressure on bellows valves = 1.0 for conventional valves, dimensionless

K_v = correction factor due to viscosity = 1.0 where the Reynolds number at the nozzle is greater than 60,000, or the viscosity at relief temperature is less than 200 SSU, dimensionless

p = relief-valve set pressure, psig

p_b = backpressure, psig

G = specific gravity of the liquid at flowing temperature, dimensionless

Vapor service:

$$A = \frac{W\sqrt{TZ}}{CKP_1k_b\sqrt{M}}$$

where:

A = effective discharge area (nozzle area in full-nozzle valves), in^2

W = flow through the valve, lb/h

T = temperature of inlet vapor, °R

Z = compressibility factor of gas, dimensionless

C = ratio of specific heats (constant pressure to constant volume) at standard conditions

K = manufacturer's coefficient of discharge (about 0.97), dimensionless

P_1 = upstream pressure, psia

K_b = capacity correction factor for backpressure (consult manufacturer's specifications), dimensionless

M = molecular weight of the gas

Steam service:

$$A = \frac{W}{50P_1K_{sh}}$$

where:

A = effective discharge area, in^2

W = flowrate, lb/h

P_1 = upstream pressure plus permissible accumulations, psia

K_{sh} = correction factor for superheat in the steam = 1.0 for saturated steam, dimensionless

No general sizing formulas can be equally applied to rupture disks, atmospheric-relief valves, or tank-venting equipment, so the manufacturer should be consulted for sizing information for each of these types.

Installation

The inlet piping on relief valves, safety valves and safety-relief valves must be at least equal in area to the valve inlet connection. Safety valves on power boilers must be mounted on nozzles directly connected to the boiler without any intervening piping. Otherwise, relief valves and safety-relief valves protecting vessels may have their inlets connected to a common piping header. The cross-sectional area of the inlet header must be at least equal to the combined inlet-connection areas of the valves. Since the valve is designed to reseat at 4% blowdown, or whenever the inlet pressure falls to 4% below set pressure, the area of the inlet piping must frequently be larger than the area of the valve inlet connection.

This means that the inlet piping must be sufficiently large to ensure that its friction loss remains below 3% of the set pressure to prevent the valve from chattering. Chattering substantially reduces the capacity of the valve and can lead to dangerous overpressuring of the protected vessel.

The valve outlet piping must be at least equal in area to the area of the valve outlet connection. The discharge piping system must be designed so that backpressure does not build up to the point where it reduces the capacity of the valve to protect the vessel. Also, system backpressure must not be permitted to rise to the point where it would prevent the opening of low-pressure valves discharging into the same system.

Pressure-relief valves should be mounted upright with the spindle in the vertical position, and in such a manner that liquid does not accumulate on the discharge side of the valve.

Legal aspects

Some codes, such as the ASME Boiler and Pressure Vessel Code, have been enacted into law, essentially verbatim, by various states and municipalities. OSHA has adopted the ASME Boiler and Pressure Vessel Code and API Standard 2000, Venting Atmospheric and Low-pressure Storage Tanks. It is likely that other codes will be included in the future. Meanwhile, it is probable that all pressure-relieving equipment is covered by OSHA's "recognizable hazard" clause.

The author

Floyd E. Anderson is Senior Instrument Engineer, Fluor Engineers and Constructors, P.O. Box 4051, Anaheim, CA 92803, where he has been continuously employed in the Instrument Engineering Dept. since joining Fluor in 1959. During the past five years, he has served as Lead Instrument Engineer on projects involving the design and construction of new process plants. Earlier assignments included responsibility for the design of several pressure-relieving systems, as well as engineering on numerous petroleum refineries and chemical and petrochemical plants.

Pressure-Relief Systems

As processing units become increasingly complex, with attendant higher levels of stored energy, the design of safe, economically sound and mechanically reliable relieving devices and disposal facilities becomes more important than ever.

MARX ISAACS, Fluor Corp.

Pressure-relief systems provide the means for protecting personnel and equipment from abnormal operation of a process. Some of the conditions that may cause excessive pressures are:

- Exposure to fire or other external sources of heat.
- Warmup or cooldown of liquid blocked between valving, or in some other closed section of the system, resulting in hydraulic expansion.
- Mechanical failure of normal safety devices, malfunctioning of control equipment, or faulty manual operation, resulting in overfilling of the equipment.
- Production of more vapor than the system can handle following an operational upset.
- Unexpected vapor generation resulting from process heat unbalance.
- Chemical reaction with heat generation and gas production in excess of system capacity.

The use of pressure-relief systems to prevent overpressurization of the plant accomplishes the objectives listed in Table I.

PRESSURE-SENSING/RELIEF DEVICES

Pressure-relief systems range from the single relief valve, or rupture disk, and piping on a single vessel to the more-complex piping systems involving many relief valves that are manifolded into a common header.

Objectives of Pressure-Relief Systems—Table I

- Compliance with local, state and national regulations, including environmental control.
- Protecting operating personnel against dangers from overpressuring of equipment.
- Minimizing material losses during and after an operational upset that has caused overpressure for a short period.
- Reducing downtime necessitated by plant overpressure.
- Preventing damage to capital equipment.
- Preventing damage to adjoining property.
- Reducing insurance premiums on capital investment.

Safety and Relief Valves

In the familiar spring-loaded relief valve, the opening increases gradually as the pressure increases above the set value. The safety valve "pops" or opens completely with only slight overpressure and remains open until the pressure drops a definite amount, then closes completely. However, such action occurs only when a compressible fluid is handled.

In the chemical process industries, the relief valves (typically used) are actually safety-relief valves that function as relief valves on liquids and as safety valves on gases and vapors. In this article, the term "pressure-relief valve" is intended to include safety valves.

A conventional safety valve (along with the nomenclature of its parts) is shown in Fig. 1a. In this valve,

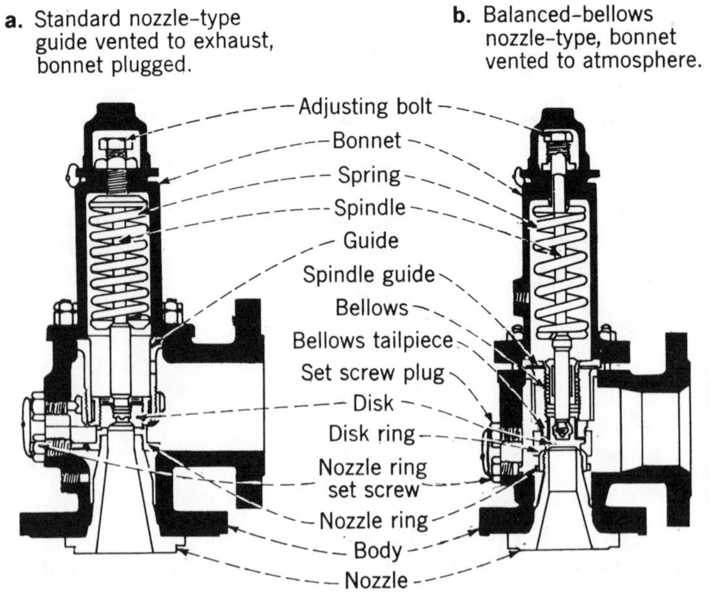

NOMENCLATURE of typical safety-relief valves—Fig. 1

Originally published February 22, 1971

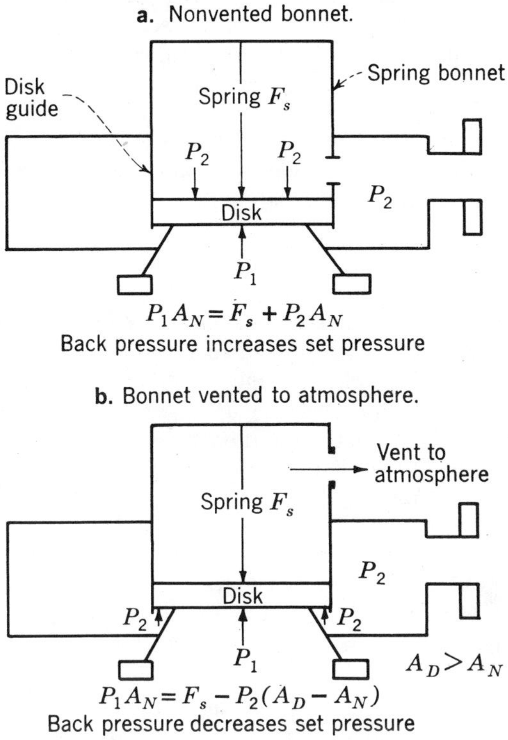

FORCES on disk of conventional relief valve—Fig. 2

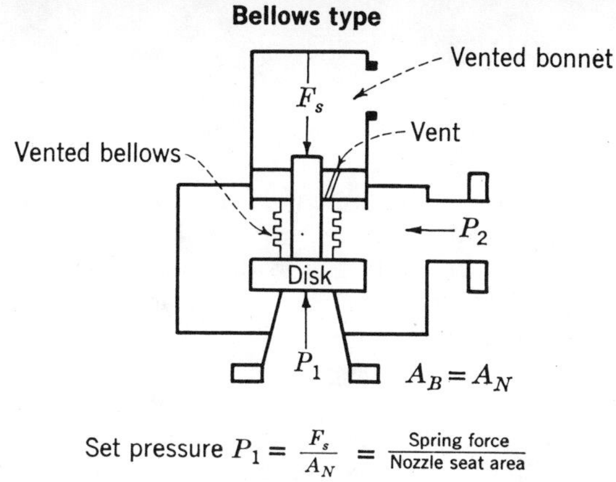

FORCES on disk of balanced-bellows valve—Fig. 3

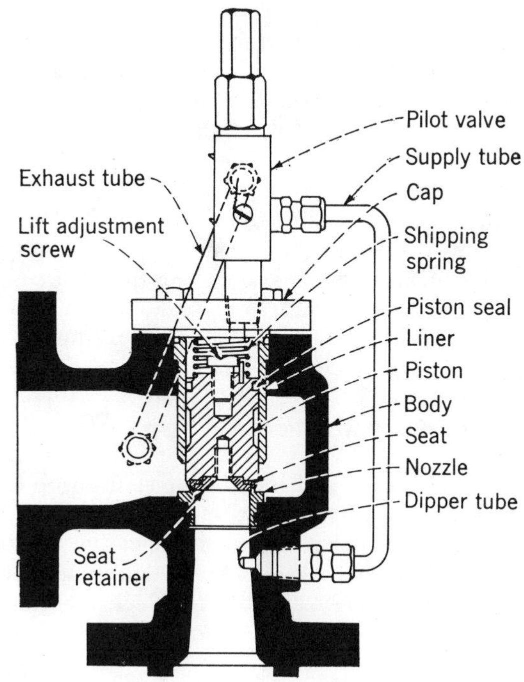

PILOT-OPERATED safety-relief valve—Fig. 4

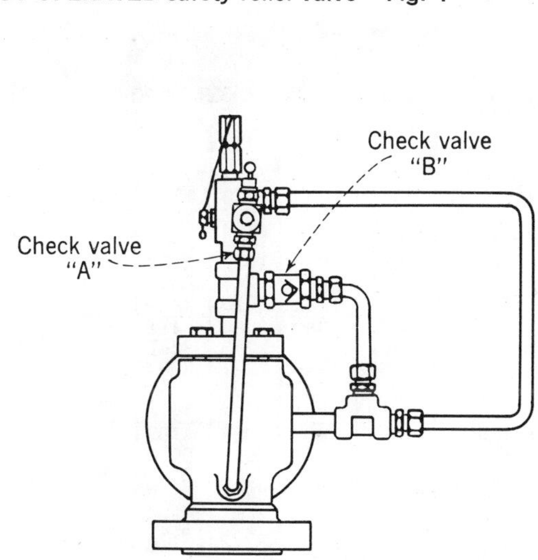

BACKFLOW arrangement prevents reverse flow—Fig. 5

the set pressure is affected by the backpressure, or pressure on the discharge side of the valve. Thus, venting the space above the disk to the discharge line yields an increase in set pressure when the backpressure increases. If vented to the atmosphere, the set pressure is decreased. With constant backpressure, the spring setting can be adjusted to compensate, and hence produce the desired set pressure. However, if the valve discharges into a header, varying backpressures are likely to be encountered, and spring adjustment is not the answer.

To handle this situation, the balanced-bellows safety-relief valve (Fig. 1b) was developed. The proportions of the bellows are such that the area open to the atmosphere is equal to the effective area of the disk, which is exposed to the process pressure. Theoretically, this eliminates any effect of backpressure on the set point. Actually, this is only approximately the case. (Manufacturing tolerance of the bellows' effective area may cause some deviations.) However, an increase in backpressure of up to 30 to 40% of the set pressure (depending on the manufacturer) can be tolerated without reducing capacity. With the standard safety-relief valve, the set pressure is increased proportionately to an increase in backpressure.

The unbalance forces across the disk of a conventional safety valve are shown in Fig. 2. In Fig. 2a, the disk guide is not plugged and the bonnet is plugged at the vent to the atmosphere. In Fig. 2b, the disk guide is plugged and the bonnet vented to atmosphere. The forces across the disk of a balanced-bellows valve operating within the design-pressure range are illustrated in Fig. 3.

The pilot-operated safety-relief valve (Fig. 4) is used for operation close to the set point. A differential piston in the main valve is loaded through an orifice by the process pressure. The spring-loaded pilot valve opens when set pressure is reached. This vents the pressure above the piston of the main valve, allowing it to open wide. When the system blowdown pressure is reached, the pilot valve closes, and full system pressure is immediately diverted to the dome of the main valve. The piston instantly moves downward to close the main valve.

When a pilot-operated valve is subjected to a backpressure, the piston may lift to cause reverse flow unless some arrangement has been made to prevent this from happening. Reverse flow may occur when several valves are manifolded into a common discharge header, with perhaps one or more exhausting while another is out of service. The possibility of reverse flow through a pilot-operated valve can be eliminated by using a backflow arrangement as in Fig. 5.

In this scheme, backpressure higher than the valve's inlet pressure is introduced into the dome through check valve B to load the piston. Check valve A prevents the backpressure from reaching the pilot supply tube and main valve inlet. In normal operation, check valve B prevents the main-valve dome pressure from bleeding into the discharge line.

Materials of Construction

Cast-steel bodies with stainless-steel trim are used on spring-loaded safety valves in process service. For special applications, valves are also made with bodies of Monel, Hastelloy, alloy steel or stainless steel, with suitable trim for the conditions to which they are to be exposed. For inlet temperatures below 450 F., carbon-steel springs are generally used. Tungsten-steel springs are required for higher temperatures. To withstand corrosion, the springs are nickel-plated or aluminized. Other alloys used for springs are Hastelloy, stainless steel and Inconel. Valves with 3.5% stainless- or nickel-steel bodies, with full stainless-steel trim including springs, are used in low-temperature applications.

Flat, metallic seating surfaces are found in most spring-loaded valves. Where leakage is difficult to prevent, an O-ring-type seat may be more suitable in applications that are subject to vibration or pulsation, or for operation close to the set point, and for fluids containing entrained solid particles. Depending on the composition of the O-ring, the maximum relieving temperature is 200 to 450 F.

Orifice Sizes

Full-nozzle safety-relief valves have standard orifice sizes. API Standard 526 was published to promote interchangeability of flanged-steel safety-relief valves.[9] However, even though the inlet flanges are in accordance with ANSI B16.5, the maximum set pressures shown for these valves are considerably lower than the pressure allowed by the flange ratings. Representative capacities for one model of valve, for steam only, and of cast-iron body only are shown in Table II for a range of pressures and for various orifice sizes.

Rupture Disks

As a protective device against overpressure, rupture disks can be used either alone or in connection with a relief valve. The disks are made of various metals and are available with a coating on one or both sides

Representative Capacities for a Relief Valve—Table II

SET PRESSURE PSIG	Orifice Designation													
	D	E	F	G	H	J	K	L	M	N	P	Q	R	T
10	141	252	395	646	1010	1653	2362	3666	4626	5577	8197	14190	20560	33410
20	202	360	564	922	1442	2361	3373	5235	6606	7964	11710	20270	29360	47710
30	262	468	733	1199	1874	3068	4384	6804	8586	10350	15210	26340	38160	
40	323	575	902	1475	2306	3776	5396	8373	10570	12740	18720	32420		
50	383	683	1070	1752	2738	4484	6407	9942	12550	15130	2223			
60	444	791	1239	2028	3170	5192	7418	11510	14530	17510	25			
70	504	899	1408	2305	3602	5899	8429	13080	16510	19900	2			
80	565	1007	1577	2581	4034	6607	9440	14650	18490	22290				
90	625	1114	1746	2858	4466	7315	10450	16220	20470	2467				
100	686	1222	1915	3134	4898	8023	11460	17790	22450	270				
110	746	1330	2084	3411	5330	8730	12470	19360	24430					
120	807	1438	2253	3687	5762	9438	13490	20920	26410					
130	867	1546	2421	3964	6194	10150	14500	22490	2839					
140	928	1653	2590	4240	6626	10850	15510	24060	30					
150	988	1761	2759	4517	7058	11560	16520	25630						
160	1049	1869	2928	4793	7490	12270	17530	2720						
170	1109	1977	3097	5070	7922	12980	18540	28						
180	1170	2085	3265	5346	8353	13690	19550							
190	1230	2192	3434	5623	8785	14390	205							
200	1291	2300	3603	5899	9217	15100								
210	1351	2408	3772	6176	9649									
220	[illegible]													

Orifice Designation	Orifice Area Sq. In.
D	0.110
E	0.196
F	0.307
G	0.503
H	0.785
J	1.287
K	1.838
L	2.853
M	3.60
N	4.340
P	6.380
Q	11.050
R	16.0
T	26.0

* Nozzle capacity in lb./hr. for 10% overpressure for steam in accordance with ASME Unfired Pressure Vessel Code.

Typical Rupture Disks and Burst Pressures—Table III

Disk		1/2	3/4	1	1½	2	3	4	6	8	10	12	14	16	18	20	24
		Size, In.															
		Burst Pressure, Psig. at 72 F.															
Aluminum, max. temp. 250 F.	Min.	65	45	34	23	15	10	8	7	5	4	3	3	3	3	3	3
	Max.	1500	1000	1000	1000	750	500	400	300	225	175	150	150	150	125	125	100
Aluminum, lead lined one side, max. temp. 250 F.	Min.	89	61	46	31	18	14	11	9	7	6	5	5	4			
	Max.	1500	1000	700	500	350	250	200	150	100	90	80	70	60			
Aluminum, vinyl coat one [illegible]	Min.	65	45	34	23	20	14	11	8	5	5	4	4	4	4	4	4
Aluminum, [illegible] coat [illegible] sides*	Min.	90	60	50	35	28	20	[illegible]	10	8	8	7	6	6	6	6	6
	Max.	1500	1000	1000	1000	750	500	400	300	225	175	150	150	150	125	125	100
Copper, max. temp. 250F.	Min.	210	140	110	65	35	28	30	26	37							
	Max.	4500	3000	3000	3000	2500	1500	1000	1000	700							
Copper, lead lined, max. temp. 250 F.	Min.	235	155	125	73	40	32	33	28								
	Max.	4500	3000	2000	1500	1000	800	600	500								
Silver, max. temp. 250F.	Min.	245	175	125	85	55	35	25	20	17							
	Max.	6000	3500	3500	3500	2500	2000	1500	1000	500							
Nickel, max. temp. 750F.	Min.	340	260	190	125	75	46	38	33	24	20	20	150	235	210	190	145
	Max.	6000	6000	6000	6000	6000	6000	6000	3600	3600	1440	1440	720	720	720	720	720
Monel, max. temp. 800F.	Min.	460	320	250	175	105	80	54	43	37	30	385	330	285	255	230	190
	Max.	6000	6000	6000	6000	6000	6000	6000	3600	3600	1440	1440	720	720	720	720	720
Inconel, max. temp. 1,000F.	Min.	760	440	410	290	180	130	100	75	40	32	27	23	20	18	16	220
	Max.	6000	6000	6000	6000	6000	6000	6000	3600	3600	1440	1440	720	720	720	720	720
Type 321 Stainless steel max. temp. 600F.	Min.	850	600	485	365	195	135	105	85	65	50	50	48	44	38	30	80
	Max.	6000	6000	6000	6000	6000	6000	6000	3600	3600	1440	1440	720	720	720	720	720

* Retainer ring must support plastic seal.

a. Prebulged **b.** Three-component

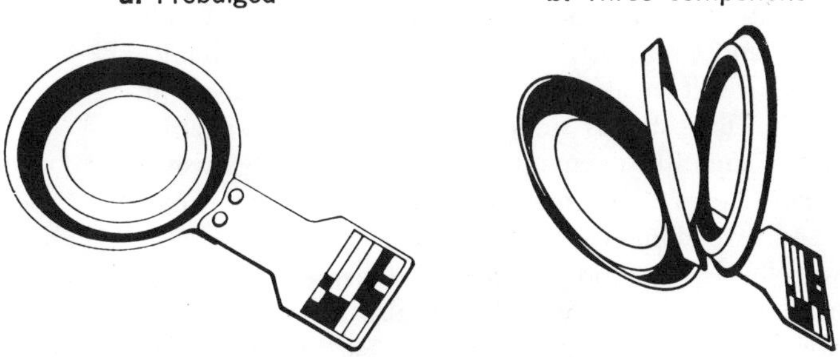

RUPTURE disks protect process systems—Fig. 6

a. Bolted-flange **b.** Union-type **c.** Screw-type

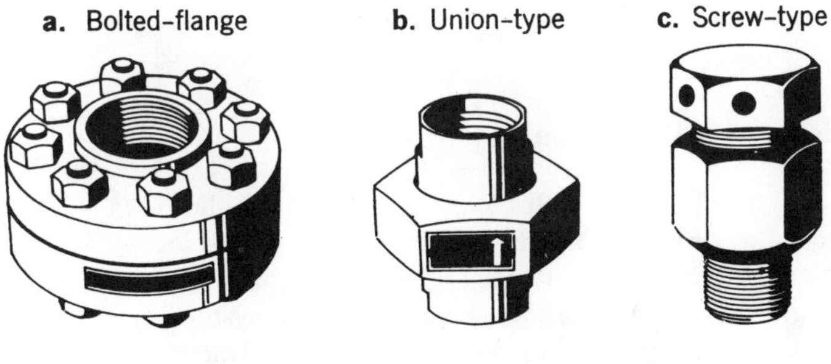

FLANGES for installing rupture disks—Fig. 7

for resistance to corrosion. Only limited use can be made of rupture disks alone because the entire contents of a processing system can be lost when the disk ruptures. However, when such disks are installed in parallel with a relief valve, additional capacity is made available for extremes of operating conditions.

Many types of rupture disks are manufactured. The prebulged disk (Fig. 6a) has gained wide acceptance. Fig. 6b shows a three-component rupture disk consisting of a vacuum support, rupture disk and retainer ring. Rupture pressure varies directly with disk thickness, and inversely with disk diameter.

The most common construction materials are: aluminum, Monel, Inconel, and austenitic stainless steel. However, disks are sometimes made of copper, silver, gold, platinum, tantalum and titanium. Metal foil, strip and sheet in the soft, annealed condition are required. The wide variety of requirements necessitates that rupture-disk manufacturers stock many increments of metal thickness ranging from approximately 0.002 to 0.060 in.

Rupture disks are readily but not exclusively available in sizes from ½ to 24 in. dia. Some manufacturers list a size range from ⅛ to 44 in. Table III shows data on representative rupture disks.

A ½-in. disk may have a minimum rupture pressure ranging from 65 to 850 psig., depending on the metal from which it is manufactured. A 24-in. disk may rupture at as low as 3 to 35 psig. Rupture pressures of 6,000 psig. are readily available on smaller sizes but rupture pressures higher than 700 psig. are rare on larger sizes.

Flanges designed to grip the disk and incorporate it into a pressure system are of many types. The more common are: the bolted type (Fig. 7a), the union type (Fig. 7b), and the screw type (Fig. 7c). These flanges are incorporated into a pressure system by welding or by threaded connections. In addition, the bolted type is available in a configuration that can be bolted to flanges existing in the system. The bolted-type assembly is used for all sizes from ½ to 24 in. in a variety of pressure ratings. The union type is limited to sizes up to 2 in., and the screw type up to 1 in.

Location of rupture disks upstream of relief valves guards against valve leakage in systems containing very toxic or extremely valuable materials. Such an application also provides added protection in a system handling materials that are likely to cause fouling or corrosion of the relief valve to the extent of making it inoperative. A vent, usually provided with an excess-flow valve, is required between the relief valve and rupture disk to prevent pressure buildup. Normally, a pressure gage is also included with the excess-flow valve to show when the disk has blown.

In some cases, the rupture disk is used downstream of the relief valve when conditions are such that upstream installation is not practical or where the mechanisms of valves connected to a header are to be protected against corrosion. With this arrangement, it is necessary to use a valve whose set point is not affected by backpressure, or a vent must be included between the relief valve and rupture disk.

In addition to better corrosion resistance, rupture disks have the following advantages over pressure-relief valves: no leakage until failure, suitable for liquid slurry service, better handling of slow explosions and detonations; affected less by high viscosities.

TRANSFER/DISPOSAL EQUIPMENT

Transfer and disposal system piping in pressure-relief systems merits at least as much mechanical-design consideration as is given to piping for handling process fluids. In fact, the design of discharge piping from pressure-relief or depressuring valves must often be more complex than for piping in the process system itself. This is likely to be the case because the relief piping is subject to a wider range of temperature, pressure, shock and corrosivity of fluid than the piping in any one segment of the process system. Moreover, the disposal system must be prepared to handle any fluid.

Thermal strains resulting from entry of hot or cold fluids cause the major stresses encountered in discharge piping of a relief system. Temperatures in some refinery-unit relief systems can range from below 0 F. to several hundred degrees above. Obviously, design of a relief piping system for flexibility under such conditions is more complicated than for a process-piping system. In the latter, temperature deviations are usually only somewhat below or above normal operating temperature, and are not generally sporadic and uncontrolled, as in relief systems.

Usually, expansion arrangements can be provided to maintain allowable stress levels over the full temperature range of a relief system. This is accomplished by means of joints, or appropriate piping configuration, or properly designed expansion employing a hot or cold spring.

It is of major importance to allow for stresses in carbon-steel piping that may possibly be cooled below the transition temperature. This can occur on entry of cold fluids or by autorefrigeration when low-boiling liquids undergo a reduction in pressure. Material specifications, allowable stresses, and impact-test requirements for carbon-steel piping to be subjected to as low as −50 F. are to be found in ANSI B31.3, "Petroleum Refinery Piping," an ASME publication. As an added precaution, stress-relieving of welded-piping systems has been found advantageous in reducing the possibility of brittle fracture of carbon steel.

For processing systems in which temperatures below −50 F. are possible, relief lines are usually constructed of materials that remain ductile at the lowest expected temperature. In choosing such a material, it is feasible to take some credit for heat gained from ambient temperature, provided a suitable transfer coefficient is used. Also, the insulating effect of ice buildup must be taken into account.

Pressure-relief valves may possibly be subjected to both thermal and mechanical stresses. This must be carefully considered when designing the associated discharge piping. Malfunction and/or leakage of valves can result from such stresses. Appropriate anchors and supports plus adequate discharge-piping flexibility are instrumental in preventing relief-valve thermal stresses from the discharge piping.

Support of discharge piping by only the outlet of the pressure-relief valve instead of by a separate support causes mechanical stresses in associated inlet piping and pressure-relief valves. Similar mechanical stresses result from forced alignment of discharge piping. The first strains sufficient to promote mechanical failure generally occur at the inlet piping. However, pressure relief-valve malfunction and leakage can be caused by moments and variable stresses at much lower levels. Reaction forces produced when a pressure-relief valve discharges may also cause stresses in disposal piping. See API RP 520, Part II, Paragraph 2.4 for a formula to compute reactive loads arising from pressure-relief-valve operation.[7]

Shock loading must also be considered as a possible source of stress in relief piping. Two of the causes of shock loading are: (1) impact of liquid slugs at points where flow changes direction, and (2) sudden release of a compressible fluid into a system with multidirectional piping.

Disposal of Discharge

For safety, refinery discharges of flammable liquids and vapors must be conducted to an appropriate location for disposal. Predominantly condensable vapor and liquid mixtures can be sent, in a closed system, to a water scrubber to remove condensable fractions. Noncondensable vapors from the scrubber are then released in some harmless and practical way, such as to the atmosphere through a vent stack or, for added safety, into a closed system leading directly to a burning pit or flare stack. The choice of disposal method is governed by space availability, local conditions, environmental regulations and economy. An extremely important safety precaution is chemical neutralization of toxic vapors or their conversion by burning to less-toxic materials.

Knockout Drums and Seals

A knockout drum immediately adjacent to each processing unit or group of units must be provided if there is a possibility of liquid discharge into the flare system. It is strongly recommended that facilities be installed for automatic transfer of accumulated liquid, through a cooler if necessary, to slop storage or other disposal. To satisfy a wide variety of possible conditions, each installation presents an individual case for which there are specific design requirements. For example, (1) water sprays may be needed for temperature control of hot vapors; (2) it may be necessary to heat light hydrocarbons to counteract the refrigerating effect of vaporization; and (3) winterizing may be required to prevent freezing of water or solidification of heavy hydrocarbons in cold climates.

Besides a knockout drum at the battery limits, a seal pot must be provided adjacent to the base of the flare stack to trap condensed liquid. The flare header should be angled so that it drains liquid back into the knockout drum.

Pipe seals are designed for filling with liquid, usually water, and placed to prevent flashback into the flare header. They generally consist of a loop or trap built into the base of the flare stack or the flare-line inlet connection. Under low-flow conditions, pipe seals may cause pulsation of gas flow to the flare. In addition, the liquid seal may be blown out of the top of the flare stack during abnormally high gas releases. However, pipe seals are usually cheaper than drum-type seals. In its simplest form, the pipe seal is appropriate where condensation with accumulation of hydrocarbons in the seal does not create a problem. Otherwise, a more complex pipe seal with a continuous, automatic skimming arrangement to remove condensed hydrocarbons is needed.

SIZING OF SYSTEM COMPONENTS

The quantity and type of fluid to be released per unit time, and the differential pressure across the valve, determine valve size. Selection of type (standard or balanced) depends on the maximum pressure built up or superimposed in the relief header in comparison with the allowable backpressure. The general practice when handling gas, vapor or steam is to calculate valve-capacity requirements at the maximum allowable working pressure of the vessel and reach full capacity at 3 or 10% above the design pressure, depending on the applicable code. The overpressure is referred to as "accumulation" and is based on the percent pressure rise from the time the valve begins to open until it has fully opened.

Liquid-relief valves reach their rated capacity at no more than 25% overpressure. The restricting area of flow is difficult to calculate because of the possible shape variations in the disk's design. Discharge coefficients also vary with design.

In calculating a valve size for gas flow at a higher or lower pressure than its calibration pressure, it must be remembered that the perfect-gas laws do not apply in determining flowrate, because of the difference in flow area at the higher or lower upstream pressure.

On discharge of safety and safety-relief valves, the flow in the throat of the nozzle reaches a constant value corresponding to the critical-flow pressure for a given upstream flow pressure. Therefore, flow through these valves, under adiabatic conditions, is directly proportional to absolute pressure.

Calculations for sizing relief valves and rupture disks are shown in Part 2 of Ref. 3. For nozzle-type valves, the following formulas apply.

For relieving steam:

$$A = \frac{W}{51.45 K_D P K_{SH}}$$

where A is nozzle area, sq. in.; W is flow capacity, lb./hr.; K_D is 90% of the actual coefficient of discharge of valves for power boilers (Section I, ASME Code) and of valves certified subsequent to 1963 for unfired pressure vessels (Section VIII, ASME Code), or 100% of the actual coefficient of discharge for valves certified prior to 1963 for unfired pressure vessels; P is set pressure (psig.) $\times$ 1.03 + atmospheric pressure for power boilers (Section I, ASME Code), or set pressure (psig.) $\times$ 1.10 + atmospheric pressure for unfired pressure vessels (Section VIII, ASME Code) except for exposure to fire, for which P is set pressure $\times$ 1.20 + atmospheric pressure; and K_{SH} is correction factor for superheat.

For gas or vapor:

$$A = \frac{W}{K_D C P}\left(\frac{TZ}{M}\right)^{1/2}$$

where A, W, K_D and P are defined as in the previous equation; T is relieving temperature, °R.; Z is compressibility factor at relieving conditions; M is molecular weight; and C is a constant for the gas or vapor, which is a function of the ratios of specific heats.

$$C = 520\left[k\left(\frac{2}{k+1}\right)^{(k+1)(k-1)}\right]^{1/2}$$

Appendix J of the ASME Code[1] contains a table of numerical values for C for various specific heat ratios.

If the ratio of specific heats k at constant pressure and constant volume is unknown, a conservative value of 315 may be used for C.

No formula for liquid relief is provided by the ASME Code. However, a relation that is derived from the general flow formula is usually accepted:

$$A = \frac{Q}{27.2 K_p K_v}\left(\frac{S}{\Delta P}\right)^{1/2}$$

where A is nozzle area, sq. in.; Q is capacity, gpm.; K_p is an overpressure factor that equals 1.0 for 25% accumulation, or 0.6 for 10% accumulation; K_v is viscosity factor; S is specific gravity at flowing temperature; and ΔP is the pressure differential across the valve, psi.

Sizing Rupture Disks

Where inlet piping is short, and discharge is direct to the atmosphere, rupture disks can be considered as sharp-edged orifices in calculating their sizes.

For liquids: $$A = \frac{Q}{23.1}\left(\frac{S}{\Delta P}\right)^{1/2}$$

For gases: $$A = \frac{W}{35.2P}\left(\frac{M}{T}\right)^{1/2}$$

where the symbols have the same meaning as in the previous formulas.

When appreciable lengths of inlet or discharge piping are involved, pressure drop (or flow) through the system must be calculated. For this, the disk assembly can be assumed equivalent to 50 dia. of straight pipe.

Discharge Piping, Knockout Drums and Seals

The basic considerations in determining line sizes in the main discharge header and laterals are:

- Maximum required vapor-relieving capacity.
- Maximum permissible backpressure in headers.
- Type of safety-relief valves to be used.
- Governing codes.
- Engineering principles involved.

After establishing these items, line sizing becomes a matter of fluid-flow calculations.

Maximum required vapor-relieving capacity is based on the valve or group of valves that may be simultaneously required to vent the maximum amount of vapor as a result of an abnormal process situation. The maximum allowable backpressure in the header system is established according to the lowest design pressure of a vessel in the system, or is based on the type of relief valves to be used and on the governing codes. Further, in a manifold discharge system, the header should be so designed as to limit backpressure to approximately 25% of set pressure of the lowest set valve. Details of line-sizing calculations are given in Ref. 8.

Knockout drums are usually sized by trial-and-error calculations. The first step is to determine the required drum size for separating the entrained liquids. Residence time of the gas or vapor in the vessel must be considered. Liquid particles separate when this time is equal to, or greater than, the time required to travel the available vertical height at the dropout velocity of the liquid particles. In addition, gas velocity must be low enough to permit the liquid droplets to fall. The available vertical height is taken as the distance from the surface of the accumulated liquid. Residence time of the gas must be such that no large slugs of liquid are allowed to reach the flare. Since the flare can handle small droplets, the allowable vertical velocity in the drum can be based upon that required for separation of droplets 150 microns or larger in dia.

The second step in sizing a knockout drum is consideration of the effect of stored liquid. Streams directed through the drum are selected according to their composition and in line with plant operational policies. Recommended liquid holdup time is 10 to 30 min.

Selection of drum size and choice of vertical or horizontal position are based on economics. Often, a horizontal drum is more economical when a large volume of liquid storage is needed and vapor velocity is high.

Seal-drum sizing requires as its first step the determination of maximum allowable backpressure in the vent header. This in turn sets the maximum distance that the inlet pipe is submerged. The ratio of cross-sectional area of the inlet pipe to vessel free-area for gas flow above the liquid surface should be at least 1 to 3 at all times, to avoid surges of gas flow to the flare. Detailed sizing calculation methods for knockout and seal drums are included in Ref. 8.

RELIEF SYSTEM DESIGN

Among the more important factors to be considered in designing a relief system are:

- Applicable codes: local, state and federal; ASME Boiler and Pressure Vessel Code, Section VIII.
- Relieving capacities of individual valves when conditions cause abnormally high pressures.
- Operating characteristics of the relief valves.
- Design pressures of process equipment requiring protection from overpressure.
- Chemical and physical properties of vapors and gases to be released through the system.

The ASME Code specifies safe practices in design, construction, inspection and repair of unfired pressure vessels. Requirements of the code are:

- All unfired pressure vessels other than steam generators are to be protected by pressure-relieving devices capable of preventing vessel pressures higher than 10% above maximum allowable working pressure, excepting where excessive pressure results from a fire or other unexpected source of heat. In the latter case, the pressure-relieving devices must prevent overpressures higher than 20% above maximum allowable working pressure, with all pressure-relieving devices operating.
- Size of the outlet pipe is to be such that any pressure existing or developing in the discharge line does not reduce the relieving capacity of the protective devices below the requirements for adequate prevention of overpressure.

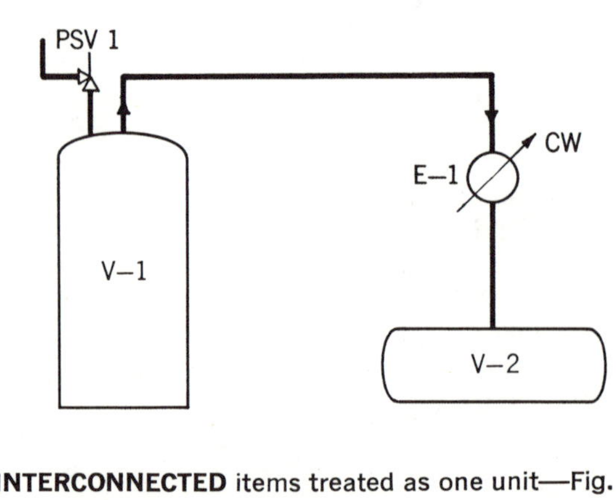

INTERCONNECTED items treated as one unit—Fig. 8

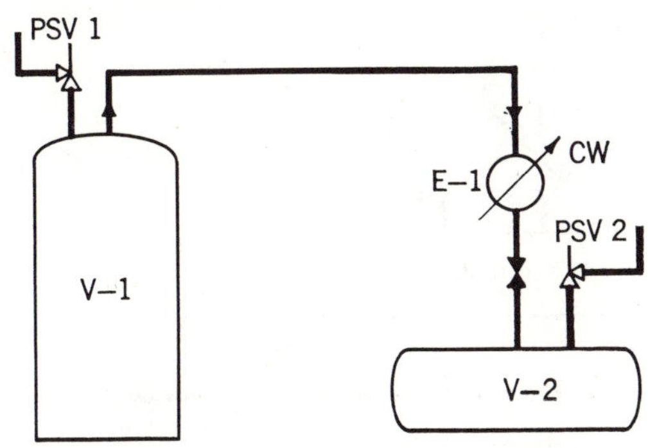

BLOCKED units require separate relief valve—Fig. 9

a. Heat exchanger protected on cold side.

b. Fired-tube heater protected on tube side.

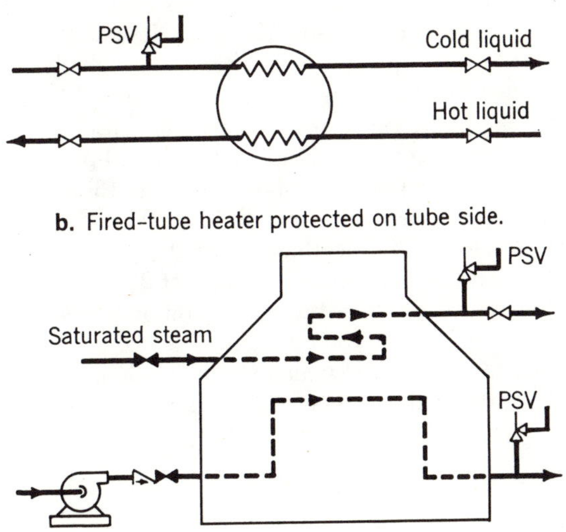

BLOCKED liquid lines need pressure relief—Fig. 10

Fire is the most prevalent danger and most devastating hazard in petroleum refineries and chemical plants. When exposed to fire, enclosed equipment containing liquid can be subjected to pressures beyond the rupture point, if adequate pressure relief is not provided. Heat exchangers, fractionating columns, storage vessels and operating vessels are liable to exposure to fire.

Interconnected pieces of equipment with a piping system that does not include valves for isolation of individual items can be considered as a single unit for applying a relief valve or valves, but set pressure should be that of the lowest designed pressure of any piece of equipment. In Fig. 8, the three interconnected items are protected by a single relief valve, PSV-1. Where items or units can be isolated by valves (as in Fig. 9), a relief valve is needed for each unit; PSV-1 is designed to relieve the unit consisting of V-1 and E-1, and PSV-2 is sized to relieve V-2 only.

Blocked Inlets and Outlets

Blocked-in sections of cold liquid-filled lines exposed to the sun must be protected by relief valves. Otherwise, the piping can be ruptured from expansion of the trapped liquid.

A relief valve is required on the cold side of a heat exchanger if the installation allows blocking-in of the liquid-filled cold side while a hot liquid is flowing through the hot side, as in Fig. 10a. If not pressure-relieved, the blocked-in cold-side liquid can be heated sufficiently to expand and rupture the shell.

If the liquid flowing in the coils of a fired heater can be blocked-in, the heater must be pressure-relieved to prevent overpressure in case heat is being applied while the liquid in the coils is entrapped. In Fig. 10b, closing block valves in the liquid feed line blocks-in the coils. This can also be done by closing the outlet valve and the check valve, to prevent reverse flow.

a. Reciprocating compressor

b. Reciprocating or gear pump

c. Turbine driver

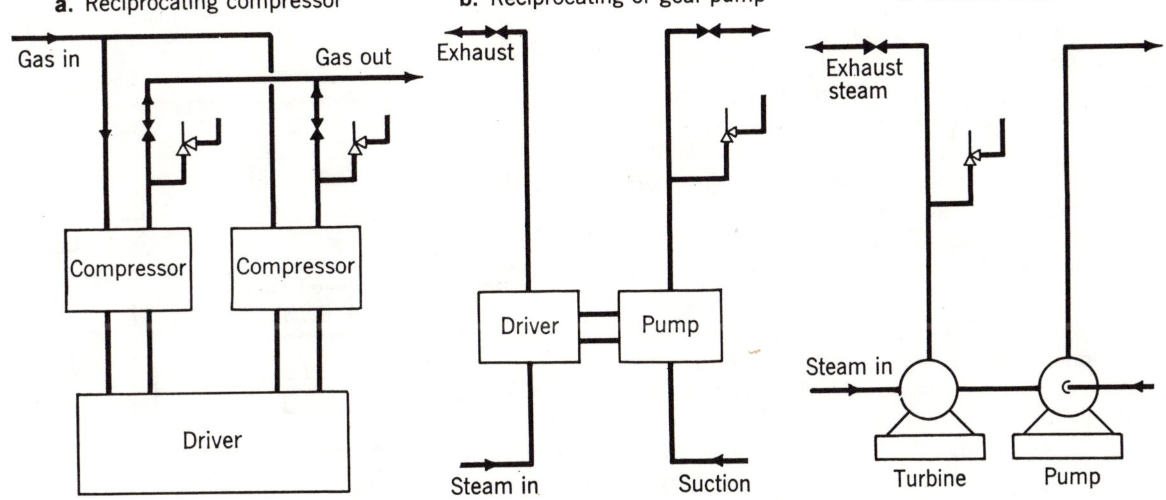

COMPRESSORS, pumps and turbines are subject to overpressure and require relief valves—Fig. 11

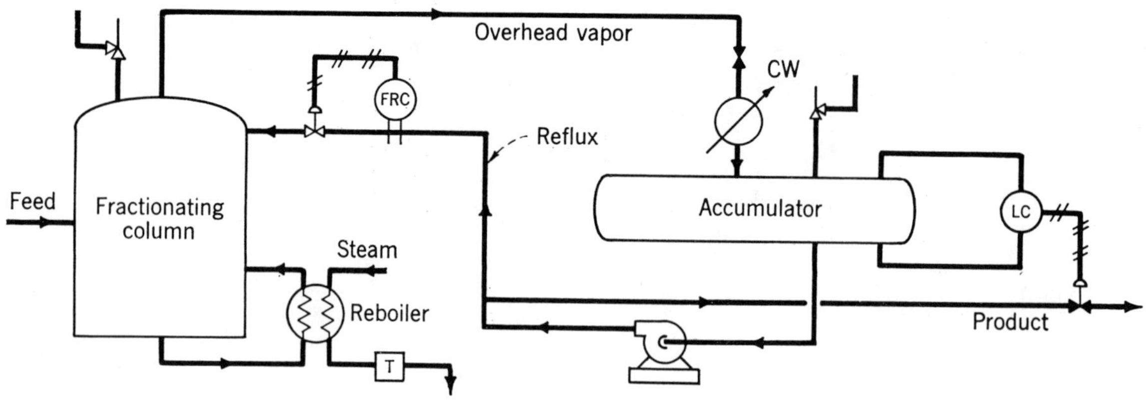

FRACTIONATING unit may have overpressure due to blocking, loss of coolant, equipment or instrument failure—Fig. 12

The relief valve should be placed on the coil outlet side, to permit flow through the coil and prevent coking in case there is a block-in.

A relief valve is also needed on the superheater steam coil of the heater in Fig. 10b, because of the possibility of starting the heater with the valves closed on the steam coils.

Reciprocating compressors and pumps, gear pumps, the steam-exhaust side of turbines, and fractionating columns are all subject to overpressure caused by blocked outlets or loss of condensing medium. Fig. 11a and 11b show relief-valve arrangements whereby the blocked fluid can discharge on the suction side.

Since the steam-exhaust side of a turbine is usually designed for a much lower pressure than the inlet side, the relief valve in Fig. 11c is placed on the exhaust side of the turbine drive of a pump. If the steam outlet should be blocked without relief, the pressure on the exhaust side would rise to that on the inlet side. However, if the exhaust block-valve is omitted, the relief valve is not required provided the downstream system is properly protected.

In a fractionating column, overpressure can develop when the flow of overhead vapors is blocked and the source of heat to the column continues. In Fig. 12, overpressure would result if the block valve in the overhead vapor line were accidentally closed and heat to the reboiler continued. So a relief valve is a necessity for safe operation of this unit.

Again referring to Fig. 12, overpressure in the column can also be caused by loss of reflux, which in turn can be due to pump or instrument failure or to loss of cooling water. The column is overheated without the reflux, which acts as a coolant under normal operating conditions.

Another important source of overpressure is failure of an instrument itself or loss of air to one or all of the related instruments on a unit. "Freezing" of a control valve in the open or closed position is a type of failure to be considered. With reference to Fig. 13a, if the maximum design pressure of the column were 60 psi., and stripping steam at 150 psi. were in use, overpressure would occur if the flow-control valve "froze" in the fully open position. However, this would not occur if the overhead condenser had been designed to condense the full capacity of the steam through the control valve. Overpressure could also be caused by closing of the block valve between the condenser and the accumulator, due to buildup of excess 150-psi. steam.

In the absorption unit shown in Fig. 13b, rich oil leaving the absorber at 450 psi. flows through a level-control valve to the deethanizer at 200 psi. The latter is designed for a maximum pressure of 250 psi. Failure of the control valve in the open position would result in excess pressure in the deethanizer, since this valve is the only "partition" between the two columns. Therefore, a relief valve is essential for safe operation.

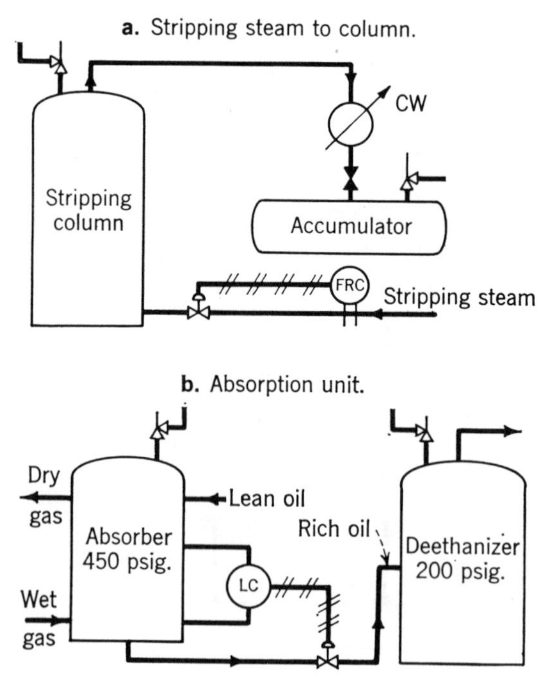

CONTROL-valve failure causes overpressure—Fig. 13

Other Causes of Overpressure

The possibility of human errors in operation must also be taken into account in designing pressure-relief systems. For example, an operator might inadvertently close the block valves on the cold side of the heat exchanger in Fig. 10a, and open the block valves on the hot side. He might shut off the block valves (Fig. 10b) on the steam coil containing water, while continuing to fire the heater. In Fig. 11a and 12, he might manually close the discharge valves on the compressor and pump. These are only a few examples of how misoperation can cause overpressure in a processing system.

The most common equipment failures causing overpressure are those of the plant cooling-water system, fractionating-column reflux pumps, and heat-exchanger tubes. Pumps can fail through mechanical breakdown or loss of power. Since exchanger tubes occasionally split, consideration should be given to the effect of high-pressure fluid on the low-pressure side. If the high pressure is on the tubeside and a tube splits, the shellside, shell piping, or associated equipment can be overpressurized. High pressure on the shellside should receive similar consideration.

Transmission of pressure waves back and forth in a liquid-filled pipe results in the condition called "water hammer." It can be caused by rapid shutoff of a pump or valve in the liquid line. Pressure thus created is sometimes sufficiently excessive to rupture piping or equipment.

For equipment not designed for underpressure, protection against negative pressure, or vacuum, must be provided if this condition is likely to occur. Many times, vessels have collapsed because of lower pressure on the inside than the atmospheric pressure on the outside. These are a few of the causes of partial or full vacuum in equipment:

- Pumping more liquid out than is entering a vessel.
- Draining liquid out of an unvented vessel, e.g., in removing water after a hydrostatic test.
- Injecting cold liquid into a vessel too soon after steaming it out.

PROTECTING AGAINST EXPLOSIONS

Water Into Hot Oil

There are no generally accepted methods of calculating the relieving requirements for the condition of water contacting hot oil, even though this is widely recognized as a potential source of overpressure. If the quantity of water is known (which is rarely the case) and the heat available in the process stream can be calculated, the relief valve can be designed as though it were a steam valve. But it would still be questionable whether such a valve could open fast enough to relieve the almost instantaneous expansion of liquid to vapor (approximately 1 to 1,400 at atmospheric pressure) that occurs under these conditions. The same statements generally apply to instantaneous vaporization of light hydrocarbons by hot oil, although the vapor-to-liquid volume ratio can be considerably lower than with water vaporization.

Although no pressure-relieving device is normally provided for this contingency, proper considerations in process design and operation can greatly reduce or virtually eliminate the possibility of explosion from this source. Some of the precautions that can be taken are double blocks and bleeds on water connections, appropriate steam-condensate traps, and avoidance of water-collecting pockets.

Internal Explosion

General rules for accurate sizing of relieving devices for internal explosion are very difficult, if not impossible, to formulate. No single method of sizing has been widely accepted, because of the thermodynamics involved, the uncertainty in determining the rates of temperature and pressure rise, and the lack of a basis for ascertaining the gaseous mixtures present. It is generally agreed that rupture disks rather than relief valves should be used for protection from internal explosion, mainly because the disks react more rapidly than relief valves to the instantaneous buildup of pressure. Moreover, limits of overpressure as prescribed by the vessel code are not necessarily applicable under these conditions of extremely rapid pressure rise.

Internal explosions cause pressure buildup in vessels as a function of:

- Pressure and composition in the vessel immediately prior to the explosion.
- Rupture-disk bursting pressure at the operating temperature.
- Rupture-disk area in relation to vessel volume.
- Timing and direction of the explosive pressure wave.

In controlled explosion tests, the potential maximum explosion pressure of typical air-hydrocarbon mixtures at ambient temperature has been shown to be about eight times the pressure in the vessel just before the explosion. In addition, the peak pressure can be considerably higher than, and is not limited by, the rating of the rupture disk.

Since there are no generally accepted design methods for protection against internal explosion, selecting the most effective means of protection is the designer's responsibility and is entirely dependent on his judgment. Here are two suggested approaches:

- Design the vessel to withstand internal explosion by applying a suitable safety factor to the normal operating pressure. However, the required safety factor might make this method impractical.
- Based on the contained volume of the vessel, a rupture disk can be sized by an arbitrary method. For protecting against an air-hydrocarbon explosion at an oil refinery, a relief area of 2 sq. ft./100 cu. ft. of vapor volume is considered adequate. A reliable, official guideline cannot be recommended until further investigations are made on the subject of internal explosion.

Chemical Reaction

Extensive knowledge of reaction kinetics and reaction rates is required in calculating the rate of pressure relief necessitated by a chemical reaction in a closed system. Considerations for chemical reactions that can approach the condition of explosion are similar to those discussed for internal explosion.

Exothermic reactions are particularly hazardous because of the:

- Acceleration of reaction rate with rising temperature, with rapid attendant production of high rates of energy release.
- Large volumes of noncondensables released—in many instances, when the temperature has risen to excessive levels, and decomposition reactions start.

Normal relief of overpressure is likely to be inadequate to cope with these conditions. To prevent possible runaway reactions, rapid pressure relief must be provided.

If a fully reliable pressure-relief system could be designed, destructive overpressures might be prevented by adding enough volatile fluid to absorb excess reaction heat, thus cooling the system to a safe level. In so doing, a calculable overpressure would be produced for which relieving capacity, though large, could be provided. Internal refrigeration from volatilization of the liquid would prevent more-serious overpressure conditions that might result in runaway chemical reactions.

Explosion-Suppressive Systems

In designing an explosion/fire protection system in conjunction with processing and storage facilities for a particular material, it is necessary to determine by appropriate tests the explosive and reactive characteristics of the material in the form or dust, vapor or gas. The fact is recognized that explosions are not instantaneous. With any combustible material, there is a measurable time-lapse of milliseconds between ignition and buildup of pressure to explosive levels. In most instances, this lapse can be used to advantage in extinguishing an explosion that has already begun, by applying one or more of the following counteractive techniques.

- Discharge of a suppressant to extinguish the explosion before appreciable pressure buildup.
- Opening of relief vents by a controlled explosive charge in time to relieve rising pressure. Suppressants are often used in conjunction with venting, to extinguish after-fires.
- Inclusion of suppression provisions downstream of the hazardous area to prevent chain explosions or spread of fire.
- Isolation of one section of a plant from another to prevent the spread of explosions and resulting fires.
- Incorporation of relays for immediately shutting down the process when an explosion occurs.

Fig. 14 is a diagram of an explosion-protection system.

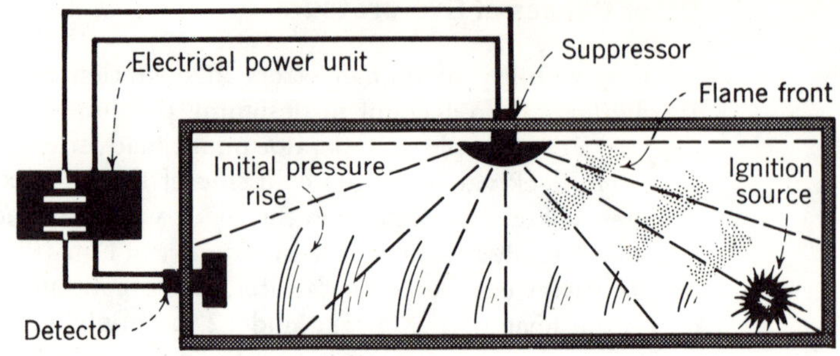

DETECTOR senses initial pressure rise—Fig. 14

TRANSFER/DISPOSAL SYSTEM DESIGN

Prior to design of the disposal section of a pressure-relief system, detailed analysis must be made of all possible conditions under which fluids will be discharged from pressure relief and/or depressuring devices in the system to determine maximum load for any emergency. This is the total of the loads of individual devices relieving at emergency conditions.

The maximum load is that which demands the greatest head loss through the system. It is not necessarily the largest number of pounds per hour the system is required to relieve. For example, a vapor having a molecular weight of 19 and flowing at a rate of 100,000 lb./hr. at 300 F. results in a greater head loss and greater load on the relieving system than does 150,000 lb./hr. at 100 F. of a vapor (molecular weight = 44).

Overpressure, requiring emergency relief, arises from a specific contingency. Since simultaneous occurrence of two or more unrelated contingencies is most unlikely, this is seldom used as a basis for calculating the maximum system load. Lines from individual relief valves are sized for maximum design flow, but sections of a main header or subheader must be sized for a specific maximum contingency.

In addition to the discharge of pressure-relief valves, the disposal system may have a load from vapor depressuring sources. The latter are auxiliary facilities for rapid reduction of pressure in equipment through release of vapors. Headers servicing relief valves often receive depressuring streams. The load from depressuring must be added to the loads from the individual relief valves in calculating the maximum load on a disposal system.

Single-Valve Systems

The single-valve system discharging to the atmosphere is the simplest type of relieving/disposal system. Excessive backpressure is avoided by using the shortest practical downstream line. This type of discharge system is frequently installed at elevated locations such as the tops of towers; however, the technique is generally undesirable, and when used is limited to light gases.

The single-valve, closed-discharge system is somewhat more complex. The discharge line is routed to a vent stack, flare or other disposal means to which no other relief-valve discharge line or depressuring line is connected. Fluids at unusual temperature and/or pressure conditions, as well as toxic or corrosive gases or vapors, are often relieved through a single-valve closed-discharge system.

Multiple-Valve Systems

The multiple-valve closed-discharge system is the most widely used pressure-relief system. After calculating the line size based on individual loads, maximum flowrates under emergency conditions, allowable backpressure and fluid discharge temperature, routing of lines becomes the next important consideration. Possible routing of main headers can be obtained by inspection. Instead of combining all discharge streams into a large header serving an entire operating unit, it may be more economical, under the particular conditions, to subdivide the protection with two headers, each servicing one-half of the area. Such decisions must be based on cost of pipe, supports, and other related materials. Final selection of a disposal method may also depend on inherent ease of maintenance.

A multiple-header system is frequently needed in a multiple-valve system serving a single unit, for economic, safety or other reasons. The need for such a system arises when it is advantageous to isolate particular relief or depressuring streams because of:

- Presence of corrosive materials.
- Considerable differences in operating pressures of equipment served by the system.
- Discharge streams that are likely to subject piping to abnormally high or low temperatures.

Subheaders of corrosion-resistant material can be used to discharge corrosive streams, instead of constructing an entire common header for all discharge streams of the more expensive alloy. Or, with only subheaders handling the corrosive streams, it might be even more economical to make them of carbon steel, with periodic replacement. In some cases, subheaders handling corrosive discharge are routed through neutralizing equipment before venting to the flare system.

Operation is at considerably different pressure ranges in many processing units. Often, a small, separate header is sufficient to handle the discharges of relieving devices on high-pressure equipment. This can be done if full advantage is taken of the built-up backpressure permissible, without restricting the capacity of the relief devices.

High-pressure sources can sometimes be routed to a system serving low-pressure equipment, by increasing the size of the header if feasible. However, there must be adequate provision against excessive backpressure on the low-pressure devices that are expected to discharge simultaneously. In this plan, balanced-type pressure-relief valves are required on the low-pressure equipment. In some cases, it may be more economical to provide separate headers and balanced relief valves than to increase size of a common header.

If extremes of temperature are expected, separate discharge headers may be more economical than increasing the size of a large common header to provide for the expansion of hot streams. Alloys to withstand very-low temperatures can also add to the cost of enlarging a common header.

Installation procedures for pressure-relief valves for gas, vapor and liquid service are given in Ref. 7. Also, pertinent data for the design, installation and operation of pressure-relief and depressuring systems are compiled in Ref. 8 (adjunct to API RP 520[6, 7]).

Acknowledgment

The author thanks E. Lochte, K. Owen and C. Teague, of Fluor's Houston Div., Instrument Engineering Dept., for their assistance in supplying the reference material for this article and for their suggestions on the content.

References

1. "ASME Boiler and Pressure Vessel Code," Unfired Pressure Vessels, Section VIII, American Soc. of Mechanical Engineers, New York. 1965.
2. Conison, J., Designing Vapor Relief Systems, *ISA (Instr. Soc. Am.) J.*, June 1964, p. 51.
3. Rearick, J. S., How to Design Pressure Relief Systems, *Hydrocarbon Process.*, Part 1, Aug. 1969, p. 104; Part 2, Sept. 1969, p. 161.
4. Jenett, E., Design Considerations for Pressure-Relieving Systems, *Chem Eng.*, July 8, 1963, p. 125.
5. Jenett, E., Components of Pressure-Relieving Systems, *Chem. Eng.*, Aug. 19, 1963, p. 151.
6. "Recommended Practice for the Design and Installation of Pressure-Relieving Systems in Refineries," Part I—Design, API RP 520, 3rd ed., American Petroleum Institute, New York, 1967.
7. "Recommended Practice for the Design and Installation of Pressure-Relieving Systems in Refineries," Part II—Installation, API RP 520, 2nd ed., American Petroleum Institute, New York, 1963.
8. "Guide for Pressure Relief and Depressuring Systems," API RP 521, 1st ed., American Petroleum Institute, New York, 1960.
9. "Flanged Steel Safety Relief Valves for Use in Petroleum Refining," API RP 526, American Petroleum Institute, New York, 1963.

Meet the Author

Marx Isaacs is a process engineer and technical writer/editor, Fluor Corp., Houston Div., 4620 N. Braeswood, Houston, Tex. 77035. After graduating from Tulane University with a B.S. in chemical engineering, he worked for Shell Oil Co., first as a chemist and later as a technologist. Before joining Fluor, he was for three years the technical editor of Corrosion, a reserach publication of the National Assn. of Corrosion Engineers.

Coping with high-pressure letdown

Designing valves and piping for letdown is an old problem that has cropped up more and more as new high-pressure coal conversion plants go onstream. Here the author describes accepted techniques for offsetting the destructive and disruptive effects of letdown.

Les Driskell, *Chemical Plants Div., Dravo Corp.*

☐ Bringing a fluid from high to low pressure under controlled conditions can be exceedingly troublesome. With gases, there may be problems with excessive noise and extremely low temperatures. With vapors (and sometimes even with gases), droplets may precipitate within the valve to threaten the surfaces they impinge upon. Pressure letdown of a liquid may be accompanied by flashing or cavitation, with attendant noise and damage to valve and piping materials. And if the liquid is corrosive or abrasive, the problem may be so severe as to necessitate changes in the process flow sheet.

Clean, dry gas

A clean, dry gas is by far the easiest fluid to handle. The high-velocity noise problem—the major difficulty in such services—can now be dealt with by using special control valves and auxiliary devices that are readily available from valve manufacturers. During the past eight years there has been a flurry of activity in the field of aerodynamic noise reduction. *Source treatment*, using special quiet valves, prevents or attenuates acoustic power at the source. *Path treatment* reduces noise by attenuating it along the transmission path from the source to the ear. Either or both of these approaches may be used, depending upon the costs [*1*].

Auto-refrigeration

Gas expansion during pressure letdown can create low temperatures by auto-refrigeration [*2*]. For example, if carbon dioxide at 20°F and 300 psig is expanded to atmospheric pressure through a valve, the temperature of the downstream gas becomes −60°F (Fig. 1). Because most steels become brittle at these temperatures, the valve body and downstream piping material must be specified carefully to allow for such temperature extremes.

When a gas flows through a throttling valve, the overall thermodynamic process is isenthalpic, but from the inlet to the vena contracta—the point of minimum cross-sectional area of the stream—the process is nearly isentropic [*2*]. For this reason, the temperature at the vena contracta can be appreciably lower than that at the valve outlet, depending on the thermodynamic properties of the fluid. Ethylene, and even superheated steam, can condense precipitously within the valve body to create droplets of high-velocity liquid. Valve construction materials must be selected to withstand the erosive scrubbing action of such streams, and measures must be taken to abate the added noise that may result.

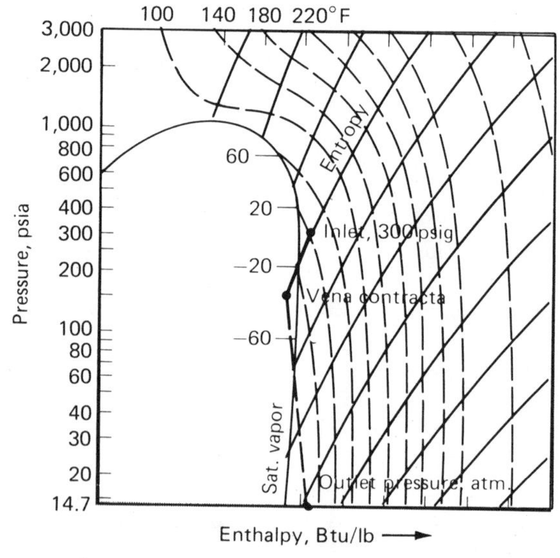

Pressure-enthalpy curves show that when CO_2 lets down from 300 psig and 20°F to atmospheric pressure, temperature drops to −60°F. Duration of subcooling is insufficient to cause droplet formation. Fig. 1

Originally published October 25, 1976

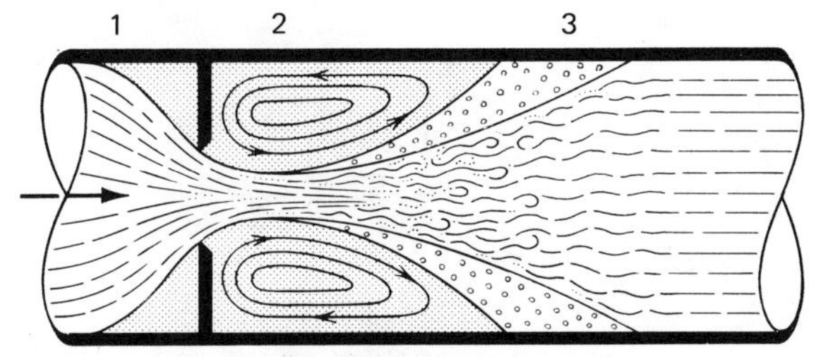

High-velocity liquid passing through restriction induces a low-pressure section at the vena contracta. Even lower pressures occur in the vortices created at the jet boundary layer. Fig. 2

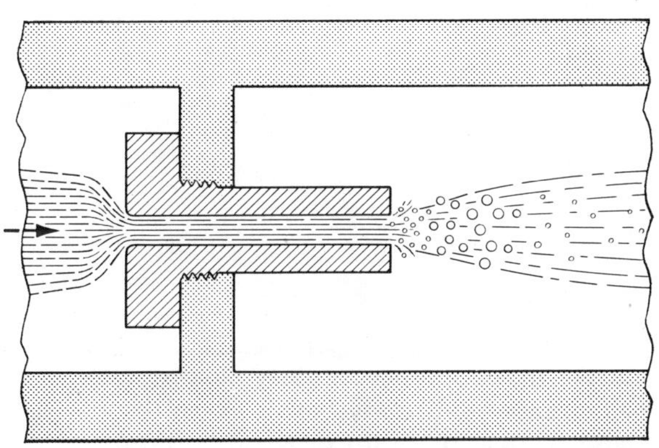

The choke is made of a hard material, and because of its shape, minimizes damaging impingement. Cavitation or flashing energy is spent in a downstream pool of liquid. Fig. 3

Solids formation

Solids in the form of ice or hydrates may be induced during pressure letdown by auto-refrigeration if water is present. In a continuous process, icing within the valve body is unlikely if nearly all of the pressure drop is taken across a single orifice, since crystals are denied sufficient holdup time to grow. But if there is substantial pressure loss in the upstream piping, or if the valve flowpath has more than one restriction in series, ice can form and accumulate in the valve. In blowdown service, where the entire upstream volume is exhausted, icing can be expected. Hydrates, a chemical combination of water with gas, can form at temperatures much higher than the freezing point of water. Hence, the tendency toward hydrate formation in such services is much greater than that toward ice formation.

Dry slurries—gas streams bearing solid particles—add another dimension to the problem of gas letdown, because of their "sandblasting" effect on valve components. This form of abrasion can be mitigated by ordering trim or facing that is substantially harder than the flowing particles. Now available are special valve designs which have been streamlined so as to reduce or prevent direct impingement on the seat and other vital parts of the trim.

Flashing liquid

Liquid letdown is also complicated by a change in fluid state. If the vapor pressure of the upstream liquid is greater than the downstream system pressure, part of the liquid will vaporize. This can produce high-velocity, mixed-phase flow in the valve and in downstream piping, resulting in both noise and erosion. If the liquid is clean (negligible particulate matter), a reduction in line velocity may diminish the problem. Velocity in the valve body can be cut by moving to an oversize body—i.e., reduced trim. A good choice of valve style is the angle-type body, with the flow tending to close the plug. This style offers an unobstructed downstream path which minimizes erosion caused by impingement.

The downstream piping is another matter. Increasing the pipe diameter in order to lower fluid velocity may not be enough to offset erosion. Slug flow of condensate can still occur, and wreaks havoc on the piping system because of the impact of liquid slugs on pipe walls wherever there are changes in direction. If at all possible, the angle-body valve should discharge directly into a vessel containing a reservoir of liquid; this configuration dissipates fluid energy and momentum into the reservoir, rather than into the piping system.

Hydrocarbons are not as aggressive as water in flashing service, in part because their density is less and their surface tension is significantly less than water. Consequently, liquid condensate "bullets" in flashing hydrocarbon lines are lighter and softer and less destructive than those in water lines.

Cavitating liquid

Much more troublesome than flashing liquid is a cavitating liquid. No engineering material can withstand the long-term effects of cavitation. But the most insidious aspect of the problem is that the conditions for cavitation are so seldom recognized. Consider the commonplace application of a vertical centrifugal pump on a cooling-tower basin. Because of the steep head-curve normally associated with this style pump, the pressure is often controlled by bypassing part of the flow back to the basin. In this service a marked degree of cavitation will be present under the following conditions, assuming that water temperature is 80°F, and that the valves are wide open.

Valve type	Upstream pressure exceeds
V-plug globe	57 psig
Contour-plug globe	26 psig
Standard ball valve	5 psig

The problem of cavitation is a baffling one. We cannot accurately predict the conditions at which a specific liquid in a specific valve will begin to cavitate. Even more importantly, we cannot predict the conditions under which mechanical damage can be expected to begin, because some valves escape damage in light cavitation. If the vapor bubbles in a cavitating stream collapse before they can reach a solid surface, they will cause no material damage to the valve or piping.

Letdown of slurries

If the liquid under pressure is a slurry, the designer has fewer options than with a clean liquid. If the fluid is corrosive, and the solid particles are abrasive, the design possibilities become severely limited.

In order to better understand the phenomenon of slurry letdown, let us consider what happens as a simple liquid passes through a throttling orifice at increasingly higher velocities (Fig. 2). As the fluid moves from point 1 to point 2, the flow lines converge and the fluid accelerates. The increased velocity up to point 2 causes a reduction in pressure in accordance with the Bernoulli equation. At the orifice edge and beyond, a separation zone is formed by the discontinuity of the jet and the boundary layer. The high-velocity gradients in this zone create small but intense vortices that generate still lower pressures at their centers.

As stream velocity rises, the local pressure in the vortices falls below the vapor pressure of the liquid, and vapor cavities begin to form at nucleation sites. These sites are believed to consist of submicroscopic nuclei of free, undissolved gases that exist in the crevices of foreign particles suspended in the liquid.

The vapor bubbles move downstream toward point 3. At the same time, deceleration in the widening jet, and the accompanying decay of the vortices, raises local pressure to the point where bubbles become unstable and collapse—thus completing the cavitation cycle.

The unusual manner in which these bubbles collapse makes the process extremely damaging to engineering materials. As a bubble collapses, its spherical shape flattens and then suddenly implodes into a toroid, ejecting a high velocity jet which concentrates fluid energy over a very small area [*3*]. It has been determined that pressures exceeding 1,000,000 psi can be generated in this manner. If the implosion occurs at a solid boundary, the surface will become pitted and destroyed by the collapse of these bubbles.

If upstream pressure is held constant, reducing downstream pressure causes the bubbles to grow still larger, and pressure recovery—with concommitant bubble collapse—takes place farther downstream. Presence of larger vapor bubbles in the stream boosts velocity in this downstream zone and promotes abrasion and/or corrosion if the fluid contains suspended particles or corrosives. When downstream pressure is less than the vapor pressure of the inlet fluid, the bubbles do not condense at all, and the condition previously described as flashing exists. With flashing, damage to piping stems from the high velocities that can be attained if the cross-sectional area of the pipe is not greatly enlarged to handle the volume of the vapor-liquid mixture.

In addition to promoting mechanical damage on metal surfaces, cavitation creates considerable noise and vibration. This vibration may become so severe that the integrity of the piping system itself may be jeopardized.

Cavitation can be prevented by reducing or eliminating pressure-recovery downstream of the valve orifice. Special valves which exploit multiple, parallel flow paths, each having many changes in direction, have been developed for this purpose. The use of this design for slurries is limited, however, because narrow flow passages are subject to plugging with solids, and because sharp turns in the channel foster abrasion.

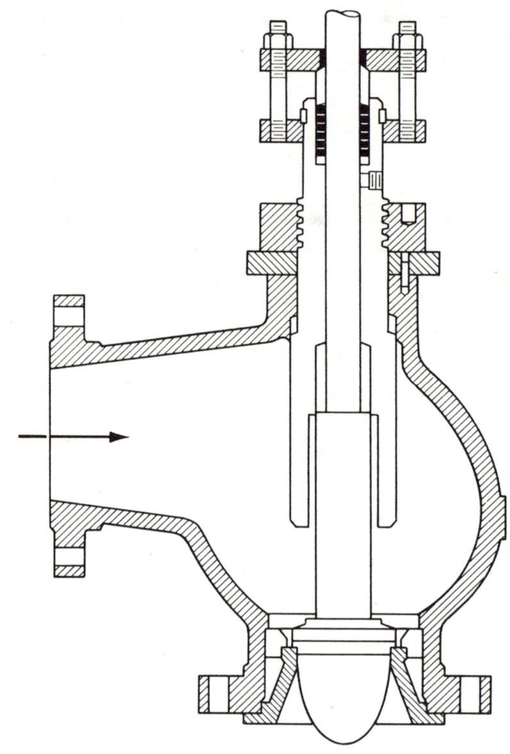

Letdown valve for moderately severe service: expanded body slows velocity; placing orifice at outlet cuts body damage from discharge stream; special two-piece seat expands radially. (*ITT Conoflow*) Fig. 4

Other ways of reducing pressure recovery include operation of two standard valves in series, or use of a multi-step valve, which has a series of throttling orifices.

If noise and vibration can be tolerated, a controlled cavitation can be permitted to occur in such a manner that the bubbles do not collapse against a vulnerable surface. The simplest device using this technique is the oil-field choke (Fig. 3). The choke is a fixed orifice of hard material with a rounded entrance and a long, straight passage. Vaporization does not take place until the stream is just beyond the outlet, where the jet begins to expand. A liquid pool downstream of the choke absorbs the energy of the jet, slows the high velocity, and allows the bubbles to collapse harmlessly.

One example of a throttling valve that has been used successfully on slurries at moderate pressure drops (400 psi) is shown in Fig. 4. Features that should be noted are: (a) an expanded body which reduces inlet velocity; (b) an orifice discharge at the valve outlet designed to aim the cavitating fluid at a downstream liquid pool; and (c) a two-piece seat that is deformed by

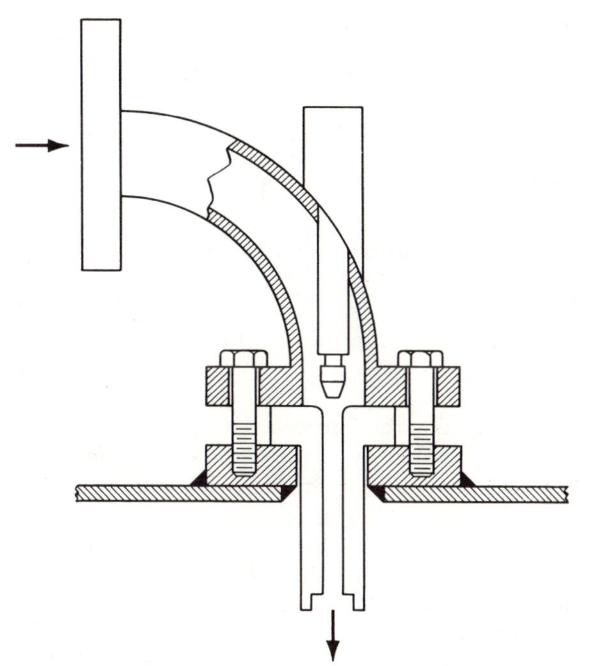

Slurry letdown valve: long-sweep entrance avoids sharp turns; long orifice delays vaporization until fluid leaves valve; recessed orifice exit forms buffer-chamber at the tip. (*ITT Hammel Dahl*) Fig. 5

the steep seating-angle of the plug in order to seal the closure more effectively and avoid wire-drawing when the valve is closed.

Another example of a throttling valve for slurry letdown is shown in Fig. 5. Note the resemblance of the valve orifice to the oil-field choke shown previously. The orifice and plug are usually made of a hard ceramic material such as alumina, which resists both abrasion and corrosion. Special features of this valve include: (a) a long sweep entrance to minimize wear due to impingement of abrasive particles; (b) an orifice designed with a rounded entrance and a long, straight choking section, which delays the onset of vaporization; (c) a recessed orifice exit that forms a buffer chamber, protecting the tip from high-velocity particle bombardment caused by explosive vaporization of the issuing stream; and (d) a discharge that can be directed into a large volume, such as a tank or expanded pipe section, for dissipation of stream energy.

General guidelines for letdown valves

Valves that are to be used in high-pressure letdown service should be carefully designed with respect to plug guiding and dynamic stability. In practice, this stipulation calls for a powerful and stiff actuator. Also, close attention should be given to selecting materials that will minimize the effects of corrosion on valve parts that are exposed to the high-velocity path of the stream.

Table I shows that cavitation alone is best dealt with by the use of labyrinth-style trim, which has little or no pressure recovery. These valves eliminate cavitation completely, and thereby avoid the noise and vibration

Hazards that are likely to occur with different gas services, and measures for avoiding or reducing damage from liquid letdown

Special problem or design feature	Gas service			Liquid service				
	Clean, dry	Wet	Solids	Abrasives	Flashing	Cavitating	Abrasive and flashing	Abrasive and cavitating
Noise	X	X	X			X		X
Auto-refrigeration	X	X	X					
Condensation	X	X						
Icing/hydrates		X						
Hard trim, no impingement				X			X	X
Labyrinth path						X		
Strait-choking nozzle with deenergizing pool						X*		X
Enlarged outlet					X		X	

* Alternative design

which are created by it. If the fluid is a slurry, however, it is unlikely that the narrow passages found in these valves will accommodate the suspended particles, unless the particle dimensions are exceedingly small. So we must use the alternative method shown in the table—installation of a choking nozzle with a de-energizing volume—and deal with the noise problem in other ways, such as path treatment.

There are two other problems that must be resolved: turndown and shut-off. As the valve plug approaches the seat, a local high-velocity zone is created, and the plug and orifice surfaces themselves become subject to the ravages of cavitation and erosion. The closer the plug approaches the seat, the more severe the problem becomes. If the seat leaks when it is completely closed, damage to seating surfaces will be severe. Therefore, it is important to design the process system to avoid operation at flowrates below the practical limits of a particular control valve. And if the seat design will not shut off tightly, an auxiliary on-off valve must be placed in series with the throttling valve; the auxiliary valve should be designed to close automatically when the control-valve opening reaches its lower operating limit.

Process changes to ease letdown

With shrewd design of the process system, it is often possible to greatly simplify the letdown problem by eliminating the need for turndown in the throttling

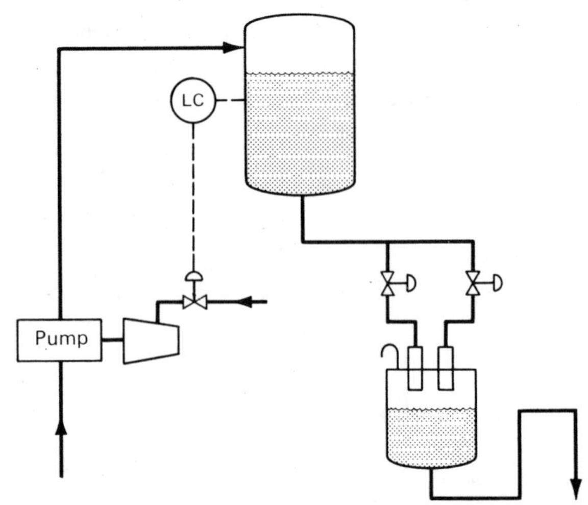

Novel design where the process has been altered to avoid throttling an abrasive slurry. Fixed chokes are used to set the throughput, and the feed stream is modulated to suit. **Fig. 6**

Example: Gas letdown — Box 1

An application requires the use of a ball valve to let down nitrogen gas under the following flow conditions: w (flow rate) = 64,500 lb/hr; P_1 (upstream pressure) = 1,000 psia; P_2 (downstream pressure) = atmospheric; T_1 (upstream temperature) = 940°F; k (ratio of specific heats) = 1.40; and γ (specific weight) = 1.87 lb/ft^3.

Solution

The ISA equation for gas flow [4] is

$$w = 63.3 C_v Y \sqrt{x P_1 \gamma_1}$$

where x is the pressure-drop ratio (pressure drop divided by P_1), Y is the expansion factor and C_v is the valve coefficient. Depending on the type of valve, x has a maximum effective value above which the flow is "choked." This limit applies when $x \geq F_k x_T$, where $F_k = k/1.40$ is a thermodynamic property of the gas, and x_T is the limiting pressure-drop ratio for air through a specific valve, as determined by physical testing. With choked flow, Y reaches its lower limit of 0.667.

The ball valve under consideration has a rated value (fully open) for x_T of 0.15, as reported in a manufacturer's catalog.

Under these circumstances $F_k = k/1.40 = 1.0$, and choked flow occurs at $x \geq F_k x_T$ or $(1.0)(0.15) = 0.15$. Actually $x = (1{,}000 - 14.7)/1{,}000 = 0.985$ and the flow is choked. Substituting in the ISA equation we may now solve for C_v.

$$C_v = \frac{64{,}500}{(63.3)(0.667)\sqrt{(0.15)(1{,}000)(1.87)}} = 91$$

The catalog lists a 2-in. valve having a C_v of 120, which is adequate.

device. Fig. 6 illustrates a simplified process in which a steam-turbine-driven feed pump is the system element that is modulated. The turbine is regulated by system demand. High-pressure letdown takes place through two fixed chokes which are activated by on-off valves. If these two chokes have different capacities, a total of three operating flow rates are available. Similarly, for n chokes, a total of $2^n - 1$ rates are available. In this service, wear is inevitable and a fixed choke is opted for because it has the longest service life; even so, replacement costs for this design are far less than for a highly specialized letdown valve.

Calculation techniques

The design engineer is naturally concerned with techniques for predicting flowrates through pressure letdown valves or orifices, and for estimating orifice size. In the case of gases flowing at high pressure drops, the technique is well established and the experimentally determined coefficients C_v and x_T (see Box 1) are available from valve manufacturers.

Liquid letdown presents an entirely different problem. The customary high velocity causes vaporization to occur in a manner which is poorly understood, and the effect on flow is difficult to predict. Almost all the experimental data has been collected using pure water as the fluid, and no useful data exists for liquid mixtures. The ISA standard [5] presents a method for pure liquids, but this method has extremely narrow limits of applicability, especially if the fluid is other than water.

The flowrate through any restriction is measured by the area and velocity of the stream at the vena contracta. When there is no change of state, there is always a fixed relationship between the overall pressure drop and the pressure drop to the vena contracta. The velocity at the vena contracta may thus be inferred from the overall pressure drop because of this fixed pressure recovery. The experimentally determined sizing factor, C_v, contains factors for both the effective orifice area and the pressure recovery.

When vaporization takes place at the vena contracta, the pressure recovery relationship no longer applies, and the pressure drop to the vena contracta cannot be inferred from the overall pressure drop. Another means must be found to predict it. A pressure recovery factor F_L can be determined by testing and is published by control-valve manufacturers. This factor converts the sizing factor, C_v, for use with pressure drop $P_1 - P_{vc}$ instead of $P_1 - P_2$, where P_{vc} is the pressure at the vena contracta. Although the equation is considered reliable, there is some difficulty in predicting P_{vc}.

Box 2 shows a widely used method for predicting the flow of water under choked conditions, based on the ISA standard [5]. If the liquid is nonaqueous, one is faced with the problem of predicting the minimum effective vena-contracta pressure without the benefit of useful experimental data.

The process stream to be let down is frequently

Example: Liquid letdown **Box 2**

The equation that is applicable to high-pressure liquid letdown, assuming that vaporization will cause choked flow, is as follows:

$$w = 63.3\, F_L C_v \sqrt{\gamma(P_1 - P_{vc})}$$

where C_v and F_L are constants experimentally determined by the valve manufacturer; w is the flowrate in lb/hr; P_1 is the upstream pressure in psia; P_{vc} is the effective vena contracta pressure in psia; and γ is specific weight in lb/ft^3.

If the liquid is pure water, P_{vc} may be estimated from the vapor pressure (P_v) using the following equation derived from the ISA standard:

$$P_{vc} \simeq 0.96 P_v - 0.005\, P_v^{\,1.5}$$

Note that the effective vena contracta pressure is always less than the vapor pressure. This implies that the liquid is superheated, or metastable at this condition.

Consider an angle valve from a manufacturer's catalog listed as having a rated C_v of 23.7 and an F_L of 0.87. We will calculate how much 320°F water can flow through the valve when dropping from 600 to 45 psia. The steam table shows $P_v = 90$ psia, therefore

$$P_{vc} \simeq 0.96(90) - 0.005(90)^{1.5} = 82$$

Substituting these values in the above equation,

$$w = 63.3(0.87)(23.7)\sqrt{56.66(600\text{-}82)}$$
$$= 224{,}000 \text{ lb/h}$$

multi-phase: gas-solid, liquid-solid, or liquid-gas. It is possible to predict the flow pattern of such a stream if there is no change in composition due to the throttling process. Flow of liquid slurries may be computed in the same manner as for a pure liquid, with adjustments for the actual density. Parameters for gas streams carrying suspended liquids or solids may be computed by methods described in references [1] or [2], where the expansion factor Y is applied to the gas fraction only. Because of the changes in composition and density, an accurate technique is not available for predicting the flow of a liquid with its own vapor through a control valve.

Calculation to predict the onset and severity of cavitation cannot be done according to present technology with any assurance of validity. A considerable body of data is available on cold-water service in specific valve styles, but even here the damage threshold is imprecise. For most other liquids, cavitation would probably start at even less severe conditions, but on the other hand, the damage threshold is likely to be much higher.

Some manufacturers publish a cavitation index number for their various valve styles. This number is called K_c and denotes the ratio $(P_1 - P_2)/(P_1 - P_v)$ at which the valve will start to choke on cold water. The ratio K_c is usually considerably greater than the ratio causing "critical" cavitation, which is defined as the point at which the noise level increases sharply, and beyond which mechanical damage is apt to occur. Typical K_c values are:

60-degree butterfly	0.3
Ball	0.22
Globe (smooth plug)	0.53–0.70
Globe (V-plug)	0.80

While these ratios are fair indications of the inception of choking, they should not be used to predict damage. For example, critical cavitation on a 60-degree butterfly valve will occur at a ratio of 0.17.

Experience has demonstrated that several factors are important in controlling damage resulting from cavitation.

Valve location—It is sometimes possible to find a point in the stream where the downstream pressure is higher, or where the temperature and vapor pressure are lower. The right combination of these variables can suppress vaporization. If not, perhaps the system can be changed to make the stream flash instead of cavitate. (Caution! If flashing is a possibility, avoid the use of multi-stage valves—orifices in series—because cavitation can then occur at an interstage orifice.)

Valve selection—Choose a valve with low pressure-recovery, such as a V-port globe, or use two valves in series to reduce the pressure drop across the last valve. A valve designed especially for this purpose may do the job, but is probably the most costly solution.

Air injection—Air or another suitable gas can be injected into the stream at the vena contracta of the valve. This works well, especially with butterfly and ball valves, because the vena contracta can be reached by drilling holes at the right point. The gas acts as a snubber to prevent the violent collapse of the cavitation bubbles. If the vapor pressure is less than atmospheric pressure, compressed air is not required. Atmospheric air will be inspirated automatically.

References

1. Hutcheson, J. W., ed., "ISA Handbook of Control Valves," Second edition, Pittsburgh, Pa., 1976, pp. 221–266.
2. Driskell, L., Control Valve Sizing with ISA Formulas, *Instrumentation Technology*, July 1974, pp. 33–48.
3. Ivany, R. D., Hammitt, F. G., and Mitchell, T. M., Cavitation Bubble Collapse Observations in a Venturi, *Transactions of ASME, Journal of Basic Engineering*, Sept. 1966, pp. 649–657.
4. Control Valve Sizing Equations for Compressible Fluids, ISA Standard S39.3, 1973.
5. Control Valve Sizing Equations for Incompressible Fluids, ISA Standard S39.1, 1972.

The author

Les Driskell is Section Chief, Chemical Plants Div. of Dravo Corp., Pittsburgh, PA 15222, where he directs the activities of the instrument and electrical departments. After doing graduate work in industrial instrumentation at the University of Louisville, he worked at Jos. E. Seagram and Sons Co. and E. I. duPont before joining Chemical Plants Div. (formerly Blaw-Knox Co.). He is a registered professional engineer in Pennsylvania, and a Fellow of the Instrument Soc. of America. He has written and lectured extensively on control valves and other subjects and is chairman of the ANSI subcommittee on control-valve sizing.

Rupture disks for gases and liquids

V. Ganapathy, Bharat Heavy Electricals Ltd., India *

☐ Rupture disks are frequently used in process plants to protect vessels from pressure surges, as well as to separate safety and relief valves from process fluids.† The codes offer precise rules for the installation of rupture disks; and most manufacturers will guarantee rupture disks that they size according to the capacities and operating conditions set forth in a purchase requisition or specification.

However, the engineer frequently needs to know these sizes long before bids are received from a manufacturer, in order to specify vessel nozzles, plan piping, etc. These nomographs might be useful for estimating such a size. They are based on conventional formulas, as follows: **

For gases and vapors:

$$d = (W/146P)^{1/2}(1/Mw)^{1/4}$$

*Bharat Heavy Electrical, Ltd., Tiruchirapalli-620014, India.

†See Liptak, Bela G., "Instrument Engineers Handbook," Chilton Pub., p. 1,006.

**See Anderson, Floyd E., Pressure relieving devices, p. 293.

For liquids:

$$d = 0.236(Q)^{1/2}(Sp)^{1/4}/P^{1/4}$$

where: d = minimum rupture-disk dia., in
w = relieving capacity, lb/h
P = relieving pressure, psia
Mw = molecular weight
Q = relieving capacity, gpm
Sp = liquid specific gravity

Example: 50,000 lb/h of hydrogen must be relieved to the atmosphere from a pressure of 80 psig. What is the diameter of the rupture disk required? Connect $P = 80$ on Fig. 1 with $W = 50{,}000$ and extend the line to its intersection with the reference line. Connect this intersection with $Mw = 2$, and read on the d-scale 1.75 in.

Example: 100 gpm of liquid with a specific gravity of 0.9 must be relieved from 200 psig to atmosphere. What diameter of rupture disk is required? On Fig. 2, connect $Sg = 0.9$ with $P = 200$, and extend to the intersection on line 1. Connect this intersection with $Q = 100$, and read diameter = 0.625 in. on the d-scale.

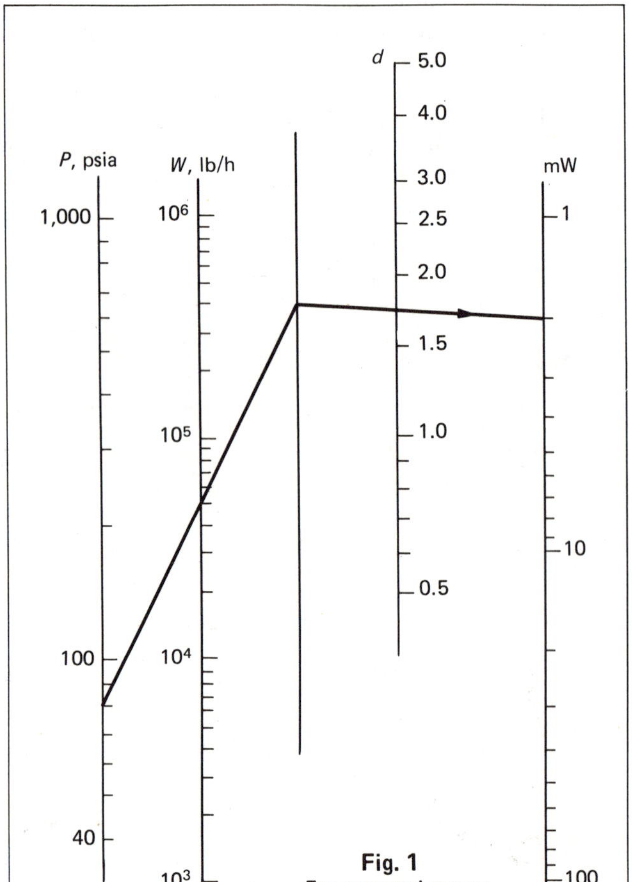

Fig. 1
For gases and vapors

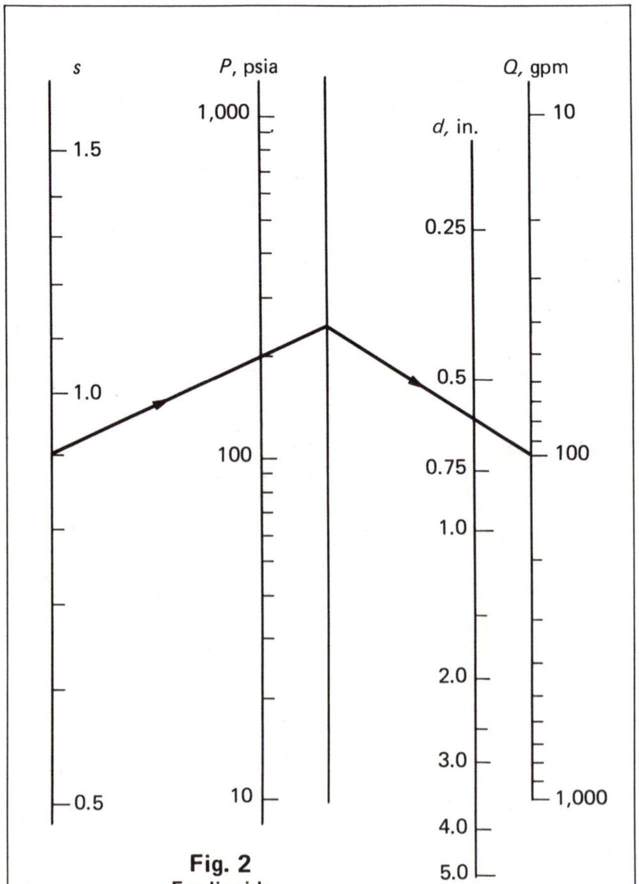

Fig. 2
For liquids

Originally published October 25, 1976

Rupture disks for low-burst pressures

Rupture disks are now being used at pressures as low as 1 psig to protect equipment and personnel in a wide range of chemical processes.

Roy J. Zook, Zook Enterprises, Inc.

☐ Low-pressure rupture disks are used as overpressure protection in a wide range of pressure and vacuum systems. They also prevent corrosion and fouling of relief valves, and provide secondary backup relief on closed storage tanks in the event of malfunction of valves or vents during loading and unloading, or during pressure changes caused by sun heating and night cooling.

In a powder-drying process, for instance, which removes used methylene chloride solvent, 4-in-dia. graphite rupture disks—rated at 5 psig—are used as primary overpressure protection. Another example is a 4-in-dia. disk—rated to burst at 15 psig—which is used as primary protection in production of dry cleaning fluid. The disk is on a CCl_4 seal pot that is used to baffle the process fluid between stages. The disk also acts as a neutralizer in the event of a blowdown and consequent venting of the product, which is highly acidic.

In this discussion, the term "low-burst pressure" will apply to pressures in the range of 1 psig to about 30 psig, but especially to pressures below 20 psig. Fig. 1 provides pressure ranges for protection devices, including conservation vents, relief valves and rupture disks.

Factors affecting low-pressure rupture disks

At low-burst pressures, six factors become critical in their effect on rupture-disk burst reliability, depending upon material of construction. Of these factors, three result from the extremely thin pressure membranes required at low ratings. Such thinness increases their sensitivity to corrosion, temperature and fatigue; graphite disks constitute an exemption because they are not affected by temperature or fatigue.

A fourth factor, which should be considered for equipment and process economy, is the allowable ratio of operating pressure to burst rating (O/B ratio), which, except for graphite, should be widened at low pressures to accommodate the increased effect of fatigue and creep.

A fifth consideration is the decreasing accuracy of all rupture disks as burst ratings become lower. Finally, the economy of vent-line diameter should be considered at low pressures, because minimum burst ratings of all disks are an inverse function of the diameter.

Rupture-disk materials

Materials used for low-pressure rupture disks can be divided into two main categories: graphite and metals, either one of which can be plastic coated or lined. The three fundamental designs are shown in Fig. 2.

Graphite and fluorocarbon-lined-graphite disks are particularly adaptable to low-pressure applications because of: their corrosion resistance; availability at low-burst ratings in small diameters at close burst tolerances; insensitivity of burst rating to temperature; nonfatiguing and creep-free characteristics; and their ability to be rated as close to operating pressures as desired. They also offer an indefinite service life, which is subject only to corrosion in the relatively few instances when this occurs.

After considering corrosion resistance, solid- and composite-metal rupture disks are quite popular for high-temperature use without insulation, as well as for partial cost recovery in the case of precious metals. Nevertheless, wherever metal rupture disks are used, the following operating characteristics apply: (1) burst rating varies inversely with temperature; (2) metal fatigue and creep limit operating flexibility and service

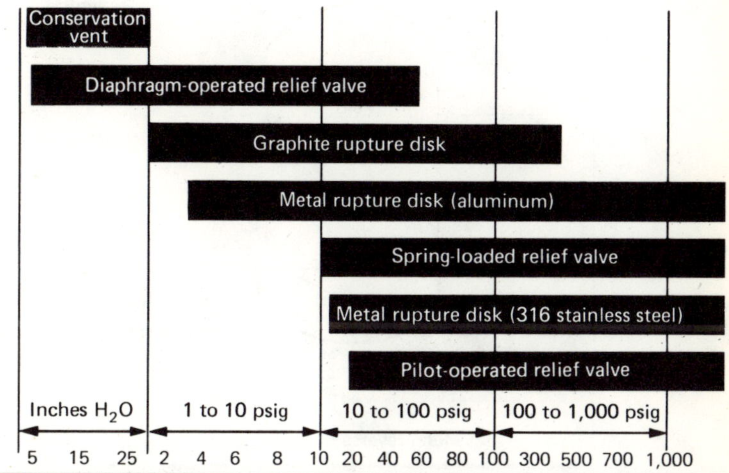

Typical ranges of several protection devices — Fig. 1

The data presented in this article have been obtained from catalogs published by major manufacturers of metal and graphite rupture disks.

Originally published March 1, 1976

Typical high-temperature disk limits	Table I
Material of construction	**Temperature, °C**
Aluminum, solid	120
Aluminum, composite	425
Graphite, unlined	170
Graphite, lined	200
Graphite, insulated	340
Monel, solid	425
Monel, composite	540
Silver, solid	120
Silver, composite	425
Stainless steel 316, solid	480
Stainless steel 316, composite	540

Higher temperatures can be used by flowing coolant over disk's vent side.

life; (3) for lower pressures, O/B ratios should be increased beyond the normally recommended 70%; (4) many metals cannot be supplied in desired vent sizes at low-burst ratings.

Plastic coatings and films

Coatings are generally too porous to block corrosion of the base material over an extended time. They cannot be recommended for corrosion resistance, particularly at low pressures, although they may be suitable to prevent material buildup that might affect burst rating. By contrast, linings—which are sufficiently impermeable for corrosive service—definitely affect rupture-disk burst rating at low pressures, and should always be supplied by the disk manufacturer. To rate such two-material systems, they must be tested as a single unit so that the unit will conform with code requirements.

No amount of corrosion is acceptable for any low-pressure disk made of any material. Obviously, the lower the pressure rating, the thinner the membrane must be, and the more critical is corrosion resistance to operating reliability. Where base-material selection cannot satisfy a corrosion resistance requirement, metal disks can be supplied with a variety of protective linings, including Kel-F, Teflon, vinyl, lead, silver, gold and platinum. Graphite disks are available with a fluorocarbon lining to act as a corrosion barrier in those instances where extreme oxidizers will attack the graphite base.

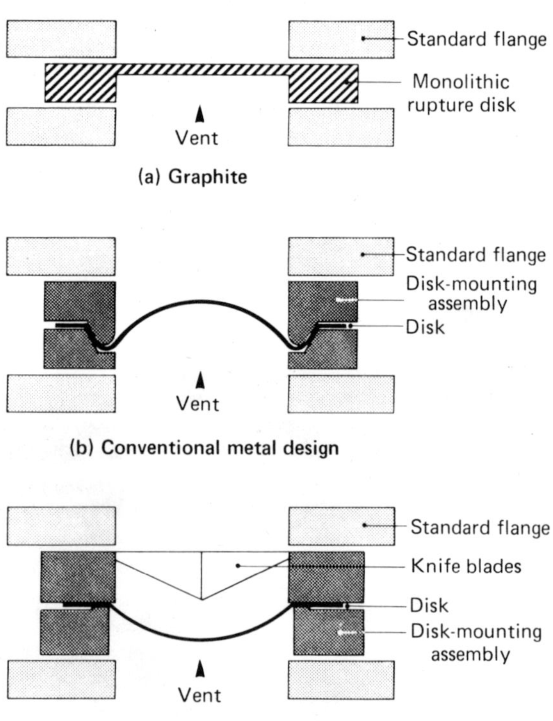

Fundamental types of rupture disks: (a) graphite; (b) conventional metal, concave to the pressure side; (c) reverse-buckling design with the metal pressure-membrane convex to the pressure side Fig. 2

Temperature limits

The upper and lower operating-temperature limits of the disk material must be considered, because it is the temperature of the disk material itself that governs its suitability, and not the process temperature. Low limits are seldom a concern. However, graphite and metal disks can provide service in the –200 to –250°C range with proper gaskets. At high limits, where disks are located some distance away from the heat source, dead space in the vent line will act as an insulator.

Metal disks accommodate high temperatures without protection. For example, Monel can be used up to 540°C. Graphite disks are used with a standard insulating assembly for temperatures above 200°C, up to 340°C. A guide to operating-temperature limits of various rupture-disk materials is provided in Table I.

Insofar as metal disks are concerned, high operating temperatures increase the tendency of the metal to creep, depending upon the metal used. As a result, metal disks must be rated for a precise operating temperature. Graphite does not creep, and is insensitive to temperature up to its operating limit.

A second consideration in the case of metal disks occurs when significant variations in operating temperatures might occur at low pressure. Metal disk-burst ratings vary inversely with temperature. As a result, increased temperature alone can conceivably rupture a low-pressure metal disk, even though the process pressure has not reached the burst rating of the disk. For instance, a 4-in aluminum disk rated to burst at 10 psig at 22°C, will burst at 8.1 psig at 120°C, according to published temperature–burst curves. In some temperature-sensitive service, this may be desirable.

Fatigue, creep and O/B ratio

Fatigue and creep eventually affect the metals' rupture-disk burst-rating by lowering the original tensile strength of the materials. To counteract this, and to achieve maximum service life, conventional metal disks in normal service are recommended for use at operating pressures not exceeding 70% of the burst rating of the disk (O/B ratio). When economic life is less critical,

Tolerances of rupture-disk designs **Table II**

Ordered burst rating, psig	Graphite design	Reverse-buckling design	Conventional design
lanufacturing design range, %			
1	None	NA*	NA*
2	"	"	"
3	"	"	+33, −33
4	"	"	+50, −25
5	"	"	+40, −20
6	"	"	+33, −17
7	"	"	+36, −21
8	"	"	+31, −19
9	"	"	+28, −17
10	"	None	+25, −15
11	"	"	+27, −18
12	"	"	+25, −17
13	"	"	+23, −15
14	"	"	+21, −14
15	"	"	+20, −13
20	"	"	+20, −10
25	"	"	+20, −12
30	"	"	+17, −10
50	"	"	+12, − 8
100	"	"	+ 9, − 5
upture tolerance, %			
1	±75	NA*	NA*
2	±38	"	"
3	±25	"	± 25
4	±19	"	±25
5	±15	"	±25
6	±13	"	±20
7	±11	"	±20
8	± 9	"	±20
9	± 8	"	±15
10	± 8	±20	±15
11	± 7	±18	±15
12	± 6	±17	±15
13	± 6	±15	±10
14	± 5	±14	±10
15	± 5	±13	±10
20	± 5	±10	± 7
25	± 5	± 8	± 7
30	± 5	± 7	± 5
50	± 5	± 4	± 5
100	± 5	± 2	± 5
ximum accumulative tolerance, psig			
1	±0.75	NA*	NA*
2	±0.75	"	"
3	±0.75	"	+1.74, −1.74
4	±0.75	"	+3.00, −2.00
5	±0.75	"	+3.25, −2.25
6	±0.75	"	+3.12, −2.22
7	±0.75	"	+3.92, −2.87
8	±0.75	"	+4.08, −3.12
9	±0.75	"	+3.87, −2.88
10	±0.75	±2.00	+4.00, −3.00
11	±0.75	±2.00	+4.62, −3.63
12	±0.75	±2.00	+4.80, −3.84
13	±0.75	±2.00	+4.29, −3.25
14	±0.75	±2.00	+4.34, −3.36
15	±0.75	±2.00	+4.50, −3.45
20	±1.00	±2.00	+5.40, −3.40
25	±1.25	±2.00	+6.75, −4.75
30	±1.50	±2.00	+6.00, −4.50
50	±2.50	±2.00	+8.50, −6.50
100	±5.00	±2.00	+14.00, −10.00

t available.

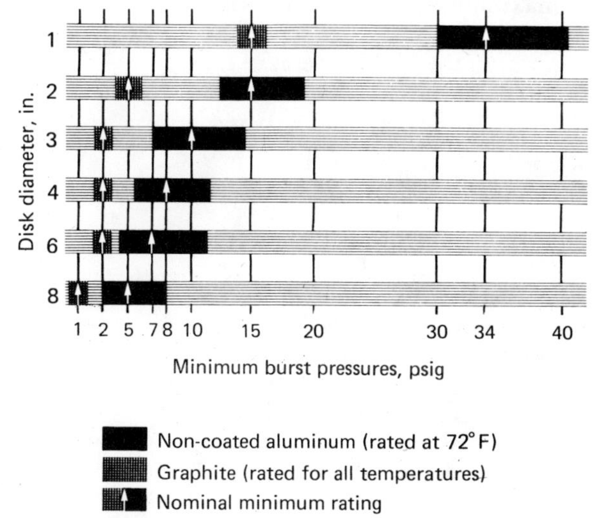

Disk accuracy at minimum burst pressure **Fig. 3**

O/B ratios up to 80% are possible. A new metal disk operating on a reverse-buckling principle is much less subject to the effects of fatigue; it therefore features a recommended maximum O/B ratio of 90%.

However, very thin membranes used at low pressures, as well as at elevated temperatures, aggravate fatigue and creep effects. A wider margin (lower O/B ratio) is therefore recommended between the operating pressure and the burst rating. Because of metal fatigue and creep, a strict metal-disk replacement schedule is normally followed, especially when the disks are used at low pressures.

Graphite is not affected by fatigue or creep, either at elevated temperatures or at low burst ratings. Graphite disks may be used at burst ratings as close to the operating pressure as is consistent with code requirements, which typically limit burst rating to equipment-design pressure.*

Decreasing accuracy at low pressures

As low-burst limits are approached, accuracy of all disk materials decrease as a percent of rated burst pressure. Depending upon disk material, design, and manufacturer, a manufacturing-design range (MDR) tolerance may be used in addition to a burst-rating tolerance. Only the reverse buckling-metal design and graphite disks are furnished as standard in all ratings without an MDR tolerance. When MDRs are not used, nameplate ratings should be identical to ordered ratings. This provides improved accuracy and better control, particularly for low pressures.

When an MDR is used, it could create two undesirable situations brought about by frequently used ordering procedures:

1—Ordering a disk at equipment-design pressure could result in nonconformance with the code, which states that normally "rupture disks shall be rated to

* According to UG 133d of the ASME Code, Div. I, Section 8.

rupture at a pressure not to exceed the maximum allowable working pressure of the vessel" (UG 133d). For instance, the MDR tolerance for aluminum is +40% and −20% at 5 psig; +25% and −15% at 10 psig; +20% and −13% at 15 psig; and +20% and −10% at 20 psig.

In each instance, odds favor a higher nameplate rating than the nominal one ordered, because the MDR tolerance is about twice as great on the high side as on the low. Obviously, if a disk is ordered at equipment-design pressure without special instructions, then a reasonable possibility exists that the nameplate rating of the actual disk supplied will be higher than allowed by the code, especially at low pressures.

2—Reordering low-pressure disks by nameplate rating appearing on the existing used disk is the second way to trip the odds towards higher-than-desired nameplate ratings. For example, an order for a 5-psig aluminum disk might reasonably use half of the high side of the +40%, −20% MDR tolerance. Thus, the new disk would be supplied at a 6-psig nameplate rating. If this same disk were reordered again at 6 psig, at half of the +40% high side MDR tolerance, the next new disk would be rated at 7.2 psig. Even a few nameplate reorders could cause a misapplication.

The normal burst-rating tolerance also becomes less accurate as a percentage of nameplate rating as low-burst limits are approached. Table II compares accuracies of representative disks by listing, percentagewise, MDR and burst-rating tolerances separately. Then the two tolerances are added to show maximum possible accumulative tolerance in psig at specified ordered ratings. For example, the maximum accumulative tolerance for a disk ordered at 10 psig would be +4, −3 psig for conventional solid metal, ±2 psig for reverse-buckling metal, and ±0.75 psig for graphite.

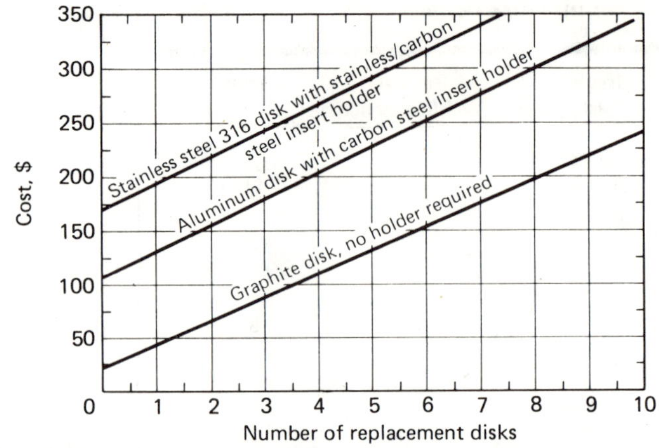

Rupture-disk cost comparison, Prices are as of September 1975, for single units for standard 4-in.-dia. disks designed for 150-lb flange ratings. Other common diameters are comparable **Fig. 4**

Minimum available burst ratings, psig **Table III**

Material	Diameter, in							
	2	1.5	2	3	4	6	8	20
Aluminum	34	23	15	10	8	7	5	3
Graphite	15	5	5	3	3	3	1	1
Monel	175	115	67	49	37	28	22	9
Silver	90	58	34	25	19	14	11	–
Stainless steel 316	320	210	120	90	68	51	40	16

Individual manufacturers may specify higher minimums.

Minimum ratings and tolerances by diameter

Minimum burst ratings are a function of diameter in all materials. Table III shows some typical, lowest burst-ratings available in standard diameters. Should the vent line diameter to be too small for a desired low-burst rating, either an alternate material can be selected in that diameter, or a larger-diameter disk can be used in the preferred material. Graphite provides the lowest possible ratings; aluminum is second.

At these minimum ratings, tolerances are at their maximum. To show this characteristic, Fig. 3 compares aluminum to graphite, assuming allowable tolerances for both materials, as derived from Table II.

Conclusions

When rupture disks are used at low-burst ratings, these factors should be considered, depending on the material selected: (1) MDR tolerance; (2) burst-rating tolerance; (3) material-temperature limit; (4) temperature effect on burst rating; (5) ratio of operating pressure to burst rating (O/B ratio); (6) fatigue and creep characteristics of the material of construction chosen; and (7) corrosion.

When metals are to be used, all of the factors listed above must be very carefully considered. Graphite, however, is subject only to: (1) burst-rating tolerance; (2) material-of-construction temperature limit; and (3) corrosion.

The author

Roy J. Zook is President of Zook Enterprises, Inc. (8309 Washington St., Chagrin Falls, OH 44020). When impervious-graphite processing equipment was developed at Union Carbide Corp , he was associated with their design and application. As founder and president of Falls Industries, Inc., manufacturers of graphite and graphite processing equipment, he was part of the group that developed graphite rupture disks. Now, he is devoting full time to improving graphite rupture-disk manufacturing techniques and new designs for broader applications. He received his B.M.E. degree from Ohio State University.

Power Actuators for Valves

Power operators for valves are widely used for their convenience, speed and safety. They are either driven electrically or by fluid pressure; both types have advantages, although electric motor operators are generally more versatile.

EDWARD LAWSON and W. J. DENKOWSKI
Limitorque Corp.

Power actuators are available for virtually all types and sizes of valves. They may be driven pneumatically, electrically or hydraulically. Whether or not an actuator should be considered in a particular instance depends on the valve's location and its role in the process.

Valves in pipelines and water supply lines, or those located in remote areas of a plant, require a power actuator for the sake of convenience alone, if no other reason. An actuator can be justified on most valves that must be turned on, off or throttled frequently. Speed is also a factor, especially on large valves that may require several minutes to open or close manually.

Safety, of course, is often the overriding consideration. In the event of fire, process upset or other emergency, a power actuator is a must to quickly shut down a line. They may be set to close automatically when process conditions exceed certain limits.

Automatic control valves are commonly operated by pneumatic diaphragm-and-spring actuators or the pressure-balanced piston type. (Control valve actuators are discussed in the preceding article.)

How to Specify

On most new valves, actuators are sized and furnished by the valve manufacturer. Whether the unit is for new or existing equipment, the buyer should supply the following information:

Valve size, type and make.
Line pressure.
Maximum differential pressure across the valve.
Stem diameter, pitch and lead, and hand of threads.
Desired closing or opening time in seconds.
Turns of yoke nut to open valve.
Phase, frequency and voltage of power.
Maximum ambient and fluid temperatures.
Type and frequency of service.
Control voltage.
Special considerations such as the need for explosionproof enclosures on electric drives.

Originally published October 11, 1971

Hydraulic Units are Simple

Hydraulic actuators may be simple devices with a minimum of mechanical parts. Air or fluid pressure acts on a duplex piston that is connected to the valve stem. The required length of stem travel and seating thrust is obtained by varying cylinder size. Units usually have hand pumps for manual operation when it is necessary to bypass the central power system.

A key advantage of hydraulic operators is their much faster opening and closing speeds compared to electric motor-gear types. They can be set to close almost instantaneously, while a motor-gear actuator is typically designed to have a closing cycle of 10 to 60 seconds or longer if necessary. Of course, hydraulic hammer in a line that is shut down too quickly can cause serious damage.

Hydraulic actuators may be preferred because compressed air accumulators can store enough energy to operate a valve in the event of a power failure. Also, speed control valves are available to vary the rate of stem travel.

To avoid valve creepage requires a tightly sealed piston and check valves in the actuator. Any leakage in these components may cause undesirable stem movement. Care should be taken to avoid leaks and loss of fluid throughout the hydraulic system. For remote electrical control, fluid flow in the actuator is controlled by solenoid valves, which must be carefully maintained to avoid false operation of the main valve.

In high-pressure valve service, a high-pressure hydraulic system is needed to keep the actuator piston at reasonable size and cost. For example, a 24-in., 600-lb. gate valve must have a seating thrust of around 130,000 lb. A ball valve under the same conditions requires about 207,000 in.-lb. torque at the ball stem.

A possible problem with hydraulic operators is that there is no reliable method of achieving controlled-thrust seating on gate valves. If the system has enough reserve power to unseat a stuck valve, this same force may be applied each time the valve is closed, causing damage to the seat.

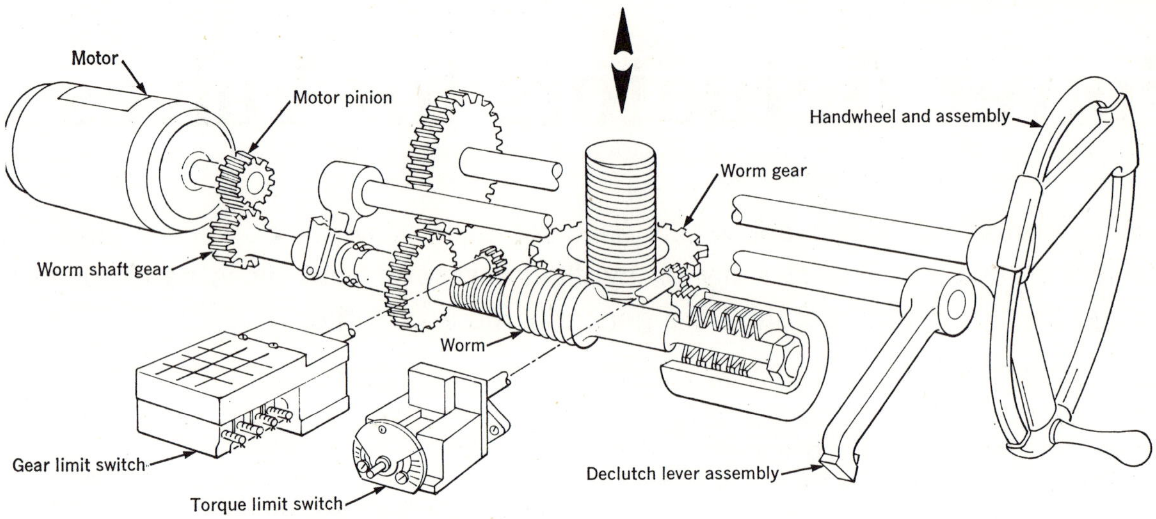

COMPONENTS of typical electric motor valve actuator—Fig. 1

Electric Actuators

Solenoids are linear actuators that consist of a soft iron core which moves within an electrical field set up by a surrounding coil. They are widely used on small on-off globe valves, particularly for emergency shutoff service. Solenoids have economic limitations on force output and length of travel.

The electric motor-gear actuator is a good choice for overall versatility. In simplest form this is made up of an electric motor connected to the valve stem through a gear box (Fig. 1). Many refinements on this basic design are available to handle practically all actuator situations.

One of the main benefits of electric actuators is that they can be equipped with a torque limit switch that controls the amount of seating force. This provides a constant seating thrust for tight closure and at the same time protects all valve parts from overloads. The torque limit switch may be adjusted to compensate for changing service conditions.

Units also have geared limit switches to govern valve disc travel while opening or closing, and position indicator lights. All electrical components are enclosed in a housing that can be weatherproof, submersible or explosionproof. The motor drive can be disengaged for manual operation with the handwheel.

Motors are specially designed for valve actuator service and normally run at 1,800 or 3,600 rpm. Valve stem travel usually varies between 4 to 12 in./min. on globe valves and 12 to 72 in./min. for gate valves. Speeds over 36 in./min. usually require a motor brake to absorb forces of inertia. Operators for ball, plug and butterfly valves typically are designed to close in 30 to 60 sec.

Electric actuators can be practical on valves as small as ¾ to 1 in. for high pressure service (above 125 psi.) and 2 to 3 in. on low pressure lines. There is really no upper limit. Commercial units are available with thrusts up to 500,000 lb. and closing torques of 60,000 ft. lb. Fig. 2 shows an installation on a 10 ft. dia. butterfly valve, with a quadrant worm gear on the actuator's output drive.

Experience has shown that these actuators are rugged and dependable. A principal advantage is that the gear train provides self locking in any position. This is important because dynamic flow conditions can have a serious windmilling effect on ball, plug and butterfly valves, and produce high torques at the stem of gate valves.

With polyphase induction motors, closing speed is almost constant. On the other hand, speed cannot be changed in the field without altering the gear ratio. Also, there is no way to close a valve automatically in the event of power failure unless direct current from batteries is available.

ELECTRIC MOTOR operates huge butterfly valve in water service—Fig. 2

Construction Materials

Selecting the proper materials for valve operator mechanisms is important to achieve the right combination of strength and durability, which is required for heavy stresses. Materials commonly used for principal components are listed in Table I.

Efficient and precise gearing is critical. The gear train of an electric valve operator consists of a reduction helical gear set, and a worm and worm-gear combination. Helical gears are heat-treated, chrome-moly alloy steel. After final hardening, they have enough capacity to absorb the high impact reversing stresses that occur during valve closing and opening.

Worms are alloy steel suitable for carburizing, hardening and grinding of the threads. Either cast or forged bronze is selected for the mating gear. Other elements, such as manual parts and the automatic clutch, must have shear strengths compatible with the gear train.

Materials also must be selected for long service life. The operator on an isolating valve, for example, may be unused for years, while it is exposed to various temperatures, pressures, humidities and corrosive environments. Then, at the push of a button in the control room, it must be able to close without hesitation.

Lubrication

For long service life with a minimum of attention, most electric operators are grease lubricated. They should be inspected every six months to a year. Some manufacturers prefer oil lubrication, particularly for special services. Oil requires more frequent inspection for leakage, and the operator's mounting position must be specified.

Lubricating greases are generally defined as semi-fluid to solid products, either petroleum or synthetic based, combined with a thickening agent. Most contain a soap, mixture of soaps or jelling agents. They have other additives to improve wear, oxidation resistance and extreme pressure performance.

The lubricant, of course, should not corrode bearings, shafts and other internal parts. It should not precipitate sediment and be stable well above the designated service temperature. Properties of some typical lubricating greases are shown in Table II. The manufacturer should be consulted for severe services at high or low temperatures.

Testing

There are three kinds of testing that are widely used for electric valve operators. These are a shock

Material Selection Chart—Table I

Components		Standard on-off	Continuous modulating	High shock (navy)	Low temp.	High temp.	Nuclear class I containment vessel
Thrust loads, bearing parts, housing, housing cover, adapters		Hi-strength cast iron class 35		Ductile iron or cast steel	Ductile iron	Hi-strength class 35	Hi-strength class 35
Helical gears		Alloy steel—heat treated					
Worm		Alloy steel carburized and hardened or induction hardened					
Worm gear		High strength bronze					
Power gear shafting		Alloy steel—heat treated					
Manual declutch parts and shifting		Carbon and/or alloy steel					
Control components gear limit and torque switch		Standard design moderate shock plastic parts		Special high shock resistant plastic parts			
Motors	Insulation	Class B	Class B, F, H	Class B	Class H	Class H	Class H
	Construction	Standard	Standard	Special housing design as required	Standard	Standard	Standard
	Duty	15 or 30 min.	Continuous				

Lubricating Greases for Valve Operators—Table II

Soap Base	Basic Properties	Texture	Water Resistance	Max. Temp., Cont. Service	Resistance to Softening
Lithium	Multipurpose High and low temperature range Good mechanical stability	Smooth	Good	200 F. melting point 350 F. max.	Good to Excellent
Calcium Complex	Multipurpose Good high & low temperature range High mechanical stability High melting point	Smooth to buttery	Excellent	300 F. melting point 400–500 F. max.	Good to Excellent
Migrogel Non-soap	Multipurpose Very high mechanical stability High melting point	Smooth	Excellent	350 F. melting point 500 F.+ max.	Excellent

and vibration test, a vibration and acceleration test and an environmental test.

The shock test (Navy MIL-S-901C) is very demanding and often requires special construction of the actuator. The unit along with its controls is mounted on a special stand and struck from beneath by a 3,000-lb. hammer. G forces can be varied by the position of the hammer. The unit must withstand up to nine blows (18 for nuclear submarine service).

Vibration testing illustrated in Fig. 3 is performed to MSL-STD-167 or specifications drawn by regulatory agencies. It usually begins with a frequency sweep test to detect natural resonances, using a sinusoidal steady-state imput. This is followed by several vibration cycles at natural frequency or at 35 Hertz.

HEAVY EQUIPMENT is used for vibration testing—Fig. 3

ACTUATORS find other applications, in this case to position blades in large mixing tank—Fig. 4

The environmental test for nuclear service is one of the most difficult to survive. In one version, the motor is heat-aged to simulate 40-years service. Then the entire unit is placed in a chamber with saturated steam for 7 days. The pressure is 105 psig. and temperature varies up to 340 F. The operator must perform normally at specified times during the test.

Testing to customer specifications is becoming more common because increasingly complicated designs make it difficult to extrapolate from past experience. Setups may be quite sophisticated to simulate special loading and environmental conditions, and are time consuming and expensive.

Special Applications

Electric valve operators may be used as electromechanical servo mechanisms. Many solid state controls can be adapted for set point control or for modulating services. For example, the output from a process controller can drive a shunt-wound operator motor, which balances the position feedback impulse from the operator against the signal from the controller.

Actuators may be custom designed to solve special problems. In one case, all components were squeezed into a slim unit only 10.5 in. wide. These were provided to the Navy to operate jet fuel manifold valves, where only a 12 in. space was available between valve stems. Special water-tight, explosion-proof actuators were developed for a petroleum tanker. In this remotely controlled system, one man can unload 650,000 bbl. from 21 tanks in less than 18 hr.

Finally, electric actuators may be useful in any situation where there is a need for controlled torque or thrust. Fig. 4 shows an unusual application in which a large operator is mounted above a mixing tank to allow automatic positioning of the mixer blades within the tank.

Meet the Authors

Edward F. Lawson is sales manager, Limitorque Corp. (affiliated with Philadelphia Gear Corp.), King of Prussia, PA 19406. He has held this position since 1962, and previously was a sales engineer for Philadelphia Gear. Mr. Lawson is a graduate of the U.S. Merchant Marine Academy and served as an engineering officer in the Merchant Marines from 1951 to 1953, and with the Navy from 1953 to 1955.

Walter J. Denkowski is chief engineer of Limitorque Corp. A native of Poland, he completed his engineering education at Polytechnic University of Warsaw. Mr. Denkowski began his career in this country in 1946 as a diesel locomotive designer at Baldwin Locomotive Works. He joined Philadelphia Gear in 1950, where he has been instrumental in developing three patented lines of valve operators.

Valve Installation, Operation and Maintenance

After the proper valve has been selected for a particular application, the engineer's job is still far from complete. Service requirements must be analyzed for proper installation, operation and maintenance.

HAROLD C. TEMPLETON, Consulting Engineer

Operation of any process plant depends upon the satisfactory functioning of the many types of valves installed in the piping system. Types of valves covered here are stop valves, i.e., valves used normally in the fully open or fully closed position; globe-type valves used for throttling control as well as on-off service; and self-actuated check valves.

Stop valves and globe-type valves are usually manually operated and are used infrequently. An increasing number are being installed with remote operation to reduce operating manpower requirements.

INSTALLATION

Flanged-end valves are the type most frequently used in most process piping systems. Systems that require very high integrity and assurance from leakage are often designed with weld ends. Small pipe sizes are frequently assembled with screwed ends, depending upon pipe and valve material and fluid to be handled. Typical contact faces for flanges are shown in Fig. 1.

Flanged ends have standard dimensions and pressure ratings for various construction materials. Flange diameters, bolt circles and drilling, and raised face diameters are covered by various American National Standards Institute (ANSI) and Manufacturers Standardization Society of the Valve and Fittings Industry (MSS) standards. Identical and similar pressure classes in different materials have the same dimensions.

When steel raised-face flanges are assembled to cast iron, bronze or stainless steel flanges, special care must be exercised to prevent breakage or bending of the lower strength material. Also, if raised-face steel flanges are joined to flat-faced flanges of other materials, recommended practice is to remove the raised face and use a full-face gasket in the joint. If the raised-faced steel flange must be used, low strength bolting is recommended to avoid damage to the weaker material. MSS SP 46, Assembly of Steel Raised Face Flanges to Cast Iron, Brass, Bronze or Stainless Steel Flanges, covers recommended practice in such cases.

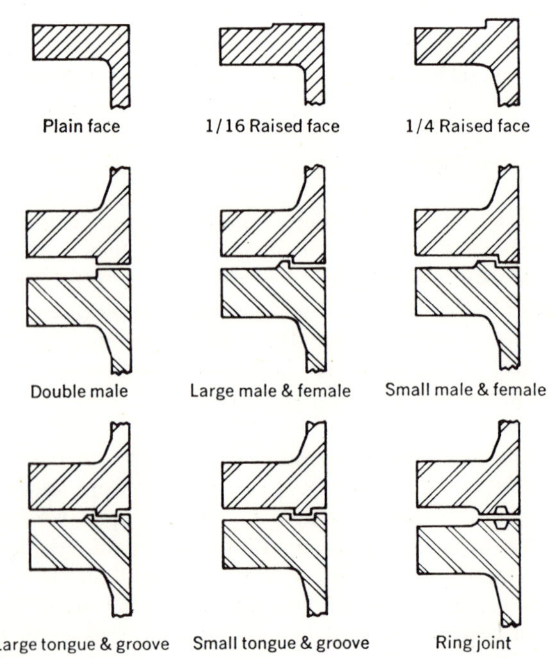

CONTACT FACES for standard flanges—Fig. 1

Originally published October 11, 1971

National Standards for Valves and Flanges—Table I

MSS SP 6	Finishes of Contact Faces of Connecting End Flanges of Ferrous Valves and Fittings
MSS SP 37	MSS 125 LB Bronze Gate Valves
MSS SP 42	MSS 150 LB Corrosion Resistant Cast Flanged Valves
MSS SP 46	Assembly of Steel Raised Face Flanges to Cast Iron, Brass, Bronze or Stainless Steel Flanges
MSS SP 65	High Pressure Chemical Industry Flanges and Threaded Stubs for use with Lens Gaskets
ANSI B 16.1	Cast Iron Pipe Flanges and Flanged Fittings
ANSI B 16.5	Steel Pipe Flanges and Flanged Fittings
ANSI B 16.10	Face to Face and End to End Dimensions of Ferrous Valves
ANSI B 16.11	Steel Socket Welding Fittings
ANSI B 16.20	Ring Joint Gaskets and Grooves for Steel Pipe Flanges
ANSI B 16.21	Non Metallic Gaskets for Pipe Flanges
ANSI B 16.24	Brass or Bronze Flanges and Flanged Fittings
ANSI B 16.25	Butt Welding Ends

Gaskets

Gaskets for valve-to-pipe flanges fall into three categories: soft, semimetallic and metallic. Each of these groups has many sub categories.

Soft gaskets include those made of elastomers; the various compressed asbestos sheet gaskets, usually with a synthetic rubber binder for corrosion resistance and sealing properties; and the fluorinated hydrocarbons such as PTFE materials, both filled and unfilled. The most widely used chemical service soft gaskets are compressed asbestos and PTFE sheet.

Semimetallic gaskets cover the metal-jacketed asbestos filled types of which there are many variations, the corrugated metal and profiled metal constructions with asbestos cemented into the concentric grooves, and spiral wound types with either asbestos paper or PTFE sheet between the windings.

Metallic gaskets are made to several standardized designs. The ring joint is the most common although the lens joint has become popular recently for high pressure service. There are many non-standard types such as corrugated metal, plain solid metal, profiled metal with concentric V grooves and other variations.[1] These non-standard metal gaskets are not widely used in pipe line flanges.

Maximum service temperature for most elastomeric gaskets rarely exceeds 200 F. PTFE gaskets are limited to 500 F., above which decomposition begins to take place. Good practice usually dictates a use temperature below this, normally 400 F., to avoid cold flow.

Compressed asbestos sheets are limited by the binder, with 800 F. the usual maximum use temperature. Specific limits for a particular gasket should be verified with the manufacturer. When these gaskets are reinforced with metal mesh they can be used at somewhat higher temperatures, with 1,000 F. an absolute maximum service temperature.

Soft gasketing requires a serrated flange finish of sufficient roughness to prevent creeping. This is particularly important with the PTFE materials, which are subject to cold flow.

Properties of Semimetallic Gaskets

Semimetallic gaskets are limited by the type of metal and asbestos. If the asbestos is confined, it can produce a satisfactory seal even though it has virtually turned to a powder with no tensile strength. Stainless steel or nickel-base alloys are used in most semimetallic gasketing for elevated temperatures, i.e., above 1,000 F.

Semimetallic gasketing will require different surface finishes depending upon the specific type. Metal-jacketed asbestos gaskets, which require the metal envelope to be seated against the flange face, and the metal core gaskets with asbestos over part of the surface, are most easily sealed against smooth finishes.

Spiral-wound gaskets can be used against different surface finishes.[2] For general service, a finish of 125 to 200 root mean square is adequate. Hazardous service and hard-to-hold fluids require a surface finish no coarser than 125 RMS. Vacuum applications should have a finish no coarser than 80 RMS.

Bolting load for spiral wound gaskets must match the compressive characteristics of the gasket. Consultation with the manufacturer will assure a proper gasket-bolt combination. Compression rings are recommended in most cases to assure that gaskets are not overcompressed.

Metallic gaskets require smooth finishes to assure sealing, with 62 RMS the maximum roughness normally acceptable. These and some semimetallic gaskets seal by deforming the gasket metal to conform with the flange surface so smoothness is essential. The gasket should be softer than the flange to obtain metal flow without deforming the flange. A hardness differential is often unattainable when joining austenitic stainless steel flanges with gaskets of the same alloy so this combination is seldom used except on ring joint and lens flanges.

Phonographic or concentric serrated finishes are preferred for use with all types of gasketing. In no case should tool marks extend radially across the flange seating surface.

Ring Joints

Ring joint nd lens are the most commonly used metallic pipe flange gaskets (Fig. 2). Ring joint facings are easy to assemble and tighten. They have the advantage of positive sealing and maintain their integrity in a fire. If the ring does not seal on the inner edge there will be a pocket that can trap line fluid.

Lens joint facings have the advantage of no crevice to trap line fluid. They are as easily made up as flat or raised-face flanges. Little spring is required in piping to install or remove a lens joint flange, while considerable spring is needed to remove a ring joint flange.

Flanged connections should be bolted to torque loadings appropriate to the material and the pressure class. Various flange and fitting standards give guidance to the bolt load.[3] This is particularly important with low strength fasteners to assure that bolts are not overstressed in making up the joint. This can cause leakage when the bolts relax.

Bolts should be tightened according to the sequence in Fig. 3. Torque wrenches give the most control, although a reasonable substitute is an accurately calibrated air wrench. Bolts and nuts should be adequately lubricated to assure reasonable reproducibility of torque loadings.

Screwed connections are the lowest cost type of joint and can be assembled by low skilled workers. However, they have many disadvantages. Leaks are difficult to seal, the threads are subject to crevice corrosion, and disassembly may be almost impossible if corrosion has occurred. Materials that gall easily, for example austenitic stainless steels, are very difficult to tighten. PTFE tape wrapped around the male thread is very useful in making up tight screwed joints.

Leaks from screwed joints may produce conditions on the exterior that will rapidly attack piping and valves. An example is handling concentrated sulfuric acid in carbon steel. Leaking acid will be diluted from atmospheric hydration and rapidly attack the outside of the carbon steel. In such cases, seal welding of the joints may be necessary.

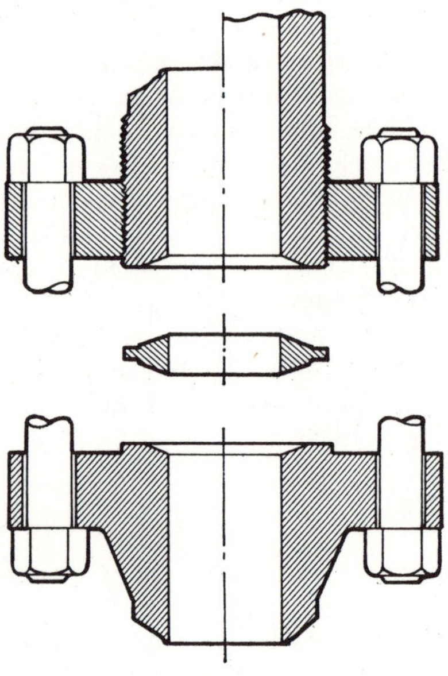

RING JOINT gaskets offer positive sealing—Fig. 2

Welded Systems

Welded piping systems give the maximum assurance of leak tightness. Socket weld ends are generally used up to 2 in. dia. and butt weld ends for larger sizes. In some cases, butt welding is selected for smaller sizes in order to avoid the crevice in the socket weld connection.

Socket weld connections are easily made up with less skill required than for butt welds. When fitting up socket weld joints, the pipe should not be bottomed in the socket before welding. This will produce very high stresses when the weld metal contracts and pulls the pipe into the socket. Failure to leave a small space when welding gate valves can distort the seats so they will not seat properly in service.

Butt-weld end valves of steel and stainless steel are machined to ANSI B 16.5, ANSI B 16.25 and ANSI B 16.10. Special weld end preparations are often called for with consumable inserts, a type not covered by B 16.5.

Care should be taken to avoid damaging valve seats with weld spatter. Metal seats should be closed during welding with protection over the disc sur-

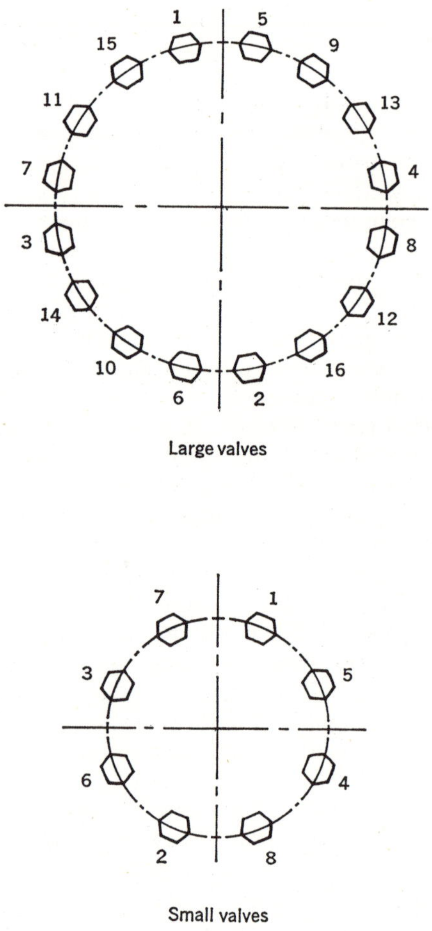

SEQUENCE for tightening flange bolts—Fig. 3

faces if there is any chance of weld spatter hitting them. Valves should be kept as cool as possible to avoid body seat distortion. This is especially critical on the shorter end-to-end socket weld valves. Ball valves with internal parts of PTFE may require removal of seals to avoid damaging them from welding heat.

Other types of end connections, such as hub and compression, are not widely used in process valving. Manufacturers should be consulted for engineering information and recommendations.

Precautions During Installation

Handling is critical especially on large valves that require mechanical lifting devices. Valves should not be hoisted by their handwheel, yoke arms or operator mechanisms. Slings should be placed under the body with provision to prevent overturning. Small valves that can be lifted by hand pose no problems in handling, of course.

When valves are received it is generally better to leave them in the original shipping container unless they are stored inside. In any case flange protectors should be left on until just before installation. After removal, the valve interior should be checked for cleanliness and any foreign material removed. Normally, valves should be stored with discs in the closed position.

Valves should be checked for operation through a full cycle: closed-open-closed. Check valves are normally shipped with interior blocking to protect disc and seat from mechanical damage. Be certain this is removed before the valve is installed in the line.

Space between pipe line flanges should be set so valves fit without springing the piping. Valves are not intended to be used as tension or compression members to pull poorly fitted piping together or to wedge it apart.

After the valve is in the line, the bonnet bolt tightness should be checked. Nuts loosen in shipment and gaskets can relax or flow to cause a leaky body-bonnet joint. Gland bolting should also be inspected for tightness. If lubrication of stem and bushing is permissible, the exposed stem threads and yoke bushing (stem nut) should be lubricated at this time. The unit should again be checked for operation after installation, through a full closed-open-closed cycle.

OPERATION

Gate valves for process service usually have wedge gates with a 10 deg. included angle between the seats. Sealing is accomplished by forcing a tapered wedge or a pair of discs down between the seats, as shown in Fig. 4. A variation uses a flexible wedge that can afford some very minor adjustment to match the angle between the seats. The parallel seat gate with metal seats requires a wedging arrangement to move the discs against the seats. This design is seldom used in process piping.

Solid wedge sealing depends on precise machining and fitting of the wedge to the mating surfaces. To obtain a continuing seal in service, the seat angle must remain unchanged without mechanical or corrosion damage to the disc and seat.

Tight seals in double-disc gate valves requires flat and properly angled seats. As long as the flatness of discs and seats is maintained, changes in seat angle from pipe line stresses will not affect the seal. If the design is one which permits discs to rotate when the valve is operated, wear will be distributed over the entire disc surface.

Corrosion resistant valves are normally made with integral seats. The same alloy at the same hardness level is used for both seat and body. Many valves are supplied with renewable seat rings, often made of a more corrosion resistant alloy than the body and bonnet. Common practice on steel valves is to maintain a hardness difference of over 50 Brinell between disc and seat to reduce the possibility of galling. Cast iron valves of the so-called "all iron" type usually have integral seats.

Throttling with a gate valve is a questionable practice and should only be attempted in an emergency. Throttling characteristics of gate valves are very poor, and severe vibration of the disc will occur if pressure drop across the seats is high. In addition, the guides may be inadequate to prevent the disc from striking the seat, which can cause severe seat and disc damage. Severe throttling may also cause wire drawing with damage to seats and discs.

Gate valves are intended for clean service, that is, process fluid with little or no solid material in it. To use gate valves in dirty service, a tapping boss with a washout connection should be installed in the bowl of the body between the seats. In this way, any collected material can be washed or blown out before the valve is closed. Where solids have collected in a bowl without a washout connection, closing the valve to the point where the disc nearly strikes the collected solids may create enough turbulence so that the solids will be entrained and carried away. Several partial-stroke cycles may be required to remove enough material so the valve can be closed.

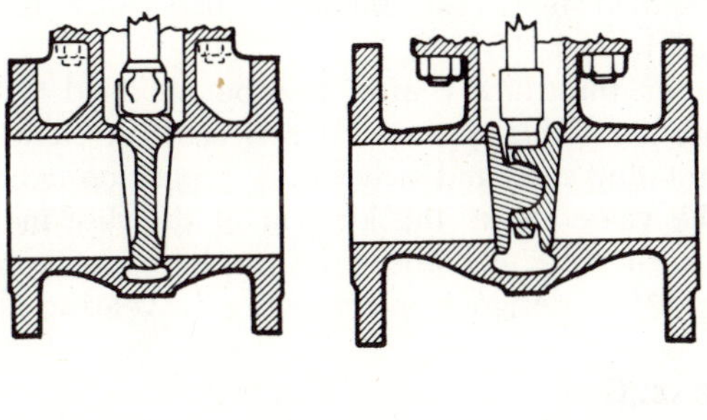

SEALING design in solid wedge (left) and double-disc gate valves—Fig. 4

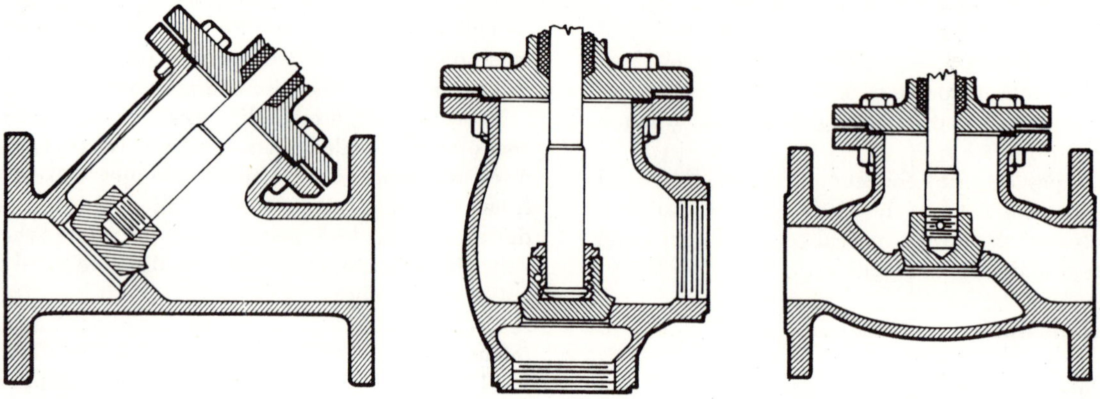

THREE TYPES of globe valves have similar seating—Fig. 5

Globe Valve Operation

Globe valves have three basic types: the right angle globe, the Y globe and the angle valve (Fig. 5). All three types function by raising or lowering a disc on the end of a stem to open or close the valve, or to control flow within certain limits. The renewable disc (RD) type is made with a flat insert of a composition material such as PTFE. The RD globe valve has poor throttling characteristics and is used primarily for tight shut-off, both on clean fluids and on those containing particles that would scratch the seat and disc of a metal seat valve.

Most globe valves are made with a swivel disc and a rotating-rising stem. Included seat angle is commonly 30 deg., with a slightly smaller angle on the disc to provide a line contact between seat and disc. The swivel disc arrangement is very satisfactory if the pressure drop across the seat is within reasonable limits. Excessive drops can cause vibration with resulting wear of the stem foot, disc and swivel nut connection with eventual failure.

In some cases, severe vibration has caused fatigue failure of the stem just below the bottom of the stuffing box. If a large pressure drop is required across a globe valve, an orifice plate should be placed upstream of the valve. Maximum pressure drop that can be handled by a conventional globe valve is best limited to 100-150 psig.

Closer throttling control can be obtained with a V-port disc. This type of valve is constructed with a non-rotating stem and swivel disc. A position indicator on the valve shows the location of the disc in relation to the seat. A higher pressure drop can be taken through this design than through a swivel-disc globe.

Gate vs. Globe

Globe valves in general can be expected to have lower leakage rates than are attainable with gate valves. A special type of disc, with a PTFE insert wafer, seals easily with very low torque on the handwheel even at high pressures. This type is also well suited for sealing at cryogenic temperatures. Valves that incorporate PTFE in the seat design are limited to maximum service temperatures of 500 F.; a more realistic limit for satisfactory service is 400 F.

Gate and globe valves have one common trait: the force of the disc against the seat can be externally controlled by the operator. This is very desirable because it is possible to cycle the valve to obtain a seal on a valve with damaged, corroded or clogged seating surfaces. In the hands of a skilled operator, many troublesome valves can be effectively closed off.

This feature of an externally variable seating force also can cause problems when unskilled and heavy handed operators are responsible for valve operations. Training to assure that valves are treated with respect is the best answer. It is not possible to build a valve that cannot be damaged by operator abuse.

Metal gate and globe valves are not limited to the pressure they can contain or seal against. They are offered over the full range of pressure classes appropriate to the construction material. Limitations on maximum allowable temperatures are set by various codes, and are largely dependant on packing materials. Most other valves have much more restricted maximum pressure or temperature limits because of their seat-seal materials.

Ball Valves

Ball valves fall into two broad categories based on the basic body design: end entry and top entry. Most are end entry and wide variety of designs are available. The body-body end design, in which the connection can be either flanged or threaded, entraps the ball and seals it between the two parts (Fig. 6). Another design uses a one-piece body with a retainer threaded on one end to hold the seats and ball. An

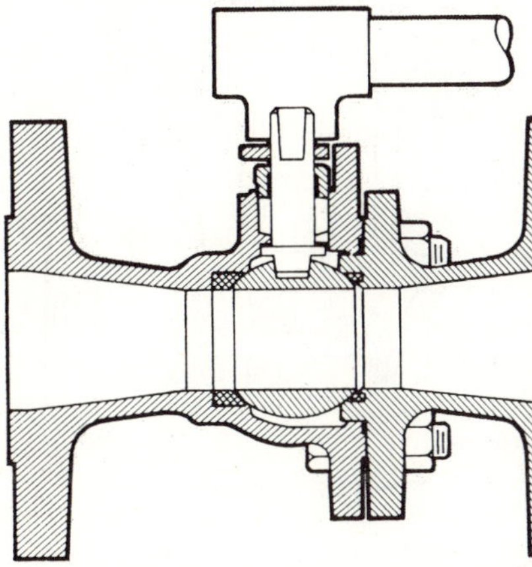

BALL VALVE with body-body end design—Fig. 6

alternate configuration makes the body in three pieces, which are either threaded or bolted together.

Top entry ball valves place the seats at an angle, somewhat akin to the seat angle of the wedge gate, and use a spring to push the ball against the seats to obtain a seal.

PTFE seat seals, both filled and unfilled, are used in most ball valves so the maximum realistic service temperature is 400 F. Pressures that can be handled by the seats are usually lower than those carried by a comparable gate or globe valve. The lower pressure-temperature capability is most marked in sizes over 2.5 in. and at temperatures above 200-250 F. Seat designs vary widely and no adequate evaluation criteria exists except service tests, which may not be truly conclusive.

Ball valves are most effective when full open or full closed because the seats are vulnerable to damage from throttling. They are best suited for clean streams because solid material may damage the seats and pack in behind the ball. Fluids that might set up or polymerize in a stagnant pocket (such as the cavity around the ball) should not be handled.

Initial assembly of ball valves usually compresses the seats against the ball. Various seal arrangements provide a spring lip, a torsion seat or other design to obtain a seal without excessive seat compression, and to keep the seal effective for the life of the valve. PTFE, filled or unfilled, will cold flow under load so that after a period of time little or none of the original assembly compression remains. When this point is reached, the valve will seal only if there is enough line pressure to force the ball against the downstream seat.

Ball valves are simple to operate with one quarter turn required from open to close. Closing of ball valves can cause severe water hammer and piping supports should anticipate this. Stem seals are simpler than those for gate or globe valves, and stuffing boxes are much shorter.

Plug Valves

Plug valves fall into two categories: lubricated and non-lubricated. Lubricated plug valves rely on the lubricant to effect a seal, and successful operation requires a regular schedule for lubrication. Operating conditions are in general limited to temperatures and process fluids for which a suitable lubricant exists.

The most severe problem with the plug valve is to operate one that has been in the same position for an extended time and has not been serviced. Some designs use a lift arrangement to unseat the plug before rotating it. This type can be used without lubrication and has been found satisfactory in elevated temperature service with considerable build-up of hard deposits.

Non-lubricated plug valves have a non-metallic sleeve between the body and plug to serve both as a seal and as an anti-friction member. The sleeve is keyed into the body by various locking arrangements and the plug is compressed down into the sleeve to produce a seal. Use temperatures are the same as for ball valves.

Operation of plug valves is a quarter turn unless they are gear operated. Remote actuation is easy to install, though operating torques are high, especially on the non-lubricated variety.

Plug valves are one of the ideal types for handling streams with considerable amounts of solids, including fairly heavy slurries. Lubricated plug valves can be used to throttle, though abrasion can occur in the presence of solids. Non-lubricated plug valves are not generally suited for throttling, except with low pressure drop, because of the danger of collapsing the sleeve.

Butterfly and Diaphragm Valves

Butterfly valves are most commonly made with an elastomer sleeve that serves as a full liner, a seat seal and the gasket surface of the flanges. Service limitations depend on the properties of the lining material and a variety of materials are used. Valves with PTFE liners require some changes in design because of the material's mechanical properties and lack of elasticity. Service temperature limits of PTFE-lined butterfly valves are roughly the same as for ball valves.

Butterfly valves are low pressure, usually 150 psi. class, and are most commonly made in the wafer or lug styles. Remote control is easily done with a simple double-acting air cylinder.

Some all-metal butterfly valves are available with various seal arrangements. These are used in elevated temperature service as well as at lower temperatures. Tight shut-off is difficult in valves of this type, but they have the assurance of reliable operation even with cyclic temperatures.

Butterfly valves can be used for throttling and are suited for a variety of services, from handling clean fluids to process streams carrying considerable solids.

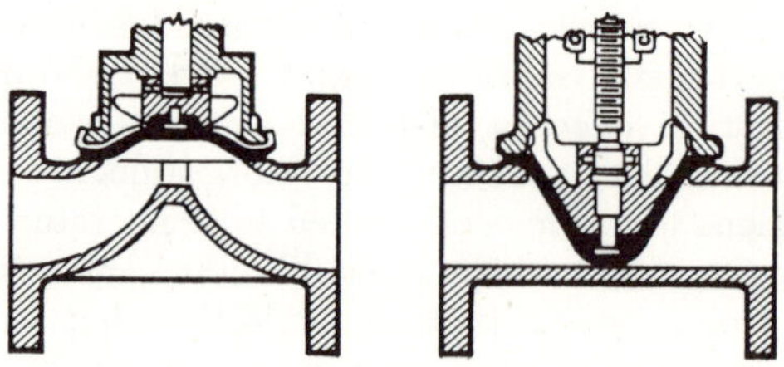

HIGH WEIR diaphragm valve (left) has higher pressure drop than flow-through design—Fig. 7

Diaphragm valves rely on an impervious flexible diaphragm to block fluid flow. Both high and low weir designs are available, as shown in Fig. 7. The high weir causes less flexing of the diaphragm, but produces higher pressure drop.

Diaphragm valves can be used for throttling and to handle solids-laden streams as well as in clean service. They have no pockets to trap material, which is an advantage of interest to some chemical plants and to the food industry.

Service conditions are limited by the diaphragm material. Pressure must be low (usually 150 psig. maximum) because the diaphragm is not able to withstand high forces. Diaphragm failure can be a major problem and can cause considerable damage when handling corrosive chemicals.

Types of Check Valves

Check valves have two basic configurations: the swing check and the piston or lift check illustrated in Fig. 8. Tilting-disc valves are similar to swing check valves, but have some distinctive features. The piston or lift check class also includes ball check valves. Any of these may be supplied with a spring-loaded disc or ball.

Check valves usually are installed only for one purpose: to prevent back flow. In general, they do not seat reliably enough with low leakage to handle high-pressure back flows. In most cases, the check valve is backed up with an appropriate stop valve for positive control over leakage.

Swing check valves are suited for installation in horizontal lines with the cover in the up position, or in vertical lines with the flow upward. Pressure drop is comparatively high when installed in vertical lines. Dash pot arrangements on some swing check valves reduce slamming of disc against seat. Normal construction is with a metal disc, although a renewable disc similar to those in globe valves is available in some pressure classes. Sealing across the seat is usually improved by a non-metallic insert.

Lift check types, including ball checks, have a higher pressure drop than swing check valves. Sealing in the lift check style, however, is somewhat better than that attainable with the swing check design. Spring loading will further improve sealing ability, but in many services the springs are subject to corrosive attack. Spring loaded valves can be used in any line position or orientation; those without springs can be installed only in horizontal lines with the cover in the up position.

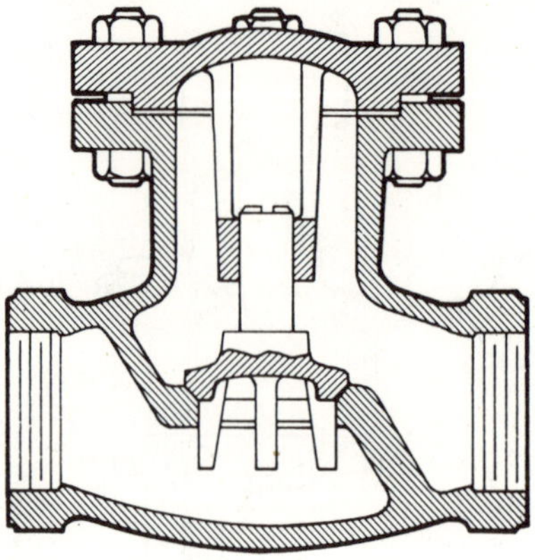

LIFT CHECK valve prevents back flow—Fig. 8

Stuffing Boxes

Sealing of the valve stem to prevent leakage requires a stuffing box appropriate to the valve's construction. Two types are used: conventional stuffing boxes and seals using O-rings.

Valves that require the stem to move in and out, with or without rotation, must have a stuffing box. These are either single (Fig. 9) or double with a lantern ring. Single-depth boxes are used on lower-pressure valves (150 psi. class) and those under 2 in. dia. Double stuffing boxes can have the lantern area tapped and plugged, or have a connection so fluid can be carried away.

Conventional stuffing boxes are suited for a wide variety of packing materials. The most common are various forms of braided asbestos containing appropriate lubricants. Wire is inserted in some packings to reduce their tendency to be extruded. An inhibitor is sometimes used to prevent corrosion by graphite lubricants.

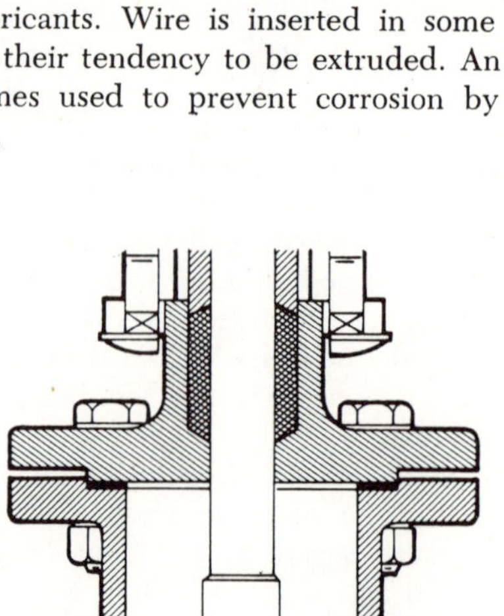

SINGLE stuffing box for valve stem—Fig. 9

PTFE packing is also widely used, usually in the form of molded "chevron" or V rings. The chevron type seals without excessive tightening of the gland bolts, thus giving easier valve operation. Braided PTFE packings are also available.

Quarter-turn valves using PTFE in the seat seals normally use a stuffing box for sealing the gland. On these valves, the stuffing box is usually not as deep as on a gate or globe valve. These are adequate in service, probably because sealing is much easier when only rotational motion is involved.

Conventional stuffing boxes can be used over the entire range of operating temperatures. Good seals are readily attained, although the glands must be quite tight for high pressure service. Scoring of stems does not always cause packing leakage, but every means possible should be used to avoid this problem.

Sealing With O-Rings

Elastomeric O-rings provide good sealing but are not suitable where there is sliding motion through the stuffing box. PTFE does not make a satisfactory dynamic O-ring seal, even with quarter-turn valves, because of its cold flow properties. O-ring seals are rarely used in high-pressure service and temperatures are limited by the material selected.

Bellows seal valves incorporate a formed metal bellows that makes a barrier between the disc and the body bonnet joint. One example is shown in Fig. 10. The bellows is more effective than a conventional stuffing box or any other gland-seal arrangement. They are normally backed up with a conventional stuffing box, and are often fitted with a leak detector to warn in the case of failure.

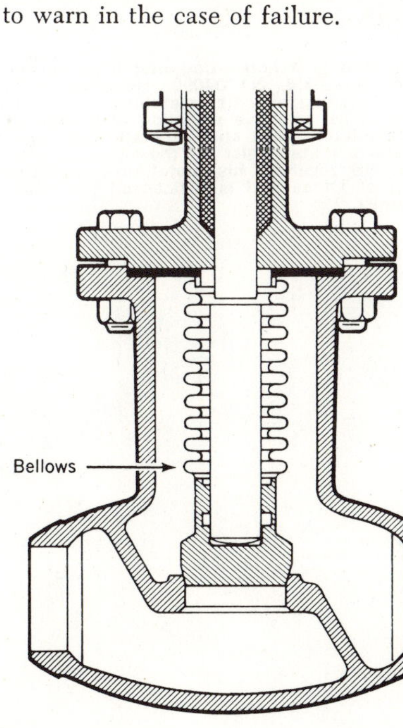

BELLOWS is the most positive process fluid isolation arrangement—Fig. 10

The bellows is the weak point of the system and service life can be quite variable. Maximum pressures are limited by the bellows design, as well as valve size, and are usually under 500 psig. Lacking any other method of providing an equivalent seal, bellows must be used; however their limitations must be recognized.

MAINTENANCE

Maintenance of valves in service is usually limited to tightening the body-bonnet joint and the gland bolting, or, in case of emergency situations, installing new packing rings. Changing the packing in a valve in service is always a hazardous operation and should be attempted only after the back seat is tightly seated against the bonnet. (Back seats are only found in gate and globe valves.)

Extensive maintenance of valves in the pipe line, even when the line is out of service, is usually only done under unusual conditions. Extent of in-line repair is limited by the valve design. It is much more satisfactory to remove a flanged valve and install a replacement, than to attempt an in-line repair, even if the valve design does permit some degree of repair without removal. Large valves may on occasion be serviced in the line because it is difficult to remove them to a shop.

Valves that can be repaired in the line, to the extent of correcting seating problems or installing new discs or seat seals, are the gate, globe, check, plug, top entry ball and diaphragm. Most ball and butterfly valves must be removed from the line to gain access to the ball and seat seals.

In-line reseating of gate and swing back check valves requires a special machine that clamps on to the body flange and cuts a new seat surface. In view of the need for both seat flatness and precise angle match with the disc, the success of such a procedure with solid-wedge gate valves is doubtful. If more than a bare minimum cut is taken off the seats, a new disc is required to obtain proper seating. Seats of globe and lift check valves can be machined from the bonnet end with good results.

Valves with PTFE seals can be reconditioned by installing new seals along with a new ball or plug if the old part is worn or corroded. However, this may be impractical if the body is also corroded in the seat or sleeve area. With diaphragm valves, the diaphragm can usually be replaced without removing the valve from the line.

Advantages of Shop Repair

Removal of valves from the line for repairs has several advantages. Plant downtime will often be less if replacement valves are available. Quality of repairs will normally be better and inspection more thorough because of access to all surfaces. Also, valves can be tested for seat tightness, which is difficult while a valve is in the line.

Shop maintenance of metal gate, globe and check valves consists of remachining seats and discs. Stems that are worn or scratched should be replaced. Discs of steel valves can be salvaged by building up seating surfaces with weld metal or hard-surface overlay. The yoke bushing (stem nut) should be replaced if it is worn, and new packing and gaskets should be installed. It is best to replace bolts and nuts since inspection may well cost more than new parts. New seat rings can be installed, but if this is necessary the repair may no longer be economic.

Shop repair of stainless steel equipment is much the same as for valves of carbon or low-alloy steel. One exception is that welding of stainless steel should be avoided because it will sensitize the metal to corrosive attack. If welding is essential, the parts should be solution annealed and water quenched to maintain corrosion resistance. If the parts are made of low-carbon stainless steel, or stabilized with columbium or titanium, welding with low-carbon or stabilized electrodes will produce a weld deposit free from sensitization.

Ball valve repair entails installation of new seat seals, a new ball and stem if necessary, and new packing, gaskets, bolts and nuts. Little remachining of ball valve components is normally required.

Lubricated plug valves can be repaired by weld build-up and remachining of the body or plug. However, the precision machining required to refit these parts is not readily available and it is doubtful if such repairs are truly economic, at least on smaller valves.

Non-lubricated plug valves require replacement of the PTFE sleeve, new packing, a new gasket, and possibly a new plug. The condition of the body under the sleeve is not always of great importance and in many cases, the body bore is not machined.

Butterfly valves are repaired by removing the stem, disc, and liner sleeve, which is the usual reason for a repair. New discs are usually not required, but O-ring seals or packing should be replaced along with stem and stem bushings if they are worn.

In Summary

Valve repairs are best done out of the pipeline, although in-line replacement of PTFE parts and some metal parts will produce satisfactory results on some valve types.

The remachining necessary on metal globe, gate and check valves requires adequate equipment and know-how on the part of the shop personnel. Most operating plants would not be justified in attempting such repairs themselves, and specialists should be considered for this type of work.

Installation of seat seals, new metal parts, liners and other miscellaneous parts might be done either in-plant, or given to a valve repair shop.

Valve repair is generally considered economic if the unit can be reconditioned for no more than 65% of the replacement cost. Repair costs average close to 50% of the replacement cost; however, many valves are not repairable because of exceeding the cost criterion above. Generally a valve is not repairable if the body cannot be reused as cost of repair will exceed the value recoverable.

Bibliography

1. "Johns-Manville Metallic Gaskets," Johns-Manville Sales Corp., 270 Madison Ave., New York, NY 10016.
2. "Flexitallic Spiral Wound Gaskets—Design Criteria," Bulletin 171, 1971 First Edition; Flexitallic Gasket Co., Inc., Camden, NJ 08101.
3. "Strength Guide," Technical Bulletin 103, H. M. Harper Co., Morton Grove, Illinois.

Meet the Author

Harold C. Templeton is a consulting metallurgical engineer at 1110 Boulevard, Westfield, NJ 07090. He obtained his valve experience while working as chief metallurgist of Aloyco, Inc., serving as a specialist in valve selection and application. Previously, Mr. Templeton was chief metallurgist for Lebanon Steel Foundry and a metallurgist for the boiler division of Babcock & Wilcox. He received his metallurgical degree at the Colorado School of Mines and is a licensed Professional Engineer in New Jersey.

Control-Valve Noise: Cause and Cure

HANS D. BAUMANN, Masoneilan International, Inc.

As part of the present great interest in bettering the environment, a new attitude has developed toward control valves. To paraphrase a familiar saying: "Control valves should be seen but not heard."

The Federal Government has added legal backing to this attitude by setting a 90-db. noise time for equipment operating in the vicinity of plant personnel (for 8-hr. exposure). As a result, process and contracting engineers alike are setting up standards to ensure adherence to this law. The obvious question to put to the equipment manufacturer is, "How noisy will the valve be in our plant?"

In order to answer this question and be able to guarantee the specific noise level, control-valve manufacturers have had to study the problem of valve noise in depth (Fig. 1).

VALVE being tested in acoustical chamber—Fig. 1

The first requirement is for a method to predict the anticipated noise level under operating conditions. The equally important second question is, "What can we do to reduce the noise level if excessive?"

One must distinguish the three distinct and different noise phenomena emanating from a control valve:

1. Noise induced by mechanical vibration of the trim.
2. Noise produced by cavitating liquids.
3. Noise caused during aerodynamic throttling.

It is quite important that these three noise sources be understood so far as their generating mechanism is concerned. Only then can an effective improvement or cure be made. Luckily, mechanical-vibration noise seldom happens simultaneously with cavitation and aerodynamic noise. However, if this happens, the cure of one is usually the cure of the other.

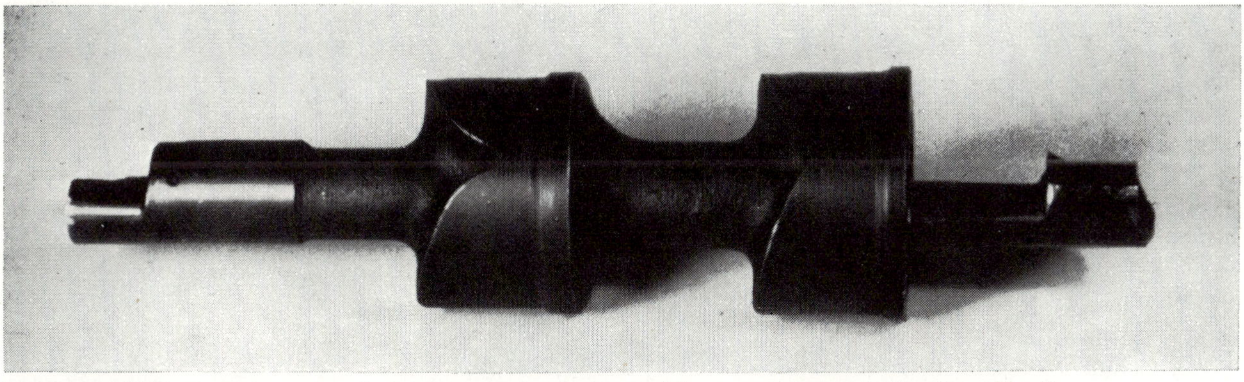

DISPLACEMENT of guide posts of stainless-steel valve caused by severe resonant vibration—Fig. 2

Originally published May 17, 1971

Noise Produced by Mechanical Vibration

Two mechanisms are involved: The first is mechanical vibration, induced by pulsation of the fluid passing through the valve. The frequency is usually low, i.e., between 50 and 500-Hz. However, if this turbulence-induced vibration of the valve trim approaches the natural frequency of the plug-stem combination, then we have the second mechanism—resonance. This resonance occurring at frequencies between 2,000 and 7,000-Hz. is most harmful, since it can lead to fatigue failure of the valve stem or plug post and can even displace solid stainless-steel parts by fractions of an inch (see Fig. 2).

The only beneficial aspect of this vibratory noise is that it warns operators that a mechanical failure is in the offing.

Luckily, the phenomenon has become less common since the introduction of top-guided single-seated valves, since they have (as a rule) less clearance in the guide bushings; also, the lower weight of a single-seated plug increases the natural frequency of the trim, making it less susceptible to fluid-induced vibration.

Possible cures for this type of noise include reduction of guided clearances, and increase in stem size (a 40% increase in stem diameter doubles the undamped natural frequency of the valve trim). Another attempt to cure this problem can be made by changing the flow or pressure conditions to which the valve is subjected. Quite often, a simple reversal of the flow direction through the valve sufficiently alters the flow pattern to shift inducing frequencies away from the trim excitation range.

Noise Produced by Cavitating Liquids

Cavitation noise should never be heard in a well-designed process plant. Hardly anything destroys a valve trim as surely as a cavitating liquid. With the introduction of special valve trims having very little pressure recovery and special valves having multiple velocity-headloss trim, there is seldom an excuse to have cavitation in a throttling valve (except, perhaps, for some rather large valve sizes where no anticavitation trim is available as yet).

With the present availability of good engineering data, it is possible to predict quite accurately whether or not a selected valve will cavitate under a given process condition. One such equation introduced by me in 1962 allows the prediction of a critical pressure drop at which a given value will cavitate,[1] i.e.,

$$\Delta p_{crit.} = C_f^2 (p_1 - p_v). \tag{1}$$

If the operating pressure drop in the plant exceeds $\Delta p_{crit.}$ then the valve is cavitating. In that case, one should solve for the required Critical Flow Factor, C_f, and then select a valve from the manufacturer's catalog that has a C_f factor equal to or higher than the one calculated by the following equation:

$$C_f = \sqrt{(p_1 - p_2)/(p_1 - p_v)} \tag{2}$$

Even though cavitation can be avoided in almost all cases, there is still interest in the prediction of cavitating noise. Our laboratory investigations indicate noise to be a function of: the amount of decrease in downstream pressure beyond the pressure that causes incipient cavitation, and the difference between downstream pressure and vapor pressure. The peak in cavitation noise can be expected where these two variables are nearly equal; i.e., the noise decreases as the difference between p_2 (actual) and p_2 (incipient cavitation) approaches zero, and if the difference between the outlet pressure and the vapor pressure approaches zero. The latter is understandable since the process of *cavitation* is converted into a process of *flashing*. The relationship just described is best illustrated in Fig. 3. Note a gradual increase in noise due to a normal fluid-borne turbulence up to the $\Delta p_{crit.}$, the point of full cavitation. The noise value reaches a peak when the excess pressure head and what we might call "cavitation energy"—i.e., the difference between p_2 and p_v—hold equal sway. Finally, the sound pressure level goes down again as the cavitation energy approaches zero with further decrease in p_2. It is of interest that this relatively small angle-valve ($C_f = 0.55$) with a C_v of only 4.25, an inlet pressure of 240 psia. and a pressure drop of only 89 psia. already exceeds the 90-db. limit of the Walsh-Healey Act.

To satisfy demands for an empirical equation to predict cavitation noise, I suggest:

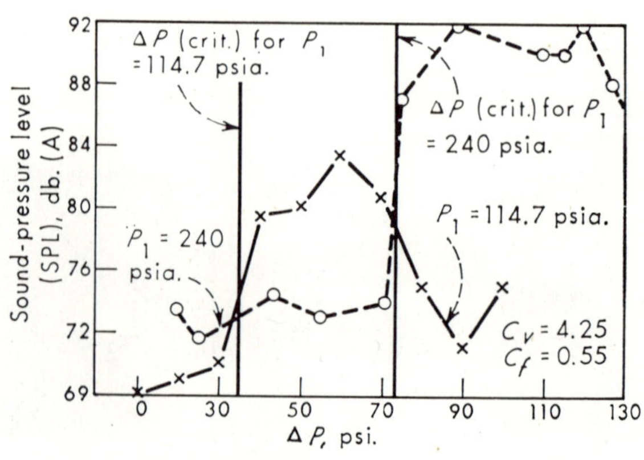

CAVITATION noise of single-seated valve at two inlet pressures (water at 60 F.)—Fig. 3

$$SPL = 10 \log (C_v C_f) + 8 \log (p_{2(crit.)} - p_2) + 20 \log (p_2 - p_v) + 33 \text{ (in dbA.)} \quad (3)$$

Where: $p_{2(crit.)} = p_1 - C_f^2 (p_1 - p_v)$

Note this equation is reasonably accurate only for water, and when using Schedule 40 downstream pipe. We expect that further research will enable extension of this equation to other liquids and let us predict the effect of differences in pipe-wall thickness.

Using this equation, one is able to predict the noise of a 6-in. ball valve reducing 100-psia. water to 25 psia. as follows:

Given: $C_v = 1{,}000$, $T = 70$F., $C_f = 0.7$, $P_v = 0.5$ psia. Hence, $p_{2crit.} = 100 - 0.7^2 (100 - 0.5) = 51.2$ psia.

$$SPL = 10 \log (700) + 8 \log (51.2 - 25) + 20 \log (25 - 0.5) + 33 = 28.5 + 11.3 + 27.8 + 33 = 100.6 \text{ db.}$$

Aerodynamic Noise

This is the most important form of acoustical annoyance so far as control valves or pressure-reducing valves are concerned. Aerodynamic noise is a byproduct of the reconversion of kinetic energy through turbulence into heat downstream of the throttling orifice. These are two basic contributory factors. One is the terminating shock front of a supersonic jet generating from the vena contracta of the valve orifice (at higher-than-critical pressure drop). The second comes from the general turbulence of the fluid boundary and is effective above, as well as below, choked flow in the valve orifice.

Unfortunately, there is no way to avoid aerodynamic noise, since we have not as yet invented a valve that can reduce pressure without causing turbulence. However, the degree of noise generation can be affected by various parameters, as will be discussed later.

The important question facing an instrument engineer when laying out a new plant is: Which control valves will be noisy, (i.e., exceed the 90 db. usually considered the upper limit)? We published in Sept., 1969, a practical equation[2] that allows the prediction of the sound-pressure level for any style of valve under any given pressure condition. This equation has been further improved and converted into a graphical form for simpler handling.[2]

The general theory behind this equation has recently been made public[3] and I would like to repeat it here in broad terms in order to promote a better understanding as to which parameters affect the sound pressure level in a valve and how they may be favorably altered.

Apparent Noise-Producing Orifices—Table I

Valve Type	n Factor
Control ball-valve	1.0
Camflex Single-port globe valve Angle valve Butterfly valve	1.4
Double-port valve 4-port cage valve	2.2

Nomenclature

C_f	Critical-flow factor of control valve, a ratio
C_v	Flow coefficient, gpm./(psi.)$^{1/2}$ (for water, 60 F.)
f	Frequency, cycles/sec. (Hz.)
g	Gravitational constant, ft./sec.^{2}
L_T	Transmission loss, db.
m	Weight of pipe wall, lb./sq.ft.
N	Strouhal number, a ratio
n	Number of apparent frequency-producing orifices (see Table I)
P	Sound pressure, lb./sq.ft.
p	Static pressure, psia.
R_{cv}	Critical-pressure ratio, p_1/p_2, at Mach 1
r	Radial distance to noise source, ft.
S_g	Gas-property correction factor, db. (see Table II)
SPL	Sound-pressure level, db. (Ref., 2×10^{-4} microbar)
V	Velocity, ft./sec.
W	Power, ft.-lb./sec.
X	Fraction of mechanical power conversion ($p_1 - p_2/0.47\, p_1$) limit to 1.
η	Acoustical efficiency, a ratio
ρ	Mean density lb./cu.ft.

Subscripts

a =	Acoustical
i =	At vena contracta
m =	Mechanical
o =	downstream
s =	Sonic
v =	Vapor
1 =	Inlet
2 =	Outlet

Theory of Aerodynamic Sound

As mentioned before, the sound pressure measured in the proximity of a throttling control valve is a result of pressure waves in the atmosphere—the pressure waves' root mean square (rms.) values are expressed in microbars. The sound-pressure level (SPL) in decibels is equivalent to $20 \times \log_{10}$ of the ratio between the rms. value and the reference value (taken as 2×10^{-4} microbars).

The acoustical power that generates these pressure waves is created by the supersonic shock front in a jet, and by turbulent boundary layers within the valve. It is directly related to the amount of mechanical energy converted in the valve. This then makes the SPL value a direct function of mass flow or C_v, since the latter is an expression of flow capacity.

The acoustical efficiency factor η describing the ratio between acoustical power and mechanical power is a function of the valve style and the pressure ratio across the valve. Here is an expression for the mechanical power converted in a valve:

$$W_m = [\rho_i v^3 \pi (2.3 \times 10^{-4}) C_v C_f]/8g \quad \text{(in ft.-lb/sec.)} \qquad (4)$$

(For the mathematical development of this and other equations see Ref. 3.)

Multiplying the mechanical power by an acoustical efficiency factor, η, will now yield the acoustical power:

$$W_a = W_m \times \eta \qquad (5)$$

Fig. 4 shows semi-empirical, acoustical efficiency factors plotted against pressure ratios across the valve. The acoustical efficiency varies with the pressure recovery characteristic of a particular valve as expressed by the Critical Flow Factor, C_f (see Table III). It shows, for example, that a streamlined angle valve, flow-to-close (C_f = 0.5) will produce relatively higher noise than a V-ported globe valve (C_f = 0.95) when below the critical-pressure ratio of approximately 2 to 1, but considerably less noise above this ratio.

Having found the acoustical power, a conversion into sound pressure can be made as follows:

$$P = (W_a \rho_o V_s / 4\pi r^2 g)^{1/2} \quad \text{(in lb./sq.ft.)} \qquad (6)$$

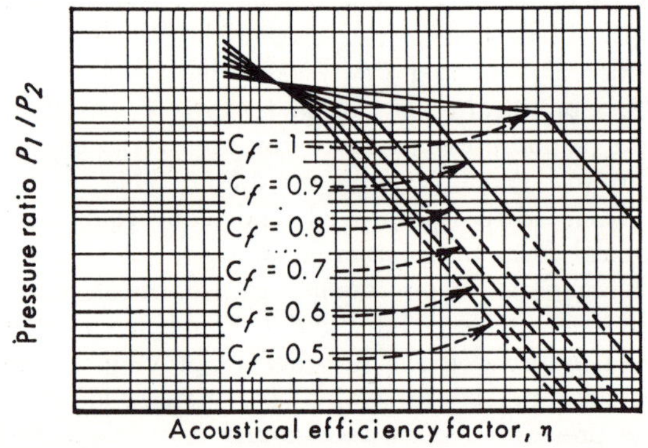

ACOUSTICAL efficiency factors of control valves for various critical-flow factors, C_f—Fig. 4

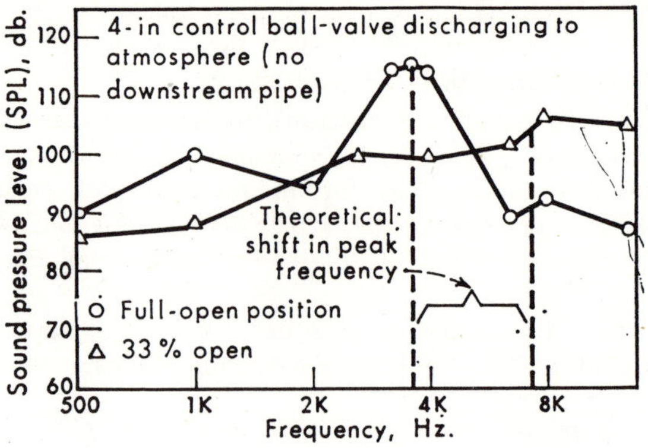

$$\text{Peak freq. } (\Delta) = \text{peak freq. } (o) \sqrt{\frac{C_v\ C_f\ (o)}{C_v\ C_f\ (\Delta)}} = 3{,}500 \sqrt{\frac{258}{65}} = 7{,}000$$

PEAK NOISE frequency shift with shift in apparent valve orifice diameter—Fig. 6

Finally, the transmission loss must be known, i.e., the amount of sound power attenuated by the downstream pipe wall. Assuming that the predominant frequency of the valve noise coming from the orifice is below the coincidence frequency or "ring frequency" of the pipe (see Ref. 4 and Fig. 5), one can relate the pipe wall attenuation to that of a flat panel, i.e.:

$$\text{Transmission loss, } L_T = 17 \log (mf) - 36 \qquad (7)$$

wher $fe = NVn/0.015\ \sqrt{C_vC_f}$.

The Strouhal number, N seems to vary between 0.1 and 0.2, depending on the pressure ratio across the vena contracta. The frequency is also a function of the apparent orifice diameter, expressed as $(1/n)$ $0.015\ \sqrt{C_vC_f}$ (see Fig. 6), wherein n is taken as a modifier to convert an irregular orifice into an equivalent circular shape (see Table I). Note also that the attenuation is a function of 17 log pipe-wall density, m, (see Fig. 7). Finally, a correction factor has to be used to make the basic equations (strictly

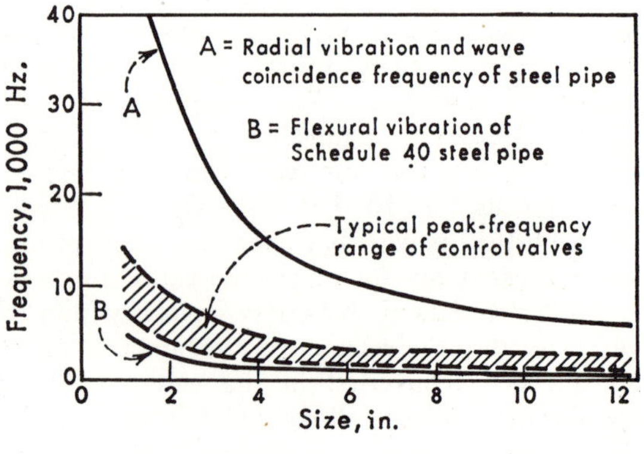

PEAK frequencies (typical) of control valves as related to vibration of adjacent piping—Fig. 5

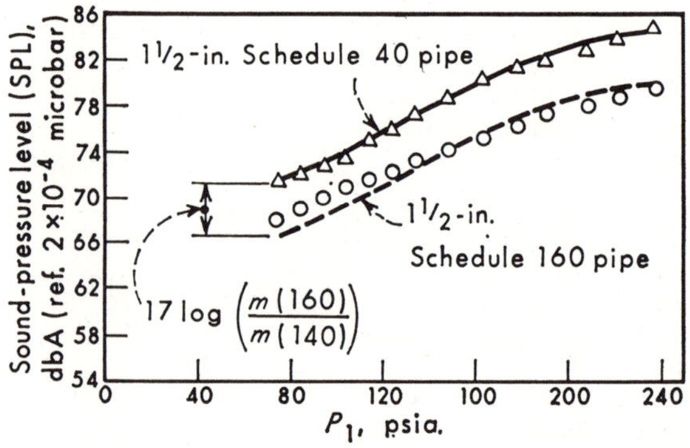

TRANSMISSION LOSS effect of pipe wall thickness. Switch to heavy pipe reduced noise by 4.5 db.—Fig. 7

Gas-Property Factor—Table II

	S_g, db.		S_g, db.
Saturated steam...	−2	Carbon dioxide.....	+1
Superheated steam	−3	Carbon monoxide..	0
Natural gas........	−1	Helium............	−6.5
Hydrogen..........	−10	Methane..........	−1
Oxygen............	+0.5	Nitrogen..........	0
Ammonia..........	−2	Propane...........	+1
Air................	0	Ethylene..........	−1
Acetylene..........	−1	Ethane............	−1

applicable only for air) suitable for other gases. This is done with the use of a gas-property factor S_g listed in Table II. Thus the final equation for the aerodynamic sound-pressure level at 3 ft. from the valve outlet and pipe wall becomes:

$$SPL = 10 \log_{10} (X \eta 10^{11} C_v C_f p_1 p_2) - L_T + S_g \quad (8)$$

This equation has been found quite accurate despite the rather complex nature of the subject. Table III shows calculated vs. tested data on valve sizes 1 to 12 in., and a wide range of pressures (up to 4,000 psi.).

What To Do About Aerodynamic Noise

One problem with the SPL equation is that in the planning stages an engineer may not know the exact style and type of valve that will be used in a given application. Since this is essentially a filtering-out stage, i.e., for isolating critical applications, a simple rule-of-thumb would be very handy. One such rule is not to worry about a valve (handling compressible media at critical pressure drop) as long as the product of absolute inlet pressure in psia. multiplied by C_v is below 1,000. All valves falling

Comparison of Calculated SPL With Laboratory and Field Test Data—Table III

Valve Size, In.	Type*	Fluid	Pipe Size, In.	Schedule	n	P_1	P_2	C_vC_f	C_f	$C_{FR/R}$†	SPL Calc.	SPL Test
1	Y style	Air	1	40	1.4	145	17.5	3.04	0.65		81.1	83
1	Y style	Air	1	40	1.4	24.7	14.7	3.04	0.65		56.8	57
1	Y style	Air	1	40	1.4	115	19.0	3.04	0.65		78.0	80
1	Y style	Sat. steam	1½	40	1.4	44.7	15.7	2.9		0.65	63.7	65
1½	S.P. globe	Air	1½	40	1.4	145	15.7	2.40	0.50		78.0	78.8
2	4-port cage	Sat. steam	2	40	2.2	71.7	21.7	31.6	0.95		89.9	89.5
2	4-port cage	Sat. steam	2	40	2.2	111	32.2	31.6	0.95		94.1	96
2	Angle	Nat. gas	10	160	1.4	4,000	2,300	25.5		0.50	110.6	111
2	Regulator	Sat. steam	8	40	1.4	170	34.7	11.8		1.0	98.3	96
2	D.P. globe	Ind. gas	4	80	2.2	530	80	14		1.0	106.6	105
2	D.P. globe	Ind. gas	4	80	2.2	530	80	17		1.0	108.1	110
3	Regulator	Sat. steam	8	40	1.4	170	34.7	19.3		1.0	99.7	102
3	D.P. globe	HC vapor	6	40	2.2	241	72	48		0.95	102	103
4	Ball	Air	4	40	1.0	150	50	180	0.76		108.7	107
4	Butterfly	Air	4	40	1.4	150	50	.180	0.62		105.3	103
4	8 port cage	Air	4	40	2.5	150	50	180	0.90		105	110
4	Ball	Air	4	40	1.0	100	72	336	0.60		103	100
4	D.P. globe	Air	4	40	2.2	150	50	180	0.86		104.6	107
4	D.P. globe	Fuel gas	8	40	2.2	255	90	97		0.89	104.5	108
4	Ball	Air	8	40	1.0	615	115	108		0.84	121.2	120
4	Ball	Air	8	40	1.0	615	315	108		0.84	113	114
4	Ball	Air	8	40	1.0	150	50	180		0.79	109.1	108
4	Ball	Air	8	40	1.0	615	75	108		0.84	123.2	122
6	S.P. globe	Nat. gas	8	40	1.4	305	91	76.7		0.80	104.4	106
6	D.P. globe	Nat. gas	8	40	2.2	270	90	100		0.90	104	105
6	Ball	Nat. gas	8	40	1.0	650	325	244		0.76	116	110
6	D.P. globe	Nat. gas	8	40	2.2	635	295	70		0.9	101	105
6	D.P. globe	Nat. gas	8	40	2.2	665	340	191		0.9	110	110
6	D.P. globe	Nat. gas	8	80	2.2	250	105	101		1.0	98.3	102
8	Regulator	Nat. gas	12	40	3.0	365	165	177		1.0	101.8	101
8	D.P. globe	Fuel gas	10	40	2.2	190	22.7	216		0.5	105.9	98
8	D.P. globe	Nat. gas	10	40	2.2	200	23	225		0.55	104.5	100
10	D.P. globe	Steam	10	40	2.2	245	130	800	0.95		110	107
12	D.P. globe	Nat. gas	18	40	2.2	165	45	.367		1.0	114.4	122

* S.P. = single port; D.P. = double port.
† Use $C_{FR/R}$ instead of C_f when valve is installed between reducers.
The weight, m, of pipe reducers is based on valve-size pipe wall. Calculated values include 3 db. for hemispherical radiation.

above this limit should be calculated and treated properly once the final specifications become available.

An important requirement in noise control is to keep the valve-outlet velocity below a certain limit, depending on the type and size of the valve, in order to prevent the occurrence of a secondary noise source that might be even worse than that produced by the valve itself. This is particularly important with valves having special "low noise" trim. To demonstrate the importance, Fig. 8 gives the SPL values of air flowing through various pipe sizes as calculated from equations suggested by Heitner.[5] This example draws attention to this often overlooked aspect and points out the need for proper valve and pipe dimensioning.

If the calculated SPL value of a reducing valve under maximum load exceeds the stated limit by only 5 to 10 db., then the following simple cures are applicable:

- Increase the pipe-wall thickness downstream (doubling the wall thickness will decrease SPL by 5 db.).
- Use acoustical insulation downstream. This will reduce the SPL number between 5 and 10 db. per inch of insulation, depending on the density of the insulating material.

If the valve noise is more than 10 db. above the selected limit, then one must choose a different approach such as the use of downstream, in-line silencers. These devices generally attenuate between 10 and 20 db., depending on the frequency range. Care should be taken to mount the silencer directly adjacent to the valve body and to make sure that the valve outlet velocity is below sonic; otherwise the silencer will act as a pressure-reducing device for which it is not suitable. One should also be aware that part of the valve throttling noise will be transmitted upstream and reradiated through the upstream pipe. This means that the valve that would have a calculated untreated SPL of 100 db. could be attenuated by a silencer to 80 db. downstream. However, the system would still radiate noise at 90 db. upstream, cancelling most of the silencer's beneficial effects.

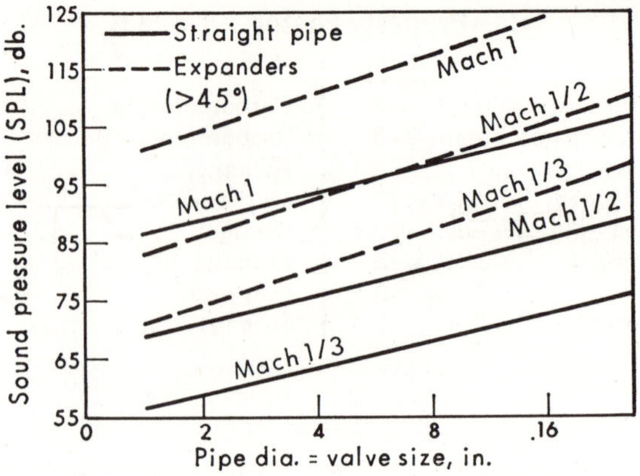

AERODYNAMIC NOISE generated by pipe velocity, including effect of downstream expanders—Fig. 8

Another approach usually recommended by my company is the use of noise-reducing expansion plates[6] downstream of a valve. The primary function of these plates is not to attenuate the valve noise, but to absorb some of the pressure reduction over the whole system. This way, the pressure ratio across the control valve can be kept below critical, with the associated benefit of having lower acoustical power generation. Noise-reducing Lo-db plates can be used if the product of absolute inlet pressure (p_1) times C_v is below 5,000 when used downstream of conventional control valves. This limit does not exist when placed downstream of multiple-step reduction valves and conventional valves used for *on-off applications* only. One major benefit of these plates is the reduction of the valve-outlet velocity to safe levels by the buildup in static pressure.

The use of multiple plates is recommended where

Sound Pressure Levels of Regular Vs. Low-Noise Valves—Table IV

Lo-db Size	Valve Type	X†	Volume, Std. Cu. Ft./Hr.	Fluid	P_1 Psia.	P_2 Psia.	Pipe Schedule	Sound-Pressure Level Lo-db Valve Calcul.*	Lo-db Valve Tested	Standard Valve
1½ × 1½	S. Seat	2	31,180	air	240	17.5	40	79.4	78.5	87
2 × 2½	S. Seat	6	1.4×10^6	gas	3,725	1,151	160	78.5	81	107
2 × 2½	S. Seat	6	$0.5 \times 10_9$	gas	4,000	1,060	160	70.5	67	96
2 × 4	S. Seat	4	2.7×10^6	gas	4,000	1,060	80	89.3	88	112
2 × 4	S. Seat	4	4.1×10^6	gas	4,000	1,060	160	89	94	110
3 × 6	S. Seat	4	2.7×10^6	gas	3,850	1,060	80	87	83	111
3 × 6	S. Seat	4	3.4×10^6	gas	3,850	1,060	80	88.8	85	113
3 × 6	S. Seat	2	4.1×10^6	gas	2,300	1,060	80	94.2	96	112
4 × 4	D. Seat	2 × 2	350,000	air	215	115	40	79.2	80	91.2
6 × 6	D. Seat	2 × 2	360,000	gas	282	138	40	75	79	90

* SPL $= 18.5 \log (C_v C_f) + 28 \log (P_1) - 13 \log (P_2) + 20 \log (4/X) + 28 - Sa + Sg$ (in dbA) † X = trim area ratio

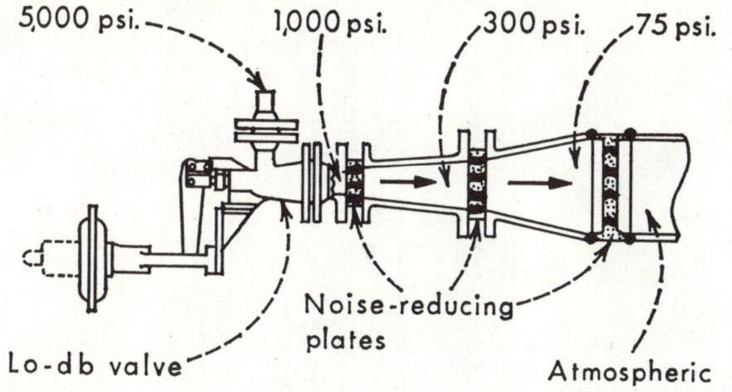

LOW-NOISE valve installation reducing from pressure of 5,000 psi. down to atmospheric—Fig. 9

the ratio between upstream pressure and ultimate downstream pressure is more than 10 to 1. Fig. 9 shows typical layout of a valve reducing from 5,000 psi. to atmospheric. Here the valve could reduce the pressure from 5,000 to 1,000 psia., the first plate from 1,000 psi. to 300 psi., the second plate from 300 psi. to 75 psi., and the last plate from 75 psi. to atmospheric. Note that these plates are installed in expanders whose increasing flow areas match the given pressure level and keep the velocity gradient through the whole system nearly constant.

Analyzing Eq. 4 and 6 will yield the conclusion that the most effective way to combat aerodynamic sound in a valve is by reducing the throttling velocity. This is indeed so, since SPL varies nearly to the 8th power of throttling velocity. Several valve styles now on the market make use of this beneficial effect, notably the Self Drag valve marketed by Control Components, Inc., and the Lo-db valve introduced by Masoneilan in 1967.[7] The latter valve uses a multistep cone trim that forces the fluid to make repeated sharp turns, each of which has a high-velocity head-loss coefficient, producing a nearly "constant adiabatic" pressure drop across the valve. The high specific head-loss change per stage means low velocity. Hence, noise level is substantially below that of conventional angle or globe valves (Table IV).

Unfortunately, these valves are more costly than their predecessors, owing to the more-expensive trim configuration and the valve size, which as a rule is double that of a conventional valve. The reason for the increased valve size is, of course, to keep the fluid velocity within limits. No such restrictions are placed on a conventional valve (when disregarding the sound pressure level).

The comment has often been made that if a valve is installed in a remote area without operating personnel being constantly present, the noise problem can safely be disregarded. This can be at times a dangerous simplification, since noise is nothing but audible *vibration.* The vibration itself can cause considerable mechanical damage. Damage to pressure gages and valve- or pipe-mounted instruments is one of the many detrimental effects. Of more-drastic consequence is that even 2,500 lb. ANSI flange bolting can vibrate loose, as I have witnessed myself. Fig. 10 shows typical acceleration values measured downstream of a 2-in. high-pressure angle valve "flow-to-close," reducing gas from 4,000 psi. to 2,300 psi. Notice that after the Lo-db valve was installed the pipe acceleration level decreased from 125 db. to 103 db. (14g. to 1g. acceleration level). The original acceleration level of 14g. was quite severe and compares to that experienced when holding a riveting hammer.[8]

In any case, elimination of vibration is an important aspect that should not be overlooked. Money spent for safety and for the reduction of maintenance costs, is money well spent.

Note: The acoustical-test results presented in this article, with the exception of Fig. 6 and 10, refer to the A-weighted scale (dbA), i.e., the frequency range corresponding roughly to the sensitivity of the human ear. Such overall frequency-weighted data are often called "Sound Levels" and usually exceed the specific SPL at a given peak frequency by 1 to 3 db.

References

1. Baumann, H. D., The Introduction of a Critical Flow Factor for Valve Sizing, *ISA Transactions,* 2, No. 2 (Apr. 1963) Paper presented at ISA Conference, Oct. 1962.
2. "Masoneilan Handbook for Control Valve Sizing," Masoneilan International, Inc., Norwood, Mass.
3. Baumann, H. D., On The Prediction of Aerodynamically Created Sound Pressure Level of Control Valves, ASME Paper WA/FE-28, Dec. 1970.
4. Sawley, R. J., While, P. H., Energy Transmission in Piping Systems and in Relation to Noise Control, ASME Paper 70-WA/PET-3, Dec. 1970.
5. Heitner, J., How to Estimate Plant Noises, *Hydrocarbon Process.,* Dec. 1968.
6. Hynes, K. M., The Development of a Low Noise Constant Area Throttling Device, ISA Paper 839-70, Oct. 1970.
7. Baumann, H. D., Multistep Valve Design Cuts Throttling Noise, *Instrum. Tech.,* Oct. 1969.
8. Harris, C. M., Crede, C. E., "Shock and Vibration Handbook," Vol. 3, pp. 44-42 and 47-53, McGraw-Hill, New York, 1961.

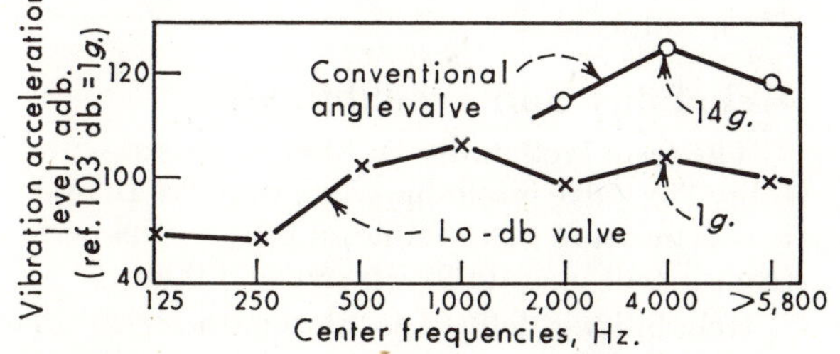

VIBRATION acceleration levels when reducing 4-million cu.ft./hr. gas from 4,000 to 2,300 psi.—Fig. 10

Meet the Author

Hans D. Baumann is Vice-President Engineering for Masoneilan International, Inc., Norwood, Mass. 02602. He received his undergraduate education in Germany, and has done graduate work at Western Reserve University, Cleveland and Northeastern University, Boston. He is a member of the American Soc. of Mechanical Engineers, and a senior member of the Instrument Soc. of America. He is a registered professional engineer in Illinois and Pennsylvania.

How to improve online control-valve performance

Reliability, fail-safe operation and economics are the essential criteria for making the final decision in selecting optimum control valves.

E. Ross Forman, *United Engineers & Constructors Inc.*

☐ The control valve is the only controlled variable-restriction in the control loop. In effect, it is a variable-area orifice that removes discrete amounts of energy from the system in order to control the process. Fig. 1 shows the place of the valve in the loop.

While pumps, reactors, boilers and compressors introduce pressure rise into a process system, the valve deliberately introduces pressure drop. Unlike equipment, such as a turbine, that generates useful work and ultimately electrical energy from a pressure drop, the control valve accomplishes control of the process variable by consuming energy. Since 30 to 50% of the total dynamic system drop may be expended in the control valve, there is always concern about its functioning reliably under continual stress.

There are more points of potential failure in a valve than in any other device in the control loop. As can be seen from the typical valve body (Fig. 2), the interior can wear due to erosion and corrosion. Furthermore, the constant need for the valve to moderate against the process medium causes packing wear, bellows fatigue, and wear of accessories such as positioners. Aging can cause diaphragms to fail. Frequent cycling to open and close the valve during operation can wear out auxiliary devices such as limit switches and solenoid valves (Fig. 3). Under such conditions, it is amazing that the control valve is as reliable as it is.

Failure analysis

There are a number of reasons why it may be necessary to account for the failure of a valve despite the most well-written specification and the finest-manufactured and best-tested valve that can be obtained.

One reason is the pure economic penalty of not having a key valve available for operation, whether its function is to control or to isolate the process. If the valve is in a crucial part of the process where a serious interruption in production can be caused by its failure, it means a loss of income. This loss is easily calculated by the hour or day, and can be postulated ahead of time. This is increasingly important in most plants because of the increasing size of processes. Major sources of income come from the larger facilities, so that any downtime is significant.

Another reason may involve the release of hazardous or toxic materials in the plant environment.

Regardless of the reason, valve failure cannot be tolerated in most circumstances, and the design phase must include appropriate analysis.

Obviously, the best way to preclude the possibility of failure is to have redundant valves. This may mean spare valves that duplicate the entire function of the original ones, or perhaps just the duplication of the components most likely to fail, such as limit switches. The task is to determine whether such redundancy is required and how much there should be. This is done by failure analysis of the process in the design phase, since no amount of quality control, field testing, or maintenance can compensate for a deficient design.

Unfortunately, it is usually easier to conceive of how a process will work than how it will fail. The human mind is basically success-oriented and looks for positive results. Conceiving negative solutions is difficult, and until recently was considered an art rather than a science. By applying proven statistical methods, the selection of a design can be success-oriented and also have a low incidence of failure that will be adequate for the plant's objective.

Reliability and availability

The term "reliability" is often confused with "availability." A valve is reliable when it works. In other words, it is a measure of the time stability of its performance (e.g., it will operate 997 h out of 1,000 h).

Reliability is defined as "the characteristic of an item expressed by the probability that it will perform a required function under stated conditions for a stated period of time." This definition of reliability mentions

Originally published June 5, 1978

only the time of operation and not the time that the item may not operate. For example, the valve involved in a failure may be in a hazardous area, and not accessible until the unit is shut down for maintenance. In this case, the valve must be reliable for this time period, no matter how long that may be. At the end of this period, the plant is shut down and repairs can be made at will, so that no further consideration is necessary.

However, let us suppose that the valve is accessible for inspection, test and repair during plant operation. Then, a more meaningful measure of the integrity of the valve will be "availability" [defined by the IEEE (Institute of Electrical and Electronics Engineers) as "the characteristic of an item expressed by the probability that it will be operational at a randomly selected future instant of time."]

When designing the system, it is necessary to (a) define the success required for the valve and its system, and (b) establish reliability goals for the valve.

Success of the valve and its system must be defined as rigorously as possible to assure that valve operation takes place as planned. A complete appreciation of the system must exist for it to be properly defined. There may be many criteria for success, and all of them will influence the number and arrangement of the valves in the process. The definition must include the environmental conditions, how long the system must function, the number of cycles of operation involved, and any other data. The following is an example of such a "definition":

"The valve and actuator must operate satisfactorily during accident and post-accident conditions at any time during the design life of the equipment. Accident conditions are: temperature = 310°F, pressure = 62 psig, relative humidity = 100%, atmosphere = saturated steam plus nitrogen. Temperature and pressure duration time is 10 h. The valve shall be capable of actuating 250 cycles/yr (full open to closed, and return)."

Reliability goals are based on several factors. In the nuclear industry, the "single failure" criterion is often used as the primary goal. Simply stated, the valve should fulfill its success definition in the event of failure of a single component. In numerical terms, the data should provide the expected system reliability or availability. Such goals are arrived at by:

1. *Risk acceptance*—What is the highest risk that can be accepted in return for the benefits of the system as designed? Risk is defined as the product of the probability of failure and the consequences of that failure. Consequences may be measured in terms of lost profits, injuries, or the release of toxic or hazardous substances.

2. *Grandfather systems*—Past experience on existing systems that have had a good history of reliability and availability.

3. *Industry-standard goals*—This is often a tentative objective based on current knowledge.

Examples illustrate analysis methods

The most convenient method for explaining the studies made on reliability and availability is based on a system where two valves are required for isolating the process fluid. For instance: How many valves are re-

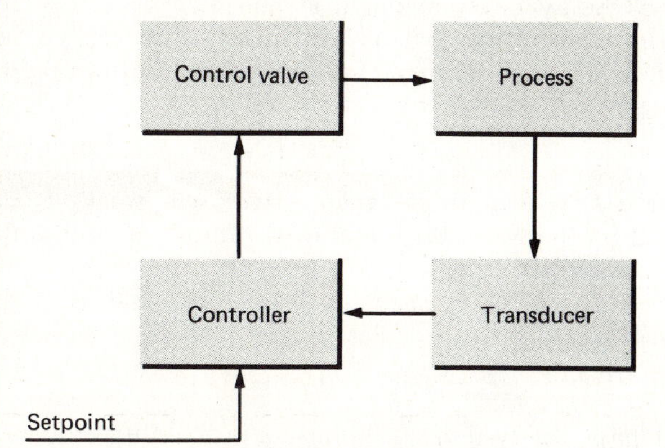

Location of control valve in control loop **Fig. 1**

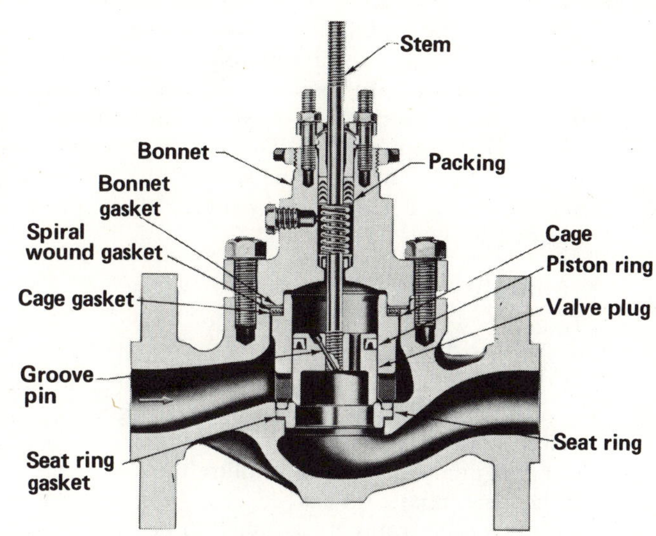

Points of potential failure in control valves **Fig. 2**

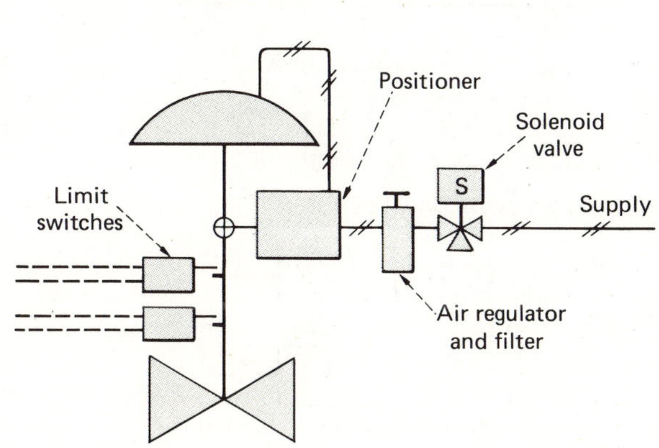

Cycling wears out auxiliary devices **Fig. 3**

quired that will meet the demands of the system, so that it will be available as much as possible?

Let us consider two valves in series, as shown in Fig. 4. The probability of successful operation of the system (i.e., closing the two valves) is:

$$P_g = P_1 \times P_2 \qquad (1)$$

where P_g = probability that system will work, P_1 = probability that Valve 1 will work, and P_2 = probability that Valve 2 will work.

If the probability of each valve working is 90%, then substituting the appropriate values into Eq. (1) yields:

$$P_g = 0.90 \times 0.90 = 0.81$$

Hence, the system would have a probability of working only 81% of the time.

If an additional valve, P_3, is added in parallel to valves P_1 and P_2, and has a probability of 90% (Fig. 4), the probability of successful operation becomes:

$$P_g' = P_g + (1 - P_g)(P_3) \qquad (2)$$

Therefore, the probability of working for this system becomes 98.1%—a substantial improvement over the original design. At this point in the design, a decision must be made whether to add redundant valves or to leave the arrangement as is. The answer depends upon the system goals. (It should be stressed that the probability for valves is considerably higher than shown in the examples.)

Arrangement	Failure rate, λ_T	Mean time between failures, h	Mean downtime, θ_T, h	Probability of operation, %
One valve 	0.01	100	10	90
Two valves in series 	0.02	50	10	81
Two valves in series with one in parallel 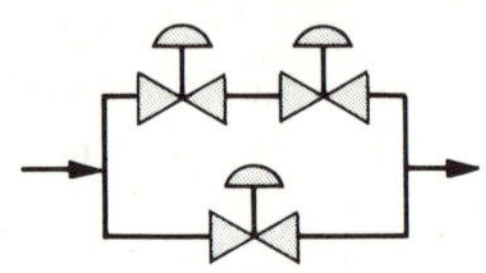	0.004	250	5	98.1
Two valves in parallel 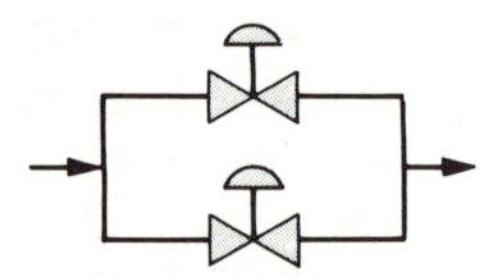	0.002	500	5	99

Redundant valves can increase system reliability **Fig. 4**

Availability analysis

In design work, we must not only analyze for failure but also for the time required to fix a failed valve so that the process can operate again. Therefore, our analysis must consider the time that it takes to get the valve repaired and back online.

For this analysis, both the failure rate ($\lambda_1, \lambda_2, \ldots$) and the mean repair time ($\theta_1, \theta_2, \ldots$) are used. For our examples, failure rates are assumed to be constant, and the repair is started as soon as failure occurs.

For two valves in series:

$$\lambda_T = \lambda_1 + \lambda_2 \qquad (3)$$

$$\theta_T = \frac{\lambda_1\theta_1 + \lambda_2\theta_2}{\lambda_1 + \lambda_2} \qquad (4)$$

For two valves in parallel:

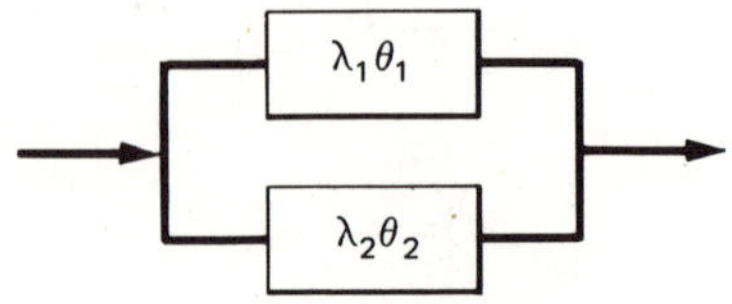

$$\lambda_T = (\lambda_1\lambda_2)(\theta_1 + \theta_2) \qquad (5)$$

$$\theta_T = \frac{\theta_1\theta_2}{\theta_1 + \theta_2} \qquad (6)$$

For our discussion, let us assume that each valve has a failure rate of 1% of its operating time, and that 10 h are needed to repair it. Thus, in a 100-h cycle, the valve could fail once and it would take 10 h to repair.

For the two valves in series, we find λ_T and θ_T by substituting into Eq. (3) and (4), respectively:

$$\lambda_T = 0.01 + 0.01 = 0.02$$

$$1/\lambda_T = 1/0.02 = 50 \text{ h}$$

$$\theta_T = \frac{(0.01)(10) + (0.01)(10)}{0.02} = 10 \text{ h}$$

Two valves in series result in a mean time between failure of 50 h, and a mean downtime of 10 h.

If the two valves are in parallel, the mean time between failure becomes 500 h, a considerable improvement. The mean downtime becomes 5 h.

For the system of two valves in series with one valve in parallel (Fig. 4), the mean time between failure becomes 250 h, and the mean downtime is 5 h.

These examples show how redundant valves can increase the success of a system. Each case must be analyzed on the specific objective of the system in order to ensure that excess equipment is not installed.

Fail-safe considerations

Valves have an additional function aside from their control assignment, in that they must fail-safe to meet operating and safety requirements. Fail-safe is defined in this discussion as the position of the valve following the loss of the operating medium, whether it be air, electronic or electrical.

The fail-safe operation depends on the type of process. In chemical processes, the essential objectives upon failure of the operating medium are to shut off the feed, eliminate the source of heat energy, and reduce operat-

Ratios of control-valve costs for 6-in. globe body **Table I**

Body construction and rating	Single-seated,* top-guided	Double-seated,† top- and bottom-guided
Iron, 125 psi, standard construction, 150 psi ΔP shutoff	1.1	1.0 (base)
Steel, 150 psi, standard construction, 150 psi ΔP shutoff	1.7	1.6
Steel, 300 psi. standard construction, 150 psi ΔP shutoff	1.7	1.6
Chrome-moly steel, 300 psi. Construction for 900° F, Extension finned-bonnet Stellited seat joint Stellited guide bushings Stellited plug-guide posts	2.9	3.0
Steel, 600 psi, 1,200 psi ΔP, 60° F, 0.5% C_v Maximum leakage Stellited seat-joint	2.1	1.8
Steel, 600 psi, 1,200 psi ΔP, 60° F, tight shutoff Stellited seat joint High-pressure piston actuator (90 psi) Not fail-safe	2.1	–
Chrome-moly steel, 600 psi, 1,200 psi ΔP, 600° F Pilot-operated, balanced design, reliable fail-safe action	2.8	–

*With diaphragm actuator of 200 in.^{2}
†With diaphragm actuator of 145 in.^{2}
Source: Hutchison, J.W., ed., "ISA Handbook of Control Valves," 2nd ed., Instrument Society of America, Pittsburgh, 1976.

ing pressure. Failure of the control valve to accomplish these objectives could cause chemical reactions to increase in rate, with resulting increases of heat and pressure. This would eventually cause loss of valuable products, or send wastes through the relief system, and possibly damage equipment by means of burnout, hot spots, coking or rupture. There is also the possibility of injury to personnel.

Fig. 5 shows various examples of fail-safe action. Note that a symbol is used that clearly shows the fail-safe position of the valve. This symbol is used on system descriptions and loop diagrams, so that the intent of the design is fully understood.

Specification techniques

While there can be no compromise with the engineering of the valve, other considerations can influence the final cost. Some of these factors must be evaluated

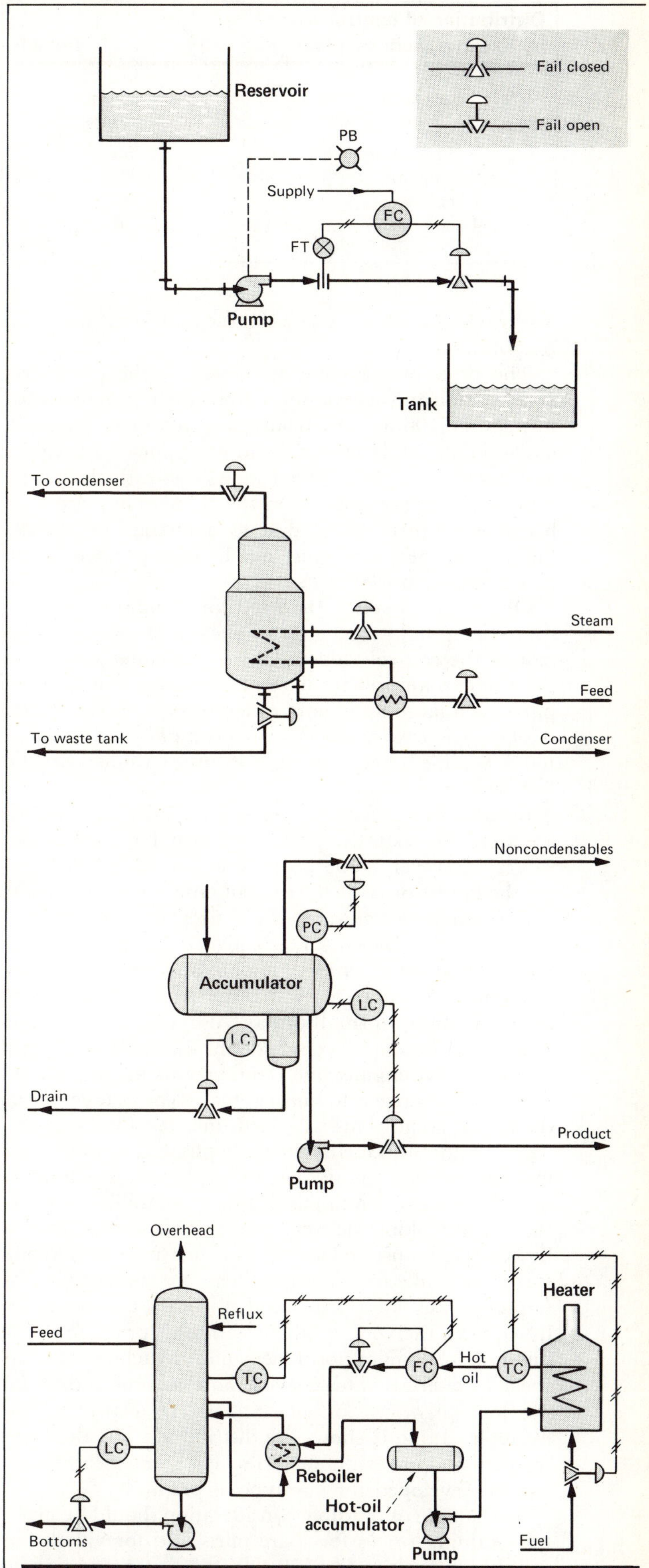

Fail-safe position for control valves must be clearly shown in flow diagram **Fig. 5**

Distribution of control valves by size in a chemical plant — **Table II**

Size range, in.	Amount, % of total
1 1/2 and under	65
2 and under	83
3 and under	91
4 and under	96

during the specification phase if the optimum valve is to be obtained.

The price of the valve is related to the type, size, construction and accessories. Valves can be purchased for less than \$100 to practically any amount—in some cases, \$250,000. However, in most applications, valves cost less than \$2,000. Table I shows cost-ratio data for a typical 6-in. globe valve, and the changes in price as it becomes more complex due to a change in service. Obviously, the valve must not be overspecified if the cost is going to match its use.

Often overlooked is the need for standardization of spare parts and maintenance. These may override first cost of the control valves. This is particularly true for the single plant, where it would be expensive to stock parts for a large number of different-type valves. When stocking (inventory) is done on a companywide basis, a more flexible policy on unique installations may be followed.

Budgets are prepared long before individual valves are sized, so that list prices are used for costing the valves. However, some procedures can be followed to get the lowest price at the time of bidding, which again may occur long before the final sizes are selected.

Competitive bidding is effective if there is an adequate specification and a full bid-analysis done. Furthermore, a vendor that wants the order will bid competitively even on small orders. Also, allow the vendor to suggest alternative constructions that best fit his line and still meet engineering requirements for the valves.

A good practice is to lump as many valves together in the bid package. This will result in quantity discounts, where available. During the early phase of a project, an inquiry and order may have to be placed on a preliminary basis in order to ensure the availability of items that require long delivery times. This means that a bid-package must be written and later amended. While this sounds like bid-guessing, there are successful techniques to use. Base prices and discounts can be established with the vendor, along with unit prices for add-ons and deletions of positioners, limit switches, trim and special materials. This is even more feasible if data on previous plant use of such valves are available. For example, Table II shows the distribution of valve sizes in chemical processing facilities. The same information can be developed for any type of plant.

At the time of bidding, consideration should be given to obtaining prices for spare parts and for valves—in suitable quantities (say, 10, 20 or 30)—in addition to those stated in the specification. Pricing based on such possible future purchases could have an influence in the selection of the vendor, and will prevent a low bidder from making it up on extras. A certain percentage of any large order should be for spare valves, in order to prevent delays at the plant when emergency replacements are required due to design changes or failures.

Generally, pricing will be competitive unless many unique features such as exotic trim, nonstandard materials, fail-safe capacity tanks, solenoid shutoff valves and valve-position switches are required. Then, the vendor may no longer be able to use mass-production techniques that save money.

What about the low bidder?

The bid analysis should fully explore all facets of the bid to make sure that the low bid is valid. To ensure that the vendor understands all of the requirements of the specification, a vendor review meeting should be held to go over all engineering requirements in detail. Also, the shop can be visited to make sure that the vendor has the manufacturing capacity for the order.

Ideally, a maintenance history would exist on some other installation so that the durability of the product can be evaluated. A complete analysis might indicate the selection of someone other than the low bidder if the facts proved that other costs really have made his low bid a high.

Valve delivery

While the delivery of a valve would seem to be automatic if a reputable bidder has been selected, there are some factors to consider. A standard valve can take six to eight weeks for delivery if it is in normal production. Special requirements such as exotic materials and the nuclear code "N" stamp can result in a delivery date of over a year. In fact, the valve can easily be the critical item in the construction schedule.

One of the first things to do is to make sure that the vendor can meet the delivery, as promised, before the order is placed. Shop loads can be checked. If the order is large, the purchaser's expeditor can follow the order through the vendor's shop to make sure the parts are not diverted to a larger order for another customer. No reputable vendor minds such expediting. In fact, substitutions can often be made that ensure faster delivery because of close coordination between the purchaser and vendor.

Lastly, once the specification is issued, it should not be changed by the purchaser in any way if the delivery dates are to be met.

The author

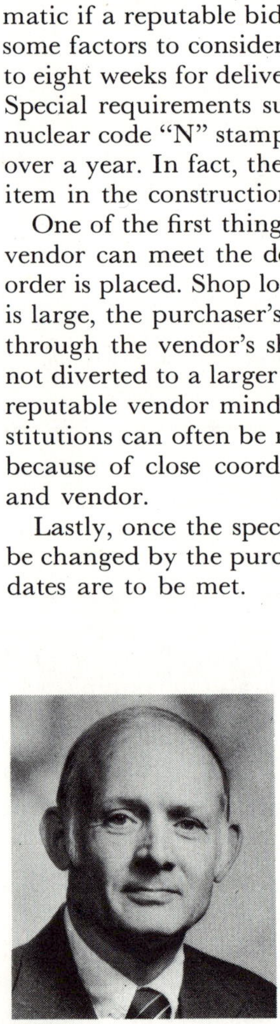

E. Ross Forman is supervisor of the power division for United Engineers & Constructors Inc., 30 S. 17 St., Philadelphia, PA 19101. He joined United in 1971 and is responsible for all phases of instrumentation and controls. Previously, he was employed for 12 years by Catalytic, Inc. as chief instrument engineer. He has a B.S. and an M.B.A. from Drexel University, and is a member of ASME and Instrument Soc. of America. He is a professional engineer in N.J., Pa. and N.C.

How To Estimate Pressure Drop Across Liquid-Control Valves

Engineers doing preliminary designs for equipment procurement must often estimate pressure drops before control valves are finally chosen. Here is a technique that is easy to use and that gives good results.

HANS D. BAUMANN, Masoneilan International, Inc.*

Pumps are usually ordered early in an engineering project—before the control valves have been specified. But in order to make an intelligent pump selection, the engineer must be able to estimate the pressure drop across the control valves in the pumped fluid's path.

Various rules-of-thumb have been proposed for this purpose, such as assigning 33% of the dynamic pipe-friction loss, or selecting a 15-psi minimum drop [*1*].

While these methods work reasonably well, they can lead to the selection of an unnecessarily large pump, thereby wasting energy in the form of excess horsepower. This article suggests an alternative method—the selection of valve pressure-drop as a function of two parameters that should already be known to the process engineer: the pipe velocity at maximum design flow; and the maximum pipe frictional loss and pump droop (reduction in pump head between zero and maximum design flow).

The total-dynamic-head change determines the relative valve capacity and, therefore, the valve style (if one wants to stay within certain pressure-drop ratios as far as installed characteristic and efficiency are concerned), while the maximum pipe velocity sets the pressure-drop across this valve. The accompanying graphs and tables make the parameters easy to determine.

Line Velocity Determines Valve Pressure-Drop

Determining maximum line-velocity has to be done early in order to specify the pipe diameter for the known maximum design-flow volume that the system will handle, and in order to calculate the pressure drop across pipe and fittings.

There seems to be no "universal" rule. While the tendency in the past 20 years has gradually been to increase liquid line-velocity for the sake of low installation cost, the reversal of this trend has recently been seen due to concern for hydrodynamic noise. For example, an 8-in Schedule-40 pipe will radiate above 90 dbA at 3 ft from the pipe, due to normal turbulence in elbows, fittings, etc., if the line velocity exceeds about 24 ft/s. As a result, one major engineering company has made a general rule not to exceed 7 ft/s as maximum line-velocity for liquids.

I believe a more flexible rule would be in order, assigning higher velocity to smaller piping systems and vice versa, since the sound pressure level seems to vary as 20 times the log of pipe diameter; i.e., for a given velocity, a 10-in pipe will tend to be 20 db louder than a 1-in pipe with the same wall thickness.

Here are some other suggested limits currently in use: Crane Technical Paper #410 recommends for boiler-

Typical C_v/D^2 Values (Nonchoked conditions) at 90% Open — Table I

Valve Type	0.4 Factor Reduced Trim*	Pipe/Valve Diameter Ratio 1	1.5	2
Single-seated globe	4	10	4.2	2.3
Split-body valve	4.3	10.8	4.6	2.5
Double-seated globe and eccentric rotary plug (Camflex)	5	12.6	5.4	3
High-capacity cage 1 to 3-in. Butterfly, 60 deg open	6.4	16	7	3.4
Y-style and angle, flow-to-close	7.2	18	7.2	3.9
Contoured ball		25	8.8	4.5
Butterfly, full-open low-torque type		32	12	5.2

*For line-size valves

*Since writing this article, the author has left Masoneilan to become a consultant—see "Meet the Author" at the end of the article.
This article is based on a paper given at the 29th Annual Symposium on Instrumentation for the Process Industries, sponsored by the Dept. of Chemical Engineering, Texas A&M University, College Station, Tex., Jan. 16–18, 1974.

Originally published April 29, 1974

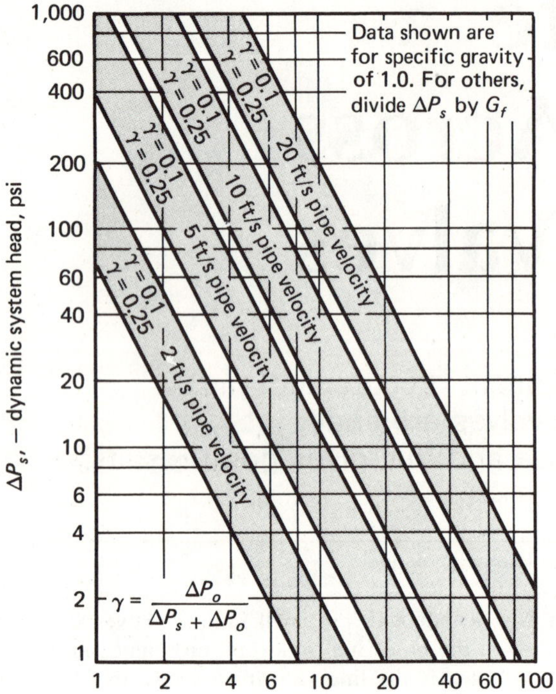

RELATIVE control-valve capacity determination—Fig. 1

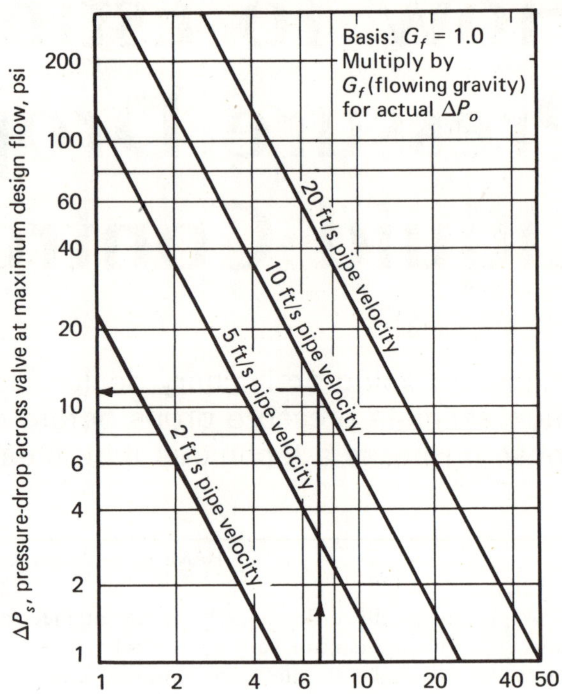

VALVE pressure-drop at maximum design flow—Fig. 2

feed application, 8 to 15 ft/s; pump suction, 4 to 7 ft/s; general service, 4 to 10 ft/s; and city water, up to 7 ft/s. Major engineering contractors specializing in paper mills allow pipe velocities up to 20 ft/s, while a well-known pump company suggests 3 ft/s maximum for suction and 5 ft/s maximum for discharge velocity.

The selected maximum pipe-velocity indirectly determines the pressure drop across the control valve once a specific valve type is selected. Rather than be concerned with the actual valve diameter and trim size, I feel that it is better at an early stage of the project to select only a style of valve with a known "relative valve capacity" (taken here as C_v/(pipe dia.)^2) in order to relate the valve capacity to the pipe size rather than the body size. The valve type may logically be chosen because of the restrictions of the system (see Fig. 1 and Table I) or may be selected on the basis of experience with similar systems. The use of such a "relative valve capacity," C_v/D^2, leaves the instrument engineer full choice later on to select line-size valves, reduced-trim valves, or full-ported valves installed between reducers.

The aforementioned relationship between line velocity, U, and valve pressure-drop, ΔP_o, (at maximum design flow) can be expressed as follows:

$$\Delta P_o = G_f V^2/C_v^2 \quad \text{(psi)} \tag{1}$$

wherein $V = 60 \times$ pipe area (ft^2) $\times U \times k$* (gpm) (2)

ΔP_o could therefore be expressed as:

$$\Delta P_o = \frac{6U^2 G_f}{(C_v/D^2)^2} \quad \text{(psi)} \tag{3}$$

For detailed development of this equation, (see box at left). Fig. 2 shows the results, indicating the dependency of valve pressure-drop on the square of the maximum pipe velocity and also the square of the relative valve capacity.

Development of Eq. 3

From the basic C_v equation (see ISA Standard SP39.1):

$$\Delta P_o = \frac{G_f V^2}{C_v^2} \text{ (psi)}$$

$$V = 60(D^2\pi/4)(Uk/144)$$

$$V = 2.45\, U\, D^2 \text{ (gpm)}$$

Therefore: $$\Delta P_o = \frac{G_f\, 2.45^2\, U^2\, D^4}{C_v^2}$$

and $$\Delta P_o = \frac{6\, U^2\, G_f}{(C_v/D^2)^2} = \text{(psi)}$$

Concern for Installed-Valve Characteristic

From the data in Fig. 2, one could quite easily select the valve pressure-drop, assuming a C_v/D^2 for a favorite style of valve from the information given in Table I [*2*]. Incidentally, the information in Table I incorporates a safety factor of 10% suggested by R. Moore [*1*], allowing for general manufacturing tolerances between the stated catalog C_v and a given valve's actual flow-capacity.

While this simplified method works, a better argument would be to select the relative valve capacity (and,

*$k = 7.48$ gal/ft^3, D = pipe dia., in.

Pressure-Drop Selection Table – Table II

Valve Type	D/d	2 Ft/s ΔP_S*	2 Ft/s ΔP_O*	5 Ft/s ΔP_S*	5 Ft/s ΔP_O*	10 Ft/s ΔP_S*	10 Ft/s ΔP_O*	20 Ft/s ΔP_S*	20 Ft/s ΔP_O*
		Pipe Velocity at Maximum Design Flow							
Single-seat globe	1	0.8	0.24	14	1.5	54	6	216	24
	1.5	12	1.3	77	8.5	308	34	1224	136
	2	42	4.7	260	29	1050	116	4180	465
Ditto with 0.4 factor trim →	1	14	1.5	86	9.5	342	38	1370	152
Split-body valve	1	1.8	0.2	12	1.3	47	5.2	189	21
	1.5	10	1.1	63	7	252	28	1010	112
	2	34	3.8	216	24	865	96	3460	385
Ditto with 0.4 factor trim →	1	10	1.1	63	7	252	28	1010	112
Double-seated globe and eccentric rotary plug valve (Camflex)	1	1.4	0.16	9	1	36	4	144	16
	1.5	7.5	0.83	47	5.2	189	21	755	84
	2	24	2.7	153	17	612	68	2450	272
Both with 0.4 factor trim →	1	23	2.5	54	6	216	24	865	96
High-capacity cage valve 1-3 in. size, butterfly 60-deg. open	1	0.9	0.10	5.4	0.6	22	2.4	87	9.6
	1.5	4.5	0.50	28	3.1	112	12.4	450	50
	2	19	2.1	117	13	468	52	1870	208
Cage valve with 0.4 factor trim →	1	5.4	0.60	33	3.7	1.35	15	540	60
Y-style and angle valve, flow-to-close	1	0.5	0.06	3.6	0.4	14	1.6	58	6.4
	1.5	4.3	0.48	27	3	108	12	432	48
	2	14	1.6	90	10	360	40	1440	160
Both with 0.4 factor trim →	1	4.2	0.47	26	2.9	108	12	432	48
Contoured ball	1	0.4	0.05	2.7	0.3	11	1.2	43	4.8
	1.5	2.7	0.30	17	1.9	69	7.6	270	30
	2	12	1.3	72	8	288	32	1150	128
Butterfly, full open	1	0.2	0.02	1.2	0.14	5.4	0.6	22	2.4
	1.5	1.4	0.16	9	1	36	4	144	16
	2	7.7	0.86	49	5.4	198	22	792	88

*Figures denote Max. ΔP_S, (i.e., droop in pump cahracteristic and pipe friction losses for $\gamma = 0.1$) and corresponding valve pressure drop ΔP_O.

thereby, indirectly the type of valve) from two other criteria, namely:

a. Conserving pump horsepower.

b. Keeping the system gain as linear as possible by limiting the distortion of the inherent valve characteristic.

If one assumes as an arbitrary limitation that under (a) the efficiency of the system will be detrimentally affected if the control-valve pressure drop exceeds 25% ($\gamma = 0.25$) of the total dynamic head (pipe friction plus pump droop plus valve pressure-drop)—and, on the other hand, assumes that the installed valve characteristic (even with a percentage plug) becomes too nonlinear if the valve pressure-drop is less than 10%* of the dynamic head loss ($\gamma = 0.10$)—then the choice of available C_v/D^2 becomes very narrow indeed, as can be seen from Fig. 1. For example, the calculated pipe-friction loss in a given piping system with fluid having a specific gravity of 1 comes to 50 psi. The droop in pump characteristic is estimated to be 28 psi, which gives a ΔP_s of 78 psi. Assuming a pipe velocity of 10 ft/s, one has a choice from Fig. 1 between a "relative valve capacity" of 8 C_v/D^2 (limitation of installed characteristic), or a C_v/D^2 of 4.7 (valve

*Except for rotary valves with high inherent rangeability (> 100:1), where a limit of 5% is suggested.

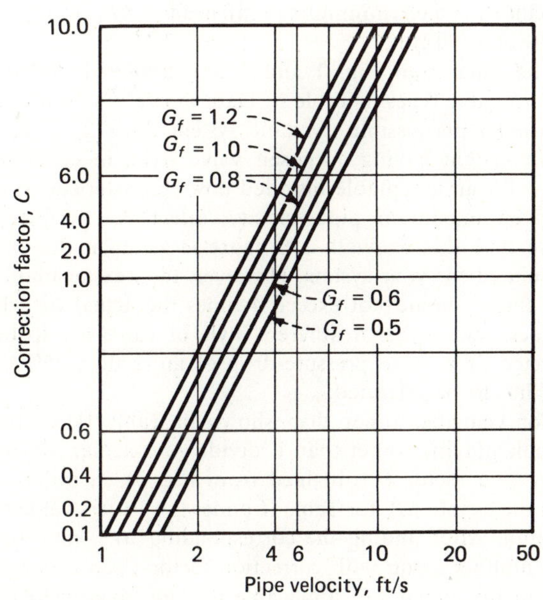

CORRECTION factors for velocity and sp. gr.—Fig. 3

Nomenclature

C	Correction factor (see Fig. 3)
C_v	U.S. gal/min per 1-psi pressure-drop across valve
D	Pipe diameter, in.
G_f	Specific gravity under flowing conditions (water at 60°F = 1)
k	Conversion factor, 7.48 gal/ft^3
P_o	Pressure drop across valve at maximum design flow (see Fig. 7), psi
P_s	Pressure drop caused by pipe friction and drop in pump head (see Fig. 7), psi
S	Actuator signal
U	Pipe velocity at maximum design flow, ft/s
V	Flowrate = $(60\pi D^2/4)(Uk/144) = 2.45\ UD^2$, gal/min
γ	Ratio of valve pressure-drop at maximum design flow to valve pressure-drop near zero flow

pressure-drop exceeds 25% of system drop below this capacity). From Table I, we have a choice between:

Camflex valve or double-seated globe with 0.4 factor trim. $C_v/D^2 = 5$	Camflex valve or double-seated globe with 1/1.5 line size. $C_v/D^2 = 5.4$
High-capacity cage valve, 0.4 factor trim. $C_v/D^2 = 6.4$	High-capacity cage valve, 1/1.5 line size $C_v/D^2 = 7$
60-deg., open butterfly, 1/1.5 line size. $C_v/D^2 = 7$	Y style or angle valve, 0.4 factor trim or 1/1.5 line size. $C_v/D^2 = 7.2$
Contoured ball valve, ½ line size. $C_v/D^2 = 4.5$	Full-open low-torque butterfly valve, ½ line size. $C_v/D^2 = 5.2$

Assuming a 60-deg open butterfly valve is selected (installed between reducers and giving a C_v/D^2 of 7), and turning to Fig. 2, the valve pressure-drop is found to be 12 psi, which can now be added to the pump specification data (minimum pump head = $\Delta P_s + \Delta P_o$ = 78 + 12 = 90 psi).

The data from Fig. 1 and 2 are tabulated directly against valve types in Table II. Data shown represent the *maximum* permissible dynamic system drop, ΔP_s, in a given system having a chosen valve style, in order to allow for an acceptable installed flow-characteristic. For a given maximum pipe-velocity, select the preferred valve type that shows a ΔP_s figure *larger* than the one calculated for your system. However, the next available ΔP_s figure should not exceed 4 times the actual ΔP_s. If it does, then either a different style of valve has to be selected or a valve pressure-drop of more than 25% of ΔP_s has to be tolerated.

For velocities other than shown in Table II, or for specific gravities other than 1, divide your actual ΔP_s by correction factor C (obtained from Fig. 3), in order to find the usable ΔP_s for Table II under the 10-ft/s velocity column. After finding the corresponding ΔP_o, multiply this number again with correction factor C, in order to obtain the actual ΔP_o applicable to your given velocity and specific gravity conditions.

Example: $G_f = 0.8$, velocity = 9 ft/s

Estimated $\Delta P_s = 80$ psi, preferred valve—single-seat globe.

1. Correction factor from Fig. 3 for $U = 9$ ft/s and G_f of 0.8 = 0.65.
2. ΔP_s for use in Table II = 80/0.65 = 123 psi.
3. Table II gives next highest ΔP_s of 308 psi for globe valve installed between reducers 1.5/1 pipe/valve-dia. ratio and the 10-ft/s reference velocity; the corresponding $\Delta P_o = 34$ psi.
4. Actual $\Delta P_o = 34 \times C = 34 \times 0.65 = 22.1$ psi.
5. Specify minimum pump head to be 80 psi (ΔP_s) + 22.1 psi (ΔP_o) plus static head if any.

The ΔP_s given in Table II is purposely limited to a maximum dynamic head loss for a γ ratio of 0.1. This really reflects the main criterion that the process engineer should shoot for, namely: What is the lowest possible pressure drop the valve will tolerate (in order to do a satisfactory control job) for the sake of reduced pump size and conservation of pumping horsepower? Modern control valves have no mechanical requirements that call for a certain minimum, internal pressure-drop for reasons of stability, hysteresis or otherwise. Therefore, the only real need for such minimum pressure-drop is the maintenance of a reasonable "installed characteristic" (the gain of the final control element in regard to the control system). The gain of the control is defined as the slope of the installed valve characteristic.

In most systems, the ideal control-valve gain should be constant—i.e., a linear relationship should exist between actuator signal and system flow. Only in loops where the transmitter input-signal is nonlinear, such as from orifice-type meters, is there a need for nonlinear control-valve gain. Nevertheless, with the limited choice of available control-valve characteristics, too great a distortion even here can be anticipated if the γ ratio (valve drop at full flow divided by the total system pressure-drop) is less than 0.1.

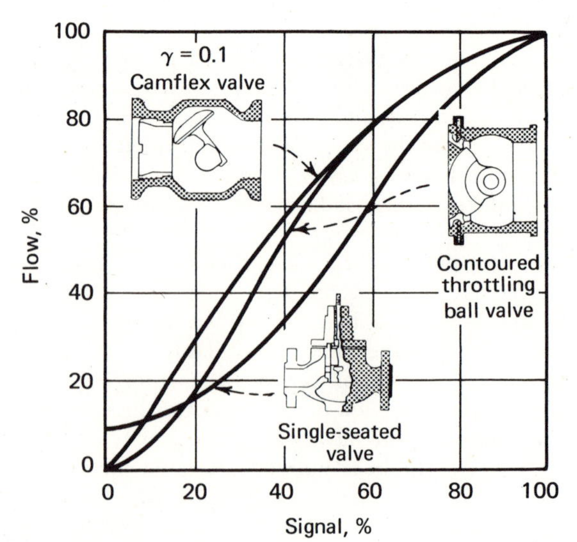

INSTALLED CHARACTERISTIC for three typical valve styles, all passing identical flowing quantities with equal percentage characteristic—Fig. 4

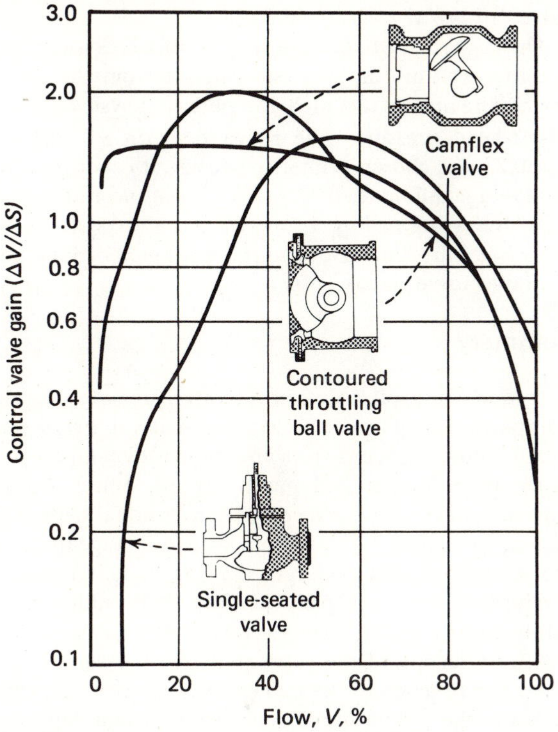

INSTALLED GAIN for three valve styles all having equal % inherent characteristic and identical design flow—Fig. 5

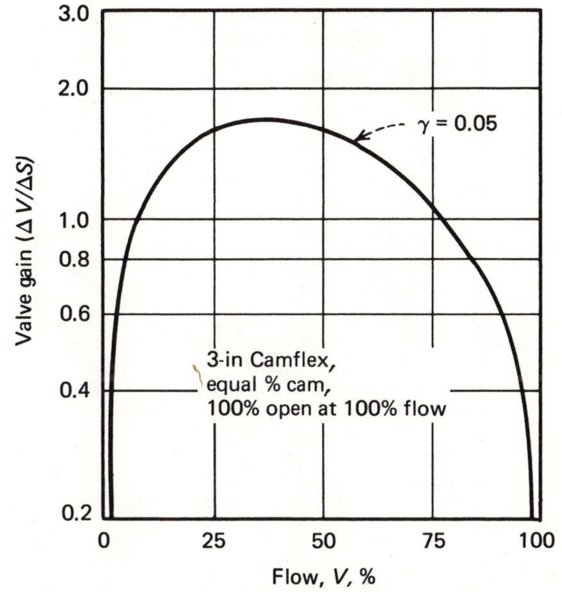

INSTALLED GAIN characteristic of 3-in Camflex valve with ratio of only 0.05 (pressure drop wide open only 5% of drop near closed position)—Fig. 6

To illustrate this point, actual installed characteristics are shown for three commonly used valve styles in Fig. 4. The valves chosen are a single-seated globe valve, at 90% open; a rotary valve with eccentric plug, 75% open; and a throttling contoured ball-valve installed between reducers 1/1.5 line size. All have an equal-percentage characteristic and are used for identical flow capacities.

As can be seen, the inherent limitation in rangeability of the globe-style control valve is very apparent and control is impossible below 10% of flow. On the other hand, modern rotary style valves do have both a superior installed characteristic and an actual usable rangeability to around 1% of system flow (due to their high *inherent* rangeability).

Perhaps a more meaningful presentation is to plot "gain" against "percent of flow" (data shown in Fig. 5). If one arbitrarily assumes that a given control system would yield stable and acceptable control with a control-valve gain varying between 0.5 and 2 (i.e., ± 2:1), then the eccentric rotating plug-valve could operate from near zero to 95% flow, the contoured ball valve from between 2% to 95% flow,* and the globe valve only from 21% to 99% flow. The data represented here indicate that, depending on the particular installation, a γ ratio of 0.1 might already be too low for typical globe valves, while modern rotary control-valves still give acceptable gain characteristics with γ ratios even down to 0.05—i.e., valve pressure drops of only 5% of total system head-change (see Fig. 6).

*The ball valve is somewhat handicapped by the added distortion of the "inherent characteristic" by adjacent pipe reducers, typical of all "high C_v" valves.

Working Example

As an example: In a typical depropanizer application, a pump has to circulate propane into the reflux tower at an elevation equivalent to 20 psi; the maximum drop through pipe and fittings at the maximum design flowrate amounts to 18 psi; the droop in pump head is estimated to be 10 psi. The total dynamic system loss, ΔP_s, is therefore 28 psi. The pipe velocity is calculated at 10 ft/s. Since the specific gravity is 0.5, enter 0.5×28 psi = 14 psi ΔP_s on the left-hand side of the graph of Fig. 1 and find that for 10 ft/s we could select C_v/D^2 between 11 and 19, while staying within the recommended limits. Since this is a relatively small installation, a single-seated globe valve is chosen.

Using a C_v/D^2 of 10 from Table I, we find, from Fig. 2, a pressure drop of 6 psi for water. Multiply by a sp. gr. of 0.5 to give an actual drop of 3 psi for the control valve. The total pump head that has to be selected is now 28-psi dynamic loss + 3-psi valve drop + 20-psi elevation, making the minimum required head 51 psi.

Including some safety factor, the next standard pump size to choose has a head of 60 psi, which then should be specified. Note that the process engineer should *add* this extra head to the valve pressure-drop in his specifications, which would then be used by the instrument engineer to select the final valve. The *actual* pressure-drop for sizing purposes (as seen by the instrument engineer for final selection) would be 60 psi minus an assumed safety factor of 10% (6 psi) and minus 48-psi dynamic loss and elevation, which equals 6 psi. This increase in actual pressure drop changes the required valve capacity by the square root of the ratio of original-to-actual pressure-drop. The required capacity therefore is 10 C_v/D^2 multiplied by $\sqrt{3/6}$ or 7.1 C_v/D^2. This means that either

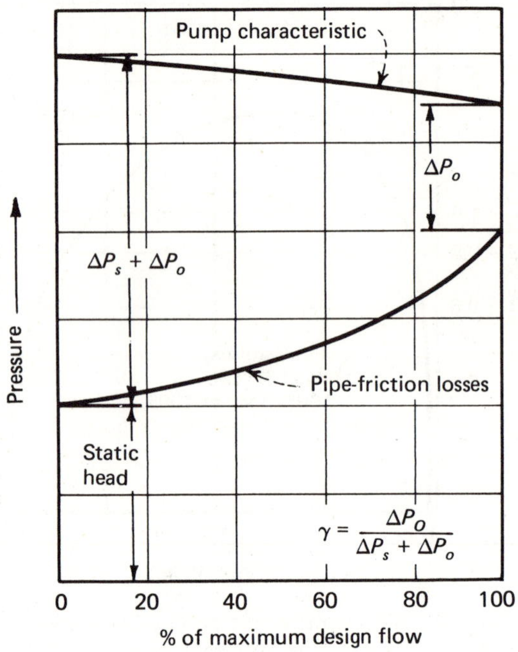

HEAD-CHANGE schematic for typical pumping system—Fig. 7

the originally selected valve will operate at 70% stroke, or a different valve style (smaller valve size) will have to be selected.

The most desirable inherent flow-characteristic is dependent upon the γ ratio. In this case, the γ ratio = $\Delta P_o/(\Delta P_s + \Delta P_o) = 6/34 = 0.177$, indicating that an equal-percentage characteristic is most desirable.

It should be noted that safety factors and "round off" on pump head always *increase* the final valve drop. This is one more reason to prudently select a γ ratio as low as possible (i.e., small valve-pressure-drop to begin with).

Economic Considerations

As explained in the foregoing, the limitation for selecting the lowest possible valve pressure-drop (ΔP_o) is concern for the installed characteristic. The upper limit is really controlled by the desire for reasonable operating expenses (pump horsepower). Simon and Roby [5] have investigated the relationship between pump operating expenses and valve size (valve ΔP) quite thoroughly, coming to the conclusion that the wide-open pressure-drop across the valve (ΔP_o) should be limited to 15 psi for pumping systems of up to 47 hp.* Above 47 hp, the drop should be lower. While this study, conducted in 1960, did not take into account the availability of newer rotary-style control valves, it should still be fairly up to date. In order to understand the economics, one must realize that it costs $0.08 per year for each gallon pumped against 1-psi pressure-drop (based on 60% pump-motor efficiency and $0.012 per kW, specific gravity = 1).

Assuming a system with 500-gpm average flow, the yearly cost for 10-psi pressure-drop is $400; the savings realized over a period of 2 years would pay for the *installed cost* difference between a 4-in and 3-in control valve (10-psi ΔP_o vs. 20-psi ΔP_o).

* At less than 240-psi head.

The suggested upper limit for ΔP_o of 25% system drop is quite arbitrary and, as one can see from the above, might already be too high for pumping systems above 60-psi-head pressure. The range between $\gamma = 0.1$ and $\gamma = 0.25$ was chosen primarily because it covers roughly the average difference in C_v/D^2 between successive valve sizes. Referring to Fig. 1, one should always start from the $\gamma = 0.1$ line and move to the left to choose the next available valve size (C_v/D^2).

Summary

To find the applicable control-valve pressure-drop for pump-selection purposes, select the basic valve style (and relative flow capacity) from the known dynamic system head (pipe friction and pump droop). Select a valve capacity near $\gamma = 0.1$ (except for conventional globe style and angle valves if the desired *installed* rangeability or "turndown" should be larger than 5:1). This results in a relatively *low* valve pressure-drop. N.B. added pump safety factors always increase final valve pressure-drop.

After selection of the valve style, the pressure drop is given as a function of the known pipe velocity and, when added to the system losses plus elevation, can determine the required pump head.

This recommendation will lead to a more economical selection of control-valve pressure-drop, particularly when utilizing the advantage in wide rangeability of modern rotary-type control valves, which work well with a valve-to-system-drop ratio down to 0.05. Finally:

■ The calculations show that not all valve styles are interchangeably applicable for given flow conditions if one desires to stay within γ ratios between 0.25 and 0.01.

■ Conventional control valves are not very suitable for conserving pump horsepower, since their poor inherent rangeability allows only a limited "turndown" range due to drastic gain changes at low γ ratios.

References

1. Moore, R. W., Allocation of Valve Pressure Drop, Paper 6.1, First ISA Final Control Elements Symposium, Wilmington, Del., Apr. 1970.
2. Liptak, Bela G., "Instrument Engineers' Handbook, Vol. II, Process Control," Chilton Book Co., Philadelphia, p. 44.
3. Boger, H. W., The Effect of Installed Flow Characteristic on Control Valve Gain, ISA Paper 68-920, Oct. 1969.
4. Buckley, P. S., Selection of Optimum Final Element Characteristics, Proceedings of 5th National Chemical & Petroleum Instrumentation Symposium, Wilmington, Del., May 1964.
5. Roby, M. A., Simon, H., Pump and Control Valve Design Determined by Minimum Cost, ISA Paper 3-SF60, May 1960.

Meet the Author

Hans D. Baumann, 29 Villa Dr., Foxboro, MA 02035, is an independent engineering consultant specializing in automatic control systems, control valves and regulators. At the time of writing this article, he was Vice President—Technology for Masoneilan International, Inc., where he headed all U.S. and worldwide engineering activities, including R&D.

He is a member of American Soc. of Mechanical Engineers and Instrument Soc. of America, and is a registered professional engineer.

Reconditioning makes old valves "new"

A good valve reconditioning shop can take your old valves and upgrade them to like-new condition at perhaps half the cost of buying new valves. Here are data on what such shops do, and some tips on what to look for in choosing one.

Michael P. Nicholas, MCC Reconditioning

☐ An attractive method of conserving cash and energy is to recondition used valves to a "like-new" configuration. More and more companies are realizing that many valves can have several life cycles. By reworking or replacing worn parts, a valve-reconditioning shop can put what might have been a scrap valve back in service in a few days or weeks, at a fraction of the cost of a new one.

The overriding reason for valve reconditioning is saving money. Not only is the purchase of a more-expensive new valve avoided, but by reusing the same valve, one can avoid the extensive piping modifications that might be required to accommodate a new valve design. In addition, the turnaround times associated with valve reconditioning are in terms of weeks rather than the many months of lead time required for many types of new valves. New-valve lead times have increased because U.S. foundry capacity has trouble in meeting the demand for castings.

Valve reconditioning also affords the valve user considerable flexibility, in that reconditioning shops can modify configurations, convert trims, install bypasses, convert flanged to weld-end styles, and perform a great variety of operations that help avoid the expenses associated with purchasing special valves.

What is involved

Valve reconditioning, as outlined in this article, implies much more than simply cleaning a valve and repainting it. Valve reconditioning is a process whereby used valves are remanufactured to achieve operating performance characteristics identical to those of a new valve. This is ensured by testing reconditioned valves in accordance with new-valve specifications.

Typically, to restore the used valve:

1. The valve is dismantled and the components are cleaned, either by chemical means or blasting.
2. Component parts are thoroughly inspected. Decisions are made as to repair or replacement of worn parts.
3. Welding and machining operations build up worn parts or produce new parts. A stores area, with casting blanks and spares, supports the production of new parts.
4. The valve components are then reassembled using new packing and, if required, new bolting.
5. The assembled valve is tested in accordance with new-valve specifications.

Many valve reconditioners will provide a warranty on their reconditioned valves for the same length of time as a new valve.

In addition to the basic reconditioning process, some valve reconditioners offer a total valve maintenance program, which includes storing the customer's used valves, maintaining inventory-control records and storing finished popular-size valves at no charge.

Cost of reconditioning

Valve reconditioning typically costs less than 50% of new-valve cost. Specific charges will vary as a function of: how worn the used valve is; if special alloys are involved; number of replacement parts required; and modifications desired by the customer. The key factors in analyzing reconditioning costs are: the valve reconditioner's charge structure and the valve mortality rate (i.e., the proportion of valves that cannot be repaired).

Reconditioning charges may be structured variously:

- Time and material (T&M).
- Base charge plus parts.
- Base charge with standard parts included.

Under the T&M method, charges are based on hours of labor on the valve and the materials used. While this sounds appealing, it is very difficult to anticipate charges, verify accurate invoicing and compare invoices under this method. Also, T&M offers no real incentive for a job to be turned around rapidly, as the longer it takes to repair the valve the greater the charges will be.

Base charge plus parts can also be deceptive. While the base charge may appear to be a very low percentage of new-valve cost, the price of new parts is very high. This

Originally published May 22, 1978

The real costs of valve reconditioning

10 used valves; new-valve cost = $100	
Shop A charges $25 per valve to recondition	Shop B charges $20 per valve to recondition
Rebuilds each valve by welding, machining and installing new parts	Switches component parts among valves with little or no machining and welding
Saves 10 of 10 valves. Cost: $250	Saves 7 of 10 valves. Cost: $140
	Customer buys 3 new valves. Cost: $300
Total cost: $250	Total cost: $440

method is akin to offering a low sticker price on a car while adding charges for headlights and brakes.

The flat charge for valve reconditioning is the most beneficial to the customer. Not only does it allow for anticipation of costs and easy comparison of invoices, but it presses the reconditioner to maximum efficiency.

Mortality rate is the other key factor, i.e. what percentage of the valves sent for reconditioning are repaired and returned. A reconditioner who charges less but saves fewer valves may cost you more (see box).

It first appears from the box that Shop B is 20% cheaper than Shop A but, overall, Shop B actually costs the customer 40% more than Shop A.

Delivery and turnaround times

Turnaround times on standard, popular-size valves are typically two to four weeks from date of receipt. Larger size valves (8 in. and greater) may take four to six weeks and very large ones (e.g., 36-in. gate valves) may take four to eight weeks. Most reconditioners, though, offer emergency or shutdown service and can turn valves around in a matter of days when overtime is authorized. A valve reconditioner can typically repair a used valve in the larger-size ranges much quicker than a new valve can be obtained. On smaller sizes, some reconditioners will maintain a stock of repaired valves in popular sizes for the customer.

Shipping times to and from the reconditioning shop may be a factor, but many shops have their own trucks to facilitate quick pickup and delivery. In addition, the nationwide reconditioners have shops located in key areas of the country, so that shipping time between customer and reconditioner is typically less than two days.

Assessing a reconditioning shop

To help in analyzing a valve-reconditioning shop, the following points should be evaluated:

What is the shop's main business?—Is the operation a machine shop that also does valve-repair work, or does the shop concentrate on valves? A critical consideration in choosing a valve remanufacturer is his ability to respond to your emergency or shutdown requirements. A shop that concentrates on valve repair handles these types of emergency situations constantly and is sensitive to the implications of delay in turning valves around. A machine shop may not be as willing to break into runs presently going through the equipment.

Does the shop also buy and sell surplus valves?—If the facility is involved exclusively in valve repair, there is no incentive for the shop to determine your valve is beyond repair, in order to sell you a valve from its stock. Look for a shop that can ensure that you will receive only your own valves back.

How is the pricing structure?—If the shop has published prices by size and pressure classifications, it is much easier to foresee costs and compare invoices. Pricing on a "time and material" basis is necessary in those cases where the work is unique, but be cautious of a shop that does all of its work on T&M. Under such pricing, it is much harder for you to anticipate costs, compare invoices and check for inflated labor and material costs. You may avoid a shop that has what appears to be a high price scale, in favor of a T&M shop, only to end up paying more and never knowing it.

Also look at how such items as freight, replacement parts and specialty items are handled. The reconditioning shops affiliated with new-valve manufacturers are at an advantage here, as they can purchase replacement parts at a lower price than can an outsider.

Personnel—Take a close look at the qualifications of the facility's superintendent and sales manager. How extensive are their backgrounds in valve reconditioning? Do they have the expertise to get you out of an emergency situation when you need them? Also, will they get out of bed at 3 a.m. when you suffer an emergency shutdown?

Shop operation and work flow—How well is the shop organized? Are valve components segregated and identified properly?

Quality-control specifications—Review the shop's quality-control specifications with the shop superintendent. What are the guidelines for accepting or rejecting piece parts? Do the testing standards conform to industry specifications? In short, how will the shop ensure that your reconditioned valve is as good as new?

Summary

Valve reconditioning offers an opportunity to realize significant savings by recycling worn assets. There are many qualified and capable valve reconditioning shops throughout the country that can tailor a program to your requirements.

It is worthwhile to spend some time appraising the reconditioning sources available. Unfortunately, some shops do less than first-class work, and deceptive pricing practices are sometimes used. Valve reconditioning has advanced to the point that the technology exists to restore valves to like-new performance characteristics. Make certain that the reconditioner you deal with is one of the many that are dedicated to ensuring safety, performance and dependability.

The author

Michael P. Nicholas is Vice-President/General Manager of MCC Reconditioning, 1900 Dempster St., Evanston, IL 60204, a Unit of Mark Controls Corp., with responsibility for its six domestic reconditioning facilities. He studied mechanical and aerospace engineering at Illinois Institute of Technology and holds an M.S. Science degree in systems engineering from that institution. He is a member of the American Soc. of Mechanical Engineers and the American Petroleum Institute.

Choosing and applying line-blind valves

When pipelines must frequently be closed with absolute certainty ("blinded"), the specialized line-blind valve is fast and economical.

***Harry E. Farley**, Fluid Control Div., FMC Corp.*

☐ Consider this: you have a line carrying a potentially destructive material (gas, solid, or liquid) that must be closed off and rendered completely leakfree. The actual physical closing of the line is not difficult. The difficulty lies in being certain that the line is *positively* shut off. In a situation where guesswork may be tantamount to courting disaster, the line-blind valve is a logical solution.

The ordinary valve, by its very nature, cannot provide assurance of positive shutoff. Structurally, there is no absolute visual proof of complete closure in an enclosed valve, since breakage, fatigue or other damaging factors may have caused a failure that, in turn, could permit leakage.

Line-blind valves use the same basic principles as the older methods that will be described later, but they provide for a faster, cleaner, easier operation. There are three basic designs: one has individual body flanges held together by large bolt assemblies; the second has an enclosed chamber, slotted for the spectacle plate; the last one uses the gate-valve principle. They are available in a wide variety of types, sizes and special metals.

Where minimum leakage can be allowed and conditions are not hazardous, simpler line-blind valves can be used. (Leakage, in this context, refers to residual drainage of product from the line.) Where no movement or expansion of the line can be permitted during blinding, and where there can be no spillage, a more sophisticated design is indicated. (Spillage is a normal consequence of operating most line blinds.)

Traditional line-blinding methods

A variety of mechanical (non-valve) approaches to positive line blinding have been devised. Among these: solid plate between flanges; blank flange and spool; and double block and bleed, using a valve as a bleed drain. Each of these approaches has one or more disadvantages for many applications (see Table I). All are expensive and time-consuming. To overcome the drawbacks of these older approaches, a specialized family of line-blind valves has been developed.

From Table I, we can conclude the following:

Standard valve: expensive, no intrinsic assurance of absolute closure, can leak. *Solid plate between flanges:* inexpensive investment, but labor-intensive, time-consuming to operate. *Blanking flanges:* highly labor-intensive, time-consuming, cumbersome, heavy. *Double block and bleed:* expensive, no intrinsic assurance of absolute closure, can leak. *Line-blind valves:* positive, safe, fast, visible indication of condition, economical when total cost picture is compared.

Various manufacturers have developed line-blinding devices, and valves in sizes from ½ to 64 in., and with pressure ratings to more than 3,000 psi.

To understand the problem of line blinding, one must understand the technique. The preference for line-blind valves over the more cumbersome and expensive methods is self-evident. Let us consider some of these methods:

Solid plate between flanges—This requires removing some of the flange bolts and nuts and spreading the flanges to insert the blanking plate, then replacing the bolts and nuts to hold and seal the plate. It takes several pipefitters with tools to blind a line in this manner, and an equal amount of time to remove the blank and rebolt the flanges to open the line (Fig. 1).

Blanking flanges—This requires the removal of all bolts on two flanges, and a pipe "spool," and the installation of a blank flange or flanges. It involves two to four pipefitters with tools and 30 to 45 min. of time (Fig. 2).

Double block and bleed—This method, consisting of a "spool" with bleeder valve installed between two valves, is often considered blinding. However, it is neither 100% positive, as one cannot see into the valves to be certain there is no leakage, nor visible, as the bleeder valves may be closed or plugged and be ineffective (Fig. 3).

Plate-between-flanges vs. the 3-bolt line blind—This is an actual case history, based upon customer application of line-blind valves.

A 10-in. line must be blinded once each day. Previously, using a solid plate between flanges for blinding took two to four pipefitters and special tools to loosen all twelve flange bolts, remove six of these, and spread the flanges to fit the plate. Then the bolts had to be replaced, and all had to be tightened. Total time required was 30 to 45 min. per operation, on the average. A 3-bolt line-blind valve (Fig. 4) was installed. One laborer can operate the valve without special tools, by using a bar. Total time required for a single operation is 5 min. If you multiply the time savings over a year, you will see how the line-blind valve has paid for itself several times over in comparison to the mechanical method.

The most basic, most popular line-blind valve is the 3-bolt model. Instead of removing several flange bolts with

Originally published October 8, 1979

Comparison of various commonly employed line-blinding techniques **Table I**

Blinding technique	Protection: Completely leakfree	Protection: Absolute shutoff	Protection: Constant visible proof of closed/open condition	Economy: Five minutes or less to operate (typical)	Economy: Requires only one man to operate	Economy: No tools required to operate	Comments
Three- and five-bolt line-blind valves	Yes	Yes	Yes	Yes	Yes	Yes	Most common types of line-blind valves.
Sliding-gate line-blind valves	Yes	Yes	Yes	Yes	Yes	Yes	May be power-operated or may use handwheel actuator.
Visible-wedge line-blind valves	Yes	Yes	Yes	Yes	Yes	Yes	No line residue spill with this type, and no line motion required for blinding.
Solid plate between flanges	Yes	Yes	Maybe	No	No	No	Typically requires at least three pipefitters 5 to 6 times as long as line-blind valves for operation.
Blank flange and spool	Yes	Yes	Maybe	No	No	No	Greater time and labor problems than solid plate, owing to extra size and weight of spool.
Block and bleed	No	No	Maybe	Maybe	Yes	Yes	Expensive. Obstructed bleed can give false indication of closure.
Standard valves	No	No	No	Yes	Yes	Yes	Nature of valve design does not allow complete assurance of absolute seal nor provide failproof visible indication of closed/open condition.

tools, three bolts are loosened, one turn each, with a bar. This operation can be done by a single laborer. The spectacle plate is then repositioned and the three nuts tightened. When the hole in the spectacle plate is visible, it is obvious that the solid portion is blinding the pipeline; when the solid portion is visible, the line is open. The valve bolts are large enough to meet pressure ratings of flanges having many more bolts. The line bolts are interchangeable. Routine maintenance is limited to occasional thread lubrication.

The sliding-gate line-blind valve (Fig. 5) was designed for remote operation, or use in any application where blinding must be completed in a hurry but still must be secure. With either powered or manual operation, it is not affected by misalignment or stresses in the piping system. The sliding-gate line-blind valve can be operated with pressure on the line up up to 50 psi. Pneumatic, electric, hydraulic or manual actuators may be used with such a valve. No bolts or other fittings need be loosened or tightened prior to blinding, or opening the line. In either the open or the blind position, the gate is seated against a ring seal for leakfree operation. Some spillage will occur when the valve is opened or closed with product in the line. Sliding-gate line-blind valves may be used in either vertical or horizontal lines.

Several line-blind valves have been designed for specialized applications. Among these are the visible-wedge line-blind valve (Fig. 6), which provides either a full-bore opening or positive blinding, inside a seat in a three-quarter-enclosed body. No line residue can spill when

Installing a solid plate between flanges **Fig. 1**

Removing a spool piece so that blanking flanges can be installed **Fig. 2**

Basic data on line-blind valves **Table II**

Item	Notes
Working pressure	Range: 150-1,500 psi. 600 psi maximum "standard." Higher ratings are "specials."
Line size	Range: ½-64 in. "Standards" usually in ½-20 in. range.
Speed of operation	Range: 0.5 s to 5+ min. Depends on valve type and line size.
Line rigidity	Most require some line motion to operate.
Size compatibility	Some are designed for same laying length as standard valves.
Special materials	Wide selection available for body, seals.
End-connections	Commonly available: butt-weld, or flanged. Socket weld and slip-on are specials.
Non-spill	Non-spill types have to be drained periodically.
Remote control	Only type available is sliding-gate line-blind valve.
Blinding with pressure	Can be done with sliding-gate type (50 psi max.).
Block and blind	Special line-blind valves are available, or use a combination of standard fittings.

wedges are being changed, and no line movement is necessary when spectacles are changed. Spectacle units are wedge-shaped, and are raised by a handwheel or geared operator suitable for air or electric actuator or impact wrench. When the spectacle is raised sufficiently to clear the body, the safety latch is released and the spectacle may be rotated to the desired position. The wedge is then lowered into the mating seat section to form a metal-to-metal seal, augmented by seal rings.

The automated "block and blind assembly" is the merger of two popular pieces of equipment, with either one or two butterfly valves and a sliding-gate line-blind valve. It is operated by an air cylinder and provides automatic-sequencing remote operation.

The "fluid saver" line-blind valve has telescoping hubs that slide one into the other with internal packing between them. This design provides a closed chamber for the spectacle plate, to prevent spillage of line fluid while the plate is being reversed.

Double-block-and-bleed piping setup **Fig. 3**

A three-bolt line-blind valve **Fig. 4**

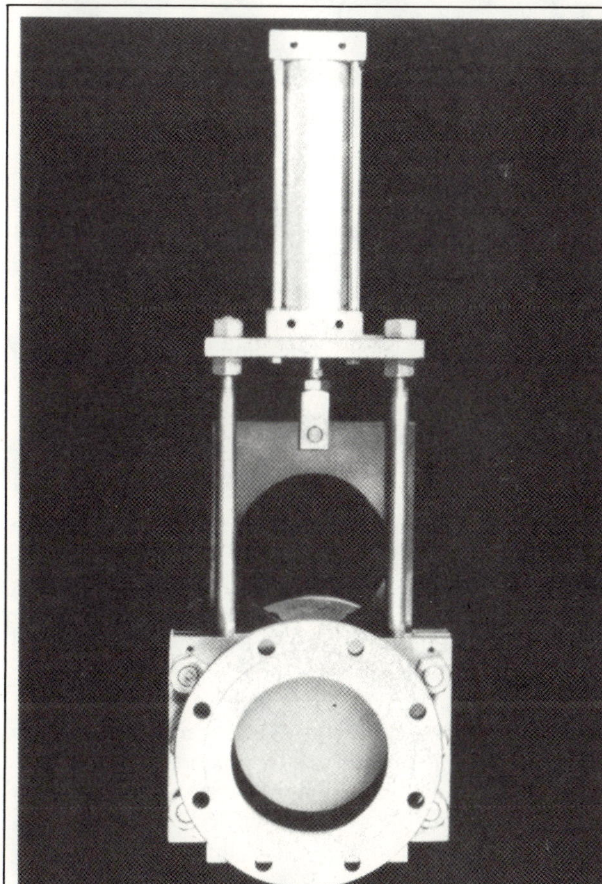

A sliding-gate line-blind valve **Fig. 5**

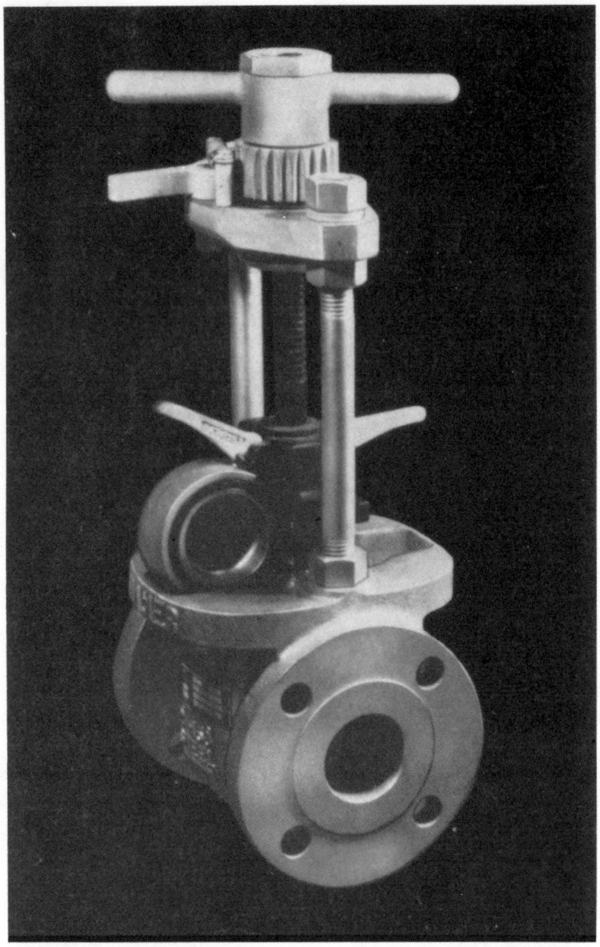

A visible-wedge line-blind valve — Fig. 6

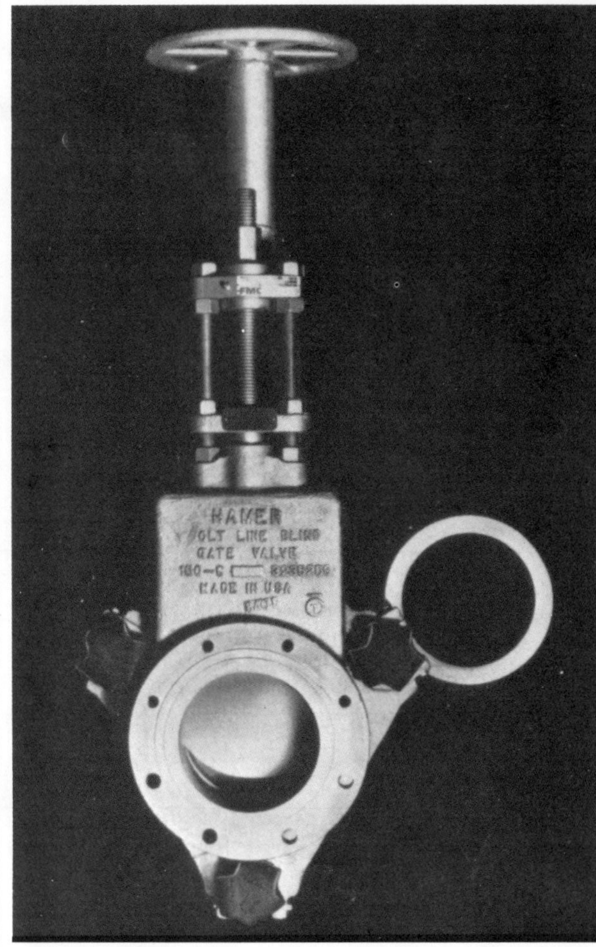

A double-block line-blind valve — Fig. 7

The line-blind gate valve is a combination of a standard 3-bolt line-blind valve with either one or two wedge-type gate valves. This enables the operator to block the line and then immediately blind it for safe, positive shutoff.

Single- or double-block line blinds consolidate the spectacle-plate features of the 3-bolt line blind and the independent wedge-gate operation into one compact unit. Single-block models shut off the line against pressure from one direction. Double-block models blind against pressure from either direction (Fig. 7).

Economy, in many instances, is the deciding factor between using line-blind valves and the older, more-time-consuming methods of blinding. Where blinding is only done infrequently, the older method may be the more cost-effective. But where frequent positive shutoff is called for, the line-blind valve is usually the better choice. Where there is the possibility of flammable or toxic liquids or gases seeping past a closure, or when there is the need to provide absolute assurance against product contamination due to commingling, line blinding is essential. Last, do not forget the economic factor, whereby seepage could cause losses of expensive product or environmental damage.

Summary

The cost of using inadequate line-blinding procedures can be very high—lost time, damage, even loss of life. Line-blind valves are not expensive. For example, an average 6-in. ball valve will cost approximately $986 compared with $465 for a comparable 3-bolt 150-psi line-blind valve. This differential increases with an increase in size, favoring line-blind valves.

In the final analysis, required frequency of operation may well be the deciding factor for selection. In a system where blinding is seldom required, cost of a line-blind valve may never be amortized. Conversely, if frequent, rapid blinding is required, it may take only a short time to gain back the initial investment.

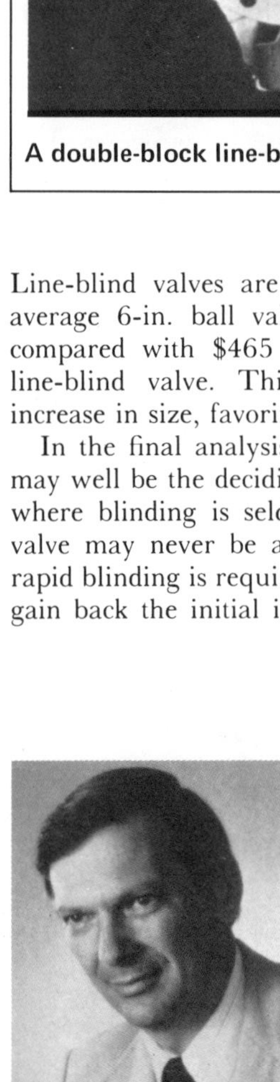

The author

Harry E. Farley is engineering manager, Fluid Control Div., FMC Corp., P.O. Box 19465-TR, Houston, TX 77024, phone: (715) 932-6362, where he is in charge of development engineering, product engineering and quality assurance. He holds a B.S. in mechanical engineering from the University of Houston, is a member of the American Soc. of Mechanical Engineers, Instrument Soc. of America and American Water Works Assn., and is a licensed professional engineer in Texas.

Section III
INSULATION

Cost-effective thermal insulation
Insulation saves energy
Guide to better insulation
The cost of missing pipe insulation
How to select insulation thickness for hot pipes
Economic pipe insulation for cold systems
Steam tracing of pipelines
Designing steam tracing
Sizing piping insulation
Electric pipe tracing
Floating balls insulate hot tanks

Cost-effective thermal insulation

With increasing energy costs, increased insulation can give a greater return on investment than increased plant capacity. Here is what you need to know about insulation selection, thickness calculations and economics.

Michael R. Harrison and *Charles M. Pelanne*
Johns-Manville

Originally published December 19, 1977

☐ It is estimated that the U.S. chemical process industries could, by the early 1980s, save 252 trillion Btu/year by increasing its use of insulation and improving system maintenance. About 200 trillion Btu of this would come from improved insulation of steam lines.

For the U.S. petrochemical sector, energy costs are some $4.75 billion, or 23% of the total operating cost.

Under today's energy costs, increased insulation frequently pays a better return on investment than increased capacity. Hence, insulation is moving up from its traditional 2% of total plant construction costs to figures ranging from 4 to 8%.

Consequently, economic calculations are more critical and insulation system design and installation are given more engineering thought. The process industries are also upgrading the types of insulation being used, to make sure that the energy savings will continue through the life of the plant.

Among the recent trends in the use of insulation in chemical process plants have been:

- Creation and use of computerized methods for the calculation of insulation thickness.
- Greater use of contract companies that are specialists at installing insulation.
- An upgrading of the insulation materials used for process plants.
- More attention to jacketing. Almost all insulation for process plants is now jacketed, to provide protection against weather, mechanical damage, and fire. Most jacketing is aluminum or, where acid-corrosion problems exist, stainless steel. Some plastics, such as polyvinylacetate, are also used because of their resistance to a wide range of chemicals.
- A long-term effort to get better values for the heat transfer properties of insulation as used. The laboratory goal now is a systems approach, giving accurate numbers for the total heat-loss of a whole piping system.
- Greater use of engineered insulations, designed to give specific combinations of insulating value, size, weight and damage resistance. Combinations—layers of different insulating materials and jackets—are also used this way.
- Newly fabricated forms of insulation designed to make easier the most difficult insulation installation jobs: around valves, flanges, and other fittings.

Particular CPI insulation requirements

Most insulations employed in plants of the chemical process industries are in the lower intermediate range, from ambient to about 500°F. Most applications are on steam lines, process vessels and process piping.

In these applications, the insulation generally should meet five criteria:

1. The material must be compatible with the chemicals being processed. It should be essentially inert regarding flash-point depression and similar properties, when chemicals may contact the insulation.
2. The material's properties, particularly thermal-transport properties, should be reproducible so that they do not vary significantly from batch to batch. This is particularly important when process materials must be kept within a specified temperature range.
3. Both the insulation and its barrier system should resist mechanical abuse over the life of the plant, since both piping and vessels in CPI plants are generally accessible to workers and are walked on, have ladders leaned against them, and tools dropped on them.
4. The barrier system has to be effective, allowing little or no water leakage. Penetration of the insulation by water seriously diminishes its insulating properties until it dries out.
5. Installation of the insulating material on valves and pipe fittings should be as easy as possible. Normally, the cost of insulating a steam valve is about three times that of insulating a similar length of straight pipe.

The material most often used in insulating process vessels and piping for intermediate temperatures is calcium silicate, because of the need for resistance to mechanical abuse. For further protection against mechanical abuse, as well as protection against weather and other damage, jacketing is almost always employed. The corrosion resistance of the jacket is also an important consideration.

For low-temperature applications (minus 150°F to chilled-water temperatures), expanded polyurethane and polystyrene, or foamed glass are most often used. Permeability to water vapor is the major problem faced in these applications. Water vapor penetrates and deposits as ice at the cold surface, progressively destroying the insulation properties of urethane or styrene, even though the warm surface appears intact. Foamed glass, which both resists moisture penetration and better withstands the periodic heating and drying necessary to counteract this moisture penetration, is growing in popularity over the plastic foams.

Fiberglass is the insulation most used for applications ranging from the temperature of chilled water up to the boiling point (212°F). Fiberglass would work as well as or even better than other insulations at colder temperatures (it is an extremely efficient insulator even in the cryogenic range) if the vapor-barrier could be maintained intact. But at below freezing temperatures, a vapor-barrier breakdown can produce a massive failure with fiberglass rather than a localized failure as foam might undergo. With fiberglass, the vapor could freeze and destroy the insulation for a whole tank, whereas with foam, because it is a series of air cells, the freezing vapor would not penetrate as far.

Lower-temperature cryogenic systems generally employ either massive solid insulations or vacuum types. In the latter, highly polished metal supporting-walls

Estimated short-term energy-conservation potential in the six largest energy-consuming industries — **Table I**

Industry	Estimated energy savings by insulation and maintenance* (10^{12} Btu/yr)	Estimated energy savings by insulation and maintenance* (10^{15} joules/yr)	Estimated total industry process energy usage† (10^{12} Btu/yr)	Estimated total industry process energy usage† (10^{15} joules/yr)
Chemical	252	266	2,662	2,810
Primary metals	500	528	5,900	6,230
Petroleum	343	362	3,044	3,210
Paper	208	219	2,563	2,700
Glass, cement, structural clay products, others	190	201	2,000	2,110
Food	47	50	1,283	1,350

*Estimated precision, ±30%.

†Process energy only; does not include feedstock energy usage. Estimated precision, ±10%.

Estimated heat-rejection in the six largest fuel-consuming industries (1972–1973) **Table II**

Industry (or process) and operation	Rejected heat, (10^{12} Btu/yr*): Radiation, convection, conduction, and unaccounted losses†	Below 100°C	100–250°C	250–800°C	800–1,800°C	Total, all ranges
Chemical						
Chlorine/caustic soda	40	254	51			
Ethylene/propylene	20	151	79	40		
Ammonia	32	115	139	40		
Ethylbenzene/styrene	1	24	28			
Carbon black	2		32			
Sodium carbonate	2	36	16			
Oxygen/nitrogen	1	171	147			
Cumene		4	4			
Phenol/acetone		8	8			
Other	150	628	290	79		
Total	**248**	**1,391**	**794**		**159**	**2,345**
Primary metals						
Steel	298	79	397	596	795	
Aluminum	119	378	79	20	79	
Other	119	397	119	40	79	
Total	**536**	**854**	**595**	**656**	**953**	**3,058**
Petroleum	318	994	497	695		2,186
Paper	199	1,391	755	119		2,265
Glass, cement, other						
Cement	119	79	40	219		
Glass	99	60	12	87	36	
Other	99	79	40	159	20	
Total	**317**	**218**	**92**	**465**	**56**	**1,049**
Food	60	795	159	119		1,073
Grand totals	**1,679**	**5,643**	**3,110**	**2,213**	**1,009**	**11,975**
Ratio to grand total Σ_T		0.47	0.26	0.18	0.09	

*To convert to 10^{15} J/yr, multiply by 1.0543. The estimated accuracy of the values is ±20%.
†Radiation, etc., losses are distributed within other losses shown in the table.

have a vacuum space, reflective foils, and various inorganic or organic materials coated with metals between them.

The high-temperature range above 1,000°F calls for more-heat-resistant materials. Eight principal types of thermal insulation are employed: 1. calcium silicate based material (on piping and vessels); 2. mineral fibers; 3. ceramic fibers; 4. oxide fibers; 5. carbon fibers; 6. rigid ceramic insulating-brick; 7. castable ceramic insulating-refractories; and 8. multiple metal-foil systems for vacuum applications (similar to the cryogenic insulation).

Insulations available

Insulation materials are normally classified according to the service-temperature ranges they are suitable for. These tend to be arbitrary. When operating temperatures reach a certain upper limit, the materials may become structurally unstable, or become uncompetitive because they have a relatively high thermal conductivity. Whenever lower-temperature limits are given for most insulations, it usually simply means that they are not cost effective in these applications, not that they are technically unable to do the job. Within each temperature range, choice among the materials available is generally based on other properties and cost.

Cryogenic range (455°F to −150°F)

Insulations fall within two types: vacuum and massive, the latter consisting of one or more solid phases distributed with a gas, such as dry air, to produce a very low thermal conductivity.

Vacuum insulation systems, consisting typically of highly polished metal supporting walls with a vacuum space between them (sometimes with multiple metal reflective foils or opacified powders inside) are usually custom-designed and installed by the insulation vendors. Performance is generally guaranteed.

Representative equipment insulations — **Table III**

Insulation type	Temperature range, °F	Conductivity, k Btu/(h)(ft^2)(°F/in.)	Density, lb/ft^3	Applications
Urethane foam	–270 to 225	0.11 to 0.14	2.0	Tanks and vessels
Fiberglass blankets	–270 to 450	0.17 to 0.60	0.60 to 3.0	Chillers, tanks (hot and cold), process equipment
Elastomeric sheets	– 40 to 220	0.25 to 0.27	4.5 to 6.0	Tanks and chillers
Fiber glass boards	Ambient to 850	0.23 to 0.36	1.6 to 6.0	Boilers, tanks, and heat exchangers
Calcium silicate boards, blocks	450 to 1,200	0.22 to 0.59	6.0 to 10.0 to	Boilers, breechings, and chimney liners
Mineral fiber blocks	to 1,900	0.36 to 0.90	13.0	Boilers and tanks

The theory behind vacuum cryogenic insulation is that of the double-walled dewar flask, which relies in part on vacuum between the walls, and in part on reflection of radiant heat. Without gas inside, heat transfer will be mainly by radiation. Coating the inside hot surface facing the evacuated area reduces heat transfer to a level proportional to the emissivity of the coating (0.01 for silver, for example).

The floating-shield approach takes advantage of the fact that thermal radiation can often be cut in half by floating a radiation shield between the cold and warm surfaces. Powders such as expanded perlite, silica aerogel, carbon black, calcium silicate, diatomaceous earth and fiber are used; the material is packed in before the air is pumped out.

The multilayer type, a series of reflective foil shields of aluminum separated by low-conductivity fillers such as fiberglass, comes in blanket form. A related type—a number of aluminum-coated layers of polyester film, crinkled to reduce heat transfer by conduction from layer to layer—withstands high acceleration loads without loss of insulating effectiveness.

Less costly than the evacuated forms described above are foam types. Foamed polyurethane and polystyrene, either foamed into flexible sheets, foamed in place or foamed in rigid insulation sections, initially showed great promise, but deterioration from permeation by water vapor and air have caused dissatisfaction. Foamed glass stands up better to drying-out processes and is less permeable. Foam types work through their microstructure of closed cells containing air or other gas. The cells, although not as small as in some newly developed insulations, are small enough to greatly reduce heat transfer by gas-phase convection, assuming they are filled with a gas other than air.

Low-temperature range (–150°F to 212°F)

Some of the evacuated types of insulation for cryogenic service are used in the lower end of this range, but foams are employed more. The main problems are moisture permeation, and fire hazards—and, at lower temperatures, heat-transfer rates that are too great to be acceptable.

Since the cost of refrigeration is greater than the cost of heating, more insulation is often justified in low-temperature applications.

Moisture problems—In the low-temperature range, moisture is by far the major insulation problem. Water permeability of four major materials is as follows:

Material	Water permeability, perm-in.
Foam glass	0.00
Urethane foam	0.3 to 0.6
Polystyrene foam	1 to 4
Fiberglass	100 to 200

Permeation by water or water vapor can dramatically increase the conductivity of the insulation. Water-vapor permeation is not a serious problem until the moisture condenses—except with foam, where the vapor replaces air in the cells. Ideally, the insulation should either absorb no moisture or readily give up any that enters, and it should resist water deterioration. Moisture barriers are extensively used, but in practice it is almost impossible to achieve the perfect barrier needed. Extra thicknesses of insulation, even beyond what would be economically dictated for cold-line applications, are sometimes employed to keep the warm surface temperature above the dewpoint of the water vapor in the air.

Moisture interferes with the insulating properties, at least temporarily, in all temperature ranges but is not a serious problem except at lower temperatures. Above ambient temperatures, the vapor pressure of water is higher at the insulation's inner surface than in the surrounding area, and this pressure tends to push moisture out. Protection from weather and spillage is what is needed.

But when the temperature of the inner surface is less than ambient, this flow reverses—the vapor is pushed in, to an extent proportional to the difference in temperatures between the insulated body and the outside air.

Even with water-impervious insulation, where water vapor may enter through unsealed joints or cracks, localized freezing can cause minor damage. When ice melts and runs along the piping to collect at low points

Secondary properties of representative insulations† **Table IV**

Property	Calcium silicate	Mineral fiber	Fiberglass	Elastomeric, closed cell	Cellular glass	Urethane foam
Alkalinity, pH	8 to 12.5	7 to 9	7.5 to 9	–	7 to 8	6.5 to 7.5
Capillarity	Will wick	Will wick	Neg	None	Neg	None
Flame-spred index	–	Less than 25 *	Less than 25	25 to 75	–	75 or less*
Smoke-density index	–	Less than 50 *	Less than 50	150 to 490*	–	450 or less *
Non-combustible	N.C.	Some N.C.	–	–	N.C.	–
Compressive strength lb/in^2 @% deformation	100* to 250 @ 5%	1 to 18 @ 10%	0.02 to 3.4 @10%	2 to 6 @ 25%	100 @ 5%	16 to 100 @ 5%
Specific heat, btu/(lb)(°F)	0.20 to 0.28	0.22	0.20	0.19 to 0.27	0.20	0.23 to 0.25

† These data extracted from "Thermal Insulation," by John F. Malloy, Van Nostrand Reinhold, 1969.
* Indicates change from referenced table.

during a shutdown, then refreezes, major damage can result.

Vapor barriers

Vapor barriers take three forms:

Structural barriers—Often prefabricated to exact dimensions needed and ready to install, these are rigid sheets of reinforced plastic, aluminum or stainless steel jacketing—flat, corrugated or embossed.

Membrane barriers—Metal foils, laminated foils and treated papers, plastic films and sheets, and coated felts and paper—these are either part of the insulation as supplied, or can be supplied separately.

Coating barriers—In fluid form as a paint (or semi-fluid of the hot-melt variety), the material may be asphaltic, resinous or polymeric. These provide a seamless coating but require time to dry.

For low-temperature insulation, the vapor barrier should perform considerably better than 1 perm. The colder the equipment, the better the resistance should be, ranging from 0.2 perm asphalt-saturated shiny glazed papers for chilled water, down to 0.01 perm laminated paper and aluminum products for below freezing.

When the virtually inevitable happens in low-temperature applications, and water vapor gets in, the insulation must be heated and dried. With polyurethane and polystyrene foam, which suffer deterioration, (polystyrene softens at about 90°C, for example), care must be exercised in this procedure.

In the higher part of this temperature range—from about −20°F to the upper limit, 212°F—a variety of organic and inorganic massive insulations are used, sometimes in loose fill forms. These include:

- Compressed and granulated cork.
- Sandwiched cellular glass and felt board.
- Glass fibers bonded with organic resins.
- Expanded and cellular forms of polystyrene.
- Polyurethane foams.
- Rubber and resin combinations.
- Vinyl chloride cellular foams.
- Wood fibers with suitable binders.
- Polyvinyl acetate.
- Cork-filled mastic.
- Expanded vermiculite and perlite.
- Aluminum foil on paper.

Selection among these insulations is based on a number of factors, generally considered in this order:

1. Form available, such as rigid board, flexible board, loose fill, etc., and how it fits the insulation design.
2. Price.
3. Availability.
4. Thermal conductivity.
5. Moisture permeability.

Intermediate range (212°F to 1,000°F)

This temperature range includes conditions encountered in most chemical and petrochemical processes and in steam systems, so it is by far the most important to the chemical process industries.

The most important type of insulation in this range is fiber-reinforced hydrous calcium silicate.

Insulation types used in this service range include:

1. Calcium silicate.
2. Diatomaceous silica.
3. Cellular glass (to about 850°F).
4. Glass fiber bonded with high-temperature binders.
5. Magnesium carbonate with asbestos or other fibers and binders (to about 600°F).
6. Rock wool or mineral-derived fibers.
7. Expanded perlite with binders.
8. Metal-sheet reflective systems.

Selection of material in this temperature range is much more strongly dictated by the value of thermal conductivity than it is in the lower temperature ranges. But other factors, such as mechanical properties, forms available and cost of installation are also strong influ-

Representative pipe insulations — **Table V**

Insulation type	Temperature range, °F	Conductivity, *k* Btu/(h)(ft²)(°F/in.)	Density, lb/ft³	Applications
Urethane foam	–300 to 300	0.11 to 0.14	1.6 to 3.0	Hot and cold piping
Cellular glass blocks	–350 to 500	0.20 to 0.75	7.0 to 9.5	Tanks and piping
Fiberglass blanket for wrapping	–120 to 550	0.15 to 0.54	0.60 to 3.0	Piping and pipe fittings
Fiberglass preformed shapes	– 60 to 450	0.22 to 0.38	0.60 to 3.0	Hot and cold piping
Fiberglass mats	150 to 700	0.21 to 0.38	0.60 to 3.0	Piping and pipe fittings
Elastomeric preformed shapes and tape	– 40 to 220	0.25 to 0.27	4.5 to 6.0	Piping and pipe fittings
Fiberglass with vapor barrier jacket	20 to 150	0.20 to 0.31	0.65 to 2.0	Refrigerant lines, dual-temperature lines, chilled-water lines, fuel-oil piping
Fiberglass without vapor barrier jacket	to 500	0.20 to 0.31	1.5 to 3.0	Hot piping
Cellular glass blocks and boards	70 to 900	0.20 to 0.75	7.0 to 9.5	Hot piping
Urethane foam blocks and boards	200 to 300	0.11 to 0.14	1.5 to 4.0	Hot piping
Mineral-fiber preformed shpaes	to 1,200	0.24 to 0.63	8.0 to 10.0	Hot piping
Mineral-fiber blankets	to 1,400	0.26 to 5.6	8.0	Hot piping
Fiberglass field applied jacket for exposed lines	500 to 800	0.21 to 0.55	2.4 to 6.0	Hot piping
Mineral-wool blocks	850 to 1,800	0.36 to 0.90	11.0 to 18.0	Hot piping
Calcium silicate blocks	1,200 to 1,800	0.33 to 0.72	10.0 to 14.0	Hot piping

ences. Various products from different manufacturers within a given generic insulation type have different upper temperature limits, depending on the manufacturing process and raw materials. Specific data should be obtained for each specific product.

Principal intermediate-range materials

Calcium silicate—This is a mixture of lime and silica reinforced with organic and inorganic fibers and molded into the shape desired. Temperature range covered is 100 to 1,500°F. Flexural strength is very good. Compressive strength is high. These are the reasons this material is so widely used in the chemical process industries. Calcium silicate meets specifications for prevention of stress-corrosion cracking of austenitic stainless steel. It is available in both half-round and quarter-round segments for pipes, as well as in the form of blocks.

Mineral fiber—For this material, rock and slag fibers are bonded together with a heat-resistant binder. Upper temperature limit is up to 1,200°F (generally, about as high as for calcium silicate). The material has a practically neutral pH. It can be made to meet requirements involving stress-corrosion cracking of austenitic stainless steel. Its compressive strength is much less than for calcium silicate. It is available in both rigid molded form for piping or vessel use, and as a flexible blanket for irregular surfaces.

Expanded silica, or perlite—This insulation is an inert siliceous volcanic rock with some combined water. The rock is expanded by heating to above 1,600°F—the water vaporizes expanding the rock volume and creating a structure of minute air cells surrounded by vitrified product. Added binders hold it all together and resist moisture penetration, and inorganic fibers reinforce the structure. The material has low shrinkage. Water absorption is low. The chloride content is low enough to eliminate its causing stress-corrosion cracking in austenitic stainless steel.

Glass—In various forms, glass products serve from the cryogenic level to around 1,200°F, with some specialty, high-purity products good up to 1,800°F. Among the forms available are flexible fiber blankets without binder, glass fiber with an organic binder, (generally a thermosetting resin good to 1,000°F or higher); semirigid fiberglass boards (for up to 850°F); molded sections for pipe; and foam or cellular form, used for low-temperature insulation applications. Glass has a low moisture absorption and can be easily dried. Upon drying, strength and thermal qualities are recovered. Its low chloride content keeps it from promoting stainless-steel stress-corrosion cracking.

Loose fill—A number of insulation materials are available in the form of loose fill: granulated insulation materials or loose fibers, diatomaceous silica, etc.

High-temperatures (1,000°F to 1,600°F)

This approaches the refractory range of materials, which begins at 1,600°F. The higher the temperature, the fewer the materials available, with major dropoff points at about 1,200°F, 1,600°F and 2,400°F.

Major materials in the high-temperature range are (listed in ascending order of temperature resistance):

1. Mineral fiber: 1,000 to 1,900°F
2. Calcium silicate: 1,200 to 2,000°F
3. Multiple-metal-foil systems for vacuum applications: to 2,500°F
4. Ceramic fibers based on the Al_2O_3-SiO_2 systems: 1,600°F to 2,600°F.
5. Castable-ceramic insulating refractories: 2,000 to 3,000°F.
6. Oxide fibers, primarily Al_2O_3 or ZrO_2: 2,800 to 3,000°F.
7. Rigid ceramic insulating brick: 2,000 to 3,200°F.
8. Carbon fibers: to 3,600°F (above this temperature, irreversible conductivity increases take place).

In high-temperature applications, service conditions can be severe. Among the abuses the insulating refractory must withstand are abrasion and erosion by molten materials, direct flame impingement, corrosive atmospheres and severe thermal shock. Although heat-transfer principles and energy economics are similar for both high-temperature and intermediate-temperature applications, use and design of refractories is an engineering art in its own right and will not be treated more fully here.

Slowing heat-transfer

Rather than thinking of insulation as a material, the chemical engineer should consider it as a strategy for slowing the transfer of heat. This point of view will guide the engineer through problems of material incompatibility, space limitations, and so on. The strategy starts with keeping in mind the three modes through which heat is transferred: radiation, conduction and convection.

Electromagnetic (infrared) radiation—This is the primary mode of heat transfer. It can operate even in a vacuum, where neither other mode can. The amount of heat transferred by radiation is governed by both the emissivity and the temperature of the surfaces involved in the emission and absorption of the heat energy, and by the relative area of these surfaces. Because radiant-energy-flow is proportional to the difference of the fourth power of the absolute temperatures of the hotter and cooler surfaces, the rate of heat transfer increases dramatically as the temperature increases.

Control of radiation heat-transfer is achieved by changing the emittance level of the surfaces involved, or by the insertion of absorbing or reflecting surfaces (sheets, fibers, particles, etc.) between the two temperature boundaries.

Conduction—In this mode, atomic or molecular motion conducts heat through any gaseous, liquid or solid material. The tighter the molecular network, the more it will conduct. Thus, steel is more conductive than air. The motion, and therefore the heat transfer, will continue until the whole network is terminally balanced.

Control of conduction heat-transfer is achieved by use of less-conductive elements (e.g., glass instead of steel) and by breaking the continuous structure (fibers instead of solid mass).

Convection or mass motion—This is a function of density and happens only in a fluid. Here, a whole mass of the material moves, carrying heat energy with it (as opposed to simply having the individual molecules move). The lighter, heated fluid will rise and be replaced by colder, and thus heavier, gas or liquid. This exchange is governed by the fluid's freedom of movement and by its density; the denser the fluid, the more effective the heat transfer. The size, shape and orientation of the cell containing the fluid are factors affecting the convective heat transfer. Control of convection is achieved by the creation of small cells within which the temperature gradients are small.

Radiation is the primary mode of heat transfer; the other two, conduction and convection, come into play only as a result of radiation. If the molecules did not absorb the radiant energy, there would be none of the difference in temperature necessary to create molecular and mass motion.

Thus, the insulation strategy is to minimize radiation transfer, minimize convective transfer, and introduce a minimum of solid conduction media (insulation has, in general, a negligible effect on air conduction). A compromise game is played between these modes to create an insulation that will work well for a given set of conditions.

As large a number of radiation absorbers as possible are placed between the temperature boundaries, while minimizing the solid conduction. Since a gas is often present, the strategy calls for creating as many small cells as possible to inhibit convection.

All of these mechanisms work together in the insulation to form a very complex system. To simplify examination of the heat-transfer mechanisms and their combined effect, the familiar term "thermal conductivity" or simply "*k*" is used.

Theoretically, there are too many variables for "*k*" to be a precise description of an intrinsic property of a homogeneous material, as it purports to be. But, generally, it is the only practical way to define the characteristics of an insulation, so it is widely used.

An important variable is the specific temperature at which the insulation is functioning. The relative importance of the various modes of heat transfer will vary considerably as the temperature conditions vary.

Even more important in insulation design is that the engineer quickly gets moved off any theoretical base into hard problems involving criteria other than optimum resistance to heat flow. Some of these problems are: temperature limitations, thermal-shock resistance, coefficient of expansion, strength, hardness, compressibility, specific heat, erosion resistance, chemical resistance, environmental resistance, space limitations and—certainly not least—cost.

The choice is generally a compromise governed by an attempt to balance the cost of the installed insulation and the amount of money that is expected to be saved by lowered energy requirements. The increasing cost of energy, of course, is bringing more and more portions of CPI plants under scrutiny for possible insulation. The uncertainty about how these energy costs will rise over future decades makes the cost/benefit engineering decisions tougher.

Insulation design theory

In making these compromises when designing the insulation system for a process plant, it is useful to know the kinds of compromises made by the designer of the insulation materials available.

The apparent thermal conductivity of mass insulations is a combination of the following heat-transfer mechanisms: thermal conductivity of air, thermal conductivity of the solid components of the insulating structure, the convective heat transfer within the pore structure, the radiation heat transfer within the structure, and the interactions of these mechanisms.

The thermal conductivity of air in low-density materials is nearly a constant, since the proportion of air to solids in most insulations ranges from 99% air in 1.6-lb/ft^3-density mineral fiber, to no less than 75%.

The thermal conductivity of the solid is related to the number of discontinuities in the network (e.g., contact between the fibers or other solid particles) and to the gradient along the fiber or structural component as influenced by the orientation of the individual fiber.

When the solid portion of the insulation is small, as it should be, convective transfer appears. With a low portion of solids, cells are formed large enough to allow natural convective air turbulence. For the phenomenon to become apparent, the cells usually must be several millimeters across. At room temperature, for example, in a 0.50-lb/ft^3-density fiberglass insulation with a uniform 4-μm-fiber-diameter distribution, the average pore size is about 1 μm. A small amount of convection will occur in slightly higher densities because of the nonuniformity of the fiber distribution.

Radiation is a significant portion of the total heat transfer at room temperature in fibrous insulation of densities below 2 lb/ft^3; as temperature increases, radiation is significant at considerably higher densities. The radiation component can be divided into two major parts: (1) that which is not absorbed but is scattered and reflected by the fibrous structure, and (2) the fiber-to-fiber reradiation, dependent on the absorptance and emittance of the fiber and the difference in temperature between two fibers within view of each other. Interaction between the air and the solid portion of the insulation is the final heat-transfer mode. Air conduction acts as a short-circuiting process, since the air contacts all radiation-absorbing surfaces.

A typical fiberglass insulation in an ambient-temperature situation illustrates how this all works. First, take the heat flux without the presence of the insulation—that is, extend the density curve to zero. The total apparent "k_a" equals 1.667, of which the air contributes 0.178, convection 0.435 and radiation 1.054. Then add fiberglass insulation into the heat-flow path, giving a rapid reduction in "k_a." The convection is eliminated at about 0.6 lb/ft^3 bulk density. The radiation transmission is being continually reduced and partially transformed into radiation conductivity as the fibers are heated and start reradiating between themselves.

Solid conduction contributes very little at densities below 2 to 2.5 lb/ft^3. The curve shows that there is no advantage in increasing the density beyond 2.5 lb/ft^3 for this set of conditions. Economically, it may be advantageous to limit the density to 0.5 lb/ft^3, which gives a "k_a" of about 18% of the original apparent "k", considering that a further increase in density to 2 lb/ft^3 would only bring it down to 12.5% of the original apparent "k" with no insulation.

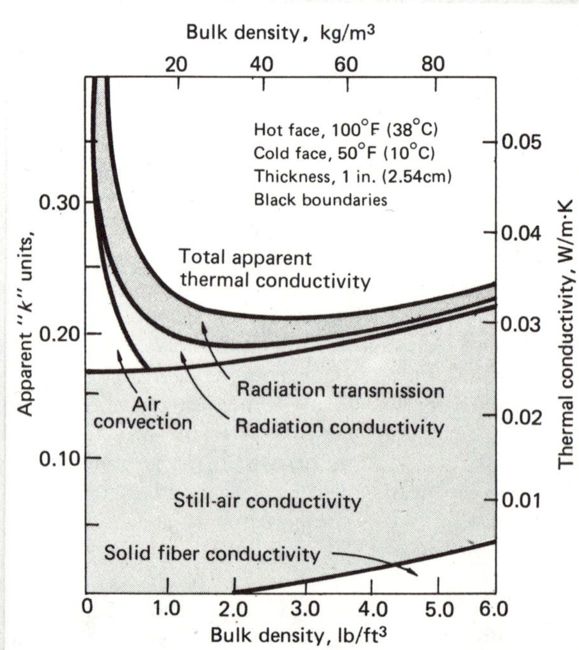

Thermal conductivity components vs. density for a typical fiberglass insulation **Fig. 1**

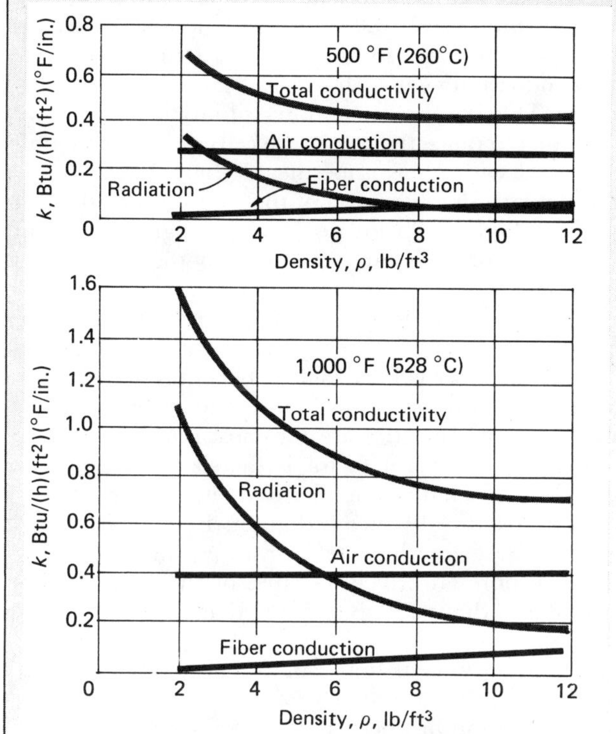

Contribution of each mode of heat transfer to the total apparent thermal conductivity **Fig. 2**

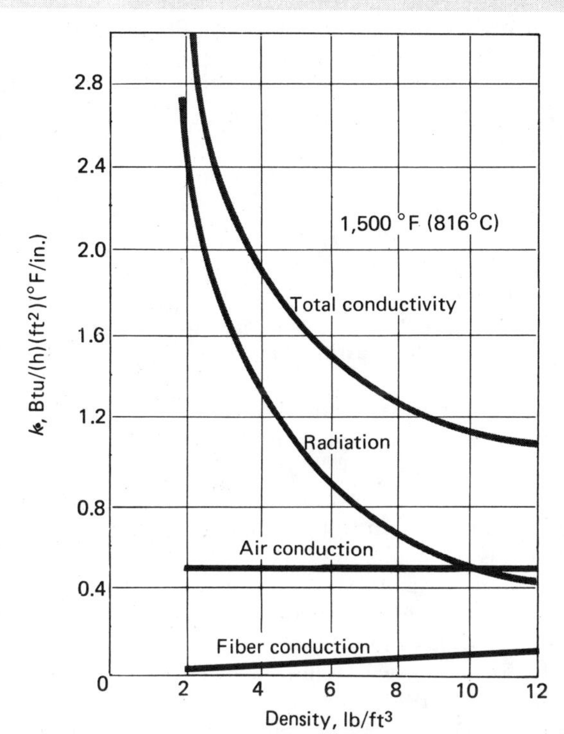

Contribution of each mode of heat transfer to the total apparent thermal conductivity at 1,500°F **Fig. 3**

Density requirements increase as the temperatures increase. With this insulation system, for example, optimum densities would be 6 lb/ft^3 at 500°F; above 12 lb/ft^3 at 1,000°F; and so on. At each temperature, a minimum apparent thermal-conductivity value occurs at a density range at which the sum of the variable components is at a minimum. The density for minimum apparent-conductivity increases with increasing temperature, because considerably greater density is required to reduce the radiation to the point where the sum of the variable components attains a minimum value. The minimum apparent thermal conductivity also increases with temperature because of the increase of air conduction, as well as the higher total of the other variable components. No definite generalization can be made concerning the superiority of higher- or lower-density material without carefully considering all of the factors in any specific application.

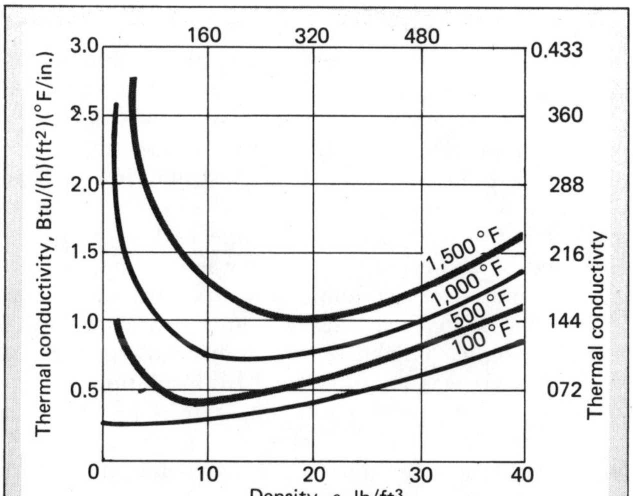

Thermal conductivity vs. density **Fig. 4**

It must be realized that all of the above is calculated in simple terms. Heat flow in actual material is much more complicated and virtually impossible to reduce to formula. Instead, theory has been correlated with experimental data by the insulation manufacturers in specific instances, giving limited density ranges.

Engineered insulations

Insulation manufacturers, in an effort to strike the best bargain among desired qualities of mechanical strength, chemical compatibility, weight, compactness, and so forth, with the least heat conductivity at the best possible cost, engineer their insulations toward particular applications.

The engineering of insulations in practice is less simple. The influences of the variables—radiation and conduction among solids and gases, convection in gases—must be considered closely.

Introduction of thermally opaque solids in the path of a radiant heat flux restricts its flow. The more solids introduced, the greater the reduction in radiant flow. But, of course, this results in an increase in solid conduction, which is influenced by the structural bonds in the network, which connect in series the solid fiber or other particles in the direction of heat flow. The practice in fiber insulation is to lay the fibers in place perpendicular to the heat flow to minimize this conduction, which tends to move along the length of the fiber.

Gas conduction within the air or other medium between the solid particles can be eliminated or minimized by evacuation or reduction of pressure, or by replacement of the air by a high-density gas such as a fluorocarbon (as in polyurethane foam). The gas conduction is approximately proportional to the square root of its molecular weight. Another method is to try to reduce the spacing between solid particles without at the same time increasing the parallel-solid conduction component.

Gas conduction is an important constant. Thus, if the air or gas is removed or made less effective, the overall "k_a" will be improved. A gradual reduction of the gas conduction can be obtained by reducing the cell spacing in the insulation to the range of the free mean path of free gas molecules. This in effect blocks off the molecules from each other, reducing molecule-to-molecule collisions. This reduces thermal energy transfer from gas molecule to gas molecule.

This principle was followed in engineering Min-K* insulation for the space program, where a highly efficient insulation with low weight and low space requirements was needed. The developers reduced the process of gas conduction by creating very small cells in which the gas does not operate as a free gas. This minimized the gas conduction, stopped convection and created a multitude of discontinuous, poorly conductive solid barriers to absorb radiation.

Other insulation systems demonstrating extraordinary thermal transmission values are: "Super" insulations, which can only function effectively in a vacuum, as in the multi-layer reflective insulation used for cryogenic purposes, and in polyurethane foams in which a low-conductivity gas such as fluorocarbon is substituted for air. This is only a temporary situation since air will ultimately replace the gas in the cells unless the foam is contained within gas-impermeable barriers.

*Min-K, a registered trademark of Johns-Manville.

Even with the reduction of the mean free path, radiation at high temperature is still a significant heat-transfer factor. This can be reduced by the use of opacifying agents.

The more-dense insulations perform better at higher temperatures. At lower temperatures, all densities converge in their "k" values, so little benefit is derived from an increase in density. For ambient-temperature applications, lower densities are the most economical.

Another important factor is the size of the fiber or particle. The smaller the fiber diameter, the greater the surface area to obstruct radiation. But, again, this is expensive. At very high temperatures, fine 1-μm fibers perform more effectively because of the shorter wavelengths of radiant energy generated as the temperatures increase.

In summary, heat flow can be effectively controlled by insulation within which:

1. Radiation has been reduced to a minimum level. It remains as the controlling effect on the lower part of the "k"-versus-density curve.
2. Solid conduction can be held to a minimum.
3. Convection is minimal, particularly at the structural density required to obtain reasonable radiation absorption.
4. Gas conduction, though present in most common insulating materials, in some instances can be reduced or eliminated.

Economic thickness calculations

There are many design criteria involved in specifying insulation, particularly in the process industries. But one, economic considerations, which until the past three years was comparatively low on the list, is steadily becoming more important.

By economic considerations, we mean the calculation of the amount of insulation that would save enough energy over a set time period to justify the investment. Economic thickness calculations, based on equations first published in 1926, have been refined and standardized over the years. New input, particularly concerning methods for developing investment return analyses, have made these calculations more useful just when they are most needed.

The method has stood up under 50 years of close scrutiny by engineers and has proven accurate. First performed by L.B. McMillan, a former Johns-Manville engineer, the calculations have been noted for their flexibility in the number of variables they have been able to accommodate.

Interest in economic thickness has increased considerably with the rise in energy costs. Previously the cookbook approach was the standard way to specify thermal-insulation thickness in chemical plants. The lower cost of energy did not justify an engineer's taking great pains in custom-designing insulation thicknesses for each pipe size and temperature range, unless the process itself demanded it. As the last item installed, it was often a matter of putting in whatever amount of insulation fit the space left in pipe racks—or it was governed by the amount of money left in the budget.

The concept of economic thickness has now been refined by the Federal Energy Administration into the ETI (Economic Thickness for Industrial Insulation) method. This was computerized by insulation manufacturers and others so plant data can be fed into the computer to get insulation-thickness specifications offering the best economics for a particular pipe in a particular plant.

These computer programs determine both the optimum thickness of insulation for piping in new plants, and the economics of retrofitting more insulation in old plants, according to energy costs and company accounting procedures.

One chemical plant was subjected to the economic thickness calculations and the company found that bringing old insulation, which had deteriorated to 75% of its original effectiveness, back to 95% effectiveness gave a return on its investment of 103%. Without such hard facts, however, the engineers would have had a hard time convincing management that this kind of investment was justified. Traditionally, investment capital just was not spent on maintaining insulation.

Such decisions are made by management after comparing the financial return on insulation to other investment opportunities, such as plant expansion, new and more efficient equipment, and so forth.

Below are four common investment analysis methods, as applied to upgrading insulation.

Simple Payback

Essentially, this is the time required to repay the capital investment with the operating savings.

Example:

	Thickness current std.	ETI	Difference
Insulation investment, $	200,000	250,000	50,000
Annual fuel cost, $	40,000	30,000	10,000

$$\text{Simple payback} = \frac{\text{Investment difference}}{\text{Annual fuel savings}} = \frac{50{,}000}{10{,}000} = 5.0 \text{ Years}$$

(While easy to calculate, and hence often referred to, this is generally used only to measure the degree of financial risk of various investment opportunities).

Discounted payback period

This indicator is defined as the length of time an investment requires to generate discounted, after-tax cash flows equal to the initial insulation cost.

Minimum annual cost

This measure annualizes all cost of insulation and the costs of heat lost through the insulation, discounting them at an assigned interest rate.

	Thickness, current standard	ETI	Comments
(1) Annual energy cost, $	40,000	30,000	
(2) After-tax energy cost, $	20,800	15,600	(1) × (1.0 − 0.48)
(3) Insulation depreciation, $	10,000		$200,000/20 Years
		12,500	$250,000/20 Years
(4) Total annual cash costs, $	10,800	3,100	(2) minus (3)
(5) Capital recovery factor* for 20 years at 10% = 1/8.514 = 0.1175.			
(6) Equivalent annual insulation cost	23,491		(200,000 × 0.1175)
		29,363	(250,000 × 0.1175)
(7) Total annual cost	34,291	32,463	(4) plus (6)

Conclusion: The lowest annual cost ($32,463) is the better of the two investment options.

*Capital recovery factor—Converts a one-time cost to an annual cost. The summation of the discount factor over a 20-year period = 8.93.

Maximum net present value

This takes the sum of the annual expenses, reduces this sum by the tax rate, and subtracts annual depreciation charges to produce an annual after-tax cash expense. The present value of the annual after-tax cash expense is found and to this is added the capital cost of the insulation.

	Thickness, current standard	ETI	Comments
(1) Annual energy cost, $	40,000	30,000	
(2) After-tax energy cost, $	20,800	15,600	(1) × (1.0 − .48)
(3) Insulation depreciation, $	10,000		$200,000/20 Years
		12,500	$250,000/20 Years
(4) Total annual cash costs, $	10,800	3,100	(2) minus (3)
(5) Present-value factor for 20 years at 10% = 8.514			
(6) Present value of annual cash flows*	91,951	26,393	(4) times (5)
(7) Present value of cash flow for insulation purchase	200,000	250,000	
(8) Net present value	291,951	276,393	(6) plus (7)

Conclusion: The ETI thickness provides the lowest net present value—$276,393 (which is the better of the two investment options.)

*Assumes no fuel-price escalation.

In performing economic thickness calculations, the plant's engineer and its insulation contractor must supply the computer with the following data (shown with sample answers, for a typical chemical process plant).

1. Type and cost of fuel (oil, 40¢/gal).
2. Heating value of fuel (140,000 Btu/gal).
3. Expected annual fuel-price increase (10%).
4. Efficiency of conversion of fuel to heat (80%).
5. Capital investment for heat plant ($2 million).
6. Cost of money to finance plant (10%).
7. Heating-equipment depreciation period (20 years).
8. Expected total annual heat production (600,000 million Btu).
9. Basis for economic thickness calculation (minimum annual cost).
10. Income tax rate (48%).
11. Insulation depreciation period (20 years).
12. Cost of money, or return on investment requirement, for insulation (12%).
13. Insulation material (fiberglass, pipe).
14. Jacket emissivity (0.2).
15. Surface resistance (0.7).
16. Installed insulation prices ($3.00 per linear ft for 1 in. through $21.86 per linear ft for 6 in.).
17. Pipe size (6 in.).
18. Piping complexity factor (1.15—for average on a scale ranging from simple to complex, as provided on table).
19. Ambient temperature (60°F).
20. Process temperature (350°F).

Industrial energy prices projected to 1990 **Table VI**

FEA Region I: New England
1977 dollars/10^6 Btu

	Industrial electricity	Industrial natural gas	Industrial residual oil	Industrial coal
1977	10.75	2.01	2.58	1.47
1978	10.95	3.34	2.70	1.57
1979	11.15	3.66	2.82	1.67
1980	11.35	3.98	2.94	1.77
1985	12.17	2.99	2.96	1.85
1990	13.71	3.87	3.06	1.90

FEA Region II: New York/New Jersey
1977 dollars/10^6 Btu

	Industrial electricity	Industrial natural gas	Industrial residual oil	Industrial coal
1977	8.92	1.76	2.60	1.30
1978	8.93	1.80	2.76	1.42
1979	8.95	1.85	2.92	1.54
1980	8.97	1.90	3.08	1.66
1985	8.92	2.45	3.10	1.74
1990	9.69	3.15	3.19	1.79

FEA Region III: Mid-Atlantic states
1977 dollars/10^6 Btu

	Industrial electricity	Industrial natural gas	Industrial residual oil	Industrial coal
1977	8.21	1.45	2.62	1.11
1978	8.30	1.58	2.81	1.26
1979	8.40	1.72	3.00	1.40
1980	8.50	1.85	3.19	1.55
1985	9.73	2.41	3.21	1.64
1990	10.90	2.87	3.31	1.68

FEA Region IV: South Atlantic states
1977 dollars/10^6 Btu

	Industrial electricity	Industrial natural gas	Industrial residual oil	Industrial coal
1977	6.73	1.11	2.38	1.28
1978	7.19	1.20	2.55	1.43
1979	7.65	1.30	2.72	1.58
1980	8.10	1.40	2.89	1.74
1985	9.02	2.13	2.91	1.83
1990	10.04	2.71	3.01	1.87

FEA Region V: Midwest
1977 dollars/10^6 Btu

	Industrial electricity	Industrial natural gas	Industrial residual oil	Industrial coal
1977	6.97	1.34	2.70	1.19
1978	7.23	1.42	2.83	1.30
1979	7.49	1.50	2.96	1.41
1980	7.75	1.58	3.09	1.52
1985	9.09	2.13	3.12	1.58
1990	10.40	3.11	3.21	1.64

FEA Region VI: Southwest
1977 dollars/10^6 Btu

	Industrial electricity	Industrial natural gas	Industrial residual oil	Industrial coal
1977	7.06	1.21	2.48	1.67
1978	8.25	1.43	2.63	1.58
1979	9.45	1.64	2.79	1.49
1980	10.65	1.86	2.94	1.40
1985	10.74	1.88	2.96	1.50
1990	12.41	3.04	3.06	1.59

FEA Region VII: Central states
1977 dollars/10^6 Btu

	Industrial electricity	Industrial natural gas	Industrial residual oil	Industrial coal
1977	6.96	1.13	2.80	1.34
1978	7.32	1.33	2.90	1.38
1979	7.68	1.52	3.00	1.41
1980	8.04	1.72	3.10	1.45
1985	9.23	3.04	3.12	1.52
1990	10.27	3.33	3.22	1.56

FEA Region VIII: North Central states
1977 dollars/10^6 Btu

	Industrial electricity	Industrial natural gas	Industrial residual oil	Industrial coal
1977	4.75	1.16	2.88	1.03
1978	5.08	1.41	2.95	1.08
1979	5.41	1.66	3.02	1.14
1980	5.74	1.91	3.08	1.20
1985	6.84	1.94	3.18	1.29
1990	8.80	3.14	3.34	1.36

FEA Region IX: Western states
1977 dollars/10^6 Btu

	Industrial electricity	Industrial natural gas	Industrial residual oil	Industrial coal
1977	8.49	1.74	2.92	1.71
1978	9.26	2.07	2.93	1.65
1979	10.02	2.40	2.94	1.58
1980	10.79	2.72	2.96	1.52
1985	10.74	2.48	2.98	1.63
1990	11.49	3.21	3.07	1.74

FEA Region X: Northwestern states
1977 dollars/10^6 Btu

	Industrial electricity	Industrial natural gas	Industrial residual oil	Industrial coal
1977	2.78	1.54	2.91	1.12
1978	3.26	1.75	2.95	1.14
1979	3.75	1.95	2.98	1.16
1980	4.24	2.16	3.01	1.19
1985	4.76	2.17	3.03	1.59
1990	5.78	3.80	3.12	1.77

21. Annual hours of operation (8,760).
22. Length of pipe (600 ft).
23. Previous thickness specified (2 in.).

Given these parameters, our Energy Reduction Analysis computer program concluded that in this case, for the 600 ft of 6 in. pipe, an insulation thickness of 5 in. would provide a surface temperature of 73°F, and would save 206 million Btu's per year over the previously specified 2 in., for a dollar saving over the life of the plant of $15,762 (present value).

Planning insulation

The thermal efficiency of the insulation chosen, and the economic thickness, are not the whole story.

Chemical engineers designing a plant's insulation must also be concerned with such things as chemical attack (from both spillage and fumes), the possibilities of physical abuse, fire safety, personnel protection, difficulties in cleaning, future requirements of gaining access to insulated pipes and equipment, and so on.

Also, in chemical processes, frequently more is involved in controlling heat losses than saving energy. Many processes have rigid temperature parameters, and a product cannot cool too much in being transferred from one part of the plant to another.

In the design phase, it must be decided whether added insulation can sufficiently control this heat loss, or if a heated tracer must be added to the pipeline. (Where tracers are to be added, a specially shaped pipe insulation, allowing space for the tracers, can be ordered.)

Similarly, insulation beyond what conventional economic analysis indicates might pay for itself by saving pumping costs and avoiding other problems in handling viscous liquids.

Design factors

In dealing with high temperatures, it is important to look at the whole system and make sure that the insulations chosen can withstand the full range of temperatures that will be encountered.

With the trend toward thicker insulations, gaps in the covering, caused by the expansion of metal pipe as it heats up, become more of a problem. Expert insulation contractors prepare joints to minimize these future gaps, but some loss of insulating value must be counted upon.

Form of insulation

Having picked the insulating material and thickness, the form of the material is the next most important selection. Getting as high a ratio of factory fabrication to jobsite fabrication is a key factor in keeping insulating costs to a minimum. Other important insulation-cost factors that should be given attention are ease of installation, and avoidance of damage to the installed insulation, particularly while other construction work continues.

Manufacturers of insulation have devised a number of ways to increase factory fabrication of their products, cutting back on field work. Reflective insulation is virtually all prefabricated to exact dimensions provided by drawings. Solid insulation—mass types such as calcium silicate—is more factory fabricated than heretofore.

Pipe insulation is generally formed in semicircles that are clamped or strapped together over the pipe. Insulation for larger pipes, over about 25 in., are available in quad segments. Calcium silicate pipe insulation is available with either aluminum jacketing or—when superior corrosion- and fire-resistance or strength are required—stainless-steel jacketing already attached.

For lower-temperature applications, fiberglass pipe insulation is available with either a water-vapor barrier or metal jacketing.

Pipe insulation is generally provided in either 2 or 3-ft lengths, which have been found to be the most advantageous sizes for field use. Such lengths are easier to handle without damage and require less cutting to adjust either for changes in pipe directions or for fittings.

For larger-diameter pipe, and other curved surfaces, with diameters in excess of about 34 in., scored blocks of calcium silicate are available for wrapping around the curves. These scored blocks or sheets, as well as flexible blankets (with or without vapor barriers), or metal mesh, or corrugated metal jacketings, are also useful in insulating irregular shapes.

A number of flexible insulation fittings for piping are available. One-piece elbows made of polyvinyl chloride over fiberglass are used on lines insulated with fiberglass or calcium silicate. Aluminum, molded plastic and other preformed fittings for elbows are also available and are widely used.

Tank insulation

There is a trend now toward insulating large tanks, even in places with comparatively mild climates, such as Texas. The energy that would otherwise be needed to reheat materials justifies it.

Generally, either polyurethane foam or fiberglass is used. Urethane works well up to about 250°F; the fiberglass up to 450°F. In some installations, urethane is avoided because of the potential fire hazard. On the other hand, it is less costly to apply.

Insulating an old, dented tank requires an expert, using proper scaffolding and welding equipment. Particularly where the insulation is to be jacketed, as it usually is, tank owners tend to bring in contract insulation units, rather than try to tackle the job with their own maintenance men.

Generally, metal-jacketed fiberglass is either attached by means of studs welded to the tank wall, or held in place by banding. Banding is used on old tanks that are corroded and have thin spots that often make welding difficult.

Jacketing

Jacketing serves several functions in process-pipe and -equipment insulation:

- Protects the insulation against mechanical damage from workmen and equipment throughout the life of the plant.
- Protects exposed insulation from weather.
- Serves as a water-vapor barrier.
- Protects against spills and leaks.
- Protects against fire.

Metal jacketing provides the best protection against mechanical damage. But the designer must be certain that the jacketing specified is adequate for the service conditions encountered. For example, we recently subjected four jacketed insulations to three common service conditions. Tested were calcium silicate with 0.016-in.-

thick jackets, and mineral wool with 0.016-in., 0.032-in., and 0.050-in. preformed aluminum jackets.

First a ladder was rested on the insulation supporting the full weight of a 190-lb man. All but the 0.016-in. jacketed calcium silicate and the 0.050-in. jacketed mineral wool deformed. Then the 190-lb man walked on the insulation. Again, only calcium silicate and the 0.050-in. jacketed mineral wool were unaffected. The 0.016-in. jacketed mineral wool was badly dented; the 0.032-in. jacketed sample was deflected, but sprung back. Similarly, a hammer was dropped from about 4 ft., which dented all three specimens, creasing the two lighter thicknesses on mineral wool.

Corrosion resistance is a particular concern in pipe jacketing and equipment insulation in chemical plants. Where there are corrosive atmospheres, spills or leaks, the jacketing material must resist such conditions, which generally means using stainless steel rather than aluminum. Some types of plastic jacketing, such as the polyfluorides, are excellent against both corrosives and weather, but others, such as PVC, which also offers fine protection against most corrosives, can run into problems in the presence of sunlight, ozone or weather. Ability of a plastic to withstand fungal or bacterial growth can also be a consideration, as in food plants.

In chemical plants, fire safety is always a prime concern. In determining the flame-spread and smoke-production ratings of an insulation system, the insulation and its jacketing should be considered as a composite unit, with the system tested as a whole, rather than by relying on tests of the components separately. Experiment and experience have shown that components do not always yield the same results when put together, as did the insulation, jacketing and cement individually.

Basic data on insulation fire performance derives from the basic ASTM E-84 tunnel test (National Fire Protection Assn. test NFPA 255 or Underwriters Labs test UL 723). Plain calcium silicate insulation normally has flame-spread, fuel-contributed and smoke-developed ratings all of zero. Mineral wool is virtually as good. Up to 25 for flame-spread and 50 for smoke-developed are generally acceptable. Urethane foams have the greatest problems in this area, with flame-spread ratings up to 75, and smoke-developed ratings up to 400, although newer varieties, such as isocyanurate foam, are rated as low as 20 for flame-spread and 30 for smoke-developed.

An insulation with poor fire ratings adds to the danger when the product handled already poses a fire hazard—a problem that is holding down use of urethane foams for insulating tanks in the petroleum and petrochemical industries. But even an insulation with excellent fire ratings can become part of the hazard if it soaks up spilled or leaked flammable materials. Good jacketing that has stayed intact and leakproof is the best answer for this on straight pipe. Valves, pumps, fittings pose special problems because of the greater likelihood of leaks. Such leaks surrounded by molded insulation can result in the soaking up of considerable flammable material without anyone realizing it.

Valves and fittings

Valves, flanges and other fittings are the main stumbling blocks for anyone insulating a piping system. Not only must potential leaks be guarded against, and provisions for access for inspection and maintenance made, but wrapping the troublesome shapes with a heat barrier equaling that on the straight pipe is the most time-consuming and expensive part of the insulation job. Insulating a valve can be three times as expensive as doing a similar length of straight pipe. Because of this, when energy was comparatively cheap, valves were frequently left uninsulated.

Under present conditions, however, it pays to insulate. A 6-in. valve under 60 psi at 70°F ambient can lose 70 Btu/ft^2/h, or 40 million Btu/yr, at a cost of $140–150 a year. An uninsulated flange can lose $216 a year worth of energy. This can mean, at a conservative figure of $120 per fitting each year, and some 250,000 such fittings in a plant, annual losses of $30 million.

Current valve and flange insulating methods have brought the cost down to under $70 for a 6-in. valve. In most cases, insulating a valve or flange will pay for itself in six months to a year. Molded insulations, shaped to fit the valve, flange or other fittings are available. More important in CPI plants—particularly those handling flammable products—are special fitting-insulation systems using refractory fiber that is bonded to corrugated stainless steel. Insulation time and cost are saved; a leak detection system can be incorporated; and the insulation can be removed for maintenance without being destroyed.

Older methods of insulating such fittings involved covering the valve or fitting with insulating cement, smoothing the surface and finishing it with a canvas jacket indoors or a wire-mesh-reinforced emulsion outdoors. In other cases, sleeves of a larger-diameter pipe insulation (for flanges), or structures built up out of block insulation, might be employed.

Reflective and encapsulated types

Reflective-type insulation, which blocks heat transfer by reflection rather than by resistance to conduction and convection through solid mass, is essential in nuclear power plants and is finding growing use in other applications where the qualities needed within a nuclear reactor containment are desirable enough to justify their much higher cost.

These qualities are:

- Resistance to contamination, moisture damage, vibration, mechanical damage.
- Removability and replaceability without destruction of the insulation. (Nuclear insulation, if it had to be destroyed, would pose the same radioactive-materials disposal problem as other nuclear wastes.)
- Protection from fire.

Reflective insulation is virtually all shop-fabricated according to specifications and drawings that are meticulously worked out. Although, in some construction-speedup situations, reflective insulation had been ordered and fabricated from plant drawings before piping was completed, this had led to generally unacceptable amounts of both wasted insulation materials and field work. Shops set up on the plantsite can alter and even build new reflective insulation to adjust for piping changes, but this work is slow and expensive. Reflective insulation installed too soon also leads to waste, both via piping and other structural changes, and through damage by following work crews.

Encapsulated insulation is used where qualities of reflective insulation (such as precision form, physical strength, removability, high degree of protection from the environment) are desired, but space or flooding conditions preclude use of reflective insulation. Any of a number of standard mass-insulation materials, including refractory- and ceramic-fiber blankets and felts, calcium silicate, Min-K, can be encapsulated in a variety of metals, including stainless steel, aluminum, titanium or Inconel, depending on the temperature and other service characteristics involved. Where less protection is required, semiencapsulated insulation can be employed.

Scheduling and installation

As the trend continues toward greater shop fabrication of insulation, more preplanning is also required for the mass types. In simplest terms, if you want 7 in. of insulation on a pipe, you have to be sure that it is 7 in. away from the next pipe, wall, piece of equipment or other obstruction (and, of course, if you want 7 in. on both pipes, they must be more than 14 in. apart). More space must be allowed at curves, bends, elbows, flanges, valves, etc. Also, some extra space must be left for the insulation installers to work.

Future developments

While research continues on devising special insulations to fit special requirements, such as Min-K, the most important work now going on is development of better data on existing insulations, such as calcium silicate, to allow the engineer to better plan the insulation of his plant. Criticism has been leveled, often justifiably, that the kind of exact performance data needed for today's economic analysis and planning of process plant insulation are not available. The "k" factor itself is an approximation, and other manufacturing, installation and service conditions throw the theoretical numbers off even further.

The research centers of leading insulation manufacturers are working to narrow the gap between theoretical data and actual performance. Therefore, insulation designers have begun planning a whole new way of calculating heat transfer of insulations.

Traditionally, such calculations have involved taking the "k" factor of an insulation and multiplying it by the inches of thickness, throwing in the temperature of the product, the size and length of the pipe, and coming up with a theoretical heat-loss number for the plant or system.

But this is not quite good enough now. Considerable work is being done with a new systems approach, based on experimentally derived data. The problem has been that variations in installation; gaps developing at insulation joints (as the pipe expands under the heat of startup); the special heat-transfer calculation problems of insulated curves, valves, pumps and other piping—all add up to a different set of numbers than the theoretical calculations do.

In the systems approach, the insulation laboratory would put together small sections of piping, including insulation gaps, curves and fittings, and then closely monitor heat loss as a measurement of the heat difference between the beginning and the end of the piping. In extending this number to the actual plant, the engineer would be able to consider not only the length of the piping system but also its relative complexity; that is, how many bends and valves it has. When this method is fully developed, far-more-reliable numbers will be available.

The rationale behind this research emphasis is that the greatest energy and money savings will result from better use of the insulation materials we now have available. Both engineers and corporate financial management, given this kind of reliable data, will be assured that more insulation than previously specified, both in new and old plants, will pay. Short of basic process changes, which means building new plants, this is how the chemical process industries will save energy. The amounts saved will be considerable.

The authors

Michael R. Harrison is a technical services engineer for the Industrial Products Div. of Johns-Manville, Greenwood Plaza, Denver, CO 80217, where he provides engineering and technical support for his division's market managers and sales force. He holds a B.Sc. in mechanical engineering from the University of Colorado, Boulder, Col.

Charles M. Pelanne is a senior research specialist in the Johns-Manville Applied Research Section of the Central Research Dept., Research & Development Center, Ken-Caryl Ranch, Denver, CO 80217. He is an expert in the development of thermal conductivity measurement equipment, particularly heat-flow meter techniques, and is also expert in the field of heat transfer through insulations.

Guidelines for Heat-Loss Calculations From Storage Tanks — Table I

Condition	Neglect:	Graphical Solution. Use:	Algebraic Solution. Calculate:
Vessel is uninsulated. Contained liquid is agitated.	Film resistance of contained liquid.	Fig. 1. Correct for wind velocity or outside surface emissivity, if required.	Heat loss through outside air film only, by using Eq. (7b) and (4b).
Vessel is uninsulated. Contained vapor condenses appreciably at wall temperature.	Film resistance of contained vapor.	Fig. 1. Correct for wind velocity or outside surface emissivity, if required. Temperature correction factor, $W = 1.0$	Heat loss through outside air film only, by using Eq. (7b) and (4b).
Vessel is unisulated. Contained liquid fouls or solidifies at wall temperature.	Film resistance of both contained liquid and vapor adjacent to fouled surface.	Fig. 1 with temperature correction factor $W = 0.4$; and wind velocity or emissivity correction as required.	Heat loss through fouling layer and adjacent air film by using Eq. (4a) or (8).
Vessel is uninsulated, and contains stagnant liquid or dry vapor.		Fig. 1 with temperature correction factor, W, obtained from table for type of contained material; and wind velocity or emissivity correction as required.	Heat loss through inside-fluid film and adjacent air film by using Eq. (6) or (6a), (7) and (4).
Vessel is insulated with calcium silicate or similar material.	Film resistance of both contained liquid and vapor adjacent to insulated surface.	Fig. 2 with wind-velocity correction factor obtained from table. Insulation thickness may be obtained from Fig. 2.	Heat loss through insulation and adjacent air film by using Eq. (7a) and (3).

Insulation Saves Energy

Shortcut graphical methods allow rapid evaluation of heat losses from uninsulated and insulated vessels. Depending on the temperature difference between vessel and ambient air, the method also sets the required thickness of outside insulation.

RICHARD HUGHES and VICTOR DEUMAGA, The Badger Co.

Accurate calculations for determining the heat flow between the contents of a vessel and its surroundings are often complex. Multiple resistances in series are usually present in the heat-flow path. In addition, conduction, convection and radiation flows may be simultaneously involved. And there may be heat flow by evaporation and condensation within the confines of the vessel.

Therefore, the particular assumptions and simplifications involved in the method used to calculate heat losses must be understood, and the accuracy of the method must be consistent with a particular requirement.

For example, a simple shortcut approximation will usually indicate whether outside insulation of a given thickness is required. On the other hand, it is often desirable to accurately check a final design (sometimes at varying ambient or internal tank conditions) to determine the average annual heat losses.

Accordingly, we will present two methods of heat-loss calculation:

1. A rapid graphical method that can be used for exploratory evaluation of heat losses.
2. A more accurate algebraic method that is suitable for a computerized solution if multiple calculations are required.

Graphical Method Is Rapid

We can obtain an approximate value for the unit heat loss from Fig. 1 or 2. The unit heat loss when multiplied by the surface area under review yields the overall hourly

Originally published May 27, 1974

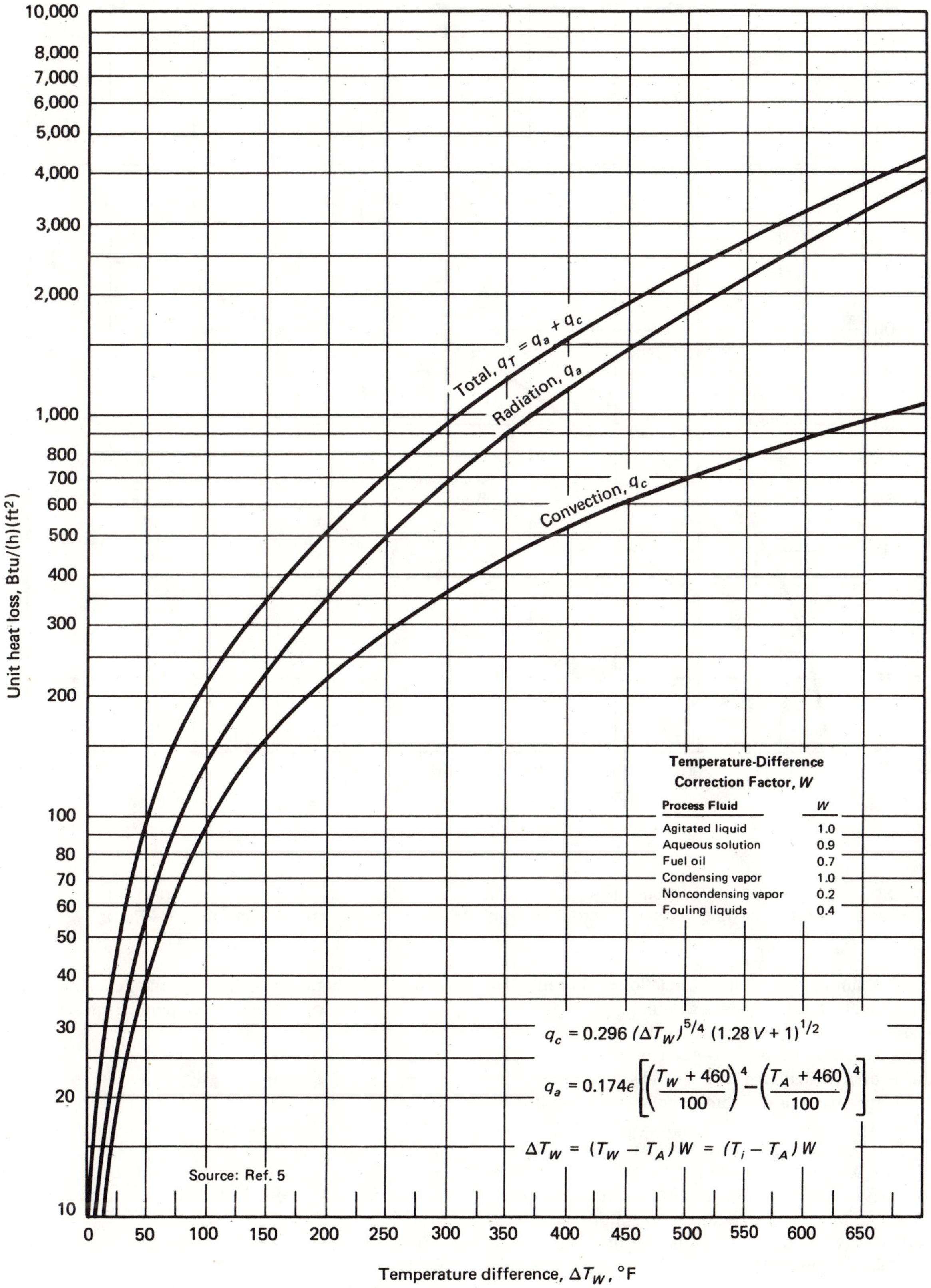

UNINSULATED tanks have heat losses that depend on nature of tank contents. Values in chart are for wind velocity of zero, surface emissivity of 0.9, and ambient air temperature of 70°F—Fig. 1

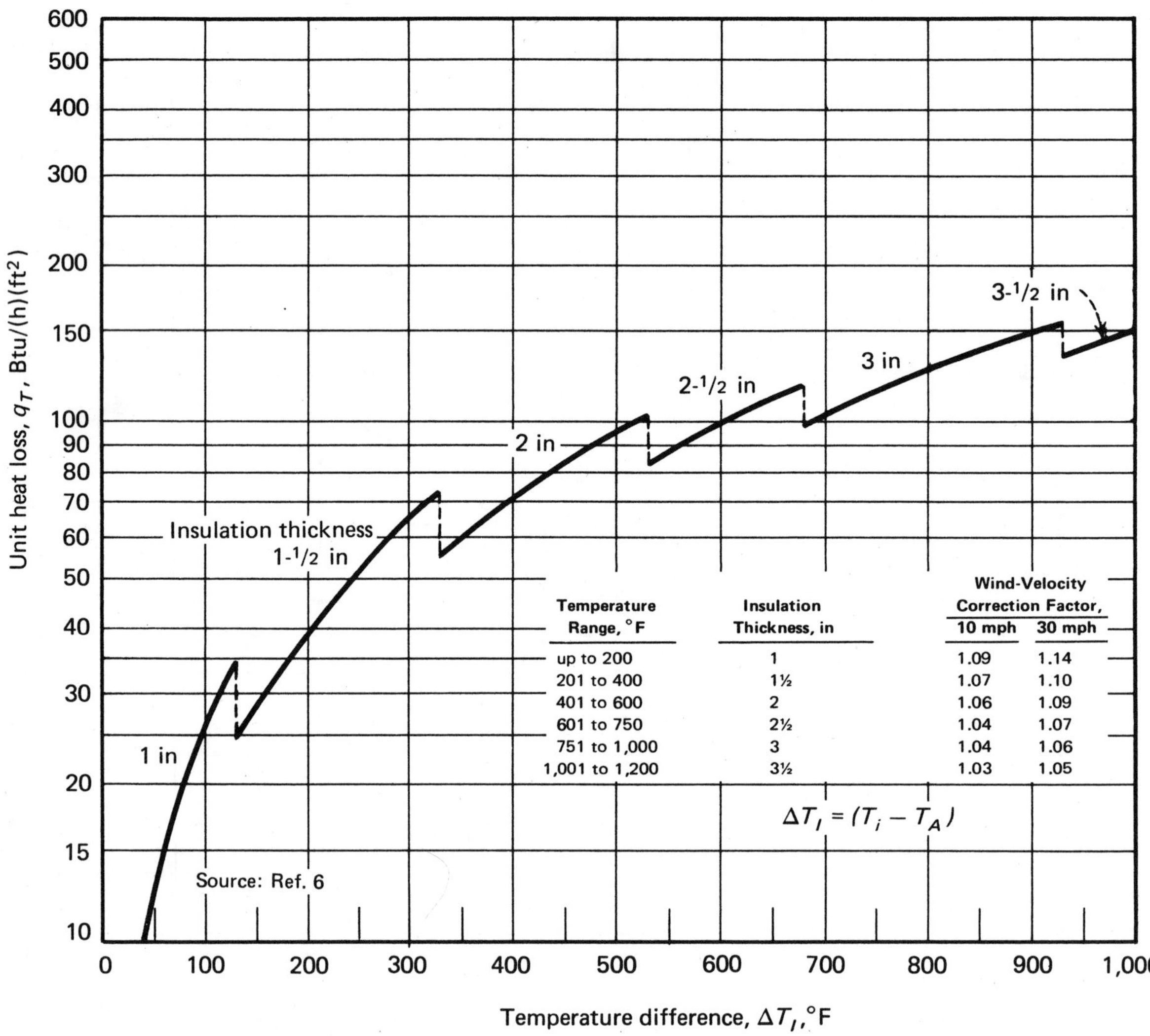

Temperature Range, °F	Insulation Thickness, in	Wind-Velocity Correction Factor, 10 mph	30 mph
up to 200	1	1.09	1.14
201 to 400	1½	1.07	1.10
401 to 600	2	1.06	1.09
601 to 750	2½	1.04	1.07
751 to 1,000	3	1.04	1.06
1,001 to 1,200	3½	1.03	1.05

INSULATED tanks, covered with calcium silicate, have heat losses based on negligible resistance to heat flow on process side. Values in chart are for wind velocity of zero, emissivity of 0.8, ambient air temperature of 70°F—Fig. 2

heat loss. A summation of the discrete losses through the tank roof and tank side, exposed to both liquid and vapor, gives the total hourly heat loss to the atmosphere. In addition, heat loss from the bottom of the tank to the ground is easily calculated from Eq. (5) to complete the heat-loss approximation. Table I summarizes conditions for which Fig. 1 and 2 are recommended.

Fig. 1 is used to approximate heat losses from uninsulated tanks. A table of correction factors for temperature difference is included to compensate for common internal resistances to heat flow. In addition, the unit heat loss is divided into a convection and a radiation component. This allows rapid, independent correction for wind velocity and surface emissivity, if required. Fig. 1 also contains a total unit-heat-loss curve at zero wind velocity and 0.9 radiation emissivity.

Fig. 2 is used to approximate heat losses from insulated tanks. A table of recommended insulation thickness for various temperature ranges is included, and wind-loss correction factors at the recommended insulation thickness are given.

Analysis of Algebraic Method

Heat losses from tanks are calculated by using the natural-convection heat-transfer relations for the sides and roof of the tank, and the conduction heat-transfer relation for the bottom where the tanks rest on the ground.

The total heat loss, Q_T, is divided into four components:

$$Q_T = Q_L + Q_V + Q_R + Q_G \tag{1}$$

where Q_L is heat loss from liquid in tank to atmosphere through sidewall; Q_V is heat loss from vapor in tank to atmosphere through sidewall; Q_R is heat loss from vapor in tank to atmosphere through roof; and Q_G is heat loss from liquid in tank to ground through the bottom of the tank.

Table I summarizes conditions for which specific equations are recommended to calculate the component heat losses.

Heat Losses: Sidewall and Top

The method for calculating side and top heat losses, Q_L, Q_V and Q_R, is essentially the same. In all cases, heat loss is calculated through two series resistances to heat transfer. These resistances are: (1) the tank inside-film or the tank insulation, and (2) the air film (when insulation is used, the inside-film resistance is neglected). Note that in all cases the tank wall itself is neglected as a separate resistance. If the tank is fabricated of heat-transfer-resistant material, then the wall itself can be treated as an insulation layer. Similarly, if appreciable fouling or solidification occurs within the tank, the fouling layer becomes equivalent to an insulation layer.

The usual individual and overall relationships for heat transfer are given by:

$$Q/A = q_a + q_c = q = U(T_i - T_A) = h_i(T_i - T_W) = (k_I/X_I)(T_W - T_I) = h_A(T_I - T_A) \quad (2)$$

When the tank surface under consideration is insulated, the unit heat loss, q, is given by:

$$q = (k_I/X_I)(T_i - T_I) = h_A(T_I - T_A) \quad (3)$$

Note that the inside-film resistance is neglected, and $T_W = T_i$. The inside liquid-film resistance can also be neglected, even in the absence of insulation, if the liquid contents are agitated. In this case, a direct resolution is possible by using Eq. (4b). Similarly, the inside vapor-film resistance can be neglected if appreciable condensation occurs in the portion of the tank's surface that is exposed to vapors.

When the tank surface under consideration is not insulated, the unit heat loss, q, is given by:

$$q = h_i(T_i - T_W) = h_A(T_W - T_A) \quad (4)$$

For either insulated or uninsulated tanks, the method of computation is the same. The air temperature, T_A, and the tank-contents temperature, T_i, are known. Individual film heat-transfer coefficients [as defined by Eq. (6) and (7)] are used, together with insulation resistance, where applicable, to calculate the remaining unknown intermediate temperature, T_I, in Eq. (3), or T_W in Eq. (4), and hence to calculate the unit heat loss. The total heat loss is then calculated as the product of the unit heat loss and tank surface area under consideration.

When the tank surface is coated internally with a fouling or solidified layer, the unit heat loss is found from a modification of Eq. (4), which is:

$$q = (1/R_i)(T_i - T_W) = h_A(T_W - T_A) \quad (4a)$$

When the inside-film resistance is small and the tank surface is insulated, the unit heat loss becomes a particular case of Eq. (4), which is:

$$q = h_A(T_i - T_A) \quad (4b)$$

Nomenclature

A	Area of heat transfer, ft^2
C	Specific heat, Btu/(lb)(°F)
D	Diameter of tank, ft
g	Gravity acceleration, 4.17×10^8 ft/h^2
H	Tank height, ft
k	Thermal conductivity, Btu/(h)(ft^2)(°F/ft)
P	Pressure, psia
Q	Heat transferred, Btu/h
q	Unit heat transferred, Btu/(h)(ft^2)
R	Fouling factor, (hr)(ft^2)(°F)/Btu
T	Temperature, °F
U	Overall heat-transfer coefficient, Btu/(h)(ft^2)(°F)
V	Wind velocity, mph
W	Temperature-difference correction factor
X	Thickness, ft
β	Coefficient of volumetric expansion, 1/°F
ϵ	Emissivity
μ	Viscosity, lb/(h)(ft)
ρ	Density, lb/ft^3
Subscripts	
A	Environment
a	Radiation
c	Convection
F	Fluid film
G	Ground
i	Tank content
I	Insulation
L	Liquid
M	Insulation mean
R	Roof
r	Fouling layer
T	Total
V	Vapor
W	Tank wall

Heat Losses: From Bottom of Tank

When a tank rests on the ground, the heat loss through the bottom, Q_G, is given by:

$$Q_G = 2Dk_G(T_L - T_G) \quad (5)$$

Eq. (5) was derived [*1*] from an equation given by McAdams [*3*]. The inside-film resistance to heat transfer is neglected.

If the tank is mounted aboveground, the bottom tank surface, A_G, is then added to the side-surface, A_L, when calculating Q_L by Eq. (3) or (4).

Finally, Q_T is found as the sum of the component heat losses from Eq. (1).

Film Coefficients and Fouling Factor

Inside liquid-film coefficient for a liquid in contact with the tank surface ($h_i = h_L$) is given by:

$$h_L = 0.45k_F\left[\frac{\rho_F^2 g\beta_F C_F(T_i - T_W)}{\mu_F k_F}\right]^{0.25} \quad (6)$$

Eq. (6) was derived [*2*] from an equation given by McAdams [*4*].

Inside vapor-film coefficient for a vapor in contact with the tank surface ($h_i = h_V$) is given by:

$$h_V = 9.686 \times 10^{-2} P^{0.5}(T_F)^{-0.1}(T_V - T_W)^{0.25} \quad (6a)$$

Eq. (6a) is a particular case of Eq. (6), and can be used for both nitrogen and air at moderate temperatures (50 to 400°F) and pressures (0 to 500 psig). When the vapor present is appreciably different from air, the vapor-film coefficient is calculated from Eq. (6) by using vapor-film physical properties. For a condensing vapor, the vapor-film resistance is neglected, and hence the coefficient is not calculated.

The outside air-film coefficient is calculated from one of the following equations.

When the tank surface is uninsulated, Eq. (7) is recommended for substitution into Eq. (4):

$$h_A = 0.296(T_W - T_A)^{1/4}(1.28V + 1)^{1/2} + 0.174\epsilon\left[\frac{\left(\frac{T_W + 460}{100}\right)^4 - \left(\frac{T_A + 460}{100}\right)^4}{T_W - T_A}\right] \quad (7)$$

Eq. (7) includes a wind-velocity correction factor and a radiation loss factor [5].

When the tank surface is insulated, Eq. (7a) is recommended for substitution into Eq. (3):

$$h_A = 0.296(T_I - T_A)^{1/4}(1.28V + 1)^{1/2} + 0.174\epsilon\left[\frac{\left(\frac{T_I + 460}{100}\right)^4 - \left(\frac{T_A + 460}{100}\right)^4}{T_I - T_A}\right] \quad (7a)$$

Eq. (7a) differs only from Eq. (7) in that the surface temperature of the insulation, T_I, is used instead of the tank-wall temperature, T_W.

In either case, Eq. (3) or (4) is then solved by trial and error for the intermediate temperature, T_I or T_W, which is then used to calculate the unit heat loss, q.

When the tank surface is uninsulated and the liquid-film or vapor-film resistance for the material contained in the tank is small, Eq. (7b) is recommended for substitution into Eq. (4b):

$$h_A = 0.296(T_i - T_A)^{1/4}(1.28V + 1)^{1/2} + 0.174\epsilon\left[\frac{\left(\frac{T_i + 460}{100}\right)^4 - \left(\frac{T_A + 460}{100}\right)^4}{T_i - T_A}\right] \quad (7b)$$

For these conditions, Eq. (4b) can be used directly to calculate the unit heat loss, q.

If the thickness of the fouling layer can be estimated, the inside fouling factor used in Eq. (4a) can be calculated from:

$$R_i = X_r/k_r \quad (8)$$

The thermal conductivity, k_r, is usually taken as the fluid thermal conductivity at the mean temperature across the fouling layer.

Temperatures and Physical Properties

Initially, we can assume that the intermediate temperature, T_W or T_I, is midway between T_i and T_A. The inside fluid properties are used at the average inside-film temperature of $(T_i + T_W)/2$. Similarly, outside properties at the average outside-film temperature of $(T_W + T_A)/2$ are used to evaluate physical properties for the film coefficients in Eq. (4).

When the tank is insulated, the average insulation temperature, T_M, becomes $(T_i + T_I)/2$. Similarly, the average outside-film temperature becomes: $(T_I + T_A)/2$. These average temperatures are used to evaluate physical properties for the film coefficients in Eq. (3).

For greater accuracy, the average film temperatures can be recalculated by using T_I or T_W, as found from Eq. (3) and (4). Hence, a new heat loss can be obtained by using physical property data at the new average film temperature.

Emissivity of Surface — Table II

Surface	Fraction of Black-Body Radiation	
	at 50° - 100°F	at 1,000°F
Black nonmetallic surfaces such as asphalt, carbon, slate, paint, paper.	0.90 - 0.98	0.90 - 0.98
Red brick and tile, concrete and stone, rusty steel and iron, dark paints (red, brown, green)	0.85 - 0.95	0.75 - 0.90
White or light-cream brick, tile, paint or paper, plaster, whitewash.	0.85 - 0.95	0.60 - 0.75
Bright aluminum paint, gilt or bronze paint.	0.40 - 0.60	–
Dull brass, copper, or aluminum; galvanized steel; polished iron.	0.20 - 0.30	0.30 - 0.50
Polished brass, copper, Monel metal	0.02 - 0.05	0.05 - 0.15
Highly polished aluminum, tin plate, nickel, chromium.	0.02 - 0.04	0.05 - 0.10

Source: "ASHRAE Guide (Heating, Ventilating, Air-Conditioning Guide)," Chap. 5, p. 52, American Soc, of Heating, Refrigerating and Air-Conditioning Engineers, New York, 1959.

Problem Illustrates Procedures

A fuel oil at 12° API at 60°F with a viscosity of 50 SSF (Saybolt seconds furol) at 122°F is stored at 300°F in a 20-ft-dia. by 30-ft-high carbon-steel tank at atmospheric pressure. The oil level in the tank is 18 ft. Air temperature is 70°F. Let us calculate heat losses to the environment for the following two cases:

Case 1—Total surface of the tank is uninsulated and black. Wind velocity is 0 mph. Surface emissivity of tank is 0.9. Thermal conductivity of ground is 0.8 Btu/(h)(ft²)(°F/ft).

We will begin the computations by finding the heat loss from the wetted inside surface of the tank. The wetted area is:

$$A_L = \pi D H_L = \pi(20)(18) = 1{,}130 \text{ ft}^2$$

$$\Delta(T_i - T_A) = (300 - 70) = 230°\text{F}$$

From Fig. 1, we find the unit heat loss to be 664 Btu/(h)(ft²) for this temperature difference. Therefore:

$$Q_L = A_L q_L = 1{,}130(664) = 750{,}320 \text{ Btu/h}$$

Next, we calculate the heat loss from the dry inside surface and roof surface of the tank:

$$A_V = \pi D H_V + \pi D^2/4$$
$$A_V = \pi(20)(30 - 18) + \pi(20)^2/4$$
$$A_V = 754 + 314 = 1{,}068 \text{ ft}^2$$
$$\Delta T_W = (T_i - T_A)W = 230(0.20) = 46°\text{F}.$$

Note that W is the temperature-difference correction factor for a noncondensing vapor (Fig. 1).

For this temperature difference, we find the unit heat loss from Fig. 1 to be: 84.3 Btu/(h)(ft^2). Hence:

$$Q_V = A_V q_V = 1{,}068(84.3) = 90{,}030 \text{ Btu/h}$$

Assuming that $T_G = T_A$, we substitute the appropriate quantities into Eq. (5) to find Q_G as:

$$Q_G = 2(20)(0.8)(300 - 70) = 7{,}360 \text{ Btu/h}$$

Therefore, the total heat loss is:

$$Q_T = Q_L + Q_V + Q_G$$
$$Q_T = 750{,}320 + 90{,}030 + 7{,}360 = 847{,}710 \text{ Btu/h}$$

Alternatively, we can find the exact solution by using the algebraic equations. After making the necessary calculations, we find that:

$$Q_T = Q_L + Q_V + Q_G$$
$$Q_T = 750{,}980 + 135{,}782 + 7{,}360 = 894{,}122 \text{ Btu/h}$$

Case 2—Roof of tank is uninsulated and is coated with aluminum paint. Side-wall is insulated with calcium silicate, or equivalent, and has a surface emissivity of 0.8. Wind velocity is 30 mph. Tank contents are not agitated. Thermal conductivity of ground is 0.8 Btu/(h)(ft^2)(°F/ft).

As in Case 1, the wetted surface of the tank, A_L, is 1,130 ft.^{2} This represents the interior circumferential area of the tank wetted by the fuel oil to a height of 18 ft.

Heat loss from the vessel wall to ambient air is a function of:

$$\Delta T_I = (T_W - T_A) = (300 - 70) = 230°\text{F}$$

From Fig. 2, we find the recommended insulation thickness to be $1\frac{1}{2}$ in, and the wind-velocity correction factor to be 1.10. Also, from Fig. 2, we find that the unit heat loss, corrected for wind velocity, becomes:

$$q_L = 46(1.10) = 50.6 \text{ Btu/(h)(ft}^2)$$

Hence: $Q_L = A_L q_L = 1{,}130(50.6) = 57{,}178$ Btu/h

Now, we calculate the heat loss from the insulated dry-side surface:

$$A_V = \pi(20)(30 - 18) = 754 \text{ ft}^2$$
$$\Delta(T_W - T_A) = 230°\text{F}$$
$$q_V = 46(1.10) = 50.6 \text{ Btu/(h)(ft}^2)$$
$$Q_V = A_V q_V = 754(50.6) = 38{,}152 \text{ Btu/h}$$

In order to calculate the heat loss from the uninsulated roof, we must use Fig. 1. We find that $A_R = \pi(20)^2/4 = 314$ ft,^{2} and that surface emissivity = 0.4 from Table II.

$$\Delta T_W = (T_i - T_A)W = 230(0.20) = 46°\text{F}$$

Using this corrected temperature difference and Fig. 1, we find that the unit heat loss for radiation and convection (correcting the radiation loss for an emissivity of 0.4 and the convection loss for the wind velocity) is:

$$q_a = 54.2(0.4) = 21.7 \text{ Btu/(h)(ft}^2)$$
$$q_c = 35.5[(1.28)(30) + 1]^{1/2} = 222.8 \text{ Btu/(h)(ft}^2)$$
$$q_R = q_a + q_c = 21.7 + 222.8 = 244.5 \text{ Btu/(h)(ft}^2)$$

Hence: $Q_R = A_R q_R = 314(244.5) = 76{,}773$ Btu/h

Heat loss to the ground is the same as Case 1, i.e. $Q_G = 7{,}360$ Btu/h.

Therefore, the total heat loss is the sum of the component parts, or:

$$Q_T = 57{,}743 + 38{,}529 + 76{,}773 + 7{,}360$$
$$Q_T = 180{,}405 \text{ Btu/h}$$

Again, we can calculate a more-exact answer by using the algebraic equations. For this case:

$$Q_T = 54{,}419 + 36{,}279 + 49{,}887 + 7{,}360$$
$$Q_T = 147{,}945 \text{ Btu/h}$$

This problem illustrates the ease with which rapid approximations to heat loss can be made.

References

1. Stuhlburg, D., How to Design Tank Heating Coils, *Petrol. Refiner*, Apr. 1959, p. 143.
2. Ibid, p. 146.
3. McAdams, W. H., "Heat Transmission," 3rd ed., p. 54, McGraw-Hill, New York, 1954.
4. Ibid, p. 172.
5. Malloy, J., "Thermal Insulation," p. 35, Van Nostrand Reinhold, New York, 1969.
6. "Heat Insulation Manual," pp. 72–79, Pabco Div., Fibreboard Corp., Emeryville, Calif.

Meet the Authors

◀ **Richard E. Hughes** is a process supervisor with The Badger Co., One Broadway, Cambridge, MA 02142. He joined Badger in 1959 and has a wide range of experience in chemical and petroleum processes. He has a B.S. in chemical engineering from the University of Witwatersrand, South Africa, and an M.S. in chemical engineering from Yale University. He is a member of AIChE.

Victor Deumaga is a process design engineer with The Badger Co., Cambridge, MA 02142. He joined Badger in 1972, and has experience in vinyl chloride, acrylonitrile and phthalic anhydride processes. He has a B.S. in chemical engineering from Pratt Institute, and an M.S. in chemical engineering from Massachusetts Institute of Technology. ▶

Guide to Better Insulation

Insulation practices developed during the years of inexpensive energy are conflicting with today's need for more-efficient thermal systems. Here is a basic guide to the insulation industry and its products.

ROBLEE B. MARTIN, Keene Corp.

The international events of the last two years that have upset decades of cheap and plentiful energy obviously have many implications for the chemical process industries (CPI). Among the many changes underway is much greater emphasis on insulation to conserve energy. Unfortunately, because of the way insulation has been regarded in the past, there are misconceptions and knowledge gaps in many engineers' understanding of the subject.

Traditionally, personnel safety has been the primary reason for applying insulation to process equipment and piping. If this statement seems surprising, consider that the majority of specifications have called for a maximum external temperature (usually 120 to 140°F) rather than a maximum allowable heat loss. As a result, insulation manufacturers have faced demands to provide more-thermally-efficient products, without regard to quality. Even the sophisticated user has seldom looked beyond the dollar cost of unit heat-flow resistance.

The insulation industry thus has developed into one that is highly volume/cost conscious, but not generally well equipped with the process controls required to turn out uniform, quality products. With insulation costs usually less than 0.5% of total facility cost, many are content to accept products that may contain a few percent of defective or off-specification material. However, the new energy environment will compel using insulation manufactured with much greater attention to quality.

The Insulation Engineer

Specifying insulation has been a low-key, secondary function in many industrial and engineering organizations. Insulation engineers are usually knowledgeable and experienced in the major business of their employers, but have gathered their facts about insulation via salesmen, outdated texts and literature. The proper primary expertise of the insulation-specifications engineer should be *insulation*. (If he makes a mistake related to the process or product, it is likely to be detected. If he makes an "insulation error," it will very probably remain undetected.)

Inhouse specifications should be developed around the requirements of the process and should be only secondarily concerned with what kind of insulation is available. That is, define the ideal or most-desirable insulation and then qualify the specification on the basis of what suppliers have to offer. Historically, the reverse has happened—engineers have examined what has been available and then composed specifications around the products that have appeared most suitable (or least unsuitable).

The immediate results of either approach are the same, but the long-range effect is greatly different. The first approach defines and continually upgrades objectives, and prods the insulation manufacturer in the proper direction. The second—and prevalent approach—maintains an unsatisfactory status quo.

Insulation manufacturers recognize that it is incredibly difficult to get on an "approved list," but once you are on such a list, it is almost as difficult to get thrown off. Obviously, both situations are wrong.

The success ratio of innovations in insulation has been dismal. A manufacturer is often reduced to simply trying to duplicate a competitor's "approved" product that, unknown to the manufacturer or user, is far from what is truly needed. The approved product was just the best thing available when the specification was written.

It is not uncommon for specifications to be an impediment to progress. Consider, for example, calcium silicate products, which are an outgrowth of the older magnesia insulations and are made in the same facilities by a similar process. The improvement represented by calcium silicate is substantial and clearly documented. Yet, it took ten years before the majority of specifications were changed. Even today, there are specs that call for magnesia, despite the fact that it is virtually unobtainable.

BASIC INSULATION PRINCIPLES

The two underlying principles that govern development of better insulation products are use of materials having low thermal conductivity, and the inclusion of pores, which reduce heat transfer and density. Although these principles appear patently obvious, they are often overlooked by the user, as shown by restrictions and compromises imposed on other properties. In addition, the actual function of pores is poorly understood; their role varies significantly depending on temperature.

Below "red heat," conduction is the primary mechanism in heat transfer from process equipment. Pores in the insulating material increase the length of the path over which heat transfer occurs. Use of very small pores,

Originally published May 12, 1975

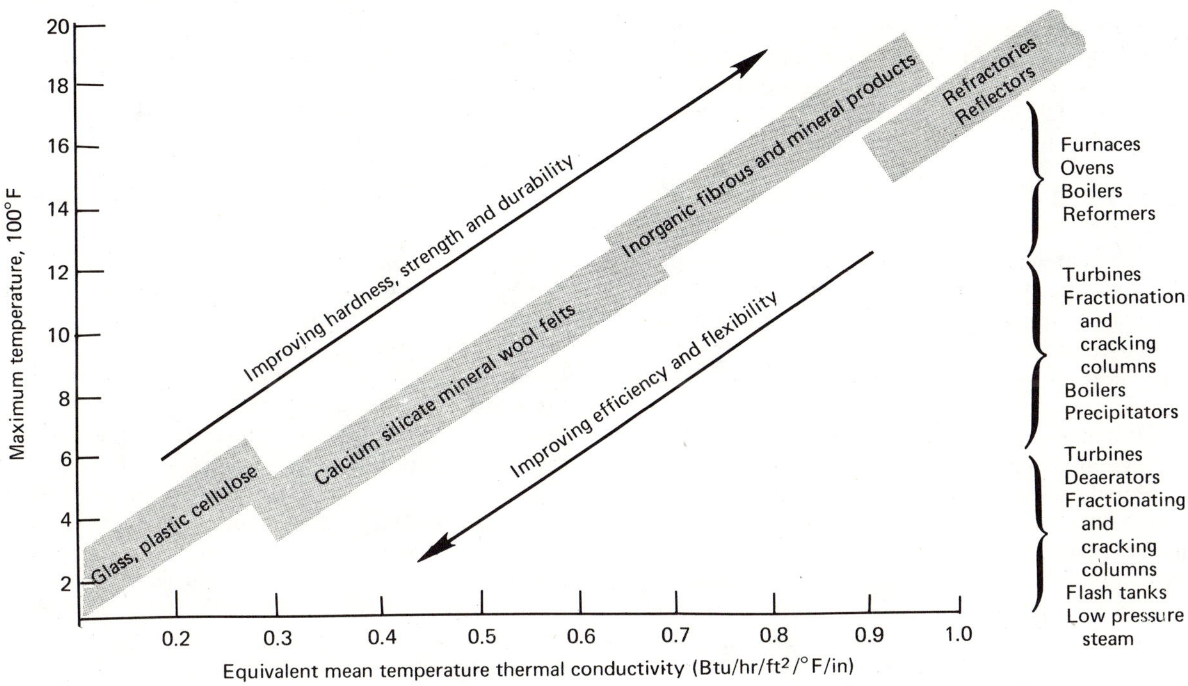

CHARACTERISTICS OF INSULATING MATERIALS. High-temperature products are strong and durable, but are less efficient than those designed for moderate and low temperatures.

or pores filled with low-pressure gas, effectively rules out convection as an important heat-transfer mechanism.

Radiation becomes progressively more important at temperatures above red heat (radiation is proportional to the fourth power of absolute temperature). However, an efficiency factor is involved each time energy is emitted, radiated and absorbed. Heat transfer by radiation across 10 pores, each having an efficiency factor of 0.6, will result in a gross apparent efficiency of 0.006. In real situations, with hundreds or even thousands of pores involved, radiation is virtually eliminated as a significant heat-transfer mechanism. The prevailing activity is conduction along a contorted and elongated path.

Efficiency vs. Temperature

The thermal conductivity of most pure materials decreases as temperature increases. But this is not the case for real-life thermal insulation. The pores or voids that improve efficiency also reverse the temperature/thermal-conductivity relationship, so that for materials having useful insulation properties, efficiency drops as temperature rises.

The adverse effect is compounded by the fact that insulation materials suitable for high-temperature applications are somewhat perverse—their thermal conductivities are intrinsically higher than those of low-temperature products. It is important that engineers fully appreciate this; otherwise they run the risk of fruitlessly searching for a mythical alchemist's stone.

As a result of the relationship between temperature and conductivity, maximum performance in insulating a very hot surface is seldom achieved with just one material. An inner layer of high-temperature product should be backed up by a layer (or layers) of material having lower temperature capability but greater efficiency. This approach requires extra installation labor, but less insulation overall. Any net increase in expense is repaid through lower operating cost.

Another point to remember is that porous materials are almost invariably less strong than monolithic ones. When insulation must have useful strength, the usual solution involves special combinations of materials or composites.

Above 2,000°F

From the user's point of view, top temperature capability is the most logical way of classifying insulation. Above 2,000°F, ceramic refractories and reflecting insulation are predominant.

Refractories range from low-performance inexpensive firebrick to chemically pure foams such as alumina beads. They generally have superior strength, poor insulation efficiency, and high installed cost. However, refractories usually are the cheapest, most practical method of insulating very hot, large, flat surfaces, particularly when structural integrity is a major consideration. Designers prefer granules or beads, which can be poured into place, for irregular or curved surfaces.

Many engineers overlook the fact that coarse sand or gravel provides an excellent, low-cost insulation layer. It is naive to expect the salesman to mention this.

Reflective insulations tend to be highly engineered systems for high-temperature applications where radiation is the predominant heat-transfer mechanism. Most installations involve piping or small vessels. Usually, material costs are high, but the labor required is low. In fact, a significant advantage of reflective insulations is that they can be rapidly removed for inspection or maintenance of equipment. Installed costs range from \$2 to \$5/(ft²)(in) usually, but can be substantially greater.

1,200 to 2,000°F

Refractory and reflective insulations also find applications in the 1,200 to 2,000°F-range. However, it is common to use more-thermally-efficient types (with an unavoidable compromise in strength). Principal products include inorganically bonded fibrous insulation, based on ceramic or special mineral-wool fibers, and denser materials that employ perlite or diatomaceous earth as a bulk filler component.

In the trade, these products are referred to as rigid, and are sold by the board foot (12 x 12 x 1 in). Mechanical strength is limited—they contribute little to the structural integrity of the system beyond being self-supporting. Compared to refractories, these materials offer better insulation efficiency at lower installed cost, generally about \$1.25 to \$3/(ft²)(in).

The engineer should make sure that high-temperature properties such as shrinkage, chemical reactivity and permanent embrittlement fit the application. Some of these products are fundamentally acid in nature; others are basic. The possibility of accidental exposure to acid or caustic environments will influence final selection.

600 to 1,200°F

Calcium silicate and resin-bonded mineral-wool felts dominate this temperature range. Aside from high efficiency, their properties are quite different.

Calcium silicate is a hard, rigid product supplied in flat blocks, or half sections of pipe covering. While having useful crushing strength, it is brittle and subject to breakage during shipment, installation, and plant operations. Calcium silicate is easily repaired, however, accepts paint well, and provides an excellent substrate for a variety of insulation cements.

Mineral-wool felts come in many forms, with densities ranging from 2.5 to 12 lb/ft³. Rigid or flexible types are available, and a variety of factory-installed facings can be ordered. Mineral wool is supplied by most manufacturers in flat form but can be converted, either by the maker or a jobshop, to pipe insulation in several proprietary designs. Installation costs tend to be lower than for calcium silicate, and damage during transit or at the jobsite is minimal.

Calcium silicate and mineral wool are used in applications such as turbines (hot side), fractionation and cracking columns, mainstream and reheat piping, boilers, ductwork, precipitators and incinerators. Installed costs range from about \$1 to \$2/(ft²)(in).

Below 600°F

Resin-bonded fibrous-glass insulation fills about 80% of the market for industrial insulation below 600°F. Like mineral-wool felts, these products are available in a variety of densities and other physical properties. They are extremely efficient and have many different factory-applied facings available at low cost.

Installation is rapid and inexpensive, although in some cases external structural support is needed. Lower-density forms are available in convenient, continuous rolls. As with other materials, performance characteristics should be carefully matched with the intended application.

A variety of other products find limited but important uses below 600°F. Cellulose (macerated paper) and plastic foam are occasionally encountered, although combustibility and thermal instability mitigates against their use above 300°F. High-cost cellular-glass products with exceptional efficiency are applied to refrigeration and cryogenic equipment, as are mineral-wool felts.

Among applications in this temperature range are turbines (cool side), deaerator tanks, condensate systems, fractionation and cracking columns, flash tanks, low-pressure steam lines, and heated viscous-liquid lines. Installed costs are usually between 70¢ and \$1.50/(ft²)(in).

Future Needs

The figure summarizes the generalizations presented here. We hope we have demonstrated that there is an important need for better inhouse expertise on insulation. With the increasing cost of energy, and considering the "thermally sensitive" nature of CPI operations, insulation engineering should begin on the first day of designing new facilities. It no longer makes any sense to design first and then worry about if and how the facility should be insulated.

Engineers need a better understanding of the total impact of insulation on operating costs. Technical expertise alone is not enough. It must be backed up by proper cost accounting. A substantially higher investment in insulation will usually yield a rapid, handsome return.

Finally, it should be stressed that selection of the insulation contractor is every bit as important as selection of the product. There are fewer than 10 major manufacturers of quality industrial insulation, but there are more than 500 insulation contractors. It follows, without derogatory intent, that there are far more incompetent contractors than incompetent manufacturers. Major producers operate their own contracting services, and buyers would do well to weigh carefully the advantages of minimizing the sources of responsibility.

Meet the Author

Roblee B. Martin is president, Insulation Div., Keene Corp., P. O. Box 145, Princeton, NJ 08540. He is a chemical engineer holding B.S. and M.S. degrees from Columbia University, where he also taught chemical engineering. Mr. Martin has held several executive positions, most recently as president of the Presidents Assn. of the American Management Assns. He joined Keene last year.

The cost of missing pipe insulation

Here's a graphical method for finding the annual expense of heat loss from uninsulated piping—at any efficiency of energy conversion, and for any cost of fuel and wind condition.

Rene Cordero, *Allied Chemical Corp.* *

☐ For piping whose surface has been left uninsulated or whose insulation has been so severely damaged as to be useless, the yearly (8,760 h) cost of the heat loss per square foot can be found from Fig. 1. Costs so determined will generally be within the accuracy of data that can be obtained in the field, and are suitable for most applications without further refinement.

Fig. 1 costs are based on a net efficiency of energy conversion of 100%, a fuel price of $1.00/million Btu, and heat transfer to still air. Corrections are necessary to arrive at annual costs for other conditions:

■ *Net efficiency of less than 100%*—To correct for a different net efficiency, divide the cost derived from Fig. 1 by the actual net efficiency of energy conversion. (For intermediate-size boilers most common in Allied plants, the net efficiency is typically 72%.)

■ *Fuel cost other than $1.00/million Btu*—To correct for a different cost, multiply the cost derived from Fig. 1 by the actual cost of the fuel.

■ *Outdoor installation exposed to moving air*—To correct for variant wind conditions, multiply the cost obtained from Fig. 1 by a correction factor from Fig. 2 for the actual wind velocity. For most of the continental U.S., 10 mi/h represents a reasonable annual average.

Assumptions made in developing the graphs

The unit costs provided by Fig. 1 are average values for a range of pipe sizes in still air. Although there is some variation for different pipe sizes and ambient temperatures, the variations are small compared to the cost of heat loss caused by low air velocities.

The heat loss due to missing insulation is the difference between the heat loss from bare pipe and that through normal insulation. Insulation standards have changed over recent years because of varying fuel and insulation costs and other economic factors. And although the loss through insulation will vary with the standard adopted for a project, such variations will be negligible compared to the heat loss from bare pipe.

In general, corrections for precise ambient conditions, pipe size and insulation thickness will not be jus-

*To meet the author, see p. 397.

Originally published February 14, 1977

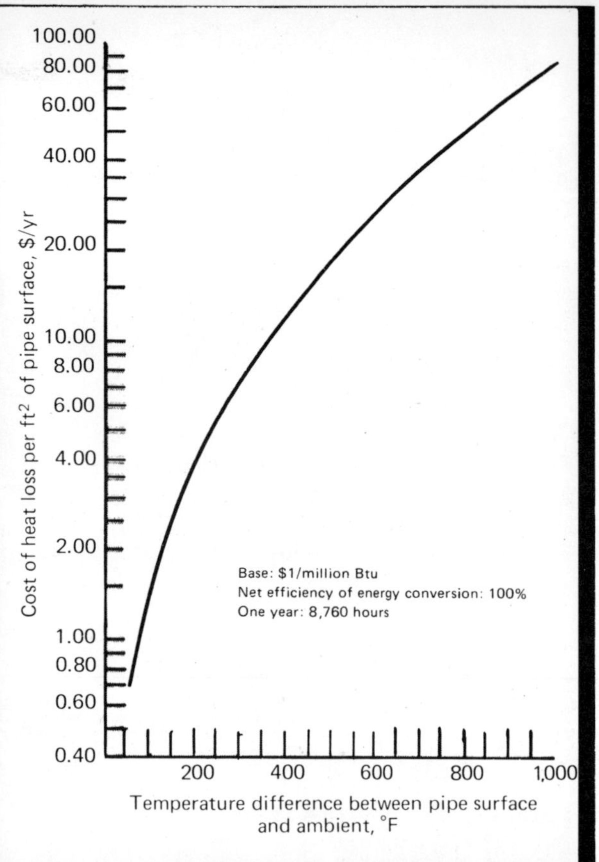

Cost of heat loss when insulation is missing **Fig. 1**

tified because they will go beyond the reasonable accuracy of field data.

Sample problems illustrate the procedure

The first problem requires estimating the annual cost for the extra energy lost from a pipe in still air without insulation:

Pipe dia., in.	6
Pipe temperature, °F	375
Average ambient temperature, °F	75
Length of uninsulated pipe, ft	10
Fuel cost, $/million Btu	1.44
Net energy conversion efficiency, %	72

For the ΔT of 300°F, Fig. 1 gives a unit economic loss of $7.60/yr/ft² of pipe surface. This figure is converted into the cost per linear foot by multiplying it by 1.734 ft²/ft, the surface area per foot of a nominal 6-in. pipe. The answer is $13.20/ft of pipe.

The annual cost for the entire 10 ft of pipe is, of course, $132.00. The corrections for net efficiency and fuel cost (132.00 × 1.44/0.72) give an annual cost for missing insulation of $264.00.

The second problem is similar to the first, except that the pipe is exposed to an average wind velocity of 10 mi/h during the year.

From Fig. 2, the wind correction factor is 2.14. Correcting the previously calculated annual cost for the uninsulated pipe in still air ($264.00 × 2.14), gives an annual cost for missing insulation of $565.00.

References

1. Crocker, S., and King, R. C., "Piping Handbook," 5th ed., McGraw-Hill, New York, 1967.
2. "Engineering Manual on Building Materials and Power Products," Johns-Manville.

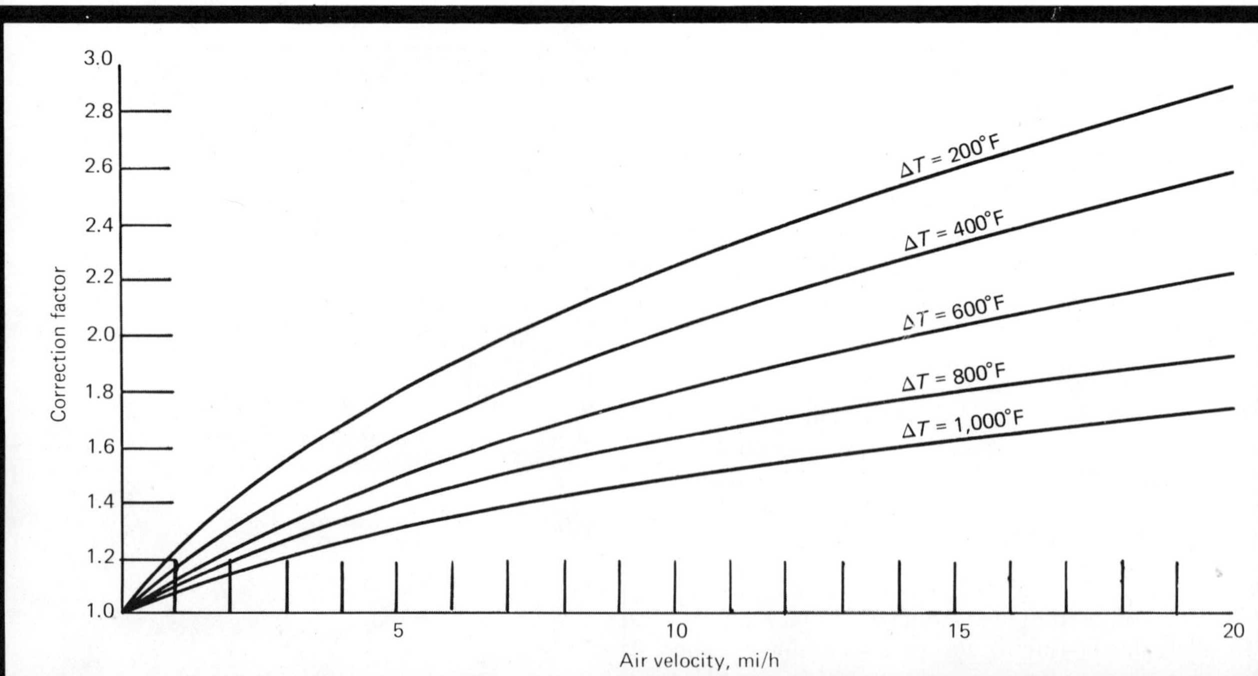

Correction for wind velocity; ΔT is difference between pipe and ambient temperature **Fig. 2**

How To Select Insulation Thickness For Hot Pipes

Rising fuel costs have made traditional insulation-thickness-selection criteria obsolete. Here is a group of nomographs that can be used to choose insulation for hot pipes that will be economic under present-day and future conditions.

JULES L. ABRAMOVITZ and RENE CORDERO,

Allied Chemical Corp.

Energy shortages and higher fuel costs have made insulation more important. Engineers must reevaluate the insulation thickness that provides optimum insulation at the most reasonable cost.

A new set of standards is desirable, because the older standards have been strained beyond reasonable extrapolation. The new standards should be simplified—by moderate sacrifices of mathematical accuracy, if necessary—and tailored to typical plant applications. Also, the standards should provide modifying factors to adjust for most conceivable variations in design conditions.

The procedures outlined in this article meet these conditions. Although these procedures have not yet been applied to an actual design, test calculations have been made for numerous applications, with results that show very favorable recovery periods for the added investment.

Only hot systems are covered in this article. (We are preparing another article on cold systems that we expect will appear in a future issue. In that situation, the new economic criteria indicate that the thicknesses needed for economic purposes are close to those needed for functional antisweat needs.)

Evaluating the Basic Economic Forces

Two major costs enter into the economics of insulation—that of the installation itself (materials and labor), and the operating expense caused by heat loss through the insulation. A direct comparison of heat-loss costs with initial costs is possible when the initial cost is expressed as an equivalent annual cost. Optimum economy is achieved when the annual cost of the capital investment plus the annual cost of the heat loss is at a minimum.

Fig. 1*a* through 1*c* picture the economic phenomena described in the preceding paragraph (with insulation-thickness presented as the independent variable). Fig. 1*c* is the sum of Fig. 1*a* and 1*b*; the theoretical thickness that achieves optimum economy corresponds to the lowest point of this curve.

Mathematical procedures, if properly applied to cor-

Originally published July 21, 1975

Photo: Johns-Manville

rect basic data, will indicate one specific value for this optimum. However, owing to the wide U-shape of the curve (Fig. 1*c*), commercially available thicknesses that are slightly greater or less than the theoretical one will yield a system whose life-economy differs only slightly from the theoretical minimum. Hence, even though the "optimum thickness" may be impractical, the overall economy is essentially the same with the next higher, or lower, commercial thickness.

Engineering Impact of New Economic Criteria

The economic thicknesses indicated by the new procedure are, in most cases, nearly double those calculated by older standards. It is apparent that the new standard cannot be applied (practically) to the repair or replacement of damaged insulation in existing installations. In most cases, existing spacing and available pipe-rack space would prohibit the change.

To test the value of the new criteria, a recent installation that was designed under the old standard was re-evaluated in light of the new criteria. The initial cost of applying the thicker insulation (at the time of construction) was estimated as $75,600. With current fuel prices, the incremental fuel saving would have been about $25,000/yr. Unfortunately, as in most cases, hindsight could not be usefully employed.

Using the Insulation Nomographs

The assumptions that have been built into the following nomographs for hot insulation are listed below:

Assumptions affecting the value of heat losses:

- Overall conversion efficiency, fuel to useful heat—72%.
- Annual hours of operation of insulated system—8,760.
- Average annual ambient temperature—45 to 75°F.
- Capital cost of heat-generating equipment, realized over a 15-yr life span—negligible.
- Effect of wind on heat loss of well-insulated outdoor pipe and equipment—negligible.

Assumptions affecting equivalent annual cost of capital investment in insulation:

- Anticipated service life—15 yr.
- Annual cost of money—10%.
- Annual cost of maintenance of insulation—negligible.

The overalll conversion efficiency (the first item noted) incorporates an 80% efficiency (based on higher heating value) of heat-generation equipment, which is typical of modern oil-burning boilers in the medium-size range, and the use of 10% of the steam produced (or the equivalent value in electrical energy) to drive auxiliaries and otherwise serve the boilers.

The annual cost of money is obviously subject to change in today's fluctuating money markets. (We will explain later how to adjust for such changes.)

User Inputs

There are five user-selected variables:

- Pipe size.
- Operating fluid temperature.
- Raw fuel cost—projected at plant's mid-life.
- Insulation material.
- Installed cost of insulation.

For the sake of simplicity, the installed cost of insulation is based on one thickness (1½ in) of jacketed insulation applied to one pipe size (2½ in). This is the only installed-cost figure needed, no matter what the actual pipe size or insulation thickness.

There is a separate nomograph of each pipe size or limited group of pipe sizes. On each, an array of curves (for various fuel costs) has been drawn for calcium silicate, and another array for fiber glass, each within the temperature range of its normal application. Where there is a preference for such application, the calcium silicate lines may be projected to temperatures below 350°F.

For mineral wool, use the fiber-glass lines. For this application they can be projected (as straight lines) to temperatures as high as 1,000°F. The projected curves can also be used with fiber-glass having higher temperature limits. (*Continued on p. 96*)

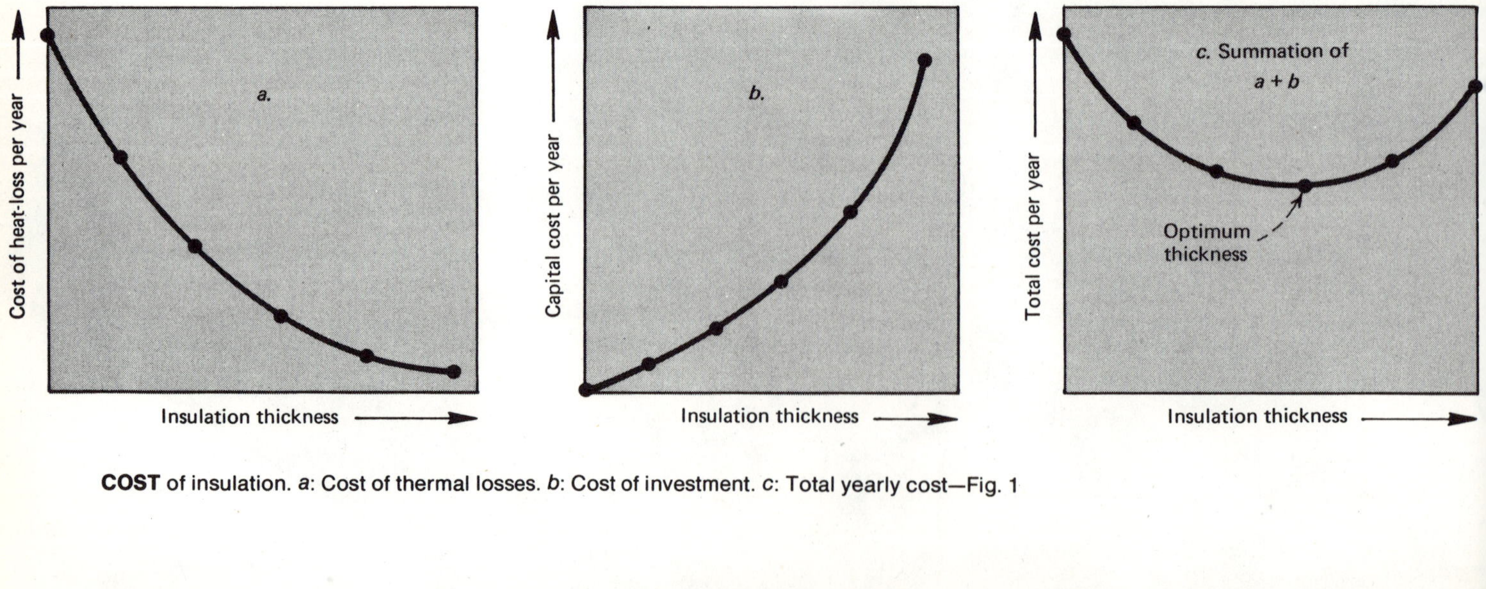

COST of insulation. *a*: Cost of thermal losses. *b*: Cost of investment. *c*: Total yearly cost—Fig. 1

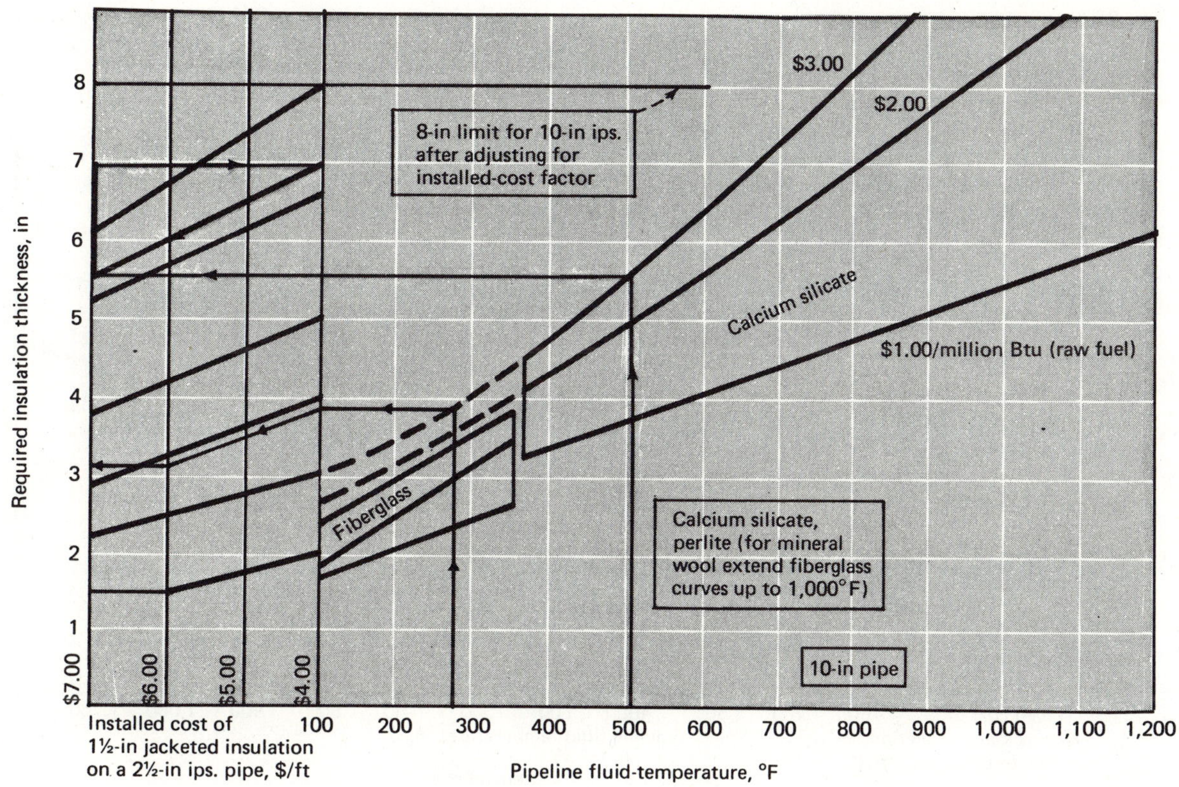

EXAMPLE for use of subsequent nomographs for determining optimum thickness of insulation—Fig. 2

Example of Selection of Insulation Thickness

For an application where calcium silicate insulation is preferred. See Fig. 2.

Given:	Solution:
a. Pipe size—10-in	a. Use 10-in pipe curves
b. Fluid temperature—275°F	b. Project CaSi curves to lower temperature region
c. Material—CaSi	c. Enter at 275°F and proceed vertically to CaSi curves
d. Raw-fuel cost-$2.85/10^6	d. Intersect $2.85 curve (by interpolation)
e. Local installed cost—$6/ft (for 1½-in on 2½-in pipe)	e. Proceed horizontally to $4 cost line and follow sloping lines to $6 cost line
f. Find economic insulation	f. Proceed horizontally and read 3.1-in

Answer: Use 3-in insulation as closely approximating the optimum.

Adjusting for Higher Fuel Costs

For fuel costs above $3/10^6 Btu, proceed as follows:

Given:	Solution:
a. Pipe size—10-in	a. Use 10-in pipe curves
b. Fluid temperature—500°F	b. Enter Fig. 2 at 500°F
c. Material—CaSi	c. Proceed vertically to CaSi curves
d. Raw-fuel cost—$4/10^6 Btu	d. Intersect $3 curve and proceed horizontally, all the way to "Insulation thickness" ordinate (5½-in). Enter Table II and read out correction factor of 1.27. Multiply insulation thickness by 1.27 and plot on nomograph (7-in). Project horizontally to right to $4 "Installed cost" line
e. Local installed cost—$6/ft (for 1½-in insulation on 2½-in pipe).	e. Follow sloping lines to $6 line
f. Find economic insulation	f. Proceed horizontally to left and read 6-in., which is an available thickness

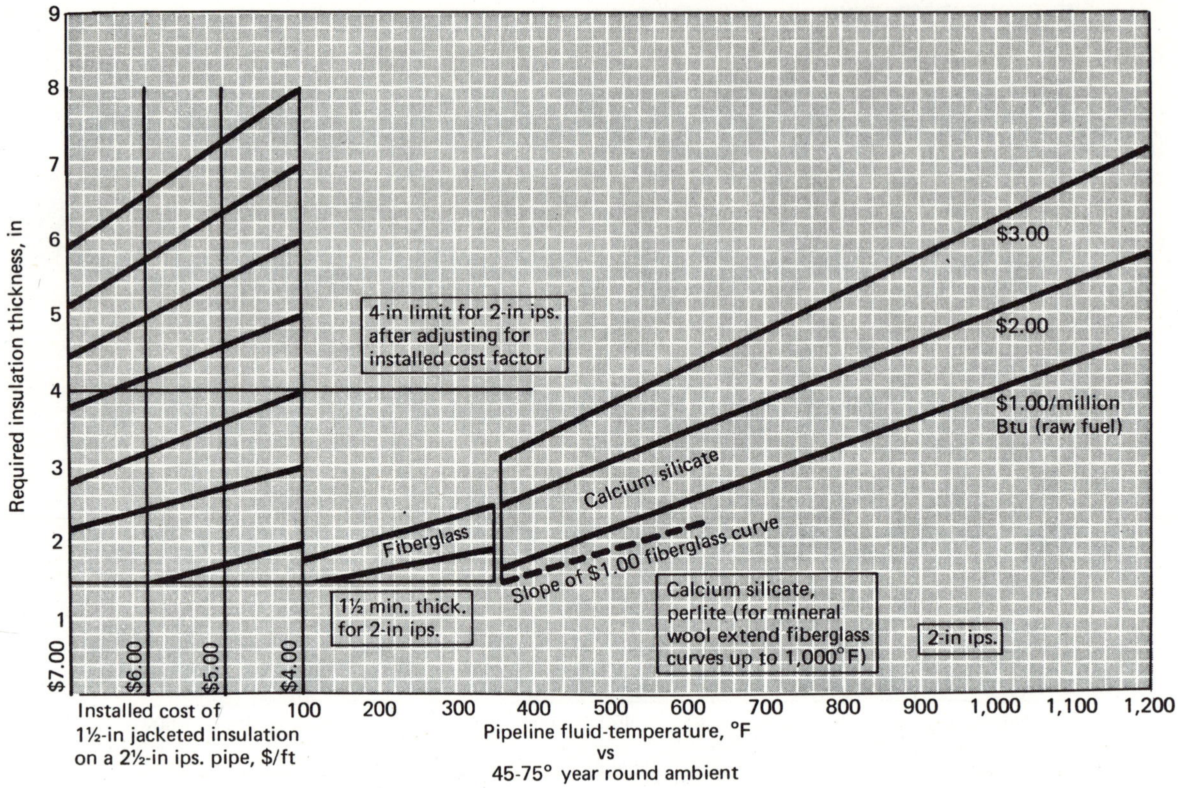

Required insulation thickness, in
4-in limit for 2-in ips. after adjusting for installed cost factor
$3.00
$2.00
$1.00/million Btu (raw fuel)
Calcium silicate
Fiberglass
Slope of $1.00 fiberglass curve
1½ min. thick. for 2-in ips.
Calcium silicate, perlite (for mineral wool extend fiberglass curves up to 1,000°F)
2-in ips.
$7.00
$6.00
$5.00
$4.00
Installed cost of 1½-in jacketed insulation on a 2½-in ips. pipe, $/ft
Pipeline fluid-temperature, °F vs 45-75° year round ambient

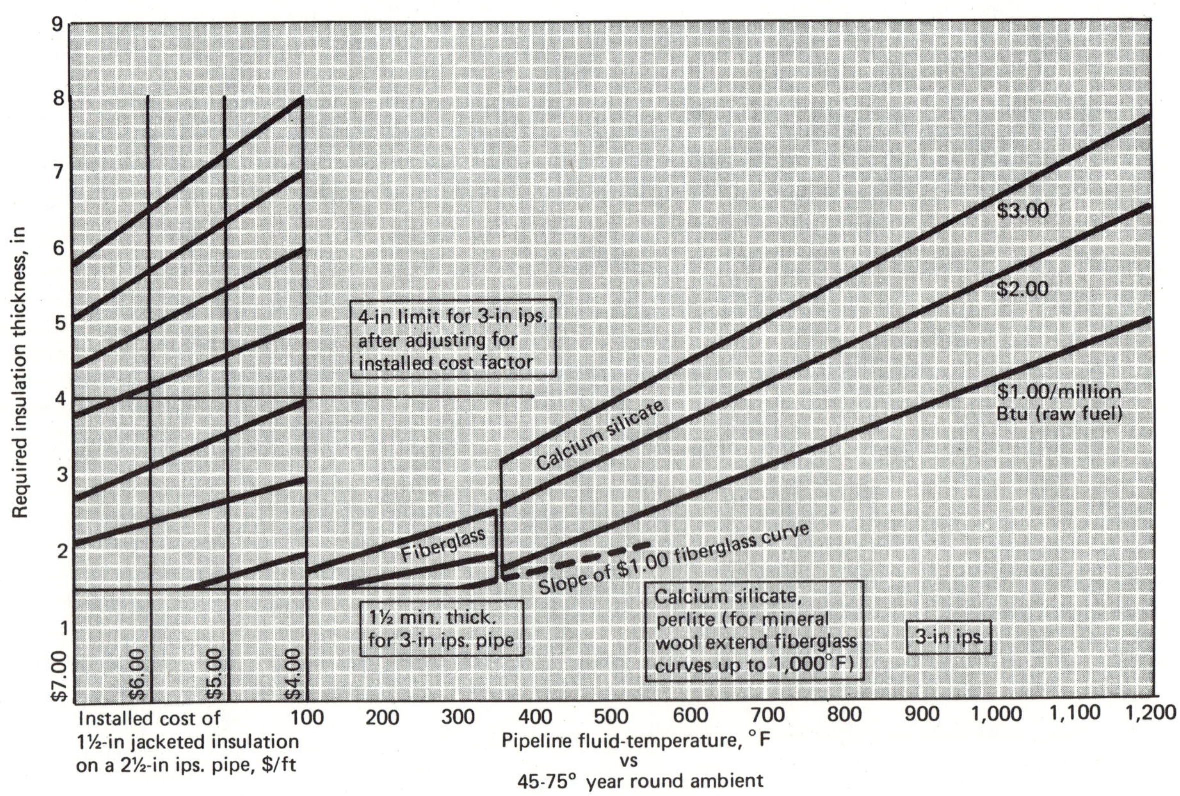

Required insulation thickness, in
4-in limit for 3-in ips. after adjusting for installed cost factor
$3.00
$2.00
$1.00/million Btu (raw fuel)
Calcium silicate
Fiberglass
Slope of $1.00 fiberglass curve
1½ min. thick. for 3-in ips. pipe
Calcium silicate, perlite (for mineral wool extend fiberglass curves up to 1,000°F)
3-in ips.
$7.00
$6.00
$5.00
$4.00
Installed cost of 1½-in jacketed insulation on a 2½-in ips. pipe, $/ft
Pipeline fluid-temperature, °F vs 45-75° year round ambient

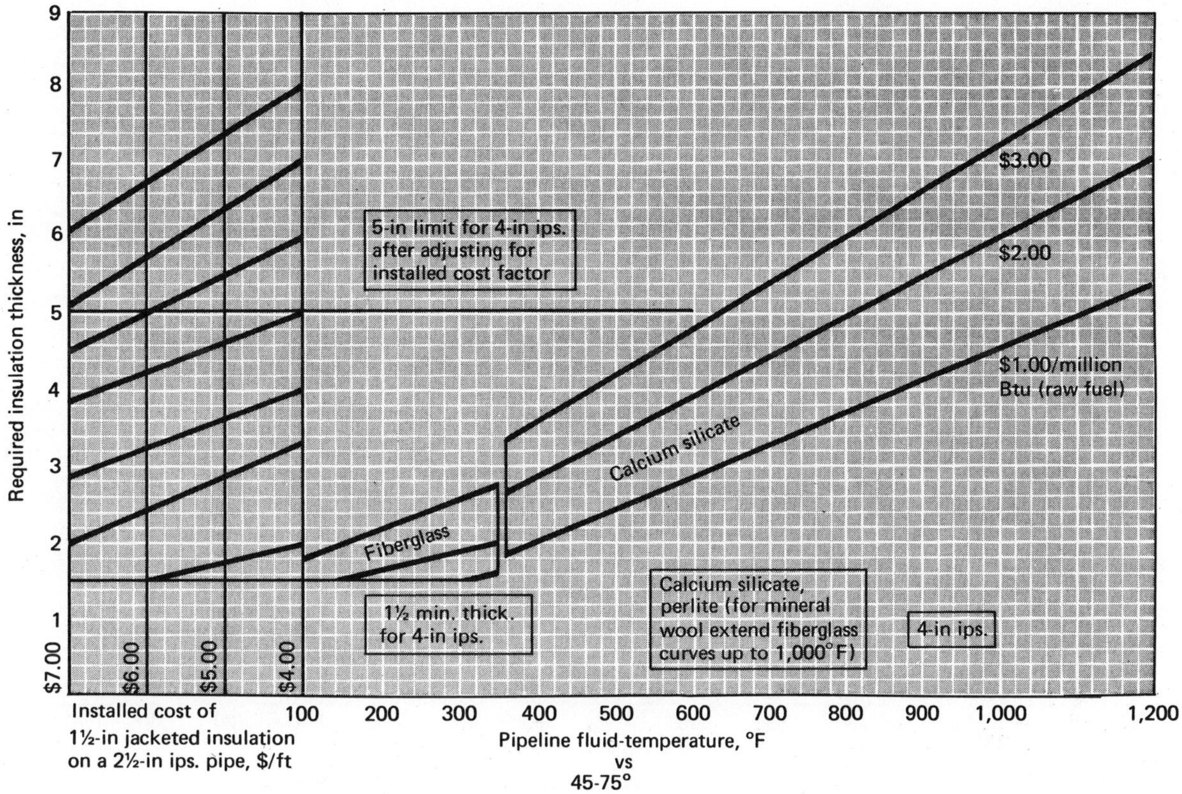
Required insulation thickness, in
9
8
7
6
5
4
3
2
1
$7.00
$6.00
$5.00
$4.00
5-in limit for 4-in ips. after adjusting for installed cost factor
$3.00
$2.00
$1.00/million Btu (raw fuel)
Calcium silicate
Fiberglass
1½ min. thick. for 4-in ips.
Calcium silicate, perlite (for mineral wool extend fiberglass curves up to 1,000°F)
4-in ips.
Installed cost of 1½-in jacketed insulation on a 2½-in ips. pipe, $/ft
100
200
300
400
500
600
700
800
900
1,000
1,200
Pipeline fluid-temperature, °F vs 45-75°

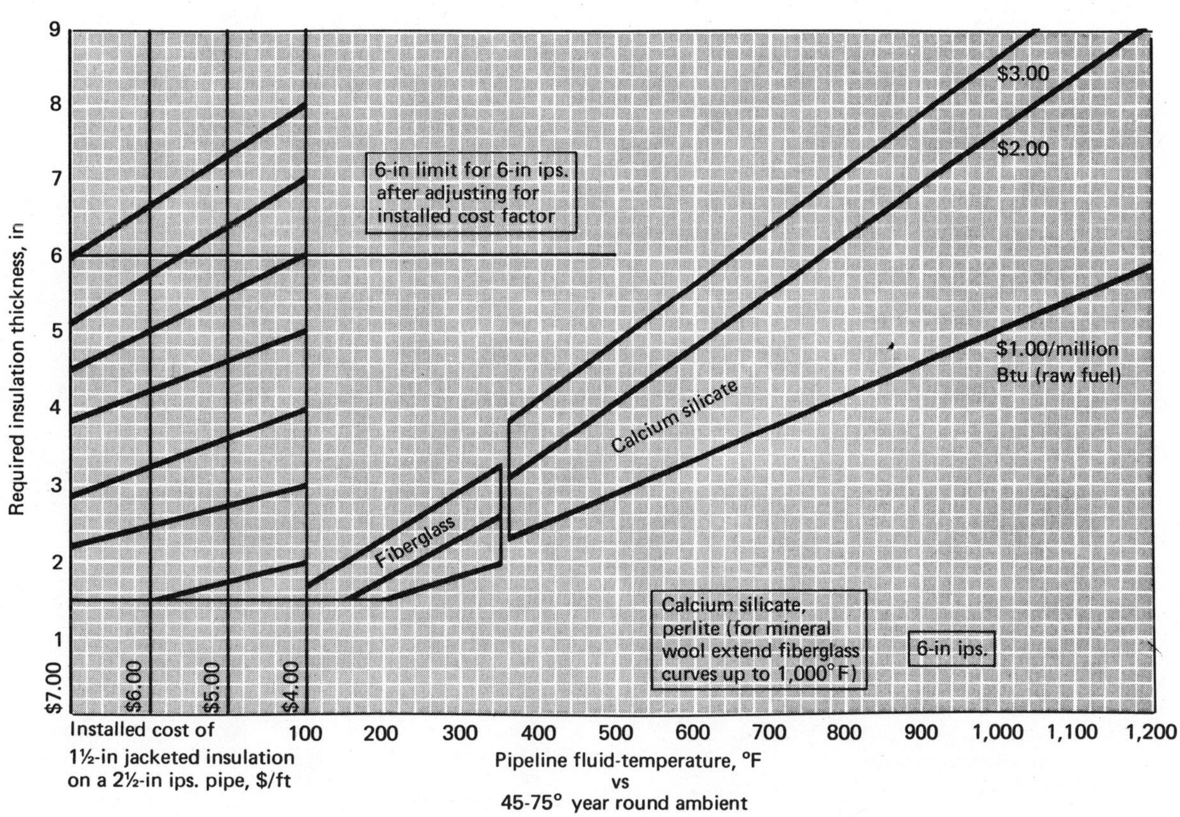
Required insulation thickness, in
9
8
7
6
5
4
3
2
1
$7.00
$6.00
$5.00
$4.00
6-in limit for 6-in ips. after adjusting for installed cost factor
$3.00
$2.00
$1.00/million Btu (raw fuel)
Calcium silicate
Fiberglass
Calcium silicate, perlite (for mineral wool extend fiberglass curves up to 1,000°F)
6-in ips.
Installed cost of 1½-in jacketed insulation on a 2½-in ips. pipe, $/ft
100
200
300
400
500
600
700
800
900
1,000
1,100
1,200
Pipeline fluid-temperature, °F vs 45-75° year round ambient

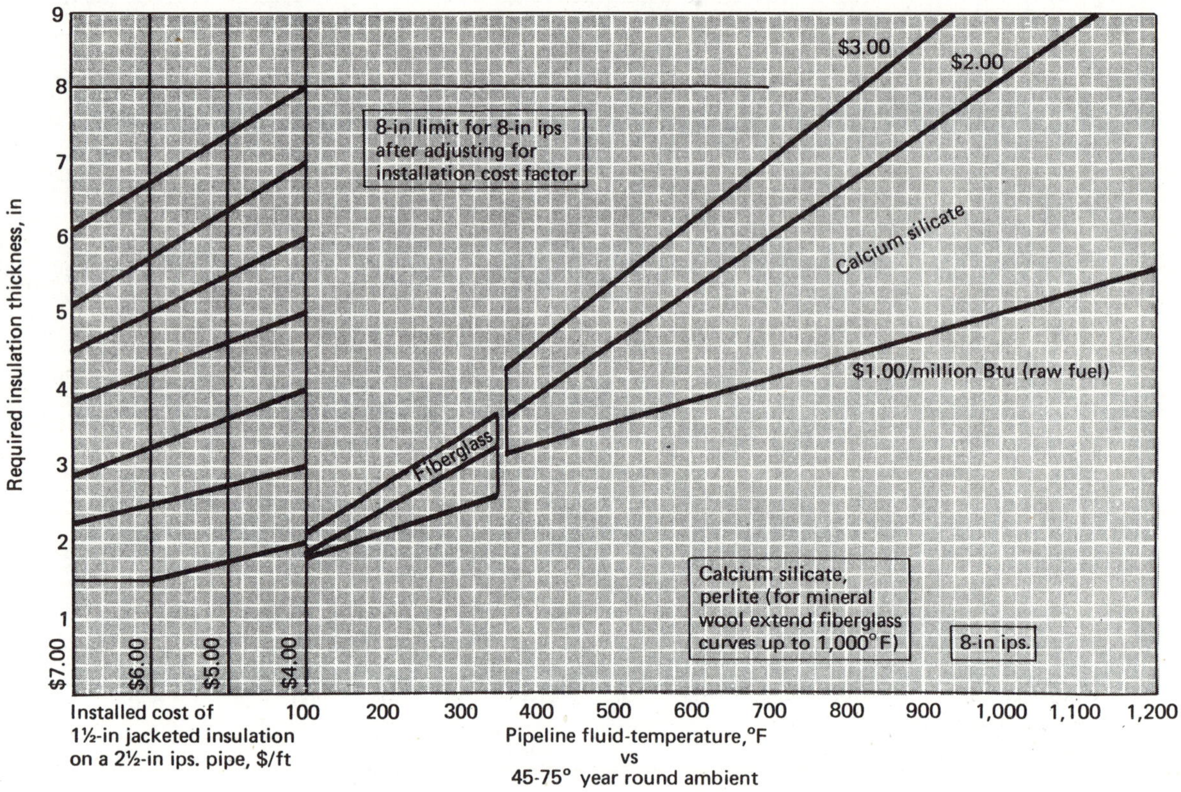

Required insulation thickness, in
9
8
7
6
5
4
3
2
1
$7.00
$6.00
$5.00
$4.00
Installed cost of
1½-in jacketed insulation
on a 2½-in ips. pipe, $/ft
100
200
300
400
500
600
700
800
900
1,000
1,100
1,200
Pipeline fluid-temperature,°F
vs
45-75° year round ambient
$3.00
$2.00
8-in limit for 8-in ips
after adjusting for
installation cost factor
Calcium silicate
$1.00/million Btu (raw fuel)
Fiberglass
Calcium silicate,
perlite (for mineral
wool extend fiberglass
curves up to 1,000°F)
8-in ips.

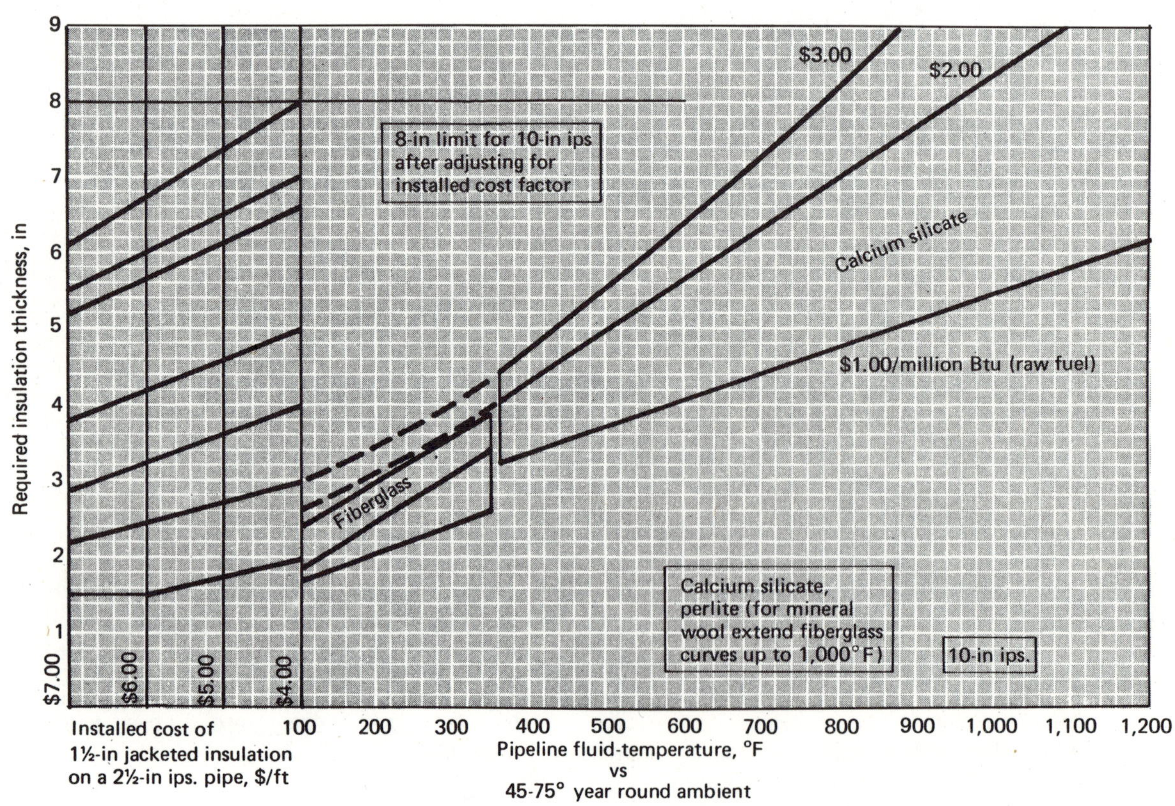

Required insulation thickness, in
9
8
7
6
5
4
3
2
1
$7.00
$6.00
$5.00
$4.00
Installed cost of
1½-in jacketed insulation
on a 2½-in ips. pipe, $/ft
100
200
300
400
500
600
700
800
900
1,000
1,100
1,200
Pipeline fluid-temperature, °F
vs
45-75° year round ambient
$3.00
$2.00
8-in limit for 10-in ips
after adjusting for
installed cost factor
Calcium silicate
$1.00/million Btu (raw fuel)
Fiberglass
Calcium silicate,
perlite (for mineral
wool extend fiberglass
curves up to 1,000°F)
10-in ips.

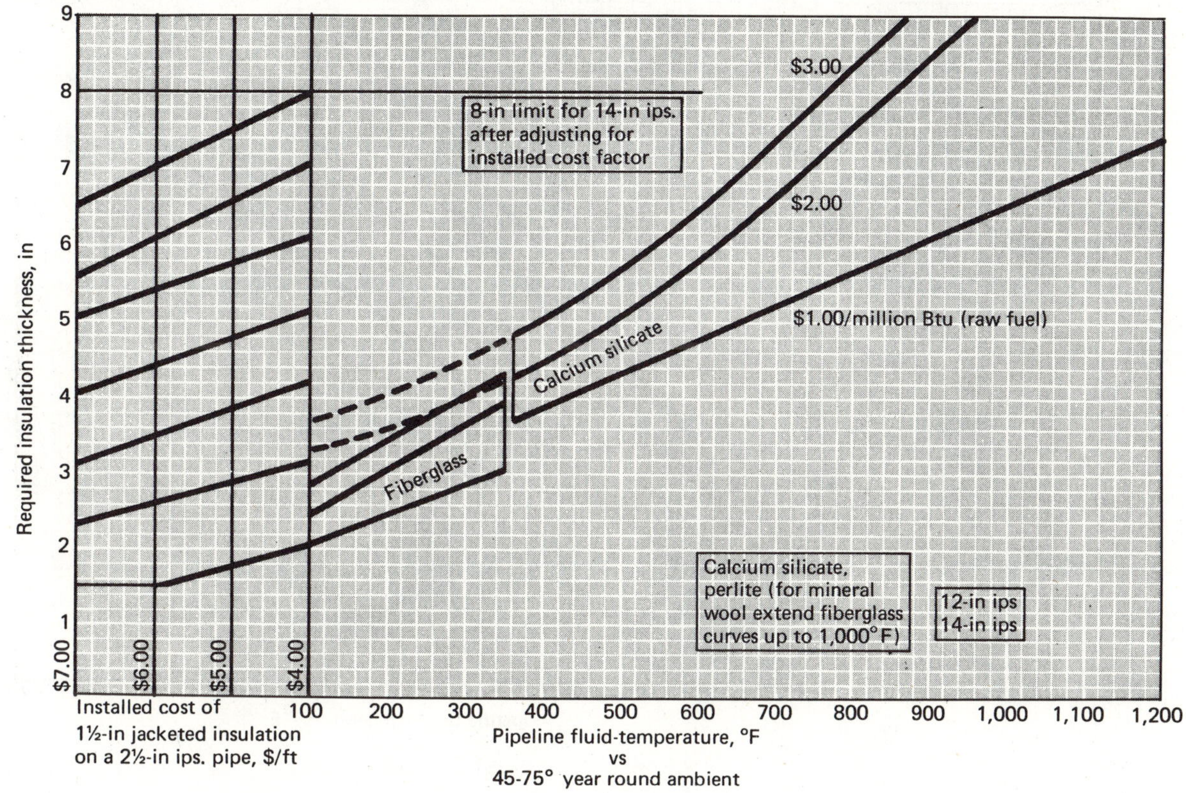
8-in limit for 14-in ips. after adjusting for installed cost factor
$3.00
$2.00
$1.00/million Btu (raw fuel)
Calcium silicate
Fiberglass
Calcium silicate, perlite (for mineral wool extend fiberglass curves up to 1,000°F)
12-in ips
14-in ips
Required insulation thickness, in
$7.00
$6.00
$5.00
$4.00
Installed cost of 1½-in jacketed insulation on a 2½-in ips. pipe, $/ft
Pipeline fluid-temperature, °F vs 45-75° year round ambient

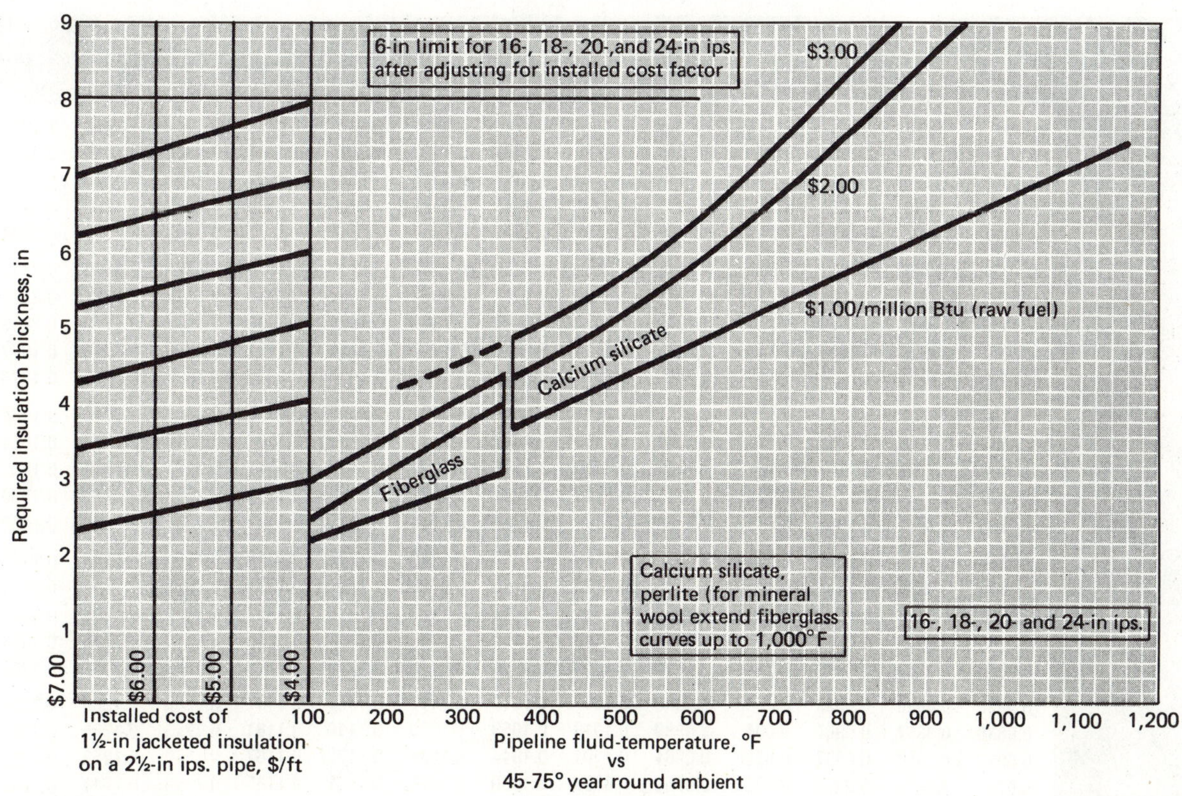
6-in limit for 16-, 18-, 20-, and 24-in ips. after adjusting for installed cost factor
$3.00
$2.00
$1.00/million Btu (raw fuel)
Calcium silicate
Fiberglass
Calcium silicate, perlite (for mineral wool extend fiberglass curves up to 1,000°F
16-, 18-, 20- and 24-in ips.
Required insulation thickness, in
$7.00
$6.00
$5.00
$4.00
Installed cost of 1½-in jacketed insulation on a 2½-in ips. pipe, $/ft
Pipeline fluid-temperature, °F vs 45-75° year round ambient

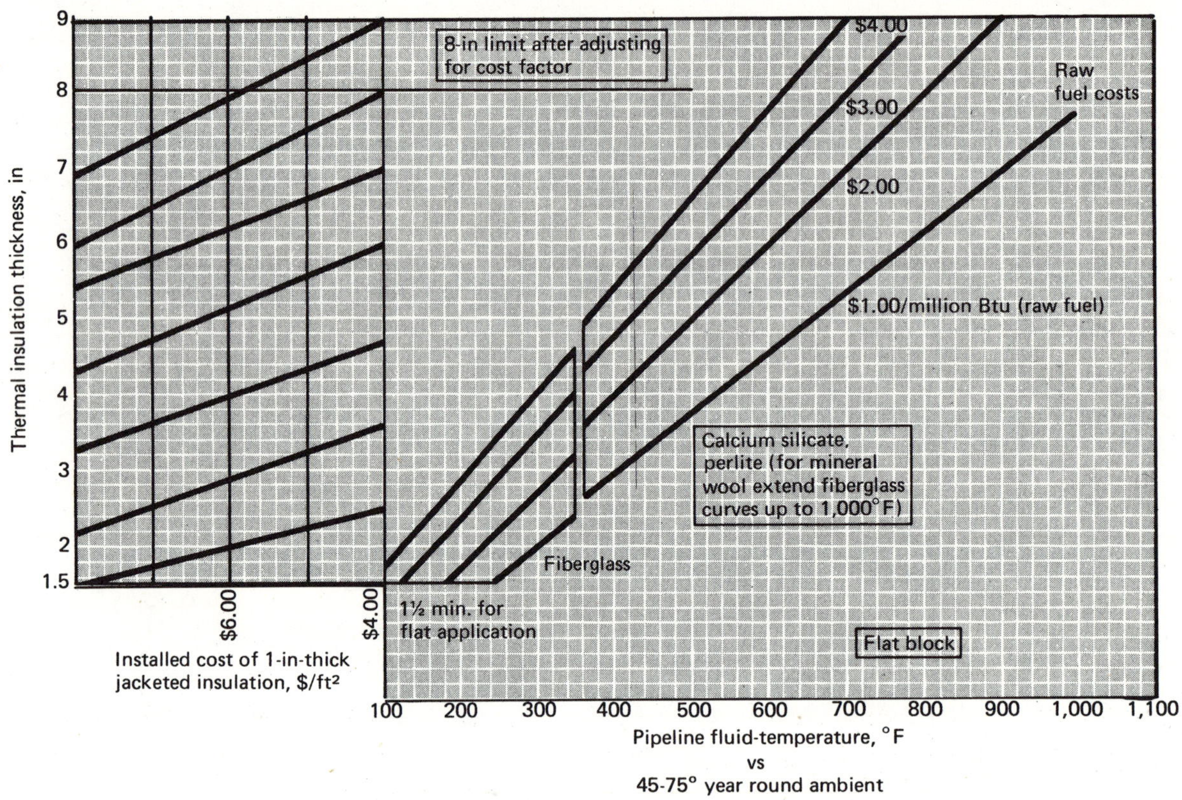

Specified Annual Interest (Factor *A*)—Table I

Amortization Period, Yr	Value of *A* for Various Interest Rate, Annual, %												
	6.0	6.5	7.0	7.5	8.0	8.5	9.0	9.5	10.0	10.5	11.0	11.5	12.0
1	1.060	1.065	1.070	1.075	1.080	1.085	1.090	1.095	1.100	1.105	1.110	1.115	1.120
2	0.5454	0.5493	0.5531	0.5569	0.5608	0.5646	0.5685	0.5723	0.5762	0.5801	0.5839	0.5878	0.5917
3	0.3741	0.3776	0.3811	0.3845	0.3880	0.3915	0.3961	0.3986	0.4021	0.4057	0.4092	0.4128	0.4163
4	0.2886	0.2919	0.2952	0.2986	0.3019	0.3053	0.3087	0.3121	0.3155	0.3189	0.3223	0.3258	0.3292
5	0.2374	0.2406	0.2439	0.2472	0.2505	0.2538	0.2571	0.2604	0.2638	0.2672	0.2706	0.2740	0.2774
6	0.2034	0.2066	0.2098	0.2130	0.2163	0.2263	0.2229	0.2263	0.2296	0.2330	0.2364	0.2398	0.2432
7	0.1791	0.1823	0.1856	0.1888	0.1921	0.1954	0.1987	0.2020	0.2054	0.2088	0.2122	0.2157	0.2191
8	0.1610	0.1642	0.1675	0.1707	0.1740	0.1773	0.1807	0.1840	0.1874	0.1909	0.1943	0.1978	0.2013
9	0.1470	0.1502	0.1535	0.1568	0.1601	0.1634	0.1668	0.1702	0.1736	0.1771	0.1806	0.1841	0.1877
10	0.1359	0.1391	0.1424	0.1457	0.1490	0.1524	0.1558	0.1593	0.1627	0.1663	0.1698	0.1733	0.1770
11	0.1268	0.1301	0.1334	0.1367	0.1401	0.1435	0.1469	0.1504	0.1540	0.1575	0.1611	0.1648	0.1684
12	0.1193	0.1226	0.1259	0.1293	0.1327	0.1362	0.1397	0.1432	0.1468	0.1504	0.1540	0.1577	0.1614
13	0.1130	0.1163	0.1197	0.1231	0.1265	0.1300	0.1336	0.1372	0.1407	0.1444	0.1482	0.1519	0.1557
14	0.1076	0.1109	0.1143	0.1178	0.1213	0.1248	0.1284	0.1321	0.1357	0.1395	0.1432	0.1470	0.1509
15	0.1030	0.1064	0.1098	0.1133	0.1169	0.1204	0.1241	0.1277	0.1315	0.1352	0.1391	0.1429	0.1468
16	0.0990	0.1024	0.1059	0.1094	0.1130	0.1166	0.1203	0.1240	0.1278	0.1316	0.1355	0.1394	0.1434
17	0.0954	0.0989	0.1024	0.1060	0.1096	0.1133	0.1170	0.1208	0.1247	0.1285	0.1325	0.1364	0.1405
18	0.0924	0.0959	0.0994	0.1030	0.1067	0.1104	0.1142	0.1180	0.1219	0.1259	0.1298	0.1339	0.1379
19	0.0896	0.0932	0.0968	0.1004	0.1041	0.1079	0.1117	0.1156	0.1195	0.1235	0.1276	0.1316	0.1358
20	0.0872	0.0908	0.0944	0.0981	0.1019	0.1057	0.1095	0.1135	0.1175	0.1215	0.1256	0.1297	0.1339
21	0.0850	0.0886	0.0923	0.0960	0.0998	0.1037	0.1076	0.1116	0.1156	0.1197	0.1238	0.1280	0.1322
22	0.0830	0.0867	0.0904	0.0942	0.0980	0.1019	0.1059	0.1099	0.1140	0.1181	0.1223	0.1265	0.1308
23	0.0813	0.0850	0.0887	0.0925	0.0964	0.1004	0.1044	0.1084	0.1126	0.1167	0.1210	0.1252	0.1285
24	0.0797	0.0834	0.0872	0.0910	0.0950	0.0990	0.1030	0.1071	0.1113	0.1155	0.1198	0.1241	0.1285
25	0.0782	0.0820	0.0858	0.0887	0.0937	0.0977	0.1018	0.1060	0.1102	0.1144	0.1187	0.1231	0.1275

Correction Factors for Higher Fuel Costs—Table II

Raw Fuel Cost, $/10^6 Btu	Pipeline-Fluid Temperatures, °F					
	100	200	300	400-599	600-799	800-1,000
	(Fiber Glass)			(Calcium Silicate)		
4	1.11	1.16	1.21	1.27	1.30	1.33
5	1.21	1.29	1.39	1.42	1.49	1.50

Observe that Fig. 2 shows a 1½-in minimum insulation thickness—this is common to all the nomographs. Several manufacturers have indicated that they are discontinuing insulation material thinner than 1½ in. in recognition of dwindling user interest. It is probable that 1½-in insulation will soon be the thinnest competitively available.

The nomographs also indicate maximum recommended insulation thicknesses. These maxima vary with the pipe size, and are in recognition of the economic impact of excessive thickness in such areas as pipe-rack capacity, pipe support problems, and the like. These maxima are based solely on our engineering judgments.

Adjusting for Variations in Assumptions

Corrections to fit conditions other than those assumed are made by adjusting one or more of the plotted variables which, when applied to the nomographs, will indicate the optimum insulation thickness.

Hours of Operation and Efficiency—To adjust for fewer than 8,760 hr/yr operation, or for overall efficiency other than 72%, multiply "Raw fuel cost per million Btu" by:

$$(\text{hours of operation}/8{,}760)\ (72\%/\%\ \text{efficiency})$$

Ambient Temperature Other Than 45-75°F—For *higher* ambients, *subtract* from "Pipeline-fluid temperature" the quantity:

$$\text{Ambient} - 60$$

For *lower* ambients, *add* to "Pipeline-fluid temperature" the quantity:

$$60 - \text{Ambient}$$

Service Life, Maintenance, Cost of Money—To adjust for a service life of other than 15 yr, a cost of money other than 10%, or appreciable insulation maintenance, multiply "Installed cost of 1½-in jacketed insulation on a 2½-in pipe" by:

$$(A + M)/0.1315$$

where A is the uniform annual payment needed to reach unit value over the expected service life, while paying a specified annual interest (read from Table I). M is the expected average annual maintenance charge, expressed as a decimal part of the installed cost. (0.1315 is the uniform annual payment at 10% for 15 years, the assumed conditions for the nomograph.)

Fuel Costs Above $3—For applications where projected raw-fuel costs exceed $3/10⁶ Btu, adjust curve values by using Table II. The resultant thicknesses will exceed, in many cases, the maxima established on the nomographs. As stated previously, the maxima are engineering judgments, and the user is advised to weigh physical space considerations against overall operating economy.

Differing Energy-Cost Levels—Applications will arise where a portion of the plant's piping owes its temperature level to conventional combustion of fossil fuels, while other streams are heated by significantly differing energy sources such as low-pressure waste steam, exhaust gases and direct-fired calciners. The user is encouraged to apply appropriate raw-fuel costs and efficiency adjustments for each distinct system to derive maximum benefit from this article.

Cylindrical Equipment—Cylindrical equipment of less than 30-in. O.D. may be treated as pipe of approximately the same O.D.

References

1. "Econ-I—How to Determine Economic Thickness of Thermal Insulation," Thermal Insulation Manufacturers Assn., 7 Kirby Plaza, Mt. Kisko, NY 10549, 1973.
2. King, Reno C., Crocker, Sabin, eds., "Piping Handbook," 5th ed., McGraw-Hill, New York, 1967.

Meet the Authors

◀ **Jules L. Abramovitz** is a Mechanical Design Engineer in the Corporate Engineering Dept. of Allied Chemical Corp., P. O. Box 2245R, Morristown, NJ 07960. He received his B.S. in mechanical engineering from Newark College of Engineering, and his M.S. in mechanical engineering from Fairleigh Dickinson University.

Rene Cordero is a Mechanical Design Engineer for Allied Chemical Corp. at Morristown, N.J. He received his B.M.E. degree from Catholic University and his M.M.A.E. from the University of Delaware. He is a registered Professional Engineer in the state of New Jersey. ▶

Economic pipe insulation for cold systems

Preventing cold-pipe sweating once usually outweighed other considerations in determining insulation thickness. Nomographs in this article are geared to selecting the thicker insulation now demanded by higher energy costs.

Jules L. Abramovitz, *Coaltek Associates* *

☐ Cold-insulation design presents fundamentally different problems from those of hot systems.* One basic difference is that capital costs for refrigeration equipment add substantially to the cost of generating refrigeration. This "capital equipment fuel surcharge" can tack 25% onto prevailing electric rates.

A second major difference is that cold-insulation specifications must satisfy two requirements: an economic one, and the functional one of preventing sweating, which damages insulation. A third difference is that in selection nomographs one independent variable must be stated in ¢/kWh, rather than in the 10^6 Btu of hot systems.

Occasional and moderate condensation upon the surface of cold insulation can be tolerated (it can never be completely eliminated), but frequent and excessive condensation damages the insulation. To prevent this destruction, cold insulation must be designed to limit surface wetting. In the past, this necessity has overshadowed economic considerations. But under current conditions, and those likely in the future, energy economics will prevail in many instances. Proper design now requires that thickness be determined for both criteria, and that the more demanding criterion be applied.

Cold systems vary in temperature of operation and in capacity of refrigeration machines, factors that materially affect the cost of equipment and operation. On the basis of 10^6 Btu/h capability, refrigeration equipment is far more expensive than heating equipment. The investment in refrigeration equipment cannot be ignored; annualized, it adds substantially to operating costs.

The nomographs presented are based on costs for a recent installation. Because installations differ, methods of adjusting for these costs and operations have been incorporated in the solutions.

Differences between refrigeration-system and average-ambient temperatures are generally moderate, usually in the range of 100°F or less. Nevertheless, the nomographs take into account the difference between average annual ambient temperature and that of the fluid inside the pipe, rather than only the temperature of the fluid (as do nomographs for hot insulation).

*To meet the author, see p. 397.

Developing a refrigeration energy audit

The basic technical data of the reference installation upon which the nomographs are based are:

Capacity, tons	1,050
Operating temperature, °F	10
Brake horsepower, full load, hp	2,300
Motor efficiency, full load, %	95
Duty, continuous, high load, h/yr	8,450

The system requires auxiliary power as follows:

Cooling-water pump, hp	145
Cooling-tower fan, hp	166
Brine-circulation pump, hp	70

The total power needed, therefore, is 2,681 hp, or 2.55 hp/ton of refrigeration.

The energy liability is therefore:

$$(2.55 \text{ hp/ton})(0.745 \text{ kW/hp})(1/0.95) = 2.0 \text{ kW/ton}$$

Or:

$$(2.0 \text{ kW/ton})(10^6/12{,}000 \text{ Btu/h/ton}) = 167 \text{ kWh}/10^6 \text{ Btu}$$

The capital equipment cost, including a part of the auxiliaries, is \$645/ton of refrigeration. Paying 9% interest on the capital financed over a 13-year economic life adds a 1.73 cost factor to the foregoing base cost. Allowing 1% per year for maintenance results in an additional cost factor of 0.13.

The total equipment cost is therefore:

$$645\ (1.73 + 0.13) = \$1{,}200/\text{ton}$$

Originally published October 25, 1976

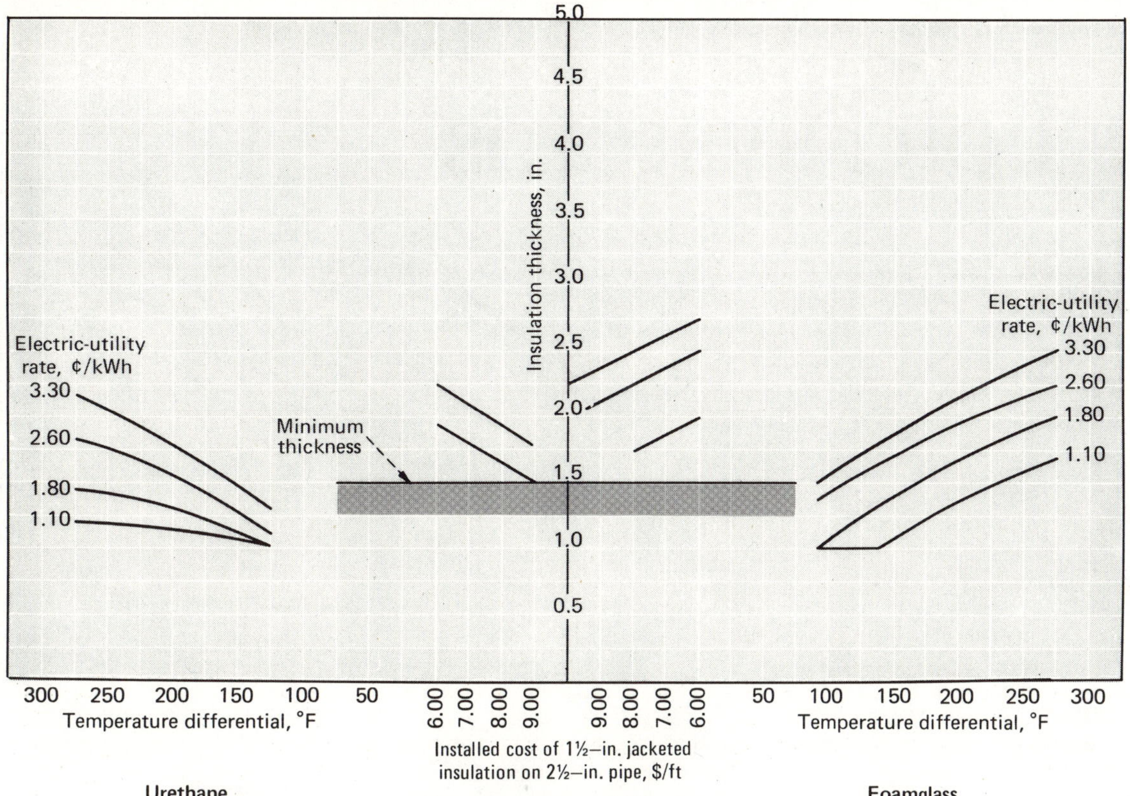

For 2-in. pipe—Fig. 1

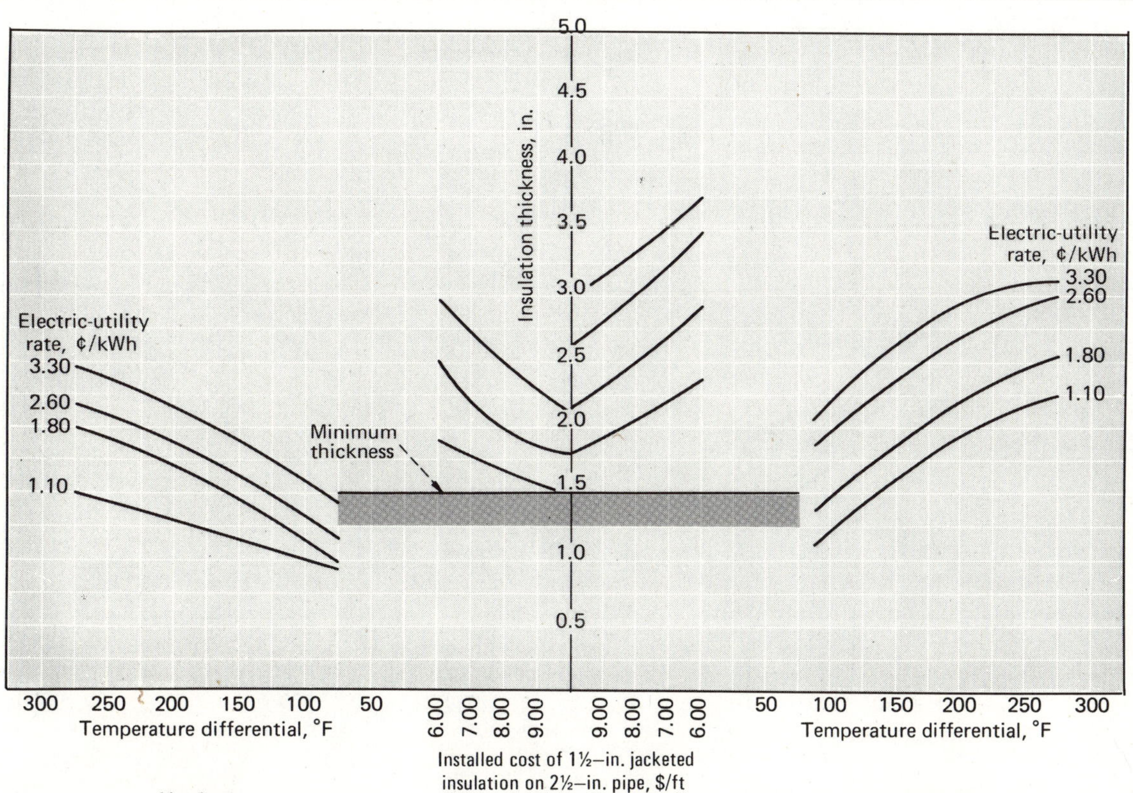

For 4-in. pipe—Fig. 2

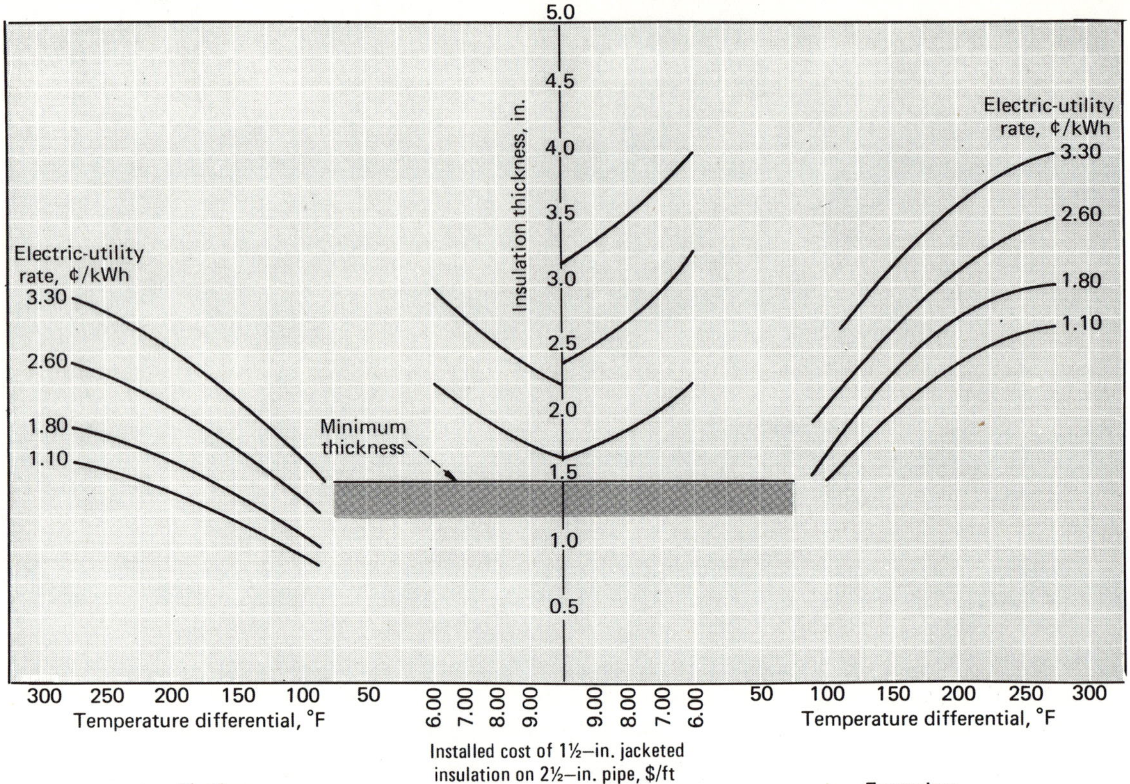

For 6-in. pipe—Fig. 3

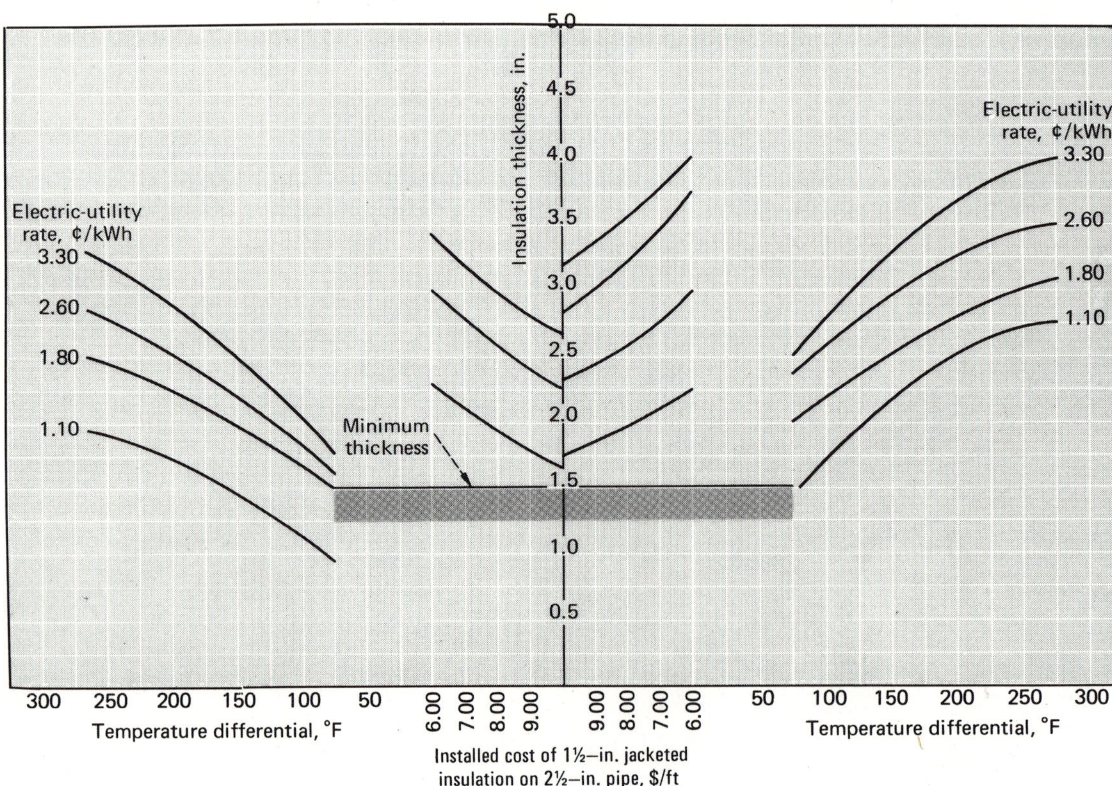

For 8-in. pipe—Fig. 4

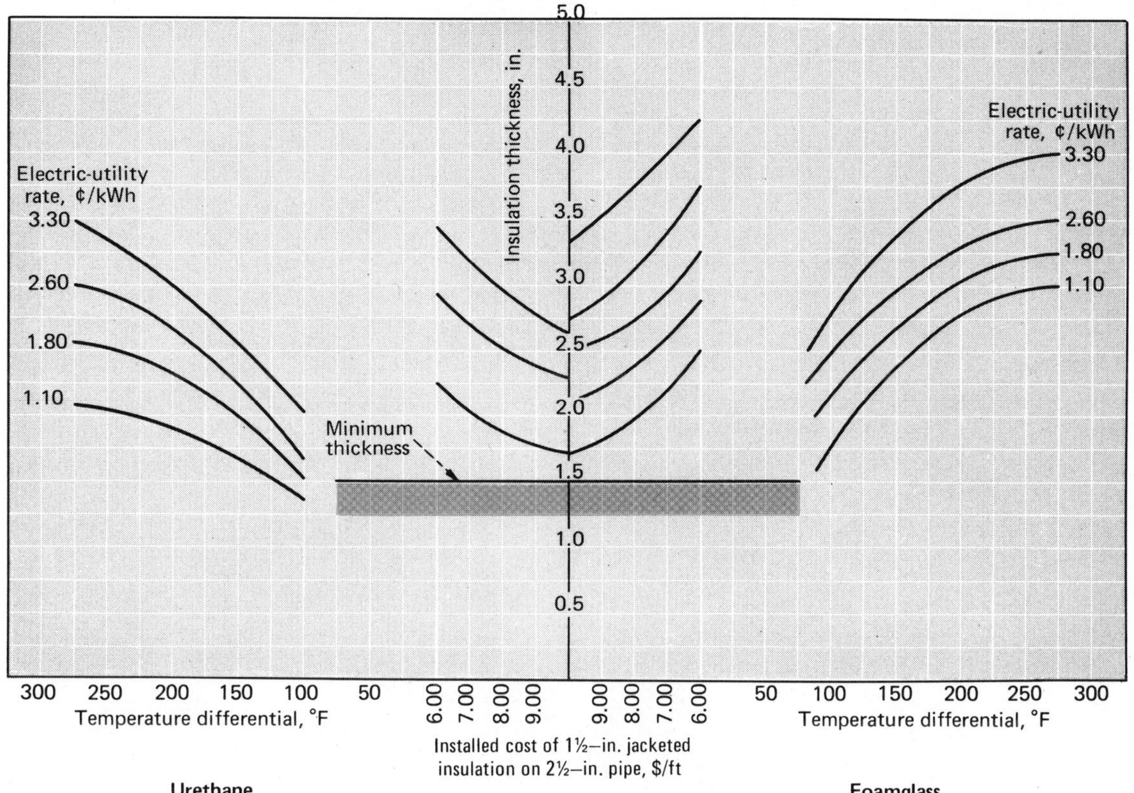

For 10-in. pipe—Fig. 5

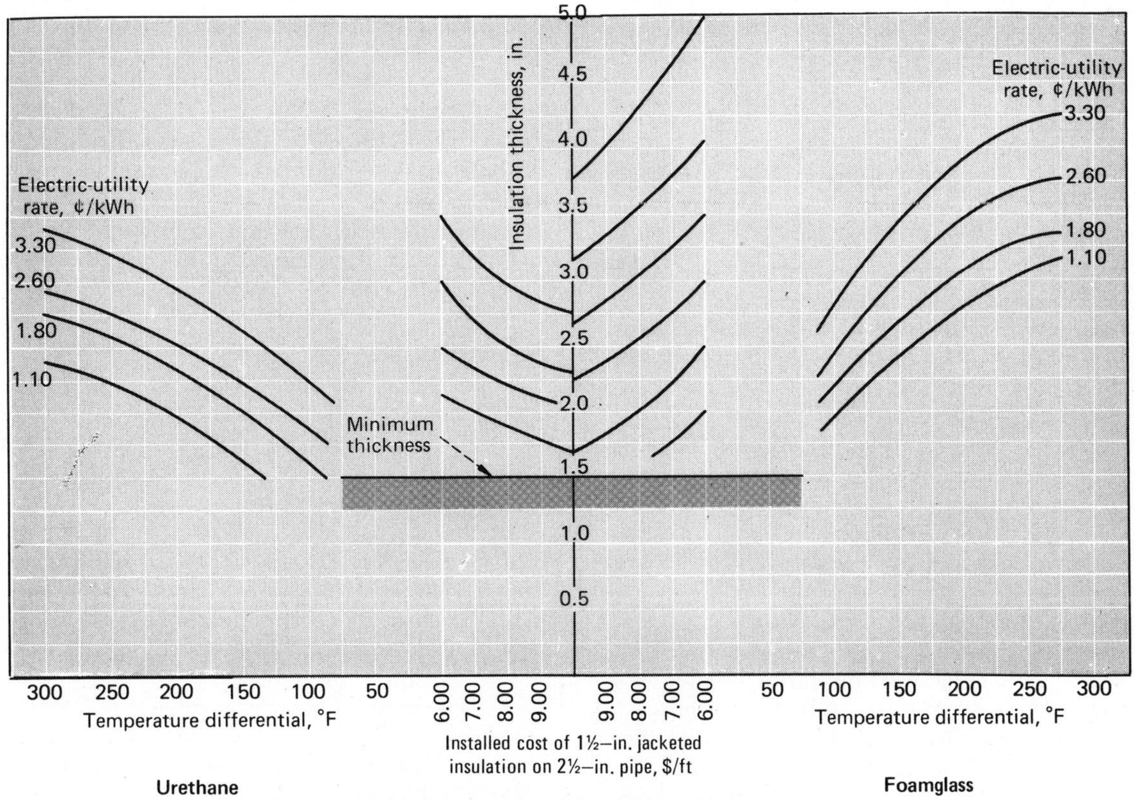

For 12-in. and 14-in. pipe—Fig. 6

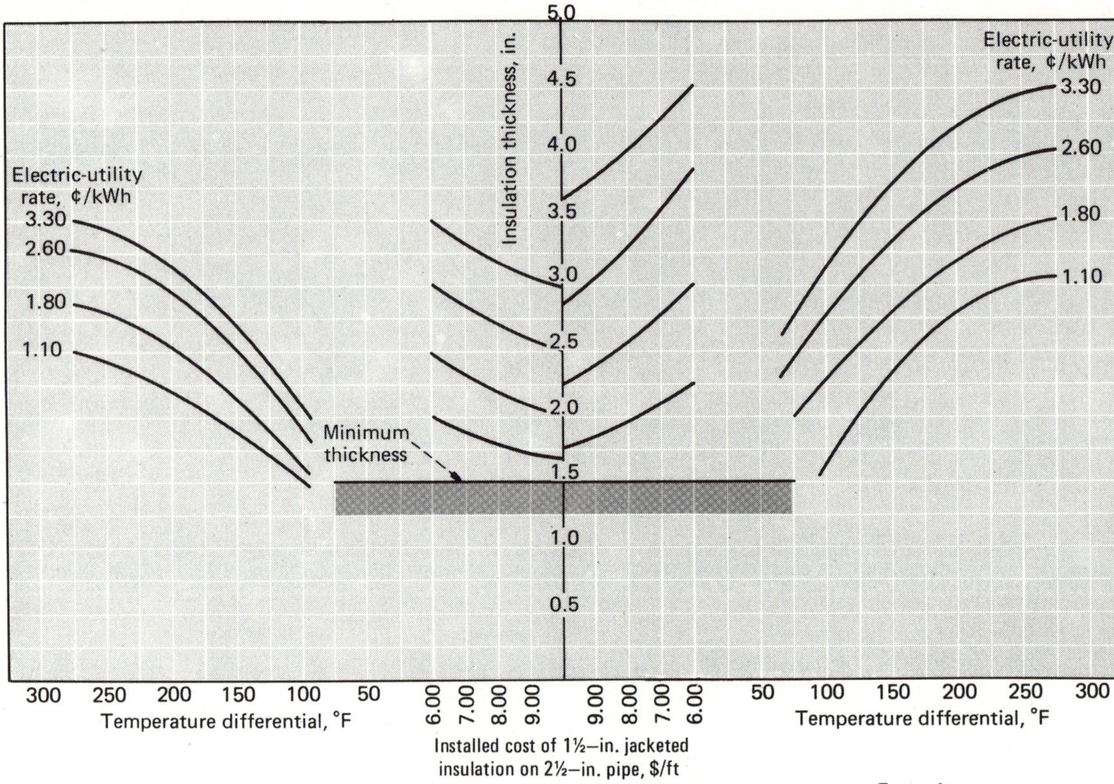

For 16-in. and 18-in. pipe—Fig. 7

Spreading this cost over 13 years (8,450 hours of operation per year) results in a capital cost ratio of:

$$\left(\frac{\$1{,}200/\text{ton}}{8{,}450\ \text{h/yr} \times 13\ \text{yr}}\right)\left(\frac{10^6}{12{,}000\ \text{Btu/h/ton}}\right) =$$

$$\$0.91/10^6\ \text{Btu, or } 91¢/10^6\ \text{Btu}$$

The foregoing values are subject to frequent fluctuation, and adjustment procedures are presented later.

Basis and use of the nomographs

Fig. 1 through 7 represent graphic solutions of the economic thickness of insulation plotted against temperature differential. Two arrays of curves are included in each figure, one for urethane, the other for foamglass, insulation. For each material, parameters of electrical rates are provided. Capital equipment cost does not appear separately; instead, its equivalent of $91¢/10^6$ Btu has been incorporated into the stated electric rates.

The following information must be known to use the nomographs: pipe size, in.; temperature difference, °F, between the pipe fluid and the average ambient; electric utility rate, ¢/kWh, projected to the midlife of the plant; insulation material; and installed cost.

For the sake of simplicity, installed cost is based on one thickness, $1\frac{1}{2}$ in., of jacketed insulation applied to a single pipe size, $2\frac{1}{2}$ in. Only this installed cost is needed, regardless of the actual pipe size or insulation thickness.

Common to all the nomographs is a minimum insulation thickness of $1\frac{1}{2}$ in. (Several manufacturers have indicated they are discontinuing making insulation thinner than $1\frac{1}{2}$ in. because of dwindling interest.)

Fig. 8 and 9 represent graphic solutions for the insulation thickness necessary to limit surface condensation. The temperature determinants here are fluid temperature (horizontal scale) and atmospheric dewpoint at design conditions. Two dewpoints are shown (75°F and 83°F); interpolation between them is linear. (In our work, "5% values"—the atmospheric conditions that are exceeded only 5% of the cooling-season hours—are used as design points.) A 90°F dry-bulb temperature is used for setting the heat-transfer rate from the pipe outward.

The thicker insulation required by the two criteria should be installed.

Adjusting for variations from basis

Adjustments are likely when (1) a derived capital equipment cost factor other than $91¢/10^6$-Btu output is necessary, all other variables remaining constant; and (2) derived capital cost and energy-consumption rate differ from the stated base.

A capital cost ratio of $91¢/10^6$-Btu net output of refrigeration was calculated for an equipment service life of 13 years. At a power consumption rate of 167 $\text{kWh}/10^6$ Btu, this cost can be expressed as an equivalent "energy surcharge" by:

$$\left(\frac{91¢}{10^6\ \text{Btu}}\right) \div \left(\frac{167\ \text{kWh}}{10^6\ \text{Btu}}\right) = 0.55\ ¢/\text{kWh}$$

(By analogy, the capital expense for a jet costing $500,000, having a service life of 2 million miles, and consuming 2 gallons of fuel per mile would be converted into an equivalent fuel surcharge as follows:

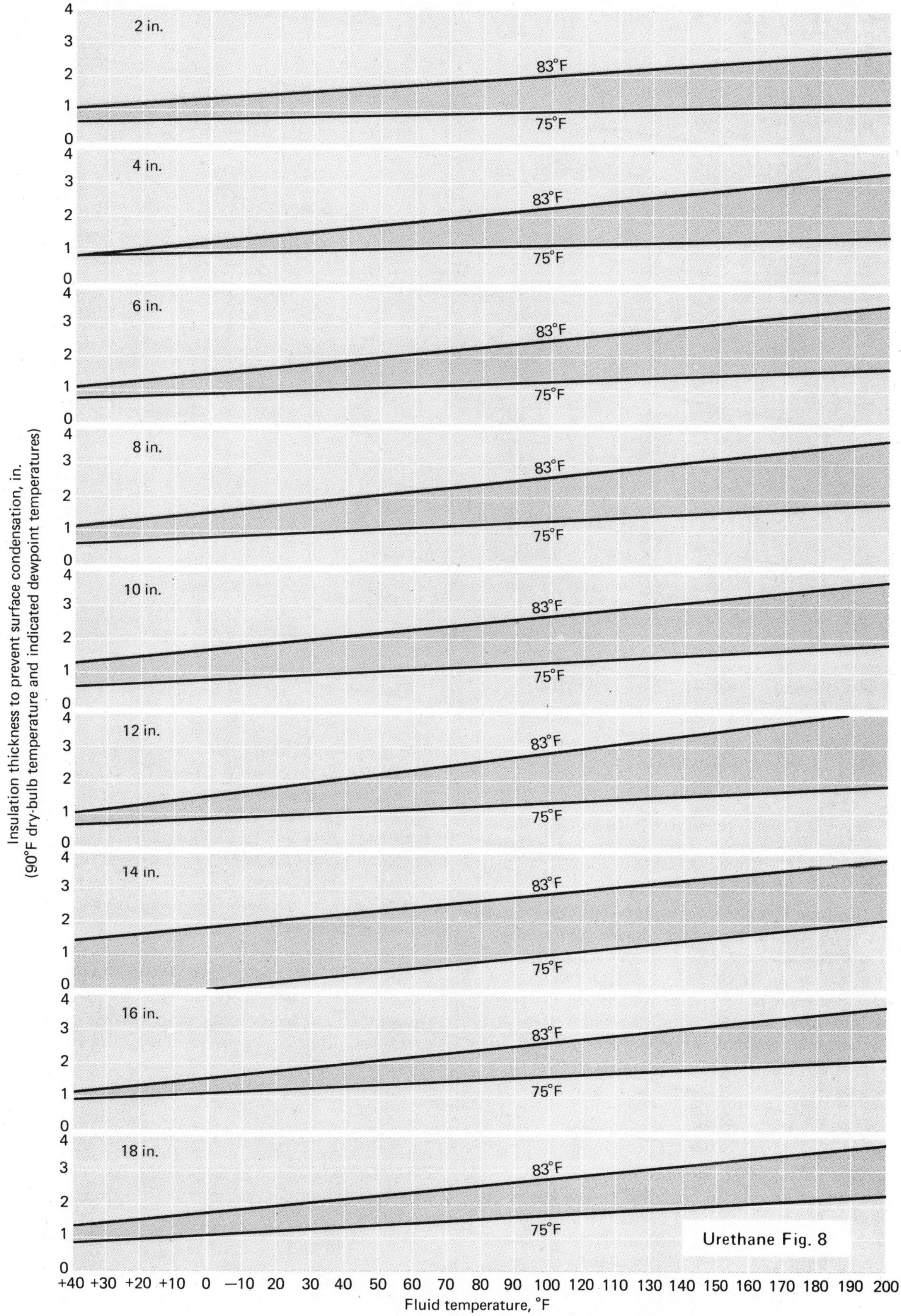

Urethane Fig. 8

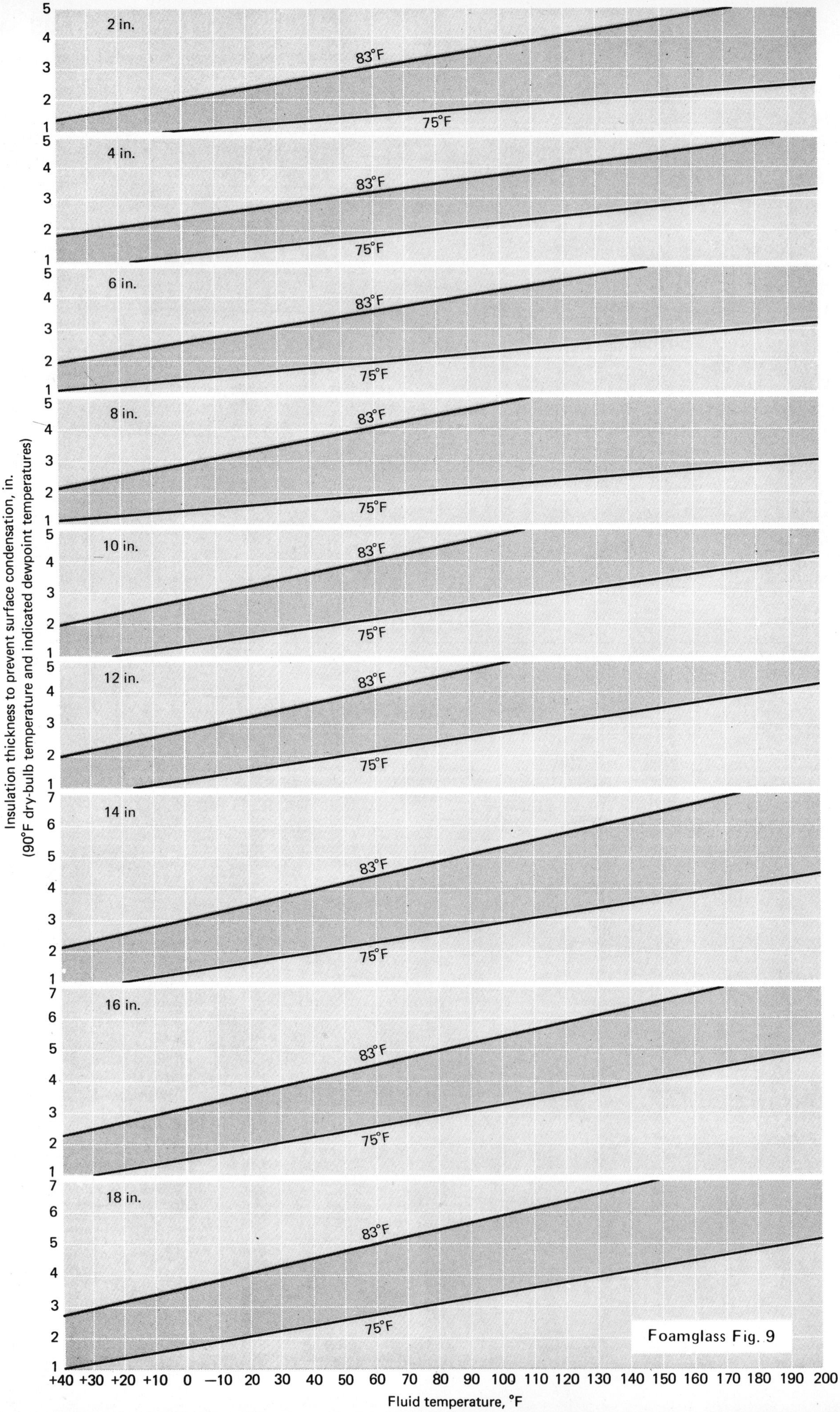

Insulation thickness to prevent surface condensation, in.
(90°F dry-bulb temperature and indicated dewpoint temperatures)
2 in.
4 in.
6 in.
8 in.
10 in.
12 in.
14 in
16 in.
18 in.
83°F
75°F
+40 +30 +20 +10 0 −10 20 30 40 50 60 70 80 90 100 110 120 130 140 150 160 170 180 190 200
Fluid temperature, °F
Foamglass Fig. 9

Amortization factors for Eq. (1) to solve cost of equipment

Amortization period	Interest rate, % 6.0	6.5	7.0	7.5	8.0	8.5	9.0	9.5	10.0	10.5	11.0	11.5	12.0
1	1.060	1.065	1.070	1.075	1.080	1.085	1.090	1.095	1.100	1.105	1.110	1.115	1.120
2	0.5454	0.5493	0.5531	0.5569	0.5608	0.5646	0.5685	0.5723	0.5762	0.5801	0.5839	0.5878	0.5917
3	0.3741	0.3776	0.3811	0.3845	0.3880	0.3915	0.3951	0.3986	0.4021	0.4057	0.4092	0.4128	0.4163
4	0.2886	0.2919	0.2952	0.2986	0.3019	0.3053	0.3087	0.3121	0.3155	0.3189	0.3223	0.3258	0.3292
5	0.2374	0.2406	0.2439	0.2472	0.2505	0.2538	0.2571	0.2604	0.2638	0.2672	0.2706	0.2740	0.2774
6	0.2034	0.2066	0.2098	0.2130	0.2163	0.2263	0.2229	0.2263	0.2296	0.2330	0.2364	0.2398	0.2432
7	0.1791	0.1823	0.1856	0.1888	0.1921	0.1954	0.1987	0.2020	0.2054	0.2088	0.2122	0.2157	0.2191
8	0.1610	0.1642	0.1675	0.1707	0.1740	0.1773	0.1807	0.1840	0.1874	0.1909	0.1943	0.1978	0.2013
9	0.1470	0.1502	0.1535	0.1568	0.1601	0.1634	0.1668	0.1702	0.1736	0.1771	0.1806	0.1841	0.1877
10	0.1359	0.1391	0.1424	0.1457	0.1490	0.1524	0.1558	0.1593	0.1627	0.1663	0.1698	0.1733	0.1770
11	0.1268	0.1301	0.1334	0.1367	0.1401	0.1435	0.1469	0.1504	0.1540	0.1575	0.1611	0.1648	0.1684
12	0.1193	0.1226	0.1259	0.1293	0.1327	0.1362	0.1397	0.1432	0.1468	0.1504	0.1540	0.1577	0.1614
13	0.1130	0.1163	0.1197	0.1231	0.1265	0.1300	0.1336	0.1372	0.1407	0.1444	0.1482	0.1519	0.1557
14	0.1076	0.1109	0.1143	0.1178	0.1213	0.1248	0.1284	0.1321	0.1357	0.1395	0.1432	0.1470	0.1509
15	0.1030	0.1064	0.1098	0.1133	0.1169	0.1204	0.1241	0.1277	0.1315	0.1352	0.1391	0.1429	0.1468
16	0.0990	0.1024	0.1059	0.1094	0.1130	0.1166	0.1203	0.1240	0.1278	0.1316	0.1355	0.1394	0.1434
17	0.0954	0.0989	0.1024	0.1060	0.1096	0.1133	0.1170	0.1208	0.1247	0.1285	0.1325	0.1364	0.1405
18	0.0924	0.0959	0.0994	0.1030	0.1067	0.1104	0.1142	0.1180	0.1219	0.1259	0.1298	0.1339	0.1379
19	0.0896	0.0932	0.0968	0.1004	0.1041	0.1079	0.1117	0.1156	0.1195	0.1235	0.1276	0.1316	0.1358
20	0.0872	0.0908	0.0944	0.0981	0.1019	0.1057	0.1095	0.1135	0.1175	0.1215	0.1256	0.1297	0.1339

$\$5 \times 10^5/2 \times 10^6$ miles ÷ 2 gal/mile = \$0.125/gal.)

For example, if the equipment capital-cost factor were \$1.50/10⁶ Btu instead of 91¢/10⁶ Btu, the equivalent "energy surcharge" would be:

$$\left(\frac{150¢}{10^6 \text{ Btu}}\right) \div \left(\frac{167 \text{ kWh}}{10^6 \text{ Btu}}\right) = 0.90¢/\text{kWh}$$

The 0.35¢/kWh *difference* between 0.90¢/kWh and 0.55¢/kWh is *added* to the applicable electric-utility rate when using the nomographs.

In Case 2, the adjustment mechanism is more involved: an adjusted electric utility rate is derived, and the nomographs are entered with this adjusted rate.

When the derived capital cost (including the financing) and energy consumption rate differ from the reference installation, the first step is to calculate a realistic capital-cost ratio, in $\$/10^6$ Btu:

$$S_b = (83.3\, S/H)(A + M) = \$/10^6 \text{ Btu} \qquad (1)$$

In Eq. (1), S = the cost of the equipment and a proportionate part of the cost of the auxiliaries, \$/ton; H = projected hours of operation per year; A = the amortization factor, which may be calculated with the formula $A = i/1 - (1 + i)^{-n}$ (in which i is the interest rate, and n is the number of payment periods), or may be found in the table above; and M = maintenance as an annual percentage of S.

The second step is to determine the adjusted electric utility rate, in ¢/kWh, for entering the nomographs:

$$R_{adj} = \left[R + \left(\frac{S_b - 0.91}{1.67}\right)\right]\frac{E}{167} = ¢/\text{kWh} \qquad (2)$$

In Eq. (2), R = any projected electric-utility rate, ¢/kWh; S_b = the capital-cost ratio derived with Eq. (1), $\$/10^6$ Btu; and E = any total-energy-consumption rate, kWh/10⁶ Btu.

Example illustrates second case

Assume the following data:

Energy consumption rate, kWh/10⁶ Btu	190
Projected life of system, yr	11
Projected annual service, h	7,800
Projected electric rate, ¢/kWh	1.95
Capital equipment cost, \$/ton	750
Maintenance, %/yr	2
Cost of capital, %/yr	10.5

The adjusted capital-cost ratio, calculated with Eq (1), is \$1.42/10⁶ Btu. The amortization factor, A, is 0.1575.

The adjusted electric-utility rate, determined with Eq (2), is 2.56¢/kWh. Enter this in nomographs.

Mechanically driven refrigeration units—The procedures discussed are not applicable to systems using mechanically driven units. Of course, the condensation criteria are the same whether the unit is electrically or mechanically driven, so Fig. 8 and 9 can be used for either.

☐ Heat tracing is used for liquid-containing pipelines to prevent the liquids from freezing or becoming too viscous. With pipelines containing gases, heat tracing prevents gas components from condensing.

Let us consider these items in more detail. Heat tracing will normally be required when:

■ The lowest ambient site temperature will be below the freezing point of the liquid carried in the pipes. An exception must be made for underground water-pipes installed below the ground frost-level. Examples of liquid lines requiring heat tracing are: phosphoric acid, molten sulfur, glacial acetic acid, benzoic acid, cresol, naphthalene, phthalic anhydride, sorbitol, *p*-xylene and water. Sometimes, heat tracing may be avoided by means of a ring main that keeps the liquid circulating through the pipework.

■ The liquid becomes highly viscous at a temperature above the ambient. Examples are: certain crude oils, fuel oils, polymeric materials, waxes, bitumen and tar residues, and caustic soda liquor.

■ The gas carried in the line has a dewpoint above the ambient temperature, and condensation of liquid from the gas is undesirable. Examples: fuel gas in oil refineries where the liquid causes trouble in the gas burners; natural gas containing moisture that may cause freezeup of control valves, or even the whole system; compressor suction lines (liquid is harmful to compressors); and H_2S/water vapor (causes corrosion on condensation).

Heat loss from pipes is reduced by insulation. Usually the thickness of insulation is based on an economic analysis both of capital costs and the costs of heat-loss from the insulated line.

In winterizing practice, water lines are insulated in an effort to avoid tracing. However, there is always a heat loss from insulated lines. If the liquid cooling resulting from this heat loss cannot be tolerated, heat tracing will be necessary.

In many cases, there is danger of line blockage, and it will not be possible to unblock the line by flushing with a solvent or by blowing in steam and air. If there is no

Originally published March 26, 1979

Steam tracing of pipelines

Here is a rundown on the uses of heat tracing, and the procedures for designing both internal and external steam-tracers.

*I. P. Kohli, Consultant**

heat tracing, the line may have to be disassembled, with associated high costs and long shutdown times. The reassembled pipe system may have to be pressure tested, and in some cases dried, before startup.

It is often said that tracing should provide for heat loss from the line—i.e., tracing should not be used for process heating. But in actual practice, facilities may have to be provided for preheating the line before startup, or for quickly thawing a frozen-up line.

The length of a traced line is quite variable. It may be a few feet in a process area, a few thousand feet between offsites and the process area, or over a hundred miles as in the case of underground lines carrying crude oil or fuel oil.

Heat for pipe tracing

Piping may be heated by using fluid heating media (steam, hot water, hot oil, or Dowtherm), or by electricity. The fluid heating systems are simple in operation; any part of the pipework may be isolated by shutting valves. However, there is a possibility of fluid leakage, leading to insulation damage and product contamination. Also, temperature control is rather poor due to the large heat capacity of the fluid-heated system. Still, steam tracing is commonly used because there is surplus low-pressure steam available in most plants.

Since steam has a high latent heat, only a small quantity is required for a given heating load. Also, steam has a high film heat-transfer-coefficient, condenses at constant temperature, and flows to the point of use without pumps.

Liquid heating media (hot water or oil) have to be pumped at a high rate to satisfy a given heating load, ensure uniform heating and provide a reasonable heat-transfer coefficient. The use of liquid and vapor media requires heated tracing pipes either inside or outside the pipeline. Vapor systems need two additional lines (vapor header and condensate return line—see Fig. 1). Liquid systems also require a return header.

Heating by electricity is a clean operation; there is no chance of fluid leakage. An electrical hazard appears only if the system is improperly installed. Temperature is readily controlled by using on/off thermostats. Thus, energy is used only when needed. Usually, electrical heating is employed when surplus steam is not available, or the temperature required would call for high-pressure steam. It is ideal for long-distance lines, because a supply of electricity is usually available along the line. (For details on electrical pipe tracing, see pp. 422–425.)

Pipeline heating using steam

Heat to pipelines may be provided by steam pipes either inside or outside the line. Steam jacketing (Fig. 2) may also be used, but owing to its high cost it is employed only for special situations involving high heating loads. Jackets (which require oversized insulation) may also be considered for short process runs.

Internal steam tracing

When internal-trace pipes are installed, all the available heat-transfer surface is utilized. The disadvantages are: (a) reduction in the equivalent internal diameter of the pipeline, (b) loss of ability to clean the pipeline by pigging or by using rotary brushes, (c) difficulty in cleaning fouled heat-transfer surfaces.

The trace pipe can only be installed in straight lengths of pipeline that are free of valves. Trace-pipe lengths have to be short, to prevent problems in supporting the trace pipe. It must enter and leave the pipeline frequently, increasing the possibility of leaks. Stresses arising due to differential expansion of the trace and the pipeline should be considered.

One possible application of the internal type of trace is for a fuel-oil unloading line, where the operating temperature is such that it is uneconomical to insulate the line, and where the line is used only intermittently. (Insulation is essential for externally traced lines.) In-

*Now working with Davy International, 8 Baker St., London, WI, U.K.

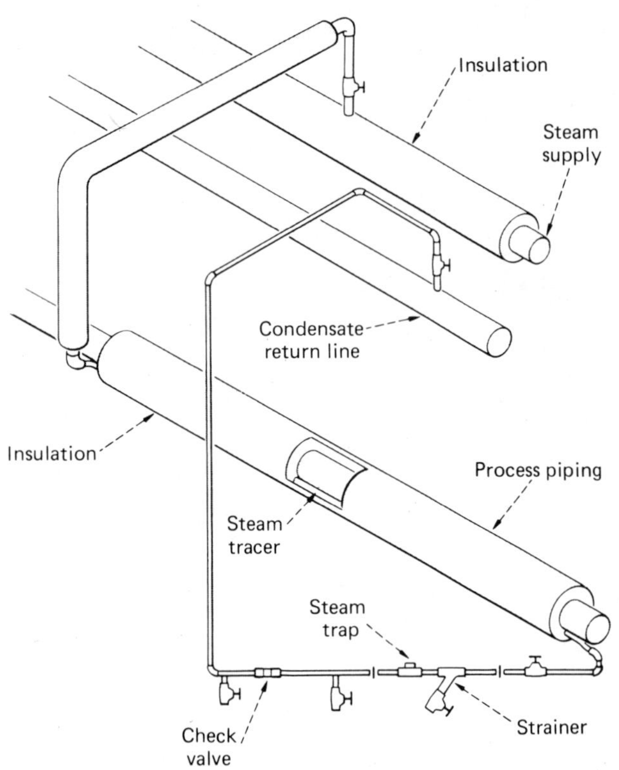

Typical steam-tracing system using external tracer Fig. 1

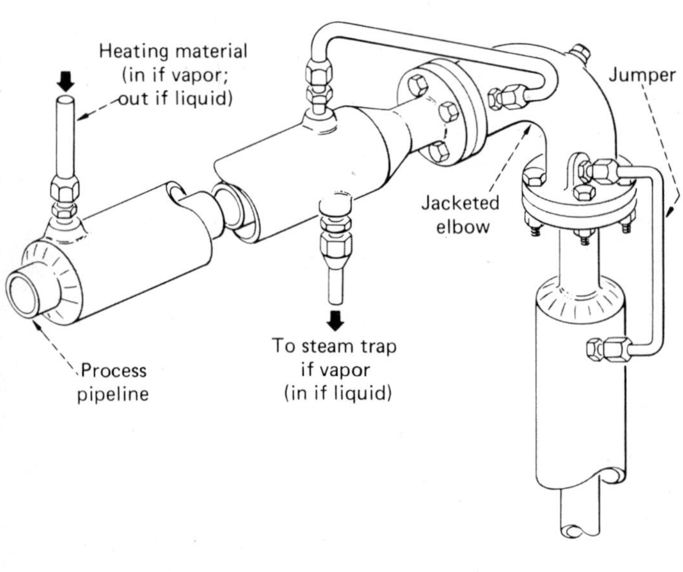

Steam jacketing is used in special situations Fig. 2

Nomenclature

A	Area of pipe, ft^2/ft of length
A_p	Outside area of tracer pipe, ft^2/ft length
A_1	Outside area of pipe lagging, ft^2/ft length
C_p	Specific heat, Btu/(lb)(° F)
D	Dia., in.
D_e	Equivalent pipe dia., ft
D_i	Inside dia., in.
D_{ip}	Pipe I.D., in.
D_o	Outside dia., in
D'_o	Outside dia. of tracer pipe, in.
d_i	I.D. of lagging, in.
d_o	O.D. of lagging, in.
G	Mass velocity, lb/(h)(ft^2)
h_a	Film coefficient to air, Btu/(h)(ft^2)
h_i	Inside film coefficient, Btu/(h)(ft^2)
h_o	Outside film coefficient, Btu/(h)(ft^2)
j_H	Colburn factor, dimensionless
k_a	Thermal conductivity, Btu/(h)(ft^2)(° F/ft)
k_1	Thermal conductivity of lagging, Btu/(h)(ft^2)(° F/ft)
L	Pipe length, ft
N_{Re}	Reynolds number, dimensionless
Q_a	Heat flux to air, Btu/(h)(ft^2)
Q_{conv}	Heat flux by convection, Btu/(h)(ft^2)
Q_p	Heat loss from pipe without tracer, Btu/(h((ft^2)
q_2	Heat loss, Btu/(h (ft of length)
t_a	Air temperature, ° F
t_p	Temperature inside pipe, ° F
t_s	Temperature, surface of lagging, ° F
t_{st}	Average steam temperature, ° F
t_w	Pipewall temperature, ° F
T_p	Pipe temperature, ° F
U_o	Heat transfer coefficient based on outside surface, Btu/(h)(ft^2)(° F)
W_1	Mass flowrate, lb/h
Greek letters	
μ	Viscosity, lb/(ft)(h)
μ_w	Viscosity at wall temperature, lb/(ft)(h)
ρ	Fluid density, lb/ft^3

ternal tracing is acceptable only if leakage of steam into the product can be tolerated.

External tracing

External tracing involves the placing of steam pipes or tubes outside the pipeline. The traced line is then insulated by using preformed sectional insulation.

Heat transfer from trace to pipeline is by conduction, convection in the air space, and radiation. The contact area between the trace and the pipeline is quite small. However, when heat-sensitive liquids are being heated, or when the pipe is plastic-lined, this contact may give rise to undesirable hot spots. In such cases, asbestos packing rings are often used to eliminate any direct contact between the trace and the pipeline.

The simplest method of external tracing is to wrap copper tube around the pipeline, and then to cover the traced pipe with insulation. This is the only way to trace around valves, pipe fittings and instruments

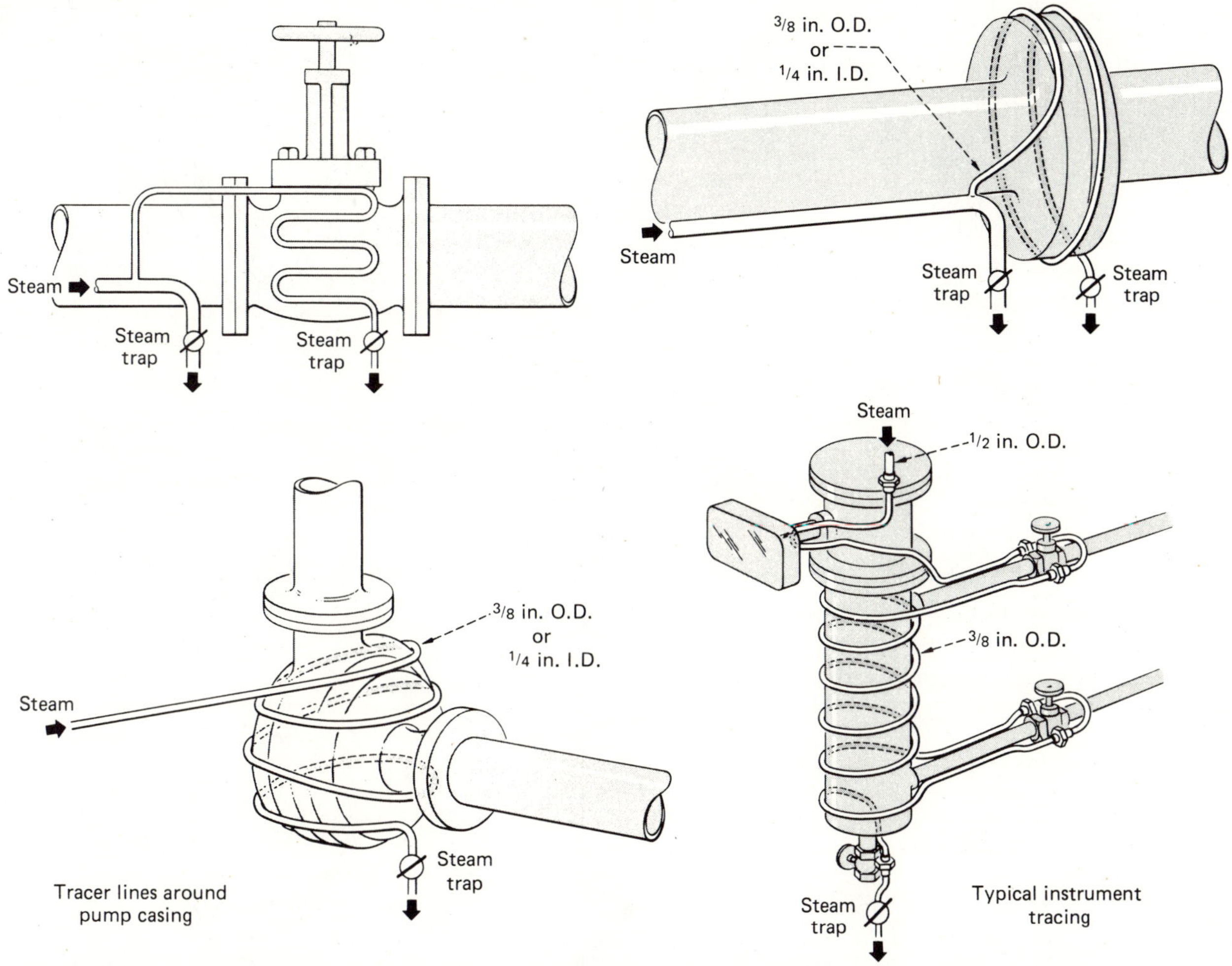

How various pipe fittings, pumps, instruments, etc., are treated when using steam tracing **Fig. 3**

(Fig. 3). This procedure is unsuitable for horizontal runs because steam condensate collects at low points and may freeze during a shutdown.

It is essential to ensure that the trace lines are self-draining. Copper tubing of 0.5-in. O.D. and 0.035-in. wall thickness is usually used with straight piping. If the tubing has to be bent into a small radius (as when tracing valves), 0.375-in.-O.D. tubing may be used.

The length of a single trace tube (from steam supply valve to steam trap) is limited by pressure drop in the trace. The trap should have a condensate drainage capacity to match the heating load. At a steam pressure of 100 psig or higher, the length of a single trace should not exceed 200 ft. If the steam pressure is lower, a tracer length of 100 ft is recommended.

To improve contact, the trace tube may be wired to the pipeline. Even then, the conductive heat-transfer rate is quite low. It may be increased by putting a layer of heat-conducting cement (graphite mixed with sodium silicate or other binders) between the trace and the pipeline. This provides much more surface for conductive heat transfer.

When higher heat loads are desirable, straight lengths of ½-in. carbon-steel pipe are clipped along the pipeline. The number of tracer pipes depends upon the heat load—large-diameter pipes carrying liquids that have a high melting point may require up to ten tracers. Steel bands, fitted by using a packing-case banding machine, help to minimize air gaps between the pipeline and tracers. At pipeline bends, the tracers are also bent.

At valves and flanges, the tracers are formed into loops that also function as expansion joints. The loops are formed in a nearly horizontal plane, to ensure self-drainage. The traced pipeline is covered with larger-sized preformed lagging.

In some cases, the traced line is wrapped with aluminum foil and then covered with shaped lagging (Fig. 4). The foil increases the radiation heat transfer. It is essential that the space between the pipeline and the tracer be kept free of particles of lagging material.

Conductive heat transfer may be enhanced by welding the trace on the pipeline. However, welding causes problems due to differential expansion of the pipeline and trace. Because horizontal pipelines are traced on the lower half, welding is a difficult operation. It is more convenient to use heat-transfer cement, as noted earlier.

Before applying the cement, it is essential that both the pipeline and tracer pipes be wire-brushed.

Internal steam tracing—design example

Basis

Size a steam-tracing system to maintain an intermittently used fuel-oil unloading line at 140°F.

Data

Average temperature	20°F
Wind speed	20 mi/h
Line length	500 ft
Size	12-in. nominal bore
Max. unloading rate	500 tons/h
Steam available at 65 lb/in^2 gage	
Specific gravity of oil	0.985
Viscosity (see Fig. 5)	
Note: Viscosity in cP × 2.42 = viscosity in lb/(ft)(h)	
Thermal conductivity	0.0778 Btu/(h)(ft^2)(°F/ft)
Specific heat	0.47 Btu/(lb)(°F)

Calculation

As the line will only be used intermittently, use internal steam tracing and an unlagged pipe.

Heat loss—Neglecting the effect of the internal tracer at this point:

$$\text{Reynolds No., } N_{Re} = \frac{6.31W}{\mu D_{ip}}$$

where W_1 = mass flowrate
$= 500 \times 2{,}240$ lb/h

μ = viscosity
$= 200$ cP at 140°F

D_{ip} = pipe dia.
$= 12$ in.

$$\therefore N_{Re} = \frac{6.31 \times 2{,}240 \times 500}{200 \times 12} = 2{,}940$$

$$L/D = 500$$

Hence, from Fig. 6:

$$\left(\frac{h_i D_{ip}}{k_a \times 12}\right)\left(\frac{C_p \cdot \mu}{k_a}\right)^{-1/3}\left(\frac{\mu}{\mu_w}\right)^{-0.14} = 8.0$$

$$\text{whence, } h_i = \left(\frac{12k_a}{D_{ip}}\right)\left(\frac{C_p\mu}{k_a}\right)^{1/3}\left(\frac{\mu}{\mu_w}\right)^{0.14} \times 8.0$$

Assuming for the first trial that $t_s = 90$°F:

$D_{ip} = 12$ in.
$\mu = 200 \times 2.42$ lb/(ft)(h) [at 140°F]
$\mu_w = 520 \times 2.42$ lb/(ft)(h) [at 90°F]
$C_p = 0.47$ Btu/(lb)(°F)
$\rho = 0.985 \times 62.4$ lb/ft^3
$k_a = 0.0778$ Btu/(h)(ft^3)(°F/ft)

$$\therefore h_i = \left(\frac{8 \times 12 \times 0.0778}{12}\right)\left(\frac{0.47 \times 200 \times 2.42}{0.0778}\right)^{1/3}\left(\frac{200 \times 2.42}{520 \times 2.42}\right)^{0.14}$$

$$= 0.621\ (2{,}920)^{1/3}\ (0.385)^{0.14}$$
$$= 0.621\ (14.3)\ (0.875)$$
$$= 7.78 \text{ Btu/(h)(ft}^2\text{)(°F)}$$

take $h = 8.0$ Btu/(h)(ft^2)(°F)

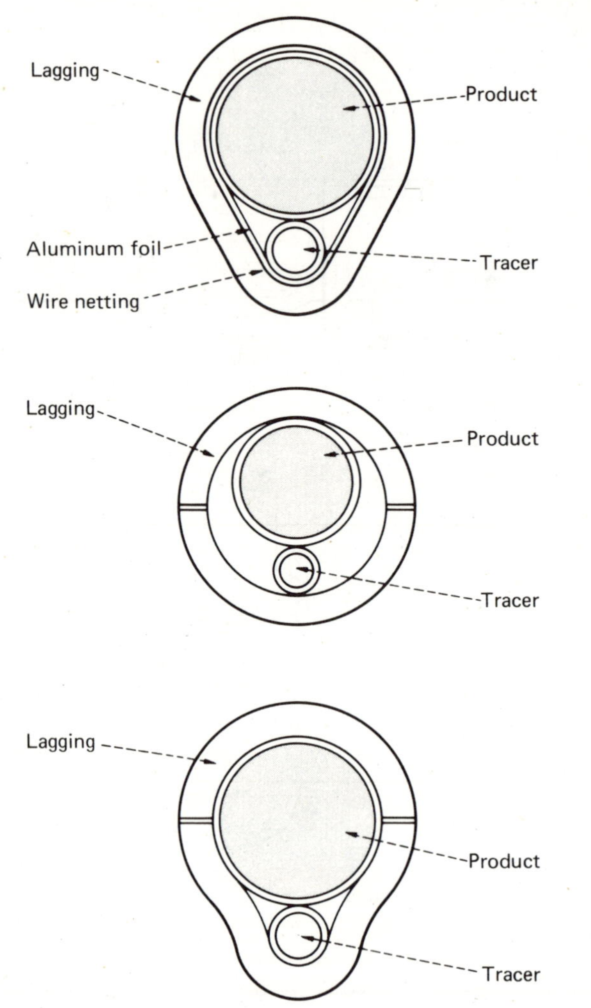

Various configurations of tracer and lagging **Fig. 4**

Correcting to O.D. of pipe:

$h_i = 8.0 \times (12.7/12) = 8.5$ Btu/(h)(ft^2)(°F)

Hence, $Q_{conv} = h_i A\ (t_p - t_s)$
$= 8.5A\ (140 - t_s)$ Btu/(h)(ft^2)

and $Q_a = h_a A\ (t_s - t_a)$

From Tables I and II,

$h_a = 1.83 \times 2.76$

$= 5.04$ Btu/(h)(ft^2)(°F)
[corrected for wind velocity]

$\therefore Q_a = 5.04A\ (t_s - 20)$ Btu/(h)(ft^2)

Since $Q_{conv} = Q_a$
$8.5A\ (140 - t_s) = 5.04A\ (t_s - 20)$
i.e., $1{,}190 - 8.5\ t_s = 5.04\ t_s - 100.8$
$\therefore t_s = 1{,}290/13.54 = 95$°F

There is an insignificant difference in μ_w between 90° and 95°F; similarly, h_a does not change.

$\therefore$ heat loss $= (8.5)\ \pi(12.7/12)(140 - 95)$
$= 1{,}290$ Btu/(h)(ft of pipe)

Due to the low temperature of the pipe, radiation heat loss has been neglected.

Determine tracer size—For first trial, assume tracer is 1.5 in. I.D. (1.9 in. O.D.)

$$\text{Equivalent diameter, } D_e = \frac{144 - (1.9)^2}{12} = 11.7 \text{ in.}$$

$$N_{Re} = D_e \frac{G}{\mu}$$

where: G = mass velocity, lb/(h)(ft²)

$$= \frac{500 \times 2{,}240}{\frac{\pi}{4}\left(1 - \frac{(1.9)^2}{144}\right)}$$

$$= 1.45 \times 10^6 \text{ lb/(h)(ft}^2)$$

$$D_e = 0.975 \text{ ft}$$

$$\mu = 200 \text{ cP at } 140°\text{F}$$

$$\therefore N_{Re} = \frac{11.7 \times 1.45 \times 10^6}{12 \times 200 \times 2.42} = 2.8 \times 10^3$$

From Fig. 6, $j_H = 9.0$; $N_{Re} = 2{,}800$; $L/D_e = 150/0.975 = 154$

Heat-transfer coefficient for heat transferred to the fuel oil based on I.D. of the 12-in. nominal-bore pipe at the steam-tracer surface where the fuel-oil viscosity is:

$$\mu_w = \frac{h_i D_e}{k_a}\left(\frac{C_p \mu}{k_a}\right)^{-1/3}\left(\frac{\mu}{\mu_w}\right)^{-0.14} = p$$

Assuming a 20-lb/in.² pressure drop along the steam tracer, the average steam pressure = 55 lb/in² gage (steam available at 65 lb/in² gage).

Average steam temperature = 303°F

$\therefore \mu_w = 1.0 \times 2.42$ lb/(ft)(h)[for fuel oil]

and heat-transfer coefficient at steam-tracer outside surface,

$$h_o = 9 \times \left(\frac{0.0778}{0.975}\right)\left(\frac{0.47 \times 200 \times 2.42}{0.0778}\right)^{1/3}\left(\frac{200}{1.0}\right)^{0.14}$$

$$= 21.6 \text{ Btu/(h)(ft}^2)(°\text{F})$$

$$\frac{1}{U_o} = \frac{1}{h_o} + 0.005 = \frac{1}{21.6} + 0.005$$

$$= 0.051 \text{ Btu/(h)(ft}^2)(°\text{F}), \text{ approx.}$$

$$\therefore U_o = 19.6 \text{ Btu/(h)(ft}^2)(°\text{F})$$

Heat transfer from steam tracer = heat lost from pipe without steam tracer:

$$Q_p = U_o A_p (t_{st} - t_p)$$

$$\therefore 1{,}290 = 19.6 A_p (303\text{-}140)$$

$$\text{Hence, } A_p = \frac{1{,}290}{19.6 \times 163}$$

$$= 0.40 \text{ ft}^2\text{/ft run of pipe}$$

Surface area of 1-in. nominal-bore pipe = 0.344 ft²/ft
Surface area of 1½-in. nominal-bore pipe = 0.498 ft²/ft
Therefore, 1½-in. nominal-bore tracing will be acceptable.

Check effect of increased fluid velocity on outside film coefficient at pipe wall (internal tracer increases the fluid velocity).

$$N_{Re} = \frac{D_e G}{\mu} = 1 \times \frac{1.45 \times 10^6}{2.42 \times 200} = 3{,}010$$

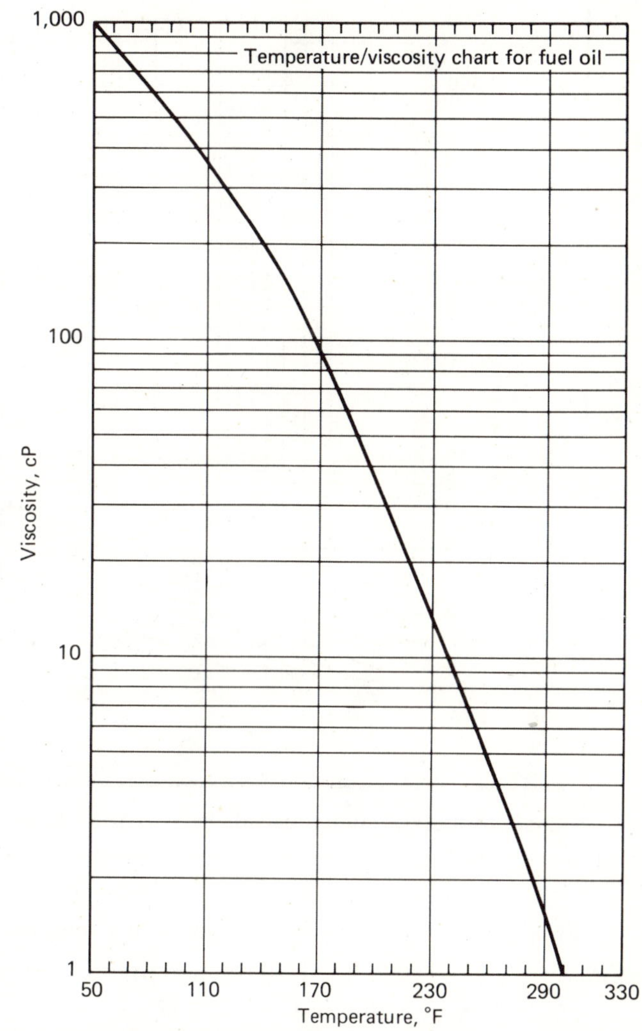

Typical temperature/viscosity chart for fuel oil **Fig. 5**

where G is based on the equivalent diameter of the pipe, and μ is at 140°F, the desired fuel-oil temperature.

$$j_H = 8.05 \text{ [from Fig. 6]}$$

There will be no significant increase in the calculated heat loss.

Check pressure-drop increase for process line

Correction factor

$$= \frac{(D_i)^{2.4}(D_i + 0.5D_o')^{1.2}}{(D_i^2 - D_o'^2)^{1.8}}\left(\frac{\mu_w}{\mu}\right)^{0.14}$$

$$= \frac{(12)^{2.4}(12 + 0.95)^{1.2}}{(144 - 3.6)^{1.8}}\left(\frac{520}{200}\right)^{0.14} = 1.33$$

Check steam tracer pressure-drop

Maximum tracer length = 150 ft
Heat load = 150 × 1,290 Btu/h
Latent heat of steam = 901 Btu/lb

$$\therefore \text{Steam load} = \frac{150 \times 1{,}290}{901}$$

$$= 215 \text{ lb/h per tracer}$$

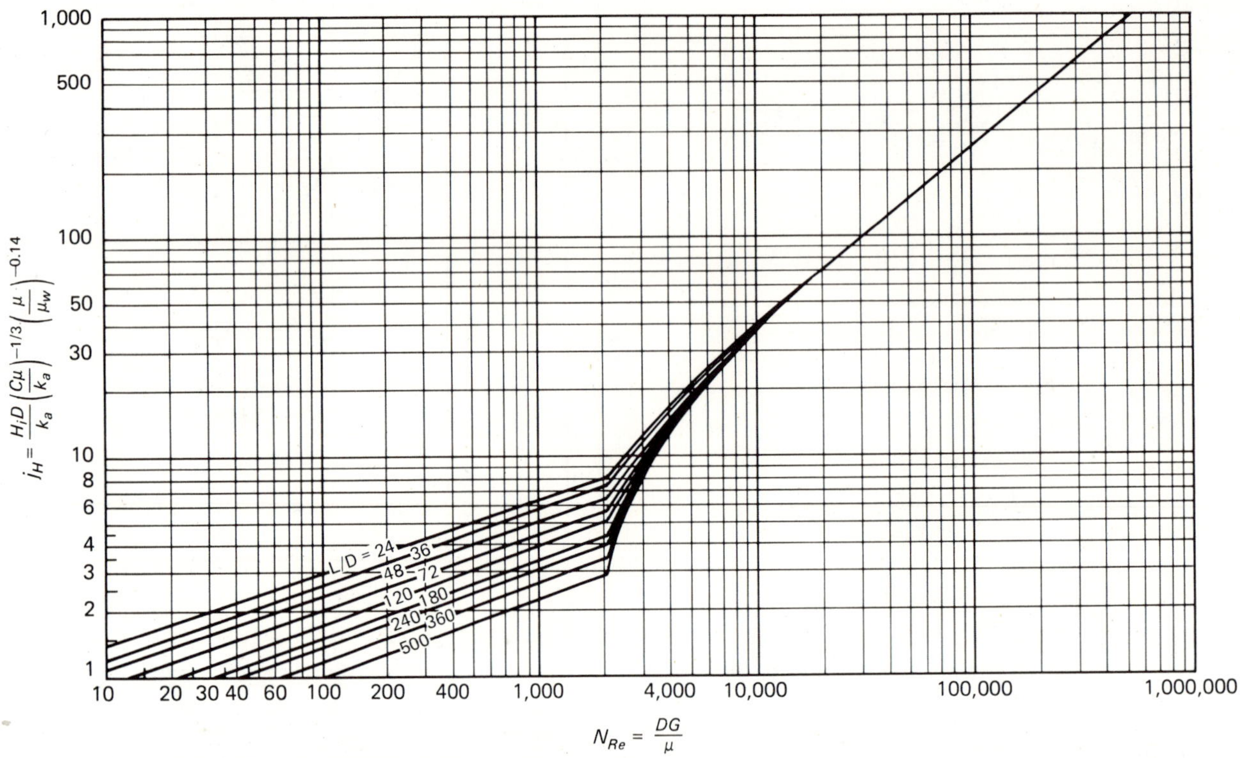

Chart for determining tube-side heat transfer, based on Reynolds number and Colburn factor **Fig. 6**

From Fig. 7, steam pressure drop at inlet conditions:

$$0.55 \text{ lb/(in}^2\text{)(100-ft run)}$$

$$\therefore \text{pressure-drop/tracer} = 1.5 \times 0.275 = 0.412 \text{ lb/in}^2$$

$\therefore$ the average steam temperature = 312°F (65 lb/in² gage)

$$U_o = 19.6 \text{ Btu/(h)(ft}^2\text{)(°F)}$$

$$\Delta T = 312 - 140 = 172\text{°F}$$

$$\therefore \text{area of tracer required} = \frac{1{,}290}{(19.6 \times 172)} = 0.383 \text{ ft}^2\text{/ft}$$

The surface area of $1\frac{1}{2}$-in. nominal-bore pipe is 0.498 ft²/ft

Designing external steam tracing

Basis

Size an external tracing system to maintain a 3-in. line at 320°F. The line will be lagged with $1\frac{1}{2}$ in. of insulation.

k_1	= 0.033 Btu/(h)(ft)(°F/ft)
Wind speed	= 20 mi/h
Minimum air temperature, T_a	= 20°F
Line length	= 100 ft
Steam available at 150 lb/in²gage	
Pipe temperature (t_p)	= 320°F

Calculation

Determine heat load—Assuming 5 lb/in.² pressure drop along the tracer, then the average steam temperature is:

$$t_{st} = \frac{366 + 363\text{°F}}{2} = 364.5\text{°F, say, } 364\text{°F}$$

$$\therefore \text{average temperature} = 0.5(t_{st} + t_p) \quad [\text{°F}] = 0.5(364 + 320) \quad [\text{°F}] = 342\text{°F}$$

Allowing for tracer, I.D. of lagging = 4 in.

$$\therefore q_2 = \frac{2\pi k_1(t_p - t_s)}{\log_e \frac{d_o}{d_i}} = h_a A_1(t_s - t_a)$$

$$d_o = 7 \text{ in.}, d_i = 4 \text{ in.}$$

Assuming for the first trial that $h_a = 4.0$ Btu/(h)(ft²)(°F):

$$q_2 = \frac{2 \times \pi \times 0.033(342 - t_s)}{2.303 \log_{10}(7.0/4.0)} = 4.0\pi \times (7.0/12) \times 1(t_s - 20)$$

$$\text{or } \frac{0.0895(342 - t_s)}{0.245} = 7.35(t_s - 20)$$

$$\text{or } 0.368(342 - t_s) = 7.35t_s - 147$$

$$\text{or } 126 - 0.368t_s = 7.35t_s - 147$$

$$\therefore t_s = \frac{273}{7.718} = 35\text{°F}$$

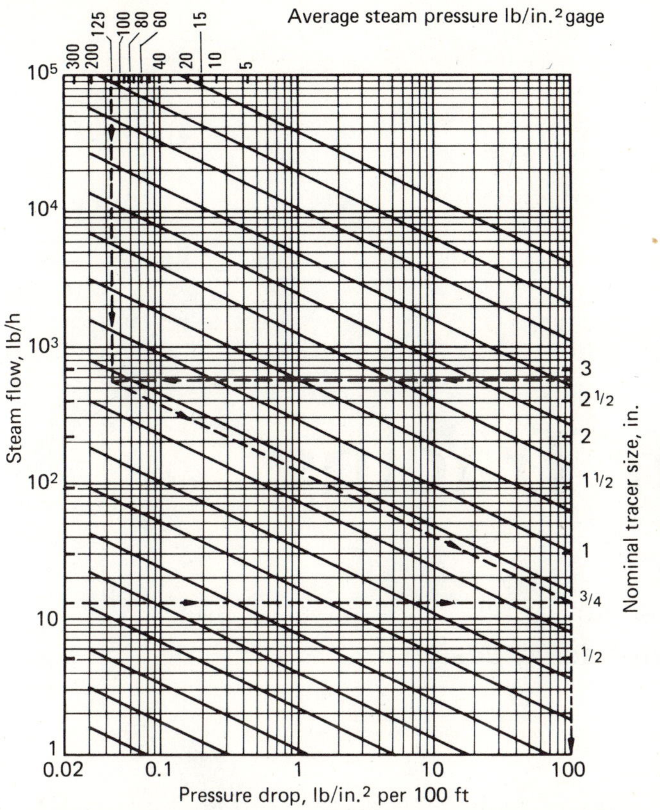

Determination of pressure drop in pipes **Fig. 7**

Tables I and II are not sufficiently accurate in this region, but the value of 4.0 taken for h_a is on the safe side, allowing for wind speed.

$$\therefore q_2 = 7.35(t_s - 20) \text{ Btu/(h)(ft run of pipe)}$$
$$= 7.35(35 - 20)$$
$$= 113 \text{ Btu/(h)(ft run of pipe)}$$

Determine the tracer size—This is done by use of thermal-conductance data on tracer to pipe: Btu/(h)(°F)(ft of pipe). This takes into account the heat-transfer coefficient and pipe surface area. These have been found by experiment.

If one ½-in. tracer without cement is used, heat transfer

$$= 0.393(t_{st} - t_p)$$
$$= 0.393(364 - 320) = 17.3 \text{ Btu/(h)(ft run)}$$

This does not meet the heat load requirement. Two ½-in. tracers without cement do not overcome the problem. If one ½-in. tracer with heat-transfer cement is installed, the heat-transfer rate is:

$$4.58(364 - 320) = 201 \text{ Btu/(h)(ft of run)}$$

A ⅜-in. tracer with cement will yield a heat-transfer rate of:

$$3.44(364 - 320) = 151 \text{ Btu/(h)(ft of run)}$$

Standard practice requires the installation of ½-in. tracer pipes. There is always some overdesign with steam tracers.

Values of h_a for pipes in still air **Table I**

Nominal pipe dia., in.	$(t_s - t_a)$, °F [For an unlagged pipe $t_s = t_w$]							
	50	100	150	200	250	300	400	500
½	2.12	2.48	2.76	3.10	3.41	3.75	4.47	5.30
1	2.03	2.38	2.65	2.98	3.29	3.62	4.33	5.16
2	1.93	2.27	2.52	2.85	3.14	3.47	4.18	4.99
4	1.84	2.16	2.41	2.75	3.01	3.33	4.02	4.83
8	1.76	2.06	2.29	2.60	2.89	3.20	3.83	4.68
12	1.71	2.01	2.24	2.54	2.82	3.12	3.83	4.61
24	1.64	1.93	2.15	2.45	2.72	3.03	3.70	4.48

h_a = Btu/(h)(ft^2)(° F,) based on still air.
From McAdams, W. H., "Heat Transmission," 3rd ed., McGraw-Hill, New York, 1954, p. 179.

Correction factor for h_a at different wind velocities **Table II**

Wind velocity, mi/h	$(t_s - t_a)$, °F [For an unlagged pipe, $t_s = t_w$]				
	100	200	300	400	500
2.5	1.46	1.43	1.40	1.36	1.32
5.0	1.74	1.69	1.64	1.59	1.53
10.0	2.16	2.10	2.02	1.93	1.84
15.0	2.50	2.42	2.33	2.27	2.08
20.0	2.76	2.69	2.58	2.45	2.30
25.0	2.98	2.89	2.78	2.64	2.49
30.0	3.15	3.06	2.94	2.81	2.66
35.0	3.30	3.21	3.10	2.97	2.81

From Thermon Data Book, Premaberg (GB) Ltd.

C_t, Thermal conductance, tracer to pipe **Table III**

Tube size	C_t with no heat-transfer cement, Btu/(h)(°F)(ft of pipe)	C_t with heat-transfer cement Btu/(h)(°F)(ft of pipe)
3/8	0.295	3.44
1/2	0.393	4.58
5/8	0.490	5.73

The author

I. P. Kohli, 118 Chadwell Heath Lane, Romford, Essex, U.K., is a consultant process engineer specializing in the design of offshore oil/gas production facilities. He has had experience in several of the chemical process industries, including pharmaceuticals, fibers, oil, gas and petrochemicals. He was graduated from Bombay University, and holds an M.Sc. in hydrocarbon chemistry and petrochemicals from Manchester University. He is a chartered engineer and a committee member of the London/South East branch of the Institution of Chemical Engineers.

Designing Steam Tracing

Use this analytical approach to determine the effect of all variables, in order to obtain the optimum steam-tracing design.

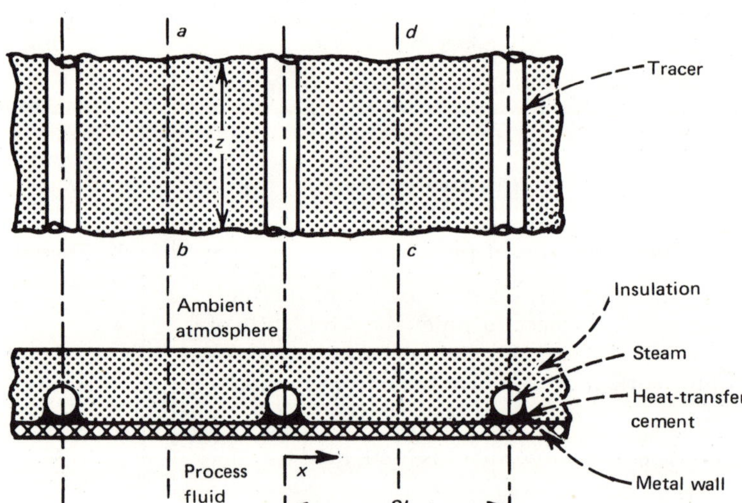

DEVELOPED SECTION of wall, showing steam tracing and insulation—Fig. 1 ▶

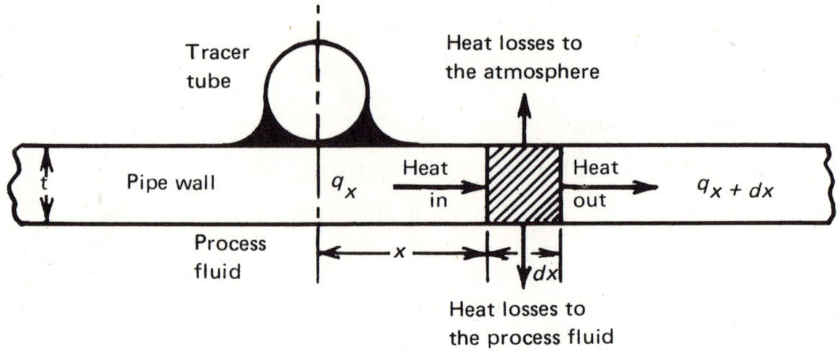

STEADY-STATE heat combustion through the wall—Fig. 2 ▶

Originally published April 3, 1972

CARL G. BERTRAM, VIKRAM J. DESAI and EDWARD INTERESS, The Badger Co.

In the chemical process industries, pipes and equipment often need to be heated. There are many reasons for this. Many materials become viscous or even solidify at ambient temperatures—phthalic anhydride solidifies at 268 F.

Some materials are relatively noncorrosive in the vapor phase but become highly corrosive when they condense—hydrogen sulfide and water vapor condense to form corrosive hydrosulfuric acid.

Although there are other ways of heating pipes and equipment, such as by electrical heating cables or by jacketing, steam tracing is becoming more widely used. An important reason for increased use has been the development of plastic heat-transfer cements.

The high thermal conductivity and good bonding characteristics of these cements, when used properly in well-designed installations, have virtually eliminated most of the disadvantages associated with steam tracing, e.g.,

- Unpredictable heat-transfer rate.
- Slow heatup.
- Uneven temperature distribution.
- Need for high temperature-differentials between tracer surface-temperature and process-fluid temperature.

To date, most installations of steam tracing that use heat-transfer cement have been designed by using either an empirical approach or one based on somewhat simplified equations that often do not accurately represent the system. Consequently, the general policy for steam-tracing design has been conservative. The cost of overdesign has been justified by considering the potential for increased operating and maintenance costs, and plant shutdowns, created by insufficient tracing (underdesign). However, steam-tracing costs may be upwards of 10% of the total piping cost of the plant. The money spent on overdesign is well worth saving.

The analytical approach to steam-tracing design presented here eliminates the over- and underdesign problems, and gives an optimum solution. A general solution can be obtained that determines the following:

1. The optimum pitch of steam-tracing. To be optimum, the tracing tubes must be spaced so that no part of the pipe wall can fall below the required temperature.
2. The temperature profile of the wall. The general solution gives the temperature profile on the wall at various distances from the tracer tube.
3. The heat transferred to the process fluid.
4. The total heat transferred by each tracer.

The parameters used in the general solution are wall thickness, wall material (thermal conductivity), insulation thickness and material, heat-transfer coefficient between wall and process fluid, process temperature, steam temperature, and ambient temperature.

Theory and Derivation

The heat balance at steady state around the differential element $t\,dx$ of the developed section of the wall in Fig. 1 and 2 is described by:

$$-Ktz\frac{dT}{dx} - \left(-Ktz\frac{dT}{dx} - Ktz\frac{d^2T}{dx^2}\,dx\right) - [h_o(T - T_{amb}) + h_i(T - T_p)]\,z\,dx = 0 \quad (1)$$

Letting:

$$A = \frac{h_o + h_i}{Kt} \text{ and } B = \frac{h_i T_p + h_o T_{amb}}{Kt},$$

Eq. (1) reduces to:

$$\frac{d^2T}{dx^2} - AT + B = 0$$

This linear differential equation with constant coefficients gives the following general solution for the wall-temperature distribution:

$$T = C_1 e^{\sqrt{A}x} + C_2 e^{-\sqrt{A}x} + (B/A) \quad (2)$$

C_1 and C_2, and hence the actual temperature distribution, are dependent on the following boundary conditions:

$$T = T_o \text{ at } x = 0$$

$$\frac{dT}{dx} = 0 \text{ at } x = L$$

The following solution results from Eq. (2):

$$\frac{T - (B/A)}{T_o - (B/A)} = \frac{e^{\sqrt{A}L\alpha}(1 + e^{2\sqrt{A}L(1-\alpha)})}{1 + e^{2\sqrt{A}(L)}} = \frac{\cosh(\sqrt{A}L(1 - \alpha))}{\cosh(\sqrt{A}L)}, \quad (3)$$

where $\alpha = x/L$.

Eq. (3) gives the temperature of the wall between the two parallel tracers, and is plotted in Fig. 3 as

$$\frac{T - (B/A)}{T_o - (B/A)} \text{ versus } \sqrt{A}(L)$$

for various values of α.

For $\alpha = 1$, the point described is midway between the two tracers. The minimum wall temperature occurs at this point. When the minimum allowable wall temperature is specified, the maximum tracer pitch, $2L$, can be calculated.

The heat transferred to the process fluid through the surface *abcd* in Fig. 1 is:

$$\frac{Q}{z} = 2\int_o^L (T - T_p)h_i\, dx \text{ Btu./(hr.)(ft.) of tracer}$$

Nomenclature

A	A parameter $= (h_o + h_i)/Kt$, dimensionless
B	A parameter $= [(h_iT_p) + (h_oT_{amb})]/Kt$, dimensionless
C_1, C_2	Constants
c_p	Heat capacity of process fluid, Btu./(lb.)(°F.)
D	Pipe diameter, ft.
f	Darcy friction factor, dimensionless
G	Mass velocity of the process fluid, lb./(hr.)(sq. ft.)
h_{air}	Heat-transfer coefficient between ambient air and insulation, Btu./(hr.)(sq. ft.)(°F.)
h_i	Process-fluid heat-transfer coefficient, Btu./(hr.)(sq. ft.)(°F.)
h_o	Overall heat-transfer coefficient between the wall and ambient air, Btu./(hr.)(sq. ft.)(°F.)
k	Thermal conductivity of the process fluid, Btu./(hr.)(sq. ft.)(°F./ft.)
K	Thermal conductivity of pipe or vessel wall, Btu./(hr.)(sq. ft.)(°F./ft.)
K_{ins}	Thermal conductivity of insulation, Btu./(hr.)(sq. ft.)(°F./in.)
L	One-half the steam tracing pitch, ft.
N_{Nu}	Nusselt number, $(h_iD)/K$, dimensionless
N_{Pr}	Prandtl number, $(C_p\mu)/k$, dimensionless
N_{Re}	Reynolds number, $(DG)/\mu$, dimensionless
Q	Heat transferred to the process fluid, Btu./hr.
Q_T	Total heat transferred by the tracer, Btu./hr.
t	Pipe or vessel wall-thickness, ft.
T	Pipe or vessel wall-temperature, °F.
T_{amb}	Ambient air temperature, °F.
T_{mid}	Midpoint temperature at $x = L$, °F.
T_o	Temperature of pipe or vessel wall at tracer, °F.
T_p	Process-fluid temperature, °F.
T_s	Steam temperature, °F.
V_v	Specific volume of vapor, cu. ft./lb.
V_L	Specific volume of liquid, cu. ft./lb.
w'	Steam consumption per unit length of tracer, Btu./(hr.)(ft.)
x	Distance along the pipe or vessel wall, ft.
X_{ins}	Insulation thickness, in.
z	Length of the tracer, equivalent ft.
ΔH	Enthalpy change of steam through tracer, Btu./lb.
ΔP	Total steam-pressure loss through tracer, psi.
α	A parameter $= x/L$
μ	Viscosity, lb./(hr.)(ft.)
ρ	Density of fluid, lb./cu. ft.

Substitution of T from Eq. 3 gives:

$$\frac{Q}{z} = 2\left(\frac{[T_o - (B/A)]h_i}{\sqrt{A}}\right) \times \tanh(\sqrt{A}L) + 2h_i[(B/A) - T_p]L \quad (4)$$

Btu./(hr.)(ft.) of tracer

The heat transferred to the process fluid, and the heat losses to the surrounding atmosphere (i.e., the total heat transferred by the tracer) are:

$$\frac{Q_T}{z} = 2\int_o^L (T - T_p)h_i\, dx + 2\int_o^L (T - T_{amb})h_o\, dx$$

$$= \frac{2(h_i + h_o)}{\sqrt{A}}[T_o - (B/A)]\tanh(\sqrt{A}L)$$

Btu./(hr.)(ft.) of tracer (5)

Application

Following are stepwise procedures for solving the two most common tracing-design problems.

Problem Type I—Maintain stagnant-process-fluid bulk temperature at a specified minimum.

Step 1—Determine h_i and h_o from physical parameters and empirical equations.

Step 2—Calculate the values of constants A and B and calculate the ratio of B/A.

Step 3—Set T_o equal to the expected lowest temperature in the tracer, specifically the saturated steam temperature at the tracer outlet. Normally, trapping distances are based on a 10% or 10-psi. pressure drop, whichever is greater.

Step 4—From Eq. (4) calculate L by using a value of Q/z of 100 Btu./(hr.)(ft.) of tracer. For certain problems, a value other than 100 may be preferable, depending on the reheat time required if the steam supply were lost. The designer must, therefore, verify that the heatup time possible meets this requirement while not resulting in an uneconomical tracing design.

Step 5—The maximum pitch is $2L$ for both spiral-wound tracing (shown below) and parallel tracing. For

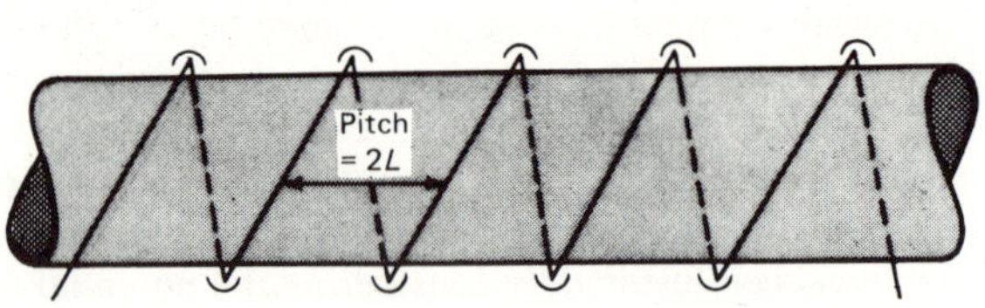

the case of multiple parallel tracers, the number of tracers required is:

$$\text{Number of tracers} = \frac{\text{Pipe circumference}}{2L} = \frac{\pi D}{2L}$$

where a fractional number of tracers is rounded off to the next integral number. These tracers are equally spaced around the circumference of the pipe.

Step 6—Determine the total heat transferred by the tracer by substituting the final value of L from Step 5 in Eq. (5).

Step 7—Calculate the steam consumption per foot of tracer from the following:

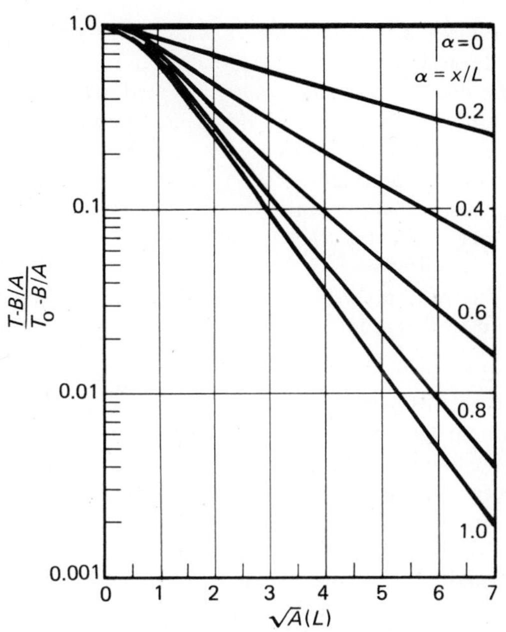

GRAPHICAL solution for steam-tracing design—Fig. 3

$$w' = \text{steam consumption (lb./(hr.)(ft.) of tracer)} = \frac{\frac{Q_T}{z}}{\Delta H}$$

where ΔH is the change in enthalpy of the steam from the tracer inlet to the outlet.

Problem Type II—Maintain wall temperature of process equipment or piping at a specific minimum.

Steps 1 to 3—Same as Problem Type I.

Step 4—Calculate:

$$\frac{T_{mid} - (B/A)}{T_o - (B/A)}$$

Step 5—From Fig. 3 (for $\alpha = 1$), determine $\sqrt{A}(L)$ and solve for L. The required tracing pitch is $2L$. The number of tracers is:

$$\frac{\text{Pipe circumference}}{2L} = \frac{\pi D}{2L}$$

A fractional number of tracers is rounded off to the next integral number.

Step 6—Substitute the value of L from Step 5 into Eq. (4) and (5) and calculate the heat transferred to the process fluid and the total heat transferred by the tracer.

Step 7—Same as Step 7 in Problem Type I.

Example—Maintaining Bulk Temperature

Problem 1—A 10-in. stainless-steel pipe is carrying phthalic anhydride at 300 F. The line is to be steam-traced using heat-transfer cement and is insulated with 1½ in. of calcium silicate insulation to maintain the bulk temperature at 300 F. The design case is for the process fluid being stagnant with an average, inside, natural-convection coefficient of 20 Btu./(hr.)(sq. ft.)(°F.). Determine the number of parallel tracers required, the heat transferred to the process fluid, and the steam consumption, using 150-psig. saturated steam.

Data:

Ambient temperature, $T_{amb} = -10$ F.
Supply steam temperature, $T_s = 366$ F.
Process fluid temperature, $T_p = 300$ F.
Thermal conductivity of stainless-steel pipe wall, $K = 9.8$ Btu./(hr.)(sq. ft.)(°F./ft.)
Thermal conductivity of insulation $K_{ins} = 0.3$ Btu./(hr.)(sq. ft.)(°F./in.).
Pipe wall thickness = 0.165 in.
Copper tracer, O.D. ≐ 0.50 in., I.D. = 0.43 in.

Assume the heat-transfer coefficient between the insulation and the ambient atmosphere, $h_{air} = 2$ Btu./(hr.)(sq. ft.)(°F).

Solution—use Problem Type I procedure.

Step 1—Calculate h_o, the overall heat-transfer coefficient between the wall and the ambient atmosphere:

$$\frac{1}{h_o} = \frac{1}{h_{air}} + \frac{X_{ins}}{K_{ins}}$$

$$\frac{1}{h_o} = \frac{1}{2.0} + \frac{1.5}{0.3}$$

Therefore, $h_o = 0.182$ Btu./(hr.)(sq. ft.)(°F.) and, given, $h_i = 20$ Btu./(hr.)(sq. ft.)(°F.)

Step 2—Determine the constants A and B.

$$A = \frac{h_o + h_i}{Kt} = \frac{0.182 + 20}{(9.8)(0.165/12)} = 150$$

$$\sqrt{A} = 12.25$$

$$B = \frac{h_i T_p + h_o T_{amb}}{Kt}$$

$$= \frac{(20)(300) + (0.182)(-10)}{(9.8)(0.165/12)} = 44{,}510$$

$$B/A = \frac{44{,}510}{150} = 296.7$$

Note: $B/A = 296.7$ F corresponds to the equilibrium temperature of the pipe wall in the absence of steam-tracing.

Step 3—Assume a 10% pressure drop for the tracer circuit. Outlet pressure = 135 psig., corresponding to a saturated steam temperature of 358 F.

Set $T_o = 358$ F.

Step 4—Design for a net heat input of 100 Btu./hr.-ft. of tracer. Using Eq. (4), solve for L assuming tanh $[\sqrt{A}(L)] = 1$.

$$L = \frac{Q/z - 2(T_o - B/A)h_i/\sqrt{A}}{2h_i(B/A - T_p)}$$

$$L = \frac{100 - 2(358 - 296.7)(20)/12.25}{2(20)(296.7 - 300)} = 0.76 \text{ ft.}$$

Checking assumption,

$$\tanh(\sqrt{A}L) = \tanh(12.25)(0.76) = \tanh 9.3 \approx 1.0.$$

Therefore assumption was valid and $L = 0.76$ is correct.

Step 5—The minimum required pitch $= 2L = 2 \times 0.76 = 1.52$ ft.

$$\text{Number of parallel tracers} = \frac{\pi D}{2L} = \frac{\pi\left(\frac{10.75}{12}\right)}{(1.52)} = 1.85$$

With 2 parallel tracers, the resultant distance between the tracers is:

$$\frac{\pi D}{2} = 1.4 \text{ ft or } 16.8 \text{ in.}$$

Step 6—The total heat transferred by the tracer is:

$$\frac{Q_T}{z} = \frac{2(20.182)}{12.25}(358 - 296.7) \tanh (12.25)(0.7)$$
$$= 202 \text{ Btu./(hr.)(ft.)}$$

Step 7—The steam consumption is:

$$w' = \frac{202}{865} = 0.234 \text{ lb./(hr.)(ft.)}$$

Note: To determine the maximum equivalent feet of tracing run per trap, corresponding to the tracer circuit pressure-drop assumed in Step 3, use the following equation:*

$$z^3 = \frac{1.48 \times 10^{11} D^5 (\Delta P)}{f(w')^2 (V_V - V_L)}$$

Substituting w' from Step 7,

$$z^3 = \frac{1.48 \times 10^{11}(0.43/12)^5(15)}{(0.012)(0.234)^2(3.02 - 0.018)}$$

$$z = 405 \text{ maximum equivalent feet}$$

*This equation is derived by integrating the Darcy equation for a fluid with a changing specific volume. No account is taken for the fact that the fluid has two phases, since the largest portion of the total pressure drop is taken where the fluid is substantially all vapor. Furthermore, the steam pressure is taken to be that at the outlet of the tracer circuit, thereby rendering a somewhat-conservative steam specific volume for the circuit.

Problem—Prevent Condensation

Problem 2—An air stream saturated with hydrogen sulfide and water vapor is flowing at 150 F. through a 28-in.-dia. carbon steel duct insulated with 1½ in. of calcium silicate insulation. It is necessary to maintain the duct wall above 150 F. to prevent condensation of hydrogen sulfide and water.

Determine the required pitch of steam-tracing and number of tracers to be applied with heat-transfer cement, using 30-psig. saturated steam at 274 F. to maintain the duct wall above 150 F. Also calculate the heat transferred to the process fluid, and the steam consumption.

Data:

Air stream flowrate = 53,500 lb./hr.
Ambient temperature, T_{amb} = 10 F.
Steam temperature, T_s = 274 F.
Process fluid temperature, T_p = 150 F.
Thermal conductivity of carbon-steel duct wall,
K = 28 Btu./(hr.)(sq. ft.)(°F./ft.)
Thermal conductivity of insulation,
K_{ins} = 0.3 Btu./(hr.)(sq. ft.)(°F./in.)
Thermal conductivity of air =
0.0168 Btu./hr. sq. ft.(°F./ft.)
Viscosity of air = 0.02 cp.
Specific heat of air = 0.25 Btu./(lb.)(°F.)
Duct wall thickness = 0.25 in.
Copper tracer, O.D. = 0.50 in., I.D. = 0.43 in.
Assume the heat-transfer coefficient between the insulation and the ambient atmosphere, h_{air} = 2 Btu./(hr.)(sq. ft.)(°F.)

Solution—Use problem Type II procedure.

Step 1—Calculate h_i, the heat-transfer coefficient between the wall and the process air stream.

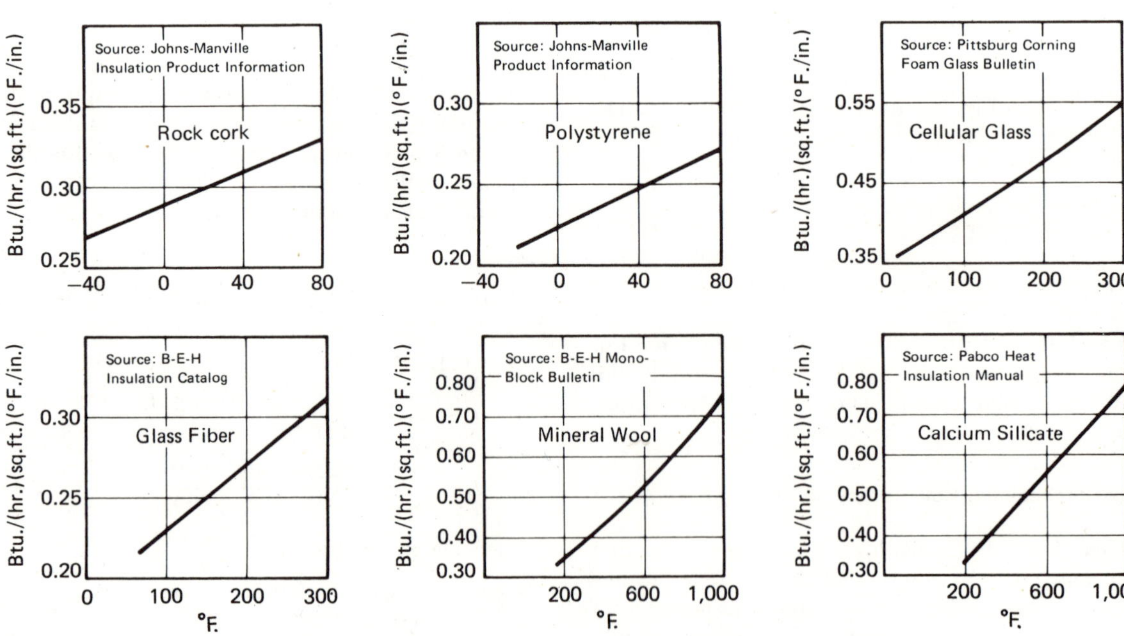

THERMAL conductivities of insulation materials—Fig. 4

Mass velocity, $G = \dfrac{53{,}500}{\frac{\pi}{4}\left(\frac{28}{12}\right)^2} = 12{,}500$ lb./(hr.)(sq. ft.)

$$N_{Re} = \frac{DG}{\mu} = \frac{(28/12)(12{,}500)}{(0.02)(2.42)} = 603{,}000.$$

$$N_{Pr} = \frac{C_p\mu}{k} = \frac{(0.25)(0.02 \times 2.42)}{0.0168} = 0.72$$

From the Dittus-Boelter equation,

$$N_{Nu} = 0.023(603{,}000)^{0.8}(0.72)^{0.4}$$

$$= 850 = \frac{h_i D}{k}$$

and

$$h_i = \frac{(850)(0.0168)}{(28/12)} = 6.12 \text{ Btu./(hr.)(sq. ft.)(°F.)}$$

Calculate h_o, the overall heat-transfer coefficient between the wall and the ambient atmosphere:

$$\frac{1}{h_o} = \frac{X_{ins}}{K_{ins}} + \frac{1}{h_{air}}$$

$$\frac{1}{h_o} = \frac{1.5}{0.3} + \frac{1}{2.0}$$

$$h_o = 0.18 \text{ Btu./(hr.)(sq. ft.)(°F.)}$$

Step 2—Determine the constants A and B:

$$A = \frac{h_o + h_i}{Kt} = \frac{0.18 + 6.12}{(28)(0.25/12)} = 10.8$$

$$\sqrt{A} = 3.29$$

$$B = \frac{h_i T_p + h_o T_{amb}}{Kt}$$

$$= \frac{(6.12)(150) + (0.180)(10)}{(28)(0.25/12)} = 1{,}575$$

and

$$B/A = 146$$

Step 3—Assume a 10-psi. pressure drop for the tracer circuit. Outlet pressure = 20 psig., corresponding to a saturated steam temperature of 260 F.

Set $T_o = 260$ F.

Step 4:

$$\frac{T_{mid} - (B/A)}{T_o - (B/A)} = \frac{150 - 146}{260 - 146} = 0.035$$

Step 5:

From Fig. 3, for $\alpha = 1$:

$$\sqrt{A}(L) = 4.05$$

$$L = 4.05/3.29 = 1.215 \text{ ft.} = 14.6 \text{ in.}$$

The maximum required pitch $2L = 2.43$ ft = 29.2 in.

For parallel tracers:

$$\text{number of tracers} = \frac{\pi(28/12)}{2.43} = 3.02.$$

Since this solution is dependent on the calculated parameter, h_i, which can vary over about a 20% range, verification shows that a larger value of h_i requires four tracers located 1.82 ft. apart for a sound design.

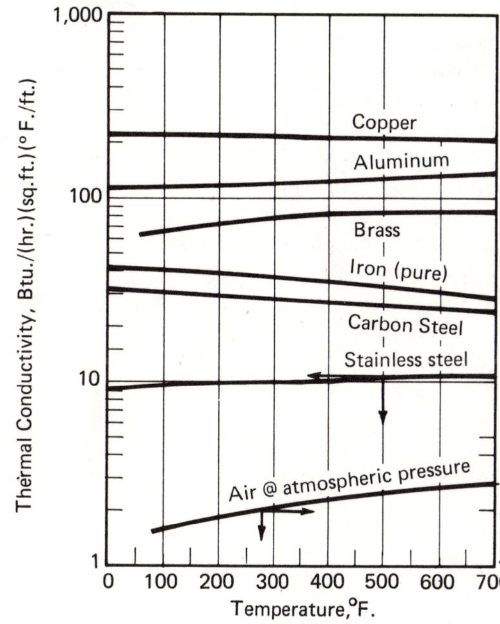

THERMAL conductivities of metals and air—Fig. 5

Step 6—The heat transferred to the process fluid is:

$$Q/z = \frac{2(260 - 146)(6.12)}{3.29} \tanh (3.29)(0.91) + 2(6.12)(146 - 150)(0.91)$$

$$= 377 \text{ Btu./(hr.)(ft.) of tracer.}$$

The total heat transferred by the tracer is:

$$\frac{Q_T}{z} = \frac{2(6.12 + 0.18)(260 - 146)}{3.29} \tanh (3.29)(0.91)$$

$$= 434 \text{ Btu./(hr.)(ft.) of tracer.}$$

Step 7—The steam consumption is:

$$w' = \frac{434}{976} = 0.445 \text{ lb./(hr.)(ft.) of tracer}$$

Note: The maximum equivalent length (feet) of tracing run per trap is:

$$z = \left[\frac{(1.48 \times 10^{11}(43/12)^5(10)}{(0.012)(0.444)^2(11.9 - 0.017)}\right]^{1/3} = 146 \text{ ft.}$$

(See Problem 1.)

Problem—Heating a Vessel Bottom

Problem 3—The bottom of a 4-ft.-dia., stainless-steel, solvent-recovery column holds a liquid that freezes at 320 F. and polymerizes at 400 F. The bottom head must be traced to keep the material fluid after a shutdown. Determine the required pitch of the tracing using 150-psig. saturated steam.

Data:

Ambient temperature = −20 F.

Supply steam temperature, $T_s = 366$ F.

Thermal conductivity of stainless steel =
9.8 Btu./(hr.)(sq. ft.)(°F./ft.)
Insulation thickness = 2 in.
Thermal conductivity of insulation =
0.3 Btu./(hr.)(sq. ft.)(°F./ft.)
Wall thickness = 0.375 in.
Assume the heat-transfer coefficient between the insulation and air is 2.0 Btu./(hr.)(sq. ft.)(°F.), and assume an inside convection coefficient of 20.

Solution—Use Problem Type II.

Step 1:

$$\frac{1}{h_o} = \frac{1}{h_{air}} + \frac{X_{ins}}{K_{ins}} = \frac{1}{2} + \frac{2}{0.3}$$

$$h_o = 0.14 \text{ Btu./(hr.)(sq. ft.)(°F.)}$$

and

$$h_i = 20 \text{ Btu./(hr.)(sq. ft.)(°F.) (assumed)}$$

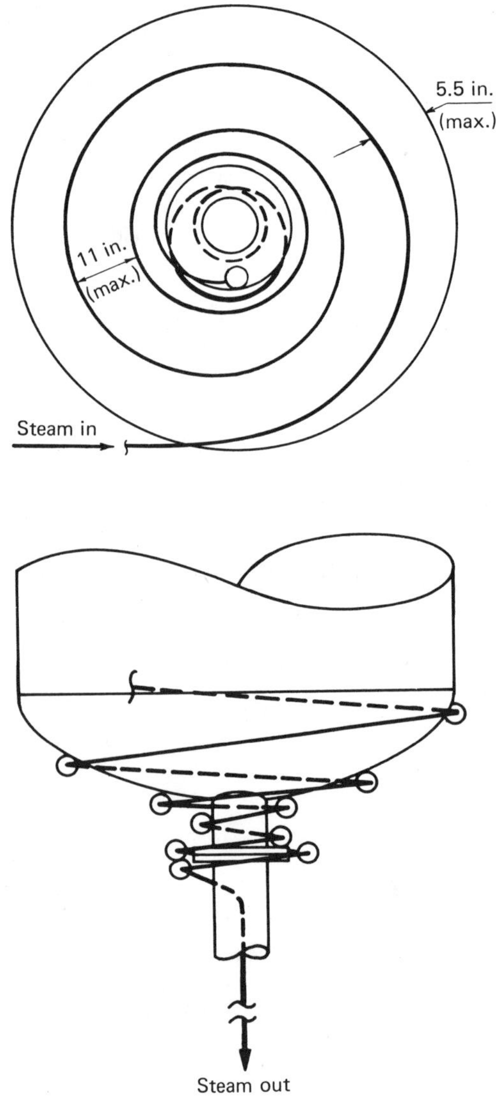

STEAM-TRACED vessel bottom (Problem 3)—Fig. 6

Step 2:

$$A = \frac{h_o + h_i}{Kt} = \frac{0.14 + 20}{(9.8)(0.375/12)} = 65.6$$

$$\sqrt{A} = 8.05$$

$$B = \frac{h_i T_p + h_o T_{amb}}{Kt} = \frac{(20)(320) + (0.14)(-20)}{(9.8)(0.375/12)} = 20{,}900$$

$$B/A = \frac{20{,}900}{65.6} = 318$$

Step 3—Using a 15-psi. pressure drop for tracer circuit, outlet pressure = 135 psig., corresponding to a saturated-steam temperature of 358 F.

Set $T_o = 358$ F.

Step 4:

$$\frac{T_{mid} - (B/A)}{T_o - (B/A)} = \frac{320 - 318}{358 - 318} = 0.05$$

Step 5—From Fig. 3, for $\alpha = 1$,

$$\sqrt{A}(L) = 3.7,$$

$$L = 3.7/8.05 = 0.46 \text{ ft.} = 5.5 \text{ in.}$$

Therefore, the maximum allowable pitch for tracing the bottom head of the column = $2L$ or 11 in. A typical tracing layout for this problem is illustrated in Fig. 6.

References

1. Desmon, L. G., Sams, E. W., Natl. Advisory Comm. Aeronaut Res. Mem. E50H23 (1950).
2. House, F. F., Pipe Tracing and Insulation, *Chem. Eng.*, June 17, 1968.
3. McAdams, W. N., "*Heat Transmission,*" 3rd ed., McGraw-Hill, New York, 1954.
4. Sieder, E. N., Tate, G. E., *Ind. Eng. Chem.*, **28**, pp. 1429–36 (1936).

Meet the Authors

Carl G. Bertram

Vikram J. Desai

Edward Interess

Carl G. Bertram is a process engineer with The Badger Co., One Broadway, Cambridge, MA 02142. He has had experience in many aspects of chemical-plant design and construction, including process design, pollution control, economic evaluations, utility systems and startup. He holds a B.S. from the University of Pennsylvania and an M.S. from Villanova University, both in chemical engineering.

Vikram J. Desai is on leave from The Badger Co. and is studying for his doctorate at Massachusetts Institute of Technology—from which he also received his S.M. He obtained his B.S. from Birla Institute of Science and Technology, India. He has previously worked for American Cyanamid Co. and Oak Ridge National Laboratories as well as several companies in India, and is a member of AIChE.

Edward Interess is a process engineer with The Badger Co. His experience has included all aspects of process design of several processes, including acrylonitrile, vinyl chloride and phthalic anhydride. He holds a B.S. Ch.E. and an M. Eng. from Rensselaer Polytechnic Institute and an M.B.A. from Boston University, and is a member of AIChE and ACS.

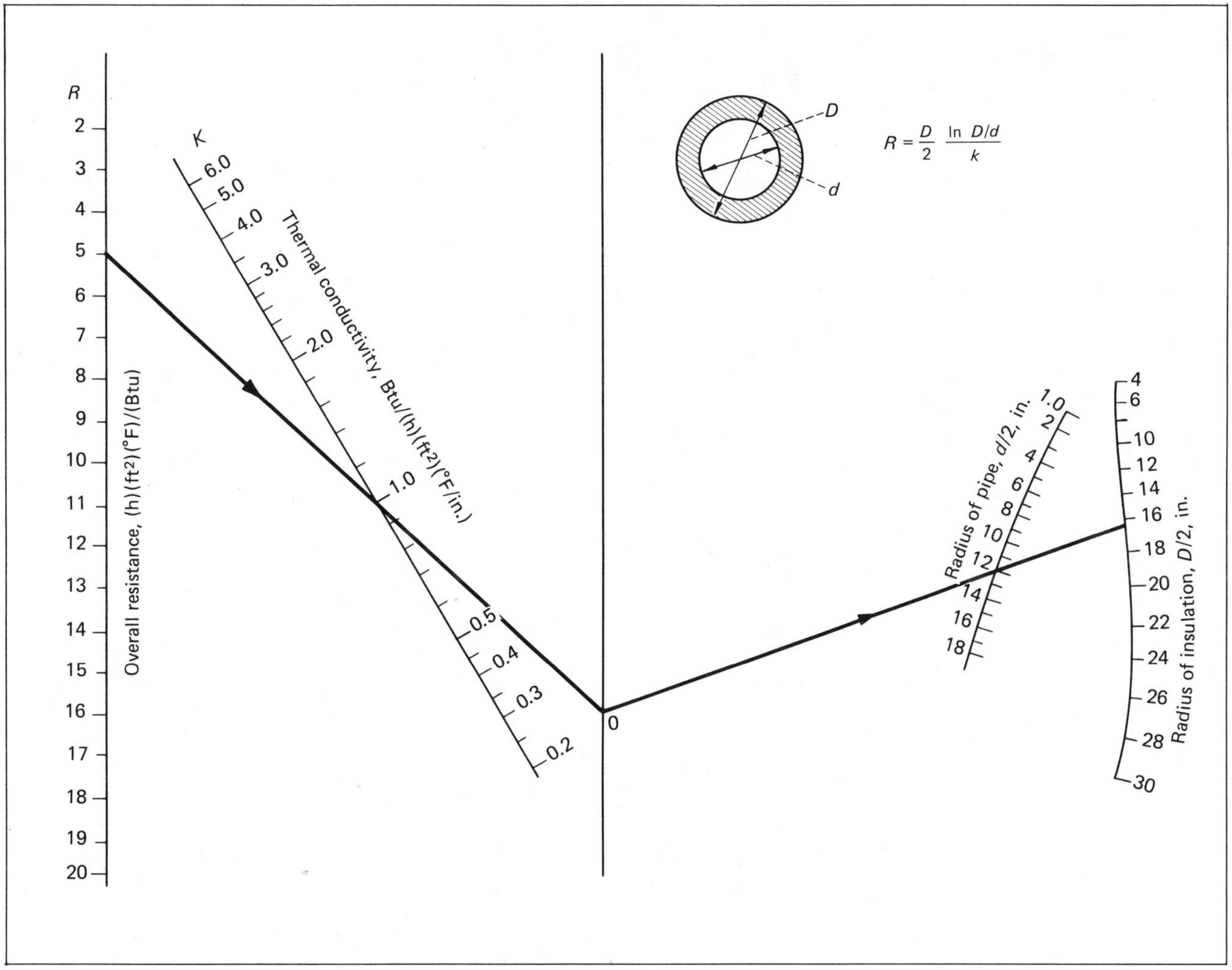

Sizing piping insulation

*V. Ganapathy, Bharat Heavy Electricals Ltd.**

☐ In many piping insulation problems, calculation of the resistance to heat transfer for the insulation becomes a tedious problem. This resistance is defined as:

$$R = D \ln (D/d)/2k$$

and the heat loss is:

$$Q = \Delta T/\Sigma(d \ln (D/d)2k) = \Delta T/\Sigma R$$

where: R = resistance to heat transfer, $(\text{h})(\text{ft}^2)(^\circ\text{F})/\text{Btu}$

D = outside diameter of insulation, in.

d = inside diameter of insulation, or diameter of the pipe, in.

k = thermal conductivity of the insulation, $\text{Btu}/(\text{h})(\text{ft}^2)(^\circ\text{F/in.})$

ΔT = the temperature difference across the insulation, °F.

This chart relates *R, D, d,* and *k*, so that any one of the four can be determined in terms of the other three. It thus eliminates a trial-and-error calculation. *Example:* If a 24-in. pipe is to be insulated with material having a thermal conductivity of 1.0 $\text{Btu}/(\text{h})(\text{ft}^2)(^\circ\text{F/in.})$ so that the overall resistance of the insulation to heat loss from the hot pipe to the ambient air is 5.0 $(\text{h})(\text{ft}^2)(^\circ\text{F})/\text{Btu}$, what are the thickness and outside diameter of the insulation? On the chart, connect $R = 5.0$ with $k = 1.0$ and extend the straight line to its intersection with the reference line. Connect this intersection with $(d/2) = 12$, and read off $(D/2) = 16.2$ in. Thus the thickness is $(16.2 - 12) = 4.2$ in.

*Product Development (Boilers), High Pressure Boiler Plant, Tiruchirapalli-620 014, India.

Originally published November 21, 1977

Electric Pipe Tracing

These guidelines should acquaint a project engineer with aspects of electric pipe tracing that are necessary to achieve desired heating results on time and on budget.

CHARLES W. BROWN, Nelson Electric, Div. of Sola Basic Industries

In the past, steam tracing of pipes maintained temperatures in process lines that would otherwise have dropped due to losses through the insulation. However, for the last several years, electric tracing has been gaining importance and capturing many of the tracing applications formerly left to steam. One reason is that today's high interest rates penalize severely for downtime.

Although steam tracing is generally more economical for large, high-temperature pipes, or when a product needs to be heated, electric heating to maintain temperatures in pipelines has these advantages over steam:

- Heat is simple to provide with automatic controls.
- Heating is more uniform.
- Operating and maintenance costs are lower.
- Heating can be applied equally to short and long process lines.

Other applications for electric pipe tracing may involve: heating of certain storage tanks and vessels; heating of subsoil under liquefied natural and synthetic natural-gas storage tanks; heating of hoppers for electrostatic precipitators; and melting of snow and ice at loading docks and ramps.

Organizing an Electric-Tracing Project

Downtime or missed startup dates caused by frozen valves or broken pipes create severe economic penalties to a plant due to high interest rates. Therefore, a major electric heat-tracing system must be well planned because it may involve as many as 300 heating circuits. With heaters, controls, junction boxes, pipe straps, tools and accessories, such circuits may contain 800 to 900 items. These must be designed, fabricated, shipped, installed, tested and calibrated.

Pipe erection may complicate matters. Pipes 2½-in. O.D. and larger are usually purchased on a spool schedule, and installed to agree with piping drawings. Pipes 2-in or smaller are frequently rerouted in the field. Consequently, a major portion of a piping system must be verified by field measurements; instrument piping is especially subject to field changes.

Even though there is a shortage of experienced heat-tracing engineers, at least a design engineer and a field supervisor must be involved in any electrical tracing system. These people must have daily access to and full cooperation from the piping group. Most cable manufacturers will train a client's personnel at no cost to ensure proper installation of equipment. Planning meetings should be held, in which process, piping, instrument and electrical personnel should all be involved. Responsibilities and authorities must then be clearly defined. The project manager will be responsible for construction per-

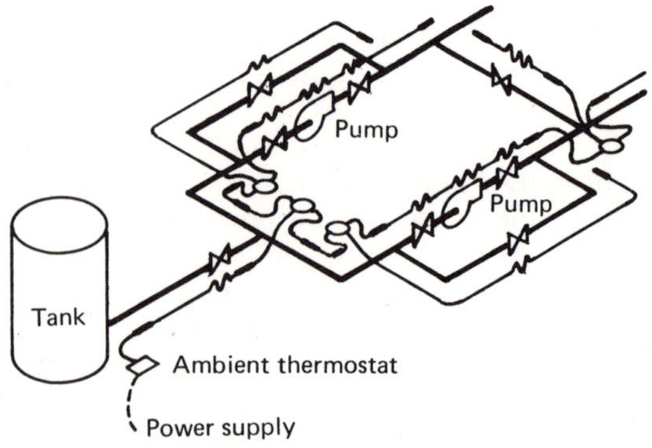

FREEZE PREVENTION for piping systems—Fig. 1

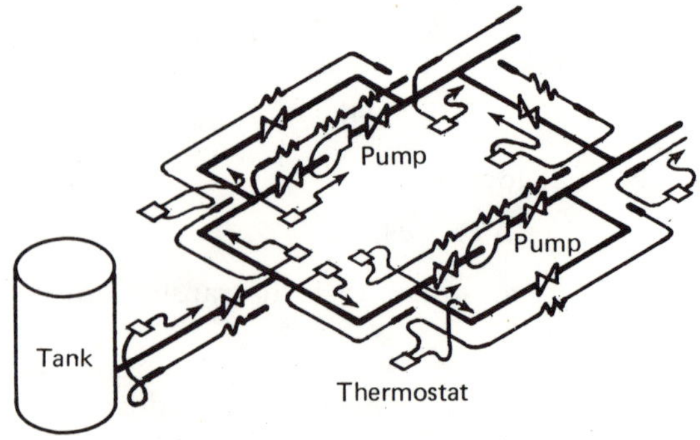

PIPE SEGMENTS with individual tracer, controller—Fig. 2

Originally published June 23, 1975

formance, while the engineer will be judged by the heating results obtained.

Determining Circuits

A task to be completed early in the project design is the determination of the heating circuits, which includes identifying the piping systems to be traced and specifying the schemes that are to be used.

Fig. 1 shows a sample piping system with the heating circuit and control added. This tracing circuit includes ten cables, connected in series, and controlled from a single, ambient-sensing thermostat. This scheme would be applicable for freeze protection of a water system, or of a noncritical system with infrequent flow.

Note that the valves and pumps are traced by the same cables used to heat the pipes. The heating cables must have additional length to replace the higher losses at valves, pumps, filters, pipe supports and wall penetrations. Cable manufacturers generally tabulate cable-length additions required for standard-size piping components. The engineer should therefore send the vendor outline drawings for motor-operated valves, pumps, filters, and all instruments not covered by standard-instrument drawings.

Separate heaters with removable insulated enclosures are recommended for: valves, pumps, or filters that require frequent access for maintenance; and for circuits requiring multiple tracers for high wattage.

Fig. 2 shows the same system with a different tracing scheme. Each pipe segment has an individual tracer and controller. This scheme would be used for processes having close temperature tolerances, changing-flow patterns, or those that are critical to plant operation or safety. Redundant heat tracings from separate power sources are sometimes used for critical circuits.

Isometric drawings similar to Fig. 1 and 2 are usually prepared by cable manufacturers. The complete drawing includes a bill of material and all pertinent reference data, including the location coordinates. These drawings, along with a manual of installation instructions, are used by field personnel to install the tracing system.

The engineer, on the other hand, must supply the manufacturer with specifications for determining the tracing circuits for various piping systems. The preparation of such specifications will require input from the process, piping, and instrumentation engineers. Construction and operating personnel should understand why tracing circuits and control requirements vary from system to system.

In addition to the specifications, the cable manufacturer should be supplied with: (1) plot plan; (2) plan and elevation drawings; (3) process-flow diagrams; (4) piping and instrument diagrams; (5) isometric piping-spool pieces; (6) company standard-instrument drawings; (7) vendor outlines for special instruments, valves, pumps, filters, etc.

Special Conditions and Situations

Normally, project managers are not involved in the calculation of heat losses, wattage applied, safety margins, etc., because the electrical engineer covers the methods to be used in the specifications.

The electrical engineer should also specify special conditions that may exist at the jobsite. Since most of this information must be gathered from operating, construction and process people, the project manager should be aware of these needs.

For instance, a thermostat that performs well in Chicago may fail in Alaska at -60°F. Also, the atmosphere in Borger, Tex., is nothing like that of Linden, N.J. A cable sheath temperature of 900°F may be acceptable in a coal gasification application, while 330°F is not acceptable for coal dust. Pipes that are periodically steam purged may require special techniques.

There may also be special situations of which the project manager should be aware. Such situations may best be solved by a redesign of the piping system or an alternative method of heating. Some such situations are the following:

Dike Penetration—Here, the preferred system is an overhead rack that provides access to the pipe and heating cable. If the pipe must go underground through a dike or roadbed, consult the cable manufacturer before making your field decision on pipe design. Access to cables may require conduit welded to the pipe, and 45-deg instead of 90-deg elbows.

Gut Tracing—Placing heating cables inside a buried pipe is recommended only as a last resort, because of special problems such as turbulence, chemical properties of the product, and cable-entry method.

Submerged Pipes—Pipes buried in a location with a high water table need special treatment. To assure that the water does not saturate the pipe insulation, a pipe-within-a-pipe construction may be used.

Temperature Controls

Another task to determine is the controls that are to be used for the various piping systems. Typically, such controls represent about one-half the cost of the heat-tracing system, and may represent 90% of the operation and maintenance problems.

Enclosure types must be specified for the particular location, such as NEMA* 1 for indoors, NEMA 4 for outdoors, and NEMA 7 for hazardous locations.

Water may enter outdoor enclosures through overhead conduit systems, if these are inadequately sealed, or moisture may be breathed into the enclosure with heat cycling. In humid climates and sea-coast locations, this moisture may corrode the thermostats or contactors to the extent that operation is impeded. Breathers, drains, and space heaters are recommended for these areas.

Thermostats—the workhorses of electric heat tracings—are used in the vast majority of control applications, because they are fairly rugged devices and are also relatively inexpensive. One drawback is that the maximum length of the capillary tube is 10 ft, or 25 ft with compensation.

If it is desirable to have the temperature probe remote from the controller, then it is necessary to use a thermocouple or resistance temperature-detector-type of con-

*National Electrical Manufacturer's Assn.

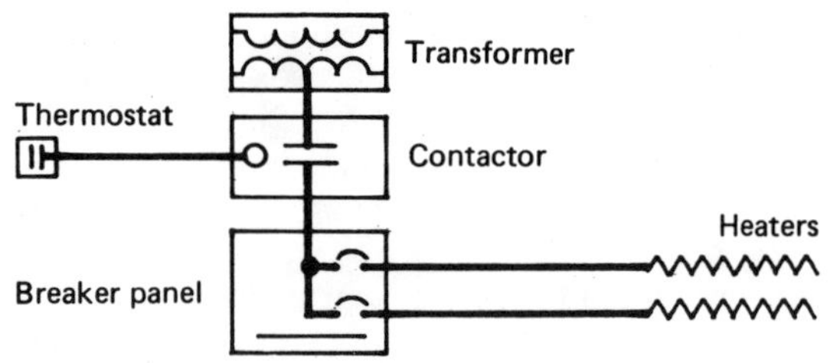

SINGLE THERMOSTAT for freeze-protection system—Fig. 3

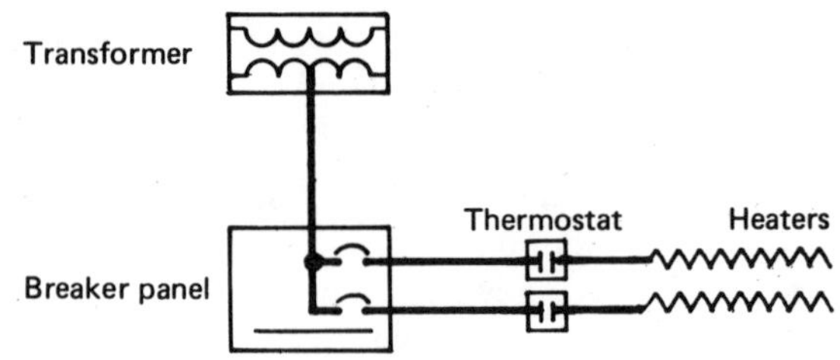

PROCESS CONTROL: one thermostat per heater—Fig. 4

trol. Since the thermostat also can only stand a limited amount of excess temperature, special analysis should be made for thermostats on a freeze-protected system, when steam purge or higher temperature is involved. The thermostat bulb is sometimes sandwiched in the insulation at a lower gradient temperature and setpoint. A thermocouple or resistance temperature-detector type of static control can be remote from a sensor probe, and can meet excess-temperature needs.

Control Schemes

In a freeze-protection system, a single thermostat may be used to control a contactor, and both in turn supply power to as many as 300 heaters (Fig. 3). This scheme offers a lower initial cost and fewer controls to maintain, although more energy is used because the heaters are on when the ambient temperature drops below the setpoint.

Fig. 4 shows a process control with an individual thermostat for each heater. This provides individual control that may be needed by changes in process flow and product requirements. Although this scheme has a higher initial cost and more controllers to maintain, less energy is required, and closer product temperatures maintained.

Sometimes it is desirable to have some indication of the status of the heat-tracing circuits. With the two previous methods shown, the first indication that something has gone wrong with the heating system is when an operator notices that a pipe is frozen or plugged.

Fig. 5 shows a thermostat control with individual controls, but with ammeters and a high-low temperature alarm added. This alarm would consist of two additional thermostats, one set higher and one lower than the control thermostats, as desired. It is common to place a red mark on the ammeter scale when the heater circuit is checked out during startup. Thus, the operator can detect any change in the heating circuit, if the ammeter indicating needle is not close to the red mark.

There are two shortcomings to this system. One is that the thermostat and heating circuits are energized only about one-third of the time. An operator seeing the ammeter's dial on zero would not be able to tell if the circuits are normal. Therefore, a bypass switch may be placed around the thermostat, so that the operator can observe the ammeter at any time. This switch should be in the anmeter panel. Since the thermostat may be remote, it will require additional wires to each heater-control circuit. The other shortcoming is that the ammeter system monitors the circuit only when the operator chooses to observe the ammeters. In a plant with 300 circuits, the ammeters are likely to be observed infrequently.

Fig. 6 shows a critical process-control system with a current monitor alarm. An auxiliary contact on the thermostat is placed in series with an electrically operated relay. If this contact does not open, a light is turned on in the annunciator panel and an alarm sounded.

The advantages of the current transformer and monitoring alarm are: (1) it is in operation 24 h/day; (2) it will detect if the heater has shorted out and tripped the supply circuit breakers; and (3) it provides a remote alarm that can be located in the operator's control room. Neither of these schemes, however, will detect a thermostat failure, such as a broken capillary tube.

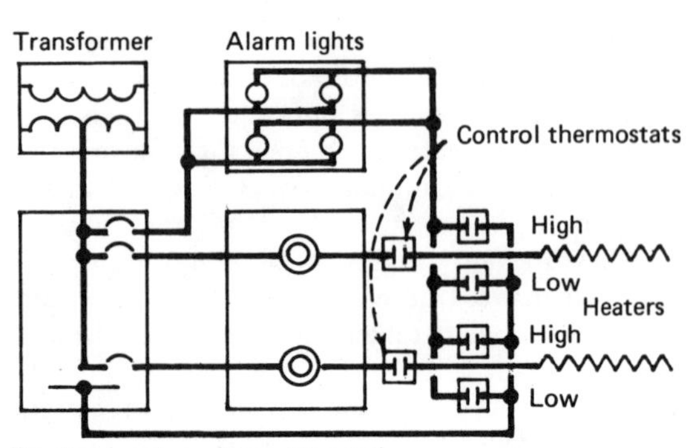

CONTROL with an ammeter and a high-low alarm—Fig. 5

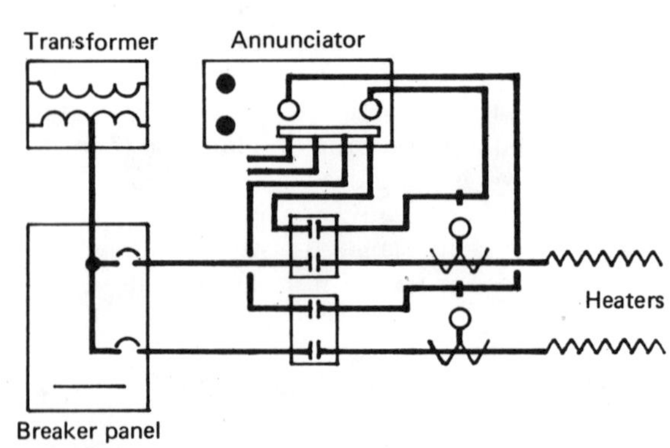

MONITOR ALARM for critical process control—Fig. 6

A minimum of 20% spare controls should be allowed for future additions and maintenance purposes. Also, a scheme should be selected that is flexible to accommodate last-minute changes, as well as load balancing in the field. The alternative is to receive the control package several weeks after the heating cables are installed.

Field Installation

Even an experienced supervisor in electric tracing needs the cooperation of other departments to accomplish his job successfully. He must be informed of all changes and of the magnitude of the project. He must also understand the specifications, and have copies of installation instructions; piping, instrument, and plant and elevation drawings; and heat-tracing isometrics. His responsibilities will include receiving, storing and issuing the hundreds of material items involved, as well as preinstallation and postinstallation testing, checkout, and record-keeping.

One common problem is the shortage of experienced electric-heat tracing electricians. A sufficient number—maybe only six to eight—should be assigned to heat tracing exclusively, and given special training.

Final cable design and fabrication should not begin until all piping is installed and verified by field measurement. This precaution is necessary because of field routing of 2-in and smaller pipes, and instrument piping. Most cable manufacturers make preliminary isometric drawings to facilitate such measurements, and provide service engineers to supervise the measurement and recording. On major projects, when the pipe installation may take several months, it is economical to train field people to make pipe measurements.

If changes are made in the piping and other piping systems are added, material requirements change. The supervisor should be authorized to requisition additional materials as required, and should make periodic status reports to the project manager to ensure that the project is on schedule and budget.

Applicable Codes and Standards

The subject of codes and standards for electric-heat tracing is indeed complex and beyond the scope of this article. The following comments should assist the project manager in determining a course of action:

Local Codes—In the majority of heat-tracing projects, local codes incorporate national codes and standards. A couple of noticeable exceptions are California and New York City. In general, cable manufacturers do not profess to meet all local codes. They will submit drawings and instructions so that the user can gain acceptance. But they reserve the right to decline the order if modifications or testing needed to gain acceptance are not economically feasible. The engineer should specify the local codes to be met and the name of the local enforcing authority.

Occupational Safety and Health Act—OSHA places the burden for compliance on the purchaser. OSHA inspectors may issue special requirements for particular installations. The engineer for the employer-purchaser should seek maximum acceptability by: (1) selecting a system that has been accepted by Underwriters' Laboratories or Factory Mutual; (2) having the system inspected by another federal agency, if no nationally recognized testing laboratory has certified the system; a state, municipal or other local authority responsible for enforcing occupational safety of the National Electric Code may also be contacted; and (3) receiving a certificate from the manufacturer that the equipment was designed and fabricated for the particular customer, where it is stated that the equipment is safe for the intended use, on the basis of test data that the employer keeps and makes available for inspection by the Assistant Secretary of Labor or his authorized representatives.

National Electrical Code—NEC is an installation guide, not a design specification. A manufacturer will supply equipment that—when installed in accordance with NEC—will be acceptable to the authority enforcing the code. As with OSHA, the burden of compliance remains with the employer-purchaser.

Underwriters' Laboratories—UL is a nationally recognized testing laboratory that certifies devices, systems and materials to have acceptable limits for life, fire and casualty hazards. The certificate is a report available to the user from the manufacturer upon request, and by application of the UL label to the equipment.

The reports include design, factory fabrication, field installation and testing. Separate test reports are available for pipe tracing, for general industrial, and for Class 1, Group D, Div. 2, hazardous locations.

A project manager should know that material and installation costs will be substantially higher for a hazardous location due to two factors: (1) all controllers must be in enclosures appropriate for the location; and (2) additional heating cable may be needed for large pipes having high wattage and high process temperatures.

National Electrical Manufacturers Assn.—NEMA defines the composition, construction, dimensions, tolerances, safety operating characteristics, performance, quality, rating, testing, and intended service of a product. A manufacturer supplies controllers that have been designed, built, and tested by NEMA standards.

References

1. Holstein, W. H., Jr., What It Costs To Steam and Electrically Trace Pipelines, *Chem. Eng. Progr.*, Mar. 1966.
2. Butz, C. H., When Is Electricity Cheaper Than Steam for Pipe Tracing?, *Chem. Eng.*, Oct. 10, 1966, p. 230.
3. Tucker, R. T., "Steam Heating Versus Electric Heating," paper presented at the Petroleum Industry Electrical Assn. and Petroleum Equipment Suppliers Assn. conference held in 1969.

Meet the Author

Charles W. Brown is Manager of Heating Products at Nelson Electric, Div. of Sola Basic Industries, P.O. Box 726, Tulsa, OK 74101. Previously, he was substation and relay engineer with the Public Service Co. of Oklahoma. Holder of a B.S.E.E. degree in electrical engineering from the University of Oklahoma, he is also a registered professional engineer in Oklahoma, and a senior member of the Institute of Electrical and Electronic Engineers.

Floating balls insulate hot tanks

*Harold E. Venable**

☐ Heated chemical tanks used for anodizing, plating and cleaning operations suffer costly heat loss through the exposed liquid surface. Such loss can be significantly reduced by insulating the surface with floating polypropylene balls. The tank loading system is the key to the success of the balls as energy savers; if the balls are continuously carried from the tank, their use is obviously impractical.

For example, our anodizing and plating tanks are loaded by mounting or hanging parts on various types of racks. The racks conduct the electric charge that is required by the process. The load is lowered through the polypropylene-ball covering and pulled out, while only dislodging a few balls.

However, our chemical cleaning tanks require parts to be loaded in a wire-mesh basket. The parts vary greatly in size and shape, so the basket is open at the top. Such a loading system would remove the balls.

In one application, we put the balls in an anodizing tank containing a hot seal solution of acetic acid, sodium hydroxide and demineralized water. The tank is 6 ft deep, 4 ft wide and 20 ft long, and holds 3,500 gal of solution at 180°F. Prior to the use of the balls, the tank energy demand cost $6,500/yr. The balls cost about $700 and reduced energy cost to $1,000/yr.

We considered two types of polypropylene balls, solid and hollow. Hollow balls were used initially because: (1) they have a higher flotation, resulting in a better seal and less heat loss, and (2) they crush readily if stepped on, making them safer for workers. The hollow balls lasted 8–9 months before punctures markedly reduced their effectiveness. Though the use of hollow balls was successful, we are replacing them with solid balls to see if these work better.

The polypropylene balls are inexpensive and easy to work with. However, to ensure troublefree operation, the operator must be willing to spend a few extra minutes both to prevent balls from being dragged out of the tank and to replace lost ones.

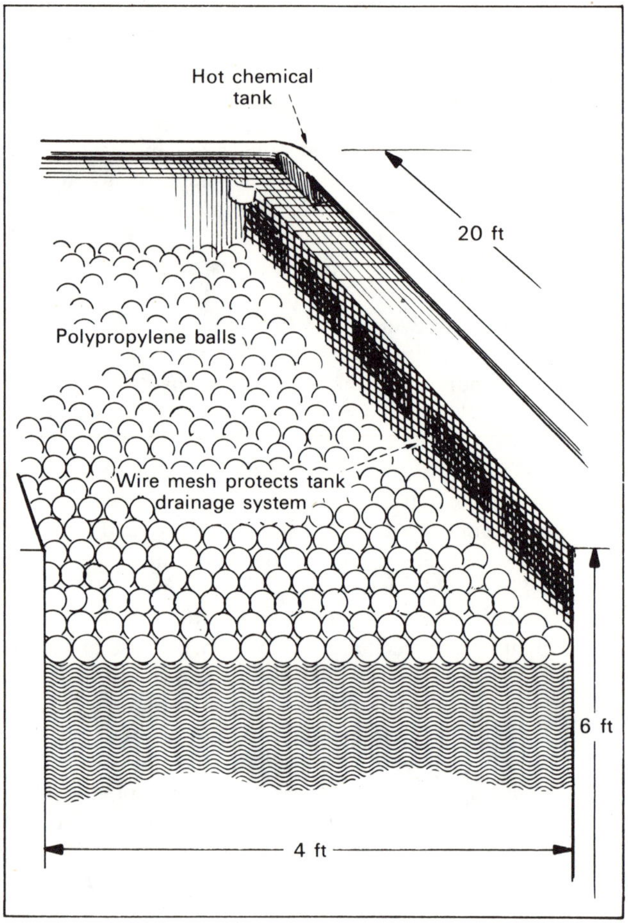

An added benefit of ball insulation is a decrease in fumes and evaporation, which reduces solution costs and bridge-crane and exhaust-system corrosion.

*Rohr Industries, Inc., P.O. Box 878, Chula Vista, CA 92012

Originally published February 12, 1979

Section IV
SLURRY HANDLING

Slurry Piping Systems: Trends, Design Methods, Guidelines

Mixtures of solids and liquids may be pumped a few hundred feet via in-plant piping, or several hundred miles via cross-country pipelines. Here is an up-to-date, practice-oriented review of both long- and short-distance systems.

T. C. AUDE, N. T. COWPER, T. L. THOMPSON and E. J. WASP, Bechtel Inc.

On Aug. 14, 1970, an underground river of coal began flowing 273 miles across the state of Arizona, and into a 1,500-Mw. power plant near Davis Dam in Nevada. Equivalent in capacity to two 160-car railroad trains per day, the unseen "river" is the 18-in.-dia. Black Mesa coal slurry pipeline, so far the largest and longest slurry pipeline in the world.

Black Mesa is the latest development in the transportation of solids by pipeline, a rapidly expanding field of growing interest to the chemical process industries. The field had its modern beginnings in the 1950's with the 108-mi., 10-in. Consolidation Coal pipeline and the 72-mi., 6-in. American Gilsonite pipeline. Since that time, considerable technical and operating knowledge has been gained. Coupled with the very broad experience that exists for short-distance slurry pipelines as well as for slurry handling in general, this knowledge has advanced the design and building of long-distance systems from an art to a maturing technology.

As for short-distance systems, these provide inplant transportation for liquid/solid mixtures in such common operations as crystallization, filtration, pulp handling, and disposal of sludge and other wastes. Chemical engineers have long been familiar with such systems but here, too, there have been some advances in methods and hardware.

This report will treat long- and short-distance systems sometimes separately and sometimes together (i.e., where the same design method or equipment concept applies to both). A minimum of theory will be presented because, for engineers deeply involved in non-Newtonian flow, the literature is quite vast, and any attempt to recapitulate it briefly would not be very meaningful. Necessarily, many of the rule-of-thumb design methods and other guidelines presented here will have exceptions, and are only meant to be an improvement over some of the rank guesses that are often used without regard to the basic phenomenon of solid-liquid systems.

Originally published June 28, 1971

SLURRY CHARACTERISTICS

Flow of mixtures of solids and liquids (i.e., slurries) in pipes differs from flow of homogeneous liquids in several important ways. With liquids, the complete range of velocities is possible, and the nature of the flow (laminar, transitional, or turbulent) is defined by the physical properties of the fluid and system. With slurries, two additional distinct flow regimes and several more physical properties are superimposed on the liquid system. The two regimes of slurry flow involve:

• *Homogeneous Slurries*—Here, the solid particles are homogeneously distributed in the liquid media, and the slurries are characterized by high solids concentrations, and fine particle sizes. Such slurries often exhibit non-Newtonian rheology (i.e., the effective viscosity is not constant, but varies with the applied rate of shearing strain). Some examples are sewage sludge, clay slurries, and cement-kiln-feed slurry.

• *Heterogeneous Slurries*—Here, concentration gradients exist along the vertical axis of a horizontal pipe even at high flowrates; i.e., the fluid phase and the solid phase retain their separate identities. Heterogeneous slurries tend to be of lower solids concentration and have larger particle sizes than homogeneous slurries. Florida phosphate rock is a good example.

Many slurries encountered commercially are of mixed character; the finer particle-size fractions join with the liquid media to form a homogeneous vehicle, while the coarser sizes act heterogeneously. Pipeline coal slurry is a prime example of the mixed characteristic.

Thus, the designer will often be faced with defining the dominant characteristic of a slurry. Unfortunately, evaluation of the complex mixed regime is outside the scope of this article. This analysis will, however, allow the designer of short systems to bracket the answer and usually achieve a practical solution.

Critical Velocity

The two types of slurries, homogeneous and heterogeneous, have entirely different critical-velocity characteristics, as seen in Fig. 1. On the chart, Curve A illustrates the deposition critical velocity typical of heterogeneous slurries. The characteristic hook in the friction-loss-vs.-velocity curve results from deposition of solids on the bottom of a horizontal pipe, as shown in the sketch. (In vertical pipes, the solids that in a horizontal pipe would be deposited, are easily transported, because their settling velocity is usually much lower than normal flow velocities.)

The deposition critical velocity is directly related to the settling velocity of the coarser particles in a heterogeneous slurry and the degree of turbulence in the pipe; it therefore increases with increasing particle size or specific gravity, and with increasing slurry concentration or viscosity. Deposition velocity generally exhibits an increase proportional to the square root of pipe diameter.

Curve B shows viscous-transition critical velocity, which is characteristic of homogeneous slurries. While design of a system for operation below the transition critical velocity is acceptable for truly homogeneous slurries, no turbulent forces exist to suspend even trace amounts of heterogeneous particles.

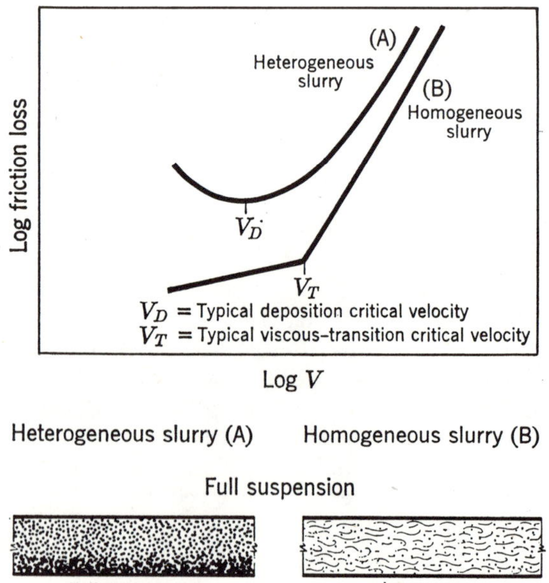

CRITICAL-VELOCITY characteristics depend on whether slurry is heterogeneous or homogeneous—Fig. 1

The transition velocity and laminar friction losses are very sensitive to the rheological properties of a homogeneous slurry. Transition velocity tends to increase with viscosity and therefore increase with solids (volume) concentration, greater quantity of fines and lower solids gravity. Transition velocity for slurries with a yield stress (Bingham plastic*) is very little affected by pipe diameter, whereas it is directly proportional to diameter for slurries with Newtonian rheological properties.

DESIGN PARAMETERS

Before making design calculations, the engineer must measure or estimate the slurry concentration, rheology and flow regime, and must select a "trial" pipe diameter.

Solids Concentration

The concentration of solids in a slurry will often be controlled by the upstream or downstream process. In this case, the dependent variables of slurry viscosity (or rheology) and specific gravity are also fixed. When addition of liquid before transportation is required, the amount of dilution may be an important factor in the downstream process, as well as in the slurry transportation.

Solids concentrations up to about 40% by volume are readily handled when the solids surface area is low (coarse slurries); however, fine materials such as clay or sewage sludge may not be fluid at even a 10% solids concentration. As a general rule, solids that are less than 25% finer than 325 mesh (0.44 microns) will be quite fluid when slurried to a 40% solids volume. For thicker and finer slurries, a knowledge of the relationship between slurry rheology and concentration may be required to select a pumpable concentration.

A good estimate of a comfortable pumping concentration can be made by reducing the static settled slurry-concentration 10 to 15 percentage points in volume concentration. Slurries approaching static settled concentration can be pumped, providing the high friction losses are tolerable (as, for instance, in short systems). The static settled concentration is determined by allowing a slurry of known concentration to settle under static conditions in a graduated cylinder until the slurry/water interface reaches equilibrium. The settled concentration is then calculated from the starting concentration and the beginning and ending volume. If the slurry does not settle, it is already too thick.

The slurry volume concentration and specific gravity are directly related, depending only on the solids specific gravity and the liquid specific gravity. Fig. 2 relates these properties for solids having a specific gravity of 2.7, and water. It is helpful to prepare a chart such as this for the slurry being evaluated.

Slurry Rheology, Viscosity

Much of the above discussion of solids concentration is based on consideration of the slurry rheology (or viscosity). The rheological properties of a slurry, for the purposes of this article, determine the "viscosity" used for friction loss calculation and for the transition critical velocity of fine, thick slurries. The following two rheological cases adequately cover most commercial slurries for short pipeline design:

- *Newtonian* slurries are described by the simple rheological property of viscosity (Fig. 3), and can be treated as true fluids, provided the flowing velocity is high enough to suspend the solids. The slurries are characteristically composed of graded materials, with few fines at moderate concentrations. Fig. 4 shows reduced viscosity as a function of slurry concentration for spheres, which is the minimum viscosity case at a given concentration.[1]
- *Bingham-plastic* slurries require a knowledge of shear stress as a function of shear rate (a rheogram, as in Fig. 3) to determine the parameters of coefficient of rigidity (η) and yield stress (τ_0) at any concentration. These data must be obtained in the laboratory with a rheometer such as the Contraves or Fann; the simple Stormer or Brookfield viscometers can be used to get a conservative estimate of these

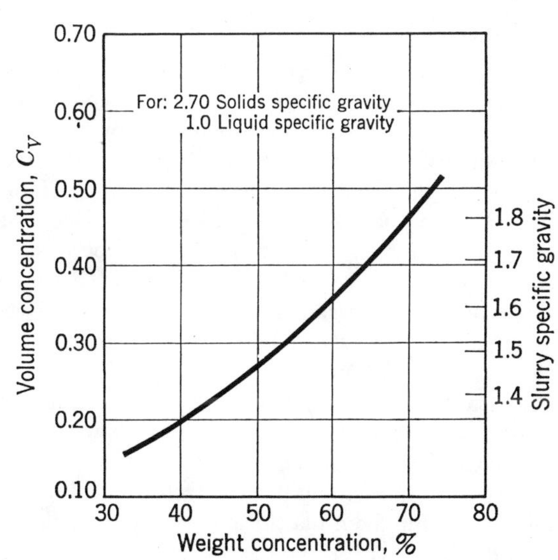

SLURRY WEIGHT concentration bears this relationship to volume concentration and specific gravity—Fig. 2

* You will recall that although Bingham-plastic fluids are similar to Newtonian fluids in that there is a linear relationship between shear stress and shear rate, Bingham plastics require a finite shear stress to initiate flow and overcome the "rigidity" of the fluid (as discussed later).

properties. Bingham-plastic slurries are characteristically composed of fine solids at higher concentrations. See Kenny[2] for information on the analysis of rheometer data, and on other rheological aspects of these slurries.

DESIGN CHARACTERISTICS

The seven basic design-calculation steps are:

1. Classify the slurry as being either homogeneous or heterogeneous.
2. Select the slurry concentration (if a variable).
3. Select a trial pipe size, based on the system's throughput requirement.
4. Calculate the critical velocity.
5. Check that the design velocity is at least 1 ft./sec. above, but not excessively above, the critical velocity. It may be necessary to select another trial pipe size and repeat the calculations until an acceptable relationship between critical velocity and design velocity is achieved.
6. Calculate the design friction losses (distinguishing between horizontal and vertical pipe for heterogeneous slurries).
7. Calculate the system pressure gradient and pump discharge pressure.

The first two steps were discussed in the design-parameters section; let us examine the remaining ones, first for homogeneous slurries, then for heterogeneous ones.

Homogeneous Slurries: General Procedure

The viscous transition from turbulent to laminar flow has been defined as the critical velocity for homogeneous slurries. As discussed earlier, the absence of turbulent energy in laminar flow allows coarser particles to settle and accumulate. Design in laminar flow for very fine (nonsettling) slurries requires an accurate knowledge of the slurry rheology, because the laminar (or viscous) friction losses are quite sensitive to that property.

Having established the design parameters and classified the slurry as homogeneous, the viscous-transition critical velocity can be calculated by using an analysis of transition velocity for Bingham-plastic materials developed from Hedstrom[5] and Hanks[6]. Hanks' curve (Fig. 6) is used to correlate the Hedstrom number (based on system and slurry properties) to the critical Reynolds number. This critical Reynolds number becomes the familiar 2,100 for slurries with Newtonian (true fluid) rheology; i.e., with no yield stress.

Once satisfied that the selected pipe size results in a design velocity above critical transition velocity, friction losses are calculated using the normal relationship between the Fanning friction factor and Reynolds number. For Bingham-plastic material, the coefficient of rigidity (η) is substituted for viscosity in Reynolds number calculations for turbulent flows—i.e., flow above the transition velocity.

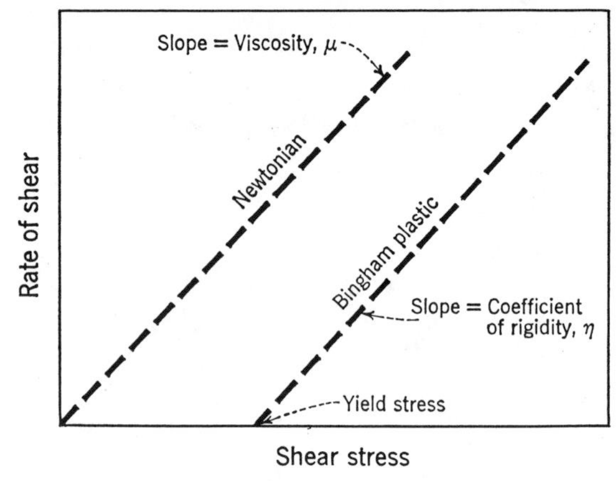

TYPICAL RHEOGRAM for two types of slurries—Fig. 3

Flow Regime

As discussed earlier, slurries are broadly classified as being in the homogeneous or heterogeneous regime, each regime having a distinctive character, particularly as to critical velocity. Fig. 5 provides a means of classifying a slurry as homogeneous or heterogeneous based on the top particle size (largest 5%) and the specific gravity of the solids. The middle area between fully homogeneous and fully heterogeneous slurries is controlled by a compounding of the properties of each type. Analysis of the compound regime is beyond the scope of this paper; however, Wasp has covered the subject in some detail.[3,4]

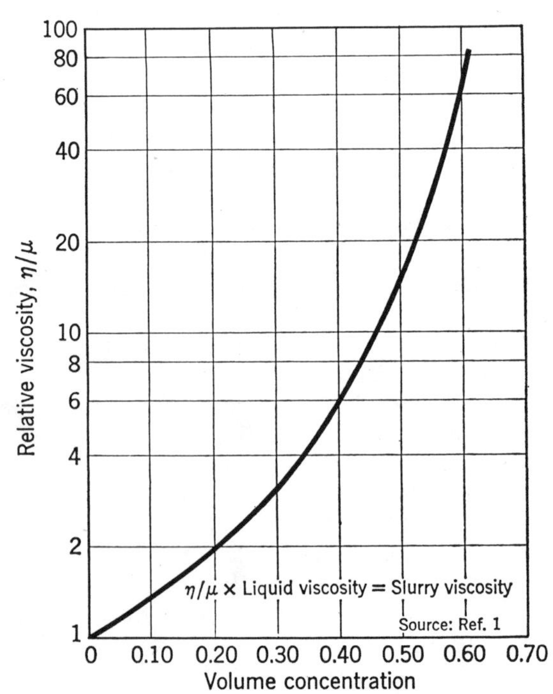

VISCOSITY for spherical-particle slurry is a function of slurry concentration—Fig. 4

Nomenclature

C_d	Drag coefficient, dimensionless
C_V	Volume concentration of solids
D	Pipe inside diameter, in.
f, f_w	Fanning friction factor
g	Gravitational constant, ft./sec.^{2}
Δh	Frictional head loss, ft. of fluid flowing/100 ft.
L_r	Length, ft.
$(N_{Fre})_c$	Critical Froude number, dimensionless
N_{He}	Hedstrom number, dimensionless
N_{Re}	Reynolds number, dimensionless
$(N_{Re})_c$	Critical Reynolds number, dimensionless
ΔP	Frictional head loss, psi./100 ft.
P_L	Plasticity number, dimensionless
V	Velocity, ft./sec.
V_d	Deposition critical velocity, ft./sec.
V_t	Viscous-transition critical velocity, ft./sec.
η, μ	Coefficient of rigidity, or viscosity for Newtonian slurries), cp.
ρ, ρ_w	Density, lb./cu. ft.
τ_0	Yield stress, dynes/sq. cm.

Design Velocity and Pipe Diameter

Because the volumetric flow is fixed by solids throughput and concentration in the slurry, and because the design velocity and pipe diameter are directly related, we can concentrate on velocity. The design velocity should of course be above critical velocity, which in turn depends on pipe diameter to a greater or lesser extent, depending on the slurry regime.

A velocity/diameter combination must be selected to enter the calculations described below. A velocity in the range of 4 to 7 ft./sec. is usually practical and economical. Velocities below 4 ft./sec. are seldom desirable, whereas, at the other end of the range, velocities above 7 ft./sec. may be necessary for strongly heterogeneous slurries. Pipe abrasion is a consideration above 8 to 10 ft./sec., and can be serious at higher velocities.

These calculations are described in detail below; also, an example is worked out on p. 80.

Viscous Transition Velocity

In Fig. 6, the critical Reynolds number is related to the Hedstrom number. The relationship between the design Reynolds number and the Hedstrom number is:

$$N_{He} = N_{Re} \times P_L \tag{1}$$

where

$$N_{Re} = DV\rho/\eta \tag{2}$$

and

$$P_L = \frac{\tau_0/\eta}{V/D} \tag{3}$$

Note that the velocity (V) actually cancels out of the Hedstrom number; however, engineers will probably find it more convenient to work with the dimensionless ratios shown. This can be done by using the system properties of velocity (V) and pipe inside diameter (D), and the slurry properties of density (ρ), coefficient of rigidity (η), and yield stress (τ_0).

The critical Reynolds number is directly proportional to the viscous-transition velocity, as follows:

$$(N_{Re})_c = (DV_t\rho)/\eta \tag{4}$$

Therefore, we can use the simple ratio:

$$\frac{V_t}{V} = \frac{(N_{Re})_c}{N_{Re}}, \quad \text{or: } V_t = V \times \frac{(N_{Re})_c}{N_{Re}} \tag{5}$$

If the design velocity (V) is not at least 1 ft./sec. greater than the transition velocity (V_t), select a smaller pipe size and recalculate the transition velocity.

Friction Losses

The Fanning friction factor (f) is found by using the Reynolds number calculated with Eq. (2) and the

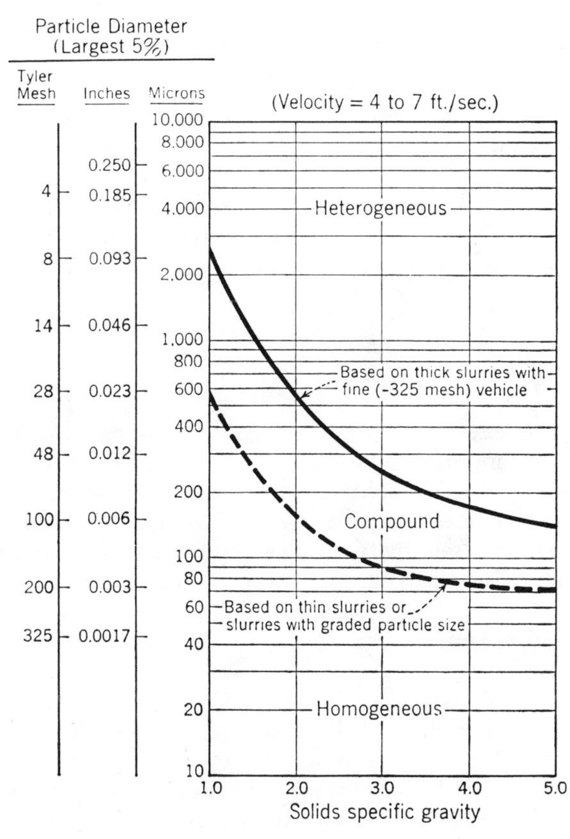

SLURRY FLOW REGIME (heterogeneous, homogeneous) is a function of solids size and specific gravity—Fig. 5

normal f vs. N_{Re} chart. Friction losses can then be found via the equation:

$$\Delta P = \frac{2f\rho V^2 L}{gD} \quad (6)$$

It is usually simpler to use some unit length, like 100 ft., for L, then extend to the system length in another step.

Selection of the friction factor for Eq. (6) requires an assumption of the pipe roughness that will exist in the operational slurry system, just as for pipes carrying liquids. This can be a major factor in longer slurry pipelines, so provision may have to be made for sending cleaning "pigs" through the pipe, and for using corrosion inhibitors during operation. In short, designing for rough pipe usually is the better solution. The often quoted Hazen Williams "C" factor of 100 will result in a pressure-loss capability about 100% greater than would be provided for new steel-pipe conditions.

Heterogeneous Slurries: General Procedure

The velocity of deposition of the coarsest size fraction on the bottom of a horizontal pipe is the critical velocity for heterogeneous slurries. Of course, this deposition will not occur, and is not a concern, in purely vertical systems. Also, friction pressure losses are usually lower in vertical pipes for heterogeneous slurries, because there can be no concentration gradient across a vertical pipe.

Calculations for both the vertical and horizontal case are described below, and a numerical example is worked out on p. 81.

Deposition Velocity

For heterogeneous slurries, for our purposes defined by Fig. 5, the volume concentration of solids in the slurry, and the solids density, are related to a critical Froude number as shown in Fig. 7:[3]

$$(N_{Fr})_c = V_d/\sqrt{gD} \quad (7)$$

therefore,

$$Vd = (N_{Fr})_c\sqrt{gD} \quad (8)$$

The effect of particle size on deposition velocity in this approach was covered in the regime selection of Fig. 5. Although the velocity of first deposition is not affected by particle size (i.e., above a certain size), the sensitivity of the system and horizontal friction losses are greater for coarser slurries.

Friction Losses

Having confirmed that the selected pipe diameter results in a velocity above critical, turbulent friction losses are calculated using the familiar f vs. N_{RE} relationship, with appropriate overpressure corrections for heterogeneous solids distribution in horizontal pipes.

Heterogeneous slurries flowing in horizontal pipes have higher friction losses due to nonuniform distribution of solids across the pipe. Durand's empirical relationship corrects the later friction factor for the solids loading as follows:

$$f = f_w\left[1 + 82\left(\frac{gD}{V^2} \cdot \frac{\rho - \rho_w}{\rho_w}\right)^{3/2} \frac{C_v}{C_d{}^{3/4}}\right] \quad (9)$$

It has been found that the overpressure correction, $(f - f_w)$, should be found for each size fraction for materials with wide granularity to completely account for the relationship of drag coefficient to particle size.[4] The volume concentration of each size fraction is used in that calculation. A quick, conservative estimate can be made by assuming the minimum value of the drag coefficient (0.44) for all sizes. See Perry's "Chemical Engineers' Handbook" for the drag-coefficient calculation.[7]

The sum of the overpressure corrections, $(f - f_w)$ for the several size or solids-gravity fractions, times the water friction factor (f_w), gives the friction factor

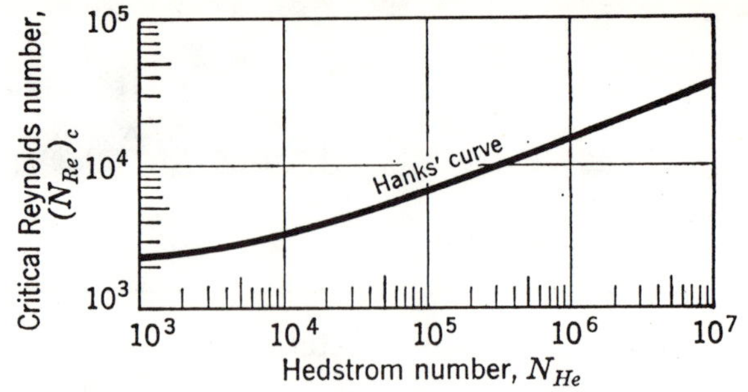

HANKS' CURVE relates critical Reynolds number to Hedstrom number for Bingham flow in pipes—Fig. 6

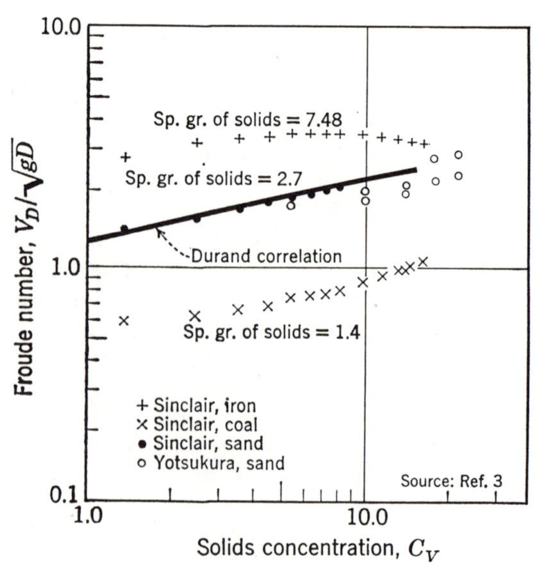

FROUDE NUMBER at deposition, for various suspensions, shows this relationship to concentration—Fig. 7

Homogeneous Slurries: Calculations for an Inplant System—Example 1

Basis for Problem

Slurry type	Fly ash in water
Solids content	60%, by weight
Solids specific gravity	2.7
Slurry volume concentration	0.356
Slurry specific gravity	1.61
Slurry density	100 lb./cu. ft.
Slurry rheology	@ 68 F.
Slurry coef. or rigidity	60 cp.*
Slurry yield stress	50 dynes/sq.cm.
System data (see sketch):	
Flow	800 gpm.
Pipe dia., trial basis	8.625-in. O.D.; 8.125-in. I.D.
Velocity	5 ft./sec.
Length (including fittings)	450 ft.
Inlet elevation	770 ft.
Discharge elevation	750 ft. (to atmosphere)

*Note: 100 cp. = 1 dyne/(sq. cm.) (sec.)

Calculating the Critical Velocity

$$N_{He} = N_{Re} \times P_L = \frac{DV\rho}{\eta} \times \frac{\tau_0/\eta}{V/D}$$

$$N_{Re} = \frac{8.125\text{ in.}}{12\text{ in./ft.}} \times \frac{5\text{ ft.}}{\text{sec.}} \times \frac{100\text{ lb.}}{\text{cu. ft.}} \times \frac{1{,}488\text{ cp./(lb./ft. sec.)}}{60\text{ cp.}} = 8{,}400$$

$$P_L = \frac{50\text{ dynes}}{\text{sq. cm.}} \times \frac{100}{60\text{ cp.}} \times \frac{8.125\text{ in.}}{5\text{ ft./sec.}} \times \frac{\text{ft.}}{12\text{ in.}} = 11.2$$

$$N_{He} = 8{,}400 \times 11.2 = 95{,}000$$

$$(N_{Re})_c = 6{,}500 \text{ (see Fig. 6)} = DV_t\rho/\eta$$

Transition velocity: $\dfrac{V_t}{V} = \dfrac{(N_{Re})_c}{N_{Re}}$

$$V_t = \frac{(N_{Re})_c}{N_{Re}} \times V = \frac{6{,}500}{8{,}400} \times 5\text{ ft./sec.} = 3.8\text{ ft./sec.}$$

Calculating Friction Losses

$f = 0.0082$

For $N_{Re} = 8{,}400$, and commercial steel-pipe roughness

$$\Delta P = \frac{2f\rho V^2 L}{gD} = 2 \times 0.0082 \times \frac{100\text{ lb.}}{\text{cu. ft.}} \times \frac{5^2\text{ ft.}^2}{\text{sec}^2} \times 100\text{ ft.} \times \frac{\text{sec.}^2}{32.2\text{ ft.} \times 8.125\text{ in.}} \times \frac{1\text{ ft.}}{12\text{ in.}}$$

$\Delta P = 1.3$ psi. per 100 ft. pipe

$\Delta h = 1.3\text{ psi.}/(0.433 \times 1.61\text{ psi./ft.})$
$= 1.88$ ft. (slurry) per 100 ft. pipe

Calculating Pump Discharge Pressure

Static press.: 770 ft. − 750 ft.	= 20.0 ft.	= 13.9 psi.
Friction press.: 1.88 ft. × 450 ft./100 ft.	= 8.5 ft.	= 5.9 psi.
Total press.	28.5 ft.	= 19.8 psi.

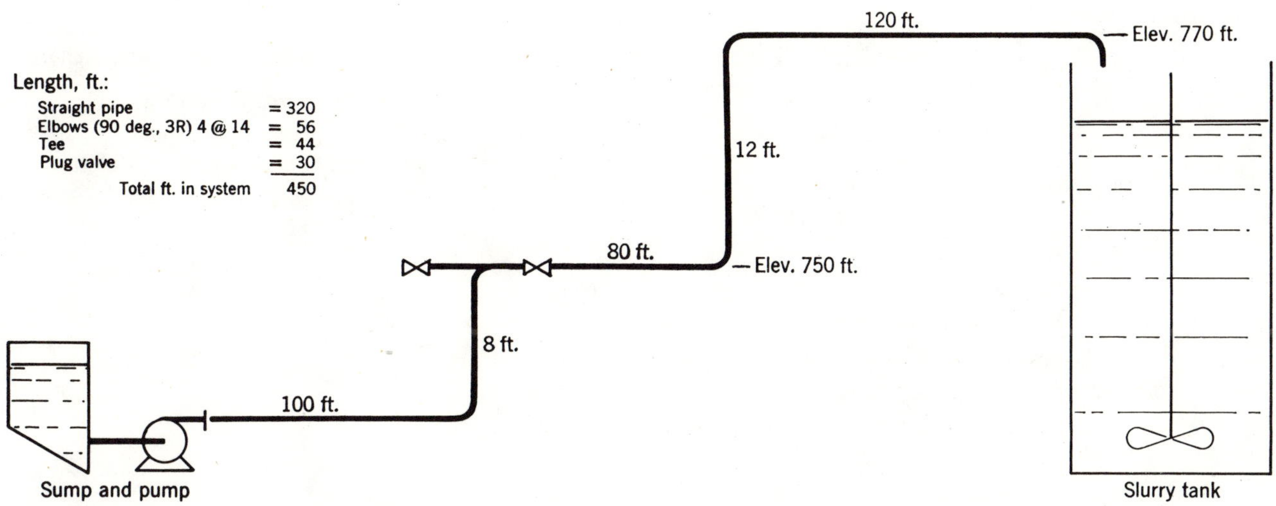

Heterogeneous Slurries: Calculations for an Inplant System—Example 2

Basis for Problem

Slurry type	Graded sand in water
Solids content	35%, by weight
Solids specific gravity	2.7
Solids size	60% is 8 × 14 mesh
	40% is 14 × 28 mesh
Slurry volume concentration	0.166
Slurry specific gravity	1.28
Slurry density	80 lb./cu. ft.
Slurry viscosity	1.8 cp. (Fig. 3)

System data
(configuration is the same as in Example 1):

Flow	1,900 gpm.
Pipe dia., trial basis	8.625-in. O.D.; 8.125-in. I.D.
Velocity	12 ft./sec.
Length (incl. fittings)	430 ft. horizontal; 20 ft. vertical
Inlet elevation	770 ft.
Discharge elevation	750 ft. (to atmosphere)

Slurry Friction Losses: Vertical Pipe

$$N_{Re}\text{ (slurry)} = DV\rho/\eta$$

$$= \frac{8.125\text{ in.}}{12\text{ in./ft.}} \times \frac{12\text{ ft.}}{\text{sec.}} \times \frac{80\text{ lb.}}{\text{cu. ft.}} \times \frac{1{,}488\text{ cp./(lb./ft. sec.)}}{1.8\text{ cp.}}$$

$$= 537{,}000$$

$$f = 0.0038$$

$$\Delta P = \frac{2 f\rho V^2 L}{gD}$$

For commercial steel pipe, $\Delta P =$

$$2 \times 0.0038 \times \frac{80\text{ lb.}}{\text{cu. ft.}} \times \frac{12^2\text{ ft.}^2}{\text{sec.}^2} \times 100\text{ ft.} \times \frac{\text{sec.}^2}{32.2\text{ ft.} \times 8.125\text{ in.}} \times \frac{1\text{ ft.}}{12\text{ in.}}$$

$$\Delta P = 2.8\text{ psi. per 100 ft. of pipe}$$

$$\Delta h = 2.8\text{ psi.}/0.433 \times 1.28\text{ psi./ft.}$$

$$= 5.0\text{ ft. (slurry) per 100 ft. of vertical pipe}$$

Slurry Friction Losses: Horizontal Pipes

$$N_{Re}\text{ (water)} = DV\rho_w/\mu$$

$$= \frac{8.125\text{ in.}}{12\text{ in./ft.}} \times \frac{12\text{ ft.}}{\text{sec.}} \times \frac{62.3\text{ lb.}}{\text{cu. ft.}} \times \frac{1{,}488\text{ cp./(lb./ft. sec.)}}{1\text{ cp.}} = 820{,}000$$

$$f_w = 0.0037$$

f = See Eq. (9) for heterogeneous slurries

$$f - f_w = 82\left(\frac{gD}{V^2} \times \frac{\rho - \rho_w}{\rho_w}\right)^{3/2} \frac{C_V f_w}{C_d^{\ 3/4}}$$

The friction factor increment $(f - f_w)$ is found for each size, and then added to f_w:

Size (Mesh)	Wt. %	C_V	C_d
14 × 28	60	0.090	0.44
28 x 48	40	0.076	0.44
		0.166	

In this example, a single calculation for f can be made, due to the single value of C_d:

$$f = 0.0037\left[1 + 82\left(\frac{32.2\text{ ft./sec.}^2}{12^2\text{ ft.}^2\text{/sec.}^2} \times \frac{8.125\text{ in.}}{12\text{ in./ft.}} \times \frac{80 - 62.3}{62.3}\right)^{3/2} \frac{0.166}{0.44^{3/4}}\right]$$

$$f = 0.0037\ (1.23) = 0.00455$$

$$\Delta P = \frac{2f\rho V^2 L}{gD}$$

$$= \frac{2 \times 0.00455 \times 80\text{ lb./cu. ft.} \times 12^2\text{ ft.}^2\text{/sec.}^2 \times 100\text{ ft.}}{32.2\text{ ft./sec.}^2 \times 8.125\text{ in.} \times 12\text{ in./ft.}}$$

$$= 3.3\text{ psi. per 100 ft. of horizontal pipe}$$

$$h = 3.3\text{ psi./}(0.433 \times 1.28\text{ psi./ft.})$$

$$= 6\text{ ft. (slurry) per 100 ft. of horizontal pipe}$$

Calculating the Critical Velocity

$$(N_{Fr})_c = 2.0 = V_d/\sqrt{gD} \qquad \text{(Fig. 7)}$$

$$V_d = 2.0\sqrt{gD} = 2.0\sqrt{\frac{32.2\text{ ft.}}{\text{sec.}^2} \times \frac{8.125\text{ in.}}{12\text{ in./ft.}}}$$

$$= 9.4\text{ ft./sec.}$$

Calculating the Pump Discharge Pressure

Static pressure, 20-ft. rise		= 11.0 psi
Friction pressure, horizontal:		
6.0 ft./100 ft. × 430 ft. = 25.8 ft.		= 14.3 psi
Friction pressure, vertical:		
5.0 ft./100 ft. × 20 ft. =	1.0 ft.	= 0.6 psi
Total pressure	46.8 ft.	= 25.9 psi

that is used in Eq. (6) to calculate horizontal-pipe, heterogeneous-slurry friction losses.*

The friction losses of many slurries are known to be dampened below the values calculated from the Fanning friction factor; however, this phenomenon is not completely understood and therefore cannot be generalized. In any event, neglecting this possible dampening leads to a conservative prediction.

In a vertical pipe, the solids are uniformly distributed, so the pressure losses are calculated as described earlier for homogeneous slurry. This takes advantage of the lower pressure losses encountered in vertical pipes for heterogeneous slurries.

Comparing the Two Cases

Pump Discharge Pressure—In both examples, the calculations are carried to a pump discharge pressure for the assumed system, to illustrate the effect of static elevation change on system pressure. Note that the threefold difference in friction losses causes only a one-third increase in system pressure, as a result of the moderate elevation rise (20 ft. in 450 ft.) in the system.

The effect of fittings on slurry friction losses is adequately compensated for by including the usual equivalent length for fittings and valves in the system length.

Having developed the pump discharge pressure, and knowing the system flow and configuration, we can proceed to the mechanical design, and to the economic evaluation of alternates. Pumping horsepower is directly related to pumping pressure and flow:

$$\text{Horsepower} = \frac{(\text{Pressure, psi.})\ (\text{Flow, gpm.})}{1{,}714e}$$

where e is the pump and drive train efficiency. The pumping pressure and system configuration define the operating and static pressures that must be contained at any point in the system.

In many cases, a preliminary design and an estimate of capital and operating costs are required before a final pipe diameter can be selected. We will discuss this more fully in the "Economic Aspects" section.

Operability—By operability, we mean sensitivity to operating upset or shutdown. Here again, the two slurry regimes have distinctive characteristics.

Homogeneous slurries tend to be slow or nonsettling under static conditions, and the settled slurry is usually soft and fluffy. Shutdown and restart of the system will probably not be a problem, even if vertical piping is involved. On the other hand, homogeneous slurries can be very sensitive to moderate concentration increases above the design concentration. It is not difficult to conceive of plugging a system by pumping enough high-concentration (high yield-stress) slurry to exceed the system's pressure capabilities.

Heterogeneous slurries settle rapidly and form a sandy or hard deposit under static conditions. The settled solids may close and plug a vertical or inclined section, but usually can be washed out of a horizontal pipe (Fig. 9). Vibrations from other equipment will greatly increase the possibility of plugging where the pipe cross-section is closed by settled solids. Frequent mechanical joints are provided in coarse slurry systems to facilitate unplugging and replacement for wear.

MECHANICAL CONSIDERATIONS

Slurries are abrasive by their very nature. They "sandblast" on impingement at high velocity; they are a "grinding compound" between moving mechanical parts; they are a "cutting tool" when throttled through a restriction; and they are "sandpaper" when dragged along the bottom of a pipe. The abrasive nature of slurries is a major consideration in design and selection of equipment for the slurry system. Selection of major mechanical components of the system—pumps, pipe, valves and fittings, and instrumentation—is discussed below.

Slurry Pumps

Slurry pumping is roughly split between slurry centrifugal and positive-displacement pumps, depending on the system pressure requirements. Various lock-hopper and surge-leg pumping systems have been considered to isolate the primary pump from the slurry.

Centrifugal Slurry Pumps—These are limited in casing pressure and efficiency, due to the nature of the material they are designed to handle. Their casing-pressure capacity is limited by the vertical-split-casing design necessary for easy replacement of the impeller and of the wear-linings of the casing. Impeller tip speed is generally limited to 4,400 ft./sec. to minimize wear on the volute. Efficiency is low because of the robust nature of the impeller design, and relatively wide throat impeller clearance.

It is generally necessary to provide seal water at a pressure above the pump suction pressure to a lantern ring on the shaft, but some mechanical shaft-sealing systems have been satisfactory on low-pressure applications.

Rubber-lined centrifugal pumps have been found very satisfactory on low-pressure applications—i.e., for fine abrasive slurries where the particles do not have enough inertia to cut the rubber lining; wear-resistant (e.g., nickel-hardened) alloys are used for coarser slurries. *(Cont. on p. 439)*

* Allen-Sherman-Hoff Pump Co. has developed an inline remixing device that, if installed at proper intervals in a horizontal line, is said to resuspend the slurry and reduce the heterogeneous overpressure.

Photo: Sun Oil Co., and Allen-Sherman-Hoff Pump Co.

SPLIT-SHELL, CENTRIFUGAL PUMPS—completely lined with replaceable molded-elastomer parts—are arranged for three-stage pumping of abrasive tailings (sand-water slurry) from the primary-oil-extraction plant of Great Canadian Oil Sands Ltd. The 12 pumps in photo have 39-in. impellers. Cylindrical enclosures contain neoprene-lined magnetic flowmeters—Fig. 8

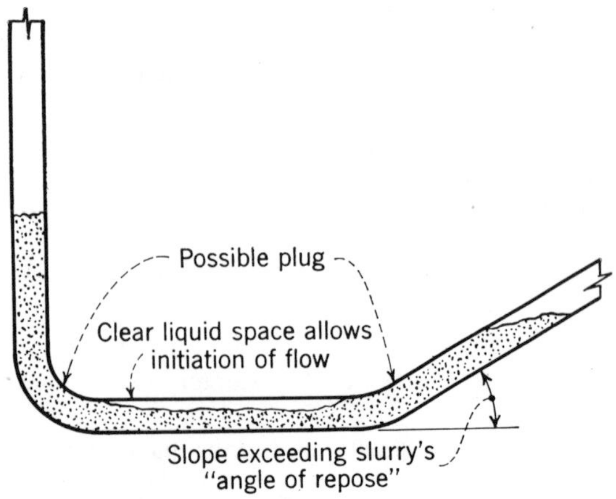

VERTICAL OR INCLINED sections may undergo plugging by heterogeneous slurries—Fig. 9

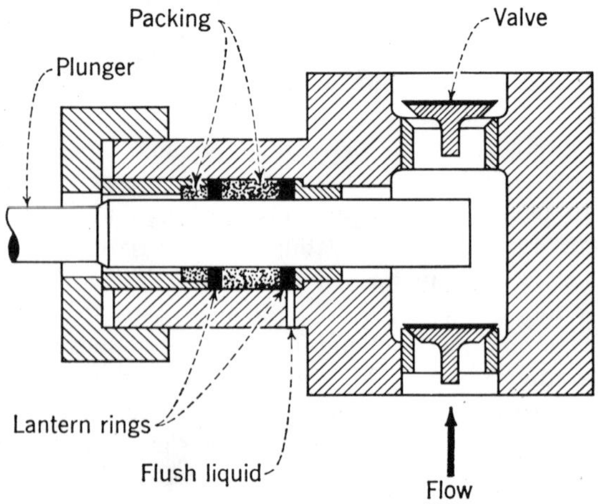

PLUNGER PUMPS with flushing arrangements are useful for very abrasive slurries—Fig. 10

Multiple pumps in series can develop final-stage discharge pressures up to about 600 psi.

Fig. 8 shows the large split-shell centrifugal pumps that handle the sand-tailings slurry in Great Canadian Oil Sands Ltd.'s Athabascan project.

Positive-Displacement Pumps—These are used where pumping pressures above 600 psi. are required. For very abrasive slurries, the plunger-type pump is used, with a flushing arrangement injecting clear liquid to keep solids away from the plunger packing. This is illustrated in the fluid-end schematic diagram Fig. 10.

Fig. 11 shows a similar diagram for a double-acting piston pump. Piston pumps have application in less-abrasive service, such as for coal slurry; they have the advantage of displacing slurry on both the in and out stroke, but depend on sealing the full differential pressure across the piston.

Positive-displacement pumps up to 1,700 hp. are now in service at flows about 2,000 gpm. Units in excess of 3,000 hp. are being considered, but large-volume and high-pressure systems will require multiple units in parallel.

A pulsation-dampening system is required when positive-displacement pumps are used. The primary elements of this system should be proper piping configuration and restraint, and gas-filled pulsation dampeners. Short, straight, suction lines are preferable; longer ones call for a centrifugal pump to charge the positive-displacement pumps. Pulsation dampeners should be fitted as close to the pump as possible on both the suction and discharge side.

Multiple-pump installations will require detailed

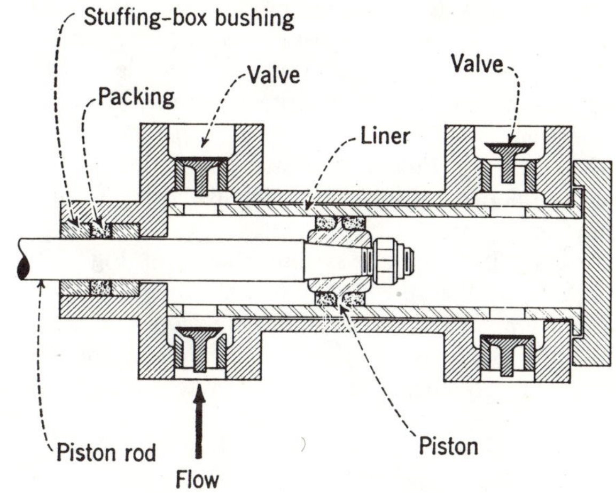

PISTON PUMPS are useful for moderately abrasive slurries such as coal—Fig. 11

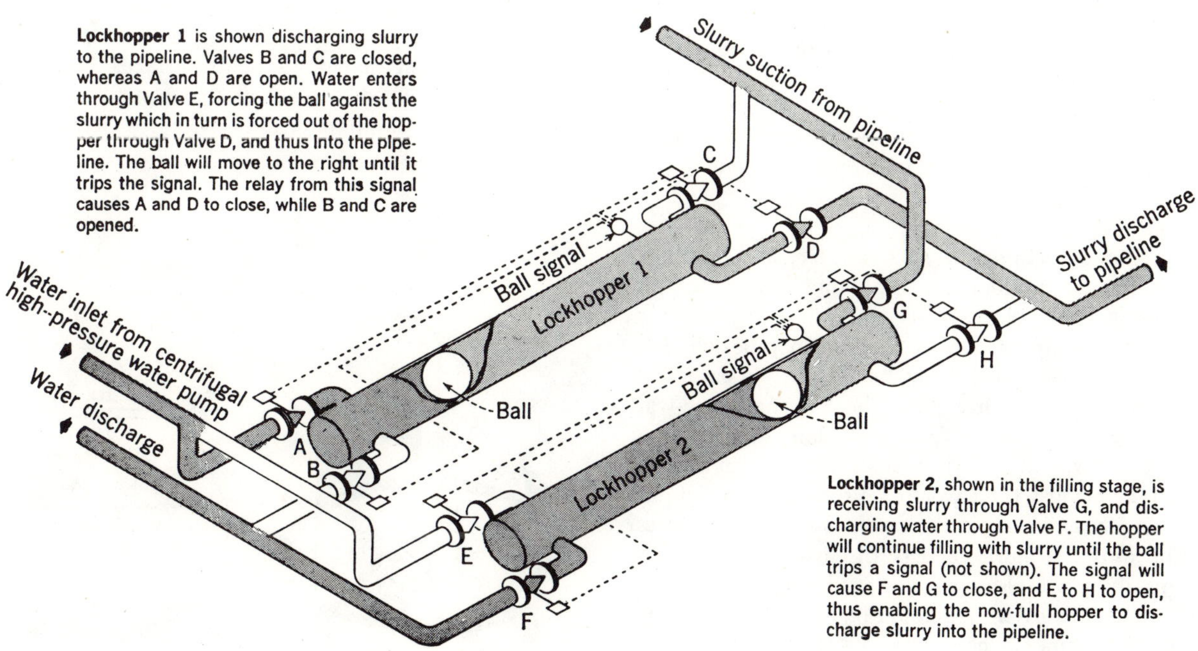

LOCKHOPPER SYSTEM is designed to allow use of centrifugal water pumps by isolating the abrasive slurry—Fig. 12

vibration analysis of the piping system. Instrument connections to positive-displacement-pump piping, such as pressure gages, are quite susceptible to vibration damage to the instrument, or failure at the line connection.

Other Slurry Pumps—Various pump designs have been developed to isolate the abrasive slurry from the pumping equipment.

1. The "lockhopper" system shown in Fig. 12 is designed to allow the use of conventional, multistage water pumps, and to let them develop high heads while the slurry is switched in and out of the lockhopper by sequenced valves. This system is now used in hydraulic hoisting from underground mines, where the pump is mounted on the surface and the lockhopper at the working level.
2. The "surge leg" pump is a special case of the lockhopper concept, where a positive-displacement pump is fitted with a chamber (surge leg) full of clear liquid between the piston and pump valves. The isolating fluid may be oil or water, and makeup fluid must be periodically added. This type of pump has been installed in some very abrasive services (by Wilson Snyder Pumps).
3. Diaphragm pumps are very useful for low-volume, low-head systems where pulsing flow is not a concern.
4. The "advancing cavity," Moyno, pump is ideal for slurries at moderate flow and pressures where steady delivery is required. It is well suited to very thick slurries.

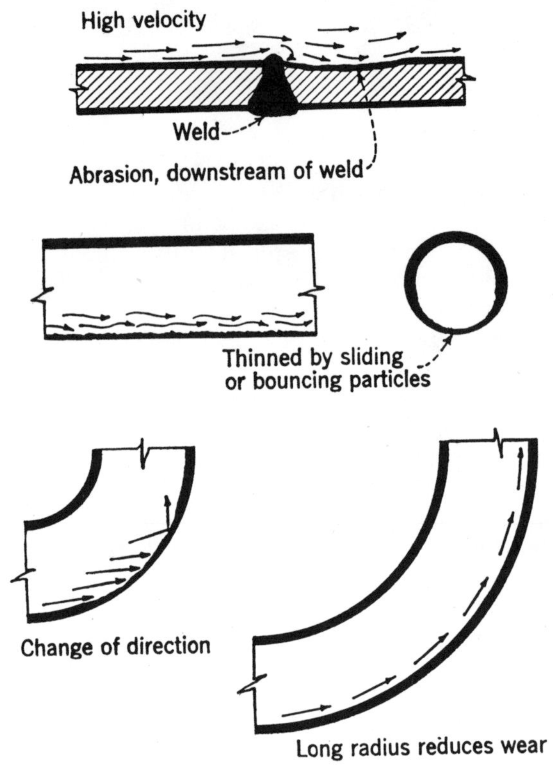

ABRASIVE WEAR can have various causes—Fig. 13

Piping System

When laying out the slurry piping system, the designer must consider:

- Flushing or draining the piping on normal or emergency shutdown.
- Replacement of wear points—near pump discharge, sharp bends, at and downstream of restrictions.
- Rotation of straight horizontal sections (very coarse slurries).
- Access for unplugging.
- Elimination of dead spaces at tees and tappings.

The above considerations apply much more strongly to heterogeneous slurries.

Corrosion may be a consideration in pipe-schedule selection, but this must be evaluated case-by-case. The designer should realize that the internal pipe-wall protection often gained from corrosion products may be eroded away in a slurry system, resulting in a much higher metal loss than would be expected in a liquid pumping system having the same chemical properties.

The abrasive slurry effects discussed under slurry pumps above are also operating in the pipeline system, although to a lesser degree (Fig. 13). General pipe abrasion becomes a consideration above about 7 ft./sec., and is a major consideration at velocities above 15 ft./sec. Rubber or concrete lining may be desirable for these high velocities. Extra-long-radius elbows are often used to minimize wear at changes of direction. Couplings that require grooving the pipe ends introduce a weak point when abrasion is a factor.

Valves, like pumps, must be designed for abrasive service. They preferably provide a full opening, do not depend on machined-metal surfaces for closure, and do not have dead pockets that can fill with solids and restrict operation. Valves with restricted or cir-

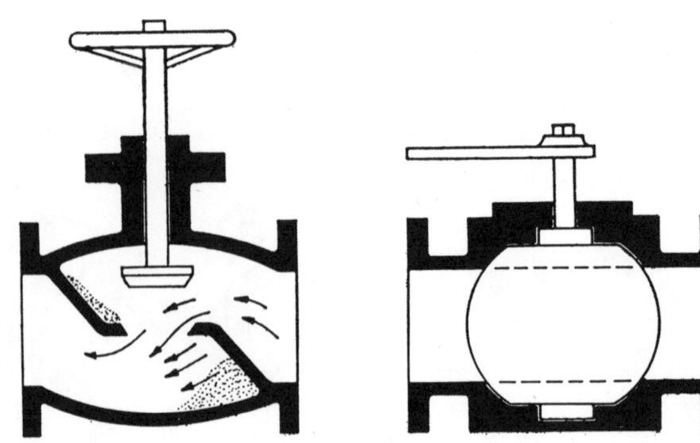

VALVES: Impingement on seat and walls will cause excessive wear (left); the full-opening flow arrangement (right) minimizes wear—Fig. 14

cuitous openings create abrasion downstream (Fig. 14). Many suitable low-pressure valves are available, often with rubber-to-rubber or metal-to-rubber sealing.

To date, high-pressure/large-diameter applications have used lubricated plug or ball valves, neither of which is ideally suited for slurry service.

In some cases, plug valves are specified with Stellite facing on the plug and internal body, to reduce wear. Most ball valves must be fitted with flush and drain connections, so that solids can be removed from the body after operation.

Instrumentation

The presence of solids in a slurry system complicates measurement of the system variables, because many conventional primary measuring elements will be worn away or plugged by the solids. Wear may affect the measuring elements themselves (such as orifices or turbine blades), or the system (i.e., due to turbulence created by the measuring element).

The need to avoid plugging of the sensing element or impulse lines is a major consideration in selecting measuring elements for slurry systems. It is a particularly troublesome problem at pressure tappings, and can disable any element with small clearances or static zones.

Segregation of a slurry in the measuring element must also be considered. A radiation density meter or magnetic flow meter will not read correctly in a horizontal pipe where solids are deposited. Fast-settling slurries will foil instruments requiring a side stream if flows are not maintained above critical velocity.

An excellent review of instrumentation for slurry systems is given by Liptak;[15] the major sensing elements are discussed briefly below.

Magnetic flowmeters have proved to be adequate, though expensive, primary flow-measuring elements; their chief advantage is that they create no flow restriction.

Where positive-displacement pumps are used, their speed provides an accurate flow-rate indication.

Radiation-density meters provide a slurry-concentration indication, again without flow restriction. In our experience, they require frequent calibration. Continuous weighing devices such as the Halliburton densometer have been used, although they present the problem of taking a side stream and returning it to the main stream.

The difficulty in pressure measurement is in keeping the pressure taps from plugging. This is best overcome by using a diaphragm close-mounted to the pipe to separate the slurry from the pressure sensor. In some cases, a continuous backflush of pressure-sensing lines may be required.

Transmitter or gages should not be mounted on the piping around positive-displacement pumps, because they will often quickly fail from the vibration. Pressure devices can be supported separately with a capillary connecting element and diaphragm.

SLURRY PIPELINE APPLICATIONS

Pipeline applications may be considered in two distinct classifications: short-distance inplant transport, and long-distance transport. The latter, for purposes of this discussion, will involve systems whose major purpose is to move the slurry to another geographical area that is closer to the point of end-use—usually distances from fifteen to several hundred miles.

Short-distance inplant or intraplant systems, on the other hand, are primarily associated with moving slurry from one point in a process to another, and usually are in the range of a few feet to thousands of feet. Pipelines for tailing and for dredging materials will also be considered as short distance, although in certain cases they can be many miles in length.

Short Distance Inplant Applications

Short-distance applications are as varied and numerous as types of chemical processes. The types of slurry pumped number in the hundreds[8] and include various foodstuffs, chemicals, chemical wastes, industrial wastes, sewage, and mine and quarry products.

One of the most interesting specialized applications is the transport of a thorium-uranium slurry for nuclear power generation; a great deal of basic theoretical work in slurry transport was done at Oak Ridge for this application.[9]

Virtually every mineral-processing installation has a tailings pipeline to transport gangue and waste from the mineral-disposal site. On the Mesabi range in Minnesota, for example, 40,000 tons of tailings are carried daily through nine 12-in.-dia. pipelines for distances up to 40,000 ft. On the island of Bougainville in the Solomons, Bougainville Copper Pty. Ltd.'s new copper facilities will include an extendable 4,000 to 15,000 ft., 36-in.-dia. tailings pipeline to move more than 30,000 gpm.

Design of tailings pipelines involves nearly all the considerations of longer pipelines, and is further complicated by the variability in screen analysis, concentration and tonnage throughput. The tailings line must be able to handle all these variables. Since the processing plants are usually modular with several lines, consideration must be given to the operating mode when one or more process lines are not operating. Due to critical-velocity considerations, the answer in many instances is a multiple-line system.

Dredging-pipelines are normally associated with the deepening of harbor channel, but could also involve mining. Probably the best-known hydraulic mining operation is in the Florida phosphate fields, where phosphate-rock systems pumping up to 25,000 gpm. are in use. Pipelines with diameters up to 20 in. and lengths up to 6 mi. exist.

Long-Distance Applications

Although materials amenable to long-distance slurry-pipeline transport are fewer than for short-distance inplant applications, the list is still quite large. Materials presently transported by long-distance slurry pipelines include: limestone and other cement raw materials, iron concentrate, coal, gilsonite, salt in brine, phosphate rock, kaolin, copper concentrate, sewage sludge, uranium-bearing slimes, sugar cane, and wood pulp.

Other materials for which the technical feasibility of long-distance slurry transport has been established include: potash, lead-zinc concentrates, sulfur, laterite (nickel ore), pyrite, coke, wood chips, solid wastes.

Several of the more-interesting applications will be briefly described below. For a more-complete list, the reader is referred to a recent literature survey by A. L. Job[10] of Canada's Dept. of Energy Mines and Resources, and to "The Transportation of Solids in Steel Pipelines" by the Colorado School of Mines Research Institute.[11]

Black Mesa Coal Pipeline: Long and Big

The longest and largest slurry pipeline in operation to date is the previously mentioned 273-mi. Black Mesa Coal Pipeline, which traverses the state of Arizona, and was commissioned in the fall of 1970. This 18-in.-dia. pipeline can transport over 5½ million tons of coal annually from the mine site in the Navajo-Hopi Indian reservation to the Mohave power plant site on the Nevada side of the Colorado River, south of Davis Dam. The coal will supply all of the fuel requirements for the two 750-Mw. generating units at Mohave.

The < 2-in. coal provided at the mine site by Peabody Coal Co. is reduced to ¾ in. by impactors, and then ground to pass 14 mesh in three parallel rod-mill lines. From the rod mills, the slurry is stored in three agitated slurry tanks that feed the initial pump station. Three additional pump stations are required along the route. Pumps are electric-powered 1700 P Wilson-Snyder double-acting duplex piston pumps, the largest of this type in existence. Each station has three of these pumps in parallel, except Station Two, which has four. The pipeline system is owned and operated by Black Mesa Pipeline Inc., a subsidiary of the Southern Pacific Transportation Co.

At the power plant, the coal discharges into one of three 24-hr. holding tanks, and is then pumped to centrifuges for dewatering. There are 20 centrifuges for each 750-Mw. generating unit. The dewatered coal, at about 25% total moisture, then goes through a mill and is blown into the boiler with heated air. See Ref. 12 for a more extensive description of the system.

Savage River Pipeline: Mountains, Valleys

The first long-distance iron-concentrate pipeline in the world is the 53-mi., 9⅝-in.-dia. slurry pipeline operated by Pickands Mather & Co. This line is in the island state of Tasmania, Australia, and traverses some exceedingly rugged terrain (Fig. 15) over its route from the minesite concentrator at Savage River to the pelletizing and shipping facilities at Port Latta. In fact, the ore deposit was known for more than 100 years but was considered inaccessible even though it was only 50 mi. from tide water. The slurry pipeline concept made development of that ore body economical. The pipeline, which has been in operation since November 1967, is designed to transport 2.25 million long-tons of iron concentrate per year.

A single pump station (consisting of four electric-motor-driven 600-hp. plunger pumps) is used for the slurry, which is transported as it is produced from the concentrator—i.e., no special processing is required. Design concentration is 60% solids by weight. The material is 100% < 100 mesh. For a detailed description, see Ref. 13.

Consolidation Coal Pipeline: A Pioneer

No discussion of slurry pipelines would be complete without acknowledging the significant contribution made by the landmark 108-mi., 10-in. coal slurry pipeline put into operation in 1957 by Consolidation Coal Co. The line extends from Cadiz, Ohio, to Eastlake, Ohio, and uses three pumping stations, each containing three 450-hp., positive-displacement, double-acting, duplex piston pumps. Flowrate is 1,100 gpm., or about one-fourth of the Black Mesa line.

This pipeline has operated very successfully for a period of six years, has experienced a 98% availability factor, and has had considerable economic impact: As a result of the pipeline, the railroads have made radical reductions not only on the 1.3 million tons/yr. of pipeline coal, but on all 5 million tons/yr. of coal that is being transported from that region in Ohio.

Other Significant Slurry Pipelines

Some of the many other slurry pipelines around the world deserve at least brief mention.

As shown in Table I, which summarizes the features of selected slurry pipelines, two copper-concentrate pipelines are presently planned. A 17-mi., 6-in.-dia. line is now under construction in Bougainville. Engineering is also underway on a project in West Irian, Indonesia, to bring concentrate from an interior minesite to a port on the Arafura Sea. The elevation rapidly drops from 8,600 ft. at the concentrator to sea level over the 70-mi. route.

American Gilsonite Co. constructed a 72-mi., 6-in.-dia. gilsonite pipeline in 1957. This line uses a single pump station to move the slurry from Bonanza Mines,

Utah, to the refinery at Grand Junction, Colo. Slurry is pumped at a concentration of 48% by weight, with a maximum particle size of 4 mesh.

Several limestone pipelines exist. The longest is a 57-mi. line in England owned by Rugby Portland Cement Co. This line moves ground chalk at the rate of 200 tons/hr. through the 10-in.-dia pipeline. A 17-mi., 7-in.-dia. limestone pipeline system will be in operation this summer in northern California.

Although the major slurry pipelines have all been constructed quite recently, it is interesting to note that in 1914, a 1,750-ft.-long line pumped coal from barges on the Thames River near London to a boiler plant.

In the future, longer and larger slurry pipelines are expected, with lengths of 1,000 mi. being practical, coupled with diameters of 40 in. or greater. This projected increase in length simply entails the addition of pump/pipe segments to the system, with due consideration given to the effect of transient phenomena, such as concentration waves. Extrapolation to larger-diameter pipes can be made with confidence—based, for instance, on recent data the authors have gathered for 18 and 24-in.-dia. pipes. As mentioned previously, positive-displacement pumps that are suitable for the volumes and pressures required for really large-diameter pipelines are now being designed.

Selected Commercial Slurry-Pipelines—Table I

Location	Material	Length, Mi.	Dia., In.	Million Tons/Yr.
Ohio	Coal	108	10	1.30
Arizona	Coal	273	18	4.8
Canada*	Coal	500	24	12.0
Utah	Gilsonite	72	6	0.38
England	Limestone	57	10	1.70
Colombia	Limestone	9	5	0.35
Trinidad	Limestone	6	8	0.57
California**	Limestone	17	7	2.0
South Africa	Gold tailings	22	6, 9	1.05
Tasmania	Iron concentrates	53	9	2.25
Japan	Copper tailings	40	8	1.0
Canada*	Sulfur/hydrocarbon	800	12–16	—
Bougainville**	Copper concentrates	17	6	1.0
West Irian†	Copper concentrates	68	3	0.3
Africa†	Phosphate	3	12	5.0
Ohio	Wastes (raw sewage)	13	12	—
New Zealand†	Magnetite	6	8, 12	1.0

* In planning phase † In design phase ** Under construction

AERIAL CROSSING enables Savage River pipeline to traverse rough terrain in Australia—Fig. 15

ECONOMIC ASPECTS

Having followed the general design procedure outlined earlier in the paper, the design engineer is usually faced with several choices that can only be decided on the basis of economics. Often, after considering critical velocity limitations, he will find that more than one pipe diameter is technically feasible. In order to select the correct diameter, the alternatives must be compared both on the basis of first cost and annual operating costs. Because operating costs are incurred over the life of the project, they must be discounted in order to realistically weigh them against capital costs. For very short distance lines, however, operating costs (power and supplies) may be quite low, which would allow the diameter decision to be made on the basis of capital costs alone.

Short-Distance Inplant Pipelines

For quick evaluations of alternatives, some very simple procedures can be used. These should be considered as initial screening techniques only. If the economic impact of the decision is large, such as would be the case in longer-distance pipelines, then more-sophisticated techniques must be employed.

The concept of capital charges is necessary for the analysis. Capital charges are simply the average annual payments necessary to support a capital investment. They include depreciation, interest, property taxes, profit, and income taxes. As a percent of initial investment, they usually range from 15% for a utility to 25% or more for more-speculative ventures. A good percentage for preliminary evaluations is 20%. This simply means that if the initial capital savings between two alternatives are \$100,000, the annual equivalent savings that may be directly compared with operating costs are \$20,000. Therefore, to support an alternative that has a \$100,000 initial cost premium, the annual operating cost savings must be more than \$20,000 per year.

The significant operating costs are usually power (e.g., electricity or fuel), and supplies such as pump parts. Normally, the operating labor for the various alternatives would be the same and therefore could be neglected. Capital expenses are largely comprised of pipeline-steel and pump costs (including installation labor). For preliminary analysis of alternatives, these two cost components will suffice; instrumentation costs usually do not have to be estimated until later.

In addition to the pipe-diameter decision, the type of pump selected often must also be based on economics. For example, in a particular application, either a series of centrifugal pumps or positive-displacement pumps in parallel might be technically feasible. The only way to properly decide is to make an economic analysis. The fact that positive-displacement pumps have a higher efficiency than centrifugal pumps should not be overlooked (85% vs. 60% is typical). In this case, the higher initial cost of positive-displacement pumps should be weighed against higher power costs for centrifugal ones. Maintenance costs should also be included in the evaluation.

In certain, very abrasive, slurries, there may be a choice of installing a more-expensive, abrasion-resistant or lined pipe, vs. a lower-cost conventional pipe that would have to be replaced during the life of the project. Assuming replacement is tolerable, then the familiar "present worth" analysis must be made at several appropriate interest rates to select the economic alternative.

Economics of Long-Distance Pipelines

For evaluation of long-distance slurry pipelines, a more-detailed analysis is required, since the economic impact of conservative design, as well as pipe-diameter and pump-type decisions, can be quite significant.[14] However, for purposes of preliminary screening, Fig. 16 can help the engineer compare a slurry system with alternate modes of transportation, to determine whether the slurry alternative merits a more complete investigation. The curve shows total transportation costs in ¢/ton-mile (short tons) as a function of the annual tonnage of solids transported for materials with various specific gravities. The transportation costs include direct operating costs and annual capital charges.

Operating Costs—Direct operating costs include electrical power costs, operating and maintenance labor, corrosion inhibitors, expendable pump parts, and other maintenance supplies—plus administrative costs and a contingency allowance. Operating labor is assumed to include a system operator for each shift, and a day man at each of the automated pump stations, to perform routine maintenance jobs and inspection.

Capital Costs—The capital costs provide for a complete pipeline transportation system—including pipe, positive-displacement pump stations, communications and supervisory-control facilities, mainteinance bases, as well as installation of all these. An allowance for indirect costs such as contingency, interest during construction, working capital, engineering, and manage-

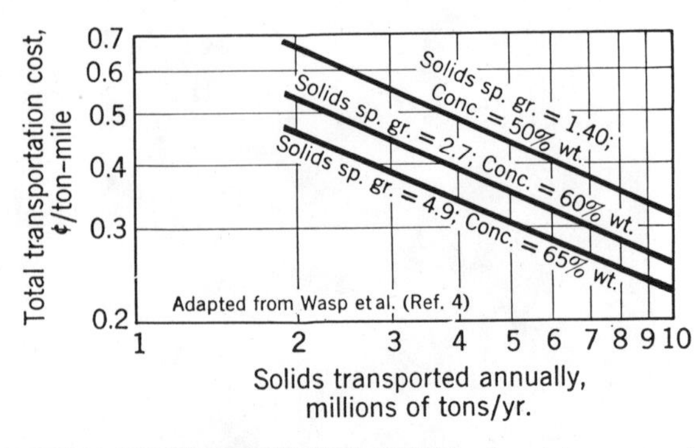

SLURRY TRANSPORTATION COSTS depend on specific gravity of the solids, and amount to be moved—Fig. 16

ment of construction is also included. Capital charges will vary depending upon the particular capital structure of the pipeline entity. For Fig. 16, an annual capital-charges allowance of 15% of capital investment has been assumed.

As shown in Fig. 16, the transportation costs (the sum of operating costs and capital charges) range from 0.68¢/ton-mile for 2 million tons/yr. of 1.4 sp.gr. material, down to 0.22¢/ton-mile for 10 million tons/yr. of 4.9 sp.gr. material.

The transportation decision is normally between rail, truck, conveyor, and slurry pipeline. Below are some comparative costs for these alternative transport modes, for solid volumes of 2 to 6 million tons/yr.:

Carrier	Transp.-Cost, ¢/Ton-Mi.	Conditions
Slurry pipeline......	0.3 to 0.7	Over 50 mi.; no slurry preparation
Rail...............	0.4 to 0.9	Unit train over 400 mi.
Truck.............	5.0 to 8.0	One-way haul with empty return
Conveyor belt......	2.0 to 6.0	Less than 15 mi.

The above again includes operating costs and capital charges (15% of investment, per year). As can be seen, slurry pipelines can often have a strong economic advantage.

References

1. Thomas, D. G., Transport Characteristics of Suspension: VIII—A Note on the Viscosity of Newtonian Suspensions of Uniform Spherical Particles, *J. Colloid Science,* **20**, No. 3, pp. 267-277 (1965).
2. Kenny, J. P., et al., FWPCA Waste Management Study, Vol. III, Technical Aspects of Pipelining of Waste Materials, Bechtel Corp. (Sept. 1969).
3. Wasp, E. J., et al., Deposition Velocities, Transition Velocities and Spatial Distribution of Solids in Slurry Pipelines, First International Conf. on the Hydraulic Transport of Solids in Pipe, Coventry, England (Sept. 1970).
4. Wasp, E. J., et al., Hetero-Homogeneous Solids-Liquid Flow in the Turbulent Regime, ASCE, Intl. Symp. on Solid-Liquid Flow, U. of Pennsylvania (Mar. 1968).
5. Hedstrom, B. O. A., Flow of Plastic Materials in Pipes, *Ind. Eng. Chem.,* **44**, pp. 651-656 (1952).
6. Hanks, R. W., Pratt, D. R., On the Flow of Bingham Plastic Slurries in Pipes and Between Parallel Plates, *Soc. of Petrol. Engrs. J.,* pp. 342-345 (Dec. 1967).
7. Perry, J. H., "Chemical Engineers' Handbook," Fourth Ed., McGraw-Hill, p. 5-59 (1963).
8. Maurer, G. W., Pipelining Can Transport Your Bulk Solids, *Mat. Hand. Eng.,* March 1966, p. 56.
9. Thomas, D. G., Non-Newtonian Suspension—Part I, *Ind. Eng. Chem.,* **55**, No. 11, pp. 18-29 (Nov. 1963).
10. Job, A. L., Transport of Solids in Pipelines, With Special Reference to Mineral Ores, Concentrates, and Unconsolidated Deposits (A Literature Survey), Dept. of Energy, Mines and Resources of Canada, Information Circular 230 (Oct. 1969).
11. "The Transportation of Solids in Steel Pipelines," Colorado School of Mines Research Foundation (1963).
12. The Longest, Largest Coal Slurry Pipeline Ever Built, *Coal Mining and Processing,* Feb. 1971, pp. 8-11.
13. McDermott, W. F., et al., Savage River Mines—The World's First Long Distance Iron Ore Slurry Pipeline, Soc. Mng., Fall Meeting Preprint 68-B-364 (Sept. 1968).
14. Wasp, E. J., Thompson, T. L., Aude, T. C., Slurry Pipeline Economics and Application, presented at First International Conf. on the Hydraulic Transport of Solids in Pipes, Coventry, Eng. (Sept. 1970).
15. Liptak, B. G., Instrumentation for Slurries and Viscous Materials, *Chem. Eng.,* Jan. 30, 1967, p. 133; Feb. 13, 1967, p. 151.

Meet the Authors

Thomas C. Aude, Development Supervisor at Bechtel Inc. (50 Beale St., San Francisco, Calif. 94119) has spent the last six years on various Bechtel slurry pipeline projects—including design, feasibility studies, laboratory evaluations, pilot-plant testing, and startup. His most recent assignment was as startup engineer for the Black Mesa coal pipeline and slurry preparation plant. Aude has an Engineer of Mines professional degree from the Colorado School of Mines.

Norman T. Cowper has 14 years of experience in design, construction, operation and maintenance of slurry pipelines, sewage disposal, and water supply systems. He joined Bechtel in 1969, having previously served as pipeline superintendent on the world's first long-distance iron-ore slurry pipeline, at Savage River, Australia. Mr. Cowper holds a bachelor's degree and engineering diploma from the University of New South Wales, Australia.

Terry L. Thompson has 11 years of diversified engineering, field and administrative experience in the pipeline industry. He is currently Chief Slurry Engineer of Bechtel's Pipeline and Production Service Div., responsible for all slurry pipeline projects. In addition to his extensive work in the Black Mesa coal slurry system, he has directed slurry studies on limestone, copper concentrates, zinc and lead concentrates, magnetite, waste materials, wood chips, phosphate and potash. Mr. Thompson has a B.S. degree in mechanical engineering from Cornell University, and a master's degree in business administration from Case-Western Reserve University.

Edward J. Wasp is manager of Slurry Systems for Bechtel's Pipeline and Production Service Div., with overall responsibility for technical development and conceptual engineering activities of the firm's slurry systems programs. Prior to joining Bechtel in 1963, Mr. Wasp was manager of process engineering and development for Consolidation Coal Co. He has B.S. and M.S. degrees in chemical engineering from Cooper Union and New York University respectively, an M.S. in mathematics from the University of Pittsburgh, and a business management certificate from the University of California at Berkeley. He has been granted 12 patents, 6 of which are involved with slurry pipelining.

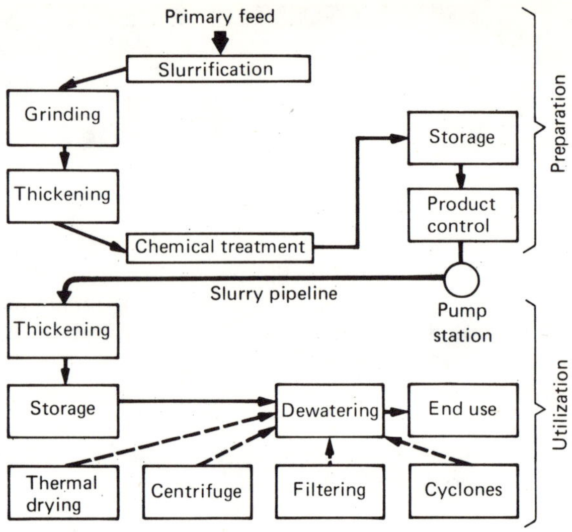

TYPICAL PROCESSING for slurry systems—Fig. 1

Processing Steps: Keys to Successful Slurry-Pipeline Systems

A slurry must be treated before and after it goes through a pipeline—important steps in the success of any slurry-transport system. Here is practical advice on equipment and design.

N. T. COWPER, T. L. THOMPSON, T. C. AUDE, E. J. WASP, Bechtel Inc.*

In the article on pp. 429-445, the authors discussed how mixtures of solids and liquids are pumped a few hundred feet via in-plant piping, or several hundred miles via cross-country pipelines[1].

In this companion article, the processing associated with such systems will be discussed. Often, these processing steps are critical to the technical and economic viability of the slurry-transport system and, therefore, deserve careful attention.

Processing may be broadly classified into slurry preparation and/or slurry utilization. Fig. 1 indicates the various processing steps that may be required. In this article, a minimum of theory and design methods will be presented: for engineers deeply involved in areas such as grinding, mixing, filtration, centrifugation, etc., the literature is quite vast, and any attempt to summarize it would not be very meaningful. The major purposes here are to indicate what processing steps are associated with the emerging slurry-transport systems; to discuss types of equipment presently used; and to note various key design parameters that will be useful to the design engineer. But first, let us define some terms:

Slurry Preparation—This covers the physical and chemical processing necessary to give the slurry characteristics required for hydraulic transport and utilization. Preparation normally involves both size reduction (crushing and grinding); and slurrification or addition of the liquid phase. Chemical treatment may also be part of slurry preparation for corrosion inhibition, thinning, and improving the characteristics of the final product.

Slurry Utilization—In slurry utilization (as well as preparation), there is virtually always a storage step involved, since it is rarely practical to close-couple the pipeline transport system with the processing plant. Therefore, various storage facilities such as tanks and

*For author biographies, see p. 445.

Originally published February 7, 1972

ponds and associated agitators, and recovery dredges must be considered. In addition, the utilization process may include facilities to change the concentration of the slurry—normally to increase the solids content via thickening, decanting, cycloning and screening. More-expensive dewatering methods may be needed, depending on the slurry and its ultimate use, such as vacuum filtration, centrifugation or thermal drying. Further, the non-used portion of the slurry, usually water, might need treatment before it returns to the environment. And additional processing steps can enhance the final product characteristics, e.g., heating to improve centrifugation, or chemical addition to enhance thickening.

SLURRY PREPARATION

Minerals extraction generally requires very fine grinding of ore; 70 to 80% of particles passing 325 mesh (44 microns) is common. Size reduction of the ore normally involves a wet process and the resulting slurries are readily handled hydraulically. Consequently, the minerals industry is ideally suited for hydraulic transportation of solids.

However, there are some hydraulic-transportation systems that do require special slurry-preparation facilities (see Table I). Coal preparation for slurry pipeline transportation is an example. In this case, a particle size specifically suited for slurry transportation must be produced.

For slurry systems, a balance has to be made between pumpability and dewatering characteristics. If sizing is too fine, pumpability may be good but the slurry may be difficult to dewater at the pipeline terminal. If the size is too coarse, then the slurry is heterogeneous and must be pumped at higher flowrates to maintain a suspension. (The cost of pumping then goes up.) The choice of particle size for a slurry depends on: (a) overall cost of preparation, pumping and use of the slurry; and (b) operability of the slurry, including shutdown/startup characteristics and critical velocity.

Commercial Slurries for Pipelines—Table I

Material	Specific Gravity	Maximum Particle Dia., Mesh	Special Preparation Required	Size Reduction Equipment Commonly Used
Gilsonite	1.05	4	Yes	High-pressure water jets, crushers
Coal	1.40	8	Yes	Impactors, cage mills, rod mills
Limestone	2.70	48	No	Impactors, ball mills
Copper	4.30	65	No	Crushers, autogenous mills, ball mills
Magnetite	4.9	150	No	Autogenous mills, ball mills

Grinding Solids

Where grinding is done specifically to prepare a material for pipelining, the step will normally involve conventional milling equipment. In some cases, the equipment may be a new application to grinding the specific material.[2] In the Black Mesa coal-preparation plant,* shown in Fig. 2, rod mills (typical of those found in ore-dressing plants) are used for the final coal-grinding stage. It was necessary to apply the known milling technology of metal-ore dressing to the grinding of coal.

In slurry preparation for pipelining, two variables become quite important (they may have only minor significance in other crushing processes). These are the control of slurry density and control of product top size.

Control of slurry density is important, primarily for the production of a consistent material fitting the hydraulic design of the pipeline. As described in the previous article,[1] slurry concentration plays an important role in both the friction loss in the pipeline and the critical velocity of the slurry. The prepara-

* A 273-mile, 18-in. dia. pipeline in Arizona, in operation for more than a year, feeding two 750 mw. electric generating units.

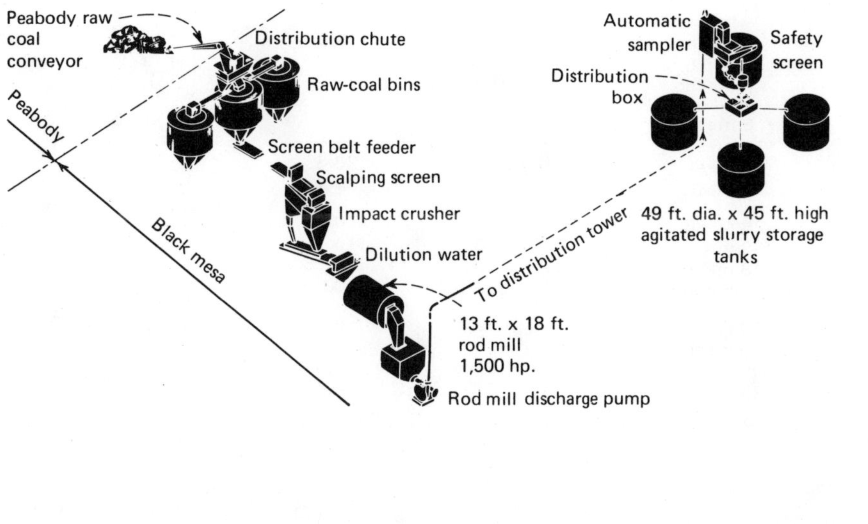

PREPARING COAL for slurry transport involves screening, crushing, grinding, and storage in mixing tanks—Fig. 2

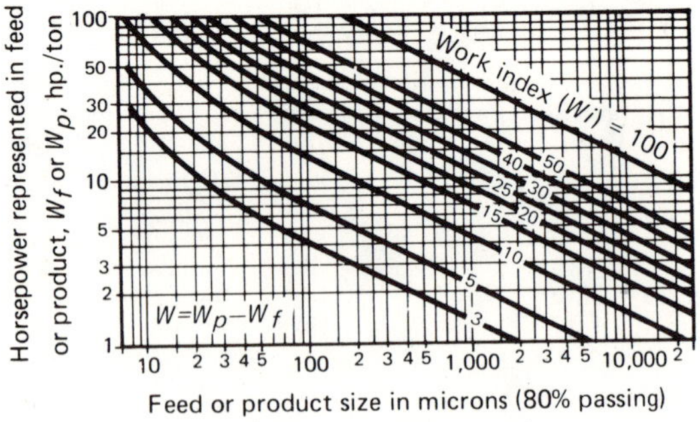

CORRELATING grinding horsepower with size—Fig. 3

tion step is usually performed at a solids concentration slightly higher than required for the pipeline with dilution control instrumentation provided downstream in the line.

Top-size control is extremely important for any slurry transported over a long distance in a pipeline. Large quantities of coarse fast-settling particles can cause pipeline plugs. To prevent this, safety screens are installed to prevent coarse particles from entering the pipeline. In systems where a minimum amount of crushing is done and where the slurry is relatively high in density and slow settling, conventional classification devices such as cyclones and screens are not satisfactory for top-size control unless the slurry can be diluted and later thickened prior to transportation.

Converting Material Sizes for Use in Bond Equation—Table II

Material Size, 99% Passing	80% Passing Equivalent	
	F or P, Microns	$(F)^{1/2}$ or $(P)^{1/2}$
1½ in.	25,000	158.0
1 in.	18,000	134.0
¾ in.	12,000	109.5
½ in.	8,500	92.2
⅜ in.	6,000	77.4
3 mesh	4,200	64.8
4 mesh	3,000	54.8
6 mesh	2,100	45.8
8 mesh	1,500	38.7
10 mesh	1,000	31.6
14 mesh	800	28.3
20 mesh	500	23.4
28 mesh	400	20.0
35 mesh	270	16.4
48 mesh	150	12.25
65 mesh	105	10.25
100 mesh	72	8.48
150 mesh	55	7.42
200 mesh	36	6.0
325 mesh	20	4.47

Available Equipment

There are many types of mills available for grinding ores. The most common include autogenous, impact cage, rod, and ball mills. Selection of the most suitable mill depends on the characteristics of the ore, particularly hardness, and the required fineness of the final product.

Grinding Power, Metal Loss

While many crushing and grinding mills are sized based on manufacturer's experience or tests, the horsepower requirements for grinding with rod or ball mills have been correlated to a theoretical Work Index (Wi) by F. C. Bond.[3] The Work Index in Bond's Third Theory is a property of the material to be ground (determined by a laboratory test or, preferably, from commercial or pilot-plant experience). Bond's approach gives excellent estimates of grinding-power requirements, and many milling installations have been sized on this basis; however, it is a big step from a fractional-horsepower laboratory mill involving a few pounds of sample to a multithousand-horsepower commercial mill processing millions of tons per year. Needless to say, great care and judgment must be used in making such extrapolations.

The basic Bond equation is:

$$W = \left[\frac{10\,Wi}{(P)^{1/2}} - \frac{10\,Wi}{(F)^{1/2}} \right] \times 1.34 \qquad (1)$$

Where: Wi Work Index.
P The size in microns that 80% of the product passes.
F The 80% passing feed size, microns.
W Grinding-power requirement, hp.-hr./short ton.

This equation is also useful for estimating the effects of variable changes on a commercial milling operation.

Work indexes from many tests are available[3, 4]; and general values for most materials can usually be located for preliminary horsepower calculations.

Table II provides a quick way to estimating the 80% passing feed or product size, based on a general knowledge of the size of the material.

Having an estimate of the Work Index and feed and product size, Fig. 3 is useful for estimating grinding-power requirements.[5] For example, what is the power required to grind coal in a wet process for a Work Index (Wi) of 11.4? The coal feed size to the mill is 99% passing ¼ in.; this corresponds to an 80% passing feed size (F) of 5,000 microns (Table II). The desired product size is 80% passing 300 microns (all passing 28 mesh). Using Fig. 3 to obtain Wp =9.0 and Wf = 2.2, then:

$$W = Wp - Wf$$
$$= 9.0 - 2.2 = 6.8 \text{ hp.-hr./ton}$$

Therefore, if it is necessary to grind 1,000 tons/hr. of this coal to 80% passing 300 microns, the minimum milling horsepower will be:

$$\text{Horsepower} = 6.8 \text{ hp.-hr./ton} \times 1{,}000 \text{ tons/hr.}$$
$$= 6{,}800$$

A significant part of grinding costs is the loss of metal through abrasion from the grinding media and mill liners. Here again, F. C. Bond has developed correlations and laboratory procedures[4] to estimate metal consumption during grinding. The abrasiveness in grinding of various materials is correlated with the abrasion index, ***Ai***.

Fig. 4 is a plot of the abrasion index against metal consumption for wet grinding in rod or ball mills. The metal loss values from Fig. 4 include both liner and grinding media.

Bond tabulates abrasion indexes from numerous tests. This tabulation was the source of the abrasion indexes used in the metal-consumption calculations shown on Table III. For example, the abrasion index for limestone, taken as 0.0238, gives a metal usage (from Fig. 4) for wet ball milling of 0.052 lb. of metal per grinding horsepower-hour. Using the previously calculated grinding power for this limestone and product size of 11.2 hp.-hr./ton (Table III), a metal consumption of 0.052 lb./hp.-hr. x 11.2 hp.-hr./ton = 0.58 lb./ton of limestone ground results.

Grinding power and metal consumption through abrasion are the two major variables in the operating costs of a grinding installation. As shown in Table III, the power costs for grinding may range from half to nearly twice the metal costs.

Relative grinding costs for coal, limestone, copper ore and magnetite ore are shown on Table III to illustrate the relative range of costs to grind a material to a size suitable for pipeline transportation. Taking coal as a base of 1.0, the relative operating costs are 1.6, 3.5, and 5.7, respectively, for limestone, copper ore and magnetite. In actual practice, only the coal grinding would be chargeable to transportation (and then only partially). The grinding illustrated for limestone, copper ore and magnetite, although sufficient for pipeline transportation, is required for the end-use.

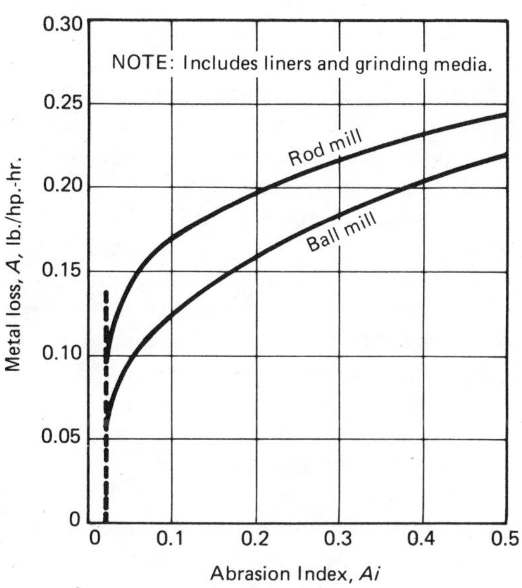

METAL LOSS in wet grinding operations—Fig. 4

Calculations for Grinding: Horsepower and Metal Consumption—Table III

Horsepower	**Coal**	**Lime-stone**	**Copper Ore**	**Magnetite**
Work Index, *Wi*[4]	11.4	12.7	11.7	13.0
Feed size, 80% passing (*F*) microns	5000	5000	5000	5000
Product size, 80% passing (*P*) microns	300	150	100	55
Wf (Fig. 3)	2.2	2.3	2.2	2.3
Wp (Fig. 3)	9.0	13.5	16.0	23.0
$W = Wp - Wf$, hp.-hr./ton	6.8	11.2	13.8	20.7
Metal consumption				
Mill, wet	**Rod**	**Ball**	**Ball**	**Ball**
Abrasion Index, *Ai*[4]	0.02	0.0238	0.1472	0.2217
A, lb./hp.-hr. (Fig. 4)	0.053	0.052	0.143	0.164
Metal consumption, *A* x *W*, lb./ton	0.36	0.58	1.97	3.4
Relative grinding costs, ¢/ton				
Power @ 1¢/kwh	5.1	8.4	10.3	15.4
Metal @ 10¢/lb	3.6	5.8	19.7	34.0
Total	8.7	14.2	30.0	49.4
Relative cost	1.0	1.6	3.5	5.7

Slurrification: Mixing Water and Solids

Slurrification normally takes the form of mixing water and solids together in a gravity feed chute, where the water assists the travel of solids through the chute. This is common practice in a wet-grinding process. The water and coarse solids feed directly into the grinding mill, that discharges a uniform slurry.

A notable exception to this is the cutter-head suction dredge. A rotating cutter head provides the shearing action to break down the bed of sand or gravel into particles that can be transported hydraulically in slurry form. The solids in the bank are sheared downwards, thus dragged down toward the dredge suction where they are picked up in the suction flow. The maximum suction lift of a slurry pump limits the density of a slurry that can be recovered by a dredge.

Slurrification is also often achieved by using hopper-shaped sumps (Fig. 5). Solids disperse over the surface of the liquid in the sump usually by gravity flow through a screen. Water is sprayed onto the surface of the screen to assist in screening, to break up conglomerates and to distribute the particles. Liquid in the sump is maintained at a constant level independent of solids addition. In the sump, the particles settle at their terminal settling velocity. As the cross-sectional area of the sump narrows (an inverted cone or pyramid shape), slurry concentration increases.

Another slurrification method requires using energy of a high-velocity stream of water to break up the solids and form a slurry. The high-velocity water stream is formed by a high-pressure nozzle. Hydraulic

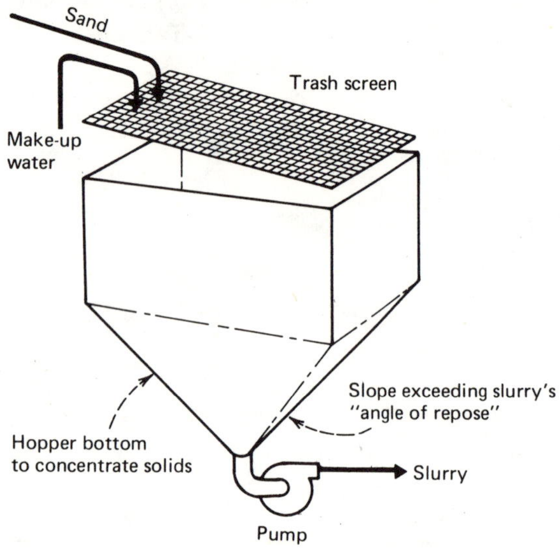

SCREEN SYSTEM for reslurrying coarse sand—Fig. 5

mining and the Marconaflo jet system[6, 7] are examples of this (Fig. 6). In this system, slowly rotating jets at about 300 psi. decompact and reslurry caked concentrate. The reslurried concentrate then flows by gravity through grates to sumps and the sump pumps.

Chemical Treatments

Corrosion in a pipeline transporting a water/solids slurry is controlled by adding corrosion inhibitors.[8, 9] Laboratory-scale tests determine the most suitable type and dosage of inhibitor to be added to a slurry. Various inhibitors may be used, depending on the corrosion mechanism taking place. For example: Sodium dichromate will form a film on the pipe wall and protect the pipe; severe attack by dissolved oxygen can be controlled by adding an oxygen scavenger such as sodium sulfite. In other systems, a high-slurry pH controls corrosion.

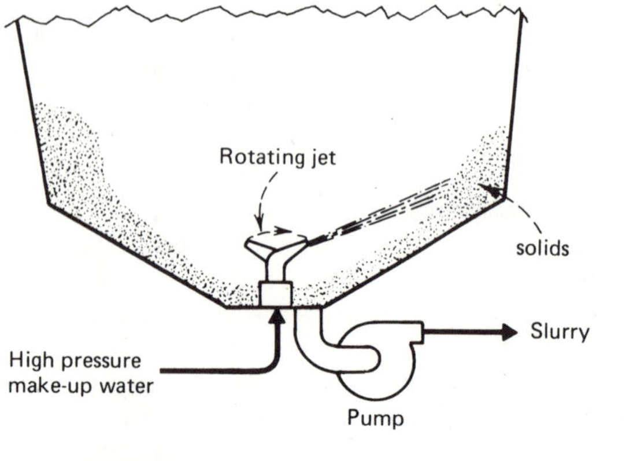

HIGH-PRESSURE JET can reslurry caked solids—Fig. 6

Normally, a solution containing inhibitor is added to the slurry by a metering pump. A corrosion meter monitors internal pipe corrosion and slurry pH can be controlled by addition of slurried lime.

Shear-thinning materials, or dispersants, such as lignin compounds and polyphosphates, are examples of materials that are effective in breaking up agglomerates. Such chemicals are used in the cement industry. These materials reduce the electric charges on the edge of the particles so that they are dispersed in the fluid trapped in the agglomerates. This fluid is now released and becomes part of the free fluid. The effect is to increase the dilution of the slurry. A surprising reduction in pressure losses is possible with low-dosage levels.

Drag reducers suppress turbulence in the pipe. These materials are usually macromolecules of very large molecular weight (large-chain polymers). The exact mechanics of how they reduce turbulence is not understood. They are quite effective in astonishingly small quantities in nonslurry systems; but they are less than effective in slurry systems because considerable suppression of turbulent energy dissipation already has taken place.[10] In addition, most slurry systems operate near the laminar-turbulent transition point where turbulent dissipation is not very high.

Controlling Product Quality

As a final control step to ensure that only specification slurry is committed to the main pipeline, a "test loop" should be included in the system. This test loop is normally a piping section about 300 ft. long, of the same diameter pipe as used in the main slurry pipeline. All slurry must pass through this section before going into the main pipeline. The test loop on the Savage River system (iron concentrates; Tasmania, Australia) is located between the feed tank and the mainline pumps. An alternative location is between the slurry storage tanks. This location, however, requires that the storage tanks be used in series, which of course reduces the effective storage volume.

Since the test loop is simply a short inline version of the main pipeline, it is useful in monitoring changes in slurry hydraulics. Pressure drop (usually by differential pressure cell) is measured in the loop along with flow, temperature and density. Any increase in the test-loop pressure drop above specified limits signifies a deleterious change in slurry properties and probably indicates that coarse fast-settling particles are forming a bed in the loop. In such an event, the off-specification slurry is either recycled or sent to the dump pond.

Slurry-Piping Considerations

The designer of slurry systems is always faced with the selection of the most economical pipe that will have an expected life consistent with the particular application. Pipe choice depends on pressure, temperature limitations, corrosiveness and abrasiveness of the slurry. Available pipe includes:

- Conventional unlined carbon steel.
- Rubber-lined carbon steel.
- Concrete-lined carbon steel.
- Special abrasion-resistant steel.
- Aluminum.

The last four types are more expensive than conventional unlined carbon steel and would only be used with a very abrasive or corrosive slurry. As noted in our previous article,[1] general pipe abrasion becomes a consideration at velocities above 7 ft./sec. and is a major consideration at velocities of 15 ft./sec. Also, pipe wear increases exponentially with velocity[11] above a certain threshold velocity. This exponential value has been reported to range from 2.1 to 2.9.

The amount of annual wear on a pipe is a function not only of the velocity but, of course, the abrasive character of the slurry. This must be evaluated on a case-by-case basis. However, some general relative values of abrasivity as measured by the Miller Number[12] on an arithmetic scale of 0 to 1,000, are as follows:

Material	Miller Number
Detergent	0
Coal A	11
Coal B	28
Fine magnetite	64
Carborundum	1,000

There is a dearth of meaningful data available for predicting absolute wear-rates for specific slurries, and the designer is often faced with only a very broad range of expected wear. In such situations, the only recourse is to make a present-worth analysis as a function of wear rate and pipe materials to study the economics of using heavy-wall unlined steel, for example, vs. a more expensive first-cost alternative. If the potential savings are great enough, it may be desirable to install the lower first-cost system and monitor the wear rate closely during the initial operation.

Corrosivity must also be evaluated on a case-by-case basis. If conventional unlined pipe is used, extra wall thickness must be added to compensate for the metal that will corrode away during the life of the system. As discussed earlier, it may also be desirable to add corrosion inhibitors.

After selecting the pipe, the designer must specify the pipe minimum yield-strength and the wall thickness required to contain the expected pressures.[1] This is done with standard formulas.

The normal economic range of wall thicknesses found in slurry systems is between 0.188 in. and 0.500 in. Heavier walls are feasible, but call for a premium both in first cost and in welding and installation costs (0.750 in. is usually the practical maximum).

On the minimum side (other than manufacturing considerations), the limiting factor for buried pipe is usually external loading. Conventional rule-of-thumb is outside dia./thickness $\leqq$ 128.

Preventing Freezing

In areas subjected to prolonged subfreezing temperatures, slurry pipelines require the same degree of protection as any water-supply piping. Although a flowing solids-water slurry will usually generate sufficient heat from wall friction to prevent freezing, a prolonged system shutdown in subzero temperatures may produce frozen pipes.

To minimize this risk, exposed piping in the immediate area of the plant can be protected by external insulation or heat tracing. For long-distance slurry pipelines, burying the pipe below the frost line is the accepted method of protecting against water-solids slurry freezing.

SLURRY UTILIZATION

In most applications, especially in the mineral and power-generation industries, it is necessary to dewater the slurry so the solids can be handled as a cake. The direct use of a slurry is a future possibility—e.g., coal-slurry firing of a boiler[13] (this has been tested) or iron concentrates used directly in the production of steel.

Storing Slurries

Size and type of storage depends on throughput of the preparation plant and slurry pipeline, the operating factor of the preparation plant, and the type of slurry being produced.

For most long-distance slurry pipelines, it is good practice, as a rule of thumb, to have a minimum storage capacity equivalent to six hours of plant throughput. The time between receiving the slurry into storage and committing it to the pipeline is necessary for laboratory analyses such as slurry percent solids and size distribution. If the slurry is outside the specification for pumping, then the retention time in storage can be used to carry out corrective measures without shutting down the system.

Agitated Storage

The calculation of mixing-power requirements has been amply covered in Ref. 14, so there is no reason to cover this subject here. Also, review articles have been published in previous issues of *CE*.

The Mohave coal tanks, with 6 million gal. in useful capacity per tank are the largest agitated slurry tanks in existence, being 125 ft. dia. by 85 ft. high. The agitator-drive horsepower is 500, the upper larger agitator blade is 34 ft. dia., and the lower agitator blade is 29 ft. dia. These blades rotate at a speed of 7 rpm.

One item the engineer should specify carefully, is what he wants in the way of performance. Two comments on this:

1. Concentration variation of the slurry on with-

drawal from the tank should be held within specified limits. There is usually little interest in the actual concentration gradient in the tank. The authors have seen the situation where a water layer was present at the top of the tank when full, and upon emptying, the tank concentration remained relatively uniform.

2. The design should avoid large quantities of solids being deposited at the edge of the tank due to insufficient turbulence.

Here are some design hints to avoid such operating problems. We have also included some useful items of "convenience."

- **Access Doors**—Access doors should be big enough so that a wheelbarrow can be rolled into the bottom of the tank for periodic removal of trash material.
- **Drains**—Drains should be at the very bottom of the tank, below the normal suction, to permit complete drainage for tank inspection.
- **Lighting**—Adequate lighting is needed at the top of the tank platform to show the slurry level.
- **Suction Piping**—Any suction piping leading to a pump should be placed a few feet above the floor of the tank to avoid drawing trash into the pump.
- **Air-Water Connections**—Suitable air-water connections at the top of the tank for lancing purposes are desirable. These connections can also be used during a power failure or to get a stalled agitator restarted.
- **Bottom Shaft-Guide**—It is not recommended that bottom bearings be used: they are difficult to maintain in the slurry. The bottom of the shaft should be fitted with a loose guide to prevent shaft whip.
- **Delivery Pipe**—The delivery pipe to the agitator tank is sometimes placed below the surface of the slurry to minimize air entrainment.

Solids Attrition in Agitated Tanks

One area of great interest in handling materials in agitated tanks involves solids attrition. Experimental data are often needed to determine the relative tendency of the material to degrade. But laboratory attrition data obtained at agitator speeds of 1,000 rpm. and higher have no relation to large commercial tanks, because such tanks have much lower rotational speeds. Note that the Mohave tanks have only a rotation speed of 7 rpm.

While the rate of shearing strain (shear level) is proportional to rotational speed, there is considerable confusion in the literature as to what is meant by "shear level," and the following brief discussion will try to clarify this point.

In a pipeline for laminar flow:

$$\text{Viscosity} = \frac{(\Delta PD/4L)}{(8V/D)} = \frac{\text{Shear stress}}{\text{Shear level}}$$

The term 8V/D, sometimes called the flow function, is really the shear level on the wall of the pipe.

In an agitated slurry tank, we may write the expression:

$$\frac{8V}{D} = \left(\frac{\text{Tank velocity parameter}}{\text{Tank distance parameter}}\right) = KN^c$$

The above units are: V, linear velocity; D, dia. of pipe; L, length of pipe; P, pressure; N, agitator speed in rev.

The last equation is discussed in Ref. 14, Chapter 10. Certain simplifications are possible: the constant c is unity for many materials. The proportionality factor K is about 10. There are interesting implications from the above formulas for mixer design. As an example, let us assume a slurry going through a pipeline at a flow rate of 4 ft./sec. with a pipe diameter of 10 in., does not produce measurable degradation of these solids. Then, by applying the above conversion, we calculate that an agitator speed of 4 rps. would give comparable shear rates in an agitator tank. This, of course, is 240 rpm., which as already noted is much higher than would exist in any commercial large-scale tank. The procedure can obviously be reversed. If one determines a critical value of N in the laboratory mixer, then the critical degrading velocity in a slurry pipe system can be calculated for a given diameter pipeline.

Static Storage

For the bulk storing of large quantities of slurry, either nonagitated tanks or ponds are usually the most economical method. Large-volume storage is only necessary in processes that operate very intermittently or in industries that require large, onsite feed stockpiles.

The bulk storing of iron-ore concentrate at marine terminals for shipment using the Marconaflo concept is an example of nonagitated storage. Another example is the storing of a coal slurry in nonagitated, emergency storage ponds for use in the event of long-term stoppage of the pipeline processes.

In operation, the slurry is discharged directly into the tank or pond, and the solids settle to their terminal concentrations. Depending on particle size of the solids, inflow slurry concentration, specific gravity of the solids, the settled slurry will remain semifluid for extended periods (provided it is not compacted by induced vibration). It is possible to reclaim this settled slurry via suitably located drawoff points. Fast-settling, high-specific-gravity slurries settle into a compact bed, and shearing action is required for reslurrification. One method now being adopted is to use water jets to reslurry the solids to a pumpable consistency. Another method involves dredge recovery.

Thickening of Slurry

The thickening of a slurry prior to pipeline pumping is necessary with some slurry pipelines. This is done for two reasons: (1) to raise the slurry density to a controllable level for pumping, and (2) to limit the amount of unnecessary water pumped into the slurry. But, if the density is too high, the slurry will be too thick for economic pumping. Therefore, the economic tradeoff must be analyzed.

At the terminal end of the pipeline, the slurry is often thickened as the first step in the dewatering process to reduce the size of the more expensive filtering equipment.

Thickeners—A thickener is a tank or basin, usually circular, in which solids settle by gravity. The primary objective of a thickener is to recover settled solids as a concentrated slurry (clarifiers produce a clear liquor from a dilute suspension). Sedimentation in a thickener is by zone or line settling. The particles in the feed slurry cohere in a floc structure, and the solids settle as a consolidated mass, leaving a sharp boundary between the settling suspension and the supernatant liquor. Thickeners usually have a raking mechanism to convey the settled solids to a discharge point.

The degree to which a slurry can be thickened depends on the settling characteristics of the solids, the physical size of the thickener (area and depth), the feedrate, and the underflow concentration.

Bench-scale batch testing of the settling characteristics of various concentrations of slurry as measured in graduates are useful for sizing thickeners as large as 300 ft. dia. In these tests, the subsidence of the slurry supernatant interface is measured as a function of time.

Data recorded from these tests are used in the Coe and Clevenger technique for sizing thickeners. The reader is referred to a recent article in *Chem. Eng.*, Aug. 23, 1971, "Batch Tests Predict Thickener Performance,"[15] by Bryant Fitch for further details.

Decanters—Thickening by decanting is helpful in dewatering a slow-settling low-specific-gravity slurry such as coal. The procedure is simply to fill a nonagitated tank or pond with slurry through a distribution pipe with evenly spaced spigots to ensure an even, uniform deposit of slurry and allow time for the solids to settle. If necessary, coagulating chemicals that are not harmful to the final process can be added to assist in clarification of the excess liquors. The clear liquid above the settled solids is then decanted.

There are usually two or more ponds or tanks in any installation: one is being filled while another is being emptied. The maximum concentration by decanting is limited by the fluid properties of the settled slurry and the ability to recoup the settled solids in slurry form.

A critical factor in the design of a decant thickening pond is the depth of settled solids. The greater the bed depth, the higher the concentration of settled solids in the pond. An advantage of this method in the dewatering of a coal slurry is illustrated in Fig. 7.

Cyclones—Cyclones have been very successfully used in dewatering of heavy beach-sand minerals, such as rutile, which is a uniformly graded, round-shaped material. For very little expenditure of energy (only 10 to 15 psi. head loss), a 25% slurry can be dewatered to a moist sand of 80 to 90% solids by weight. The simplicity of the system makes it ideal for this type of mineral operation.

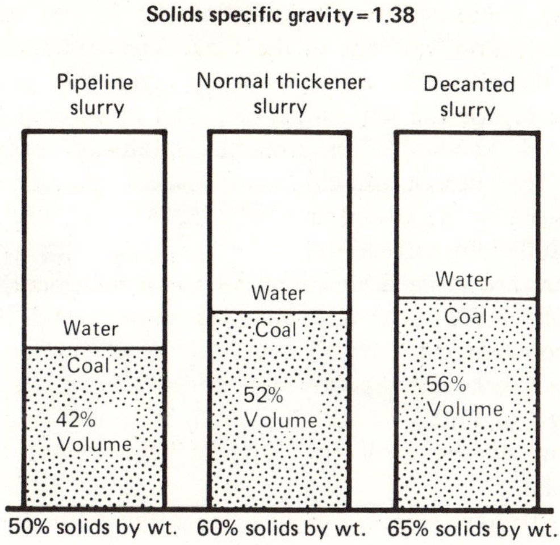

BENEFITS of decanting bituminous-coal slurry—Fig. 7

The sand discharging from the cyclone underflow will stand in its moist state at an angle in excess of the angle of repose of dry sand. Therefore, the cyclone need only be supported over a stockpile area on a lightweight structure. A network of perforated subsoil drains located under the stockpile drain the moist sand to obtain a residual moisture of less than 5%. Sand can be rapidly stockpiled at a minimum cost using this type of installation.

Screens—Dewatering screens also thicken certain selected slurries. Solids, retained in the slurry, form

Classifying Dewatering Equipment—Table IV

Major Function	Operation	General Equipment Classification	Equipment Subclassification
Recover solids	Continuous	Solid-bowl centrifuge	Cylinder-conical bowl (vertical, horizontal); solid/screen bowl combination.
		Centrifugal filter	Conical screen (helix conveyor, oscillator); cylinder screen (pusher, conveyor).
		Vacuum filter	Rotary belt/drum, horizontel belt, horizontal pan.
		Other	Various wet screens, cyclones, special filters and centrifuges, settling tanks.
	Batch automatic	Centrifugal filter	Vertical perforated basket, constant speed; horizontal basket, variable speed.
		Screen-basket centrifuge	Vertical basket, constant speed; horizontal basket, variable speed.
	Batch	Pressure leaf-filter	Plate and frame; pressure leaf, vertical/horizontal leaf.
		Settling tank	

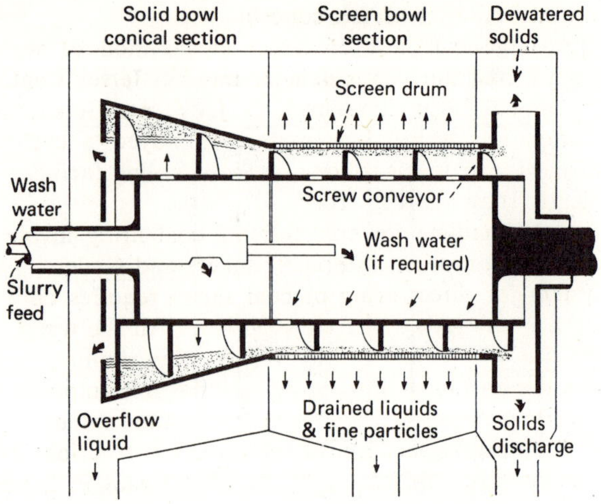

PRINCIPLES of solid-bowl screen centrifuge—Fig. 8.

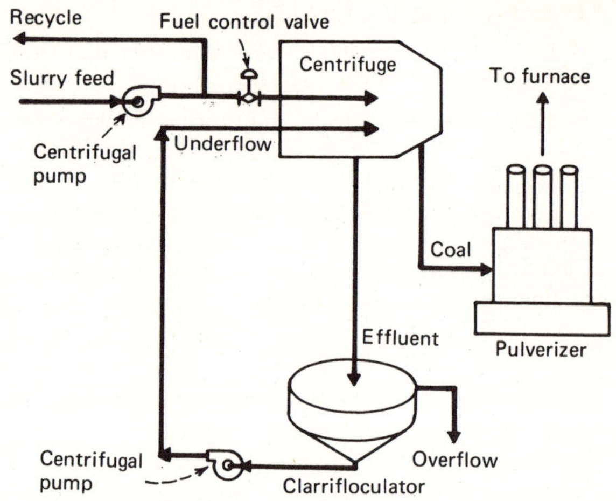

TYPICAL SYSTEM for dewatering coal slurry—Fig. 9

the screen oversize and water is the screen undersize. This is especially effective on slurries that have a gapped size distribution.

Dewatering a Slurry

The principal aim in dewatering a slurry is to recover the solids with a minimum moisture content. Clarification of the liquid phase of the slurry is usually of subsidiary importance. Table IV shows a selection of available dewatering equipment for solids recovery. In most cases, a centrifuge or a filter is necessary.

Bench-scale testing is very useful for determining the difficulty of dewatering a slurry. Simple clinical centrifuges, or Buchner funnel experiments will sometimes indicate the practicality of dewatering of a slurry by centrifuge or filter. Filter-leaf tests and laboratory basket-centrifuge tests on a sample of the proposed slurry provide valuable data for projecting plant cake moistures.

Vacuum Filters—Continuous vacuum filters are popular devices for dewatering slurries. In most of these situations, primary thickening operates upstream of the filter.

A typical vacuum filter incorporates a filtering cloth or surface attached to a moving frame. The frame moves through the thickened slurry, and the solids are caked onto the filter through the action of a vacuum. The frame continues its journey to a point where the cake is removed by mechanical and pneumatic methods. This cake goes onto a conveyor and the filtrate flows to a clariflocculator. Filter frame then moves back into the thickened slurry to repeat the cycle.

For a practical guide on equipment selection and a better understanding of the subject of filtration, see *Chem. Eng.*, June 20, 1966, by J. E. Flood, H. F. Porter, and F. W. Rennie.[16]

Dewatering of slurry in Consolidation Coal Co.'s coal-slurry pipeline in Ohio in 1957 was achieved by thickening, vacuum filtration, followed by thermal drying. The coal from this system was used to feed the conventionally coal-fired Eastlake Power Plant. (This plant originally operated using rail-delivered coal.)

Centrifuges—Continuous centrifuges are often used for dewatering slurries.

The solid-bowl centrifuge (dewatering coal slurry at Mohave[17]) consists of a rotating bowl and a rotating-screw-conveyor section that revolves concentrically within the bowl. The bowl "conveyor" conforms to the inside of the bowl. Slurry enters via a stationary feedpipe concentrically located within the hollow conveyor shaft. This feedpipe extends inward to a rotating chamber located within the screw conveyor. Here the slurry is accelerated and moves radially outward to the bowl. The liquid portion of the feed tends to form a cylinder of revolution within the bowl. Liquid depth is controlled by adjustable weir plates at the outboard end of the cylindrical section of the bowl. A rotating-screw conveyor keeps the solids moving up the conical slope. At the discharge ports, the solids fall into a discharge hopper.

Recent tests have shown that, with a screen-bowl centrifuge, cake moisture content for the material tested could be reduced by 20 to 30% as compared with the cake moisture from a solid-bowl centrifuge. A screen-bowl centrifuge is divided axially in a conical section followed by a screen-drum section. (Fig. 8). The drained solids in the conical section go via a rotating-screw conveyor to the screen section for final draining. The small quantity of solids passing through the screen is recovered in a thickener loop and recycled to the feed.

Dewatering-System Example

A typical slurry-dewatering facility is the system now in operation at Mohave Generating Station, Clark County, Nev. This system converts water-coal slurry into a coal cake containing 25% moisture. The cake feeds directly to pulverizers and the pulverized coal is direct-fired into the furnace.

Dewatering of the slurry requires two stages: centrifuges and then thermal drying.

The first-stage process schematic is shown in Fig. 9. Slurry feed, drawn directly from an agitated-slurry storage tank, is pumped to the centrifuges. In a centrifuge, the coal cake discharges directly to the pulverizer.

Centrifuge effluent contains a small percentage of very fine coal particles (minus 325 mesh). This coal is recovered in a thickening loop, where the fines are chemically flocculated and recovered in a thickener. The thickener underflow is deposited on the centrifuge "beach" to reclaim agglomerated fines.

In the second stage of the dewatering process, the cake discharged from the centrifuges goes to the mill grinding chamber for pulverizing and drying. Preheated primary air, at approximately 680 F., thermally drys the cake and transports the pulverized coal into the furnace.

Advantages of Direct Evaporation

Most materials are easier to handle in liquid form than in solid form, so it is advantageous to keep the material in this form as long as possible. The convenience of handling liquids often outweighs the disadvantage of greater heat required for evaporation. In addition, this procedure can eliminate intermediate separation steps.

A good example of direct evaporation is in the cement industry, where slurry is direct-fired and the liquid carrier evaporated in the kiln without any intermediate dewatering. The savings in power requirements of wet grinding of limestone vs. dry grinding pay for added heat requirements. Spray drying of chemicals is another example of this concept.

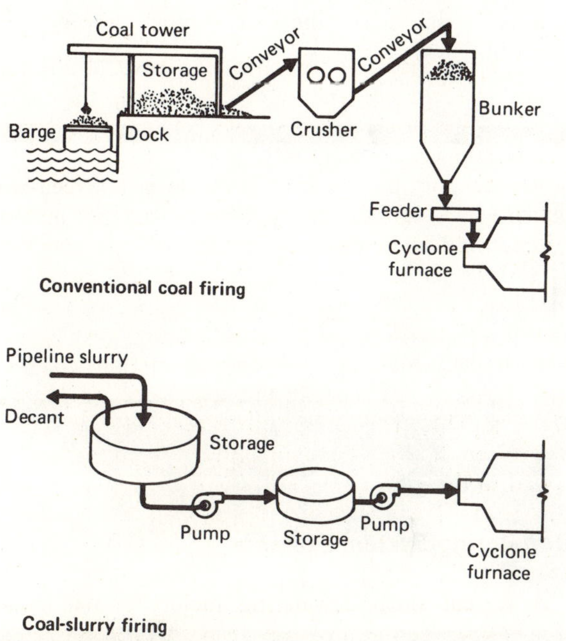

CONVENTIONAL firing vs. coal-slurry firing—Fig. 10

A large-scale test of direct evaporation of water from a coal slurry was demonstrated in Jersey Central Power and Light Company's E. H. Werner Generating Station in 1961.[13] The system employed a cyclone furnace, and the simplicity of the handling facilities is clearly shown in Fig. 10.

Coal handling for conventional coal firing involves conveyors, crushers, storage bunkers and feeders. The system for direct firing of coal slurry requires only a storage tank and a recycle pipe loop. All the water in the slurry vaporizes in the furnace. The resultant heat loss due to the latent heat of vaporization amounts to approximately 5% loss of the coal's Btu. value.

A unique application of the direct-evaporation concept involves preheating of slurries. Steam in a heat exchanger preheats a slurry prior to dewatering by centrifuges. Tests have shown that preheating (to 170 F., for example) prior to centrifuging can reduce total surface moisture by 20-30%. Total thermal load in any subsequent drying step is thus reduced significantly.

References

1. Aude, T. C., Cowper, N. T., Thompson, T. L., Wasp, E. J., "Slurry Piping Systems: Trends, Design Methods, Guidelines," p. 429.
2. Levene, H. D., "The Longest, Largest Coal Slurry Pipeline Ever Built," *Coal Mining and Proc.*, Feb. 1971.
3. Bond, F. C., "Crushing and Grinding Calculations, Part I and II," published by Allis-Chalmers (revised Jan. 1961), reprint from *British Chem. Eng.*
4. Bond, F. C., "Metal Wear in Crushing and Grinding," paper at 56th Annual Meeting of AIChE., Houston, Tex., Dec. 1-5, 1963.
5. "Rotary Grinding Mills—Selection and Capacity Determination," Allis-Chalmers, PM 3.2, Aug. 30, 1963.
6. "Marconaflo—The System and the Concept," *Eng. & Mining Jour.*, May 1970.
7. "Slurry Handling Techniques Will Lower Ore Transport Cost," *Ship-Building and Shipping Record*, Dec. 4, 1970.
8. Bomberger, D. R., "Hexavalent Chromium Reduces Corrosion in Coal-Water Slurry Pipeline," *Materials Protection*, Jan. 1965.
9. Swan, J. P., et al., "Corrosion Control Achieved in Coal Slurry Pipeline," *Materials Protection*, Sept. 1963.
10. Zandi, I., "Decreased Head Losses in Raw Water Conduits," *Jour. Amer. Water Works Assn.*, Feb. 1967.
11. Link, James M., Tuason, C. E., "Pipe Wear in Hydraulic Transport of Solids," Presented at the American Mining Congress, Las Vegas, Nev., Oct. 14, 1971.
12. Prudhomme, R. J., Rizzone, M. L., Schmieman, T. H., Miller, J. E., "Reciprocating Pumps for Long Distance Slurry Pipelines," Presented to First International Conference on the Hydraulic Transport of Solids in Pipes, Coventry, England. Sept. 1-4, 1970.
13. Taylor, D. M., "Liquefied Coal Piped Directly into Boiler," *Pipeline Industry*, Dec. 1961.
14. Holland, F., Chapman, F., "Liquid Mixing and Processing in Stirred Tanks," Reinhold, N. Y. (1966).
15. Fitch, B., "Batch Tests Predict Thickener Performance," *Chem. Eng.*, Aug. 23, 1971.
16. Flood, J. E., Porter, H. F., Rennie, F. W., "Filtration Practice Today, Centrifugation Equipment," *Chem. Eng.*, June 20, 1966.
17. Coffey, R. C., Lyons, H. G., Oakes, A. C., "Mohave Generating Station, Design Features," Presented at American Power Conference, Chicago, April 1969.

Additional Reading

McDermott, W. F., et al., "Savage River Mines—The World's First Long Distance Iron Ore Slurry Pipeline," ASME, Fall Meeting, Preprint 68-B-364, Sept. 1968.

Ellis, J. L., Bacchetti, P., "Pipeline Transport of Liquid Coal," Lignite Symposium, Bismarck, N. D., May 1971.

Walker, G. H., Wasp, E. J., "Experience and Prospects in Economic Transportation of Coal in Pipelines," Sixth World Power Conference, Melbourne, Australia, Oct. 1962.

Wasp, E. J., Thompson, T. L., Aude, T. C., "Slurry Pipeline Economics and Application," Presented at First International Conference on the 70th National Meeting of AIChE, Atlantic City, N. J., Aug. 1971.

Slurry pump selection and application

Slurry handling is becoming more common in process plants (e.g., many pollution-control projects involve pumping slurries). This article will help in selecting a slurry pump to meet your requirements.

J. Ingemar Dalstad, *Morris Pumps, Inc.*

☐ Most engineers seldom have the need to select slurry pumps. Unfortunately, the improper selection of a pump for handling slurries can have costly and dramatic results.

This article will discuss when a slurry pump should be used (it is not always obvious), the design features that are most desirable, and the other criteria and pitfalls involved in selection and application processes.

When to use a slurry pump

Much of the confusion in deciding when to specify a slurry pump arises from the lack of agreement on the meaning of the word "slurry." This is due in large part to the nearly infinite number of solid-liquid mixes. In place of the many academic slurry definitions, this broader, more functional definition if offered: A slurry is any mixture of liquid and solids capable of causing significant pump abrasion, clogging, or mechanical failure due to high loads or impact shocks.

Under some circumstances, it may seem superfluous to consider the "when" of slurry pump selection. Obviously a pump employed to move "deliberate" slurries such as mine tailings or chemical concentrates must be designed and constructed with exceptional strength, abrasion resistance, and solids-passage ability.

But what about a pump employed to supply large quantities of water from a sandy river for cooling purposes? In such "accidental" slurries, the transport of liquid is the prime purpose—the presence of solid materials is not intended (or, sometimes, even recognized). Nevertheless, failure to use a slurry pump for this type of application can frequently result in excessive maintenance, parts, and down-time costs.

The "when" of slurry pump selection might best be answered by a rule of thumb which says that whenever the fluid to be pumped contains more solids than are found in potable water, at least consider the use of a slurry pump.

Typical slurry pump features

What are the features of a slurry pump (Fig. 1) that set it apart from a standard centrifugal? Outwardly there are few differences, although the slurry pump is usually larger in size. Internally, however, there are many characteristics that make the slurry pump a very specialized breed.

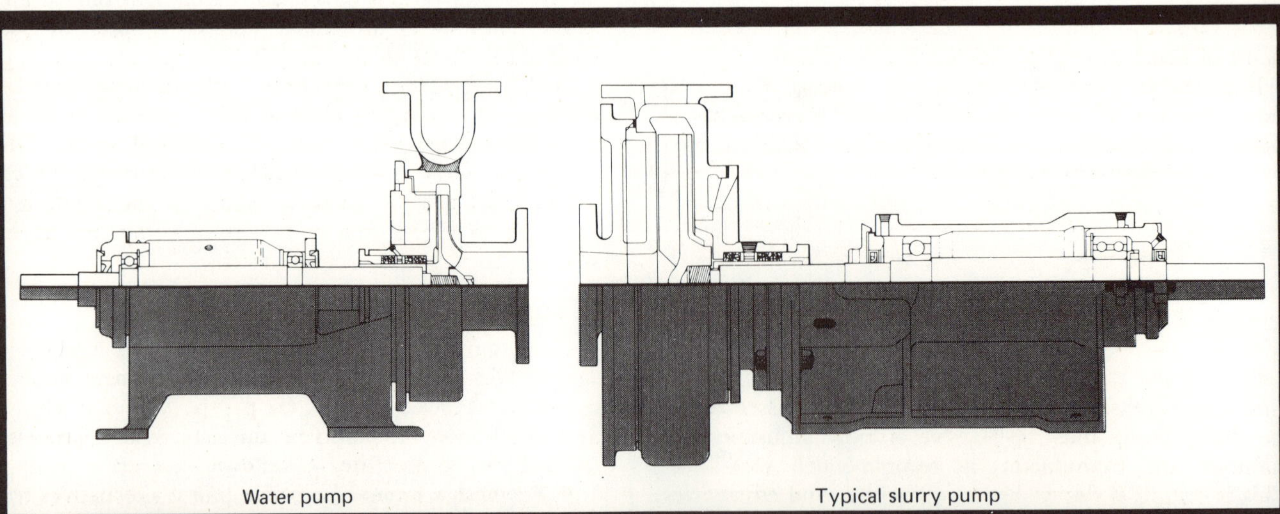

A slurry pump has many features that set it apart from the typical centrifugal pump used for clear liquids — Fig. 1

Originally published April 25, 1977

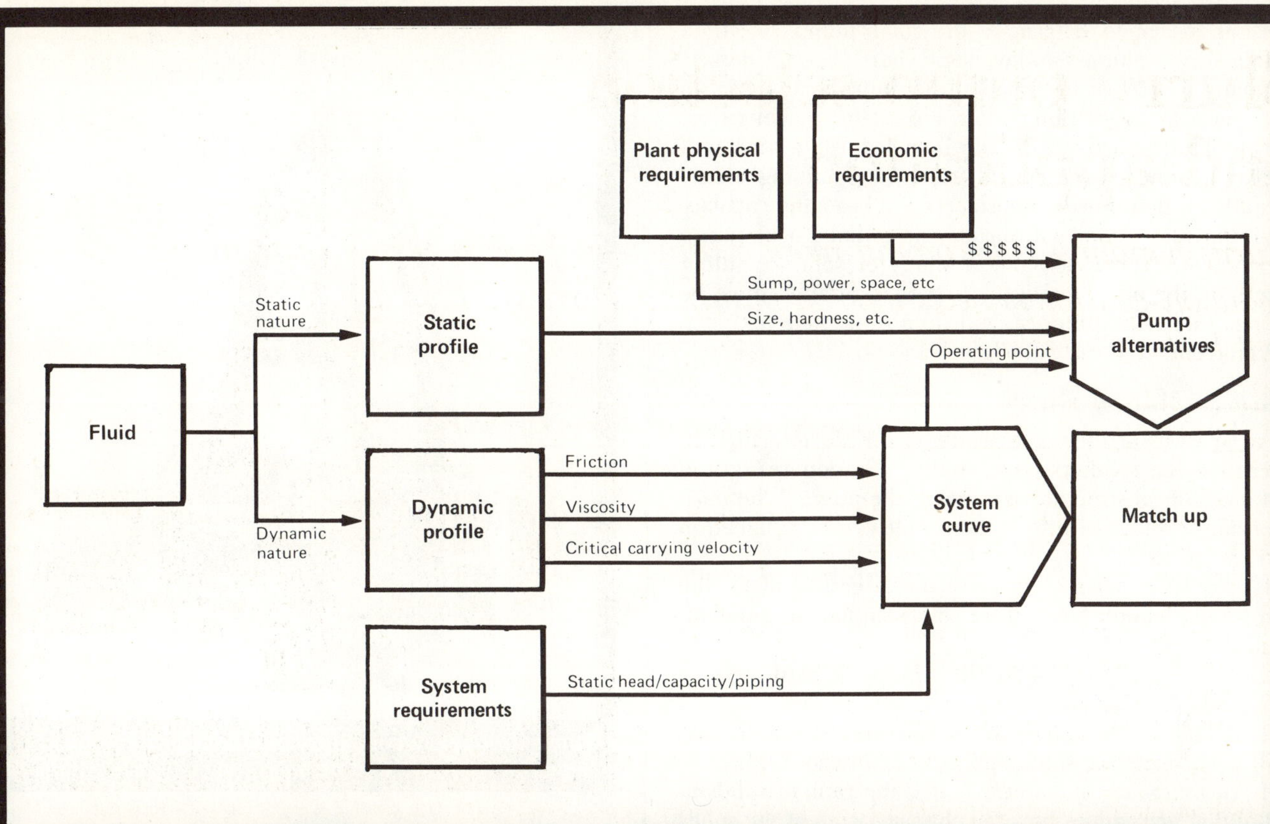

Selection-logic chart for slurry pumps shows the many kinds of information that must be considered Fig. 2

Wall thicknesses of wetted-end parts (casing, impeller, etc.) are greater than what is used in conventional centrifugals. The cutwater, or volute tongue (the point on the casing at which the discharge nozzle diverges from the casing) in the casing is generally less pronounced in order to minimize the effects of abrasion. Flow passages through both the casing and impeller are large enough to permit solids to pass without clogging the pump. Slurry pumps are available in a variety of materials of construction, to best handle the abrasion, corrosion, and impact requirements of nearly any solids-handling application.

Replaceable liners are used in critical areas of wear to reduce the cost of parts replacement. On many light-to-medium-duty pumps, the liners are used only on the suction side. A slurry pump for severely abrasive duty will use an additional liner on the hub side of the casing and, frequently, a liner in the suction nozzle as well. Another approach is the use of a complete shell lining—a sort of casing within a casing. Rubber, ceramic and other synthetic linings are used in many specialized applications.

Because the gap between the impeller face and suction liner will increase as wear occurs, the rotating assembly of the slurry pump must be capable of axial adjustment to maintain the manufacturer's recommended clearance. This is critical if design heads, capacities, and efficiencies are to continue. Other specialized features include extra-large stuffing boxes, replaceable shaft sleeves, and impeller back vanes that act to keep solids away from the pump's stuffing box.

Both radial and axial-thrust bearings on the slurry pump are generally heavier than for standard centrifugals, owing to the demands imposed by slurries of high specific-gravity. Although impeller back-vanes (used to lower stuffing-box pressures) do actually reduce axial thrust, these vanes can wear considerably in abrasive services. Consequently, the bearings must be of ample capacity to handle thrust loads by themselves. Balancing holes through the impeller should not be used to reduce axial thrust since they can either clog or initiate excessive localized impeller wear.

Nearly all slurry pumps have larger-diameter impellers than units for pumping clear liquids, to enable heads and capacities to be met at reduced rotational speed. Low-speed operation is one of the most important wear-reducing features of a slurry pump. In fact, experience shows that abrasive wear on any given pump rises at least with the third power of rpm increase.

Which pump should be selected?

Once having determined that a pump specifically designed for solids handling is required, the next steps are to select and properly apply the pump. Although selection and application are complex and interrelated processes, the task can be simplified as follows:

- Establish a group of possible pump alternatives from the many units on the market.
- Plot a system curve depicting required pump heads at various capacity levels.
- Match pump-performance curves with the system

curve in order to determine the final pump selection.

The slurry-pump-selection logic chart (Fig. 2) depicts the many sources of information that must be analyzed to arrive at a group of pump alternatives, the system curve, and the final match-up. (If the intended application will employ a standard centrifugal pump for handling water, virtually all that must be considered are head and capacity requirements, which will dictate the initial list of pump alternatives.) When selecting a slurry or solids-handling pump, on the other hand, the first consideration must be the material to be pumped, both in a static state and as a dynamic entity.

Static profile of slurry

An analysis of the static profile of the slurry will help determine the solids-passage ability, abrasion resistance, and mechanical strength required of the pump. The most important elements in the static profile can be assigned to four categories:

1. *Size of the solids*—what are the largest particles the pump must handle? Are these solids similar, or random, in size?
2. *Nature of the solids*—Are they pulpy or hard, light or dense, round or jagged, or abrasive, or corrosive?
3. *Nature of the liquid*—How corrosive is the liquid? Will it lubricate the solids and reduce abrasion?
4. *Concentration of the solids*—It is the ratio of solids to liquid that determines how the characteristics of the solids will influence the slurry as a whole.

These four static characteristics create unique demands, requiring specific pump design and construction features.

For example, Fig. 3 shows a pump designed to handle wastes, light slurries, and random large solids. Unlike the slurry pump shown in Fig. 1, this unit does not use wear liners. The emphasis here is on very large flow-passages through the casing and impeller. Because such units are generally used for pumping sewage and relatively non-abrasive industrial wastes, certain wear-reducing design features can be compromised to increase hydraulic efficiency.

Other applications involve not only large solids, but high levels of abrasion and impact shock as well. Fig. 4 shows how added protection is supplied by the extensive use of replaceable liners. Units such as these are used in the pumping of furnace slag, high-concentration ash slurries, and extremely abrasive chemical wastes.

When chemical sludges or wastes containing large solids must be pumped, the vortex pump (Fig. 5) is often the best answer. Because its impeller is fully recessed into the rear of the casing, a relatively small pump can be used to handle liquids containing very large solids.

Still other slurries may exchange the problems of large solids for the equally difficult pumping idiosyncracies associated with high concentrations of small solids. More often than not, such slurries present extreme abrasion problems. Typical are those associated with lime-slurry pumping, the handling of ore concentrates, kaolin clay, or cement slurries. Fig. 6 shows an extremely heavy-duty slurry pump ideal for such applications.

Materials of construction

In addition to pump design itself, the many available materials of construction present a wide spectrum of choice to meet the challenges found in slurry pumping. Cast iron is the basic pump material, being both low in cost and well suited for mildly abrasive applications. Cast iron with 3% nickel is sometimes used when protection against mild corrosion is needed. Ductile-iron and cast-steel offer extra strength in high-pressure applciations or when the pump is exposed to high impact shock from large, hard solids.

As abrasion increases, materials such as Ni-Hard or heat-treated high-chromium iron, with Brinnel hardness ratings in the 550 to 600 range, offer extended parts life. Chromium iron is recommended when mild corrosion will

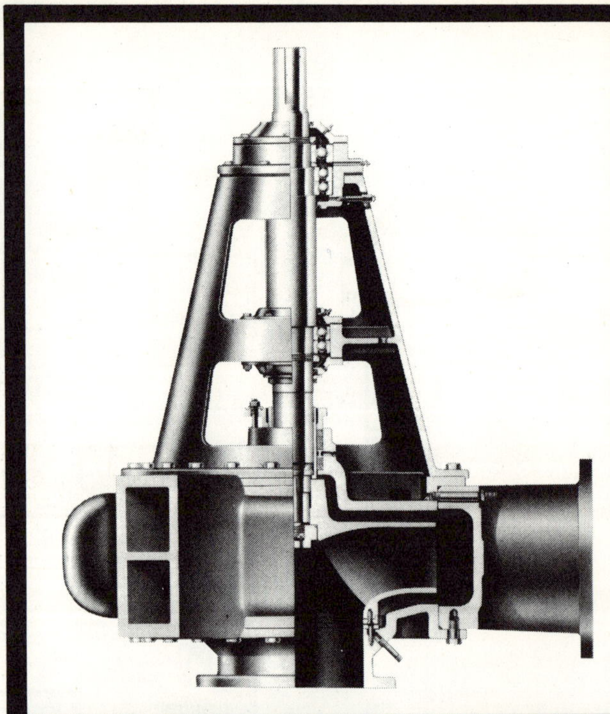

Pump for light slurries and random large solids has large flow passages and no wear liners Fig. 3

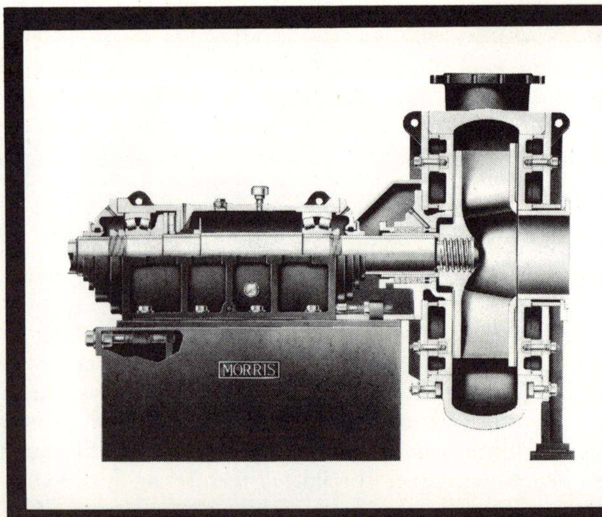

Replaceable liners characterize pump designed to handle extremely abrasive slurries Fig. 4

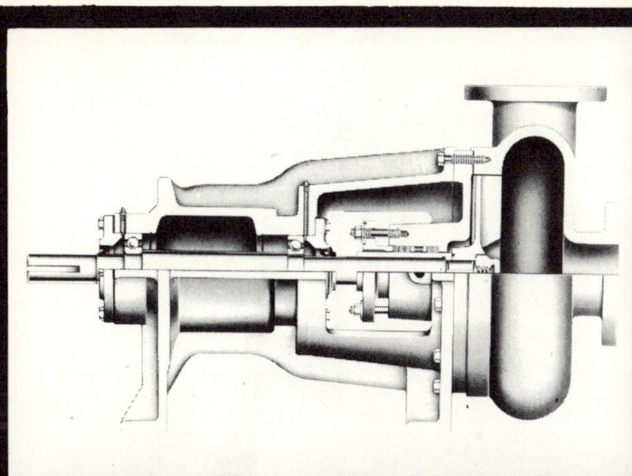

Vortex pump for large-solids slurries has impeller fully recessed in the rear of pump casing Fig. 5

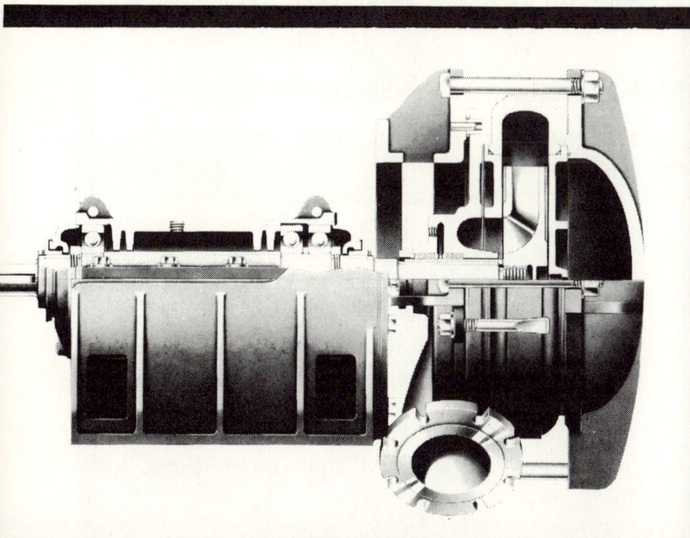

Very-heavy-duty pump for slurries with large concentrations of fine particles, with concomitant abrasion Fig. 6

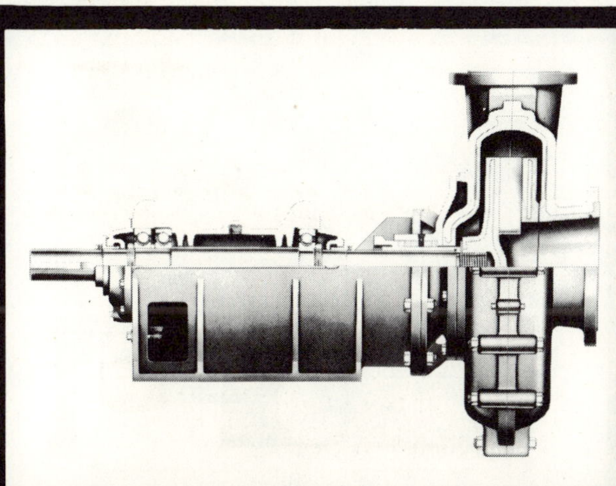

Rubber-lined pump for fine, highly abrasive and severely corrosive slurries Fig. 7

accompany abrasion. Rubber lining is frequently used in pumps for handling fine, highly abrasive, and more severly corrosive slurries (Fig. 7).

Finally, in highly corrosive situations, a number of alloys of the stainless steel and high-nickel groups are available. Ceramics and a number of plastic materials also offer protection from severe corrosion and are being used with increasing frequency to reduce costs.

Referring back to the pump-selection logic diagram (Fig. 2), it can be seen that the many static characteristics of the pumpage, combined with the physical and economic requirements of the pumping job, allow the choice of a number of pump alternatives from the many units available. This information alone, however, is not enough to enable the design engineer to make a good slurry-pump selection.

Dynamic profile, and system curves

Still to be examined are the system requirements and the dynamic profile of the material to be pumped. These two information sources will yield the system curve, the working basis for final pump selection and application. Following is a brief review of the system curve and some common errors that lead to its misuse.

The system curve (Fig. 8) is a plot of the head that must be met by the pump at any given capacity. This is entirely a function of the piping and in no way determined by the pump itself. The vertical coordinate, expressed in this example in feet of liquid, is the sum of two factors: (1) *static head*, or the actual difference in liquid levels which the pump must overcome, and (2) *friction head*, the expression of energy needed to overcome friction in the piping system. Pipe friction is a function of flowrate, increasing as flowrate increases. Friction values for pipe and fittings can be taken from tables, slurry laboratory tests, or field tests, once the slurry and the type, length, and diameter of the pipe are known. The curve in Fig. 8 describes the behavior of a Newtonian slurry. As shown, pipe friction increases by a constant percentage as capacity is increased, thereby producing a curve with a parabolic shape.

Not all slurries will exhibit Newtonian behavior. This is illustrated by Fig. 9, which shows curves for two sand slurries with different solid concentrations. At a concentration of 60%, the sand slurry is Newtonian and will yield a predictable, parabolic curve. But, when the concentration is stepped up to 70%, the curve is drastically different. The slurry becomes non-Newtonian (in this case, plastic) and the system curve cannot be plotted using normally available friction tables. Instead, pipe friction must be arrived at by actual testing in the laboratory or plant.

Fig. 10 demonstrates what will happen if the non-Newtonian nature of the 70% slurry is not recognized. The system curve (A) is plotted as if the slurry were Newtonian, and a pump with a performance curve that will intersect the system curve at a desired 600 gpm is selected. It can be seen, however, from the actual curve (B) that the intersection will never take place. If applied this way, the pump would not pump.

Another factor that will become important is demonstrated in Fig. 9—the concept of critical carrying velocity. This is the velocity at which the material suspended in the

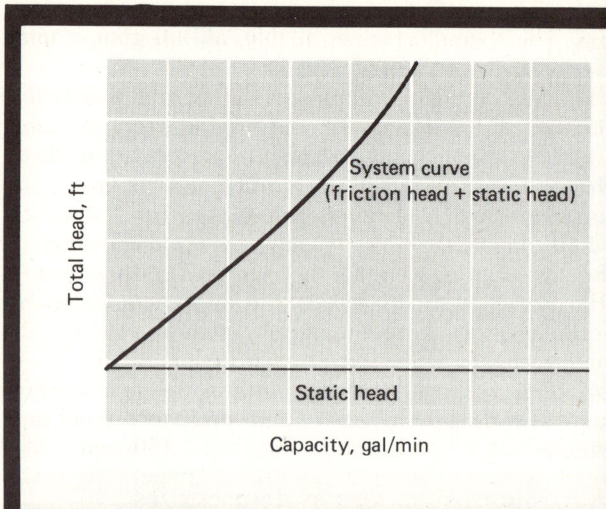

Head/capacity curve for a Newtonian slurry Fig. 8

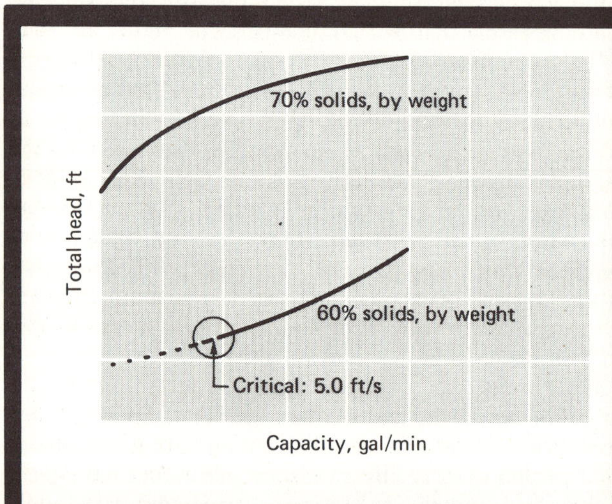

Sand slurry can be either Newtonian or Non-Newtonian depending upon concentration of sand in liquid Fig. 9

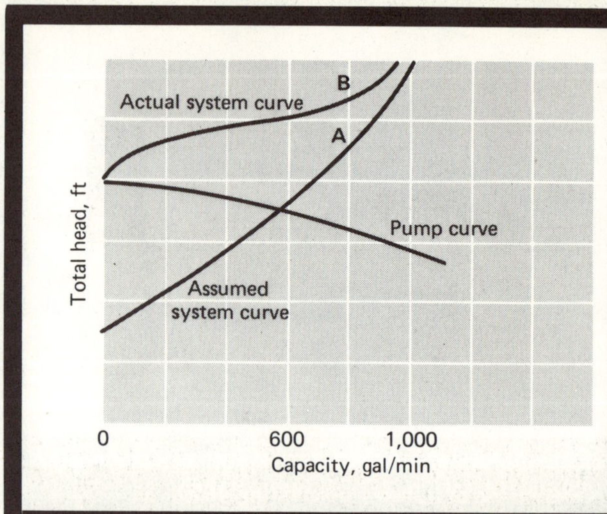

Assuming a Newtonian slurry can lead to problems if the slurry is actually non-Newtonian Fig. 10

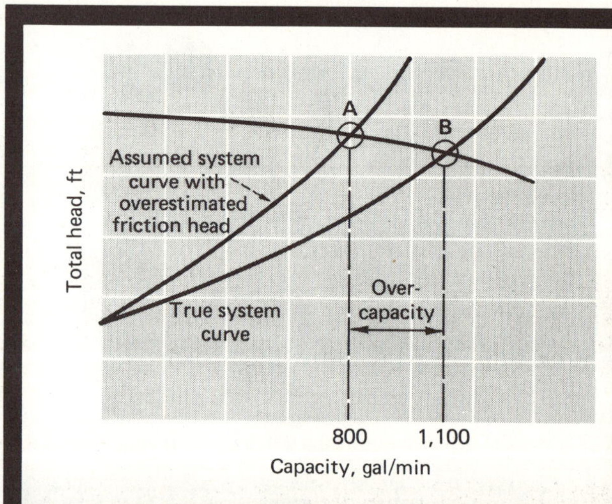

Overestimating pipe friction leads to pump overcapacity Fig. 11

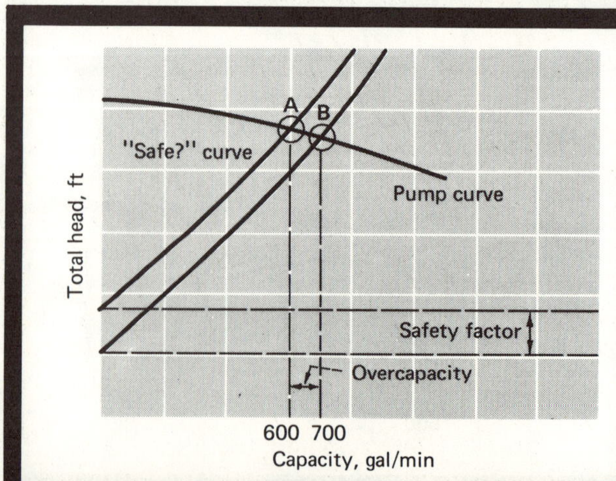

Adding a "safety factor" can lead to choosing an overcapacity pump Fig. 12

liquid will begin to settle out in the piping. The operating point of the pump must be selected so that pipeline velocity exceeds this critical carrying velocity. If this results in too great a flowrate, a smaller pipe size must be used. Critical carrying velocity can only be determined with any degree of accuracy by actual testing, although results from a test with a single pipe diameter may be extrapolated to other pipe sizes.

There are other ways to go astray in the development and use of a system curve. One relates to the overestimation of pipe friction. In Fig. 11, two system curves are plotted. The curve with the steeper slope intersects the pump performance curve at a desired capacity of 800 gpm (Point A). If, however, pipe friction has been overestimated, the true system curve might intersect the pump curve at, for example, 1,100 gpm (Point B). The effect of this error is overcapacity, which results in excess velocity in the system. This mistake will cause increased pump wear and the expenditure of excess horsepower.

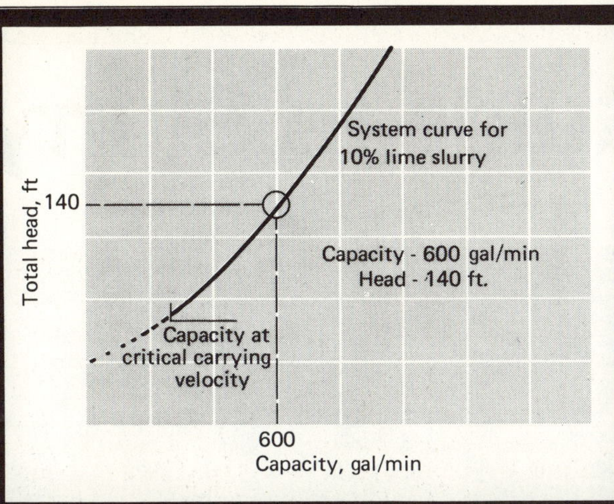

System curve of a slurry containing 10% of lime for which a pump is to be chosen Fig. 13

Depending upon just how far out on the curve the pump runs, cavitation and even mechanical failure can also result.

Possibly an even more common error is made in the name of "safety." In Fig. 12, a correct assessment of the friction has been made, and the slurry has been properly identified as Newtonian. This time, the concern is about the ability of the pump to do the job, so a few feet are added to the static head ". . . just to be sure." As a result the slope is true but the intersection planned for Point A actually takes place at Point B. Overcapacity with all its problems enters the picture again.

The match-up of system and pump

Fig. 13 and 14 demonstrate how a correctly executed system curve can help the engineer to select and properly apply a pump. In this example, the material to be pumped is a 10%-concentration lime slurry. The required capacity is 600 gpm and the critical carrying velocity is determined

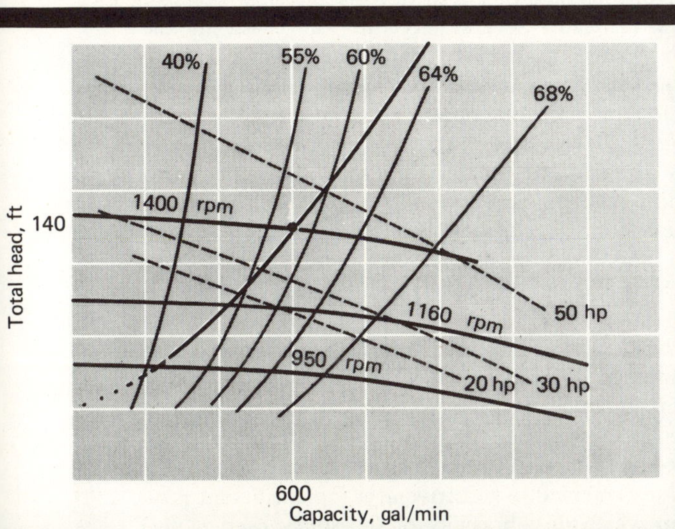

Pump-performance curves matched against system curve Fig. 14

to be well below the fluid velocity at the planned operating point. The system curve shows that, at 600 gpm, a total head of 140 ft is required.

By taking the data compiled in the examination of the static nature of the slurry and of the economic and physical requirements of the plant, and comparing them to various manufacturers' specifications, one pump in particular is generally found to have the required design and also to be available in materials of construction to meet the requirements of the job. When the pump's performance curves are matched to the system curve, an intersection occurs at a pump speed low enough to minimize wear. The curves shown in Fig. 12 are for a 4-in. horizontal slurry pump with a 14-in. impeller. Intersecting the system curve at the prescribed point is a pump curve for 1,450 rpm. In this case, 1,450 rpm is low enough to minimize wear. The selected operating point falls between 30 and 50 hp, and efficiency is just over 55%, which is quite acceptable for a relatively small slurry pump.

Driving the slurry pump

It has been pointed out that low-rpm operation is of the utmost importance to wear reduction. For highly abrasive applications, then, operating points should *not* be met by varying the impeller diameter. Instead, pump speed should be altered and a maximum-size impeller used. In addition to reduced parts wear, money will be saved by the slightly higher pump efficiency gained. Yet another advantage of using only full-diameter impellers is that the pump user does not need to carry as many replacement impellers in his inventory, because of the fact that slurry pump impellers made of Ni-Hard or chromium iron are most readily available from pump manufacturers in full sizes.

The most common way of driving a slurry pump is by using a V-belt drive with an electric motor. However, for applications above 300 hp, gear reducers are more practical. By using V-belts drives or gear reducers, high-speed motors can be used, with resulting cost savings over more-expensive, lower-speed units.

Because of the difficulty in determining friction values—and thus total pump head—for certain slurries for which test data are not readily available, it is sometimes advisable to use V-belt sheaves with adjustable pitch-diameters. Without increasing the initial purchase cost to a great extent, these sheaves simplify balancing the system at startup and allow the pump to meet future changes in operating conditions.

The author

J. Ingemar Dalstad is Sales Manager-International, for Morris Pumps, Inc., Baldwinsville, NY 13027, where he is responsible for the establishment and management of the company's many overseas representatives and licensed manufacturers. He was formerly Chief Application Engineer for the company. He holds a B.S. in mechanical engineering from the Technical Institute of Malmo, Malmo, Sweden, and has done graduate work at the Royal Institute of Technology in Stockholm.

Miller Number: Guide to Slurry Abrasion

TEST MACHINE for determining Miller numbers—Fig. 1

Weight-loss data from running wear-block tests in slurries form the basis for the Miller number, a relative measure of a slurry's abrasivity and attrition. The number is useful for estimating the service life of pumps and valves.

JOHN E. MILLER, Consultant

Even though slurry pumping is a common operation in the chemical process industries, there is a lack of information on slurry abrasivity, which may be the controlling factor in service life of pumps, valves and pipelines. The Miller number, which we have developed over the last six years, is a useful tool for determining the relative abrasivity of slurries as well as the rate at which abrasivity changes with time (attrition).

The Miller number consists of two values. The first represents abrasivity and is obtained from computer analysis of weight loss in a wear-block test. The second value is a measure of attrition—that is, the loss or gain in abrasivity as particles break down during the test period. A typical Miller number might be 74 – 7.

We include a constant in the derivation of the abrasivity value, so that 1,000 is approximately equal to a slurry of 220-mesh corundum, which has reproducible abrasivity. Slurries of most industrial materials have wide variation in their abrasivity, depending on composition and particle size and shape. This can be easily seen from the typical Miller numbers in Table I.

Developed Out of Necessity

Our work on the Miller number originated through necessity during a full-scale loop test of a 560-hp reciprocating pump handling a magnetite slurry. To evaluate the life of expendable fluid-end parts, we had to consider the amount of attrition in a recirculated slurry. We expected that attrition would cause the slurry to become less abrasive during the latter part of the test and thus give erroneous data on parts life.

We investigated several methods for measuring the slurry's change in abrasivity, but none seemed to fully duplicate the type of back-and-forth rubbing action found in a reciprocating pump. Then we recalled a "rubbing block" device used some years ago to determine the quality of hard-facing materials. This was an attempt to suspend a small amount of abrasive particles in a viscous, colloidal mud. Results were erratic because it was difficult to maintain a consistent slurry concentration under the wear block, and the reciprocating speed was too slow to provide good agitation.

We built a crude machine designed to overcome the shortcomings of the earlier device. We solved the problem of maintaining a consistent slurry by having the wear block run at a higher speed in a V-shaped tray that confined the slurry to the path of the block.

This test enabled us to arrive at a replenishment program in which the abrasivity of the slurry used in the pump study was held constant by adding "fresh" magnetite on a regular basis. Results of several years operation

Originally published July 22, 1974

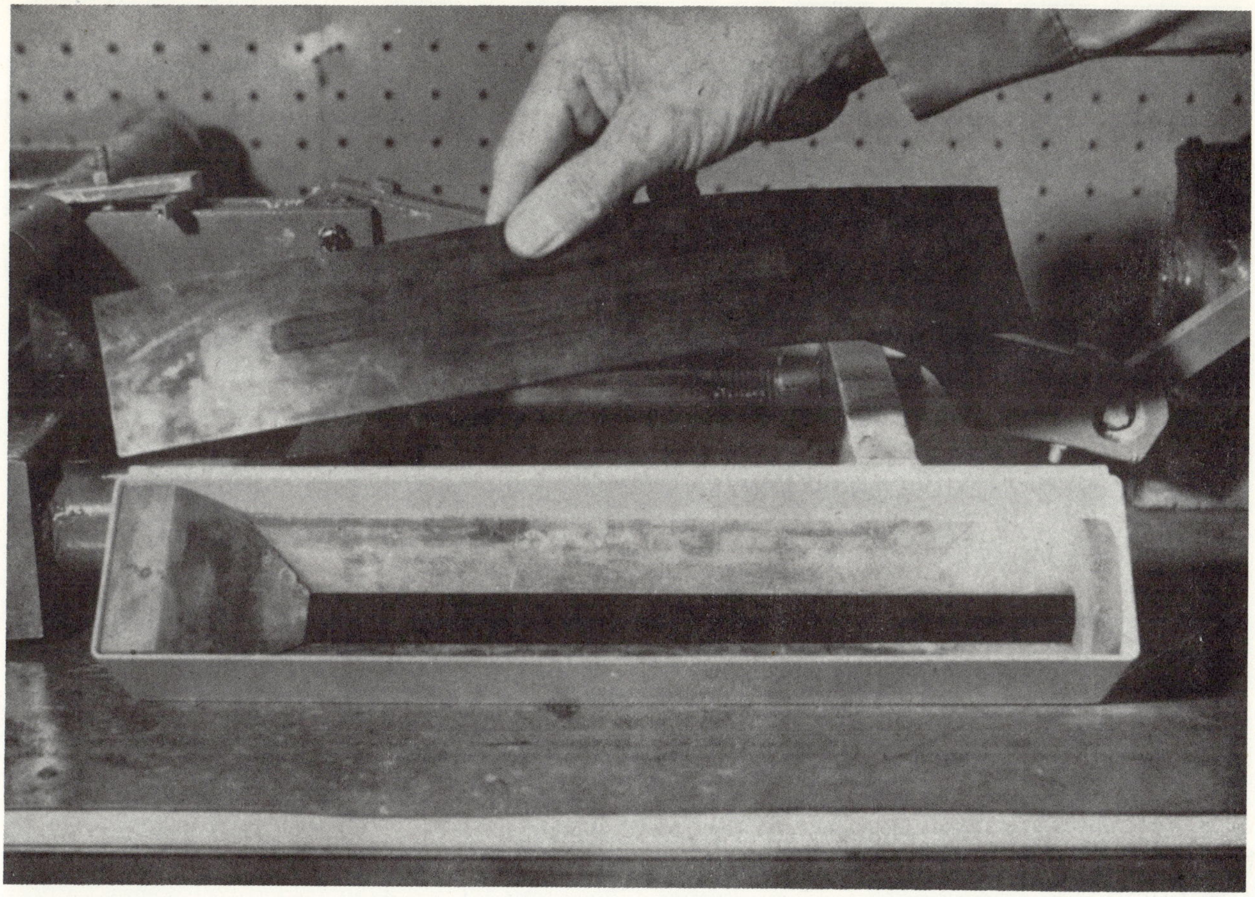

TRAY ASSEMBLY in test machine is designed to confine slurry particles in the path of the wear block—Fig. 2

of the pump confirm that our test procedure accurately predicted the wear rate of fluid-end parts.

Our success in measuring slurry abrasivity in the pump evaluation suggested that the procedure could have much wider application in handling slurries. For example, the relative abrasivity of a slurry dictates whether a piston or plunger type pump is required and whether flushing the packing is necessary. Valve selection is also influenced by the abrasivity and particle size of the solids in the slurry.

More-Elaborate Machine

To extend the usefulness of the method, we constructed a more-elaborate testing machine (Fig. 1). This consists of a standard metal wear block (1.27 x 2.54 cm) that is driven at 48 strokes/min over a distance of 20 cm. As shown in Fig. 2, the V-shaped tray that holds the wear block confines the slurry particles to the block's path.

For each test, the bottom of the tray is fitted with a new piece of neoprene rubber. Then a 50% by weight slurry of the solids to be tested is placed in the tray. (Some tests are run with actual slurries as supplied.)

A 2.3-kg weight is added to the wear block, and the machine is set in motion. At the end of each stroke, a cam lifts the block 0.32 cm to allow a flow of fresh slurry under the block. Both the block holder and the tray are made of plastic to prevent electrolysis.

Our procedure calls for weighing the wear block to 0.1 mg, after it has been scrubbed in detergent and dried for 15 min under a heat lamp. After 4 h, the block is removed from the machine, washed and weighed again. This is repeated four times, while duplicate samples are run simultaneously to verify the results.

Originally, we selected a wear block made of Colmonoy No. 6 alloy because the plunger in the test pump was constructed of this material. When using replacement blocks, we found that the method of applying the surface coating yielded variable results. The popularity of 27% chrome iron in pump liners for drilling-mud service led to picking this reproducible and readily available material as a standard. Many of the early tests had been rerun with 27% chrome iron blocks in an attempt to correlate the wear resistance of the two metals. We had enough data on hand to retain the same Miller number for all of the older tests.

Analyzing Results

After running a series of tests on available slurries, and ones made from pure minerals, we noted, as expected, that the abrasivity was not the same for each incremental run. This meant there had been a change in the characteristics of the particles, which apparently was a function of the materials' friability.

We reasoned that the rate of metal loss early in the

```
                    ABRASIVITY TEST RESULTS
               FØR DETERMINATIØN ØF MILLER NUMBER


TEST NUMBER - M126
TEST DATE   -   6/14/72
ABRASIVE MATERIAL-                    IRØN ØRE 50 PCT BY
WEAR BLØCK MTL.  - 27 PCT CHRØME IRØN        WT SLURRY

                           WEAR    BLØCK             A= 6.8577E+00

                            1        2               B= 9.7193E-01

                         ---------  ---------     * WL=(A)(HØURS)**B
IDENTIFICATIØN NUMBER       3          4
                                                 AVE MILLIGRAMS LØSS
WEIGHT IN GRAMS                                  -------------------
                                                  ACTUAL  *PREDICTED
   INITIAL                15.3535   12.9685      --------  ---------

   AFTER  4 HØURS         15.3258   12.9472        24.5       26.4

   AFTER  8 HØURS         15.2962   12.9192        53.3       51.8

   AFTER 12 HØURS         15.2716   12.8957        77.4       76.7

   AFTER 16 HØURS         15.2477   12.8728       100.8      101.5


                        ABRASIVITY  ATTRITIØN    LAP MATERIAL WEAR

MILLER NUMBER             121.7       -1.4           TRACE


                      100-----------------------------------------1
                         -         -         -         -         **
  PERCENT              90-         -         -         -       ** 2
                         -         -         -         -     **   -
    ØF                 80-         -         -         -   **     -
                         -----------------------------1**----------
  105.8                70-         -         -       **2          -
                         -         -         -     **  -          -
MILLIGRAMS             60-         -         -   **    -          -
                         -         -         1 **      -          -
   LØSS                50-------------------**--------------------
                         -         -       **2         -          -
                       40-         -     **  -         -          -
1-WEAR BLØCK 1           -         -   **    -         -          -
                       30-         - **      -         -          -
2-WEAR BLØCK 2           ---------*1*-------------------------------
                       20-      ** 2         -         -          -
3-WEAR BLØCK 1&2         -    **   -         -         -          -
                       10-  **     -         -         -          -
*-CURVE FIT LØSS         -**       -         -         -          -
                       03-----------------------------------------
                                   4         8        12         16
                                            HØURS
```

TYPICAL printout of abrasivity-test results shows a curve of the fitted test data and gives calculated Miller number—Fig. 3

Typical Miller Numbers for Slurries – Table I

Alundum 400 mesh	241+21
Alundum 200 mesh	1058–15
Aragonite	7+8
Bauxite 16 mesh	9–35, 22+6, 33–23
Calcium carbonate	14–2
Carborundum 220 mesh	1284–15
Clay	34–1, 36–9
Clay residue	66–17, 184–18, 226–17
Coal	6–26, 9+11, 12–17, 21–7, 28–17, 47–0, 57–3
Copper conc.	19–8, 37–11, 58–23, 68–3, 111–13, 128–5
Detergent	0–0, 5–22
Gilsonite	10+6
Iron ore	28–7, 37–13, 64+1, 79–4, 122–1, 157–11
Lignite	14–6
Limestone	22–2, 27+5, 29–6, 30+11, 33–5, 39–0, 43+6, 46–1
Limonite	113–3
Microsphorite	76–13
Magnetite	64–1, 67–4, 71–3, 134–1
Mud, drilling	10–14, 10–4
Nickel	31–7
Phosphate	74–7, 84+1, 134–12
Potash	0–0, 10+1, 11+2
Pyrite	194–4
Rutile	10–11
Sand & sand fill	51–10, 59–6, 68,9 85,16, 93–2, 246–9
Sea bottom	11–0
Shale	53–0, 59–3
Sodium sulfate	4–38
Sewage (digested)	15–7
Sewage (raw)	25–8
Serpentine	134+5
Sulphur	1–38, 1–12
Tailings	24–8, 58–8, 76–10, 91–5, 159–11, 217–15, 480–21, 644–14

test would be more indicative of what occurs when a slurry passes through a pump or valve for the first time. We decided to calculate abrasivity at a point two hours into the test to override the initial effect of wear on the block's finish and obtain a wear rate before attrition started. First, the weight-loss data are fitted to the equation:

$$W = At^B \tag{1}$$

where W is weight loss in g, and t is time in h. Using the least-squares linear-taylor differential-correction technique, the values of A and B are determined for a particular series of test data. Abrasivity is then defined as the partial derivative of Eq. (1) at t equal to 2 h. As mentioned earlier, we add a constant so that the abrasivity of 220-mesh corundum is about 1,000. Thus we have:

$$\text{Abrasivity} = C(dW/dt) = (C)(A)(B)(t)^{B-1}$$

Finally, attrition is calculated by taking the second derivative of Eq. (1) at t equal to 2 h.

$$\text{Attrition} = \frac{d^2W/dt^2}{dW/dt} \times 100$$

where $d^2W/dt^2 = (A)(B)(B-1)(t)^{B-2}$

We have prepared a computer program so that we can easily obtain a complete report (Fig. 3) showing the calculated Miller number and a curve of the fitted test data. The closeness of the actual weight loss to that predicted by Eq. (1) demonstrates that the wear-block weight loss is a true mathematical function.

Applications

Experience shows that slurries with a Miller number below 50 are not abrasive in the sense they can be pumped with double-acting piston-type units. Materials such as coal, limestone and sulfur are usually in this range.

When slurry abrasivity is above 50, one should consider pumps other than piston types. The first choice would be a plunger pump in which the packing and plunger can be protected by a barrier of clean fluid that is injected into the stuffing box. Also, remember that in any slurry pump the valves cannot be isolated from the fluid, and they degrade at a rate proportional to the Miller number.

We should emphasize that the number should only be used to compare the life of parts in equipment of similar design. For example, a duplex piston pump handling coal with an abrasivity of 40 would have half the parts life of the same pump moving coal of number 20. However, this experience could not be used to estimate the life of a plunger pump that can be flushed continuously.

In general, the finer the material, the less abrasive it is. Corundum at 220 mesh is about four times more abrasive than 400 mesh. Although a considerable reduction in particle size is necessary to get a significant difference in abrasion, the savings in expendable pump parts could sometimes warrant the expense of milling to a finer particle size.

At present, the usefulness of the attrition factor in the Miller number is limited. As more data are accumulated, it may become important in predicting rheological changes such as those that occur in a long pipeline.

Meet the Author

John E. Miller is a consulting engineer at 9850 Mercer Drive, Dallas, TX 75228. Formerly he was a development engineer with the Oilwell Div. of U.S. Steel, where he had worked since 1933 except for a five-year leave during which he was self-employed. Mr. Miller has a degree in petroleum and natural-gas engineering from Pennsylvania State University and is a registered professional engineer in Texas.

Instrumentation for slurry systems

Problems of fouling and solids buildup call for specialized instrumentation for making slurry measurements.

Nicholas J. Paulis *and* ***David Silvermetz,*** *National Sonics/Envirotech*

☐ As slurry handling technology has improved, the design of efficient and energy-saving systems has increased the need for automated and continuous monitoring and control of the level and density of process slurries.

Recent introduction of ultrasonic instrumentation has created a third alternative (along with photoelectric and nuclear systems) for the engineering of these controls—one which has distinct advantages in certain applications.

In the past, labor-intensive, almost haphazard, techniques were sufficient for controlling industrial slurries. Manual sampling, batch control, and even crude eyeballing were adequate to solve the problems of maintaining process flow conditions. When scale on critical parts built up to intolerable limits, the operation was simply shut down for maintenance.

But today's higher labor costs, new sophistication, better instrumentation, and closer tolerances all along the production chain have created a new role for the operator. Judgmental decisions have yielded to process controls, and measurement parameters such as density or percent suspended solids (%SS), liquid levels and blanket levels have come within the province of instrumentation.

Before we take a closer look at slurry-control techniques, let us define some measurements:

- *Slurry density* is the %SS in a slurry solution (Fig. 2a), and is similar to, but not the same as, specific gravity, since it measures neither dissolved solids nor the density of clear (or carrier) liquids. Thus, a true specific-gravity measurement usually is somewhat greater than %SS, depending on the slurry mixture.
- *Liquid level* (Fig. 2b) is the most basic measurement of industrial slurry and waste handling, and refers simply to the height of a liquid in a vessel.
- *Blanket level* (Fig. 2c) is the buildup of *settled* suspended solids in a "blanket," as measured from the bottom of a tank or other reference point. The rising "edge" of this blanket is predetermined by the sharp gradient in %SS in the solution.

Review of measurement techniques

As slurry handling became more critical, rudimentary sampling techniques simply were not adequate to provide the kind of ongoing control required.

It was easy enough to get an indication of a liquid level, but gaging blanket levels required tedious manual sampling at various points.

Early attempts at more-accurate monitoring usually employed a bypass technique that diverted part of the flow to a smaller pipe, where an inferential measurement based on weight gave an approximation of density levels. The

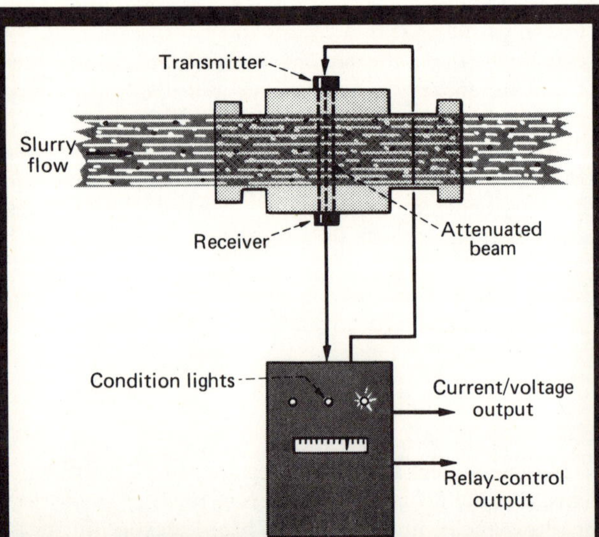

Typical "beam attenuation" control system Fig. 1

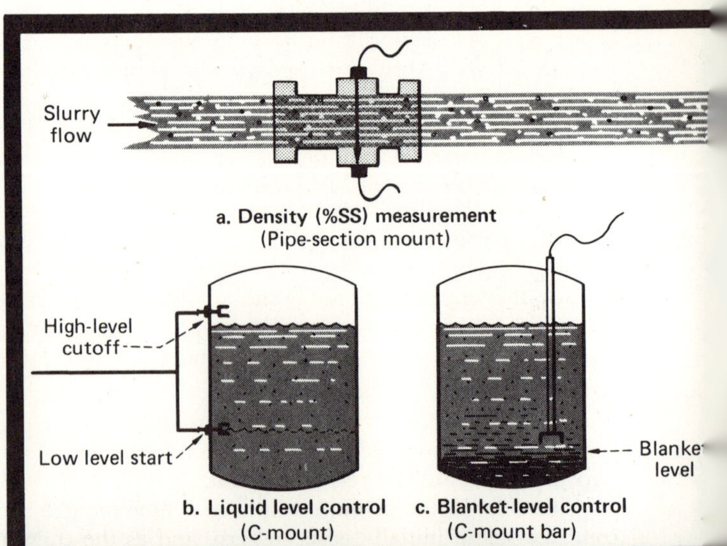

Slurry density and level measurements Fig. 2

Originally published April 25, 1977

Slurry-flow measurement

Several types of flowmeters are available for the measurement of slurries. Long-term reliability is difficult to achieve because of the nature of the material to be processed.

Mechanical flowmeters—Such flowmeters, which include propeller and turbine types, are not obstructionless. In some applications, debris in the slurry will wrap around rotating blades, solids will cause premature bearing wear, or chemical constituents will attack materials of construction.

Differential-pressure flowmeters—Venturi tubes and orifice plates are commonly used to produce differential pressure. Flow is proportional to the square root of the pressure differential measured by a manometer or transducer connected to taps in the entrance and throat of the venturi, or on the upstream and downstream sides of the orifice plate. Unless diaphragms or continuous water purges are used on pressure taps, clogging will occur in many applications. Accuracy is difficult to maintain with isolation or purging arrangements.

Electromagnetic flowmeters—Flow elements operating on the principle of electromagnetic induction are truly obstructionless. Although somewhat expensive and requiring care in installation, these flowmeters function well unless conductive or nonconductive pipewall deposits build up. Coatings affect the pickup of voltage induced by the conductive stream moving through a magnetic field. Ordinarily, the signal voltage is directly proportional to liquid velocity. Pipewall coatings upset the signal balance, producing a zero shift, and calibration span shift as well. Electromagnetic flowmeters can be fitted with ultrasonic cleaners and heated tubes to alleviate coating problems, but this means more equipment to maintain, and the required electrical power increases operating costs.

Vortex-shedding flowmeters—These meters have a uniquely shaped obstruction in the pipe, to produce secondary flow swirls or eddies having a precession frequency proportional to the liquid velocity. This vortex frequency is sensed to produce a pulse output rate that is linearly related to velocity. In slurries, solids buildup, abrasion, corrosion, or coating on the bluff body used to produce eddies, will alter calibration or cause a loss of signal.

Ultrasonic flowmeters—Flowmeters operating on several ultrasonic principles have been developed. The concept of measuring flow with an ultrasonic beam is attractive because it affords the opportunity to sense remotely with no flow obstructions and without electrical contact with the process stream. Some ultrasonic flowmeters have been designed so that they are sensitive to process liquid variations and must be dismissed from serious consideration. In other instances, a basic limitation has been a difficulty in measuring liquids containing high concentrations of solids and air. Air in particular is opaque to ultrahigh-frequency vibrations, causing a loss of signal and consequent velocity-measurement

smaller-diameter pipe tended to clog, however, and such systems often distorted true conditions. Also, they still required an operator "within the loop."

Then, a decade or so ago, instrumentation was introduced in the form of beam attenuation devices. A constant beam of particles or rays—visible light, infrared or gamma rays—is projected between a transmitter and receiver. When the beam is partially interrupted, the reduction in the received signal can be calibrated to indicate the amount of interference and, by extension, densities and levels of liquids and suspended solids.

Using ultrasonics

Ultrasonic instrumentation, a more recent addition to the range of slurry-density controls, operates on the same beam attenuation principle. Here, a pair of piezoelectric transducers are flush-mounted face-to-face in a pipe section or on a C-Bar extension. A highly directive ultrasonic signal is transmitted between them, and a control unit provides electronic excitation to the transducer pair and processes the received signal.

With clear liquid as the medium in a pipe or tank, the control readout initially is set to zero, and as the concentration of suspended solids between the transducers increases, attenuation of the beam causes a measurable loss of signal. This can then be translated into a %SS readout and/or set to start or stop various process steps in slurry handling.

As with most beam attenuation devices, ultrasonic operation is fully automatic and does not require operator attention. Most models include condition lights (indicating a light or heavy concentration of slurry) as well as a graduated-meter readout, and current and/or voltage output for more-precise readings and optional automatic control.

With fully automatic slurry-measurement control, efficiency—and, hence, plant capacity—is improved, and operating costs are significantly reduced. The ultrasonic alternative was proven first in wastewater-sludge density and blanket control-systems, and after years of successful testing and operation was applied to the handling of industrial slurries. Its advantages over other systems have made ultrasonic hardware the choice in many industrial settings.

Advantages of other systems

No one system is the panacea for slurry control problems, and the actual specification of nuclear, photoelectric or ultrasonic controls should be based on specific parameters that need to be defined.

In certain areas, one of these types may be distinctly superior; for example, in monitoring extremely heavy,

errors. In addition, suspended solids attenuate the signal until a point is reached where measurement is impossible.

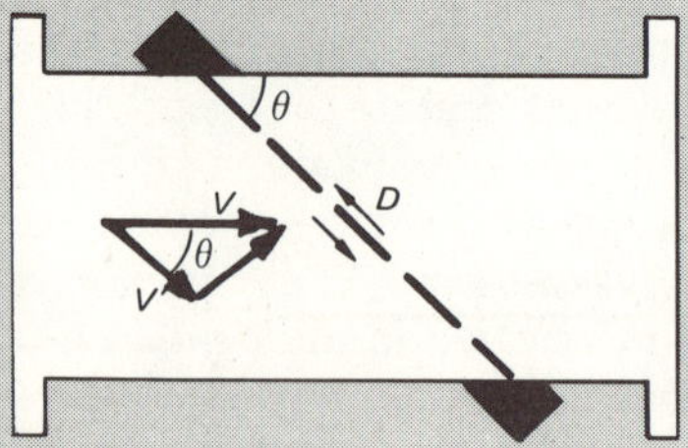

Obstructionless flowmeter—The figure depicts a new obstructionless flowmeter. Hydraulically, the flowmeter is equivalent to a pipe spool-piece.

The flowmeter alternately measures the time interval required to send a very high frequency vibrational energy pulse in the same direction as the flowing liquid, and then measures the time interval in the reverse direction (against the flowing liquid). The figure diagrams the pulse path, which is at an angle to the flow and is the same for each direction.

The transit time upstream and downstream can be expressed by:

$$T_1 = D/(C + V') \text{ and } T_2 = D/(C - V') \quad (1)$$

where: T_1 = upstream pulse transit time; T_2 = downstream pulse transit time; C = velocity of vibrational pulse propagation in the liquid; D = path length; V' = component of liquid velocity along the pulse path (see vector diagram in figure).

If the path length, D, and the velocity of the vibrational pulse through the liquid, C, are assumed to be constant, the velocity component, V', could be measured by subtracting the pulse transit times. However, the velocity of the vibrational pulse is not a constant, but varies on the order of 0.2%/°C over the usual water-temperature range. A flowmeter that does not compensate for this factor will have an error of about 2% for a 10°C temperature change of the liquid. To overcome this difficulty, the transit times are converted to frequencies, cancelling out C when the frequencies are subtracted:

$$F_1 = 1/T_1 = (C + V')/D \quad (2)$$

$$F_2 = 1/T_2 = (C - V')/D \quad (3)$$

$$F_1 - F_2 = (C + V')/D - (C - V')/D = 2V'/D \quad (4)$$

As shown in the figure:

$$V' = V \cos\theta$$
$$V = V'/\cos\theta \quad (5)$$

where V = average liquid velocity; θ = beam angle.

Substituting for V', the difference frequency is then linearly related to the average liquid velocity:

$$F_1 - F_2 = 2V/(D \cos\theta) \quad (6)$$

The flowmeter converts the transit times in each direction to a frequency, and subtracts these frequencies to produce a low pulse-rate output proportional to the liquid velocity.

Adapted and condensed from a paper by David R. Stuettig, Vice President and Technical Director, Sparling Div. of Envirotech Corp., presented at the WWEMA (Water and Wastewater Equipment Manufacturers Assn.) 4th Annual Industrial Pollution Control Conference, Houston, Tex., Mar. 30, 1976.

opaque, abrasive solids, such as cement, which is a strong point of nuclear systems—or the very low %SS or highly aerated liquids such as those found in water reclamation systems that are a strong point of photoelectric types.

Comparing ultrasonic with nuclear systems in slurry control, for example, we see that ultrasonic is better with thin slurries having a smaller %SS, while nuclear excels for heavier concentrations. Although nuclear systems are better established in industrial applications, they are about 20—50% more costly, overall, than ultrasonic ones. In addition, a whole mystique surrounds the use of cesium, cobalt or radium (the most common nuclear beam sources), which partly explains why nuclear systems do not dominate the field. Many engineers still are concerned about handling radioactive substances (and, indeed, a Nuclear Regulatory Commission license or permit is required, as well as are certified repairmen), and this same concern extends to workers and management. Also, semiannual leak-inspection tests are mandatory.

Nuclear systems simply are not applicable to blanket-level control systems because of requirements for side-mounting and shielding and because they are too unwieldy and costly for measurements of this kind.

Finally, there is the problem of radiation-source half-life, which implies that nuclear systems are subject to source decay and calibration drift.

Some applications of ultrasonic instruments

The broad capacity and flexibility of ultrasonic controls has led to their successful use in a wide range of applications. Here is one example of a recent installation.

In a sulfur dioxide (SO_2) scrubbing system for stack gas (Fig. 3), the existing installation was modified with two ultrasonic controls—a C-type sensor to monitor density in a lime-slurry holding tank and maintain a steady delivery of %SS to the scrubber to prevent scaling, and a pipe-mounted unit to control the density of recycled slurry back to the holding tank and divert part of the flow to waste storage for lime recovery. This improved system uses lime with better efficiency than was previously possible and has significantly reduced maintenance costs.

Photoelectric types for aeration

Photoelectric systems, on the other hand, are at their best in highly aerated liquid media, where ultrasonic and nuclear sensors can become "confused" by the profusion of air bubbles.

However, photoelectric systems are suitable only for very thin slurries, since their penetration of individual solid particles is almost nil, and they reach saturation very quickly. Finally, photoelectrics (whether infrared or visible light) are far more vulnerable to external sources of

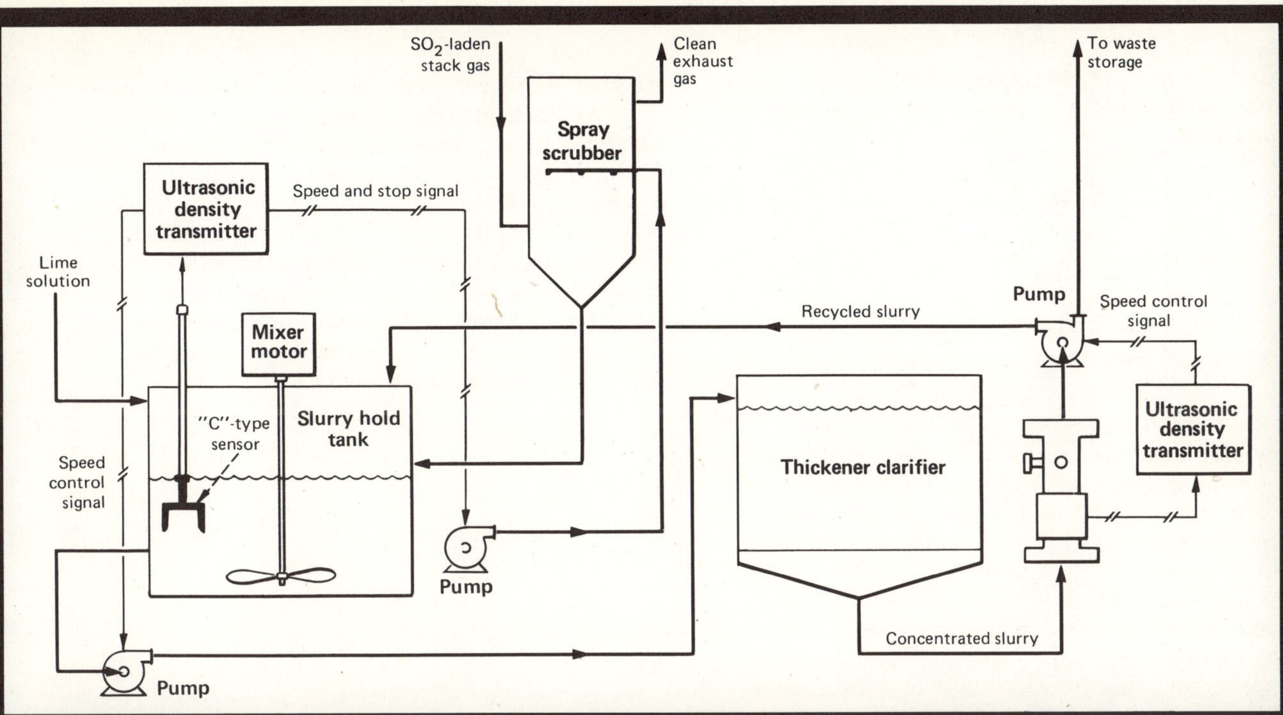

Slurry density control for SO_2 scrubbing system — Fig. 3

interference—scale, coating buildup, stray light, and colored slurry media can all affect their performance.

An abundance of answers

In all, today's specifying engineer has a number of sources to draw on in controlling industrial slurries without resorting to the crude and costly methods of a few years ago. Each of the three systems—nuclear, photo-lectric and ultrasonic—excels in certain applications, and the system designer should have all three, with their relative strengths and weaknesses, in mind when setting up job specifications.

He should be particularly aware of slurry handling requirements with regard to:

- Type of measurement (level, density).
- Degree of accuracy required.
- Type of solids.
- Range of %SS to be measured.
- Operating conditions (aeration, temperature, ease of access to controls for recalibration, etc.).

Only if he keeps these parameters firmly in mind can the engineer establish the best and most cost effective system of instrumentation for handling a particular process problem.

The authors

Nicholas J. Poulis is vice-president and director of marketing for National Sonics/Envirotech, 250 Marcus Blvd., Hauppauge, NY 11787. Formerly, he was responsible for communications-equipment sales for Republic Aviation Corp. He holds a bachelor's degree in electrical engineering from the College of the City of New York, and has attended the Baruch School of Business for additional study in industrial management. He is a member of the Institute of Electrical and Electronics Engineers.

David Silvermetz is director of engineering for National Sonics/Envirotech. He received his bachelor's degree in electrical engineering from the College of the City of New York and his masters (also in electrical engineering) from Polytechnic Institute of Brooklyn. He is a member of the Instrument Soc. of America, the Institute of Electrical and Electronics Engineers, Tau Beta Pi and Eta Kappa Nu. He is a licensed professional engineer in the state of New York.

INDEX

Field Survey Book from the Arthur G. Tracey Collection
(Historical Society of Carroll County)

James Kelly's Book of Land Courses
Aug. 15th 1851

"Parishes Lot" 4
"Buck Harbour Res." 7
"Pettycoats Lot" 9
"Peters Neglect" 10
"Creigh's Addition Enlarged" 12
"Carr's Chance" 16
"Murray's Expedition" 16
"Wilmot's Retirement" 18
"Stoddard's Delight" 19
"Joseph's Beginning" 22
"Christophers Fair Play" 2[illegible]
"Stansbury's Grove Res." 27
"Poullen" 40
"Keeths Correction 41
"Christophers Luck 48
"Deer Park" 59
"Spears Deer park" 61
Bushy Spot 62
"Chance" 63
"Merrymans Chance" 63
"Workers Prospect" 64
Good Luck 65
Georges Venture 6
"Prospect" 66

Table of Contents

Preface

The present work was initially undertaken because of interest in the land holdings of the Hoffman family of papermakers, founded by Frederick Wilhelm Hoffman, a native born German, who settled near Hoffmanville, Baltimore County, Maryland. In the year 1775 Wilhelm, or William, Hoffman built the Clipper Paper Mill on the western prong of the Great Gunpowder Falls at Hoffmanville. The Clipper Mill was the first operating paper mill in the State of Maryland. May Albright Seitz, my Grandmother, researched the early history of the Hoffman mills (See May A. Seitz; History of the Hoffman Paper Mills, Baltimore, Md., 1941) making frequent reference to many of the early land holdings of William Hoffman and of his children. May Seitz left many of these early land patents and deeds, many on paper bearing the watermarks of the early Hoffman mills, in a collection at the Maryland Historical Society.

When an attempt was made to work up a map of these early Hoffman land holdings it was soon apparent that the map was in fact a big puzzle with many bits and pieces that were lacking. In the process of researching the missing parts of the map I had to enlarge the area of interest and discovered records of other families which were also of interest. After some time and effort I delineated the area covered by two maps: Map One includes the northwest corner of present day Baltimore County (from Middletown to Hoffmanville) and Map Two includes the northeast corner of Carroll County (from Roller to Maple Grove just east of Manchester).

This has been an enjoyable project with an element of pursuit and discovery which often led to new questions and interesting new historical connections. I hope that other people who are interested in the early history and the families of northern Baltimore and Carroll Counties will find it of use.

I am indebted to the librarians at the Maryland Archives in Annapolis, the Baltimore County Historical Society and the Historical Society of Carroll County. They were always helpful and patient. I would also like to thank Mr. George J. Horvath, Jr. for the encouragement and help which he provided in getting me started and Mr. Jay A. Graybeal of the Historical Society of Carroll County for his assistance in locating materials in their collections.

Introduction

The areas covered by these two maps were originally both part of Baltimore County. After 1836 the area to the west (Map No. 2) became Carroll County, but this does not have any immediate bearing on this project since almost all the patents, warrants and surveys were prior to 1836. In the usual order a warrant was first issued by Lord Baltimore or the State of Maryland for a given parcel of land as described in the warrant. A common warrant was issued for the survey of a certain amount of land in acres that was never previously cultivated or developed. A special warrant was issued for the resurvey of tract of land that had already been surveyed and or patented. (See Skordas et. al.) In order to execute the warrants the land had to be surveyed by an official county surveyor to determine the exact boundaries of the tract of land. The survey was then registered with the State which then issued a certificate of survey to the individual. Finally, the individual applied for a patent which the State could grant only after a holding period of six months from the date of the certificate of survey (Earlier patents, pre 1770, were often issued under the same date as the survey). Upon "granting" by the State the land became the patentees' property subject to the payment of 'quit rents'. He could then deed the land as a sale to another individual or retain ownership of the land as long as he paid 'caution money' to the State. Occassionally a survey will state that caution money was not paid so a patent lapsed and the land was then 'vacated' and a new party then applied for a warrant to resurvey and the process started all over. Some of the tracts of land in the map area underwent as many as four resurveys and patents. This process could be repeated multiple times so one could easily be confused as to where all these parcels were located until one realized they were in fact a succession of surveys on an 'original' tract (i.e., Whispaw and Coutz Lott). Each time the tract was resurveyed the boundaries of the tract would usually change. Sometimes later surveys would include all or part of numerous earlier 'elder surveys' in which case one has to pay particular attention to the details of the survey. Land not included in an elder survey will be called a 'vacancy' area and would have its own set of courses. Only one of the series of successive surveys could be selected for use in the final map but often earlier or later surveys which are included within the boundaries of the selected survey will be depicted by dotted lines.

Most of the relevant land records can be found in the Maryland State Archives in Annapolis, Maryland since they were made under the authority of the State government. After the land was granted a patent by Lord Baltimore or (after 1776) the State the owner could then execute a deed of sale under the authority of the local county Government. This present work deals exclusively with the Provincial or State records. These early records are available to be studied from the original individual patents or from microfilm copies of the patent series records kept by the State in large folio sized books which are now in the State Archives. The microfilm records are also available through the genealogical libraries of the Church of Later Day Saints.

In addition, field notebooks kept by the surveyors working from the original surveys can be found in the State Archives as well as in the Arthur G. Tracey Collection in the Historical Society of Carroll County, the later of which proved of especial value in the present work.

Map Number One covers the area from Middletown west to Hoffmanville in Baltimore County and from the Maryland/ Pennsylvania Line to the north down to Beckleysville to the south. Map Number Two covers the area from Roller in Carroll County west to Maple Grove near Manchester and from the Maryland/Pennsylvania Line to the north to somewhat south of Maple Grove.

The measure of distance commonly found in all the early surveys was the perch, equal to 16.5 feet. Two scales most often used in the early surveys were 50 perches per inch and 100 perches per inch. The author plotted all the individual surveys and mapped them using a scale of 50 perches per inch. These 'master' maps were then reduced by 50% to a scale of 100 perches per inch for this publication. Similar maps of early land patents in the State of Pennsylvania just to the north of the mapped areas have ben drafted by Rev. Neall Hively (See References) using a scale of 50 perches per inch. Mr. George Horvath has mapped the area to the east of Map Number One using a scale of one inch equal to 1200 feet.

This present work is intended to aid in locating early land patents and when used inconjuction with current tax maps should, when diffences of scale are taken into account, allow one to locate the original patents in reference to present day landmarks to within 5 or 10 perches.

To make the maps the most recent tax maps of the area were reduced to a scale of 50 perches per inch. Tracings at a scale of 50 perches per inch of the individual surveys were then placed over the tax maps. Many of the lines from the original surveys can still be found in current property lines. Sometimes the outlines of whole tracts are clearly visible (i.e., California and Davis's Farm), but more often only parts of the original tracts still exist. Most original surveys site one or more points or courses which the survey shares in common with adjoining tracts; i.e., (from the survey for Nace's Tavern) "Beginning at the end of 3 perches on the third line of a tract called 'Murphy's Retirement' etc." Using such information about common boundaries one can proceed to fit together many surveys. Then using the information in the surveys which define the locations of boundaries (or courses) relative to land marks such as

Field Survey Book from the Arthur G. Tracey Collection
(Historical Society of Carroll County)

22

thence bounding on the given line of said land — S 88 W 39 p thence N — W 4¾ p untill it intersect the first line of the afsd tract of land called the Three Brothers and thence to the beg. for 68½ acres —

"Calafornia" granted Richd Richards the 5th of March 1765 Beg. N 44 E 40 p. distant from a bd white oak and a bounded red oak standing on a hill on the South side of a small branch descending into Georges Run a draught of the great falls of Gunpowder said trees being the beginning trees of the orig survey and running thence —

N 11 W 60 p
N 73 E 50 p
N 29 W 48 p
N 8 W 50 p

23

N 47 E 198 p
N 14 W 62 p
N 85 W 132 p
8 S 13 W 71 p
9 South 76 p
10 S 9 E 60 p
11 — N 65 W 76 p
12 — N 77 W 21 p
13 — S 49 W 122 p
14 — S 23 W 87 p
15 — N 87 E 11 p
16 — S 33 E 90 p
17 — S 46 E 26 p
18 — S 36 E 110 p
19 — S 20 W 86 p
20 — S 73 E 23 p
21 — N 22 E 54 p
22 — N 68 E 20 p
23 — N 33 E 70 p
24 — N 22 E 50 p and thence to the beg. for 490 acres —

The orig. called Brittania June 10th 1762
by scale N 12½ W 62 p

Being a resurvey on Brittania which is now found to contain clear of elder surveys — 41½
Vacancy added — 448¾
490

Ex'd with pat.

Richd Richards to Stophel Mire — Deed dated May 19th 1770

roads, streams, stone boundary markers or old mill sites one can tie points in the individual surveys, and in some cases whole tracts, to landmarks and survey boundaries on the present day tax map. By adjusting their positions and shapes the surveys can be combined together into a "best fit" pattern or a map.

The early surveys were only accurate to within one or two degrees bearing and a percent or so in distance measures. This they often stated within the text of the survey. In addition the early surveys often varied between themselves so adjoining tracts might share a common boundary but have different bearings and course lengths for the common boundary. Early surveys and later surveys of the same tract of land might also show variations in the courses. For anyone who is interested in the exact shapes of the original survey the author expects to make available at a later time a computer printout of all the detailed survey data collected for the individual tracts. This printout will be placed in the Baltimore and Carroll County Historical Societies.

The author has used the positions of the present day streams and rivers as shown on the present day tax maps except for the course of the Gunpowder River in the Prettyboy Reservoir area which was located using the information in the original surveys. The road locations found on these two maps are taken directly from the modern tax maps. Since road positions have changed over the past 200 years they will not be exactly the same as all the road locations found in the original surveys. However, with few exceptions the survey reference points along the old roads are in close agreement with the present day locations.

One of the difficulties in constructing a map of early land records is that of chronology. Ideally one would like the map to be composed of surveys all completed at the same time so the map would be a record frozen in one point in time. In actual fact, however, many of the land tracts underwent a succession of surveys and patents and resurveys that were continuously changing the shapes of the tracts over a period of 50 years or more. Thus some of the patents selected for the map may date to 1750 while others could be 1820 or later. The maps are therefore intended as reference maps only and are not intended to depict how large areas appeared at any one time.

It was also a common practice in the early and mid 1700s to lease land. In this case a survey was often made and the date of the actual lease agreement and the name of the lessee are often available. Many of the early leases, prior to 1775, can be found in two folio volumes in the Maryland State Archives that were transcribed by George G. Brewer about 1830 and are designated as "Propietary Leaseholds": Volumes GGB A and GGB B. In most cases these leased tracts will be found to be identical to a tract of the same name, land area and survey date which can be found as an unpatented certificate (The surveys of the unpatented certificate Series can only be found as individual surveys in folders in the Archives; they were never recorded in the book records). In numerous instances a later survey will site only the name of a lease and/or the name of the lessee and no other information can be found since the original surveys have been lost.

In organizing the successive surveys for the following work the author has used the term "prior survey for" to designate not only a survey named in a later patent as an original survey but also any other survey whose area falls within, either wholly or partially, the boundaries of the selected patent and which is prior in time to the later survey. Other terms used to designate such prior surveys are earlier survey and elder survey.

The author has chosen to provide straight line boundaries along the south, east and west sides of the map areas similar to Rev. Hively's maps and unlike those of the Horvath maps. Thus many of the tracts mapped along the boundaries will only be partially shown on the map. For information about the shape of the whole tract one should consult the original sources or the computer printout the author will provide.

The rights of the warrants, certificates of survey or the patents could be assigned by the individual(s) holding them to other individual(s). When such an assignment was made the agreement was often recorded on the outside of the patent or certificate. In the following list of tracts the assignment dates (when provided), the names of the individuals involved and is some select cases the names of witnesses to the assignment are included when available.

The information in the booklet is arranged in an alphabetical listing of all patents, unpatented certificates, leases and surveys, each with its own tract number. The exact location of a tract inside the mapped area is determined by using the map key. Each map is divided into nine sectors (See Sector Locator Map, page 15). Thus a key "Map 2-SW" would be the southwest sector of Map 2. The county (Baltimore or Carroll) in which the tract is located is given on the first line. The name of the patent, sometimes with a variation, is then given along with the acreage in acres, square rods (one square rod equals one quarter of an acre) and square perches (i.e., 10 a., 5 r., 23 p.). Next the prior (elder or original) surveys are given as "Resurvey of ..." and any successive or later surveys are designated by "Prior survey for.." Next any warrants that were granted together with the date, acreage and name(s) of the holder(s) are given. Sometimes the warrants were renewed at six month intervals to prevent the warrant from lapsing. Then the date and name of the holder of the certificate of survey is given and then the date of the patent and name of the patentee is given. If there were any assignments of the warrants and/or surveys to other parties, the dates (if known) and the names of the new assignees are given. Then the book references for the patents are given, i.e.: Series BC & GS, Book 50, folio 132 will be given as 'BC & GS 50/132.' Finally the patent certificate (or unpatented certificate) number used by the Maryland State Archives to reference the individual patent record (or unpatented certificate) is given. No patent date or book listing will be found if an unpatented certificate was issued. All but a few of the patent certificates in this book are found in Volume 114 for Baltimore County, in the State Archives in Annapolis, which is an alphabetical list of all patents and unpatented certificates. There are a few patents registered in Frederick, Anne Arundel and Carroll Counties. They are noted as such.

One may notice that the areas of the warrants may or may not be the same as that of the surveys. The surveyor may have used only part of one warrant or he may have used several warrants or their parts in order to construct the survey. One has to consult the individual surveys to determine such details. Also the areas given in the patent often varied slightly from those areas found in later resurveys of the same tract. Similarly the boundaries of two such surveys successive surveys usually vary.

A number of patents found in the listing fall completely outside of the mapped areas but share common boundaries of a tract of land that lies partially within the mapped areas. They are included because the adjoining surveys often provide information about the tracts that are shown. In such cases instead of a map key a short remark such as "South of Map 1" will be provided and a reference tract or other prior or later surveys will be named next to which the unmapped tract is located.

The spellings of patent names and names of individual patent holders could vary widely. This often can prevent one from readily locating the patent records of interest. For instance the name Baublitz is found: Boblet(s), Boblitz, Bouvellet and (in a 1796 tax record) Poplets. The tract Curfman's Choice is also found in the records as Kirkman's Choice and Courseman's Choice. Many of the early land owners were not literate and used a mark instead of a signature. The clerks and officials recording the information had to rely on their ear which in the case of German names could present a problem to the mainly Anglican clerks. I have referenced many of the variations of names one finds in the early records, but I have not attempted to include them all.

Landmark features used in the early surveys are of particular interest. These include Rivers as well as creeks and streams called variously "branches", "drafts" or "runs". The main river in the area was the Great Falls of the Gunpowder River divided in into two main branches: the first is the Northern Branch also called the "Little Falls" which originates near the Maryland/Pennsylvania line on the east side of Map No. 1 and the second is the Western Prong or Branch which originates in the northwest corner of Map No. 2. It flows southeast, diagonally across the whole map area to the southeast corner of Map No. 1. All of the mapped areas fall within the drainage basin of the Gunpowder River except for several tracts at the very western edge of Map No. 2 which are at the head of runs descending into the Potapsco River via Pipe Creek. The exact locations and courses of all the branches and runs as well as the Gunpowder River itself were determined from the modern day tax maps, except for the southeast corner of Map No. 1 where the original course of the Western Prong of the Gunpowder River now falls within the Prettyboy Reservoir area. To determine the original course of the river in the reservoir points from the original survey data which were located on the banks of the original Gunpowder Falls were used. All the streams in the mapped area can be grouped into three regions: those descending into the west side of the Western Prong of the Gunpowder Falls; those descending into the east side of the Western Prong and those descending into the west side of the Northern Prong of the Gunpowder Falls. Occasionally the name of the run or branch found in the old surveys was different from that commonly in use today. For a list of the names of the runs in the map area which can be found in the original survey data see page 13 following the introduction.

In the colonial period there were a number of roads which led from Baltimore to York, Pennsylvania. Two of these are found in the map area. The road which followed the ridge separating the streams draining into the Northern and Western Prongs of the Gunpowder and which led up to Stiltz, Pennsylvania was known as the Baltimore(town) to York(town) Waggon Road or the Middle York Road. It is now called the Middletown Road (Middletown was one half the distance between York and Baltimore). The road leading north from Reisterstown to Manchester then northerly to Conowego (Falls), Pennsylvania was known as the Baltimore(town) to Conowego (or Yorktown) Waggon Road. Since colonial era waggon roads were used to transport goods in Conestoga type wagons over long distances between major shipping points, the Conowego referred to would have been Conowego Falls on the Susquehanna River in York County, Pennsylvania. There was a natural barrier to barge traffic at the Falls until a canal was constructed in the 19th century so that goods had to move by land to this point on the River. This old road is now called the Hanover or Manchester Turnpike up to Manchester, after which it turned north-east along what is now Watertank Road and then York Road until it reached Codorus Township in York County, Pennsylvania. These two "waggon" roads were frequently used as reference points in many of the early surveys. The exact positions of all roads on Maps 1 and 2 were determined from the modern day tax maps. One should keep in mind that the locations of the roads two hundred years ago would not be exactly the same as the modern day roads.

The northern boundary of the maps is the Maryland/ Pennsylvania boundary or the Mason-Dixon Line. Before the 1760s when the Mason-Dixon Line was drawn, many of the surveys in the area just north of the line were considered to be in Baltimore County. The Mason-Dixon Line does not run exactly east-west but is tilted at a slight angle: North 89 degrees, 45 minutes and 2 seconds East or about 1/4 degree of a degree tilt.

Historically most of the land in the mapped area was originally owned by the Calvert family, the Lords Baltimore, and was so designated in the early surveys by "land in the reserve" or by "land reserved for His Lordships". By the time of the Revolutionary War much of the land in the area was leased or patented by other individuals. However, some of the land was still owned by the Calverts at the time of the War and these lands were confiscated by the Federal Government at the outset of the War. In Baltimore County, Henry Harford, the illegitimate son of the last Lord Baltimore was unable to reclaim title to these lands after the war. Such was the land that William Hoffman first leased when he built his paper mill at Hoffmanville. The "Reserve" itself was a large area lying in the lower and easterly regions of these two maps. The exact boundaries of the Reserve have never been exactly determined but Mr. George Horvath has located a number of surveys which give some of the boundaries of the Reserve.

Another feature referred to in the surveys is the "Barrens", a large area running across the middle of the map area that was either burned by the Indians for better hunting or was due to a naturally caused forest fire which left the lanscape desolate.

In the north-west section of Map No. 2, southwest of Lineboro, Carroll County there is a topographic feature which is labelled Dugg Hill Ridge on modern maps. This is a prominant feature and was the name of a very early tract of leased land. Most patents in the area start out giving the relative position, i. e.: "Beginning on the south side of Dugg Hill ...etc". The author found only one early source on Dugg Hill which is in the will of William Wright (Baltimore County,1740) and as yet no exact survey information has been found.

There were many mills built along the numerous runs and branches of the Gunpowder Falls as well as on the Falls itself. Often the early surveys would use the mills or the mill dams as reference points. For example Paper Mill Hills located at Hoffmanville (282 acres) gives the South 58 degrees East 213 1/2 perch line as "ends on the south side of the paper mill dam and 18 perches from the dam". The Whessler Mill, built by Ulrich Whessler, was located in the very north-west corner of Map No. 2 almost at the source of the western prong and was one of the earliest mills in the area. The Abraham Shaver Mill on George's Run was also very early. The Whessler and Shaver Mills were both grist mills. There were an abundance of early paper mills in the area, most built by the Hoffman family of paper makers. The clean plentiful waters of the upper Gunpowder were ideal for milling and the manufacture of paper. The Clipper Mill at Hoffmanville on the Western Prong was the first Paper Mill in the State of Maryland, built in 1775 and 1776. It was fallowed by the Gunpowder, Rockdale, Pitt and Hoffacker Paper Mills. The Gist Mill on Graves run also made paper as did the Gore Mill on Graves Run at the Carroll County Line. By the mid 1850's there were so many paper mills in the sixth district of Baltimore County that the manufacture of paper was the main industry in the district. But by the end of the century most of the paper mills had ceased to operate. Many of the early mill locations are indicated on the map; however, no effort has been made to include all of them.

The early surveyors become familiar names as one researches early land records. Each county had several official surveyors and in Baltimore county one finds the names Thomas Gist, James Calder, George Goldsmith Presbury and Darby Ensor among others.

For people who are used to working with early tax lists of Baltimore County prior to 1813 most of the area in these two maps falls in the North Hundred and some in the Pipe Creek Hundred. Later on the area was changed to Tax District No. 3 and later still, after Carroll County was formed (1836), to Districts No. 5 and 6 (Baltimore County) and the Manchester and Millers districts (Carroll County).

In order for these maps to be used to locate easily the positions of individual patents, the author has created the border so that the positions of the grids used in the modern tax maps can be easily determined. Each tax map has 24 grids or squares numbered consecutively from the top left corner to the lower right corner (6 across, 4 down). The Grid Locator Maps (pages 16 & 17) give the positions of the individual tax maps which cover the mapped area. Using the border divisions and the grid locator map one can locate the desired map number and grid number(s) which cover the area of interest. By using a suitable copier with the ability to enlarge and reduce one can bring about a correspondance of scales and create an exact overlay of the grid areas of interest.

The two U.S. Coast and Geodetic Topographic Quarangle Maps which cover the mapped area are: (1) (Md.-Pa.) Parkton Quadrangle and (2) (Md.-Pa.) Westminster Quadrangle.

LIST OF RUNS FOUND IN THE MAPPED AREA

Note: Indentations indicate smaller branches which enter larger branches before reaching the Gunpowder River. Those branches to the south of the mapped area are designated as such.

When the modern names vary from the early names the modern name will be given in parentheses.

I. Runs descending into the west side of the West Prong of the Great Gunpowder Falls from the headwaters north of Lineboro, Carroll County south-east across the map area to Prettyboy Reservoir.

- Gunpowder Falls Branch (West Manheim, Pa.)
- Middle Branch (also South Branch) of the Great Gunpowder Falls
- South Branch
    - Lick Glade Branch
- George's Cabin Branch
- Timber Branch
    - Price's Spring Branch (also Bear's Branch)
- Locust Bottom Branch
- Widow Markey's Spring Branch
- Poplar Branch
- * Graves Run
    - Indian Run (also Stansbury's Run)
        - Stansbury Run
        - Sandy Bank Run
- George's Run
    - Murphy's Run
    - Peggy's Run (south of map area)
        - Clark's Cabin Spring Branch
    - Compass Run
    - Hooker's Cabin Branch (south of map area)
- Prettyboy Run (south of map area)
    - Mirey Branch
- Bush Cabin Run (south of Map area)
    - Boring Branch
- Mingo Branch (south of map area)
- Panther Branch (south of map area)

* Another name used for Graves Run below the fork with Indian Run was "Indian Graves Run".

II. Runs descending into the east side of the Western Prong of the Great Gunpowder Falls from the headwaters north of Lineboro, Carroll County south-east across the map area to Prettyboy Reservoir.

Wolfgang Run
Broad Meadow Run
Long Meadow Run, Long Run, Long Glade (Muddy Run)
  Wildcat Run
Red Oak Run
Fishing or Fish Creek (Walker Run)
  Roaring Branch
Thomas Roger's Spring Branch (Silver Run)
Wheeler Spring Branch
Cullins or Thomas Cullin's Spring Branch (Also called Hilly Spring Branch in one survey)
Tracey Spring Branch
  Back Branch
Keillor Branch (Frog Hallow Run)
  Hoshall Spring Branch
Solomon (or John) Cross Spring Branch

III. Runs descending into the Northern Prong of the Great Gunpowder Falls also called the Little Falls starting with the source south-east of Stiltz, Penna.

Dixon Spring Branch (enters from the east side)
Cattail Marsh Branch
Morehead Spring Branch (enters from east side)
Dorsey's Spring Branch
Hunter's Spring Branch (enters from east side)
Granny Cross's Spring Branch
Jacob Markey's Spring Branch
Owl Branch (South of map area)
Bee Tree Run (South of map area)

SECTOR LOCATOR MAP

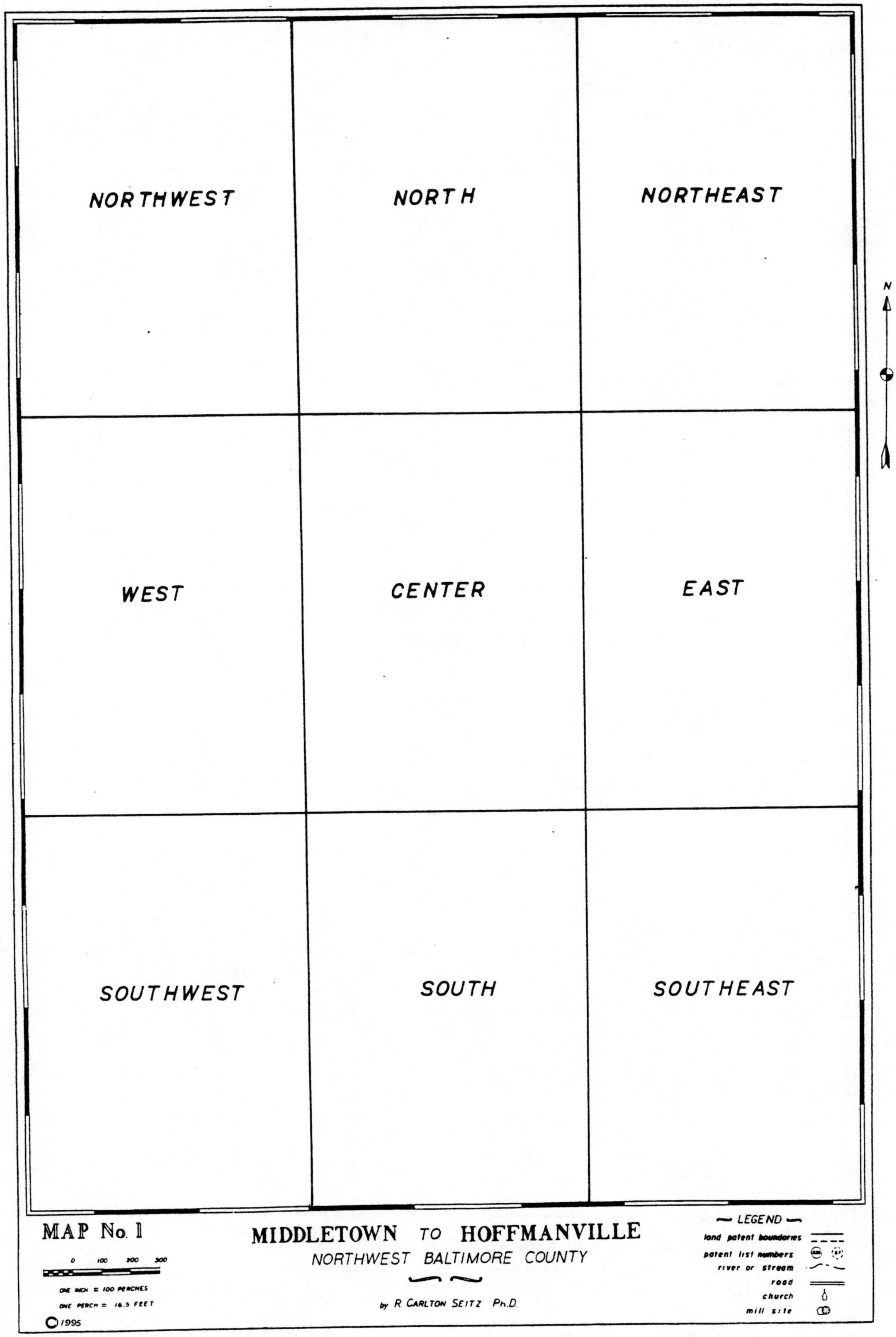

GRID LOCATOR MAP (Map No. 1) - Baltimore County

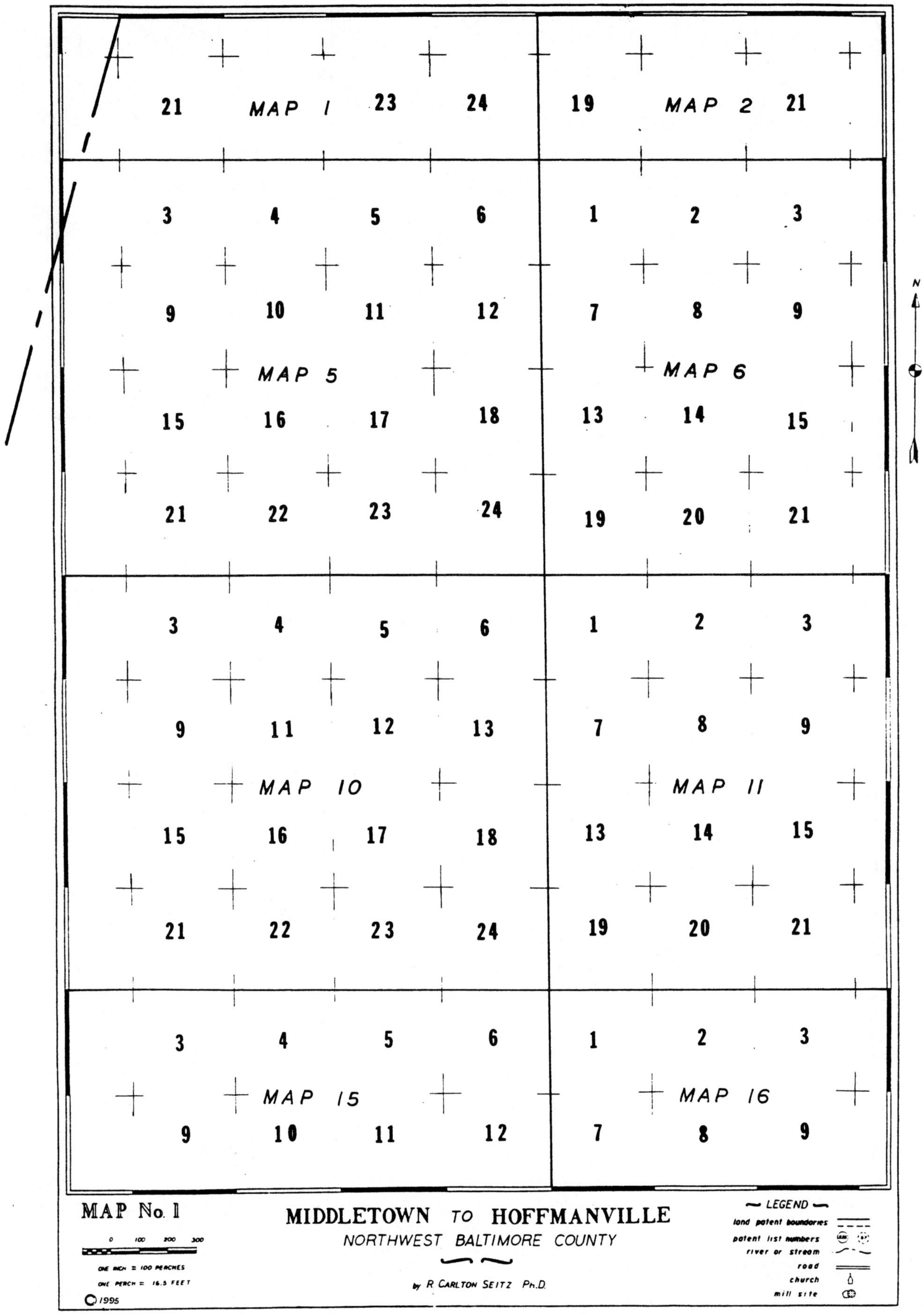

GRID LOCATOR MAP (Map No. 2) - Carroll County

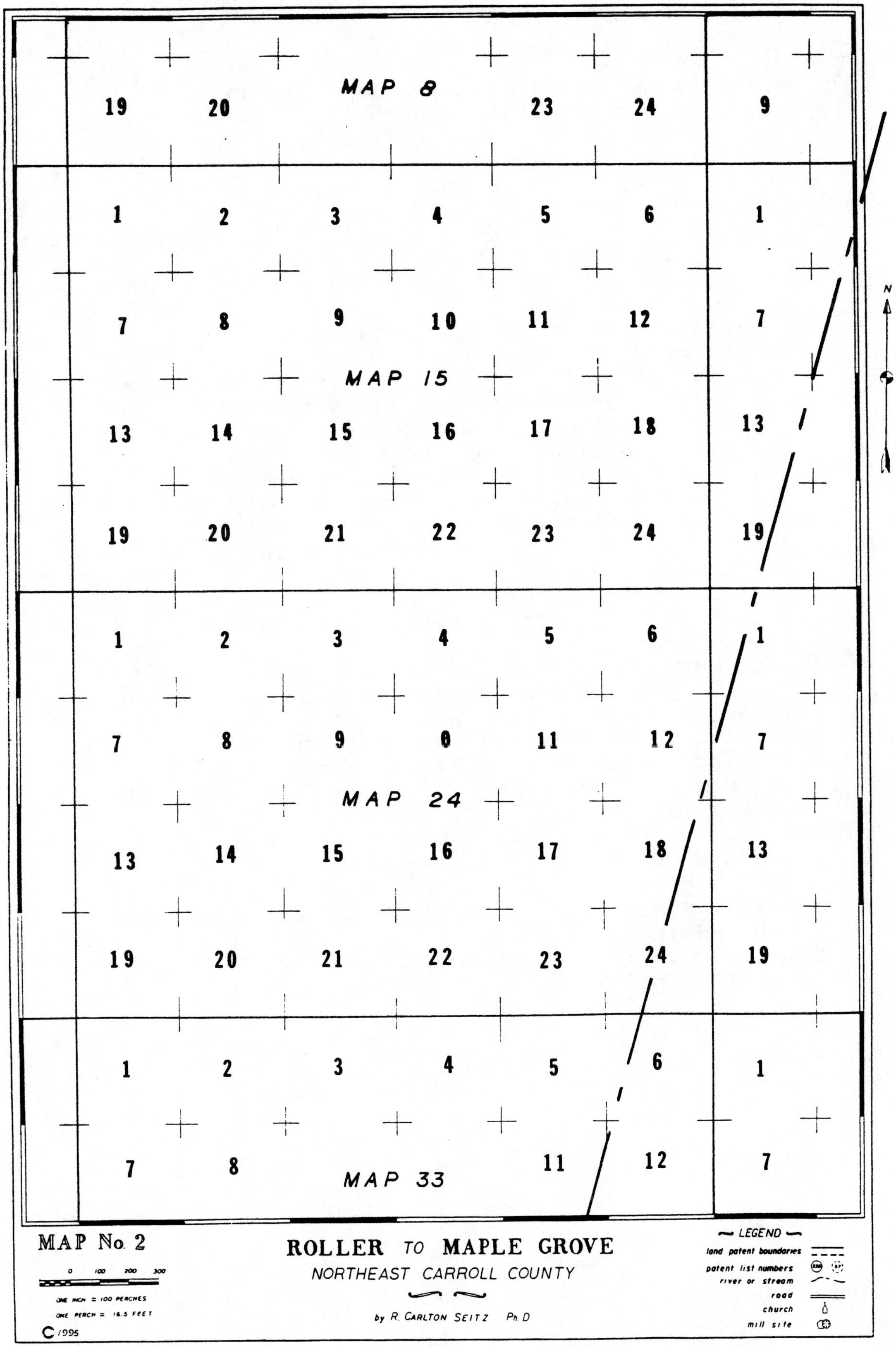

LIST OF SOURCES

1. Maryland State Archives, Annapolis Maryland
    a. Individual patent records
    b. Microfilms of patent series books
    c. Volume 114 - numbered list of patents and warrant holders in Baltimore County.
    d. Records of early warrants
    e. Proprietary leases (1707-1771) (Baltimore County: Volumes GGB A and GGB B)
    f. Baltimore County Surveys (CR 35711): Field survey books of William Smith (1764-1768 & 1782-1790)
2. Historical Society of Carroll County
   Dr. Arthur G. Tracey Collection
    a. Card Files (by patent name and by patentee name)
    b. Collection of survey field notebooks
    c. Collection of copies of the survey plats for Baltimore and Carroll Counties.
3. Maryland Historical Society; William Hoffman Collection (Ms. 1743) (Contains collection of early land records near Hoffmanville donated by May A. Seitz)
4. Haartsook, Elizabeth and Gust Skordas: Land Office and Prerogative Court Records of Colonial Maryland; Hall of Records, Commission No. 4, Annapolis, Md.
5. Horvath, George J., Jr.: From Rayville to Maryland Line, 1986. (map)
6. Hively, Rev. Neal Otto: Index and Summary to Connected Draft Maps - Codorus Township, York County, Penn., 1988 (maps and booklet with name index)
7. Department of Assessments and Taxation, Maryland State; Baltimore County and Carroll County maps. (tax maps)
8. Seitz, May A.: The History of the Hoffman Paper Mills in Maryland; Towson, Md., 1946
9. U. S. Coast and Geodetic Topographic Maps: (1) (Md.-Pa.) Parkton Quadrangle and (2) (Md.-Pa.) Westminster Quadrangle.
10. Hopkins, G.M.: Atlas of Baltimore County, Maryland; Philadelphia, 1877.
11. Hopkins, G.M.: Atlas of Carroll County, Maryland; Philadelphia, 1877.

LAND TRACT INFORMATION

Tract 1 Map 1-SW Baltimore County
Abraham's New Design 13 1/2 acres
Prior Survey for New Purchase (355 3/4 a.)
Sp. Warr. 14 Feb., 1794 (13 1/2 a.) Abraham Shaver
Survey 19 Dec., 1794 Abraham Shaver
Patent 26 Jan., 1796 Abraham Shaver
IC O/74 and IC G/653
Pat. Cert. 14

Tract 2 (West of Map 2) Carroll County
Adam's Contrivance 113 acres
(Reference Tract - Everthing Needful Corrected)
Resurvey on Russ Truss (25 a. - Conrad Prost)
Sp. Warr. 18 May, 1769 Adam Burns
Survey 14 June, 1769 Adam Burns
Patent 20 May, 1772 Adam Burns
BC & GS 45/179
Pat. Cert. 27

Tract 3 Map 2-W Carroll County
Adam's Neighbour 2 1/4 acres
Sp. Warr. 22 Nov., 1792 John Sherman
Survey 10 Dec., 1792 John Sherman
Patent 13 June, 1798 John Sherman
IC O/429 and IC M/439
Pat. Cert. 34

Tract 4 Map 1-SW Baltimore County
Adam's New Design 27 1/4 acres
Prior Survey for New Purchase (355 3/4 a.)
Warrant 20 July, 1792 (42 a.) John Adam Shaver
Survey 21 Nov., 1792 John Adam Shaver
Assignment (by terms of John Adam Shaver's will land deeded to his brother Abraham)
Patent 26 Jan., 1796 Abraham Shaver
IC O/75 and IC G/654
Pat. Cert. 35

Tract 5 (not mapped) Baltimore County
Adam Smith's Benefit 16 1/4 acres
Resurvey of Gain's Rest Enlarged (64 1/4 a.) and Gist's Purchase (465 a.)
Sp. Warr. 20 Aug., 1802 (5 a.) William Hoffman
Assigned 16 Nov., 1802 to Gottlieb Hoffman
Survey 18 Nov., 1802 Gottlieb Hoffman
Patent 8 July, 1803 Gottlieb Hoffman
IC R/241 and IC S/73
Pat. Cert. 36

Tract 6 (not mapped) Baltimore County
Addition 23 3/4 acres
(Reference tract - Crooked Billet (12 a.))

(Addition (23 3/4 acres) - continued)
Resurvey of Hurst Hills (462 3/4 a.)
Comm. Warr. 6 Feb., 1802 (140 acres) Frances Thwaites Talbot
Survey 17 Nov., 1802 Frances Thwaites Talbot
Assigned 5 May, 1804 to Col. Thomas Gist
Patent 18 May, 1804 Col. Thomas Gist
IC P/539 and IC R/265
Pat. Cert. 37

Tract 7 Map 2-SE Baltimore County
<u>Addition</u> 50 acres
Comm. Warr. 8 July, 1769 (50 a.) Charles Ridgely
Survey 20 July, 1769 Charles Ridgely
Patent 31 July, 1770 Capt. Charles Ridgely
BC & GS 39/510 and BC & GS 41/80
Pat. Cert. 52

Tract 8 Map 2-N Carroll County
<u>Addition</u> 18 acres
Prior Survey for Addition to Curpoy (158 a.)
Warrant 13 Sept., 1763 (18 a.) Christian Shaver
Survey 6 Mar., 1764 Christian Shaver
Patent 6 Mar., 1764 Christian Shaver
BC & GS 27/451 and BC & GS 28/207
Pat. Cert. 42

Tract 9 Map 2-SW Carroll County
<u>Addition</u> 13 1/2 acres
Prior Survey for John Caltrider's Fancy (229+ a.)
Comm. Warr. 6 Sept., 1766 (36 a.) Thomas Hooker,Jr.
Assigned 20 Jan., 1767 (17 a.) George Myers
Survey 17 Feb., 1767 George Myers
Patent 17 Feb., 1767 George Myers
BC & GS 34/305 and BC & GS 33/131
Pat. Cert. (?? not located in archives)

Tract 10 Map 1-SE Baltimore County
<u>Addition to Bear's Rock</u> 50 acres
Com. Warr. 14 Nov., 1793 (50 a.) Christian Singery
Survey 24 Oct., 1794 Christian Singery
Patent 17 Oct., 1797 Christian Singery
IC O/306 and IC M/300
Pat. Cert. 65

Tract 11 Map 1-S Baltimore County
<u>Addition to Borings Habitation Rock</u> 97 1/2 acres
Resurvey of Singery's Trouting Stream (575 a.)
Comm. Warr. 10 Nov., 1794 (35 a.) Ezekiel Boring
Comm. Warr. 1 Jan., 1795 (50 a.) Ezekiel Boring
Survey 1 Sept., 1795 Ezekiel Boring
Patent 28 Sept., 1796 Ezekiel Boring
IC O/20 and IC L/161
Pat. Cert. 72

Tract 12 — Map 2-SW — Carroll County

Addition to Boring's Meadow — 25 acres

Survey 20 Apr., 1755 — John Boring
Patent 20 Apr., 1755 — John Boring
BC & GS 4/267 and BC & GS 2/23
Pat. Cert. 70

Tract 13 — Map 2-SW — Carroll County

Addition to Bridgeland — 10 acres

Comm. Warr. 13 Sept., 1759 (10 a.) David Idelman
Survey 10 Mar., 1760 — David Idelman
Patent 18 Mar., 1760 — David Idelman
BC & GS 12/297 and BC & GS 13/491
Pat. Cert. 79

Tract 14 — (South of Map 1) — Baltimore County

Addition to Britton's Range — 109 1/4 acres

(Reference Tract - Britton's Range (295 a.))
Survey 5 June, 1790 — Richard Britton
Patent 21 May, 1798 — Richard Britton
IC O/439 and IC L/465
Pat. Cert. 76

Tract 15 — (South of Map 1) — Baltimore County

Addition to Buck's Range — 40 1/4 acres

(Reference Tract - Buck's Range (544 a.))
Prior Survey for Jones' First Purchase (370 a.)
Sp. Warr. 16 May, 1796 (8 A.) Nicholas Foster
Assigned 26 Nov., 1796 to John Foster
Survey 1 Mar., 1797 — John Foster
Patent 16 Oct., 1797 — John Foster
IC O/297 and IC L/378
Pat. Cert. 81

Tract 16 — Map 1-E — Baltimore County

Addition to Bushy Thicket — 9 1/2 a., 0 r., 5 p.

Resurvey on Minorsborough (54 a.)
Sp. Warr. 18 Mar., 1807 (5 a.) Thomas Gist
Survey 27 Jan., 1808 — Thomas Gist
Unpat. Cert. 23

Tract 17 — (South of Map 2) — Carroll County

Addition to Cox's Pleasant Meadow — 100 a., 0 r., 36 p.

(Reference Tract - Penelope Charcilla and Thomas Cockeys Deyesburg (1030 a.))
Resurvey of Merryman's Choice (88 a.) and Dear Bought (15 a.)
Sp. Warr. 13 Apr., 1828 — Elisha Cox
Survey 9 Apr., 1829 — Elisha Cox
Patent 2 Mar., 1832 — Elisha Cox
GGB 2/697
Pat. Cert. 91

Tract 18 Map 2-N Carroll County

Addition to Cross's Whim 18 acres

Comm. Warr. 15 July, 1765 (18 a.) Nicholas Britton
Renewed 9 Apr., 1766 Nicholas Britton
Assigned 1 Feb., 1767 to Joseph Cross
Survey 22 June, 1767 Joseph Cross
Patent 14 Nov., 1767 Joseph Cross
BC & GS 34/303 and BC & GS 32/225
Pat. Cert. 93

Tract 19 Map 2-N Carroll County

Addition to Curpoy 158 Acres

Resurvey of Level Ground (4 3/4 a.), Minor Cursor (25 a.), Addition (18 a.), Big Woods (10 1/2 a.) Lamott's Middle of the World (77 1/4 a.) and Curpoy (50 a.)
Sp. Warr. 11 Nov., 1803 Melchor Varner
Survey 2 May, 1804 Melchor Varner
Patent 23 Dec., 1805 Melchor Varner (Warner)
IC Q/639 and IC S/287
Pat. Cert. 95

Tract 20 Map 1-NE Baltimore County

Addition to Davis's Chance 9 1/2 acres

Comm. Warr. 2 Jan., 1794 (20 a.) Abraham Smith
Assigned 30 Jan., 1794 to Benjamin Davis
Survey 2 April, 1794 Benjamin Davis
Unpat Cert. 32

Tract 21 Map 1-SE Baltimore County

Addition to Deer Park 313 3/4 acres

Resurvey of May's Mountain (97 3/4 a.)
Sp. Warr. 26 Dec., 1792 Thomas Gist
Survey 11 Dec., 1793 Thomas Gist
Patent 28 Oct., 1794 Thomas Gist
IC I/634 and IC H/554
Pat. Cert. 97

Tract 22 (South of Map 1) Baltimore County

Addition to Edward's Choice 16 1/2 acres

(Reference Tract - Edward's Choice (102 1/2))
Comm. Warrant 16 Oct., 1797 (10 A.) Hny. Edwards
Survey 11 Oct., 1798 Henry Edwards
Patent 9 Dec., 1805 Henry Edwards
IC Q/642 and IC S/286
Pat. Cert. 101

Tract 23 Map 1-C Baltimore County

Addition to Fair Play 23 3/4 acres

Sp. Warr. 27 Mar., 1793 (4 a.) Jonathan Cullender
Assigned 17 Oct., 1793 to William Dorsey
Assigned 4 Jan., 1794 to John Tracey
Survey 14 Jan., 1794 John Tracey
Unpat Cert. 40

Tract 24 Map 1-W Baltimore County

Addition to Grist Mill Hills 7 a. , 20 p.

Sp. Warr. 13 Jan., 1812 Peter Hoffman
Assigned to Christian Hoffman
Survey 12 Jan., 1813 Christian Hoffman
Unpat. Cert. 43

Tract 25 (not mapped) Baltimore County

Addition to Hair's Retirement 12 1/2 acres

Prior Survey for Hair's Beautiful Prospect (327+ a.)
Comm. Warr. 10 Nov., 1794 (30 a.) Jacob Hair
Survey 23 Oct., 1795 Jacob Hair
Patent 13 Feb., 1811 Jacob Hair
IC N/549 and IC H/550 and IC L/626
Pat. Cert. 128

Tract 26 Map 2-N Carroll County

Addition to Hobson's Choice 1 acres

Prior Survey for Sugar Valley (387 a.)
Sp. Warr. 10 Nov., 1794 (1 a.) George Fair
Survey 21 Oct., 1795 George Fair
Patent 12 Nov., 1801 George Fair (Fahr)
IC N/637 and IC L/684
Pat. Cert 139

Tract 27 Map 1-SE Baltimore County

Addition to Hurst Hills 121 acres

Resurvey of Hurst's Delight (34 1/2)
Sp. Warr. 20 Apr., 1793 Bennett Hurst
Survey 25 Apr., 1793 Bennett Hurst
Unpat. Cert. 48

Tract 28 Map 1-S Baltimore County

Addition to Markey's Chance 16 1/4 acres

Comm. Warr. 10 Nov., 1794 (25 a.) Jacob Markey
Survey 2 Apr., 1795 Jacob Markey
Patent 14 Nov., 1809 Jacob Markey
JK U/225 and JK T/193
Pat. Cert. 168

Tract 29 Map 2-N Carroll County

Addition to Mattigan 25 acres

Comm. Warr. 9 Apr., 1774 (25 a.) Henry Ficht
Survey 21 July, 1774 Henry Ficht
Assigned by estate of Henry Ficht (who died 1 Jan., 1788 leaving no will)- heirs of Henry Ficht: Peter Fight, Henry Fight (Jr.), Mary w/o Herman Herkinger, Elizabeth w/o George Reinecker, Eve w/o Conrad Reinecker, Jacob Fight, Andrew Fight (who since died leaving heirs - Polly, John, Andrew and Elizabeth)
Patent 11 May, 1798 Peter Fight et. al.
IC O/449 and L/460
Pat. Cert. 171

Tract 30 (South of Map 2) Carroll County

Addition to Merryman's Grotto 90 Acres

Prior Survey for Murray's Ridge (221 1/2 a.))
Sp. Warr. 2 May, 1750 (90 a.) Jabez Murray
Survey 20 Sept., 1750 Jabez Murray
Patent 20 Sept., 1750 Jabez Murray
BY & GS 5/456 and BY & GS 3/270
Pat. Cert. 3237

Tract 31 Map 2-NW Carroll County

Addition to Michael's Beginning 57 acres

Resurvey of Michael's Beginning (25)
Prior Survey for Peter Druschel's Management (136+)
Sp. Warr. (25 a.) Michael Heming
Renewed 3 Oct., 1770 Michael Heming
Survey 27 Mar., 1771 Michael Heming
Assigned 16 May, 1787 to Martin Rub (Ruhl)
Patent 17 May, 1787 Martin Rub
IC B/694 and IC C/155
Pat. Cert. 175

Tract 32 Map 2-W Carroll County

Addition to (New) Germany 50 acres

Prior Survey for Joshua Boring son of John his Forest (220+ a.)
Warrant 12 May, 1773 (50 a.) Joshua Boring
Survey 1 Nov., 1773 Joshua Boring
Patent 20 Jan., 1785 Joshua Boring
IC B/251 and IC A/417
Pat. Cert. 112

Tract 33 Map 2-W Carroll County

Addition to Nottenstot 25 acres

Prior Survey for Nottenstot Resurvey (155 a.)
Comm. Warr. 26 Sept., 1764 Nicholas Sluthover
Renewed 12 Mar., 1765 Nicholas Sluthover
Renewed 11 Sept., 1765 Nicholas Sluthover
Assigned 11 Aug., 1766 to Anthony Noll
Survey 1 Jan., 1767 Anthony Noll
Patent 18 Feb., 1767 Anthony Noll
BC & GS 34/315 and BC & GS 32/229
Pat. Cert. 189

Tract 34 Map 2-W Carroll County

Addition to Nottestot 17 acres

Prior Survey for Nottenstot Resurvey (155 a.)
Comm. Warr. 28 Mar., 1774 (15 a.) Anthony Null
Comm. Warr. 28 Mar., 1774 (12 a.) Anthony Null
Survey 11 Apr., 1774 Anthony Noll
Patent 14 July, 1775 Anthony Noll
BC & GS 51/125 and BC & GS 48/466
Pat. Cert. 190

Tract 35 (not mapped) Baltimore County

Addition to Number Six 85 3/4 acres

(Not precisely located - in area of Hoffmanville)
Sp. Warr. 31 May, 1782 William Hoffman
Survey 6 June, 1782 William Hoffman
(never patented - survey in MHS collection of Hoffman papers)

Tract 36 Map 1-W Baltimore County

Addition to Paper Mill Hills 8 3/4 acres

Prior Survey for Christiana Creek (102 1/2 a.)
Comm. Warr. 13 Jan., 1812 (10 a.) Peter Hoffman
Survey 15 May, 1812 Peter Hoffman
Patent 29 Nov., 1816 Peter Hoffman
IB D/451 and IB C/496
Pat. Cert. 197

Tract 37 Map 2-S Carroll County

Addition to Penelope Charchilla and Thomas Cockey Deyesburg 5 3/4 a., 0 r., 16 p.

Resurvey of Fairplay (61 a.)
Sp. Warr. 9 May, 1797 (90 a.) Thomas Cockey Deye
Survey 26 Apr., 1798 Thomas Cockey Deye
Patent 16 Dec., 1799 Thomas Cockey Deye
IC O/677 and IC L/581
Pat. Cert. 199

Tract 38 Map 2-S Carroll County

Addition to Penelope Charchilla and Thomas Cockey Deyesburg 5 1/4 acres

Sp. Warr. 16 Aug., 1792 (10 a.) Thomas Cockey Deye
Survey 22 July, 1793 Thomas Cockey Deye
Patent 23 May, 1794 Thomas Cockey Deye
IC I/618 and IC G/395
Pat. Cert. 200

Tract 39 Map 2-S Carroll County

Addition to Penelope Charchilla and Thomas Cockey Deyesburg 44 1/4 acres

Resurvey of Foster's Hunting Ground (1439 a.)
Sp. Warr. 9 May, 1797 (90 acres) Thomas Cockey Deye
Survey 21 Apr., 1798 Thomas Cockey Deye
Patent 16 Dec., 1799 Thomas Cockey Deye
IC O/679 and IC M/664
Pat. Cert. 201

Tract 40 Map 1-C Baltimore County

Addition to Plesant Valley 10 acres

Resurvey of Wind Mill Hills (15 a.)
Comm. Warr. 20 Sept., 1807 (10 a.) Aaron Campbell
Survey 7 June, 1808 Aaron Campbell
Assigned 23 Aug., 1832 to Simon Markey
Patent 4 Mar., 1833 Simon Markey
GCB 2/351
Pat. Cert 207

Tract 41 Map 1-SE Baltimore County
Addition to Ragged Raiment 43 3/4 acres
Resurvey on Ragged Raiment (26 a.)
Sp. Warr. 29 May, 1812 William Matthews
Survey 2 Mar., 1813 William Matthews
Un Pat. Cert. 65

Tract 42 Map 2-SW Carroll County
Addition to Shillings Lott 8 1/2 acres
Comm. Warr. 12 Sept., 1766 (5 a.) George Myers
Comm. Warr. 9 May, 1766 Thomas Hooker, Jr.
Renewed 6 Sept., 1766 (36 a.) Thomas Hooker, Jr.
Assigned 20 Jan., 1767 (17 a.) to George Myers
Survey 17 Feb., 1767 George Myers
Patent 17 Feb., 1767 George Myers (Miers)
BC & GS 34/313 and BC & GS 32/244
Pat. Cert. 229

Tract 43 (not mapped) Baltimore County
Addition to Talbot's Lot
(Reference tract - Davis's Farm (182 1/2 a.))
(no further information - not in archives)

Tract 44 Map 2-N Carroll County
Addition to Tilley's (Dilley's) Meadow 27 acres
Prior Survey for Sugar Valley (387 a.)
Comm. Warr. 18 May, 1769 (27 a.) John Telly
Survey 17 June, 1769 John Telly
Patent 11 June, 1770 John Telly
BC & GS 41/93 and BC & GS 48/61
Pat. Cert. 241

Tract 45 Map 1-NE Baltimore County
Adventure 41 1/2 acres
Resurvey of Standing Stone (41 1/2 a.)
Sp. Warr. 31 Mar., 1796 Samuel Goodman of Anne Arundel County
Survey 20 Apr., 1796 Samuel Goodman
Patent 2 Feb., 1797 Samuel Goodman
Assigned 2 Feb., 1797 to Andrew Horn
IC O/262
Pat. Cert. 290

Tract 46 Map 1-SW Baltimore County
All That's Left 26 1/4 acres
Sp. Warr. 2 May, 1796 (161 1/4 a.) Joshua Stevenson
Survey 21 Apr., 1797 Joshua Stevenson
Patent 17 May, 1798 Joshua Stevenson
IC O/442 and IC L/463
Pat. Cert. 315

Tract 47 (South of Map 1) Baltimore County
All That's Left For Poor Little Me 0 a., 1 r., 35 p.
(Reference Tract - Singary's Prospect (26 a.))
Comm. Warr. 13 June, 1825 (2 a.) Jacob Markey of Samuel

(All That's Left for Poor Little Me - continued)
Survey 10 Jan., 1826 Jacob Markey
Unpat. Cert. 92

Tract 48 (South of Map 1) Baltimore County
<u>All That's Left for Poor Me</u> 3 acres
(Reference Tract - Hard Looks (27 a.))
Comm. Warr. 30 Nov., 1815 (3 a.) Samuel Markey
Survey 24 Nov., 1815 Samuel Markey
Patent 20 Dec., 1820 Jacob Markey
IB G/498
Pat. Cert. 317

Tract 49 Map 2-C Carroll County
<u>All That We Could Get</u> 9 a. + 3 R. + 34 p.
Comm. Warr. 13 Dec., 1816 (10 a.) John Keeth
Survey 18 Mar., 1817 John Keith
Assigned to John Hofacker (Kofacker)
(See Colt's Pasture (1+ a.))
Patent 6 Dec., 1821 John Hoffacker
IB G/594 and IB E/547
Pat. Cert. 316

Tract 50 Map 2-N Carroll County
<u>All Together Too Late</u> 47 1/4 acres
Resurvey on Rather too Late (50 a.)
Sp. Warr. 11 Nov., 1794 Conrad Kerlinger, Jr.
Survey 6 June, 1795 Conrad Kerlinger
Assigned 7 Nov., 1795 to Henry Lamott, Jr. & Sr.
Patent 15 June, 1802 Henry Lamott, Jr. & Sr.
IC R/45 and IC L/731
Pat. Cert. 314

Tract 51 Map 1-S Baltimore County
<u>Allen's Folly</u> 26 1/2 acres
Survey 13 Mar., 1787 John Allen
Assigned 25 Mar., 1809 to Benedict Hurst
Patent 5 Apr., 1809 Benedict Hurst
JK U/89 and JK T/65
Pat. Cert. 308

Tract 52 Map 2-NE Carroll County
<u>Ammon's (Almon's) Forest</u> 70 acres
Prior Survey for Bite the Biter (295 1/2 a.)
Comm. Warr. 15 Apr., 1774 (90 a.) Jacob Ammon
Survey 5 May, 1774 Jacob Ammon (Amon)
Bond entered into 15 Mar., 1780: Jacob Amon and his wife Elizabeth (of York County) entered into a bond with Jacob Messoner where they agreed to assign him all such land that they might own.
Jacob Messoner then assigned his rights to Ammon's Garden and Ammon's Forest under the terms of the bond to John Shaulk

(Ammon's Forest (70 acres) - continued)
Assigned 24 Sept., 1785 by the heirs of Jacob Amon who where Elizabeth, his wife, and Michael Jones, his step-son to John Shaulk
Patent 16 Feb., 1787 John Shaulk (Shank)
IC B/625 and IC C/74
Pat. Cert. 321

Tract 53 Map 2-N Caroll County
Ammon's Garden 20 acres
Prior Survey for Inclosure (112 1/4)
Comm. Warr. 15 Apr., 1774 (90 a.) Jacob Ammon
Survey 5 May, 1774 Jacob Ammon
Bond entered into 15 March, 1780: Jacob Almon and his wife Elizabeth (of York County) entered into a bond with Jacob Messoner where they agreed to assign him all such land they might own
Jacob Messoner then assigned his rights to Ammon's Garden and Ammon's Forest under the terms of the bond to John Shauck
Assigned 24 Sept., 1785 by the heirs of Jacob Ammon who were Elizabeth, his wife, and Michael Jones, his step-son to John Shauck
Patent 16 Feb., 1787 John Shauck
IC B/624 and IC C/72
Pat. Cert. 322

Tract 54 (West of Map 2) Carroll County
Amsterdam 180 acres
(Reference Tract - Troy (1634))
Prior Survey for Mesopotamia (1958 a.)
Warrant 29 Nov., 1751 (2000 a.) Adam Lynn
Survey 10 Aug., 1752 Adam Lynn
(No Pat. Cert. in Archives; possibly leased land)

Tract 55 Map 2-SE Baltimore County
Ann Alcock from Herefordshire in Old England, Her Gift 4 1/2 acres
Resurvey of Wilmot's Lot (217 a.)
Comm. Warr. 6 Sept., 1790 (10 a.) Thomas Cockey Deye
Survey 15 Oct., 1790 Thomas Cockey Deye
Patent 24 May, 1793 Thomas Cockey Deye
IC I/296
Pat. Cert. 343

Tract 56 (East of Map 1) Baltimore County
Ann's Best Way 6 1/4 a., 0 r., 15 p
(Reference Tract - Enoch's Hills (15 1/4 a.))
Comm. Warr. 8 Dec., 1814 (6 1/4 a. + 15 p.) Uriah Freeland
Survey 17 Feb., 1815 Uriah Freeland
Unpat. Cert. 110

Tract 57 Map 2-W Carroll County

Anthony's Hope 24 acres

Prior Survey for Nottenstot (155 a.)
Comm. Warr. 5 Nov., 1764 Christiam Bowen
Assigned 6 Mar., 1765 to Anthony Noll
Survey 7 Mar., 1765 Anthony Null
Patent 7 Mar., 1765 Anthony Hull
BC & GS 27/487 and BC & GS 28/75
Pat. Cert. 358

Tract 58 Map 1-SW Baltimore County

Any Name You Please 10 acres

Prior Survey for New Purchase (355 3/4 a.)
Sp. Warr. 24 Sept., 1795 Abraham Shaver
Survey 24 Sept., 1795 Abraham Shaver
Patent 5 Oct., 1798 Abraham Shaver (Shavell)
IC O/402 and IC M/467
Pat. Cert. 360

Tract 59 Map 1-SW Baltimore County

Anything at All 13 acres

Prior Survey for Rabbit's Hallow (287 a.)
Comm. Warr. Feb., 1792 (28 1/2 a.) Ezekiel Boring
Assigned 1 Feb., 1792 to John Boring
Survey 19 Jan., 1793 John Boring
Patent 2 July, 1794 John Boring
IC I/617 and IC G/401
Pat. Cert. 364

Tract 60 (not mapped) Carroll County

Archibald's Level 48 1/2 acres

Prior Survey for Minorsborough (54 a.)
Leased 20 Apr., 1761 Archibald Tracey
(No other information found)

Tract 61 (not mapped) Carroll County

Bad Enough 25 acres

Survey 1757 (Charles Boblitz ?)
(Courses given in Tracey card file - HSCC)
(On Indian Grave Run - probably a prior survey for Great Profit (25 a. - Charles Baublitz))

Tract 62 Map 1-NW Baltimore County

Baker's Stoney Mountain 55 Acres

Purchased 12 Sept., 1785 (55 + 5 a.) Wm. Baker
Survey 27 Jan., 1797 William Baker
Assigned 13 Apr., 1815 Barbary Baker
Patent 30 Jan., 1818 Barbary Baker
IB E/452 and IC D/710
Pat. Cert. 452

Tract 63 (South of Map 1) Baltimore County

Barney's Lott 150 acres

(Reference Tract - Ford's Choice (157 a.))
Survey 5 May, 1787 Barney Ford

(Barney's Lott (150 acres) - continued)

Assigned 17 Sept., 1789 to Mordecai Ford
Patent 16 Sept., 1806 Mordecai Ford
IC R/414 and IC S/396
Pat. Cert. 479

Tract 64 Map 1-E Baltimore County
<u>Barren Hill</u> 37 3/4 acres
Sp. Warr. 28 Feb., 1795 William Dorsey
Survey 7 May, 1795 William Dorsey
Unpat. Cert. 149
Assigned 10 Feb., 1796 to Henry Wilhelm
(Witness: Moses Freeland)

Tract 65 (not mapped) Baltimore County
<u>Barren Hills</u> 16 1/2 Acres
Prior Survey for No. 2 (40 3/4 a.) and Grist Mill Hills (257 a.)
Survey 9 May, 1776 William Hoffman
Leased 7 Aug., 1777 William Hoffman
(never precisely located - site of Clipper Paper Mill built in 1775 and 1776.)
(survey is in Hoffman collection of papers in the MHS.)

Tract 66 Map 2-W Carroll County
<u>Baughman's Goodwill (Resurvey)</u> 154 3/4 acres
Resurvey of Belfast (50 a.)
Survey 1 May, 1805 John Sherman
Unpat. Cert. 160

Tract 67 Map 1-S Baltimore County
<u>Bear's Rock</u> 177 3/4 acres
Survey 4 Jan., 1790 Christian Singery
Patent 22 May, 1794 Christian Singery
IC I/635 and IC G/394
Pat. Cert. 528

Tract 68 Map 1-NE Baltimore County
<u>Beard's Branch</u> 294 3/4 acres
Prior Survey for George's Wisdom (332 1/2 a.)
Survey 2 Dec., 1786 Ulrich Freeland
Assigned 3 Nov., 1789 to George Nunnemaker
Patent 16 May, 1791 George Nunnemaker, Sr.
IC F/392
Pat. Cert. 523

Tract 69 Map 2-NE Carroll County
<u>Beaver's Bottom</u> 24 acres
Prior Survey for Beaver Mint (138 a.)
Comm. Warr. 19 May, 1761 (100 a.) Philip Idleman
Assigned 4 Mar., 1767 (80 a.) to John Beaver
Survey 10 Aug., 1767 John Beaver
Patent 10 Aug., 1767 John Beaver

(Beavers Bottom - continued)
BC & GS 34/277 and BC & GS 32/243
Pat. Cert. 531

Tract 70 Map 2-NE Carroll County
Beaver's Bottom Hills 251 acres
Resurvey of Beaver's Bottom (24 a.)
Prior Survey for Beaver's Mint (138 a.)
Sp. Warr. 27 May, 1773 John Beaver
Survey 5 July, 1773 John Beaver
Unpat. Cert. 164

Tract 71 Map 2-NE Carroll County
Beaver Mint 138 acres
Resurvey of Beaver's Bottom (24 a.) and Beaver's Bottom Hills (251 a.)
Sp. Warr. 27 May, 1773 John Beaver
Survey 5 July, 1773 John Beaver
Patent 10 Mar., 1788 John Beaver
IC D/142 and IC C/333
Pat. Cert. 534

Tract 72 Map 2-NE Carroll County
Beaver's (Baver's) New Invention 170 acres
Resurvey of Beaver's Retirement (20) and Wilmott's Desire (136 a.)
Prior Survey for Hobson's Choice (106 1/4 a.)
Sp. Warr. 6 Apr., 1770 John Beaver
Renewed 26 Mar., 1771 John Beaver
Survey 6 Nov., 1772 John Beaver
Patent 12 Feb., 1773 John Beaver (Baver)
BG & GS 47/305 and BG & GS 42/465
Pat. Cert. 511

Tract 73 Map 2-NE Carroll County
Beaver's Retirement 20 acres
Prior Survey for Beaver's New Invention (170 a.)
Comm. Warr. 19 May, 1762 (100 a.) Philip Idleman
Assigned 4 Mar., 1767 to John Beaver
Survey 10 Aug., 1767 John Beaver
Patent 10 Aug., 1767 John Beaver
BC & GS 34/302 and BC & GS 32/232
Pat. Cert. 535

Tract 74 Map 2-NE Carroll County
Beaver Trap 609 acres
Resurvey of Beaver's New Invention (170 a.), Beaver's Mint (138 a.) and Happiness (39 a.)
Prior Survey for Lebanon (15 a.) and Double Trouble (628 a.)
Sp. Warr. 3 May, 1790 Henry Lamotte
Survey 5 Oct., 1790 Henry Lamotte
Patent 15 Jan., 1795 Henry Lamotte
IC I/571 and IC H/612
Pat. Cert. 536

Tract 75 (East of Map 1) Baltimore County

<u>Becker's Mistake</u> 4 a., 1 r., 10 p.

(Reference Tracts - Northampton (242 a.) and Mount Pleasant (25 1/2 a.))

Sp. Warr. 7 Mar., 1860 (4 a.) William McAbee
Survey 20 Apr., 1860 William McAbee
Patent 15 Apr., 1865 William McAbee
GLLD 1/98
Pat. Cert. 537

Tract 76 (West of Map 2) Carroll County

<u>Belfast</u> 50 acres

Prior Survey for Belfast Enlarged (400 a.) and Powell's Discovery (148 a.)

Comm. Warr. (50 a.) Andrew McGill
Sp. Warr. 21 Jan., 1749 Andrew McGill
Survey 25 July, 1750 Andrew McGill
Patent 25 July, 1750 Andrew McGill
BY & GS 5/476 and BY & GS 4/134
Pat. Cert. 548

Tract 77 (West of Map 2) Carroll County

<u>Belfast Enlarged</u> 400 acres

(Reference Tract - Powell's Discovery (148 a.))

Resurvey of Belfast (50 a.)
Prior Survey for Baughman's Goodwill (154 3/4 a.)
Sp. Warr. 9 Nov., 1762 Peter Stump
Renewed 9 May, 1763 Peter Stump
Renewed 21 Oct., 1763 Peter Stump
Renewed 7 Mar., 1764 Peter Stump
Renewed 4 Oct., 1764 Peter Stump
Survey 1 Oct., 1764 Peter Stump
Unpat. Cert. 173

Tract 78 Map 1-NE Baltimore County

<u>Bender's Plains</u> 186 acres

Prior Survey for Dorsey's Plains (655 1/2 a.)
Survey 9 Nov., 1786 Philip Bender
Assigned 30 Oct., 1789 to Uriah Freeman (Freeland)
Assigned 24 Dec., 1789 to Thomas Boring
Patent 30 Sept., 1790 Thomas Boring
IC F/181 and IC G/4
Pat. Cert. 567

Tract 79 Map 1-SE Baltimore County

<u>Benjamin's Beginning</u> 174 1/2 acres

Survey 8 Mar., 1787 Solomon Cross
Assigned 17 Aug., 179- to Samuel C. Price
Patent 2 Apr., 1794 Samuel C. Price
IC I/615 and IC H/465
Pat. Cert. 571

Tract 80 Map 1-NE Baltimore County

<u>Betsey's Best Way</u> 0 a., 1/4 r., 31 p.

Comm. Warr. 13 Dec., 1814 (7 a., 13 p.) Wm. Dorsey

(Bestsey's Best Way - continued)
Survey 2 July, 1815 — William Dorsey
Unpat. Cert. 182

Tract 81 — Map 1-SE — Baltimore County
<u>Betsey's Choice</u> — 98 acres
Prior Survey for Betsey's Choice Resurvey (141+ a.)
Survey 26 Oct., 1787 — John Cross
Assigned 10 Dec., 1794 to John Cooper
Patent 13 Mar., 1797 — John Cooper
IC M/154 and IC O/250
Pat. Cert. 601

Tract 82 — Map 1-SE — Baltimore County
<u>Betsey's Choice Resurvey</u> — 141 3/4 acres
Resurvey of Betsey's Choice (98 a.)
Sp. Warr. 2 Mar., 1798 — John Cooper
Survey 28 Mar., 1798 — John Cooper
Patent 9 Feb., 1854 — John Cooper
GGB 7-B/26 and GGB 8/259
Pat. Cert. 602

Tract 83 — Map 2-SE — Carroll County
<u>Betsey's Contented</u> — 45 3/4 acres
Sp. Warr. 23 Mar., 1792 (60 a.) Charles Boblets
Assigned 16 Nov., 1792 to Peter Boblets
Survey 19 Nov., 1792 — Peter Boblitz
Patent 17 Apr., 1794 — Peter Boblitz
IC H/481 and IC I/627
Pat. Cert. 603

Tract 84 — Map 2-C — Carroll County
<u>Big Owl Hill</u> — 245 1/2 acres
Resurvey of Little Meadow Enlarged (157 1/2 a.), Borings Triangle (15 a.), Stoney Spring (20 a.) Cross's Meadow (70 a.) and Stoney Hill (1291)
Prior Survey for Charles Trouble (245+ a.)
Sp. Warr. 18 Feb., 1800 — John Miller
Survey 30 Jan., 1801 — John Miller
Patent 8 Dec., 1817 — John Miller
IB D/604 and IB E/76
Pat. Cert. 625

Tract 85 — Map 2-NW — Carroll County
<u>Big Wood(s)</u> — 10 1/2 acres
Prior Survey for Addition to Curpoy (158 a.)
Comm. Warr. 15 Apr., 1774 (46 a.) Ulrich Fritz
Assigned 20 July, 1774 (4 1/2 a.) to C. Shaver
Survey 8 Sept., 1774 — Christian Shaver
Assigned by terms of will of Christian Shaver who died circa 1785: All land to son John Adam Shaver when he reaches 21 years of age.
Patent 25 Oct., 1791 — John Adam Shaver(s)
IC F/447 and IC G/90
Pat. Cert. 626

Tract 86 Map 2-NE Carroll County

<u>Bite-Bit-The-Biter</u> 2 a., 3 r., 20 p.

Comm. Warr. 21 Nov., 1803 (5 a.) George Walker
Survey 10 Jan., 1804 George Walker
Unpat. Cert. 192

Tract 87 Map 2-NE Carroll County

<u>Bite-the-Biter</u> 295 1/2 acres

Resurvey of Amon's Garden (20 a.)
Sp. Warr. 5 Feb., 1789 John Shauck of Pennsylvania
Survey 12 Dec., 1789 John Shauck
Patent 31 July, 1790 John Shauck (Shank(s))
IC F/1 and IC E/506
Pat. Cert. 633

Tract 88 Map 2-N Carroll County

<u>Bol(e) Gat's Harbour</u> 12 3/4 acres

Sp. Warr. 16 June, 1803 (5 a) Valentine Fritz
Assigned 2 Sept., 1803 to George Fair
Survey 5 Sept., 1803 George Fair
Unpat. Cert. 204

Tract ***

Bold Contrivance (See Good Contrivance)

Tract 89 Map 2-C Carroll County

<u>Bollinger's Living</u> (Luck) 178 1/2 acres

Resurvey of Wallet Resurvey (212 3/4 a.), Stoney Hills (1291 a.), Cross's Meadow (70) and Little Meadow Enlarged (157 1/2 a.)
Sp. Warr. 6 Dec., 1802 Joseph and Matthias Bollinger
Survey 14 Sept., 1803 Joseph and Matthias Bollinger
Patent 22 Oct., 1812 Joseph and Matthias Bollinger
IB B/403 and IB A/254
Pat. Cert. 655

Tract 90 Map 2-SW Carroll County

<u>Boring's Garden</u> 18 acres

Resurvey of Ely's Rock (25 a.)
Prior Survey for Penelope Charcilla and Thomas Cockey Deyesburg (236 3/4 a.)
Sp. Warr. 24 Oct., 1776 (180 a.) Thomas Stansbury and Henry Wareheim
Survey 14 Feb., 1777 Thomas Stansbury and Henry Wareheim
Patent 30 May, 1788 Thomas Stansbury and Henry Wareheim
IC D/178 and IC C/433
Pat. Cert. 707

Tract 91 Map 1-S Baltimore County

Boring's Habitation Rock 300 acres

Resurvey of Singery's Trouting Stream (575 a.)
Sp. Warr. 30 Nov., 1793 (120 a.) Ezekiel Boring
Sp. Warr. 10 Dec., 1793 (180 a.) Ezekiel Boring
Survey 8 Oct., 1794 Ezekiel Boring
Patent 24 Apr., 1795 Ezekiel Boring
IC K/183 and IC G/520
Pat. Cert. 709

Tract 92 Map 1-NE Baltimore County

Borings Hall 314 3/4 acres

Resurvey of Bender's Plains (186 a.)
Sp. Warr. 3 Dec., 1793 Thomas Boring
Survey 28 Mar., 1794 Thomas Boring
Patent 8 Sept., 1795 Thomas Boring
IC K/165 and IC H/743
Pat. Cert. 710

Tract 93 Map 1-NE Baltimore County

Borings Hall 10 acres

Prior Survey for Borings Hall Resurvey (16 1/2)
Com. Warr. 30 Nov., 1790 (10 a.) Thomas Boring
Survey 28 Sept., 1791 Thomas Boring
Patent 8 Sept., 1795 Thomas Boring
IC K/199 and IC H/741
Pat. Cert. 711

Tract 94 Map 1-NE Baltimore County

Borings Hall Corrected 261 acres 2 Rods

Resurvey of Borings Hall (314 3/4 a.)
Sp. Warr. 26 Dec., 1831 Joseph Kaufman
Survey 11 Jan., 1832 Joseph Kaufman
Patent 16 July, 1833 Jesse Burquebile
GGB 2/397
Pat. Cert. 712

Tract 95 Map 1-NE Baltimore County

Borings Hall Resurvey 16 1/2 acres, 23 ps.

Resurvey of Borings Hall (10 a.)
Sp. Warr. 1 Feb., 1802 Uriah Freeland
Survey 7 Apr., 1802 Uriah Freeland
Patent 31 Dec., 1812 Uriah Freeland
IB B/410 and JK T/531
Pat. Cert. 713

Tract 96 Map 1-SW Baltimore County

Borings Hazard 164 acres

Resurvey of Stansbury's Contrivances (14 a.)
Prior Survey for Three Elbows (263 3/4 a.)
Survey 22 Mar., 1787 Ezekiel Boring
Patent 24 Oct., 1791 Ezekiel Boring
IC F/484 and IC E/799
Pat. Cert. 714

Tract 97 Map 2-SW Carroll County

<u>Boring's (Boarings) Meadow</u> 100 acres

Prior Survey for John Caltrider's Fancy (229+ a.)
Warrant 1 Apr., 1744 (100 a.) John Boring
Survey 4 Sept., 1744 John Boring
Patent 4 Sept., 1744 John Boring
PT 2/74
Pat. Cert. 653

Tract 98 Map 2-W Carroll County

<u>Boring's Prospect</u> 20 acres

Prior Survey for Joshua Boring son of John his Forest (220+ a.)
Comm. Warr. 11 Apr., 1791 Joshua Boring
Survey 8 Sept., 1791 Joshua Boring
Patent 12 Dec., 1799 Joshua Boring
IC O/676 and IC L/580
Pat. Cert. 715

Tract 99 Map 2-S Carroll County

<u>Borings Range</u> 1375 acres

Resurvey of Ely's Rock (25 a.), Tipton's Fancy (50)
Prior Survey for Foster's Hunting Grounds (1439), Hard Times (53 a.), No Name (1 1/2 + a.), Penelope Charchilla and Thomas Cockey Deyesburg (340 1/2 a.)
Warrant 10 Apr., 1760 James Boring
Renewed 21 Sept., 1761 James Boring
Survey 24 Feb., 1762 James Boring
Patent 1 Apr., 1762 James Boring
BC & GS 14/696 1/2 and BC & GS 17/225
Pat. Cert. 716

Tract 100 Map 2-C Carroll County

<u>Borings Struggle</u> 50 acres

Prior Survey for Grandfather's Gift (343+ a.)
Comm. Warr. 20 Sept., 1759 (50 a.) Thomas Boring
Survey 25 Feb., 1760 Thomas Boring
Patent 25 Feb. 1760 Thomas Boring
BC & GS 12/265 and BC & GS 10/448
Pat. Cert. 717

Tract 101 Map 2-C Carroll County

<u>Boring's Triangle</u> 15 acres

Prior Survey for Big Owl Hill (245 1/2 a.)
Comm. Warr. 2 Apr., 1794 (15 a.) Thomas Boring
Survey 19 Mar., 1795 Thomas Boring
Assigned by terms of Thomas Boring's will - lands deeded to wife Philasander and children: Ezekiel Boring (executor), Mary Kelly, Ruth Tipton and Malinda Tracey
Patent 17 May, 1798 Ezekiel Boring et. al.
IC O/442 and IC M/421
Pat. Cert. 718

Tract 102 (not mapped) Carroll County

<u>Bowlinger's Contrivance</u>

Pat. Cert 753

(File in Archives is empty - see Joseph Bollinger's Contrivance (Pat. Cert. 2691))

Tract 103 Map 2-W Carroll County

<u>Bower's Delight</u> 45 acres

Warrant 2 June, 1773 (108 a.) Ulrich Eklar
Assigned 5 June, 1773 (31 a.) to Christian Bower
Warrant 2 June, 1773 (40 a.) Ulrich Eklar
Assigned 2 June, 1773 (14 a.) to Christian Bower
Survey 13 Nov., 1773 Christian Bower(s) of Frederick County
Patent 1 Dec., 1774 Christian Bower(s)
BC & GS 50/83 and BC & GS 49/122
Pat. Cert. 752

Tract 104 (South of Map 2) Baltimore County

<u>Bratton's Desire</u> 26 1/2 acres

(Reference tract - Wilmott's Lot)
Prior Survey for Tipton's Trouble (27 a.)
Survey 28 Oct., 1758 Andrew Bratton
Leased 28 Oct., 1758 Andrew Bratton
Assigned 17 Nov., 1804 to Sylvester Tipton and John Michael Shaffer
Leaseholds GGB A/69

Tract 105 Map 1-SW Baltimore County

<u>Bravois (Braves) Range</u> 173 1/4 acres

Survey 8 Mar., 1787 Thomas Price
Patent 24 Feb., 1797 Thomas Price
IC O/248 and IC M/150
Pat. Cert. 762

Tract 106 Map 2-SW Carroll County

<u>Bridgeland</u> 150 acres

Prior Survey for Nace's Tavern (478 + a.)
Comm. Warr. 21 Sept., 1753 (150 a.) Philip Idleman
Survey 28 Feb., 1754 Philip Idleman
Patent 28 Feb., 1754 Philip Idleman
BC & GS 1/221 and BC & GS 2/60
Pat. Cert. 776

Tract 107 Map 2-SW Carroll County

<u>Brittania</u> 48 1/4 acres

Prior Survey for California (490 a.)
Comm. Warr. 11 Feb., 1762 (42 a.) Arthur Chenoweth
Assigned 6 June, 1762 to Richard Richards
Survey 10 June, 1762 Richard Richards
Patent 10 June, 1762 Richard Richards
BC & GS 21/352 and BC & GS 22/469
Pat. Cert. 782

Tract 108 Map 1-SW Baltimore County

<u>Britton's Range</u> 295 acres

Purchase of 295 a. in the Reserve: Nicholas Britton
Survey 5 Dec., 1770 (399 a.) Nicholas Britton
Certificate examined in 1775 and never returned
Assigned: By terms of will of Nicholas Britton who died about 1785 to his only son Richard
Certificate corrected from 399 to 295 acres
* Survey 5 June, 1790 (295 a.) Richard Britton
Patent 29 July, 1790 Richard Britton
IC F/24 and IC E/502
Pat. Cert. 785

Tract 109 Map 2-E Carroll County

<u>Brother to Hard Stone</u> 9 1/2 a., 0 r., 32 p

Sp. Warr. 4 Dec., 1815 (Unexecuted part of a Warrant for 129 1/4 a., 32 p.) Wm. Hoffman, Jr.
Survey 1 May, 1816 William Hoffman, Jr.
Assigned by the terms of the will of William Hoffman, Jr. to his heirs: William J. Hoffman, Henry B. Hoffman, Susannah Hoffacker, Mary Markey, Elizabeth Hoffacker, Ann Gore and Catherine Miller
Patent 27 July, 1850 William J. Hoffman et. al.
GGB 7A/276 and GGB 5/553
Pat. Cert. 803

Tract 110 Map 1-NW Baltimore County

<u>Brother to Laurel Hills</u> 47 1/4 acres

Warrant 15 Dec., 1800 (5 a.) Henry Hoffman
Assigned 11 Mar., 1801 to Peter and Caleb Hoffman
Survey 12 Mar., 1801 Peter and Caleb Hoffman
Patent 14 Jan., 1802 Peter and Caleb Hoffman
IC R/51
Pat. Cert. 804

Tract 111 Map 2-E Baltimore County

<u>Brother to New Beginning</u> (Corrected) 25 a., 3 r, 16 p

Sp. Warr. 4 Dec., 1815 (129 1/4 a. and 32 p.) William Hoffman, Jr.
Survey 4 Dec., 1815 William Hoffman, Jr.
Assigned by the terms of the will of William Hoffman, Jr. to his hiers: William J. Hoffman, Henry B. Hoffman, Susannah Hoffacker, Mary Markey, Elizabeth Hoffacker, Ann Gore and Catherine Miller
(Certificate corrected)
Patent 27 July, 1850 William J. Hoffman et. al.
GGB 7/275 and GGB 5/547
Pat. Cert. 805

Tract 112 (South of Map 1) Baltimore County

<u>Brother to Walnut Hills</u> 30 1/2 acres

(Reference Tract - Wild Cat Hallow (54 3/4 a.))
Comm. Warr. 22 June, 1801 (60 a.) William Hoffman
Survey 24 May, 1802 William Hoffman

(Brother to Walnut Hills - continued)
Patent 4 Aug., 1803 William Hoffman (Jr.)
IC R/216 and IC S/89
Pat. Cert. 806

Tract 113 (South of Map 2) Baltimore County
<u>Bryan's Prospect</u> 25 acres
(Reference Tract - Britton's Range (295 a.))
Warrant 21 Aug., 1787 (25 a.) Bryan Tipton
Survey 3 July, 1788 Bryan Tipton
Patent 11 June, 1789 Bryan Tipton
IC E/60 and IC D/469
Pat. Cert. 834

Tract 114 Map 1-SW Baltimore County
<u>Buck Range</u> 50 acres
Prior Survey for Towson's Grove (367 a.)
Comm. Warr. 21 Apr., 1762 (50 a.) Thomas Towson
Survey 13 Sept., 1762 Thomas Towson
Patent 13 Sept., 1762 Thomas Towson
BC & GS 21/27 and BC & GS 18/245
Pat. Cert. 861

Tract 115 Map 1-SE Baltimore County
<u>Buck Range</u> 544 acres
Prior Survey for Jones' First Purchase (370 a.)
Survey 11 June, 1771 Solomon Cross
Patent 6 Apr., 1774 Solomon Cross
BC & GS 50/49 and BC & GS 44/415
Pat. Cert. 857

Tract 116 Map 1-C Baltimore County
<u>Bushy Hill</u> 10 3/4 acres
Sp. Warr. 2 May, 1796 (161 1/4 a.) Jos. Stevenson
Survey 21 Apr., 1797 Joshua Stevenson
Patent 17 May, 1798 Joshua Stevenson
IC O/429
Pat. Cert. 889

Tract 117 Map 1-S Baltimore County
<u>Bushy Ridge</u> 85 acres
Special Warrant 6 Jan., 1791 Jacob Markey
Survey 30 Sept., 1791 Jacob Markey
Patent 6 Dec., 1792 Jacob Markey
IC I/112 and IC G/205
Pat. Cert. 891

Tract 118 Map 1-E Baltimore County
<u>Bushy Thicket</u> 17 acres
Resurvey of Dorsey's Plains (655 1/2 a.)
Sp. Warr. 8 Oct., 1793 (20 a.) Benjamin Vaughn
Survey 20 Mar., 1794 Benjamin Vaughn
Patent 18 Feb., 1796 Benjamin Vaughn
IC O/71 and IC L/75
Pat. Cert. 892

Tract 119 Map 1-C Baltimore County

(The) Butcher's Cleaver 25 Acres

Sp. Warr. 6 Jan., 1791 (112 a.) Jacob Markey
Survey 4 Aug., 1791 Jacob Markey
Patent 17 Oct., 1792 Jacob Markey
IC I/119 and IC G/184
Pat. Cert. 893

Tract 120 Map 2-SW Carroll County

California (Calafornia) 490 acres

Resurvey of Brittania (48 1/4 a.)
Sp. Warr. 21 Mar., 1763 Richard Richards
Renewed 14 Sept., 1763 Richard Richards
Renewed 12 Mar., 1764 Richard Richards
Renewed 11 Sept., 1764 Richard Richards
Survey 5 Mar., 1765 Richard Richards
Patent 26 Mar., 1765 Richard Richards
BC & GS 27/513 and BC & GS 28/406
Pat. Cert. 910

Tract 121 Map 1-W Baltimore County

Campbell's Chance 17 acres

Comm. Warr. 2 Dec., 1793 (2 acres) James Campbell
Assigned by the will of James Campbell to his children and heirs
Assigned 7 July, 1797 (by Deed): Mary, wife of Basil Tracey, Ann Campbell and Elizabeth Campbell, all children of James Campbell, assign their interest in Campbell's Chance to their brothers Moses, Henry, John, Aaron and William Campbell
Survey 14 Jan., 1794 Moses, John, Aaron, Henry and William Campbell
Patent 27 Dec., 1803 Moses, John, Aaron, Henry and William Campbell
IC Q/340 and IC P/501
Pat. Cert. 917

Tract 122 Map 1-N Baltimore County

Campbell's Pursuit 130 acres

Survey 14 Mar., 1787 Wm. Dhyis (Davis), Edmund Stansbury and Jacob Marshall
Assigned 6 July, 1789 to James Campbell
Assigned by the will of James Campbell to his children and heirs
Assigned by a deed dated 7 July, 1797: Mary, wife of Basil Tracey, Ann Campbell and Elizabeth Campbell, all children of James Campbell deed their interest in Campbell's Pursuit to their brothers named therein (Moses, Henry, John, Aaron and William)
Assigned 24 June, 1805 (Deed): Moses, Henry, John and William Campbell assign their interest in Campbell's Pursuit to their brother Aaron Campbell
Patent 26 Oct., 1809 Aaron Campbell
JK U/221 and JK T/177
Pat. Cert. 918

Tract 123 Map 1-C Baltimore County

Campbell's Trouble 50 acres

Survey 1 Dec., 1789 James Campbell

Assigned by the terms of the will (unsigned) of James Campbell to his children and heirs

Assigned by a deed dated 27 July, 1797: Elizabeth and Ann Campbell, Basil Tracey and his wife Mary all assigned their interest in Campbell's Trouble to their brothers, Moses, John, Henry, Aaron and William Campbell

Patent 28 May, 1804 Moses, John, Aaron, Henry and William Campbell

IC R/258

Pat. Cert. 914

Tract 124 Map 1-S Baltimore County

Can't You be easy and let Me alone 44 acres

Sp. Warr. 5 Jan., 1797 (25 a.) Joshua Stevenson

Survey 20 Dec., 1797 Joshua Stevenson

Patent 9 May, 1799 Joshua Stevenson

IC O/571 and IC M/587

Pat. Cert. 928

Tract 125 (South of Map 2) Carroll County

Canton Bearn (Bearer) 633 acres

(Reference tracts - Everthing Needful Corrected (1755+ a.) and Petersburg Resurvey 1003 a.))

Reservey of Upper Agreton - Henry Salbocker

Sp. Warr. 17 Nov., 1791 Henry Salbocker

Survey 21 Sept., 1792 Henry Salbocker

Patent 1 May, 1793 Henry Salbocker

IC H/301

Pat. Cert. 926

Tract 126 Map 1-SE Baltimore County

Captain Little John 56 acres

Comm. Warr. 9 Mar., 1801 (30 a.) John Cross

Comm. Warr. 16 Dec., 1801 (36 a.) John Cross

Survey 8 Jan., 1802 John Cross

Patent 17 Oct., 1803 John Cross

IC Q/258 and IC S/120

Pat. Cert. 934

Tract 127 Map 1-SE Baltimore County

Captain Little John Hold Fast your Sword 12 a., 1 R., 20 p.

Sp. Warr. 1 Sept., 1815 (3 a.) Henry Shue

Comm. Warr. 23 Sept., 1815 (10 a.) Henry Shue

Survey 27 Feb., 1816 Henry Shue

Assigned 3 Dec., 1819 to Jacob Lamotte (witness to assignment - Joseph Cooper)

Patent 21 Dec., 1819 Jacob LaMotte

IB G/365

Pat. Cert. 933

Tract 128 Map 2-S Carroll County

<u>Carr's Ingenuity</u> 256 acres

Prior Survey for Penelope Charchilla and Thomas Cockey Deyesburg (1030 a.)
Sp. Warr. 11 Sept., 1772 (1400 a.) John Wilmott
Assigned 2 Dec., 1772 to Aquilla Carr
Survey 11 Jan., 1773 Aquilla Carr
Assigned 20 Sept., 1773 to John Wilmot
Patent 17 Nov., 1773 John Wilmot
BC & GS 47/298 and BC & GS 44/316
Pat. Cert. 950

Tract 129 Map 2-SE Baltimore County

<u>Catey's Delight</u> 154 1/2, 4 p.

Resurvey of Sideling Hill (50 a.), Catey's Hill (31 1/2 a.), Christopher's Fair Play (18 a.), Stansbury's Grove (1448 a.), New London (250+) and Christopher's Luck (327 3/4 a.)
Sp. Warr. 5 Feb., 1817 Christopher Boblets
Survey 10 Apr., 1817 Christopher Boblets
Patent 8 Dec., 1817 Christopher Boblets
IB D/601 and IB E/73
Pat. Cert. 969

Tract 130 Map 2-SE Baltimore County

<u>Catey's Hill</u> 31 1/2 acres

Prior Survey for Catey's Delight (153 1/2 a.)
Comm. Warr. 23 Dec., 1793 (14 a.) Michael Boblet
Survey 8 May, 1794 Christopher Boblitz
Patent 5 Oct., 1798 Christopher Boblitz
IC O/462 and IC M/466
Pat. Cert. 972

Tract 131 (East of Map 1) Baltimore County

<u>Chalk's (Chaulk's) Meadow</u> 13 acres

(Reference Tract - Stoney Level (78 1/2 a.))
Prior Survey for Expedition (337 a.)
Comm. Warr. 9 Oct., 1794 William Dorsey
Survey 13 Nov., 1794 William Dorsey
Patent 24 June, 1796 Peter Shock
IC O/37 and IC M/12
Pat. Cert. 980

Tract 132 Map 2-SW Carroll County

<u>Chaney's Ramble</u> 62 1/4 acres

Prior Survey for Penelope Charchilla and Thomas Cockey Deyesburg (340 1/2 a.)
Warrant 21 June, 1771 (62 1/4 a.) James Boring
Survey 29 Oct., 1771 James Boring
Patent 19 May, 1772 James Boring
BC & GS 45/194 and BC & GS 42/283
Pat. Cert. 998

Tract 133 Map 1-NW Baltimore County

<u>Charles Delight</u> 138 acres

Survey 7 Aug., 1787 George Mire(s)
Assigned 21 May, 1790 to Michael Erhart
Patent 1 May, 1790 Michael Erhart
IC F/8 and IC E/455
Pat. Cert. 1005

Tract 134 Map 1-NW Baltimore County

(Charles Delight) 10 3/4 acres

Survey (later survey only)
(Original patent never located)
(In a later Baltimore County Deed (29 July, 1874; J.B. Vol. 88, Folio 421) this tract is part of a whole tract called "Charles Delight" which is the same as Charles Delight (138 acres) with this tract simply added. No prior information could be found.)

Tract 135 Map 2-SE Baltimore County

<u>Charles' Flower Knot</u> 50 acres

Prior Survey for New London (250 3/4 a.)
Comm. Warr. 9 Apr., 1774 (50 a.) Hny. Fitchet (Fite)
Assigned 8 Oct., 1774 to Charles Bouvellet (Boblet)
Survey 8 Oct., 1774 Charles Boblet(s)
Patent 2 May, 1787 Charles Boblet(s)
IC B/685 and IC C/128
Pat. Cert. 1009

Tract 136 Map 2-SE Baltimore County

<u>Charles' Hill</u> 29 3/4 acres

Prior Survey for New London (250 3/4 a.)
Sp. Warr. 8 Apr., 1790 (30 a.) Charles Boblets
Survey 30 Mar., 1791 Charles Boblitz
Patent 28 Feb., 1792 Charles Boblitz
IC F/459 and IC H/20
Pat. Cert. 1010

Tract 137 Map 2-SE Baltimore County

<u>Charles' Rue</u> 30 acres

Prior Survey for New London (250 3/4 a.)
Comm. Warr. 27 Jan., 1786 (30 a.) Charles Boblets
Survey 21 Mar., 1786 Charles Boblitz
Patent 7 Aug., 1787 Charles Boblitz
IC C/201 and IC D/25
Pat. Cert. 1018

Tract 138 Map 2-C Carroll County

<u>Charles' Trouble</u> 480a., 1 r., 26 p.

Resurvey of Stansbury's Grove (1448 a.), Borings Triangle (15 a.), Big Owl Hill (245 1/2 a.), Hope Restored (20 a.) Happiness (39 a.), Price's Glory (50 a.) and Forelorn Hope (23 +)
Survey 29 June, 1829 Wm. P. Mills, Jr.
(apparently never patented - no further records)

Tract **
Chaulk's Meadow (See Chalk's Meadow)

Tract 139 Map 1-W Baltimore County
<u>Chestnut Hills</u> 58 1/4 acres
Comm. Warr. 22 June, 1801 (5 a.) Gottlieb Hoffman
Survey 19 May, 1802 Gottlieb Hoffman
Assigned 17 June, 1802 to William Hoffman
Patent 19 May, 1803 William Hoffman
IC R/233 and IC P/462
Pat. Cert. 1035

Tract 140 Map 1-NW Baltimore County
<u>Chestnut Ridge</u> 75 acres
Survey 30 Jan., 1787 Fredrick Willhelm
Assigned 26 Dec., 1789 to Henry Caufman
Patent 27 Dec., 1790 Henry Caufman
IC F/177 and IC G/23
Pat. Cert. 1040

Tract 141 Map 2-W Carroll County
<u>Chestnut Ridge</u> 17 3/4 acres +23 p
Prior Survey for Manheim (161 1/2 a.)
Sp. Warr. 29 Aug., 1791 (20 a.) Jacob Ploucher
Survey 25 Oct., 1791 Jacob Plocher
Patent 17 Aug., 1801 Jacob Plucker
IC N/629 and IC P/188
Pat. Cert. 1041

Tract 142 Map 2-E Baltimore County
<u>Christiana Creek</u> 102 1/2 acres
Resurvey of Locust Bottom (91 a.) and Addition to Paper Mill Hills (8 3/4 a.)
Survey 14 Mar., 1788 William Hoffman
Unpat. Cert. 314

Tract 143 (not mapped) Carroll County
<u>Christian's Chance</u> (Choice) 166 acres
(Reference Tracts - Pleasant Springs Res. (444 + a.) and Carr's Inginuity (256 a.))
Resurvey of Merryman's Grotto (50 a.)
Prior Survey for Green's Visenburg (171 3/4 a.)
Warrant 13 Feb., 1771 (1500 a.) John Wilmot
Assigned 12 Aug., 1772 (165 a.) to Chris. Shilling
Survey 21 Aug., 1772 Christian Shilling
Patent 20 Apr., 1773 Christian Shilling
BC & GS 47/341 and BC & GS 42/515
Pat. Cert. 1063

Tract 144 Map 2-NE Carroll County
<u>Christian's Lott</u> 5 acres
Sp. Warr. 13 Mar., 1794 (15 a.)
John and Christian Baker
Survey 25 Mar., 1794 John and Christian Baker

(Christian's Lott - continued)
Assigned 4 Oct., 1822 by Barbary Baker (widow of John Baker) 1/2 interest to Christian Baker.
Patent 10 Oct., 1822 Christian Baker
IB and GGB G/722 and IB and GGB F/185
Pat. Cert. 1065

Tract 145 Map 2-SE Baltimore County
<u>Christopher's Fair Play</u> 18 acres
Prior Survey for Catey's Delight (153 1/2 a.)
Spec. Warr. 23 Apr., 1792 (24 a.) Christopher Boblet
Survey 17 Nov., 1792 Christopher Boblitz
Patent 5 Oct., 1798 Christopher Boblitz
IC O/462 and IC M/467
Pat. Cert. 1061

Tract 146 Map 2-E Carroll County
<u>Christopher's Luck</u> 327 3/4 acres
Resurvey of Foster's Neglect (104 a.) and Stansbury's Grove (1448 a.)
Sp. Warrant 24 Mar., 1791 Christopher Hair
Survey 19 Mar., 1792 Christopher Hair
Patent 19 Nov., 1801 Christopher Hair
IC N/632 and IC P/232
Pat. Cert. 1062

Tract 147 Map 1-W Baltimore County
<u>Church Lot</u> 13 acres
Comm. Warr. 12 Feb., 1794 (5 a.) Willaim Hoffman
Comm. Warr. 13 Nov., 1793 (12 a.) William Hoffman
Survey 25 Apr., 1794 William Hoffman
Patent 30 Mar., 1798 William Hoffman
IC O/454 and IC M/397
Pat. Cert. 1071

Tract 148 Map 1-SW Baltimore County
<u>Clapperstone's Range</u> (Ridge) 25 acres
Prior Survey of Hair's Beautiful Prospect (327+)
Comm. Warr. 14 June, 1792 (25 a.) Jacob Hair
Survey 25 Nov., 1792 Jacob Hair
Patent 19 Nov., 1794 Jacob Hair
IC I/629 and IC H/571
Pat. Cert. 1074

Tract 149 Map 2-S Carroll County
<u>Coal Pit</u> 27 acres
Prior Survey for Petersburg Resurvey (1003 a.)
Comm. Warr. 1 May, 1755 (27 a.) Jacob Shilling
Survey 15 May, 1755 Jacob Shilling
Patent 15 May, 1755 Jacob Shilling
BC & GS 4/277 and BC & GS 2/245
Pat. Cert. 1126

Tract 150 (South of Map 2) Carroll County
<u>Colchester</u> 33 acres
(Reference Tract - Petersburg Resurvey (1003 a.))
Prior Survey for Petersburg Resurvey (1003 a.)
Warrant 12 June, 1759 (33 a.) Jacob Shilling
Survey 20 June, 1759 Jacob Shilling
Patent 20 June, 1759 Jacob Shilling
BC & GS 12/21 and BC & GS 8/480
Pat. Cert. 1116

Tract 151 (not mapped) Baltimore County
<u>Cole's Clearance</u> 170 acres
Survey 24 April, 1766 Joseph Cole
(Never patented)
(Located next to Cross's Beginning (27 a.))
(See field survey books of William Smith (p. 54) in the Maryland State Archives)

Tract **
Cole's Forest (See Cole's Frolick)

Tract 152 Map 1-E Baltimore County
<u>Cole's Frolick</u> 55 acres
Prior Survey for Phillips Industry (154 3/4 a.) and The Three Divisions (2519 a.)
Leased 11 Apr., 1751 Joseph Cole
Leaseholds GCB A/301
Survey 11 April, 1751 Joseph Cole
Unpat. Cert 329

Tract 153 Map 1-E Baltimore County
<u>Coles Meadows</u> 152 1/2 acres
Prior Survey for Dorsey's Plains (655 1/2 a.)
Leased 22 June, 1757 Joseph Cole
Leaseholds GGB A/305

Tract 154 Map 2-C Carroll County
<u>Colt's Pasture</u> 1 a. + 0 R + 19 p.
Comm. Warr. 2 Jan., 1817 (8 a.) John Keeth
Survey 19 Mar., 1817 John Keeth
Patent 6 Dec., 1821 John Keith
IB G/595 and IB E/547
Pat. Cert. 1150
Assigned (no date) rights to Colt's Pasture and All That We Could Get to John Hoffacker

Tract 155 (East of Map 1) Baltimore County
<u>Conrod's Chance</u> 99 acres
Prior Survey for Conrad's Chance Resurvey (142+)
Lands purchased in the Reserve by John Hand(s)
Survey 15 May, 1787 John Hand(s)
Assigned 15 Sept., 1787 to Peter Conrad
Patent 8 Dec., 1789 Peter Conrad
IC D/625 and IC E/210
Pat. Cert. 1181

Tract 156 (East of Map 1) Baltimore County
Conrod's Chance Resurvey 142 1/2 acres
Resurvey of Conrad's Chance (99 a.)
Prior Survey for Vaughn's Disappointment (504+)
Sp. War. 19 Feb., 1802 John Free
Survey 20 Feb., 1803 John Free
Patent 11 Mar., 1812 John Free
IB B/94 and IB A/157
Pat. Cert. 1182

Tract 157 (East of Map 1) Baltimore County
Contention 1 a., 1 r., 16 p.
(Reference Tract - Vaughn's Disappointment (504+ a.))
Sp. Warr. 20 Dec., 1816 (3 a.) John Cooper
Survey 15 Nov., 1817 John Cooper
Unpat. Cert. 362

Tract 158 Map 2-W Carroll County
Continuance (Contrivance) 30 acres
Sp. Warr. 4 Mar., 1767 Henry Feeder
Renewed 2 Feb., 1768 Henry Feeder
Survey 21 Sept., 1768 Henry Feeder
Patent 26 June, 1772 Henry Feeder (Feather)
BC & GS 45/176 and BC & GS 42/317
Pat. Cert. 1192

Tract 159 Map 1-SE Baltimore County
Cooper's Discovery 77+ acres
Resurvey of Joshua's Trust (38 1/2 a.)
Sp. Warr. 12 Apr., 1837 John Cooper
Survey 1 Dec., 1837 John Cooper
Patent 31 Dec., 1838 John Cooper
GGB 4/252
Pat. Cert. 1209

Tract **
Cooper's Grove (See Cross's Grove)

Tract 160 (South of Map 1) Baltimore County
Cooper's Hill 9 a., 2 r., 15 p.
(Reference Tract - Cooper's Discovery)
Sp. Warr. 13 Mar., 1841 Micajah Cooper
Survey 7 Dec., 1841 Micajah Cooper
Patent 9 Jan., 1860 Micajah Cooper
WL & WS 3/296 and IM 1/379
Pat. Cert. 1210

Tract 161 Map 2-NW Carroll County
Coutz Lott Resurvey (2 Parts) (1) 339/1/20 acres
(2) 2/1/20 acres
Resurvey of Whispaw (283 1/2 a.)
Warrant Michael Kontz (Contz)
Assigned to heirs of Michael Kontz (who died
13 Nov., 1785): Peter, George, John, Henry
and Michael (Jr.) Kontz.

(Coutz Lott Resurvey - continued)

Survey 5 Nov., 1787 — Peter Coutz et. al.
Patent 25 Mar. 1817 — Peter, George, John, Michael and Henry Kountz
IC D/563
Pat. Cert. 1226

Tract 162 — Map 2-SE — Baltimore County
<u>Coutz (Couts) Meadow</u> — 27 acres
Comm. Warr. 20 Nov., 1760 (27 a.) Michael Countz
Survey 10 Apr., 1761 — Michael Countz
Patent 10 Apr., 1761 — Michael Countz (Cootz)
BC & GS 14/136 and BC & GS 15/100
Pat. Cert. 1227

Tract 163 — (not mapped) — Baltimore County
<u>Cox's Park</u> — 85 acres
Prior Survey for Stiltz Deer Park (606 1/2 a.)
Leased 10 Aug., 1752 — John Cox
Leaseholds GGB A/229
Survey 10 Aug., 1752 — John Cox
Unpat. Cert. 372
(not precisely located - at head of 'Roaring' Run)

Tract 164 — (South of Map 2) — Carroll County
<u>Coxe's Plesant Meadow</u> — 100 acres
(Reference Tract - Penelope Charcilla and Thomas Cockey Deyesburg (1030 a.))
Comm. Warr. 18 Mar., 1757 (60 a.) Jacob Cox
Survey 29 Sept., 1757 — Jacob Cox(e)
Patent 29 Sept., 1757 — Jacob Cox(e)
BC & GS 9/353 and BC & GS 10/9
Pat. Cert. 1251

Tract 165 — Map 1-SE — Baltimore County
<u>Crooked Billet</u> — 12 acres
Resurvey of Hurst Hills (462 3/4 a.)
Sp. Warr. 29 Oct., 1794 — James Marsh
Survey 30 Sept., 1795 — James Marsh
Unpat. Cert. 377
(Assigned by sale 31 Aug., 1802 to Jesse Hoshall)

Tract 166 — (not mapped) — Carroll County
Cross's Addition — (? acres)
William Burgubine
(Located next to The Wallet (118 1/2 a.) but no survey or other information found)

Tract 167 — (East of Map 1) — Baltimore County
<u>Cross's Beginning</u> — 27 acres
(Reference Tract - Cross's Beginning (129 a.))
Leased 1 May, 1746 — Solomon Cross
Leaseholds GGB A/272
Survey 1 May, 1746 — Solomon Cross

(Crosses Beginning (27 acres) - continued)
Unpat. Cert. 380
(not precisely located - probably prior survey for Cross's Beginning (129 a.))

Tract 168 (East of Map 1) Baltimore County
<u>Cross's Beginning</u> 129 acres
(Reference Tract - Cross's Grove Enl. (271 3/4 a.))
Survey 15 Nov., 1786 Jesse Cullinson
Assigned 12 Oct., 1789 to Uriah Freeland
Patent 12 Apr., 1809 Uriah Freeland
JK U/94 and JK T/66
Pat. Cert. 1286

Tract 169 Map 1-S Baltimore County
<u>Cross's Boot</u> 86 acres
Prior Survey for Cross's Boot Enlarged (338 a.)
Survey 27 Apr., 1761 Solomon Cross
(Leased Land - no record)

Tract 170 Map 1-S Baltimore County
<u>Crosses Boot Enlarged</u> 338 Acres
Resurvey of Cross's Boot (86 a.) and Cross's Hunting Ground (38 a.)
Prior Survey for Markey's Luck (341 1/4 a.)
Purchased land in the Reserve Barbary Markey
Survey 18 Jan., 1787 Barbary Markey
Patent 4 Aug., 1789 Barbary Markey
IC D/535 and IC E/94
Pat. Cert. 1287

Tract 171 (not mapped) Baltimore County
<u>Cross's Delight</u> (? acres)
Survey 31 May, 1766 Richard Cross
(never patented)
(not precisely located - south of Back Branch)
(See field survey books of William Smith (p.57) in the Maryland State Archives)

Tract 172 Map 1-SE Baltimore County
<u>Cross's Folly</u> 31 a.
Reservey of Solomon's Reserve (80 a.)
Sp. Warr. 26 Nov., 1796 (40 a.) John Cross
Survey 2 Nov., 1797 John Cross of Solomon
Patent 13 Nov., 1800 John Cross of Solomon
IC P/48 and IC N/487
Pat. Cert. 1291

Tract 173 Map 1-E Baltimore County
<u>Cross's Grove</u> 70 acres
Prior Survey for Cross's Grove (271 3/4 a.)
Leased 21 June, 1757 Zachariah Cross
Leaseholds GGB A/233

Tract 174 Map 1-E Baltimore County

Cross's Grove 271 3/4 acres

Resurvey of Cross's Grove (70 a.)
Prior Survey for Expedition (337 a.) and
Vaughn's Disapointment (504 1/2)
Survey 14 Nov., 1786 Jacob Marshall
Patent 22 Nov., 1791 Jacob Marshall
IC F/ 538 and IC E/821
Pat. Cert. 1292

Tract 175 Map 2-N Carroll County

Cross's Grudge 32 acres

Comm. Warr. 20 Aug., 1765 (32 a.) John Teeley
Survey 4 Dec., 1765 John Teeley, Jr.
Patent 4 Dec., 1765 John Teeley, Jr.
BC & GS 30/305 and BC & GS 31/204
Pat. Cert. 1293

Tract 176 Map 1-S Baltimore County

Cross's Hunting Ground 38 acres

Prior Survey for Cross's Boot Enlarged (338 a.)
(probably leased land - no records in archives)

Tract 177 Map 2-C Carroll County

Cross's Meadow(s) 70 acres

Prior Survey for Bollinger's Living (178 1/2 a.)
and Big Owl Hill (245 1/2 a.)
Warrant 11 Dec., 1749 (70 a.) William Cross
Survey 15 May, 1750 William Cross
Patent 15 May, 1750 William Cross
BY & GS 1/528 and TI 4/634
Pat. Cert. 1296

Tract 178 Map 1-E Baltimore County

Cross's Park (Part) 50 acres

Comm. Warr. 24 Nov., 1742 (100 a.) Nicholas Haile
Assigned 26 Feb., 1742 (50 a.) to John Cross
Survey 12 Mar., 1742 John Cross
Patent 30 Aug., 1743 John Cross
L.G. E/150 and EI 6/657
Pat. Cert. 1297

Tract 179 Map 2-N Carroll County

Cross's Whimm 18 acres

Comm. Warr. 11 Sept., 1759 (18 a.) John Miller
Survey 3 Mar., 1760 John Miller
Patent 3 Mar., 1760 John Miller
BC & GS 12/295 and BC & GS 13/204
Pat. Cert. 1298

Tract 180 Map 2-C Carroll County

Cuckold Point 50 acres

Warrant 25 Apr., 1751 (50 a.) Thomas Boring, Jr.
Survey 24 Oct., 1751 Thomas Boring (Jr.)

(Cuckold Point - continued)
Patent 29 May, 1752 Thomas Boring (Jr.)
Y & S 7/36 and BY & GS 4/449
Pat. Cert. 1308

Tract 181 Map 1-NE Baltimore County
Culley's Valley 78 Acres
Survey 8 Nov., 1786 Phillip Benders
Assigned 30 Oct., 1789 to Uriah Freeland
Assigned 3 Nov., 1789 to Nicholas Sigolar of York Co., Pa.
Patent 30 Sept., 1790 Nicholas Segolar (Zeigler)
IC F/182 and IC E/539
Pat. Cert. 1314

Tract 182 Map 1-C Baltimore County
Cullings (Cullins) Lott 128 1/4 acres
Resurvey of The Thre Divisions (2591 a.)
Survey 22 Mar., 1788 John Cummings
Patent 4 Dec., 1800 Thomas Cullins
IC N/488
Pat. Cert. 1310

Tract 183 (East of Map 1) Baltimore County
Cullinson's Old Aim 109 1/4 acres
(Reference Tract - Vaughn's Disappointment (504+ a.))
Prior Survey for Price's Hardship (160 a.)
Survey 16 Nov., 1786 Nathaniel John Davidge Bell
Assigned 30 May 1789 to Thomas Miller
(witnesses: Mary Bell and Jemima Walker)
Patent 3 Jan., 1794 Thomas Miller
IC I/435 and IC G/336
Pat. Cert. 1312

Tract 184 Map 2-NE Carroll County
Curfman's (Courseman's) Choice 27 acres
Prior Survey for Beaver Trap (609 a.)
Leased 10 Jan., 1759 Daniel Curfman (Kirkman)
(witnesses: William Lux and William Smith)
Leaseholds GGB A/212

Tract 185 (East of Map 1) Baltimore County
Curfman's Folly 106 3/4 acres
(Reference Tract - Cross's Park (50 a.))
Prior Survey for Curfman's Folly (84 a.)
Survey 21 Sept., 1790 Bennett Hurst
Patent 29 Dec., 1790 Bennett Hurst
IB D/26
Pat. Cert. 1319

Tract 186 (East of Map 1) Baltimore County
Curfman's Folly (2 parts) 84 acres
(Part I - 51 acres; Part II - 33 acres)
(Reference Tract - Cross's Park (50 a.))

(Curfman's Folly (84 acres) - continued)
Resurvey of Curfman's Folly (106 3/4 a.)
Survey 19 Dec., 1792 Bennett Hurst
Assigned 15 Jan., 1812 by heirs of Bennett Hurst:
Bennnett (Jr.), Benedict, Shadrack Bazel and Meshack Hurst, Elender Mash (w/o James), Athey Mash (w/o Elijah) Mary Hoshall (w/o Jesse)
to Jacob Hurst
Patent 2 Feb., 1813 Jacob Hurst
IB D/27, JK T/536 (Part I) and JK T/537 (Part II)
Pat. Cert. (could not locate)

Tract 187 Map 2-N Carroll County
Curpoy 50 acres
Prior Survey for Addition to Curpoy (158 a.)
Warrant 1 Feb., 1759 (50 a.) Christian Shafer
Survey 28 Mar., 1759 Christian Shaver
Patent 28 Mar., 1759 Christian Shaver
BC & GS 12/4 and BC & GS 8/589
Pat. Cert. 1323

Tract 188 Map 1-E Baltimore County
Davis's Chance 316 3/4 acres
Survey 3 Nov., 1786 John Davis
Unpat. Cert. 395

Tract 189 Map 1-N Baltimore County
Davis's Farm 182 1/2 acres
Purchased part of the Reserve - William Davis, Edmund Stansbury and Jacob Marshall
Survey 16 Mar., 1787 William Davis, Edmund Stansbury and Jacob Marshall
Assigned 25 May, 1799 to James Davis
Witness: William Matthews
Patent 17 Mar., 1812 James Davis
IB B/93 and IB A/156
Pat. Cert. 1353

Tract 190 Map 1-NE Baltimore County
Davis's Folly 57 acres
Comm. Warr. 2 Jan., 1794 (20 a.) Abraham Smith
Assigned 17 Jan., 1794 to Benjamin Davis
Survey 27 Nov., 1794 Benjamin Davis
Unpat. Cert. 396

Tract 191 (South of Map 2) Carroll County
Dear Bought 15 acres
(Reference Tract - Penelope Charcilla and Thomas Cockey Deyesburg (1030 a.) and Addition to Cox's Pleasant Meadow (100 a.))
Comm. Warr. 25 Feb., 1775 Daniel Pennington
Renewed 23 Aug., 1775 Daniel Pennington
Assigned 20 Oct., 1775 to Jacob Cox

(Dear Bought - continued)
Survey 15 Feb., 1776 — Jacob Cox
Patent 23 Apr., 1792 — Jacob Cox
IC F/650
Pat. Cert. 1366

Tract 192 — Map 2-E — Baltimore County
<u>Deer Park</u> — 21 3/4 acres
Comm. Warr. 1 Feb., 1792 (50 a.) Ezekiel Boring
Survey 22 Nov., 1792 — Ezekiel Boring
Patent 2 June, 1794 — Ezekiel Boring
IC I/622 and IC G/401
Pat. Cert. 1380

Tract 193 — Map 1-E — Baltimore County
<u>Deer Park</u> — 560 acres
Prior Survey for Deer Park Completed (935 1/2 a.)
Survey 2 Mar., 1787 — Thomas Gist (Jr.)
Patent 4 Jan., 1790 — Thomas Gist, Jr.
IC F/7 and IC E/258
Pat. Cert. 1382

Tract 194 — Map 1-E — Baltimore County
<u>Deer Park Completed</u> — 935 1/2 acres
Resurvey of Deer Park (560 a.) Addition to Deer Park (313 3/4 a.) and Addition (23 a.)
Sp. Warr. 2 Dec., 1807 — Thomas Gist
Survey 10 Nov., 1808 — Thomas Gist
Patent 12 Mar., 1810 — Thomas Gist (Jr.)
JK U/305 and JK T/258
Pat. Cert. 1384

Tract 195 — Map 2-NE — Carroll County
<u>Dehoos's (Dehoof's) Pleasure</u> — 14 1/4 acres
Warrant 7 May, 1782 (15 a.) George Dehoss
Survey 28 May, 1782 — George Dehoos (Dehoof)
Assigned 15 Mar., 1793 to John Sowers
Patent 13 Mar., 1794 — John Sowers
IC I/420 and IC G/362
Pat. Cert. 1388

Tract 196 — Map 1-NW — Baltimore County
<u>Delectable Hill(s)</u> — 9 3/4 acres
Sp. Warr. 13 Mar., 1794 (15. a.) Christian and John Baker
Survey 25 Mar., 1794 — Christian and John Baker
Patent 4 Jan., 1797 — Christian and John Baker
Assigned 5 Mar., 1797 to Nicholas Ehrhart
IC O/245 and IC M/138
Pat. Cert 1389

Tract 197 — Map 2-W — Carrol County
<u>Deyes Chance</u> — 50 acres
Prior Survey for Deyes Chance Enlarged (68 1/2)
Comm Warr. 28 Feb., 1755 (150 a.) Thos. C. Deye

(Deyes Chance - continued)

Survey 10 June, 1755 — Thomas Cockey Deye
Patent 10 June, 1755 — Thomas Cockey Deye
BC & GS 4/286 and BC & GS 3/196
Pat. Cert. 1409

Tract 198 — Map 2-W — Carroll County
<u>Deye's Chance Enlarged</u> — 68 1/2 acres
Resurvey of Deye's Chance (50 a.)
Sp. Warr. 8 Mar., 1791 — Thomas Cockey Deye
Survey 8 Sept., 1791 — Thomas Cockey Deye
Patent 28 Mar., 1793 — Thomas Cockey Dey(e)
IC I/154 and IC H/257
Pat. Cert. 1410

Tract 199 — Map 2-N — Carroll County
<u>Dilley's (Tilley's) Meadow</u> — 27 acres
Prior Survey for Hobson's Choice (106 1/4)
Comm. Warr. 5 Nov., 1761 (27 a.) John Nearoff of Anne Arundel Co.
Renewed 4 June, 1765 — John Nearoff
Assigned 2 Dec., 1765 to John Dilly
Survey 3 Dec., 1765 — John Dilly
Patent 3 Dec., 1765 — John Dilly
BC & GS 30/324 and BC & GS 32/109
Pat. Cert. 1420

Tract 200 — Map 2-W — Carroll County
<u>Dilley's Venture</u> — 49 1/2 acres
Comm. Warr. 7 June, 1758 (49 1/2 a.) John Dilly
Survey 28 Nov., 1758 — John Dilly
Patent 29 Nov., 1759 — John Dilly
BC & GS 12/1 and BC & GS 10/246
Pat. Cert. 1421

Tract 201 — Map 1-SE — Baltimore County
<u>Dinah's (Dina's) Delight</u> — 88 acres
Prior Survey for Schieming Defied Res. (194 3/4)
Leased 2 Apr., 1746 — Charles Cole
Leaseholds GGB A/294
Survey 23 July, 1746 — Dinah Cole
Unpat. Cert. 430

Tract 202 — Map 1-N — Baltimore County
<u>Dixon's Marsh(es)</u> — 53 acres
Survey 21 Mar., 1787 — Jacob Dixon
Patent 22 Apr., 1790 — Jacob Dixon
IC F/26 and IC E/327
Pat. Cert. 1440

Tract 203 — Map 1-NE — Baltimore County
<u>Dorsey's Chance</u> — 21 acres
Sp. Warr. 1 Sept., 1802 — William Dorsey
Survey 4 Dec., 1801 — William Dorsey

(Dorsey's Chance - continued)
Patent 17 Oct., 1803 William Dorsey
IC Q/257 and IC S/121
Pat. Cert. 1453

Tract 204 Map 1-NE Baltimore County
Dorsey's Delay 3 a., 0 r., 30 p.
Resurvey of Myer's Luck (12 1/2 a.)
Comm. Warr. 13 Jan., 1812 (5 A.) Christian Hoffman of Baltimore County
Assigned 5 Jan., 1813 to Frederick Dicks
Survey 6 Jan., 1813 Frederick Dicks
Patent 9 Jan., 1822 Frederick Dicks
IB G/664 and IB F/141
Pat. Cert. 1461

Tract 205 Map 1-E Baltimore County
Dorsey's Plains 655 1/2 acres
Resurvey of Cole's Meadows (152 1/2 a.)
Survey 7 Mar., 1787 Elisha Dorsey
Patent 15 Oct., 1798 Elisha Dorsey
IC O/464 and IC L/498
Pat. Cert. 1475

Tract 206 Map 2-NE Carroll County
Double Trouble 628 acres
Resurvey of Beaver's New Invention (170 a.), Vaughn's Meadows (27 a.) and Beaver Trap (609 a.)
Sp. Warr. 3 Dec., 1789 John Beaver
Assigned 16 Sept., 1790 to George Walker and Joseph Curtis
Survey 19 Oct., 1790 George Walker & Joseph Curtis
Unpat. Cert. 448
(nearly the same tract as Beaver Trap (609 a.))

Tract 207 (East of Map 1) Baltimore County
Dreary Wilderness 148 acres
(Reference Tract - Solomon's Wisdom (156 1/4 a.))
Survey 14 Nov., 1786 Samuel Morris
Patent 15 Dec., 1794 Samuel Morris
IC I/612 and IC H/589
Pat. Cert. 1490

Tract 208 (West of Map 2) Carroll County
Dugg Hill (? acres)
(Reference tracts - Overprone (24 1/2 a.), Mount Pleasant (17 a.), Mt. Hendrix (48 a.), Richards Chance (50 a.), Whispaw (189 a.) and Belfast (50 a.))
Very early tract (pre 1741) owned by William Wright and sited in many other early surveys. It seems to be located just to the west of Map No. 2. No survey information found. See will of William Wright (23 Aug., 1740), Baltimore County Wills: Volume 8 folio 103.

Tract 209 Map 2-W Carroll County
<u>Easenburgh</u> 15 acres
Comm. Warr. 25 Feb., 1761 John McHard
Assigned 2 June, 1761 to Henry Feather (Feeder)
Survey 24 Aug., 1761 Henry Feather
Patent 24 Aug., 1761 Henry Fetter
BC & GS 14/661 and BC & GS 15/501
Pat. Cert. 1523

Tract 210 (South of Map 2) Carroll County
<u>Eastern Branch</u> 178 acres
(Reference Tract - Petersburg Resurvey (1003 a.))
Prior Survey for Easter Branch Resurvey (275 a.)
Purchased 10 Feb., 1752 (178 a.) Chris. Thornback
Survey 25 Mar., 1752 Christopher Thornback
Patent 14 Jan., 1754 Christopher Thornback
BC & GS 1/240 and BC & GS 8/427
Pat. Cert. 1524

Tract 211 (South of Map 2) Carroll County
<u>Eastern Branch Resurvey</u> 275 acres
(Reference Tract - Petersburg Resurvey (1003 a.))
Resurvey of Easter Branch (178 a.)
Sp. Warr. 9 Oct., 1756 Christopher Thornback
Survey 25 March, 1757 Christopher Thornback
Patent 25 Oct., 1758 Christopher Thornback
BC & GS 9/374
Pat. Cert. 1525

Tract 212 Map 1-SW Baltimore County
<u>Edward's Choice</u> 102 1/2 acres
Survey 16 Nov., 1789 John Edwards
Patent 13 Mar., 1792 John Edwards
IC F/749 and IC H/32
Pat. Cert. 1535

Tract 213 Map 2-N Carroll County
<u>Eight Sisters</u> 480 3/4 acres
Resurvey of Stoney Hill (1291 a.) and
Remnant (9 1/4 a.)
Sp. Warr. 19 Nov., 1804 Alexis Lemon
Survey 20 Apr., 1805 Alexis Lemon
Patent 24 Feb., 1807 Alexis Lemmon
IC R/513 and IC S/473
Pat. Cert. 1548

Tract 214 Map 2-SW Carroll County
<u>Ely's Rock</u> 25 acres
Prior Survey for Borings Range (1375 a.)
Warrant 23 Apr., 1754 (25 a.) James Boring
Survey 15 Oct., 1754 James Boring
Patent 15 Oct., 1754 James Boring
BC & GS 4/276 and BC & GS 2/18
Pat. Cert. 1581

Tract 215 Map 1-E Baltimore County
<u>Enoch's Hill</u> 15 1/4 acres
Prior survey for Mount Pleasant (28+ a.)
Sp. Warr. 12 Feb., 1794 (21 a.) William Dorsey
Survey 21 Mar., 1794 William Dorsey
Assigned 25 Oct., 1794 to Moses Lemon
Patent 15 Oct., 1798 Moses Lemon
IC O/465 and IC L/497
Pat. Cert. 1593

Tract 216 Map 2-C Carroll County
<u>Errick</u> 17 acres
Sp. Warr. 24 Oct., 1776 (180 a.) Thomas Stansbury and Henry Wareheim
Survey 14 Feb., 1777 Thomas Stansbury and Henry Wareheim
Patent 30 May, 1788 Thomas Stansbury and Henry Wareheim
IC D/177 and IC C/431
Pat. Cert. 1604

Tract 217 Map 1-NW Baltimore County
<u>Eve</u> 64 acres
Survey 12 Mar., 1787 William Hoffman
Patent 2 Apr., 1789 William Hoffman
IC D/412 and IC C/645
Pat. Cert. 1616

Tract 218 Map 2-SE Baltimore County
<u>Eve's Garden</u> 5 1/2 acres
Prior Survey for New London (250 3/4 a.)
Sp. Warr. 29 Feb., 1792 (5 a.) Charles Boblets
Survey 20 Oct., 1792 Charles Boblitz
Patent 8 Nov., 1793 Charles Boblitz
IC I/434 and IC H/381
Pat. Cert. 1617

Tract 219 Map 2-NW Carroll County
<u>Everybodies Land</u> 390 1/2 acres
(Parcel #13 only in map area)
Resurvey of George's Right (25 a.)
Prior Survey for Utz'z Inheretance (426 a.)
Sp Warr. 1 Mar., 1786 (176 3/4 a.) Charles Ridgely and Andrew Stiger
(omitted to pay caution money, land vacated)
Sp. Warr. 13 Oct., 1790 (376 a.) Henry Lamott
Survey 23 Aug., 1791 Henry Lamotte
(13 parcels of vacant land)
Patent 27 Sept., 1792 Henry Lamotte
IC I/120 and IC H/155
Pat. Cert. 1623

Tract 220 Map 2-SW Carroll County

Everthing Needful 2922 acres

Resurvey of Red Oak Land (2 a.)
Prior Survey for Everthing Needful Resurvey (1646 a)
Sp. Warr. 25 May, 1761 Patrick Irons
Renewed 21 Nov., 1761 Patrick Irons
Renewed 19 May, 1762 Patrick Irons
Renewed 23 Sept., 1762 Patrick Irons
Assigned 22 Mar., 1763 to Capt. Richard Richards
Survey 25 May, 1764 Capt. Richard Richards
Unpat. Cert. 499

Tract 221 Map 2-SW Carroll County

Everything Needful Resurvey 1646 acres

Resurvey of Red Oak Land (2 a.) and
Everything Needful (2922 a.)
Prior Survey for Everything Needful Corr. (1755+)
* Survey 16 May, 1763 Richard Richards
Assigned 6 Sept., 1765 to Andrew Steiger,
William Winchester and Samuel Owings
Comm. Warr. 8 Feb., 1763 (30 a.) Samuel Owings, Jr.
Comm. Warr. 7 Dec., 1764 (100 a.) Andrew Stiger
Comm. Warr. 11 Dec., 1761 (50 a.) Samuel Lane
Assigned 25 Sept., 1764 to Samuel Owings, Jr.,
(witness: Charles Kelley)
Comm. Warr. 17 Nov., 1761 (40 1/2) George Greenwood
Assigned 10 May, 1764 to John Chrussonan
(witness: Henry Piper)
Assigned 17 Sept., 1764 to William Winchester
Comm. Warr. 22 Feb., 1763 (50 a.) Martin Winter
Assigned 14 Sept., 1765 (22 a.) to Wm. Winchester
Comm. Warr. 26 Sept. 1765 (50 a.) Samuel Owings,Jr.
Comm. Warr. 17 Aug., 1746 (25 a.) Edward Ford
Assigned 26 Sept., 1765 (16 a) to Sam. Owings, Jr.
* Patent 29 Sept., 1765 William Winchester, Andrew
Stiger and Samuel Owings, Jr.
BC & GS 27/536
Pat. Cert. 1624

Tract 222 Map 2-SW Carroll County

Everything Needful Corrected 1755 a., 2 r., 32 p.

Resurvey of Everything Needful Corrected (1646 a.)
Sp. Warr. 25 Nov., 1785 Samuel Owings
Survey 14 Nov., 1786 Samuel Owings
Patent 17 Oct., 1789 Samuel Owings
IC D/633 and IC E/119
Pat. Cert. 1625

Tract 223 Map 2-SE Baltimore County

Everywhere 34 acres

Sp. Warr. 23 Mar., 1792 (15 a.) Charles Boblitz
Survey 17 Nov., 1792 Charles Boblitz
Patent 28 Dec., 1793 Charles Boblitz
IC I/452 and IC H/431
Pat. Cert. 1622

Tract 224 — Map 1-E — Baltimore County

<u>Expedition</u> — 337 acres

Resurvey of Stoney Level (78 1/4 a.), Chaulk's Meadow (13 a.), Narrow Passage (16 a.), North Spring (11 a.) and Vaughn's Disappointment (504+)

Sp. Warr. 28 Dec., 1809 — Peter Shock

Survey 1 Dec., 1810 — Peter Shock (Shank)

Unpat. Cert. 501

Tract 225 — Map 2-S — Carroll County

<u>Fairplay</u> — 61 acres

Comm. Warr. 21 Jan., 1767 (40 a.) Peter Fowble

Assigned 30 Mar., 1767 to Conrad Swhink

Comm. Warr. 19 Feb., 1767 (21 a.) Conrad Swhink

Survey 22 June, 1767 — Conrad Swhink (Shwink)

Patent 2 Nov., 1767 — Conrad Swinks

BC & GS 34/272 and BC & GS 32/264

Pat. Cert. 1633

Tract 226 — Map 1-C — Baltimore County

<u>Fairplay</u> — 50 acres

Survey 20 Mar., 1788 — John Tracy

Unpat. Cert. 507

Tract ***

Fairplay (See Christopher's Fairplay)

Tract 227 — (South of Map 2) — Carroll County

<u>Fancy</u> — 34 acres

(Reference Tract - Pleasant Springs (285 a.))

Sp. Warr. 26 May, 1774 (168 1/4 a.) Thos. Gist, Jr.

Assigned 1 July, 1774 (60 1/4 a.) to C. Singery

Survey 22 Oct., 1774 — Christian Singery

Assigned 1 Nov., 1775 to Thomas Gist, Jr.

Patent 1 Dec., 1775 — Thomas Gist, Jr.

BC & GS 51/19 and BC & GS 52/82

Pat. Cert. 1638

Tract 228 — Map 1-C — Baltimore County

<u>Farmer's Lot</u> — 287 1/4 acres

Prior Survey for Talbot's Plains (638 1/2 a.)

Survey 18 Mar., 1799 — William Davis, Edmund Stansbury and Jacob Marshall

Unpat. Cert. 511

Tract 229 — (South of Map 2) — Carroll County

<u>Fertile Meadows</u> — 50 acres

(Reference Tract - Petersburg Resurvey (1003 a.))

Prior Survey for Fertile Meadows (146 a.)

Sp. Warr. 17 Mar., 1761 (50 a.) John McLain

Assigned 9 Sept., 1761 to Richard Richards

Survey 12 Sept., 1761 — Richard Richards

Patent 12 Sept., 1761 — Richard Richards

BC & GS 14/665 and BC & GS 17/285

Pat. Cert. 1665

Tract 230 (South of Map 2) Carroll County
Fertile Meadows Revised 146 acres
(Reference Tract - Petersburg Resurvey (1003 a.))
Resurvey of Fertile Meadows (50 a.)
Prior Survey for Petersburg Resurvey (1003 a.)
Sp. Warr. 19 Aug., 1767 Richard Richards
Survey 13 Feb., 1768 Richard Richards
Patent 13 Feb., 1768 Richard Richards
BC & GS 40/206 and BC & GS 36/267
Pat. Cert. 1666

Tract 231 Map 1-NE Baltimore County
Fills (Phills) Hunting Ground 67 1/2 acres
Survey 31 Jan., 1787 Philip Stiltz
Patent 18 June, 1790 Philip Stiltz (Steltz)
IC F/5 and IC C/726
Pat. Cert. 1672

Tract 232 Map 1-NE Baltimore County
First and Second Parts of George's Mistake 58 3/4 a.
Comm. Warr. 4 Mar., 1791 (21 3/4 a.)
John Tracey of John
Sp. Warr. 7 Oct., 1791 (25 a.) John Tracey
Assigned 18 June, 1796 (46 3/4 a.) to Moses Freeland
Sp. Warr. 1 June, 1795 Moses Freeland
Survey 11 May, 1796 Moses Freeland
Patent 17 Mar., 1814 Moses Freeland
IB A/508 and IB B/633
Pat. Cert. 1680

Tract 233 Map 1-NE Baltimore County
Fisher's Hill 7 1/4 acres
Comm. Warr. 16 Aug., 1802 (12 a.) William Dorsey
Survey 24 Jan., 1803 William Dorsey
Assigned 24 Feb., 1803 to Jacob Hershinger
Patent 17 Jan., 1808 Jacob Hershinger
IC R/667 and IC S/599
Pat. Cert. 1686

Tract 234 (South of Map 1) Baltimore County
Fishmuckle Harry 10 acres
(Reference Tract - Hard Looks (27 a.))
Comm. Warr. 31 June, 1803 Basil Tracey
Survey 26 Nov., 1803 Basil Tracey
Assigned 1 Jan., 1804 to John Ruhl
Patent 30 Jan., 1810 John Ruhl
JK U/298 and JK T/234

Tract 235 Map 1-SW Baltimore County
Ford's Choice 157 acres
(Only a small bit of this tract in the map area)
Survey 18 Nov., 1762 Mordecai Ford
Patent 7 Dec., 1774 Mordecai Ford
BC & GS 50/31 and BC & GS 48/244
Pat. Cert. 1709

Tract 236 Map 1-S Baltimore County

<u>Ford's Enlargement</u> (? acres)

(Only a little bit of this large tract is inside the map area next to Ford's Choice and south of Singery's Trouting Stream (575 a.))

Resurvey of Horatio's Lot (314 a.), Thomas's Delight (25 a.), Little Profit (27 1/4 a.) and Great Profit (87 a.)

A Drawing of the plat and the name are given in Carroll County Envelope # 38 - (HSCC). No other information found.

Tract **

Ford's Fancy (Thomas Towson - 66 acres)

(Incorrectly referenced in survey (Towson's Grove))

(Not in this region, see Selsed in Limestone Valley)

Tract 237 Map 2-C Carroll County

<u>Forelorn Hope</u> 23 1/2 acres

Prior Survey for Neilson's Puzzle (256 1/4 a.)

Comm. Warr. 3 Apr., 1792 (108 3/4 a.) Charles Stansbury

(Certificate correction of Charles's Choice - unexecuted)

Comm. Warr. 24 Oct., 1792 (54 a., 1 r., 3 p.) Charles Stansbury

Survey 6 Aug., 1794 Charles Stansbury

Patent 24 Oct., 1795 Charles Stansbury

IC K/189 and IC G/589

Pat. Cert. 1716

Tract 238 Map 2-NW Carroll County

<u>Foreman's Run</u> 20 acres

Prior Survey for Troy (1634 a.)

Comm. Warr. 21 Nov., 1760 George Kiteyear

Survey 17 Jan., 1761 George Kiteyear

Patent 17 Jan., 1761 George Kiteyear (Riteyear)

BC & GS 14/144 and BC & GS 16/390

Pat. Cert. 1715

Tract **

Forster's Chance (See Foster's Chance)

Tract 239 (South of Map 2) Carroll County

<u>Foster's Beginning</u> 50 acres

Prior Survey for Foster's Meadow Enlarged (120)

Comm. Warr. 1 June, 1761 (100 a.) John Foster

Survey 8 June, 1761 John Foster

Patent 8 June, 1761 John Foster

BC & GS 14/156 and BC & GS 15/162

Pat. Cert. 1731

Tract 240 Map 2-SW Carroll County

<u>Foster's Chance</u> 50 acres

Prior Survey for John Caltrider's Fancy (229+ a.)

(Foster's Chance - continued)
Comm. Warr. 1 June, 1761 (150 a.) John Foster
Renewed 31 Mar., 1762 John Foster
Survey 27 Sept., 1762 John Foster
Patent 8 Aug., 1771 John Foster (Forster)
BC & GS 45/70 and BC & GS 43/159
Pat. Cert. 1723

Tract 241 Map 2-S Carroll County
<u>Foster's Hunting Ground</u> 1439 acres
Resurvey of Borings Range (1375 a.)
Prior Survey for Middlesex (8 1/4 a.) and
General Bonaparte (20 a.)
Sp. Warr. 8 Mar., 1771 John Foster
Renewed 9 Feb., 1773 (630 a.) John Foster
Survey 15 Feb., 1773 John Foster
Patent 14 June, 1773 John Foster
BC & GS 47/313 and BC & GS 46/108
Pat. Cert. 1733

Tract 242 (South of Map 2) Carroll County
<u>Foster's Lot</u> 30 acres
(Reference Tracts - Foster's Pleasant Meadow (259 +)
and Carr's Ingineuity (256 a.))
Prior Survey for Foster's Pleasant Meadow (259+ a.)
Comm. Warr. 30 Oct., 1763 (30 a.) John Foster
Survey 25 Apr., 1764 John Foster
Patent 25 Apr., 1764 John Foster
BC & GS 27/432 and BC & GS 26/350
Pat. Cert. 1734

Tract 243 (South of Map 2) Carroll County
<u>Foster's Meadow Enlarged</u> 120 acres
(Reference Tract - Foster's Pleasant Meadow (259+))
Resurvey of Foster's Beginning (50 a.)
Prior Survey for Foster's Pleasant Meadow (259+)
Sp. Warr. 9 Dec., 1761 John Foster
Survey 28 May, 1762 John Foster
Patent 29 Sept., 1763 John Foster
BC & GS 21/356 and BC & GS 20/673
Pat. Cert. 1735

Tract 244 Map 2-SE Carroll County
<u>Foster's (Forster's) Neglect</u> 104 acres
Prior Survey for Christopher's Luck (327 3/4 a.)
Comm. Warr. 9 Apr., 1762 Thomas Stansbury, Jr.
Renewed 15 Sept., 1763 Thomas Stansbury, Jr.
Renewed 10 Mar., 1763 Thomas Stansbury, Jr.
Renewed 29 Aug., 1763 Thomas Stansbury, Jr.
Assigned 23 Feb., 1764 to Thomas Stansbury III
Survey 24 Feb., 1764 Thoams Stansbury III
Patent 24 Feb., 1764 Thomas Stansbury III
BC & GS 27/494 and BC & GS 28/350
Pat. Cert. 1724

Tract 245 (South of Map 2) Carroll County

<u>Foster's Pleasant Meadow (Resurvey)</u> 259 1/2 acres

(Reference Tract - Towson's Grove (367 a.))

Resurvey of Foster's Lot (30 a.), Foster's Meadow Enlarged (120 a.), Polygon (340 a.), Carr's Inginuity (256 a.) and Christian's Chance (166 a.)

Sp. Warr. 28 July, 1789 John Foster

Survey 19 Apr., 1790 John Foster

Patent 17 Aug., 1791 John Foster

IC F/427 and IC E/739

Pat. Cert. 1736

Tract 246 (South of Map 2) Carroll County

<u>Fowble's (Foible's) Barren Hill</u> 35 1/2 acres

(Reference Tract - Pleasant Spring Resurvey (444+))

Warrant 27 May, 1773 (36 a.) Peter Fowple (Foible)

Survey 25 Nov., 1773 Peter Fowple

Patent 10 Nov., 1774 Peter Fowble

BC & GS 50/23 and BC & GS 49/100

Pat. Cert. 1699

Tract 247 (South of Map 2) Carroll County

<u>Fowble's (Fowple's) Lott</u> 240 acres

(Reference Tract - Petersburg Resurvey (1003 a.))

Resurvey of Shillings Lot (50 a.)

Prior Survey for Petersburg (1003 a.)

Sp. Warr. Peter Fowple

Renewed 16 Mar., 1765 Peter Fowple

Survey 20 May, 1765 Peter Fowple

Patent 27 July, 1765 Peter Fowple (Fowble)

BC & GS 27/527 and BC & GS 25/76

Pat. Cert. 1748

Tract 248 Map 2-S Carroll County

<u>Fowble's New Germany</u> 41 acres

Prior Survey for Petersburg (1003 a.)

Warrant 19 Dec., 1767 (41 a.) Peter Fowple

Survey 27 June, 1767 Peter Fowple

Patent 2 Nov., 1767 Peter Fowple

BC & GS 34/294 and BC & GS 33/136

Pat. Cert. 1749

Tract 249 (South of Map 2) Carroll County

<u>Fowble's New Town</u> 40 acres

(Reference Tract - Petersburg Resurvey (1003 a.))

Prior Survey for Petersburg (1003 a.)

Warrant 19 Feb., 1768 (60 a.) George Key

Assigned 18 Mar., 1768 (60 a.) to Peter Fowble

Survey 11 Apr., 1768 Peter Fowble

Patent 5 Nov., 1768 Peter Fowble

BC & GS 37/300 and BC & GS 33/514

Pat. Cert. 1744

Tract 250 Map 2-N Carroll County
Fox Range 25 acres
Comm. Warr. 3 Feb., 1790 (15 a.) Henry Lamott
Comm. Warr. 3 Feb., 1790 (10 a.) Henry Lamott
Survey 26 Jan., 1791 Henry Lamott
Patent 2 Apr., 1794 Henry Lamotte
IC I/616 and IC H/464
Pat. Cert. 1758

Tract 251 Map 2-SE Baltimore County
Frances Thwaitsburgh 496 3/4 acres
(Parcel 1 - 471 3/4 acres & Parcel 2 - 25 acres)
Resurvey of Petersburg (300 a.)
Prior Survey for Francis Thwaitsburg (542 a.)
Warrant 22 Aug., 1786 Thomas Cockey Deye
Survey 10 Aug., 1787 Thomas Cockey Deye
Patent 27 May, 1788 Thomas Cockey Deye
IC D/181 and IC C/427
Pat. Cert. 1764

Tract 252 Map 2-SE Baltimore County
Frances Thwaitsburgh Resurveyed 542 acres
(Parcel 1 - 511 1/2 acres & Parcel 2 - 30 1/2 acres)
Resurvey of Francis Thwaitsburg (496 3/4 a.)
Sp. Warr. 21 Dec., 1792 Thomas Cockey Deye
Survey 10 Dec., 1793 Thomas Cockey Deye
Patent 2 Oct., 1794 Thomas Cockey Deye
IC I/476 and IC H/544
Pat. Cert. 1765

Tract 253 Map 2-NE Carroll County
Fredericks Conquest 40 acres
Prior Survey for Hosbson's Choice (106 1/4 a.)
Comm. Warr. 3 Oct., 1770 (50 a.) Frederick Woolpack
Renewed 3 Oct., 1770 (40 a.) Frederick Woolpack
Survey 27 Mar., 1771 Frederick Woolpack
Patent 7 Jan., 1773 Frederick Woolpack
BC & GS 47/328 and BC & GS 43/515
Pat. Cert. 1780

Tract 254 Map 1-S Baltimore County
Friday 32 acres
Comm. Warr. 13 Jan., 1794 (32 a.) William Keith
Survey 24 Oct., 1794 William Keith
Assigned 26 May, 1807 to John Keith
Patent 14 Dec., 1810 John Keith
JK U/446 and JK T/420
Pat. Cert. 1820

Tract 255 Map 1-NW Baltimore County
Friendship 14 1/4 acres
Resurvey of Poplar Run (122 1/2 a.)
Sp. Warr. 15 Oct., 1809 William Hoffman
Survey 4 May, 1810 William Hoffman
Unpat. Cert. 557

Tract 256 Map 1-E Baltimore County
<u>Frosty Hill</u> 46 acres
Prior Survey for Pessel Tree (96 3/4 a.)
Com. Warr. 1 Mar., 1792 Abraham Collins
Com. Warr. 7 Nov., 1793 (6 A.) William Dorsey
Com. Warr. 20 Aug., 1794 (16 A.) John Tracey
Survey 1 Sept., 1794 William Dorsey
Patent 1 Sept., 1801 William Dorsey
IC N/631 and IC P/192
Pat. Cert. 1828

Tract 257 Map 1-NW Baltimore County
<u>Fruitful Garden</u> 45 1/4 acres
Survey 19 Mar., 1787 John Leahy of Penna.
Assigned 2 Sept., 1809 by estate of John Leahy of Baltimore County to his children and heirs: Mary Sharp, Catherine Siegly (Zeigler), Elizabeth Hanny, John Leahy (Jr.), Judith Leahy, Barbara Franklin, William Leahy and Susannah Leahy.
Patent 2 Sept., 1809 Mary Sharp et. al.
JK U/99 and JK T/160
Pat. Cert. 1830

Tract 258 (South of Map 2) Carroll County
<u>Gain's Delight</u> 50 acres
Prior Survey for Penelope Charchilla and Thomas Cockey Deyesburg (1030 a.)
Comm. Warr. 8 Sept., 1760 (50 a.) William Gain
Survey 20 Sept., 1760 William Gain
Patent 20 Sept., 1760 William Gain
BC & GS 12/499
Pat. Cert. 1834

Tract 259 Map 1-N Baltimore County
<u>Gains Rest</u> 28 1/2 acres
Prior Survey for Gain's Rest Enlarged (64 1/4 a.)
Leased land April, 1761 Richard Sampson
(no record in Maryland State Archives)

Tract 260 Map 1-N Baltimore County
<u>Gains Rest Enlarged</u> 64 1/4 acres
Resurvey of Gain's Rest (28 1/2 a.)
Prior Survey for Simmerman's Venture (79 1/2 a.)
Survey 12 Apr., 1787 Richard Sampson
Assigned 22 Feb., 1791 to Adam Smith
Patent 9 May, 1804 Adam Smith
IC R/277 and IC S/163
Pat. Cert. 1835

Tract 261 Map 2-S Carroll County
<u>General Bonaparte the Friend of Liberty</u> 20 acres
Resurvey of Foster's Hunting Ground (1439 a.) and Borings Range (1375 a.)
Sp. Warr. 9 May, 1797 (12 a.) Thomas Cockey Deye
Survey 20 Apr., 1798 Thomas Cockey Deye

(General Bonaparte the Friend of Liberty - continued)
Patent 16 Dec., 1799 Thomas Cockey Deye
IC O/678 and IC M/663
Pat. Cert. 1870

Tract 262 Map 1-SW Baltimore County
General Rouchambeau the Friend of General Washington Therefore the Friend of Universal Liberty
160 acres
Comm. Warr. 15 Aug., 1791 (215 a.) Thos. C. Deye
Comm. Warr. 10 Aug., 1792 (66 3/4 a.) Thos. C. Deye
Survey 13 Aug., 1792 Thomas Cockey Deye
Patent 28 Aug., 1793 Thomas Cockey Deye
IC I/133 and IC H/258
Pat. Cert. 1871

Tract 263 Map 1-N Baltimore County
George's Goodwill 5 acres
Prior Survey for Simmerman's Venture (79 1/2 a.)
Comm. Warr. 2 Dec., 1793 (5 a.) James Campbell
Assigned 8 Oct., 1794 to Adam Smith
Survey 8 Oct., 1794 Adam Smith
Patent 10 May, 1804 Adam Smith
IC R/278 and IC S/164
Pat. Cert. 1984

Tract 264 (West of Map 2) Carroll County
George's Right 25 acres
Prior Survey for Everybodies Land (390 1/2 a.) and I will take it if I can (26 1/2 a.)
Comm. Warr. 11 Nov., 1762 (25 a.) George Kiteyear
Survey 10 Mar., 1763 George Kittyear
Patent 10 Mar., 1763 George Kittyear
BC & GS 21/339 and BC & GS 22/364
Pat. Cert. 1880

Tract 265 Map 1-NE Baltimore County
George's Wisdom 332 1/2 acres
Resurvey of Beard's Branch (294 3/4 a.)
Sp. Warr. 17 May, 1791 George Nunemaker
Survey 12 May, 1792 George Nunemaker
Unpat. Cert. 586

Tract 266 (West of Map 2) Carroll County
German Church 25 Acres
(Reference Tracts - Worms (25 a.) and Everything Needful Corrected (1755+ a.))
Sp. Warr. 23 Oct., 1758 (25 a.)
Survey 20 Dec., 1758 Jacob Shilling, Philip Edelman, Jacob Utz, Michael Burn
Patent 20 Dec., 1758 Jacob Shilling, Philip Edelman, Jacob Utz, Michael Burn
BC & GS 12/4 and BC & GS 8/607
Pat. Cert. 1887

Tract 267 Map 1-C Baltimore County

<u>Gist's Delight</u> 10 acres

Prior Survey for Gist's Purchase (465 a.)
Comm. Warr. 27 Aug., 1793 (10 A.) Thomas Dorsey
Survey 17 Jan., 1794 George Gist
Patent 4 Jan., 1797 George Gist
IC O/244 and IC M/139
Pat. Cert. 1917

Tract 268 Map 1-W Baltimore County

<u>Gist's Disappointment</u> 4 1/4 a. + 2 p.

Com. Warr. 21 Dec., 1802 (10 a.) William Hoffman,Jr.
Assigned 16 Nov., 1803 to Peter Hoffman
Survey 17 Nov., 1803 Peter Hoffman
Patent 21 Feb., 1805 Peter Hoffman
IC R/328 and IC P/618
Pat. Cert. 1920

Tract 269 (South of Map 2) Carroll County

<u>Gist's Meadow(s)</u> 75 acres

(Reference Tract - Petersburg Resurvey (1003 a.))
Sp. Warr. 26 May, 1774 (168 1/2 a.) Thomas Gist, Jr.
Survey 10 Oct., 1774 Thomas Gist, Jr.
Patent 7 Sept., 1775 Thomas Gist, Jr.
BC & GS 50/241 and BC & GS 52/19
Pat. Cert. 1927

Tract 270 (South of Map 2) Carroll County

<u>Gist's Prospect</u> 49 1/2 acres

(Reference Tracts - Bratton's Desire (26 1/2 a.) and Wilmott's Lott (217 a.))
Warrant 19 Apr., 1775 (186 1/2 a.) Christian Singery
Assigned 10 May, 1775 to Thomas Gist
Assigned 6 Oct., 1775 (35 a.) to David Gist
Warrant 19 Oct., 1774 (46 a.) Peter Fowble
Renewal 8 May, 1775 (46 a.) Peter Fowble
Assigned 10 Oct., 1775 to Abraham Green
Assigned 12 Oct., 1775 (14 1/2 a.) to David Gist
Survey 10 Oct., 1775 David Gist
Patent 15 Feb., 1783 David Gist
IC A/79 and BC & GS 50/409
Pat. Cert. 1928

Tract 271 Map 1-NW Baltimore County

<u>Gist's Purchase</u> 465 acres

Resurvey of Gist's Delight (10 a.) and Mary's Delight (270 a.)
Sp. Warr. 8 Apr., 1795 George Gist
Survey 13 May, 1795 George Gist
Unpat. Cert 592

Tract 272 Map 1-NW Baltimore County

<u>Gist's Second Purchase</u> 159 1/2 acres

Sp. Warr. 2 June, 1796 (150 acres) George Gist

(Gist's Second Purchase - continued)
Survey 19 May, 1797 George Gist
Patent 27 June, 1798 George Gist
IC O/434 and IC M/439
Pat. Cert. 1931

Tract 273 Map 2-SW Carroll County
Gittinger's (Kittinger's) Glade 24 acres
Prior Survey for John Caltrider's Fancy (229+ a.)
Warrant 27 May, 1773 (24 a.) John Gittinger
Survey 10 Nov., 1773 John Kittinger
Patent 3 Dec., 1774 John Kittinger
BC & GS 50/28 and BC & GS 49/124
Pat. Cert. 1936

Tract 274 Map 2-NE Carroll County
Give it a Name 30 acres
Sp. Warr. 13 Dec., 1792 (30 a.) John Sours
Survey 28 Dec., 1792 John Sours
Patent 29 Mar., 1796 John Sours
IC O/59 and IC G/715
Pat. Cert. 1937

Tract 275 (not mapped) Baltimore County
Glen View
(Reference Tracts - Beard's Branch (294 3/4) and Adventure (41 3/4))
Named in deed of conveyance: George F. Miller and wife Leah Miller to Alfred K. Baker (Baltimore County Land Records Vol. NBM 250, Folio 214).
No other information found

Tract 276 Map 1-SE Baltimore County
Goat's Pasture 17 1/4 a., 29 p.
Comm. Warr. 25 June, 1806 (25 a., 0 r, 119 p.) Thomas Gist
Renewed 29 May, 1807 Thomas Gist
Survey 20 May, 1808 Thomas Gist
Patent 10 Mar., 1810 Thomas Gist
JK U/302 and JK T/250
Pat. Cert. 1940

Tract 277 Map 2-SW Carroll County
Good (Bold) Contrivance 15 acres
Sp. Warr. 15 Aug., 1767 (15 a.) George Myers
Survey 28 July, 1767 George Myers
Patent 27 July, 1768 George Myers
BC & GS 37/328 and BC & GS 36/12
Pat. Cert. 1949

Tract 278 (West of Map 2) Carroll County
Good Luck 524 1/2 acres
(Reference Tracts - Powell's Discovery Resurvey (148) and Everything Needful Corrected (1646 a.))
Resurvey of Winchester's Lot (972 a.) and others

(Good Luck - continued)

Sp. Warr. 30 Jan., 1793 — John Shower
Survey 20 Apr., 1793 — Capt. John Showers
Patent 8 Apr., 1795 — Capt. John Showers
IC H/658 and IC K/163
Pat. Cert. 1966

Tract 279 — Map 2-NE — Carroll County
<u>Good Neighbor</u> — 26 1/2 acres
Sp. Warr. 11 Oct., 1790 (5 a.) Henry Lamott
Survey 5 Nov., 1790 — Henry Lamott
Assigned 11 June, 1794 to George Philip Dehoof
Patent 13 June, 1794 — George Philip Dehoof
IC G/405 and IC I/609
Pat. Cert. 1972

Tract 280 — Map 1-W — Baltimore County
<u>Good Spring</u> — 19 1/2 acres
Sp. Warr. 17 Dec., 1799 — William Hoffman
Survey 4 Dec., 1800 — William Hoffman
Patent 22 Oct., 1801 — William Hoffman
IC N/621 and IC P/221
Pat. Cert. 1977

Tract **
Goodwill (See George's Goodwill)

Tract 281 — Map 1-SE — Baltimore County
<u>Got By Chance</u> — 6 a., 3 r., 24 p.
Resurvey of Hurst's Hills (462 3/4 a.)
Sp. Warr. 16 May, 1803 (16 a.) Jesse Hoshall
Survey 10 Apr., 1804 — Jesse Hoshall
Unpat. Cert. 622

Tract 282 — Map 1-N — Baltimore County
<u>Gottlieb's Lot</u> — 17 1/4 acres
Comm. Warr. 2 May, 1802 (14 a.) Gottlieb Hoffman
Survey 9 Nov., 1802 — Gottlieb Hoffman
Patent 7 July, 1803 — Gottlieb Hoffman
IC R/243 and IC S/73
Pat. Cert. 2017

Tract 283 — Map 2-C — Carroll County
<u>Grandfather's Gift</u> — 343 a., 2 r., 19 p.
Resurvey of Borings Struggle (50 a.), Stoney Hills (1291 a.) and Stansbury Hills (32 a.)
Sp. Warr. 4 Feb., 1807 — Thomas Sater
Survey 4 Feb., 1807 — Thomas Sater
Patent 2 Feb., 1808 — Thomas Sater
IB D/62 and IB C/152
Pat. Cert. 2032

Tract 284 — Map 2-C — Carroll County
<u>Grape Springs</u> — 22 1/2 acres
Resurvey of Jerico (100 a.)

(Grape Springs - continued)
Prior Survey for Stoney Hills (1291 a.)
Comm. Warr. 16 Nov., 1759 (22 a.) Philip Harren
Survey 10 Mar., 1760 Philip Harren
Patent 10 Mar., 1760 Philip Harren (Farrar)
BC & GS 12/308 and BC & GS 13/220
Pat. Cert. 2034

Tract 285 Map 2-SE Baltimore County
Great Profit 25 acres
Resurvey of Montagle (122 a.)
Prior Survey for New London (250 3/4 a.)
Warrant 29 May, 1789 (25 a.) Charles Boblett
Survey 30 June, 1789 Charles Boblitz
Patent 9 Apr., 1790 Charles Boblitz
IC F/33 and IC E/321
Pat. Cert. 2048

Tract 286 (South of Map 1) Baltimore County
Great Profit 87 acres
(Reference Tracts - Horatio's Lot (314 a.),
Singery's Trouting Streams (575 a.))
Prior Survey for Ford's Enlargement (--- a.)
Sp. Warr. 8 Nov., 1794 (54 of 67 a.) John Keith
Comm. Warr. 10 Nov., 1794 (3 of 40 a.) John Keith
Sp. Warr. 25 Nov., 1794 (12 a.) Shadrack Hurst
Assigned 4 Sept., 1795 (23 a.) to John Keith
Survey 4 Sept., 1795 John Keith
Comm. Warr. 10 Nov., 1795 John Keith
Assigned 30 Sept., 1795 (23 a.) to Benjamin Thomas
Patent 18 Sept., 1815 Benjamin Thomas
IB D/464 and IB C/249
Pat. Cert. 2049

Tract 287 Map 1-S Baltimore County
Green's Chance 22 acres
Prior Survey for Rabbit's Hallow (287 a.)
Comm. Warr. 10 Apr., 1775 Henry Surbouger
Renewed 26 Sept., 1776 (20 a) Henry Surbouger
Assigned 4 July, 1776 to Abraham Green
Survey 8 Jan., 1777 Abraham Green
Patent 15 Feb., 1783 Abraham Green
BC & GS 50/408 and IC A/79
Pat. Cert. 2059

Tract 288 Map 1-SW Baltimore County
Green's Meadows 138 1/2 acres
Original Survey for Rabbit's Hallow (287 a.)
Surveyed 6 Mar., 1786 Abraham Green
Patent 25 May, 1798 Abraham Green
IC D/467 and IC E/36
Pat. Cert. 2069

Tract 289 (South of Map 2) Carroll County
<u>Green's Visenburg</u> (Gruen's Wisenburg) 171 3/4 acres
3 Parts - (29, 4 1/2 and 138 acres)
(Reference Tracts - Carr's Ingenuity (256 a.) and
Pleasant Spring Resurvey (444 1/4 a.))
Resurvey of Christian's Chance (166 a.)
Sp. Warr. 16 Mar., 1791 George Michael Shafer
Survey 1 June, 1791 George Michael Shafer
Patent 30 May, 1795 George Michael Shaffer
IC I/414 and IC H/303
Pat. Cert. 2111

Tract 290 (East of Map 1) Baltimore County
<u>Greenspring</u> 118 acres
(Reference Tracts - George's Wisdom (332 1/2 a.) and
Snaggy Hill (50 a.))
Sp. Warr. 27 Mar., 1793 (26 a.) Moses Freeland
Comm. Warr. 11 Oct., 1793 (92 a.) Moses Freeland
Survey 18 Oct., 1793 Moses Freeland
Patent 29 Oct., 1794 Moses Freeland
IC I/619 and IC G/446
Pat. Cert. 2074

Tract **
Greiss's (Griss's) Lott (See Kreis Lot)

Tract 291 Map 1-W Baltimore County
<u>Grist Mill Hills</u> 257 acres
Resurvey of Barren Hills (16 1/2 a.) & No. 2 (40+ a)
Prior Survey for Timber and Stone (7 1/4 a.)
Survey 12 Mar., 1787 William Hoffman
Patent 2 Apr., 1789 William Hoffman
IC D/414 and IC C/649
Pat. Cert. 2095

Tract 292 Map 2-NE Carroll County
<u>Grogg's Delight</u> 10 acres
Comm. Warr. 13 Mar., 1794 (10 a.) Jacob Grogg
Survey 24 Mar., 1794 Jacob Grogg
Assigned by will of Jacob Jacob Grogg, Sr. to his
children and heirs: Jacob Grogg (Jr.), John Grogg,
Molly Rinehart, Catherine Denis, Lewis Dehoof and
Fanny Haney
Patent 2 Feb., 1813 Jacob Grogg (Jr.) et. al.
IB D/25 and IB A/325
Pat. Cert. 2096

Tract 293 Map 2-NW Carroll County
<u>Ground Oak Hill</u> 50 acres
Comm. Warr. 6 June, 1787 (50 a.) John Derivester
of York County, Pennsylvania
Survey 15 June, 1787 John Derivester
Patent 12 July, 1792 John Derivester
IC I/21
Pat. Cert. 2102

Tract 294 (not mapped) Baltimore County

<u>Ground Oak Hills</u> 5 1/2 acres

Prior Survey for Ground Oak Hills Res. (52 1/2)
Warrant 13 Nov., 1793 (10 a.) William Hoffman
Surveyed 25 Apr., 1794 William Hoffman
Patented 13 Nov., 1798 William Hoffman
IC O/455 and IC M/399 and IC P/222
Pat. Cert. 2100

Tract 295 Map 2-E Baltimore County

<u>Ground Oak Hills Resurveyed</u> 52 1/2 acres

Resurvey of Ground Oak Hills (5 1/2 a.)
Sp. Warr. 20 Feb., 1802 William Hoffman
Survey 19 May, 1802 William Hoffman
Patent 12 May, 1812 William Hoffman
IC R/232 and IC P/463
Pat. Cert. 2103

Tract 296 Map 1-NW Baltimore County

<u>Grubby Thicket</u> (Scrubby Thicket) 134 acres

Resurvey of Venus's Arbour (202 a.), Peter's Lot (25 a.) and Charles Delight (138 a.)
Survey 16 Aug., 1788 William Hoffman
Patent 2 Apr., 1789 William Hoffman
(Patent never executed)
* Certificate of Survey adjusted 23 Oct., 1793 (Second set of courses)
* Patent Recorded 31 Mar., 1795 William Hoffman
IC D/413 and IC C/646
Pat. Cert. 2110

Tract 297 Map 1-S Baltimore County

<u>Guard House</u> 82 acres

Prior Survey for Markey's Water Rights (3+ a.)
Sp. Warr. 2 May, 1796 (161 1/4 a.) Joshua Stevenson of Frederick County
Survey 27 Apr., 1797 Joshua Stevenson
Patent 9 May, 1799 Joshua Stevenson
IC O/572 and IC L/541
Pat. Cert. 2112

Tract 298 (not mapped) Baltimore County

<u>Gunpowder Meadows</u> 43 a., 0 r., 20 p.

(Reference tracts - Hurst Ill Will (10 a.) and Addition to Markey's Chance (16 1/4 a.))
Sp. Warr. 22 Jan., 1855 (40 a.) Isaac Hoshall
Survey 12 Dec., 1855 Isaac Hoshall
Patent 8 Jan., 1857 Isaac Hoshall
GGB 8/509
Pat. Cert. 2117

Tract 299 (South of Map 1) Baltimore County

(The) <u>Habitation</u> (Rocks) 39 1/2 acres

(Reference Tracts - bears on Pine Grove (52 a.) and close to Wildcat Hallow (54 3/4 a.))

(Habitation - continued)

Survey 20 Oct., 1789 John Brown
Assigned by terms of will of John Brown to widow Elie Matthews and children : Rachel, Elizabeth, Thomas and William Matthews.
Patent 13 Feb., 1792 Rachel, Elizabeth, Thomas and William Matthews
IC F/499 and IC H/15
Pat. Cert. 2120

Tract 300 Map 1-SW Baltimore County
<u>Hair's (Heir's) Beautiful Prospect</u> 329 3/4 acres
Resurvey of Hair's Retirement (175 1/4 a.), Clapperstone's Range (25 a.), Jacob's Poor Chance (27 1/2 a.), Sedgey Hill (18 a.), Hair's Patch (7 a.) and Addition to Hair's Retirement (12 1/2 a.)
Sp. Warr. 14 Feb., 1801 Jacob Hair
Survey 21 Dec., 1801 Jacob Hair
Patent 7 Dec., 1808 Jacob Hair (Hare)
JK U/30 and IC S/710
Pat. Cert. 2136

Tract 301 (not mapped) Baltimore County
<u>Hair's Palete</u> 7 acres
Prior Survey for Hair's Beautiful Prospect (327+)
Com. Warr. 29 Mar., 1790 (25 acres) Jacob Hair
Survey 7 Mar., 1791 Jacob Hair
Patent 13 Nov., 1801 Jacob Hair
IC N/638 and IC P/238
Pat. Cert. 2138

Tract **
Hair's Patch (See Hair's Palete)

Tract 302 (not mapped) Baltimore County
<u>Hair's Retirement</u> 175 1/4 acres
Resurvey of Hair's Retirement (15 a.)
Prior Survey for Hair's Beautiful Prospect (327+)
Survey 13 Mar., 1788 Jacob Hair
Patent 19 Nov., 1794 Jacob Hair
IC I/629 and IC H/571
Pat. Cert. 2139

Tract 303 (not mapped) Carroll County
<u>Hair's Retirement</u> 15 acres
Prior Survey for Hair's Beautiful Prospect (327+ a.)
Warrant 3 Mar., 1791 (15 a.) Jacob Hair
Survey 17 Feb., 1792 Jacob Hair
Patent 13 Nov., 1801 Jacob Hair
IC N/638 and IC P/238
Pat. Cert. 2140

Tract 304 Map 1-E Baltimore County
<u>Hampshire's Bargain</u> 10 acres
Prior survey for Mount Pleasant (28+ a.)
Com. Warr. 25 Mar., 1800 Benedict Hurst
Survey 13 Mar., 1801 Benedict Hurst
Assigned 19 Mar., 1801 to George Hamshire
Patent 16 Aug., 1802 George Hampshire
IC R/46 and IC L/740
Pat. Cert. 2176

Tract 305 (not mapped) Carroll County
<u>Happiness</u> 39 acres
Prior Survey for Neilson's Puzzle (256 1/4 a.), Number Three (15 1/4 a.) and Beaver Trap (609 a.)
Comm. Warr. 10 Mar., 1763 Thomas Stansbury, Jr.
Renewed 10 Sept., 1763 (100 a.) Thos. Stansbury, Jr.
Survey 25 Feb., 1764 Thomas Stansbury, Jr.
Patent 25 Feb., 1764 Thomas Stansbury, Jr.
BC & GS 27/483 and BC & GS 28/190
Pat. Cert. 2186

Tract 306 (South of Map 1) Baltimore County
<u>Hard Fare</u> 250 acres
(Reference Tract - Betsey's Choice Resurvey (141+))
Survey 26 Oct., 1786 John Foster
Assigned 17 Nov., 1789 to Christopher Walker
Patent 27 Mar., 1794 Christopher Walker
IC I/608 and IC H/457
Pat. Cert. 2191

Tract 307 (South of Map 1) Baltimore County
<u>Hard Fare</u> 37 1/4 acres
(Reference Tract - not precisely located, near Hard Fare (250 a.))
Survey 19 Sept., 1787 Christopher Walker
Assigned 17 Nov., 1789 to John Foster
Patent 16 Aug., 1791 John Foster
IC F/425 and IC E/745
Pat. Cert. 2193

Tract 308 Map 1-SE Baltimore County
<u>Hard Looks</u> 27 acres
Resurvey of Harry Mucklefish (10 a.)
Sp. Warr. 6 Sept., 1803 Samuel Markey
Survey 27 Feb., 1804 Samuel Markey
Patent 30 Dec., 1820 Samuel Markey
IB G/499
Pat. Cert. 2196

Tract 309 Map 1-E Baltimore County
<u>Hard Scuffle</u> 34 acres
Prior Survey for Take All (56 1/4 a.) and The Three Divisions (2591 a.)
Leased 16 Sept., 1751 Teague Tracey

(Hard Scuffle (34 acres) - continued)
Leaseholds GGB B/601
Survey 16 Sept., 1751 — Trego Tracy
Unpat. Cert. 662

Tract 310 (East of Map 1) Baltimore County
Hard Scuffle — 10 acres
(Reference Tract - Snaggy Hill (50 a.))
Comm. Warr. 2 Feb., 1797 (41 1/2 a.) Andrew Horn
Assigned to Moses Freeland
Survey 29 Jan., 1798 — Moses Freeland (Franklin)
Assigned 15 May, 1809 to John Free
Patent 20 June, 1809 — John Free
JK U/103 and JK T/118
Pat. Cert. 2201

Tract 311 Map 2-C Carroll County
Hard Times — 53 acres
Resurvey of Borings Range (1375 a.)
Sp. Warr. 24 Oct., 1776 (180 a.) Thomas Stansbury and Henry Wareheim
Survey 14 Feb., 1777 — Thomas Stansbury and Henry Wareheim
Patent 30 May, 1788 — Thomas Stansbury and Henry Wareheim
IC D/176 and IC C/432
Pat. Cert. 2203

Tract 312 Map 1-S Baltimore County
Hard Times — 1 1/2 a., 0 r., 10 p.
Sp. Warr. 16 Apr., 1812 (2 a.) Charles Gore
Survey 4 Mar., 1813 — Charles Gore (Goor)
Unpat. Cert. 663

Tract **
Hard Stone (See Hardstone)

Tract 313 (East of Map 1) Baltimore County
Hardesty's Chance — 120 acres
(Reference Tract - Schieming Defied (104 1/2 a.))
Prior Survey for Hardesty's Chance (132 a.)
Leased 1 Jan. 1761 — Henry Hardesty
(witnesses : William Smith and J. Ross)
Leaseholds GGB A/578
Survey 1 Jan., 1761 — Henry Hardesty
Unpat. Cert. 661

Tract 314 (East of Map 1) Baltimore County
Hardesty's Chance — 132 acres
(Reference Tract - Schieming Defied Enl. (194 3/4))
Resurvey of Hardesty's Chance (120 a.)
Prior Survey for Kidd's Design (9 1/4 a.) and Kidd's Design Enlarged (58 1/4 a.)
Survey 15 Nov., 1786 — Jacob Markey
Patent 18 June, 1788 — Jacob Markey

(Hardesty's Chance (132 a.) - continued)
IC D/162 and IC C/463
Pat. Cert. 2205

Tract 315 Map 2-E Baltimore County
<u>Hardstone</u> 22 acres
Comm. Warr. 8 May, 1797 (50 a.) William Hoffman
Survey 22 Mar., 1798 William Hoffman
Patent 26 Dec., 1799 William Hoffman
IC O/675 and IC M/677
Pat. Cert. 2202

Tract 316 Map 1-SE Baltimore County
<u>Harry Fishmuckle</u> 10 acres
Prior Survey for Hard Looks (27 a.)
Comm. Warr. 2 Dec., 1801 (10 a.) Christian Singery
Survey 9 Nov., 1802 Christian Singery
Patent 5 Sept., 1803 Samuel Markey
IC R/220 and IC S/97
Pat. Cert. 2228

Tract 317 Map 2-E Baltimore County
<u>Heavy Ground</u> 80 1/2 acres
Sp. Warr. 3 Apr., 1797 (12 a.) William Hoffman
Comm. Warr. 8 May, 1797 (50 a.) William Hoffman
Survey 23 Mar., 1798 William Hoffman
Patent 26 Dec., 1799 William Hoffman
IC O/675 and IC M/677
Pat. Cert. 2261

Tract 318 Map 2-W Carroll County
<u>Heidleburgh</u> 54 1/2 acres
Comm. Warr. 5 Mar., 1774 (54 1/2 a.) Conrad Creepaum
Survey 11 Apr., 1774 Conrad Creepaum
Patent 17 Feb., 1775 Conrad Creepaum
BC & GS 51/69 and BC & GS 48/312
Pat. Cert. 2266

Tract 319 (West of Map 2) Carroll County
<u>Heidleburg</u> 50 acres
Prior Survey for Heidlesburgh Resurvey (160 a.)
Survey 10 Aug., 1752 Adam Lynn
(Original Patent never paid for)

Tract 320 (West of Map 2) Carroll County
<u>Heidlesburgh Resurvey</u> 160 acres
(Reference Tracts - Troy (1634 a.) and Everybodies Land (Parcel # 1) on Md/Pa. line))
Resurvey of Heidlebug (50 a.)
Sp. Warr. 13 Nov., 1761 Elias Horage
Survey 22 Feb., 1762 Elias Horage
Patent 29 Sept., 1763 Elias Horage
BC & GS 21/359 and BC & GS 20/532
Pat. Cert. 2267

Tract 321 Map 1-SE Baltimore County

Henry's Chance (Change) 27 acres

Sp. Warr. 13 Mar., 1797 (2 a.) Henry Wilhelm
Comm. Warr. 13 Mar., 1797 (1 a.) Henry Wilhelm
Patent 28 Jan., 1798 Henry Wilhelm
Survey 29 Mar., 1799 Henry Wilhelm
IC O/590 and IC M/570
Pat. Cert. 2279

Tract 322 Map 2-SE Baltimore County

Henry's Hope 100 acres

Prior Survey for New London (250 3/4 a.)
Warrant 20 Nov., 1760 (100 a.) Henry Crottinger
Survey 11 May, 1761 Henry Crottinger
Patent 11 May, 1761 Henry Crottinger
BC & GS 14/164 and BC & GS 10/688
Pat. Cert. 2287

Tract 323 Map 1-S Baltimore County

Here is Life without care and Love without Fear

51 1/4 acres

Sp. Warr. 15 Aug., 1797 (270 a.) Joshua Stevenson of Frederick County
Survey 21 Dec., 1797 Joshua Stevenson
Patent 9 May, 1799 Joshua Stevenson
IC M/584
Pat. Cert. 2298

Tract 324 Map 1-NW Baltimore County

Hickman's (Heckman's) Pipes 12 acres

Comm Warr. (12 a.) George Gist
Assigned 9 Oct., 1794 to Sebastian Heckman, Sr.
Survey 15 May, 1795 Sebastian Heckman, Sr.
Assigned 19 Oct., 1811 by deed to heirs of Sebastain Heckman, Sr.: Christian and George Heckman, Susannah Heckman (w/o Henry Masemore), Elizabeth Heckman (w/o Peter Coopersmith)
Patent 5 April, 1814 Christian and George Heckman, Susannah Masemore, and Elizabeth Coopersmith
IB B/626 and IB C/51
Pat. Cert. 2263

Tract 325 (not mapped) Baltimore County

Hickory Bottom 10 acres

(Common boundary - Hair's Retirement (175 1/4 a.))
Survey 4 Feb., 1796 Thomas Gist
(Field Survey Book No. 242, page 94 (Tracey Collection - HSCC) No further information)

Tract 326 (East of Map 1) Baltimore County

High Rock 60 acres

(Reference Tract - Pessel Tree (96 3/4 a.))
Comm. Warr. 5 June, 1790 (60 a.) John Free

(High Rock - continued)
Survey 10 Sept., 1790 John Free(s)
Patent 17 Mar., 1812 John Free(s)
IB B/90 and IC A/155
Pat. Cert. 2314

Tract 327 (not mapped) Baltimore County
<u>Hilly Run</u> 188 acres
(Survey was 162 1/2 acres - deficient by 25 7/8 a.)
Prior survey for Hilly Run Resurvey (197 1/4 a.)
Lease 9 Apr., 1761 John Kilty Collins
Leaseholds GGB A/216
Survey 9 Apr., 1761 John Kilty Collins
Unpat. Cert. 707

Tract 328 (not mapped) Baltimore County
<u>Hilly (Shilly) Run Resurveyed</u> 197 1/4 acres
Resurvey of Hilly Run (188 a.)
Prior Survey for Pleasant Valley (290 a.)
Survey 5 May, 1787 James Campbell
Patent 12 Feb., 1794 James Campbell
IC I/421 and IC G/358
Pat. Cert. 2323

Tract 329 Map 2-N Carroll County
<u>Hit or Miss</u> 13 3/15 acres
Comm. Warr. 9 Sept., 1793 (13 3/15 a.)
Thomas Stansbury
Survey 6 Aug., 1794 Thomas Stansbury
Unpat. Cert. 713

Tract 330 Map 1-C Baltimore County
<u>Hobson's Choice</u> 51 1/4 acres
Amendment to Benjamin Roger's land, "Pitt" (475 a.)
Prior Survey for Pleasant Valley (290 a.)
Survey 14 May, 1771 John Hale (Hall)
Patent 13 Oct., 1774 John Hale (Hall)
BC & GS 50/34 and BC & GS 49/70
Pat. Cert. 2335

Tract 331 (South of Map 1) Baltimore County
<u>Hobson's Choice</u> 15 acres
(Reference Tract - Hard Looks (27 a.))
Comm. Warr. 14 June, 1793 (15 a.) Christian Singery
Survey 13 May, 1794 Christian Singery
Patent 5 May, 1799 Christian Singery
IC O/579 and IC M/602
Pat. Cert. 2334

Tract 332 Map 2-NE Carroll County
<u>Hobson's Choice</u> 106 1/4 acres
Resurvey of Dilley's Meadow (27 a.), Frederick's Conquest (40 a.), Beaver's New Invention (170 a.) and Vaughn's Meadows (27 a.)
Prior Survey for Sugar Valley (387 a.)

(Hobson's Choice (106 1/4 acres) - continued)
Sp. Warr. 6 Jan., 1791 George Fair
Survey 10 Feb., 1791 George Fair
Patent 12 Nov., 1801 George Fair (Fahr)
IC N/635 and IC L/684
Pat. Cert. 2332

Tract 333 (not mapped) Baltimore County
<u>Hobson's Choice</u> 1/2 a., 0 r., 20 p.
(Reference Tracts - Davis's Chance (316 a.) and Lexington (106 1/2 a.))
Warrant 14 Dec., 1803 (14 a.) Aaron Campbell
Comm. Warr. 17 June, 1808 (4 a.) Aaron Campbell
Survey 19 Apr., 1809 Aaron Campbell
Unpat. Cert. 717

Tract 334 Map 2-S Carroll County
<u>Hog Island</u> 50 acres
Warrant 2 Apr., 1750 (50 a.) Richard Richards
Survey 16 Sept., 1751 Richard Richards
Patent 30 May, 1752 Richard Richards
Y & S &/25 and BY and GS 4/415
Pat. Cert. 2343

Tract 335 Map 1-SE Baltimore County
<u>Honorificabilitudinity</u> 79 acres
Survey 29 Sept., 1790 Bennett Hurst
Patent 26 Nov., 1796 Bennett Hurst
IC O/27 and IC M/97
Pat. Cert. 2359

Tract 336 Map 2-C Carroll County
<u>Hope for Better Times</u> 14 acres
Sp. Warr. 24 Oct., 1776 Thomas Stansbury and Henry Wareheim
Survey 14 Feb., 1777 Thomas Stansbury and Henry Wareheim
Patent 30 May, 1788 Thomas Stansbury and Henry Wareheim
IC D/177 and IC C/431
Pat. Cert. 2386

Tract 337 Map 2-C Carroll County
<u>Hope Restored</u> 20 acres
Prior Survey for Charles Trouble (480+ a.)
Warrant 24 Oct., 1793 (54 1/4 a.) Charles Stansbury (unexecuted)
Comm. Warr. 20 Oct., 1794 (30 a. + 63 p.) Charles Stansbury
Survey 19 Oct., 1795 Charles Stansbury
Patent 18 Oct., 1797 Charles Stansbury
IC O/312 and IC L/383
Pat. Cert. 2387

Tract 338 (South of Map 1) Baltimore County
Horatio's Chance 20 acres
(Reference Tracts - Ford's Enlargement (--- a.)
Horatio's Lot (314 a.)) and Walnut Hills (813+))
Comm. Warr. 30 July, 1790 (20 a.) Mordecai Ford
Assigned 22 June, 1791 to Horatio Ford
Survey 22 June, 1791 Horatio Ford
Assigned 25 Oct., 1794 to Joshua Price (Sr.)
(witnesses: Joshua Lemmon and Edw. Bosman, Jr.)
Assigned to heirs of Joshua Price: wife Martha who was deceased at time of patent and his children: Benjamin, John, Joshua, Jr., Eleanor, Hannah, and Temperance Price
Patent 7 Jan., 1805 Benjamin, John, Joshua, Jr., Eleanor, Hannah and Temperance Price
IC B/327 and IC R/327
Pat. Cert. 2397

Tract 339 (South of Map 1) Baltimore County
Horatio's Lot 314 acres
(Reference tracts - Singery's Trouting Stream (575) and Borings Habitation Rock (300 a.))
Resurvey of Singery's Trouting Stream (575 a.)
Prior Survey for Horatio's Enlargement (--- a.)
Survey 22 June, 1791 Mordecai Ford
Assigned 14 Nov., 1791 to Horatio Ford
Patent 9 Jan., 1794 Horatio Ford
IC I/454 and IC G/340
Pat. Cert. 2398

Tract 340 Map 1-SW Baltimore County
Huckleberry Hill 100 acres
Prior Survey for Neighbor's Goodwill (289 a.)
Comm. Warr. 26 June, 1775 (100 a.) Edward Stevenson
Survey 20 July, 1775 Edward Stevenson
Patent 6 Aug., 1776 Edward Stevenson
BC & GS 51/320 and BC & GS 52/274
Pat. Cert. 2428

Tract 341 Map 1-S Baltimore County
Hug Me Snug 15 1/2 acres
Prior Survey for Markey's Chance Enl. (62 3/4 a.)
Sp. Warr. 15 Aug., 1787 (270 a.) Joshua Stevenson of Frederick County
Survey 24 Dec., 1797 Joshua Stevenson
Patent 9 May, 1799 Joshua Stevenson
IC M/585
Pat. Cert. 2442

Tract 342 Map 1-SE Baltimore County
Hurst Delight 34 1/2 acres
Prior Survey for Addition to Hurst Hills (121 a.)
Survey 29 Sept., 1790 Bennett Hurst
Patent 19 Apr., 1793 Bennett Hurst

(Hurst's Delight - continued)
IC I/155 and IC H/260
Pat. Cert. 2455

Tract 343 Map 1-SE Baltimore County
Hurst's (Hust's) Goodwill 44 acres
Survey 4 Dec., 1789 Samuel Price of John
Assigned 8 Nov., 1789 to John Taylor
Patent 24 May, 1796 John Taylor
IC O/27 and IC L/107
Pat. Cert. 2459

Tract 344 (South of Map 1) Baltimore County
Hurst's Hackle 26 1/2 acres
(Reference Tract - Cooper's Discovery (77+ a.))
Prior Survey for Smith's Grief (34 1/4 a.)
Sp. Warr. 13 Oct., 1793 Peter Stevely
Assigned 9 Sept., 1794 to Capt. Bennett Hurst
Survey 8 Oct., 1794 Capt. Bennett Hurst
Unpat. Cert. 751

Tract 345 Map 1-S Baltimore County
Hurst Hills 462 3/4 acres
Resurvey of Hurst's Venture (39 3/4 a.)
Prior Survey for Hurst's Ill Will (10), Crooked Billet (12 a.), Addition (23 3/4 a.) and Gunpowder Meadows (43+ a.)
Survey 16 Jan., 1771 (408 1/2 a.) Bennett Hurst
Land purchased by Bennett Hurst from the Commissioners under the Propietary Government
* Survey 11 Sept., 1794 (462 3/4 a.) Bennett Hurst
(No Patent Certificate in Md. Archives)
(See Field Note Books in Tracey Collection, CCHS: Book No. 268, p. 3 and Book No. 106, p. 41. Book No. 106 refers to Col. (Thomas) Gist's Book.)

Tract 346 Map 1-S Baltimore County
Hurst's Ill Will 10 acres
Sp. Warr. 6 Mar., 1793 (10 a.) John Taylor
Survey 13 Feb., 1794 (10 a.) John Taylor
Assigned 1 Aug., 1801 to Basil Hurst
Patent 19 Jan., 1802 Basil Hurst
IC R/50 and IC L/699
Pat. Cert. 2456

Tract 347 Map 1-E Baltimore County
Hurst's (Hust's) Lot 68 1/2 acres
Purchase of rerserved land: Bennett Hurst, Jesse Hoshall and John Hurst
Survey 20 May, 1788 Benedict Hurst
Patent 9 Aug., 1798 Benedict Hurst
IC O/430 and IC M/457
Pat. Cert. 2460

Tract 348 Map 1-C Baltimore County
<u>Hurst Marsh</u> 45 acres
Prior Survey for Pitt (475 a.)
Leased 20 June, 1750 Benjamin Hurst
Leaseholds GGB H/572
Survey 20 June, 1750 Benjamin (Bennett) Hurst
Unpat. Cert. 752

Tract 349 Map 1-S Baltimore County
<u>Hurst's Venture</u> 39 3/4 acres
Survey 21 Sept., 1790 Bennett Hurst
Patent 3 Dec., 1793 Bennett Hurst
IC I/428 and IC H/402
Pat. Cert. 2457

Tract 350 (West of Map 2) Carroll County
<u>I Will Take It If I Can</u> 26 1/2 acres
(Reference Tract - Everybodies Land (390 1/2 a.)
Resurvey of George's Right (25 a.)
Sp. Warr. 12 Nov., 1794 Henry Lamott
Survey 7 Jan., 1795 Henry Lamott
Unpat. Cert. 767

Tract 351 Map 2-SW Carroll County
<u>Idlesburgh</u> 25 acres
Prior Survey for Schieming Defeated (186 a.)
Comm. Warr. 30 Aug., 1757 (25 a.) Philip Idleman
Survey 30 Oct., 1757 Philip Idleman
Patent 30 Oct., 1757 Philip Idelman
BC & GS 9/352 and BC & GS 10/163
Pat. Cert. 2467

Tract ***
Idleman's Addition (See Addition to Bridgeland)

Tract 352 Map 2-NW Carroll County
<u>Ill Neighborhood</u> 10 acres
Comm. Warr. 15 Apr., 1774 Barbara Hoffman
Assigned 11 May, 1774 to Henry Lamuth
Survey 10 June, 1774 Henry Lamott
Patent 24 June, 1790 Henry Lamott
IC F/4 and IC C/728
Pat. Cert. 2472

Tract 353 Map 2-N Carroll County
<u>Inclosure</u> (Enclosure) 112 1/4 acres
Resurvey of Ammon's Garden (20 a.)
Sp. Warr. 1 Nov., 1793 John Showk
Survey 21 Oct., 1794 John Showk
Patent 24 Feb., 1807 John Showk (Shauck)
IC R/511 and IC S/470
Pat. Cert. 2477

Tract 354 Map 2-W Carroll County
Jabet 2 acres
Comm. Warr. 9 Feb., 1841 (2 a.) William Crumrine and Samuel B. Fuhrman
Survey 12 Mar., 1841 William Crumrine and Samuel B. Fuhman
Patent 7 Mar., 1843 William Crumrine and Samuel B. Fuhrman
GGB 6/238 and GGB 5/123
Pat. Cert. 11 (Carroll County Patent Series)

Tract 355 Map 1-N Baltimore County
Jacob's Lot (Plains) 23 1/2 a.
Purchased land in The Reserve - Philip Stiltz
Survey 19 Mar., 1787 (Jacob's Plains) Philip Stiltz
Assigned 26 May, 1789 to Jacob Keillor
Patent 9 Dec., 1789 Jacob Keillor (Keller)
IC E/211 and IC D/624
Pat. Cert. 2552

Tract 356 Map 1-E Baltimore County
Jacob's Lot 20 acres
Warrant 5 Feb., 1791 (20 a.) Jacob Marshall
Survey 28 Jan., 1792 Jacob Marshall
Patent 29 Oct., 1794 Jacob Marshall
IC I/609 and IC G/445
Pat. Cert. 2546

Tract **
Jacob's Plains (See Jacob's Lot (23 1/2 a.))

Tract 357 Map 1-SW Baltimore County
Jacob's Poor Chance 27 1/2 acres
Prior Survey for Hair's Beautiful Prospect (327+)
Comm. Warr. 10 Nov., 1794 (30 a.) Jacob Hare
Survey 29 Oct., 1795 Jacob Hair
Patent 13 Feb., 1801 Jacob Hair
IC N/550 and IC L/625
Pat. Cert. 2553

Tract 358 (South of Map 2) Baltimore County
James's Lot 35 acres
Prior Survey for Penelope Charchilla and Thomas Cockey Deyesburg (1030 a.)
Comm. Warr. 28 Sept., 1762 (35 a.) Thomas Towson
Survey 30 Sept., 1762 Thomas Towson
Patent 30 Sept., 1762 Thomas Towson
BC & GS 21/345 and BC & GS 22/198
Pat. Cert. 2571

Tract 359 (East of Map 1) Baltimore County
Jarrett's Intention 427 acres
(Reference Tract - Expedition (337 a.))
Correction 5 Mar., 1792 Jesse Jarrett
Survey 1 Aug., 1793 Jesse Jarrett

(Jarrett's Intention - continued)
Patent 25 Oct., 1800 Jesse Jarrett
IC N/487 and IC P/40
Pat. Cert. 2578

Tract **
Jemima's Ketch (See Jimmy's Ketch)

Tract 360 Map 1-N Baltimore County
<u>Jemimy (Jimmy) and Rachel's Farm</u> 14 1/4 acres
Sp. Warr. 16 Dec., 1802 Aaron Campbell
Survey 24 Dec., 1803 Aaron Campbell
Patent 8 June, 1809 Aaron Campbell
JK U/83 and JK T/106
Pat. Cert. 2599

Tract 361 Map 1-SE Baltimore County
<u>Jenny's Patch</u> 2 acres
Warrant 5 Apr., 1795 (2 a.) John Taylor
Survey 7 May, 1795 John Taylor
Patent 1 Feb., 1802 William Dorsey
IC R/71 and IC P/279
Pat. Cert. 2588

Tract 362 Map 2-C Carroll County
<u>Jerico</u> 100 acres
Prior Survey for Stoney Hills (1291 a.)
Survey 12 June, 1745 Thomas Stansbury, Jr.
Patent 12 June, 1745 Thomas Stansbury, Jr.
LG E/504 and PT 2/338
Pat. Cert. 2594

Tract 363 Map 2-NE Carroll County
<u>Jerusalem</u> 7 3/4 acres
Sp. Warr. 11 Nov., 1802 (15 acres) Jacob Grog
Survey 30 Aug., 1803 Jacob Grog (Sr.)
Patent 2 Feb., 1813 Heirs of Jacob Grogg, Sr.:
Jacob (Jr.) and John Grog, Molly Rinehart,
Catherine Denis, Louis Dehoof and Fanny Haney
IB D/24 and IB A/323
Pat. Cert. 2596

Tract 364 Map 1-N Baltimore County
<u>Jimmy's (Jemima's) Ketch</u> 3 1/2 a., 0 r., 26 p
Comm. Warr. 28 Sept., 1807 (1 a.) Aaron Campbell
Survey 7 June, 1808 Aaron Campbell
Patent 8 June, 1809 Aaron Campbell
JK T/107 and JK U/87
Pat. Cert. 2600

Tract 365 Map 2-SW Carroll County
<u>John Caltrider's Fancy</u> 229 a., 2 r., 33 p.
Resurvey of Borings Meadow (100 a.), Foster's
Chance (50 a.), Addition (13 1/2 a.) and
Kittinger's Glade (24 a.)

(John Caltrider's Fancy - continued)
Sp. Warr. 6 Nov., 1800 John Caltrider
Survey 26 Oct., 1801 John Caltrider
Patent 15 Dec., 1813 John Caltrider (Cowlrider)
IC Q/327 and IC P/491
Pat. Cert. 2630

Tract 366 Map 1-NW Baltimore County
John is it Mine? 32 acres
Sp. Warr. 3 Dec., 1793 (10 a.) William Hoffman
Survey 24 Apr., 1794 William Hoffman
Patent 30 Mar., 1798 William Hoffman
IC O/456 and IC M/399
Pat. Cert. 2644

Tract 367 Map 2-N Carroll County
John Tracey and Feltey's wife Reconciled (7 1/2 a.)
Resurvey of The Wallet Resurvey (212 3/4 a.)
Sp. Warr. 7 June, 1803 (5 a.) John Tracy
Survey 3 Sept., 1803 John Tracey
Patent 15 Nov., 1806 John Tracey
IC R/433 and IC S/433
Pat. Cert. 2660

Tract 368 Map 2-N Carroll County
John's Adventure 50 acres
Comm. Warr. 1 Feb., 1754 (150 a.) Richard Richards
Assigned 15 Oct., 1754 to John Cross of Joseph
Survey 15 Nov., 1754 John Cross
Patent 15 Nov., 1754 John Cross
BC & GS 4/272 and BC & GS 3/719
Pat. Cert. 2617

Tract 369 Map 2-W Carroll County
John's Gift 11 acres
Sp. Warr. 19 Mar., 1794 (11 a.) Richard Stansbury
Survey 6 Aug., 1794 Richard Stansbury
Patent 1 Nov., 1804 Richard Stansbury
IC P/589 and IC R/293
Pat. Cert. 2638

Tract 370 (West of Map 2) Carroll County
Johnsborough Resurvey 130 acres
(Reference Tract - Baughman's Goodwill (154 3/4))
Resurvey of Phillipsburg Resurvey (131 a.)
Prior survey for Matter's Choice Resurvey (162 a.)
Special proclamation to resurvey Phillipsburg:
(Warrant) 1 Mar., 1761 (125 a.) John Shrimplin
Survey 20 May, 1761 John Shrimplin
Assigned 29 Oct., 1762 to John Hendricks
Patent 29 Sept., 1763 John Hendricks
BC & GS 21/3 and BC & GS 17/691
Pat. Cert. 2625

Tract 371 Map 2-W Carroll County
<u>Join All</u> 10 acres
Prior Survey for Nottenstott Resurvey (155 a.)
Comm. Warr. 28 Mar., 1774 (12 a.) Anthony Noll
Survey 11 Apr., 1774 Anthony Noll
Patent 12 July, 1775 Anthony Noll
BC & GS 51/123 and BC & GS 49/257
Pat. Cert. 2661

Tract 372 (South of Map 1) Baltimore County
<u>Jones' First Purchase</u> 370 acres
(Reference Tract - Buck Range (544 a.))
Resurvey of Buck Range (544 a.) and Addition to Buck Range (40 1/4 a.)
Sp. Warr. 7 May, 1802 John Foster
Survey 10 Nov., 1802 John Foster
Assigned: 26 May, 1811 to Joseph Jones
Patent 3 Sept., 1811 Joseph Jones
IB B/60 and IB A/21
Pat. Cert. 2671

Tract 373 Map 2-NW Carroll County
<u>Joseph Bollinger's Contrivance</u> 348 acres
Resurvey of New Prussia (24 1/4 a.), Rome (36 1/4) and Plymouth (434 a.)
Sp. Warr. 20 Jan., 1790 Joseph Bollinger
Survey 29 Oct., 1790 Joseph Bollinger
Patent 30 July, 1793 Joseph Bollinger (Bowlinger)
IC I/464 and IC H/331
Pat. Cert. 2691

Tract 374 Map 2-N Carroll County
<u>Joseph's Folly</u> 27 Acres
Comm Warr. 29 Oct., 1761 Ulrich Fight
Assigned 26 Apr., 1762 to John Miller
Survey 26 Apr., 1762 John Miller
Unpat. Cert. 837

Tract 375 Map 2-SW Carroll County
<u>Joshua Boring son of John, His Forest</u> 220+ acres
Resurvey of Boring's Prospect (20 a.), New Germany (178 a.) and Addition to New Germany (50)
Sp. Warr. 12 Dec., 1799 Joshua Boring
Survey 17 Oct., 1800 Joshua Boring
Patent 14 Feb., 1803 Joshua Boring
IC Q/324 and IC P/4986
Pat. Cert. 2700

Tract 376 Map 1-SE Baltimore County
<u>Joshua's Trust</u> 38 1/2 acres
Prior Survey for Cooper's Discovery (77+ a.)
Survey 27 Oct., 1787 Joshua Smith
Patent 8 Dec., 1789 Joshua Smith
IC D/623 and IC E/247
Pat. Cert. 2712

Tract 377 (not mapped) Carroll County
<u>Keith's (Keeth's) Correction</u> 210 acres
Resurvey of Stansbury's Grove (1448 a.) and Catey's Delight (154 1/2 a.)
Sp. Warr. 26 Feb., 1835 John Keith
Survey 7 Nov., 1825 John Keith (Keeth)
Unpat. Cert. 848

Tract 378 Map 1-S Baltimore County
<u>Keith's Ketch</u> 17 acres
Resurvey of Keith's Long Mountain (154 a.)
Sp. Warr. 6 Jan., 1793 (17 a.) William Keith
Survey 1 Jan., 1794 William Keith
Patent 2 Dec., 1794 William Keith
IC I/612 and IC Q/460
Pat. Cert. 2719

Tract 379 Map 1-S Baltimore County
<u>Keith's Long Mountain</u> 154 acres
Prior Survey for Keith's Ketch (17 a.)
Survey 21 Mar., 1787 William Keith
Patent 10 Dec., 1789 William Keith
IC D/621 and IC E/248
Pat. Cert. 2720

Tract 380 Map 2-E Carroll County
<u>Kentucky</u> 100 acres
Sp. Warr. 1 Aug., 1797 (100 a.) Henry Lamott, Jr.
Survey 17 Aug., 1798 Henry Lamott, Jr.
Patent 6 Dec., 1804 Henry Lamott, Jr.
IC R/293 and IC P/587
Pat. Cert. 2727

Tract 381 (East of Map 1) Baltimore County
<u>Kidd's Design</u> 9 1/4 acres
Resurvey Hardesty's Chance (132 a.)
Prior Survey for Kidd'd Design Enl. (58 1/4 a.)
Survey 27 Mar., 1787 John Kidd
Assigned by estate of John Kidd to James Malcomb
Patent 5 Apr., 1850 James Malcomb
GGB 7-A/205 and GGB 8/2
Pat. Cert. 2729

Tract 382 (East of Map 1) Baltimore County
<u>Kidd's Design Enlarged</u> 58 1/4 acres
(Reference Tract - Buck Range (544 a.))
Resurvey of Kidd's Design (9 1/4 a.) and Hardesty's Chance (132 a.)
Sp. Warr. 20 Dec., 1816 John Kidd
Survey 17 Jan., 1817 John Kidd
Assigned by estate of John Kidd to James Malcomb
(See Schieming Defied Resurvey (GGB 7-A/208))
Patent 5 Apr., 1850 James Malcom
GGB 7-A/206 and GGB 8/3
Pat. Cert. 2730

Tract 383 Map 2-S Carroll County

Kingbury (Kings) Point 50 acres

Prior Survey for Petersburg (300 a.)

Comm. Warr. 7 Aug., 1758 (50 a.) Peter Rigo

Survey 30 Jan., 1759 Peter Rigo

Patent 30 Jan., 1759 Peter Rigo

BC & GS 12/2 and BC & GS 8/570

Pat. Cert. 2743

Tract **

Kittinger's Glade (See Gittinger's Glade)

Tract 384 Map 2-SW Carroll County

Kittinger's Orchard 9 acres

Comm. Warr. 11 June, 1768 (9 a.) John Kittinger

Survey 25 Nov., 1768 John Kittinger

Patent 25 Nov., 1768 John Kittinger

BC & GS 40/176 and BC & GS 38/108

Pat. Cert. 2754

Tract 385 (West of Map 2) Carroll County

Kittinger's Tavern 64 acres

(Reference Tract - Bridgeland (150 a.) and Nace's Tavern (478 1/2 a.))

Comm. Warr. 19 Aug., 1761 (60 a.) Richard Richards

Assigned 4 Feb., 1762 (10 a.) to John Kittinear

Warrant 11 Dec., 1761 (20 a.) George (Sater ?)

Assigned 24 Dec., 1761 (14 a.) to John Kittinear

Survey 4 Feb., 1762 John Kittinear

Patent 4 Feb., 1762 John Kittinear

BC & GS 14/705 and BC & GS 16/639

Pat. Cert. 2755

Tract 386 Map 2-W Carroll County

Kreis (Krees) Lott 9 a., 0 r., 36 p

Sp. Warr. 9 Apr., 1812 (20 a.) Henry Griss

Survey 10 Feb., 1813 Henry Kreis (Griss)

(Will of Henry Kreis names heirs: wife Margaret and George, Catherine (w/o Jacob Kinney), Elizabeth and Elias Kreis, all children of Henry.)

Assigned by court to Elias Kreis

Patent 22 Mar., 1827 Elias Kreis

IB GGB H/320 and IB GGB F/480

Pat. Cert. 2094

Tract 387 MAP 2-S Carroll County

Lady de la Lafayette 2 1/4 acres

Sp. Warr. 9 May, 1797 (90 a.) Thomas Cockey Deye

Survey 26 Apr., 1798 Thomas Cockey Deye

Patent 16 Dec., 1799 Thomas Cockey Deye

IC O/679 and IC M/664

Pat. Cert. 2775

Tract 388 Map 2-N Carroll County
<u>Lamotte's (Lamuth's) Defeat</u> 15 1/2 acres
Warrant 28 Apr., 1774 (25 a.) Henry Fitcht
Survey 7 Oct., 1774 Henry Fitcht
Patent 11 July, 1775 Henry Fitcht (Fites)
BC & GS 51/120 and BC & GS 49/254
Pat. Cert. 2777

Tract 389 Map 2-N Carroll County
<u>Lamotte's Delight</u> 20 1/4 acres
Resurvey of White Oak Bottom (61 3/4 a.)
Comm. Warr. 5 Mar., 1776 (50 a.) Henry Lamott
Survey 5 Sept., 1776 Henry Lamott
Assigned by terms of will of Henry Lamott, Sr.
Executors: Jacob Lamott and Christian Hoopman
Patent 5 Feb., 1802 Henry (Jr.), John, Jacob, Daniel, Barbary, Moses, Levy, Joshua and Catherine Lamott and Mary Walker
IC R/71 and IC P/280
Pat. Cert. 2778

Tract 390 Map 2-N Carroll County
<u>Lamuth's Middle of the World</u> 77 1/4 acres
Prior Survey for Addition to Curpoy (158 a.)
Sp. Warr. 13 June, 1774 (50 a.) Henry Lamotte
Comm. Warr. 30 May, 1774 (15 a.) Henry Lamotte
Survey 2 July, 1774 Henry Lamotte
Patent 12 July, 1775 Henry Lamotte (Lamuth)
BC & GS 51/122 and BC & GS 48/461
Pat. Cert. 2781

Tract 391 (East of Map 1) Baltimore County
(The) <u>Last Remnant</u> 5 1/2 a. + 17 p.
(Reference Tract - George's Mistake (Part I))
Sp. Warr. 11 Mar., 1814 (2 a) Moses Freeland
Survey 20 Dec., 1814 Moses Freeland
Patent 4 Dec., 1817 Moses Freeland
IB D/607 and IB E/65
Pat. Cert. 2808

Tract 392 Map 2-NE Carroll County
<u>Laurel Hills</u> 57 acres
Survey 15 Mar., 1787 William Hoffman
Patent 2 Apr., 1789 William Hoffman
IC D/413 and IC C/647
Pat. Cert. 2810

Tract 393 Map 1-NW Baltimore County
<u>Lebanon</u> 16 acres
Sp. Warr. 30 Mar., 1794 (15 a.) John and Chr. Baker
Survey 25 Mar., 1794 John and Christian Baker
Patent 5 Jan., 1797 John and Christian Baker
IC O/344 and IC L/236
Pat. Cert. 2825
Assigned 5 Mar., 1797 to Nicholas Ehrhart

Tract 394 Map 2-C Carroll County

Lebanon 15 acres

Prior survey of Sugar Valley (387 a.)
Com. Warr. 2 Apr., 1791 (15 a.) Frederick Fowble
Survey 20 Mar., 1795 Frederick Fowble
Patent 12 Nov., 1801 Alexis Lemmon
IC N/637 and IC P/235
Pat. Cert. 2824

Tract 395 Map 1-NE Baltimore County

Leibs Level 153 1/2 acres

Survey 20 Mar., 1787 Jacob Leib
(Jacob Leib given from both York Co., Pa. and Baltimore Co., Md.)
Patent 30 July, 1790 Jacob Leib
IC F/30 and IC E/504
Pat. Cert. 2831

Tract 396 Map 2-NW Carroll County

Level Ground 4 3/4 acres

Comm. Warr. 15 Apr., 1774 (36 a.) Ulrich Fritz
Assigned 20 July, 1774 (14 a.) to Christian Shaver
Survey 8 Oct., 1774 Christian Shaver
Assigned by terms of will of Christian Shaver who died circa 1785: All land to son John Adam Shaver when he reaches 21 years of age.
Patent 25 Oct., 1791 John Adam Shaver
IC F/537 and IC G/90
Pat. Cert. 3850

Tract 397 Map 1-C Baltimore County

Lexington 106 1/2 acres

Resurvey of The Three Divisions (2591 a.)
Survey 14 Mar., 1787 William Davis, Edmund Stansbury and Jacob Marshall
Patent 25 May, 1798 William Davis, Edward Stansbury and Jacob Marshall
IC O/438 and IC M/430
Pat. Cert. 2857

Tract 398 Map 2-W Carroll County

Little Britain 66 acres

Prior Survey for Little Britain Enl. (242 a.)
Comm. Warr. 2 Nov., 1761 (66 a.) Wm. Winchester
Survey 30 Nov., 1751 William Winchester
Patent 1 Oct., 1752 William Winchester
BY & GS 7/421 and BY & GS 6/326
Pat. Cert. 2882

Tract 399 Map 2-W Carroll County

Little Britain Enlarged 242 acres

Resurvey of Little Britain (66 a.)
Prior Survey of Little Britain Enl. Res. (212 a.)
Sp. Warr. 28 June, 1759 William Winchester
Survey 30 July, 1759 William Winchester

(Little Britain Enlarged - continued)
Patent 25 Mar., 1765 William Winchester
BC & GS 27/506 and BC & GS 28/86
Pat. Cert. 2885

Tract 400 Map 2-W Carroll County
<u>Little Britain Enlarged Resurvey</u> 212 acres
Resurvey of Little Britain Enlarged (242 a.)
Sp. Warr. 20 Dec., 1791 John Sherman
Survey 16 Aug., 1792 John Sherman
Patent 17 Dec., 1794 John Sherman
IC I/630 and IC H/589
Pat. Cert. 2884

Tract 401 Map 2-C Carroll County
<u>Little Meadow</u> 32 acres
Prior Survey of Little Meadow Enlarged (157 1/2)
Warrant 10 Mar., 1763 (100 a.) Thomas Stansbury
Survey 24 Feb., 1764 Thomas Stansbury
Patent 24 Feb., 1764 Thomas Stansbury
BC & GS 27/470 and BC & GS 25/210
Pat. Cert. 2896

Tract 402 Map 2-C Carroll County
<u>Little Meadow Enlarged</u> 157 1/2 acres
Resurvey of Litle Meadow (32 a.)
Prior Survey for Big Owl Hill (245 1/2 a.)
and Bollinger's Living (178 1/2 a.)
Sp. Warr. 22 Dec., 1774 Thomas Stansbury
Survey 15 Apr., 1775 Thomas Stansbury
Patent 14 Mar., 1786 Thomas Stansbury
IC B/401 and IC A/644
Pat. Cert. 2898

Tract 403 Map 2-SE Baltimore County
<u>Little Profit</u> 10 acres
Resurvey of Montagle (122 a.)
Prior Survey for New London (250 3/4 a.)
Warrant 7 Aug., 1787 (10 a.) Charles Boblitz
Survey 4 Aug., 1788 Charles Boblitz
Patent 28 May, 1789 Charles Boblitz
IC D/466 and IC E/43
Pat. Cert. 2908

Tract 404 (South of Map 1) Baltimore County
<u>Little Profit</u> 27 1/4 acres
(Reference Tract- Singery's Trouting Stream (575 a))
Prior Survey for Ford's Enlargement (--- a.)
Sp. Warr. 8 Nov., 1794 (16 a.) John Keith
Survey 3 Sept., 1795 John Keith
Assigned 30 Sept., 1795 to Benjamin Thomas
Patent 18 Sept., 1815 Benjamin Thomas
IB D/461
Pat. Cert. 2909

Tract 405 Map 2-N Carroll County

Little Rock 41 acres

Sp. Warr. 20 Nov., 1773 (105 a.) John Gill
Assigned 16 Apr., 1774 (76 a.) to Henry Lamotte
Renewed 29 Apr., 1774 (76 a.) Henry Lamotte
(Renewed with liberty to execute warrant on land next to Plymouth(434 a.))
Survey 10 May, 1774 Henry Lamotte
Patent 12 July, 1775 Henry Lamotte
BC & GS 51/121 and BC & GS 49/255
Pat. Cert. 2911

Tract 406 Map 2-SW Carroll County

Little Room 30 acres

Comm. Warr. 24 Apr., 1775 (31 a.) Fr. Addlesparger
Survey 27 Apr., 1775 Frances Addlesparger
Patent 20 Aug., 1794 Frances Audlesperger
IC I/619 and IC G/419
Pat. Cert. 2912

Tract 407 Map 1-W Baltimore County

Locust Bottom 91 acres

Prior Survey for Christiana Creek (102 1/2 a.)
Survey 10 Mar., 1787 Thomas Stansbury
Patent 14 Oct., 1788 Thomas Stansbury
IC D/321 and IC C/517
Pat. Cert. 2935

Tract 408 Map 1-NE Baltimore County

Long Dream 46 acres

Comm. Warr. 2 Dec., 1790 (22 a.) John Free
Assigned 21 Nov., 1791 Christian Crouse
Sp. Warr. 16 May, 1791 (24 a.) Martin Freedman
Assigned 28 Sept., 1791 Christian Crouse
Survey 23 Nov., 1791 Christian Crouse
Patent 19 Feb., 1802 Christian Crouse
IC R/43 and IC L/709
Pat. Cert. 2951

Tract 409 (not mapped) (Carroll) County

Long Marsh 18 acres

Leased 6 Oct., 1759 John Beavor (Beaver)
Leaseholds GGB A/150
(not precisely located - on north side of west prong of the Gunpowder Falls, probably on Long (Meadow) Run.)

Tract 410 (South of Map 1) Baltimore County

Long Slipe 21 acres

(Reference Tract - Buck Range 544 a.))
Comm. Warr. 11 Dec., 1794 (20 a.) John Cooper
Survey 3 Nov., 1795 John Cooper
Patent 13 Nov., 1800 John Cooper
IC N/488 and IC L/614
Pat. Cert. 2966

Tract 411 Map 2-N Carroll County
Loretta 51 acres
Resurvey of Wooley String (35 a.)
Prior Survey for Loretta (51 a. granted 1809)
Warrant 23 June, 1774 (125 a.) Thomas West
Assigned 15 Nov., 1774 (66 a.) to Christian Singery
Survey 20 Nov., 1774 Christian Singery
Patent 21 Oct., 1785 Christian Singery
IC A/552 and IC B/406
Pat. Cert. 2977

Tract 412 Map 2-N Carroll County
Loretta Resurveyed 51 acres
Resurvey of Loretta (51 a. granted 21 Oct., 1785)
Assigned 4 Dec., 1786 to Valentine Feitz (Foitz)
(Bond of conveyance)
Assigned 12 Apr., 1790 to John Shaulk
(Bond assigned to "Petitioner" - John Shaulk)
Original Patent vacated due to error in the 7th course which reads south instead of north.
Patent 30 June, 1809 Christian Singery
(New patent issued)
Assigned 30 June, 1809 to John Shaulk
JK U/97 and JK T/125 (IC O/403)
Pat. Cert. 2978

Tract 413 Map 2-W Carroll County
Lott 30 3/4 acres
Prior Survey of Manheim (161 1/2 a.)
Comm. Warr. 29 Aug., 1790 Michael Koutz
Survey 10 Oct., 1792 Michael Coutz
Patent 18 Aug., 1801 Jacob Ploucher
IC N/630 and IC P/188
Pat. Cert. 2981

Tract 414 Map 1-SW Baltimore County
Louis 16th King of the French 100 3/4 acres
Com. Warr. 15 Aug., 1791 (215 1/4 a.) T. C. Deye
Survey 4 Aug., 1792 Thomas Cockey Deye
Patent 29 Apr., 1793 Thomas Cockey Deye
IC I/135 and IC G/240
Pat. Cert. 3046

Tract 415 Map 1-NW Baltimore County
Lovely Plains 90 acres
Prior Survey for Lovely Plains Enl. (96 a.)
Survey 20 Feb., 1787 Christopher Ring
Patent 20 Oct., 1795 Christopher Ring (King)
IC K/199 and IC L/10
Pat. Cert. 3049

Tract 416 Map 1-NW Baltimore County
Lovely Plains Enlarged 96 3/4 a., 34 p.
Resurvey of Lovely Plains (95 a.)
Sp. Warr. 3 June, 1815 Jacob Dennis

(Lovely Plains Enlarged - continued)
Survey 17 Nov., 1815 Jacob Dennis
Patent 29 Nov., 1819 Jacob Dennis
IB G/351 and IB E/343
Pat. Cert. 3050

Tract 417 Map 1-SW Baltimore County
Madden's New Design 86 1/2 acres
Prior Survey of New Purchase (355 3/4 a.)
Survey 7 Mar., 1787 Abraham Green
Patent 25 May, 1789 Abraham Green
IC D/468 and IC E/36
Pat. Cert. 3104

Tract 418 Map 2-NW Carroll County
Magdalen Hall 14 acres
Prior Survey for Spring Run Enlarged (130 3/4 a.)
Sp. Warr. 1 Dec., 1775 (100 a.) Richard Richards and Christopher Shrode
Survey 28 May, 1776 Richard Richards and Christopher Shrode
Assigned 15 June, 1776 to Charles Smith
Patent 20 Mar., 1777 Charles Smith
BC & GS 51/455 and BC & GS 52/489
Pat. Cert. 3105

Tract 419 Map 2-W Carroll County
Manheim 161 1/2 acres
Resurvey on Lott (30 3/4 a.), Three Brothers (300), Stoney Hills (1291 a.), Chestnut Ridge (17 3/4) and Oldfields (2+ a.)
Sp. Warr. 4 Apr., 1801 Jacob Plocher
Survey 26 May, 1801 Jacob Plocher
Patent 31 May, 1802 Jacob Plocher
IC R/66 and IC PP/313
Pat. Cert. 3117

Tract 420 (not mapped) Baltimore County
Markey's Chance 10 acres
Comm. Warr. 24 Oct., 1791 (10 a.) Jacob Markey
Survey 2 Oct., 1792 Jacob Markey
Unpat. Cert. 982
(not located precisely - on north side of west prong of Gunpowder Falls probably near Markey's Chance (60 a.))

Tract 421 Map 1-S Baltimore County
Markey's Chance 60 acres
Prior Survey for Markey's Chance Enl. (62 3/4 a.)
Com. Warr. 24 Nov., 1790 (8 A.) John Hurst
Assigned 12 May, 1791 to Jacob Markey
Com. Warr. 4 Dec., 1790 (50 A.) Jacob Markey
Sp. Warr. 6 Jan., 1791 (112 A.) Jacob Markey
Survey 5 Aug., 1791 Jacob Markey
Patent 20 Oct., 1795 Jacob Markey

(Markey's Chance (60 acres) - continued)
IC K/186 and IC G/582
Pat. Cert. 3127

Tract 422 Map 1-S Baltimore County
<u>Markey's Chance Enlarged</u> 62 3/4 acres
Resurvey of Markey's Chance (60 a.) and Hug Me Snug (15 1/2 a.)
Sp. Warr. 4 Apr., 1823 Amos Markey
Survey 28 May, 1823 Amos Markey
Unpat. Cert. 983

Tract 423 (South of Map 1) Baltimore County
<u>Markey's Fortune</u> 15 1/4 a., 30 p.
(Reference Tracts - Singery's Trouting Stream (575) and Borings Habitation Rock (300 a.))
Survey 7 Aug., 1815 Jacob Markey of Samuel
(No Patent Certificate - See Field Survey Book No. 237, Page 21 (CCHS))

Tract 424 Map 1-S Baltimore County
<u>Markey's Garden</u> 0 a., 2 r., 29 p.
(Reference Tracts - Can't you be easy and let me alone (44 a.) and Singery's Prospect (26 a.))
Comm. Warr. 13 June, 1825 (2 a.) Jacob Markey
Survey 9 Jan., 1826 Jacob Markey
Unpat. Cert. 984

Tract 425 (South of Map 1) Baltimore County
<u>Markey's Hackle</u> 11 acres
(Reference Tract - Bear's Rock (177 3/4 a.))
Comm. Warr. 12 Feb., 1794 Samuel Markey
Survey 23 Oct., 1794 Samuel Markey
Patent 18 Dec., 1830 Samuel Markey
IB G/497
Pat. Cert. 3128

Tract 426 (South of Map 1) Baltimore County
<u>Markey's Hill</u> 29 acres
(Reference Tract - Bear's Rock (177 3/4 a.))
Comm. Warr. 12 Feb., 1794 (29 a.) Samuel Markey
Survey 24 Oct., 1794 Samuel Markey
Patent 20 Dec., 1820 Samuel Markey
IB G/498
Pat. Cert. 3129

Tract 427 Map 1-W Baltimore County
<u>Markey's Ketch</u> 15 acres
Comm. Warr. 17 Sept., 1794 Henry Markey
Survey 1 Sept., 1795 Henry Markey
Patent 16 Oct., 1797 Henry Markey
IC O/309 and IC L/383
Pat. Cert. 3130

Tract 428 Map 1-S Baltimore County

<u>Markey's Luck</u> 341 1/4 acres

Resurvey of Cross's Boot Enlarged (338 a.)
Sp. Warr. 29 Mar., 1779 Jacob Markey
Survey 23 Oct., 1799 Jacob Markey
Patent 18 Nov., 1800 Jacob Markey
IC N/497 and IC P/54
Pat. Cert. 3131

Tract 429 Map 1-S Baltimore County

<u>Markey's Water Rights</u> 3 a., 0 r., 31 p.

Resurvey of Guard House (25 a.)
Sp. Warr. 2 Dec., 1854 Jacob Markey
Survey 15 Jan., 1855 Jacob Markey
Unpat. Cert. 985

Tract 430 Map 1-SW Baltimore County

<u>Marquis de la Lafayette, the Friend of of Religious Liberty</u> 18 Acres

Comm. Warr. 15 Aug., 1791 (215+) Thomas Cockey Deye
Survey 4 Aug., 1792 Thomas Cockey Deye
Patent 27 Mar., 1793 Thomas Cockey Deye
IC I/134 and IC H/257
Pat. Cert. 3132

Tract 431 Map 1-N Baltimore County

<u>Marshall's Folly</u> 254 1/2 acres

Resurvey of The Three Divisions (2591 a.)
Survey 13 Mar., 1787 Thomas Marshall
Patent 26 Apr., 1790 Thomas Marshall
IC F/27 and IC E/328
Pat. Cert. 3133

Tract 432 Map 1-E Baltimore County

<u>Marshall's Marsh</u> 122 acres

Resurvey of The Three Divisions (2591)
Survey 20 Mar., 1787 William Davis, Edmund Stansbury and Jacob Marshall
Assigned 7 May, 1788 to Henry Wilhelm
Patent 11 Mar., 1812 Henry Wiehelm (Wilhelm)
IB B/96 and IC A/160
Pat. Cert. 3134

Tract 433 Map 1-E Baltimore County

<u>Marshall's Marsh</u> 7 3/4 acres

Land Purchased in the Reserve 13 Dec., 1785 Jacob Marshall
Survey 5 Mar., 1787 Jacob Marshall
Assigned 2 June, 1809 to Henry Wilhelm
Patent 10 June, 1809 Henry Wilhelm
JK T/108 and JK U/73
Pat. Cert. 3135
(On the reverse side of the certificate of survey: 25 May, 1798 William Davis, Edmund Stansbury and Jacob Marshall purchased 809 acres of land.)

Tract 434 Map 1-NW Baltimore County
<u>Marshall's Second Attempt</u> 8 acres
Comm. Warr. 20 Oct., 1795 (50 a.) Joseph Vaughn
Assigned 12 Apr., 1796 to Thomas Marshall
Survey 11 Oct., 1796 Thomas Marshall
Patent 7 June, 1808 Thomas Marshall
IC R/729 and IC S/447
Pat. Cert. 3136

Tract 435 Map 1-N Baltimore County
<u>Marshall's Struggle</u> 122 1/2 acres
Resurvey of The Three Divisions (2591)
Survey 17 Mar., 1787 William Davis, Edmund Stansbury and Jacob Marshall
Assigned 6 July, 1789 to William Marshall
(witnesses: Thomas Gist and James Campbell)
Patent 4 Sept., 1809 William Marshall
JK U/101 and JK T/162
Pat. Cert. 3137

Tract 436 Map 1-NW Baltimore County
<u>Mary's Delight</u> 270 acres
Prior Survey for Gist's Purchase (465 a.)
Survey 11 Nov., 1786 George Gist
Patent 8 Apr., 1795 George Gist
IC K/182 and IC G/514
Pat. Cert 3154

Tract 437 Map 1-SW Baltimore County
<u>Mason's Hall</u> 12 acres
Sp. Warr. 18 June, 1792 (12 a.) Philip Myers
Survey 17 June, 1793 Philip Myers
Patent 22 Apr., 1830 Philip Myers
GGB 2/87 and GGB 1/185
Pat. Cert. 3168

Tract 438 (West of Map 2) Carroll County
<u>Matter's Choice Resurvey</u> 162 acres
Resurvey of McGill's Choice (50 a.), Richard's Chance (50 a.) and Johnsborough (15 a.)
Sp. Warr. 30 May, 1785 George Matter of York Co.
Survey 30 Dec., 1785 George Matter
Assigned by deed 2 Dec., 1786 to Jacob Matter
Patent 22 Oct., 1787 Jacob Matter
IC D/108 and IC C/223
Pat. Cert. 3173

Tract 439 Map 1-SE Baltimore County
<u>Matthew's Choice</u> 145 1/2 acres
Survey 12 Mar., 1789 William Davis, Edmund Stansbury and Jacob Marshall
Assigned 26 May, 1789 to William Matthews
Patent 24 May, 1799 William Matthews
IC O/575 and IC M/598
Pat. Cert. 3177

Tract 440 Map 2-NW Carroll County
<u>Mattigan</u> 437 acres
Resurvey of Plymouth (434 a.) (excluding land claimed by the heirs of Jacob Hoffman)
Warrant 23 Mar., 1765 Henry Fites
Renewed 20 Sept., 1765 Henry Fites
Renewed 17 Mar., 1766 Henry Fites
Renewed 12 Sept., 1766 Henry Fites
Survey 14 Feb., 1767 Henry Fites
Patent 2 May, 1774 Henry Fites
BC & GS 50/93 and BC & GS 44/478
Pat. Cert. 3190

Tract 441 (not mapped) Baltimore County
<u>Mays (Mayes) Mountain</u> 100 acres
Prior Survey for Addition to Deer Park (313 3/4)
Survey 8 Mar., 1787 James Calder
Assigned 18 Oct., 1787 to Thomas Gist
Patent 19 Dec., 1789 Thomas Gist
IC D/623 and IC E/253
Pat. Cert. 3202

Tract 442 Map 1-SE Baltimore County
<u>McCallister's (McAllister's) Folly</u> 353 1/2 acres
Resurvey of Widow's Habitation (185 1/2 a.)
Sp. Warr. 25 Feb., 1791 Richard McAllister and Samuel Hornish of Penna.
Survey 4 Jan., 1792 Richard McAllister and Samuel Hornish of Penna.
Patent 5 Oct., 1793 Richard McAllister and Samuel Hornish of Penna.
IC H/359 and IC I/445
Pat. Cert. 3089

Tract 443 (West of Map 2) Carroll County
<u>McGill's Choice</u> 50 acres
(Reference tract - Good Luck (524 1/2 a.))
Resurvey of Richard's Chance (50 a.)
Prior Survey for McGill's Choice Resurvey (243)
Warrant 4 May, 1744 (50 a.) Andrew McGill
Survey 12 June, 1744 Andrew McGill
Patent 12 June, 1744 Andrew McGill
LG E/448 and PT 1/211
Pat. Cert. 3095

Tract 444 (West of Map 2) Carroll County
<u>McGill's Choice Resurvey</u> 243 acres
(Reference Tract - Good Luck (524 1/2 a.))
Resurvey of McGill's Choice (50 a.)
Prior Survey for Matter's Choice (162 a.)
Survey 26 Dec., 1749 Andrew McGill
Patent 12 June, 1754 Andrew McGill
BC & GS 4/482 and BC & GS 3/272
Pat. Cert. 3094

Tract 445 Map 2-E Carroll County

<u>Merryman's Chance</u> 56 acres

Prior Survey for Stansbury's Grove (444 a.)
Warrant 8 June, 1751 (50 a.) John Ridgely
Assigned to Samuel Merryman
Survey 23 Nov., 1751 Samuel Merryman
Patent 9 Nov., 1752 Samuel Merryman
BY & GS 5/643 and BY & GS 4/410
Pat. Cert. 3229

Tract 446 (South of Map 2) Carroll County

<u>Merryman's Choice</u> 88 acres

Prior Survey for Merryman's Choice (granted 1797)
Leased Land 12 July, 1743 Charles Merryman
Leaseholds GGB B/52
Survey 12 July, 1743 Charles Merryman
Unpat. Cert. 1013

Tract 447 (South of Map 2) Carroll County

<u>Merryman's Choice</u> 88 acres

(Reference Tract - Penelope Charchilla and Thomas Cockeys Deyesburg (1030 a.))
Resurvey of Merryman's Choice (leased 1743)
Survey 17 Nov., 1796 Jacob Cox
Assigned 14 Feb., 1797 Elisha Cox
Patent 15 Mar., 1797 Elisha Cox
IC O/273 and IC L/280
Pat. Cert. 3231

Tract 448 (South of Map 2) Carroll County

<u>Merryman's Grotto</u> 50 acres

(Reference Tract - Murray's Ridge (221 1/2 a.))
Prior survey for Addition to Merryman's Grotto (90)
Warrant 31 Dec., 1744 Thomas Stansbury
Assigned to William Merryman
Survey 12 June, 1745 William Merryman
Patent 12 June, 1745 William Merryman
LG E/513 and PT 2/325
Pat. Cert. 3238

Tract 449 Map 1-SW Baltimore County

<u>Merryman's Mountain</u> 83 acres

Prior Survey for Singery's Trouting Stream (178)
Survey 4 Feb., 1762 George Merryman
(No Pat. Cert. in Archives - probably leased land)

Tract 450 Map 2-E Carroll County

<u>Merryman's Neighbor</u> 24 1/2 acres

Warrant 13 Nov., 1793 (10 a.) William Hoffman
Warrant 13 Nov., 1793 (12 a.) William Hoffman
Warrant 13 Nov., 1793 (16 a.) William Hoffman
Survey 25 Apr., 1794 William Hoffman
Patent 30 Mar., 1798 William Hoffman
IC O/455 and IC M/400
Pat. Cert. 3248

Tract 451 Map 1-W Baltimore County
Merryman's Rock 70 1/2 acres
Survey 15 March, 1787 William Hoffman
Patent 2 April, 1789 William Hoffman
IC D/414 and IC C/647
Pat. Cert. 3251

Tract 452 (not mapped) Carroll County
Mesopotamia (Missopotamia) 1958 acres
Resurvey of Amsterdam (180 a.)
Prior Survey for Troy (1634 a.)
Warrant 29 Nov., 1751 (2000 a.) Adam Lynn
(warrant expired)
Sp. Warr. 14 Mar., 1762 Andrew Stiger of Baltimore City and William Winchester of Frederick Co.
Renewed 2 Mar., 1763 (same)
Renewed 1 Aug., 1763 (same)
Renewed 30 Jan., 1764 (same)
Survey 7 Apr., 1764 Andrew Stiger of Baltimore City and William Winchester of Frederick Co.
Assigned 5 July, 1764 to Andrew stiger and Charles Ridgely (Jr.)
Patent 16 Oct., 1764 Charles Ridgely, Jr. and Andrew Stiger
BC & GS 21/378 and BC & GS 18/473
Pat. Cert. 3314

Tract 453 (not mapped) Carroll County
Michael's Beginning 16 acres
Prior Survey for Soldier's Delight (20 a.)
Survey 16 June, 1784 Michael Deeds
(no records in State Archives)
(See field survey book No. 26, page 33 in Arthur G. Tracey Colection (HSCC))

Tract 454 Map 2-SE Carroll County
Michael's Beginning 20 acres
Warr. 13 Mar., 1791 (20 a.) Michael Boblets
Survey 18 Nov., 1791 Michael Boblets
Patent 16 Apr., 1794 Michael Boblets
IC I/632 and IC H/482
Pat. Cert. 3259
(one half in Baltimore County)

Tract 455 Map 2-NW Carroll County
Michael's Beginning 25 acres
Prior Survey for Addition to Michael's Beginning (57 a.)
Comm. Warr. 14 Mar., 1755 (25 a.) Michael Hanna
Survey 16 Apr., 1755 Michael Hanna
Patent 16 Apr., 1755 Michael Hanna
BC & GS 4/270
Pat. Cert. 3256

Tract 456 Map 2-S Carroll County
Middlesex 8 1/4 acres
Resurvey of Foster's Hunting ground (1439 a.)
Sp. Warr. 9 May, 1797 (8 1/4 a.) T. C. Deye
Survey 25 Apr., 1798 Thomas Cockey Deye
Patent 16 Dec., 1799 Thomas Cockey Deye
IC O/680 and IC L/580
Pat. Cert. 3272

Tract 457 Map 2-N Carroll County
Miner Courser 25 acres
Comm. Warr. 4 June, 1760 (25 a.) Christian Shafer
Survey 30 Nov., 1760 Christian Shafer
Patent 30 Nov., 1760 Christian Shafer
BC & GS 14/708
Pat. Cert. 3304

Tract 458 Map 1-E Baltimore County
Minersborough 54 acres
Resurvey of Archibald's Level (48 1/2 a.) and
The Three Divisions (2591 a.)
Survey 12 Mar., 1799 Edward Stansbury
Assigned 22 June, 1799 to Ann Bond and Thos. S. Bond
Assigned 4 June, 1800 to Thomas Talbot (Jr.)
Patent 5 July, 1808 Thomas Talbot (Jr.)
IC R/716 and IC O/581
Pat. Cert. 3309

Tract 459 (not mapped) Carroll County
Montagle 122 acres
(Reference Tract - Henry's Hope (100 a.))
Prior Survey for Great Profit (25 a.), Little
Profit (10 a.) and Francis Thwaitsburgh (542 a.)
Sp. Warr. 11 Sept., 1772 (1400 a.) John Wilmott
Survey 10 Jan., 1773 John Wilmott
Patent 22 Sept., 1773 John Wilmott
BC & GS 47/416 and BC & GS 46/214
Pat. Cert. 3360

Tract 460 (South of Map 1) Baltimore County
Mordecai's Addition 33 3/4 acres
(Reference Tract - Ford's Choice (157 a.), Singary's
Trouting Stream (178 a.) and Merryman's Mt. (83))
Survey 26 May, 1787 Mordecai Ford
Patent 6 Nov., 1794 Mordecai Ford
IC I/611 and IC H/558
Pat. Cert. 3340

Tract 461 (not mapped) Baltimore County
Morehead's Relief 35 acres
(Reference Tract - Cole's Meadows (152 1/2 a.)
on Cattail Marsh Branch)
Leased 2 Apr., 1760 Samuel Morehead
Leaseholds GGB B/151
(lease made out but (Morehead) neglected to take it)

Tract **
Motter's Choice (See Matter's Choice)

Tract 462 Map 1-SW Carroll County
<u>Mount Gilboa</u> (Gibbon) 29 3/4 acres
Survey 4 Dec., 1792 John McComsky
Patent 16 Nov., 1793 John McComsky heir of Daniel
IC G/313 and IC I/458
Pat. Cert. 3371

Tract 463 (West of Map 2) Carroll County
<u>Mount Hendrix Resurveyed</u> 48 acres
(Reference Tract - Baughman's Goodwill (154 3/4 a.))
Resurvey of Mount Pleasant (17 a.) and
McGill's Choice (243 a.)
Sp. Warr. 6 Apr., 1767 James Hendrix
Renewed 6 Oct., 1767 James Hendrix
Survey 3 Mar., 1768 James Hendrix
Patent 3 Oct., 1769 James Hendrix
BC & GS 40/203
Pat. Cert. 3375

Tract 464 (West of Map 2) Carroll County
<u>Mount Pleasant</u> 17 acres
(Reference Tract - Baughman's Goodwill (154 3/4 a.))
Prior Survey for Mt. Hendrix Resurvey (48 a.)
Survey 20 May, 1761 James Hendricks
Patent 20 May, 1761 James Hendricks
BC & GS 14/150 and BC & GS 15/78
Pat. Cert. 3399

Tract 465 Map 1-E Baltimore County
<u>Mount Pleasant</u> 28 a. + 6 p.
Resurvey of Hampshire's Bargain (10 a.) and
Enoch's Hill (15 1/4 a.)
Sp. Warr. 4 Dec., 1809 George Hampshire
Survey 2 May, 1810 George Hampshire
Unpat. Cert. 1069

Tract 466 Map 1-E Baltimore County
<u>Mount Pleasant</u> 135 acres
Resurvey of The Three Divisions (2591 a.)
Prior Survey for Mount Pleasant (138 a.)
Survey 13 Mar., 1787 William Richardson
Assigned 18 Feb., 1792 to William Matthews
Patent 24 May, 1799 William Matthews
IC M/599
Pat. Cert. 3393

Tract 467 Map 1-E Baltimore County
<u>Mount Pleasant</u> 138 acres
Resurvey of Mount Pleasant (135 a.)
Survey 7 Oct., 1843 John Michaels
(no records in State Archives - survey in
envelope #60 in Tracey collection - (HSCC))

Tract 468 (East of Map 1) Baltimore County

Mount Pleasant 25 1/2 acres

(Reference Tract - Northampton (242 a.))

Survey 18 Nov., 1786 John Lowe of York Co., Pa.

Patent 30 Aug., 1788 John Lowe

IC D/297

Pat. Cert. 3383

Tract 469 Map 1-SE Baltimore County

Mount Pleasant 16 acres

Sp. Warr. 11 Mar., 1795 (16 a.) Samuel Cauhey

Survey 2 Mar., 1796 Samuel Cauhey

Patent 31 Jan., 1803 Samuel Cauhey

IC R/239 and IC S/18

Pat. Cert. 3392

Tract 470 Map 2-SW Carroll County

Murphy's Retirement 60 acres

Prior Survey for Phillipsburg (164 a.)

Warrant 12 Feb., 1744 (50 a.) William Murphy

Survey 5 June, 1745 William Murphy

Patent 5 June, 1745 William Murphy

LG E/516 and PT 1/340

Pat. Cert. 3417

(Sold and conveyed to Philip Idleman)

Tract 471 (South of Map 2) Carroll County

Murray's Ridge 221 1/2 acres

(Reference Tracts - Christian's Choice (166 a.) and Pleasant Spring Resurvey (444 1/4 a.))

Resurvey of Merryman's Grotto (50 a.) and Addition to Merryman's Grotto (90 a.)

Sp. Warr. 28 May, 1762 William Murray

Survey 7 June, 1762 William Murray

Patent 24 Apr., 1764 William Murray

BC & GS 21/315 and BC & GS 22/36

Pat. Cert. 3430

Tract 472 Map 1-NW Baltimore County

Murse's (Mash's, Marsh's) Garden 137 1/2 acres

Survey 1 Feb., 1787 Thomas Smith

Assigned by terms of will to heirs of Thomas Smith: Eleanor, Elizabeth and John Smith, Catherine Zimmerman and Mary Dungan

Patent 24 Dec., 1812 Eleanor, Elizabeth and John Smith, Catherine Zimmerman and Mary Dungan

IB B/413 and IB A/296

Pat. Cert 3433

Tract 473 Map 1-N Baltimore County

My Lady's Benefit 39 1/4 acres

Survey 19 Mar., 1787 Michael Howman

Patent 30 Dec., 1794 Michael Howman

IC I/613 and IC G/476

Pat. Cert. 3438

Tract 474 Map 1-S Baltimore County

<u>My Sweet Girl, My Friend, My Pitcher</u> 62 acres

Sp. Warr. 11 May, 1797 (42 1/2 a.) Joshua Stevenson of Frederick County
Comm. Warr. 15 May, 1798 (37 3/4 a.) Christian Singery
Assigned (20 a.) to Joshua Stevenson
Survey 19 Dec., 1797 Joshua Stevenson
Patent 9 May, 1799 Joshua Stevenson
IC O/571 and IC M/586
Pat. Cert. 3441

Tract 475 Map 1-NE Baltimore County

<u>Myer's Luck</u> 12 1/2 acres

Prior Survey for Dorsey's Delay (3+ a.)
Warrant 27 May, 1797 (70 a.) Adam Hendrick
Survey Adam Hendrick
Certificate vacated and reassigned to John Garrett
Assigned (56 1/2 a.) to William Dorsey
Assigned (6 1/4 a.) to Andrew Myers
Warrant 30 July, 1801 (40 a.) Stephan Freeland
Assigned (partial) to William Dorsey
Sp. Warr. 1 Sept., 1801 (26 A.) William Dorsey
* Survey 3 Dec., 1801 William Dorsey
Assigned 12 Aug., 1802 to Andrew Miers
Patent 17 Mar., 1812 Andrew Miers (Myers)
IB B/91 and IB A/154
Pat. Cert. 3435

Tract 476 Map 2-SW Carroll County

<u>Nace's (Neace's) Tavern</u> 478 1/2 acres

Resurvey of Schieming Defeated (186 a.), Shillings Lot (50 a.), Shillings Folly (269 a.), Bridgeland (150 a.), Phillipsburg Res. (164) and Addition to Shillings Lot (8 1/2 a.)
Sp. Warr. 20 May, 1786 George Nace
Survey 1 Mar., 1787 George Nace
Patent 23 Dec., 1788 Genrge Nace (Neace)
IC D/378 and IC M/610
PPat. Cert. 3461

Tract 477 Map 1-SW Baltimore County

<u>Naked Hills</u> 60 acres

Comm. Warr. 6 Jan., 1791 (96 a.) Samuel Markey
Survey 30 Sept., 1791 Samuel Markey
Patent 27 Dec., 1799 Samuel Markey
IC O/674 and IC M/676
Pat. Cert. 3442

Tract 478 (East of Map 1) Baltimore County

<u>Narrow Passage</u> 16 acres

(Reference Tract - Expedition (337 a.))
Prior Survey for Expedition (337 a.)

(Narrow Passage - continued)
Comm. Warr. 1 May, 1797 (16 a.) Peter Shock
Survey 3 May, 1798 Peter Shock
Unpat. Cert. 1089

Tract 479 Map 1-SW Baltimore County
Neighbor's Goodwill 289 acres
Resurvey of Huckleberry Hill (100 a.)
Prior Survey for New Purchase (355 3/4 a.)
Sp. Warr. 20 July, 1791 Christian Singery
Survey 29 June, 1792 Christian Singery
Patent 22 Oct., 1795 Christian Singery
IC K/173 and IC L/14
Pat. Cert. 3472

Tract 480 (not mapped) Carroll County
Neilson's (Nellson's) Puzzle 256 1/4 acres
Resurvey of Forelorn Hope (23 1/2 a.), Stansbury's Grove (1448 a.), Price's Glory (50 a.) Happiness (39 a.) and Hope Restored (20 a.)
Sp. Warr. 6 Aug., 1833 Thomas N. Neilson of Baltimore City
Survey 6 Mar., 1834 Thomas N. Neilson
Patent 15 May, 1837 Thomas N. Neilson
GGB 4/10 and GGB 3/149
Pat. Cert. 3484

Tract 481 Map 2-E Baltimore County
(The) New Beginning 15 acres
Comm. Warr. 17 Dec., 1799 (15 a.) William Hoffman
Survey 2 Dec., 1800 William Hoffman (Sr.)
Assigned 8 Dec., 1800 to William Hoffman, Jr.
Patent 22 June, 1801 William Hoffman, Jr.
IC N/545 and IC L/647
Pat. Cert. 3487

Tract 482 Map 2-W Carroll County
New Germany 178 acres
Prior Survey for Joshua Boring son of John his Forest (220+ a.)
Comm. Warr. 22 May, 1754 (50 a.) Philip Idleman
Assigned 23 May, 1754 to Jacob Shilling, Jr.
Comm. Warr. 27 May, 1754 (128) Jacob Shilling, Jr.
Survey 30 Aug., 1754 Jacob Shilling (Jr.)
Assigned 17 Oct., 1754 to Philip Idleman
Patent 17 Oct., 1754 Philip Idleman
BC & GS 1/230 and BC & GS 2/20
Pat. Cert. 3498

Tract 483 Map 2-SE Baltimore County
New London 250 3/4 acres
Resurvey of Little Profit (9 3/4 a.), Great Profit (25 a.), Henry's Hope (100 a.), Eve's Garden (5 1/2 a.), Charles Rue (23 3/4 a.), Charles Hill (29 3/4) and Charles Flower Knot (49+)

(New London - continued)

Sp. Warr. 8 Nov., 1793 — Charles Boblitz
Survey 3 May, 1794 — Charles Boblitz
Patent 28 Nov., 1798 — Charles Boblitz
IC O/470 and IC M/507
Pat. Cert. 3508

Tract 484 (West of Map 2) Carroll County
New Market 67 acres
(Reference Tracts - Everything Needful Corr. (1755+) and Adam's Contrivance Resurvey (113 a.))
Comm. Warr. 12 Mar., 1764 (67 a.) Richard Richards
Renewed 11 Sept., 1764 — Richard Richards
Survey 5 Mar., 1765 — Richard Richards
Patent 5 Mar., 1765 — Richard Richards
BC & GS 27/468 and BC & GS 26/290
Pat. Cert. 3509

Tract 485 Map 2-NW Carroll County
New Prussia 24 1/2 acres
Prior Survey of Jos. Bollinger's Contriv. (348)
Comm. Warr. 14 Sept., 1774 (50 a.) Henry Lamuth
Renewed 14 Mar., 1775 — Henry Lamuth
Survey 9 Aug., 1775 — Henry Lamuth
Assigned 3 June, 1786 to Joseph Bollinger
Patent 10 Apr., 1788 — Joseph Bollinger
IC D/163 and IC C/343
Pat. Cert. 3514

Tract 486 Map 1-SW Baltimore County
New Purchase 355 3/4 acres
Resurvey of Madden's New Design (86 1/2 a.), Piney Hill (175 1/2 a.), Tempy's Spring (36 a.) Abraham's New Design (13 1/2 a.), Adam's New Design(27 1/4 a.), Any Name You Please (10) and Neighbor's Goodwill (289 a.)
Sp. Warr. 17 Feb., 1800 — Abraham Shaver
Survey 6 Feb., 1801 — Abraham Shaver
Patent 23 Nov., 1801 — Abraham Shaver
IC N/622 and IC P/236
Pat. Cert. 3513

Tract 487 (West of Map 2) Carroll County
Niagra 15 acres
(Reference Tracts - Baughman's Goodwill (154 3/4 a.) and Troy (1634 a.))
Comm. Warr. 13 Feb., 1760 (15 a.) Philip Pillar
Survey 25 Mar., 1760 — Phillip Pillar
Patent 25 Mar., 1760 — Phillip Pillar
BC & GS 12/315 and BC & GS 10/525
Pat. Cert. 3521

Tract 488 Map 2-C Carroll County
No Name 1 1/2 a. + 8 p.
Resurvey of Borings Range (1375 a.)

(No Name - continued)
Sp. Warr. 7 Apr., 1807 (10 a.) Absalom Foster
Survey 22 Jan., 1808 Absalom Foster
Assigned 3 Oct., 1812 to John Hoffacker
Patent 5 Nov., 1812 John Hoffacker
IB B/408 and IB A/271
Pat. Cert. 3532

Tract 489 Map 2-W Carroll County
North 37 acres
Sp. warr. 24 Oct., 1776 (180 a.) Thomas Stansbury and Henry Wareheim
Survey 14 Feb., 1777 Thomas Stansbury and Henry Wareheim
Patent 30 May, 1788 Thomas Stansbury and Henry Wareheim
IC C/433 and IC D/175
Pat. Cert. 3540

Tract 490 Map 1-NE Baltimore County
North Hampton (Northampton) 242 acres
Survey 14 Nov., 1786 Joseph Morris
Patent 4 Aug., 1790 Joseph Morris
IC F/26 and IC C/707
Pat. Cert. 3541

Tract 491 (East of Map 1) Baltimore County
North Spring 11 acres
(Reference Tract - Expedition (337 a.))
Prior Survey for Expedition (337 a.)
Comm. Warr. 21 May, 1793 (7 a.) Jesse Jarrett
Survey 1 Aug., 1793 Jessee Jarrett of Harford Co.
Unpat. Cert. 1174

Tract 492 Map 2-C Carroll County
Not Enough 15 acres
Comm. Warr. 19 May, 1795 (15 a.) Christopher Fair
Survey 29 May, 1795 Christopher Fair
Patent 7 Dec., 1805 Christopher Fair
IC Q/642 and IC S/286
Pat. Cert. 3561

Tract 493 (South of Map 1) Baltimore County
Nothing Ventured Nothing Win 20 acres
(Reference Tract - Buck Range (544 a.) and Hard Fare (250 a.))
Sp. Warr. 1 May, 1795 (20 a.) George Taylor
Survey 30 June, 1795 George Taylor
Unpat. Cert. 1178

Tract 494 Map 2-W Carroll County
Nottenstot 50 acres
Prior Survey for Nottenstot (66 a.)
Comm. Warr. 24 Jan., 1756 (50 a.) Anthony Noll
Survey 20 July, 1756 Anthony Noll

(Nottenstot (50 acres) - continued)
Patent 20 July, 1756 Anthony Noll
BC & GS 9/100 and BC & GS 7/329
Pat. Cert. 3568

Tract 495 Map 2-W Carroll County
Nottenstot 66 acres
Resurvey of Nottenstot (50 a.)
Prior Survey for Nottenstot Resurvey (155 a.)
Sp. Warr. 20 Feb., 1759 Anthony Noll
Survey 30 Apr., 1759 Anthony Noll
Patent 7 Feb., 1761 Anthony Noll
BC & GS 14/157 and BC & GS 16/209
Pat. Cert. 3566

Tract 496 Map 2-W Carroll County
Nottenstot Resurvey 155 acres
Resurvey of Anthony's Hope (24 a.), Addition to to Nottenstot (25 a.), Addition to Nottenstot (17 a.), Nottenstott (66 a.) and Join All (10 a.)
Sp. Warr. 9 Feb., 1790 Anthony Noll
Survey 4 Nov., 1790 Anthony Noll
Patent 21 May, 1792 Anthony Noll
IC F/645 and IC H/81
Pat. Cert. 3567

Tract 497 Map 1-W Baltimore County
Number One 123 acres
Sp. Warr. 24 June, 1800 (14 a.) William Hoffman, Jr.
Survey 3 Dec., 1800 William Hoofman, Jr.
Assigned by will of William Hoffman, Jr. to his children and heirs: (See 'Number Three')
Patent 22 June, 1801 William J. Hoffman et. al.
IC N/545 and IC P/176
Pat. Cert. 3575

Tract 497 A (not mapped) Baltimore County
Number 1 21 1/2 acres
Prior survey for Paper Mill Hills (282 a.)
Sp. Warr. 6 May 1782 (15 a.) William Hoffman
Survey 20 May, 1782 William Hoffman
(Site of the Gunpowder Paper Mill - among the surveys in the MHS collection of Hoffman papers.)

Tract 498 (not mapped) Baltimore County
Number 2 40 3/4 acres
Resurvey of Barren Hills (16 1/2 a.)
Prior survey for Grist Mill Hills (257 a.)
Sp. Warr. 7 May, 1782 (20 a.) William Hoofman
Survey 17 May, 1782 William Hoofman
(Site of Clipper Paper Mill and 'new' saw mill. On line #7 of Grist Mill Hills. Survey in (MHS) collection of Hoffman Papers.)

Tract 498A Map 2-E Carroll County

<u>Number Three</u> 15 1/4 acres

Resurvey of Happiness (39 a.)
Comm. Warr. 24 June, 1800 William Hoffman, Jr.
Survey 22 June, 1801 William Hoffman, Jr.
Assigned by will of William Hoffman, Jr. to his Children and heirs: William J. Hoffman, Henry B. Hoffman, Susannah Hoffman (m. Samuel Hoffacker), Mary Hoffman (m. Jacob Markey), Elizabeth Hoffman (m. Henry M. Hoffacker), Ann Hoffman (m. John Gore) and Catherine Hoffman (m. George Miller) (See GGB 7/277)
Patent 27 July, 1850 William J. Hoffman et. al.
GGB 7/273
Pat. Cert. 3577

Tract 499 (not mapped) Baltimore County

<u>Number 3</u> 10 acres

Sp. Warr. 18 May, 1782 William Hoffman (Sr.)
Survey 18 May, 1782 William Hoffman (Sr.)
(not precisely located - near Hoffmanville, Md.; survey in Hoffman Collection (MHS))

Tract 500 Map 1-W Baltimore County

<u>Number Four</u> 91 1/4 acres

Sp. Warr. 15 Dec., 1800 (10 a.) Wm. Hoffman, Jr.
Survey 10 Nov., 1801 William Hoffman, Jr.
Assigned by will of William Hoffman, Jr. to his children and heirs: (See 'Number Three')
Patent 27 July, 1850 William J. Hoffman et. al.
GGB 5/549 and GCB 7/274
Pat. Cert. 3579

Tract 501 (not mapped) Baltimore County

<u>Number 4</u> 5 acres

Comm. Warr. 6 May, 1782 (5 a.) William Hoffman (Sr.)
Survey 18 May, 1782 William Hofman (Sr.)
(not precisely located - near Hoffmanville, Md.; survey in Hoffman Collection (MHS))

Tract 502 (not mapped) Baltimore County

<u>Number 5</u> 5 acres

(same as Tract 501 but different survey courses)

Tract 503 (not mapped) Baltimore County

<u>Number Six</u> 5 acres

Comm. Warr. 6 May, 1782 (5 a.) William Hoffman (Sr.)
Survey 18 May, 1782 William Hoffman (Sr.)
Unpat. Cert. 1182
(not Precisely located - Near Hoffmanville, Md.)

Tract 504 (not mapped) Baltimore County

<u>Number 7</u> 5 acres

(same as Tract 501 but different survey courses)

Tract 505 (not mapped) Baltimore County
<u>Number 8</u> 5 acres
(same as Tract 501 but different survey courses)

Tract 506 (not mapped) Baltimore County
<u>Number 9</u> 5 acres
(same as Tract 501 but different survey courses)

Tract 507 (not mapped) Baltimore County
<u>Number 10</u> 8 acres
Sp. Warr. 31 May, 1782 William Hoffman (Sr.)
Survey 6 June, 1782 William Hoffman (Sr.)
(not precisely located - near Hoffmanville, Md.; survey in Hoffman Collection (MHS))

Tract 508 (not mapped) Baltimore County
<u>Number 11</u> 10 acres
(same as Tract 507 but different survey courses)

Tract **
Oates Inheretance (See Utz'z Inheretance (226 a.))

Tract 509 Map 2-W Carroll County
<u>Oldfields</u> 2 acres, 17 perches
Prior Survey for Manheim (161 1/2 a.)
Sp. Warr. 29 Aug., 1791 Jacob Ploucher
Survey 25 Oct., 1791 Jacob Ploucher
Patent 17 Aug., 1801 Jacob Ploucher (Bloucher)
IC N/629 and IC P/187
Pat. Cert. 3597

Tract 510 Map 2-NW Carroll County
<u>Overpronn</u> (Overprone) 24 1/2 acres
Comm. Warr. 2 June, 1773 (25 a.) Peter Miller
Survey 13 Nov., 1773 Peter Miller
Patent 20 June, 1796 Peter Miller
IC O/39 and IC M/10
Pat. Cert. 3645

Tract 511 Map 1-W Baltimore County
<u>Paper Mill Hills</u> 282 acres
Prior Survey for Paper Mill Hills (460 1/2 a.)
Survey 13 Mar., 1787 William Hoffman (Sr.)
Patent 2 Apr., 1789 William Hoffman (Hoofman)
IC D/415 and IC C/649
Pat. Cert. 3683

Tract 512 Map 1-NW Baltimore County
<u>Paper Mill Hills</u> 460 1/2 acres
Resurvey of Paper Mill Hills (282 a.)
Sp. Warr. 17 Mar., 1791 William Hoffman
Survey 12 Feb., 1792 William Hoffman
Patent 17 Nov., 1794 William Hoffman
IC I/586 and IC H/572
Pat. Cert. 3684

Tract 513 (South of Map 2) Carroll County
(The) Parade (Perade) 32 acres
(Reference Tract - Petersburg Resurvey (1003 a.))
Comm. Warr. 25 Feb., 1761 (22 a.) John Hendricks
Assigned 1 June, 1761 to Richard Richards
Survey 23 Aug., 1761 Richard Richards
Patent 23 Aug., 1761 Richard Richards
BC & GS 14/658 and BC & GS 17/327
Pat. Cert. 3778

Tract 514 Map 2-NE Carroll County
Partnership 47 acres
Warrant 21 May, 1760 (47 a.) Nicholas Merryman
Renewed 12 June, 1761 Nicholas Merryman
Renewed 22 Dec., 1762 Nicholas Merryman
Renewed 13 Nov., 1764 Nicholas Merryman
Assigned 18 Nov., 1764 to Solomon Cross
Survey 12 May, 1765 Solomon Cross
Patent 20 Nov., 1772 Solomon Cross
BC & GS 45/183
Pat. Cert. 3716

Tract 515 Map 2-N Carroll County
Paul's Delight 15 acres
Prior Survey Ulrich's Delight (95 a.)
Warrant 14 Mar., 1755 (15 a.) Paul Blaystone
Survey 29 Apr., 1755 Paul Blaystone
Patent 29 Apr., 1755 Paul Blaystone
BC & GS 4/269 and BC & GS 2/131
Pat. Cert. 3748

Tract 516 Map 2-NW Carroll County
Peace With England 153 acres
Sp. Warr. 1 Dec., 1775 (100 a.) Christopher Shrode
and Richard Richards
Comm. Warr. (36 a.) Richard Richards
Renewed 16 Feb., 1776 (36 a.) Richard Richards
Assigned 21 May, 1776 (18 a.) to Christopher Shrode
Comm. Warr. (80 a.) Christopher Shrode
Renewed 16 Feb., 1776 Christopher Shrode
Survey 21 May, 1776 Christopher Shrode
Assigned 21 May, 1776 (40. a) to Richard Richards
Patent 20 Mar., 1777 Richard Richards and
Christopher Shrode
BC & GS 51/456 and BC & GS 52/491
Pat. Cert. 2953 (Frederick County Patent Series)

Tract 517 Map 2-SE Carroll County
Penelope Charchilla and Thomas Cockey's Deyesburg
(In eight parts) 1030 acres
Resurvey of Towson's Grove (367 a.),
Jame's Lot (35), Carr's Inginuity (256 a.),
Polygon (340 a.), Rachel's Retirement (42 a.),
Gain's Delight (50 a.)
Warrant 15 Aug., 1789 Thomas Cockey Deye

(Penelope Charchilla and Thomas Cockey Deyesburg (1030 acres) - continued)

Survey 5 July, 1790 — Thomas Cockey Deye
Patent 20 Apr., 1791 — Thomas Cockey Deye
IC F/400 and IC E/687
Pat. Cert. 3773

Tract 518 — Map 2-SW — Carroll County

<u>Penelope Charchilla and Thomas Cockey's Deyesburg</u> — 340 1/2 acres

Resurvey of Borings Range (1375 a.), Chaney's Ramble (62 1/4 a.) and Shillings Folly (269 a.)
Warrant 30 Apr., 1790 — Thomas Cockey Deye
Survey 31 Mar., 1791 — Thomas Cockey Deye
Patent 19 Dec., 1791 — Thomas Cockey Deye
IC I/3 and IC G/123
Pat. Cert. 3774

Tract 519 — (South of Map 2) — Carroll County

<u>People(s) Will</u> — 37 3/4 acres

(Reference Tract - Petersburg Resurvey (1003 a.))
Sp. Warr. 28 Oct., 1796 (40 a.) Peter Fowble
Survey 4 Nov., 1796 — Peter Fowble
Unpat. Cert. 1229

Tract 520 — Map 1-E — Baltimore County

<u>Pessel Tree</u> — 96 3/4 acres

Resurvey of Frosty Hill (46 a.)
Sp. Warr. 2 Sept., 1801 — William Dorsey
Survey 5 Dec., 1801 — William Dorsey
Patent 21 Jan., 1831 — William Dorsey
GGB 2/179 and GGB 1/265
Pat. Cert. 3786

Tract 521 — Map 2-W — Carroll County

<u>Peter Druschel's Management</u> — 136 1/2 acres

Resurvey of Addition to Michael's Beginning (57 a.)
Sp. Warr. 6 Jan., 1791 — Peter Druschel
Survey 10 Feb., 1791 — Peter Druschel
Assigned 20 Dec., 1791 to Melchiore Warner
Patent 9 Mar., 1797 — Melchiore Warner
IC L/281 and IC O/248
Pat. Cert. 3795

Tract 522 — (South of Map 2) — Carroll County

<u>Peter's Addition</u> — 27 acres

(Reference Tract - Pleasant Spring Resurvey (444+))
Prior Survey for Petersburg (1003 a.)
Comm. Warr. 27 July, 1765 (27 a.) Peter Fowble
Survey 3 Dec., 1765 — Peter Fowble
Patent 3 Dec., 1765 — Peter Fowble
BC & GS 30/316 and BC & GS 32/151
Pat. Cert. 3787

Tract 523 Map 2-SE Carroll County

Peter's Beginning 14 acres

Comm. Warr. 3 Mar., 1791 (17 a.) Peter Boblets
Survey 18 Nov., 1791 Peter Baublitz
Patent 16 Apr., 1794 Peter Baublitz
IC I/627 and IC H/480
Pat. Cert. 3788

Tract 524 Map 2-SE Carroll County

Peter's Delight 50 acres

Comm. Warr. 27 Jan., 1786 (50 a.) Peter Boblitz
Survey 21 Mar., 1786 Peter Boblits
Patent 2 May, 1787 Peter Boblitz (Boblets)
IC B/689 and IC C/129
Pat. Cert. 3796

Tract 525 Map 1-NW Baltimore County

Peter's Lot 25 acres

Prior Survey for Grubby Thicket (134 a.)
Survey 2 Feb., 1787 George Mire(s) (Wire)
Assigned 21 Sept., 1789 to Peter Costallow
Assigned 1 Oct., 1789 to Michael Erhard
Patent 21 May, 1790 Michael Erhard
IC F/11 and IC E/454
Pat. Cert. 3805

Tract 526 Map 2-S Carroll County

Peter's Plain(s) 10 acres

Sp. Warr. 23 Mar., 1792 (60 a.) Charles Boblets
Assigned 16 Nov., 1792 to Peter Baublitz
Survey 19 Nov., 1792 Peter Baublitz
Patent 16 Apr., 1794 Peter Baublitz
IC I/611 and IC H/480
Pat. Cert. 3809

Tract 527 Map 2-S Carroll County

Petersburg 100 acres

Prior Survey for Petersburg (1003 a.)
Comm. Warr. 4 June, 1765 (100 a.) Peter Fowpin
Survey 2 Dec., 1765 Peter Fowpin (Fowble)
Patent 2 Dec., 1765 Peter Fowpin (Fowble)
BC & GS 30/315 and BC & GS 29/460
Pat. Cert. 3791

Tract 528 Map 2-SW Carroll County

Petersburg Resurvey 1003 acres

Resurvey of Fowbles Lot (240 a.), Fowble's New Germany (41 a.), Fowble's New Town (40 a.), Petersburg (100 a.), Peter's Addition (27 a.), Colchester (33 a.), Coal Pit (27 a.), Fertile Meadows (146 a.) and Shillings Lot (50 a.)
Sp. Warr. 10 Apr., 1775 Peter Fowble
Renewed 26 Sept., 1776 Peter Fowble
Survey 17 Mar., 1777 Peter Fowble
Patent 17 May, 1784 Peter Fowble

(Petersburg Resurvey (1003 acres) - continued)
IC B/30 and IC A/317
Pat. Cert. 3790

Tract 529 Map 2-SE Carroll County
<u>Petersburg</u> 300 acres
Resurvey of Kingbury Point (50 a.)
Prior Survey for Francis Thwaitsburg (421 3/4 a.)
Sp. Warr. 26 Jan., 1767 Peter Igow
Renewed 23 July, 1767 Peter Igow
Renewed 14 Jan., 1768 Peter Igow
Renewed 6 July, 1768 Peter Igow
Survey 16 Dec., 1768 Peter Igow
Patent 18 Nov., 1769 Peter Igow
BC & GS 40/190 and BC & GS 38/77
Pat. Cert 3789

Tract 530 (West of Map 2) Carroll County
<u>Petticoat's (Peddicote's) Choice</u> 13 acres
(Reference Tract - Everthing Needful Corr. (1646))
Warrant 14 Oct., 1786 (10 a.) Daniel Goodwin
Assigned 6 Oct., 1787 to Dorsey Peddicote
Warrant 1 Mar., 1787 (4 a.) Daniel Goodwin
Assigned 6 Oct., 1787 Dorsey Peddicote
Survey 10 Oct., 1787 Dorsey Petticoat
Assigned 27 Oct., 1789 to Gerrard Murray Stansbury
Patent 21 Oct., 1790 Gerrard Murray Stansbury
IC F/186 and IC E/543
Pat. Cert. 3768

Tract 531 Map 1-S Baltimore County
<u>Petticoats (Pedicoats) Loose</u> 91 1/2 acres
Prior Survey for Singery's Trouting Stream (178)
(Survey given in Pat. Cert 4482 for Singery's
Trouting Stream (178 a.) - no other information)

Tract 532 (West of Map 2) Carroll County
<u>Philipsburg</u> 6 acres
Prior Survey for Philipsburg (131 a.)
Sp. Warr. 5 Aug., 1758 (6 a.) Philip Crease
Survey 23 Sept., 1758 Philip Crease
Patent 29 Aug., 1758 Philip Crease (Crouse)
BC & GS 11/113 and BC & GS 8/149
Pat. Cert. 3824

Tract 533 (West of Map 2) Carroll County
<u>Philipsburg Resurvey</u> 131 acres
(Reference Tracts - Baughman's Goodwill (154 3/4 a.)
Johnsborough (130 a.) and Troy (1634 a.))
Resurvey of Philipsburg (6 a.)
Sp. Warr. 21 Feb., 1759 (125 a.) Philip Crouse
Survey 19 May, 1759 Philip Crease
Unpat. Cert. 1241

Tract 534 Map 1-NW Baltimore County
Philip's Discovery 4 1/4 a., 0 r., 4 p.
Comm. Warr. 19 Mar., 1807 (6 a.) Philip Stiltz
Survey 19 May, 1807 Philip Stiltz
Unpat. Cert. 1243

Tract 535 Map 1-E Baltimore County
Philip's Industry 154 3/4 acres
Resurvey of Cole's Frolick (55 a.) and
The Three Divisions (2591 a.)
Purchase of land in The Reserve: Robert Cummings
Assigned 3 Jan., 1786 to Philip Stiltz
Survey 8 Mar., 1787 Philip Stiltz
Patent 27 Feb., 1796 Philip Stiltz
IC G/664 and IC O/69
Pat. Cert. 3831

Tract 536 (not mapped) Baltimore County
Philip's Lot 1 a., 0 r., 2 p.
(Reference Tract - What's Left Resurvey (146+ a.))
Sp. Warr. 6 Dec., 1808 (15 a.) Philip Myers
Survey 13 May, 1809 Philip Myers
Patent 20 Apr., 1830 Philip Myers
GGB 2/86 and GGB 1/183
Pat. Cert. 3833

Tract 537 Map 1-SW Baltimore County
Philip's Spot 4 a., 0 r., 6 p.
Sp. Warr. 6 Dec., 1808 (5 a.) Philip Myers
Survey 30 May, 1809 Philip Myers
Patent 22 Apr., 1830 Philip Myers
GGB 1/184 and GGB 2/86
Pat. Cert. 3837

Tract 538 Map 2-SW Carroll County
Phillipsburgh 164 acres
Resurvey of Murphy's Retirement (60 a.)
Prior Survey for Nace's Tavern (478 1/2 a.)
Sp. Warr. 25 Aug., 1749 Philip Idleman
Survey 28 Jan., 1749 Philip Idleman
Patent 15 July, 1751 Philip Idleman
BY & GS 5/430 and BY & GS 4/34
Pat. Cert. 3825

Tract 539 Map 1-S Baltimore County
Picket's Guard 9 acres
Comm. Warr. 12 Jan., 1790 (35 a.) Chris. Singery
Survey 10 Dec., 1790 Christian Singery
Unpat. Cert. 1245

Tract 540 (West of Map 2) Carroll County
Piddle's Delight 8 acres
(Reference Tract - Troy (1634 a.) and Peace
with England (153 a.))

(Piddle's Delight - continued)
Sp. Warr. 20 Mar., 1820 (2 a.) William Roberts
Survey 28 Feb., 1821 William Roberts
(Copy of survey - Envelope #45 in Tracey Collection (HSCC))

Tract 541 (South of Map 1) Baltimore County
<u>Pine Grove</u> 52 acres
(Reference Tracts - Wildcat Hallow (53 3/4 a.) and Brother to Walnut Hills (30 1/2 a.))
Comm. Warr. 16 Dec., 1800 (52 a.) Samuel Markey
Survey 10 Mar., 1801 Samuel Markey
Patent 18 Dec., 1820 Samuel Markey
IB G/496
Pat. Cert. 3849

Tract 542 Map 1-SW Baltimore County
<u>Piney Hill</u> 175 1/4 (180) acres
Prior Survey for New Purchase (355 3/4 a.)
Sp. Warr. 11 Sept., 1772 (1400 a.) John Wilmott
Survey 2 Jan., 1773 John Wilmott
Patent 22 Sept., 1773 John Wilmott (Wilmoth)
BC & GS 47/349
Pat. Cert. 3852

Tract 543 Map 1-C Baltimore County
<u>Pitt</u> 475 acres
Resurvey of Hurst's Marsh (45 a.) and Roger's Reserve (60 a.)
Survey 25 Oct., 1770 Benjamin Rogers
Patent 6 Aug., 1773 Benjamin Rogers
BC & GS 47/391 and BC & GS 46/144
Pat. Cert. 3859

Tract 544 Map 2-SE Baltimore County
<u>Plains of Susa</u> 57 acres
Prior Survey for Wilmott's Lot (217 a.)
Comm. Warr. 21 Mar., 1769 (57 a.) Benjamin Hammond
Assigned 1 Aug., 1769 to Thomas Hammond
Survey 5 Sept., 1769 Thomas Hammond
Patent 11 June, 1770 Thomas Hammond
BC & GS 41/76
Pat. Cert. 3864

Tract 545 Map 2-SE Baltimore County
<u>Plains of Susa</u> 57 acres
(Folder - only the assignment from Thomas Hammond)
Assigned 19 May, 1798 to Philip Hammond
Unpat. Cert 1261

Tract 546 Map 2-SE Baltimore County
<u>Plains of Susa</u> 60 1/2 acres
Resurvey of Plains of Susa (57 a.)
Survey 19 June, 1799 Philip Hammond
Unpat. Cert. 1263

Tract 547 (not mapped) Baltimore County
<u>Pleasant Garden</u> 4 3/4 acres
Prior Survey for Walnut Hills (813 1/2 a.)
Comm. Warr. 12 Jan., 1790 (35 a.) Christian Singery
Survey 18 Nov., 1790 Christian Singery
Patent 13 Feb., 1792 Christian Singery
IC F/479
Pat. Cert. 3868

Tract 548 Map 2-E Baltimore County
<u>Pleasant Mount</u> 44 3/4 acres
Prior Survey for Pleasant Mount Corrected (136 a.)
Survey 13 Apr., 1787 William Hoffman
Patent 2 Apr., 1789 William Hoffman
IC D/415 and IC C/645
Pat. Cert. 3888

Tract 549 Map 1-W Baltimore County
<u>Pleasant Mount Corrected</u> 136 acres
Resurvey of Pleasant Mount (44 3/4 a.)
Sp. Warr. 12 Feb., 1794 William Hoffman
Survey 6 Apr., 1794 William Hoffman
Patent 30 Mar., 1798 William Hoffman
IC O/454 and IC M/398
Pat. Cert. 3889

Tract 550 (not mapped) Carroll County
<u>Pleasant Spring</u> 285 acres
Prior Survey for Pleasant Spring Res. (444 1/4 a.)
Comm. Warr. 31 Mar., 1774 (524 1/2 a.) Thomas Jones
Renewed 7 Sept., 1774 Thomas Jones
Assigned 23 Jan., 1775 (430 a.) to Thomas Gist, Jr.
Assigned 24 Jan., 1775 (430 a.) to Chris. Singery
Survey 30 Jan., 1775 Christian Singery
Assigned 5 Sept., 1775 to Thomas Gist, Jr.
Patent 11 Sept., 1775 Thomas Gist
BC & GS 51/181 and BC & GS 49/406
Pat. Cert. 3894

Tract 551 Map 2-S Carroll County
<u>Pleasant Spring Resurvey</u> 444 1/4 acres
Resurvey of Pleasant Spring (285 a.), Shilling's Folly (269 a.), Boring's Range (1375 a.), Foster's Hunting Ground (1439 a.), Fowble's Barren Hill (35 1/2 a.), Colchester (33 a.), New Germany (178 a.) and Spiretown (51 a.)
Sp. Warr. 5 June, 1789 Thomas Gist
Survey 23 Apr., 1790 Thomas Gist
Patent 1 June, 1791 Thomas Gist
IC F/387 and IC G/63
Pat. Cert. 3895

Tract 552 Map 1-C Baltimore County
<u>Pleasant Valley</u> 290 acres
Resurvey of Hilly Run Resurvey (197 1/4 a.) and Hobson's Choice (51 1/4 a.)
Sp. Warr. 12 Feb., 1794 James Campbell
Survey 28 Nov., 1794 James Campbell
Unpat. Cert. 1279

Tract 553 Map 1-E Baltimore County
<u>Pleasant Valley</u> 8 acres
Comm. Warr. 4 Dec., 1793 (8 a.) William Dorsey
Survey 20 Mar., 1794 William Dorsey
Assigned 17 Sept., 1794 to John Taylor
Patent 24 May, 1796 John Taylor
IC O/57 and IC L/107
Pat. Cert. 3901

Tract 554 Map 1-NE Baltimore County
<u>Plowman's Prospect</u> (? acres)
Land held by Jonathan Plowman
(Beard's Branch (293 3/4 a.) line #22 to beginning of Plowman's Prospect and Dorsey's Plains (655+ a.) line #14 to beginning of Plowman's Prospect and near Morehead's Marsh - no other information found)

Tract 555 Map 2-NW Carroll County
<u>Plymouth</u> 50 acres
Prior Survey for Plymouth (350 a.)
Sp. Warr. 31 July, 1746 Henry Hendrickson
Renewed 18 May, 1747 (25 a.) Henry Hendricks
Comm. Warr. 19 Oct., 1747 (100 a.) Nicholas R. Gay
Assigned 4 Nov., 1747 to Henry Hendricks
Survey 20 Nov., 1747 Henry Hendricks
Assigned 7 June, 1749 to Ulrich Whisler
Patent 7 June, 1749 Ulrich Whisler
BY & GS 1/241 and BC & GS 2/288
Pat. Cert. 3907

Tract 556 Map 2-NW Carroll County
<u>Plymouth</u> 350 acres
Resurvey of Plymouth (50 acres)
Prior Survey for Plymouth (434 a.)
Sp. Warr. 4 June, 1749 Ulrich Whisler
Survey 25 Nov., 1749 Ulrich Whisler
Patent 5 Apr., 1751 Ulrich Whisler (Whessler)
BY & GS 5/475
Pat. Cert. 3908

Tract 557 Map 2-NW Carroll County
<u>Plymouth</u> 434 acres
Resurvey of Plymouth (350 a.)
Prior Survey for Joseph Bollinger's Contr. (348 a.)
Sp. Warr. 11 May, 1757 Ulrich Whisler
Survey 10 Nov., 1757 Ulrich Whisler

(Plymouth (434 acres) - continued)
Patent 25 Mar., 1760 — Ulrich Whisler
BC & GS 12/278 and BC & GS 13/309
Pat. Cert. 3906

Tract 558 — Map 1-NW — Carroll County
<u>Point</u> — 1 a., 1 r., 26 p. .
Sp. Warr. 5 Aug., 1853 — Jacob Hoff
Survey 13 Mar., 1854 — Jacob Hoff
Patent 22 Mar., 1855 — Jacob Hoff
GGB 7-B/150 and GGB 8/375
Pat. Cert. 3921
(part of this tract lies in Baltimore County)

Tract 559 — (South of Map 2) — Carroll County
<u>Polygon</u> — 340 acres
(Reference Tracts - Foster's Pleasant Meadow (259+) and Penelope Charcilla and Thomas Cockeys Deyesburg (1030 a.))
Prior Survey for Penelope Charchilla and Thomas Cockey Deyesburg (1030 a.)
Sp. Warr. 11 Sept., 1772 (1400 a.) John Wilmott
Survey 6 Jan., 1773 — John Wilmott
Assigned 19 Mar., 1773 to Patrick Duncan (Dunkin)
Patent 23 Sept., 1773 — Patrick Duncan (Dunker)
BC & GS 47/346 and BC & GS 44/272
Pat. Cert. 3932

Tract 560 — (West of Map 2) — Carroll County
<u>Pomerania</u> — 50 acres
(Reference Tracts - Johnsborough (130 a.), Matter's Choice (162 a.) and Troy (1634 a.))
Comm. Warr. 8 July, 1754 — John Crady
Assigned 7 Jan., 1755 to William Winchester
Survey 8 Jan., 1755 — William Winchester
Patent 8 Jan., 1755 — William Winchester
BC & GS 4/281 and BC & GS 3/199
Pat. Cert. 3936

Tract 561 — Map 1-S — Baltimore County
<u>Poor Dependence</u> — 47 1/2 acres
Survey 22 Dec., 1789 — William Francis
Patent 1 Mar., 1792 — William Francis
IC F/458 and IC H/24
Pat. Cert. 3941

Tract 562 — (West of Map 2) — Carroll County
<u>Poor Man's Barrack</u> — 100 acres
(Reference Tract - Peace with England (153 a.))
(Warrant) 10 Feb., 1752 — Nicholas Sluthover - Single, German and Protestant
Survey 25 Mar., 1752 — Nicholas Sluthover
Patent 25 May., 1756 — Nicholas Sluthover
BC & GS 5/315 and BC & GS 7/406
Pat. Cert. 3943

Tract 563 Map 1-NW Baltimore County

<u>Poplar Run</u> 122 1/2 acres

Survey 11 Nov., 1786 John Wilay (Hilary)
Assigned 9 Feb., 1790 to John Willhelm
Patent 6 June, 1791 John Willhelm
IC F/390 and IC G/66
Pat. Cert. 3950

Tract 564 Map 2-W Carroll County

<u>Powell's Discovery Resurvey</u> 148 acres

Resurvey of Belfast Enlarged (400 a.)
Sp. Warr. 9 Nov., 1762 Peter Stump
(Resurvey on Belfast - Warrant expired)
Sp. Warr. 10 Nov., 1764 (350 a.) Philip Powell
Survey 29 Nov., 1764 (350 a.) Philip Powell
Assigned 7 Dec., 1764 to Henry Feeder
(Certificate for 350 a. was never patented)
* Survey 10 June, 1765 Henry Feeder
Patent 29 Sept., 1766 Henry Feeder
BC & GS 30/302 and BC & GS 31/350
Pat. Cert. 3970

Tract 565 Map 2-W Carroll County

<u>Powell's Discovery Resurvey</u> 60 acres

Sp. Warr. 10 Nov., 1787 Henry Crottinger
Survey 25 Sept., 1788 Henry Crottinger
Assigned 22 Dec., 1804 by the terms of the will of Henry Crottinger to John Jacob Crottinger
Patent 24 Dec., 1804 John Jacob Crottinger
IC R/296 and IC P/586
Pat. Cert. 3971

Tract 566 (South of Map 1) Baltimore County

<u>Prettyboy Hills</u> 56 3/4 acres

(Reference Tract - Bear's Rock (177 3/4 a.) and Walnut Hills (813 1/2 a.))
Comm. Warr. 27 June, 1805 (50 a.) Thomas Gist
Survey 2 May, 1806 Thomas Gist
Patent 23 Dec., 1807 Thomas Gist
IC R/665 and IC S/588
Pat. Cert. 3982

Tract 567 Map 1-C Baltimore County

<u>Prices's Chance</u> 70 acres

Prior Survey for Talbot's Purchase (66 3/4 a.) and The Three Divisions (2591 a.)
Survey 13 Apr., 17-- John Price
(Probably leased land - See Talbot's Purcahse)

Tract 568 Map 2-C Carroll County

<u>Price's Glory</u> 50 acres

Prior Survey for Neilson's Puzzle (256 1/4 a.)
Comm. Warr. 9 Oct., 1751 John Price, Jr. of Garrison Ridge
Survey 4 Apr., 1752 John Price, Jr.

(Price's Glory - continued)
Patent 10 Oct., 1752 — John Price, Jr.
BY & GS 7/9 and BC & GS 4/361
Pat. Cert. 3992

Tract 569 (East of Map 1) Baltimore County
<u>Price's Hardship</u> 160 acres
(Reference Tract - Vaughn's Disappointment (504+ a))
Resurvey of Cullinson's Old Aim (109 1/4 a.)
Sp. Warr. 28 Feb., 1795 — Samuel Price
Survey 12 Feb., 1796 — Samuel Price
Assigned by court to heirs of Samuel Price
Patent 29 Mar., 1814 — Solomon and John Price
IB B/627
Pat. Cert. 3995

Tract 570 (West of Map 2) Carroll County
<u>Prickum</u> 36 acres
(Reference Tracts - Troy (1634 a.) and Everybodies Land (Parcel #3))
Comm. Warr. 4 June, 1760 (36 a.) Peter Chepp
Survey 30 Nov., 1760 — Peter Chepp (Zepp)
Patent 30 Nov., 1760 — Peter Chepp (Zepp)
BC & GS 30/320 and BC & GS 31/192
Pat. Cert. 4005

Tract ***
Prospect (See Gist's Prospect)
Prospect (See Bryan's prospect)

Tract 571 Map 1-SW Baltimore County
<u>Rabbits Hollow</u> 287 acres
Resurvey of Green's Chance (22 a.), Green's Meadows (138 1/2 a.) and Anything at All (13 a.)
Sp. Warr. 10 Nov., 1794 — George Keller
Survey 23 Sept., 1795 — George Keller
Patent 27 Sept., 1796 — George Keller
IC C/23 and IC L/159
Pat. Cert. 4052

Tract 572 Map 2-E Carroll County
<u>Rabbit's (Rabet's) Play</u> 12 acres
Sp. Warr. 3 Mar., 1791 (20 a.) Charles Boblets
Survey 16 April, 1794 — Charles Boblitz
Patent 8 Nov., 1796 — Charles Boblitz
IC I/631 and IC H/481
Pat. Cert. 4053

Tract 573 Map 2-NE Carroll County
<u>Raccoon Harbour</u> 46 1/2 acres
Comm. Warr. 16 Dec., 1800 (15 a.) William Hoffman
Survey 10 Mar., 1801 — William Hoffman
Patent 3 May, 1802 — William Hoffman
IC R/65 and IC L/723
Pat. Cert. 4055

Tract 574 (not mapped) Baltimore County

<u>Race Ground</u> 20 3/4 acres

Resurvey of Solomon's Reserve (80 a.)
Sp Warr. 29 Mar., 1799 John Coone
Survey 27 Mar., 1800 John Coone (Kone)
Unpat. Cert. 1332
(Not precisely located - next to Benjamin's Beginning (174 1/2 a.), see Balto. Co. Deeds AWB Vol. 461, folio 129; 3 June, 1851)

Tract 575 Map 2-S Carroll County

<u>Rachel's Retirement</u> 42 acres

Resurvey of Tipton's Hope (60 a.) and Carr's Inginuity (256 a.)
Prior Survey of Penelope Charchilla and Thomas Cockey Deyesburg (1030 a.)
Comm. Warr. 13 Feb., 1773 (60 a.) Rachel Thorp
Survey 28 July, 1773 Rachel Thorp (Throp)
Assigned 1 Sept., 1773 to John Wilmot
Patent 9 Dec., 1774 John Wilmot
BC & GS 50/75 and BC & GS 48/248
Pat. Cert. 4060

Tract 576 Map 1-SE Baltimore County

<u>Ragged Raiment (Remnant)</u> 26 acres

Prior Survey for Addition to Ragged Raiment (43+ a.)
Comm. Warr. 1 Mar., 1792 (15 a.) William Matthews
Comm. Warr. 16 Oct., 1792 (11 a.) William Matthews
Survey 22 Nov., 1792 William Matthews
Unpat. Cert. 1336

Tract 577 Map 2-N Carroll County

<u>Rather Too Late</u> 50 acres

Prior Survey for All Together to Late (47 1/4 a.)
Survey 1 Oct., 1794 Henry Lamotte Sr. & Jr.
Unpat. Cert. 1346

Tract 578 Map 1-S Baltimore County

<u>Raven Mountains</u> 13 1/2 acres

Comm. Warr. 13 June, 1801 (8 a.) Benedict Hurst
Survey 10 Apr., 1802 Benedict Hurst
Assigned 18 June, 1802 to James Patterson
Patent 31 Jan., 1803 James Patterson
IC R/238 and IC S/19
Pat. Cert. 4074

Tract 579 Map 2-E Baltimore County

<u>Red Hallow</u> 30 acres

Sp. Warr. 24 June, 1796 (20 a.) Solomon Flowers
Assigned to Uriah Freeman
Comm. Warr. 2 Feb., 1797 Andrew Harman
Assigned to Uriah Freeman
Comm. Warr. 2 Feb., 1797 Uriah Freeman
Survey 29 Jan., 1795 Uriah Freeman
Patent 2 Feb., 1813 Uriah Freeman

(Red Hallow - continued)
IB D/29 and JK T/535
Pat. Cert. 4083

Tract 580 (West of Map 2) Carroll County
Red Oak Land 2 acres
Prior Survey for Everything Needful (1646 a.)
Comm. Warr. 24 Dec., 1760 (27 a.) Patrick Irons
Survey 25 Mar., 1761 Patrick Irons
Patent 25 Mar., 1761 Patrick Irons
BC & GS 14/116
Pat. Cert. 4081

Tract 581 Map 2-C Carroll County
Remnant 9 1/4 acres
Prior Survey for Eight Sisters (480 3/4 a.)
Comm. Warr. 20 Oct., 1794 Charles Stansbury
Assigned 9 Oct., 1795 to Alexis Lemon
Survey 20 Oct., 1795 Alexis Lemon
Patent 12 Nov., 1801 Alexis Lemmon
IC N/637 and IC P/233
Pat. Cert. 4094

Tract 582 (West of Map 2) Carroll County
Richard's Chance 50 acres
Prior Survey for McGill's Choice (50 a.)
Warrant 15 Nov., 1749 (50 a.) Richard Richards
Survey 1 Jan., 1749 Richard Richards
Patent 1 Jan., 1749 Richard Richards
BY & GS 1/532
Pat. Cert. 4110

Tract 583 (East of Map 1) Baltimore County
Richardson's Hills and Dales 168 1/2 acres
(Reference Tract - Northampton (242 a.))
Purchase of land in "Reserve" - Zachariah Richardson
Survey 18 Nov., 1786 Zachariah Richardson
Assigned 17 Sept., 1788 to John Low(e)
Patent 24 Nov., 1789 John Low(e)
IC D/622 and IC E/189
Pat. Cert. 4125

Tract 584 Map 1-NE Baltimore County
Roberts Venture 136 1/2 acres
Survey 10 Nov., 1786 Zachariah Roberts
Assigned 12 June, 1790 Christian Crous(e)
Patent 19 Feb., 1802 Christian Crous(e)
IC R/40 and IC P/284
Pat. Cert. 4182

Tract 585 Map 1-W Baltimore County
Rock (Rockey) Point 4 a., 1 r., 30 p.
Sp. Warr. 10 Mar., 1860 (3 a.) William J. Hoffman
Survey 25 Oct., 1860 William J. Hoffman
Unpat. Cert. 1407

Tract 586 Map 2-N Carroll County

Rocky Hills 129 1/2 acres

Resurvey of Ulrich's Delight (95 a.)
Sp. Warr. 3 Nov., 1794 Ulrick Fites
Survey 22 Oct., 1795 Ulrick Fites
Patent 8 Feb., 1797 Ulrick Fites
IC O/246 and IC L/260
Pat. Cert. 4207

Tract 587 Map 1-W Baltimore County

Rockey Point 30 acres

Comm. Warr. 3 Mar., 1791 (30 a.) Henry Markey
Survey 20 Feb., 1792 Henry Markey
Patent 18 May, 1798 Henry Markey
IC O/442 and IC L/465
Pat. Cert. 4211

Tract **

Rocky Point (See Rock Point (4+ a.))

Tract 588 Map 1-C Baltimore County

Roger's Reserve 60 acres

Prior Survey for Pitt (475 a.)
Leased 10 Aug., 1752 William Rogers
Leaseholds GGB B/386
Survey 10 Aug., 1752 William Rogers
Unpat. Cert. 1418

Tract 589 Map 2-NW Carroll County

Rome 36 1/4 acres

Prior Survey for Jos. Bollinger's Contr. (348 a.)
Sp. Warr. 20 Nov., 1773 (105 a.) John Gill
Renewed 29 Apr., 1774 John Gill
(with liberty to apply warrant to land next to Plymouth (434 a.))
Assigned 29 Apr., 1774 (76 a.) to Henry Lamuth
Survey 10 May, 1774 Henry Lamuth
Patent 12 July, 1775 Henry Lamuth (Lamotte)
BC & GS 51/122 and BC & GS 48/460
Pat. Cert. 4243

Tract 590 Map 1-C Baltimore County

Round about Flat Top 35 acres

Prior Survey for Round About Flat Top (68 a.)
Leased 6 Oct., 1752 Teague Tracey
Leaseholds GGB B/603
Survey 10 Oct., 1752 Teaque Tracey
Unpat. Cert. 1426

Tract 591 Map 1-C Baltimore County

Round about Flat Top (Resurvey) 68 acres

Resurvey of Round about Flat Top (35)
Prior Survey for Talbot's Plains (638 1/2 a.)
Land purchased for heirs of Thomas Talbot of Harford County (68 a.) by Edward Stansbury

(Round about Flat Top (Resurvey (68 acres) - continued)
Survey 13 Mar., 1799 heirs of Thomas Talbot
Assigned to Thomas S. Bond and Ann Bond
Assigned 4 June, 1800 to Thomas Talbot (Jr.)
(witnessed by Jesse Jarrett)
Unpat. Cert. 1425

Tract 592 (not mapped) (? County)
Round Meadow 38 acres
Survey 2 July, 1744 James Calder
(not precisely located - on east side of Indian Grave Run: See card files in Tracey Collection (HSCC) - no other information)

Tract 593 Map 2-SE Baltimore County
Royals Defeat(ed) 150 acres
Prior Survey for Royals Defeat (192 1/4 a.)
Survey 8 Mar., 1787 Mordecai Price
Assigned 5 Nov., 1788 to Thomas Gist
Patent 4 Jan., 1790 Thomas Gist
IC F/4 and IC E/253
Pat. Cert. 4266

Tract 594 Map 2-SE Baltimore County
Royals Defeat(ed) Resurvey 192 1/4 acres
Resurvey of Royals Defeat (150 a.) and Stoney Mountain (19 1/4 a.)
Sp. Warr. 29 Jan., 1805 Thomas Gist
Survey 9 Dec., 1805 Thomas Gist
Patent 25 June, 1806 Thomas Gist
IC R/375 and IC S/367
Pat. Cert. 4265

Tract 595 Map 1-NE Baltimore County
Ruhl's Bargain 363 1/2 acres
Resurvey of United Friendship (260 1/2 a.) and The Three Diviions (2591 a.)
Sp. Warr. 8 Sept., 1795 George Ruhl
Survey 10 Dec., 1795 George Ruhl
Patent 14 Nov., 1809 George Ruhl (Rule)
JK U/227 and JK T/194
Pat. Cert. 4275

Tract 596 Map 1-NW Baltimore County
Sampson's Folly 118 acres
Prior Survey for Try For It (94 1/2 a.)
Survey 10 Aug., 1752 Richard Sampson
Unpat. Cert. 1445

Tract 597 Map 1-SE Baltimore County
Sam's Grief 15 acres
Comm. Warr. 22 June, 1801 (60 a.) Wm. Hoffman, Jr.
Survey 22 May, 1802 William Hoffman, Jr.

(Sam's Grief - continued)

Assigned by will of William Hoffman, Jr. to his children and heirs: William J. Hoffman, Henry B. Hoffman, Susannah Hoffman (m. Jacob Markey), Elizabeth Hoffman (m. Henry M. Hoffacker), Ann Hoffman (m. John Gore) and Catherine Hoffman (m. George Miller)
Patent 27 July, 1850 William J. Hoffman et. al.
GGB 7/272
Pat. Cert. 4314

Tract 598 Map 1-SE Baltimore County
Sam's Puzzle 94 acres
Survey 21 Nov., 1787 Warnell Tracey
Assigned 8 Apr., 1795 to Samuel Cauckay (Cauhey)
Patent 21 Nov., 1795 Samuel Cauckay (Cauhey)
IC O/20 and IC M/33
Pat. Cert. 4317

Tract 599 Map 2-W Carroll County
Sarah's Addition 4 a., 1 r., 9 p.
Resurvey of Sherman's Addition (15 3/4 a.)
Sp. Warr. 1 Apr., 1864 (6 a.) Sarah Rorhbaugh
Survey 7 June, 1864 Sarah Rohrbaugh
Patent 23 Sept., 1765 Sarah Rohrbaugh
WL & WS 3/559 and WL & WS 2/173
Pat. Cert. 18 (Carroll County patent series)

Tract 600 Map 2-SW Carroll County
Schieming Defeated 186 acres
Resurvey of Idlesburg (25 a.)
Prior Survey for Nace's Tavern (478 1/2 a.)
Sp. Warr. 10 Sept., 1765 George Miers
Renewed 18 July, 1767 George Miers
Survey 23 June, 1767 George Miers
Patent 19 May, 1772 George Miers (Myers)
BC & GS 43/259 and BC & GS 45/193
Pat. Cert. 4340

Tract 601 Map 1-SE Baltimore County
Schieming Defied 104 1/2 acres
Prior Survey for Scheeming Defied Enl. (194 3/4)
Survey 27 Mar., 1787 John Kidd
Patent 20 Aug., 1788 John Kidd
IC D/288 and IC C/495
Pat. Cert. 4342

Tract 602 Map 1-SE Baltimore County
Schieming (Scheeming) Defied Enlarged 194 3/4 a.,18 p
Resurvey of Schieming Defied (104 1/2), Dinah's Delight (88 a.) and Wolf Den (50 a.)
Sp. Warr. 20 Dec., 1816 John Kidd
Survey 3 May, 1817 John Kidd
Assigned by estate of John Kidd to John Malcomb

(Schieming Defied Enlarged (194 3/4 acres) - continued)
Patent 6 Apr., 1850 John Malcomb
GGB 7-A/208
Pat. Cert. 4485

Tract 603 Map 1-S Baltimore County
<u>Scratch and Tear</u> 10 acres
Comm. Warr. 2 Dec., 1801 (20 a.) Christian Singery
Survey 10 Nov., 1802 Christian Singery
Patent 6 Sept., 1803 Christian Singery
IC R/219 and IC P/477
Pat. Cert. 4354

Tract **
Scrubby Thicket (See Grubby Thicket)

Tract 604 Map 1-S Baltimore County
<u>Second Adventure</u> 10 acres
Comm. Warr. 1 Mar., 1792 (10 a.) John Tracey
Survey 22 Oct., 1792 John Tracey of John
Assigned 13 Mar., 1812 to Henry Wilhelm
Patent 31 Dec., 1812 Henry Wilhelm
IB B/409 and IB A/300
Pat. Cert. 4363

Tract 605 Map 1-SW Baltimore County
<u>Sedgey Hill</u> 18 acres
Prior Survey for Hair's Beautiful Prospect (327+ a.)
Comm. Warr. 29 Mar., 1790 (25 a.) Jacob Hair
Survey 7 Mar., 1791 Jacob Hair
Patent 15 Feb., 1795 Jacob Hair (Hare)
IC K/138 and IC H/628
Pat. Cert. 4381

Tract 606 Map 2-N Carroll County
<u>Shaver's (Shaffer's) Folly</u> 52 acres
Comm. Warr. 13 Sept., 1763 (18 a.) Henry Grothinger
Assigned 1 Nov., 1764 to Jacob Shaver
Renewed 5 Nov., 1764 Jacob Shaver
Comm. Warr. 28 Nov., 1764 (34 a.) Jacob Shaver
Survey 7 Mar., 1765 Jacob Shaver
Patent 7 Mar., 1765 Jacob Shaver (Shafer)
BC & GS 27/468 and BC & GS 28/205
Pat. Cert. 4411

Tract 607 (not mapped) Baltimore County
<u>Shaver's New Design</u> 36 acres
Comm. Warr. 26 July, 1774 (36 a.) Christian Singery
Assigned 1 Aug., 1774 to Christian Shaver
Survey 4 Aug., 1774 Christian Shaver
Unpat Cert. 1476
(not precisely located - on south side of Poplar Branch)

Tract 608 Map 2-W Carroll County
<u>Sherman's Addition</u> 15 3/4 acres
Resurvey of Notenstott (66 a.)
Sp. Warr. 20 Dec., 1791 (8 a.) John Sherman
Survey 17 Aug., 1792 John Sherman
Patent 13 June, 1798 John Sherman
IC O/437 and IC L/473
Pat. Cert. 4433

Tract 609 (not mapped) Carroll County
<u>Sherman's Goodwill</u> 7 1/2 acres
(Reference Tract - Nottenstot (66 a.))
Sp. Warr. 13 Aug., 1792 (8 a.) Adam Feather
Survey 18 Aug., 1792 Adam Feather
Patent 21 Nov., 1794 Adam Feather
IC I/617 and IC H/581
Pat. Cert. 4434

Tract 610 Map 1-NW Baltimore County
<u>Shieres' Goodluck</u> 64 1/4 acres
Survey 16 Jan., 1787 Martin Shierers
Assigned 16 Dec., 1800 to Frederick Hoviss
Patent 16 Dec., 1800 Frederick Hoviss
IC N/496 and IC P/77
Pat. Cert. 4436

Tract 611 Map 2-SW Carroll County
<u>Shillings Folly</u> 269 acres
Resurvey of Shillings Lot (50 a.)
Prior Survey for Nace's Tavern (478 1/2 a.)
Sp. Warr. 29 May, 1752 (40 a.) Jacob Shilling
Renewed 27 Apr., 1753 Jacob Shilling
Survey 20 May, 1753 Jacob Shilling
Patent 23 Apr., 1754 Jacob Shilling
BC & GS 1/244 and BC & GS 2/52
Pat. Cert. 4437

Tract 612 (South of Map 1) Carroll County
<u>Shillings Lott</u> 40 acres
(Reference Tract - Shilling's Folly (269 a.))
Prior Survey for Nace's Tavern (478 1/2 a.),
Fowble's Lot (240 a.), Shilling's Folly (269 a.)
and Petersburg (1003 a.)
Warrant 27 Aug., 1751 (54 a.) Jacob Shilling
Survey 3 Oct., 1751 Jacob Shilling
Patent 29 May, 1752 Jacob Shilling
Y & S 7/26 and BY & GS 4/440
Pat. Cert. 4438

Tract 613 Map 2-SE Baltimore County
<u>Sideling Hill</u> 50 acres
Prior Survey for Catey's Delight (153 1/2 a.)
Warrant 29 May, 1789 (50 a.) Charles Boblitz, Jr.
Survey 1 July, 1789 Charles Boblitz, Jr.
Patent 9 Apr., 1790 Charles Boblitz, Jr.

(Sideling Hill - continued)
IC F/34 and IC E/320
Pat. Cert. 4469

Tract 614 Map 1-N Baltimore County
<u>Simmerman's (Simmon's) Venture</u> 79 1/2 acres
Resurvey of Gain's Rest Enlarged (64 1/4 a.),
George's Goodwill (5 a.) and Smith's Folly (8 a.)
Sp. Warr. 8 Jan., 1812 John Simmerman
Survey 14 May, 1812 John Simmerman (Simmons)
Unpat. Cert. 1488

Tract 615 Map 2-SW Carroll County
<u>Singery's Chance</u> 153 acres
Prior Survey for Singery's Chance Resurvey (114+)
Comm. Warr. 2 Dec., 1775 (50 a.) Christian Singery
Survey 25 Dec., 1775 Christian Singery
Patent 8 Oct., 1782 Christian Singery
BC & GS 50/322 and IC A/4
Pat. Cert. 4477

Tract 616 Map 2-SW Carroll County
<u>Singery's Chance Resurvey</u> 114 1/4 acres
Reservey of Singery's Chance (153 a.)
Sp. Warr. 9 May, 1797 Christian Singery
Survey 1 May, 1798 Christian Singery
Patent 13 Apr., 1799 Christian Singery
IC O/595 and IC M/575
Pat. Cert. 4478

Tract **
Singery's Hunting Stream (See Singerys Trouting
Stream (196 1/2 a.))

Tract 617 Map 1-SE Baltimore County
<u>Singery's Level(s)</u> 21 1/4 acres
Comm. Warr. 12 Jan., 1790 (35 a.) Christian Singery
Survey 19 Nov., 1790 Christian Singery
Patent 18 Oct., 1797 Christian Singery
IC O/307 and IC M/301
Pat. Cert. 4779

Tract 618 Map 1-S Baltimore County
<u>Singery's Prospect</u> 26 acres
Comm. Warr. 20 July, 1792 (26 a.) Christian Singery
Assigned 29 Nov., 1792 to Jacob Singery
Survey 25 May, 1793 Jacob Singery
Patent 18 Oct., 1796 Jacob Singery
IC O/42 and IC M/61
Pat. Cert. 4480

Tract 619 Map 1-S Baltimore County
<u>Singery's Spot</u> (Lot) 25 acres
Comm. Warr. 13 July, 1790 Christian Singery
Survey 23 June, 1791 Christian Singery

(Singery's Spot (25 acres) - continued)
Assigned 28 Nov., 1792 to Jacob Singery
Patent 12 Mar., 1799 Jacob Singery
IC O/598 and IC L/532
Pat. Cert. 4481

Tract 620 Map 1-SW Baltimore County
Singery's Trouting Stream 178 acres
Resurvey of Merryman's Mountain (83 a.) and
Petticoats Loose (91 1/2 a.))
Prior Survey for Singery's Trouting Stream (575 a.)
Survey 30 Sept., 1770 Christian Singery
Patent 20 Apr., 1775 Christian Singery
BC & GS 51/68 and BC & GS 49/211
Pat. Cert. 4482

Tract 621 Map 1-S Baltimore County
Singery's Trouting Stream Resurvey 575 acres
Resurvey of Singery's Trouting Stream (178 a.)
Prior survey for Singery's Hunting Stream (196+)
Sp. Warr. 3 Jan., 1775 Christian Singery
Survey 11 Apr., 1792 Christian Singery
Caveat Filed 12 Apr., 1793 Ezekiel Boring and
William Cooke
Assigned by deed 7 May, 1804 from Christian Singery
to Elijah Foster
(Baltimore County Deeds WG 8/254)
Assigned by deed 11 Sept., 1817 to Henry Hoover
and Anthony Sanger
(Baltimore County Deeds WG 143/139)
(No patent certificate granted - see next tract)

Tract 622 Map 1-S Baltimore County
Singery's Trouting (Hunting) Stream 196 1/2 acres
Resurvey of Singery's Trouting Stream Res. (575 a.)
Caveat filed 12 Apr., 1793 to resurvey Singery's
Trouting Stream (575 a.): ordered by Ezekiel
Boring and William Cooke.
Returned 30 Nov., 1793
Caveat ruled good 2 June, 1794
Chancellor's order to correct the certificate of
survey renewed 6 Sept., 1825
Survey corrected 29 Sept., 1825 Henry Hoover
and Anthony Sanger(y)
Patent 11 Sept., 1826 Henry Hoover
and Anthony Sanger(y)
IB & GGB E/614 and IB & GGB H/285
Pat. Cert. 4483

Tract 623 Map 1-NW Baltimore County
Skin a Man 78 1/4 acres
Survey 17 Mar., 1787 Andrew Kerrick
Patent 10 July, 1795 Andrew Kerrick (Korrick)
IC K/186 and IC H/712
Pat. Cert. 4486

Tract 624 Map 2-SE Baltimore County

Slabbing Town 100 acres

Comm. Warr. 9 Aug., 1757 (50 a.) Michael Konk
Comm. Warr. 6 Sept., 1757 (50 a.) Michael Konk
Survey 30 Nov., 1757 Michael Konk
Patent 30 Nov., 1757 Michael Konk
BC & GS 9/352 and BC & GS 8/214
Pat. Cert. 4488

Tract 625 Map 1-SW Baltimore County

Slightly Ketch (Kitch) 4 1/4 acres

Sp. Warr. 24 Feb., 1824 (2 a.) Philip Hair of Jacob
Survey 30 Mar., 1824 Philip Hair
Assigned 28 June, 1824 to Abraham Hair of Jacob
Witness: Henry Kidd
Patent 28 Feb., 1826 Abraham Hair of Jacob
IB & GGB H/222 and IB F/422
Pat. Cert. 4490

Tract 626 Map 2-NW Carroll County

Sluthover's Choice 25 1/2 acres

Comm. Warr. 2 June, 1758 (25 1/2 a.) N. Sluthover
Survey 29 Nov., 1758 Nicholas Sluthover
Patent 29 Nov., 1758 Nicholas Sluthover
BC & GS 12/8 and BC & GS 8/692
Pat. Cert. 4494

Tract 627 Map 2-NW Carroll County

Smith's Field (Smithfield) 132 acres

Resurvey of Smith's Folly (22 1/2 a.), Whispaw (140)
Warrant 12 Sept., 1766 Charles Smith
Renewed 11 Mar., 1767 (20 a.) Charles Smith
Survey 22 June, 1767 Charles Smith
Patent 26 June, 1772 Charles Smith
BC & GS 43/292 and BC & GS 45/177
Pat. Cert. 4520

Tract 628 (not mapped) Carroll County

Smith's Folly 22 1/2 acres

Prior Survey for Smith's Field (132 a.)
Comm. Warr. 7 June, 1758 (22 1/2 a.) Charles Smith
Survey 28 Nov., 1758 Charles Smith
Patent 28 Nov., 1758 Charles Smith
BC & GS 12/17 and BC & GS 10/262
Pat. Cert. 4522

Tract 629 Map 1-N Baltimore County

Smith's Folly 8 acres

Prior Survey for Simmerman's Venture (79 1/2 a.)
Comm. Warr. 3 Dec., 1793 (8 a.) Samuel Price
Assigned 17 Jan., 1794 to Adam Smith
Survey 17 Jan., 1794 Adam Smith
Patent 10 May, 1804 Adam Smith
IC R/279 and IC S/165
Pat. Cert. 4521

Tract 630 (South of Map 1) Baltimore County

Smith's Grief 34 1/4 acres

(Reference Tract - Cooper's Discovery (77+ a.))
Resurvey of Hurst's Hackle (26 1/2 a.)
Sp. Warr. 26 Mar., 1795 (15 a.) John Cooper
Survey 25 Feb., 1796 John Cooper
Patent 9 Feb., 1854 John Cooper
GGB 7B/24 and GGB 8/257
Pat. Cert. 4525

Tract 631 Map 1-NE Baltimore County

Smith's Prospect 150 acres

Sp. Warr. 15 Apr., 1790 (150 a.) William Smith
Survey 25 Aug., 1790 William Smith
Patent 5 Dec., 1791 William Smith
IC F/450 and IC G/124
Pat. Cert. 4527

Tract 632 Map 2-SE Baltimore County

Snaggy Hill 8 acres

Comm. Warr. 3 Mar., 1791 (20 a.) Charles Boblets
Survey 18 Nov., 1791 Charles Boblitz
Patent 2 Aug., 1803 Charles Boblitz
IC R/215 and IC S/84
Pat. Cert. 4533

Tract 633 Map 1-E Baltimore County

Snaggy Hill 50 acres

Comm. Warr. 17 Feb., 1796 (50 a.) Moses Freeman
Survey 10 Jan., 1797 Moses Freeman
Patent 14 Mar., 1814 Moses Freeman
IB B/636
Pat. Cert. 4532

Tract 634 Map 1-E Baltimore County

Snowy Hill 17 acres

Comm. Warr. 28 Feb., 1795 (2 a.) William Dorsey
Survey 24 Feb., 1796 William Dorsey
Assigned 17 May, 1799 to John Tracey of John
Assigned 23 Apr., 1803 to William Dorsey
Patent 17 May, 1803 William Dorsey
IC R/227 and IC P/467
Pat. Cert. 4539

Tract 635 Map 2-S Carroll County

Soldier's Delight 20 acres

Resurvey of Michael's Beginning (16 a.)
Sp. Warr. 29 Mar., 1790 (20 acres)
Survey 22 Apr., 1790 Philip Deeds
Patent 14 Oct., 1799 Philip Deeds
IC O/669 and IC M/628
Pat. Cert. 4542

Tract 636 Map 1-SE Baltimore County

Solomon's Reserve 80 acres

Prior Survey for Race Ground (20 3/4 a.) and Cross's Folly (31 a.)

Leased 10 Aug., 1752 Solomon Cross

Leaseholds GGB A/274

Survey 10 Aug., 1752 Solomon Cross

Unpat. Cert. 1512

(Location not certain but apparently has a common line (north 84 perches) on Buck Range (544 a.))

Tract 637 Map 1-NE Baltimore County

Solomon's Wisdom 156 1/4 acres

Survey 16 Nov., 1786 Solomon Nunamecher of York Co., Pennsylvania

Patent 16 Dec., 1790 Solomon Nunamecker

IC F/175 and IC E/589

Pat. Cert. 4549

Tract 638 Map 2-E Carroll County

Sower's Bargain 21 1/4 acres

Survey 7 Oct., 1789 John Sowers

Assigned 24 Mar., 1791 to Henry Lamotte

Witness: John Lamott

Patent 12 Nov.,1795 Henry Lamotte

IC K/200 and IC L/24

Pat. Cert. 4559

Tract 639 (South of Map 2) Carroll County

Spiretown 51 acres

(Reference Tract - Pleasant Springs Res. (444+ a.))

Prior Survey for Pleasant Spring Res. (444 1/4 a.)

Comm. Warr. 6 Mar., 1769 (51 a.) Peter Fowble

Survey 22 Mar., 1769 Peter Fowble

Patent 18 July, 1770 Peter Fowble (Foeble)

BC & GS 41/88 and BC & GS 42/320

Pat. Cert. 4574

Tract 640 Map 2-S Carroll County

Spring Garden 50 acres

Sp. Warr. 2 Mar., 1762 (50 a.) John Tipton

Survey 6 May, 1762 John Tipton

Patent 29 Sept., 1763 John Tipton

BC & GS 27/451

Pat. Cert. 4597

Tract 641 Map 2-NW Carroll County

Spring Run 77 acres

Prior Survey for Spring Run Enlarged (130 3/4 a.)

Comm. Warr. 14 Apr., 1762 Michael Dummer

Assigned 22 July, 1762 (52 a.) to Peter Mostallow

Comm. Warr. 3 Feb., 1762 (25 a.) Geo. Kittyear

Assigned 15 July, 1762 to Peter Mostallow

Survey 29 July, 1762 Peter Mostallow

(Spring Run - continued)

Patent 29 July, 1762 — Peter Mostallow
BC & GS 27/422 and BC & GS 26/32
Pat. Cert. 4605

Tract 642 — Map 2-NW — Carroll County
<u>Spring Run Enlarged</u> — 130 3/4 acres
Resurvey of Spring Run (77 a.), Foreman's Run (20), Madgalen Hall (14 a.) and Tuscuhem (26 a.)
Sp. Warr. 9 Aug., 1787 — Barton Stonebraker of Frederick County
Survey 25 Mar., 1798 — Barton Stonebraker
Unpat. Cert. 1526

Tract 643 — Map 1-W — Baltimore County
<u>Springfield</u> — 102 acres
Purchased land in the Reserve - William Hoofman
Survey 16 Mar., 1787 — William Hoofman
Patent 2 Apr., 1789 — William Hoofman
IC D/412 and IC E/26
Pat. Cert. 4584

Tract 644 — Map 2-E — Baltimore County
<u>Squirrel's Nest</u> — 10 acres
Comm. Warr. 18 Mar., 1791 (10 a.) William Hoffman
Survey 18 Feb., 1792 — William Hoffman
Patent 17 Feb., 1794 — William Hoffman
IC I/421 and IC G/357
Pat. Cert. 4612

Tract 645 — Map 1-NW — Baltimore County
<u>Stagg's Range</u> — 54 1/2 acres
Survey 6 Aug., 1787 — John Rice
Assigned 6 Nov., 1789 to William Ruhl
Patent 29 July, 1790 — William Ruhl (Rule)
IC F/29 and IC C/735
Pat. Cert. 4615

Tract 646 — Map 1-E — Baltimore County
<u>Standing Stone</u> — 41 1/2 acres
Prior survey for Adventure (41 1/2 a.)
Comm. Warr. 3 Dec., 1793 (36 a.) Andrew Horn
Survey 11 Apr., 1794 — Andrew Horn
Unpat. Cert. 1536

Tract 647 — (not mapped) — Baltimore County
<u>Stansbury's Contrivances</u> — 14 acres
Prior Survey for Boring's Hazard (164 a.)
Leased 10 Nov., 1746 — Thomas Stansbury, Jr.
Leaseholds GGB B/545
Survey 18 Nov., 1746 — Thomas Stansbury, Jr.
Unpat. Cert. 1539

Tract 648 Map 2-E Carroll County

Stansbury's Grove 444 acres

Resurvey of Merryman's Chance (56 a.)
Prior Survey for Stansbury's Grove Res. (1448 a.)
Sp. Warr. 6 June, 1763 Thomas Stansbury
Survey 26 Nov., 1763 Thomas Stansbury
Patent 29 Sept., 1765 Thomas Stansbury
BC & GS 27/522 and BC & GS 23/468
Pat. Cert. 4634

Tract 649 Map 2-E Carroll County

Stansbury's Grove 1448 acres

Resurvey of Stansbury's Grove (448 a.)
Prior Survey for Catey's Delight (153 1/2 a.), Charles' Trouble (480+ a.), Neilson's Puzzle (256 1/4 a.) and Keeth's Correction (210)
Sp. Warr. Thomas Stansbury, Jr.
Renewal 15 Nov., 1771 Thomas Stansbury, Jr.
Survey 4 May, 1772 Thomas Stansbury (Jr.)
Patent 23 Mar., 1773 Thomas Stansbury (Jr.)
BC & GS 47/317 and BC & GS 44/89
Pat. Cert. 4635

Tract 650 Map 2-C Carroll County

Stansbury's Hills 32 acres

Prior Survey for Grandfather's Gift (343+ a.)
Sp. Warr. (150 a.) Thomas Stansbury
Renewed 21 Dec., 1774 (31 1/2 of 150 a.) (by same)
Survey 8 April, 1775 Thomas Stansbury
Patent 15 March, 1786 Thomas Stansbury
IC B/402 and IC A/644
Pat. Cert. 4636

Tract 651 Map 1-W Baltimore County

Stansbury's Ramble 346 1/2 acres

Survey 9 Mar., 1787 Thomas Stansbury
Patent 14 Oct., 1788 Thomas Stansbury
IC D/321 and IC C/517
Pat. Cert. 4649

Tract 652 (West of Map 2) Carroll County

Steven's Hope 150 acres

(Reference Tract - Powell's Discovery (148 a.))
Warr. 16 May, 1737 Samuel Stevens
Survey 30 Oct., 1737 Samuel Stevens
Patent 8 Sept., 1739 Samuel Stevens
EI 5/432 and LG B/102
Pat. Cert. 4668

Tract 653 Map 1-S Baltimore County

Stevenson's Cow Pasture 181 1/2 acres

Sp. Warr. 5 Jan., 1797 (50 a. + 25 a.)
Joshua Stevenson of Frederick County
Sp. Warr. 15 Aug., 1797 (270 a.) Joshua Stevenson
Survey 23 Dec., 1797 Joshua Stevenson

(Stevenson's Cow Pasture - continued)
Patent 9 May, 1799 — Joshua Stevenson
IC O/570 and IC M/587
Pat. Cert. 4672

Tract 654 — Map 1-S — Baltimore County
<u>Still Worse</u> — 25 acres
Survey 22 Dec., 1789 — William Francis
Patent 1 Mar., 1792 — William Francis
IC F/458 and IC H/24
Pat. Cert. 4687

Tract 655 — Map 1-N — Baltimore County
<u>Stiltz's (Steltz's) Deer Park</u> — 606 1/2 acres
Resurvey of Cox's Park (85 a.), Stiltz Park (10 a.) and The Three Divisions (2591 a.)
Purchased land in "Reserve" - Philip Stiltz
Survey 13 Jan., 1787 — Philip Stiltz
Patent 8 Dec., 1789 — Philip Stiltz
IC D/626 and IC E/212
Pat. Cert. 4661

Tract 656 — (not mapped) — Baltimore County
<u>Stiltz Park</u> — 10 acres
Prior Survey for Stiltz Deer Park (606 1/2 a.)
Comm. Warr. 25 Apr., 1782 (10 a.) Philip Stiltz
Survey 1 May, 1782 — Philip Stiltz
Unpat. Cert. 1550
(not precisely located - begins at head of 'Roring' Branch)

Tract 657 — Map 2-C — Carroll County
<u>Stoney Hill</u> — 1291 acres
Resurvey of Jerico (100 a.)
Prior Survey for Grandfather's Gift (343+ a.) and Eight Sisters (480 3/4 a.)
Sp. Warr. 3 Mar., 1761 — Thomas Stansbury
Renewed 23 Sept., 1761 — Thomas Stansbury
Renewed 22 Mar., 1762 — Thomas Stansbury
Renewed 10 Sept., 1762 — Thomas Stansbury
Survey 3 Mar., 1763 — Thomas Stansbury
Patent 29 Sept., 1763 — Thomas Stansbury
BC & GS 21/376
Pat. Cert. 4718

Tract 658 — Map 2-C — Carroll County
<u>Stoney Hill(s)</u> — 1291 acres
Resurvey of Jerico (100 a.)
Sp. Warr. 3 Mar., 1761 — Thomas Stansbury
Renewed 23 Sept., 1761 — Thomas Stansbury
Renewed 22 Mar., 1762 — Thomas Stansbury
Survey 3 Mar., 1763 — Thomas Stansbury
Unpat. Cert. 1556

Tract 659 Map 1-C Baltimore County

Stoney Hill 16 acres

Sp. Warr. 26 Aug., 1791 (16 a.) Jonathan Cullins
Survey 6 Aug., 1792 Jonathan Cullen(s)
Assigned 20 Feb., 1801 to George Palmer, Sr.
Patent 9 Mar., 1801 George Palmer, Sr.
IC N/562
Pat. Cert. 4724

Tract 660 Map 1-N Baltimore County

Stoney Hills 86 1/2 acres

Survey 26 Aug., 1789 Nicholas Henry
Assigned 11 Sept., 1789 to Dewalt Snyder (Snider)
Patent 28 July, 1790 Dewalt Snyder
IC F/27 and IC E/500
Pat. Cert. 4719

Tract 661 Map 1-E Baltimore County

Stoney Level 78 1/4 acres

Prior Survey for Expedition (337 a.)
Survey 6 Mar., 1787 Elisha Dorsey
Unpat. Cert. 1558
Assigned 31 July, 1790 to Michael and Peter Shock

Tract 662 Map 2-SE Baltimore County

Stoney Mountain 19 1/4 acres

Sp. Warrant 22 June, 1789 Benjamin Price
Survey 15 June, 1790 Benjamin Price
Patent 3 June, 1791 Benjamin Price
IC F/391 and IC G/67
Pat. Cert. 4726

Tract 663 (South of Map 2) Carroll County

Stoney Point 33 3/4 acres

Comm. Warr. 21 May, 1787 (33 3/4 a.) B. G. Vaughn
Survey 15 May, 1788 Benjamin Gist Vaughn
Patent 28 Feb., 1789 Benjamin Gist Vaughn
IC D/400 and IC C/629
Pat. Cert. 4729

Tract 664 Map 1-SE Baltimore County

Stoney Point 19 acres

Sp. Warr. 27 Feb., 1804 (150 a.) Wm. Hoffman, Jr.
Survey 25 Apr., 1804 William Hoffman, Jr.
Assigned by the will of William Hoffman, Jr. to his children and heirs: William J. Hoffman, Henry B. Hoffman, Susannah Hoffman (m. Samuel Hoffacker), Mary Hoffman (m. Jacob Markey), Elizabeth Hoffman (m. Henry M. Hoffacher), Ann Hoffman (m. John Gore) and Catherine Hoffman (m. George Miller)
Patent 27 July, 1850 William J. Hoffman et.al.
GGB 7/270
Pat. Cert. 4727

Tract 665 Map 2-NW Carroll County
<u>Stoney Point</u> 74 acres
Sp. Warr. 10 Apr., 1787 Conrad Kerlinger
Survey 15 June, 1787 Conrad Kerlinger
Unpat. Cert. 1559

Tract 666 (not mapped) Baltimore County
<u>Stoney Rock</u> 19 acres
Comm. Warr. 29 Apr., 1782 (10 a.) Philip Stiltz
Survey 1 May, 1782 Philip Stiltz
Patent 29 May, 1807 Philip Stiltz
IC R/524 and JK T/47
Pat. Cert. 4736
(not precisely located - west side of Middletown Road, approx. location of Marshall's Struggle)

Tract 667 Map 2-C Carroll County
<u>Stoney Spring</u> 20 acres
Prior Survey for Big Owl Hill (245 1/2 a.)
Warrant 28 Oct., 1767 (20 a.) Andrew Burgubine
Survey 9 Apr., 1768 Andrew Burgubine
Patent 24 Sept., 1768 Andrew Burgubine
BC & GS 37/294 and BC & GS 36/18
Pat. Cert. 4738

Tract 668 Map 1-W Baltimore County
<u>Stoney Ups and Downs</u> 35 acres
Survey 14 May, 1771 John Hall son of Joshua
Patent 13 Oct., 1774 John Hall
BC & GS 48/170 and BC & GS 50/29
Pat. Cert. 4739

Tract 669 (West of Map 2) Carroll County
(The) <u>String</u> 2 1/8 acres
(Reference Tracts - Bridgeland (150 a.), California (490 a.) and Worms (25 a.))
Sp. Warr. 25 Dec., 1830 (2 a.) Philip Gore
Survey 15 Sept., 1838 Philip Gore
Patent 1 Feb., 1840 Philip Gore
GGB 4/473
Pat. Cert. 22 (Carroll County patent series)

Tract 670 (East of Map 1) Baltimore County
<u>Struggle Hard</u> 25 acres
Prior Survey for Stuggle Hard Completed (55 a.)
Sp. Warr. 28 July, 1790 (25 a.) Daniel Curfman
Survey 30 Sept., 1790 Daniel Curfman
Patent 18 May, 1792 Daniel Curfman
IC F/647 and IC H/82
Pat. Cert. 4750

Tract 671 (East of Map 1) Baltimore County
<u>Struggle Hard Completed</u> 55 acres
(Reference Tracts - Cross's Park (50 a.) and Curfman's Folly (84 a.))

(Struggle Hard Completed (55 acres) - continued)
Resurvey of Struggle Hard (25 a.)
Sp. Warr. 14 Mar., 1825 Jacob Hurst
Survey 22 Feb., 1826 Jacob Hurst
Patent 13 Jan., 1856 Jacob Hurst
GGB 7B/213 and GGB 8/437
Pat. Cert. 4751

Tract 672 Map 2-NW Carroll County
Stumps's Lot 22 1/2 acres
Comm. Warr. 17 Mar., 1764 Peter Stump
Renewed 4 Oct., 1764 (22 1/2 a.) Peter Stump
Survey 7 Mar., 1765 Peter Stump
Patent 7 Mar., 1765 Peter Stump
BC & GS 27/444
Pat. Cert. 4756

Tract 673 Map 2-NE Carroll County
Sugar Valley 387 acres
Resurvey of Hobson's Choice (106 1/4 a.), Addition to Hobson's Choice (1 a.), Addition to Tilley's Meadow (27 a.) and Lebanon (15 a.)
Sp. Warr. 5 Aug., 1797 George Fair
Survey 20 Sept., 1798 George Fair
Patent 12 Nov., 1801 George Fair
IC N/634 and IC P/230
Pat. Cert. 4763

Tract 674 (not mapped) Baltimore County
Sugar Valley 10 acres
(Reference Tract - Singery's Level (21 1/4 a.))
Comm. Warr. 15 Dec., 1800 William Hoffman, Jr.
Survey 25 Mar., 1801 William Hoffman, Jr.
Patent 21 June, 1802 William Hoffman, Jr.
IC R/47 and IC P/329
Pat. Cert. 4762

Tract 675 Map 1-E Baltimore County
Take All 56 1/4 acres
Resurvey of Hard Scuffle (34 a.) and The Three Divisions (2591 a.)
Sp. Warr. 29 May, 1807 (20 a.) Henry Wilhelm(s)
Survey 25 May, 1808 Henry Wilhelm(s)
Unpat. Cert. 1585

Tract 676 Map 1-N Baltimore County
Talbot's Lot Resurvey 164 1/2 acres
Resurvey of Wheeler's Garden Spot (25) and Wheeler's Lot (139 1/4 a.)
Prior Survey for Talbot's Plains (638 1/2 a.)
Warrant 10 Aug., 1770 (164 1/4 a.) Thomas Talbot
Survey 17 Oct., 1770 Thomas Talbot
Patent 11 May, 1774 Thomas Talbot
BC & GS 50/83
Pat. Cert. 4792

Tract 677 Map 1-C Baltimore County

Talbot's Plains 638 1/2 acres

Resurvey of Farmer's Lot (287 1/4 a.), Talbot's Lot (164 1/2 a.), Talbot's Purchase (66 3/4 a.), Round about Flat Top (68 a.) and The Three Divisions (2591 a.)

Sp. Warr. 5 July, 1808 Thomas Talbot (Jr.)

Survey 1 June, 1809 Thomas Talbot (Jr.)

Unpat. Cert. 1588

Tract 678 Map 1-C Baltimore County

Talbot's Purchase 66 3/4 acres

Resurvey of Price's Chance (70 a.)

Purchased for heirs of Thomas Talbot, together with other tracts by Edward Stansbury

Survey 15 March, 1799 heirs of Thomas Talbot

Assigned 22 June, 1799 by Edward Stansbury for the heirs of Thomas Talbot to Ann Bond. (Witness: Thomas Talbot (Jr.))

Assigned 4 June, 1800 by Ann Bond and Thomas S. Bond to Thomas Talbot (Jr.) (Witness: Jessee Jarrett)

Unpat. Cert. 1589

Tract 679 (South of Map 1) Baltimore County

Tear and Scratch 12 1/8 acres

(Reference Tract - Addition to Bear's Rock (50 a.))

Sp. Warr. 19 Sept., 1810 (20 a.) Richard Hooker

Assigned 5 Jan., 1811 to William Murray

Survey 6 June, 1811 William Murray

Assigned 2 June, 1817 to Richard Hooker of Thomas

Patent 17 Nov., 1818 Richard Hooker

IC G/108

Pat. Cert. 4800

Tract 680 Map 1-SW Baltimore County

Tempy's Sping 36 acres

Prior Survey for New Purchase (355 3/4 a.)

Comm. Warr. 6 Jan., 1791 (96 a.) Samuel Markey

Survey 8 Dec., 1791 Samuel Markey

Patent 27 Dec., 1799 Samuel Markey

IC O/760 and IC M/676

Pat. Cert. 4839

Tract 681 (South of Map 1) Baltimore County

Thomas Delight 25 acres

(Reference Tracts - Great profit (87 a.) and Little Profit (27 1/4 a.))

Prior Survey for Ford's Enlargement (--)

Comm. Warr. 25 Aug., 1759 (25 a.) Thomas Vaughn

Survey 7 Feb., 1760 Thomas Vaughn

Patent 7 Feb., 1760 Thomas Vaughn

BC & GS 12/301 and BC & GS 13/172

Pat. Cert. 4867

Tract 682 Map 2-W Carroll County
Three Brothers 300 acres
Prior Survey for Manheim (161 1/2 a.)
Warrant Feb., 1745 (300 a.) John, William and Reuben Boring
Survey 2 Aug., 1746 John, William and Reuben Boring
Patent 2 Aug., 1746 John, William and Reuben Boring
LG E/544 and LG 2/279
Pat. Cert. 4886

Tract 683 Map 2-W Carroll County
Three Brother's Neighbour 29 acres
Comm. Warr. 10 May, 1763 (29 a.) Henry Feeder
Survey 6 July, 1763 Henry Feeder
Patent 6 July, 1763 Henry Feeder
BC & GS 21/415 and BC & GS 22/270
Pat. Cert. 4888

Tract 684 Map 1-NW Baltimore County
(The) Three Cornered Piece 13 acres
Comm. Warr. 22 June, 1801 (5 a.) Peter Hoffman
Survey 20 May, 1802 Peter Hoffman
Patent 2 May, 1803 Peter Hoffman
IC R/233 and IC P/461
Pat. Cert. 4889

Tract 685 (not mapped) Baltimore County
(The) Three Divisions 2591 acres
Resurvey on Price's Chance (70 a.), Coles Forest (55), Hard Scuffle (34 a.) and Wheeler's Lot (139)
Prior Survey for Marshall's Marsh (122 a.) Talbot's Plains (638 1/2 a.), Hilly Run Resurveyed (197 3/4 a.), Take All (56 1/4 a.), Minersborough (54 a.), Phillips Industry (154+), Mt. Pleasant (135 a.), Davis's Farm (182 1/2 a.), Lexington (106 1/2), Cullins Lot (128 1/4 a.), Dorsey's Plains (655 1/2 a.) and others
Survey 24 Oct., 1770 Jonathan Plowman
Unpat. Cert. 1622

Tract 686 Map 1-C Baltimore County
Three Elbows 263 3/4 acres
Resurvey of Boring's Hazard (164 a.)
Sp. Warr. 17 Dec., 1791 Ezekial Boring
Survey 22 Nov., 1792 Ezekial Boring
Patent 3 June, 1794 Ezekial Boring
IC I/609 and IC G/402
Pat. Cert. 4890

Tract 687 Map 1-W Baltimore County
Timber and Stone 7 1/4 acres
Resurvey of Grist Mill Hills (257 a.)
Comm. Warr. 22 June, 1802 (5 a.) Henry Hoffman
Assigned 13 Nov., 1802 to William Hoffman
Survey 13 Nov., 1802 William Hoffman

(Timber and Stone - continued)
Patent 6 July, 1803 — William Hoffman
IC R/241 and IC S/74
Pat. Cert. 4915

Tract 688 (South of Map 2) Carroll County
<u>Tipton's Delight</u> 18 acres
(Reference Tract - Penelope Charchilla and Thomas Cockey Deyesburg (1030 a.))
Sp. Warr. 5 Aug., 1796 (25 a.) Brian Tipton
Survey 29 July, 1797 — Brian Tipton
Patent 2 Feb., 1798 — Brian Tipton
IC O/457 and IC L/425
Pat. Cert. 4925

Tract 689 Map 2-S Carroll County
<u>Tipton's Fancy</u> 50 acres
Prior Survey for Borings 's Range (1375 a.)
Warrant 4 Apr., 1760 (50 a.) Edward Tipton
Survey 20 Sept., 1760 — Edward Tipton
Patent 20 Sept., 1760 — Edward Tipton
BC & GS 12/500
Pat. Cert. 4927

Tract 690 Map 2-S Carroll County
<u>Tipton's Hope</u> 60 acres
Prior Survey for Rachel's Retirement (42 a.)
Comm. Warr. 21 Jan., 1767 (60 a.) Peter Fowble
Assigned 30 Mar., 1767 to Edward Tipton
Survey 30 Mar., 1767 — Edward Tipton
Patent 30 Mar., 1767 — Edward Tipton
BC & GS 34/276 and BC & GS 32/341
Pat. Cert. 4930

Tract 691 (South of Map 1) Baltimore County
<u>Tipton's Hunting Grounds</u> 34 1/2 acres
(Reference Tract - Britton's Range (295 a.))
Sp. Warr. 20 Mar., 1804 — Brian Tipton
Survey 18 Oct., 1804 — Bryan Tipton
Assigned 5 Jan., 1819 to Edward Lawson
Witness: Richard Hooker
Patent 2 Mar., 1827 — Edward Lawson
IB & GGB H/318 and IB & GGB F/473
Pat. Cert. 4931

Tract 692 (South of Map 2) Carroll County
<u>Tipton's Lot</u> 25 acres
(Reference Tract - Petersburg Res. (1003 a.))
Warrant 17 Oct., 1754 (25 a.) Edward Tipton
Survey 10 Apr., 1755 — Edward Tipton
Patent 10 Apr., 1755 — Edward Tipton
BC & GS 2/326 and BC & GS 4/248
Pat. Cert. 4932

Tract 693 (South of Map 2) Baltimore County

<u>Tipton's Trouble</u> 27 acres

(Reference Tract - Wilmott's Lot (217 a.))
Resurvey of Bratton's Desire (26 1/2 a.)
Sp. Warr. 2 Sept., 1803 (24 a.) Silvester Tipton
Assigned 17 Nov., 1804 to George M. Shaffer and John M. Shaffer
Survey 14 Aug., 1804 George M. Shaffer and John M. Shaffer
Patent 4 Aug., 1805 George M. Shaffer and John M. Shaffer
IC Q/562 and IC P/683
Pat. Cert. 4934

Tract 694 Map 2-SE Baltimore County

<u>Towson's Grove</u> 367 acres

Resurvey of Buck Range (50 a.)
Prior Survey for Penelope Charchilla and Thomas Cockey Deyesburg (1030 a.)
Sp. Warr. 22 Feb., 1763 Thomas Cockey Deye
Survey 10 Feb., 1765 Thomas Cockey Deye
Patent 7 Mar., 1765 Thomas Cockey Deye
BC & GS 27/511
Pat. Cert. 4958

Tract 695 Map 1-S Baltimore County

<u>Tracey's Beginning</u> 73 1/4 acres

Comm. Warr. 4 Mar., 1791 (40 a.) John Tracey
Comm. Warr. 4 Mar., 1791 (45 a.) John Tracey
Sp. Warr. 7 Oct., 1791 (10 a.) John Tracey
Survey 18 Feb., 1792 John Tracey
Patent 20 May, 1799 John Tracey
IC M/579 and IC L/550
Pat. Cert. 4960

Tract 696 (not mapped) Baltimore County

<u>Tracey's Chance</u> 63 1/2 acres

Survey 30 May, 1766 Teague Tracey
(never patented)
(not precisely located - west of Cole's Frolick)
(See field survey book of William Smith (p. 58) in the Maryland State Archives)

Tract 697 (South of Map 2) Carroll County

<u>Tracey's Hazard</u> 55 3/4 a., 0 r., 25 p.

(Reference Tract - Penelope Charcilla and Thomas Cockey Deyesburg (1030 a.))
Prior Survey for Tracey's Pleasant Ridge (73 a.)
Comm. Warr. 14 Mar., 1767 (54 a.) Basil Tracey
Survey 10 Aug., 1767 Basil Tracey
Patent 16 Nov., 1767 Basil Tracey
BC & GS 34/294
Pat. Cert. 4962

Tract 698 (South of Map 2) Carroll County

<u>Tracey's Pleasant Ridge</u> 73 acres

(Reference Tract - Penelope Charchilla and Thomas Cockeys Deyesburg (1030 a.))

Resurvey on Tracey's Hazard (55 3/4 a.)

Sp. Warr. 24 Apr., 1790 Basil Tracey

Assigned 10 Mar., 1791 to Michael Shafer

Survey 10 Mar., 1791 Michael Shafer

Patent 9 Dec., 1806 Michael Shafer

IC R/437 and IC S/425

Pat. Cert. 4965

Tract 699 (South of Map 2) Carroll County

<u>Transylvania</u> 50 acres

Prior Survey for Transylvania (355 a.)

Warrant 15 Mar., 1748 (50 a.) Richard Richards

Survey 30 Mar., 1749 Richard Richards

Patent 30 Mar., 1749 Richard Richards

BY & GS 1/256 and BY & GS 2/175

Pat. Cert. 4968

Tract 700 (South of Map 2) Carroll County

<u>Transylvania</u> 355 acres

(Reference Tracts - Petersburg Resurvey (1003 a.) and Gist's Meadow (75 a.))

Resurvey of Transylvania (50 a.)

Sp. Warr. 5 Feb., 1750 Richard Richards

Survey 10 June, 1751 Richard Richards

Patent 14 Oct., 1752 Richard Richards

BY & GS 5/647 and BY & GS 4/426

Pat. Cert. 4969

Tract 701 Map 2-NW Carroll County

<u>Troy</u> 1755 3/4 acres

Resurvey of Mesopotamia (1958 a.)

Prior Survey for Troy (1634 a.)

Survey 8 May, 1756 Charles Ridgely

Unpat. Cert. 1653

(Oversize folder in Archives was empty)

Tract 702 Map 2-NW Carroll County

<u>Troy</u> 1634 acres

Resurvey of Troy (1755 3/4 a.)

Sp. Warr. 18 Oct., 1764 Charles Ridgely and Andrew Stiger

Renewed 28 Apr., 1766 A. Stiger and Chas. Ridgely

Survey 26 Sept., 1766 A. Stiger and Chas. Ridgely

Patent 26 Sept., 1766 A. Stiger and Chas. Ridgely

BC & GS 34/316 and BC & GS 33/167

Pat. Cert. 4990

Tract 703 Map 1-NW Baltimore County

<u>Try for It</u> 94 1/2 acres

Resurvey of Sampson's Folly (118 a.)

(Try For It - continued)
Sp. Warr. 4 June, 1805 (5 a.) Philip Stiltz
Survey 21 Feb., 1806 Philip Stiltz
Patent 18 Mar., 1807 Philip Stiltz
IC R/521 and JK T/39
Pat. Cert. 4995

Tract 704 Map 2-NE Carroll County
Tumbling Down the Hill 60 acres
Sp. Warr. 7 Dec., 1797 (60 a.) Henry Lamott
Survey 6 Dec., 1798 Henry Lamott
Patent 6 Dec., 1804 Henry Lamott
IC R/291 and IC P/573
Pat. Cert. 4999

Tract 705 Map 2-NW Carroll County
Tuschem 26 acres
Prior survey for Spring Run Enlarged (130 3/4 a.)
Reservey of Troy (1634 a.)
Sp. Warr. 1 Dec., 1775 (100 a.) Richard Richards and Christopher Shrode
Survey 28 May, 1776 Richard Richards and Christopher Shrode
Unpat. Cert. 1656
Assigned 15 June, 1776 to Charles Smith

Tract 706 Map 1-W Baltimore County
Two Brothers 170 1/2 acres
Survey 6 Mar., 1787 Capt. John Showers
Patent 4 Apr., 1788 Capt. John Showers (Shrivers)
IC D/168 and IC C/345
Pat. Cert. 5011

Tract 707 Map 2-NE Carroll County
Two Butter Merchants 64 acres
Sp. Warr. 11 Nov., 1802 (15 a.) Jacob Grogg, Sr.
Survey 2 Oct., 1804 Jacob Grogg, Sr.
Assigned by estate of Jacob Grogg, Sr. to his heirs: (children) Jacob Grogg, Jr., John Grogg, Molly Rinehart, Catherine Denis, Louis Dehoss and grandchild Fanny Haney.
Patent 2 Feb., 1813 Jacob Grogg, Jr., John Grogg, Molly Rinehart, Catherine Denis, Louis Dehoss and Fanny Haney
IB D/21 and IB A/320
Pat. Cert. 5013

Tract 708 Map 1-W Baltimore County
Two Down at Once 16 acres
Comm. Warr. 22 June, 1802 (5 a.) Jacob Fitzer
Assigned 12 Nov., 1802 to Peter Hoffman
Survey 13 Nov., 1802 Peter Hoffman
Patent 8 July, 1803

(Two Down at Once - continued)
IC R/242 and IC S/74
Pat. Cert. 5014

Tract 709 Map 1-W Baltimore County
<u>Two Springs</u> 50 acres
Com. Warr. 6 Oct., 1795 (50 a.) Ezekial Boring
Survey 27 Sept., 1796 Ezekiel Boring
Patent 16 Oct., 1797 Ezekiel Boring
IC O/294
Pat. Cert. 5016

Tract 710 Map 2-N Carroll County
<u>Ulrich's Delight</u> 95 acres
Resurvey of Paul's Delight (15 a.)
Prior Survey for Rockey Hills (129 1/2 a.)
Sp. Warr. 14 Apr., 1762 Ulrich Fites
Survey 8 Oct., 1762 Ulrich Fites
Patent 29 Sept., 1764 Ulrick Fites
BC & GS 27/500 and BC & GS 28/84
Pat. Cert. 5019

Tract 711 Map 1-W Baltimore County
<u>Unexpected Discovery</u> 262 acres
Comm. Warr. 15 Aug., 1797 (187 1/2 a.)
Joshua Stevenson of Frederick Co.
Sp. Warr. 15 Aug., 1797 (74 3/4 a.) Joshua Stevenson
Survey 22 Dec., 1797 Joshua Stevenson
Patent 9 May, 1799 Joshua Stevenson
IC O/568 and IC L/542
Pat. Cert. 5022

Tract 712 (not mapped) Baltimore County
<u>United Friendship</u> 181 acres
Prior Survey for United Frienship Enl. (260 1/2 a.)
Survey 8 Nov., 1786 William Nash
Patent 11 May, 1792 William Nash
IC F/653 and IC H/180
Pat. Cert. 5030

Tract 713 Map 1-NE Baltimore County
<u>United Frienship Enlarged</u> 260 1/2 acres
Resurvey of United Friendship (181 a.)
Prior Survey for Ruhl's Bargain (363 1/2 a.)
Sp. Warr. 12 May, 1792 William Nash
Survey 8 May, 1793 William Nash
Assigned 28 Feb., 1795 to George Ruhl
Patent 3 Mar., 1795 George Ruhl
IC K/134 and IC H/639
Pat. Cert. 5032

Tract 714 (West of Map 2) Carroll County
<u>Utz's (Oates) Inheretance</u> 226 acres
(Reference Tracts - Troy (1634 a.) and
Winchester's Lot (973 a.))

(Utz's Inheretance (226 acres) - continued)
Sp. Warr. 2 Feb., 1770 (200 a.) Jacob Oates
Survey 3 Feb., 1770 Jacob Oates
Patent 2 July, 1771 Jacob Oates (Utz)
BC & GS 47/347 and BC & GS 44/327
Patent Cert. 3583

Tract 715 (West of Map 2) Carroll County
<u>Utz's Inheretance</u> 426 1/2 acres
(Reference Tracts - Winchester's Lot (972 a.), Troy (1634 a.) and Everybodies Land (390 1/2 a.))
Resurvey of Winchester's Lot (972 a.), Troy (1634), Hogstadt (314 a.), Oxford (14 3/4 a.) and Everybodies Land (390 1/2 a.)
Sp. Warr. 19 Nov., 1804 Peter Utz
Survey 4 May, 1805 Peter Utz
Patent 28 June, 1808 Peter Utz
IC R/701
Pat. Cert. 5041

Tract 716 Map 2-N Carroll County
<u>Vance's Delight</u> 25 acres
Prior Survey for Vance's Enlargement (120 a.)
Sp. Warr. 17 Oct., 1754 Jacob Shilling, Jr.
Survey 16 Mar., 1755 Jacob Shilling Jr.
Assigned 15 Oct., 1755 to Valentine Vance
Patent 29 Sept., 1757 Valentine Vance
BC & GS 9/370 and BC & GS 8/178
Pat. Cert. 5047

Tract 717 Map 2-N Carroll County
<u>Vance's Enlargement</u> 120 acres
Resurvey of Vance's Delight (25 a.)
Prior Survey for Vance's Garden (121 3/4 a.)
Warrant 1 Oct., 1765 (18 a.) Betty Vance
(not executed - applied to Vance's Enlargement)
Sp. Warr. 30 May, 1774 (25 a.) Valentine Vance
Survey 11 June, 1774 Valentine Vance
Patent 8 May, 1775 Valentine Vance
BC & GS 51/88 and BC & GS 49/218
Pat. Cert. 5048

Tract 718 Map 2-N Carroll County
<u>Vance's Garden</u> 121 3/4 acres
Resurvey of Vance's Enlargement (120 a.)
Sp. Warr. 7 Aug., 1792 Valentine Vance
Survey 24 July, 1793 Valentine Vance
Patent 3 Nov., 1794 Valentine Vance (Vantz)
IC I/618 and IC H/562
Pat. Cert. 5049

Tract 719 Map 1-E Baltimore County
<u>Vaughn's Disappointment</u> 504 1/2 acres
Resurvey of Conrad's Chance Resurveyed (142 1/2 a.) and Cross's Grove (271 3/4 a.)

(Vaughn's Disappointment - continued)
Prior Survey for Expedition (337 a.)
Sp. Warr. 1 Dec., 1794 Jacob Marshall
Survey 19 Nov., 1795 Jacob Marshall
Patent 9 Mar., 1797 Jacob Marshall
IC O/251 and IC M/155
Pat. Cert. 5052

Tract 720 Map 2-NE Carroll County
Vaughn's Meadows 27 acres
Prior Survey for Sugar Valley (387 a.), Beaver's New Invention (170 a.) and Hobson's Choice (106+)
Comm. Warr. 27 Apr., 1763 Thomas Stansbury
Assigned 26 Oct., 1763 to Richard Vaughn
Survey 26 Oct., 1763 Richard Vaughn
Patent 26 Oct., 1763 Richard Vaughn
BC & GS 27/486 and BC & GS 26/162
Pat. Cert. 5055

Tract 721 Map 1-NW Carroll County
Venus's Arbour 202 acres
Chancellor's Order 8 Dec., 1797 to correct Venus's Arbour to exclude the part called Grubby Thicket
Survey 13 Mar., 1798 Joseph Walker (Jr.)
Patent 27 Mar., 1799 Joseph Walker (Jr.)
IC O/587 and IC L/537
Pat. Cert. 5060
(part of this tract is in Baltimore County)

Tract 722 Map 2-NE Carroll County
Walker's Folly 6 acres
Comm. Warr. 11 Oct., 1797 (6 a.) George Walker
Survey 2 Oct., 1798 George Walker
Unpat. Cert. 1683

Tract 723 Map 2-NE Carroll County
Walker's Goodwill 26 1/8 acres
Survey 13 Apr., 1787 William Hoffman
Patent 2 Apr., 1789 William Hoffman
IC D/411 and IC C/648
Pat. Cert. 5077

Tract 724 Map 2-N Carroll County
(The) Wallet 118 1/2 acres
Prior Survey for Wallet Resurveyed (212 3/4 a.)
Sp. Warr. 21 Mar., 1774 (118 1/2 of 150 a.)
Thomas Stansbury, Jr.
Survey 13 Apr., 1774 Thomas Stansbury, Jr.
Patent 14 Mar., 1776 Thomas Stansbury, Jr.
BC & GS 51/254 and BC & GS 52/168
Pat. Cert. 5083

Tract 725 Map 2-N Carroll County
Wallet Resurveyed 212 3/4 acres
Resurvey of The Wallet (188 1/2 a.)

(Wallet Resurveyed (212 3/4 acres) - continued)
Prior Survey for John Tracey and Feltey's Wife Reconciled (7 1/2 a.)
Sp. Warr. 4 Jan., 1790 — Christopher Fair
Survey 16 Dec., 1790 — Christopher Fair
Patent 2 Dec., 1801 — Christopher Fair
IC N/627 and IC P/254
Pat. Cert. 5084

Tract 726 — Map 1-C — Baltimore County
<u>Walls of Derry</u> — 81 3/4 acres
Sp. Warr. 27 Aug., 1793 (10 a.) William Dorsey
Assigned 4 Jan., 1794 — Jonathan Cullings
Survey 13 Jan., 1794 — Jonathan Culling(s)
Patent 12 June, 1809 — Jonathan Culling(s)
JK U/73 and JK T/109
Pat. Cert. 5088

Tract 727 — Map 1-W — Baltimore County
<u>Walnut Hills</u> — 44 1/4 acres
Warrant 15 Dec., 1800 (5 a.) Henry Hoffman
Survey 13 Mar., 1801 — Henry Hoffman
Patent 14 Jan., 1802 — Henry Hoffman
IC R/60 and IC P/278
Pat. Cert. 5092

Tract 728 — (South of Map 1) — Baltimore County
<u>Walnut Hills</u> — 813 1/2 acres
(Reference Tracts - Bear's Rock (177 3/4 a.), Prettyboy Hills (56 3/4 a.) and Pine Grove (52))
Resurvey of Pleasant Garden (4 3/4 a.)
Sp. Warr. 28 Nov., 1791 — Christian Singery
Survey 15 Nov., 1792 — Christian Singery
Patent 31 Dec., 1793 — Christian Singery
IC I/454 and IC H/432
Pat. Cert. 5093

Tract 729 — Map 2-E — Carroll County
<u>We Found the Beginning</u> — 40 acres
Sp. Warr. 5 Oct., 1796 (40 a.) Henry Lamott
Survey 21 Sept., 1797 — Henry Lamott, Jr.
Patent 6 Dec., 1804 — Henry Lamott, Jr.
IC P/581 and IC R/292
Pat. Cert. 5144

Tract 730 — Map 2-N — Carroll County
<u>West Lotham</u> — 18 acres
Comm. Warr. 20 Nov., 1764 (18 a.) John Dilley
Survey 7 Mar., 1765 — John Dilly
Patent 7 Mar., 1765 — John Dilly (Jr.)
BC &GS 27/445 and BC & GS 26/255
Pat. Cert. 5168

Tract 731 — Map 1-S — Baltimore County
<u>Whartleberry Garden</u> — 6 acres

(Whartelberry Garden - continued)
Comm. Warr. 26 Mar., 1795 Jacob Markey
Survey 2 Apr., 1795 Jacob Markey
Unpat. Cert. 1706

Tract 732 Map 2-W Carroll County
<u>What You Please</u> 10 acres
Comm. Warr. 21 Jan., 1790 (10 a.) Joseph Bollinger
Survey 2 Nov., 1790 Joseph Bollinger
Patent 30 July, 1793 Joseph Bollinger
IC H/332 and IC I/416
Pat. Cert. 5176

Tract 733 Map 1-SW Baltimore County
<u>What's Left</u> 112 1/2 acres
Prior Survey for What's Left Resurvey (146 3/4)
Survey 7 Mar., 1787 Laurance Myers
Patent 4 Aug., 1790 Laurance Myers
IC F/30 and IC C/738
Pat. Cert. 5170

Tract 734 Map 1-SW Baltimore County
<u>What's Left Resurveyed</u> 146 3/4 acres
Resurvey of What's Left (112 1/2 a.)
Sp. Warr. 16 July, 1791 Philip Myers
Survey 22 June, 1792 Philip Myers
Patent 22 April, 1830 Philip Myers
GGB 2/88
Pat. Cert. 5173

Tract 735 Map 1-C Baltimore County
<u>Wheeler's Garden Spot</u> 25 acres
Prior Survey for Talbot's Lot (164 1/2 a.)
Leased Land 1 July, 1746 John Wheeler
Leaseholds GGB B/654
Survey 1 July, 1746 John Wheeler
UnPat. Cert. 1713

Tract 736 Map 1-C Baltimore County
<u>Wheeler's Lot</u> 139 1/4 acres
Prior Survey for Talbot's Lot (164 1/2 a.) and
The Three Divisions (2591 a.)
Leased Land 16 Apr., 1761 John Wheeler
Leaseholds GGB B/652
Survey 10 Aug., 1761 John Wheeler
Unpat. Cert. 1714

Tract 737 Map 2-NW Carroll County
<u>Whispaw</u> 140 acres
Prior Survey for Whispaw (189 a.)
Sp. Warr. 16 July, 1752 (140 a.) Ulrich Whessler
Survey 30 Nov., 1752 Ulrich Whessler
Patent 10 Aug., 1753 Ulrich Whisler
BC & GS 1/245 and Y & S 6/425
Pat. Cert. 5202

Tract 738 Map 2-NW Carroll County

Whispaw Resurvey 189 acres

Resurvey of Whispaw (140 a.)
Prior Survey for Whispaw (283 1/2 a.)
Sp. Warr. 21 Jan., 1767 Michael Couts
Renewed 20 July, 1767 Michael Couts
Renewed 20 Jan., 1768 (120 a.) Michael Couts
Survey 1 Mar., 1768 Michael Courts
Patent 5 Sept., 1776 Michael Couts (Coutz)
BC & GS 41/416 and BC & GS 42/377
Pat. Cert. 5201

Tract 739 Map 2-NW Carroll County

Whispaw Resurvey Resurvey 283 1/2 acres

Resurvey of Whispaw Resurvey (189 a.)
Prior Survey for Couts Lot Resurvey (241+ a.)
Sp. Warr. 13 May, 1784 Michael Coutz
Survey 4 May, 1785 Michael Coutz
Patent 24 Oct., 1787 Michael Coutz
IC D/106 and IC C/229
Pat. Cert. 5203

Tract 740 Map 2-N Carroll County

White Oak Bottom 61 3/4 acres

Prior Survey for Lamott's Delight (20 1/4 a.)
Sp. Warr. 10 Dec., 1798 (20 a.) Henry Lamott
Survey 12 Apr., 1799 Henry Lamott
Patent 6 Dec., 1804 Henry Lamott
IC R/290 and IC P/579
Pat. Cert. 5217

Tract 741 Map 1-NW Baltimore County

Wibling's (Weibling's) Chance 13 acres

Sp. Warr. 11 Mar., 1844 (10 a.) John Weibling
Survey 25 Mar., 1844 John Weibling
Unpat. Cert. 1693

Tract 742 Map 2-C Carroll County

Widow's Favor 20 acres

Sp. Warr. 19 Mar., 1794 (20 a.) Alexis Lemmon
Assigned 5 Aug., 1794 to Hannah Sater
Survey 5 Aug., 1794 Hannah Sater
Patent 24 Oct., 1795 Hannah Sater
IC G/590 and IC K/187
Pat. Cert. 5230

Tract 743 Map 1-SE Baltimore County

Widow's Habitation 185 1/2 acres

Prior Survey for McAllister's Folly (353 1/2 a.)
Survey 11 May, 1789 Delilah Wheeler
Assigned 5 June, 1789 to Richard McCallister and Samuel Harnash of Pennsylvania
Patent 11 Feb., 1791 Richard McCallister and Samuel Harnash of Pennsylvania

(Widow's Habitation - continued)
IC F/397 and IC E/645
Pat Cert. 5231

Tract 744 Map 1-SE Baltimore County
Wildcat Hallow 54 3/4 acres
Sp. Warr. 27 Feb., 1804 Wm. Hoffman, Jr.
Survey 24 Apr., 1805 Wm. Hoffman, Jr.
Assigned by the will of William Hoffman, Jr. to his children and heirs: William J. Hoffman, Henry B. Hoffman, Susannah Hoffacker (m. Samuel Hoffacker) Mary Hoffman (m. Jacob Markey), Elizabeth Hoffman (m. Henry M. Hoffacker), Ann Hoffman (m. John Gore) and Catherine Hoffman (m. George Miller)
Patent 27 July, 1850 William J. Hoffman et. al.
GGB 7/277
Pat. Cert. 5240

Tract 745 Map 2-NE Carroll County
Wildcat Harbour 56 acres
Comm. Warr. 19 May, 1762 (100 a.) Philip Idleman
Assigned 4 Mar., 1767 (56 a.) to John Beaver
Survey 10 Aug., 1767 John Beaver
Patent 10 Aug., 1767 John Beaver
BC & GS 34/305 and BC & GS 32/235
Pat. Cert. 5239

Tract 746 (East of Map 1) Baltimore County
(The) Wilderness 11 1/2 acres
(Reference Tract - Northampton (242 a.))
Comm. Warr. 4 Aug., 1790 (10 a.) Thomas Wantland
Survey 3 Aug., 1790 Thomas Wantland
Patent 6 Apr., 1802 Thomas Wantland
IC R/43 and IC L/721
Pat. Cert. 5242

Tract 747 Map 1-C Baltimore County
Wiley's Ketch 54 3/4 acres
Survey 9 Feb., 1787 Peter Costeloe
Assigned 23 Sept., 1793 to Thomas Smith
Assigned 24 Oct., 1793 to Ezekiel Boring
Patent 24 Apr., 1795 Ezekiel Boring
IC K/193 and IC G/520
Pat. Cert. 5244

Tract 748 (South of Map 1) Baltimore County
Williams Lott 60 1/2 acres
(Reference Tract - Britton's Range (295 a.))
Survey 11 Apr., 1787 Jonathan Tipton, Sr.
Assigned by estate of Jonathan Tipton, Sr. to his children and heirs: Jonathan, Jr., William, Bryan, James and Silvester Tipton, Lydia Price (wife of Benjamin), Eleanor Wheeler, (wife of William), Rachel Tracey, Eleanor Tracey, Jonathan Tracey, William Tracey

(Williams Lott - continued)
and Basil Tracey (all children of William
and Kazial Tipton Tracey)
Patent 19 Mar., 1804 Jonathan Tipton, Jr. et. al.
IC P/526 and IC R/249
Pat. Cert. 5269

Tract 748 A Map 1-N Baltimore County
Williams Triple (Trifle) 6 acres
Comm. Warr. 2 Dec., 1793 (2 a.) James Campbell
Assigned 9 Oct., 1794 to William Marshall
Survey 9 Oct., 1794 William Marshall
Unpat. Cert. 1758

Tract 749 Map 2-NE Carroll County
Wilmott's Desire (Pleasure) 136 acres
Prior Survey for Beaver's New Invention (170 a.)
Sp. Warr. 11 Sept., 1772 (1400 a.) John Wilmot
(for land south of Map No. 2)
Survey 27 Oct., 1772 John Wilmot
Patent 22 Sept., 1773 John Wilmot
BC & GS 47/351
Pat. Cert. 5290

Tract 750 Map 2-SE Baltimore County
Wilmott's Lott 217 acres
Resurvey of the Plains of Susa (57 a.)
Sp. Warr. 11 Sept., 1772 (1400 a.) John Willmott
Survey 20 Nov., 1772 John Wilmotte
Assigned 18 Mar., 1773 to John Ford
Patent 7 Aug., 1773 John Ford
BC & GS 47/336
Pat. Cert. 5295

Tract 751 (West of Map 2) Carroll County
Winchester's Lott 50 acres
Prior Survey for Winchester's Lot (972 a.)
Comm. Warr. 21 June, 1744 (50 a.) Wm. Winchester
Survey 29 Oct., 1744 William Winchester
Patent 18 Mar., 1746 William Winchester
BC & GS 1/238
Pat. Cert. 5326

Tract 752 (West of Map 2) Carroll County
Winchester's Lott Resurvey 972 acres
(Reference Tracts - Powell's Discovery (148 a.) and
McGill's Choice Resurvey (243 a.))
Resurvey of Winchester's Lot (50 a.)
Prior Survey for Good Luck (524 1/2 a.)
Sp. Warr. 5 Sept., 1750 William Winchester
Renewed 15 Aug., 1751 William Winchester
Survey 23 Oct., 1751 William Winchester
Patent 1 Sept., 1752 William Winchester
BY & GS 4/225 and BY & GS 7/7
Pat. Cert. 5327

Tract 753 Map 1-E Baltimore County
<u>Wind Mill Hills</u> 15 acres
Prior survey for Addition to Pleasant Valley (10 a.)
Comm. Warr. 22 Mar., 1793 (15 a.) George Walker and George Curtis
Assigned 1 Nov., 1793 to John Beaver
Survey 22 Jan., 1794 John Beaver
Unpat. Cert. 1769

Tract 754 Map 1-NW Baltimore County
<u>Windsor</u> (Winsor) 10 acres
Comm. Warr. 25 Mar., 1794 (10 a.) Christopher Ring
Survey 27 Mar., 1794 Christopher Ring
Patent 7 June, 1794 Christopher Ring
IC O/253 and IC L/333
Pat. Cert. 5330

Tract 755 Map 1-S Baltimore County
<u>With Little I am Content</u> 22 acres
Sp. Warr. 5 Jan., 1797 (25 a.) Joshua Stevenson of Frederick County
Survey 19 Dec., 1797 Joshua Stevenson
Patent 9 May, 1799 Joshua Stevenson
IC O/574 and IC M/585
Pat. Cert. 5340

Tract 756 Map 1-SE Baltimore County
<u>Wolf Den</u> 50 acres
Prior Survey for Schieming Defied (194 3/4 a.)
(lease land (?) named in several patents and later deeds but no other information)

Tract 757 Map 1-SE Baltimore County
<u>Woodlot</u> 12 acres
Comm. Warr. 28 Oct., 1790 Jesse Orshall (Hoshall)
Survey 22 Oct., 1791 Jesse Orshell (Hoshall)
Unpat. Cert. 1776

Tract 758 Map 2-N Carroll County
<u>Wooley String</u> 35 acres
Prior Survey for Loretta Resurvey (51 a.)
Comm. Warr. 14 Sept., 1774 (35 a.) Wooley Foits
Survey 28 Feb., 1775 Ulrich Fite
Patent 9 Nov., 1795 Ulrich Foit (Fites)
IC K/193 and IC L/22
Pat. Cert. 5368

Tract 759 Map 2-W Carroll County
<u>Worms</u> 25 acres
Comm. Warr. 4 Mar., 1761 (25 a.) Stephen Long
Survey 27 June, 1761 Stephen Long
Patent 27 June, 1761 Stephen Long
BC & GS 14/707 and BC & GS 17/451
Pat. Cert. 5372

Tract 760 — Map 1-NE — Baltimore County

<u>Yardley</u> — 10 acres

Comm. Warr. 27 July, 1790 — Thomas Arms
Survey 10 Sept., 1790 — Thomas Arms
Patent 13 Mar., 1800 — Thomas Arms
IC N/381 and IC M/712
Pat. Cert. 5382

Tract 761 — Map 1-E — Baltimore County

<u>Yoder's Park</u> (Part) — 158 1/2 acres

Survey 6 Mar., 1787 — Yoder Maesoner (Masimore)
Assigned 17 Feb., 1796 to Henry Masimore
Patent 12 May, 1800 — Henry Masimore
IC N/389 and IC L/597
Pat. Cert. 5387

Tract 762 — Map 2-NE — Carroll County

<u>Young Man's Delight</u> — 141 1/2 acres

Survey 16 Mar., 1787 — Jacob Grogg (Sr.)
Assigned by will of Jacob Gross, Sr. to heirs: Jacob Grogg, Jr., Molly Rhinehart, Catherine Denis, John Grogg, Louis Dehoff his only children and grandchild Fanny Haney.
Patent 6 June, 1833 — Jacob Grogg, Jr. et. al.
GCB 2/385 and GGB 1/468
Pat. Cert. 5401
(part of this tract is in Baltimore County)

Tract **

Zimmerman's Venture (See Simmerman's Venture)

NAME INDEX